Land Use Changes and the Corresponding Ecological Risks

Land Use Changes and the Corresponding Ecological Risks

Editors

Wei Song
Hualin Xie

MDPI

Basel • Beijing • Wuhan • Barcelona • Belgrade • Novi Sad • Cluj • Manchester

Editors
Wei Song
Institute of Geographic Sciences and Natural Resources Research, Chinese Academy of Sciences
Beijing, China

Hualin Xie
Jiangxi University of Finance and Economics
Nanchang, China

Editorial Office
MDPI
St. Alban-Anlage 66
4052 Basel, Switzerland

This is a reprint of articles from the Special Issue published online in the open access journal *International Journal of Environmental Research and Public Health* (ISSN 1660-4601) (available at: https://www.mdpi.com/journal/ijerph/special_issues/LUCCER).

For citation purposes, cite each article independently as indicated on the article page online and as indicated below:

Lastname, A.A.; Lastname, B.B. Article Title. *Journal Name* **Year**, *Volume Number*, Page Range.

ISBN 978-3-0365-8842-1 (Hbk)
ISBN 978-3-0365-8843-8 (PDF)
doi.org/10.3390/books978-3-0365-8843-8

Contents

About the Editors

Wei Song

Wei Song, Ph.D., is an associate research fellow of Institute of Geographic Sciences and Natural Resources Research, Chinese Academy of Sciences. He obtained a PhD in Science (2009) from the University of Chinese Academy of Sciences, and a Master's Degree in Management (2006) from China Agricultural University. His specific research interests cover four aspects: (1) land use change monitoring using remote sensing technology; (2) land use change modelling, such as the development of System Dynamics Models and Multiagent Models; (3) the effects of land use change on ecosystem services; and (4) land use policies. He has published more than 140 academic papers in English and Chinese journals. His published papers have been cited in Web of Science 3,449 times, with an H-Index of 30, and in Google Scholar 5,348 times, with an H-Index of 40. Four papers in which he was the first author/corresponding author have been included in ESI's highly cited papers. His research results have been granted 7 provincial and ministerial awards; the 14th National Youth Geography Science and Technology Award; and 9 software copyrights.

Hualin Xie

Hualin Xie, Ph.D., is a Professor for Land Resource Management in the Institute of Ecological Civilization at Jiangxi University of Finance and Economics (Nanchang, China). He obtained a PhD in Science (2006) from Beijing Normal University, and a Master's Degree in Management (2003) from China Agricultural University. He has been the Director of the Institute of Ecological Civilization since 2016. His scientific interests are centered around land use, territorial space optimization and ecological product value realization. He is the author of over 20 scientific monographs and advanced textbooks and over 200 publications in scientific journals and proceedings. He was named as a Highly Cited Chinese Researcher by Elsevier in 2019, 2020 and 2021.

Preface

The global population has been increasing dramatically since the 1950s. High population pressure is not only linked with an unprecedented scale of resource consumption, but also aggravates the exploitation of the ecological environment by humans. As an important link between human activities and ecosystems, land use mode and intensity are not only an important driving factor of global ecological environmental changes but also a result of these environmental changes, and there is a close interdependent relationship between land use and global environmental changes. At global and regional scales, humans are increasingly realizing that a functional ecological environment is an important basis for the sustainable development of a social economy and for improving human wellbeing. In recent years, numerous studies have focused on land use changes and the corresponding ecological responses, including the identification of regional land use change processes based on remote sensing technology, the discussion of landscape effects on land use changes, and the construction of a series of ecosystem service evaluation models.

However, before the concept of ecological management can be integrated into studies of land use resources, it is necessary to systematically (1) reveal typical land use change processes, such as the reclamation and abandonment of marginal arable land, in ecologically fragile areas; (2) assess ecosystem vulnerability; (3) identify important ecological sources and key ecological corridors; (4) clarify the concept and connotation of land use ecological risks; (5) reveal the quantitative relationship between land use change and ecological risks; and (6) evaluate the ecological responses and potential risks of regional land use changes. Such results can provide a scientific basis for the establishment of adequate policies.

We appreciate the academic editors and reviewers for their valuable time and insightful comments on the published 30 research papers and 3 review papers in the Special Issue book "Land Use Changes and the Corresponding Ecological Risks". We also thank all the authors of this book for their contributions. The findings of the book will support scientific bases and policy implications for sustainable land management and ecological protection construction.

This book was supported by the Strategic Priority Research Program of Chinese Academy of Sciences (Grant No. XDA20040201).

Wei Song and Hualin Xie

Editors

International Journal of
Environmental Research and Public Health

MDPI

Article

The Spatio-Temporal Patterns and Driving Forces of Land Use in the Context of Urbanization in China: Evidence from Nanchang City

Yuxi Liu [1], Cheng Huang [1,2,*] and Lvshui Zhang [1,*]

1 School of Forestry, Jiangxi Agricultural University, Nanchang 330045, China
2 Shanghai Key Lab for Urban Ecological Processes and Eco-Restoration, School of Ecological and Environmental Sciences, East China Normal University, Shanghai 200241, China
* Correspondence: chuang@jxau.edu.cn (C.H.); zhanglvshui@jxau.edu.cn (L.Z.)

Abstract: Land use change has been one of the common problems in the context of urbanization in China. Social economy and land use interact with each other, and it is especially important for human society to adhere to sustainable development, and to deal with the contradictory relationship between the social–economic needs and land use change. The objectives of this study are: (1) Obtain time-series land-use classification data and its spatial distribution in Nanchang City; (2) Identify the characteristics and driving force of spatial–temporal land use changes in Nanchang City from 2000 to 2020; (3) Discuss the relationship between the urban expansion and social economy in Nanchang City. The results show that the spatial distribution of land use in Nanchang City has changed significantly from 2000 to 2020, and the largest area of land-use type in Nanchang City has been cropland. The cropland has continuously declined, and the urban area has increased significantly. A lot of cropland has been transformed into urban areas, and land use degree in Nanchang City has significantly increased. The spatial pattern of land use has greatly changed, and the city spatial pattern has become more aggregated, while the spatial distribution of cropland, forest and grassland has become more fragmented. Moreover, there has been an obvious correlation between social-economic development and the level of land use, and GDP has been the main driver of land use change. The central urban area of Nanchang city has been the main hotspot of land use change.

Keywords: land use change; drivers; urban expansion; landscape

Citation: Liu, Y.; Huang, C.; Zhang, L. The Spatio-Temporal Patterns and Driving Forces of Land Use in the Context of Urbanization in China: Evidence from Nanchang City. *Int. J. Environ. Res. Public Health* **2023**, *20*, 2330. https://doi.org/10.3390/ijerph20032330

Academic Editor: Paul B. Tchounwou

Received: 28 November 2022
Revised: 20 January 2023
Accepted: 26 January 2023
Published: 28 January 2023

1. Introduction

Land use is a complex system containing natural elements such as geology, hydrology, soil, vegetation and human activities [1]. It is closely related to social–economic development [2]. Land use change is a reflection of human activities and a key factor influencing climate change [3,4]. Land-use change mechanisms, monitoring and modeling have become the focus of the scientific community [5,6]. Land use surveys, land resource assessment and utilization planning are the application fields of traditional land-use research [7]. Socio-economic, cultural and spatial land management make land use show significant regional and periodic differences [8,9].

Since China officially became a member of the World Trade Organization (WTO) in 2001, China's economy and urbanization processes have entered a stage of rapid development. Urbanization is considered to have a significant impact on land use patterns and the environment [10]. So far, more than 54% of the world's total population (about 8 billion) lives in urban areas and this figure is expected to exceed 67% in 2050. Urbanization is placing an increasing demand on urban land, yet the demand for land resources for urbanization cannot be met indefinitely as land resources are scarce resources [11]. How to use land resources efficiently and intensively has become one of the issues of sustainable development in human society [12]. Clarifying the mechanisms of interaction between land use change and social economy is important for the formulation of urban sustainable development strategies and the sustainable use of land resources.

Urbanization is one of the most important factors of LULC change, and it also brings up a series of ecological problems [13], such as the heat island effect [14], environmental pollution [15], habitat destruction [16] and loss of cropland [17]. Urban expansion is one of the essential parts of land use research. Since the 1820s, researchers have focused mainly on urban morphology and spatial structure in land use [18]. Suggested theories for urban expansion include the concentric model, sector theory and multi-core model [19]. With the continuous development of remote sensing and geographic information science (GIS) technology, the research on urban expansion based on the perspective of land use has developed into quantitative analysis, dynamic monitoring, spatiotemporal characteristic analysis and driving mechanism research of urban expansion [18].

Due to the difficulty of obtaining time-series spatial data, the early land use research mainly discussed the area change of land use from a statistical perspective, so the spatial pattern change of land use was not fully studied. The development of remote sensing technology and computer technology has provided rich monitoring methods and efficient data-processing methods for land use research. Additionally, the application of remote sensing technology not only provides rich and diverse spatial monitoring data but also greatly saves human and financial costs. Liu [20] showed 13 cities in China that urban expansion is influenced by a combination of population, monetary and land policies. Danilo [21] used an object-oriented classification to analyze aerial photographs and obtain land use situations in the Langhe region of Italy from 1954 to 2000. He also further analyzed changes in the spatial–temporal pattern of land use. Although land use data have good spatial resolution and effectively promote the study of urban space, it is mainly based on Landsat data interpretation, which relies on traditional methods that are time-consuming and labor-intensive. With the progress of Big Data technology and the abundance of remote sensing data, the rapid and low-cost land use data acquisition of time series is possible, and the joint research of remote sensing data and socio-economic time series is gradually enriched [22].

To solve the problem of discontinuous land-use data and obtain urban spatial distribution data quickly, researchers have turned their attention to night-time lighting data from meteorological satellites [23–25]. However, the inherent disadvantages of night-time light data, such as low resolution (1 km), determining that the data are suitable for the extraction of urban boundary information and lack more detail, make the data unsuitable for comprehensive land-use analysis and monitoring.

Advances in remote sensing and artificial intelligence technologies have dramatically changed the way we look at the Earth [12]. The Remote Sensing Big Data Platform (RSBDP) provides a powerful and data-rich platform for remote sensing analysis for researchers and provides remote sensing analysis functions, such as GEE [26]. Currently, RSBDP has been widely used in a variety of fields such as land cover mapping, and Hanberry [27] noted that RSBDP and machine learning methods performed well in wildfire identification and classification. Some researchers found that Landsat images and GEE were suitable for mapping flood areas, with an overall accuracy of over 90% [28]. The researchers found that by using machine learning methods and Landsat data, artificial targets in heterogeneous environments could be accurately detected [29]. The researchers found that GEE and random forest classifier can effectively extract vegetation distribution information [30]. RSBDP-based land-use classification methods and statistical characterization are hot issues in land use research [31].

In this study, we propose to use spatiotemporal Big Data and time-series analysis methods to explore the driving force of land-use pattern change in Nanchang from the perspective of spatiotemporal change, so as to provide decision-making support for sustainable use of land resources and ecological environmental protection. A large number of existing studies of land use cannot quantify the correlation between land use change and social economy using statistical methods due to the lack of continuous time-series data on land use. The innovations of this study are to quantify the correlation between land use change and social economy using continuous time-series land use data and statistical methods. In addition, the correlation among comprehensive land use index, per capita GDP, population and fixed asset investment was analyzed in this study.

2. Materials and Methods

2.1. Research Areas

Nanchang is the capital city of Jiangxi Province and an important city in the urban agglomeration in the middle reaches of the Yangtze River in China. It is also the central city of the Poyang Lake Ecological Economic Zone. Located in the north-central of Jiangxi Province (Figure 1), it covers an area of 7194.98 km^2. As of 2020, Nanchang has a resident population of 6,437,500 and a GDP of 574,551 million CNY. In addition, Nanchang is one of the first low-carbon pilot cities in China, a national innovative city and an international garden city. In recent years, the economy of central China was growing rapidly at a rate of about 8%, leading to increased urbanization and rapid land use changes, which have had an impact on urban spatial form and the ecological environment. Therefore, this study analyzes the relationship between land use patterns and social–economic development in Nanchang City over the past 20 years and provides decision support for urban green development and sustainable development.

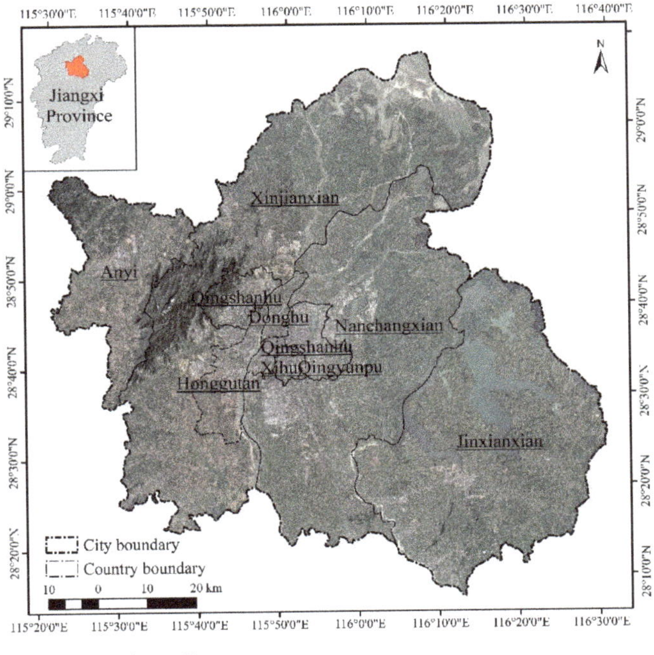

Figure 1. Nanchang City.

2.2. Materials

This study mainly uses statistical data and spatial data. Among them, the statistical data were obtained from the Nanchang Statistical Yearbook (2001–2021), including population, GDP and fixed investment. Spatial data include land use and administrative division data. Among them, land use data were obtained by using remote sensing Big Data platform, selecting Landsat impact data and using a supervised classification method and machine learning algorithm to classify remote sensing images. Benefiting from the application of the RSBDP and machine learning algorithm, this study can quickly obtain continuous time-series land-use data and carry out innovative research. The administrative division data were obtained from the China Geographic Information Bureau and were produced in 2015. This study classifies land use in Nanchang into seven types: cropland, forest, shrub, grassland, water, barren and urban.

2.3. Methods

The research framework is shown in Figure 2. Previous studies on urban land use and drivers were limited by the lack of spatial data on land use and mainly used discontinuous land use data every 5 years for analysis. Therefore, previous studies could not fully utilize the statistical data of continuous time series to mine the relationship between land use and the social economy, nor could they construct statistical models of social economy and land use. They could only analyze the social-economic impact on land use in the datum year and the target year through LMDI or other methods. In recent years, thanks to the maturity of remote sensing Big Data and remote sensing cloud-computing platforms, the acquisition and analysis of spatial–temporal data based on the remote sensing Big Data platform becomes a reality, and the spatial–temporal data of continuous time series can be quickly acquired and applied to practice research. As a result, integrated research of spatial–temporal change and social economy based on time series can be carried out smoothly.

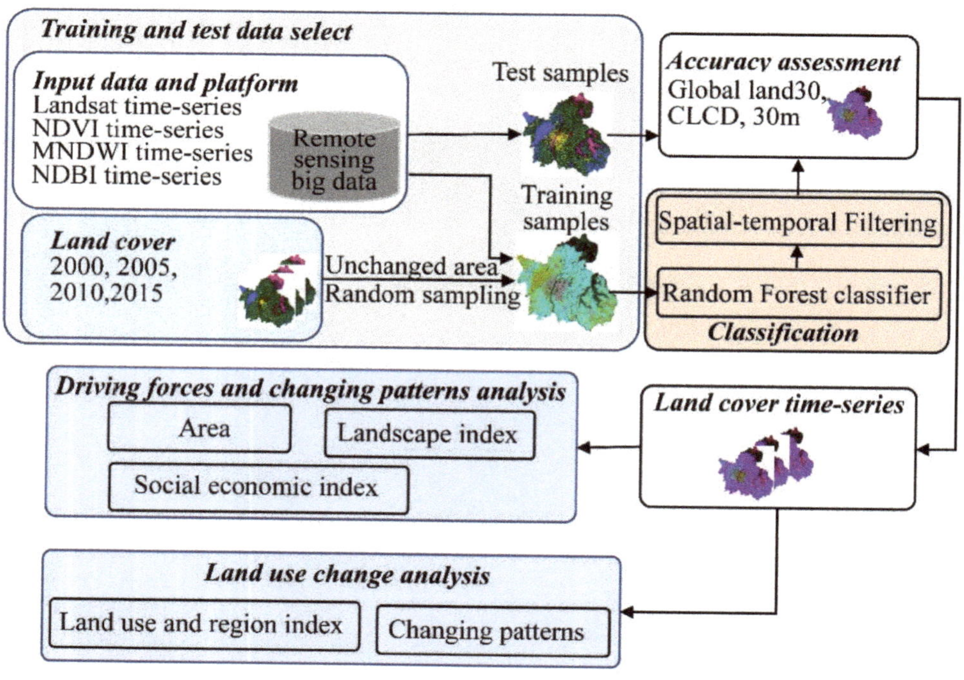

Figure 2. Research framework diagram.

2.3.1. Land Use Classification

This study uses remote sensing Big Data analysis platform to obtain time series Landsat satellite images, and then supervised and classified them by combining random forest and transfer learning methods to obtain the land use data of Nanchang City from 2000 to 2020. The land use classification process is shown in Figure 2. Firstly, we used the 4 periods of land use data (2000, 2005, 2010, 2015) released by the Institute of Geographical Sciences and Resources, Chinese Academy of Sciences (https://www.resdc.cn/Default.aspx, accessed on 27 March 2021) to randomly collect unchanged stable regions form the 4 periods of land use data as training samples for supervised classification, and made them into sample data labels. Then, we chose seven types of land use samples: cropland, forest, grassland, water, barren and urban. Secondly, we screened the Landsat dataset of 2000–2020 using RSBDP, calculated NDVI, MNDWI and NDBI and used the results as the three bands. We then exported them as a new dataset. After that, the sample data labels were imported into the RSBDP platform and were extracted for the sample dataset to train the Random Forest classifier. We used the trained Random Forest classifier to supervise and classify the Landsat dataset from 2000 to 2020. Finally, we implemented spatial–temporal filtering and accuracy assessment on the classification results, and if the classification results did not meet the accuracy requirements, the sample labels were modified. The classification operation was repeated until the classification accuracy was more than 80%, and then the classification was finished. In this study, the classification method was the same as the existing literature, and more detail can be found in the literature [32].

2.3.2. Land Use Comprehensive Index

The quantitative study of land use degree is based on its limit, that is, when land resource utilization reaches the limit, human beings will not able to continue developing land resources. The lower limit of land resource use is the starting point for human exploitation of land resources [33]. The magnitude of the composite index reflects the level of land use. In order to quantitatively evaluate the degree of land use, the existing research classifies the ideal state of land use into four levels, and gives corresponding weights to the four levels of land use, so as to achieve a quantitative evaluation of the land use degree. This study uses weight assignment of land use degree from the existing studies to calculate the comprehensive index of land use in Nanchang (Table 1).

Table 1. The classification values of land use degree.

Type of Land Use	Barren	Forest, Grassland, Water	Cropland	Urban
Classification index	1	2	3	4

The comprehensive index of land use is a Weaver index, and its calculation model is shown below [33]:

$$L_i = 100 \times \sum_{j=1}^{n} A_j \times P_j \quad (1)$$

where L_i is the comprehensive index of land use with values in the range (100, 400); A_j is the graded index of high land use at level j and P_j is the percentage of the graded area of land use degree at level j.

2.3.3. Relative Change Rate of Land Use

In order to quantitatively analyze the different characteristics of land use change in each district and county of Nanchang City, we introduced the index of land-use relative change rate. The relative change rate of land-use type change is built on the basis of the change rate index. It is the ratio of the change rate of land use type in target areas to the change rate of land use type change in the study area, which is a hot spot area for analyzing differences in specific land use type change and specific land use type change within the study area [34]. The calculation model of the relative change rate of land use is as follows:

$$\mathrm{K} = \frac{|K_{t+1} - K_t| \times C_t}{K_t \times |C_{t+1} - C_t|} \quad (2)$$

where K_{t+1}, K_t are the area of a particular land use type in the target areas at moment t+1 and t, respectively; C_{t+1}, C_t are the area of the land use type in the study area at moment t+1 and t, respectively. The significance of the absolute values is to avoid confusion caused by the direction of land use change and to facilitate comparison between target areas. The significance of the relative change rate is to reveal differences in the land use change of study areas [35].

2.3.4. Landscape Index Analysis

In order to deeply understand the spatial variation characteristics of land use, we selected four landscape-scale indicators, Shannon's Evenness Index (SHEI), Patch Density (PD), Contagion Index (CONTAG) and Number of Patches (PD) for the study. We also chose Total Landscape Area (CA), Largest Patch Index (LPI), Patch Cohesion Index (COHESION) and Splitting Index (SPLIT) to quantitatively evaluate the spatial pattern of land use in Nanchang City. The above indicators were calculated using Fragstats version 4.2 (Gen. Tech. Rep. PNW-GTR-351. Portland), a landscape pattern analysis software.

3. Results and Discussion

3.1. Land Use Change

The results showed the overall accuracy of land use was 88.36%, and the Kappa index was 42.49%. From 2000 to 2020, the spatial distribution of land use in Nanchang City has changed significantly (Figure 3). Among them, urban expansion has been significant, especially in the central city of Nanchang and Honggutan. Cropland has always been the largest land use type in Nanchang.

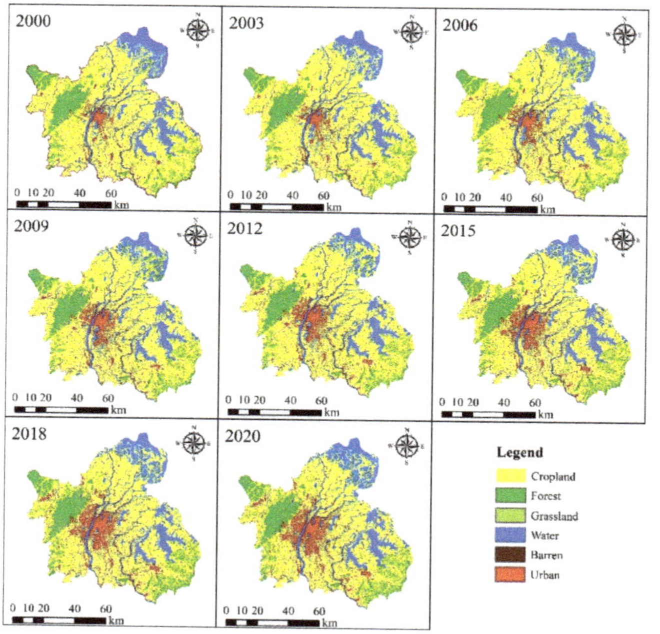

Figure 3. Nanchang City Land Use Map 2000–2020.

In the last 21 years, the area of cropland in Nanchang has decreased by 2.94%, from 66.74% to 63.80% (Figure 4a). The loss of cropland area has mainly devolved into urban (51.49%), forest (28.86%) and water (19.15%) areas (Figure 4b). In order to protect the ecological environment, the Chinese government has implemented the project of "returning farmland to lakes" in the middle and lower reaches of the Yangtze River. Therefore, part of the cultivated land around Poyang Lake has become wetland. Urban expansion was the main factor leading to the loss of cropland, which was consistent with the conclusions of existing studies that the urbanization process is the main factor leading to the massive loss of cropland. From 2000 to 2020, the urban area has increased from 379.15 km^2 to 722.66 km^2, an increase of 90.60% (Figure 4a; Table 2). Of this increase, 87.25% of the urban area from cropland to urban, 8.09% from water areas and 3.10% from forest (Figure 4b). It is worth noting that the forest area of Nanchang City has increased from 802.25 km^2 in 2000 to 842.67 km^2 in 2020, an increase of 5.04%. In addition, Grassland, Water and Barren areas have decreased by 4.58 km^2 (48.82%), 160.52 km^2 (12.75%) and 1.22 km^2 (10.40%), respectively (Table 2). Urban expansion has been the main reason for the decrease in grassland area, with 2.81 km^2 of grassland converted to urban. On the other hand, Barren areas have decreased by 1.22 km^2 from 2000 to 2020, of which 0.91 km^2 has been converted to Grassland. Most of it has transformed into grassland in urban parks. A total of 2.90 km^2 of Barren areas have been converted into urban. Therefore, urban expansion has made dramatic changes in Barren areas.

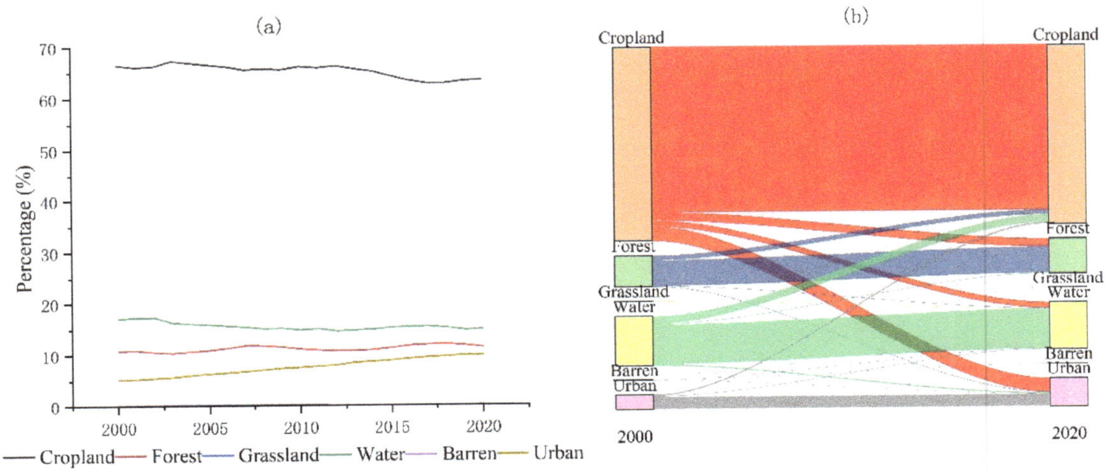

Figure 4. Land use statistics and change map for Nanchang City 2000–2020. (**a**) Summary of the area of six types of land use; (**b**) The 2000–2020 land use transfer map.

Table 2. Land use transfer matrix (unit: square kilometers).

2000 \ 2020	Cropland	Forest	Grassland	Water	Barren	Urban
Cropland	4320.89	122.25	1.96	266.18	0.96	9.90
Forest	178.63	654.66	0.13	8.87	0.00	0.37
Grassland	1.10	1.27	0.65	0.80	0.91	0.07
Water	118.48	8.10	3.50	952.10	4.93	11.26
Barren	2.02	4.63	0.33	1.39	2.06	0.09
Urban	318.62	11.33	2.81	29.54	2.90	357.46

Figure 5 shows the variation trend of urban, cropland and forest areas in the nine administrative districts of Nanchang. The urban expansion in Honggutan was the most significant, increasing from 13.48 km^2 to 63.19 km^2, an increase of 368.75% (Figure 5a), followed by Qingshanhu and Xinjian County, with an increased urban area of 92.76% and 74.02%, respectively. Honggutan has been under construction since 2000 and has experienced a high growth period from 2013 to 2016, with an average annual expansion of 5.18 km^2. The district with the smallest urban area expansion was Donghu (15.06%), followed by Qingyunpu (44.39%), Jinxian (51.68%) and Xihu (54.68%). Donghu, Qingyunpu and Xihu were the central urban areas of Nanchang City with a high degree of land development, and therefore the urban area has increased less in the past 21 years. Jinxian was a suburban administrative district of Nanchang City and was located at the periphery of the central urban area, with slower social and economic development, hence its urban area was expanding slowly. The largest urban area of Nanchang in 2020 was Nanchang county (208.47 km^2), followed by Xinjian county (133.34 km^2) and Qingshanhu (111.87 km^2). The smallest urban area was Donghu (21.82 km^2), Xihu (24.79 km^2) and Qingyunpu (27.50 km^2). In the last 21 years, the highest rate of cropland loss has been found in Xihu (51.37%), followed by Qingyunpu (49.68%) and Qingshanhu (34.04%) (Figure 5b). Among the nine districts and counties in Nanchang, only Xinjian County has increased its cropland by 4.58%, while the other eight districts and counties all have lost cropland. In addition, the water and grassland areas of Xinjian have decreased by 20.02% and 70.78%, respectively. Urban expansion was the main factor in cropland and ecological land loss. From 2000 to 2020, the forest area of Qingyunpu decreased by 97.33%, followed by Donghu (95.00%) and Xihu (54.72%) (Figure 5c). At the end of 2020, the total forest area of Qingyunpu, Donghu and Xihu was less than 0.01 km^2, except for Jinxian county and Anyi county, whose forest area increased by 34.04% and 9.29%, respectively; the forest areas of the other seven districts and counties all decreased to different degrees (Figure 5c).

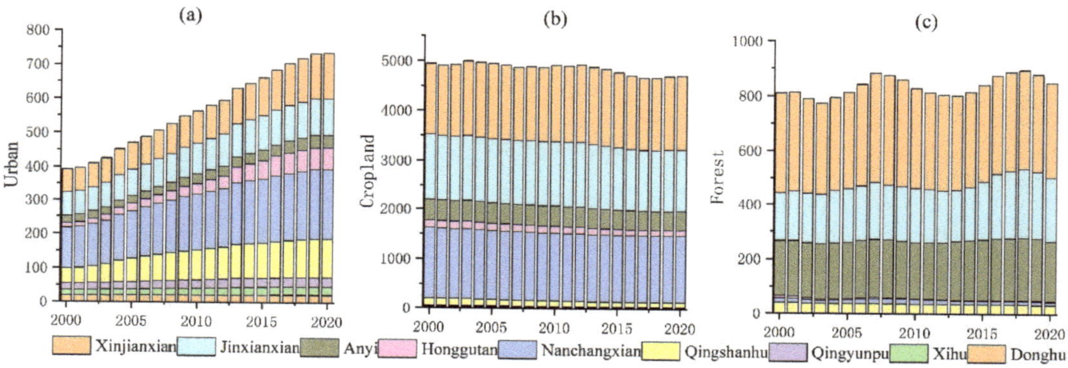

Figure 5. Area of urban, cropland, forest changing patterns (unit: km^2). (**a**) changing patterns of urban area from nine counties; (**b**) changing patterns of cropland area from nine counties; (**c**) changing patterns of forest area from nine counties.

In the last 21 years, cropland, forest and urban land have occupied the largest areas in Nanchang city. Among them, urban land has an average annual growth rate of 3.21%, which was significantly higher than cropland (−0.22%) and forest (0.27%) (Figure 6a,b). Among them, the largest annual change rate of Nanchang was in 2004 (6.48%), followed by 2013 (5.95%) and the smallest annual change rate was in 2020 (0.16%) (Figure 6a). The annual growth rate of urban land has been generally higher than forest and cropland from 2000 to 2020. It has a small fluctuation, and the city showed a stable increasing trend (Figure 6b). The abnormal value is the urban increase rate from 2003 to 2004, during which Nanchang City had the largest growth rate of fixed assets' investment (67.9%). Thus, the rapid growth of fixed assets investment may be one of the important factors for the rapid urban expansion (Figure 6b). It is worth noting that the distribution of the annual change rate of the forest area has large dispersion and high fluctuation. This may be due to the uncertainty of the land use classification method and classification accuracy. The median line of the annual change rate in cropland is negative and less volatile, indicating that cropland change is dominated by loss and has a stable trend.

From 2000 to 2020, the comprehensive index of land use in Nanchang City has increased from 276.86 to 283.19, indicating a significant increase in land use degree in Nanchang City (Figure 7a). Compared with the non-time series, time series land use shows more details of land use change. For example, the comprehensive index of land use in

Nanchang City decreased from 2013 to 2017, and land use was in a period of decline. The comprehensive index of land use increased slowly from 2005 to 2007, indicating that the land use of Nanchang was in the adjustment period. The Pearson Correlation Analysis was employed in this study, and the results show that the correlation coefficients of the comprehensive index of land use and GDP per capita, population and fixed assets investment were 0.92, 0.90 and 0.85, respectively, at 0.01 significance level (Figure 7b). This indicates that the change in the degree of land use is influenced by GDP per capita, population and fixed assets investment. GDP per capita is an important indicator of social affluence, and GDP per capita is an important influencing factor for the increase in land use degree.

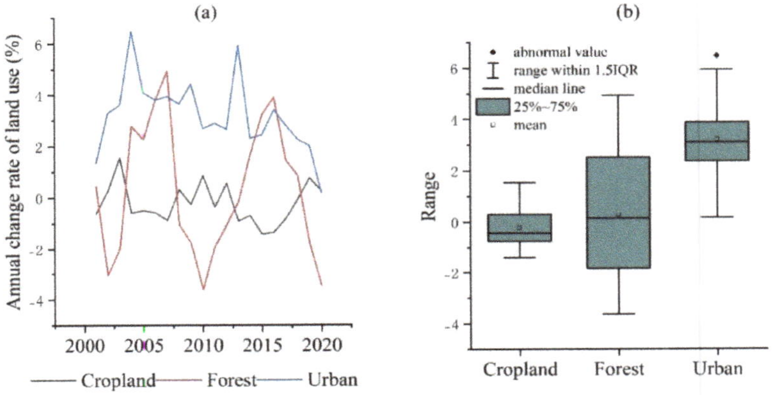

Figure 6. Cropland, forest, urban annual rate of change of land use (%). (**a**) Cropland, forest, urban annual change rate (%); (**b**) Box plot of the annual change rate.

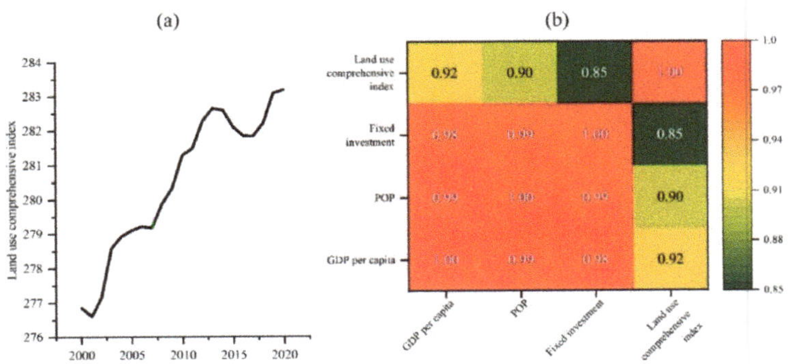

Figure 7. Comprehensive index of land use and its social economic-correlation. (**a**) Comprehensive index of land use from 2000 to 2020; (**b**) Correlation coefficient between the comprehensive index of land use and social-economic indicators, all the $p < 0.01$.

After normalizing the data, this study established a multiple regression model of land use composite index and GDP, and the results showed that the increase in per capita GDP and population was an important factor to promote land intensive use (Table 3). At the same time, the rapid increase in GDP and fixed asset investment will lead to extensive development of land use. In addition, the R squared of the regression model is 0.94.

Cropland was the largest land-use type and urban land was the land use type with the largest change rate in Nanchang. Then, cropland and urban land were employed to discover the regional differences in the changes. The results show that there were significant regional differences between the change in cropland and urban areas in the nine districts and counties of Nanchang (Figure 8). The relative change rate of cropland in Qingyunpu was the largest in 2008 (28.07), and the relative change rate fluctuated widely around the average, followed by Xihu (Figure 8a). The relative change rates of cropland in Qingyunpu and Xihu districts were significantly higher than that in the other seven districts and counties, indicating that the relative change of cropland in Qingyunpu and Xihu districts was drastic, which reflected the drastic loss of cropland and the acuity of land resource development. Moreover, cropland loss in Xihu and Qingyunpu was 51.37% and 49.68, respectively. The urban relative-change rate in Honggutan was largest in 2004 (3.81), followed by 2013 (3.67) and 2015 (3.66), and the urban relative-change rate in Honggutan was much greater than the other eight districts and counties (Figure 8b). From 2000 to 2005, the urban relative-change rate of Qingshanhu has been greater than two, second only to Honggutan and since then the urban relative-change rate of Qingshanhu has been regionally stable and less than 1.5 (Figure 8b). The urban relative-change rate indicated that Honggutan had experienced continuous and high-intensity development of land resources from 2000 to 2016, and its development intensity had been significantly higher than that of other districts and counties. Meanwhile, Qingshanhu has experienced high-intensity development of land resources and urban expansion from 2000 to 2005. The relative change rates of urban land in Anyi County and Xinjian County have increased gradually in the past 20 years, and the intensity of urban expansion has been strengthened. Xinjian County, on the other hand, began experiencing high-intensity development of land resources in 2017, but its development intensity was still lower than that of the earlier development intensity in Honggutan. Therefore, Honggutan was the key area of Nanchang's urban expansion in the past 20 years.

Table 3. Regression coefficient.

	Regression Coefficient	Standard Error	Significance
GDP	−2.06	2.06	0.00
GDP per capita	3.26	1.19	0.00
POP	0.39	0.54	0.00
Fixed investment	−0.77	0.98	0.00
Intercept	0.08	0.05	0.01

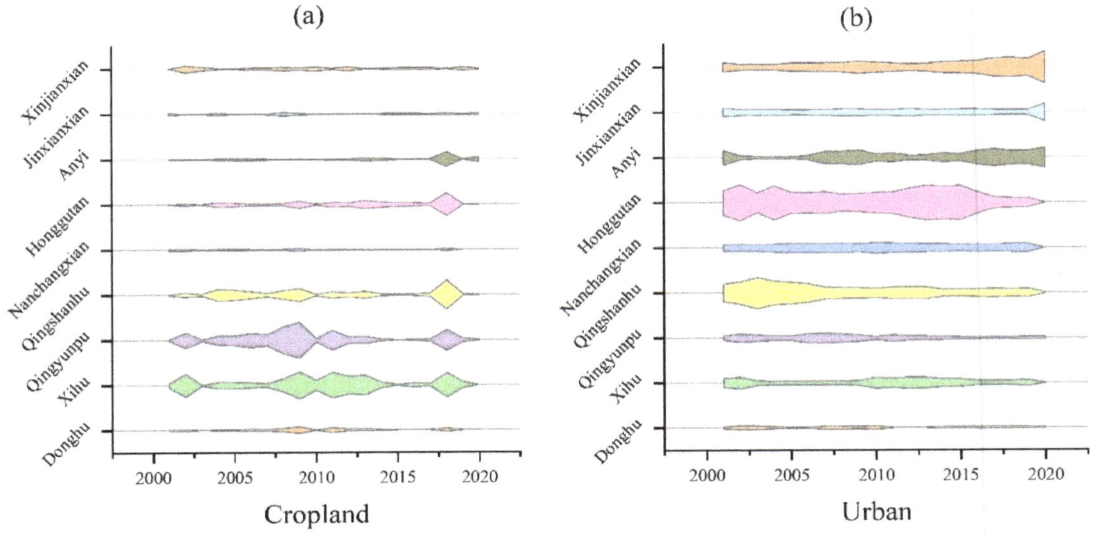

Figure 8. Kite diagram of relative change rates in cropland and urban land use areas. (**a**) the relative change rates in cropland; (**b**) relative change rates in urban.

3.2. *Characteristics of Land Use Pattern Change*

The Shannon's Evenness Index (SHEI) at the landscape scale of land use in Nanchang increased by 7.56%, showing a fluctuating increasing trend (Figure 9a). It indicates that the diversity of land use in Nanchang City has increased and the patches' distribution uniformity of each land use type has also increased. The Patch Density (PD) index has gone through two stages of increasing from 2000 to 2008 and decreasing from 2009 to 2020, and the PD in 2020 is 9.20% higher than that in 2000 (Figure 9b). It indicates that the process of land use change in Nanchang City has gone through two stages of increasing and then decreasing in landscape heterogeneity, which also shows the increasing and decreasing process of human activities. Contagion Index (CONTAG) has decreased by 4.66% from 2000 to 2020, with an average annual growth rate of −0.24%, this means that urban expansion leads to an increase in landscape fragmentation and the deterioration of landscape continuity (Figures 8a and 9c). The number of patches (NP) has increased by 9.20% from 2000 to 2020, indicating an increase in the number of patches of land use in Nanchang City and an increase in landscape fragmentation. In summary, land use in Nanchang City has gone through dramatic changes in the past 21 years. Urbanization and intensified human activities have accelerated the landscape fragmentation of land use spatial distribution and the spatial heterogeneity has significantly increased.

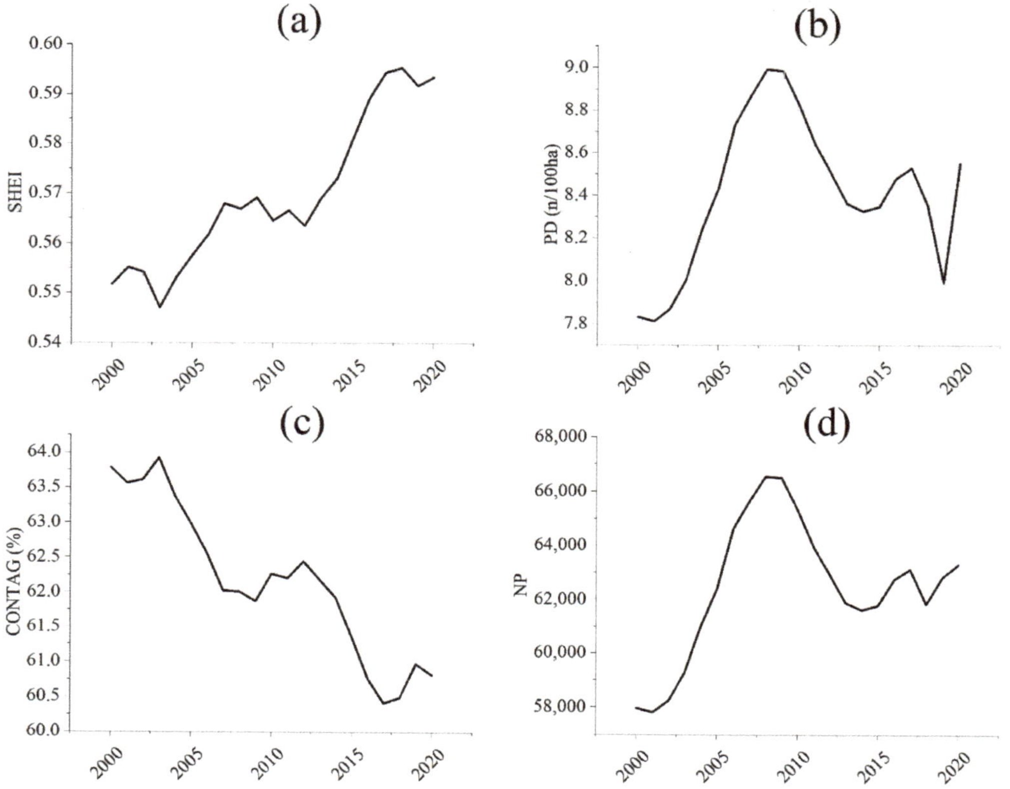

Figure 9. Landscape index change of land use in Nanchang, 2000–2020. (**a**) the Shannon's Evenness Index (SHEI) of Nanchang from 2000 to 2020; (**b**) the Patch Density (PD) index of Nanchang from 2000 to 2020; (**c**) the Contagion Index (CONTAG) of Nanchang from 2000 to 2020; (**d**) the number of patches (NP) of Nanchang from 2000 to 2020.

With the social economy developing, urbanization and increased human activities have become the main factors of dramatic land-use changes. Urban land was the only type of land use that has increased substantially among the six land-use types. Therefore, in order to deeply understand the variation of urban land-use types, this study analyzed the change characteristics of urban land in Nanchang. Existing studies are limited by the difficulty of obtaining land use data and untimely updates and are only able to analyze land use at discontinuous times. Unlike existing studies, this study used spatial–temporal Big Data technology of remote sensing to obtain continuous time data of land use rapidly and reliably, and combined it with time series statistics for comprehensive analysis to explore the change characteristics and driving mechanisms of land use.

The land use with continuous time series of Nanchang City indicated that the urban area has expanded by 87.81% with an average annual growth rate of 3.21% (Figure 10a). The Largest Path Index (LPI) of urban patches increased by 186.49% with an average annual growth rate of 5.48% (Figure 10b), and the largest patch was the central city of Nanchang, indicating a significant trend of urban expansion in the central city. It is noteworthy that the largest LPI growth rate was in 2002 and 2008 at 15.49% and 15.89%, respectively, which implied significant urban expansion. The fixed assets' investment in Nanchang City increased significantly in 2002 and 2008, 41.40% and 34.00% higher than in 2003 and 2009, respectively, and continued to grow at a high rate for four and three years from then on.

Thus, we have reason to believe that the increase in fixed assets investment contributed to the rapid increase in LPI of the urban patches and urban expansion (Table 4). Existing studies also show that economic development is directly related to the urban landscape pattern [36,37]. The regression coefficient of LPI with GDP, GDP per capita and fixed investment was calculated, and the total R square is 0.99 (Table 4). This shows that POP and fixed investments are the key factors influencing the change of LPI.

Table 4. Regression coefficient of LPI with GDP, GDP per capita and fixed investment (R square is 0.99).

	Regression Coefficient	Standard Error	Significance
Intercept	0.01	0.02	0.43
GDP	−3.79	0.67	0.00
GDP per capita	3.50	0.39	0.00
POP	0.66	0.18	0.00
Fixed investment	0.61	0.32	0.01

The Patch Cohesion Index (COHESION) reflects the degree of connection between urban patches, and the larger the value, the stronger the spatial connection between patches. The COHESION of Nanchang City has increased by 3.30% from 2000 to 2020, with an average annual growth rate of 0.16%. Among them, the average annual growth rate from 2000 to 2008 was 0.33% and from 2009 to 2020 was 0.05% (Figure 10c). This indicates that urban areas have expanded rapidly from 2000 to 2008, and there was a strong spatial connection of urban expansion. This implies that the direction of urban expansion is closely related to spatial factors such as the distance of existing urban land space. The spatial relationship between urban expansion and the existing urban area has weakened in 2009–2020, implying that urban expansion was mainly punctiform and there was no strong spatial linkage between patches. Moreover, the Splitting Index (SPLIT) is one of the important indicators to measure the degree of landscape separation, and the larger the value, the higher the separation degree between landscape patches and the more fragmented the spatial distribution. The SPLIT of urban land-use types in Nanchang City has decreased by 90.48% from 2000 to 2020, indicating that the separation degree of urban land in Nanchang City has decreased and the fragmentation degree of urban spatial distribution has also decreased, which means that the aggregation degree of urban spatial distribution has increased (Figure 10d).

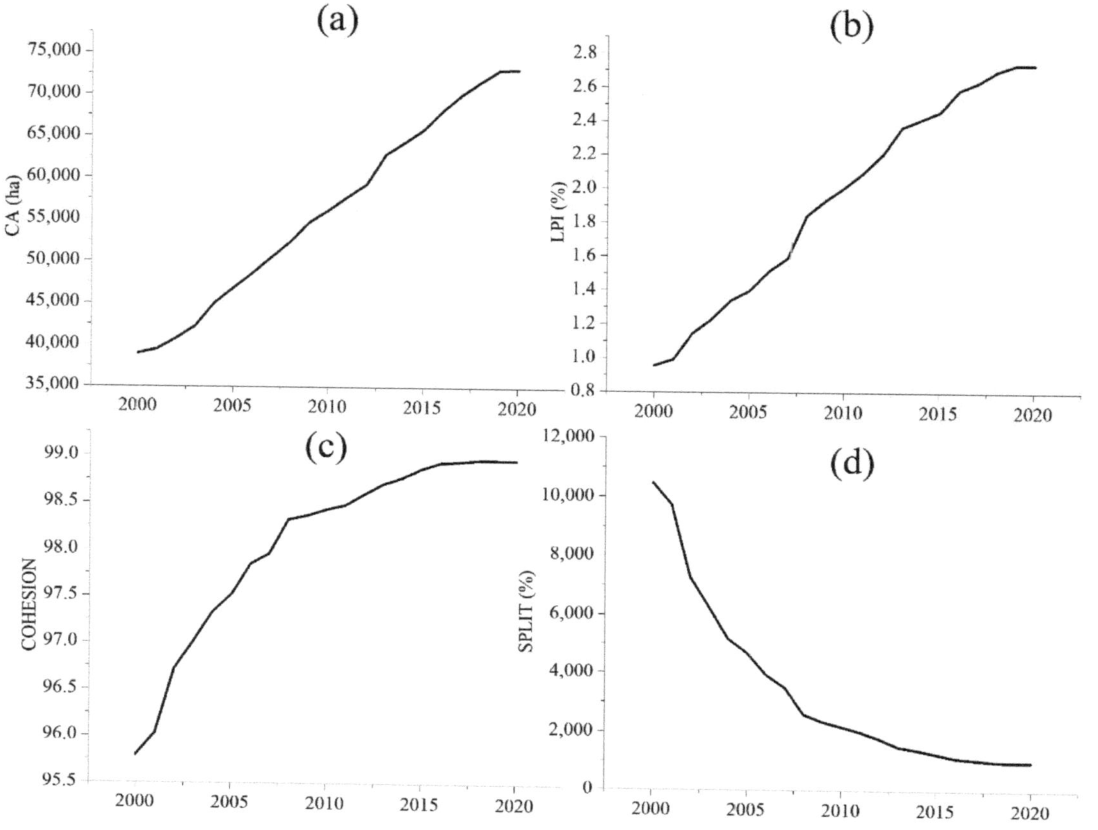

Figure 10. Changes in urban landscape pattern index. (**a**) urban landscape area; (**b**) the Largest Path Index (LPI) of urban patches; (**c**) the Patch Cohesion Index (COHESION) of urban; (**d**) Splitting Index (SPLIT) of urban.

There were four hot spots of land use change (HSLUC) in Nanchang, which were located in the northeast, northwest, central and southeast of Nanchang (Figure 11). Only the HSLUC in northeast China was formed because the project of converting cropland to lakes transformed cropland into wetland. Urban expansion is the main reason for the formation of the other three HSLUC. Among them, the central urban area of Nanchang city was the main HSLUC, its urban area expansion was the largest, and the land use change was the most drastic.

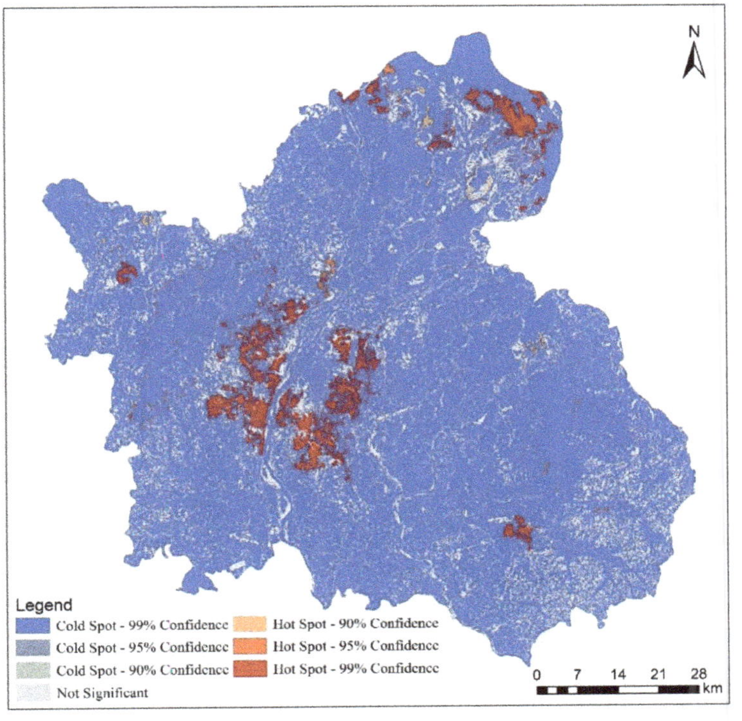

Figure 11. Hot spot of land use change from 2000 to 2020.

3.3. Uncertainty

This study used remote sensing spatial–temporal Big Data and machine learning methods to obtain continuous time-series data of land use and analyzed the spatial–temporal pattern change of land use in Nanchang City with time series statistical data. We bridged the gap of time series land-use research based on a statistical perspective. However, there are still uncertainties in the results of this study, mainly concerning the uncertainty of the method and the uncertainty of the data accuracy.

One of the main uncertainties of the method is the limitations in the selection of training datasets, which cannot extract all features and use them for model training. As a result, there is uncertainty in the accuracy of the model. Furthermore, the land use data have their uncertainties of accuracy as the third party of sample data, and we cannot guarantee that the precision of the sample data is completely accurate. Data errors may be passed on to the classification data of land use in this study, leading to uncertainty in the results of this study. In the subsequent study, we can improve the classification method of land use and select land use data with higher accuracy and credibility as the training sample data to reduce the uncertainty of the study results.

4. Conclusions

Land use in Nanchang City has changed dramatically from 2000 to 2020. Among them, the urban area has increased enormously, and the area of cropland, on the other hand, has decreased significantly. Meanwhile, most of the reduced cropland area has been transformed into urban land. This study provides more evidence for the loss of cropland caused by urban expansion. In addition, there were significant differences in land use changes among the nine administrative units in Nanchang. Among them, Honggutan showed dominance in urban expansion, while the old city showed a significant loss of forest areas which were converted into urban use. Over the past 21 years, land use changes in Nanchang have been dominated by urban expansion, which led to the loss of cropland and forest. The spatial pattern of land use has changed significantly, and the urban spatial pattern has become more aggregated, while the spatial distribution of cropland, forest and grassland has become more fragmented. The landscape pattern of land use has been more fragmented, and the city has expanded rapidly. Moreover, there was a significant correlation between social–economic development and land use level, and the social economy was obviously correlated with the landscape patterns of land use. Population, GDP and fixed assets investment were the three main drivers of land use change. They affected the area change of land use types, as well as the form and pattern of land use spatial patterns, and further affected the development of the city. Therefore, understanding the characteristics of land use change can help to clarify the driving mechanisms of land use change and provide support for government decisions and the rational and efficient use of land resources.

Author Contributions: Y.L.: Formal analysis, writing-original draft, methodology, data curation; C.H.: Conceptualization, writing–review and editing; L.Z.: Methodology, writing–review. All authors have read and agreed to the published version of the manuscript.

Funding: This study was supported by the National Natural Science Foundation of China (grant number 32260418), Social Science Foundation of Jiangxi Province (grant number 22YS08), Science and Technology Project of Education Department of Jiangxi Province (grant number GJJ200412), Shanghai Key Lab for Urban Ecological Processes and Eco-Restoration Open Fund and the Basic Scientific Research Funds of Central Universities funded (SHUES2022A10).

Institutional Review Board Statement: Not applicable.

Informed Consent Statement: Not applicable.

Data Availability Statement: The data are from the Nanchang Statistical Yearbook, the China Geographic Information Bureau, the Resource and Environment Science and Data Center, and Google Earth Engine platform.

Conflicts of Interest: The authors declare no conflict of interest.

References

1. Nasiri, V.; Deljouei, A.; Moradi, F.; Sadeghi, S.M.M.; Borz, S.A. Land use and land cover mapping using sentinel-2, landsat-8 satellite images, and google earth engine: A comparison of two composition methods. *Remote Sens.* **2022**, *14*, 1977. [CrossRef]
2. Aquilué, N.; De Cáceres, M.; Fortin, M.; Fall, A.; Brotons, L. A spatial allocation procedure to model land-use/land-cover changes: Accounting for occurrence and spread processes. *Ecol. Modell.* **2017**, *344*, 73–86. [CrossRef]
3. Vlad Sandru, M.I.; Iatu, C.; Andru, D.; Cimbru, D. Approaching land cover-land use changes using statistical data validation for urban policies improvement. *J. Settl. Spat. Plan.* **2017**, *8*, 119–129. [CrossRef]
4. Wang, Q.; Wang, H.; Chang, R.; Zeng, H.; Bai, X. Dynamic simulation patterns and spatiotemporal analysis of land-use/land-cover changes in the Wuhan metropolitan area, China. *Ecol. Model.* **2022**, *464*, 109850. [CrossRef]
5. Purvaja, R.; Ramesh, R.; Glavovic, B.; Ittekkot, V.; Samseth, J. Regional initiatives for interlinking global coastal scientific research projects. *Environ. Dev.* **2015**, *14*, 66–68. [CrossRef]
6. Wang, S.; Huang, S.; Budd, W.W. Integrated ecosystem model for simulating land use allocation. *Ecol. Model.* **2012**, *227*, 46–55. [CrossRef]
7. Tarawally, M.; Wenbo, X.; Weiming, H.; Mushore, T.D.; Kursah, M.B. Land use/land cover change evaluation using land change modeller: A comparative analysis between two main cities in Sierra Leone. *Remote Sens. Appl. Soc. Environ.* **2019**, *16*, 100262. [CrossRef]
8. Lagrosa, J.J.; Zipperer, W.C.; Andreu, M.G. Projecting land-use and land cover change in a subtropical urban watershed. *Urban Sci.* **2018**, *2*, 11. [CrossRef]
9. Sİyavuş, A.E. Changes in land use and land cover of Düzce province (1990–2018). *Coğraf. Derg.* **2021**, *42*, 121–138. [CrossRef]
10. Alberti, M.; Marzluff, J.M.; Musacchio, L.R.; Wu, J. Ecological resilience in urban ecosystems: Linking urban patterns to human and ecological functions. *Urban Ecosyst.* **2004**, *7*, 241–265. [CrossRef]
11. Zhang, L.; Lei, J.; Li, X.; Gao, C.; Zeng, W. The features and influencing factors of urban expansion in China during 1997–2007. *Prog. Geogr.* **2011**, *30*, 607–614. [CrossRef]
12. Ah Koo, K.; Uk Park, S. The effect of interplays among climate change, land-use change, and dispersal capacity on plant redistribution. *Ecol. Indic.* **2022**, *142*, 109192. [CrossRef]

13. Barlow, C.M.; Pellatt, M.G.; Kohfeld, K.E. Garry oak ecosystem stand history in southwest British Columbia, Canada: Implications of environmental change and indigenous land use for ecological restoration and population recovery. *Biodivers. Conserv.* **2021**, *30*, 1655–1672. [CrossRef]
14. Vojvodikova, B.; Ticha, I.; Fojtik, R.; Jupova, K. Land use changes and effects on heat islands in the city. *IOP Conf. Ser. Earth Environ. Sci.* **2020**, *444*, 12056. [CrossRef]
15. Chen, L.; Yang, X.; Kang, J. *A Case Study of Land Use Planning Environmental Assessment Based On the Air Pollution Analysis*; Lee, G., Ed.; Springer: Berlin/Heidelberg, Germany, 2012; pp. 319–328.
16. Gomes, E.; Inácio, M.; Bogdzevič, K.; Kalinauskas, M.; Karnauskaitė, D.; Pereira, P. Future scenarios impact on land use change and habitat quality in Lithuania. *Environ. Res.* **2021**, *197*, 111101. [CrossRef] [PubMed]
17. Qiu, B.; Li, H.; Tang, Z.; Chen, C.; Berry, J. How cropland losses shaped by unbalanced urbanization process? *Land Use Policy* **2020**, *96*, 104715. [CrossRef]
18. Liu, Y.; Zou, R.; Dong, Y. Analysis of temporal and spatial characteristics of urban expansion in Xiaonan district from 1990 to 2020 using time series landsat imagery. *Remote Sens.* **2021**, *13*, 4299. [CrossRef]
19. Dai, M. A review of urban spatial expansion at home and abroad. *Mark. Res.* **2020**, *5*, 16–18. [CrossRef]
20. Liu, J.; Zhan, J.; Deng, X. Spatio-temporal patterns and driving forces of urban land expansion in china during the economic reform era. *AMBIO J. Hum. Environ.* **2005**, *34*, 450–455. [CrossRef]
21. Godone, D.; Garbarino, M.; Sibona, E.; Garnero, G.; Godone, F. Progressive fragmentation of a traditional Mediterranean landscape by hazelnut plantations: The impact of cap over time in the Langhe region (nw Italy). *Land Use Policy* **2014**, *36*, 259–266. [CrossRef]
22. Xie, Y.; Weng, Q. Updating urban extents with nighttime light imagery by using an object-based thresholding method. *Remote Sens. Environ.* **2016**, *187*, 1–13. [CrossRef]
23. Huang, C.; Zhuang, Q.; Meng, X.; Guo, H.; Han, J. An improved nightlight threshold method for revealing the spatiotemporal dynamics and driving forces of urban expansion in China. *J. Environ. Manag.* **2021**, *289*, 112574. [CrossRef] [PubMed]
24. Stathakis, D.; Tselios, V.; Faraslis, I. Urbanization in European regions based on night lights. *Remote Sens. Appl. Soc. Environ.* **2015**, *2*, 26–34. [CrossRef]
25. Zhou, Y.; Smith, S.J.; Elvidge, C.D.; Zhao, K.; Thomson, A.; Imhoff, M. A cluster-based method to map urban area from DMSP/OLS nightlights. *Remote Sens. Environ.* **2014**, *147*, 173–185. [CrossRef]
26. Oszczak, B.; Matuszak, J.; Cwiklak, J. Decoding of GPS data for single point positioning computation by using python programming language. *Int. Multidiscip. Sci. GeoConf. SGEM* **2017**, *17*, 193–202. [CrossRef]
27. Hanberry, B.B. Classifying large wildfires in the united states by land cover. *Remote Sens.* **2020**, *12*, 2966. [CrossRef]
28. Inman, V.L.; Lyons, M.B. Automated inundation mapping over large areas using landsat data and google earth engine. *Remote Sens.* **2020**, *12*, 1348. [CrossRef]
29. Owen, H.J.F.; Duncan, C.; Pettorelli, N. Testing the water: Detecting artificial water points using freely available satellite data and open source software. *Remote Sens. Ecol. Conserv.* **2015**, *1*, 61–72. [CrossRef]
30. Li, J.; Wang, J.; Zhang, J.; Liu, C.; He, S.; Liu, L. Growing-season vegetation coverage patterns and driving factors in the China-Myanmar economic corridor based on Google Earth engine and geographic detector. *Ecol. Indic.* **2022**, *136*, 108620. [CrossRef]
31. Becker, W.R.; Ló, T.B.; Johann, J.A.; Mercante, E. Statistical features for land use and land cover classification in Google Earth engine. *Remote Sens. Appl. Soc. Environ.* **2021**, *21*, 100459. [CrossRef]
32. Yang, J.; Huang, X. The 30 m annual land cover dataset and its dynamics in China from 1990 to 2019. *Earth Syst. Sci. Data* **2021**, *13*, 3907–3925. [CrossRef]
33. Zhuang, D.; Liu, J. Study on the model of regional differentiation of land use degree in China. *J. Nat. Resour.* **1997**, *12*, 10–16. [CrossRef]
34. Zhu, H.; Li, X.; He, S.; Zhang, M. Spatial-temporal change of land use in Bohai rim. *Acta Geogr. Sin.* **2001**, *56*, 253–260. [CrossRef]
35. Zhu, H.; Li, X. Discussion on the index method of regional land use change. *Acta Geogr. Sin.* **2003**, *58*, 643–650. [CrossRef]
36. Mahtta, R.; Fragkias, M.; Güneralp, B.; Mahendra, A.; Reba, M.; Wentz, E.A.; Seto, K.C. Urban land expansion: The role of population and economic growth for 300+ cities. *Npj Urban Sustain.* **2022**, *2*, 5. [CrossRef]
37. Nong, D.H.; Lepczyk, C.A.; Miura, T.; Fox, J.M.; Wilson, R.K. Quantifying urban growth patterns in Hanoi using landscape expansion modes and time series spatial metrics. *PLoS ONE* **2018**, *13*, e196940. [CrossRef]

International Journal of
Environmental Research and Public Health

MDPI

Article

Spatiotemporal Dynamics of Wetland in Dongting Lake Based on Multi-Source Satellite Observation Data during Last Two Decades

Liwei Xing [1], Liang Chi [1], Shuqing Han [1], Jianzhai Wu [1], Jing Zhang [1], Cuicui Jiao [2] and Xiangyang Zhou [1,*]

1 Key Laboratory of Agricultural Blockchain Application, Ministry of Agriculture and Rural Affairs/Agricultural Information Institute, Chinese Academy of Agricultural Sciences, Beijing 100081, China
2 College of Economics, Sichuan University of Science & Engineering, Zigong 643000, China
* Correspondence: zhouxiangyang01@caas.cn

Abstract: Monitoring the dynamics of wetland resources has practical value for wetland protection, restoration and sustainable utilization. Dongting Lake wetland reserves are well known for both their intra-annual and inter-annual dynamic changes due to the effects of natural or human factors. However, most wetland monitoring research has failed to consider the seasonal wetlands, which is the most fragile wetland type, requiring more attention. In this study, we used multi-source time series remote sensing data to monitor three Dongting Lake wetland reserves between 2000 and 2020, and the seasonal wetlands were separated from permanent wetlands. Multispectral and indices time series were generated at 30 m resolution using a two-month composition strategy; the optimal features were then selected using the extension of the Jeffries–Matusita distance (J_{Bh}) and random forest (RF) importance score; yearly wetland maps were identified using the optimal features and the RF classifier. Results showed that (1) the yearly wetland maps had good accuracy, and the overall accuracy and kappa coefficients of all wetland maps from 2000 to 2020 were above 89.6% and 0.86, respectively. Optimal features selected by J_{Bh} can improve both computational efficiency and classification accuracy. (2) The acreage of seasonal wetlands varies greatly among multiple years due to inter-annual differences in precipitation and evaporation. (3) Although the total wetland area of the three Dongting Lake wetland reserves remained relatively stable between 2000 and 2020, the acreage of the natural wetland types still decreased by 197.0 km^2, and the change from natural wetland to human-made wetland (paddy field) contributed the most to this decrease. From the perspective of the ecological community, the human-made wetland has lower ecological function value than natural wetlands, so the balance between economic development and ecological protection in the three Dongting Lake wetland reserves requires further evaluation. The outcomes of this study could improve the understanding of the trends and driving mechanisms of wetland dynamics, which has important scientific significance and application value for the protection and restoration of Dongting Lake wetland reserves.

Keywords: wetland; remote sensing; dynamic; random forest; J_{Bh} extension

Citation: Xing, L.; Chi, L.; Han, S.; Wu, J.; Zhang, J.; Jiao, C.; Zhou, X. Spatiotemporal Dynamics of Wetland in Dongting Lake Based on Multi-Source Satellite Observation Data during Last Two Decades. *Int. J. Environ. Res. Public Health* **2022**, *19*, 14180. https://doi.org/10.3390/ijerph192114180

Academic Editor: Paul B. Tchounwou

Received: 28 September 2022
Accepted: 27 October 2022
Published: 30 October 2022

1. Introduction

Land cover and land use dynamics are some of the most important international research topics in Earth systems, with profound implications for natural ecosystems and human society [1,2]. Wetlands are vulnerable and sensitive to climate change, rapid population increase and fast economic development, and have been the major hotspots of monitoring in terms of land cover and land use change [3,4]. Studying the spatial distribution characteristics of wetland types and the complexity of their spatial–temporal dynamics can help us to understand the mutual influence between wetland ecosystems and natural or human factors, including global warming, environmental pollution and urban sprawl, which is conducive to the better protection and management of wetlands [5,6].

Monitoring long-term wetland dynamics by field investigations may be challenging due to poor site access and their labor-intensive and time-consuming nature [7–9]. Remote sensing images have been widely used to monitor large-scale wetland changes, with the advantage of obtaining ground information promptly and efficiently [10–12]. A series of global land cover products have mapped wetland distributions using remote sensing data, such as GlobalLand30 [13], FROM-GLC BUMODIS [14], GLCNMO2013 [15], DISCover [16] and GLC2000 [17]. However, these products do not include seasonal wetlands and cannot describe the inner-annual wetland changes.

Time series remote sensing data have shown an advantage in monitoring seasonal changes in vegetation and water [18–20]. The Advanced Very High-Resolution Radiometer (AVHRR) and Moderate-resolution Imaging Spectroradiometer (MODIS), with high temporal resolution, are always used to monitor the inner-annual dynamics of water [21–24], but the coarse spatial resolution causes misclassification and they cannot identify smaller wetlands precisely. Meanwhile, Landsat and Sentinel-2 data have a better temporal resolution, and previous research has demonstrated the potential of Landsat or Sentinel data for distinguishing wetland types [25–28]. However, there are only a few studies in which Landsat data are merged with Sentinel data to obtain an equal interval time series at 30 m spatial resolution for wetland monitoring.

Dongting Lake, a typical river-connecting lake, interacts with the Yangtze River [29]. In recent years, Dongting Lake wetlands have been changed and destroyed due to human activities and the dynamics of the hydrological regime, especially the operation of the Three Gorges Dam (TGD) [30]. There are a large number of studies monitoring wetland changes in Dongting Lake with remote sensing data [29,31]. However, most of the research is focused on flood area and water body changes, and any long-term dynamic study involving Dongting Lake wetland monitoring is usually based on a single date to identify wetlands and ignores the seasonal changes in wetlands [10,32], which leads to a lack of comparison among wetland classification results over multiple years. Dongting Lake, with smooth relief, tends to be influenced by seasonal precipitation, and may undergo significant variation among different seasons [33]. There is an urgent requirement to monitor the wetlands' intra-annual and inter-annual dynamics in Dongting Lake using satellite series time at a high spatial resolution. Therefore, the main objectives of this study are (1) to generate yearly wetland maps in the three Dongting Lake wetland reserves at 30 m spatial resolution, and (2) to analyze the seasonal and inter-annual dynamics of wetlands in Dongting Lake wetland reserves from 2000 to 2020 and reveal the motivation and regulation of these dynamics.

2. Study Area and Materials

2.1. Study Area

Dongting Lake is located in the north of Hunan Province (28°30′~30°20′ N, 110°40′~113°10′ E), and is the second-largest freshwater lake in China [29]. The lake has a subtropical monsoon climate that is characterized by a rainy season between April and September and a dry season between October and March. Dongting Lake, with smooth relief, tends to be influenced by seasonal precipitation, and may undergo significant changes over several days [34]. During the rainy season, the lake provides storage for river flood waters. During the dry season, it provides water to the river to allow river transportation to continue without significant interruption. Seasonal mudflats provide rare habitats and spawning grounds for migratory birds and fish [35]. High-frequency measurements are, therefore, needed to study short- and long-term fluctuations in the inundation areas of lakes. Dongting Lake includes three national wetland nature reserves (Figure 1), the East Dongting Lake wetland reserve, the South Dongting Lake wetland reserve and the West Dongting Lake wetland reserve, which were included in the list of Ramsar Sites in 1992 and 2002 [7]. Dongting Lake wetland reserves include various types of land cover. In this study, the classification system uses a two-tier hierarchical structure. Level 1 comprises two categories: wetlands and uplands. Level 2 comprises 7 categories: four types of natural wetlands (permanent water, permanent marsh, flooded wetland and seasonal marsh), an artificial wetland (paddy field), one type of natural upland (forest) and one type of artificial upland (construction land).

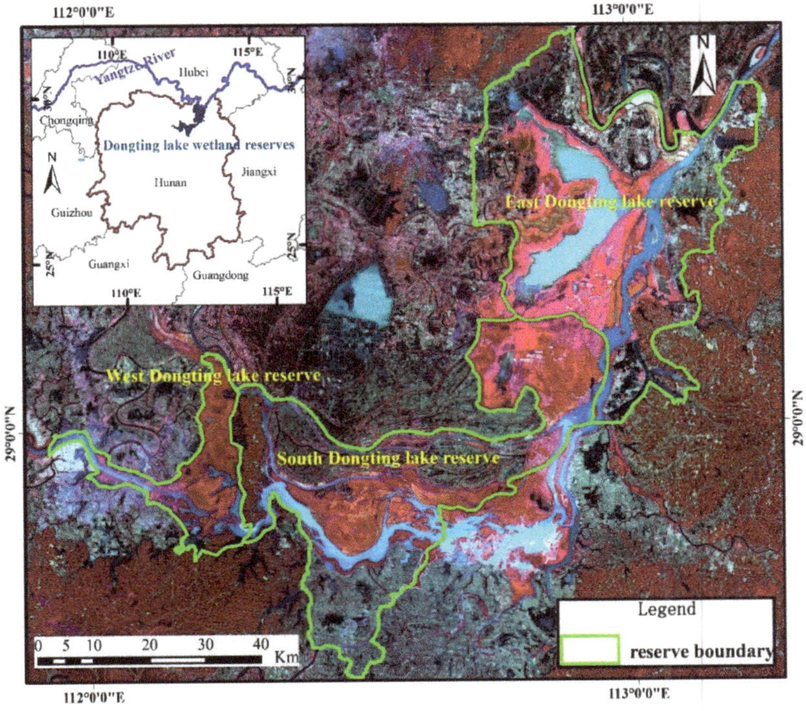

Figure 1. Location of study area.

2.2. Materials

2.2.1. Remote Sensing Data

Remote sensing data with higher temporal resolution are of key importance for mapping seasonal wetlands. A series of medium-resolution satellites at 20–30 m spatial resolution enable us to map and monitor seasonal wetlands. In this study, all Landsat-5, Landsat-7, Landsat-8 and Sentinel-2 data between 2000 and 2020 (Table 1) were prepared and preprocessed on Google Earth Engine [36,37]. In addition, the top of atmosphere (TOA) reflectance of both Landsat 5/7/8 and Sentinel-2 was utilized because surface reflectance (SR) data were not available on GEE before 2019; this is appropriate as existing studies have shown that the correlations of Landsat and Sentinel-2 for TOA reflectance data are higher than those for SR data [38,39]. Finally, 3235 images were used in total, including 551 Landsat-5 TM images, 987 Landsat-7 ETM+ images, 425 Landsat-8 OLI images and 1272 Sentinel-2 MSI images (Table 1). All data were re-sampled to 30 m, and quality assurance (QA) bands were used to mask cloud pixels. Afterwards, two-month image time series were generated using the two-month median composition for the multi-spectral bands. We used a two-month composition strategy in this study because it is the best choice to compensate for the missing data and retain the seasonal variation in the land surface. Finally, we acquired the normalized difference water index (NDWI) [40], normalized difference moisture index (NDMI) [41] and normal difference vegetation index (NDVI) [42] time series from the two-month composited images. These indices were selected because NDWI time series have been successfully used for detecting permanent and seasonal water (Equation (1)); NDMI time series have been proven to increase the separability among wetland plant varieties (Equation (2)), and they can determine the phenological parameters of wetland plants (Equation (3)).

$$\text{NDWI} = \frac{\rho(\text{Green}) - \rho(\text{NIR})}{\rho(\text{Green}) + \rho(\text{NIR})} \tag{1}$$

$$\text{NDMI} = \frac{\rho(\text{NIR}) - \rho(\text{SWIR1})}{\rho(\text{NIR}) + \rho(\text{SWIR1})} \tag{2}$$

$$\text{NDVI} = \frac{\rho(\text{NIR}) - \rho(\text{Red})}{\rho(\text{NIR}) + \rho(\text{Red})} \tag{3}$$

where ρ(Green), ρ(Red), ρ(NIR) and ρ(SWIR1) denote the TOA reflectance of the Green, Red, NIR and SWIR1 bands, respectively, which are Band2, Band3, Band4 and Band5 for Landsat-5 and Landsat-7 data; Band3, Band4, Band5 and Band6 for Landsat-8; and Band3, Band4, Band8 and Band11 for Sentinel-2 (Table 2).

Table 1. The number of images used in our study.

Number	Landsat-5	Landsat-7	Landsat-8	Sentinel-2	SUM
2000	37	44	0	0	81
2001	52	54	0	0	106
2002	38	52	0	0	90
2003	42	37	0	0	79
2004	58	47	0	0	105
2005	49	51	0	0	100
2006	50	50	0	0	100
2007	33	48	0	0	81
2008	55	55	0	0	110
2009	58	54	0	0	112
2010	38	45	0	0	83
2011	41	49	0	0	90
2012	0	38	0	0	38
2013	0	52	58	0	110
2014	0	40	47	0	87
2015	0	45	60	0	105
2016	0	48	53	0	101
2017	0	42	51	0	93
2018	0	45	58	30	133
2019	0	43	52	623	718
2020	0	48	46	619	713
SUM	551	987	425	1272	

Table 2. Introduction to bands.

Generic Name	Landsat-5	Landsat-7	Landsat-8	Sentinel-2
Blue	1 (450–520)	1 (450–520)	2 (450–510)	2 (458–522)
Green	2 (520–600)	2 (520–600)	3 (530–590)	3 (543–578)
Red	3 (630–690)	3 (630–690)	4 (640–670)	4 (650–680)
Near-Infra-Red (NIR)	4 (760–900)	4 (770–900)	5 (850–880)	8 (785–900)
Short-Wave Infra-Red 1 (SWIR1)	5 (1550–1750)	5 (1550–1750)	6 (1570–1650)	11 (1565–1655)
Short-Wave Infra-Red 2 (SWIR2)	7 (2080–2350)	7 (2090–2350)	7 (2110–2290)	12 (2100–2280)

2.2.2. Training and Validation Samples

In this study, training and validation samples were used to train the classifier and evaluate the accuracy of the classification results, respectively. Initially, 2000 geolocation sampling points were randomly generated in the study region using ArcGIS (ESRI, Redlands, SC, USA) for each year. Subsequently, the samples were visually interpreted as permanent water, permanent marsh, flooded wetland, seasonal marsh, paddy field, forest and construction land based on remote sensing images with high spatial resolution from Google Earth (http://earth.google.com, accessed on 5 January 2022) and the characteristics of the reconstructed two-month composition time series data set (NDVI, NDWI and NDMI) for each land cover in the same year. Only the samples with 120 m × 120 m pure pixels were retained (Figure 2). Finally, these samples in each year were divided into two parts for training and validation; two thirds of these samples were randomly selected and used as training samples to classify, and the remaining one third of the samples were used as validation samples to verify the accuracy of the classification results.

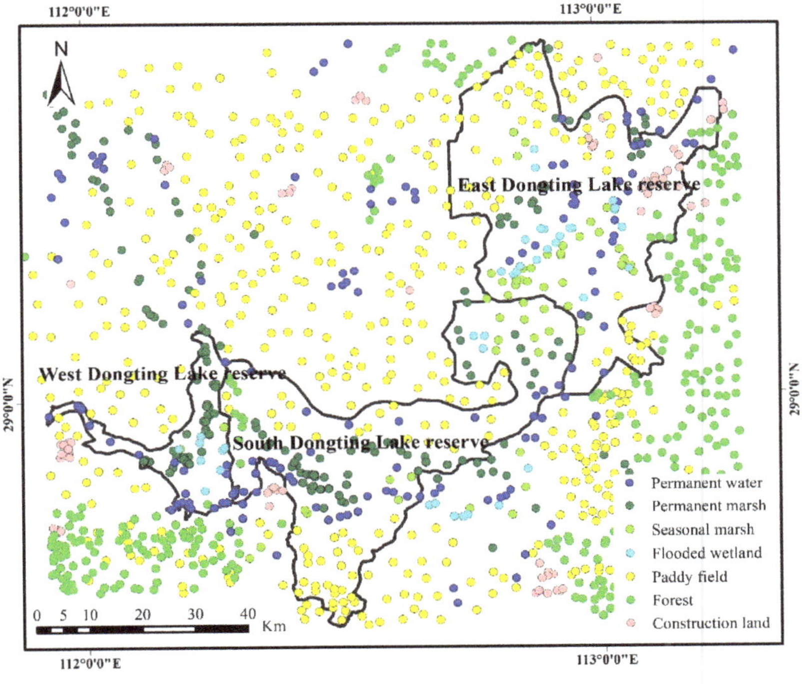

Figure 2. Distribution of samples for each land cover type in 2020.

3. Methodology

3.1. Flowchart

The framework for mapping seasonal wetlands developed in this paper is shown in Figure 3. We generated two-month time series with 54 features, including multi-spectral bands and indices (NDWI, NDMI and NDVI) for all time phases. To select the optimal features and improve the classification efficiency, importance scores for all these features were calculated based on the random forest algorithm using the training samples, and the ranking of importance scores for all 54 features were acquired. Afterwards, we calculated the separability among all wetland types using the extension of the Jeffries–Matusita (JM) distance (J_{Bh}); the number of features used increased from 1 to 54, and the sequence of adding features was based on the descending order of the importance score. In other words, the features with a higher importance score were added to the feature set, which was used for separability calculation, in higher priority. The separability among all wetland types increased with the increase in the number of features, and it was saturated when the features reached a certain amount; the features before the saturation points were then used as optimal features. Next, the RF algorithm was used to classify the wetland types based on the optimal features, and the classification accuracy was calculated to measure the classification performance. Finally, we analyzed the dynamics of wetland areas in the three Dongting Lake wetland reserves from 2000 to 2020 based on the wetland maps that we generated.

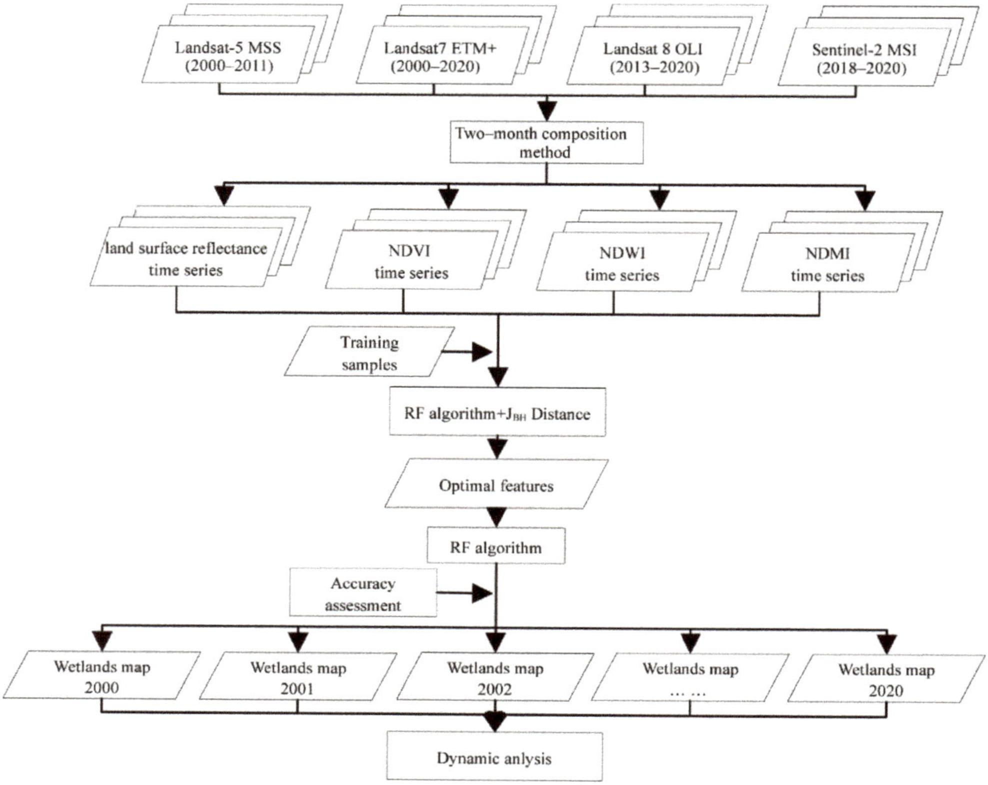

Figure 3. Flowchart of the methodology followed in this study.

3.2. Random Forest

The most commonly used classification algorithms for wetlands are maximum likelihood (ML, [43]), decision tree (DT, [44]), random forest (RF, [44]), and support vector machine (SVM, [45]). In our study, the RF algorithm was chosen for wetland mapping. It has high efficiency and flexibility and is suitable for large datasets. RF is an ensemble-based machine learning algorithm that combines a set of Classification and Regression Trees (CARTs). Two thirds of the samples were used to train each tree. The remaining one third of samples was then used to validate the classification result, with an error called the "out-of-bag (OOB) error". Next, the final output was determined by majority voting on all classification results obtained by each tree. Two parameters needed to be set in the RF algorithm: the number of decision trees to be generated (Ntree) and the number of features to best split each node (Mtry). Another function of RF is to derive the importance of each feature. In this study, the RF package for R was used to calculate the importance scores of features and classify wetland types. The Ntree was set to 1000 to ensure that the OOB errors stabilized and reached convergence. The Mtry was set to the square root of the number of input features.

3.3. J_{Bh} Extention Method

The Jeffries–Matusita (JM) distance was selected to estimate the separability of wetland types, as many studies showed that the JM distance can measure the separability more accurately than other distance measures [46]. The JM distance between two classes was given by Equation (4):

$$\mathrm{JM}(c_i, c_j) = \int_x \left[\sqrt{p(X/c_i)} - \sqrt{p(X/c_j)} \right]^2 dX \tag{4}$$

where c_i and c_j represented the two wetland classes. Under normality assumptions, Equation (4) was reduced to JM = $2 \cdot (1 - e^{-B})$, where

$$B = \frac{1}{8}(u_i - u_j)\left(\frac{C_i + C_j}{2}\right)^T (u_i - u_j) + \frac{1}{2}\ln\left(\left|\frac{|C_i + C_j|}{2\sqrt{|C_i| * |C_j|}}\right|\right) \tag{5}$$

and C_i and C_j represented the covariance matrices of classes i and j, respectively. $|C_i|$ and $|C_j|$ denoted the determinants of C_i and C_j, respectively. The range of JM distance was from 0 to 2. A high value indicated high separability between the two classes [47].

When considering the separability of multiple classes, the extension of the JM distance (J_{Bh}) was used in this study. The J_{Bh} can be calculated as in Equation (6) based on Bhattacharyya bounds [48].

$$J_{Bh} = \sum_{i=1}^{N} \sum_{j>i}^{N} \sqrt{p(w_i) * p(w_j)} * JM^2(i, j) \tag{6}$$

where N was the number of classes, and $p(w_i)$ and $p(w_j)$ were the prior probability of class i and class j calculated using training samples.

3.4. Accuracy Evaluation

In this study, the accuracy of the wetland classification results was evaluated as the overall accuracy (OA) (Equation (7)), user's accuracy (UA) (Equation (8)), producer's accuracy (PA) (Equation (9)) and kappa coefficient (Equation (10)). These parameters can be calculated based on the confusion matrix [49].

$$p_o = \sum_{i=1}^{r} p_{ii}/N \tag{7}$$

$$p_{u_i} = p_{ii}/p_{i+} \tag{8}$$

$$p_{A_i} = p_{ii}/p_{+i} \tag{9}$$

$$Kappa = \frac{N * \sum_{i=1}^{r} p_{ii} - \sum_{i=1}^{r}(p_{i+} * P_{+i})}{N^2 - \sum_{i=1}^{r}(p_{i+} * p_{+i})} \tag{10}$$

where p_o is the overall accuracy; p_{u_i} is the user's accuracy; p_{A_i} is the producer's accuracy; p_{ii} is the number of correctly classified samples of class i; r is the number of classes; N is the number of training samples; p_{i+} is the number of classified samples of class i; and p_{+i} is the number of training samples of class i.

4. Results and Discussion

4.1. Classification Accuracy

Validation samples were used to verify the wetland classification results generated in this study. OA, PA, UA and the kappa coefficient calculated from the confusion matrix were used to evaluate the wetland classification accuracy. The results showed that the wetland maps in the study had high accuracy. The OA ranged from 89.6% to 95.6%, with an average value of 93.1%. The kappa coefficient ranged from 0.86 to 0.94, with an average value of 0.91 (Figure 4). The producer's and user's accuracies for each land cover type were above 72.9% and 73.9%, respectively (Table 3). Both PA and UA for permanent water, permanent marsh, paddy field and seasonal marsh were high and stable for each year, with averages of 96.7%, 95.8%, 95.3%, 91.7% and 95.3%, 95.8%, 93.9%, 94.4%, respectively. The

PA and UA only for flooded wetland were below 90%, with averages of 88.3% and 89.7%, respectively. The main omissions or commissions of flooded wetland were attributed to its smaller area, with the problem of mixed pixels. In general, the classification results in our study are reliable and acceptable for monitoring wetland changes in Dongting Lake wetland reserves.

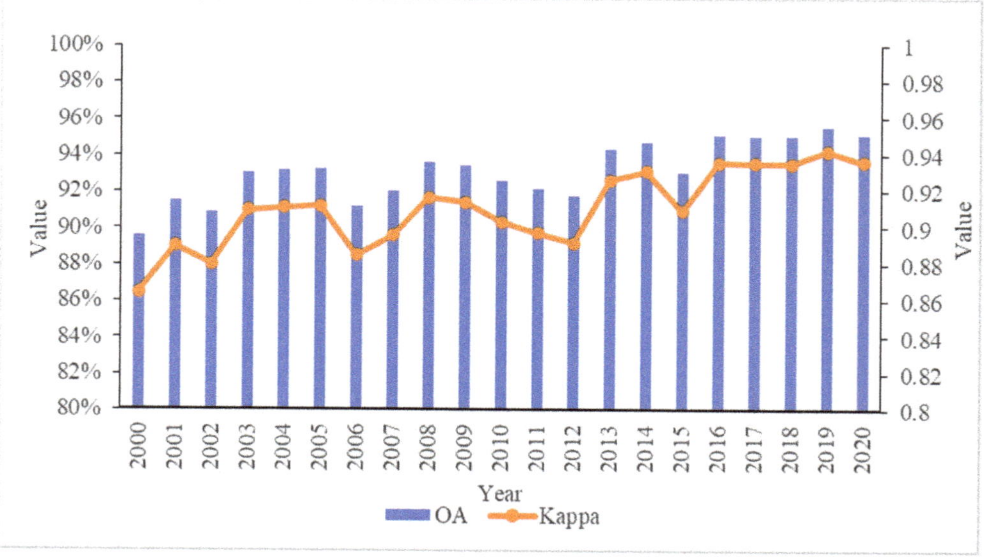

Figure 4. Overall accuracy (OA) and kappa coefficient for each year.

Table 3. Classification accuracy for each year.

Year	Permanent Water	Permanent Marsh	Flooded Wetland	Seasonal Marsh	Paddy Field	Forest	Construction Land
	PA/UA	PA/UA	PA/UA	PA/UA	PA/UA	PA/UA	PA/UA
2000	97.44%/95%	95.02%/94.17%	89.13%/93.18%	96.77%/96.77%	92.4%/91.07%	79.07%/81.93%	72.86%/73.91%
2001	94.85%/97.87%	93.67%/92.83%	82.14%/82.14%	94.44%/91.89%	93.57%/93.02%	88.95%/88.44%	78.57%/82.09%
2002	96.67%/95.6%	96.24%/95.88%	81.25%/83.87%	82.35%/93.33%	95.32%/87.87%	77.91%/85.9%	81.43%/95%
2003	97.06%/97.06%	92.48%/96.09%	84.62%/81.48%	83.33%/94.59%	95.61%/95.61%	95.35%/82.41%	80%/98.25%
2004	96.34%/95.18%	98.12%/96.67%	81.25%/86.67%	81.08%/93.75%	94.74%/92.84%	91.86%/90.8%	78.57%/87.3%
2005	90.91%/88.24%	95.49%/96.58%	85.96%/89.09%	93.33%/90.32%	94.15%/95.55%	93.02%/86.49%	88.57%/96.88%
2006	98.63%/97.3%	92.48%/90.44%	94.29%/97.06%	96.97%/94.12%	93.86%/92.51%	80.23%/84.66%	88.57%/92.54%
2007	97.3%/96%	99.25%/96.7%	86.84%/94.29%	90.63%/93.55%	93.57%/88.89%	79.65%/88.39%	85.71%/92.31%
2008	97.14%/93.15%	95.11%/93.7%	80.56%/93.55%	93.94%/91.18%	93.57%/92.22%	94.77%/95.88%	88.57%/96.88%
2009	95.83%/98.57%	95.49%/95.49%	96.97%/88.89%	97.06%/94.29%	95.03%/93.12%	87.79%/92.07%	85.71%/86.96%
2010	93.55%/87.88%	99.25%/94.29%	82.69%/91.49%	90.63%/93.55%	93.57%/96.1%	88.37%/89.41%	80%/81.16%
2011	97.22%/97.22%	94.36%/94.36%	91.43%/94.12%	92.59%/96.15%	93.86%/91.98%	87.21%/89.29%	82.86%/84.06%
2012	97.22%/89.74%	96.99%/95.56%	82.05%/88.89%	87.5%/96.55%	95.32%/91.57%	81.4%/88.05%	81.43%/87.69%
2013	97.5%/95.12%	95.49%/97.69%	88.24%/88.24%	94.87%/97.37%	97.37%/95.42%	90.7%/91.23%	82.86%/84.06%
2014	97.33%/97.33%	96.62%/99.23%	93.33%/87.5%	95.35%/95.35%	97.95%/95.71%	89.53%/91.67%	81.43%/80.28%
2015	98.67%/96.1%	95.49%/94.42%	85.29%/85.29%	88.24%/93.75%	95.32%/94.49%	88.95%/93.29%	82.86%/80.56%
2016	98.67%/97.37%	98.5%/98.5%	94.12%/91.43%	94.12%/96.97%	96.78%/96.78%	90.12%/91.18%	82.86%/81.69%
2017	96%/96%	96.62%/98.47%	92.86%/88.64%	94.44%/94.44%	97.95%/97.1%	91.28%/91.81%	84.29%/94.12%
2018	97.22%/97.22%	93.61%/96.14%	97.14%/91.89%	97.67%/97.67%	95.91%/96.19%	94.77%/90.56%	91.43%/94.12%
2019	97.18%/95.83%	96.99%/98.85%	94.12%/91.43%	94.74%/97.3%	98.25%/97.67%	91.81%/91.28%	85.71%/84.51%
2020	98.84%/97.7%	95.13%/96.58%	91.3%/95.45%	86.54%/90%	97.95%/96.82%	93.02%/92.49%	88.57%/87.32%

Note: UA, user accuracy; PA, producer accuracy.

4.2. Wetland Types and Their Distributions

The inter-annual Dongting Lake wetland classification results were obtained using optimal features and the RF algorithm. The results showed that the total area of wetlands in Dongting Lake was 3619.67 km^2 in 2020 (Figure 5), accounting for more than 93% of the total area of the three wetland reserves. Natural wetlands, including permanent water (624.05 km^2), permanent marsh (1013.56 km^2), seasonal marsh (535.57 km^2) and flooded wetlands (218.42 km^2), accounted for 61.99% of the total area of the three wetland reserves. The paddy field area in 2020 was 1228.08 km^2, accounting for 31.83% of the total area of the three wetland reserves (Figure 6).

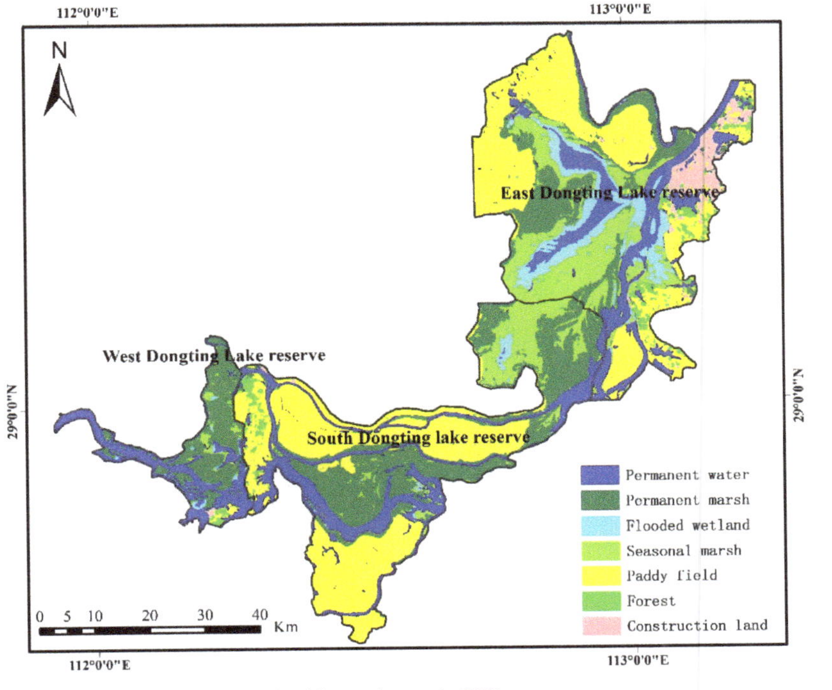

Figure 5. The classification result of the study area in 2020.

Figure 7 shows the spatial distribution of wetland types in these three reserves. Almost all natural wetlands (Figure 7A–D) and human-made wetlands (Figure 7E) were located in the East Dongting Lake wetland reserve and South Dongting Lake wetland reserve, accounting for 99% and 86% of their respective total areas. More than 50% of permanent water (Figure 7A), with an area of 318.13 km^2, was located in the East Dongting Lake wetland reserve, and 82% of flooded wetland (Figure 7C) was also located in the East Dongting Lake wetland reserve, as the central district of Dongting Lake is located in the East Dongting Lake wetland reserve and the flooded wetland was generally distributed around permanent water. Permanent marsh (Figure 7B) was mainly located in the South Dongting Lake wetland reserve, accounting for 49% of the total area of permanent marsh, while seasonal marsh (Figure 7D) was mainly located in the East Dongting Lake wetland reserve.

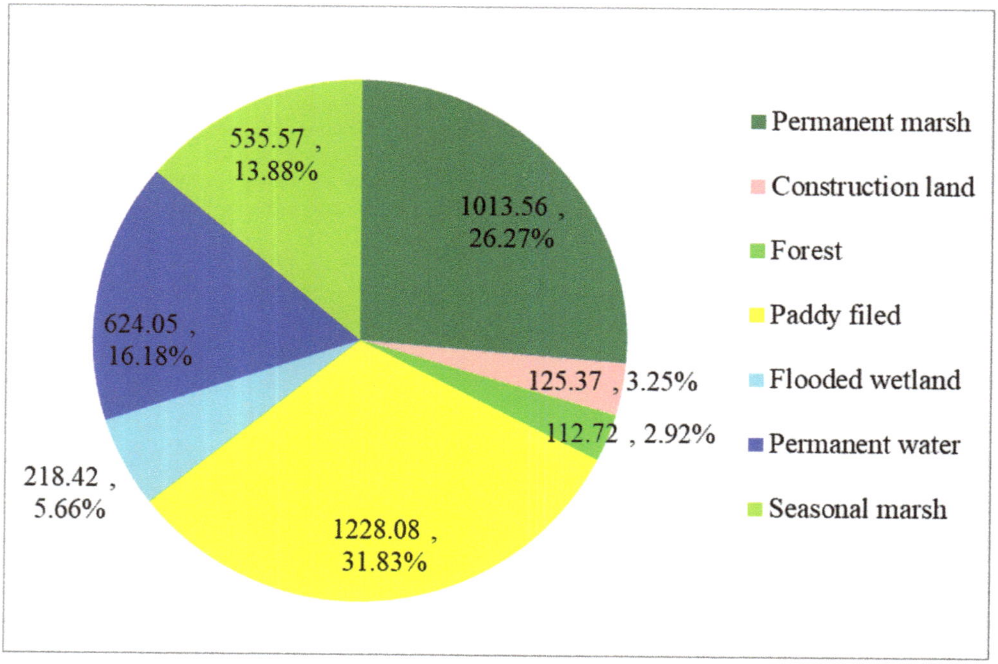

Figure 6. Area ratio of each land cover type in 2020.

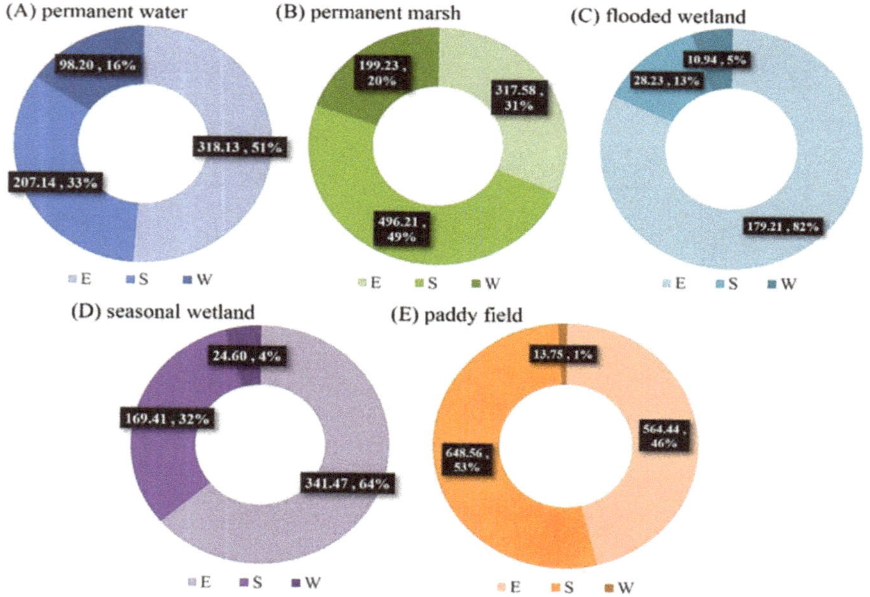

Figure 7. Spatial characteristics of different wetlands in the three Dongting Lake wetland reserves. Note: E represents the East Dongting Lake wetland reserve; S represents the South Dongting Lake wetland reserve; W represents the West Dongting Lake wetland reserve.

Seasonal wetlands, with high heterogeneity of temporal and spatial changes, have more abundant ecological functions and higher biodiversity than permanent wetlands [50,51]. Figure 8 shows the spatiotemporal dynamic changes in the seasonal wetlands at the three Dongting Lake wetland reserves, by overlaying the mapping results of the seasonal wetlands from 2000 to 2020. The seasonal wetlands were always distributed between permanent water and permanent marsh. The overall inter-annual change characteristics are that seasonal wetlands occur less frequently near permanent water and permanent marsh. However, the occurrence frequency of seasonal wetlands increases with the increase in distance, and decreases after a certain distance. In the East Dongting Lake wetland reserve, the spatial distribution of seasonal wetlands remained stable during 2000 and 2020 (frequency = 21); in the South Dongting Lake wetland and West Dongting Lake wetland reserves, most pixels that were identified as seasonal wetlands were classified as other land cover types in other years. The total area of land that was covered by seasonal wetlands ($0 < \text{frequency} \leq 21$) during the 21 years was 2079.3 km^2, accounting for more than 53.9% of the total area of the three Dongting Lake wetland reserves, while the acreage of the land dominated by seasonal wetlands in all 21 years (frequency = 21) was 252.9 km^2, accounting for only 6.6% of the total area of the three Dongting Lake wetland reserves. Therefore, seasonal wetlands are seriously affected by inter-annual fluctuations, such as rainfall, temperature, and human activities, and more attention should be paid to this specific wetland type considering its ecological value and vulnerability [33].

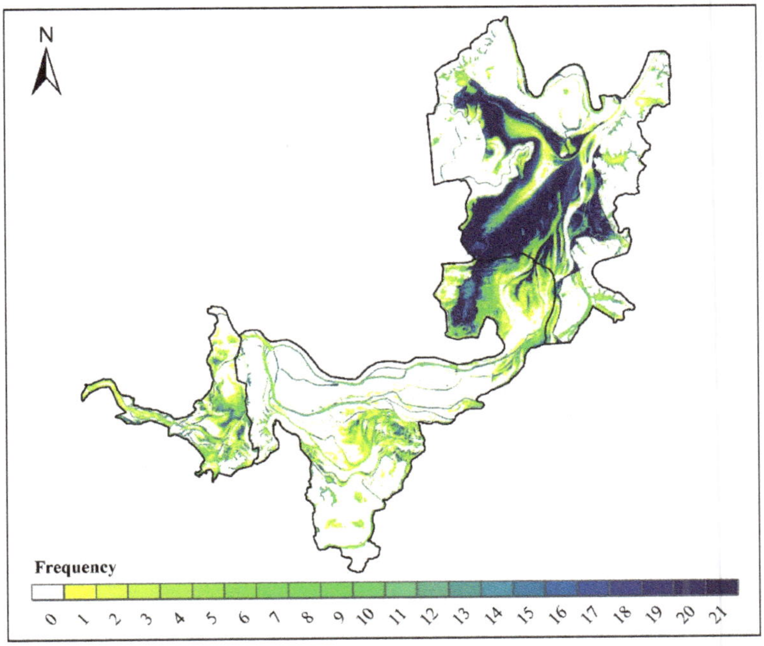

Figure 8. Spatiotemporal dynamic changes in seasonal wetlands. Note: The numbering 0–21 reflects the number of seasonal wetland occurrences in the same location from 2000 to 2020.

As a whole, the acreage of seasonal wetlands ($0 < \text{frequency} \leq 21$) in the East Dongting Lake wetland reserve, the South Dongting Lake wetland reserve and the West Dongting Lake wetland reserve was 1190.3 km^2, 668.7 km^2 and 220.2 km^2, accounting for 62.8%, 41.7% and 61.8% of each wetland reserve, respectively. Meanwhile, there were still some differences in the spatial characteristics of these three wetland reserves (Figure 9). In the East Dongting wetland reserve, the high-frequency seasonal wetland region ($19 \leq$ frequency ≤ 21) accounted for the largest proportion of the seasonal wetlands ($0 <$ frequency ≤ 21) area (34%). However, the low-frequency seasonal wetland region ($1 \leq$ frequency ≤ 3) accounted for the largest proportion in both the South Dongting Lake wetland reserve (33%) and the West Dongting Lake wetland reserve (37%) (Figure 10). Thus, the inter-annual variation in seasonal wetlands is small in the East Dongting Lake wetland reserve, but large in the South/West Dongting Lake wetland reserve.

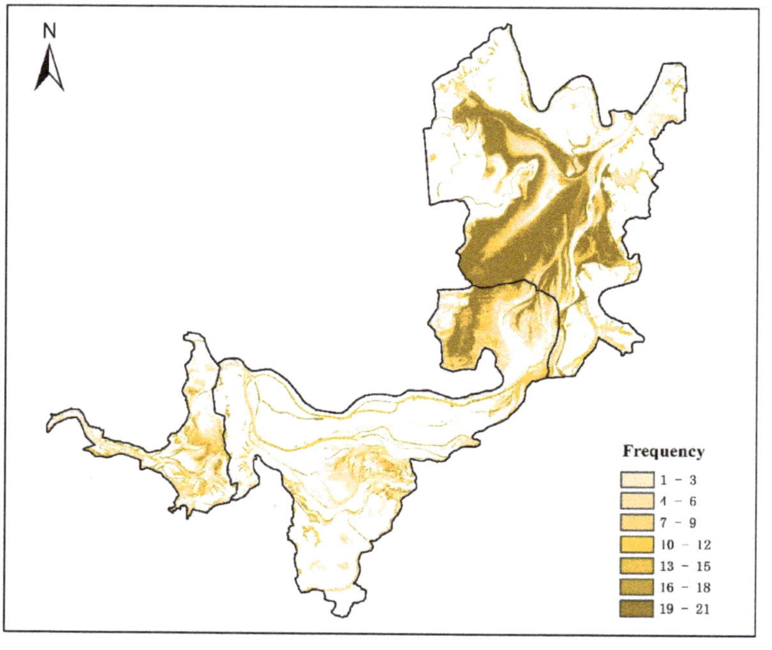

Figure 9. The spatial characteristics of seven levels' frequencies.

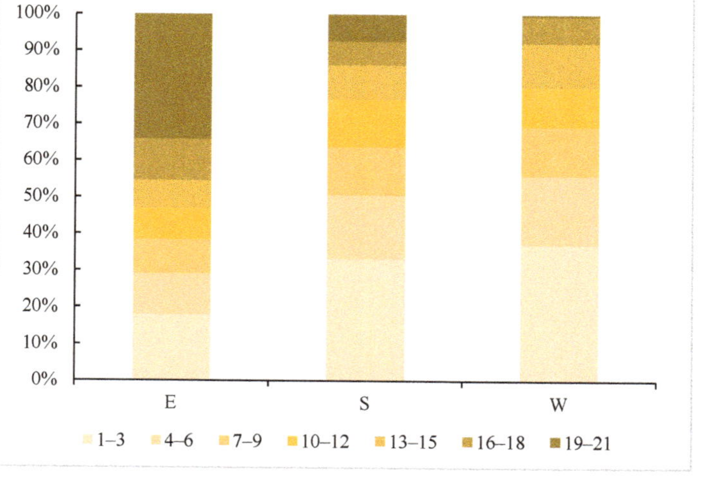

Figure 10. The proportion of the seven levels' frequencies in the three wetland reserves.

4.3. Wetland Changes

Figure 11 shows the yearly classification results of the three Dongting Lake wetland reserves between 2000 and 2020. The total wetland area of the three Dongting Lake wetland reserves slightly increased by 63.1 km^2 from 2000 to 2020, accounting for 1.6% of the total area of the three reserves. This means that the wetland area of these three reserves remained stable. However, the acreage of natural wetland (permanent water, permanent marsh, flooded wetland and seasonal marsh) decreased year by year, due to the area decline of permanent marsh, flooded wetland and seasonal marsh. The area of natural wetlands decreased by 197.0 km^2 from 2000 to 2020 (Figure 12). The natural wetland area in 2000 was 2588.6 km^2, and it decreased to 2480.8 km^2 in 2010. From 2010 to 2020, the natural wetland area decreased to 2391.6 km^2. Human-made wetlands (paddy fields) increased by 260.0 km^2 during the two decades, but the main function of human-made wetlands is to create economic value. Some researchers have found that the ecological functions of natural wetlands are far greater than the economic benefits [52], and the expansion of paddy fields also places pressure on water resources. Therefore, it is crucial to balance the agriculture development and ecology functions and implement some management measures to protect the natural wetlands in the three Dongting Lake wetland reserves.

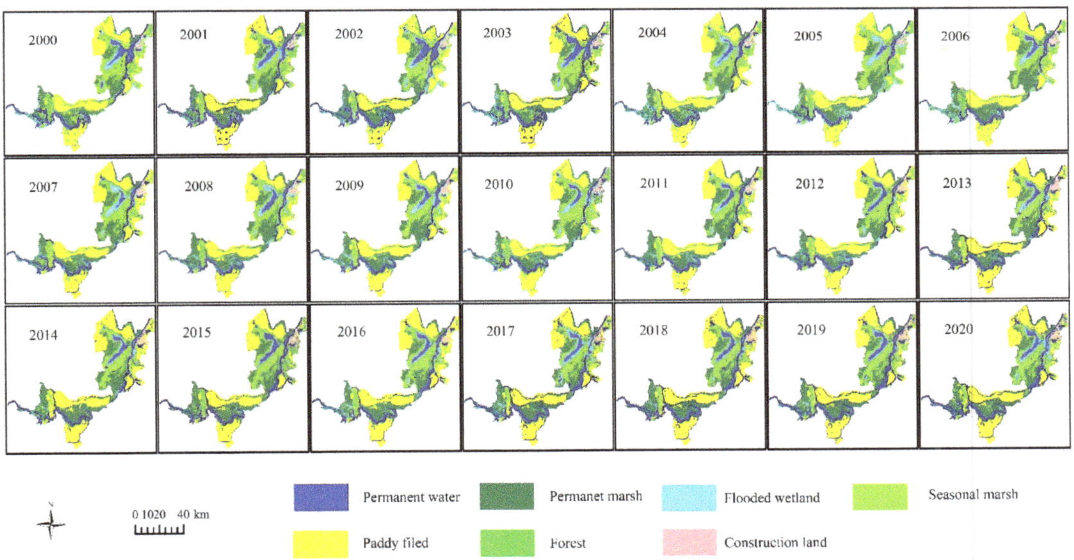

Figure 11. Spatiotemporal distribution of classification results of the three Dongting Lake wetland reserves from 2000 to 2020.

4.4. Influencing Factor Analysis

There were a number of factors affecting the surface water acreage of Dongting Lake, such as precipitation, evaporation, runoff, and infiltration. Monthly precipitation and evaporation data from 2000 to 2020 were acquired from the GLDAS Noah Land Surface Model L4 monthly 0.25 $\times$ 0.25 degree V2.1 [53], which were downloaded from the NASA Earth Data platform (https://www.earthdata.nasa.gov/, accessed on 6 April 2022). In our study, effective monthly precipitation equals the monthly precipitation minus the monthly evaporation. Then, the annual effective precipitation is obtained by summing the monthly effective precipitation. We used the annual effective precipitation to analyze the potential factors involved in the annual maximum surface water acreage variation. The annual maximum surface water area was the total acreage of seasonal marsh, flooded wetlands and permanent water bodies. Results showed a high positive correlation between the annual maximum surface water and annual effective precipitation (correlation coefficient r = 0.7, $p < 0.001$), which indicates that the annual maximum surface water area of Dongting Lake increased as the annual effective precipitation increased, and it deceased as the annual effective precipitation decreased (Figure 13). Therefore, the effective precipitation was one of the most important factors that influenced the dynamic changes in the surface water area of Dongting Lake.

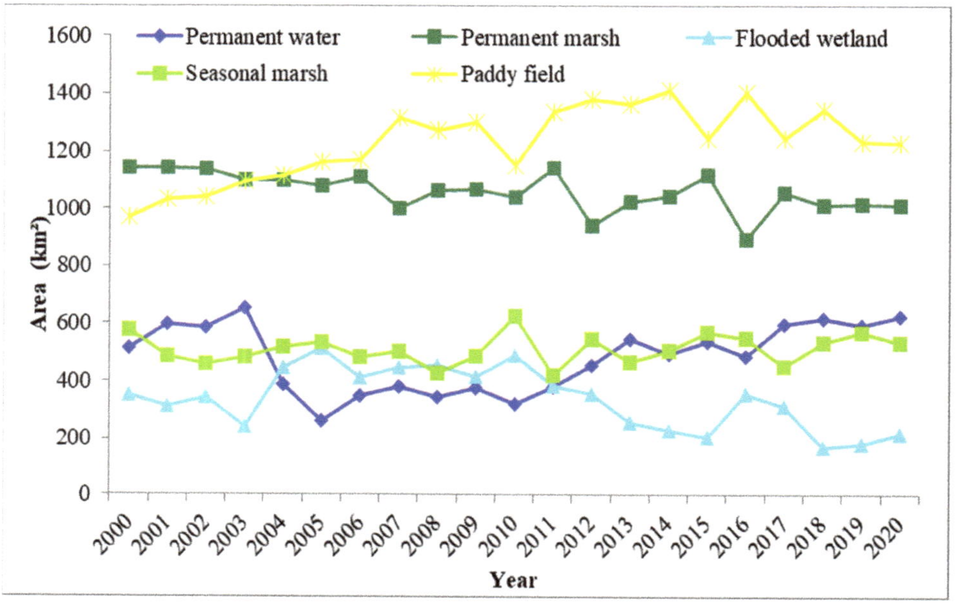

Figure 12. Dynamic changes in the three Dongting Lake wetland reserves from 2000 to 2020.

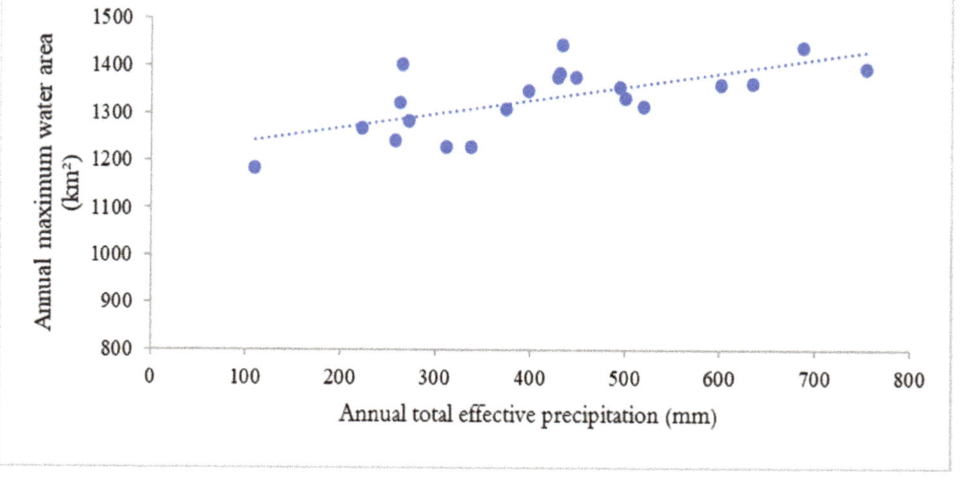

Figure 13. Linear regression analysis of the annual total effective precipitation and annual maximum water area of the three Dongting Lake wetland reserves from 2000 to 2020.

5. Conclusions

In this study, we used all available satellite images at 10~30 m spatial resolution to generate yearly wetland maps of the three Dongting Lake wetland reserves between 2000 and 2020, and monitored the wetland dynamics throughout the two decades; the classification categories included seasonal wetlands, which have high ecological value but are often omitted in existing studies. The main conclusions are as follows:

(1) Using a two-month composition to construct time series data, we effectively eliminated the influence of clouds and strips. The reconstructed two-month composition time series data set (NDVI, NDWI and NDMI) could effectively reflect the information on wetland phenology and water inundation. The results showed that the two-month composition strategy had good potential to be used as basic data for yearly wetland distribution mapping, and this strategy effectively improves the utilization of multi-source remote sensing data.

(2) The use of optimal features and the random forest classifier achieved good wetland identification accuracies, as the OA and kappa coefficients of our classification results were above 89.6% and 0.86, respectively. The PA and UA for all land cover types were above 72.9% and 73.9%, respectively. Feature optimization not only reduces data redundancy and improves operation efficiency, but also achieves wetland identification. However, due to the different characteristics of wetland vegetation in different regions, the optimal features are different.

(3) The total area of wetlands (including natural and human-made wetlands) in these three Dongting Lake wetland reserves essentially remained stable between 2000 and 2020. Although human-made wetlands (paddy fields) increased by 260.0 km^2, the area of natural wetlands decreased by 197.0 km^2. The acreage of seasonal wetlands decreased by 176.8 km^2, which was affected by both human factors (farmland expansion) and natural factors (precipitation and evaporation).

Therefore, more attention should be focused on the fragile seasonal wetlands, which play a vital role in promoting species diversity and supporting their survival. In our study, we only analyzed the correlation between the annual maximum surface water acreage and the annual effective precipitation. As more meteorological, human and economic data are obtained, the detailed reasons for the annual dynamics of these three Dongting Lake reserves will be further explored in the future.

Author Contributions: Data curation, L.X.; Formal analysis, L.X.; Methodology, L.X.; Writing—original draft, L.X.; Writing—review and editing, L.C., S.H., J.W., J.Z., C.J. and X.Z. All authors have read and agreed to the published version of the manuscript.

Funding: This research was funded by the National Natural Science Foundation of China (42271313) and the Central Public-Interest Scientific Institution Basal Research Fund, China (JBYW-AII-2022-06, JBYW-AII-2022-21, JBYW-AII-2022-40, JBYW-AII-2022-36 and JBYW-AII-2022-13).

Institutional Review Board Statement: Not applicable.

Informed Consent Statement: Not applicable.

Data Availability Statement: Not applicable.

Conflicts of Interest: The authors declare no conflict of interest.

References

1. Turner, B.L.; Lambin, E.F.; Reenberg, A. The emergence of land change science for global environmental change and sustainability. *Proc. Natl. Acad. Sci. USA* **2007**, *104*, 20666–20671. [CrossRef] [PubMed]
2. Song, X.-P.; Hansen, M.C.; Stehman, S.V.; Potapov, P.V.; Tyukavina, A.; Vermote, E.F.; Townshend, J.R. Global land change from 1982 to 2016. *Nature* **2018**, *560*, 639–643. [CrossRef] [PubMed]
3. Wang, Y.-S.; Gu, J.-D. Ecological responses, adaptation and mechanisms of mangrove wetland ecosystem to global climate change and anthropogenic activities. *Int. Biodeterior. Biodegrad.* **2021**, *162*, 105248. [CrossRef]
4. Fu, J.; Liu, J.; Wang, X.; Zhang, M.; Chen, W.; Chen, B. Ecological risk assessment of wetland vegetation under projected climate scenarios in the Sanjiang Plain, China. *J. Environ. Manag.* **2020**, *273*, 111108. [CrossRef] [PubMed]

5. Zhong, Y.; Xue, Z.; Davis, C.C.; Moreno-Mateos, D.; Jiang, M.; Liu, B.; Wang, G. Shrinking Habitats and Native Species Loss Under Climate Change: A Multifactorial Risk Assessment of China's Inland Wetlands. *Earth's Future* **2022**, *10*, 6. [CrossRef]
6. Kåresdotter, E.; Destouni, G.; Ghajarnia, N.; Hugelius, G.; Kalantari, Z. Mapping the vulnerability of arctic wetlands to global warming. *Earth's Future* **2021**, *9*, e2020EF001858. [CrossRef]
7. Xing, L.; Niu, Z.; Xu, P.; Li, D. Wetlands classification and assessment of Ramsar sites in China based on time series Moderate Resolution Imaging Spectroradiometer (MODIS) imagery. *Mar. Freshw. Res.* **2017**, *69*, 658–668. [CrossRef]
8. Töyrä, J. Pietroniro, Alain, Towards operational monitoring of a northern wetland using geomatics-based techniques. *Remote Sens. Environ.* **2005**, *97*, 174–191. [CrossRef]
9. Rebelo, L.-M.; Finlayson, C. Max, Nagabhatla, Nidhi, Remote sensing and GIS for wetland inventory, mapping and change analysis. *J. Environ. Manage.* **2009**, *90*, 2144–2153. [CrossRef]
10. Jing, L.; Zhou, Y.; Zeng, Q.; Liu, S.; Lei, G.; Lu, C.; Wen, L. Exploring wetland dynamics in large river floodplain systems with unsupervised machine learning: A case study of the Dongting Lake, China. *Remote Sens.* **2020**, *12*, 2995. [CrossRef]
11. Jefriza; Yusoff, I.M.; Abir, I.A.; Syahreza, S.; Lateh, H. The applications of InSAR technique for natural hazard detection in smart society. *J. Phys. Conf. Ser.* **2020**, *1572*, 012067. [CrossRef]
12. Orhan, O. Monitoring of land subsidence due to excessive groundwater extraction using small baseline subset technique in Konya, Turkey. *Environ. Monit. Assess.* **2021**, *193*, 4. [CrossRef]
13. Chen, J.; Chen, J.; Liao, A.; Cao, X.; Chen, L.; Chen, X.; He, C.; Han, G.; Peng, S.; Lu, M. Global land cover mapping at 30 m resolution: A POK-based operational approach. *ISPRS J. Photogramm. Remote Sens.* **2015**, *103*, 7–27. [CrossRef]
14. Friedl, M.A.; McIver, D.K.; Hodges, J.C.; Zhang, X.Y.; Muchoney, D.; Strahler, A.H.; Woodcock, C.E.; Gopal, S.; Schneider, A.; Cooper, A. Global land cover mapping from MODIS: Algorithms and early results. *Remote Sens. Environ.* **2002**, *83*, 287–302. [CrossRef]
15. Kobayashi, T.; Tateishi, R.; Alsaaideh, B.; Sharma, R.C.; Wakaizumi, T.; Miyamoto, D.; Bai, X.; Long, B.D.; Gegentana, G.; Maitiniyazi, A. Production of global land cover data–GLCNMO2013. *J. Geogr. Geol.* **2017**, *9*, 1. [CrossRef]
16. Loveland, T.R.; Reed, B.C.; Brown, J.F.; Ohlen, D.O.; Zhu, Z.; Yang, L.; Merchant, J.W. Development of a global land cover characteristics database and IGBP DISCover from 1 km AVHRR data. *Int. J. Remote Sens.* **2000**, *21*, 1303–1330. [CrossRef]
17. Bartholome, E.; Belward, A.S. GLC2000: A new approach to global land cover mapping from Earth observation data. *Int. J. Remote Sens.* **2005**, *26*, 1959–1977. [CrossRef]
18. Hamunyela, E.; Rosca, S.; Mirt, A.; Engle, E.; Herold, M.; Gieseke, F.; Verbesselt, J. Implementation of BFASTmonitor algorithm on google earth engine to support large-area and sub-annual change monitoring using earth observation data. *Remote Sens.* **2020**, *12*, 2953. [CrossRef]
19. Praticò, S.; Solano, F.; Di Fazio, S.; Modica, G. Machine learning classification of mediterranean forest habitats in google earth engine based on seasonal sentinel-2 time-series and input image composition optimisation. *Remote Sens.* **2021**, *13*, 586. [CrossRef]
20. Pickens, A.H.; Hansen, M.C.; Hancher, M.; Stehman, S.V.; Tyukavina, A.; Potapov, P.; Marroquin, B.; Sherani, Z. Mapping and sampling to characterize global inland water dynamics from 1999 to 2018 with full Landsat time-series. *Remote Sens. Environ.* **2020**, *243*, 111792. [CrossRef]
21. Feng, L.; Hu, C.; Chen, X.; Cai, X.; Tian, L.; Gan, W. Assessment of inundation changes of Poyang Lake using MODIS observations between 2000 and 2010. *Remote Sens. Environ.* **2012**, *121*, 80–92. [CrossRef]
22. Jain, S.K.; Saraf, A.K.; Goswami, A.; Ahmad, T. Flood inundation mapping using NOAA AVHRR data. *Water Resour. Manag.* **2006**, *20*, 949–959. [CrossRef]
23. Kang, S.; Hong, S.Y. Assessing Seasonal and Inter-Annual Variations of Lake Surface Areas in Mongolia during 2000–2011 Using Minimum Composite MODIS NDVI. *PLoS ONE* **2016**, *11*, e0151395.
24. Klein, I.; Gessner, U.; Dietz, A.J.; Kuenzer, C. Global WaterPack–A 250 m resolution dataset revealing the daily dynamics of global inland water bodies. *Remote Sens. Environ.* **2017**, *198*, 345–362. [CrossRef]
25. Xing, L.; Niu, Z.; Jiao, C.; Zhang, J.; Han, S.; Cheng, G.; Wu, J. A Novel Workflow for Seasonal Wetland Identification Using Bi-Weekly Multiple Remote Sensing Data. *Remote Sens.* **2022**, *14*, 1037. [CrossRef]
26. Wang, X.; Xiao, X.; Zou, Z.; Hou, L.; Qin, Y.; Dong, J.; Doughty, R.B.; Chen, B.; Zhang, X.; Chen, Y. Mapping coastal wetlands of China using time series Landsat images in 2018 and Google Earth Engine. *ISPRS J. Photogramm. Remote Sens.* **2020**, *163*, 312–326. [CrossRef]
27. Mao, D.; Wang, Z.; Du, B.; Li, L.; Tian, Y.; Jia, M.; Zeng, Y.; Song, K.; Jiang, M.; Wang, Y. National wetland mapping in China: A new product resulting from object-based and hierarchical classification of Landsat 8 OLI images. *ISPRS J. Photogramm. Remote Sens.* **2020**, *164*, 11–25. [CrossRef]
28. Dang, K.B.; Nguyen, M.H.; Nguyen, D.A.; Phan, T.T.H.; Giang, T.L.; Pham, H.H.; Nguyen, T.N.; Tran, T.T.V.; Bui, D.T. Coastal wetland classification with deep u-net convolutional networks and sentinel-2 imagery: A case study at the tien yen estuary of vietnam. *Remote Sens.* **2020**, *12*, 3270. [CrossRef]
29. Ding, X.; Li, X. Monitoring of the water-area variations of Lake Dongting in China with ENVISAT ASAR images. *Int. J. Appl. Earth Obs. Geoinf.* **2011**, *13*, 894–901. [CrossRef]
30. Li, J.; Dai, Y.; Ou, C.; Peng, P.; Deng, C. Effects of store water application of the three gorges reservoir on Yangtze River on water and sediment characteristics in the Dongting Lake. *J. Soil Water Conserv* **2011**, *25*, 15.

31. Hu, Y.; Huang, J.; Du, Y.; Han, P.; Huang, W. Monitoring spatial and temporal dynamics of flood regimes and their relation to wetland landscape patterns in Dongting Lake from MODIS time-series imagery. *Remote Sens.* **2015**, *7*, 7494–7520. [CrossRef]
32. Yang, L.; Wang, L.; Yu, D.; Yao, R.; He, Q.; Wang, S.; Wang, L. Four decades of wetland changes in Dongting Lake using Landsat observations during 1978–2018. *J. Hydrol.* **2020**, *587*, 124954. [CrossRef]
33. Xing, L.; Tang, X.; Wang, H.; Fan, W.; Wang, G. Monitoring monthly surface water dynamics of Dongting Lake using Sentinel-1 data at 10 m. *PeerJ* **2018**, *6*, e4992. [CrossRef] [PubMed]
34. Zhang, J.; Xu, K.; Yang, Y.; Qi, L.; Hayashi, S.; Watanabe, M. Measuring water storage fluctuations in Lake Dongting, China, by Topex/Poseidon satellite altimetry. *Environ. Monit. Assess.* **2006**, *115*, 23–37. [CrossRef]
35. Li, F.; Huang, J.; Zeng, G.; Yuan, X.; Li, X.; Liang, J.; Wang, X.; Tang, X.; Bai, B. Spatial risk assessment and sources identification of heavy metals in surface sediments from the Dongting Lake, Middle China. *J. Geochem. Explor.* **2013**, *132*, 75–83. [CrossRef]
36. Flood, N. Continuity of Reflectance Data between Landsat-7 ETM+ and Landsat-8 OLI, for both top-of-atmosphere and surface reflectance: A Study in the Australian landscape. *Remote Sens.* **2014**, *6*, 7952–7970. [CrossRef]
37. Griffiths, P.; Nendel, C.; Hostert, P. Intra-annual reflectance composites from Sentinel-2 and Landsat for national-scale crop and land cover mapping. *Remote Sens. Environ.* **2019**, *220*, 135–151. [CrossRef]
38. Hao, P.; Chen, Z.; Tang, H.; Li, D.; Li, H. New Workflow of Plastic-Mulched Farmland Mapping Using Multi-Temporal Sentinel-2 Data. *Remote Sens.* **2019**, *11*, 1353. [CrossRef]
39. Harald, V.; Freek, V. Sentinel-2A MSI and Landsat 8 OLI Provide Data Continuity for Geological Remote Sensing. *Remote Sens.* **2016**, *8*, 883.
40. McFeeters, S.K. The use of the Normalized Difference Water Index (NDWI) in the delineation of open water features. *Int. J. Remote Sens.* **1996**, *17*, 1425–1432. [CrossRef]
41. Jin, S.; Sader, S.A. Comparison of time series tasseled cap wetness and the normalized difference moisture index in detecting forest disturbances. *Remote Sens. Environ.* **2005**, *94*, 364–372. [CrossRef]
42. DeFries, R.S.; Townshend, J. NDVI-derived land cover classifications at a global scale. *Int. J. Remote Sens.* **1994**, *15*, 3567–3586. [CrossRef]
43. John, A. RA Fisher and the making of maximum likelihood 1912–1922. *Stat. Sci.* **1997**, *12*, 162–176.
44. Breiman, L. Random forests. *Mach. Learn.* **2001**, *45*, 5–32. [CrossRef]
45. Cortes, C.; Vapnik, V. Support-vector networks. *Mach. Learn.* **1995**, *20*, 273–297. [CrossRef]
46. Van Niel, T.G.; McVicar, T.R.; Datt, B. On the relationship between training sample size and data dimensionality: Monte Carlo analysis of broadband multi-temporal classification. *Remote Sens. Environ.* **2005**, *98*, 468–480. [CrossRef]
47. Adam, E.; Mutanga, O. Spectral discrimination of papyrus vegetation (*Cyperus papyrus* L.) in swamp wetlands using field spectrometry. *ISPRS J. Photogramm. Remote Sens.* **2009**, *64*, 612–620. [CrossRef]
48. Bruzzone, L.; Roli, F.; Serpico, S.B. An extension of the Jeffreys-Matusita distance to multiclass cases for feature selection. *IEEE Trans. Geosci. Remote Sens.* **1995**, *33*, 1318–1321. [CrossRef]
49. Foody, G.M. Status of land cover classification accuracy assessment. *Remote Sens. Environ.* **2002**, *80*, 185–201. [CrossRef]
50. Chen, M.; Liu, J. Historical trends of wetland areas in the agriculture and pasture interlaced zone: A case study of the Huangqihai Lake Basin in northern China. *Ecol. Model.* **2015**, *318*, 168–176. [CrossRef]
51. Gxokwe, S.; Dube, T.; Mazvimavi, D. Leveraging Google Earth Engine platform to characterize and map small seasonal wetlands in the semi-arid environments of South Africa. *Sci. Total Environ.* **2022**, *803*, 150139. [CrossRef] [PubMed]
52. Qiu, P.; Songjun, X.; Genzong, X.; Ying, F. Comparisons of natural wetland, Semi-constructed wetland and engineered wetland. *J. Hainan Norm. Univ. (Nat. Sci.)* **2010**, *23*, 209–213, 231.
53. Rodell, M.; Houser, P.; Jambor, U.; Gottschalck, J.; Mitchell, K.; Meng, C.-J.; Arsenault, K.; Cosgrove, B.; Radakovich, J.; Bosilovich, M. The global land data assimilation system. *Bull. Am. Meteorol. Soc.* **2004**, *85*, 381–394. [CrossRef]

International Journal of
Environmental Research and Public Health

MDPI

Article

Assessing the Landscape Ecological Risks of Land-Use Change

He Gao [1,2] and Wei Song [1,3,*]

1 Key Laboratory of Land Surface Pattern and Simulation, Institute of Geographic Sciences and Natural Resources Research, Chinese Academy of Sciences, Beijing 100101, China
2 School of Geosciences, Yangtze University, Wuhan 430100, China
3 Hebei Collaborative Innovation Center for Urban-Rural Integration Development, Shijiazhuang 050061, China
* Correspondence: songw@igsnrr.ac.cn

Abstract: In recent years, a changing global climate and the continuous expansion of the intensity and scope of human activities have led to regional differentiation in the surface landscape. This has caused numerous ecological risks under multiple pressure sources, gradually becoming an important factor restricting the sustainable development of economic and social health. With the continuous development of the social economy, land use and associated ecological risks will inevitably change. According to the forest transformation theory and the environmental Kuznets curve, we put forward the theoretical framework of ecological risk transformation of land-use change and took Zhangjiachuan County (China) as an example to verify it. Therefore, on the basis of Landsat satellite data, this paper used landscape structures to calculate an ecological risk index, and evaluated the ecological risk of land-use changes through pattern index analyses. The results show that, from 2000 to 2020, the ecological risk index of land-use change in Zhangjiachuan County exhibited an increasing and then decreasing trend, showing an overall "inverted U-shaped" trend of change consistent with the transformation theoretical framework of ecological risks of land use change. Secondly, in terms of patterns, the ecological risk of land-use change in Zhangjiachuan County showed a distribution feature of high in the west and low in the east. In 2000, high-risk areas were mainly concentrated in the central and northern areas, while low-risk areas were mainly concentrated in the eastern areas. From 2000 to 2015, the medium-risk areas expanded to the west and midwest, and the geographic centers of the risk areas were slightly offset. From 2015 to 2020, the overall pattern of ecological risk areas was basically the same as that of the previous stage, but the medium-risk areas were slightly reduced. In terms of quantity, from 2000 to 2015, the areas of the lowest risk level and low risk level decreased, while the areas of medium risk level, high risk level, and the highest risk level increased; from 2015 to 2020, the areas of the lowest risk level and low risk level increased, and the areas of medium risk level, high risk level, and highest risk level decreased. Lastly, the spatial aggregation of ecological risks in Zhangjiachuan County weakened slightly from 2000 to 2005, gradually increased from 2005 to 2015, and then slightly weakened from 2015 to 2020.

Keywords: land-use change; landscape pattern; ecological risk; theoretical framework of ecological risk transformation of land-use change; ecological fragile area; Zhangjiachuan county; China

Citation: Gao, H.; Song, W. Assessing the Landscape Ecological Risks of Land-Use Change. *Int. J. Environ. Res. Public Health* **2022**, *19*, 13945. https://doi.org/10.3390/ijerph192113945

Academic Editor: Richard A. Lord

Received: 9 October 2022
Accepted: 24 October 2022
Published: 27 October 2022

1. Introduction

Over the centuries, the multiple effects of human activities and natural succession have led to dramatic changes in land use and global ecology, inevitably triggering various conflicts between economic growth and sustainable development [1,2]. The ecological risks caused by land-use changes are currently among the hottest issues of concern in developed and developing countries worldwide [3]. The extensive influence of land-use changes and ecological changes has led to changes in ecosystem structures and landscape patterns, leading to a large number of ecological problems, such as ecological function degradation, soil erosion, land desertification, environmental pollution, and biodiversity reduction [4,5]. The ecological environment is constantly affected by these problems, greatly increasing the risk posed to ecosystems [6] and seriously threatening human well-being [5,7,8]. Assessing the ecological risks of land-use changes is critical for establishing early warning systems of ecological risks, with an accurate and effective control of ecological risks, and it is of great significance to the formulation of regional ecological protection policies and for adopting ecological protection measures.

Risk assessment began in 1980 and initially consisted of toxicology research on chemical pollutants and risk research on human health. It then became used as a kind of management tool to assess multiple chemical pollutants and various events that may cause environmental pollution and was finally extended to consider the assessment of ecological risks caused by human activities. The term "ecological risk assessment" was first introduced by the US Environmental Protection Agency (EPA), which defined ecological risk as the possibility of exposure to adverse ecological effects caused by or likely to result from one or more stressors [9], and the framework was subsequently expanded and revised to form the basic guide for the current risk assessment [10]. Since 1990, in the context of increasingly prominent ecological problems, the focus of risk assessments has gradually shifted from human health assessments to ecological risk assessments [11], and risk receptors have expanded to populations, communities, and whole ecosystems. From the late 1990s to the beginning of the 21st century, with the continuous improvement and maturation of ecological risk assessment systems, the field of ecological risk assessment gradually expanded and entered the stage of regional ecological risk assessments, and the risk receptors extended to the watershed and regional landscape scales.

Landscape ecological risk assessment is an important branch of ecological risk assessment [12], which tends to quantitatively identify and directly assess ecological risks from the perspective of spatial landscape patterns caused by land-use changes. Since land use is regarded as a comprehensive reflection of the direct impact of human economic and social activities on surface resources and natural environment [13], the spatiotemporal heterogeneity of land use is affected by regional terrains and geomorphic features closely related to landscape patterns and landscape ecological risks [14]. At present, the research on the ecological risks of land-use changes has mainly focused on analyzing long-term historical evolution characteristics [15,16], revealing the driving mechanisms of ecological risks caused by land-use changes [17], and simulating future evolution trends [18,19]. On the one hand, through the analysis of the long-term historical evolution characteristics of the ecological risks of land-use changes in a region, we can obtain reliable evidence for the long-term changes of ecosystems in the region, reveal its evolution trajectory and mechanisms, and provide key scientific information for the formulation of regional ecological restoration paths and objectives. For example, Recanatesi et al. [20] analyzed the changing characteristics of soil vulnerability and landscape degradation and their spatial distribution during the period from 1960 to 2010 in Italian agricultural and forestry areas using four thematic indicators of environmentally sensitive areas and the comprehensive index of desertification risk. Krajewski et al. [21] used the landscape change index to analyze the driving mechanisms of forest resource changes and forest transformation in landscape parks on the basis of 140 years (1863–2013) of map data. On the other hand, ecological risk simulations of regional land-use changes allow us to predict the possible impact or harm caused by human activities on the ecosystem and formulate ecological protection and risk prevention policies, which are of great significance for promoting regional sustainable development. For example, Li et al. [22] used the conversion of land use and its effects at a small region extent (CLUE-S) model to simulate and analyze the future land-use changes of the Luanhe River basin from 2010 to 2030 under three scenarios of trend, rapid economic growth, and ecological security, and then assessed the spatial distribution characteristics of the landscape ecological risks caused by it.

With the rapid development of geographical information system (GIS) and remote sensing (RS) technologies and the wide application of landscape ecology theories and methods [23], landscape ecological risk assessment based on land-use changes has become one of the most effective methods in ecological risk assessment [24]. At present, the ecological risk assessments have mainly focused on cities, river basins, coastal areas, administrative regions, and nature reserves. The urban populations of cities are excessively dense, and their spatial structure expands rapidly, with various land-use types being frequently transformed into each other, and ecological problems being increasingly prominent [25,26]. The watershed area ecology is fragile, and the natural endowment conditions such as water resources are poor. The ecological improvement and deterioration of landscapes coexist, with the overall deterioration trend being greater than the improvement one, and the degree of ecological deterioration of a landscape is constantly increasing [27,28]. In the process of development in coastal areas, the unrestricted urban expansion during the early stages of economic development has led to landscape fragmentation, and the overall ecological risk has increased, becoming more spatially clustered. With the suppression of disorderly expansion, the overall ecological risk decreases [29,30]. During the early stages of the development planning of administrative regions, the purpose of regional economic development is usually achieved by sacrificing agricultural lands. During this period, land use changes drastically, resulting in increased ecological risks. Subsequently, due to the gradual enhancement of residents' awareness of environmental protection and the implementation of targeted ecological prevention and control and protection policies, ecological problems caused by human activities have been alleviated and ecological risks have been reduced [31–33]. Nature reserves can be divided into core areas, buffer zones, and experimental zones according to their functions. Human activities are common in the experimental areas, the ecological risk of land-use changes is the most serious, and the land-use changes in the buffer zone and the core area are relatively stable. Therefore, the ecological risk of nature reserves is greatest in the experimental areas, followed by the buffer areas, and finally the core areas [34,35].

Compared with traditional ecological risks, the landscape ecological risk assessment pays more attention to spatial heterogeneity, and its key lies in establishing an evaluation system and selecting appropriate indicators. At present, there are two widely used methods: one is the ecological risk assessment system based on the "pressure–receptor–response" model and failure mechanism [25]. The ecological risk assessment system consists of identifying the risk source intensity, receptor exposure, and risk effect. The evaluation method is a comprehensive relative risk model (RRM) [36]. The evaluation system focused on stressors and habitats of concern in the study area. For example, Muditha and Heenkenda et al. [37,38] used this evaluation system to rank and classify the stressors and habitats within a region and modeled the interaction between the two through exposure and effect filters, revealing the spatial and temporal distribution of ecological risk in ports. However, this model is only suitable for large-scale areas that need to focus on multiple stressors and is often used to assess the ecological risks of a specific stressor or disturbance source, which has certain limitations. The second approach uses the deviation from the optimal mode as a risk source, evaluating the ecological risk, and regarding the whole system as a receptor. The landscape pattern index is commonly used to assess the ecological risks of the study subjects as a function of the ecological changes throughout a region, and the most representative method is the landscape loss model. Shi et al. [39] used the Markov model and landscape index analysis to construct an ecological risk assessment model, which revealed that the ecological risks in Huaibei, a typical resource-based city in China, were affected by land-use changes. In addition to assessing the landscape ecological risks alone, some scholars have also studied the ecosystem services associated with landscape ecological risks. For example, Gong et al. [40] revealed the spatiotemporal changes in the grain yield of cultivated land, carbon storage, water yield, biodiversity index, and ecological risks in the Bailongjiang River basin of China by introducing ecosystem services and landscape ecological risks into the formulation of ecological policies and governance of ecological problems. It is difficult to obtain RRM assessment data and completely uniform assessment criteria during the comparative analysis of different time series in the same study region. Therefore, with the support of a landscape ecology theory, a landscape loss model based on land-use changes can both quantitatively describe the landscape structure [41,42] and explain the evolutionary mechanisms of landscape ecological risks from the perspective of spatial landscape pattern changes. This model becomes an important tool to analyze and reveal the spatial and temporal characteristics of landscape ecological risks.

The ecological risk of land-use change has obvious stages. Analyzing and studying the evolution characteristics of the ecological risks of land-use changes on medium and long timescales can provide a scientific and reasonable basis for the planning of regional ecological protection schemes. Earlier research on this topic mainly focused on the construction of ecological risk models and spatial analysis [29], and relatively little theoretical analysis has been conducted on the ecological risks of land-use changes on medium to long timescales. We previously studied the ecological risk changes of land-use changes in other countries and found that there is indeed a close relationship between the pattern evolution of ecological risks and land-use changes. Therefore, according to the theories of forest transformation and an environmental Kuznets curve, this paper reconstructed a theoretical framework for the ecological risk transformation of land-use changes and put forward a theoretical hypothesis. Considering Zhangjiachuan County, Gansu Province, China as the verification case area, this paper constructs an ecological risk assessment model through landscape disturbance and landscape vulnerability indices by taking into account land-use changes in order to comprehensively reveal the evolution characteristics of the overall spatial and temporal patterns of ecological risks in Zhangjiachuan County. Specifically, the research objectives of this paper were to (1) analyze the spatiotemporal land-use change characteristics in Zhangjiachuan County from 2000 to 2020, (2) reveal the spatiotemporal evolution patterns of the ecological risks of land-use changes, and (3) study the dynamic characteristics of ecological risks at the landscape level through spatial autocorrelation analyses. The remainder of this paper is arranged as follows: in Sections 2 and 3, brief descriptions of the study area, datasets, and methods are provided; Section 4 introduces the results of land use and landscape ecological risk changes in Zhangjiachuan County; the discussion and conclusions are then provided in Sections 5 and 6, respectively.

2. Overview of the Study Area and Data Sources

2.1. Overview of the Study Area

Zhangjiachuan County (105°54′–106°35′, 34°44′–35°11′) is located in the southeast of the Gansu Province, the northeast of Tianshui City, at the foot of the west side of Longshan Mountain. Its northern part is linked with Zhuanglang County and Huaxiang County in Pingliang City. To the south is Tianshui City, Qingshui County, and it is adjacent to Qin'an County, Tianshui City to the west, and Long County, Shaanxi Province to the east. Zhangjiachuan County is located in the transition zone of Liupanshan trough and Longxi Loess Plateau, belonging to the loess hilly and gully region in the middle reaches of the Yellow River. The terrain is uneven in most areas, with alternating ridges and ravines, small plots, steep slopes, and the terrain slopes from northeast to southwest. The county is 62 km long from east to west and 48 km wide from north to south, with a total area of 1311.8 km^2 (Figure 1).

Zhangjiachuan County is a traditional dry farming county, with deep mountains and fragmented terrains. The average elevation is 2011.4 m, with the highest point at 2659 m and the lowest at 1486 m. The climate is warm temperate semi-humid, with an average temperature of 7.6 °C, highs of 31.7 °C, and lows of 20.6 °C. The annual average rainfall is 599.8 mm, and the appropriate amount of precipitation varies greatly in time and space, with the months of July, August, and September accounting for 59.3% of the annual rainfall. The rainfall in Zhangjiachuan County falls mostly through heavy rainstorms which last for a long time and are of high intensity, which can easily cause soil erosion. In recent years, with the enhancement of human activities, the natural vegetation has declined sharply with soil erosion becoming serious, and the land has been seriously degraded. In addition, the geographical environment in the study area is complex, and the climate is volatile, leading to frequent natural disasters such as droughts, hails, rainstorms, and floods. Overall, the ecosystem in the study area is relatively fragile and the sustainable development of the ecosystem has become threatened [43].

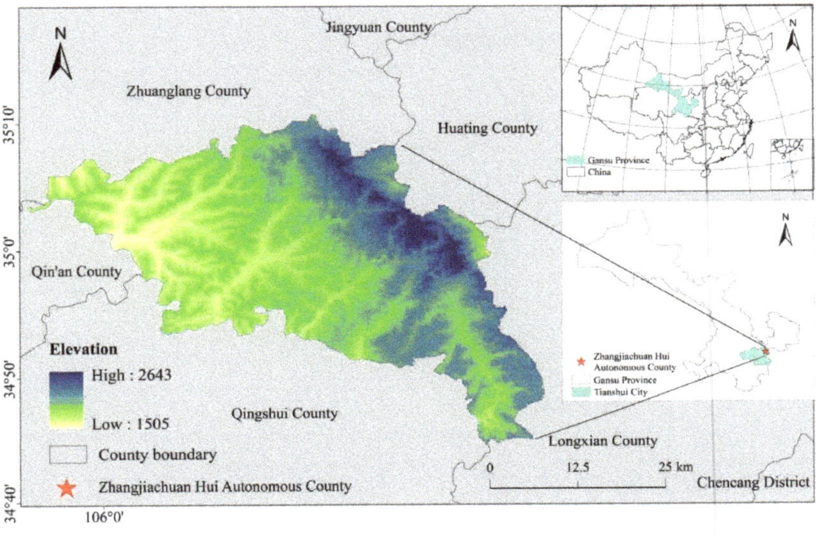

Figure 1. Geographical location of Zhangjiachuan County.

2.2. Data Source

This paper used two types of data to conduct a land-use classification and landscape ecological risk assessment:

(1) Landsat image data. The primary data source used in this study was Landsat satellite images. The satellite images included Landsat TM (2000–2011), Landsat ETM + (2012), and Landsat OLI (2013–2020). The images were filtered using a series of functions in the Google Earth engine (GEE) filter, with a time standard of April–September per year and a cloud coverage standard of less than 10%. The available remote sensing images were generated following cloud removal, image mosaicking, and cropping processing. The specific data sources are shown in Table 1.

(2) Other data. Using the administrative boundary data of the National Catalog Service for Geographic Information [44], we combined the 30 m resolution digital elevation model (DEM) data of the geospatial data cloud platform to serve as the classification basis [45]. The analyses were performed using Google historical image data (2000–2020) from 91 Bitmap Assistant [46] with a resolution of 0.52 m.

Table 1. Landsat image information.

Data Source	Landsat Image Set ID	Year
Landsat 5	LANDSAT/LT05/C01/T1_TOA	2000–2011
Landsat 7	LANDSAT/LE07/C01/T1_TOA	2012
Landsat 8	LANDSAT/LC08/C01/T1_TOA	2013–2020

3. Research Method

The main technical steps to assess the landscape ecological risk in Zhangjiachuan County are shown in Figure 2. First, the images were preprocessed through the GEE online editor, and the image data from 2000 to 2020 were classified using the random forest classifier. Then, the landscape ecological risks were evaluated, and the spatial and temporal change characteristics were explored. Finally, the spatial autocorrelation of the ecological risk index was analyzed using the Moran's I index and local spatial autocorrelation analysis methods.

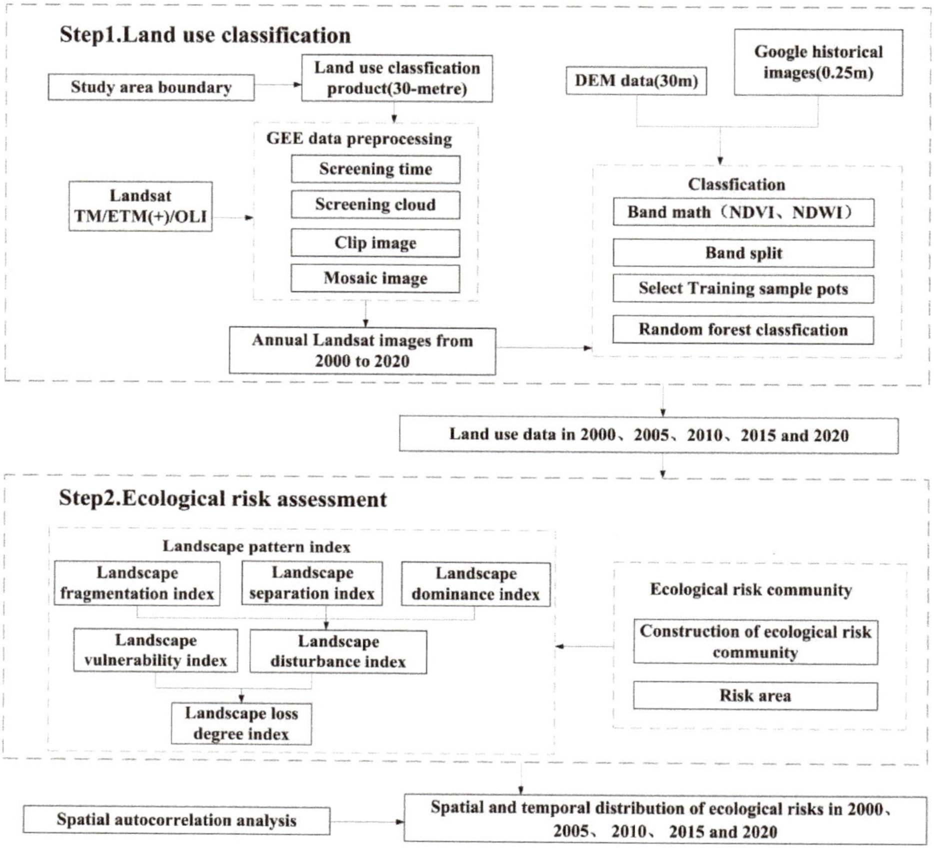

Figure 2. Ecological risk assessment process of land-use change.

3.1. Theoretical Framework of the Ecological Risk Transformation of Land-Use Changes

The main basic theories of the ecological risk transformation theory framework of land-use changes include the forest transformation theory and environmental Kuznets curve. Among them, Mather [47] divided the forest area into two stages of a "U-shaped" change process, from a reduction to an increase. According to the Kuznets curve proposed by Simon, Panayotou [48] divided the change relationship between environmental quality and per capita income into two stages, namely, the "inverted U-shaped" change process, in which environmental quality first increases and then decreases with the per capita income. In the process of forest transformation, the process of abandoned logging land being reclaimed and developed into agricultural lands by farmers can reflect the change in land use, while the economic, social, and ecological issues, among others, which are brought about by social and economic development, can also be fed back to land use. Therefore, Mather's forest transformation theory and the Panayotou environmental Kuznets curve are also applicable for the ecological risk change process of land-use changes and show the characteristics of stages.

The root of ecological risk change lies in the change in land use, and its essence depends on the mutual feedback of land-use changes and ecological risks under transmission mechanisms. It not only reflects the consistent change characteristics of the similarity attribute and the transformation theory in different social development stages, but also exhibits a trend of inconsistent change rates in different development stages under the combined effect of natural factors and human interferences. Lastly, the ecological risk, accompanied by the land-use changes, presents an "inverted U-shaped" phase change difference, forming a "land-use change ecological risk" relationship curve (Figure 3).

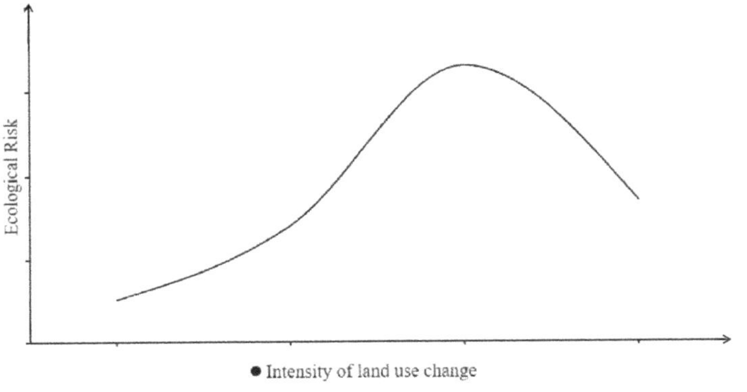

Figure 3. Relationship curve showing the ecological risk of land-use change.

Ecological risks present an "inverted U-shaped" change trend with land-use changes. Land-use changes are dynamic and characterized by large interannual changes and inconsistent change rates. Therefore, the change rate of the ecological risks of land-use changes fluctuates at different stages. At the initial stage of social development when the economy is in its infancy, the process of urbanization is slow. Driven by the national macro policies, the construction of large-scale development zones has led to an increase in the area of urban and rural land and industrial and mining construction land. At the same time, the incentive policies issued in urban construction, land management, and other major infrastructure construction have led to an increase in construction land. In order to meet the demands for the expansion of construction land, the occupation of cultivated and ecologically significant lands has gradually increased, resulting in the fragmentation of ecological landscapes, the destruction of vegetation and biological natural habitats to a certain extent, and the continuous reduction in the ecological carrying capacity and environmental capacity, whilst affecting the ecological quality.

In the process of land use, human beings pursue a highly efficient economic society, accelerate the demand for land resources, and ignore the nonrenewable nature of land resources. In the middle period of social development, with accelerated economic development, industrialization and urbanization promote each other; the proportion of nonagricultural industries and nonagricultural populations increases sharply, urban spaces begin to spread to the surrounding areas, and cultivated and ecological lands are largely occupied, resulting in a sharp deterioration of surface natural conditions, a sharp decline in green space areas, and a large consumption of natural resources, thereby compromising the productivity and production functions of certain ecosystems. On the other hand, the remote coupling theory, involving the relaxation of population mobility controls and population migration, typically represented by migrant workers, promotes land-use changes in different ways. Population losses and village abandonment have become a common phenomenon, accompanied by the marginalization and even abandonment of agricultural lands, causing further damage to the landscape connectivity. During this period, land use changed dramatically, causing a sharp increase in ecological risks.

Following economic maturation, the ecological risk reaches its peak, and the increasingly serious ecological problems have gradually aroused great attention of the country. According to the "win–win sustainable development strategy" demonstrated by Gao [49], the ecological risks caused by land-use changes can be mitigated through the combined effects of economic and ecological benefits. The intensification of land use should be strengthened by fully considering the bearing capacity of environment, resources, and ecology. It is necessary to divide the area according to the nature of the city, environmental conditions, and functions, reasonably adjust the industrial structure and construction layout, make the evolution pattern of construction land more compact, and continuously reduce its fragmentation and the separation of individual patches. The intensification of construction lands triggered during this social development stage is beneficial to the improvement of the regional ecological stability. At the same time, the land ecological project, the project of returning cultivated land to forest and grass, land reclamation construction, and regional environmental management, such as water and soil loss management, can be carried out to classify and use scarce land resources, so that forest coverage and species richness can gradually increase. Furthermore, the quantity and quality of grass, wetland, and other ecological land spaces can be improved and abandoned lands can be reused, thereby increasing the connectivity of the landscape. At this stage of development, land use intensity tends to slow down, ecological damage is improved, and ecological risks are gradually reduced.

3.2. Land-Use Classification

The image interpretation was mainly carried out in GEE. Following the spatiotemporal filtering and cloud removal processing of the Landsat images, functional methods were used to stitch the images, and the classification accuracy of vegetation and buildings was improved by adding the normalized difference vegetation index (NDVI) and normalized difference building index (NDBI) together. The NDVI values were calculated from the crop growth period (screening time from April to September). The NDVI and NDBI indices were calculated using the following formulas [50]:

$$NDVI = (NIR - R)/(NIR + R), \tag{1}$$

$$NDBI = (SWIR - NIR)/(SWIR + NIR), \tag{2}$$

where NIR, R, and $SWIR$ are the surface reflectivity of the near-infrared, red-band, and short-wave infrared, respectively.

The B2, B3, B4, B5, B6, and B7 bands of Landsat OLI images (six bands corresponding to Landsat TM/ETM+ images) were combined, cut, and spliced, and six auxiliary classification spectral indices (Table 2) were combined to improve the classification accuracy. The quality of the training samples directly affected the final classification results. This paper combined digital elevation model (DEM) data and Google historical images to provide a visual interpretation for drawing training samples [51–53]. According to the land-use classification standard and the actual situation of the research area, the land-use types in Zhangjiachuan County were divided into five categories: cultivated land (623 sample rectangles), forest (503 sample rectangles), grass (516 sample rectangles), construction land (218 sample rectangles), and water areas (183 sample rectangles). In order to ensure the accuracy of the land use classification results, the Landsat image set and Google historical imagery were superimposed and compared to extract the land parcels that have remained unchanged for several years. Combined with the random points generated by Arc GIS, these random points were used as a reference for the selection of the samples, so as to ensure that the selected year-by-year classified samples were evenly distributed in the entire study area. The time iteration and cyclic classification were performed in the GEE program using a combination of a supervised classification and manual visual interpretation to obtain land-use data for 2000, 2005, 2010, 2015, and 2020, and the classification accuracy was verified using Google historical imagery.

Table 2. Auxiliary classification spectral index.

Spectral Index	Calculation Formula
MNDWI [54]	$MNDWI = \frac{(G-MIR)}{(G+MIR)}$
RVI [55]	$RVI = \frac{NIR}{R}$
DVI [56]	$DVI = NIR - R$
SAVI [57]	$SAVI = \frac{1.5*(NIR-R)}{(NIR+R+0.5)}$
NDMI [58]	$NDMI = \frac{(G-SWIR)}{(G+SWIR)}$
EVI [59]	$EVI = \frac{2.5*(NIR-R)}{(NIR+6*R-7.5*B+1)}$

Note: (MNDWI: modified normalized difference water index; RVI: ratio vegetation index; DVI: difference vegetable index; SAVI: soil-adjusted vegetation index; NDMI: normalized dry matter index; EVI: enhanced vegetation index).

A random forest classifier was selected to plot the land-use situation. The random forest method is a comprehensive multi-decision tree-based comprehensive classifier trained and predicted by Breman which consists of a decision-tree classification using the bagging strategy and in-house algorithms on a GEE platform [51]. Due to its high classification accuracy, relatively robust performance, the ability to include more variables, fast prediction speed, good multisource remote sensing data processing capabilities, etc., it has been widely used in land-use classification [60]. The images in the study area were classified by running a random forest in the GEE, using 70% of the data as training samples and 30% as validation samples. To improve the accuracy of the classification results, two parameters were set for the random forest classifier; the number of decision trees was set to 100, and the number of feature variables was the square root of the total number of feature variables [61,62].

The selected validation samples were used as true values to analyze the accuracy of land-use classification from 2000 to 2020. The overall accuracy and Kappa coefficient could be calculated using the confusion matrix tool in the GEE platform. The total classification accuracy is equal to the total number of correctly classified pixels divided by the total number of pixels. The Kappa coefficient and overall accuracy can be measured according to the following equations [63]:

$$P_e = \frac{\sum_{i=1}^{m} a_i * b_i}{n^2}, \quad (3)$$

$$k = \frac{p_o - p_e}{1 - p_e}, \quad (4)$$

where p_o is the overall classification accuracy, a_i is the true number of samples for each land-use type, b_i is the number of predicted samples for each land-use type, m is the total number of land-use types, n is the total number of samples, and k is the Kappa coefficient. The precision assessment showed that the overall accuracies of land-use classification in 2000, 2005, 2010, 2015, and 2020, were 87.18%, 85.50%, 88.58%, 91.29%, and 89.05%, respectively, meeting the overall requirements of landscape research [64].

3.3. Construction of an Ecological Risk Cell

To fully demonstrate the spatial differentiation of landscape indicators and ecological risks in Zhangjiachuan County, this study combined the actual situation and divided it into 1378 ecological risk assessment units consisting of squares with a side length of 0.1 km (Figure 4). The ecological risk index was calculated in each cell, and the results were assigned to the central point of the assessment unit [34].

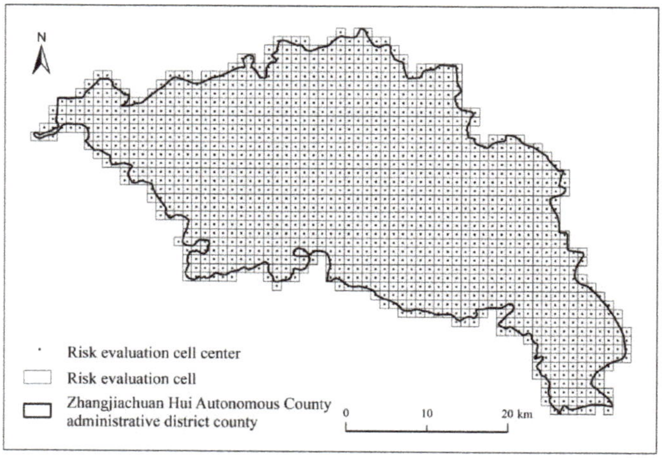

Figure 4. Ecological risk assessment unit of Zhangjiachuan County.

3.4. Construction of Landscape Ecological Risk Indices

The extent of an ecological risk depends on the strength of the regional ecosystem subjected to external interferences and the size of the internal resistance. Different landscape types have different roles in protecting species, maintaining biodiversity, improving the overall structure and function, and promoting the natural succession of landscape structures; moreover, different landscape types have different resistances to external interferences [65]. In this paper, the ecological risk index was calculated to estimate the ecological risks of Zhangjiachuan County in 2000, 2005, 2010, 2015, and 2020, and to reflect on the relationship between the landscape patterns of land-use changes and the ecological risks associated with those. The calculation formula was as follows [66]:

$$ERI_i = \sum_{i=1}^{N} \frac{A_{ki}}{A_k} R_i, \tag{5}$$

where ERI_i is the ecological risk index of the i-th risk cell, A_{ki} is the area of class-i landscape of the k-th risk cell, A_k is the area of the k-th risk cell, and R_i is the landscape loss index [67] of class-i landscape.

$$R_i = E_i * F_i, \tag{6}$$

where F_i is the ecological fragility index, which refers to the fragility of the ecosystem under a strong external disturbance of human beings. A smaller fragility denotes a greater resistance and less risk to the ecosystem. However, the variability of different landscape types in response to external interference is related to the stage of natural succession applied [66,68]. The ecological risk orders of different land-use types from low to high were construction lands, forests, grasses, cultivated lands, and water areas. After normalization, the F_i values for the five land-use types were 0, 0.25, 0.5, 0.75, and 1, respectively [69–72]. S_i is the index of landscape disturbance degree for the category i land-use type. The calculation formula is as follows [73]:

$$E_i = aC_i + bN_i + cD_i, \tag{7}$$

where C_i is the landscape fragmentation degree, N_i is the landscape separation degree, D_i is the landscape dominance degree, and a, b, and c reflect the influence of human interference on the ecosystem, representing the weights of C_i, N_i, and D_i, respectively; $a + b + c = 1$.

C_i reflects the changes in the landscape structure, function, and ecological processes; its calculation method is as follows [32]:

$$C_i = \frac{n_i}{A_i}, \tag{8}$$

where n_i is the number of patches of landscape type i, and A_i is the total area of landscape type i.

N_i is the degree of separation of individual patches in a given landscape, expressed as follows [74]:

$$N_i = \frac{A}{2A_i}\sqrt{\frac{n_i}{A}}, \tag{9}$$

where A is the total landscape area, n_i is the number of patches of landscape type i, and A_i is the total area of landscape type i.

D_i describes the advantages of patches in a given land-use type, calculated as follows [75]:

$$D_i = \frac{Q_i + M_i}{4} + \frac{Li}{2}, \tag{10}$$

where Q_i is the number of sample squares/total number of squares in plaque i, M_i is the number of plaque i/total number of plaques, and L_i is the area of plaque i/total area of quadrat.

According to relevant references and previous studies, the importance from high to low included landscape fragmentation (C_i), landscape separation (N_i), and landscape dominance (D_i), with weights of 0.5, 0.3, and 0.2 assigned to the three indicators a, b, and c, respectively. [28,65,66].

The spatial distribution of the ecological risk index was analyzed through the natural breakpoint classification. The ecological risk index was divided into five categories: lowest risk (ERI $\leq$ 0.1), low risk (0.1 $\leq$ ERI $\leq$ 0.15), medium risk (0.15 $\leq$ ERI $\leq$ 0.35), high risk (0.35 $\leq$ ERI $\leq$ 0.45), and highest risk (ERI > 0.45).

3.5. Spatial Autocorrelation Analysis

Spatial autocorrelation analysis is used to present the spatial correlation features of the spatial reference unit and adjacent unit as a function of the attribute values (ERI in the study). Global spatial and local spatial autocorrelation analyses were performed by using the GeoDa software to describe the spatial distribution of landscape ecological risks in Zhangjiachuan County. The Moran index (Moran's I) was applied to measure the global spatial autocorrelation of the ecological risk, and the numerical values represent the clustered distribution, discrete distribution, and random distribution of landscape ecological risk in space [76]. Moran's I can be calculated according to the following formula [77]:

$$I = \frac{\sum_{i=1}^{n}\sum_{i=1}^{m} w_{ij}(x_i - \overline{x})(x_j - \overline{x})}{S^2\sum_{i=1}^{n}\sum_{i=1}^{m} W_{ij}}, \tag{11}$$

$$S^2 = \frac{1}{n}\sum_{i=1}^{n}(x_i - x)^2, \tag{12}$$

$$\overline{x} = \frac{1}{n}\sum_{i=1}^{n} x_i, \tag{13}$$

where x_i and x_j represent the ERI of the reference cell i and the adjacent cell j, respectively, n and m represent the number of cells i and j, respectively, and W_{ij} is a binary matrix of adjacent spaces. When region i is adjacent to region j, W_{ij} is equal to the value 1; otherwise, W_{ij} is equal to the value 0. S^2 denotes the mean variance ($i = 1, 2, \ldots, n; j = 1, 2, \ldots, m$).

The local indicators of spatial autocorrelation (LISA) index reflects the degree to which a geographical phenomenon (or attribute values of a local unit in the entire region) is related to the same phenomenon or attribute values of adjacent local units. In general, Moran's I terms are usually decomposed and represented on different area units to form a LISA cluster diagram. The LISA cluster maps are then analyzed to generate high–high aggregation "hotspots" and low–low aggregation "cold spots" in the local space to evaluate any abnormal local spatial feature [78,79]. The formula for calculating the LISA index is as follows:

$$LISA = z_i \sum\nolimits_{j=1}^{n} w_{ij} z_{ij} (i \neq j), \tag{14}$$

where z_i and z_j represent the standardization of ERI in cells i and j, respectively, and w_i is a spatial weight matrix. If $I_i > 0$, the spatial difference between cell i and its neighboring cell j is very small, with a high–high cluster (HH, highest value in a high value neighborhoods) or low–low cluster (LL, lowest value in a low value neighborhood); if $I_i < 0$, there is a significant spatial difference in ecological risk, with high–low outliers (HL, high values in low value neighborhoods) and low–high outliers (LH, low values in high-value neighborhoods) [80,81].

4. Results

4.1. Spatiotemporal Characteristics of Land-Use Changes

The land types used in Zhangjiachuan County include cultivated land, forest, grass, water body, and construction land (Figure 5). In 2000, cultivated land was the most widely distributed land-use type in Zhangjiachuan County, with an area accounting for 48.76% of the total area, followed by grass and forest, accounting for 33.45% and 16.58%, respectively. The construction land and water bodies accounted for 1.15% and 0.06%, respectively. Cultivated land, construction land, and grass are widely distributed throughout the central and western parts of Zhangjiachuan County, while forests are mainly distributed in the east.

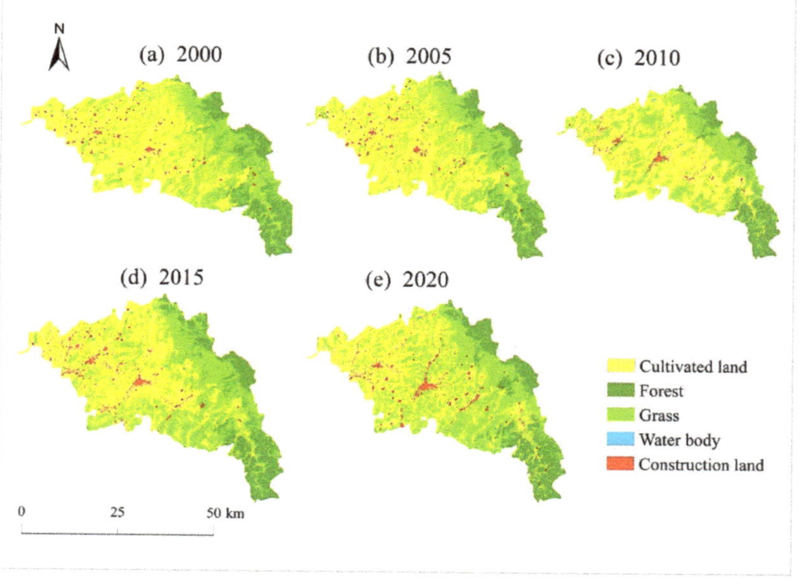

Figure 5. Spatial distribution of land use in Zhangjiachuan County from 2000 to 2020.

From 2000 to 2020, the land-use changes in Zhangjiachuan County were mainly characterized by a decrease in cultivated land areas and an increase in grass and construction land areas. The proportions of areas with water or forest changed less. From 2000 to 2020, the cultivated land areas in Zhangjiachuan County decreased by 35.22% (164.66 km^2), areas with construction land increased by 35.23% (10.86 km^2), and grass areas increased by 14.66% (74.50 km^2). In turn, areas with forest or water body changed relatively little, increasing only 6.52% (14.99 km^2) and 10.96% (0.25 km^2), respectively (Table 3). Among the different time periods, the cultivated land areas decreased the most from 2000 to 2005, decreasing by 13.76% (76.48 km^2). Areas with grass saw the largest increase rate from 2015 to 2020, up by 8.26% (42 km^2), while those with construction land increased the most from 2010 to 2015, increasing by 18.07% (5.20 km^2).

Table 3. Land-use transfer matrix from 2000 to 2020.

km^2		(Year) 2020				
		Cultivated Land	Forest	Grass	Water Body	Construction Land
2000	Cultivated land	345.79	8.78	209.33	1.01	18.05
	Forest	4.76	180.17	29.34	0.01	0.23
	Grass	109.18	47.48	267.11	0.77	8.74
	Water body	0.13	0.03	0.06	0.4	0.1
	Construction land	7.21	0.05	1.9	0.11	5.7

From 2000 to 2020, the areas with cultivated land constantly decreased in Zhangjiachuan County, prompting the expansion of construction land and grass areas. From 2000 to 2005, around 173.62 km^2 of cultivated land was converted for different land uses, of which 94.14% was converted into grass and 4.17% was converted into construction land. The transferred area of grass was 193.19 km^2, 84.61% of which came from cultivated land, and the transferred area of construction land was 10.61 km^2, 68.22% of which came from cultivated land. From 2005 to 2010, the transferred area of cultivated land was 168.37 km^2, of which 89.09% was converted into grass and 8.16% was converted into construction land. The transferred area of grass was 191.67 km^2, 78.61% of which was formerly cultivated land, while the transferred area of construction lands amounted to 18.72 km^2, 73.38% of which was previously cultivated land. From 2010 to 2015, the area of cultivated land transferred out was 180.18 km^2, of which 85.78% was converted into grass and 6.93% was converted into construction land. The transferred area of grass amounted to 194.53 km^2, 79.45% of which came from cultivated land, and the transferred area of construction land was 19.43 km^2, 64.29% of which previously consisted of cultivated land. From 2015 to 2020, the area of cultivated land transferred out was 181.94 km^2, of which 87.70% was converted into grass and 9.04% was converted into construction land. The transferred area of grass was 189.41 km^2, 84.24% of which formerly consisted of cultivated land, and the transferred area of construction land was 20.55 km^2, with 80.04% having previously consisted of cultivated land (Figure 6).

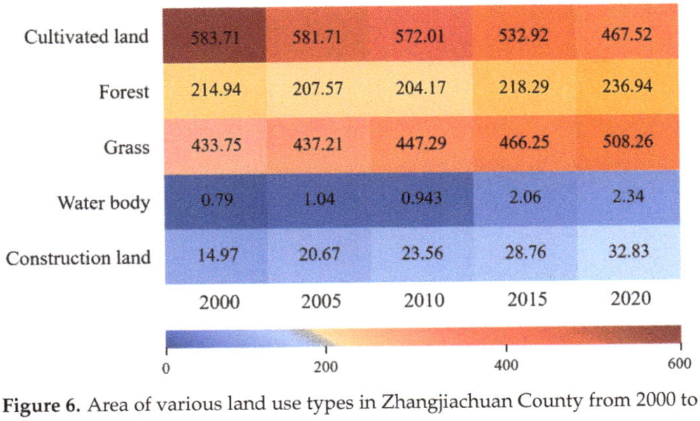

Figure 6. Area of various land use types in Zhangjiachuan County from 2000 to 2020 (km^2).

4.2. Spatiotemporal Changes of Landscape Ecological Risks

In 2000, the landscape ecological risk area of Zhangjiachuan County was mainly concentrated near the Shixiakou Reservoir in the north and the Dongxiakou Reservoir in the middle; the low-risk area was mainly concentrated in the Longshan mountainous area in the east. On the whole, the ecological risk of land-use changes in Zhangjiachuan County presented a trend of being high in the west to low in the east but being mainly low (Figure 7). From 2000 to 2020, the ecological risks of land-use changes in Zhangjiachuan County were mainly the transformation between low risk and medium risk, with the lowest-risk area changing very little. From 2000 to 2005, the moderate-risk-level regions gradually emerged in the central and western regions of Zhangjiachuan County, and the low-risk areas basically remained unchanged. From 2005 to 2010, the change pattern of medium-risk regions was similar to that of the previous stage, further expanding in the central and western regions. From 2010 to 2015, the medium-risk areas further increased, the geographic center was shifted back to the midwest from the west of the previous stage, and the risk area distribution was relatively concentrated. Lastly, from 2015 to 2020, the ecological risk in Zhangjiachuan County improved, and the overall pattern was basically the same as that of the previous stage, with the areas prone to medium-level risks decreasing.

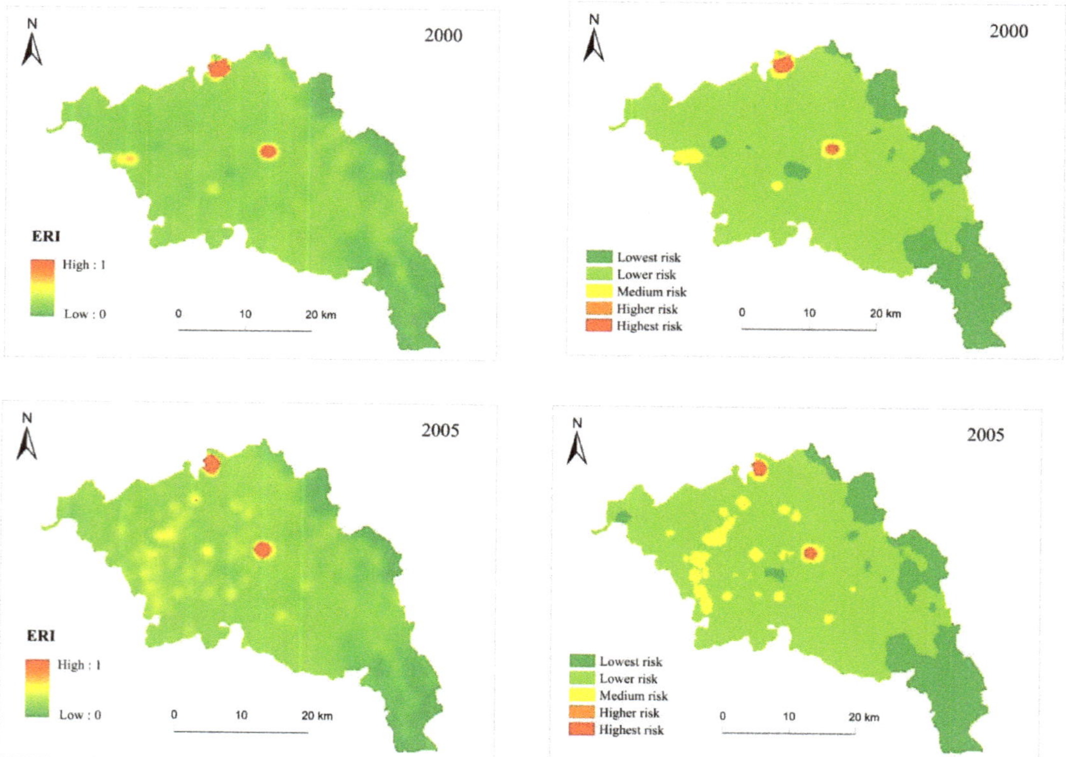

Figure 7. *Cont.*

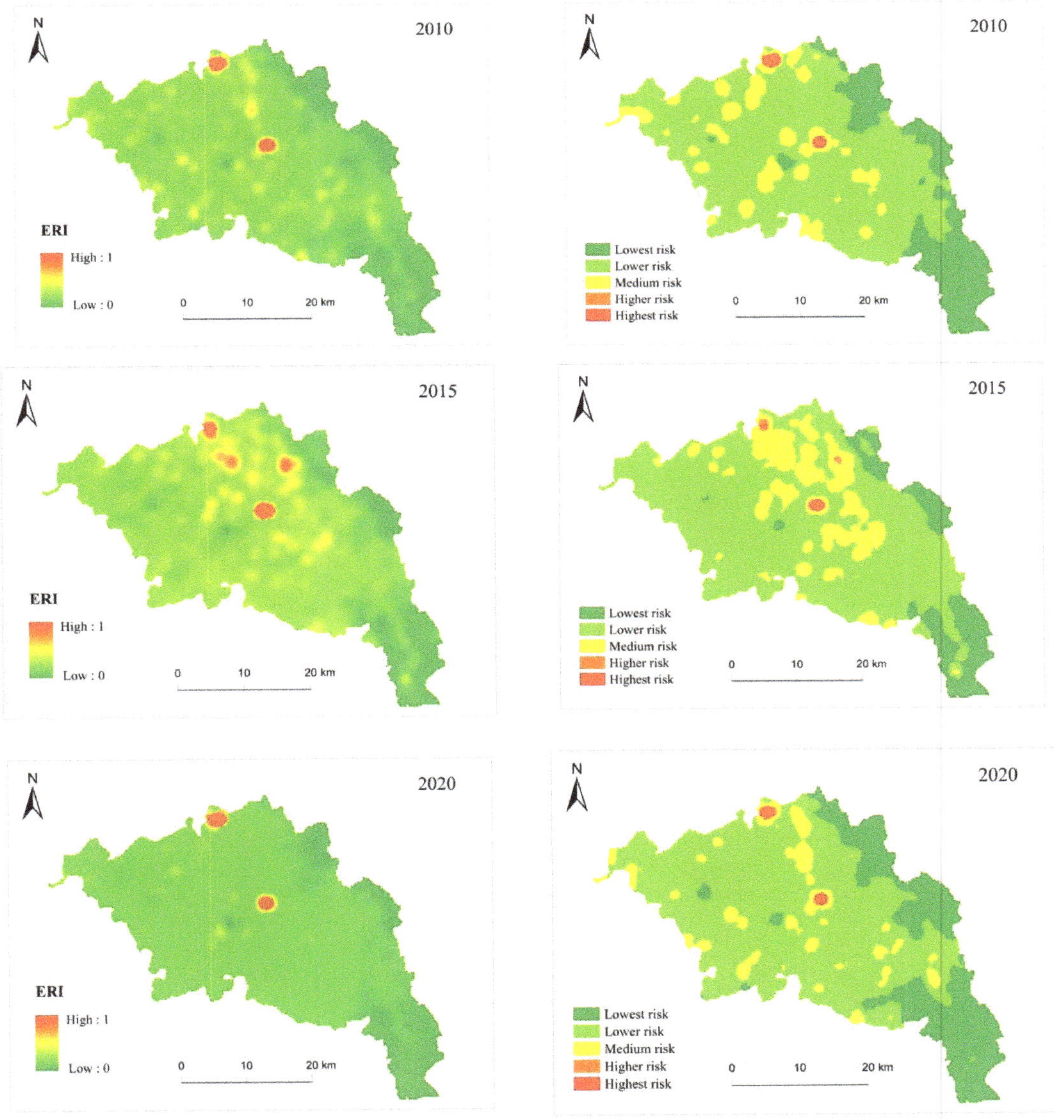

Figure 7. Spatial distribution of ecological risk index (**left**) and grade (**right**) in Zhangjiachuan County from 2000 to 2020.

From 2000 to 2020, the ecological risk value of each ecological risk cell increased and decreased, and the overall ecological risk value showed a trend of increasing first and decreasing later (Figure 8). However, the lowest-risk areas and low-risk areas showed a trend of decreasing and then increasing, while the areas of medium, high, and highest risk first increased and then decreased. From 2000 to 2015, the areas of the lowest risk and low risk levels decreased by 17.14% (38.76 km^2) and 17.22% (140.40 km^2), respectively; the areas of medium risk, high risk, and highest risk increased by 89.53% (181.2 km^2), 21.50% (0.92 km^2), and 11.27% (0.64 km^2), respectively. The ecological risks associated with land-use changes increased during this period. From 2015 to 2020, the areas associated with the lowest risk and low risk levels were expanded by 26.42% (81.24 km^2) and 2.27% (18.92 km^2), and the areas of medium risk, high risk, and highest risk decreased by 67.49% (81.56 km^2), 59.70% (1.6 km^2), and 32.63% (1.24 km^2), respectively. The ecological risks of land-use changes decreased during this period.

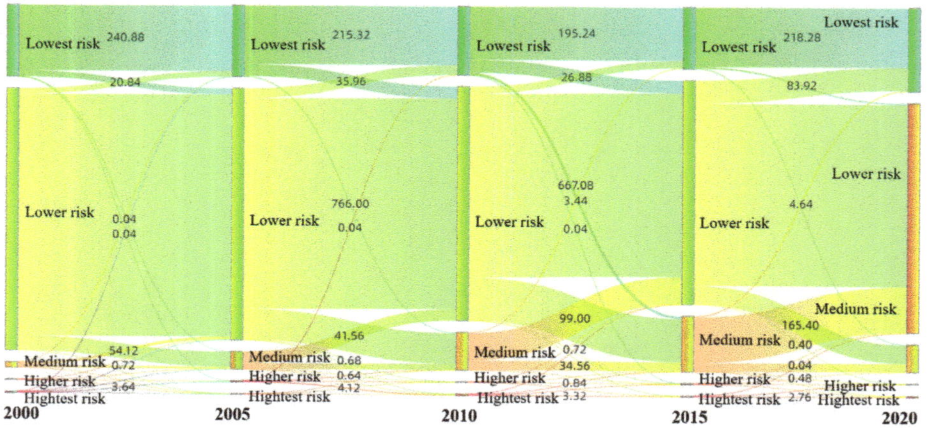

Figure 8. Ecological risk transfer in Zhangjiachuan County from 2000 to 2020 (km^2).

4.3. Spatial Autocorrelation Analysis

The spatial autocorrelation of the ecological risks of land-use changes is closely related to spatial gain. Figure 9 shows the calculated results of Moran's I at different spatial distances varying from 0.1 to 0.5 km. All Moran's I values were greater than 0, indicating that the ecological risk values of adjacent cells were similar in space, and the ecological risk index of Zhangjiachuan County had a significant positive spatial correlation within 20 years ($p < 0.05$). Moran's I with a space distance of 0.1 km was the largest and decreased with increasing distance. When the distance ranged from 0.1 to 0.15 km, Moran's I showed consistent change trends in 2000, 2005, 2010, 2015, and 2020. When equal distance Moran's I values were considered, the value was the lowest in 2005 and the highest in 2015, showing a general change trend of decline from 2000 to 2005, increase from 2005 to 2015, and decline from 2015 to 2020. This suggests that the spatial concentration of ecological over the studied time period decreased slightly from 2000 to 2005, increased gradually from 2005 to 2015, and decreased slightly from 2015 to 2020.

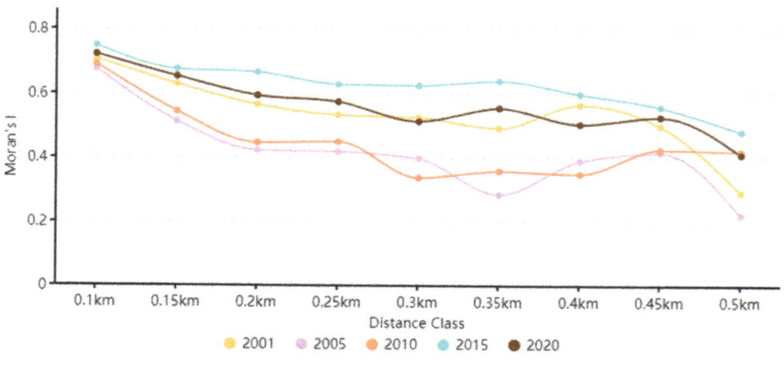

Figure 9. Moran's I values for the ecological risk index (ERI) using different distances.

The Moran index can often be used to study the overall distribution and spatial aggregation of a region, but it cannot show the spatial correlations within it; therefore, local autocorrelation LISA analysis was used to research the correlation degree of ecological risks in the study area and to study whether it exhibits some spatial aggregation (Figure 10). Four kinds of significant autocorrelations—high–high (HH), low–low (LL), low–high (LH), and high–low (HL)—appeared in the study area. The HH areas were mainly found in the center and northwest of Zhangjiachuan County, the LL areas were mainly distributed in the east and southeast, and the HL areas mainly occurred in the northwest and southwest. The ecological risks in the midwestern regions were relatively high, while those in the east and southeast regions were relatively low.

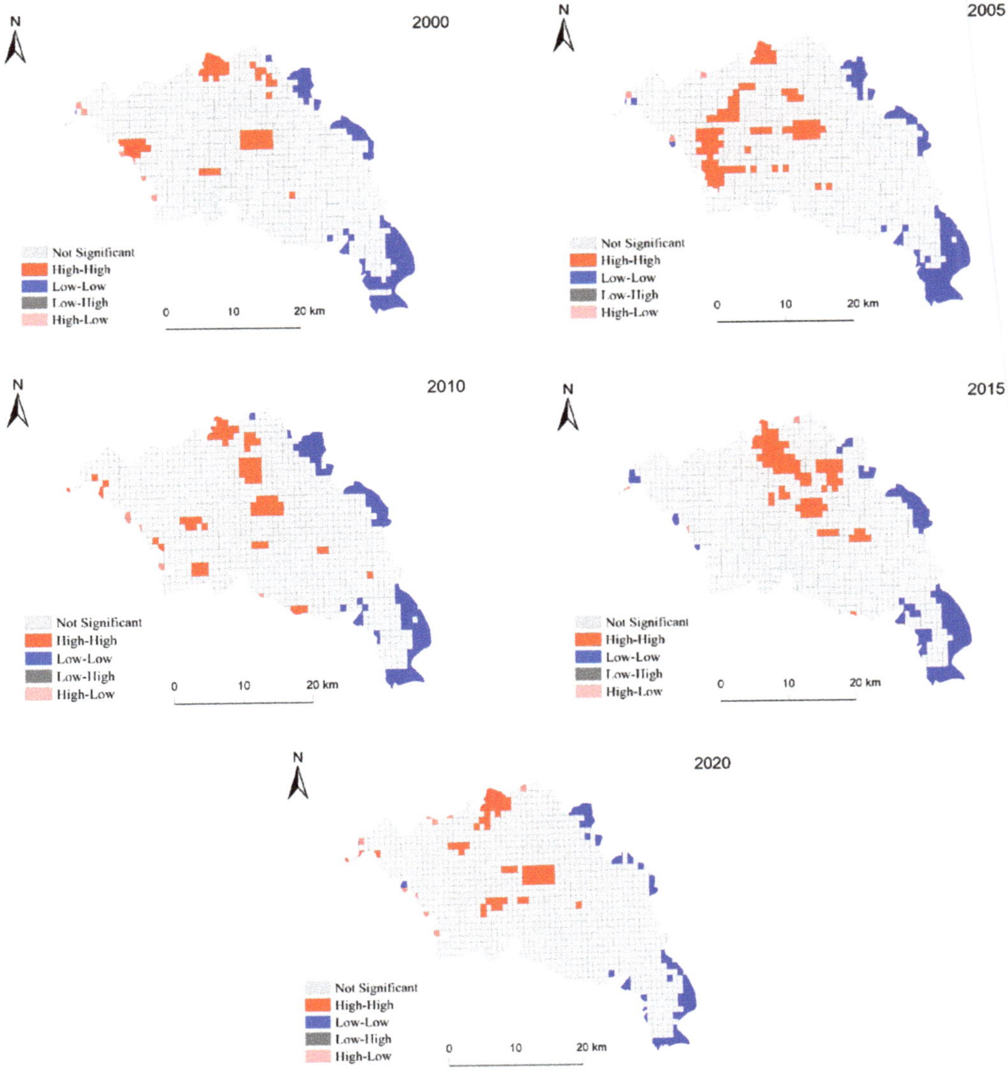

Figure 10. Local indicators of spatial autocorrelation (LISA) map of the local spatial auto-orrelation of ecological risks in Zhangjiachuan County from 2000 to 2020.

4.4. Verification of the Theoretical Framework of Ecological Risk Transformation of Land-Use Change

In general, the ecological risk index of land-use changes in Zhangjiachuan County showed a trend of increasing first and then decreasing (Figure 11). Over 2000–2010, the change was relatively slow. The risk index increased from 146.8 in 2000 to 153.3 in 2010 by a total increase of 4.2% (6.5). Over 2010–2015, the ecological risk value increased sharply to 172.6 in 2015, a total increase of 11.2% (19.3). From 2015 to 2020, the ecological risk value decreased to 157.5 in 2020, with a reduction rate of 8.7% (15.1). The value of the ecological risk index showed an "inverted U-shaped" change trend in the area over the 20 studied years. Therefore, the temporal and spatial change characteristics of ecological risks of land-use changes in Zhangjiachuan County are consistent with the ecological risk change trend of the transformation theoretical framework of ecological risks of land-use changes.

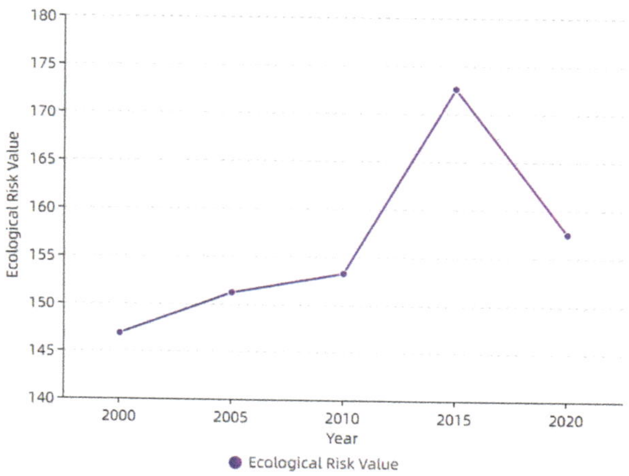

Figure 11. Overall landscape ecological risk values of Zhangjiachuan County from 2000 to 2020.

5. Discussion

5.1. Formation Mechanism of the Spatial Differentiation of Ecological Risks

Land is the carrier of the main social and economic activities, and it is one of the most intuitive forms of human development and utilization of the natural geographical environment [82]. Changes in the structures and patterns of lands are strongly related to the spatial and temporal distribution and dynamics of landscape ecological risks. According to the landscape ecological risk assessment of spatial patterns and analyses of the influence of the number, functions, and combinations of landscape elements on ecological risk, land-use changes can affect the structure and function of landscapes to varying degrees [11]. The ecological risks of land-use changes in Zhangjiachuan County are high in the west and low in the east, and the overall risk is low. Longshan Mountain is located In the east of the study area, with high terrains, as well as a cold and humid climate, consisting mostly of forested lands with a low human impact; the ecological risk in the east is, therefore, low. The midwestern regions are characterized by frequent droughts, large surface water runoff, rare vegetation, and serious water and soil loss; most areas consists of cultivated and construction lands with intense human influences, thereby making the ecological risk in this region high. The areas associated with the most serious risks were located close to the Shixiakou Reservoir in the north and Dongxiakou Reservoir in the middle. The construction of the reservoir has changed the original flood plain of the river, damaged the habitat of organisms, flooded the land vegetation, and reduced the extent of the wetland areas, greatly changing the type of land use and triggering ecological risks.

The ecological risks of land-use changes in Zhangjiachuan County generally showed a trend of increasing first and then decreasing. The risk increased from 2000 to 2015 and decreased slightly from 2015 to 2020. The west of Zhangjiachuan County is located in the hilly and gully area of the Loess Plateau, with a low terrain, poor vegetation, and the abandonment of cultivated lands; increased landscape fragmentation was observed from 2000 to 2005. At the same time, the expansion of Longshan Town in the west also introduced certain risks to the ecology of the studied area. Over 2005–2010, the main reason for the expansion of medium-risk areas in the midwestern regions was that the expansion of construction land occupied cultivated land, destroyed the landscape connectivity of the original cultivated lands, and deteriorated the landscape ecology. From 2010 to 2015, the implementation of the transformation project of medium- and low-yield fields in the comprehensive agricultural development of the Dayang Township in the west improved the ecological problems in the west of Zhangjiachuan County. However, due to the influence of national policies, some farmers in Zhangjiachuan County migrated to Xinjiang and other places, which aggravated the loss of cultivated land areas in the east. At the same time, the town of Zhangjiachuan expanded rapidly, which made the landscape more fragmented and dramatically increased the ecological risks in the area. In general, from 2000 to 2015, on the one hand, Zhangjiachuan County had a complex geographical environment, undulating terrain, changeable climate, and frequent natural disasters, resulting in certain ecological risks. On the other hand, the remoteness of the area and difficulties associated with transportation both hindered the improvement of agricultural mechanization; therefore, the area could not meet the requirements of high-intensity farming processes [83]. At the same time, the economic benefits of agricultural planting were not obvious, which led to the low enthusiasm of some farmers for planting, along with the growing phenomenon of large areas of idle cultivated land, turning cultivated land into wasteland [84,85]. At the same time, with the expansion of the population and the improvement of the per capita consumption level, the continuous development and expansion of construction lands led to a more severe loss and fragmentation of cultivated lands, and frequent human activities and urbanization fragmented and complicated the landscape, thereby increasing the ecological risk.

With the rapid development of modern society and the economy from 2015 to 2020, increasingly serious ecological problems have attracted great attention from the state, and a series of deployments have been made; for example, the land greening action was launched to promote the comprehensive control of desertification, rocky desertification, and soil erosion. During the 13th Five-Year Plan period, Zhangjiachuan County planned and implemented four small watershed comprehensive treatment projects, including the 2016 National Comprehensive Agricultural Development Project, Zhangjiachuan Maguan Project, and Area Comprehensive Control Project, which further promoted the comprehensive control level of soil erosion and ecological quality of Zhangjiachuan County [86]. Therefore, from 2015 to 2020, the ecological risk decreased as a result of the abovementioned measures.

According to the theoretical framework of ecological risk transformation of land-use changes and a field verification of Zhangjiachuan County, the construction land areas in Zhangjiachuan County began to expand on a small scale during the early stages of the study, occupying part of the cultivated lands and grasslands; at the same time, a phenomenon of abandoned cultivated lands also prevailed, which damaged the landscape connectivity, resulting in a relatively slow increase of ecological risks from 2000 to 2010. From 2010 to 2015, Zhangjiachuan Town continued to expand outward, construction land started to emerge in a disorderly manner, and the loss of cultivated land intensified, leading to a sharp increase in ecological risks during this period. The implementation of the ecological environment governance project improved the ecological quality of the area and reduced the ecological risks over 2015–2020. Therefore, the overall ecological risk of land-use changes in Zhangjiachuan County from 2000 to 2020 presented an "inverted U-shaped" change trend consistent with the theoretical framework of the ecological risk transformation of land-use changes.

5.2. Comparison of the Ecological Risk Results and Other Studies

This study put forward a theoretical framework for the transformation of the ecological risks of land-use changes. It is believed that there was a transformation of the ecological risk of land-use changes in Zhangjiachuan County, showing an "inverted U-shaped" change. The relationship between land-use changes and ecological risks is complex. According to the forest transformation theory and an environmental Kuznets curve, this paper further improved the research process and content of the ecological risk of land-use changes. It provides an important theoretical reference for ascertaining the ecological risks associated with the processes of regional land-use transformation, expanding the research vision of regional ecological risks and forming a new perspective. At the same time, it upholds the concept of theory guiding practice, fills the shortcomings of the relative lack of previous research theories in this field, and provides theoretical support for the ecological risk research of land-use changes in other regions worldwide.

The reasons of the ecological risk of land-use change in this study area are described below. On the one hand, Zhangjiachuan County is experiencing ecological problems due to changes in its geographical environment and climate. On the other hand, strong human activities have caused the loss of cultivated lands and expansion of construction lands, leading to a more fragmented landscape and increasing the intensity of the ecological risks in the area. The relationship between land-use changes and landscape ecological risk is complex, and different modes of action can cause different degrees of ecological risk. Song et al. [87] studied the impact of land-use changes in mining areas on the ecological environment and showed that the land use/coverage of the mining area often underwent great changes with changes in mining activities, while the vegetation, soil, and water would also suffer different degrees of damage. Kayumba [88] studied the wetlands in Bayanbulk and found that human activities such as urbanization, overgrazing, and tourism seriously changed the ecosystem of the region, causing ecological deterioration and ecological risks.

According to the analysis of the landscape ecological risks of land-use changes in Zhangjiachuan County, intense human activities have sharply reduced the natural vegetation in the area, and the landscape has become more fragmented, resulting in a series of ecological problems and increasing the associated ecological risks. However, these risks are reversible. The ecological problems could be remedied through certain countermeasures, thereby reducing the ecological risks. This is consistent with the research conclusions of Liang [67] on the dynamic changes of land use and ecological risks in the Three Gorges Reservoir Area of China and those of Liu [13] on the Shaanxi Province of China. However, the observations made in this work differ from the research conclusions of Peng [11] on the ecological risks of the Yongjiang River basin in Zhejiang Province, those of Yan [27] on typical areas of the Yellow River basin in China, and those of Hassan Omar [3] who studied land0use changes in Zanzibar (Tanzania). The main reason for this difference is the certain blindness in carrying out ecological protection and restoration in some areas, with a lack of systematic and comprehensive arrangements, resulting in few ecological restoration effects. Therefore, different land-use control strategies should be developed to handle ecological problems in different regions, and ecological restoration interventions should be conducted to further realize the sustainable utilization of land resources.

5.3. Policy Enlightenment

Although the problem of cultivated land fragmentation still existed in the study area from 2015 to 2020, and the areas of construction lands continued to expand, and Zhangjiachuan County effectively improved the landscape ecology of the area by conducting a series of comprehensive management projects. This paper showed that the implementation of governance projects can improve the regional ecology of a region and plays a positive role in reducing ecological risks. For example, following the completion of a dam in 2009, the Three Gorges Reservoir Area implemented a number of measures such as a project of returning cultivated lands to forests and developing orchards to promote the management of the Three Gorges Reservoir area and reduce ecological risks. Moreover, the ecological water transmission project in the Tarim River basin has increased the natural vegetation area and the riverbank vegetation coverage area, and the overall natural ecology of the whole region has been improved [89]. At present, the ecological governance projects have achieved some results, but the factors that cause ecological risks still exist and have not been effectively addressed. Therefore, Zhangjiachuan County still faces a certain degree of ecological risks in the future. According to our results, future policy formulations should focus on the aspects described below.

First, the ecological restoration of the reservoir should be strengthened. On the one hand, the protection of the ecological environment around the reservoir should be enhanced. We should promote the cultivation of water conservation forests, protect the existing forests and grasslands in the reservoir area, follow appropriate measures to return cultivated lands to forests, effectively prevent water and soil loss, ensure water supplies, and purify and reduce water pollution. On the other hand, the water environment in the reservoir area should be comprehensively improved. By improving the county government infrastructure, we should increase the construction of sewage treatment systems, adjust agricultural production structures, improve planting methods, advocate for ecological agriculture, effectively control agricultural non-point-source pollution, and minimize the ecological risk of the reservoir.

Secondly, the agricultural operation mode should be innovated. On the one hand, rural land reforms should be promoted. To some extent, owning fragmentated cultivated lands is a key factor limiting the enthusiasm of farmers for production and leading to a low efficiency of cultivated land utilization. Under the background of the transfer of the rural labor force in Zhangjiachuan County, the rural land transfer system should be improved, the transfer behavior should be standardized, and cultivated lands should be centrally managed. At the same time, the land management department should strengthen the supervision and management of the transfer of land management rights to protect the interests of farmers. On the other hand, the mode of agricultural development should be changed. The development of the "party building + enterprises + cooperatives + production bases + farmers" should be explored and established, an industrial chain of "farmers planting + cooperative purchasing + enterprise production" should be formed, and a stable increase in the income of farmers should be ensured. At the same time, according to the climate, terrain, environment, land resources, and other characteristics of all townships (towns) and villages in the county, timely and reasonably adjustments should be made by planting different types of crops according to local conditions, selecting good varieties, vigorously promoting good varieties and advanced applicable technologies, increasing the economic benefits of agricultural planting, and improving the enthusiasm of farmers.

6. Conclusions

Using the forest transformation theory and an environmental Kuznets curve, this paper analyzed the close relationship between land-use changes and the ecological risks associated with these; the theoretical framework of the ecological risk transformation of land-use changes was considered for Zhangjiachuan County, Tianshui City, Gansu Province as an example. The spatial and temporal pattern changes of land use in Zhangjiachuan County from 2000 to 2020 were analyzed, and the spatial and temporal characteristics of landscape ecological risks in the study area were evaluated. Our results show that the land-use changes in Zhangjiachuan County from 2000 to 2020 mainly included a decrease in the area of cultivated lands and the expansion of grasslands and construction lands. The ecological risks increased slowly from 2000 to 2010, increased sharply from 2010 to 2015, and decreased from 2015 to 2020. The overall "inverted U-shaped" trend is consistent with the transformation theoretical framework of the ecological risks of land-use changes. In terms of patterns, the ecological risk of land-use changes in Zhangjiachuan County was basically reflected in the distribution trend of high in the west and low in the east. In terms of areas, those with the lowest risk and low risk levels decreased from 2000 to 2015, and areas with medium risk, high risk, and highest risk levels increased, thereby aggravating the ecological risks of the area. From 2015 to 2020, the areas with the lowest risk and low risk levels increased, while the areas with medium, high, and highest risk levels decreased, reducing the ecological risks during this period. By comparing the ecological risk change of land-use change with that of some countries, systematic and comprehensive arrangements should be made when carrying out regional ecological restoration projects, and differentiated land-use control strategies should be formulated according to local conditions to achieve the sustainable use of land resources.

There were some limitations to this study. The spatiotemporal evolution of ecological risks is a comprehensive and complex process, which is affected by the population, economy, and production, thus necessitating further research and analysis. At the same time, longer time series should be studied to clarify the impact of land-use changes on ecological risks, and the direct relationship between landscape patterns and ecological risks also requires further research. In summary, future ecological risk research on land-use changes should consider highly variable social, economic, and environmental factors. The relationship between land-use changes and ecological risks would also need to be confirmed through more empirical research. While ecological protection is not a linear decision-making process, it requires a dynamic adaptive response to changing land-use types. Land-use changes in areas with high ecological risks should be paid attention to, land remediation efforts should be multiplied, and the resilience to ecological risks should be improved.

Author Contributions: Conceptualization, W.S.; methodology, H.G. and W.S.; formal analysis, H.G. and W.S.; investigation, W.S. and H.G.; resources, W.S.; writing—original draft preparation, H.G. and W.S.; writing—review and editing, W.S.; supervision, W.S. All authors have read and agreed to the published version of the manuscript.

Funding: This research was funded by the Strategic Priority Research Program of Chinese Academy of Sciences (grant number XDA20040201), the Second Tibetan Plateau Scientific Expedition and Research (grant number 2019QZKK0603), and the Project of National Natural Science Foundation of China (grant number 42071233).

Institutional Review Board Statement: Not applicable.

Informed Consent Statement: Not applicable.

Data Availability Statement: All relevant datasets in this study are described in the manuscript.

Conflicts of Interest: The authors declare no conflict of interest.

References

1. Foley, J.A.; DeFries, R.; Asner, G.P.; Barford, C.; Bonan, G.; Carpenter, S.R.; Chapin, F.S.; Coe, M.T.; Daily, G.C.; Gibbs, H.K. Global consequences of land use. *Science* **2005**, *309*, 570–574. [CrossRef] [PubMed]
2. García-Nieto, A.P.; Geijzendorffer, I.R.; Baró, F.; Roche, P.K.; Bondeau, A.; Cramer, W. Impacts of urbanization around Mediterranean cities: Changes in ecosystem service supply. *Ecol. Indic.* **2018**, *91*, 589–606. [CrossRef]
3. Omar, H.; Cabral, P. Ecological risk assessment based on land cover changes: A case of Zanzibar (Tanzania). *Remote Sens.* **2020**, *12*, 3114. [CrossRef]
4. Wu, X.; Hu, F. Analysis of ecological carrying capacity using a fuzzy comprehensive evaluation method. *Ecol. Indic.* **2020**, *113*, 106243. [CrossRef]
5. Bryan, B.A.; Gao, L.; Ye, Y.; Sun, X.; Connor, J.D.; Crossman, N.D.; Stafford-Smith, M.; Wu, J.; He, C.; Yu, D. China's response to a national land-system sustainability emergency. *Nature* **2018**, *559*, 193–204. [CrossRef] [PubMed]
6. Qiu, M.; Zuo, Q.; Wu, Q.; Yang, Z.; Zhang, J. Water ecological security assessment and spatial autocorrelation analysis of prefectural regions involved in the Yellow River Basin. *Sci. Rep.* **2022**, *12*, 1–15. [CrossRef] [PubMed]
7. Cao, Q.; Zhang, X.; Lei, D.; Guo, L.; Sun, X.; Wu, J. Multi-scenario simulation of landscape ecological risk probability to facilitate different decision-making preferences. *J. Clean. Prod.* **2019**, *227*, 325–335. [CrossRef]
8. Chen, J.; Dong, B.; Li, H.; Zhang, S.; Peng, L.; Fang, L.; Zhang, C.; Li, S. Study on landscape ecological risk assessment of Hooded Crane breeding and overwintering habitat. *Environ. Res.* **2020**, *187*, 109649. [CrossRef] [PubMed]

9. United States. Environmental Protection Agency. Report on the Ecological Risk Assessment Guidelines Strategic Planning Workshop. In *Risk Assessment Forum*; US Environmental Protection Agency: Washington, DC, USA, 1992.
10. United States. Environmental Protection Agency. Guidelines for ecological risk assessment. In *Risk Assessment Forum*; US Environmental Protection Agency: Washington, DC, USA, 1998.
11. Tian, P.; Li, J.; Gong, H.; Pu, R.; Cao, L.; Shao, S.; Shi, Z.; Feng, X.; Wang, L.; Liu, R. Research on land use changes and ecological risk assessment in Yongjiang River Basin in Zhejiang Province, China. *Sustainability* **2019**, *11*, 2817. [CrossRef]
12. Suter, G.W. Endpoints for regional ecological risk assessments. *Environ. Manag.* **1990**, *14*, 9–23. [CrossRef]
13. Liu, D.; Chen, H.; Zhang, H.; Geng, T.; Shi, Q. Spatiotemporal evolution of landscape ecological risk based on geomorphological regionalization during 1980–2017: A case study of Shaanxi Province, China. *Sustainability* **2020**, *12*, 941. [CrossRef]
14. Xu, X.; Yang, G.; Tan, Y.; Zhuang, Q.; Li, H.; Wan, R.; Su, W.; Zhang, J. Ecological risk assessment of ecosystem services in the Taihu Lake Basin of China from 1985 to 2020. *Sci. Total Environ.* **2016**, *554*, 7–16. [CrossRef]
15. Cui, L.; Zhao, Y.; Liu, J.; Han, L.; Ao, Y.; Yin, S. Landscape ecological risk assessment in Qinling Mountain. *Geol. J.* **2018**, *53*, 342–351. [CrossRef]
16. Carlson, M.; Browne, D.; Callaghan, C. Application of land-use simulation to protected area selection for efficient avoidance of biodiversity loss in Canada's western boreal region. *Land Use Policy* **2019**, *82*, 821–831. [CrossRef]
17. Xie, L.; Wang, H.; Liu, S. The ecosystem service values simulation and driving force analysis based on land use/land cover: A case study in inland rivers in arid areas of the Aksu River Basin, China. *Ecol. Indic.* **2022**, *138*, 108828. [CrossRef]
18. Tian, P.; Cao, L.; Li, J.; Pu, R.; Gong, H.; Li, C. Landscape characteristics and ecological risk assessment based on multi-scenario simulations: A case study of Yancheng Coastal Wetland, China. *Sustainability* **2020**, *13*, 149. [CrossRef]
19. Liang, Y.; Song, W. Integrating potential ecosystem services losses into ecological risk assessment of land use changes: A case study on the Qinghai-Tibet Plateau. *J. Environ. Manag.* **2022**, *318*, 115607. [CrossRef]
20. Salvati, L.; Tombolini, I. A diachronic classification of peri-urban forest land based on vulnerability to desertification. *Int. J. Environ. Res.* **2014**, *8*, 279–284.
21. Ayre, K.K.; Landis, W.G. A Bayesian approach to landscape ecological risk assessment applied to the Upper Grande Ronde Watershed, Oregon. *Hum. Ecol. Risk Assess. Int. J.* **2012**, *18*, 946–970. [CrossRef]
22. Li, Y.; Huang, S. Landscape ecological risk responses to land use change in the Luanhe River Basin, China. *Sustainability* **2015**, *7*, 16631–16652. [CrossRef]
23. Li, C.; Zhang, J.; Philbin, S.P.; Yang, X.; Dong, Z.; Hong, J.; Ballesteros-Pérez, P. Evaluating the impact of highway construction projects on landscape ecological risks in high altitude plateaus. *Sci. Rep.* **2022**, *12*, 5170. [CrossRef] [PubMed]
24. Jin, X.; Jin, Y.; Mao, X. Ecological risk assessment of cities on the Tibetan Plateau based on land use/land cover changes–Case study of Delingha City. *Ecol. Indic.* **2019**, *101*, 185–191. [CrossRef]
25. Wang, D.; Ji, X.; Li, C.; Gong, Y. Spatiotemporal variations of landscape ecological risks in a resource-based city under transformation. *Sustainability* **2021**, *13*, 5297. [CrossRef]
26. Mondal, B.; Sharma, P.; Kundu, D.; Bansal, S. Spatio-temporal assessment of landscape ecological risk and associated drivers: A case study of Delhi. *Environ. Urban. ASIA* **2021**, *12* (Suppl. 1), S85–S106. [CrossRef]
27. Qu, Y.; Zong, H.; Su, D.; Ping, Z.; Guan, M. Land Use Change and Its Impact on Landscape Ecological Risk in Typical Areas of the Yellow River Basin in China. *Int. J. Environ. Res. Public Health* **2021**, *18*, 11301. [CrossRef]
28. Mo, W.; Zhao, Y.; Yang, N.; Xu, Z.; Zhao, W.; Li, F. Effects of Climate and Land Use/Land Cover Changes on Water Yield Services in the Dongjiang Lake Basin. *ISPRS Int. J. Geo-Inf.* **2021**, *10*, 466. [CrossRef]
29. Zhang, W.; Chang, W.J.; Zhu, Z.C.; Hui, Z. Landscape ecological risk assessment of Chinese coastal cities based on land use change. *Appl. Geogr.* **2020**, *117*, 102174. [CrossRef]
30. Li, J.; Pu, R.; Gong, H.; Luo, X.; Ye, M.; Feng, B. Evolution characteristics of landscape ecological risk patterns in coastal zones in Zhejiang Province, China. *Sustainability* **2017**, *9*, 584. [CrossRef]
31. Yang, Y.; Chen, J.; Lan, Y.; Zhou, G.; You, H.; Han, X.; Wang, Y.; Shi, X. Landscape Pattern and Ecological Risk Assessment in Guangxi Based on Land Use Change. *Int. J. Environ. Res. Public Health* **2022**, *19*, 1595. [CrossRef]
32. Ji, Y.; Bai, Z.; Hui, J. Landscape Ecological Risk Assessment Based on LUCC—A Case Study of Chaoyang County, China. *Forests* **2021**, *12*, 1157. [CrossRef]
33. Zhao, Y.; Huang, C.M.; Zhang, H.J. Ecological risk assessment of provincial land-use overall planning in China. *Hum. Ecol. Risk Assess. Int. J.* **2014**, *20*, 1491–1506. [CrossRef]
34. Wang, H.; Liu, X.; Zhao, C.; Chang, Y.; Liu, Y.; Zang, F. Spatial-temporal pattern analysis of landscape ecological risk assessment based on land use/land cover change in Baishuijiang National nature reserve in Gansu Province, China. *Ecol. Indic.* **2021**, *124*, 107454. [CrossRef]
35. Peng, L.; Dong, B.; Wang, P.; Sheng, S.; Sun, L.; Fang, L.; Li, H.; Liu, L. Research on ecological risk assessment in land use model of Shengjin Lake in Anhui province, China. *Environ. Geochem. Health* **2019**, *41*, 2665–2679. [CrossRef]
36. Ai, J.; Yu, K.; Zeng, Z.; Yang, L.; Liu, Y.; Liu, J. Assessing the dynamic landscape ecological risk and its driving forces in an island city based on optimal spatial scales: Haitan Island, China. *Ecol. Indic.* **2022**, *137*, 108771. [CrossRef]
37. Heenkenda, M.K.; Bartolo, R. Regional ecological risk assessment using a relative risk model: A case study of the Darwin Harbour, Darwin, Australia. *Hum. Ecol. Risk Assess. Int. J.* **2016**, *22*, 401–423. [CrossRef]

38. Kanwar, P.; Bowden, W.B.; Greenhalgh, S. A regional ecological risk assessment of the Kaipara Harbour, New Zealand, using a relative risk model. *Hum. Ecol. Risk Assess. Int. J.* **2015**, *21*, 1123–1146. [CrossRef]
39. Zhou, S.; Chang, J.; Hu, T.; Luo, P.; Zhou, H. Spatiotemporal Variations of Land Use and Landscape Ecological Risk in a Resource-Based City, from Rapid Development to Recession. *Pol. J. Environ. Stud.* **2019**, *29*, 475–490. [CrossRef]
40. Gong, J.; Cao, E.; Xie, Y.; Xu, C.; Li, H.; Yan, L. Integrating ecosystem services and landscape ecological risk into adaptive management: Insights from a western mountain-basin area, China. *J. Environ. Manag.* **2021**, *281*, 111817. [CrossRef]
41. Liu, J.; Chen, J.; Qin, Q.; You, H.; Han, X.; Zhou, G. Patch pattern and ecological risk assessment of alpine grassland in the source region of the Yellow River. *Remote Sens.* **2020**, *12*, 3460. [CrossRef]
42. Qiao, F.; Bai, Y.; Xie, L.; Yang, X.; Sun, S. Spatio-Temporal Characteristics of Landscape Ecological Risks in the Ecological Functional Zone of the Upper Yellow River, China. *Int. J. Environ. Res. Public Health* **2021**, *18*, 12943. [CrossRef] [PubMed]
43. Zhang, Y.; Li, Y.; Lv, J.; Wang, J.; Wu, Y. Scenario simulation of ecological risk based on land use/cover change—A case study of the Jinghe county, China. *Ecol. Indic.* **2021**, *131*, 108176. [CrossRef]
44. Huete, A.; Didan, K.; Miura, T.; Rodriguez, E.P.; Gao, X.; Ferreira, L.G. Overview of the radiometric and biophysical performance of the MODIS vegetation indices. *Remote Sens. Environ.* **2002**, *83*, 195–213. [CrossRef]
45. Xiao, C.; Li, P.; Feng, Z. Monitoring annual dynamics of mature rubber plantations in Xishuangbanna during 1987-2018 using Landsat time series data: A multiple normalization approach. *Int. J. Appl. Earth Obs. Geoinf.* **2019**, *77*, 30–41. [CrossRef]
46. Rondeaux, G.; Steven, M.; Baret, F. Optimization of soil-adjusted vegetation indices. *Remote Sens. Environ.* **1996**, *55*, 95–107. [CrossRef]
47. Mather, A.S. The forest Transition. *Area* **1992**, *24*, 367–379. Available online: http://www.jstor.org/stable/20003181 (accessed on 13 February 2022).
48. Panayotou, T. *Empirical Tests and Policy Analysis of Environmental Degradation at Different Stages of Economic Development*; International Labour Office: Geneva, Switzerland, 1993. [CrossRef]
49. Qun, G.; Hanying, M. Ecological restoration, social-economic changes and sustainable development in the Three Gorges Reservoir area: A case study in Yunyang, Chongqing Municipality. *Int. J. Sustain. Dev. World Ecol.* **2007**, *14*, 174–181. [CrossRef]
50. Zha, Y.; Gao, J.; Ni, S. Use of normalized difference built-up index in automatically mapping urban areas from TM imagery. *Int. J. Remote Sens.* **2003**, *24*, 583–594. [CrossRef]
51. Xie, S.; Liu, L.; Zhang, X.; Yang, J.; Chen, X.; Gao, Y. Automatic land-cover mapping using landsat time-series data based on google earth engine. *Remote Sens.* **2019**, *11*, 3023. [CrossRef]
52. Shen, Z.; Wang, Y.; Su, H.; He, Y.; Li, S. A bi-directional strategy to detect land use function change using time-series Landsat imagery on Google Earth Engine: A case study of Huangshui River Basin in China. *Sci. Remote Sens.* **2022**, *5*, 100039. [CrossRef]
53. Long, J.; Liu, Y.; Xing, S.; Qiu, L.; Huang, Q.; Zhou, B.; Shen, J.; Zhang, L. Effects of sampling density on interpolation accuracy for farmland soil organic matter concentration in a large region of complex topography. *Ecol. Indic.* **2018**, *93*, 562–571. [CrossRef]
54. Xu, H. Modification of normalised difference water index (NDWI) to enhance open water features in remotely sensed imagery. *Int. J. Remote Sens.* **2006**, *27*, 3025–3033. [CrossRef]
55. Arii, M.; van Zyl, J.J.; Kim, Y. A general characterization for polarimetric scattering from vegetation canopies. *IEEE Trans. Geosci. Remote Sens.* **2010**, *48*, 3349–3357. [CrossRef]
56. Purnamasari, E.; Kamal, M.; Wicaksono, P. Comparison of vegetation indices for estimating above-ground mangrove carbon stocks using PlanetScope image. *Reg. Stud. Mar. Sci.* **2021**, *44*, 101730. [CrossRef]
57. Huete, A.R. A soil-adjusted vegetation index (SAVI). *Remote Sens. Environ.* **1988**, *25*, 295–309. [CrossRef]
58. Klemas, V.; Smart, R. The influence of soil salinity, growth form, and leaf moisture on-the spectral radiance of spartina-alterniflora canopies. *Photogramm. Eng. Remote Sens.* **1983**, *49*, 77–83.
59. Liu, H.Q.; Huete, A. A feedback based modification of the NDVI to minimize canopy background and atmospheric noise. *IEEE Trans. Geosci. Remote Sens.* **1995**, *33*, 457–465. [CrossRef]
60. Belgiu, M.; Drăguţ, L. Random forest in remote sensing: A review of applications and future directions. *ISPRS J. Photogramm. Remote Sens.* **2016**, *114*, 24–31. [CrossRef]
61. Teluguntla, P.; Thenkabail, P.S.; Oliphant, A.; Xiong, J.; Gumma, M.K.; Congalton, R.G.; Yadav, K.; Huete, A. A 30-m landsat-derived cropland extent product of Australia and China using random forest machine learning algorithm on Google Earth Engine cloud computing platform. *ISPRS J. Photogramm. Remote Sens.* **2018**, *144*, 325–340. [CrossRef]
62. Wu, R.; Wang, J.; Zhang, D.; Wang, S. Identifying different types of urban land use dynamics using Point-of-interest (POI) and Random Forest algorithm: The case of Huizhou, China. *Cities* **2021**, *114*, 103202. [CrossRef]
63. Congalton, R.G. A review of assessing the accuracy of classifications of remotely sensed data. *Remote Sens. Environ.* **1991**, *37*, 35–46. [CrossRef]
64. Liu, F.; Qin, T.; Girma, A.; Wang, H.; Weng, B.; Yu, Z.; Wang, Z. Dynamics of land-use and vegetation change using NDVI and transfer matrix: A case study of the Huaihe River Basin. *Pol. J. Environ. Stud.* **2018**, *28*, 213–223. [CrossRef]
65. Lin, W.; Cen, J.; Xu, D.; Du, S.; Gao, J. Wetland landscape pattern changes over a period of rapid development (1985–2015) in the ZhouShan Islands of Zhejiang province, China. *Estuar. Coast. Shelf Sci.* **2018**, *213*, 148–159. [CrossRef]
66. Li, C.; Chen, J.; Liao, M.; Chen, G.; Zhou, Q. Ecological Risk Assessment of Shan Xin Mining Area Based on Remote Sensing and Geography Information System Technology. *J. Geogr. Inf. Syst.* **2018**, *10*, 234–246. [CrossRef]

67. Wu, P.; Zhan, W.; Cheng, N.; Yang, H.; Wu, Y. A Framework to Calculate Annual Landscape Ecological Risk Index Based on Land Use/Land Cover Changes: A Case Study on Shengjin Lake Wetland. *IEEE J. Sel. Top. Appl. Earth Obs. Remote Sens.* **2021**, *14*, 11926–11935. [CrossRef]
68. Xu, X.G.; Lin, H.P.; Fu, Z.Y.; Bu, R.C. Regional ecological risk assessment of wetland in the Huanghe River Delta. *Acta Sci. Nat. -Univ. Pekinensis.* **2001**, *37*, 111–120. [CrossRef]
69. Fan, J.; Wang, Y.; Zhou, Z.; You, N.; Meng, J. Dynamic ecological risk assessment and management of land use in the middle reaches of the Heihe River based on landscape patterns and spatial statistics. *Sustainability* **2016**, *8*, 536. [CrossRef]
70. Shi, H.; Yang, Z.; Han, F.; Shi, T.; Li, D. Assessing Landscape Ecological Risk for a World Natural Heritage Site: A Case Study of Bayanbulak in China. *Pol. J. Environ. Stud.* **2015**, *24*, 269–283. [CrossRef]
71. Peng, J.; Zong, M.; Hu, Y.n.; Liu, Y.; Wu, J. Assessing landscape ecological risk in a mining city: A case study in Liaoyuan City, China. *Sustainability* **2015**, *7*, 8312–8334. [CrossRef]
72. Mo, W.; Wang, Y.; Zhang, Y.; Zhuang, D. Impacts of road network expansion on landscape ecological risk in a megacity, China: A case study of Beijing. *Sci. Total Environ.* **2017**, *574*, 1000–1011. [CrossRef]
73. Genxu, W.; Guodong, C. Study on the landscape pattern of a desert-oasis ecological system: A spatial grid method and its application. *Arid Zone Res.* **1999**, *16*, 6–11. [CrossRef]
74. Kong, X.; Fu, M.; Zhao, X.; Wang, J.; Jiang, P. Ecological effects of land-use change on two sides of the Hu Huanyong Line in China. *Land Use Policy* **2022**, *113*, 105895. [CrossRef]
75. Ran, P.; Hu, S.; Frazier, A.E.; Qu, S.; Yu, D.; Tong, L. Exploring changes in landscape ecological risk in the Yangtze River Economic Belt from a spatiotemporal perspective. *Ecol. Indic.* **2022**, *137*, 108744. [CrossRef]
76. Ghulam, A.; Ghulam, O.; Maimaitijiang, M.; Freeman, K.; Porton, I.; Maimaitiyiming, M. Remote sensing based spatial statistics to document tropical rainforest transition pathways. *Remote Sens.* **2015**, *7*, 6257–6279. [CrossRef]
77. Darand, M.; Dostkamyan, M.; Rehmani, M.I.A. Spatial autocorrelation analysis of extreme precipitation in Iran. *Russ. Meteorol. Hydrol.* **2017**, *42*, 415–424. [CrossRef]
78. Alexander, L.V.; Bador, M.; Roca, R.; Contractor, S.; Donat, M.G.; Nguyen, P.L. Intercomparison of annual precipitation indices and extremes over global land areas from in situ, space-based and reanalysis products. *Environ. Res. Lett.* **2020**, *15*, 055002. [CrossRef]
79. Zhang, F.; Yushanjiang, A.; Wang, D. Ecological risk assessment due to land use/cover changes (LUCC) in Jinghe County, Xinjiang, China from 1990 to 2014 based on landscape patterns and spatial statistics. *Environ. Earth Sci.* **2018**, *77*, 491. [CrossRef]
80. Batlle, J.R.M.; Van Der Hoek, Y. Clusters of high abundance of plants detected from local indicators of spatial association (LISA) in a semi-deciduous tropical forest. *PLoS ONE* **2018**, *13*, e0208780. [CrossRef]
81. Yu, T.; Bao, A.; Xu, W.; Guo, H.; Jiang, L.; Zheng, G.; Yuan, Y.; Nzabarinda, V. Exploring variability in landscape ecological risk and quantifying its driving factors in the Amu Darya Delta. *Int. J. Environ. Res. Public Health.* **2020**, *17*, 79. [CrossRef]
82. Liu, S.L.; Cui, B.; Dong, S.; Yang, Z.; Yang, M.; Holt, K. Evaluating the influence of road networks on landscape and regional ecological risk—A case study in Lancang River Valley of Southwest China. *Ecol. Eng.* **2008**, *34*, 91–99. [CrossRef]
83. Liang, X.; Li, Y.; Zhao, Y. Coupling Land Use Analysis and Ecological Risk Assessment: A Study of the Three Gorges Reservoir Area, China. *Mt. Res. Dev.* **2020**, *40*, R1. [CrossRef]
84. Zhang, Y.; Li, X.; Song, W. Determinants of cropland abandonment at the parcel, household and village levels in mountain areas of China: A multi-level analysis. *Land Use Policy* **2014**, *41*, 186–192. [CrossRef]
85. Song, W.; Pijanowski, B.C. The effects of China's cultivated land balance program on potential land productivity at a national scale. *Appl. Geogr.* **2014**, *46*, 158–170. [CrossRef]
86. Zhang, G. An analysis of the relationship between the construction of small watershed comprehensive management project and ecological protection in Zhangjiachuan county during the 13th Five Year Plan period. *Gansu Agric.* **2022**, *02*, 67–69.
87. Song, W.; Song, W.; Gu, H.; Li, F. Progress in the remote sensing monitoring of the ecological environment in mining areas. *Int. J. Environ. Res. Public Health* **2020**, *17*, 1846. [CrossRef] [PubMed]
88. Kayumba, P.M.; Chen, Y.; Mind'je, R.; Mindje, M.; Li, X.; Maniraho, A.P.; Umugwaneza, A.; Uwamahoro, S. Geospatial land surface-based thermal scenarios for wetland ecological risk assessment and its landscape dynamics simulation in Bayanbulak Wetland, Northwestern China. *Landsc. Ecol.* **2021**, *36*, 1699–1723. [CrossRef]
89. Mamat, Z.; Halik, Ü.; Keyimu, M.; Keram, A.; Nurmamat, K. Variation of the floodplain forest ecosystem service value in the lower reaches of Tarim River, China. *Land Degrad. Dev.* **2018**, *29*, 47–57. [CrossRef]

International Journal of *Environmental Research and Public Health*

MDPI

Article

Ecosystem Health Responses of Urban Agglomerations in Central Yunnan Based on Land Use Change

Binpin Gao [1], Yingmei Wu [1,*], Chen Li [1], Kejun Zheng [1,2] and Yan Wu [1]

1 Faculty of Geography, Yunnan Normal University, Kunming 650500, China
2 Yunnan Academy of Social Sciences, Kunming 650000, China
* Correspondence: wuyingmei@hotmail.com

Abstract: Land use change in urban agglomerations is gradually becoming a major cause and a key factor of global environmental change. As a consequence of the interaction between land use and ecological processes, the transformation in natural ecosystem structure and function with human activity disturbances demands a systematic assessment of ecosystem health. Taking the Central Yunnan urban agglomeration, undergoing transition and development, as an example, the current study reveals the typical land use change processes and then emphasizes the importance of spatial heterogeneity of ecosystem services in health assessment. The InVEST model-based ecosystem service assessment is incorporated into the ecosystem health evaluation, and hotspot analysis is performed to quantitatively measure the ecosystem health response degree to land use according to spatial latitude. The study had three major findings: First, the urban land expansion in the urban agglomeration of central Yunnan between 1990 and 2020 is the most significant. Further, the rate of the dynamic change of urban land is 16.86%, which is the highest among all land types. Second, the ecosystem health of the central Yunnan urban agglomeration is improving but with obvious spatial differences, showing a trend of increasing from urban areas to surrounding areas, with the lowest ecosystem health level and significant clustering in the areas where the towns are located. The ecosystem health level is mainly dominated by the two classes of ordinary and well grades, and the sum of the two accounts for 63.35% of the total area. Third, the process of land transfer, mutual transfer between forest and grassland, and conversion from cropland to forest land contributed the most to the improvement of ecosystem health across the study area. Furthermore, the conversion from cropland and grassland to urban land is an important cause of the sustained exacerbation of ecosystem health. Significantly, the study provides a scientific reference for maintaining ecosystem health and formulating policies for macro-control of land in the urban agglomerations of the mountain plateau.

Keywords: land use change; ecosystem health; ecosystem services; InVEST model; urban agglomerations

Citation: Gao, B.; Wu, Y.; Li, C.; Zheng, K.; Wu, Y. Ecosystem Health Responses of Urban Agglomerations in Central Yunnan Based on Land Use Change. *Int. J. Environ. Res. Public Health* **2022**, *19*, 12399. https://doi.org/10.3390/ijerph191912399

Academic Editors: Paul B. Tchounwou and Xiaowei Chuai

Received: 6 August 2022
Accepted: 26 September 2022
Published: 29 September 2022

1. Introduction

With the process of globalization and integration of the world economy, international competition and cooperation relations have expanded from single urban cities to urban agglomerations [1]. The spatial structure of China's economic development has also profoundly changed, and urban agglomerations have become the main spatial form for bearing development factors and gradually become an important spatial unit for urbanization construction in China [2]. In recent years, China's major urban agglomerations, such as the Yangtze River Delta, the Pearl River Delta, and the Beijing–Tianjin–Hebei urban agglomerations, have been developing rapidly. The 19 national urban agglomerations that have been formed carry more than 75% of the urban population and contribute more than 80% of the country's GDP, but they have also been sensitive areas where ecological and environmental problems are highly concentrated and exacerbated. Urban agglomerations in China face serious problems caused by the imbalance between urbanization and ecosystem interaction processes [3,4], which is directly manifested by the dramatic changes in land use patterns, resulting in the degradation of ecosystems and their service functions, significantly altering the health and integrity of natural ecosystems. In the face of the current situation, countries around the world are paying more and more attention to the far-reaching effects of land use changes caused by rapid urbanization on global ecosystems [5,6], it is important to assess the ecological effects of land use changes from the perspective of ecosystem health [7–9] to coordinate the relationship between economic development and ecological conservation in urban agglomerations areas.

Ecosystem health refers to maintaining the integrity, stability, and sustainability of ecosystem structure and function with disturbances due to human activity and is considered the ultimate goal of environmental management [10–12]. A healthy ecosystem should be active, maintainable in its organizational structure, and able to recover itself under stress, and it is the core guarantee of sustainable human development [13,14]. Most existing studies assessing ecosystem health used either the indicator system approach [15–17], the pressure-state-response (PSR) framework [18–20], or the vigor-organization-resilience (VOR) framework [21,22]. However, to evaluate ecosystem health from the landscape viewpoint, exploring the interaction between ecosystem service change and different land use types is important, besides considering the changes in landscape structure due to human activities [23]. Based on this, Xiao et al., Ge et al., Xu et al. and Wang et al. [24–27] all explored the effects of land use change from an ecosystem health perspective and explained the ecological effects of regional land use change from different aspects.

Unlike economically developed urban agglomerations in China, the Central Yunnan urban agglomerations, as one of the important urban agglomerations in Southwest China, is at the stage of transition from early development to maturity. Although the urban agglomerations in central Yunnan have rich biological resources and innate advantages for development, their ecosystem is more sensitive. In recent decades, with constant frictions between natural ecosystems and urban construction, makes it an ideal research area for studying land use change and ecosystem health responses. In addition to incorporating the InVEST (Integrated Valuation of Ecosystem Services and Tradeoffs) model-based assessment of ecosystem service into the ecosystem health framework, this study further explores spatially the response of two trends of ecosystem health-deterioration and improvement of the study area to land use change. The key objectives of the study include: (1) To study the features of land use change of the Central Yunnan Urban Agglomeration between 1990 and 2020. (2) To assess the ecosystem health in the central Yunnan urban agglomerations and analyze its spatial and temporal evolutionary characteristics. (3) To quantitatively measure land use change responses in the deteriorating and improving ecosystem health areas. The study significantly contributes to government policies and decision-making in formulating land management plans, balanced development, and meeting conservation needs. Meanwhile, it also provides a reference for the enrichment of the ecosystem health evaluation system for mountainous urban agglomerations.

2. Materials and Methods

2.1. Study Area

The central Yunnan urban agglomeration is situated in the central Yunnan Province in southwest China (24°58′ N–25°09′ N, 100°43′ E–104°49′ E) (Figure 1). It is the most economically developed region in Yunnan Province, including 49 counties, cities, and districts, with a land area of 111,356.04 km^2, accounting for 28.26% of the province's land area, of which only 11.84% is plains. The study area belongs to the lake basin landscape of the Zhongshan plateau, with karst landforms developed in the east. The overall mountain and inter-mountain basin topography are dominant, with a large vertical height difference (between 116~4282 m above sea level), which is a typical plateau mountainous urban agglomeration. Due to the complex topographic fragmentation and fragile ecological environment in the region, it is among the urban agglomerations of western China that are more seriously constrained by topographic structures, resources, and environment. The main climate type is low latitude plateau mountain monsoon climate with a small annual temperature difference but significant vertical differences in climate, rainfall increases with altitude, good lighting conditions, and rich biological resources.

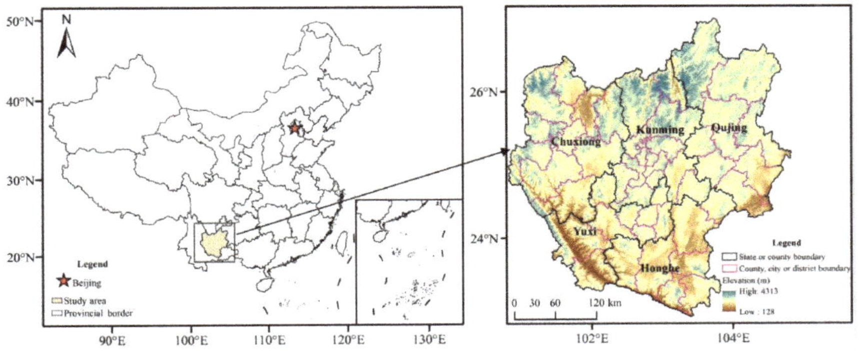

Figure 1. Location of central Yunnan urban agglomeration in China.

As a major growth pole of China's Yangtze River Economic Belt, the central Yunnan urban agglomeration is in a transitional stage from the nurturing phase of development to the mature phase. With a population of over 21 million, accounting for 46.5% of the province's population, and a total GDP exceeding 1500 billion yuan, accounting for 61.47% of the province's GDP. The dense transportation facilities, rapid population, and economic and social development of the cluster area have caused the large-scale transformation of ecological land into construction land. Additionally, due to the high-intensity economic development, the ecological and environmental problems brought about by the continuous expansion of the urban scale have become more and more prominent.

2.2. Data Sources

The research adopts 30 m resolution land use data of 1990, 2000, 2010, and 2020 as the basis for the study. The dataset comes from the Resource and Environmental Science Database of the Chinese Academy of Sciences (http://www.resdc.cn) (accessed on 6 October 2021), with Landsat5/7/8, GF2 and other satellite remote sensing images as the main information sources. And according to the research objectives and the actual situation of the surface, the land use types are divided into seven categories, which included cultivated land, forest land, grassland, water area, urban land, rural land, and unutilized land. Other physical, geographic, and economic data include elevation, precipitation, soil depth, potential evapotranspiration, normalized vegetation index, net primary productivity, grain yield, and sown area. Table 1 shows the specific sources for each of these data types used in the research.

2.3. Methodological Steps

Firstly, the spatial and temporal characteristics of land use types in the central Yunnan urban agglomeration area from 1900 to 2100 were analyzed by land use change measurement. Second, the InVEST model was used to assess changes in grain production, water conservation, carbon storage, soil conservation, habitat quality, and provide aesthetic landscape. This was applied to the physical health level assessment of ecosystems quantified by the vigor–organization–resilience (VOR) model, which together form a framework for ecosystem health assessment. Finally, the exploratory spatial data analysis (ESDA) method was used to explore the "cold spots" and "hot spots" of ecosystem health changes in the whole central Yunnan urban agglomeration. The response of ecosystem health to land use change was further revealed. The specific flowchart is as follows (Figure 2):

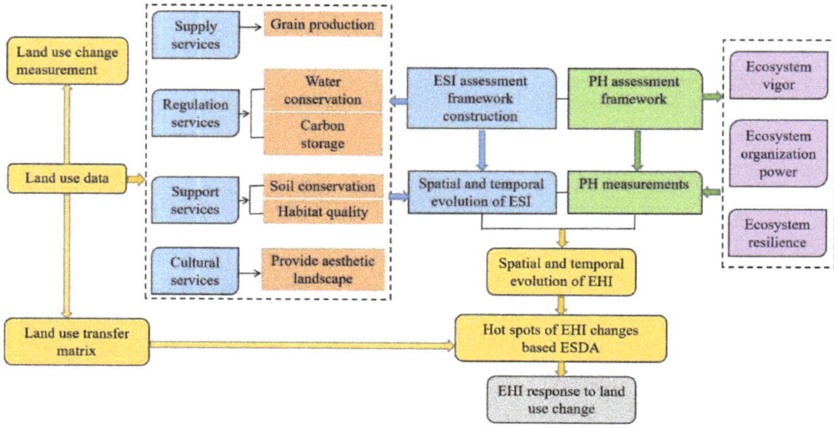

Figure 2. Flowchart of methodological steps. ESI (ecosystem service index), PH (physical health index), EHI (ecosystem health index).

Table 1. Study data sources.

Data Type	Data Sources
Land use data	Resource and Environmental Science Data Center of the Chinese Academy of Sciences (https://www.resdc.cn/) (accessed on 6 October 2021).
Elevation data	Geospatial Data Cloud Platform (http://www.gscloud.cn) (accessed on 6 October 2021).
Precipitation data	National Earth System Science Data Center (http://www.geodata.cn) (accessed on 25 January 2022).
Potential evapotranspiration	National Earth System Science Data Center (http://www.geodata.cn) (accessed on 25 January 2022).
Soil depth	World Soil Database (HWSD) China Soil Dataset (v1.1) (http://www.cryosphere.csdb.cn) (accessed on 25 January 2022).
Normalized vegetation index (NDVI)	The NDVI data in 2000, 2010 and 2020 were generated based on the MODIS vegetation index products with a spatial resolution of 250 × 250 m and a 16-day temporal resolution obtained by the NASA Earth Observation System, using the annual maximum synthesis method. The NDVI data in 1990 was obtained from the national Earth System Science Data Center.
Net Primary Productivity (NPP)	The spatial resolution of the MODIS data product is 500 m, the band is cut and stitched, and the pixel value is multiplied by a scale factor of 0.0001 to calculate. The 1990 NPP data was gathered from Chen Pengfei, "Monthly net primary productivity 1 km raster dataset of Chinese terrestrial ecosystems north of 18° N (1985–2015)" [J/DB/OL]. Electronic Journal of Global Change Data Warehousing, 2019 [28]. (http://www.geodoi.ac.cn/) (accessed on 23 December 2021).
Grain production and sown area	According to"Yunnan Provincial Statistical Yearbook" to obtain the grain production and sown area of 49 (district) counties in each year.
Grain prices	National compilation of agricultural cost-benefit information.

2.3.1. Land Use Change Measurement

(1) The land use dynamics degree. It can quantitatively describe the quantity change in a certain land use type at a certain time. The land use change process across the study

area can be observed by calculating the dynamic degree of every land use type, using the following calculation formula:

$$D_i = \left(\frac{S_i^{t2} - S_i^{t1}}{S_i^{t1}}\right) \times \frac{1}{\Delta t} \times 100\%$$

where D represents the dynamic attitude in land type ith, S_i^{t2} suggests the area in land type i at the latter moment, S_i^{t1} means the area in land type i at the initial moment, Δt refers to the time interval, and if the time interval is measured in years then D denotes the average annual change rate in land type i.

(2) Land use type transfer. The land use data across the study area was calculated by two-by-two superposition through the raster calculation in ArcGIS 10.8 to conclude the land use type transfer and spatial distribution across the study area in two phases with the following formula:

$$C_{i*j} = A_{i*j} \times 10 + B_{i*j} \quad (\text{Applicable when land use type } < 10)$$

where C_{i*j} refers to the land use change from period A to period B, and A_{i*j} and B_{i*j} are the land use types of any two periods.

2.3.2. Ecosystem Health Assessment Framework

Costanza proposed that understanding ecosystem health requires the recognition of humans as an important component of ecosystems [11]. Traditional ecosystem health assessments explore the sustainability of spatial unit patterns and ecological processes only in terms of the ecosystem itself, ignoring the benefits that humans derive from a properly functioning ecosystem [29], whereas ecosystem services can precisely link ecosystem processes, functions, and human well-being.

Therefore, this study follows the ecosystem health assessment framework proposed by Costanza and draws from the "ecosystem vigor-ecosystem organization-ecosystem resilience-ecosystem services" integrated ecosystem health assessment framework constructed by Peng et al., Pan et al. and Chen et al. [30–32]. To further emphasize the importance of spatial heterogeneity of ecosystem services in health assessment, we incorporated ecosystem service indicators quantified using the InVEST model into the assessment framework in an attempt to enhance the richness of ecosystem health assessment indicators and further improve the scope and precision of the assessment perspective [14]. The evaluation of the ecosystem health in the central Yunnan urban agglomeration comprises two components: the physical health level in the ecosystem and the integrated ecosystem service index. The specific formula is as follows:

$$EHI = \sqrt{PH \times ESI}$$

where EHI means the ecosystem health in the assessed area; PH denotes the physical health index in the assessed area; ESI is the integrated ecosystem service index of the assessed area.

2.3.3. Selection and Assessment of Integrated Ecosystem Service Indicators

The UN Millennium Ecosystem Assessment categorized the ecosystem service function classification system into four major functions: product supply service, regulating service, supporting service, and cultural service. Among them, product supply service function is the function of the ecosystem to produce or provide products; regulating service function is the function of the ecosystem to regulate human ecological environment including water production capacity and carbon storage capacity; supporting service function is the basic function necessary to ensure the provision of all other ecosystem service functions, including soil conservation, habitat quality, and cultural service function is the aesthetic landscape experience, non-material benefits from the ecosystem [33].

For assessment, six typical ecosystem service functions, namely, habitat quality, carbon storage, soil conservation, water conservation, grain production, and provide aesthetic landscape, were selected from the four major services, including ecosystem supply services, regulation services, support services, and cultural services. These ecosystem service functions are consistent with the typical features of the central Yunnan urban agglomeration and reflect the integrated water, food, soil, atmospheric, and overall ecological information of the habitat. The details of the specific process are shown in Table 2.

Table 2. Principles and methods for assessing the ecosystem service functions of each ecosystem in the central Yunnan urban agglomeration.

Ecosystem Services	Ecosystem Functions	Fundamentals	Measurement Formula
Supply Services	Grain Production	Based on the linear correlation between grain yield and NDVI, grain yield was assigned according to the ratio of raster NDVI values to total NDVI values of cultivated land based on land use type [34].	$G_i = G_{sum} \times \frac{NDVI_i}{NDVI_{sum}}$ where G_i is the grain yield of arable raster i, G_{sum} suggests the total grain yield in the study unit, $NDVI_i$ means the $NDVI$ value for arable raster i, and $NDVI_{sum}$ indicates the sum of $NDVI$ values in the study unit.
Regulation Services	Water Conservation	According to the water cycle principle, the water yield is obtained by calculating parameters, including precipitation, plant transpiration, surface evaporation, root depth as well as soil depth [35]. Afterward the runoff path topography index is measured using the DEM and the runoff residence time on the grate is measured with soil permeability and surface runoff flow coefficient. The water yield is corrected to obtain the water content [36].	$Retention = min\left(1, \frac{249}{Velocity}\right) \times min\left(1, \frac{0.9 \times TI}{3}\right) \times min\left(1, \frac{Ksat}{300}\right) \times Yield$ (1) where *Retention* means the water content (mm); *Velocity* refers to the flow rate coefficient; *TI* denotes the topographic index measured using Equation (2); *Ksat* indicates the soil saturation hydraulic conductivity (cm/d), measured with Equation (3); *Yield* denotes the water yield, measured using Equation (4). $TI = \lg\left(\frac{Drainage\_Area}{Soil\_Depth \times Percent\_Slope}\right)$ (2) where *Drainage_Area* represents the number of grids of the catchment area (dimensionless); *Soil_Depth* suggests the soil depth (mm); *Percent_Slope* indicates the percentage slope. $In(Ksat) = 20.62 - 0.96 \times In(Clay) - 0.66 \times In(Sand) - 0.46 \times In(OC) - 8.43 \times BD$ (3) where *Ksat* denotes soil saturated hydraulic conductivity (cm/d), *Clay* indicates soil clay content (%), *Sand* represents soil sand content (%), *OC* refers to soil organic carbon content (%), *BD* means soil bulk weight (g/cm^3). $Y_{jx} = \left(1 - \frac{AET_{xj}}{P_x}\right) \times P_x$ (4) where Y_{jx} represents the annual water yield; P_x means the average annual rainfall in raster cell x; AET_{xj} refers to the average annual evapotranspiration in raster cell x in land use type j.

Table 2. *Cont.*

Ecosystem Services	Ecosystem Functions	Fundamentals	Measurement Formula
	Carbon Storage	The average carbon density for above-ground carbon pool, below-ground carbon pool, soil carbon pool and dead organic carbon pool were calculated separately for different land types and were summed by multiplying the area in every land type by the corresponding carbon density.	$C_i = C_{i,\ above} + C_{i,\ below} + C_{i,\ soil} + C_{i,\ dead}$ $C_{total} = \sum_{i=1}^{n} C_i \times S_i$ where C_i is the ith land use type; $C_{i,above}$ signifies the above-ground carbon density in land use type i (t/hm^2); $C_{i,below}$ suggests the below-ground biological carbon density in land use type i (t/hm^2); $C_{i,soil}$ signifies the soil carbon density in land use type i (t/hm^2); $C_{i,dead}$ denotes the carbon density of dead organic matter in land use type i (t/hm^2), C_{total} means the total carbon stock in the ecosystem (t); S_i indicates the area in land use type i (hm^2); n denotes the number of land use types, and n is 7 in this paper.
Support Services	Soil Conservation	Soil retention is obtained by measuring the difference between potential erosion and real erosion and adding it to the sediment holding capacity.	$SM_x = RKLS_x - USLE_x + SDR_x$ $RKLS_x = R_x \times K_x \times LS_x$ $ULSEx = RKLS \times C_x \times P_x$ $SR_x = SE_x \sum_{y=1}^{x-1} USLE_y \prod_{z=y+1}^{x-1} (1 - SE_x)$ where SM_x means the soil retention of raster x, SDR_x means the sediment retention for raster x, and SE_x represents the sediment retention efficiency for raster x. $PKLS_x$ suggests the potential soil loss for raster x, and $USLE_x$ and $USLE_y$ stand for the real erosion of raster x and its upslope raster y, i.e., soil erosion under vegetation cover and soil and water conservation measures, respectively. R_x, K_x, LS_x, C_x, and P_x denote the rainfall erosion force factor, soil erodibility factor, topography factor, vegetation cover factor, and soil and water conservation measure factor for raster x, respectively.
	Habitat Quality	Generate habitat quality maps by the Habitat Quality module under the InVEST model, combining information on land cover and biodiversity threat factors.	$Q_{xj} = H_j \left[1 - \left(\frac{D_{xj}^z}{D_{xj}^z + k^z}\right)\right]$ where Q_{xj} means the habitat quality index for raster x in land use type j; H_j suggests the habitat suitability in land use type j, with the value set to [0, 1]; D_{xj} indicates the degradation of habitat for raster x in land use type j; k signifies the half-saturation constant, which takes half of the maximum degradation 0.056 (system default 0.5).
Cultural Services	Provide Aesthetic Landscape	The sown area, yield, and average price for three main crops (rice, wheat, and corn) in 49 (district) counties were used as the base data to calculate the economic value of crops per unit area. Combined with the base equivalence table of ecosystem services per unit area in the research by Xie et al. [37], the ecosystem service values of aesthetic landscapes were calculated and expressed spatially based on grid division.	$E = \frac{1}{7} \sum_{i=1}^{m} \frac{O_i P_i Q_i}{M}$ where E is the economic value of crop production per unit area of the study area; i means the crop type; O_i, P_i and Q_i represent the sown area, yield per unit area and average price of i crops, respectively; M is the total area of three crops (rice, wheat, and corn) of the study area. $ESV = \sum(A_i \cdot VC_i)$ where ESV refers to the ecosystem service value of aesthetic landscape; A_i means the area in land type ith; VC_i represents the ESV coefficient of the aesthetic landscape in land type ith.

To comprehensively measure the ecosystem services in the central Yunnan urban agglomeration, the evaluation results for the above six ecosystem service functions were normalized in ArcGIS. After eliminating the effects by different magnitudes, they were superimposed and calculated according to the mean weights. The integrated Ecosystem Service Index (ESI) in the central Yunnan urban agglomeration in 1990, 2000, 2010, and 2020 were each calculated as follows.

$$ESI = \sum_{i=1}^{n} W \times ES'$$

where *ESI* is the integrated ecosystem service index for the study area; *W* means the weight coefficient for various ecosystem service types; *ES′* is the standardized function of different types of ecosystems, and *n* is the type of ecosystem services in this study ($n = 6$).

2.3.4. Ecosystem Physical Health Indicator Selection and Assessment

The physical health of the ecosystem can be assessed on the basis of three criteria: the activity of the ecosystem to provide energy, the structure of the ecosystem to maintain health under stress, and the ability of the ecosystem to self-regulate and recover. This study is based on the results of Das, Manob et al. [38] and Pan et al. [32] to characterize the ecosystem's physical health with three indicators of health level: vigor (V), organization (O), and recovery (R).

(1) Ecosystem vigor (V): The NPP bands of 2000, 2010, and 2020 were cut and spliced by obtaining MODIS data products to calculate the annual true values of NPP for the four periods of the central Yunnan urban agglomeration. As the NPP reflects more vegetation vigor in terrestrial ecosystems, and the highland lakes in the study area play a key role in ecosystem vigor, the watershed was set to 1 in the normalization process according to the actual situation in the study area [39,40].

(2) Ecosystem organization power (O): The Weighted Mean Fractional Dimension (AWMPFD), Shannon Diversity Index (SDI), and Simpson Diversity Index (MSDI) were selected to measure the landscape heterogeneity (LH). Sprawl (CONTAG), separation (SPLIT), and connectivity (CONNECT) were then used to characterize landscape connectivity. In addition, woodlands and watersheds bear vital ecological functions of the central Yunnan urban agglomeration and should be protected as priority landscape types. Thus, the separation (SPLIT) and connectivity (CONNECT) of important patches were included as important landscape connectivity (ILC) separately in the calculation of ecosystem organization power. For the weight setting, landscape heterogeneity, and important landscape connectivity, weights were determined with reference to the previous studies. After each index was obtained through Fragstates software, the ecosystem organization power was calculated and normalized [23].

$$\begin{gathered} EO = 0.35 \times LC + 0.35 \times LH + 0.3 \times ILC \\ = 0.1 \times AWMPED + 0.15 \times SDI + 0.1 \times MSDI + 0.1 \times CONTAG + 0.25 \times SPLIT \\ +0.1 \times (SPLIT_1 + SPLIT_2) + 0.05 \times (CONNECT_1 + CONNECT_2) \end{gathered}$$

where *EO* means the ecosystem organization force coefficient, *LH* represents landscape heterogeneity, *LC* denotes landscape connectivity, *ILC* is important to landscape connectivity, $SPLIT_1$ and $SPLIT_2$ are the separation degree of watershed and woodland, respectively, and $CONNECT_1$ and $CONNECT_2$ are the connectivity degree of watershed and woodland, respectively.

Ecosystem resilience (R): Referring to the study by Liu et al. and Peng et al. [7,15], the resilience and resistance in various land use types were assigned. The resilience coefficients in various types of sites were corrected by combining NDVI data of the study area (Formula (1)). In the correction process, the study considered that water bodies do not have obvious vegetation reflection characteristics; therefore, water bodies were not included in the correction process to guarantee the accuracy of the results. The central

Yunnan urban agglomeration is an area of intensive human activities, and the high-intensity human activities and rapid economic development have caused damage to the ecosystem caused by disturbances from outside, which have exceeded the ecosystem's regulation capacity. Therefore, the resilience weighting should be higher than the resistance, which is set at 0.6 and 0.4, respectively (Formula (2)). The regional ecosystem resilience was calculated as follows.

$$RC_i = \frac{NDVI_i}{NDVI\_mean_j} \times RC_j \quad (1)$$

where RC_i denotes the resilience coefficient of raster i, $NDVI_i$ denotes the $NDVI$ value of raster i, $NDVI\_mean_j$ denotes the average $NDVI$ value of class j at raster i, and RC_j means the resilience assignment in land use class j (Table 3).

$$ER = 0.6 \times RC_i + 0.4 \times RT_j \quad (2)$$

where ER is the ecosystem resilience index for the central Yunnan urban agglomeration, RC_i represents the resilience coefficient for raster i, and RT_j is the resistance assignment of raster j of the land use type.

Table 3. Principles and methods for assessing the ecosystem service functions of each ecosystem in the central Yunnan urban agglomeration.

Type of Land Use	Cultivated Land	Forest Land	Grassland	Water Area	Urban Land	Rural Land	Unutilized Land
Ecosystem resilience	0.4	0.6	0.8	0.7	0.2	0.5	1
Ecosystem resistance	0.6	1	0.7	0.5	0.3	0.4	0.2

Assessment of the level of physical health of ecosystems: According to the definition of ecosystem health by Costanza, the level of physical health was calculated as follows [22]:

$$PH = \sqrt[3]{V \times O \times R}$$

where PH represents the ecosystem health index; V, O, and R mean ecosystem vigor, organization, and resilience, respectively.

2.3.5. Hot Spots Analysis

Hot spots analysis (Getis-ord Gi*) is performed to recognize the distribution of hot spots and cold spots in the local space of the study area [41,42]. In order to analyze the ecosystem health response to land use change of the central Yunnan urban agglomeration from spatial latitude, the research explored the spatial clustering of ecosystem health changes and identified the hot spot areas and cold spot areas. With each stage of change, the hot spot areas are referred to as ecosystem health improvement areas and cold spot areas as ecosystem health deterioration areas.

2.3.6. Measuring the Impact of Land Use Change on Ecosystem Health Hot Spots Analysis

The study overlays land use change mapping and ecosystem health change hotspot analysis mapping to analyze land use change of cold hotspot areas for ecosystem health change from 1990 to 2020. The specific method is as follows: the improvement and degradation areas for each stage of ecosystem health change are calculated and determined separately by ArcGIS 10.8. The resulting area was used as the extent of land use data extraction [43] to generate the land use transfer matrix, and the impact of each transfer type on ecosystem health is calculated. The degree of effect of land use transfer on the ecosystem was calculated using the contribution indicator [44,45] with the following equation:

$$LEI = \frac{(LE_t - LE_0)LA}{TA}$$

where *LEI* is the contribution of ecosystem health caused by a specific type of land use shift in the study area; LE_0 and LE_t represent the ecosystem health indices for land use type at the beginning and the end of the change, respectively; *LA* means the total area of the land use type; *TA* denotes the total area of all land types in the study area.

3. Results

3.1. Characteristics of Land Use Change in the Central Yunnan Urban Agglomeration

Forest land was the predominant land use type during the study period, occupying over 49% of the total study area (Table 4). This is followed by grassland, with over 26% of the total area, and cultivated land, with over 20% of the total area. Finally, urban land makes up the least land use type, with less than 2% of the total area. During our course of study, we observed some general features of land use change across the area. These include, firstly, a continuous increase in the urban land and water area, secondly, a continuous decrease in the cultivated land area, and lastly, first an increase and then a decrease in the forest land and rural land area, while a decrease followed by an increase in the grassland area, whereas the area of unutilized land remained basically the same for a long time.

Table 4. Area and proportion of land use types in central Yunnan urban agglomeration, 1990–2020.

Type of Land Use	1900		2000		2010		2020	
	Area (km^2)	Proportion (%)	Area (km^2)	Proportion (%)	Area (km^2)	Proportion (%)	Area (km^2)	Proportion (%)
cultivated land	23,442.81	21.05	23,041.63	20.69	22,963.36	20.62	22,528.40	20.23
forest land	54,721.12	49.14	54,945.99	49.34	54,857.48	49.26	54,649.97	49.08
grassland	30,625.01	27.50	30,633.62	27.51	30,331.00	27.24	29,893.95	26.85
water area	1282.37	1.15	1294.87	1.16	1326.63	1.19	1461.55	1.31
urban land	284.97	0.26	411.78	0.37	739.61	0.66	1726.02	1.55
rural land	844.17	0.76	872.45	0.78	976.49	0.88	935.60	0.84
unutilized land	155.59	0.14	155.58	0.14	161.47	0.15	160.54	0.14

The urban land area shows a strikingly significant increase from 284.97 km^2 in 1990 to 1726.02 km^2 in 2020, with the proportion increasing from 0.26% to 1.55%. According to the dynamic attitude analysis, urban land in the central Yunnan urban agglomeration has the maximum dynamic rate of change in comparison to other land types. The rate of change was 4.45%, 7.96%, and 13.34% during 1990–2000, 2000–2010, and 2010–2020, respectively, and the highest dynamic rate of change was 16.86% during 1990–2020. Although, the cultivated land and grassland area changed with higher intensity, the dynamic rate of change was significantly lower than that of urban land.

According to the land use transfer matrix (Figure 3), the largest area converted to urban land among all land types during 1990–2020 is cultivated land and grassland, with 818.20 km2 and 357.49 km2, respectively, accounting for 23.28% and 9.50% of the total converted area. As shown by the urban land sources at different stages, a large part of cultivated land converted, with 119.98 km^2, 258.71 km^2, and 476.41 km^2 of cultivated land occupied by urban land in the three stages of 1990–2000, 2000–2010, and 2010–2020, respectively.

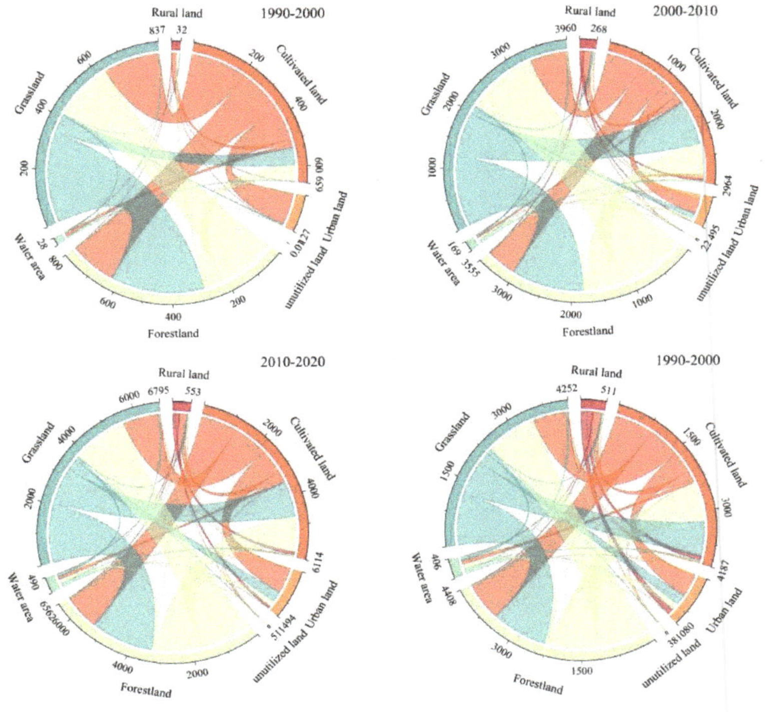

Figure 3. Land use change chord map of central Yunnan urban agglomeration, 1990–2020.

3.2. Spatial and Temporal Evolutionary Features of Ecosystem Service Functions of the Central Yunnan Urban Agglomeration

For our research, we observed six ecosystem service functions: grain production, water content, carbon storage, soil conservation, habitat quality, and aesthetic landscape of the central Yunnan urban agglomeration. These functions were overlaid to determine the spatial distribution of the ecosystem service index (ESI) of the central Yunnan urban agglomeration in 1990, 2000, 2010, and 2020 (Figure 4). It can be seen from the figure that the ecosystem service functions of the central Yunnan urban agglomeration in 1990–2020 first decreased, then increased, and finally decreased. In 1990, the ESI values of the central Yunnan urban agglomeration were 0–0.711, with a mean value of 0.327 and an extensive distribution of low and medium values. This indicates that the ecosystem service function of the central Yunnan urban agglomeration should be promoted. The areas with high values of ESI are small and scattered, mostly in Honghe Prefecture, Yuxi and Chuxiong. For example, Jianshui County (0.362), Eshan County (0.378), and Dayao County (0.379) in these regions. The range of ESI values for the central Yunnan urban agglomeration in 2020 was 0~0.682, with a mean value of 0.3253. Although the range of high-value areas was higher compared to 1990, the frequent transformation of land use structures affected the stability of the overall ecosystem function and showed the spatial characteristics of the interactive distribution of high-value areas and low-value areas. Further, the low-value areas are primarily distributed in central Qujing City, central and southern Kunming City, central and eastern Honghe Prefecture and eastern Yuxi City. For example, Qilin District (0.269), Chenggong District (0.198), Lusi County (0.290) and Jiangchuan District (0.287) in these regions.

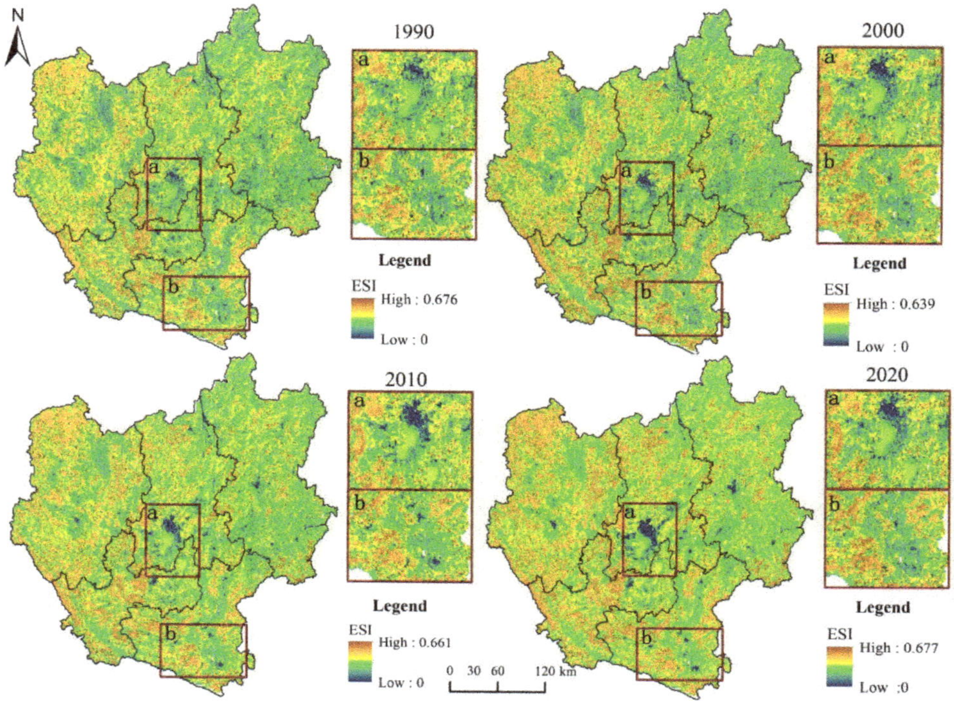

Figure 4. Spatial distribution pattern of ESI of central Yunnan urban agglomeration, 1990–2020.

3.3. Spatial and Temporal Evolutionary Characteristics of Ecosystem Health in the Central Yunnan Urban Agglomeration

To effectively characterize the spatial differentiation of the ecosystem health index (EHI) of the study area, the natural breakpoint method was applied. Within this, the EHI of the central Yunnan urban agglomeration was divided into five classes from low to high, and the spatial distribution maps of EHI classes were obtained (Figure 5). Evidently, in the central study area, there has been a low EHI value area for a long time, i.e., the central urban area of Kunming city. Among the five levels of ecosystem health, the relatively weak and ordinary levels are mostly located in areas covered by arable land and grassland. Due to the constraints of the mountainous terrain of the study area, the arable land and grassland patches are fragmented. The land use is relatively homogeneous in these areas, making the ecosystem health of the region relatively low. The areas with high ecosystem health levels are mainly located in areas with high cover of forestland with an intensive natural ecological background.

3.4. Effects of Land Use Change on the Ecosystem Health of the Central Yunnan Urban Agglomeration

We observed two trends of ecosystem health changes in the central Yunnan urban agglomeration: improvement (hot spot clustering) and deterioration (cold spot clustering). After overlaying the land use change mapping with the EHI change hot spot map, the land use shifts in the colder hot spot areas in the study site between 1990 and 2020 were studied. The types of land use changes with the most significant impact on ecosystem health were obtained using the contribution ratio.

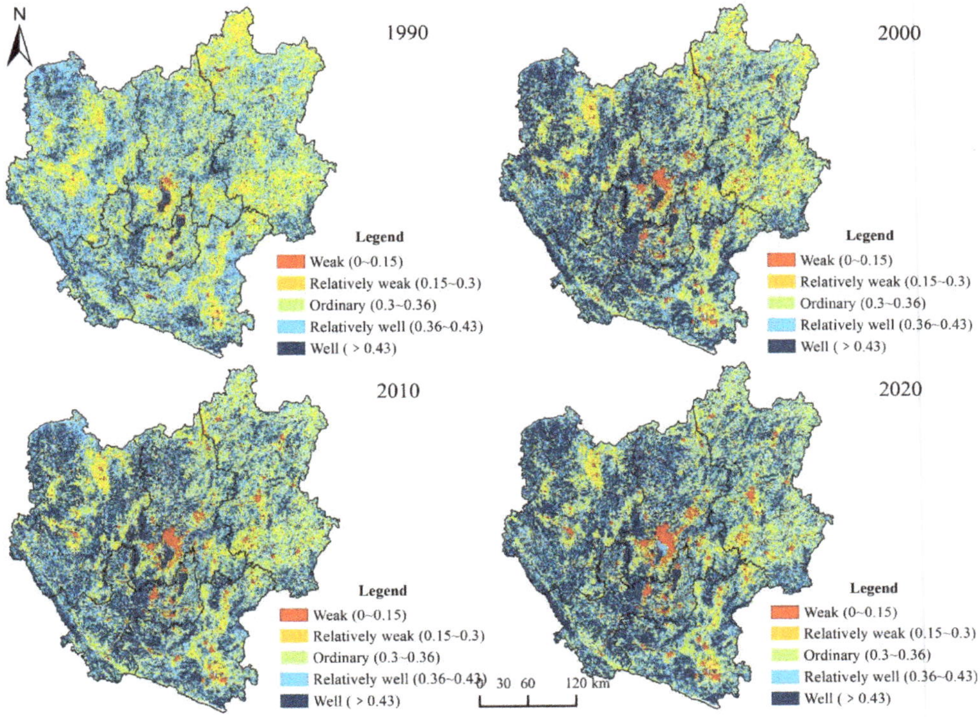

Figure 5. Spatial distribution of EHI by class in central Yunnan urban agglomeration, 1990–2020.

In terms of the area ratio of ecosystem health at different levels (Table 5), in 1990, the ordinary health level accounted for 29.7% of the total area of the study area, while the relatively well health level accounted for 33.65%. The areas with well ecosystem health started to expand in 2000, wherein they accounted for 38.15% in 2020, which is an increase of 19.39% compared to 1990. Simultaneously, the areas with weak ecosystem health also expanded by 3.91%. In the future, attention needs to be paid to the possible impact on ecosystem health of a certain increase in the area of low health level.

Table 5. Percentage area and amount of change of each class of EHI in central Yunnan urban group, 1990–2020.

Ecosystem Health Rating	1990	2000	2010	2020	1990–2000	2000–2010	2010–2020	1990–2020
Weak	0.68%	4.83%	4.25%	4.59%	4.15%	−0.59%	0.34%	3.91%
Relative weak	17.01%	11.42%	11.26%	9.85%	−5.59%	−0.16%	−1.14%	−7.16%
Ordinary	29.70%	27.38%	26.31%	24.09%	−2.32%	−1.07%	−2.23%	−5.61%
Relatively well	33.65%	23.13%	24.93%	23.12%	−10.52%	1.81%	−1.81%	−10.53%
Well	18.96%	33.24%	33.25%	38.35%	14.27%	0.01%	5.11%	19.39%

The spatial agglomeration state of EHI changes during 1990–2020 was identified by Getis-Ord G* statistics. Red areas (Gi z-score $\geq$ 1.65) are hot spots (ecosystem health improvement agglomerations); blue areas (Gi z-score ≤ -1.65) are cold spots (ecosystem health deterioration agglomerations); gray areas (1.65 > Gi z-score > -1.65) are non-significant zones of change (Figure 6).

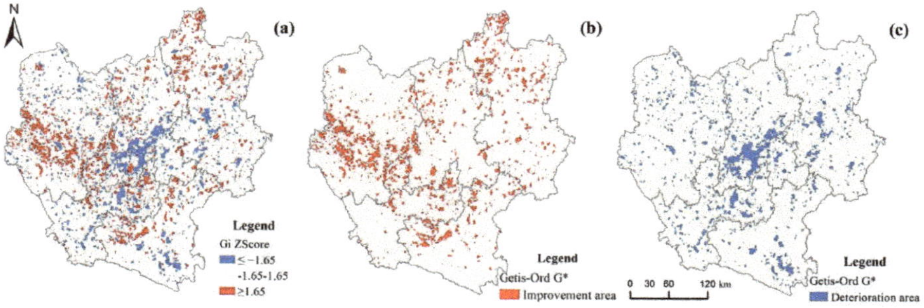

Figure 6. (**a**) Hot and cold spots areas of change in ecosystem health levels from 1990–2020. (**b**) Ecosystem health improvement areas. (**c**) Ecosystem health deterioration areas.

As seen in Figures 7 and 8, the types of land use changes that caused the improvement of ecosystem health from 1990 to 2020 were smaller in scale and relatively scattered. Additionally, the exchanges between forest land and grassland and the conversion from cultivated land to forest land was found to be the dominant type of land conversion across this area (a–e). The improvement in ecosystem health was driven primarily by the conversion from grassland to forest land, the conversion from cultivated land to forest land, and the conversion of grassland to the water area, with contribution indices of 0.020, 0.018, and 0.017, respectively. The shift from grassland and cultivated land to forest land provides the impetus for the continued improvement of ecosystem health.

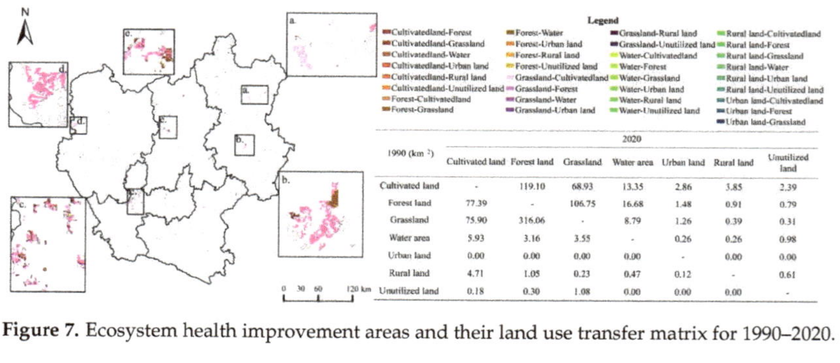

1990 (km ^{2})	2020						
	Cultivated land	Forest land	Grassland	Water area	Urban land	Rural land	Unutilized land
Cultivated land	-	119.10	68.93	13.35	2.86	3.85	2.39
Forest land	77.39	-	106.75	16.68	1.48	0.91	0.79
Grassland	75.90	316.06	-	8.79	1.26	0.39	0.31
Water area	5.93	3.16	3.55	-	0.26	0.26	0.98
Urban land	0.00	0.00	0.00	0.00	-	0.00	0.00
Rural land	4.71	1.05	0.23	0.47	0.12	-	0.61
Unutilized land	0.18	0.30	1.08	0.00	0.00	0.00	-

Figure 7. Ecosystem health improvement areas and their land use transfer matrix for 1990–2020.

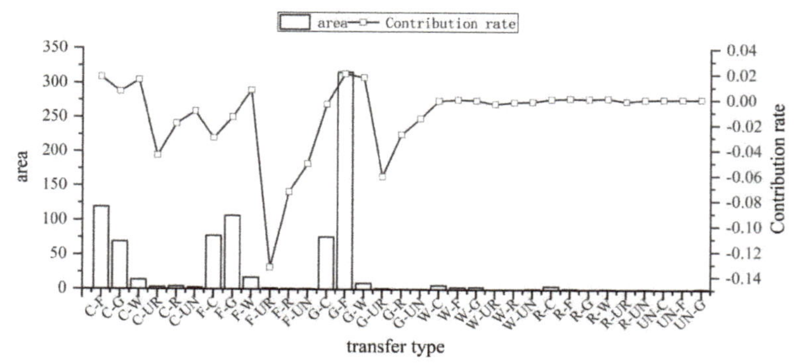

Figure 8. Contribution of various types of land use changes in ecosystem health improvement areas. C (cultivated land), F (forest land), G (grassland), W (water area), UR (urban land), R (rural land), UN (unutilized land).

As seen in Figures 9 and 10, the types of land use changes that caused deterioration of ecosystem health from 1990 to 2020 are relatively spatially clustered. The land use changes in the deterioration area were primarily found from the main urban area of Kunming, and the conversion from arable land and grassland to urban land was the dominant type of land use change across this area (f–j). Deterioration of ecosystem health occurred mainly due to the conversion from cultivated land to urban land, the conversion from cultivated land to rural land, and the conversion from cultivated land to unutilized land, with contribution indices of −1.04, −0.43, and −0.22, respectively. Our study determined that the deterioration of ecosystem health in the central Yunnan urban agglomeration is significantly related to land use changes due to human activities. It has caused an increasing amount of ecological land to be converted into cultivated land and urban land. It, thus, becomes a major cause of the deterioration of ecosystem health in the region.

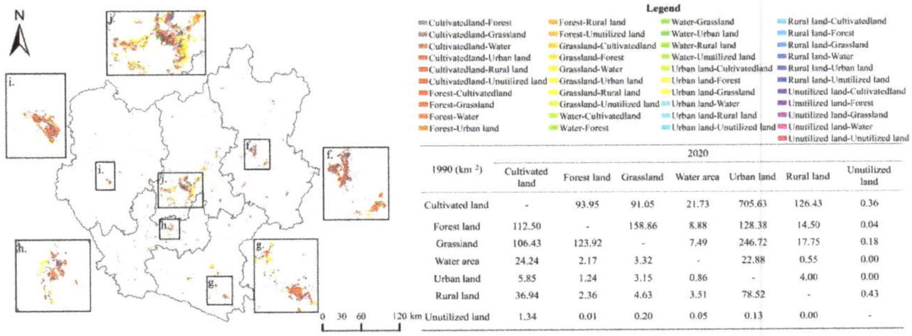

1990 (km 2)	2020 Cultivated land	Forest land	Grassland	Water area	Urban land	Rural land	Unutilized land
Cultivated land	-	93.95	91.05	21.73	705.63	126.43	0.36
Forest land	112.50	-	158.86	8.88	128.38	14.50	0.04
Grassland	106.43	123.92	-	7.49	246.72	17.75	0.18
Water area	24.24	2.17	3.32	-	22.88	0.55	0.00
Urban land	5.85	1.24	3.15	0.86	-	4.00	0.00
Rural land	36.94	2.36	4.63	3.51	78.52	-	0.43
Unutilized land	1.34	0.01	0.20	0.05	0.13	0.00	-

Figure 9. Ecosystem health deterioration areas and their land use transfer matrix for 1990−2020.

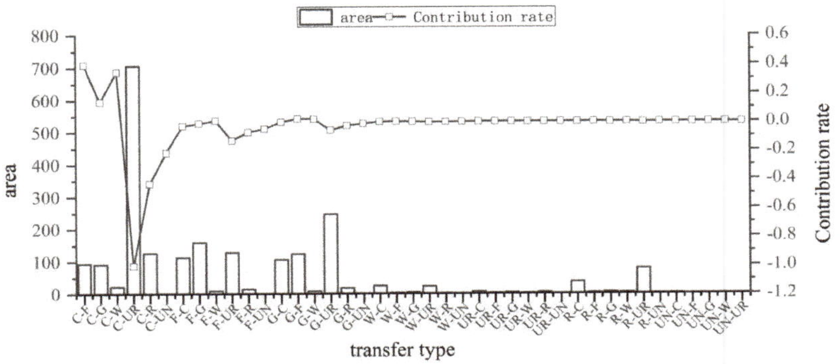

Figure 10. Contribution of various types of land use changes in ecosystem health deterioration areas. C (cultivated land), F (forest land), G (grassland), W (water area), UR (urban land), R (rural land), UN (unutilized land).

4. Discussion

Our research explores the impact of land use change on ecosystem health and established that it is a key factor affecting the spatial variability of ecosystem health. We showed through our analysis that combining land use change mapping with ecosystem health hotspot analysis can effectively reveal the spatial variability of ecosystem health responses while revealing the transition patterns among various types of land uses.

4.1. Interpretation of Land Use Changes in Central Yunnan Urban Agglomeration

Land resources form the basic conditions and key components in the sustainable development of urban agglomerations [46]. Our study site, the urban land in the central

Yunnan urban agglomeration, has seen rapid expansion since 1990. The rates of change in urban land reached 16.86% in the last 30 years, which became the most important factor perturbing the overall land structure. The central Yunnan urban agglomeration developed rapidly between 2010–2020. Further, because of the accelerated economic globalization and regional integration, four cities of Central Yunnan (Kunming, Qujing, Yuxi, and Chuxiong) signed a cooperation framework for integrated development. It resulted in the rapid economic growth, enhanced functions of the central cities, and rapid population concentration. The growth in the urbanization level, i.e., from 32.05% in 2010 to 58.94% in 2019. The conversion of forest land, grassland, and rural land into urban land has increased significantly. All four predominant cities of the central Yunnan urban agglomeration are located in the mountainous basin, and urban development is mostly concentrated in the dam area. This limits the land resources accessible to the central urban area.

Predictably, the contradictions between high concentration of population, industry, and limited land resources of the central Yunnan urban agglomeration will become the weakest link in future development. For future work on land development and management of the urban agglomeration, we should focus on characteristics like the unique natural conditions and landscape patterns of the mountainous areas, strengthening the macro control of land, improving land conservation and intensive use, and promoting land use efficiency. With guided demands and regulated supply, it will be possible to reasonably determine the new urban land scale, control the disorderly expansion of urban industrial and mining land, and guide the adjustment of the internal structure of urban land. Simultaneously, this will lead to strict control over all kinds of non-agricultural construction land occupying cultivated land, the protection of the functioning in production, ecological, and landscape isolation zones of cultivated land of the study area.

4.2. Ecosystem Health Level Analysis

Our study shows an improvement in the ecosystem health in the central Yunnan urban agglomeration, but the ecological health level is spatially unevenly distributed. The central and eastern regions show poorer characteristics of ecosystem health while the western region is relatively better. Among them, the areas with high values of ecosystem vigor are distributed in the areas where the lakes and waters are located, as well as the forest areas within Chuxiong, Yuxi, and Honghe (Figure S7). Driven by rapid urbanization, the spatial pattern of ecosystem health shows an increasing trend from the urban areas to the surrounding areas. This characteristic is also reflected in the sub-indices, where ecosystem organizational power shows a decreasing distribution from the central high-value area to the peripheral areas, reaching the lowest value at the edge of the study area (Figure S8). Spatial variation in ecosystem resilience coefficients is closely related to anthropogenic interventions. The lowest level of ecosystem resilience is found in areas with urban development, and the state of low-value clustering is apparent (Figure S9). This indicates that the environmental problems and landscape fragmentation, brought about by the concentration of urban population and high-intensity development, are key factors influencing ecosystem health [47].

Although the ecosystem health of the Central Yunnan urban agglomeration is relatively well, the degree of aggregation becomes more pronounced in areas of weak health. The major factors causing this include the integrated development within urban agglomerations, the amplified gravitational and spatial radiation capacity between cities, the deepening degree of interaction, and the frequent transitions between land use intensity and land use types [48]. Therefore, in the process of ecosystem health maintenance, attention should be paid to monitoring the ecological health of low-value agglomeration areas, reducing the interference of large-scale human activities on the ecosystem, overcoming the negative effects caused by the agglomeration effect, and enhancing the self-regulatory ability of the area.

4.3. Relationship between Land Use Change and Ecosystem Health

The ecosystem health improvements and deteriorations are majorly driven by interconversions among land use types of the central Yunnan urban agglomeration. Among them, the land change types with the most significant impact on ecosystem health improvement were the conversions from grassland to forest land, the conversions from cultivated land to forest land, and the conversions from grassland to water areas. Thus, the most effective means to maintain the ecological health and stability of the central Yunnan urban agglomeration would include accelerating the protection and restoration of the ecological backgrounds of the improvement area, enhancing vegetation cover and biodiversity in the area, and maintaining the ecosystem vigor of forest land and water areas. On the contrary, the types of land changes that had the most significant impact on the deterioration of ecosystem health were the conversions from cultivated land to urban land, with a contribution index of -1.04, and the area of occupied cultivated land is 818.20 km^2. However, food security is the basis of national security, and in the face of the large-scale occupation of arable land by urban land, efforts must be made to restrict land development intensity. In the future, government decisions should encourage the development of a gradient development model for the gradient development mode of mountain towns, make rational use of land resources, strengthen intensive land use, determine the optimal land use structure, and guarantee the development of urban construction while achieving effective protection of the ecological function of arable land.

It has also been found in previous studies that urban sprawl can, directly and indirectly, lead to the degradation of ecological services and contributes to the decline in the value of global ecosystem services [49–51]. Rural–urban migration, economic growth, all intensified urban expansion and land depletion. During this period, the degraded areas of urban ecosystem service function in central Yunnan were also mainly concentrated in the urban expansion areas (Figures S1–S6). Therefore, in order to promote ecosystem health, policy makers should pay more attention to the contradiction between the scarcity of urban land in the study area and ecological conservation in mountainous areas. On the one hand, it implements the construction of major ecological projects such as "natural forest protection", "return of cultivated land to forest and grass", soil and water conservation, and rock desertification control [42], which actively carries out the restoration of degraded terrestrial ecosystems. In addition, according to the "Yunnan Province Ecological Function Zoning Plan" [52], the focus is on strengthening the ecological function protection and restoration of the middle reaches of the Jinsha River Basin Soil and Water Conservation Zone, the Red River Basin Soil and Water Conservation Zone, and the Pearl River Headwaters Water Conservation Zone. For areas with weaker ecosystem health, local governments must increase investment in environmental protection to improve water connotation and soil conservation capacity. This will improve the purification and self-regulation capacity of the ecosystem and reduce the vulnerability of the ecological environment in mountainous areas.

4.4. Limitations and Future Research

In the current research, the ecosystem services assessed, based on the InVEST model, were incorporated into the ecosystem health evaluation, which better reflects the ecological processes compared to the value-accounted ecosystem services. This methodology, to some extent, compensates for the study of land use change in the ecosystem material-energy cycle. However, it is still unclear how ecological processes triggered by land use change, and their cascading effects can simultaneously contribute to changes in ecosystem health levels [53]. Moreover, the study only focused on the unidirectional processes between land use change and ecosystem health effects, while the quantitative integration between natural feedback mechanisms and the land use change model needs further attention in the future [54].

As the urban agglomeration of central Yunnan gradually moves from development to maturity, the intensity of anthropogenic disturbances further increases. Based on the limited land resources in mountainous areas, the perspective of ecosystem health research on land use changes should pay more attention to the rational allocation of land resources. In addition, we should continue to explore the factors driving and influencing mechanisms of ecological health. These would provide the scientific basis for policymakers and implementers to establish suitable conservation measures.

5. Conclusions

The research was conducted on mountainous urban agglomerations in the southwest plateau of China. It aims at exploring the inherent correlation of regional ecosystem health with land use patterns and incorporates the InVEST model-based ecosystem service assessment into the ecosystem health evaluation. This methodology further emphasizes the interaction process between land use and natural ecosystems. Additionally, the study explored the responses of two trends of ecosystem health deterioration and improvement of the study area to land use change by combining land use change mapping and hotspot analysis. The findings of the study are shown below:

(1) Forest land was the predominant land use type during the study period. The transformation of cultivated and grassland to urban land is the most significant in the process of land type transformation. The rapid expansion of urban land became the most important factor in disrupting the overall land use structure change in the study area.

(2) The spatial variability of ecosystem health level is significant, with the central and eastern regions being worse and the western regions being relatively good. The areas with the lowest levels of ecosystem health are urban development areas.

(3) Ecosystem health is influenced by land use shifts. The improvement of health levels is closely related to the mutual transfer between forest land and grassland, and the conversion from cultivated land to forest land. The fast expansion of urban land caused by urbanization and the conversion from cultivated land and grassland to urban land are important reasons for the deterioration of ecosystem health.

Supplementary Materials: The following supporting information can be downloaded at: https://www.mdpi.com/article/10.3390/ijerph191912399/s1.

Author Contributions: Conceptualization, B.G.; Data curation, B.G.; Formal analysis, B.G. and C.L.; Funding acquisition, Y.W. (Yingmei Wu); Investigation, K.Z. and Y.W. (Yan Wu); Supervision, Y.W. (Yingmei Wu); Validation, C.L.; Writing—original draft, B.G. and Y.W. (Yingmei Wu); Writing—review and editing, C.L., K.Z. and Y.W. (Yan Wu). All authors have read and agreed to the published version of the manuscript.

Funding: This research was supported by the National Natural Science Foundation of China: 41761037; 42101278. Yunnan Normal University Graduate Research Innovation Fund Project: YJSJJ22-A20; YJSJJ22-B106.

Institutional Review Board Statement: Not applicable.

Informed Consent Statement: Not applicable.

Data Availability Statement: The data that support the findings of this study are available from the corresponding author upon reasonable request.

Conflicts of Interest: The authors declare no conflict of interest.

References

1. Fang, C.; Yu, D. Urban agglomeration: An evolving concept of an emerging phenomenon. *Landsc. Urban Plan.* **2017**, *162*, 126–136. [CrossRef]
2. He, Y.; Lin, Y.; Zhou, G.; Zhu, Y.; Tang, K. Spatial pattern and drivers of urbanization in china's mid-level developing urban agglomeration: A case study of chang-zhu-tan. *Reg. Sustain.* **2021**, *2*, 83–97. [CrossRef]
3. Wang, Z.; Liang, L.; Sun, Z.; Wang, X. Spatiotemporal differentiation and the factors influencing urbanization and ecological environment synergistic effects within the beijing-tianjin-hebei urban agglomeration. *J. Environ. Manag.* **2019**, *243*, 227–239. [CrossRef] [PubMed]
4. Fang, C.; Zhou, C.; Gu, C.; Chen, L.; Li, S. A proposal for the theoretical analysis of the interactive coupled effects between urbanization and the eco-environment in mega-urban agglomerations. *J. Geogr. Sci.* **2017**, *27*, 1431–1449. [CrossRef]

5. Patra, S.; Sahoo, S.; Mishra, P.; Mahapatra, S.C. Impacts of urbanization on land use/cover changses and its probable implications on local climate and groundwater level. *J. Urban Manag.* **2018**, *7*, 70–84. [CrossRef]
6. Sharma, S.; Nahid, S.; Sharma, M.; Sannigrahi, S.; Anees, M.M.; Sharma, R.; Shekhar, R.; Basu, A.S.; Pilla, F.; Basu, B.; et al. A long-term and comprehensive assessment of urbanization-induced impacts on ecosystem services in the capital city of india. *City Environ. Interact.* **2020**, *7*, 100047. [CrossRef]
7. Peng, J.; Liu, Y.; Li, T.; Wu, J. Regional ecosystem health response to rural land use change: A case study in lijiang city, china. *Ecol. Indic.* **2017**, *72*, 399–410. [CrossRef]
8. He, J.; Shi, X.; Fu, Y.; Yuan, Y. Evaluation and simulation of the impact of land use change on ecosystem services trade-offs in ecological restoration areas, china. *Land Use Policy* **2020**, *99*, 105020. [CrossRef]
9. Li, B.; Shi, X.; Chen, Y.; Jiang, Y. Quantitative assessment of the ecological effects of land use/cover change in the arid region of northwest china. *Environ. Monit Assess* **2019**, *191*, 704. [CrossRef]
10. Rapport, D.J. What constitutes ecosystem health? *Perspect. Biol. Med.* **1989**, *33*, 120–132. [CrossRef]
11. Costanza, R. Ecosystem health and ecological engineering. *Ecol. Eng.* **2012**, *45*, 24–29. [CrossRef]
12. He, J.; Pan, Z.; Liu, D.; Guo, X. Exploring the regional differences of ecosystem health and its driving factors in china. *Sci. Total Environ.* **2019**, *673*, 553–564. [CrossRef] [PubMed]
13. Rapport, D.J.; Costanza, R.; McMichael, A.J. Assessing ecosystem health. *Trends Ecol. Evol.* **1998**, *13*, 397–402. [CrossRef]
14. Liu, Y.; Peng, J.; Wan, A.; Xie, P.; Han, Y. New research progress and trends in ecosystem health. *Acta Ecol. Sin.* **2015**, *35*, 5920–5930.
15. Ma, K.; Kong, H.; Guan, W.; Fu, B. Ecosystem health assessment:Methods and directions. *Acta Ecol. Sin.* **2001**, *21*, 2106–2116.
16. Xie, X.; Pu, L. Assessment of urban ecosystem health based on matter element analysis: A case study of 13 cities in jiangsu province, china. *Int. J. Environ. Res. Public Health* **2017**, *14*, 940. [CrossRef]
17. Meng, L.; Huang, J.; Dong, J. Assessment of rural ecosystem health and type classification in jiangsu province, china. *Sci. Total Environ.* **2018**, *615*, 1218–1228. [CrossRef]
18. Sun, J.; Li, Y.P.; Gao, P.P.; Xia, B.C. A mamdani fuzzy inference approach for assessing ecological security in the pearl river delta urban agglomeration, china. *Ecol. Indic.* **2018**, *94*, 386–396. [CrossRef]
19. Wang, Q.; Li, S.; Li, R. Evaluating water resource sustainability in beijing, china: Combining psr model and matter-element extension method. *J. Clean. Prod.* **2019**, *206*, 171–179. [CrossRef]
20. Xue, L.; Wang, J.; Zhang, L.; Wei, G.; Zhu, B. Spatiotemporal analysis of ecological vulnerability and management in the tarim river basin, china. *Sci. Total Environ.* **2019**, *649*, 876–888. [CrossRef]
21. Costanza, R. Toward an Operational Definition of Ecosystem Health. In *Ecosystem Health: New Goals for Environmental Management*; Island Press: Washington, DC, USA, 1992; pp. 239–256.
22. Kang, P.; Chen, W.; Hou, Y.; Li, Y. Linking ecosystem services and ecosystem health to ecological risk assessment: A case study of the beijing-tianjin-hebei urban agglomeration. *Sci. Total Environ.* **2018**, *636*, 1442–1454. [CrossRef] [PubMed]
23. Peng, J.; Liu, Y.; Wu, J.; Lv, H.; Hu, X. Linking ecosystem services and landscape patterns to assess urban ecosystem health: A case study in shenzhen city, china. *Landsc. Urban Plan.* **2015**, *143*, 56–68. [CrossRef]
24. Xiao, Y.; Guo, L.; Sang, W. Impact of fast urbanization on ecosystem health in mountainous regions of southwest china. *Int. J. Environ. Res. Public Health* **2020**, *17*, 826. [CrossRef] [PubMed]
25. Ge, F.; Tang, G.; Zhong, M.; Zhang, Y.; Xiao, J.; Li, J.; Ge, F. Assessment of ecosystem health and its key determinants in the middle reaches of the yangtze river urban agglomeration, china. *Int. J. Environ. Res. Public Health* **2022**, *19*, 771. [CrossRef]
26. Xu, D.; Cai, Z.; Xu, D.; Lin, W.; Gao, J.; Li, L. Land use change and ecosystem health assessment on shanghai–hangzhou bay, eastern china. *Land* **2022**, *11*, 867. [CrossRef]
27. Wang, Z.; Liu, Y.; Li, Y.; Su, Y. Response of ecosystem health to land use changes and landscape patterns in the karst mountainous regions of southwest china. *Int. J. Environ. Res. Public Health* **2022**, *19*, 3273. [CrossRef]
28. Chen, P. Monthly npp dataset covering china's terrestrial ecosystems at north of 18°n (1985–2015). *J. Glob. Chang. Data Discov.* **2019**, *3*, 34–41. [CrossRef]
29. Costanza, R.; d'Arge, R.; de Groot, R.; Farber, S.; Grasso, M.; Hannon, B.; Limburg, K.; Naeem, S.; Robert, V.O.; Paruelo, J.; et al. The value of the world's ecosystem services and natural capital. *Nature* **1997**, *387*, 253–260. [CrossRef]
30. Chen, W.; Zhao, X.; Zhong, M.; Li, J.; Zeng, J. Spatiotemporal evolution patterns of ecosystem health in the middle reaches of the yangtze river urban agglomerations. *Acta Ecol. Sin.* **2022**, *42*, 138–149. [CrossRef]
31. Peng, J.; Wang, Y.; Wu, J.; Zhang, Y. Evaluation for regional ecosystem health: Methodology and research progress. *Acta Ecol. Sin.* **2007**, *27*, 4877–4885. [CrossRef]
32. Pan, J.; Wang, Y. Ecological security evaluation and ecological pattern optimization in taolai river basin based on cvor and circuit theory. *Acta Ecol. Sin.* **2021**, *41*, 2582–2595.
33. Millennium ecosystem assessment. *Ecosystems and Human Well-Being: Biodiversity Synthesis*; Island Press: Washington, DC, USA, 2005.
34. Zhao, X.; Du, Y.; Li, H.; Wang, W. Spatio-temporal changes of the coupling relationship between urbanization and ecosystem services in the middle yellow river. *J. Nat. Resour.* **2021**, *36*, 131–147. [CrossRef]
35. Qiu, W.; Zhang, Y.; Yu, J.; Zhang, C.; Zheng, C.; Yu, S. Feasibility verification with field application of a water conservation module using the invest model. *J. Zhejiang AF Univ.* **2018**, *35*, 810–817. [CrossRef]
36. Liu, Y.; Liu, X.; Zhang, B.; Li, M. Spatial features analysis of water conservation function in the hilly areas of the loess plateau based on invest model. *Acta Ecol. Sin.* **2020**, *40*, 6161–6170.

37. Xie, G.; Zhang, C.; Zhang, L.; Chen, W.; Li, S. Improvement of the evaluation method for ecosystem service value based on per unit area. *J. Nat. Resour.* **2015**, *30*, 1243–1254. [CrossRef]
38. Das, M.; Das, A.; Pereira, P.; Mandal, A. Exploring the spatio-temporal dynamics of ecosystem health: A study on a rapidly urbanizing metropolitan area of lower gangetic plain, india. *Ecol. Indic.* **2021**, *125*, 107584. [CrossRef]
39. Zheng, K.; Li, C.; Wu, Y.; Gao, B.; Wu, Y.; Li, C. Spatio-temporal evolution of habitat quality and its influencing factors in ecological conservation area in sichuan-yunnan provinces based on value assessment. *J. Ecol. Rural Environ.* **2022**. *accepted*. [CrossRef]
40. Peng, J.; Xu, F. Effect of grid size on habitat quality assessment: A case study of huangshan city. *J. Geo-Inf. Sci.* **2019**, *21*, 887–897. [CrossRef]
41. Shi, Y.; Shi, D.; Zhou, L.; Fang, R. Identification of ecosystem services supply and demand areas and simulation of ecosystem service flows in shanghai. *Ecol. Indic.* **2020**, *115*, 106418. [CrossRef]
42. Li, C.; Wu, Y.; Gao, B.; Zheng, K.; Wu, Y.; Li, C. Multi-scenario simulation of ecosystem service value for optimization of land use in the sichuan-yunnan ecological barrier, china. *Ecol. Indic.* **2021**, *132*, 108328. [CrossRef]
43. Chen, Y.; Wang, J.; Xiong, N.; Sun, L.; Xu, J. Impacts of land use changes on net primary productivity in urban agglomerations under multi-scenarios simulation. *Remote Sens.* **2022**, *14*, 1755. [CrossRef]
44. Guo, C.; Gao, S.; Zhou, B.; Gao, J. Effects of land use change on ecosystem service value in funiu mountain based upon a grid square. *Acta Ecol. Sin.* **2019**, *39*, 3482–3493.
45. Cui, G.; Zhang, Y.; Chao, Y.; Zhao, Y.; Pan, B. Land use change and eco-environmental effects in qinling mountains in recent 40 years. *Res. Soil Water Conserv.* **2023**, *30*. *accepted*. [CrossRef]
46. Lu, X.; Ke, S. Evaluating the effectiveness of sustainable urban land use in china from the perspective of sustainable urbanization. *Habitat Int.* **2018**, *77*, 90–98. [CrossRef]
47. Yang, Y.; Jun, Z.; Sui, X.; He, X. Study of the spatial connection between urbanization and the ecosystem-a case study of central yunnan (China). *PLoS ONE* **2020**, *15*, e0238192. [CrossRef] [PubMed]
48. Shu, H.; Xiao, C.; Ma, T.; Sang, W. Ecological health assessment of chinese national parks based on landscape pattern: A case study in shennongjia national park. *Int. J. Environ. Res. Public Health* **2021**, *18*, 11487. [CrossRef]
49. Zank, B.; Bagstad, K.J.; Voigt, B.; Villa, F. Modeling the effects of urban expansion on natural capital stocks and ecosystem service flows: A case study in the puget sound, washington, USA. *Landsc. Urban Plan.* **2016**, *149*, 31–42. [CrossRef]
50. Anaya-Romero, M.; Muñoz-Rojas, M.; Ibáñez, B.; Marañón, T. Evaluation of forest ecosystem services in mediterranean areas. A regional case study in south spain. *Ecosyst. Serv.* **2016**, *20*, 82–90. [CrossRef]
51. Tolessa, T.; Senbeta, F.; Kidane, M. The impact of land use/land cover change on ecosystem services in the central highlands of ethiopia. *Ecosyst. Serv.* **2017**, *23*, 47–54. [CrossRef]
52. The People's Government of Yunnan Province. Development Plan of Central Yunnan Urban Agglomerate. 2020. Available online: http://www.yn.gov.cn (accessed on 25 January 2022).
53. Liu, C.; Deng, C.; Li, Z.; Liu, Y.; Wang, S. Optimization of spatial pattern of land use: Progress, frontiers, and prospects. *Int. J. Environ. Res. Public Health* **2022**, *19*, 5805. [CrossRef]
54. Wang, X.; Meng, J. Research progress on the environmental-ecological impacts of land use change. *Acta Sci. Nat. Univ. Pekin.* **2014**, *50*, 1133–1140. [CrossRef]

International Journal of
Environmental Research and Public Health

MDPI

Article

How Does Topography Affect the Value of Ecosystem Services? An Empirical Study from the Qihe Watershed

Li Li [1,*,†], Yonghui Li [2,†], Lan Yang [1], Ying Liang [1], Wenliang Zhao [3] and Guanyu Chen [3]

1 School of Government, Beijing Normal University, Beijing 100875, China
2 Henan Provincial General Institute of Urban and Rural Planning and Design, Zhengzhou 450044, China
3 School of Surveying and Planning, Shangqiu Normal University, Shangqiu 476000, China
* Correspondence: lili19960217@163.com
† Co-first author: These authors, Li Li and Yonghui Li, contributed equally to this work.

Abstract: Topographic position indices (TPIs) measure essential impacts on ecosystem service supply capacity. The identification of changes in ecosystem services and value metrics under varying TPIs has become a topical subject of global change research. Multidimensional changes in spatiotemporal and geographical aspects of ecosystem service values (ESVs) are assessed in this article using land cover/use data from 2000–2015. Effects of land-use/cover changes and topographic indices on ESVs are explored using the Chinese terrestrial unit area ecosystem service value equivalence table combined with topographic factors. A sensitivity index is introduced to quantify the robustness of total ESV to land-use/cover and topographic indices. The results show that: (1) The total ESV in the Qihe watershed declined with a change in land-use/cover during the period 2000–2015. The maximum ESV was CNY 1.984 billion in 2005 and the minimum was CNY 1.940 billion in 2010; (2) The response of ESV to land/use cover varied greatly across TPIs, with the most significant change in ESV occurring in the 0.6–0.8 TPI range and the greatest change in a single ecosystem service occurred in water areas; (3) The sensitivity indices of ESVs are all less than 1. The sensitivity indices of unused land and water tended to zero. Woodland sensitivity indices were the highest at 0.53, followed by those of arable land and grassland, owing to the large proportion of arable land and grassland areas in the overall area of land-use categories.

Keywords: ecosystem service value; land-use/cover change; topographic position index; sensitivity index; Qihe watershed

Citation: Li, L.; Li, Y.; Yang, L.; Liang, Y.; Zhao, W.; Chen, G. How Does Topography Affect the Value of Ecosystem Services? An Empirical Study from the Qihe Watershed. *Int. J. Environ. Res. Public Health* **2022**, *19*, 11958. https://doi.org/10.3390/ijerph191911958

Academic Editor: Paul A. Sandifer

Received: 4 August 2022
Accepted: 15 September 2022
Published: 22 September 2022

1. Introduction

Complex topography typically offers a variety of ecosystem services and significant spatial heterogeneity across watersheds. These services often include biodiversity conservation, water supply, food production, and soil conservation [1–5]. Nonetheless, the integrated measurement of ecosystem service capacity and regional differences between watersheds has presented a research challenge for geographers, ecologists, and economists [6–10].

Ecosystem Service Value (ESV) research was pioneered by Constanza in 1997 [11–14]. Ouyang et al. [15,16] and Xie et al. [17,18], among other prominent Chinese scholars, quickly followed up with an assessment of the capacity and value of ecosystem services in China. Using the global ecosystem services assessment by Constanza, Xie et al. [19] established a Chinese terrestrial ecosystem services assessment system. The value of five ecosystem service functions (1. Preserving the equilibrium of O_2 and CO_2 in the atmosphere; 2. Aiding water conservation; 3. Conserving total organic matter; 4. Providing nutrient storage and cycling; 5. Providing a purifying effect on the environment) in China was estimated by Ouyang et al. [20] using alternative engineering, shadow pricing, and profit and loss analysis.

The Millennium Ecosystem Assessment (MEA) demonstrated that the capacity and the value of global ecosystem services are largely underestimated, and that an accurate estimate of the capacity and value of ecosystem services can improve land-use science [21]. Simultaneously, the MEA proposed that ecosystem services research should evolve from the current single static value assessment to the assessment of the ecosystem services' impact on human well-being, including concepts such as regional variability, multi-scale ecosystem services, and the dynamic evolution of ecosystem services [22,23].

Owing to these developments, the valuation of ecosystem services has become a high priority topic in ecosystem services research, especially research focusing on the impact of changes in ecosystem services in the context of global change and including the consideration of human activities on regional sustainable development [24,25]. In this context, many ecosystem service payment projects have been implemented in watersheds around China and are providing a basis for government policies on ecological protection [26,27]. Quantitative investigations of anthropogenic influences on the ESV, focusing on land-use/cover change, are becoming popular [28–31]. Nonetheless, there remains a lack of scientific standards for applying scientific rigor to ecosystem services in regional development and ecological conservation, making it difficult to operationalize in regional development planning.

Yang et al. analyzed the trade-offs between ecological health and socioeconomic development in 2040 under different land-use scenarios, by using multi-temporal, high-resolution (0.5 m) remote sensing satellite imagery and biophysical models, setting a precedent for the practical application of ecosystem service analysis [32]. Regional variability and dynamic changes in ecosystem services are defined by human activities with land-use at their center, dramatically influencing the structure, processes, and function of the ecosystem. Understanding the multidimensional patterns of ecosystem service changes and influencing variables at the local watershed scale bear practical implications for land resource management and human well-being enhancement [33–36].

At present, most research has focused on the quantitative analysis of a single time node and a single type of service capacity in a region, whereas the trade-off synergies and geographical and temporal differences of numerous service capabilities have received insufficient attention [37]. The equivalent factor method has the advantage of visualizing changes in ecosystem services and requires fewer parameters, so it is often used to estimate the value of ecosystem services [18]. Studies have been conducted to assess the ESV at different scales such as for provincial scales [38], mountainous regions [39], and watersheds [40], and to estimate the ESV from different land-use types such as glaciers [41], forests [42], grasslands [43], and wetlands [44]. However, little research has been conducted on multidimensional variations in ESV at the small watershed scale in combination with topographic features [9,45].

Small watersheds are basic and complete natural geographical units, and their complex geomorphological types render them capable of a variety of ecosystem service functions (biodiversity conservation, water supply, production, regulation, etc.). It is vital to research the spatial–temporal variability of ESV in small watersheds for human well-being. However, there are few studies that incorporate the multidimensional analysis of spatial–temporal variability of ecosystem service in small watersheds with topographic gradient effects. Consequently, determining ways to evaluate the spatial–temporal variability of ESV in connection with topographic features has emerged as a critical issue in this study. Given that ecosystem services are characterized by regional heterogeneity and dynamic changes, especially resulting from human activities centered on land use, the structure, processes, and function of an ecosystem can change significantly. Thus, it is of practical significance to understand the multidimensional change patterns and influencing factors of ecosystem services at small watershed scales to facilitate the rational use of land resources and the improvement of human well-being.

The Qihe watershed is located in the transition area between the second and third steps in China, between the Taihang Mountains and the North China plain. This watershed serves an important water-conservation role, and the terrace transition zone is distinguished by its peculiar geographic relief. From 2000 to 2015, the ESV in this region was evaluated using land-use/cover data and a coefficient-corrected terrestrial ecosystem unit area scale was developed. Within the watershed, the total ESV, individual ESVs, and sensitivity indices were calculated. The topographic position of individual ESVs was also evaluated to further investigate the impacts of land-use/cover and identify spatial differentiation patterns on ESV in a small watershed.

The main objective of this study is to reveal how topography affects the spatial and temporal distribution of ESV in a mountain-plain transition zone. We have two specific questions: (1) What are the spatial–temporal characteristics of ecosystem service values in the Qihe watershed? (2) How do topographic features affect the ESVs? To answer these questions, we first corrected China's terrestrial value ecosystem service equivalence table using grain prices and production in Henan Province. Secondly, we analyzed the differences in spatial–temporal ecosystem services. Finally, we used the topographic position index (TPI) and a sensitivity index to investigate the characteristics of the regional ESVs.

2. Data Sources and Research Methods

2.1. Study Area

The Qihe watershed (35°32′–36°15′ N, 113°15′–114°23′ E) is located between the southwestern part of the North China Plain and the southern part of the Taihang Mountains. The Qihe River originates from the Fangnaoling mountains in Lingchuan county, Shanxi Province. It then flows through the Henan Province into the Weihe River, a tributary of the Haihe River. The watershed area is 2227 km^2, and the elevation trend is from high in the west to low in the east (Figure 1). The main climate type is a warm temperate semi-humid continental monsoon climate, with an average annual precipitation of 574 mm and an average annual temperature of 11.9 °C. Complex topography renders the ecological environment of the Qihe watershed fragile, and diverse landform types present complex variations in ecosystem services within the basin.

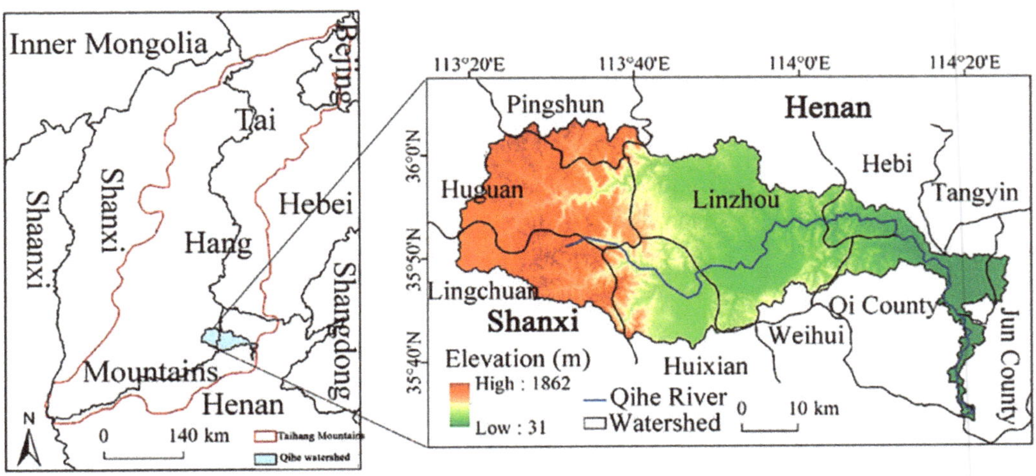

Figure 1. Location and elevation of the study area.

2.2. Data Sources and Initial Data Processing

Land cover/use data from the Qihe watershed (2000–2010) were obtained from the China Earth System Data Sharing Platform-Middle and Lower Yellow River Scientific Data Center (http://www.geodata.cn/). Based on LANDSAT multi-band remote sensing images (from Geospatial Data Cloud, http://www.gscloud.cn/), 2015 land-use data were interpreted visually using human–machine interaction and surveyed in the field employing historical land-use maps of the study area, with a kappa coefficient of 86%. Digital elevation model (DEM) data were obtained from the Geospatial Data Cloud (http://www.gscloud.cn/). The land-use data were all in the form of 1:100,000 vector data and the raster data were in a uniform grid format with a spatial resolution of 30 m. The geographic coordinate system used was WGS_1984_Albers. The socio-economic data used in the study were obtained from the Henan Provincial Statistical Yearbook (2000–2015) and the China Statistical Yearbook (2000–2015).

We referred to the research methods of Xie et al. [17–19] who excluded construction land in this study area from their estimation of ESV. A coefficient correction of the Chinese terrestrial ESV per unit area scale was performed using food production and arable prices in Henan province. This model was used to estimate the value and change trend of five major ecosystem services categories (arable land, woodland, grassland, water area, and unused land) from 2000 to 2015.

A TPI was used to evaluate the shift in total and individual ESV. A 5 km × 5 km grid was constructed in ArcGIS 10.3 and the different land-use types on the grid were multiplied by ESV coefficients, and then divided by the grid-cell area to obtain ESV densities. Changes in total ESV and individual ESV in relation to the TPI were calculated separately using land-use type area changes. The ESVs in relation to TPIs were calculated and spatially differentiated by utilizing a fishing-net function [46].

2.3. Methodology

The ESV of the Qihe watershed was investigated using the research framework for the study of ESV multidimensional changes (spatial, temporal and TPI) summarized in Figure 2. The framework consisted of three main components:

(1) Data preparation: In 2015, land-use data from the Qihe watershed were obtained using human–computer interactive visual interpretation and field survey of remote sensing images, based on reference to land-use maps of previous years. Additionally, land-use data for 2000, 2005, and 2010 were downloaded from the China Earth System Data Sharing Platform—Middle and Lower Yellow River Scientific Data Center. Socio-economic data and other relevant data were extracted from the China Statistical Yearbook (2000–2015) and Henan Provincial Statistical Yearbook (2000–2015). DEM data were obtained from the Geospatial Data Cloud Platform.

(2) Ecosystem service value accounting: The existing China terrestrial ecosystem services table could not be directly applied to the calculation of regional ESV. Consequently, its parameters were corrected using grain production and prices from Henan Province. Further integration of historical land-use data was then performed to estimate the value of ecosystem services in the years 2000, 2005, 2010, and 2015.

(3) Multi-dimensional change analysis of ESV: A comprehensive analysis of the changes in the total ESV and individual ESVs in three dimensions (i.e., spatial, temporal, and TPI) was performed.

2.3.1. Estimating the Value of Ecosystem Services

As noted above, the value coefficients per unit area of the terrestrial ecosystem in China were modified in this study. Grassland, forest, cropland, desert, and watershed in the new system correspond to grassland, woodland, arable land, unused, and water in the original system, respectively [7–9]. The Qihe watershed area in Henan Province spans 1424 km^2, accounting for 64% of the total area. The average grain yield of 5305.24 kg/hm^2 and the grain price of 1.36 CNY/kg in Henan Province from 2000 to 2015 were used to correct the table of the terrestrial ESVs [46].

The value of food production per unit area is given by the formula:

$$Va = \frac{1}{7}\sum_{m=1}^{n} \frac{a_m p_m q_m}{A} \tag{1}$$

where m refers to the type of crop, with $m = (1, 2, 3, \cdots, n)$; Va denotes the economic value of food production function per unit area of the arable ecosystem; p_m is the average price of the m food crop; q_m is the yield per unit area of the m crop; a_m is the area of the m crop cultivation; A is the area of food cultivation. V_{ij} is defined as follows:

$$V_{ij} = e_{ij} v_a \tag{2}$$

where V_{ij} is the service value of ecosystem service i in ecosystem j per unit area; e_{ij} is the equivalent factor of the service value of ecosystem service i in ecosystem j in the study area; i is the ecosystem service type, with $i = (1, 2, 3, \cdots, n)$, and j is the type of ecosystem.

According to Equation (1), at constant prices in 2015, the ecosystem service value of food production per unit area of arable land was calculated as 1030.73 CNY/ha. We refer to Xie et al. [18,47] and Ouyang et al. [15,16] for other land-use types (Table 1).

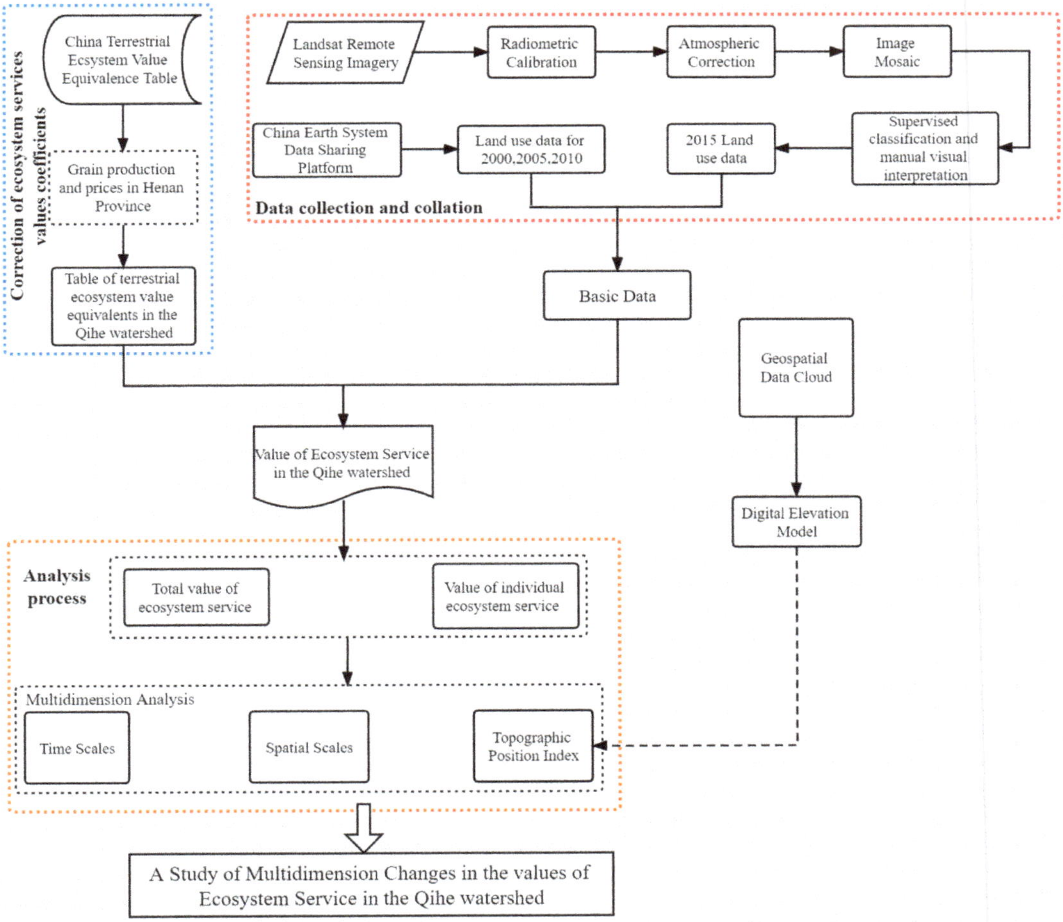

Figure 2. The research framework used for studying multi-dimensional (spatial, temporal, and TPI) changes in the ESV of the Qihe watershed.

Table 1. Table of ecological service value equivalents per unit area of terrestrial ecosystems in the Qihe watershed (CNY/ha) as used in this study area.

Type I	Type II	Arable Land	Woodland	Grassland	Water Area	Unused Land
Adjustment Services	Gas Regulation (GR)	515.37	2196.22	88.60	0.00	0.00
Support Services	Climate Regulation (CR)	917.35	1770.13	219.42	407.00	0.00
	Water Conservation (WC)	618.44	2078.02	220.25	18,033.2	26.50
	Soil formation and conservation (SFC)	1504.87	2575.16	353.56	8.80	17.70
	Waste Disposal (WD)	1690.40	1419.60	1227.92	16,086.60	8.80
	Biodiversity Conservation (BC)	731.82	2195.01	580.43	2203.30	300.80
Supply Services	Food Production (FP)	1030.73	462.67	802.33	88.50	8.80
	Raw Materials (RM)	103.07	1601.40	4.97	8.50	0.00
Cultural Services	Entertainment Culture (EC)	10.31	833.94	93.56	3840.20	8.80
	Total	7122.36	15,132.15	3591.04	40,676.10	371.40

2.3.2. Single Land-Use Dynamic Approach

The single land-use dynamic approach was introduced to measure the quantitative change characteristics of a land-use type over a set time horizon in the watershed [48]. The calculation formula used is as follows:

$$K = \frac{U_b - U_a}{U_a} \times \frac{1}{F} \times 100\% \tag{3}$$

where U_a is the area of land-use type a at the beginning of the period; U_b is the area of the same land-use type at the end of the period; F denotes the study period; and K is the annual rate of change during the study period.

2.3.3. Topographic Position Index (TPI)

The TPI was introduced to reflect the multidimensional changes in the ESV within the watershed along topographic gradients, and to characterize the spatial pattern distribution of ESV [49]. The calculation formula is as follows:

$$T = \log_{10}\left(\left|\frac{E}{E_{mean}} + 1\right| \times \left|\frac{D}{D_{mean}} + 1\right|\right) \tag{4}$$

where E is the elevation value of the raster; E_{mean} is the average elevation value of the raster; D is the slope value of the raster; D_{mean} is the average slope value in the raster; T is the topographic position index. The magnitude of T is affected by both the elevation value and the slope of the study area. If the elevation is larger and the slope is steeper, T is larger, and vice versa.

2.3.4. Sensitivity Analysis of Ecosystem Service Values

In this paper, a Coefficient of Sensitivity (CS) was introduced to test the effects of land-use change on ESV, and to discern the dependence of ESV on the value coefficients derived from 2000 to 2015 [9]. If CS > 1, this reveals that the change in the ESV coefficient of one land-use type has a significant impact on the total ESV. If CS < 1, this can indicate that the change in the ESV in one land-use type does not have a significant impact on the ESV in the entire study area per unit area. The CS was defined as follows:

$$CS = \left|\frac{(ESV_j - ESV_i)/ESV_i}{\left(VC_{jk} - VC_{ik}\right)/VC_{ik}}\right| \tag{5}$$

where ESV is the total ESV of the study area (CNY); VC is the ESV coefficient of each land-use type (CNY/hm^2); subscript k refers to the land-use type; subscripts i and j refer to before and after the adjustment of the ESV coefficient, respectively.

3. Results

3.1. Land-Use/Cover Changes in the Qihe Watershed

The land-use/cover types in the Qihe watershed are mainly grassland, arable land, and woodland, with a smaller area of watershed and unused land. Land-use/cover changed significantly during the study period (Table 2). The area of construction land, watershed, and unused land increased. Forest land area first increased and then decreased, grassland area decreased and then increased, while arable land area continued to decrease. The land-use single dynamic approach demonstrated that unused land was the highest, followed by watershed, and woodland was the smallest.

Table 2. Land-use/cover change (ha) and percentage (%) in Qihe watershed from 2000 to 2005.

Time	Category	Arable Land	Woodland	Grassland	Waters Area	Construction Land	Unused Land
2000	Area (ha)	73,417	68,287.3	69,438.7	3596.49	7893.35	10.9705
2005		68,306.28	72,657.67	69,415.12	3603.19	8737.88	10.971
2010		66,183.28	67,930.32	73,540.46	4063.92	10,932.70	79.67
2015		60,445.36	66,183.25	78,293.08	5595.32	12,128.45	80.32
2000–2005	Area change (ha)	−5110.72	4370.37	−23.58	6.70	844.53	0.00
2005–2010		−2123	−4727.35	4125.34	460.73	2194.82	67.61
2010–2015		−5737.92	−1747.06	4752.61	1531.4	1195.75	0.69
2000	Percentage (%)	32.97	30.67	31.21	1.62	3.52	0.01
2005		32.62	30.67	31.16	1.62	3.92	0.01
2010		29.71	30.5	33.02	1.82	4.91	0.04
2015		27.14	29.72	35.15	2.51	5.44	0.04
2000–2005	Single-motion Attitude (%)	−1.39	1.28	−0.01	0.01	2.14	0.01
2005–2010		−0.62	−1.30	1.19	2.56	5.02	123.2
2010–2015		−1.73	−0.51	1.29	7.54	2.19	0.17

3.2. Changes in the Value of Ecosystem Services in the Qihe Watershed

3.2.1. Temporal Change

Each land-use type ESV was obtained by multiplying various land-use type areas at different periods with the corresponding ESV coefficients. The highest share of woodland ESV in total (53%) from 2000–2015 is shown in Figure 3. It is clear that the grassland areas account for 36% of the total (Figure 3a), but the ESV percentage is less than 14% (Figure 3b). The reason for this is that the ESV coefficients of both water and woodland land-use are greater than the ESV coefficient of other land-use types [18].

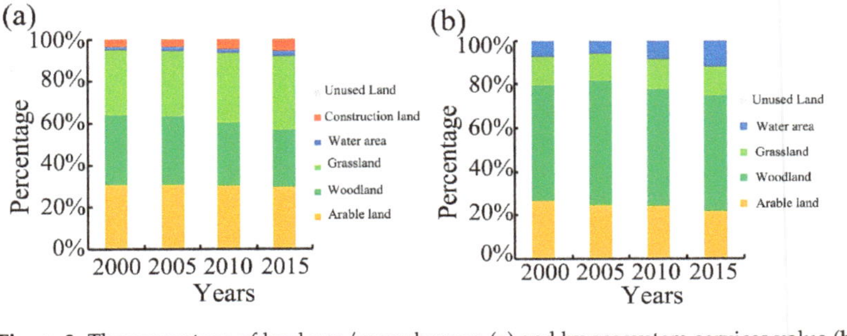

Figure 3. The percentage of land-use/cover by area (**a**) and by ecosystem services value (**b**) in the Qihe watershed.

The total ESV in 2000 was CNY 1.954 billion (Table 3), and mainly composed of arable land, woodland, and grassland ESVs. The total ESV of the study area increased by CNY 1.981 billion in 2005. The CNY 0.66 million increase in woodland ESV accounted for the major part of the gain and compensated for the CNY 0.36 million decrease in arable land ESV.

Table 3. Change amount (CNY) and change rate (%) of ecological service value in the Qihe watershed from 2000 to 2015.

Land-Use	ESV/1 × 10^8 CNY				2000–2005		2005–2010		2010–2015	
/Cover Type	2000	2005	2010	2015	Change Amount/ 1 × 10^8 CNY	Change Rate/%	Change Amount/ 1 × 10^8 CNY	Change Rate/%	Change Amount/ 1 × 10^8 CNY	Change Rate/%
Cultivated land	5.23	4.87	4.71	4.31	−0.36	−7	−0.15	−3.12	−0.41	−8.67
Forest land	10.33	10.99	10.28	10.01	0.66	6	−0.71	−6.51	−0.26	−2.57
Grassland	2.49	2.5	2.64	2.81	−0.01	−0.4	0.14	5.6	0.17	6.43
Waters	1.46	1.47	1.65	2.28	0.0027	0.2	0.19	12.79	0.62	37.68
Unused land	0.00	0.00	0.00	0.00	0.00	0.00	0.00	0.00	0.00	0.00
Construction Land	0.00	0.00	0.00	0.00	0.00	0.00	0.00	0.00	0.00	0.00
Total	19.54	19.81	19.29	19.4	0.27	1.38	−0.52	−2.62	0.11	0.57

Between 2000 and 2005, the land-use type with the most significant decrease in ESV was arable land (5110.72 ha), with the largest increase in woodland (4370.37 ha). The total ESV decreased by CNY 25 million between 2000 and 2010, because woodland and arable land were converted to other land-use types. The increase in water and grassland areas compensated for the decrease in total value. The largest increase in land-use type was in water area. Compared to the period 2000–2005, the total ESV of the water land-use type decreased at a high rate of change (1.9%), and the ESV showed an increasing and then decreasing trend.

During the 2000–2015 period, the total ESV decreased to CNY 1.942 billion, with the decrease in area of arable land and woodland being the main reason for the decrease in the total ESV. The total ESV underwent an increase of CNY 27 million from 2000 to 2005, a decrease of CNY 52 million yuan from 2005 to 2010, and an increase of CNY 11 million from 2010 to 2015. The decrease in the Qihe watershed total ESV is mainly attributed to the decrease in areas of arable land and woodland, and increase in the area of construction land-use.

Each ESV and its contribution rate from 2000 to 2015 were summarized using the secondary type value coefficients multiplied by the corresponding land-use type for each calendar year (Table 4). The different ecosystem function ESVs exhibited small variations, with the highest contribution of 18.83% from WD and the smallest contribution from EC (3.99%). The ranking of the individual ESVs is as follows: WD > SFC > WC > BC > CR > GR > FP > RM > EC.

3.2.2. Spatial Variation

Based on land-use/cover data from 2000, 2005, 2010, and 2015, different ESV land-use types in the four years were calculated for each grid cell, as well as the region's total ESV [37]. The value density was classified into five classes (0–1000 CNY/km^2, 1000–2000 CNY/km^2, 2000–4000 CNY/km^2, 4000–7000 CNY/km^2 and >7000 CNY/km^2) by referring to the study of Xu et al. [48]. The amount of ESV density change was divided into six categories (<−4000 CNY/km^2, −4000 to −1000 CNY/km^2, −1000–0 CNY/km^2, 0–2000 CNY/km^2), displaying a clear reflection of the difference in spatial distribution and ESV change trend [50].

Table 4. Ecosystem service value (1×10^8 CNY) and the contribution rate (%) of the Qihe watershed.

Type I	Type II	2000		2005		2010		2015		Grade
		ESV	Contribution Rate	ESV	Contribution Rate	ESV	Contribution Rate	ESV	Contribution Rate	
Adjustment Services	Gas Regulation (GR)	1.94	9.92	2.01	10.15	1.9	9.85	1.83	9.43	6
	Climate Regulation (CR)	2.05	10.49	2.08	10.5	1.99	10.32	1.92	9.9	5
	Water Conservation (WC)	2.67	13.66	2.73	13.78	2.72	14.11	2.93	15.1	3
Support Services	Soil formation And conservation (SFC)	3.11	15.91	3.14	15.85	3.01	15.6	2.89	14.9	2
	Waste Disposal (WD)	3.64	18.62	3.61	18.22	3.63	18.82	3.82	19.69	1
	Biodiversity Conservation (BC)	2.52	12.89	2.58	13.02	2.49	12.91	2.47	12.73	4
Supply Services	Food Production (FP)	1.66	8.49	1.6	8.07	1.59	8.24	1.56	8.04	7
	Raw Materials (RM)	1.17	5.98	1.24	6.26	1.16	6.01	1.12	5.77	8
Cultural Services	Entertainment Culture (EC)	0.78	3.99	0.82	4.14	0.8	4.14	0.85	4.38	9
	Total	19.54	100	19.81	100	19.29	100	19.4	100	-

The overall ESV in the Qihe watershed was high in the southwest and low in the northeast (Figure 4). ESV densities > 7000 CNY/km^2 were mainly distributed in areas covered by woodlands and grasslands in the upper reaches of the watershed. Densities of 4000–7000 CNY/km^2 were mainly distributed in the central part of the watershed covered by cropland and grassland. The regional ESV density of grassland cover was between 2000–4000 CNY/km^2 and was the most widely distributed, while the ESV density in the middle and lower reaches of the watershed was <2000 CNY/km^2, displaying a fragmented distribution. During the period spanning 2000–2005, the spatial ESV distribution density was diminished in the upper reaches of the watershed and increased in the middle and lower reaches. The most significant decrease in ESV in the upper reaches was caused by the rapid expansion of woodland reclamation into arable land and construction land. At the same time, the expansion of the water area caused an increase in ESV density in the middle and lower reaches, leading to a gradual improvement in habitat quality in the middle and lower reaches of the basin [51]. The most obvious change in ESV density between 2005 and 2010 was in the lower reaches due to the growth in construction land area and reduction in grassland and arable land area. As different land-use types correspond to various ESV coefficients, a land-use type shift in the watershed will cause a corresponding change in its ESV. It is clear that during the study period, the ESV in the Qihe watershed was in a dynamic process of change. The decrease in woodland and grassland areas, and the rapid expansion of construction land explained the most obvious changes. Overall, the total ESV showed a decreasing trend.

3.3. Analysis of the TPI of ESVs

Referring to research by Chen et al. [44], the TPI was classified into six levels (0–0.2, 0.2–0.4, 0.4–0.6, 0.6–0.8, 0.8–0.1 and >1). The TPI is high in the east and low in the west (Figure 5a). TPI values from 0–0.2 are mainly distributed in the lower reaches of the watershed; 0.4–0.6 TPI are distributed in the middle reaches, and 0.8–1 TPI are found in the upper reaches of the basin. Figure 5b demonstrates that land-use area is mainly distributed on 0.2–0.4 and 0.6–0.8 TPI, which account for 27% and 25%, respectively. Values of TPI > 1 have the least distributed area (0.22 km^2) and the smallest ratio (0.01%). Overall, the TPI < 1 is distributed most widely in the Qihe watershed, accounting for 99.9% of the area.

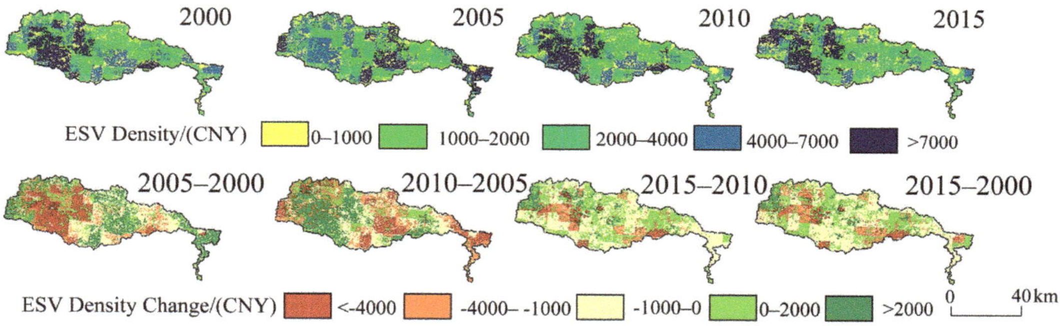

Figure 4. Spatial distribution and change in ecosystem service value density in the Qihe watershed from 2000 to 2015.

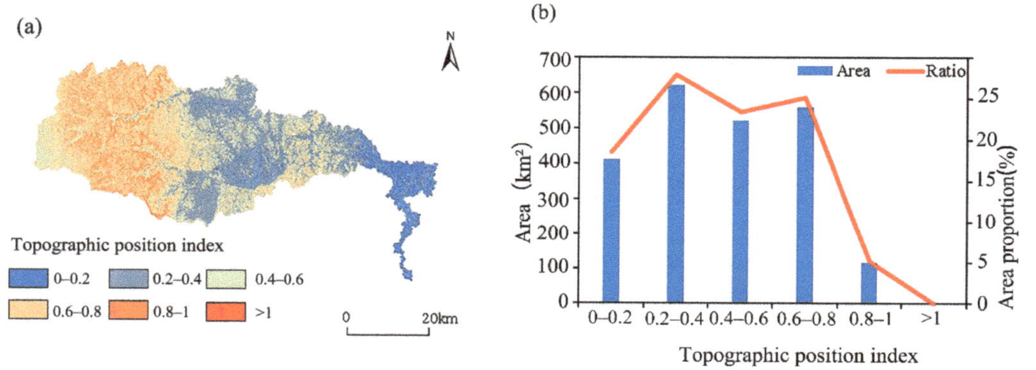

Figure 5. Topographic position index distribution map (**a**) and land-use type area by different topographic position index (**b**).

3.3.1. Topographic Factor Analysis of ESV Change

The land-use/cover in 2000 and 2015 were used to analyze the change in ESV and individual ESVs in relation to TPI (Figure 6a). The most significant decrease in arable land ESV (CNY 39.52 million) occurred within the 0–0.2 TPI range, while grassland and woodland ESV increased by CNY 12.57 million and CNY 2.34 million, respectively, mainly caused by low-value TPI areas being highly influenced by human activities [29].

A significant increase in watershed ESV (CNY 25.22 million) and an increase in woodland ESV of CNY 6.85 million occurred within TPIs of 0.2–0.4. The high ESV coefficient of water areas was the major factor behind the significant increase in water ESV, while arable land and grassland ESV decreased by CNY 12.5 million CNY and CNY 3.2 million CNY. Small overall changes in the ESV of areas with TPI from 0.4 to 0.6 were due to increases in water, grassland, and woodland ESVs and decreases in arable land ESV. Woodland ESV decreased in areas with high TPI (i.e., TPI > 0.6), with the largest reduction being within the 0.6–0.8 TPI range (CNY 25.89 million).

As shown in Figure 6b, the individual ESVs vary across TPI values. For example, the CR ESV decreased by CNY 21.14 million, FP ESV decreased by CNY 19.25 million, BC ESV decreased by CNY 16.14 million, GS ESV decreased by CNY 14.13 million, and WC decreased by CNY 9.01 million, which were mainly due to the largest reduction in a woodland area during this interval. The WC ESV increased by CNY 10.98 million and CNY 9.07 million within the 0.2–0.4 TPI and 0.4–0.6 TPI intervals, respectively. Tables 1 and 2 indicate that the large water area and high WC ESV coefficient are the main reasons for the watershed ESV increase.

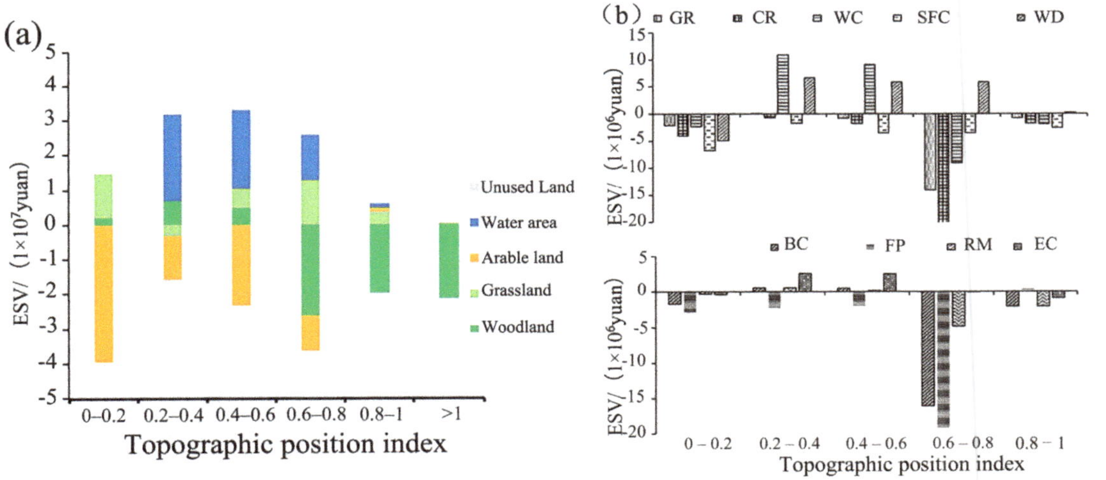

Figure 6. Changes in total ESV (**a**) and individual ESVs (**b**) in the Qihe watershed by topographic position index from 2000 to 2015.

3.3.2. Spatial Characteristics of TPI of Ecosystem Service Value Change

Here, we use the land-use/cover data from the two years 2000 and 2015 and combine them with the TPI analysis to investigate the dynamic change process of ESV. By referring to the work of Li et al. [22], the ESV was divided into <6000 CNY, 6000–10,000 CNY, 10,000–30,000 CNY, and >30,000 CNY in total, with topographic position indices of 0–0.2, 0.2–0.4, 0.4–0.6, 0.4–0.8, 0.8–1 and >1 (Figure 7). The results suggest that ESV greater than CNY 30,000 in 2000–2015 was distributed over a large area and concentrated within TPIs of 0.2–0.4, 0.6–0.8, and 0.8–1. ESV < 6000 CNY was mainly distributed within 0–0.2 and 0.4–0.6 TPI grading, suggesting that ESV is higher and widely distributed in 0.2–0.4, 0.6–8, and 0.8–1 TPI, primarily due to the wide distribution area of grassland and woodland and the higher ESV coefficients of these two types. The TPI > 1 accounts for a small proportion of the area, and the distribution of grassland and woodland in this zone was small, thus the ESV distribution is not significant and located within two intervals of CNY 6000–10,000 and CNY 10,000–30,000. There was a small change in land-use types resulting in the change in ESV from 2000 to 2015. The construction land area increase had no direct effect on the total ESV, while the increase in water area and grassland by 1998.83 ha and 8854.38 ha, respectively, compensated to some extent for the total ESV loss caused by the arable land area decrease.

3.4. *Sensitivity Analysis of the ESVs in the Qihe Watershed*

The modified ESV coefficients of the Qihe watershed were adjusted up and down by 50%, respectively, to calculate the total ESV for all years, and to estimate the sensitivity of the results to this value (Table 5) [9]. The calculated results of the adjusted ESV coefficients for each land-use type indicate a sensitivity index of less than 1. The CS of unused land and water tends to zero, reflecting the inelasticity of the total ESV concerning the service value coefficient, demonstrating the reliability of the results in this paper. The large ESV coefficient of woodland land led to the highest sensitivity index (about 0.53), followed by arable land and grassland, owing chiefly to the large proportion of arable land and grassland area in the total. Both the small area of unused land and low ESV coefficient resulted in the lowest sensitivity index for unused land (0.0005). During the period 2000–2015, the CS of grassland, water, and unused land showed a stable and then increasing trend. The CS of arable land displayed a gradual decrease, the woodland CS displayed an increase and then a decrease, and their CS changes were consistent with the changes in their respective adjusted areas. Overall, the ESV sensitivity index indicates that the ESV coefficients of various land-use types still bear many uncertainties, but the total ESV in the Qihe watershed remains in a stable state.

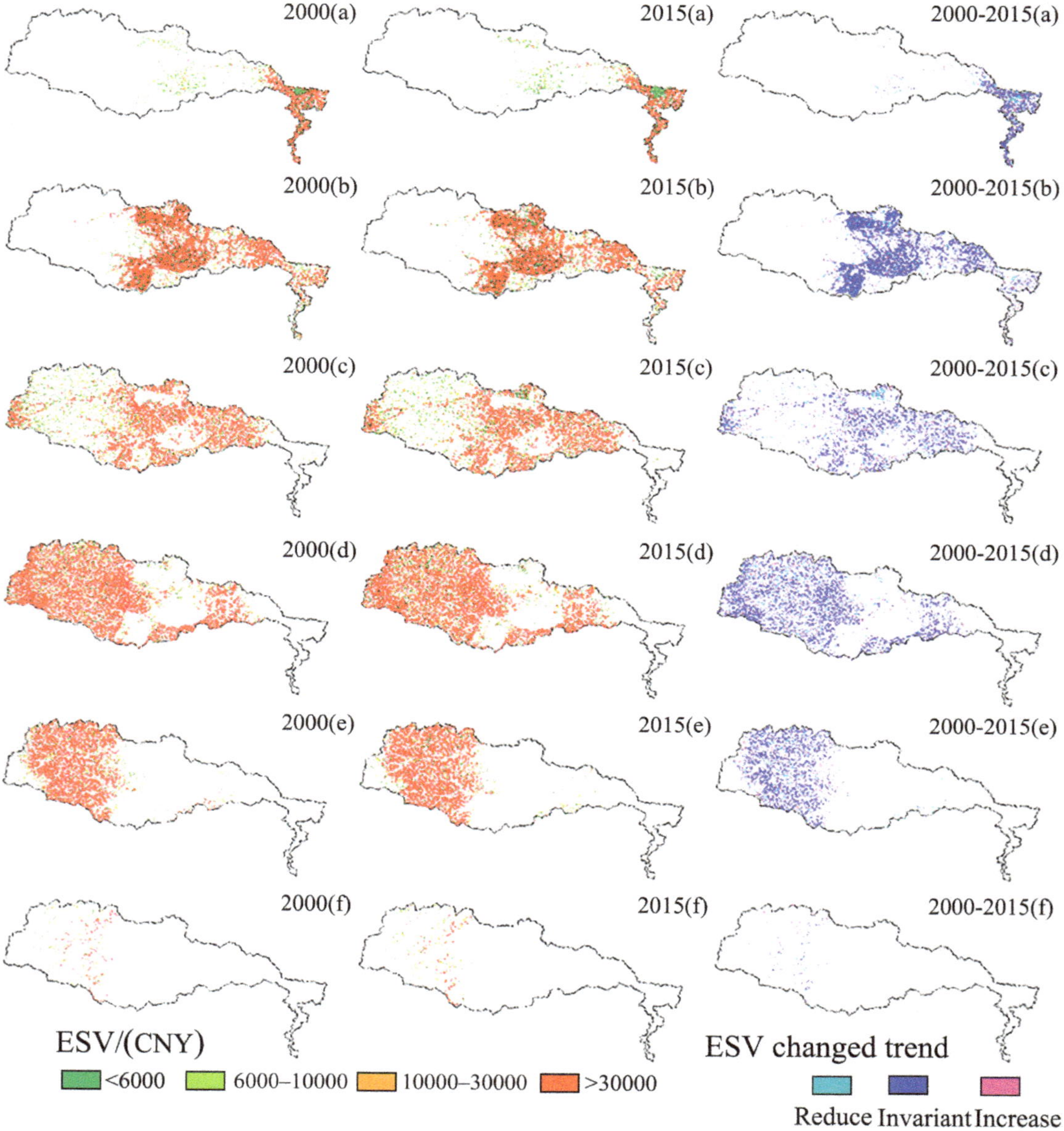

Figure 7. Characteristics of change in ecosystem service values of different topographic position indices (Note: a, b, c, d, e and f indicate 0–0.2, 0.2–0.4, 0.4–0.6, 0.6–0.8, 0.8–1, TPI > 1) in the Qihe watershed from 2000–2015.

Table 5. Changes in the total value (CNY Billion), amount of change (CNY Billion) and sensitivity index of ecosystem services in Qihe watershed after adjustment.

	ESV/(CNY Billion)				Amount of Change/(CNY Billion)				Sensitivity Index (CS)			
Value Factor	**2000**	**2005**	**2010**	**2015**	**2000–2005**	**2005–2010**	**2010–2015**	**2000–2015**	**2000**	**2005**	**2010**	**2015**
Cultivated land VC + 50%	22.13	22.25	21.64	21.56	0.12	−0.61	−0.08	−0.57	0.27	0.25	0.24	0.22
Cultivated land VC-50%	16.90	17.39	16.93	17.26	0.49	−0.46	0.32	0.36				
Forestland VC + 50%	24.69	25.32	24.43	24.42	0.63	−0.89	−0.01	−0.27	0.53	0.55	0.53	0.52
Forestland VC-50%	14.35	14.32	14.15	14.40	−0.03	−0.17	0.25	0.05				
Grassland VC + 50%	20.77	21.06	20.61	20.81	0.29	−0.45	0.21	0.04	0.13	0.13	0.14	0.14
Grassland VC + 50%	18.27	18.57	17.97	18.00	0.30	−0.60	0.04	−0.27				
Water VC + 50%	20.25	20.55	20.11	20.55	0.30	−0.44	0.43	0.30	0.07	0.07	0.09	0.12
Water VC-50%	18.79	19.09	18.46	18.27	0.30	−0.63	−0.19	−0.52				
Unused land VC + 50%	19.52	19.82	19.29	19.41	0.30	−0.53	0.12	−0.11	0.0005	0.0005	0.001	0.001
Unused land VC-50%	19.52	19.82	19.29	19.41	0.30	−0.53	0.12	−0.11				

4. Discussion

The equivalent factor method used here to estimate the ESV in the Qihe watershed can visually reflect the change in ESV. The advantage of the equivalent factor method is its lower data demand compared with the price per unit area of the service function method, which is suitable for the study of ESV at regional and global scales [20]. The changes in land-use during the research period had a profound impact on the ESV, with the changes acting as a guide for adjusting the land-use structure and optimizing the land-use pattern. Topographic elements have a significant impact on regional land-use patterns and spatial structure. Therefore, investigating the dual response of ESV to land-use/cover and topographic factors can be a useful method for assessing the quality of the ecological environment in a watershed. Analyzing the interaction between individual ESV and topographic factors plays an important role in enhancing human well-being and building harmonious habitat relationships. The purpose of this paper is to provide a reference for small watershed-scale ecosystem service research and ecological environment construction.

5. Conclusions

The study conclusions are as follows:

(1) The land-use types in the Qihe watershed from 2000 to 2015 were mainly arable land, forest land, and grassland, the sum of which accounts for more than 90% of the total area. The land-use/cover changes were obvious as the areas of cultivated land and forest land decreased by 12,971.61 ha and 2104.05 ha, respectively, and the areas of grassland and water increased by 8854.38 ha and 43,234.8 ha, respectively.

(2) The ESV in the Qihe watershed decreased by CNY 0.14 billion from 2000 to 2015. During the study period, the total ESV increased, then decreased, and then increased again. The highest ESV occurred in 2015, with a value of CNY 1.981 billion. The contribution level of each individual ESV remained stable, with waste treatment exhibiting the highest contribution level of 18.84%, followed by soil formation and protection.

(3) There was a significant influence of topography on the ESV. The largest decrease of CNY 39.52 million in cropland ESV and the largest increase of CNY 12.56 million in grassland ESV occurred within the 0–0.2 TPI range. The largest increase in the 0.2–0.4 TPI range was that of water ESV (CNY 25.19 million) and the largest decrease in the 0.6–0.8 TPI range was that of grass ESV (CNY 25.89 million). The largest reductions in individual ESVs were observed in the 0.6–0.8 TPI range. The ESV of

water supply increased by CNY 10.98 million and 9.07 million within the areas of TPI in the 0.2–0.4 and 0.4–0.6 intervals, respectively.

(4) The sensitivity index of the ESV in the Qihe watershed is less than 1. This implies a certain lack of elasticity for the value coefficient and characterizes the robustness of the research results in this paper.

The ESV of the Qihe watershed was estimated using the Chinese terrestrial unit area ecosystem service value scale. Additionally, the spatial and temporal evolution characteristics from the period 2000–2015 were analyzed. Quantitative studies of the ESV in this watershed bear insufficient explanatory power for the trade-offs and synergistic relationships between ecosystem service functions. In future research, the trade-offs and synergistic relationships of ecosystem service functions within the Qihe watershed will be refined based on this paper.

Author Contributions: Conceptualization, L.L. and Y.L. (Yonghui Li); methodology, L.L.; software, Y.L. (Yonghui Li); validation, L.L., Y.L. (Ying Liang) and L.Y.; formal analysis, L.L.; investigation, Y.L. (Yonghui Li); resources, W.Z.; data curation, Y.L. (Ying Liang); writing—original draft preparation, Y.L. (Ying Liang); writing—review and editing, L.L.; visualization, G.C.; supervision, Y.L (Yonghui Li). All authors have read and agreed to the published version of the manuscript.

Funding: The research was supported by the Beijing Normal University Interdisciplinary Research Foundation for First-Year Doctoral Candidates (Grant BNUXKJC 2124).

Institutional Review Board Statement: Not applicable.

Informed Consent Statement: Not applicable.

Data Availability Statement: Not applicable.

Conflicts of Interest: The authors declare no conflict of interest.

References

1. Gamfeldt, L.; Snäll, T.; Bagchi, R.; Jonsson, M.; Gustafsson, L.; Kjellander, P.; Ruizjaen, M.C.; Fröberg, M.; Stendall, J.; Philipson, C.D.; et al. Higher levels of multiple ecosystem services are found in forests with more tree species. *Nat. Commun.* **2013**, *4*, 1340–1348. [CrossRef] [PubMed]
2. Zhong, L.; Wang, J.; Zang, X.; Ying, L.X. Effects of agricultural land consolidation on ecosystem services: Trade-offs and synergies. *J. Clean. Prod.* **2020**, *264*, 121412. [CrossRef]
3. Polasky, S.; Tallis, H.; Reyers, B. Setting the bar: Standards for ecosystem services. *Proc. Natl. Acad. Sci. USA* **2015**, *112*, 7356–7361. [CrossRef] [PubMed]
4. Liang, Y.; Song, W. Integrating potential ecosystem services losses into ecological risk assessment of land use changes: A case study on the Qinghai-Tibet Plateau. *J. Environ. Manag.* **2022**, *318*, 115607. [CrossRef] [PubMed]
5. Toman, M.; Speclal, S. Forum on valuation of Ecosystem Services: Why not to calculate the value of the world's ecosystem services and natural capital. *Ecol. Econ.* **2004**, *25*, 57–60. [CrossRef]
6. Salzman, J.; Bennett, G.; Carroll, N.; Goldstein, A.; Jenkins, M. The global status and trends of Payments for Ecosystem Services. *Nat. Sustain.* **2018**, *1*, 136–144. [CrossRef]
7. Ren, H.; Zhang, J.J.; Zhu, W.B.; Wang, L.Y.; Zhang, L.J.; Zhu, L.Q. Impact of land use change on habitat in the Qihe River Basin of Taihang Mountains. *Prog. Geogr.* **2018**, *37*, 1693–1704. (In Chinese)
8. He, S.S.; Ye, L.P.; Zhu, W.B.; Cui, Y.P.; Zhu, L.Q. Soil erosion and water supply change in Qihe watershed of Taihang Mountains from 2000 to 2015. *Geogr. Res.* **2018**, *37*, 1775–1787. (In Chinese)
9. Li, Y.S.; Zhun, G.J.; Liang, T.; Wu, L.X.; Liu, W.J. Study of land use change on the gains and losses of ecosystem service function values of Chaohu Lake Basin. *Geogr. Res.* **2009**, *28*, 1656–1664. (In Chinese)
10. Song, W.; Deng, X.Z. Land-use/land-cover change and ecosystem service provision in China. *Sci. Total Environ.* **2017**, *576*, 705–719. [CrossRef]
11. Costanza, R.; D'Arge, R.; De Groot, R.; Farber, S.; Grasso, M.; Hannon, B.; Limburg, K.; Naeem, S.; O'Neill, R.V.; Paruelo, J.; et al. The value of the world's ecosystem services and natural capital. *Nature* **1997**, *387*, 253–260. [CrossRef]
12. Sutton, P.C.; Costanza, R. Global estimates of market and non-market values derived from nighttime satellite imagery, land cover and ecosystem service valuation. *Ecol. Econ.* **2002**, *41*, 509–527. [CrossRef]
13. Costanza, R.; Groot, R.D.; Sutton, P.; Ploeg, S.; Anderson, S.J.; Kubiszewski, I.; Farber, S.; Turner, R.K. Changes in the global value of ecosystem services. *Glob. Environ. Chang.* **2014**, *26*, 152–158. [CrossRef]

14. Costanza, R.; Groot, R.D.; Braat, L.; Kubiszewski, I.; Fioramonti, L.; Sutton, P.; Farber, S.; Grasso, M. Twenty years of ecosystem services: How far have we come and how far do we still need to go? *Ecosyst. Serv.* **2017**, *28*, 1–16. [CrossRef]
15. Ouyang, Z.Y.; Zhu, C.Q.; Yang, G.B.; Xu, W.H.; Zheng, H.; Zhang, Y.; Xiao, Y. Gross ecosystem product: Concept, Accounting framework and case study. *Acta Ecol. Sin.* **2013**, *33*, 6747–6761. (In Chinese) [CrossRef]
16. Ouyang, Z.Y.; Zheng, H.; Xiao, Y.; Polasky, S.; Liu, J.G.; Xu, W.H.; Wang, Q.; Zhang, L.; Xiao, Y.; Rao, E.; et al. Improvements in ecosystem services from investments in natural capital. *Science* **2016**, *352*, 1455–1459. [CrossRef]
17. Xie, G.D.; Zhang, C.; Zhang, L.M.; Xiao, Y.; Lu, C.X. Improvement of the evaluation method for ecosystem service value based on Per Unit Area. *J. Nat. Resour.* **2015**, *30*, 1243–1254. (In Chinese)
18. Xie, G.D.; Lu, C.X.; Leng, Y.F.; Zheng, D.; Li, S.C. Ecological assets valuation of the Tibetan Plateau. *J. Nat. Resour.* **2003**, *18*, 189–196. (In Chinese)
19. Xie, G.D.; Zhen, L.; Lu, C.X.; Xiao, Y.; Chen, C. Expert knowledge based valuation method of ecosystem services in China. *J. Nat. Resour.* **2008**, *23*, 911–919. (In Chinese)
20. Ouyang, Z.Y.; Wang, R.S.; Zhao, J.Z. Ecosystem services and their economic valuation. *Chin. J. Appl. Ecol.* **1999**, *10*, 635–640. (In Chinese)
21. Assessment, M.E. *Ecosystems and Human Well-Being*; Island Press: Washington, DC, USA, 2005.
22. Li, Z.H.; Dong, S.C.; Li, F.J.; Li, Y.; Wan, Y.K. Scenario analysis of ecological effects of urbanization in Wuwei oasis, China. *J. Desert Res.* **2012**, *33*, 937–942.
23. Power, M. The planetary piggy bank. *Nature* **1997**, *388*, 529–530. [CrossRef]
24. Hossain, M.S.; Dearing, J.A.; Rahman, M.M.; Salehin, R. Recent changes in ecosystem services and human well-being in the coastal areas of Bangladesh. *Reg. Environ. Chang.* **2016**, *16*, 429–443. [CrossRef]
25. Boyd, J.; Banzhaf, S. What are ecosystem services? The need for standardized environmental accounting units. *Ecol. Econ.* **2007**, *63*, 616–626. [CrossRef]
26. Zhu, W.B.; Li, S.C. A diagnostic framework of payments for ecosystem services and associated case studies. *Acta Ecol. Sin.* **2014**, *34*, 2460–2469. (In Chinese)
27. Fu, Q.; Hou, Y.; Wang, B.; Bi, X.; Li, B.; Zhang, X.S. Scenario analysis of ecosystem service changes and interactions in a mountain-oasis-desert system: A case study in Altay Prefecture, China. *Sci. Rep.* **2018**, *8*, 12939–12952. [CrossRef]
28. Li, J.L.; Shi, X.Y.; Zhu, Y.C. Environmental impact assessment of Puyang land use planning based on theory of ecosystem services value. *Popul. Resour. Environ. China* **2015**, *25*, 255–258. (In Chinese)
29. Liu, Y.Q.; Liao, L.W.; Long, H.L.; Qin, J.X. Effects of land use transitions on ecosystem services value: A case study of Hunan province. *Geogr. Res.* **2015**, *34*, 691–700. (In Chinese)
30. Li, X.W.; Fang, J.Y.; Piao, S.L. Land use changes and its implication to the ecological consequences in lower Yangtze Region. *Acta Geogr. Sin.* **2003**, *58*, 659–667. (In Chinese)
31. Rietveld, J.; Schilling, M.A.; Bellavitis, C. Platform strategy: Managing ecosystem value through selective promotion of complements. *Organ. Sci.* **2019**, *30*, 1232–1251. [CrossRef]
32. Bai, Y.; Wong, C.P.; Jiang, B.; Hughes, A.C.; Wang, M.; Wang, Q. Developing China's Ecological Redline Policy using ecosystem services assessments for land use planning. *Nat. Commun.* **2018**, *9*, 3034–3047. [CrossRef]
33. Renard, D.; Rhenmtulla, J.M.; Bennett, E.M. Historical dynamics in ecosystem service bundles. *Proc. Natl. Acad. Sci. USA* **2015**, *112*, 13411–13416. [CrossRef] [PubMed]
34. Zhang, Z.; Xia, F.; Yang, D.; Huo, J.; Wang, G.; Chen, H. Spatiotemporal characteristics in ecosystem service value and its interaction with human activities in Xinjiang, China. *Ecol. Indic.* **2020**, *110*, 105826. [CrossRef]
35. Han, R.; Feng, C.C.; Xu, N.; Guo, L. Spatial heterogeneous relationship between ecosystem services and human disturbances: A case study in Chuandong, China. *Sci. Total Environ.* **2020**, *721*, 137818. [CrossRef] [PubMed]
36. Fang, L.; Wang, L.; Chen, W.; Sun, J.; Cao, Q.; Wang, S.; Wang, L. Identifying the impacts of natural and human factors on ecosystem service in the Yangtze and Yellow River Basins. *J. Clean. Prod.* **2021**, *314*, 127995. [CrossRef]
37. Liu, S.; Sun, Y.; Wu, X.; Li, W.; Liu, Y.; Tran, L. Driving Factor Analysis of Ecosystem Service Balance for Watershed Management in the Lancang River Valley, Southwest China. *Land* **2021**, *10*, 522. [CrossRef]
38. He, C.; Shao, H.; Xian, W. Spatiotemporal Variation and Driving Forces Analysis of Eco-System Service Values: A Case Study of Sichuan Province, China. *Int. J. Environ. Res. Public Health* **2022**, *19*, 8595. [CrossRef]
39. Xu, N.Y.; Guo, L.; Xue, D.Y.; Sun, S.Q. Land use structure and the dynamic evolution of ecosystem service value in Gannan region, China. *Acta Ecol. Sin.* **2019**, *39*, 1969–1978. (In Chinese)
40. Mueller, H.; Hamilton, D.P.; Doole, G.J. Evaluating services and damage costs of degradation of a major lake ecosystem. *Ecosyst. Serv.* **2016**, *22*, 370–380. [CrossRef]
41. Zhang, Z.Y.; He, X.L.; Liu, L.; Li, Z.Q.; Wang, P.Y. Ecological service functions and value estimation of glaciers in the Tianshan Mountains, China. *Acta Geogr. Sin.* **2018**, *73*, 856–867. (In Chinese)
42. Chen, W.J.; Jan, J.F.; Chung, C.H.; Liaw, S.C. Resident Willingness to Pay for Ecosystem Services in Hillside Forests. *Int. J. Environ. Res. Public Health* **2022**, *19*, 6193. [CrossRef] [PubMed]
43. Richter, F.; Jan, P.; El, B.N.; Lüscher, A.; Buchmann, N.; Klaus, V.H. A guide to assess and value ecosystem services of grasslands. *Ecosyst. Serv.* **2021**, *52*, 101376. [CrossRef]

44. Chen, C.; Liu, X.A.; Yan, L.L.; Wang, J.; Peng, P.H. Evaluation on Ecosystem Service Values of Sichuan Nanhe National Wetland Park. *Wetl. Sci.* **2018**, *16*, 238–244.
45. Zhang, R.; Wang, Y.M.; Chang, J.X.; Li, Y.Y. Response of land use change to human activities in the Yellow River Basin based on water resources division. *J. Nat. Resour.* **2019**, *34*, 274–287. (In Chinese) [CrossRef]
46. Li, L.; Zhu, W.B.; Li, Y.H.; Zhu, L.Q.; Xu, S.B.; Feng, X.X. Ecosystem Service Value Gains and Losses of Qihe River Basin Based on Topographic Gradient Characteristics. *Res. Soil Water Conserv.* **2019**, *26*, 287–295. (In Chinese)
47. Xie, G.D.; Zhang, Y.L.; Lu, C.X.; Zheng, D.; Cheng, S.K. Study on valuation of rangeland ecosystem services of China. *J. Nat. Resour.* **2001**, *16*, 47–53. (In Chinese)
48. Xu, L.M.; Li, H.; Chen, D.H.; Ye, L.Z.; Li, J.G.; Zhao, J.P. Spatial-temporal variability and driving force of ecosystem service: A case study of Bozhou in Xinjiang. *J. Subtrop. Resour. Environ.* **2018**, *13*, 66–74. (In Chinese)
49. Liu, G.L.; Zhang, L.C.; Zhang, Q. Spatial and temporal dynamics of land use and its influence on ecosystem service value in Yangtze River Delta. *Acta Ecol. Sin.* **2014**, *34*, 3311–3319. (In Chinese)
50. Zhang, J.J.; Zhu, W.B.; Zhao, F.; Zhu, L.Q.; Li, M.J.; Zhu, M.; Zhang, X.D. Spatial variations of terrain and their impacts on landscape patterns in the transition zone from mountains to plains—A case study of Qihe River Basin in the Taihang Mountains. *Sci. China Earth Sci.* **2018**, *48*, 476–486. (In Chinese) [CrossRef]
51. Chen, Z.; Huang, Y.B.; Zhu, Z.P.; Zheng, Q.Q.; Que, C.X.; Dong, J.W. Landscape pattern evolution along terrain gradient in Fuzhou City, Fujian Province, China. *Chin. J. Appl. Ecol.* **2018**, *29*, 4135–4144. (In Chinese)

International Journal of
Environmental Research and Public Health

MDPI

Article

Metacoupling of Water Transfer: The Interaction of Ecological Environment in the Middle Route of China's South-North Project

Qingmu Su [1], Hsueh-Sheng Chang [2], Xiang Chen [2,3,*] and Jingjing Xiao [4]

1 School of Architecture and Planning, Fujian University of Technology, Fuzhou 350118, China
2 Department of Urban Planning, National Cheng Kung University, Tainan 70101, Taiwan
3 Institute of Urban-Rural Planning & Design, Xiamen University, Xiamen 361001, China
4 Department of Geology, National Taiwan University, Taibei 106216, Taiwan
* Correspondence: p28073038@ncku.edu.tw

Abstract: At present, nearly half of the population of China live in water-deficient areas where water needs to be transferred from surrounding or remote water sources to meet local water demands. Although the water transfer project has alleviated the demands for water in the water-deficient areas, and brought water-supply income to water source regions, it has also posed some cross-regional negative impacts, including the changes in the original ecology within the water source, the impacts on the downstream water demands, and the risk of biological invasion in the distant water receiving areas. Therefore, it can be seen that the impact of water transfer is complicated and will be manifested in various aspects. The Middle Route of China's South–North Water Transfer Project (SNWTP-MR), as the world's largest cross-watershed water transfer project, exerts particularly important effects on regional sustainable development; however, it also produces complex interactions within the ecological environment itself, downstream and in the distant water receiving cities. Thus, this work attempts to apply a metacoupling analysis framework of water transfer to explore the ecological interaction of water transfer in SNWTP-MR on each system. The metacoupling framework can be divided into intracoupling, pericoupling and telecoupling. This study focuses on the analysis of the causes and effects of the intracoupling, pericoupling and telecoupling of SNWTP-MR from the perspective of ecological values and ecological risks. We found that the coupling of water transfer brings about 23 billion yuan of ecological service value to the water source annually, but also increases the internal ecological risk index by 9.31%, through the calculation of changes in land use; secondly, the power generation benefit significantly increases, and the flood control standards have shifted from once-in-20 years to once-in-a-century. However, the ecological risks are also significant, such as poor water quality, eutrophication of water resources, competition for water between industry and agriculture, deterioration of waterway shipping, and threats to biodiversity, etc. Considering only water supply, the population carrying capacity of the water resource in distant water receiving cities is increased by 16.42 million people, which enhances the value of water resources and creates a cross-regional green ecological landscape belt. Nevertheless, the biological invasion and water pollution have greatly affected the safety of water supply. It can be seen that the cross-regional water transfer does not always damage the interests of the sending system and the spillover system while benefiting the receiving system; its impacts are complex and variable. Through this paper, it is hoped to provide a reference for the analysis of the ecological compensation, resource development and allocation in SNWTP-MR by revealing the metacoupling relationship of SNWTP-MR. This paper will provide new ideas for researching the metacoupling relationship, thereby offering valuable reference for the study of the interaction generated by large-scale water transfer.

Keywords: water transfer; metacoupling; ecological environment; Middle Route of China's South–North Water Transfer Project (SNWTP-MR); sustainable development

Citation: Su, Q.; Chang, H.-S.; Chen, X.; Xiao, J. Metacoupling of Water Transfer: The Interaction of Ecological Environment in the Middle Route of China's South-North Project. *Int. J. Environ. Res. Public Health* **2022**, *19*, 10555. https://doi.org/10.3390/ijerph191710555

Academic Editor: Paul B. Tchounwou

Received: 29 July 2022
Accepted: 22 August 2022
Published: 24 August 2022

1. Introduction

As sustainable development is a major challenge around the world, how to achieve the sustainability of urban water supply systems is one of the global concerns [1,2]. The world population reached 7.7 billion in May 2019 and is estimated to increase to 11.2 billion in 2100 [3]. However, the rapid urbanization and population growth will lead to a series of severe challenges, especially in terms of urban water security. The water sources in large cities only account for 1% of the world's total water sources, but the urban population consumes 41% of the world's water resources [4,5]. Therefore, large cities or places where the population gathers need a large amount of water transferred from surrounding areas or remote distant areas to meet the local water demands. China's water resources are estimated to total 2800 billion m^3, ranking fourth in the world. However, affected by natural factors and human activities, the water resources are unevenly distributed in different regions. For example, the water resources in the northern areas are unable to support the sustained and sound development of the economy, resulting in increasingly significant predicaments. To solve this problem, the Chinese government has implemented regional water transfer projects through dam construction, water diversion, etc., among which the most influential is China's South-North Water Transfer Project (SNWTP). However, there is no systematic answer as to whether other impacts will be caused if the water is transferred; considering the large scale of the project, it is necessary to have a comprehensive analytical framework to evaluate the impact of the water transfer project on river ecology and urban ecology [6,7].

China's South-North Water Transfer Project (SNWTP) is by far the world's largest cross-watershed water transfer project, and the Middle Route of China's South–North Water Transfer (SNWTP-MR) can alleviate the problem of insufficient water in the north to a large extent, especially in Beijing, Tianjin, Hebei, and Henan where high water risks exist [8]. However, under multiple influences such as the unique natural regional unit of the water source (at the junction of three provinces and the north-south boundary of China), the human activities (regional cultivation and reservoir expansion) and policy implementation (water withdrawal and immigration in the middle route), SNWTP-MR has led to complex environmental changes and variations in hydrological processes. Moreover, the resulting evolution and migration of the ecological environment system directly affects the security of the ecological environment and socio-economic sustainable development. At the same time, stakeholders often lack cross-sectoral coordination [9]. Therefore, it is necessary to study the interaction caused by SNWTP-MR on water source areas, downstream and in distant water receiving cities to reveal the systematic coupling and sustainable development of the ecological environment.

Many scholars have explored these issues from different aspects. There are studies focusing on the interactions within the coupling system, or concentrating on the social and economic interactions between coupling systems [10,11]. Su et al. (2021) divided the metacoupling of the water transfer project into intracoupling, pericoupling and telecoupling, opened the internal structure of the traditional data envelopment analysis (DEA) system, and analyzed the efficiency of water transfer [12]. Chang et. al. (2021) used inter-regional land exchange as the basis for fair distribution, and analyzed the cross-regional impact of stormwater flow through local coupling and telecoupling [13]. In general, the conceptual framework of telecoupling provides a much-needed comprehensive method for systematic research which clearly examines the interaction of the coupling of human and natural systems on time and space scales; it is an effective framework to solve the sustainability of urban water resources, and is also an important reference framework for this paper [6,14].The framework consists of five interrelated components: the coupling of human and natural system (system); the material, information and energy flow within the system (flow); agent promoting the flow (agent); driving causes of flow (cause); and the effects of the flow (effects). The direction of the flow determines that the system can be considered as a sending system (for example, the sending system in this study mainly refers to the watershed range of the Danjiangkou Reservoir), the receiving system (for example, the distant cities that receive water from the sending system and the downstream of the

Danjiangkou Reservoir), and the spillover system (for example, the counties and cities along the SNWTP-MR) [15]. This study abandons the "single dimension" mode in the traditional research on coupling, and combines with the conceptual framework of telecoupling to build the metacoupling framework of water transfer in SNWTP-MR, including the intracoupling within the system, and the pericoupling and telecoupling between two systems.

This study attempts to answer the following questions under the framework of exploring the metacoupling of water transfer from SNWTP-MR: How can we systematically analyze the interaction among the water source area, downstream areas, and the distant water receiving cities in SNWTP-MR? How can the ecological value and uncertain risks be studied based on technologies such as GIS? How should we further quantitatively reveal the coupling relationship among various systems? In this study, these problems and related issues are investigated under the new metacoupling framework to attempt to illustrate the ecological coupling of SNWTP-MR with limited information.

2. Materials and Analytical Framework

2.1. Study Area

China's SNWTP-MR is 1430 km long and provides domestic water to 155 billion m^2 of land, including Beijing, Tianjin, Hebei and Henan (Figure 1A). Since the route was officially put into use in December 2014, as of 12 December 2018, the total amount of water transfer was 22.2 billion m^3 [16]. The water source is the Danjiangkou Reservoir on the Han River (the tributary of the Yangtze River) and its upstream area, including 95,200 km^2 of the drainage divides in 43 counties in Henan, Hubei and Shaanxi Province (Figure 1B). The downstream watershed of the Han River covers an area of 43,800 m^2 (Figure 1C), including 30 counties in Henan and Hubei provinces.

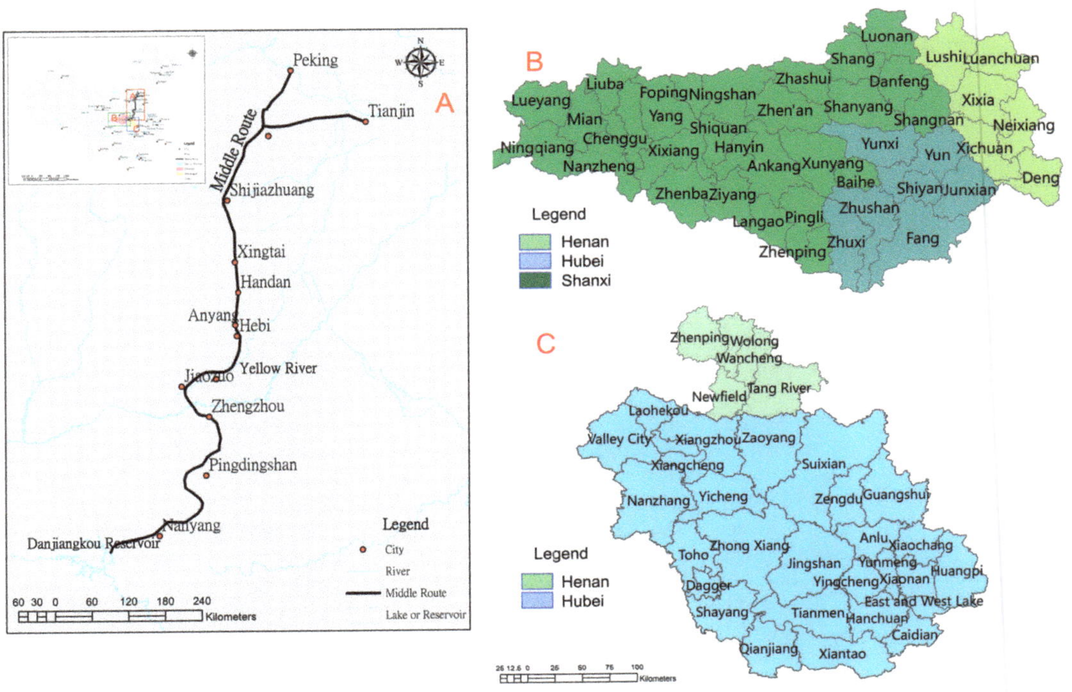

Figure 1. Range of influence of SNWTP-MR. ((**A**) is the whole process of water transfer in the SNWTP-MR; (**B**) is the scope of the water source area for water transfer; (**C**) is the downstream of the water source area).

2.2. Data Sources

First, remote sensing map data were obtained from the United States Geological Survey (https://earthexplorer.usgs.gov/, accessed on 1 October 2019), and 10 image data of 30 m × 30 m of Landsat in 2013 and 2019 were downloaded, respectively; they were used for evaluating the vulnerability of the ecological environment of the SNWTP-MR water source area. SNWTP-MR was officially put into use in December 2014. Secondly, the information collected from the SNWTP-MR website, Beijing Statistical Yearbook, Hebei Statistical Yearbook, Henan Statistical Yearbook, Danjiangkou Reservoir data, etc., was used as the main source of data in this study.

2.3. Analytical Framework

The metacoupling system is a set of two or more coupled systems that interact internally as well as nearby and far away, facilitated by agents affected by various causes with various effects [17]. The entire metacoupling system consists of a sending system, receiving system, and spillover system (Figure 2), among which the sending system provides materials and information for the entire coupling; the receiving system obtains materials and information from the sending system and provides energy and capital for the sending system; while the spillover system is the area affected by the transfer of materials and information, and is the medium for the transfer between the receiving system and the sending system. Further, the metacoupling framework can be divided into three parts: intracoupling, pericoupling and telecoupling [17] (Figure 2).

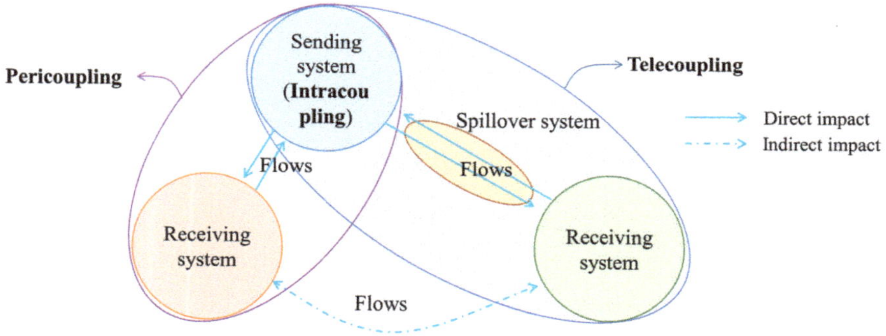

Figure 2. Metacoupling relationship.

Intracoupling refers to the internal mutual influence of the sending system. For example, the sending system of SNWTP-MR is Danjiangkou Reservoir, and intracoupling studies the relationship of ecological changes in the water source area's own system.

Pericoupling refers to the mutual influence between the sending system and the receiving system, emphasizing the influence of the sending system on the surroundings, and there is no need for conduction between them through other media. For example, the pericoupling of SNWTP-MR refers to ecological coupling in the same watershed.

Telecoupling is similar to pericoupling, but it emphasizes the cross-regional relationships. The sending system and the receiving system need to pass through the spillover system to have a relationship. For example, the Danjiangkou Reservoir of SNWTP-MR needs to be connected to the northern water-receiving city through the construction of a canal.

Due to the complex impact of the SNWTP-MR project, this article uses a system perspective to decompose the impact, which will help to present the problem and simplify the internal complexity of SNWTP-MR. We segmented metacoupling to facilitate the understanding of the interactions among them, so as to assess the sustainable development of the ecological environment of SNWTP as a whole. However, the metacoupling framework can integrate the impacts that may not be quantifiable or measurable, and expand the existing methods. For quantifiable impacts, the framework can provide the quantitative results of the coupling process, and for non-quantifiable impacts, it can improve the qualitative understanding of the relationship between agent and flow [18].

The metacoupling framework (Figure 3) begins with the analysis of the ecological environment in intracoupling, pericoupling, and telecoupling, including the main influences, the causes and ways of coupling, to clarify the ecological values and ecological risks brought by these potential factors. Intracoupling evaluates the vulnerability of the ecological environment in the water source area of SNWTP-MR according to the data in typical years selected, analyzes the temporal and spatial pattern and evolution trend of the ecological environment, and pre-processes the data on Plowland, Vegetation, Water, Artificial surface, and Bare Soil of the water source area with the help of remote sensing and geographic information system (GIS), so as to reveal the spatial coupling of Danjiangkou waters and the ecological effects of the construction of SNWTP-MR. These effects include impact on the ecological value (the economic value of land type) and ecological risk (improvement of ecological risk index), which are then measured by economic evaluation of ecosystem services and intensity coefficients of ecological risk theory. The Danjiangkou Reservoir is the boundary to divide the sending system (upstream water source area) and the receiving system (downstream area). Pericoupling reveals that the ecological value and ecological risk of the two systems are not always mutually matching and benignly complementary under the coupling of environmental ecological effects. For example, a situation may occur where migration may mitigate the carrying capacity of ecological environment in upstream, but poses a burden to the environmental carrying capacity downstream. We explored the ecological values from the aspects of power generation benefits, seasonal water transfer, water quality, flood control benefits, and the amount of water abandonment of floods. We also investigated the ecological risks from issues such as eutrophication of water, the deterioration of shipping capacity, the competition for water between industry and agriculture, and biological threats. A clear understanding of the interaction between the upstream and downstream aspects of the Danjiangkou Reservoir is beneficial to dealing with the contradictory relationship between them; telecoupling divides the sending system, the spillover system and the receiving system. The remote water transfer will affect the safety of water supply, population carrying capacity, water quality and groundwater level in water receiving areas, and produce ecological risks such as water source pollution and biological intrusion. The spillover system is directly and indirectly affected by water pollution and agricultural irrigation, and it also exerts pressure on the water quality protection and water development of the sending system. By analyzing the telecoupling relationship, we can understand the interaction among them in terms of ecological value and ecological risk. The framework encompasses with the environmental interaction of metacoupling, and based on the interaction of ecological environments in the coupling relationship, we can reveal how to promote the sustainable development under this interaction of coupling.

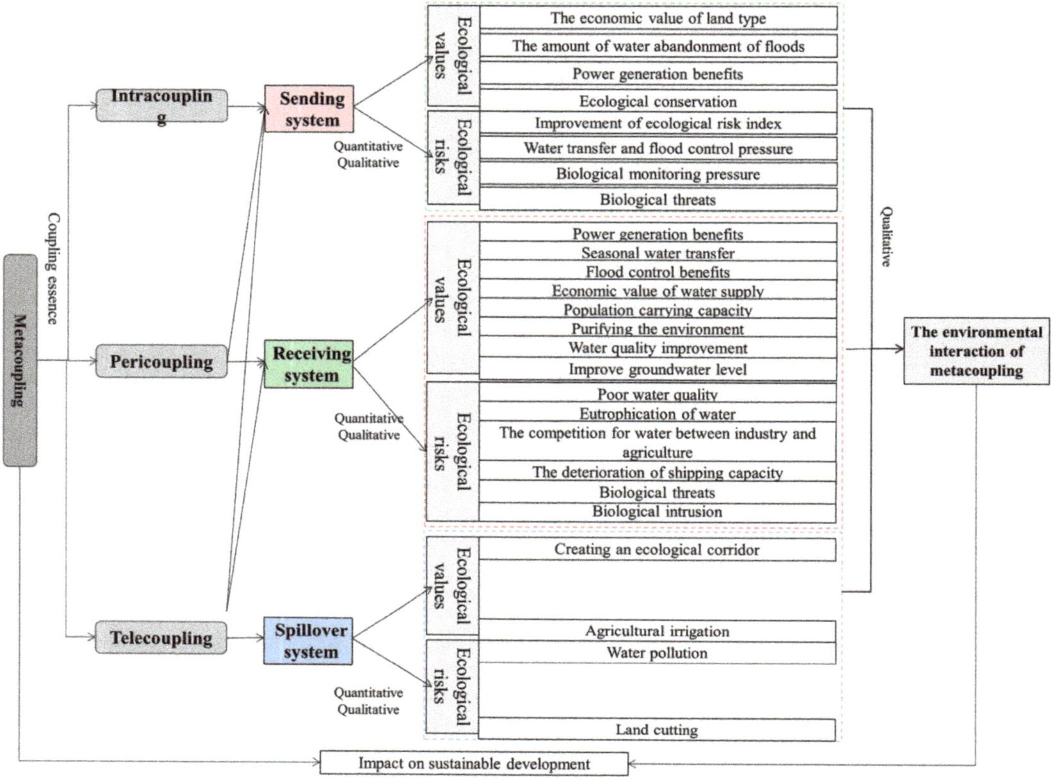

Figure 3. Research framework of SNWTP-MR metacoupling.

3. Application of the Metacoupling Framework in SNWTP-MR

3.1. Metacoupling Framework Analysis

Under the framework of metacoupling, we redefine the multi-system coupling relationship of SNWTP-MR as follows, focusing mainly on system, agent, flow, cause and effect.

3.1.1. System

Intracoupling is mainly the interaction among different types of land use in Danjiangkou water source, emphasizing the mutual transformation among Plowland, Vegetation, Water, Artificial Surface and Bare Soil. Since pericoupling is in the cross-boundary watershed, the different demands of upstream and downstream will affect the impact of water transfer in the watershed. Therefore, we use the sending system and the receiving system to analyze the pericoupling relationship between upstream and downstream caused by SNWTP-MR. In telecoupling, the watershed range of Danjiangkou Reservoir is the sending system, the distant water receiving city (Province) (Henan, Hebei, Tianjin, Beijing) is the receiving system, and the counties and cities along the route constitute the spillover system.

3.1.2. Agent

The metacoupling of SNWTP-MR mainly involves the participation of local and central government, enterprises and farmers in the construction of the reservoir, the transfer of water, the protection of the environment and the coordination among the various systems. To help overcome the project's technological and engineering challenges, thousands of experts—representing government agencies, corporations, banks, and non-governmental organizations—have been recruited both domestically and internationally, in addition to millions of laborers. At the central government level, there is the SNWTP Construction Committee Office under the State Council. At the local and regional levels, there are construction committees in each affected city and province [15].

3.1.3. Flow

The flow of metacoupling involved in water transfer is mainly the volume of water transfer, for example, the multi-year average volume of water transfer is 9.7 billion m^3 per year, of which the volume of water transfer to Henan, Hebei, Beijing, and Tianjin is 3.71, 3.47, 1.24, and 1.02 billion m^3, respectively, and the necessary volume of runoff in downstream areas [19]. Secondly, there is a spatial flow of industrial and agricultural pollutants with the water transfer. In addition, the water delivery at the pump station also leads to the flow of energy. Thirdly, there is a spatial transfer of population in the areas affected by the project construction. Finally, the construction and operation of the SNWTP-MR also results in the flow of funds. The main flows of intracoupling include water, money, and population; the main flows of pericoupling are water, money, and energy; and the main flows of telecoupling are water, money, pollution, and energy.

3.1.4. Cause

The reason for the intracoupling of the whole SNWTP-MR is mainly attributed to the change in the nature of the land which was caused by the construction of Danjiangkou Reservoir, and the development and utilization of land due to the spatial transfer of population and local economic development. In addition, land development will also be restricted due to the strengthened government focuson ecological protection. From the perspective of the dam itself, the reason for pericoupling is that because the dam body of the Danjiangkou Reservoir is raised, a large part of the water storage is used for the downstream water supply, so that the water for use in the downstream area is reduced. From the perspective of the water receiving area and downstream area, the reason for pericoupling is that the receiving water area and the downstream area have different purposes for the use of water resources, and there is a competitive relationship between the two. From the perspective of upstream and downstream, the interests of upstream and downstream users are different, resulting in conflicts and contradictions in water allocation. Under these multiple influences, the services and the willingness for development in the watershed will be inevitably affected. The potential reasons for telecoupling include four aspects: ecology, economy, politics and technology. From the perspective of ecology, there are excessive exploitation of groundwater and serious water pollution in the north, so the spatial water transfer is conducive to improving the carrying capacity of ecological environmental resources. From the point of view of the economy, there is a large distribution of agriculture and industry in the north, accompanied by a lack of water resources, so in order to achieve the sustainable development of the economy, it is a feasible option to seek alternative water resources. From a political perspective, the distribution of water resources and population in the north and the south is uneven; the remote water transfer in SNWTP-MR can increase taxes for local government and provide employment opportunities. From a technical point of view, the terrain along the route is high in the south and low in the north, which means that natural runoff can be achieved through the influence of gravity in most parts of the route, and there are not many places where a pumping station is needed for water delivery [20]; thus, water transfer is feasible from the aspects of geographical elevation and energy.

3.1.5. Effects

Intracoupling affects the hydrogeomorphic characteristics, causing the pressure of population transfer, and limiting the development of industry and agriculture to a certain extent. Obviously, the result of coupling will also promote ecological conservation (see

Table 1 below for details). The impact of pericoupling is that the expansion of the Danjiangkou Reservoir brings both opportunities and challenges to the ecological and economic use of water resources, and also influences the transformation of regional industry and society (see Table 2 below for details). The impact of telecoupling is manifested in multiple aspects, including economy, society, ecology, etc. Water transfer can put pressure on the finances of different systems or increase their income, and also will threaten or promote the environmental carrying capacity of each system. At the same time, water transfer causes a threat to biological survival and the pressure on migration (see Table 3 below for details). In this paper, we focus on the ecological value and ecological risk generated by SNWTP-MR. Therefore, a description of detailed quantitative and qualitative effects is conducted below from the perspective of ecological value and ecological risk, in order to better present the impact of metacoupling.

Table 1. Analysis of the intracoupling framework of SNWTP.

	Sending System
System description	The watershed range of Danjiangkou Reservoir
Causes	Construction of Danjiangkou Reservoir, spatial transfer of population, local economic development, government's interest in ecological protection
Agents	Central and local government, enterprises, farmers
Flows	Water, money and population
Effects (−)	The reservoir region immigration and resettlement cause ecological damage (relocation of 300,000 people). The hydrogeomorphic features have been changed, which increases the ecological risk. Industrial, agricultural and fishery development is restricted, thus farmers' incomes are reduced.
Effects (+)	The water storage in the reservoir is increased, which improves the utilization of water resources. Local awareness of ecological protection has been aroused, and the ecological destruction rate in the water source is effectively contained.

Table 2. Pericoupling of SNWTP.

	Sending System	**Receiving System**
System description	The watershed range of Danjiangkou Reservoir	Counties and cities in the downstream of Han River
Causes	For the SNWTP-MR, the dam body of the Danjiangkou Reservoir is raised, which reduces downstream water for use. The water receiving areas have different purposes for water consumption from the downstream areas, and there is a competitive relationship between the two. The interests of upstream and downstream users are different.	
Agents	Central and local government, enterprises, farmers	
Flows	Water, money and energy	
Effects (−)	The pressure on flood control is increased; the transfer and resettlement of the population from the reservoir-inundated region brings huge burden; the hydrological and geomorphological characteristics have been changed.	Water flow is reduced; water quality is decreased; flow rate is slowed, shipping capacity is reduced; biobalance is destroyed, and biodiversity is threatened; water consumption in industry and agriculture is affected.
Effects (+)	The water storage in reservoir is increased; total power generation is improved; the value of tourism development is increased; the utilization efficiency of water resource is increased.	The regional electricity guarantee is increased; the flood control pressure is reduced; the seasonal water allocation can reduce the threat to water supply caused by extreme weather; industries with high demands of water can carry out industrial transformation and upgrading.

Table 3. Telecoupling of SNWTP.

	Sending System	Receiving System	Spillover System
System description	The watershed range of Danjiangkou Reservoir	Henan, Hebei, Tianjin, Beijing	Counties and cities along the route
Causes	Geographical elevation allows water transfer, the terrain along the route is high in the south and low in the north, and naturalrunoff can be realized in most parts of the route; There is an uneven distribution of water resources and population in the north and south. There is a large distribution of agriculture and industry in the north, accompanied by the lack of water resources; There is excessive groundwater exploitation and serious water pollution in the north.		Spatial transfer of water
Agents	Central and local government, enterprises, farmers		
Flows	Water, money, pollution, energy		
Effects (−)	Migration out of the reservoir region causes environmental burden and brain drain; threats to species' survival.	Biological invasion; elevated water, nitrogen and phosphorus nutrients; water use risk; human and financial input; payment for water resources.	Water quality decline; the need for water resources protection along the route increases financial pressure.
Effects (+)	Income from water transfer; dam power generation; ecological conservation; tourism development. A large amount of water can be transferred out during the flood season to reduce the threat of floods.	Agricultural irrigation water; domestic water; reduced groundwater exploitation; increased population capacity and economic growth; water quality improvement;	Ecological corridor; tourism development; use of corridor to achieve water transfer balance in small range.

3.2. From the Perspective of Ecological Value

The intracoupling of SNWTP-MR refers to the interactions between different types of land use in Danjiangkou water source areas, so it is necessary to analyze their patterns of spatial and temporal change pattern. Based on the satellite remote sensing images of the Danjiangkou water source in 2013 and 2019, we classified the land use type by the Threshold method of eCognition 8.7 software (Developed by Trmible, Sunnyvale, CA, USA) and the Pretreatment method of ENVI 5.1 software (Developed by Exelis Visual Information Solutio, Boulder, CO, USA) (including radiation calibration, atmospheric correction, image mosaic, image cropping). According to the theories of Costanza et al. and Yang Guoqing [21,22], when calculating the ecological service value and ecological risk, land use is mainly divided into five categories, and this study also continues this classification. The categories include: Plowland (paddy field, dry land), Vegetation (woodland, bush, open woodland, other woodland, high coverage grass, medium coverage grass, low coverage grass), Water (canals, lake, reservoir pond), Artificial Surface (urban land, rural settlement), and Bare Soil (sandy ground, Gobi saline-alkali land, bare earth, bare rock texture) (Figures 4 and 5). The "accuracy assessment" analysis index was used for accuracy verification. The specific method is to compare and test the drawings analyzed in this study using the vector files of the identified land types on the software. The verification result is: the accuracy of the classification in 2013 is about 87%, about 85% in 2019, so the accuracy is high. This can be seen from the changes in land use (Figure 6). From the red circles A and B in Figure 6, it is also found that there is obvious land development; at the same time, the expansion of the reservoir increases the water area significantly (the red circle C in Figure 6), which will provide a solid foundation for remote water transfer.

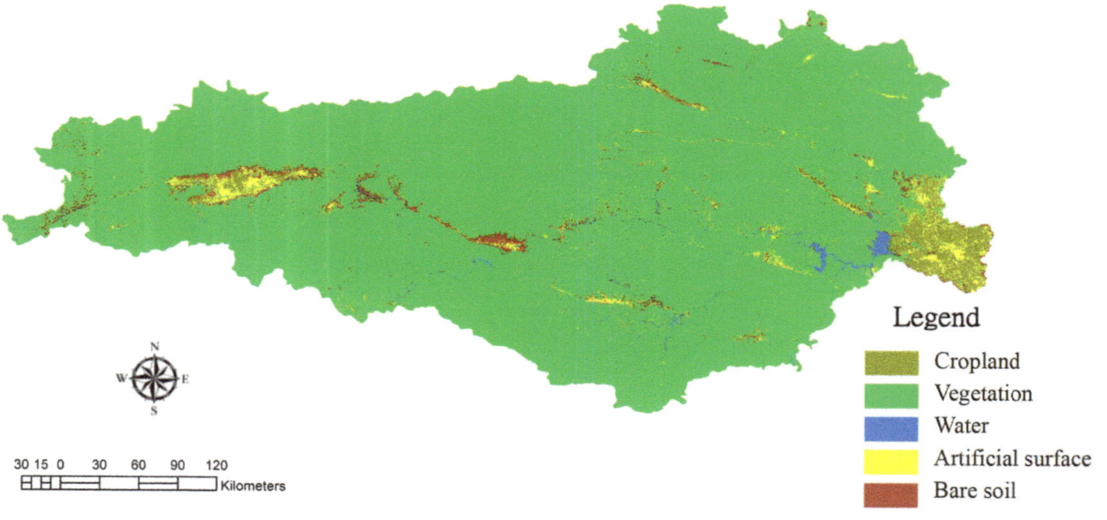

Figure 4. Land use of Danjiangkou water source in 2013.

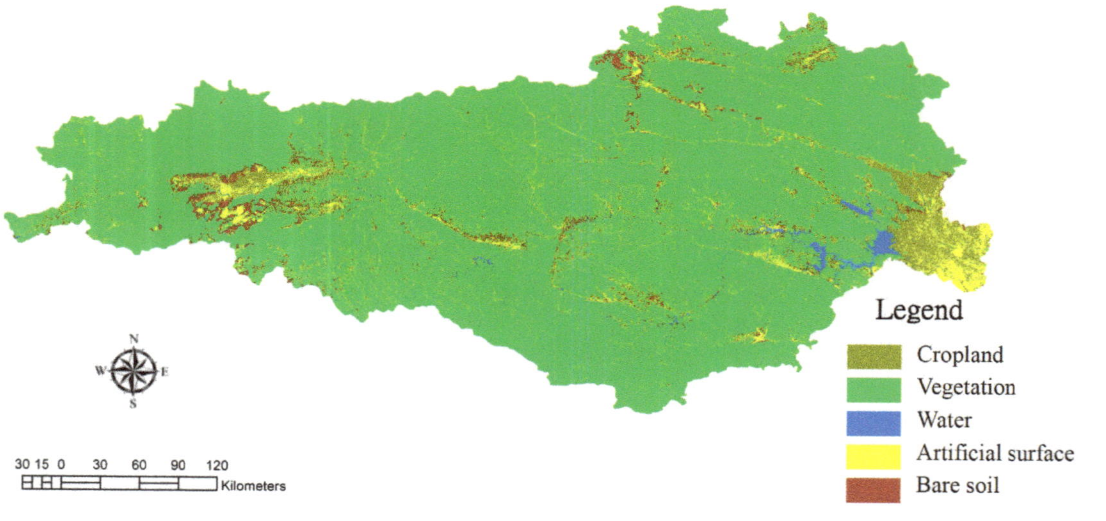

Figure 5. Land use of Danjiangkou water source in 2019.

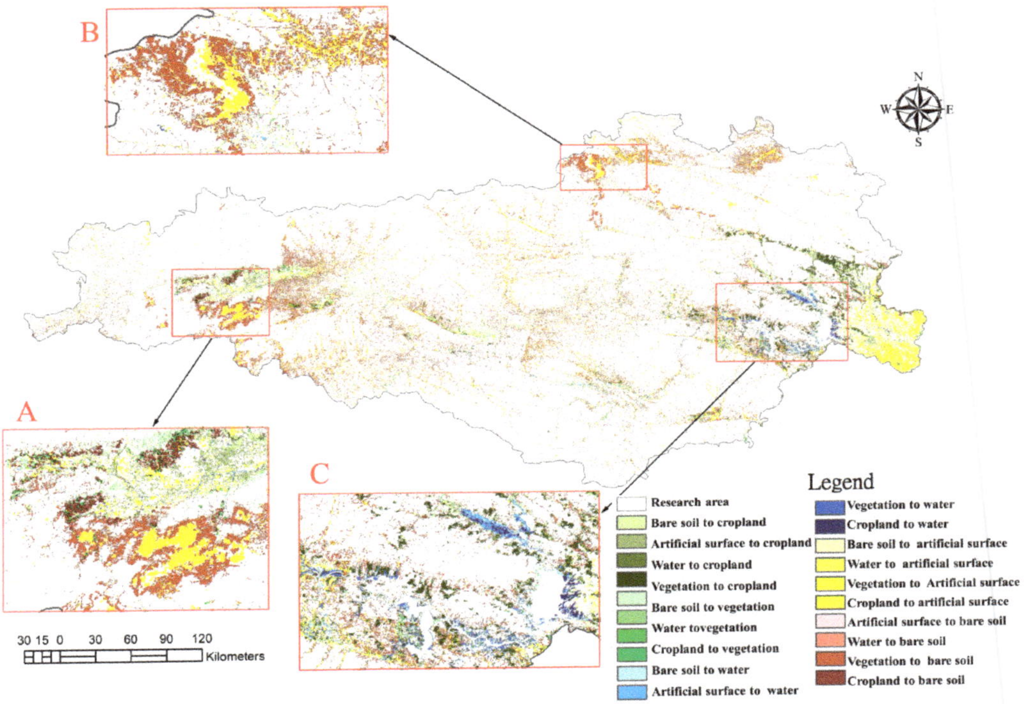

Figure 6. Total changes in land use from 2013 to 2019. (The circle (**C**) represents the increased area of the reservoir. Circle (**A**) and circle (**B**) indicate artificial surfaces added in 2019 over 2013).

Tables 3 and 4 show the changes in land use. We use the spatial dynamic degree of change in land use to reflect the speed of land use change under coupling. The spatial dynamic degree refers to the ratio of the product of the translation of this type into the sum of other types and the translation of other types into the sum of this type to the number of certain land types at the end of the period, which is used to reflect the rate of change [23]. It can be seen from Table 5 that the water transport by SNWTP-MR leads to an increase in the water level of the Danjiangkou Reservoir, forcing the population migrated due to the Reservoir to develop more of the highland. This can also be indirectly reflected by the increase in the area of the Plowland and the Artificial Surfaces. At the same time, the area of Vegetation has not changed much in the past six years, which also demonstrates that the protection of the water source of Danjiangkou Reservoir has exerted certain effects. The changes in intracoupling will also affect the ecological value and ecological risk in the area.

Table 4. Matrix of the land change and transfer in the water source of Danjiangkou Reservoir in 2013–2019.

2013–2019	Plowland (km^2)	Vegetation (km^2)	Water (km^2)	Artificial Surface (km^2)	Bare Soil (km^2)	Total	Reduction (km^2) (ΔU_{re})
Plowland (km^2)	2305.81	142.07	84.96	886.73	183.38	3602.95	1297.14
Vegetation (km^2)	1161.04	88,234.62	253.31	2565.63	7672.90	99,887.50	11,652.89
Water (km^2)	40.66	186.96	640.01	109.25	96.69	1073.58	433.57
Building (km^2)	562.75	286.15	125.23	1063.31	237.32	2274.75	1211.44
Bare Soil (km^2)	308.12	1773.17	91.38	953.67	1362.08	4488.42	3126.34

Table 4. *Cont.*

2013–2019	Plowland (km^2)	Vegetation (km^2)	Water (km^2)	Artificial Surface (km^2)	Bare Soil (km^2)	Total	Reduction (km^2) (ΔU_{re})
Total	4378.38	90,622.96	1194.89	5578.60	9552.37		
Increments $\left(km^2\right)(\Delta U_{in})$	2072.57	2388.34	554.88	4515.29	8190.29		
Change (km^2)	775.43	−9264.54	121.31	3303.85	5063.95		

Table 5. Spatial dynamic degree of land change in the water source of Danjiangkou Reservoir in 2013–2019.

Name	Formula	Plowland	Vegetation	Water	Buildings	Bare Soil
spatial dynamic degree	$C = \frac{\Delta U_{in} \times \Delta U_{re}}{U}$ C : Rate of change over a period of time; ΔU_{in}: Other types translate to the sum of this type; ΔU_{re}: This type translates to the sum of other types; U: The number of certain land types at the end of the period	0.77	0.15	0.83	1.03	1.18

Based on the above-mentioned changes in land use, the ecological service value of the water source of SNWTP-MR can be measured. In terms of the measurement of the ecological value of intracoupling, Costanza proposed the theoretical framework of economic evaluation of ecosystem services, which provides a perspective for the measurement of the relative ecological value of land use types [21,24]. Costanza et al. classified land use into 17 types, and the ecological services of each type include four major categories, namely, supply service, regulation service, support service and cultural service, and nine sub-services. Specifically, the relative value of each unit of arable land is linked to the market price of its output value, and the value of other types of land depends on its relative importance to arable land (equivalence coefficient) [25]. Built-up land is considered incapable of providing ecosystem services [26,27]. For example, the output value per hectare in 2013 was 25,837 yuan, while in 1995 it was 13,350 yuan, so the relative increase was 1.94 times. We can calculate the ecological service value of the year according to the corresponding ratio. According to the theory proposed by Costanza et al. and the average grain price per hectare for the current year, Plowland, Vegetation, Water, and Bare Soil of 1 hm^2 can provide the ecological value of 4.7279, 1.8774, 10.7772, and 3.303 million yuan, respectively. Therefore, the increase in the water area due to the expansion of the Danjiangkou Reservoir can create an ecological service value of 1.307 billion yuan; on the whole (Table 6), the coupling of the internal land use types has increased the value of 206.86 billion yuan provided in 2013 to the ecological services value of 229.937 billion yuan, an increase of 23 billion yuan. Therefore, although the construction of SNWTP-MR has changed various land types to some extent, as a whole, it is still conducive to providing ecological value in water sources. (The ecological value emphasized in this article generally refers to the services that ecology can provide, including quantifiable and non-quantifiable values, such as the value of ecological services. Ecological service value refers to the ecological value that can be measured by money. Benefits refer to effects and incomes, which can be economic, social and environmental benefits, and include the expansion of ecological value, such as flood control benefits.)

Table 6. Ecological service value of intracoupling.

Name	Formula	Plowland (km^2)	Vegetation (km^2)	Water (km^2)	Artificial Surface (km^2)	Bare Soil (km^2)	Total		Sources
Value of ecological services	$V_S = \sum_{i=1}^{m} T_i \times P_i$ T_i: Area of type i land type; P_i: Ecological value of each land type	207.01	1701.36	128.78	0.00	31.55	2068.69	2013 year (100 million yuan)	[22]
		382.83	1610.93	224.99	0.00	80.62	2299.37	2019 year (100 million yuan)	

Note: The units of money used in this study are expressed in Renminbi (RMB).

In terms of pericoupling, first it is found that the power generation benefit of water resources and seasonal water transfer have been improved. For example, the total length of the Danjiangkou dam is 2.5 km, the maximum dam height of the project is 97 m, the installed capacity is 900,000 kilowatts, and the multi-year average annual power generation is 4 billion kilowatt hours, which can bring an income of 2.292 billion yuan to the water source of Danjiangkou Reservoir every year if calculated at the lowest electricity price of 0.573 yuan in the downstream Wuhan city. During the main flood season from July to September each year, the water level can be regulated to below the flood control level, and seasonal water transfer can be realized. At the end of the flood season in October each year, the required electricity can be generated and water can be stored; at the same time, the stored water can also be used to maintain seasonal normal runoff [28,29]. Second, the amount of water loss (abandonment) due to the flood is reduced. Suppose we subtract the downstream water consumption from the water output of the reservoir and divide the result by the downstream water consumption as the water abandonment rate. The maximum water consumption of the downstream cities of 8.17 billion m^3 is selected as the water consumption in the lower reaches of the Han River [30]; the water output of Danjiangkou Reservoir in 2013 before the water transfer was 26.71 billion m^3. After the water transfer in 2014, the water output was 15.45 billion m^3, so the water abandonment rate was reduced by 137.8%, and the utilization of water resources has been significantly promoted. Third, flood control benefits: After the Danjiangkou Reservoir was raised in height, the reserved flood control capacity of the Reservoir was increased by 3.28 billion m^3 (in summer flood season) and 2.51 billion m^3 (in autumn flood season). Meanwhile, by combining the total flood diversion capacity of the downstream flood diversion areas, the flood control standard in the lower reaches of the Han River can be changed from once-in-20-years to once-in-a-century (controlling the once-in-a-century flood as in 1935) to eliminate the flood threats to more than 700,000 people downstream [31].

In terms of telecoupling: First, in this study, it is found that water supply can create more economic value. The water transfer increases the income of the sending system, and at the same time guarantees the water use of the receiving system and alleviates the shortage of water. From Table 7, we can see that SNWTP-MR can bring the water supply value of 9.85 billion yuan to the water source every year, which will make up for the cost on engineering development and the loss of economic value caused by water source protection. Second, the system creates an ecological corridor. The water in Danjiang River passes through sluice gates, canals and ecological forests on both sides, building a green corridor with a length of more than 1430 km and a width of several tens of meters to form a cross-regional green ecological landscape belt. The spillover system can use the corridor to bring excess water into the corridor for use in water-deficient areas, so as to achieve the balance of water transfer in a small area. The creation of the ecological corridor will enhance the greening level and ecological benefits in various systems. Third, (3) the water transfer improves the population carrying capacity of water resources. According to the research results of the Sustainable Development Strategy Research Group of China's 21st-century Agenda Management Center, with the coefficient of water supply capacity adopted as 0.39 (technical parameter), the calculation results (Table 7) show that the water transfer from SNWTP-MR will increase the population carrying capacity of Henan, Hebei,

Beijing and Tianjin by 5.99, 5.57, 2.67 and 2.18 million people, respectively, in 2019. Fourth, the telecoupling promotes the benefits of flood control, humidity increase, temperature reduction and environmental purification. According to the relevant research, the value of the absorption of sulfur dioxide and nitrogen oxides and of dust retention by forest is determined to be 215.6, 6.0, 21.7 kg/hm^2, respectively. The sewage charges are 1.2, 0.6, 0.2 yuan/kg, respectively, and the width of the SNWTP-MR greenway is 50 m [32,33]. It can be seen from Table 7 that the benefit of environmental purification in Henan, Hebei, Beijing, and Tianjin is increased to 97, 79, 10, and 3 million yuan, respectively. Fifth, the water quality of urban living water and rivers and lakes is improved. The water inflow from SNWTP-MR is blended with the water in northern counties and cities, which reduces the hardness of the water. The hardness of the tap water in Beijing is reduced from 380 mg·L^{-1} to 120–130 mg·L^{-1} [34]. At the same time, excess water can be used to supplement the water volume of rivers and lakes. For example, Hebei Province uses the water from SNWTP-MR to replenish 0.07 billion m^3 of water to the Hutuo River and Qili River. Sixth, there is an increase the agricultural water and the area of waters. When providing both domestic and industrial water, SNWTP-MR also takes into account the water for ecological environment and agriculture along the route; at the same time, SNWTP-MR can replenish water in rivers and lakes and water sources in areas along the route, which is conducive to increasing the area of waters. For example, SNWTP-MR has supplemented an accumulation of 171 million m^3 of water to Beijing Miyun Reservoir, effectively inhibiting the decline of water in Miyun Reservoir [35]. Finally, the system alleviates the problem of over-exploitation of groundwater. The distant water receiving cities (receiving system) effectively alleviate the deterioration of groundwater ecology by replacing the local groundwater source with water brought by SNWTP-MR. The groundwater level in the water-supplement areas has increased by different degrees. For example, in 2017, the average buried depth of groundwater in Beijing increased by 0.53 m year-on-year, and the groundwater level in Tianjin increased by 0.17 m [34].

Table 7. Benefits created by SNWTP-MR for water receiving provinces and cities.

	Index **Province (City)**	**Henan**	**Hebei**	**Beijing**	**Tianjin**	**Total**	**Sources**
Data	Multi-year average (billion m^3)	37.7	34.7	12.4	10.2	95	[19]
	Water price (yuan/m^3)	0.37	0.97	2.33	2.16		[36]
	Per capita water consumption (m^3/person)	245	243	181	182		[37]
	Pipe length (km)	731	596	80	25	1432	
Benefit	Formula						
Benefits of water supply	$VQ = P \times Q$ (billion m^3) Q: Total water transfer	13.95	33.66	28.89	22.03	98.53	[38]
Benefits of population carrying capacity of water resources	$W = \frac{\alpha \times Q}{C}$ (Ten thousand people) α: Coefficient of water supply capacity C:The amount of water used per capita	599	557	267	218	1642	[39]
Benefit environmental purification	$Ve = L \times W \times \sum_{i=1}^{n=3} (Qi \times Pi)$ (Billion yuan) Q_i: Respectively indicate sulfur dioxide, nitrogen oxides and dust retention P_i: Sewage charge price W: Width of SNWTP-MR greenway	0.97	0.79	0.1	0.03	1.89	[32,33]

Summary. The benefits brought by water transfer in SNWTP-MR are reflected in multiple aspects (Figure 7). The expansion of the Danjiangkou Reservoir in the water source changed the original land type, and the relative ecological value of the land use type has improved. At the same time, the expansion and water transfer of the Danjiangkou Reservoir broke the original spatial and temporal configuration pattern of water resources, which brings some additional ecological values to the upstream and downstream areas, including the increase in total power generation caused by the increase in reservoir water storage, the enhanced ability of seasonal water transfer, and the effective coordination of flood control in the region. These, in turn, promote the reduction of the water abandonment rate of floods in the sending system (water source), so that the power generation benefit can be improved. SNWTP-MR focuses on solving the problem of water shortage in the north.The transfer through the spillover system can provide economic value of water supply, improve the water quality of urban domestic water and rivers and lakes, alleviate the problem of over-exploitation of groundwater, improve the population carrying capacity of water resources and promote the benefits of flood control, carbon fixation and oxygen release, and air purification. These further bring economic value of water supply to the water source, raise the awareness of ecological conservation, and also create ecological corridor for the spillover system and alleviate the shortage of agricultural irrigation water. The analysis of the ecological value brought by SNWTP-MR through the metacoupling framework is helpful for us to understand the coordinated development among various systems.

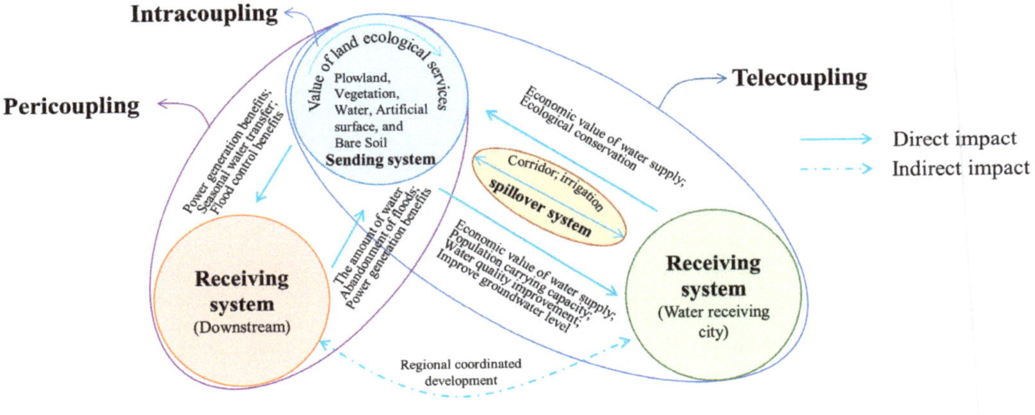

Figure 7. Schematic diagram of SNWTP-MR metacoupling ecological value.

3.3. From the Perspective of Ecological Risk

Based on the above-mentioned changes in land use type, the changes in the ecological risk of the SNWTP-MR water source can be measured. In terms of intracoupling, the ecological risk caused by the intensity of the changes in different land types is cumulative. Therefore, it is necessary to judge the accumulated risks from the perspective of the overall system and establish an empirical connection between land use structure and ecological risk. The researches in this aspect are mainly conducted through establishing the intensity coefficient of ecological risk to transform the land use types into ecological risk variables [22]. As can be seen from Table 8, the ecological risk index of the Danjiangkou Reservoir water source was 0.165 in 2013, and 0.18 in 2019, an increase of 9.31%. Therefore, the construction of SNWTP-MR has increased the level of ecological risk in the water source, and the ecological vulnerability has been destroyed to some extent. At the same time, the population transfer and resettlement from the submerged area of the reservoir (300,000 people) has caused a large amount of ecological land upstream of the water source to be used for the construction of new towns, which will lead to further ecological damage, and the disturbance of human activities will exacerbate ecological fragility.

Table 8. The ecological risk index of intracoupling.

Name	Formula	Plowland (km^2)	Vegetation (km^2)	Water (km^2)	Artificial Surface (km^2)	Bare Soil (km^2)	Total	Year
ecological risk index	$E = \sum_{i=1}^{m} \frac{T_i \times \delta_i}{T}$ δ_i: Intensity coefficient of ecological risk of the i-th land type; T: area of Total land; T_i: Area of type i land type	0.010	0.135	0.002	0.014	0.004	0.165	2013
		0.012	0.122	0.002	0.035	0.009	0.180	2019

In terms of pericoupling:First, the reduction of the downstream water flow causes ecological problems. From 2012 to 2016, the outflow water of Danjiangkou Reservoir was 36.27, 26.71, 15.45, 28.19, and 15.01 billion m^3, respectively, showing a decreasing trend, which caused the water level in the lower reaches of the Han River to drop by between 0.3 m and 1 m [40]. The reduction of water volume also resulted in the eutrophication of water resources and the gradual deterioration of water quality in the downstream of the Han River. The reduction of downstream water flow has given rise to the worse overall shipping capacity. After the water transfer, the average annual navigable time is only 121 days, and the navigation guarantee rate is significantly reduced to 33% [41]. At the same time, the decreased water level will result in the deterioration of downstream water quality. As the amount of inflow is reduced, the flow rate will slow down, which makes the self-purification capacity of the water body in the lower reaches of the Han River drop drastically. At the same time, industry and agriculture pump a large amount of water, and a large part of this water directly enters the river, further polluting the river, while the urgent need for the improvement of water quality in the downstream will also put more pressure on the water quality protection in the upstream water source. In addition, the shortage of water intensifies the competition for water between industry and agriculture, and the rate of average water supply guarantee of each water plant has dropped by 34.7% [40]. For the downstream watershed, the reduction of water volume and the increase of the dam body have led to the destruction of the ecological balance of aquatic organisms in the Han River, and threat to biodiversity. Second, biodiversity and fish stocks are threatened. After the water transfer, the number of fish spawning grounds has been reduced by 25, the spawning time is delayed, and the total amount of fish is reduced by 1/4 or more, among which the production of wild fish is reduced by 50%, greatly destroying fish diversity and output value. Besides, the variety of hygrophyte in the lower reaches of the Han River will be affected to a certain extent, and the area of swamp vegetation, such as Aceh Weed Swamp, Valerian Swamp, Reed Marsh, and Calamus Marsh will decrease [42,43].

In terms of telecoupling: First, the primary impact is the biological invasion, which threatens species survival. The construction of water transfer canals has created channels for biological exchanges. Some fish in the Danjiangkou Reservoir may enter other reservoirs and watershed areas along the SNWTP-MR, which will cause changes in the number of fish species in other reservoirs and watersheds. Secondly, the water resources along the route are polluted, threatening the safety of water supply. For distant water receiving cities, SNWTP-MR adopts open channels for water delivery under direct sunlight, with long routes, shallow water and low water flow rate. The water temperature changes with air temperature (from 20 °C in September to May to above 30 °C in July and August), which will be conducive to algae reproduction and lead to a further increase in algal density and the risk of algae odor [35]. There are clams in the Danjiangkou Reservoir, with a short reproductive cycle, and the larvae belong to plankton, while the adults grow on the hard substrate, attached by the foot. They often block the water pipeline and the water plant filter, threatening the safety of water use. Meanwhile, clams are tolerant to the temperature in the north, so there is a risk that they will migrate northward along the SNWTP-MR [44,45].

Summary From the perspective of ecological risk (Figure 8), the expansion of the Danjiangkou Reservoir in the water source has changed the original land type, which increases the ecological risk index and further enhances the ecological vulnerability. At the same time, the water transfer to the north will inevitably lead to a significant reduction in downstream water flow, which will then result in poorer water quality, eutrophication of water resources, competition between industry and agriculture for water, deterioration of conditions for waterway shipping, threats to the diversity of aquatic organisms and fish survival, further intensifying the ecological conflicts between upstream and downstream. Moreover, the most direct ecological impact of remote water transfer on the distant water receiving city is the problem of biological invasion and water pollution, which in turn forces the water source to strengthen the monitoring of biological invasion. The conflict in water use between the downstream area and the distant water receiving city also aggravates the conflicts in the coordination of water resources in the water source. By clarifying the ecological risk of the water transfer in SNWTP-MR via a metacoupling framework, we can take necessary measures to prevent and mitigate its negative effects.

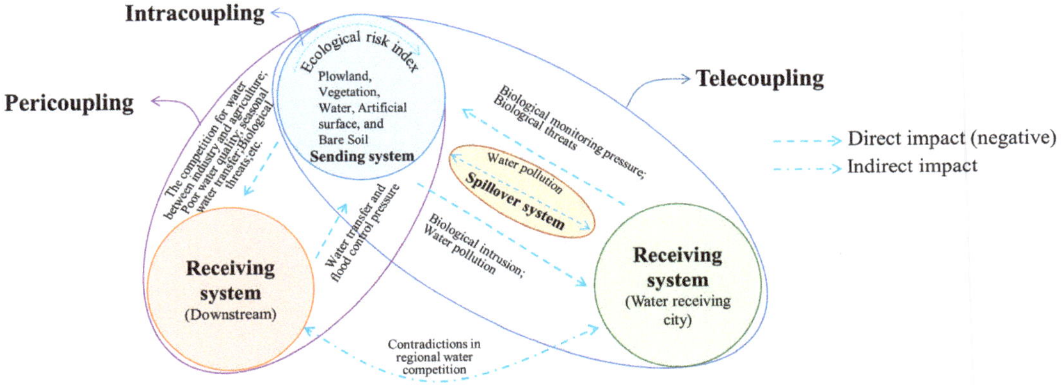

Figure 8. Ecological risk of the metacoupling of SNWTP-MR.

4. Discussion—The Impact of the Metacoupling on the Sustainability of SNWTP-MR

SNWTP-MR brings risks to the ecological environment while creating benefits for the ecological environment. It is difficult to say who should take the responsibility and who can get the benefits. The metacoupling framework can help us to better understand and manage the sustainability of regional water transfer, identify gaps between systems and hidden problems based on the direct and indirect positive and negative feedback, as illustrated in Figure 9. The coupling of water transfer does not always damage the benefits of the sending system and the spillover system while benefitting the receiving system, but presents a more complex and varied situation. It is known from the previous sections that the intracoupling can bring about 23 billion yuan of ecological service value. The pericoupling significantly reduces the flood control pressure in the downstream, and the telecoupling can compensate the water volume for the spillover system, create a green corridor and bring growth of domestic water, population carrying capacity and economy to the receiving system. The metacoupling of SNWTP-MR can fully present the interrelationship between systems. When we clarify the value and risks it brings and try to overcome the risks, a water transfer project such as SNWTP-MR can bring win-win benefits to each system.

The systematic illustration of the metacoupling of the SNWTP-MR provides a framework for the reasonable allocation of water resources and promotes the sustainability of water transfer. Long-distance water transfer, which reduces downstream water inflow, coupled with different interests of upstream, downstream and the distant water receiving city, affects the willingness of all parties to protect rivers and pay for the ecological services, and intensifies the conflicts among all the parties in the competition for water [6]. For example, the distant water receiving cities consumed 9.7 billion m^3 of water in the Danjiangkou Reservoir, almost one-third of the water volume in the Danjiangkou water source, which causes the changes in the land use type of the water source to produce a chain effect, and imposes tremendous pressure on the ecosystem services [46,47]. The annual decline in the downstream water volume has threatened the agricultural land and food production, and resulted in the deterioration of river channels to some extent.The role of the feedback and negative feedback of the metacoupling relationship provides useful information for long-distance water transfer, and offers a good opportunity to correct water allocation in annual water supply, so that the water allocation can achieve a positive balance to support social and economic development while protecting and improving the future environment. Applying the metacoupling framework to understand the sustainability issues can yield valuable insights to identify both beneficiaries and losers; thus, the overexploitation and unreasonable allocation of resources can be avoided [16].

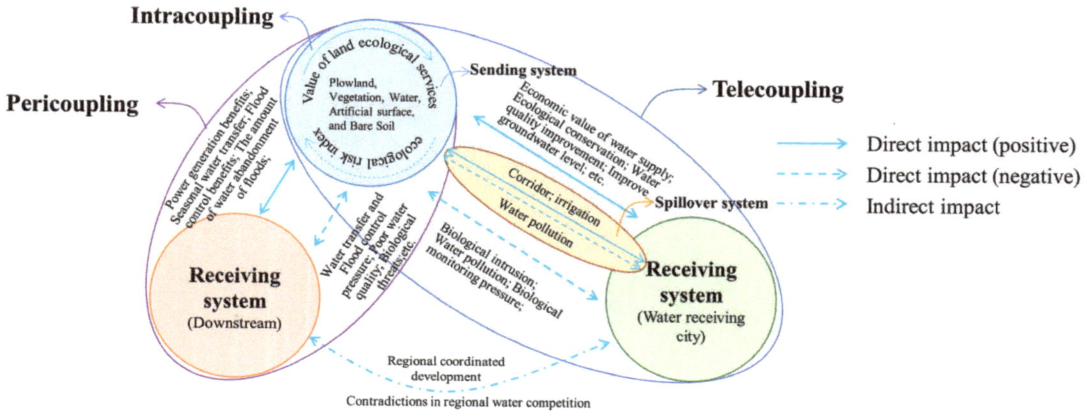

Figure 9. The ecological benefit feedback of the metacoupling of SNWTP-MR.

The analysis of the benefits and disadvantages to various systems in SNWTP-MR, brought by the metacoupling framework, provides a theoretical reference for who should take greater responsibility for ecological compensation. We know that SNWTP involves not only resettlement but also ecological and economic costs. Migration resettlement and food demand may lead to reclamation of agricultural land, degradation of forests and surrounding grasslands, which has been verified in our intracoupling framework. The reduction in downstream water volume also causes real pressure; although the water crisis can be alleviated in water receiving city, the increase in water use will also reduce the efficiency of water consumption. In this aspect, the metacoupling framework can help us to estimate and determine who will compensate, what to pay and how much to pay [25]. From the perspective of fairness and justice, ecological compensation can offset the social and economic benefits given up by water delivering areas. The analysis of the causes and impacts of coupling can help us to further understand the social and equity issues related to water use and coordinate the conflict between future land use and ecological protection, so as to promote the sustainability of water transfer [18]. This study is helpful for diagnosing which links have gone wrong in the sustainability of water transfer, which needs to be adjusted, and what linkage effects will be produced after the adjustment. Therefore, this study is conducive to the sustainability of long-term water transfer.

The analysis results of the metacoupling of SNWTP-MR helps to clarify the problems and challenges faced by different organizations, systems and agents, and can help them to take separate or joint actions. We know that SNWTP-MR has changed the economic structure and the method for ecological treatment in water sources, and also promoted the transformation of regional industry and society. The complexity of water transfer has transcended the water trading and investment itself. Therefore, the metacoupling framework is conducive to uncovering the complex interactions between water resources and environmental sustainability. In advancing the sustainable development of SNWTP-MR, joint actions can be taken to reduce costs, improve water quality, reduce pressure on water use and environment, promote social technological innovation and the implementation of the payments for watershed service plans, and enhance the complementarity of water resources [48].

Previous studies have mainly discussed SNWTP-MR ecological payment, water transfer policies, water environment governance, and water market incentive coordination studies. Few studies have been carried out on the metacoupling impact of the entire water transfer from a global perspective [49–51]. Therefore, the innovation of this research lies in solving the problem of the interaction of ecological environment of the pericoupling and telecoupling of water transfer and sustainable development under the new integrated framework (metacoupling), providing a new research idea for coupling relationships. At the same time, it reveals the metacoupling relationship of SNWTP-MR and the mutual influence of the ecological environment, which can provide a reference for SNWTP-MR, and also provide a valuable reference for other large-scale water transfer interactions. For example, Australia is considering sending large amounts of water over long distances from the north to the south of the country [52]. The cross-regional coupling relationship is more complex than the local coupling and requires a lot of resource integration. Metacoupling can find many research gaps and can conduct a comprehensive and in-depth analysis [17,53]. Therefore, the metacoupling framework can also be used for transnational and trans-regional research on ecosystem services, food trade, information dissemination, energy and species invasion.

This research still faces some research limitations. First, we are aware of the fact that it is challenging to study the interaction among multiple systems at the same time. Although some conclusions have been drawn by systematically analyzing the metacupling of SNWTP-MR with limited data, there is still more work to do to clarify the intricate relationship in reality. The biggest challenge is that the data are not easy to be established and acquired, which means that many environmental and socioeconomic impacts are not yet quantitatively measured. In the future, more efforts need to be put into predicting the value and risk brought by the metacoupling relationship of SNWTP-MR. Second, the Metacoupling framework is still a conceptual framework. The unified evaluation of the integration of intracoupling, pericoupling and telecoupling is still a problem. Subsequent research can introduce some quantitative methods to conduct a unified evaluation of the metacoupling systems, such as data envelopment analysis or multi-region input and output models. Third, Differences in the time period taken by the remote sensing map will lead to a certain deviation between the changes in water volume and other data and the real measurement data. Therefore, this study has a certain error range, to which readers should pay attention.

5. Conclusions

This paper uses the metacoupling analysis framework of water transfer to reveal the influence of the interaction of ecological environment of SNWTP-MR. Thus, the following conclusions can be drawn:

(1) By analyzing intracoupling, it can be shnown that water transfer has caused spatial and temporal changes in the land use of the Danjiangkou Reservoir, which affects the ecological value that it provides and exacerbates the internal ecological risk. Considering only land changes, the construction of SNWTP-MR has created an ecological service value of 23 billion yuan for the water source of Danjiangkou Reservoir, but also increased the original ecological risk index by 9.31%, and the ecological vulnerability has been changed to some extent.

(2) By analyzing pericoupling, it can be found that the upstream and downstream have different purposes for water use, mainly because the expansion of the reservoir produces different ecological value and ecological risk for the upstream and downstream. The expansion of Danjiangkou Reservoir has increased the total power generation of the Reservoir, increasing the revenue of generation by 2.292 billion yuan; at the same

time, the ability for seasonal water transfer has also been greatly enhanced, so that the downstream flood control standards have been improved to once-in-100-years, and the threat of flooding has been greatly reduced. However, the ecological risks are also present in many aspects, such as poor water quality, eutrophication of water resources, water competition between industry and agriculture, deterioration of waterway shipping conditions, and threats to biodiversity.

(3) By analyzing telecoupling, the interaction among the sending system, the receiving system and the spillover system can be seen. From the perspective of ecological value, the sending system obtains a large amount of income from water transfer; the receiving system has greatly improved the urban domestic water consumption. For example, the water volume for consumption has been increased by 9.5 billion m^3, and the hardness of tap water has also dropped from 380 $mg \cdot L^{-1}$ to 120–130 $mg \cdot L^{-1}$, and the problem of over-exploitation of groundwater has been alleviated. The population carrying capacity of the water resources has increased by 16.42 million people, and the spillover system has created a cross-regional green ecological landscape belt that can create a value of 189 million yuan of environmental purification benefits. From the perspective of ecological risk, both biological invasion and water pollution have become problems that all systems need to face together.

The metacoupling framework systematically discusses the values and risks brought by various systems of the SNWTP-MR, and provides a theoretical reference for who should take more responsibility for the ecological compensation. This may help avoid excessive development and unreasonable allocation of resources for this purpose.

Author Contributions: Q.S.: Conceptualization, Methodology, Software, Data curation, Writing—Original draft preparation, Visualization, Writing—Reviewing and Editing, Financial aid; H.-S.C.: Conceptualization, Methodology, Writing—Reviewing. X.C.: Conceptualization, Writing—Reviewing, Financial aid; J.X.: Visualization; Data curation. All authors have read and agreed to the published version of the manuscript.

Funding: The research start-up fund project of Fujian University of Technology (No. GY-Z22039).

Informed Consent Statement: Informed consent was obtained from all subjects involved in the study.

Data Availability Statement: All data generated or analyzed during this study are included in this article.

Conflicts of Interest: The authors declare no conflict of interest.

References

1. Hogeboom, R.J.; Kamphuis, I.; Hoekstra, A.Y. Water sustainability of investors: Development and application of an assessment framework. *J. Clean. Prod.* **2018**, *202*, 642–648. [CrossRef]
2. Zhou, X.-Y. Spatial explicit management for the water sustainability of coupled human and natural systems. *Environ. Pollut.* **2019**, *251*, 292–301. [CrossRef] [PubMed]
3. UN DESA. *World Population Prospects 2019-Population Division*; UN DESA: New York, NY, USA, 2019.
4. McDonald, R.I.; Weber, K.; Padowski, J.; Flörke, M.; Schneider, C.; Green, P.A.; Gleeson, T.; Eckman, S.; Lehner, B.; Balk, D.; et al. Water on an urban planet: Urbanization and the reach of urban water infrastructure. *Glob. Environ. Chang.* **2014**, *27*, 96–105. [CrossRef]
5. Swyngedouw, E. UN Water Report 2012: Depoliticizing Water. *Dev. Chang.* **2013**, *44*, 823–835. [CrossRef]
6. Deines, J.M.; Liu, X.; Liu, J. Telecoupling in urban water systems: An examination of Beijing's imported water supply. *Water Int.* **2015**, *41*, 251–270. [CrossRef]
7. Fitzhugh, T.W.; Richter, B.D. Quenching Urban Thirst: Growing Cities and Their Impacts on Freshwater Ecosystems. *BioScience* **2004**, *54*, 741–754. [CrossRef]
8. Gassert, F.; Reig, P.; Shiao, T.; Luck, M.; ISciences LLC. *Aqueduct Global Maps 2.1: Constructing Decision-Relevant Global Water Risk Indicators*; World Resources Institute: Washington, DC, USA, 2014; p. 31.
9. Su, Q.; Zhang, W. Cross-sector collaboration in the process of urban planning reform in China—A case of 'multiple-plan coordination' work in Xiamen city, China. *J. Asian Public Policy* **2022**, 1–18. [CrossRef]
10. Liu, J.; Hull, V.; Luo, J.; Yang, W.; Liu, W.; Viña, A.; Vogt, C.; Xu, Z.; Yang, H.; Zhang, J.; et al. Multiple telecouplings and their complex interrelationships. *Ecol. Soc.* **2015**, *20*, 44. [CrossRef]

11. McCord, P.; Tonini, F.; Liu, J. The Telecoupling GeoApp: A Web-GIS application to systematically analyze telecouplings and sustainable development. *Appl. Geogr.* **2018**, *96*, 16–28. [CrossRef]
12. Su, Q.; Chen, X. Efficiency analysis of metacoupling of water transfer based on the parallel data envelopment analysis model: A case of the South–North Water Transfer Project-Middle Route in China. *J. Clean. Prod.* **2021**, *313*, 127952. [CrossRef]
13. Chang, H.-S.; Su, Q. Exploring the coupling relationship of stormwater runoff distribution in watershed from the perspective of fairness. *Urban Clim.* **2021**, *36*, 100792. [CrossRef]
14. Tonini, F.; Liu, J. Telecoupling Toolbox: Spatially explicit tools for studying telecoupled human and natural systems. *Ecol. Soc.* **2017**, *22*, 11. [CrossRef]
15. Liu, J.; Yang, W. Integrated assessments of payments for ecosystem services programs. *Proc. Natl. Acad. Sci. USA* **2013**, *110*, 16297–16298. [CrossRef]
16. Britt, C.-M.; Michael, W. Of maps and eating bitterness: The politics of scaling in China's South-North Water Transfer Project. *Political Geogr.* **2017**, *61*, 19–30.
17. Liu, J. Integration across a metacoupled world. *Ecol. Soc.* **2017**, *22*, 29. [CrossRef]
18. Yang, W.; Hyndman, D.W.; Winkler, J.A.; Viña, A.; Deines, J.M.; Lupi, F.; Luo, L.; Li, Y.; Basso, B.; Zheng, C.; et al. Urban water sustainability: Framework and application. *Ecol. Soc.* **2016**, *21*, 4. [CrossRef]
19. Zhong, Z.; Liu, G.; Wu, Z. Analysis and practices of water regulation in the Middle Route of South-to-North Water Transfer Project. *South-North Water Transf. Water Sci. Technol.* **2018**, *18*, 95–99.
20. Sheng, J.; Webber, M. Incentive-compatible payments for watershed services along the Eastern Route of China's South-North Water Transfer Project. *Ecosyst. Serv.* **2017**, *25*, 213–226. [CrossRef]
21. Costanza, R.; d'Arge, R.; de Groot, R.; Farber, S.; Grasso, M.; Hannon, B.; Limburg, K.; Naeem, S.; O'Neill, R.V.; Paruelo, J.; et al. The value of the world's ecosystem services and natural capital. *Nature* **1997**, *387*, 253–260. [CrossRef]
22. Yang, G. *Coupling Relationship between Land Use/Cover Chang and its Eco-Environment Effect in the Three Gorges Reservoir Region of Chongwing*; Chongqing Normal University: Chongqing, China, 2011.
23. Luo, G.; Zhou, C.; Chen, Y. Process of Land Use/Land Cover Change in the Oasis of Arid Rigion. *Acta Geogr. Sin.* **2003**, *58*, 63–72.
24. Song, W.; Deng, X. Land-use/land-cover change and ecosystem service provision in China. *Sci. Total Environ.* **2016**, *576*, 705–719. [CrossRef] [PubMed]
25. Chen, Y.; Li, X.; Liu, X.; Zhang, Y.; Huang, M. Quantifying the teleconnections between local consumption and domestic land uses in China. *Landsc. Urban Plan.* **2019**, *187*, 60–69. [CrossRef]
26. Lu, X.; Shi, Y.; Chen, C.; Yu, M. Monitoring cropland transition and its impact on ecosystem services value in developed regions of China: A case study of Jiangsu Province. *Land Use Policy* **2017**, *69*, 25–40. [CrossRef]
27. Garcia, L.; Celette, F.; Gary, C.; Ripoche, A.; Valdés-Gómez, H.; Metay, A. Management of service crops for the provision of ecosystem services in vineyards: A review. *Agric. Ecosyst. Environ.* **2018**, *251*, 158–170. [CrossRef]
28. Wei, Y.; Tang, D.; Ding, Y.; Agoramoorthy, G. Incorporating water consumption into crop water footprint: A case study of China's South–North Water Diversion Project. *Sci. Total Environ.* **2016**, *545–546*, 601–608. [CrossRef]
29. Su, Q. Long-term flood risk assessment of watersheds under climate change based on the game cross-efficiency DEA. *Nat. Hazards* **2020**, *104*, 2213–2237. [CrossRef]
30. Dou, M.; Lu, Y.; Yang, H.Z.; Lu, Y.-L. Research on the effects of Water Resources Availability in the Lower reachs of the Hanjiang River. *China Rural. Water Hydropower* **2016**, *42*, 34–37.
31. He, Y.; Li, W. Influence of the Middle Route of South-to-North Water Transfer Project on Flood Control of the Middle and Lower Reaches of Hanjiang River and Countermeasures. *Water Dev. Res.* **2015**, 54–63. [CrossRef]
32. Tang, J.; Cao, H.; Chen, J. Evaluation of ecosystem service value for water sources region of middle route project of South-to-North Water Diversion. *Yangtze River* **2018**, *49*, 29–34.
33. Wang, J.; Qi, Y.; Sun, X. Effects of Excessive Grazing on Grassland Eco-system Services Valuation. *J. Nat. Resour.* **2006**, *21*, 109–117.
34. Liu, H. Constructing the ecological barrier of the Middle Route of South-to-North Water Transfer. *South-North Water Transf. Track* **2017**, *13*, 22–23.
35. Lin, M.; Zhang, Q.; Li, Z.; Zhang, G.; Zhang, Z.; Yang, Z.; Si, S.; Niu, H.; Sun, J.; Fan, H. Characteristics of the variance of the water uality and quantity in the middle route of South-to-North Water Diversion Project and corresponding. *Water Wastewater Enchineering* **2016**, *42*, 9–13.
36. National Development and Reform Commission. *Notice on the Water Supply Price Policy at the Initial Stage of the Main Project of the First Phase of the South-to-North Water Transfer Project*; National Development and Reform Commission: Beijing, China, 2014.
37. National Development and Reform Resources. *China Water Resources Bulletin*; Ministry of Water Resources: Beijing, China, 2018.
38. Yang, L.; Zhu, Q.; Sun, J.; Yong, D.U.; Shen, B. Water supply benefit evaluation of Middle Route Project of South-to-North Water Diversion in Beijing City. *Tangtze River* **2017**, *48*, 44–46.
39. Wang, D.; Deng, S. The Influence Degree Evaluation of South Water Inflow Transfer Project to North Intake Area in Hebei. *J. Hebei Univ. Econ. Bus.* **2015**, *36*, 125–128.
40. Song, Q. Suggestions on Strengthening the Environmental Protection of the Middle and Lower Reaches of the Hanjiang River after the South-to-North Water Diversion Project. *China Dev.* **2017**, *17*, 88–89.
41. Liu, G. Discussion on countermeasures for water transport development in the middle and lower reaches of Hanjiang River after mid-line water transfer. *J. Jingchu Univ. Technol.* **2013**, *27*, 51–55.

42. Rui, X.; Zhang, Y.; Wang, L.; Zhang, Y.; Dou, M.; Qiao, Y.; Zhang, M. Characteristics Identification of Multiple Influencing Factors on Hanjiang River Algal Bloom. *Res. Environ. Sci.* **2019**, *33*, 911–920.
43. Wang, Z.; Liu, J.Z.; Liu, Y.Y.; Fan, H. Research on the Cumulative Superposition Effect of Water Diversion Project on Ecological Environment in the Middle and Lower Reaches of Hanjiang River. *China Rural. Water Hydropower* **2018**, *3*, 29–36.
44. Yan, B. *Safety Assessment of Water Quality in the Water Sourse of the Middle Route Project of the South to North Water Transfcr Project in Henan Province*; North China University of Water Resouses and Electric Power: Zhengzhou, China, 2017.
45. Pan, L. *Management and Adjustment of Beijing Water Supply in the South-to North Water Benefited Area 2016*; China University of Mining & Technology: Beijing, China, 2016.
46. Ke, X.; van Vliet, J.; Zhou, T.; Verburg, P.H.; Zheng, W.; Liu, X. Direct and indirect loss of natural habitat due to built-up area expansion: A model-based analysis for the city of Wuhan, China. *Land Use Policy* **2018**, *74*, 231–239. [CrossRef]
47. van Vliet, J.; Eitelberg, D.A.; Verburg, P.H. A global analysis of land take in cropland areas and production displacement from urbanization. *Glob. Environ. Chang.* **2017**, *43*, 107–115. [CrossRef]
48. El-Nwsany, R.I.; Maarouf, I.; El-Aal, W.A. Water management as a vital factor for a sustainable school. *Alex. Eng. J.* **2019**, *58*, 303–313. [CrossRef]
49. Sheng, J.; Wang, H.; Qiu, W. Water quality and incentive coordination in water markets: The eastern route of China's South-North Water Transfer Project. *J. Hydrol.* **2022**, *607*, 127526. [CrossRef]
50. Qu, X.; Chen, Y.; Liu, H.; Xia, W.; Lu, Y.; Gang, D.-D.; Lin, L.-S. A holistic assessment of water quality condition and spatiotemporal patterns in impounded lakes along the eastern route of China's South-to-North water diversion project. *Water Res.* **2020**, *185*, 116275. [CrossRef] [PubMed]
51. Zhang, X.; Wang, G.; Tan, Z.; Wang, Y.; Li, Q. Effects of ecological protection and restoration on phytoplankton diversity in impounded lakes along the eastern route of China's South-to-North Water Diversion Project. *Sci. Total Environ.* **2021**, *795*, 148870. [CrossRef] [PubMed]
52. Zhao, Z.-Y.; Zuo, J.; Zillante, G. Transformation of water resource management: A case study of the South-to-North Water Diversion project. *J. Clean. Prod.* **2017**, *163*, 136–145. [CrossRef]
53. Su, Q.; Chang, H.-S.; Pai, S.-E. A Comparative Study of the Resilience of Urban and Rural Areas under Climate Change. *Int. J. Environ. Res. Public Health* **2022**, *19*, 8911. [CrossRef]

International Journal of
Environmental Research and Public Health

MDPI

Article

Land Use Dynamic Evolution and Driving Factors of Typical Open-Pit Coal Mines in Inner Mongolia

Lijia Zhang [1,2], Zhenqi Hu [1,*], Dazhi Yang [3,4], Huanhuan Li [5], Bo Liu [6], He Gao [7], Congjie Cao [7], Yan Zhou [2], Junfang Li [8] and Shuchang Li [3,4]

1 School of Geosciences & Surveying Engineering, China University of Mining and Technology, Beijing 100083, China
2 Land Consolidation and Rehabilitation Center, Ministry of Natural Resources, Beijing 100035, China
3 Key Laboratory of Land Surface Pattern and Simulation, Institute of Geographic Sciences and Natural Resources Research, Chinese Academy of Sciences, Beijing 100101, China
4 College of Resources and Environment, University of Chinese Academy of Sciences, Beijing 100049, China
5 School of Land and Tourism, Luoyang Normal University, Luoyang 471000, China
6 School of Geomatics, Liaoning Technical University, Fuxin 123000, China
7 School of Earth Sciences, Yangtze University, Wuhan 430100, China
8 School of Earth Sciences and Resources, Chang'an University, Xi'an 710064, China
* Correspondence: huzq@cumtb.edu.cn

Abstract: Although coal is difficult to replace in the short term, the large-scale production and consumption of coal have significant impacts on the ecological environment. The severe disturbances, such as land excavation and occupation, that accompany the mining of mineral resources have caused dramatic changes in land cover and a significant pressure on the sensitive and fragile ecological environment. To analyze the temporal and spatial evolution trends and the differences in land use in different typical mining areas in Inner Mongolia, as well as the evaluation system and driving mechanisms of land use evolution, this study takes the typical open-pit coal mines in Inner Mongolia as the research objects and, based on the Google Earth Engine (GEE) platform, analyzes the dynamic evolution characteristics and driving factors of land use in typical open-pit coal mines in Inner Mongolia from 2001 to 2020. The change trend of land use in typical open-pit mining areas in Inner Mongolia for the past 20 years is obvious, with the highest fluctuations for grassland, mining land, cropland, and residential/industrial land. Land use in the open-pit coal mining area is greatly affected by mining factors. From the perspective of spatial variation, the most important driving factor is the distance from national roads and railways, followed by the annual average temperature and annual average precipitation and topographical conditions, such as elevation. In terms of policy, land reclamation and ecological restoration in mining areas have a positive impact on land use change. Improving the mechanism for environmental compensation in mining areas can promote the efficient and rational use of mining areas and the protection of ecosystems.

Keywords: open-pit coal mines; GEE; land use; dynamic degree; driving factors; Inner Mongolia

Citation: Zhang, L.; Hu, Z.; Yang, D.; Li, H.; Liu, B.; Gao, H.; Cao, C.; Zhou, Y.; Li, J.; Li, S. Land Use Dynamic Evolution and Driving Factors of Typical Open-Pit Coal Mines in Inner Mongolia. *Int. J. Environ. Res. Public Health* **2022**, *19*, 9723. https://doi.org/10.3390/ijerph19159723

Academic Editor: Yongli Cai

Received: 28 June 2022
Accepted: 3 August 2022
Published: 7 August 2022

1. Introduction

Coal is the world's largest and most widely distributed non-renewable energy, and plays an irreplaceable role in the development of the national economy [1]. Against the background of climate change, since the 21st century, greenhouse gas emission reduction, carbon neutrality, and the adjustment of energy structure have received significant attention, and numerous developed countries have progressively adopted clean energy to replace coal energy consumption [2,3]. However, in developing countries, especially China, India, South Africa, and Indonesia, among others, coal resources are still the most important energy sources [4–6]. China's coal production far exceeds that of other countries, accounting for about 51% of the global coal production [7], and the mining industry has brought improved infrastructure, economic development, and elevated living standards for locals [8,9]. In most countries, coal resources are still the most important energy source for power and heat generation [1], and the principal raw material for various daily necessities, such as dyes, fertilizers, and pesticides [10].

However, the large-scale production and consumption of coal have seriously negatively impacted the ecological environment, with an emphasis on land resources, by directly destroying the surface soil layer and original vegetation [11–13]. Wastewater from mining is generally discharged into rivers, resulting in the death of aquatic animals and plants and the destruction of river ecosystems [14]. The infiltration of acidic wastewater into the ground also leads to the deterioration of groundwater quality and the destruction of the vegetation. In addition, excessive mining activities produce large amounts of carcinogenic heavy metals that are difficult to degrade in the natural state, such as chromium, nickel, cadmium, with negative impacts on human and animal health [15,16].

Open-pit mining is the most commonly used method of coal mining because of its low cost and convenience; two thirds of the world's mineral resources are extracted via open-pit mining [17,18]. More than 50% of coal resources in the United States, Australia, Spain and other countries are mined in open-pit mines [19,20]. However, because of the "stripping-mining-transportation-disposal-land making" in the mining area, the original ecosystem is predominantly degraded, with deteriorated surface water and groundwater and a diminished carbon storage capacity [21–23]. According to the data released by the National Research Council (NRC) of the United States, open-pit mining of 124 billion tons of coal in the United States will destroy about 4 million hm^2 of land. Under the same mining volume, China will destroy 2.728 million hm^2 of land [24–26]. The severe disturbances, such as land excavation, compaction and occupation during the mining of mineral resources, have caused dramatic changes in the land cover in this area [27], resulting in altered ecosystem types, patterns and processes and, ultimately, in changes in ecosystem services [28]. In this sense, studies on land use change in open-pit coal mining areas are valuable to assess the evolution of the ecosystems in such areas [29].

In China, more than 90% of large open-pit coal mines are located in arid and semi-arid areas with a fragile ecological environment [30–32]. For example, in Inner Mongolia and Xinjiang [33], land use changes caused by resource over-exploitation have largely changed ecological processes [34]. In this context, investigating land use changes in open-pit coal mining areas can help optimize reclamation planning in such areas, adjust land use structure [35,36], and provide an important basis for the development of adequate management strategies and a sustainable coal mining policy.

As early as in the 1960s, studies on land use monitoring in mining areas have been carried out. In 1969, the land protection department in the United States monitored local mine environments and disasters, using remote sensing technology to monitor the land reclamation in coal mining areas, thereby providing a basis for the development of land reclamation strategies [37]. Brink et al. [38] took sub-Saharan Africa as the study area and, based on the high-scoring earth observation data, monitored and analyzed the changes in regional land use types during the period from 1975 to 2000. However, China's remote sensing and geographic information system technology started late and is in a relatively undeveloped stage. The use of remote sensing satellites to monitor land use in mining areas was gradually developed after the 1980s. Since then, the remote sensing technology has gradually been developed, providing a certain amount of data for research and analysis.

Because of the late start of the monitoring technology, immature monitoring methods and low-accuracy monitoring results are frequent [39,40]. However, this phenomenon gradually decreases with the improvement of the research methods.

Globally, the application of remote sensing and GIS technology has gradually matured; various high-precision satellites, such as QuickBird, Landsat, Spot, and Sentinel, were born. The monitoring of land use in mining areas is performed with accurate data and technical support, which has resulted in a large number of studies. For instance, Raval et al. [41] used traditional remote sensing technology to monitor and quantitatively analyze land use change in kaolin mining areas in India from 2000 to 2009, providing technical support for the rapid mapping of land use changes in these areas. Sonter et al. [42] considered the mining area as a separate land use type for the classification of remote sensing images, described the land use change process in the Brazilian mining area over a period of time, and compared it with that of the surrounding non-mining areas, with the aim to analyze the differences and the underlying reasons. Using all archived Landsat imagery between 2000 and 2015, Wohlfart et al. [43] calculated the temporal and textual measures of spatially continuous spectra based on dense Landsat time series for each year to obtain values related to mining, agriculture, forestry and urbanization in the Yellow River Basin Zoning land cover change map. Using Landsat image data from 2013 to 2016, Padmanaban et al. [44] studied the land use change in a reclaimed mining area in Kirchheller Heide, Germany, using the vegetation cover index (NDVI) to analyze the changes in vegetation productivity and to determine the geological and surface environment changes that may occur in the mining area. Several methods of land use monitoring in mining areas have been developed, among which remote sensing and GIS technology are the most important ones and can be applied in high-precision and long-term land use monitoring in mining areas. However, the image processing flow of mining areas needs to be optimized, and the efficiency needs to be improved.

The application of remote sensing technology in the monitoring and management of land use change has gradually intensified. For most researchers interested in land-use change monitoring, the acquisition and processing of long-term remote sensing data are time-consuming and labor-intensive. When traditional image-processing software (ENVI, ERDAS) is used for land use change monitoring, original image data from specific channels need to be downloaded, and complex steps are required, such as image data correction, registration, splicing, and cropping. Processing power and storage space require researchers to have good theoretical knowledge and adequate image-processing skills. In this context, Google Earth Engine (GEE) [45] has become an important tool for geography and space-related research, providing powerful computing resources and massive online data. By invoking a large number of published geographic data products collected by the GEE platform and combining the algorithms provided by the researchers, online computing can be performed, which greatly reduces the workload of data acquisition and processing. More and more scholars use the GEE platform for land use monitoring research. For example, Hamud et al. [46] used the GEE platform to monitor land use cover changes in Somalia. Lin et al. [47] monitored land cover change on a rapidly urbanizing island using the GEE. This approach can greatly expand the time and space scale of their original research and provide national and even global research results [48–50]. The GEE platform makes up for the deficiency of traditional image-processing software and enriches the technical methods of land use monitoring research in mining areas.

Judging from the current global research progress, most of the current technologies are applied in small mining areas and are dominated by algorithm models. There are few studies on long-term, rapid, accurate and continuous land use classification in open-pit mining areas. In addition, most of the research is concentrated in a single mining area, and investigations on multiple mining areas of a specific mining area type are scarce, and the explanation of the driving factors behind land use evolution is insufficient. In this study, seven types of land use are investigated, namely cropland, forest, grassland, water body, mining land, residential/industrial land, and unused land, according to the present situation of land use in the open-pit mining areas in Inner Mongolia. Based on the emerging GEE platform, it solves the problems of difficult data collection, large data volumes, and low interpretation efficiency in long-term large-scale analyses. This study regards typical open-pit coal mines in Inner Mongolia as the research unit, and analyzes the dynamic evolution characteristics and driving factors of land use from 2001 to 2020. The main objectives are as follows: (i) to gain an in-depth understanding of the dynamic change in land use in open-pit coal mining areas in Inner Mongolia; (ii) to identify the causes of spatial changes in land use in typical open-pit coal mining areas in Inner Mongolia; (iii) to put forward policy suggestions on land exploitation and remediation in mining areas.

2. Study Area

The Inner Mongolia Autonomous Region is located in northern China, at 37°24′–53°23′ north latitude and 97°12′–126°04′ east longitude (Figure 1), with a large plateau area, a large distance from the ocean, and mountains along the edge. The region has the richest mineral resources in China, with 17 kinds of mineral reserves in the forefront. However, high-intensity resource exploitation has a great impact on fragile ecosystems in the arid and semi-arid areas of the Inner Mongolia Autonomous Region. Open-pit mining has a more significant impact on the environment, such as ecosystem destruction and land resource degradation. Mining areas located in arid and semi-arid areas are particularly sensitive to this impact. Therefore, the land use change and ecological processes in the mining area are more complex and diverse, and its pattern change characteristics and laws are more representative.

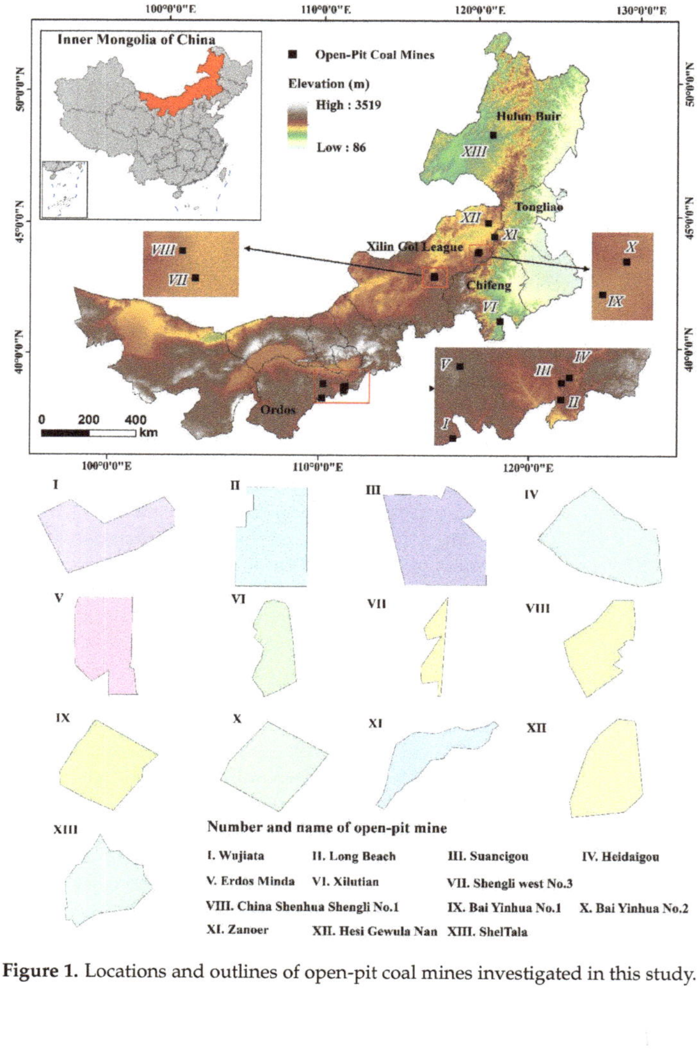

Figure 1. Locations and outlines of open-pit coal mines investigated in this study.

The precise mining area is defined according to the mining rights, so we can only consider selecting open-pit coal mines according to the mining license information issued by the Ministry of Natural Resources. Moreover, to ensure a wider representation of mining areas, we need to ensure that open-pit coal mines that cover large, medium and small areas are covered. Therefore, we selected 13 open-pit coal mines as typical study areas in mining areas with relatively complete information (Figure 1). By connecting the original registered nodes of each mining area in sequence, the boundary mining area can be delineated. The open-pit coal mines are mainly distributed in Erdos, Xilin Gol League, Hulun beier and Chifeng, covering a total area of 391 km^2 (the largest open-pit coal mine, Changtan, is 66.99 km^2, whereas the smallest one, Shengli West No. 3, only covers 1.55 km^2).

3. Data Sources and Methods

3.1. Data Sources

Generally speaking, 2001–2020 is an important period for the rapid development of China's ecological protection and restoration and the innovation and transformation of its system and mechanism. This study used the 2001–2020 image data of Landsat on the GEE platform. Among them, mainly from April to September, GEE synthesized the image data and used them as a remote sensing data source combined with DEM data, the Chinese Academy of Sciences Resources and Environment Data Center [51] vegetation type data, vegetation zoning data, and meteorological data and the random forest model was used for land use classification.

According to the characteristics of the mining area, the land use types were divided into six categories, namely cropland, forest, grassland, water body, residential/industrial square land, mining land, unused land. Residential/industrial square land refers to residential and living ancillary facilities, industrial plants, and large industrial construction. Mining land refers to the mining, quarrying, sand mining (sand) fields, brick kilns and other ground production land and tailings dumps that are independent of residential areas. Unused land refers to tailing stacking land, bare land, bare rocks, and sand areas. The social and economic statistics of raw coal output used in this study were derived from the *Inner Mongolia Statistical Yearbook* [52], the *Regulations of Inner Mongolia Autonomous Region on Mineral Resources Management* [53], and the *Statistical Bulletin of National Economic and Social Development of Inner Mongolia Autonomous Region* [54].

3.2. Methods

3.2.1. Dynamic Degree of Land Use

The dynamic degree of land use is based on the magnitude of land use change and represents the results of various types of area changes during the study period. It can directly reflect the change speed of different land use types and can be used to compare and analyze the change differences among various types [55]. In this paper, the dynamic degree model that reflects the absolute amount of land use change was used to monitor the speed change in each land use type in the study area, using the following equation:

$$k_j = \frac{u_b - u_a}{u_a} \times \frac{1}{\mathrm{T}} \times 100\% \quad (1)$$

where k_j represents the dynamic degree of a certain land use type during the research period; u_a represents the quantity of a certain land use type at the early stage of the research; u_b represents the quantity of a certain land use type at the end of the research period; T represents the length of the research time.

The study of land use dynamic changes is an important approach to arrive at a deep understanding of the process of urban land use change, and is the main method to comprehend the evolution process and pattern of land use [56,57]. To deeply explore the land use change dynamic of typical open-pit mining areas in Inner Mongolia, the single land use dynamic degree method was used.

3.2.2. Geographical Detector Model (GDM)

The GDM (geographical detector model) is an important method to detect the spatial pattern and genesis of geographic elements and is widely used in studies on land use driving mechanisms and climate change [58]. When importing the input data of GDM, discrete classification processing of driving factors is required in ArcGIS, and through sampling, discrete data of the dependent variables and result variables are obtained, and are finally imported into the GDM for factor analysis. The specific calculation method is as follows:

$$q = 1 - \frac{1}{n\sigma^2} \sum_{i=1}^{m} n_i \times \sigma^2 \quad (2)$$

where q is the explanatory power of the driving factor for the expansion of construction land; n is the total amount of driving factors, and σ^2 is the sample variance. The value range of q is (0, 1) and the larger the value, the stronger the explanatory power of the factor to land use change will be.

4. Results

4.1. Dynamic Evolution of Land Use in the Typical Open-Pit Coal Mine Area

4.1.1. Land Use Pattern

By establishing polygon training samples of land use classification in GEE, a sample set of each corresponding land class was formed. Then, all the samples were fused into a sample set. Part of them were selected as training samples to participate in classification, and part of them were used as verification samples for precision verification. We used the random forest classifier in GEE, took training samples and images as input, and carried out supervised classification to obtain the raster data of land cover. After classification, the overall accuracy, kappa coefficient, and transfer matrix were calculated by using the verification data set and the classification results, and the accuracy was evaluated. When all the land use classification results meet the accuracy requirements (Table 1), the results were retained.

Table 1. Overall accuracy and kappa coefficient values of land use classification.

Year	Kappa	Overall Accuracy	Year	Kappa	Overall Accuracy
2001	0.863	0.921	2011	0.861	0.921
2002	0.849	0.914	2012	0.852	0.915
2003	0.857	0.917	2013	0.845	0.912
2004	0.839	0.909	2014	0.868	0.924
2005	0.841	0.910	2015	0.866	0.923
2006	0.837	0.908	2016	0.903	0.944
2007	0.847	0.913	2017	0.884	0.933
2008	0.833	0.906	2018	0.849	0.914
2009	0.852	0.916	2019	0.859	0.919
2010	0.828	0.903	2020	0.854	0.916

From 2001 to 2020, the change trend of land use in open-pit coal mines was obvious (Figure 2), with large grassland and cropland areas and a considerable change range. The forest, water body, and unused land areas were small, and the fluctuation was relatively stable. The areas of mining land and residential/industrial square showed a fluctuating increase throughout the research period, whereas the grassland area showed a fluctuating decrease. The cropland area showed a downward trend from 2001 to 2015, followed by an increase after 2016. The mining land increased rapidly from 2006 to 2012, with a slower growth thereafter.

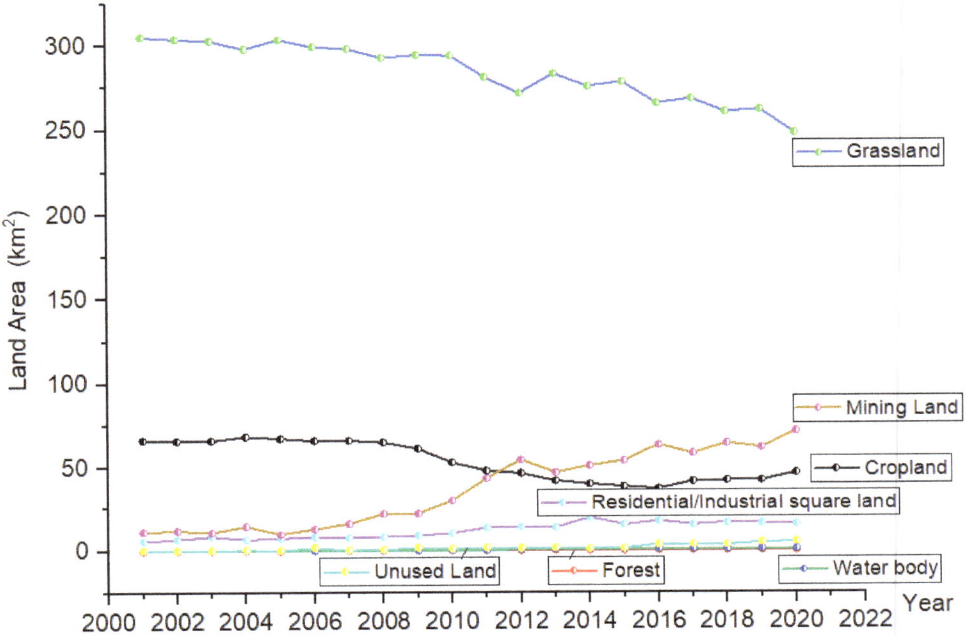

Figure 2. Land use types and areas in open-pit coal mines from 2001–2020.

4.1.2. Dynamic Degree of Land Use

According to the dynamic degree of land use in the study area (Table 2), from 2001 to 2020, the dynamic degree of unused land and mining land changed most significantly, accounting for 93.16% and 27.11%, respectively. Among them, from 2001 to 2005, the land with the largest change was residential/industrial square land, with a total transfer area of 1.68 km^2 and a dynamic degree of 5.85%, followed by mining land with a transfer area of −1.42 km^2 and a dynamic degree of −2.54%. From 2005 to 2010, the mining land showed the largest dynamic change, which was 16.97 km^2, and the dynamic change degree was 26.53%, followed by cropland, whose change amount was −13.16 km^2, with a dynamic degree of −3.98%. Compared with 2001–2005 and 2005–2010, in 2010–2015, the water body significantly changed the most, and the dynamic degree was 39.13%. From 2015 to 2020, the transfer of cropland kept increasing, with a change of 9.63 km^2, and the dynamic degree increased to 5.17%; in contrast, the residential/industrial square land decreased by −1.72 km^2, with a dynamic degree of −1.98%.

Table 2. Dynamic degree of land use in open-pit coal mines (2001–2020).

Land Use Type		Cropland	Forest	Grassland	Water Body	Residential/ Industrial Square Land	Mining Land	Unused Land
2001–2005	Variation (km^2)	1.12	0.00	−1.38	0.00	1.68	−1.42	0.00
	Dynamic Degree k(%)	0.34	0.00	−0.09	−0.34	5.85	−2.54	0.07

Table 2. *Cont.*

Land Use Type		Cropland	Forest	Grassland	Water Body	Residential/ Industrial Square Land	Mining Land	Unused Land
2005–2010	Variation (km^2)	−13.16	0.00	−5.20	0.11	2.09	16.97	−0.82
	Dynamic Degree k(%)	−3.98	−2.50	−0.35	14.73	5.04	26.53	−8.33
2010–2015	Variation (km^2)	−9.39	0.02	−2.90	0.65	1.64	10.45	−0.47
	Dynamic Degree k(%)	−3.91	30.77	−0.21	39.13	2.42	4.79	−5.56
2015–2020	Variation (km^2)	9.63	0.02	−17.44	−0.11	−1.72	8.03	1.59
	Dynamic Degree k(%)	5.17	13.79	−1.31	−2.28	−1.98	2.53	8.70
2001–2020	Variation (km^2)	−19.48	0.02	−56.51	0.72	9.88	60.42	4.98
	Dynamic Degree k(%)	−1.47	5.21	−0.93	22.60	8.58	27.11	93.16
	Annual Change (km^2)	−0.97	0.00	−2.83	0.04	0.49	3.02	0.25

Note: The data in the table were calculated according to the interpreted land use data and Formula (1).

4.2. Spatial Driving Factor Analysis of Land Use Change in Typical Open-Pit Coal Mining Areas in Inner Mongolia

Identifying the causes of spatial changes in land use in typical open-pit coal mining areas in Inner Mongolia is of great significance for exploring the landscape ecological trends of land use changes, adjusting the industrial structure of mining areas, and arriving at sustainable land development [59,60]. From the perspective of spatial heterogeneity, this study uses the ArcGIS spatial analysis function to sample the spatial location and driving factors of land use change in the open-pit mining areas from 2001 to 2020 and used the GDM for q-value detection. From the perspective of time, this paper analyzes the impact of mining and land reclamation on land use change, as well as the impact of large-scale mining, reclamation and other activities on this change.

4.2.1. Analysis of the Factors Influencing Mining and Reclamation in Mining Activities

In the past 20 years, Inner Mongolia has witnessed large-scale mining activities. According to the Statistical Yearbook of Inner Mongolia and the Statistical Bulletin of National Economic and Social Development of Inner Mongolia Autonomous Region, from 2001 to 2020, the output of raw coal in the Inner Mongolia Autonomous Region showed a rapid growth, from 81.63 to 1025.51 million tons (Figure 3).

Regarding the entire study period, the area of mining land showed a growth trend, which is closely related to the development of mining activities. From 2005 to 2012, the area of open field/unused land increased significantly, whereas that of cropland decreased greatly. After 2013, the area of mining land showed a fluctuating decrease, and some cropland recovered rapidly. This is due to the implementation of a number of land reclamation policies and measures in the study area from 2008 to 2020, which significantly improved the local land use structure. For example, in 2008 and 2013, the measures of the Inner Mongolia Autonomous Region for the management of mining and mineral deposits were issued and revised successively, and in 2009, the implementation plan for the management of mining the geological environment was issued. From 2009 to 2015, the Inner Mongolia Autonomous Region vigorously carried out the geological and ecological environment treatment in mining areas. In addition, according to the regulations on land reclamation issued by the State Council in 2011, the basic national policy of promoting mining enterprises to make rational use of land and implement cropland protection is one of the reasons for the overall increase in cropland after 2011. According to the Xinhua news agency, since 2007, the Inner Mongolia Autonomous Region has received a total of 53 national land consolidation projects, with a capital of CNY 350 million, for the transformation of wasteland, sandy land, and low-yield fields, resulting in 7713.3 hectares of new cropland. A total of 78 economic development zones were abolished in Inner Mongolia, and nearly 1333.3 hectares of land were restored. According to the Inner Mongolia Bureau of Statistics, from 2012 to 2018, the cultivated area increased by 163,000 hectares, with an average annual growth of about 0.3%. At the same time, in the grassland mining area, the land damage caused by mining has been effectively controlled (Figure 4). Therefore, mining activities and land reclamation can be regarded as the main reasons for land use changes in mining areas.

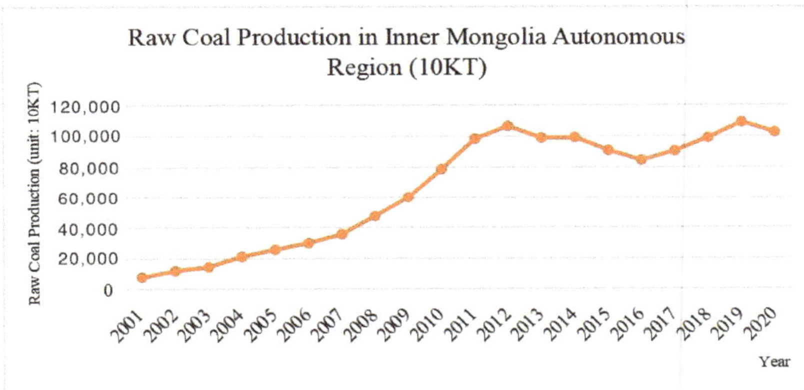

Figure 3. 2001–2020 Raw coal production in Inner Mongolia Autonomous Region.

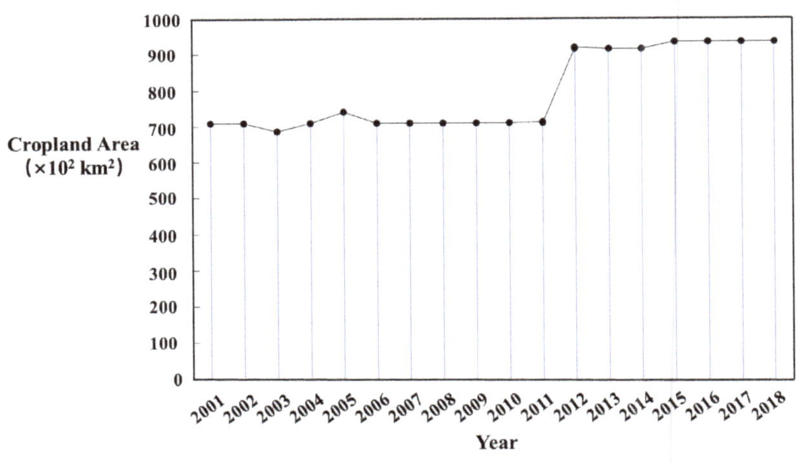

Figure 4. Changes in cropland area from 2001 to 2018 in the Inner Mongolia Autonomous Region.

4.2.2. Geographical Detector Model-Based Analysis of Natural and Geographic Drivers

To deeply explore the driving mechanisms of land use change in the open-pit coal mining area, the GDM was used to analyze the q-value and p-value of the open-pit coal mine, thereby determining the strength of the driving force. The results are shown in Table 3. The p-value of the eight spatial driving factors are all below 0.001, and the p-values of the distance from the urban road and the distance from the rural road are 0.06 and

0.81, respectively; the correlation between land use change in the open-pit coal mining area and the distance from urban roads and rural roads was weak. The distance from the national highway most significantly explained the land use change in the open-pit coal mining area, with a q-value of 0.19. The second most important factors were distance from the railway, the average annual temperature, average annual precipitation, distance from the county road and the elevation, and the q-values are 0.18, 0.16, 0.138, 0.12, and 0.12, respectively (Table 2).

Table 3. q-values of land use changes in open-pit coal mines.

Detection Type	x_1	x_2	x_3	x_4	x_5	x_6	x_7	x_8	x_9	x_{10}
q value	0.02	0.12	0.19	0.12	0.14	0.08	0.18	0.16	0.138	0.01
p value	0.06	≤0.001	≤0.001	≤0.001	≤0.001	≤0.001	≤0.001	≤0.001	≤0.001	0.81

Note: x_1 represents the distance from the city road; x_2 represents the elevation; x_3 represents the distance from the national highway; x_4 represents the average annual precipitation; x_5 represents the distance from the provincial road; x_6 represents the slope; x_7 represents the distance from the railway; x_8 represents the annual average temperature; x_9 represents the distance from the county road, and x_{10} represents the distance from the township road.

By analyzing the driving factors of land use change in open-pit coal mines, we found that the most important factor for the spatial change in land use in open-pit coal mines is traffic location conditions, followed by climatic conditions and topographical conditions. Human activities have an important impact on land use changes and landscape patterns in mining areas. However, in recent years, the government has attached great importance to the monitoring and evaluation of the environmental impacts of mining, with the development of new technologies and optimized mining planning and design. Such efforts have resulted in the alleviation of the environmental destruction via open-pit coal mining, also showing that human factors can play a role in environmental protection through policy formulation and publicity [61]. In view of the land problem of open-pit coal mines, open-pit combined mining can be carried out in conditional mining areas, or the coordinated land-saving technology can be advocated, and the production management in the pre-mining planning stage, mining disturbance stage and layout recovery stage of the mining area should be strictly controlled to reduce the damage to the original landscape.

5. Discussion

5.1. Uncertainties

Land use classification based on remote sensing data is afflicted with certain errors due to various reasons, such as differences in evaluation accuracy. Therefore, in future research, the combination of remote sensing, RTK, UAV, and 3D laser scanning technology should be strengthened to improve the accuracy and quality of data extraction. This study conducted multi-party comparisons and on-the-spot investigations and tests to repeatedly demonstrate the compatibility of classification and calculation results with local conditions, with the aim to minimize the degree of error. This can ensure that the calculation results are credible and in line with natural, economic and social trends. At the same time, a variety of sampling methods (stratified sampling, probability statistics) can be explored to provide a test paradigm for the future research accuracy of remote sensing estimation of ecological assets in open-pit coal mining areas and to offer a technical basis for the formulation of ecological restoration goals in mining areas [62].

5.2. Comparison of the Mining Activities

Coal mining has led to major changes in land use in mining areas, and the ecological environment of mining areas has been affected and destroyed. Scholars from China and around the world have conducted numerous studies on land use identification, land space planning and reclamation, and ecosystem services in mining areas. For example, He et al. [16] proposed an improved model for identifying coal mine areas, which can monitor coal mining conditions in the mining area at any time. Gao et al. [15] studied the conflict of land use in the production-living ecological space of large-scale open-pit coal mines and proposed spatial planning optimization and land reclamation measures. Bian et al. [63] analyzed the change in ecosystem service value and characteristics on the basis of analyzing land use change in the mining area. With the rapid economic and social development, the irreplaceability of resource extraction will continue for a certain period [64]. These authors determined the impact of the mining and reclamation of metal mines, oil and gas fields, and coal bed methane and other mineral resources on land use and ecological effects and compared the evolution of different types of mineral mining on land use and landscape patterns. This paper also explores how the mining of different minerals affects the ecosystem, providing a scientific reference for the formulation of ecological restoration policies in mining areas. Future studies on metal mines (especially related to pollution), oil and gas fields, and coal bed methane (occupying large amounts of land) can provide a powerful reference for the coordinated development of the regional economy and the transformation of resource-exhausted cities.

5.3. Policy Significance

This paper studies the dynamic evolution of land use in open-pit coal mines from 2001 to 2020. The time span is large, reflecting the impact of open-pit coal mining on the structure and function of land use. However, some mining areas have not been fully exploited, especially some open-pit coal mines that were issued licenses and put into operation after 2017 and 2018. Due to the progress of the corresponding procedures in the mining areas, coal price adjustments in recent years, and even the impact of the epidemic, the mining progress has not been carried out as scheduled. Therefore, the life cycle scale research based on open-pit coal mining has certain limitations. Mineral resource development policies have a significant impact on land use changes in mining areas, and unified planning and management should be carried out in the following three stages: pre-mining planning, mining disturbance, and post-mining recovery. In future research, the entire life cycle of open-pit coal mines from the infrastructure construction period, mining period, reclamation period, stabilization (underground mining) period to the management and protection period can be explored. Carrying out comparative research at different scales, such as regional and mining site scales, can provide technical support for the development and use of mining areas and the formulation of ecological restoration policies in the later stage. Even after the cessation of mining activities, land resource degradation will still occur. In this sense, carrying out a comprehensive renovation of the whole life cycle of the mining area and improving the mechanism of mining environment compensation can be applied to achieve efficient and rational use of mining area land and to protect ecological integrity. We suggest that the coal mining subsidence areas and abandoned sand pits should be subjected to slope cutting and platform building, land leveling, and vegetation restoration, and measures such as the establishment of comprehensive enclosures and the sowing of grass seeds should be taken to restore the ecological environment of the mining area. At the same time, strengthening the management and protection work in the later stage and promoting the follow-up survival guarantee measures for shrubs, grass, shelter forests, and seedlings are important steps. In this context, it is crucial to investigate the environmental and geographical characteristics of the mining area and analyze the specific issues.

6. Conclusions

Based on remote sensing images, this study used the land use dynamics analysis and geographic detector model to explore the temporal evolution trend and driving factors of land use dynamics in typical open-pit coal mining areas in Inner Mongolia. In particular, we analyzed the dynamic change process of land use in typical open-pit coal mines from 2001 to 2020, identified the reasons for the spatial changes in land use, and put forward policy recommendations for the optimization of land mining and reclamation in mining areas. From 2001 to 2020, grassland, mining land, cropland, and residential/industrial square land dynamics significantly fluctuated the most, whereas the areas of forest, water body and unused land remained relatively stable. Mining activities and land reclamation were the main reasons for land use changes in the study area. The land use in the open-pit coal mining area is greatly affected by mining factors. From the perspective of spatial variation, the most important driving factor is the traffic location condition, followed by the climatic and topographical conditions. Land reclamation and ecological restoration in mining areas have a positive impact on land use change.

Multi-mineral, multi-scale, and long-term comprehensive studies on mining areas need to be performed in the future. Strengthening the comprehensive analysis of various methods, performing real dynamic simulation, and revealing the characteristics and internal mechanisms of the land use changes in the past can provide the theoretical foundation for future land use changes. At the same time, emphasis should be placed on strengthening the research on soil, vegetation restoration, and reconstruction methods in mining areas. Land reclamation and ecological reconstruction in mining areas should be guided by new technologies to promote the development of social, economic, and environmental benefits. By focusing on the improvement of ecological quality and ecological economic construction, measures such as land consolidation, forest restoration on abandoned mining land, and conservation forest planting in water resource areas can promote the comprehensive management of water and mining land ecosystems and improve regional ecosystem functions. Appropriate vegetation allocation modes should be selected for the configuration, planting, management, and protection of plants at different site types.

Author Contributions: L.Z.: Methodology, Investigation, Resources, Data curation and Writing—original draft; Z.H.: Conceptualization, Writing—review and editing. D.Y., H.L., B.L., H.G., C.C., J.L., Y.Z. and S.L.: Data curation, Writing—original draft, Investigation and Validation. All authors have read and agreed to the published version of the manuscript.

Funding: This research was funded by National Key Research and Development Program (Grant No.2020YFC1806505).

Institutional Review Board Statement: Not applicable.

Informed Consent Statement: Not applicable.

Data Availability Statement: All relevant data sets in this study are described in the manuscript.

Conflicts of Interest: The authors declare no conflict of interest.

References

1. Dai, G.S.; Ulgiati, S.; Zhang, Y.S.; Yu, B.H.; Kang, M.Y.; Jin, Y.; Dong, X.B.; Zhang, X.S. The false promises of coal exploitation: How mining affects herdsmen well-being in the grassland ecosystems of Inner Mongolia. *Energy Policy* **2014**, *67*, 146–153. [CrossRef]
2. Goncalves, L.C.; Sebastiao, P.; Souto, N.; Correia, A. 5G mobile challenges: A feasibility study on achieving carbon neutrality. In Proceedings of the International Conference on Telecommunications, Thessaloniki, Greece, 16–18 May 2016; pp. 1–5.
3. Li, X.; Gao, J.; Zhang, J.; Wang, R.; Jin, L.; Zhou, H. Adaptive strategies to overcome challenges in vegetation restoration to coalmine wasteland in a frigid alpine setting. *Catena* **2019**, *182*, 104142. [CrossRef]
4. Areendran, G.; Rao, P.; Raj, K.; Mazumdar, S.; Puri, K. Land use/land cover change dynamics analysis in mining areas of Singrauli district in Madhya Pradesh, India. *Trop. Ecol.* **2013**, *54*, 239–250.
5. Bandyopadhyay, S.; Maiti, S.K. Application of statistical and machine learning approach for prediction of soil quality index formulated to evaluate trajectory of ecosystem recovery in coal mine degraded land. *Ecol. Eng.* **2021**, *170*, 106351. [CrossRef]
6. Rademeyer, M.C. Investigating the outcome for South African coal supply to the domestic market when faced with declining demand for exported coal. *Miner. Econ.* **2021**, *34*, 441–453. [CrossRef]
7. Thakur, T.K.; Dutta, J.; Upadhyay, P.; Patel, D.K.; Thakur, A.; Kumar, M.; Kumar, A. Assessment of land degradation and restoration in coal mines of central India: A time series analysis. *Ecol. Eng.* **2022**, *175*, 106493. [CrossRef]
8. Garai, D.; Narayana, A.C. Land use/land cover changes in the mining area of Godavari coal fields of southern India. *Egypt. J. Remote Sens. Space Sci.* **2018**, *21*, 375–381. [CrossRef]
9. Huang, Y. Drivers of rising global energy demand: The importance of spatial lag and error dependence. *Energy* **2014**, *76*, 254–263. [CrossRef]
10. Li, Q. The view of technological innovation in coal industry under the vision of carbon neutralization. *Int. J. Coal Sci. Technol.* **2021**, *8*, 1197–1207. [CrossRef]

11. Bell, F.G.; Bullock, S.; Hälbich, T.; Lindsay, P. Environmental impacts associated with an abandoned mine in the Witbank Coalfield, South Africa. *Int. J. Coal Geol.* **2001**, *45*, 195–216. [CrossRef]
12. Kumar, A.; Subrahmanyam, G.; Mondal, R.; Cabral-Pinto, M.M.S.; Shabnam, A.A.; Jigyasu, D.K.; Malyan, S.K.; Fagodiya, R.K.; Khan, S.A.; Kumar, A.; et al. Bio-remediation approaches for alleviation of cadmium contamination in natural resources. *Chemosphere* **2021**, *268*, 128855. [CrossRef]
13. Zhang, M.; Wang, J.; Feng, Y. Temporal and spatial change of land use in a large-scale opencast coal mine area: A complex network approach. *Land Use Policy* **2019**, *86*, 375–386. [CrossRef]
14. Li, S.; Wang, J.; Zhang, M.; Tang, Q. Characterizing and attributing the vegetation coverage changes in North Shanxi coal base of China from 1987 to 2020. *Resour. Policy* **2021**, *74*, 102331. [CrossRef]
15. Gao, Y.; Wang, J.; Zhang, M.; Li, S. Measurement and prediction of land use conflict in an opencast mining area. *Resour. Policy* **2021**, *71*, 101999. [CrossRef]
16. He, D.; Le, B.T.; Xiao, D.; Mao, Y.; Shan, F.; Ha, T.T.L. Coal mine area monitoring method by machine learning and multispectral remote sensing images. *Infrared Phys. Technol.* **2019**, *103*, 103070. [CrossRef]
17. Chen, J.; Li, K.; Chang, K.-J.; Sofia, G.; Tarolli, P. Open-pit mining geomorphic feature characterisation. *Int. J. Appl. Earth Obs. Geoinf.* **2015**, *42*, 76–86. [CrossRef]
18. Loupasakis, C. Contradictive mining–induced geocatastrophic events at open pit coal mines: The case of Amintaio coal mine, West Macedonia, Greece. *Arab. J. Geosci.* **2020**, *13*, 582. [CrossRef]
19. Amichev, B.Y.; Burger, J.A.; Rodrigue, J.A. Carbon sequestration by forests and soils on mined land in the Midwestern and Appalachian coalfields of the U.S. *For. Ecol. Manag.* **2008**, *256*, 1949–1959. [CrossRef]
20. Emerson, P.; Skousen, J.; Ziemkiewicz, P. Survival and growth of hardwoods in brown versus gray sandstone on a surface mine in West Virginia. *J. Environ. Qual.* **2009**, *38*, 1821–1829. [CrossRef]
21. Kumar, S.; Singh, A.K.; Ghosh, P. Distribution of soil organic carbon and glomalin related soil protein in reclaimed coal mine-land chronosequence under tropical condition. *Sci. Total Environ.* **2018**, *625*, 1341–1350. [CrossRef]
22. Yuan, Y.; Zhao, Z.; Niu, S.; Bai, Z. The reclaimed coal mine ecosystem diverges from the surrounding ecosystem and reaches a new self-sustaining state after 20-23 years of succession in the Loess Plateau area, China. *Sci. Total Environ.* **2020**, *727*, 138739. [CrossRef] [PubMed]
23. Zhao, Z.; Wang, L.; Bai, Z.; Pan, Z.; Wang, Y. Development of population structure and spatial distribution patterns of a restored forest during 17-year succession (1993–2010) in Pingshuo opencast mine spoil, China. *Environ. Monit. Assess.* **2015**, *187*, 431. [CrossRef] [PubMed]
24. Mukherjee, S.; Pahari, D.P. Underground and Opencast Coal Mining Methods in India: A Comparative Assessment. *Space Cult. India* **2019**, *7*, 39–55. [CrossRef]
25. Ahirwal, J.; Maiti, S.K. Assessment of soil properties of different land uses generated due to surface coal mining activities in tropical Sal (Shorea robusta) forest, India. *Catena* **2016**, *140*, 155–163. [CrossRef]
26. Wang, J.; Qin, Q.; Bai, Z. Characterizing the effects of opencast coal-mining and land reclamation on soil macropore distribution characteristics using 3D CT scanning. *Catena* **2018**, *171*, 212–221. [CrossRef]
27. Polasky, S.; Nelson, E.; Pennington, D.; Johnson, K.A. The Impact of Land-Use Change on Ecosystem Services, Biodiversity and Returns to Landowners: A Case Study in the State of Minnesota. *Environ. Resour. Econ.* **2010**, *48*, 219–242. [CrossRef]
28. Nelson, E.; Sander, H.; Hawthorne, P.; Conte, M.; Ennaanay, D.; Wolny, S.; Manson, S.; Polasky, S. Projecting global land-use change and its effect on ecosystem service provision and biodiversity with simple models. *PLoS ONE* **2010**, *5*, e14327. [CrossRef]
29. Wardrop, D.H.; Glasmeier, A.K.; Peterson-Smith, J.; Eckles, D.; Ingram, H.; Brooks, R.P. Wetland ecosystem services and coupled socioeconomic benefits through conservation practices in the Appalachian Region. *Ecol. Appl.* **2011**, *21*, S93–S115. [CrossRef]
30. Wu, Z.; Lei, S.; Qingqing, L.; Bian, Z.-F. Impacts of Large-Scale Open-Pit Coal Base on the Landscape Ecological Health of Semi-Arid Grasslands. *Remote Sens.* **2019**, *11*, 1820. [CrossRef]
31. Xia, M.; Dong, S.; Chen, Y.; Liu, H. Study on evolution of groundwater-lake system in typical prairie open-pit coal mine area. *Environ. Geochem. Health* **2021**, *43*, 4075–4087. [CrossRef]
32. Li, X.; Lei, S.; Liu, Y.; Chen, H.; Zhao, Y.; Gong, C.; Bian, Z.; Lu, X. Evaluation of Ecological Stability in Semi-Arid Open-Pit Coal Mining Area Based on Structure and Function Coupling during 2002–2017. *Remote Sens.* **2021**, *13*, 5040. [CrossRef]
33. Yang, Z.; Li, J.; Zipper, C.E.; Shen, Y.; Miao, H.; Donovan, P.F. Identification of the disturbance and trajectory types in mining areas using multitemporal remote sensing images. *Sci. Total Environ.* **2018**, *644*, 916–927. [CrossRef]
34. Tang, Q.; Wang, J.; Jing, Z.; Yan, Y.; Niu, H. Response of ecological vulnerability to land use change in a resource-based city, China. *Resour. Policy* **2021**, *74*, 102324. [CrossRef]
35. Shrestha, R.K.; Lal, R. Carbon and nitrogen pools in reclaimed land under forest and pasture ecosystems in Ohio, USA. *Geoderma* **2010**, *157*, 196–205. [CrossRef]
36. Zhao, Z.; Liu, F.; Xie, X.; Liu, X.; Tang, Z. Asymptotic stability of bidirectional associative memory neural networks with time-varying delays via delta operator approach. *Neurocomputing* **2013**, *117*, 40–46. [CrossRef]
37. Matschullat, J.; Borba, R.P.; Deschamps, E.; Figueiredo, B.R.; Gabrio, T.; Schwenk, M. Human and environmental contamination in the Iron Quadrangle, Brazil. *Appl. Geochem.* **2000**, *15*, 181–190. [CrossRef]
38. Brink, A.B.; Eva, H.D. Monitoring 25 years of land cover change dynamics in Africa: A sample based remote sensing approach. *Appl. Geogr.* **2009**, *29*, 501–512. [CrossRef]

39. Wang, L.; Yang, L.; Wang, W.; Chen, B.; Sun, X. Monitoring Mining Activities Using Sentinel-1A InSAR Coherence in Open-Pit Coal Mines. *Remote Sens.* **2021**, *13*, 4485. [CrossRef]
40. Ma, B.; Yang, X.; Yu, Y.; Shu, Y.; Che, D. Investigation of Vegetation Changes in Different Mining Areas in Liaoning Province, China, Using Multisource Remote Sensing Data. *Remote Sens.* **2021**, *13*, 5168. [CrossRef]
41. Raval, S.; Shamsoddini, A. A monitoring framework for land use around kaolin mining areas through Landsat TM images. *Earth Sci. Inform.* **2014**, *7*, 153–163. [CrossRef]
42. Sonter, L.J.; Moran, C.J.; Barrett, D.J.; Soares-Filho, B.S. Processes of land use change in mining regions. *J. Clean. Prod.* **2014**, *84*, 494–501. [CrossRef]
43. Wohlfart, C.; Mack, B.; Liu, G.; Kuenzer, C. Multi-faceted land cover and land use change analyses in the Yellow River Basin based on dense Landsat time series: Exemplary analysis in mining, agriculture, forest, and urban areas. *Appl. Geogr.* **2017**, *85*, 73–88. [CrossRef]
44. Padmanaban, R.; Bhowmik, A.; Cabral, P. A Remote Sensing Approach to Environmental Monitoring in a Reclaimed Mine Area. *ISPRS Int. J. Geo-Inf.* **2017**, *6*, 401. [CrossRef]
45. Gorelick, N.; Hancher, M.; Dixon, M.; Ilyushchenko, S.; Thau, D.; Moore, R. Google Earth Engine: Planetary-scale geospatial analysis for everyone. *Remote Sens. Environ.* **2017**, *202*, 18–27. [CrossRef]
46. Hamud, A.M.; Shafri, H.Z.M.; Shaharum, N.S.N. Monitoring Urban Expansion And Land Use/Land Cover Changes In Banadir, Somalia Using Google Earth Engine (GEE). *IOP Conf. Ser. Earth Environ. Sci.* **2021**, *767*, 012041. [CrossRef]
47. Lin, L.; Hao, Z.; Post, C.J.; Mikhailova, E.A.; Yu, K.; Yang, L.; Liu, J. Monitoring Land Cover Change on a Rapidly Urbanizing Island Using Google Earth Engine. *Appl. Sci.* **2020**, *10*, 7336. [CrossRef]
48. Bar, S.; Parida, B.R.; Pandey, A.C. Landsat-8 and Sentinel-2 based Forest fire burn area mapping using machine learning algorithms on GEE cloud platform over Uttarakhand, Western Himalaya. *Remote Sens. Appl. Soc. Environ.* **2020**, *18*, 100324. [CrossRef]
49. Liang, J.; Xie, Y.; Sha, Z.; Zhou, A. Modeling urban growth sustainability in the cloud by augmenting Google Earth Engine (GEE). *Comput. Environ. Urban Syst.* **2020**, *84*, 101542. [CrossRef]
50. Zhou, B.; Okin, G.S.; Zhang, J. Leveraging Google Earth Engine (GEE) and machine learning algorithms to incorporate in situ measurement from different times for rangelands monitoring. *Remote Sens. Environ.* **2020**, *236*, 111521. [CrossRef]
51. Freeman, S. The Current Global Reality: Poverty and Income Inequality. *Semin. Pediatric Neurol.* **2018**, *27*, 1–9. [CrossRef]
52. Inner Mongolia Autonomous Region Statistics Bureau. Inner Mongolia Statistical Yearbook. Available online: http://tj.nmg.gov.cn/tjyw/jpsj/ (accessed on 1 August 2021).
53. Inner Mongolia Autonomous Region People's Government. Regulations of Inner Mongolia Autonomous Region on Mineral Resources Management. Available online: https://www.nmg.gov.cn/zwgk/zfgb/1999n_5236/199909/199909/t19990901_308952.html (accessed on 21 August 2021).
54. Inner Mongolia Autonomous Region People's Government. Statistical Bulletin of National Economic and Social Development of Inner Mongolia Autonomous Region. Available online: https://www.nmg.gov.cn/tjsj/sjfb/tjsj/tjgb/202202/t20220228_2010485.html (accessed on 30 August 2021).
55. Li, P.; Zuo, D.; Xu, Z.; Zhang, R.; Han, Y.; Sun, W.; Pang, B.; Ban, C.; Kan, G.; Yang, H. Dynamic changes of land use/cover and landscape pattern in a typical alpine river basin of the Qinghai-Tibet Plateau, China. *Land Degrad. Dev.* **2021**, *32*, 4327–4339. [CrossRef]
56. Kafy, A.-A.; Naim, M.N.H.; Subramanyam, G.; Faisal, A.-A.; Ahmed, N.U.; Rakib, A.A.; Kona, M.A.; Sattar, G.S. Cellular Automata approach in dynamic modelling of land cover changes using RapidEye images in Dhaka, Bangladesh. *Environ. Chall.* **2021**, *4*, 100084. [CrossRef]
57. Wang, Z.; Li, X.; Mao, Y.; Li, L.; Wang, X.; Lin, Q. Dynamic simulation of land use change and assessment of carbon storage based on climate change scenarios at the city level: A case study of Bortala, China. *Ecol. Indic.* **2022**, *134*, 108499. [CrossRef]
58. Duan, Q.; Tan, M. Using a geographical detector to identify the key factors that influence urban forest spatial differences within China. *Urban For. Urban Green.* **2020**, *49*, 126623. [CrossRef]
59. Deslatte, A.; Szmigiel-Rawska, K.; Tavares, A.F.; Ślawska, J.; Karsznia, I.; Łukomska, J. Land use institutions and social-ecological systems: A spatial analysis of local landscape changes in Poland. *Land Use Policy* **2022**, *114*, 105937. [CrossRef]
60. Liu, J.; Wang, J.; Zhai, T.; Li, Z.; Huang, L.; Yuan, S. Gradient characteristics of China's land use patterns and identification of the east-west natural-socio-economic transitional zone for national spatial planning. *Land Use Policy* **2021**, *109*, 105671. [CrossRef]
61. Shi, X.; Zhou, F.; Wang, Z. Research on optimization of ecological service function and planning control of land resources planning based on ecological protection and restoration. *Environ. Technol. Innov.* **2021**, *24*, 101904. [CrossRef]
62. Shao, G.; Wu, J. On the accuracy of landscape pattern analysis using remote sensing data. *Landsc. Ecol.* **2008**, *23*, 505–511. [CrossRef]
63. Bian, Z.; Lu, Q. Ecological effects analysis of land use change in coal mining area based on ecosystem service valuing: A case study in Jiawang. *Environ. Earth Sci.* **2012**, *68*, 1619–1630. [CrossRef]
64. Chen, Z.; An, C.; Tan, Q.; Tian, X.; Li, G.; Zhou, Y. Spatiotemporal analysis of land use pattern and stream water quality in southern Alberta, Canada. *J. Contam. Hydrol.* **2021**, *242*, 103852. [CrossRef]

International Journal of
Environmental Research and Public Health

MDPI

Article

Construction and Optimization Strategy of an Ecological Network in Mountainous Areas: A Case Study in Southwestern Hubei Province, China

Qian Zuo [1,2], Yong Zhou [1,2,*] and Jingyi Liu [1,2]

1 Key Laboratory for Geographical Process Analysis & Simulation of Hubei Province, Central China Normal University, Wuhan 430079, China
2 College of Urban and Environmental Sciences, Central China Normal University, Wuhan 430079, China
* Correspondence: yzhou@ccnu.edu.cn

Abstract: High-intensity urban development and economic exploitation have led to the fragmentation and isolation of regional habitat patches, and biodiversity is under serious threat. Scientific identification and effective optimization of ecological networks are essential for maintaining and restoring regional ecosystem connectivity and guiding sustainable socio-economic development. Taking the mountainous areas of southwest Hubei Province (MASHP) in central China as an example, this study first developed a new integrated approach to identify ecological sources based on a quantitative assessment of ecosystem services and the morphological spatial pattern analysis (MSPA) method; it then used the Linkage Mapper tool to extract ecological corridors, applied the principle of hydrological analysis to identify ecological nodes, evaluated each ecological element to quantify its importance, and finally constructed the ecological network and further proposed some optimization countermeasures. The results show that the ecological network in the MASHP is dominated by ecological resources composed of forestland. Connectivity in the central region is significantly better than in other regions, including 49 ecological sources with an area of 3837.92 km^2, 125 ecological corridors with a total length of 2014.61 km, and 46 ecological nodes. According to the spatial distribution of crucial ecological landscape elements, a complete and systematic ecological framework of "two verticals, three belts, three groups, and multiple nodes" was proposed. The internal optimization of the ecological network in mountainous areas should focus on improving ecological flow, and strategies such as enhancing the internal connectivity of ecosystems, unblocking ecological corridors, and dividing ecological functional zones can be adopted. Based on the above analyses, this study also made recommendations for ecological protection and development and construction planning in mountainous areas. This study can provide realistic paths and scientific guidelines for ecological security and high-quality development in the MASHP, and it can also have implications for the construction of ecological networks and comprehensive ecological management in other mountainous areas.

Keywords: ecological network; ecosystem services; morphological spatial pattern analysis; landscape connectivity; ecological function zones

Citation: Zuo, Q.; Zhou, Y.; Liu, J. Construction and Optimization Strategy of an Ecological Network in Mountainous Areas: A Case Study in Southwestern Hubei Province, China. *Int. J. Environ. Res. Public Health* **2022**, *19*, 9582. https://doi.org/10.3390/ijerph19159582

Academic Editors: Wei Song and Hualin Xie

Received: 5 July 2022
Accepted: 1 August 2022
Published: 4 August 2022

1. Introduction

The continuous socio-economic development and the increasing demand for urban construction land have resulted in the encroachment of a large amount of ecological land by construction land and the constant compression of ecological space, which has divided the originally continuous natural habitat into a mosaic of mixed patches with a high degree of fragmentation and has seriously affected the functioning of the ecosystem [1–3]. This has led to the destruction of ecological processes and serious degradation of the ecological environment, which to a certain extent has intensified the fragmentation of the regional landscape; meanwhile, the connectivity between habitat patches is decreasing, the habitat quality of species continues to decline, and the circulation between different ecological streams is hindered, all of which have posed serious challenges to the sustainable development of the region [4–6]. As a result, the concept of ecological security, which is based on maintaining the integrity, health, and sustainability of ecological processes, has gradually received widespread attention from governments and academics, and the theory has provided strong support for governments to seek a balance between regional ecological conservation and sustainable economic development [7,8]. After decades of development, ecological security has become one of the essential concepts to alleviate the contradiction between natural ecological conservation and human social evolution [9,10]. More and more scholars consider constructing an ecological network as a suitable method to solve the above thorny problems [11–14]. Ecological networks are derived from the theories and methods of landscape ecology and further enriched and developed on this basis. They specifically refer to a potential spatial pattern of the ecosystem to maintain the normal biodiversity, ecosystem health, and sustainable supply of ecosystem services in a specific region [15,16], which can characterize the integrity and health of the current natural ecosystem [17] and enhance the connectivity between different habitats through "point-line-surface", to effectively resist the lasting adverse effects of habitat fragmentation on biodiversity [18,19].

At present, ecological network research is becoming more and more mature, and the mainstream paradigm of "sources identification-resistance surface construction-corridors extraction and nodes discrimination" has been formed [20–22]. As critical ecological patches, ecological sources can promote ecological processes and maintain the integrity and stability of the ecosystem [23], and they are identified mainly by the direct identification method, minimum area threshold method, and ecosystem service overlay analysis method. The first one is the direct identification method, which refers to the method of identifying specific areas of the research object, such as the green areas, forest parks, scenic spots and nature reserves with large concentrated and continuous areas and important recreational functions as ecological sources [24,25]. This method is simple and straightforward, but the screening criteria are relatively single and subjective. The second is the minimum area threshold method. For the ecological sources that have been roughly selected, data reduction is directly carried out according to previous experience, and the operation process is simplified. The choice of some thresholds remains controversial [26,27]. The third method is the ecosystem service overlay analysis, which evaluates the sensitivity and importance of ecosystems based on several important or typical ecosystem service capabilities, and uses them as the basis for identifying ecological sources. The method is widely used because of the comprehensive and relatively scientific nature of its quantitative indicators, but these ecosystem service functions are mainly related to human well-being and are not closely related to species migration and dispersal processes. Integrated approaches to identifying ecological sources have become a new research trend. The Integrated Valuation of Ecosystem Services and Tradeoffs (InVEST) model is a powerful tool that can be used to reduce subjectivity to quantitatively assess ecosystem services [28]. Morphological spatial pattern analysis (MSPA) is based on graphical principles and combines a raster algorithm to identify ecological sources at the pixel level scale [29,30], emphasizing structural connectivity and thereby increasing the scientific validity of ecological source selection [31]. Based on the goal of comprehensively enhancing ecosystem services and landscape connectivity, the combined use of the above two methods will allow for a more scientific and objective identification of ecological sources, better reflecting biological conservation needs and the suitability of species habitats. The creation of ecological resistance surfaces is another core element of ecological network establishment. The evaluation framework has developed from simply assigning values to land use types to considering human interference activities, and has become multi-faceted and more objective. Ecological corridors, as bridges between ecological patches, are linear or banded landscape elements different from the substrate on both sides and channels for exchanging materials, energy, and information between ecological patches. The minimum cumulative resistance (MCR) model proposed by a Dutch ecologist [32] has been optimized by Yu [33] and has been widely used in the extraction of ecological corridors. The model reflects the ease of "source" dispersal by the magnitude of cumulative resistance, which can effectively reflect the potential possibility and trend of species and energy flow and dispersal in the region. Circuit theory, which applies the random travel of electrons in a circuit to simulate the diffusion of species or ecosystem services within a region [34], can reflect the relative importance of ecosystem services in patches and corridors and identify key nodes in the flow process, and is also widely used in related studies [35–37]. Others have used the Linkage Mapper tool integrating the least-cost path (LCP) method and Circuit theory to extract ecological corridors [38,39]. The Linkage Mapper tool first identifies adjacent ecological sources, then constructs a network between ecological sources by adjacency and distance from each other, then calculates the cost-weighted distances and least-cost paths between different source locations, and finally combines the least-cost corridor links that need to be connected into a map [40]. However, any resulting corridors should not only be a conceptual connecting path, but should also have width. Effective corridors are those that can effectively link ecological space of wildlife with land use policies [41]. The corridor width setting level should be combined with the movement characteristics of the research object in order to achieve a balance between the purpose of protecting the ecological environment and avoiding a large conflict with economic development [17,42,43]. Ecological nodes are critical and strategic nodes that occupy important locations on ecological corridors, generally located at the weakest part of the corridor [44], and they play an important role in improving the connectivity of the existing ecological network and promoting healthy ecosystem operation [45,46].

In general, the current studies on ecological network mainly focus on different scales such as provincial areas, urban clusters, cities, and counties [47,48] and typical objects such as mining areas, plateaus, plains, and watersheds [49–51], while mountainous areas are less studied. Nearly 65% of China's land area is mountainous [52]. Mountainous areas have complex topography, which affects the transfer of materials and energy and constitutes a unique ecological environment with important ecological security maintenance functions but a fragile ecological substrate, and is thus a representative area for spatial control studies. The mountainous areas of Southwestern Hubei Province (MASHP) are located in the hinterland of Hubei Province. It is a superimposed zone of Hubei Province and even China's key ecological function zone, ecological water conservation zone, biological resource enrichment zone, key link zone between poverty alleviation and rural revitalization, and key ecological tourism development zone. The Three Gorges project is located in Sandouping Town, Yichang City, in the MASHP. The MASHP is a typical mountainous area with important ecological location, fragile ecological environment, and intertwined ecological problems and poverty problems, and the contradiction between economic and social development and ecological environmental protection is increasingly prominent [53]. The Nineteenth National Congress of China Report (2017) clearly states that the quality and stability of ecosystems can be effectively improved by building ecological corridors and biodiversity conservation networks. However, with the comprehensive promotion of ecological civilization in China, under the constraint of laws and regulations such as the Yangtze River Protection Law of the People's Republic of China and the implementation of strategies such as "to step up conservation of the Yangtze River and stop its over development" and "Green Enshi" [54], the governments in the MASHP have made substantial efforts to rectify the ecological problems and promote the implementation of joint prevention and control mechanisms and co-management and co-construction models between various regions, and the ecological environmental protection has achieved phased results. However, the contradiction between ecological environment and economic development is still prominent, and the situation of ecological and environmental security is still serious. Therefore, it is typical and urgent to study the construction and optimization of ecological networks and their protection and restoration in the MASHP.

The MASHP in central China was selected as the study area in this study. The specific research objectives are as follows: (i) based on the integration of ecosystem service hotspots and the MSPA method, the minimum area threshold is set scientifically and combined with landscape connectivity analysis to propose a new comprehensive method for identifying ecological sources; (ii) the ecological corridors are extracted based on the least-cost paths, and the intersection of the maximum paths and the minimum paths are identified as the ecological nodes; (iii) ecological sources, ecological corridors and ecological nodes of different importance levels together constitute the final ecological network; (iv) the internal optimization measures of the ecological network are proposed. The research results can provide the scientific basis for ecosystem restoration, ecological security, territorial spatial planning and high-quality economic development in the MASHP and provide reference for the construction of ecological network and comprehensive ecological restoration and management in other mountainous areas in China.

2. Materials and Methods

2.1. Study Area

The MASHP is located in the southwest hinterland of Hubei Province, China, east of the Yunnan–Guizhou Plateau and west of the Jianghan Plain, at 108°22′ E–112°05′ E, 29°08′ N–31°35′ N, including Yichang City (with 5 districts, 3 cities and 5 counties) and Enshi Tujia and Miao Autonomous Prefecture (Enshi Prefecture, with 2 cities and 6 counties) (Figure 1). The MASHP covers a total area of about 45,113 km^2 and a population of 7,921,200 (in 2020). The mountainous area is undulating and dominated by high and mid-alpine mountains, with an average elevation of over 900 m and a maximum of over 2900 m. The MASHP has a humid subtropical monsoon climate with abundant rainfall. The mountains are crisscrossed by rivers and deep river valleys, with the Yangtze River and Qingjiang as the main water systems. The area is rich in forest resources; vegetation types are coniferous forest, evergreen broad-leaved forest and shrub; rare plant and animal resources are very rich. The MASHP is a very important ecological refuge and gene pool for plants and animals in Hubei Province and even the whole central region of China. In recent years, with the rapid socio-economic development, the expansion of construction, green land crowding, tourism development and large-scale rural deforestation, dominant agricultural development, as well as the construction of dams and other water conservancy facilities in the MASHP, have caused more serious soil erosion, environmental pollution and ecological degradation, and ecological and environmental problems are increasingly prominent.

2.2. Data Sources and Processing

The data information used in this study is shown in Table 1, which mainly includes land use data, digital elevation model (DEM), traffic and river data, meteorological data, soil data and normalized difference vegetation index (NDVI), and the above data were processed to obtain the indirect data needed to be used in this study. ArcGIS 10.2 software (ESRI,2013,10.2) (Environmental Systems Research Institute, Redlands, CA, USA) was used to project, pre-process, map and analyze the data in this study. All data were unified in the Albers coordinate system, and the grid of the raster data was unified at 30 m × 30 m. Among them, the land use data were classified into 6 primary types and 25 secondary types according to the land use/land cover changes classification standard, which are defined in Table S1 in Supplementary Materials.

2.3. Method

Certain flow paths (corridors), critical junctions (nodes) and localities (sources) or their spatial combinations together constitute the ecological network structure, which is divided into four steps: (1) identifying the ecological sources; (2) constructing the resistance surface; (3) extracting the ecological corridors and ecological nodes; (4) establishing the ecological network and optimization. The research framework is shown in Figure 2.

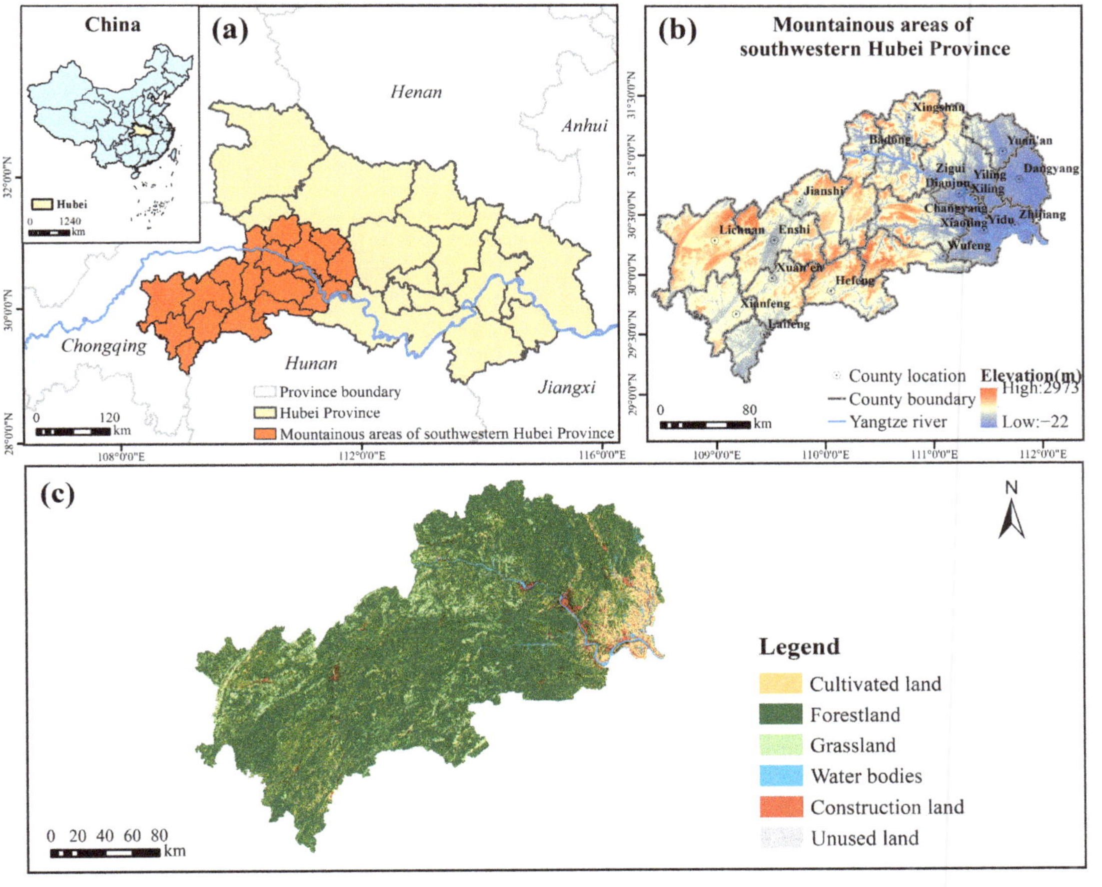

Figure 1. Location of the study area: (**a**) location in Hubei Province, China; (**b**) digital elevation map (DEM); (**c**) land cover.

Table 1. Data information table.

Data Types	Format	Data Sources
Land use data	Grids at 30 m resolution in 2020	Resource and Environment Science and Data Center (https://www.resdc.c, (accessed on 28 December 2021))
Digital elevation model (DEM)	Grids at 30 m resolution	Geospatial Data Cloud site (http://www.gscloud.cn (accessed on 30 March 2021))
Traffic road and river data	Lines in 2020	National Catalogue Service for Geographic Information (https://www.webmap.cn); OpenStreetMap (https://www.openstreetmap.org (accessed on 30 March 2022))
Meteorological data	Grids at 1 km resolution in 2020	National Meteorological Information Center (https://data.cma.cn/ (accessed on 30 March 2022))
Soil attributes	Grids at 1 km resolution	Harmonized World Soil Database v 1.2 from Food and Agriculture Organization of the United Nations (https://www.fao.org/soils-portal/soil-survey/soil-maps-and-databases/harmonized-world-soil-database-v12/en (accessed on 30 March 2022))
Normalized difference vegetation index (NDVI)	Grids at 1 km resolution in 2019	Resource and Environment Science and Data Center (https://www.resdc.cn (accessed on 30 March 2022))

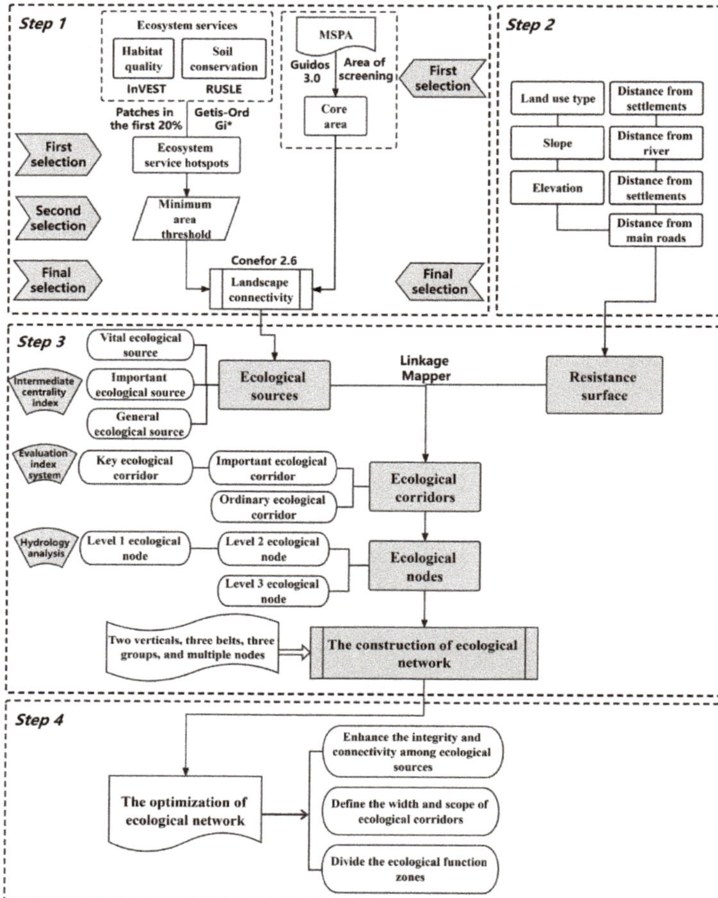

Figure 2. Framework of this study.

2.3.1. Identification of Ecological Sources

In this study, the final ecological sources were identified by integrating the assessment of two ecosystem service hotspots, namely habitat quality and soil conservation, with the MSPA method, setting the minimum area threshold scientifically and analyzing the landscape connectivity. The specific steps are:

(1) Assessment of ecosystem services

For the development positioning of the MASHP as an important ecological barrier in the middle and upper reaches of Yangtze River, as well as its ecological background situation and related regional management objectives (mainly for soil erosion), two ecosystem services, habitat quality and soil conservation, were selected for quantitative evaluation, and the top 20% patches of each ecosystem service were selected and taken as the first part of alternative ecological sources after merging.

Habitat quality indicates the potential of the ecosystem to provide an environment for species to survive and thrive, and it is positively correlated with biodiversity, thus characterizing the richness of biodiversity to some extent. In the Habitat Quality module of the InVEST model (https://naturalcapitalproject.stanford.edu/software/invest, (accessed on 31 March 2022)), the habitat quality in the MASHP is quantitatively assessed in terms of external threat factors and habitat sensitivity [55–57]. This study was based on reference studies of similar areas [58,59], determined the threat factors of habitats, the impact range and weight of each threat factor, the suitability of different types of habitats and their relative sensitivity to different threat factors (Tables S2 and S3). The closer the habitat quality is to 1, the higher the level of habitat quality is indicated, and the calculation formula is as follows:

$$Q_{xj} = H_j \left[- \left(\frac{D_{xj}^z}{D_{xj}^z + K^z} \right) \right] \quad (1)$$

where Q_{xj} is habitat quality of raster x in land use type j, H_j is the habitat suitability of land use type j, D_{xj} is the habitat degradation of raster x in land use type j, K is the half-saturation constant, which is generally taken as half of the maximum value of D_{xj}, and Z is the normalization constant, which is the default parameter of the system and usually takes the value of 2.5.

The modified universal soil loss equation (RUSLE) [60] was used to estimate the potential soil loss without considering vegetation cover factors and soil and water conservation measures and the actual soil loss considering the factors mentioned above. The difference between the two was taken as the soil conservation amount of the ecosystem [61,62]. The calculation formula is as follows:

$$SC = R \times K \times L \times S \times (1 - C \times P) \quad (2)$$

where SC is the amount of soil conservation ($t \cdot hm^{-2} \cdot a^{-1}$), R is the rainfall erosion force factor ($MJ \cdot mm \cdot hm^{-2} \cdot h^{-1} \cdot a^{-1}$), K is the soil erodibility factor ($t \cdot h \cdot MJ^{-1} \cdot mm^{-1}$), L and S are the topography factors (L is the factor of slope length, S is the slope factor), C is the surface vegetation cover factor, and P is the soil conservation measure factor.

(2) Application of hotspots analysis

The spatial distribution of cold and hot areas for ecosystem services reflects the strength of ecosystem services. The hotspot analysis method is based on the Getis-Ord Gi* statistical method [63], which analyzes the spatial clustering of high or low values of ecosystem services. Patches that are fragmented and do not have concentrated contiguity are continuously eliminated as the distance threshold increases, and close and connected patches are clustered to form larger patches. Based on the evaluation results of habitat quality and soil conservation, the top 20% patches of each ecosystem service were selected to select ecological service hot areas according to extreme confidence hot areas, and the first round of selection of the first part of alternative ecological sources was completed.

(3) MSPA

This study identified and extracted the core area in the MASHP based on the MSPA method as the second part of alternative ecological sources [30]. Considering the geographical conditions of the MASHP, forestland was set as foreground and other land use types were set as background. There is an obvious scale effect in the MSPA method, and the identification results are not consistent with different threshold values [64]. For example, if the threshold is too large, elements with smaller areas will disappear or be classified under other elements, and small core areas will be classified as isolated islands, etc. Therefore, the threshold value was set to 100 m × 100 m by considering the effects of the area, data, and scale effects of the study area. Finally, the data were binarized, rasterized, and analyzed in Guidos Toolbox 3.0 analysis software (https://forest.jrc.ec.europa.eu/en/activities/lpa/gtb/, (accessed on 31 March 2022)). The data were analyzed using the eight-neighborhood analysis method to obtain seven types of non-overlapping landscapes, namely, Core, Islet, Perforation, Edge, Loop, Bridge, Branch and Background (see Guidos Toolbox user guider for detailed meanings of landscape types and color symbols) [65]. Finally, core areas that were important for maintaining connectivity were extracted as landscape elements for subsequent analysis. In order to avoid fragmented patches from degrading the main function of the core area and generating redundant ecological corridors, the patches with an area of not less than 10 km^2 were selected for the subsequent patch connectivity calculation with reference to the studies on the threshold setting of relevant core area by MSPA [28,66]. The first round of selection of the second part of alternative ecological sources was completed at this time.

(4) Evaluation of landscape connectivity

Landscape connectivity reflects the degree to which the landscape facilitates or impedes ecological flows, has a positive impact on species richness and is key to maintaining ecosystem stability and integrity [67,68]. Based on the Conefor Sensinode 2.6 software (Jenness Enterprises, Flagstaff, AZ, USA) (http://www.conefor.org/coneforsensinode.html (accessed on 31 March 2022)), starting from the probability of connectivity (*PC*) and the integral index of connectivity (*IIC*) reflecting the connectivity of each patch to the landscape, the delta values for the *PC* index (*dPC*) and the Delta Values for the *IIC* index (*dIIC*) of alternative ecological source patches were calculated. The average value (*dI*) of the two was taken to obtain a more accurate degree of ecological patch importance [69]. A higher index means a higher degree of patch connection. The node files and distance files required for connectivity analysis in Conefor Sensinode 2.6 were generated by the plug-in for Conefor in ArcGIS 10.2 software and were calculated as follows:

$$PC = \left(\sum_{i=1}^{n} \sum_{j=1}^{n} a_i \times a_j \times P_{ij}^* \right) / A_L^2 \tag{3}$$

$$IIC = \sum_{i=1}^{n} \sum_{j=1}^{n} \left[\left(a_i a_j / \left(1 + nl_{ij} \right) \right) \right] / A_L^2 \tag{4}$$

$$dPC = (PC - PC_{remove}) / PC \times 100\% \tag{5}$$

$$dIIC = (IIC - IIC_{remove}) / IIC \times 100\% \tag{6}$$

$$dI = 0.5 \times dPC + 0.5 \times dIIC \tag{7}$$

where n is the total number of patches in the study area, a_i and a_j are the areas of patch i and patch j, respectively, P_{ij} is the maximum probability of species dispersal between patches I and j, A_L is the total landscape area in the study area, l_{ij} is the shortest path from patch I to patch j, PC_{remove} is the *PC* after removing an element in the landscape, and IIC_{remove} is the *IIC* after removing an element in the landscape.

(5) Final confirmation and optimization of ecological sources

The first part of alternative ecological sources should be of a specific size to ensure the stability of ecosystem service provision and the positive aggregation effect [70]. This study quantitatively determined the minimum area threshold of ecological sources by exploring the variation of the number of patches and the total area of patches with the selected minimum area threshold, and completed the second round of selection of the first part of alternative ecological sources. The two previously obtained parts of alternative ecological sources were combined after removing the small overlapping parts, and the *dI* values of the patches were calculated. Finally, the patches with *dI* greater than 2 were identified as the final ecological sources, and the final round of selection of the first and second parts of alternative ecological sources was completed at this time.

The intermediary centrality index ($0 < Q_i \leq 1$, the closer Q_i is to 1, the greater the importance) was calculated to define the role of ecological sources, and the results were classified qualitatively: $Q_i \geq 0.5$ is classified as vital ecological source; $0.2 < Q_i < 0.5$ is classified as important ecological source; $0 \leq Q_i \leq 0.2$ is classified as general ecological source; the formula is as follows:

$$Q_i = \frac{dPC_{connector}}{dPC} \tag{8}$$

In the equation, $dPC_{connector}$ quantifies the importance of patch i in maintaining the overall effective connection in the ecological network, and dPC quantifies the maximum flux through patch i in the complete landscape diffusion process.

Considering the increased urban expansion and demand for ecosystem services in the future, the current ecological resources may not be sufficient to meet the needs of future development, so the ecological sources need to have a wider radiative power for maintaining the ecological processes and natural succession within the sources and reducing the impact of anthropogenic disturbances in the external landscape [71]. Since the MASHP is rich in forestland resources, forestland with the strongest ecosystem services was selected as the main factor influencing the spread of ecological sources [72]. In this study, multiple ring buffer zones (100, 200, 300, 400, 500, 600, 700, 800, 900, and 1000 m) were established for ecological sources with poor landscape connectivity ($2 \leq dI < 3$), and the change in forestland growth rate in each buffer zone with the increase in buffer distance was analyzed. Finally, the optimal diffusion distance of these sources was obtained by integrating the distribution of other land use types.

2.3.2. Determination of Resistance Surface

When determining ecological resistance, referring to the experience of previous studies [46,58,73], a five-level system was used to define the resistance magnitude of a single factor, with higher scores implying greater species dispersal resistance. The AHP method was used to determine the resistance factor weights. The weights and coefficients of the resistance factors are shown in Table 2.

Table 2. Weights and coefficients of resistance factors.

Resistance Factor	Resistance Coefficient					Weight
	10	30	50	70	90	
Land use type	Forestland/Grassland	Water bodies	Cultivated land	Unused land	Construction land	0.5299
Slope (°)	<8	[8–20)	[20–30)	[30–40)	≥40	0.0636
Elevation (m)	<374	[374–755)	[755–1085)	[1085–1441)	≥1441	0.0636
Distance from river (km)	<2	[2–5)	[5–8)	[8–10)	≥10	0.1034
Distance from settlements (km)	≥25	[15–25)	[9–15)	[4–9)	<4	0.1273
Distance from main roads (km)	≥75	[55–75)	[35–55)	[15–35)	<15	0.1122

2.3.3. Extraction of Ecological Corridors and Ecological Nodes

In this study, the Linkage Mapper tool (https://linkagemapper.org/, (accessed on 5 April 2022).) in ArcGIS 10.2 software was used to extract ecological corridors. First, the cost-weighted distances (CWDs) from each pixel on the ecological resistance surface to the ecological sources were calculated; second, the least-cost paths (LCPs) between all ecological sources were determined; finally, the cut-off distances were set to generate ecological corridors, so that each source was connected and formed a network loop. In view of the identified corridors, an evaluation index system was established to evaluate the importance of corridors by comprehensively considering their functional importance and their own conditions [74]. The functional importance of the corridors included the area of the source patch connected by the corridor and the landscape connectivity of the main source connected by the corridor; the corridor conditions mainly considered the corridor length and corridor quality. Using the AHP method to determine the weights of indicators at all levels, all indicators were divided into three levels using the natural breaks method in ArcGIS 10.2, and the importance was assigned to 5, 3 and 1 from the largest to the smallest, respectively (Table 3). Finally, the ecological corridors were classified into key ecological corridor, important ecological corridor and ordinary ecological corridor according to the calculation results, from the largest to the smallest.

Table 3. Evaluation index system of ecological corridor importance.

Target Layer	Criteria Layer	Solution Layer	Grading Evaluation	Description
Ecological corridor importance	Corridor function importance (0.62)	Total area of patches connected by corridors (0.22)	5 3 1	The sum of the area of connected ecological sources by the corridor, the larger the area, the more important the corridor.
		Landscape connectivity of main connected ecological sources (0.40)	5 3 1	The higher the *dI* of main connected ecological source, the more important the corridor.
	Corridor condition (0.38)	Corridor length (0.10)	5 3 1	The longer the corridor, the greater the risk of breakage.
		Corridor quality I (0.10)	5 3 1	The ratio of CWD to the Euclidean distance represents the ease of animal migration between sources. When the value is larger, the corridor quality is poorer
		Corridor quality II (0.18)	5 3 1	The ratio of CWD to the least-cost path length (LCPL) can study the quality of corridors. When the CWD/LCPL value is larger, species suffer greater resistance to migration or dispersal through this corridor, and the corridor quality is poorer.

Strengthening the ecological environment of ecological nodes is conducive to reducing the consumption cost of ecological corridors and enhancing the ecological service function of the regional ecological network. In this study, the ecological corridors were the minimum cost distance channels between adjacent sources, namely, the minimum paths. The "ridge lines" of the minimum cumulative resistance surface were obtained in ArcGIS 10.2 using hydrology analysis module, namely, the maximum paths. In the hydrological analysis module of the spatial analysis tool of ArcGIS 10.2 software, the maximum threshold value that blocked ecological flow and species dispersal was extracted based on the minimum cumulative resistance surface, and then vectorized and smoothed the vectorized lines to obtain the "ridge lines" of the resistance surface [75,76]. This study identified the intersection of the maximum and minimum paths as ecological nodes. According to their intersection, the ecological nodes were divided into three levels according to their intersection with corridors of different importance (Table 4).

Table 4. Classification criteria of ecological nodes.

Levels of Ecological Nodes	Classification Criteria
1	Located at the intersection of key ecological corridors and the "ridge lines".
2	Located at the intersection of important ecological corridors and the "ridge lines".
3	Located at the intersection of ordinary ecological corridors and the "ridge lines".

3. Results

3.1. Spatial Patterns of Ecological Sources

3.1.1. Assessment of Ecosystem Services and Analysis of Hotspots

The spatial patterns of the two ecosystem services in the MASHP were obtained by quantitatively evaluating habitat quality and soil conservation ecosystem services. As can be seen from Figure 3, the average habitat quality index in the MASHP is 0.75, which is at a moderate to a high level. Because of the serious fragmentation of the study area, the areas with various levels of habitat quality show a staggered distribution throughout the study area. The areas with higher habitat quality are spread, mostly covered with forest vegetation, and have a good natural environment. They are less disturbed by human activities such as urban land expansion and construction, so the habitat quality indices are high. In terms of soil conservation, the values in the southwest are generally higher than those in the northeast, and the low value areas are mainly located in low elevation areas with less topographic relief, mainly including Zhijiang City and Dangyang City, where the land use types are mainly cultivated land and construction land, with sparse vegetation and high intensity of human activities. In general, the natural background state of the MASHP is good, and the high-value areas of the two ecosystem services mostly overlap with forestland and grassland.

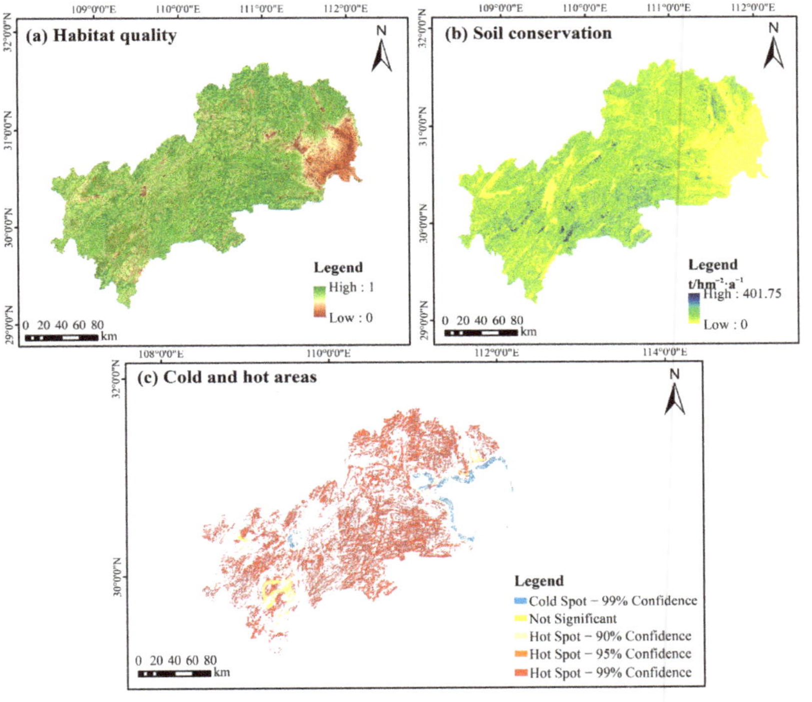

Figure 3. Spatial pattern of ecosystem services and identification of cold and hot areas of ecosystem.

The spatial distribution of cold and hot areas of ecosystem services reflected the strength of ecosystem services. The evaluation results of habitat quality and soil conservation are integrated, the top 20% ecological patches of each ecosystem service are selected, and the hot areas of ecosystem services in the MASHP are obtained based on ArcGIS 10.2 software after taking the merged set and using the hotspots analysis tool in the spatial statistics module, with extreme confidence hot areas as the selection criteria (Figure 3c). The hot areas are mainly located in the northern, central, and southwestern regions, while the cold areas are concentrated in the eastern region. For the ecological sources selected by hotspots analysis, there are problems with the excessive number of ecological source patches and high fragmentation, so it is necessary to set the minimum area threshold manually to correct them further to avoid the fragmented patches from reducing the ecosystem service function of the sources [48,71,77]. As the minimum patch area threshold of ecological source patches increases from 1 to 15 km^2 in steps of 1 km^2, the number and area of patches first show a sharp decline, and then there is a steady and apparent decrease at the threshold of 10 km^2. At the inflection point, the downward trend of the curve is flat, and there is no longer a vast fluctuation (Figure 4). The fluctuation of the above data indicates that it is better to use the threshold value of 10 km^2 to eliminate small and finely fragmented patches. Although there are a large number of removed patches, they are small in individual area, finely fragmented and scattered, and have little impact on the regional ecological environment. Therefore, patches with areas larger than 10 km^2 were selected. 68 patches become alternative ecological sources, and most of these alternative ecological sources are concentrated in the northern and central parts of the study area, and scattered in the southwestern part (Figure S1).

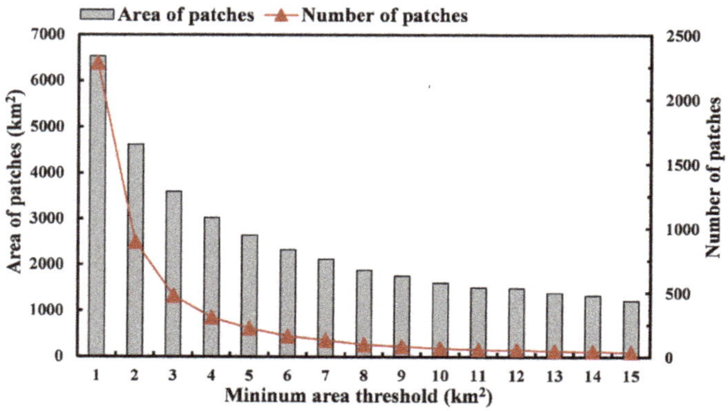

Figure 4. Effect of the minimum area threshold on the number of patches and area of patches from 1 to 15 km^2.

3.1.2. Identification of Core Ecological Patches by MSPA

The MSPA patterns distribution of the study area is shown in Figure 5. Among the seven landscape types in the foreground, the core area has the largest area of 16,750.84 km^2, accounting for 49.98% of the foreground area. Meanwhile, due to the serious fragmentation of patches in the core area, the selection of too many ecological sources will lead to excessive overlapping and redundancy of subsequent ecological corridors, which will increase management costs. Combining with the actual situation of the study area, 103 patches with an area of more than 10 km^2 in the core area were selected as the alternative ecological sources (Figure S2).

In order to avoid the phenomenon of patch overlap, the two previously obtained alternative ecological source patches were combined after removing the small overlapping parts. Finally, 171 alternative ecological source patches were obtained. Based on the results of previous studies and Equations (3)–(7), the landscape connectivity of 171 patches was assessed using Conefor Sensinode 2.6 software, setting the patch connectivity distance threshold to 1500 m and the probability of connectivity to 0.5, with 68 patches based on ecosystem service evaluation, hotspots analysis and area screening and 103 patches based on MSPA and area screening. The evaluation results show that the patches with *dI* values less than 2 are isolated and small compared to other patches, and the landscape connectivity is poor, so they should not be identified as ecological sources. Therefore, 49 patches with *dI* values greater than 2 were identified as the final ecological sources. The total area of ecological sources is 3837.92 km^2, accounting for 8.51% of the total area of the MASHP, and the land use types are mainly forestland. Overall, most of the ecological sources are concentrated in the central region, the county boundary areas in the north and south, and scattered in the southwest, with the number of sources gradually decreasing from the center to the two ends, and the landscape connectivity in the southwest is poor (see Figure 6 and Figure S3). Some of the larger ecological sources (e.g., number 22, 38) are close to the study area boundary, while other relatively small ecological sources (e.g., number 9, 42) are mainly in the central part. From the spatial distribution of ecological sources by county, Changyang Tujia Autonomous County has the largest proportion of ecological sources in the total county area (19.95%), followed by Wufeng Tujia Autonomous County (10.43%). In contrast, Xiling District, Wujiagang District, Xiaoting District and Zhijiang City have no ecological source distribution. As shown by the overlay analysis with various national nature reserves of the MASHP (http://www.papc.cn/html/folder/1-1.htm (accessed on 1 May 2022)), most of the ecological sources obtained involve the core areas of existing nature reserves and forest parks, such as Xingdoushan National Nature Reserve (in Lichuan), Mulinzi National Nature Reserve (in Hefeng), Houhe National Nature Reserve and Chaibuxi National Forest Park (in Wufeng), indicating that the selected sources are scientific.

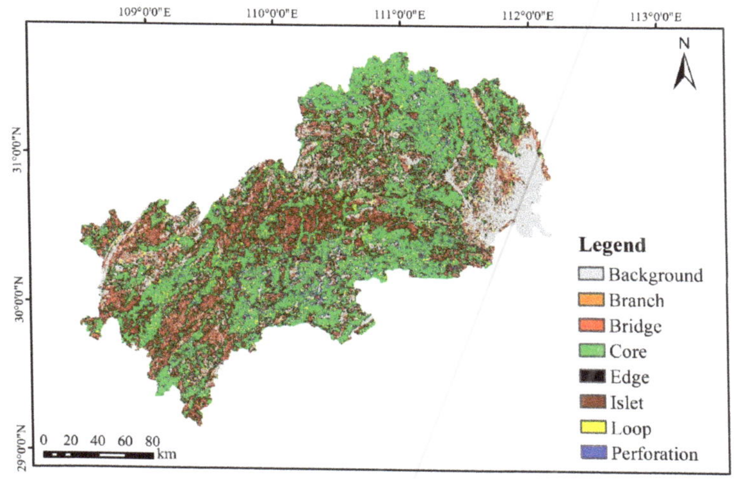

Figure 5. MSPA-based landscape feature type map.

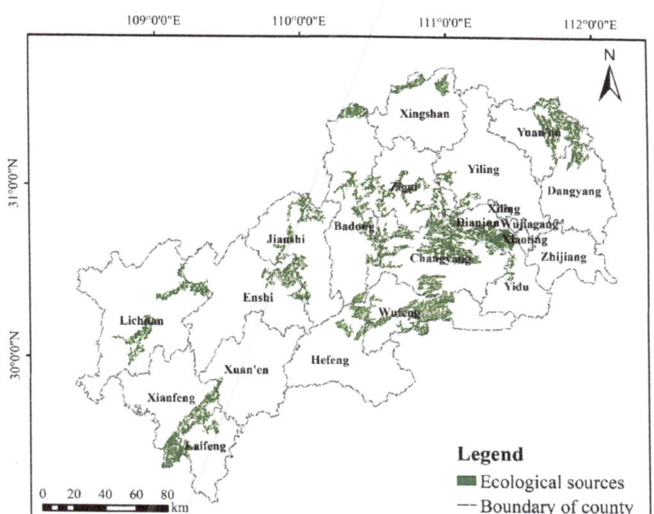

Figure 6. Ecological sources in the study area.

Based on Equation (8), Q_i was calculated to evaluate the importance of each ecological source patches, so as to determine the connotation of ecological source patches in regional potential ecological network. Finally, there are 14 vital ecological sources, 12 important ecological sources and 23 general ecological sources.

3.2. Analysis of Comprehensive Resistance Surface

According to the resistance factors and weights in Table 2, ArcGIS 10.2 was used to establish the comprehensive resistance surface of ecological factors. The minimum cumulative resistance surface was generated based on ecological sources and comprehensive resistance surface to lay the foundation for the subsequent delineation of function zones. The average resistance value is 34.52, and the comprehensive ecological resistance surface shows an overall staggered distribution of high and low resistance values, with a certain aggregation in some areas (Figure 7a). The high-value ecological resistance areas show the characteristics concentrated in the northeast and scattered in the rest of the area, basically located in the areas with the high level of urban development, high traffic density and frequent human activities. Although cropland has vegetation with good growth, its resistance value rises due to the influence of human agricultural activities and artificial control around it. Topographic obstruction factors also influence the high value of resistance areas distributed in Enshi Prefecture, and the high terrain makes the spread of ecological species narrow. The low-value ecological resistance areas are located in the north, south and southwest, where the surface vegetation coverage is high, there are rivers and lakes, and the anthropogenic interference is less than that in the urban concentration. From Figure 7b, it can be seen that some areas in the eastern, southern and western parts of the study area can have a more significant obstructive effect on the migration dispersal and material flow of species.

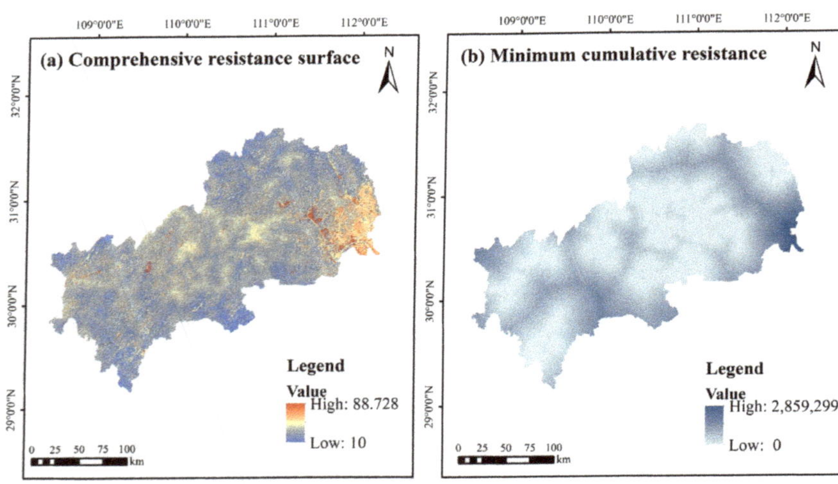

Figure 7. Spatial distribution of comprehensive resistance surface and minimum cumulative resistance.

3.3. Analysis of Ecological Corridors and Nodes

The Linkage Mapper tool was used in ArcGIS 10.2 to extract the LCPs to complete the ecological corridor identification based on the ecological sources and comprehensive resistance surface. In this study, 125 ecological corridors were identified, with a total length of about 2014.61 km. The ecological corridors are connected along the northeast–southwest direction to form a network and run through the whole area, with a dense distribution in the central part, indicating that the central source areas are better connected. Most of the corridors spreading from the east and west ends to the middle are long-distance corridors.

Affected by the barrier of urban built-up areas, there is a potential connection trend between some sources separated by distance, but no connected corridor has been formed (Figure 8). After analyzing the statistics, it is found that the corridor length distribution in three regions, Changyang Tujia Autonomous County, Xuan'en County and Enshi City, shows obvious advantages, with a total of 640.11 km, accounting for about 31.77% of the total length of ecological corridors. The corridors show an extensive spanning range in these areas. There are no ecological corridors distributed in Xiaoting District, Zhijiang City and Laifeng County, mainly for three reasons: firstly, most of the construction land in the above counties is centrally distributed, with high ecological resistance values and large barriers to communication between ecological sources; secondly, regional construction land is distributed sporadically among ecological patches, and ecological patch fracture cannot form a good ecological concentration surface; lastly, there are regions with large areas of contiguous cultivated land, without ecological source land coverage. According to Table 3, the importance of each ecological corridor was calculated. Finally, there are 26 key ecological corridors, 29 important ecological corridors, and 70 ordinary ecological corridors with total lengths of 543.80 km, 536.56 km, and 934 km, specifically. Key and important ecological corridors are important paths for the diffusion of ecosystem services from the sources to the outside, ensuring the fundamental ecological processes. Additionally, some corridors are also the only corridors for circulation between some ecological sources, with a wide range of radiation. However, because of the long length of some corridors, they are susceptible to anthropogenic impacts such as urban expansion, excessive agricultural settlement and vegetation destruction, so there is an urgent need to optimize the corridors in terms of adjusting the spatial layout of functions and increasing the density of regional vegetation cover. Ordinary ecological corridors are supplementary to other level corridors, forming a north–south coherent ecological network.

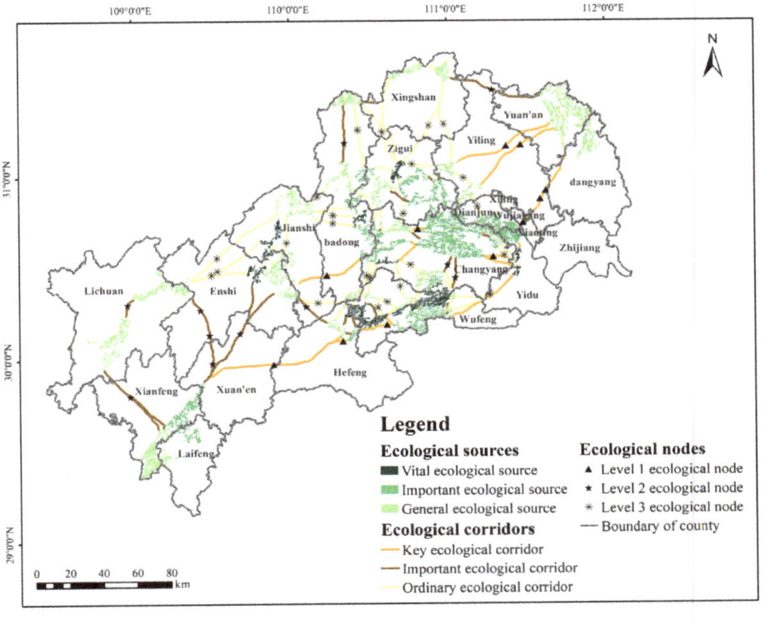

Figure 8. Ecological network in the study area.

The number, quality, and distribution of ecological nodes affect species' migration timing and success probability. A total of 46 potential ecological nodes were identified in the study area, including 11 level 1 ecological nodes, 11 level 2 ecological nodes, and

24 level 3 ecological nodes (Figure 8). These nodes mainly play the role of connections and hubs, located at the weak point of ecological material and capacity exchange, which are crucial for the mobility efficiency of the ecological corridor and greatly influenced by human activities [17]. The ecological nodes identified are relatively evenly distributed spatially, in total, on 42 ecological corridors. Still, no ecological nodes are distributed in Zhijiang City, Dangyang City, and Yuan'an County in the east and Laifeng County at the southern end. The landscape type of most ecological nodes is forestland, but any two points are far apart and poorly connected between nodes, which also indicates the need to increase ecological nodes in areas with dense source distribution to promote connectivity and strengthen the construction and protection of ecological nodes. Level 1 ecological nodes are scattered and relatively few in number, and there is no distribution in the north. They play an important role in connecting ecological sources and assume the responsibility for smooth operation of ecological flow and maintenance of ecological stability. Therefore, the preservation and optimization of the original green belt and water patch with high ecological function in the area where the nodes are located should be strengthened. Level 2 ecological nodes are more distributed in the southwest, and some of them are also in the transition zone between urban construction land and cultivated land. Landscape transformation of such nodes has high cost and difficulty, so local governments should focus on improving the ecological efficiency of green land and enhancing the ability to prevent pollution. Level 3 ecological nodes are the most numerous, mostly covering the central and northern parts of the study area. Three of them are cultivated land, vulnerable to human development activities. Moreover, the spread of agricultural non-point source pollution can greatly interfere with the circulation of ecological flow. In the future, the regional government can consider constructing "stepping stones" at ecological nodes by building small ecological parks or planting forestland to improve ecological network connectivity and provide landing points for biological migration.

3.4. Construction of Ecological Network

Ecological corridors are interwoven with ecological nodes together with ecological sources to form ecological network. The ecological network in the MASHP includes 49 ecological sources, 125 ecological corridors, and 46 ecological nodes, all of which have different levels of importance. Ultimately, these elements form a sustainable, composite, multi-level and interconnected ecological network (Figure 8). In this ecological network, the coverage of ecological sources is ideal, ecological corridors can effectively link the sources, and ecological nodes play an important role in promoting the flow of energy and material in the region.

Combining the spatial distribution characteristics of elements, the ecological spatial framework of the MASHP can be summarized as "two verticals, three belts, three groups, and multiple nodes" (Figure 9). The "two verticals, three belts" represent the general direction of ecological corridors of a certain scale connecting all ecological sources and nodes, mainly the vertical ecological corridors starting from Badong County and Xingshan County and the longer horizontal ecological corridors in the northeast–southwest direction. The ecological corridors expressed as vertical axes maintain the connectivity between the north and the central and southern parts. In contrast, the ecological corridors expressed as horizontal axes link the ecological sources at the northeast and southwest ends with the central sources, ensuring the integrity and continuity of ecological processes from north to south and from east to west, which are the best paths for species migration and ecological information flow. The "three groups" represent the northeastern, central, and southwestern ecological groups, respectively. The northeastern ecological source group mainly includes ecological sources distributed in the northern end of Badong County, Xingshan County and Yuan'an County, mainly at the edge of the county boundary. This group is responsible for important ecological functions such as soil conservation and biodiversity maintenance northeast of MASHP. The central ecological group includes ecological sources mainly in Changyang Tujia Autonomous County and Wufeng Tujia Autonomous County, which is the core group. The ecological sources in this group are concentrated, crossed by several ecological corridors, rich in species and with good natural substrates. The southwest mainly includes ecological sources in Lichuan City, Xianfeng County and Laifeng County, which is the main ecological source group in Enshi Prefecture. The source patches are large in area but small in number, which guarantee the ecosystem security and stability in the southwestern part of MASHP. The "multiple nodes" represent the ecological nodes of different levels in the network, which are located at the weakest part of the corridor function. The connectivity degree and cross-structure features of the ecological network in the MASHP have noticeable spatial differences, so further details can be improved and optimized from the perspective of improving the stability and liquidity of the overall ecosystem.

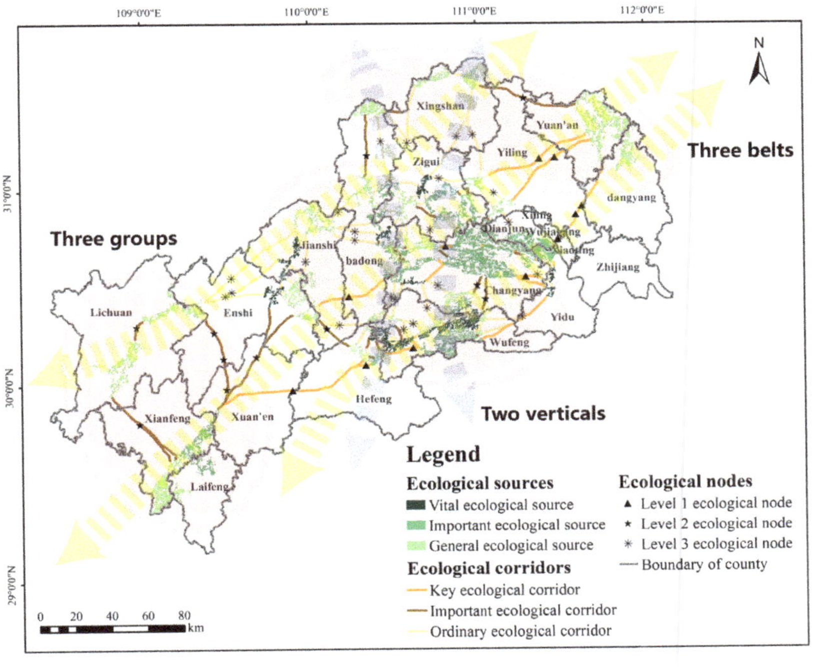

Figure 9. Ecological framework in the study area.

4. Discussion

4.1. Optimization of Ecological Network

The current problems of ecological network in the MASHP created in this study are reflected in three aspects. Firstly, the fragmentation of the landscape in the study area is serious, causing fragmentation of ecological resources, resulting in the uneven spatial distribution of ecological sources and insufficient diversity of ecological elements, for example, some small patches in the central ecological source group are isolated and not well connected. Therefore, according to the spatial structure and distribution characteristics of the existing ecological sources, the related management departments should focus on maintaining and effectively improving the ecological quality and ecological benefits of the ecological sources, enhancing the connectivity between the sources, and dividing ecological protection buffer zones to avoid damage to the ecological sources. Secondly, the existing ecological corridors need to be further widened, and only corridors with a certain width can assume the function of ecological element connectivity and communication. The width of ecological corridors in ecological network optimization can enhance the ecological effect of corridors and the connectivity between small ecological corridors. Therefore, a certain width of ecological corridors should be given to activate the ecological service function of the corridor [68]. Finally, the existing ecological network elements are in conflict with other production and living service elements, and the planned new infrastructure is also prone to damage the ecological network; for example, the completion of new national highways will affect several ecological corridors. Therefore, different ecological functional zones can be divided to set up targeted management and protection measures to effectively coordinate the value balance of agricultural production, development and construction and ecological services.

To further enhance the integrity and connectivity among ecological sources, buffer zones were set for ecological sources with poor landscape connectivity ($2 \leq dI < 3$), and the growth rate of forestland within each buffer would be observed to change as the buffer distance becomes larger, and finally, the optimal dispersal distance of these sources were obtained after integrating the distribution of other land use types (Table S4 and Figure S4). The analysis of the evolution of the buffer zones of ecological sources in the MASHP shows the following: when the buffer zones of some sources reach a certain distance, they overlap with the surrounding construction land, so this phenomenon should be avoided as much as possible; when the buffer zones of some sources are too large, the buffer zones of ecological source patches overlap each other and start to crowd the space of other lands, which is not consistent with the reality of intensive land use; when the buffer zones of some sources are increased to the corresponding distance, the connectivity with the surrounding sources enhance, but the optimal diffusion distance need to be determined after comprehensive consideration. In summary, the buffer zones are set up for ecological source patches with poor landscape connectivity. As the ecological sources spread outward, some ecological sources will form effective connections, thus enhancing the connectivity between ecological sources and the stability within the patches [71,72]. For instance, patches numbered 14 and 12 and patches numbered 9 and 16 in the central eco-logical source group have connected after installing buffer zones to realize the expansion of outward radiation (Figure S5).

The connectivity of ecological network is ensured by dredging ecological corridors. The animal species in the MASHP are mainly medium-sized mammals. According to Zhu et al. [78], buffer zones of 30, 60, 100, 400, 600, 800, and 1000 m were set for existing ecological corridors, and the area of each land use type in different corridor widths was statistically analyzed. As shown in Table 5 and Figure 10, it can be seen that the distribution area of each land use type in different corridor widths varies greatly. Forestland is the primary type in the corridor. As the corridor width increased from 30 m to 1000 m, the area of all land use types shows a continuously increasing trend. Although the proportion of forestland area to the total area of corridor area decreases continuously, the proportion is always more than 80%. The proportion of cultivated land area increases less, and after 400 m corridor width, the proportion of cultivated land area to the total area of corridor area is second only to forestland area. The proportion of grassland area increases, then decreases and then increases again but remains stable at 7% or less, with the inflection points occurring at 60 m and 600 m corridor widths, respectively. The proportion of water bodies area increases and then decreases, with the inflection point and the highest point occurring at the 600 m corridor width. The proportion of construction land area keeps rising, but it is always below 1%, and the increase rate increases significantly to 400 m. It is worth noting that the large construction land area is not conducive to the migration and diffusion of organisms between ecological sources. It also increases the difficulty of constructing ecological corridors. Taking all factors into consideration, the corridor width of 100–400 m has the least encroachment on cultivated land, weak impact on construction land, and the area share of three major ecological land use types, namely, forestland, grassland and water bodies, is at a high level, which is also the width required for migration and conservation of small- and medium-sized mammals [79]. Therefore, the optimal width range of the MASHP corridor is determined to be 100–400 m. In the future construction of the corridors, based on the natural advantages of forestland and grassland in the study area, the stability of the corridors should be enhanced by forming a vegetation community with a composite structure of trees, shrubs, and grasses according to the actual situation and by strengthening the configuration of ecologically strong and stable vegetation. In particular, when building long and important ecological corridors, the distance between core patches can be shortened by adding "stepping stones" and buffer zones to prevent the rupture of corridors. For corridors that may cross water bodies (for example, some vertical ecological corridors starting from the northern part of the study area will span the Yangtze River and its tributaries), attention should be paid to the construction of coastal vegetation buffer zones of water bodies where the corridors are located. A reasonable vegetation buffer zone can effectively filter surface pollutants to improve water quality and create conditions for the migration of inland organisms in the corridors. For corridors with cultivated land distributed inside, the focus should be on protecting semi-natural habitats such as pond wetlands, ecological ditches, farmland ridges and scrubs, and moderate conversion of farmland back to forests to maintain the ecological stability of the small regional environment. For corridors with a large distribution of internal construction land, a buffer zone of a certain width can be set outside the corridor to minimize the interference of large human construction activities in the areas around the corridor. A portion of ordinary ecological corridors exist between compact ecological sources with short lengths, and such corridors can facilitate further species circulation by enhancing the closeness between ecological sources [74]. Of course, the ecological network is a resilient ecological conservation space in which the width of the corridors can be appropriately contracted and expanded according to the practical problems faced during construction [17].

Table 5. Area of different land use types in corridors with different widths (km^2).

Land Use Type	Corridor Width (m)						
	30	60	100	400	600	800	1000
Cultivated land	2.49	6.15	15.55	126.66	213.37	295.95	383.74
Forestland	96.85	175.34	311.20	1161.49	1734.31	2305.15	2900.60
Grassland	7.89	14.44	25.42	97.13	145.43	194.68	247.85
Water bodies	1.38	2.76	5.37	25.46	38.69	51.49	64.21
Construction land	0.17	0.39	0.96	6.44	10.97	15.75	21.04

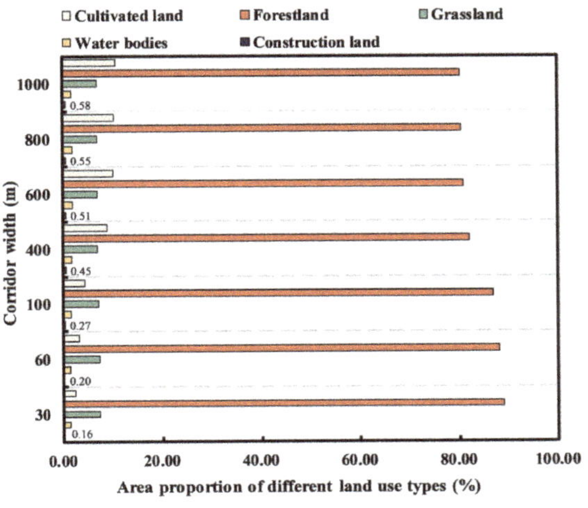

Figure 10. Area proportion of different land use types in corridors with different widths.

By dividing the ecological function zones, the precision and efficiency of ecological security construction can be ensured. The minimum cumulative resistance surface constructed based on ecological sources is the basis of ecological function zoning [76]. The standard deviation classification method was used to classify the resistance values. The mutations in the number of grids corresponding to different cumulative resistance numbers were selected as the basis for determining the resistance thresholds. When species pass through these mutation points, the minimum cumulative resistance value will increase sharply, indicating that when the ecological land area increases to a certain extent, the conservation significance of increasing the area again will decrease suddenly. According to the standard deviation of the minimum cumulative resistance, the resistance values of ecological land protection were initially divided into 17 categories, named C1–C17, and each category was separated by one-quarter variance. The number of grids in each category and the correspondence of the image element value (minimum cumulative resistance value) were counted (Table S5). As shown in Figure 11, from C1–C2, there is a significant decrease in the number of grids, and the difference in the number of grids accounts for 8.02% of the total number of grids in the MASHP; from C3–C7, the number of grids again undergoes a relatively large abrupt change, and the change is second only to that from C1 to C2, accounting for 3.18% of the total number of grids; starting from C8 onward, the number of grids tends to stabilize, remaining at a level below 6%. Therefore, according to the above threshold determination method, the minimum cumulative resistance values of 38,183, 131,896 and 600,462 were selected as the critical threshold functional zoning in the MASHP. At the same time, the spatial distribution patterns of ecological sources, ecological corridors and ecological nodes in the study area were comprehensively considered. The corresponding division results were ecological conservation zones (C1, ECZs), ecological buffer zones (C2, EBZs), ecological transition zones (C3–C7, ETZs), and ecological available zones (C8–C17, EAZs), with the objectives of protection and conservation, stability maintenance, conflict mitigation and production development, respectively. In addition, ecological function zoning also needed to consider the spatial distribution pattern of ecological corridors. The primary function of ecological corridors was to provide species migration and dispersal channels. Therefore, based on the composition of land use types in ecological corridors, the corridor restoration zones (CRZs) with the greatest resistance to species migration and dispersal in the corridor with a maximum width of 400 m were extracted from the results of the previous analysis and superimposed on the divided functional zones. The final ecological function zones were obtained by superimposing them on the functional zones already delineated (Figure 12). Different measures should be developed to enhance ecological maintenance and management according to the characteristics and objectives of each ecological function zone.

According to the zoning results, the ETZs occupy the largest proportion, accounting for 4204% of the total area of the study area. In contrast, the EBZs account for the most minor proportion of the whole area, 13.25%. The ECZs occupy 21.28% of the total area of the study area, and all ecological sources are included. The zones slightly increase in area and connectivity compared to the ecological sources and serve as a protective barrier for ecological sources, a potential area for the expansion and succession of ecological sources. The zones are the core area of ecological protection, development and occupation should be strictly restricted, conservation should be the focus, a reasonable ecological layout should be adopted, natural grassland, ecological woods and other ecological resources restoration projects should be actively implemented (e.g., the "greener mountains" strategic decision implemented by Enshi Prefecture, the various projects around the ecology of the Yangtze River implemented by Yichang City), and ecological protection of various types of cultured forests should be established. For the ECZs with urban construction land areas, the existing parks and cultural scenic spots can be combined to construct and protect urban green corridors and strengthen green space planning to improve the public green space and ecological functions of the urban ECZs. The EBZs are distributed close to ecological sources and are an extension of ecological sources. In the process of the ECZs playing its ecological service function, the EBZs provide enough buffer area to ensure the normal operation of the ECZs and reduce the interference and impact of external human activities on the ECZs.

The zones should maintain the dominant landscape stability and focus on optimizing the existing layout to mitigate various land use conflicts. The ETZs are located between the EBZs and EAZs with the largest scale, and the area distribution of cultivated land and construction land in the region is evident. It is a region with prominent contradiction between agricultural production, urban construction and ecological protection. The development and utilization of various kinds of land need to consider the balance between the three and carry out urban construction, tourism development and infrastructure expansion activities appropriately and selectively. Agricultural production should be carried out within reasonable limits, cultivated land should be improved, and the use of chemical fertilizers and pesticides and the discharge of wastewater and waste should be controlled to ensure soil ecological safety. The EAZs account for 23.07% of the total area of the study area and are located outside the ETZs, far away from the ecological sources, and are the zones where urbanization construction and agricultural development are prioritized. The zones should actively expand modernized specialty agriculture based on not affecting or destroying the ecological environment, such as the characteristic citrus, navel orange and tea in Yiling District and Zhijiang City, and white grapefruit in Xuan'en County, etc. These counties can focus on particular products to carry out industrial transformation and upgrading and continuously expand agricultural development. Countries with an excellent rustic foundation can comprehensively promote the large-scale agricultural operation, mechanized production and smart agriculture. When overlaying the ETZs and EAZs layers and layers of cultivated land and construction land, it is found that 74.49% of cultivated land and 84.30% of construction land are within these two zones, indicating that zoning results are more realistic and reliable with the real situation, and they have certain guidance. Within the CRZs, the corridors will be subject to frequent human interference during construction and maintenance, which can be maintained by setting up isolation zones and other measures to strictly control the uncontrolled growth of construction land within the corridors.

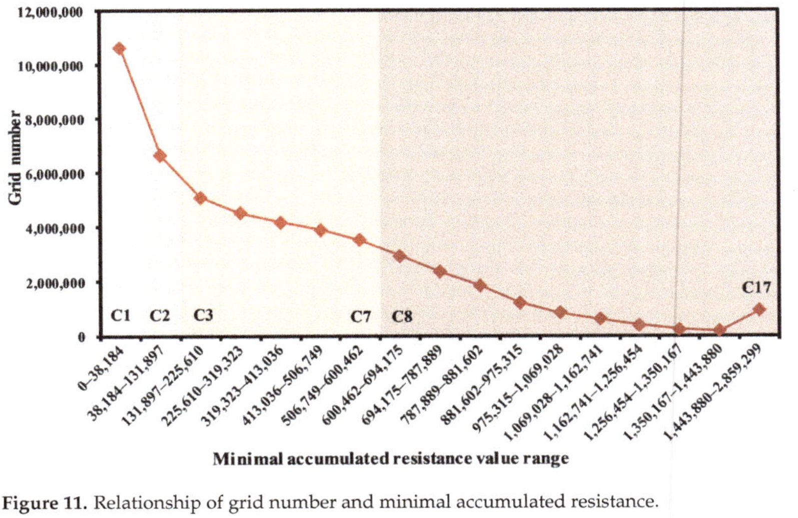

Figure 11. Relationship of grid number and minimal accumulated resistance.

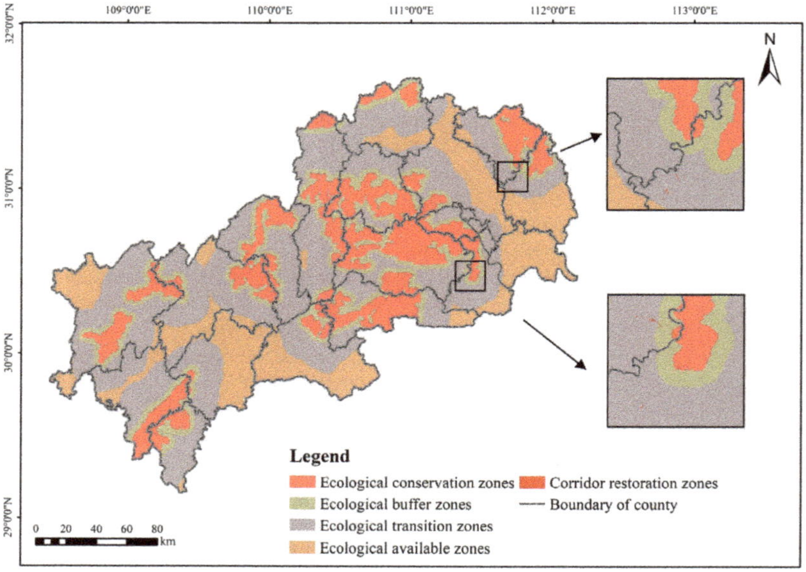

Figure 12. Ecological function zoning based on ecological network.

4.2. Methodological Advantages and Limitations

Currently, the construction and maintenance of ecological networks are essential to ensure the structural integrity of ecosystems in the region and to promote the cyclic flow of ecological streams. The basis of constructing an ecological network is to select suitable ecological sources [72,80]. There are various methods for identifying ecological sources for different conservation objectives and development priorities. Most previous studies have focused only on assessing the supply capacity of ecosystem services or only using the MSPA method alone, with integrated approaches still to be discovered [81]. Compared with other studies that also used an integrated approach to identify ecological sources [28,43,46,49], this study not only proposed a new method to integrate the identification of ecological sources, but also innovated in the subsequent quantification of the importance of each component of ecological network. Firstly, this study demonstrated that the ecosystem service hotspot analysis and MSPA method can effectively help to screen ecological sources in the first stage, and the minimum area threshold and landscape connectivity indices in the later stage can further make the ecological source identification results more accurate. The identified ecological sources provide a good supply of ecosystem services and are critical patches with internal homogeneity and the ability to spread to the surrounding area. Secondly, compared to the traditional method of distinguishing the importance of corridors, this study did not rely on the attractiveness between sources (e.g., only using gravity model) in grading, but considered the functional importance of ecological corridors and the conditions of the corridors themselves, which can provide a certain scientific basis for the priority of corridor protection and restoration. Finally, in this study, the spatial boundaries of the ecological functional zones delineated in the subsequent optimization are clear and in good agreement with the actual situation, providing an important basis for the implementation of ecological spatial protection in mountainous areas.

The study has two limitations. Firstly, this study only identified the ecosystem service capacity of the study area in terms of habitat quality and soil conservation, which was an empirical judgment based on the actual situation of the study area. The combined effects of other ecosystem service functions (e.g., carbon storage, water production) still need to be tested and improved. Secondly, this study did not construct an ecological network based on the living characteristics of specific species in the region, and subsequent studies could be improved to deepen the scientific validity of the results.

5. Conclusions

The construction of an ecological network is conducive to realizing the goal of co-development of regional ecological protection and regional development. In this study, taking the MASHP, a typical mountainous area in central China, as the study area, an integrated approach was developed to identify ecological sources based on a quantitative assessment of ecosystem services and MSPA method; an ecological resistance surface that met the characteristics of the regional environment was constructed by considering natural environmental and socioeconomic factors; ecological corridors were extracted according to the least-cost path using the Linkage Mapper tool, and ecological nodes were identified by applying the principle of hydrological analysis; finally, ecological sources, ecological corridors and ecological nodes with different levels of importance together formed the final ecological network in the MASHP, and corresponding optimization measures were proposed.

The main conclusions of this study can be drawn as follows:

(1) This study proposed a new integrated method for identifying ecological sources, considering ecosystem and landscape patch functions and innovates in quantifying the importance of ecological corridors and ecological nodes. The results show that the methods used are feasible and practical.

(2) A total of 49 ecological sources were identified, with an area of 3837.92 km^2, accounting for 8.51% of the total area in the MASHP. Most of the ecological sources are concentrated in the central region, the county boundary zones in the north and south, and scattered in the southwest, with forestland as the main land use type. A total of 125 ecological corridors were constructed, with a total length of about 2014.61 km. The overall network is connected along the northeast–southwest direction and ran through the whole area, with a dense distribution in the central part. A total of 46 ecological nodes were identified. These nodes are more evenly distributed spatially, only on 42 ecological corridors in total. Ecological sources, ecological corridors and ecological nodes were graded with different levels of importance, ultimately forming a sustainable, composite, multi-level and interconnected ecological network.

(3) Combining the spatial distribution characteristics of each element of the ecological network, the overall layout of ecological network spatial structure is "two verticals, three belts, three groups, and multiple nodes". To further improve and optimize ecological network in the MASHP, the direction of optimization can be focused on enhancing the connectivity of existing ecological sources with low connectivity, clarifying the width range of ecological corridors and delineating ecological functional zones. The multiple ring buffer analysis was used to optimize the connectivity of ecological sources and improve the effective connection between sources; the optimal range for corridor construction width is 100–400 m, based on local species and landscape structure; the study area was divided into the ecological conservation zones, ecological buffer zones, ecological transition zones, ecological available zones based on ecological sources and minimum cumulative resistance surface. Finally, the corridor restoration zones were added to form the ultimate division result.

(4) Based on the study results, according to the level, spatial structure, distribution characteristics of the components of the ecological network in mountainous areas and the different development positioning of the divided ecological functional zones, this study put forward the recommendations for ecological protection and development and construction planning in mountainous areas. In brief, each ecological source group should focus on forestland protection and water conservation and actively implement ecological resources restoration projects such as natural grassland and ecological forests (e.g., the "greener mountains" strategic decision implemented by Enshi Prefecture, the various projects around the ecology of the Yangtze River im-

plemented by Yichang City), and control anthropogenic activities and construction land boundaries so that each ecological group can better perform its ecological service functions. Ecological corridors are mainly to maintain the ecological stability of the regional microenvironment. Managers may consider setting up isolation zones around ecological corridors of longer lengths (e.g., multiple horizontal axis corridors in the southwest) and setting up buffers by building and protecting artificial green spaces to safeguard the integrity of the ecological network. Meanwhile, different protection measures should be implemented for corridors passing through water and traffic land according to local conditions. It is necessary to strengthen the ecological function of the zone where the ecological nodes are located so that the nodes are less disturbed. On the one hand, it is important to strengthen the nodes within the area of the high original ecological function of green space and water patches and optimization of the plaque. On the other hand, "stepping stone" construction can be carried out through afforestation and ecological park building to improve the connectivity of the overall ecological corridor.

Supplementary Materials: The following supporting information can be downloaded at: https://www.mdpi.com/article/10.3390/ijerph19159582/s1, Table S1: Land use classification system in the study area; Table S2: Threat factors and their stress intensity; Table S3: Habitat suitability and its relative sensitivity to different threat sources; Figure S1: Alternative ecological source patches based on ecosystem service assessment, hotspots analysis and area screening; Figure S2: Alternative ecological source patches based on MSPA analysis and area screening.; Figure S3: Numbered ecological sources in the study area; Table S4: Optimal diffusion distances; Figure S4: Change curve of the rate of forestland increment with the increase in ecological sources buffer distance in the study area; Figure S5: Spatial distribution of optimized ecological sources; Table S5: Classification based on the standard deviation of minimum cumulative resistance.

Author Contributions: Q.Z.: conceptualization, methodology, data curation, writing—original draft, writing—review and editing. Y.Z.: conceptualization, funding acquisition, writing—review and editing. J.L.: writing—review and editing. All authors have read and agreed to the published version of the manuscript.

Funding: This research was funded by the National Natural Science Foundation of China (No. 42171061) and Special Foundation for National Science and Technology Basic Research Program of China (No. 2021FY100505).

Institutional Review Board Statement: Not applicable.

Informed Consent Statement: Not applicable.

Data Availability Statement: Not applicable.

Conflicts of Interest: The authors declare no conflict of interest.

References

1. Estoque, R.C.; Murayama, Y. A Worldwide Country-Based Assessment of Social-Ecological Status (c. 2010) Using the Social-Ecological Status Index. *Ecol. Indic.* **2017**, *72*, 605–614. [CrossRef]
2. Wang, C.; Yu, C.; Chen, T.; Feng, Z.; Hu, Y.; Wu, K. Can the Establishment of Ecological Security Patterns Improve Ecological Protection? An Example of Nanchang, China. *Sci. Total Environ.* **2020**, *740*, 140051. [CrossRef] [PubMed]
3. Xiao, L.; Cui, L.; Jiang, Q.; Wang, M.; Xu, L.; Yan, H. Spatial Structure of a Potential Ecological Network in Nanping, China, Based on Ecosystem Service Functions. *Land* **2020**, *9*, 376. [CrossRef]
4. Newbold, T.; Hudson, L.N.; Hill, S.L.L.; Contu, S.; Lysenko, I.; Senior, R.A.; Boerger, L.; Bennett, D.J.; Choimes, A.; Collen, B.; et al. Global Effects of Land Use on Local Terrestrial Biodiversity. *Nature* **2015**, *520*, 45–50. [CrossRef]
5. Da Encarnação Paiva, A.C.; Nascimento, N.; Rodriguez, D.A.; Tomasella, J.; Carriello, F.; Rezende, F.S. Urban Expansion and Its Impact on Water Security: The Case of the Paraíba Do Sul River Basin, São Paulo, Brazil. *Sci. Total Environ.* **2020**, *720*, 137509. [CrossRef]
6. Dong, J.; Jiang, H.; Gu, T.; Liu, Y.; Peng, J. Sustainable Landscape Pattern: A Landscape Approach to Serving Spatial Planning. *Landsc. Ecol.* **2022**, *37*, 31–42. [CrossRef]
7. Fábos, J.G.; Ryan, R.L. International Greenway Planning: An Introduction. *Landsc. Urban Plan.* **2004**, *68*, 143–146. [CrossRef]

8. Ma, L.; Bo, J.; Li, X.; Fang, F.; Cheng, W. Identifying Key Landscape Pattern Indices Influencing the Ecological Security of Inland River Basin: The Middle and Lower Reaches of Shule River Basin as an Example. *Sci. Total Environ.* **2019**, *674*, 424–438. [CrossRef]
9. Steffen, W.; Richardson, K.; Rockstrom, J.; Cornell, S.E.; Fetzer, I.; Bennett, E.M.; Biggs, R.; Carpenter, S.R.; de Vries, W.; de Wit, C.A.; et al. Planetary Boundaries: Guiding Human Development on a Changing Planet. *Science* **2015**, *347*, 1259855. [CrossRef]
10. Huang, Q.; Peng, B.; Wei, G.; Wan, A. Dynamic Assessment and Early Warning of Ecological Security: A Case Study of the Yangtze River Urban Agglomeration. *Nat. Hazards* **2021**, *107*, 2441–2461. [CrossRef]
11. Harvey, E.; Gounand, I.; Ward, C.L.; Altermatt, F. Bridging Ecology and Conservation: From Ecological Networks to Ecosystem Function. *J. Appl. Ecol.* **2017**, *54*, 371–379. [CrossRef]
12. Xiao, S.; Wu, W.; Guo, J.; Ou, M.; Pueppke, S.G.; Ou, W.; Tao, Y. An Evaluation Framework for Designing Ecological Security Patterns and Prioritizing Ecological Corridors: Application in Jiangsu Province, China. *Landsc. Ecol.* **2020**, *35*, 2517–2534. [CrossRef]
13. Luo, Y.; Wu, J. Linking the Minimum Spanning Tree and Edge Betweenness to Understand Arterial Corridors in an Ecological Network. *Landsc. Ecol.* **2021**, *36*, 1549–1565. [CrossRef]
14. Yang, X.; Li, S.; Zhu, C.; Dong, B.; Xu, H. Simulating Urban Expansion Based on Ecological Security Pattern—A Case Study of Hangzhou, China. *Int. J. Environ. Res. Public. Health* **2021**, *19*, 301. [CrossRef]
15. Liu, D.; Chang, Q. Ecological Security Research Progress in China. *Acta Ecol. Sin.* **2015**, *35*, 111–121. [CrossRef]
16. D'Aloia, C.C.; Naujokaitis-Lewis, I.; Blackford, C.; Chu, C.; Curtis, J.M.R.; Darling, E.; Guichard, F.; Leroux, S.J.; Martensen, A.C.; Rayfield, B.; et al. Coupled Networks of Permanent Protected Areas and Dynamic Conservation Areas for Biodiversity Conservation Under Climate Change. *Front. Ecol. Evol.* **2019**, *7*, 27. [CrossRef]
17. Peng, J.; Zhao, S.; Dong, J.; Liu, Y.; Meersmans, J.; Li, H.; Wu, J. Applying Ant Colony Algorithm to Identify Ecological Security Patterns in Megacities. *Environ. Model. Softw.* **2019**, *117*, 214–222. [CrossRef]
18. Dong, J.; Peng, J.; Xu, Z.; Liu, Y.; Wang, X.; Li, B. Integrating Regional and Interregional Approaches to Identify Ecological Security Patterns. *Landsc. Ecol.* **2021**, *36*, 2151–2164. [CrossRef]
19. Zhao, X.; Yue, Q.; Pei, J.; Pu, J.; Huang, P.; Wang, Q. Ecological Security Pattern Construction in Karst Area Based on Ant Algorithm. *Int. J. Environ. Res. Public. Health* **2021**, *18*, 6863. [CrossRef]
20. Weber, T.; Sloan, A.; Wolf, J. Maryland's Green Infrastructure Assessment: Development of a Comprehensive Approach to Land Conservation. *Landsc. Urban Plan.* **2006**, *77*, 94–110. [CrossRef]
21. Yin, H.; Kong, F.; Hu, Y.; James, P.; Xu, F.; Yu, L. Assessing Growth Scenarios for Their Landscape Ecological Security Impact Using the SLEUTH Urban Growth Model. *J. Urban Plan. Dev.* **2016**, *142*, 05015006. [CrossRef]
22. Zhai, T.; Huang, L. Linking MSPA and Circuit Theory to Identify the Spatial Range of Ecological Networks and Its Priority Areas for Conservation and Restoration in Urban Agglomeration. *Front. Ecol. Evol.* **2022**, *10*, 828979. [CrossRef]
23. Peng, J.; Pan, Y.; Liu, Y.; Zhao, H.; Wang, Y. Linking Ecological Degradation Risk to Identify Ecological Security Patterns in a Rapidly Urbanizing Landscape. *Habitat Int.* **2018**, *71*, 110–124. [CrossRef]
24. Zhao, S.; Ma, Y.; Wang, J.; You, X. Landscape Pattern Analysis and Ecological Network Planning of Tianjin City. *Urban For. Urban Green.* **2019**, *46*, 126479. [CrossRef]
25. Yang, J.; Zeng, C.; Cheng, Y. Spatial Influence of Ecological Networks on Land Use Intensity. *Sci. Total Environ.* **2020**, *717*, 137151. [CrossRef]
26. Kong, F.; Yin, H.; Nakagoshi, N.; Zong, Y. Urban Green Space Network Development for Biodiversity Conservation: Identification Based on Graph Theory and Gravity Modeling. *Landsc. Urban Plan.* **2010**, *95*, 16–27. [CrossRef]
27. Su, J.; Yin, H.; Kong, F. Ecological Networks in Response to Climate Change and the Human Footprint in the Yangtze River Delta Urban Agglomeration, China. *Landsc. Ecol.* **2021**, *36*, 2095–2112. [CrossRef]
28. Wang, F.; Yuan, X.; Zhou, L.; Zhang, M. Integrating Ecosystem Services and Landscape Connectivity to Construct and Optimize Ecological Security Patterns: A Case Study in the Central Urban Area Chongqing Municipality, China. *Environ. Sci. Pollut. Res.* **2022**, *29*, 43138–43154. [CrossRef] [PubMed]
29. Vogt, P.; Ferrari, J.R.; Lookingbill, T.R.; Gardner, R.H.; Riitters, K.H.; Ostapowicz, K. Mapping Functional Connectivity. *Ecol. Indic.* **2009**, *9*, 64–71. [CrossRef]
30. Soille, P.; Vogt, P. Morphological Segmentation of Binary Patterns. *Pattern Recognit. Lett.* **2009**, *30*, 456–459. [CrossRef]
31. Carlier, J.; Moran, J. Landscape Typology and Ecological Connectivity Assessment to Inform Greenway Design. *Sci. Total Environ.* **2019**, *651*, 3241–3252. [CrossRef]
32. Knaapen, J.P.; Scheffer, M.; Harms, B. Estimating Habitat Isolation in Landscape Planning. *Landsc. Urban Plan.* **1992**, *23*, 1–16. [CrossRef]
33. Yu, K. Security Patterns and Surface Model in Landscape Ecological Planning. *Landsc. Urban Plan.* **1996**, *36*, 1–17. [CrossRef]
34. McRae, B.H.; Beier, P. Circuit Theory Predicts Gene Flow in Plant and Animal Populations. *Proc. Natl. Acad. Sci. USA* **2007**, *104*, 19885–19890. [CrossRef] [PubMed]
35. Peng, J.; Yang, Y.; Liu, Y.; Hu, Y.; Du, Y.; Meersmans, J.; Qiu, S. Linking Ecosystem Services and Circuit Theory to Identify Ecological Security Patterns. *Sci. Total Environ.* **2018**, *644*, 781–790. [CrossRef]
36. Huang, J.; Hu, Y.; Zheng, F. Research on Recognition and Protection of Ecological Security Patterns Based on Circuit Theory: A Case Study of Jinan City. Environ. *Sci. Pollut. Res.* **2020**, *27*, 12414–12427. [CrossRef] [PubMed]

37. Dai, L.; Liu, Y.; Luo, X. Integrating the MCR and DOI Models to Construct an Ecological Security Network for the Urban Agglomeration around Poyang Lake, China. *Sci. Total Environ.* **2021**, *754*, 141868. [CrossRef] [PubMed]
38. Feng, H.; Li, Y.; Li, Y.; Li, N.; Li, Y.; Hu, Y.; Yu, J.; Luo, H. Identifying and Evaluating the Ecological Network of Siberian Roe Deer (Capreolus Pygargus) in Tieli Forestry Bureau, Northeast China. *Glob. Ecol. Conserv.* **2021**, *26*, e01477. [CrossRef]
39. Wang, Z.; Luo, K.; Zhao, Y.; Lechner, A.M.; Wu, J.; Zhu, Q.; Sha, W.; Wang, Y. Modelling Regional Ecological Security Pattern and Restoration Priorities after Long-Term Intensive Open-Pit Coal Mining. *Sci. Total Environ.* **2022**, *835*, 155491. [CrossRef]
40. McRae, B.H.; Dickson, B.G.; Keitt, T.H.; Shah, V.B. Using Circuit Theory to Model Connectivity in Ecology, Evolution, and Conservation. *Ecology* **2008**, *89*, 2712–2724. [CrossRef]
41. Ford, A.T.; Sunter, E.J.; Fauvelle, C.; Bradshaw, J.L.; Ford, B.; Hutchen, J.; Phillipow, N.; Teichman, K.J. Effective Corridor Width: Linking the Spatial Ecology of Wildlife with Land Use Policy. *Eur. J. Wildl. Res.* **2020**, *66*, 69. [CrossRef]
42. Dong, J.; Peng, J.; Liu, Y.; Qiu, S.; Han, Y. Integrating Spatial Continuous Wavelet Transform and Kernel Density Estimation to Identify Ecological Corridors in Megacities. *Landsc. Urban Plan.* **2020**, *199*, 103815. [CrossRef]
43. Lin, Q.; Eladawy, A.; Sha, J.; Li, X.; Wang, J.; Kurbanov, E.; Thomas, A. Remotely Sensed Ecological Protection Redline and Security Pattern Construction: A Comparative Analysis of Pingtan (China) and Durban (South Africa). *Remote Sens.* **2021**, *13*, 2865. [CrossRef]
44. Luo, Y.; Wu, J.; Wang, X.; Wang, Z.; Zhao, Y. Can Policy Maintain Habitat Connectivity under Landscape Fragmentation? A Case Study of Shenzhen, China. *Sci. Total Environ.* **2020**, *715*, 136829. [CrossRef]
45. Yu, Q.; Yue, D.; Wang, Y.; Kai, S.; Fang, M.; Ma, H.; Zhang, Q.; Huang, Y. Optimization of Ecological Node Layout and Stability Analysis of Ecological Network in Desert Oasis: A Typical Case Study of Ecological Fragile Zone Located at Deng Kou County (Inner Mongolia). *Ecol. Indic.* **2018**, *84*, 304–318. [CrossRef]
46. Zhou, D.; Song, W. Identifying Ecological Corridors and Networks in Mountainous Areas. *Int. J. Environ. Res. Public. Health* **2021**, *18*, 4797. [CrossRef]
47. Guo, X.; Zhang, X.; Du, S.; Li, C.; Siu, Y.L.; Rong, Y.; Yang, H. The Impact of Onshore Wind Power Projects on Ecological Corridors and Landscape Connectivity in Shanxi, China. *J. Clean. Prod.* **2020**, *254*, 120075. [CrossRef]
48. Jiang, H.; Peng, J.; Dong, J.; Zhang, Z.; Xu, Z.; Meersmans, J. Linking Ecological Background and Demand to Identify Ecological Security Patterns across the Guangdong-Hong Kong-Macao Greater Bay Area in China. *Landsc. Ecol.* **2021**, *36*, 2135–2150. [CrossRef]
49. Li, S.; Xiao, W.; Zhao, Y.; Lv, X. Incorporating Ecological Risk Index in the Multi-Process MCRE Model to Optimize the Ecological Security Pattern in a Semi-Arid Area with Intensive Coal Mining: A Case Study in Northern China. *J. Clean. Prod.* **2020**, *247*, 119143. [CrossRef]
50. Hou, Q.; Du, Y.; Dong, W.; Zeng, Z.; Zhang, L.; Duan, Y.; Hou, X. Smart City Oriented Ecological Corridor Layout of Sanshui River Basin in Arid Area of Loess Plateau. Sustain. *Energy Technol. Assess.* **2021**, *44*, 100993. [CrossRef]
51. Yang, R.; Bai, Z.; Shi, Z. Linking Morphological Spatial Pattern Analysis and Circuit Theory to Identify Ecological Security Pattern in the Loess Plateau: Taking Shuozhou City as an Example. *Land* **2021**, *10*, 907. [CrossRef]
52. Peng, L.; Chen, T.; Liu, S. Spatiotemporal Dynamics and Drivers of Farmland Changes in Panxi Mountainous Region, China. *Sustainability* **2016**, *8*, 1209. [CrossRef]
53. Wang, L.; Wen, C. Traditional Villages in Forest Areas: Exploring the Spatiotemporal Dynamics of Land Use and Landscape Patterns in Enshi Prefecture, China. *Forests* **2021**, *12*, 65. [CrossRef]
54. Cui, J.; Kong, X.; Chen, J.; Sun, J.; Zhu, Y. Spatially Explicit Evaluation and Driving Factor Identification of Land Use Conflict in Yangtze River Economic Belt. *Land* **2021**, *10*, 43. [CrossRef]
55. Gong, J.; Xie, Y.; Cao, E.; Huang, Q.; Li, H. Integration of InVEST-Habitat Quality Model with Landscape Pattern Indexes to Assess Mountain Plant Biodiversity Change: A Case Study of Bailongjiang Watershed in Gansu Province. *J. Geogr. Sci.* **2019**, *29*, 1193–1210. [CrossRef]
56. Mengist, W.; Soromessa, T.; Feyisa, G.L. Landscape Change Effects on Habitat Quality in a Forest Biosphere Reserve: Implications for the Conservation of Native Habitats. *J. Clean. Prod.* **2021**, *329*, 129778. [CrossRef]
57. Yohannes, H.; Soromessa, T.; Argaw, M.; Dewan, A. Spatio-Temporal Changes in Habitat Quality and Linkage with Landscape Characteristics in the Beressa Watershed, Blue Nile Basin of Ethiopian Highlands. *J. Environ. Manag.* **2021**, *281*, 111885. [CrossRef]
58. Li, M.; Zhou, Y.; Xiao, P.; Tian, Y.; Huang, H.; Xiao, L. Evolution of Habitat Quality and Its Topographic Gradient Effect in Northwest Hubei Province from 2000 to 2020 Based on the InVEST Model. *Land* **2021**, *10*, 857. [CrossRef]
59. Dai, Y. Identifying the Ecological Security Patterns of the Three Gorges Reservoir Region, China. *Environ. Sci. Pollut. Res.* **2022**, *29*, 45837–45847. [CrossRef] [PubMed]
60. Kouli, M.; Soupios, P.; Vallianatos, F. Soil Erosion Prediction Using the Revised Universal Soil Loss Equation (RUSLE) in a GIS Framework, Chania, Northwestern Crete, Greece. *Environ. Geol.* **2009**, *57*, 483–497. [CrossRef]
61. Fang, H.; Fan, Z. Assessment of Soil Erosion at Multiple Spatial Scales Following Land Use Changes in 1980–2017 in the Black Soil Region, (NE) China. *Int. J. Environ. Res. Public. Health* **2020**, *17*, 7378. [CrossRef]
62. Liu, L.; Chen, X.; Chen, W.; Ye, X. Identifying the Impact of Landscape Pattern on Ecosystem Services in the Middle Reaches of the Yangtze River Urban Agglomerations, China. *Int. J. Environ. Res. Public. Health* **2020**, *17*, 5063. [CrossRef]
63. Getis, A.; Ord, J.K. The Analysis of Spatial Association by Use of Distance Statistics. *Geogr. Anal.* **1992**, *24*, 189–206. [CrossRef]

64. Ostapowicz, K.; Vogt, P.; Riitters, K.H.; Kozak, J.; Estreguil, C. Impact of Scale on Morphological Spatial Pattern of Forest. *Landsc. Ecol.* **2008**, *23*, 1107–1117. [CrossRef]
65. Vogt, P.; Riitters, K. GuidosToolbox: Universal Digital Image Object Analysis. *Eur. J. Remote Sens.* **2017**, *50*, 352–361. [CrossRef]
66. Ma, B.; Chen, Z.; Wei, X.; Li, X.; Zhang, L. Comparative Ecological Network Pattern Analysis: A Case of Nanchang. *Environ. Sci. Pollut. Res.* **2022**, *29*, 37423–37434. [CrossRef]
67. Taylor, P.D.; Fahrig, L.; Henein, K.; Merriam, G. Connectivity Is a Vital Element of Landscape Structure. *Oikos* **1993**, *68*, 571–573. [CrossRef]
68. Nie, W.; Shi, Y.; Siaw, M.J.; Yang, F.; Wu, R.; Wu, X.; Zheng, X.; Bao, Z. Constructing and Optimizing Ecological Network at County and Town Scale: The Case of Anji County, China. *Ecol. Indic.* **2021**, *132*, 108294. [CrossRef]
69. Bodin, Ö.; Saura, S. Ranking Individual Habitat Patches as Connectivity Providers: Integrating Network Analysis and Patch Removal Experiments. *Ecol. Model.* **2010**, *221*, 2393–2405. [CrossRef]
70. Fan, F.; Liu, Y.; Chen, J.; Dong, J. Scenario-Based Ecological Security Patterns to Indicate Landscape Sustainability: A Case Study on the Qinghai-Tibet Plateau. *Landsc. Ecol.* **2021**, *36*, 2175–2188. [CrossRef]
71. Ding, M.; Liu, W.; Xiao, L.; Zhong, F.; Lu, N.; Zhang, J.; Zhang, Z.; Xu, X.; Wang, K. Construction and Optimization Strategy of Ecological Security Pattern in a Rapidly Urbanizing Region: A Case Study in Central-South China. *Ecol. Indic.* **2022**, *136*, 108604. [CrossRef]
72. Fu, Y.; Shi, X.; He, J.; Yuan, Y.; Qu, L. Identification and Optimization Strategy of County Ecological Security Pattern: A Case Study in the Loess Plateau, China. *Ecol. Indic.* **2020**, *112*, 106030. [CrossRef]
73. Zhang, Y.-Z.; Jiang, Z.-Y.; Li, Y.-Y.; Yang, Z.-G.; Wang, X.-H.; Li, X.-B. Construction and Optimization of an Urban Ecological Security Pattern Based on Habitat Quality Assessment and the Minimum Cumulative Resistance Model in Shenzhen City, China. *Forests* **2021**, *12*, 847. [CrossRef]
74. Yuan, Y.; Bai, Z.; Zhang, J.; Xu, C. Increasing Urban Ecological Resilience Based on Ecological Security Pattern: A Case Study in a Resource-Based City. *Ecol. Eng.* **2022**, *175*, 106486. [CrossRef]
75. Chen, C.; Shi, L.; Lu, Y.; Yang, S.; Liu, S. The Optimization of Urban Ecological Network Planning Based on the Minimum Cumulative Resistance Model and Granularity Reverse Method: A Case Study of Haikou, China. *IEEE Access* **2020**, *8*, 43592–43605. [CrossRef]
76. Guan, D.; Jiang, Y.; Cheng, L. How Can the Landscape Ecological Security Pattern Be Quantitatively Optimized and Effectively Evaluated? An Integrated Analysis with the Granularity Inverse Method and Landscape Indicators. *Environ. Sci. Pollut. Res.* **2022**, *29*, 41590–41616. [CrossRef]
77. Li, S.; Zhao, Y.; Xiao, W.; Yue, W.; Wu, T. Optimizing Ecological Security Pattern in the Coal Resource-Based City: A Case Study in Shuozhou City, China. *Ecol. Indic.* **2021**, *130*, 108026. [CrossRef]
78. Zhu, Q.; Yu, K.; Li, D. The width of ecological corridor in landscape planning. *Acta Ecol. Sin.* **2005**, *25*, 2406–2412. (In Chinese)
79. Tang, F.; Zhou, X.; Wang, L.; Zhang, Y.; Fu, M.; Zhang, P. Linking Ecosystem Service and MSPA to Construct Landscape Ecological Network of the Huaiyang Section of the Grand Canal. *Land* **2021**, *10*, 919. [CrossRef]
80. Burkhard, B.; Kroll, F.; Nedkov, S.; Müller, F. Mapping Ecosystem Service Supply, Demand and Budgets. *Ecol. Indic.* **2012**, *21*, 17–29. [CrossRef]
81. Zhang, L.; Peng, J.; Liu, Y.; Wu, J. Coupling Ecosystem Services Supply and Human Ecological Demand to Identify Landscape Ecological Security Pattern: A Case Study in Beijing–Tianjin–Hebei Region, China. *Urban Ecosyst.* **2017**, *20*, 701–714. [CrossRef]

International Journal of
Environmental Research and Public Health

MDPI

Article

Landscape Ecological Risk Assessment Based on Land Use Change in the Yellow River Basin of Shaanxi, China

Zhiyuan Zhu [1,2], Zhikun Mei [1,2], Xiyang Xu [1,2], Yongzhong Feng [1,2] and Guangxin Ren [1,2,*]

1 College of Agronomy, Northwest A & F University, Xianyang 712100, China; zhuzhiyuan@nwafu.edu.cn (Z.Z.); zyzhu@nwafu.edu.cn (Z.M.); 2020050039@nwafu.edu.cn (X.X.); fengyz@nwsuaf.edu.cn (Y.F.)
2 The Research Center of Recycle Agricultural Engineering and Technology of Shaanxi Province, Xianyang 712100, China
* Correspondence: rengx@nwsuaf.edu.cn

Abstract: The Yellow River Basin in Shaanxi (YRBS) has a relatively fragile ecological environment, with severe soil erosion and a high incidence of natural and geological disasters. In this study, a river basin landscape ecological risk assessment model was constructed using landscape ecology principles to investigate the temporal and spatial evolution, as well as the spatial autocorrelation characteristics of landscape ecological risks in the YRBS over a 20-year period. The main findings from the YRBS were that the land use types changed significantly over the span of 20 years, there was spatial heterogeneity of the landscape pattern, and the ecological risk value was positively correlated. The threat of landscape ecological risks in YRBS is easing, but the pressure on the ecological environment is considerable. This study provides theoretical support administrative policies for future ecological risk assessment and protection, restoration measures, and control in the Yellow River Basin of Shaanxi Province.

Keywords: Yellow River Basin; ecological risk assessment; Shaanxi Province; land use

Citation: Zhu, Z.; Mei, Z.; Xu, X.; Feng, Y.; Ren, G. Landscape Ecological Risk Assessment Based on Land Use Change in the Yellow River Basin of Shaanxi, China. *Int. J. Environ. Res. Public Health* **2022**, *19*, 9547. https://doi.org/10.3390/ijerph19159547

Academic Editors: Wei Song and Hualin Xie

Received: 14 July 2022
Accepted: 2 August 2022
Published: 3 August 2022

1. Introduction

The Yellow River is the second largest river in China and the sixth largest river in the world; it is considered the mother river of China. Thus, the protection and development of the Yellow River Basin is important for the peace and prosperity of the people [1–3]. Presently, with the rapid development of urbanization, the interference of human activities on the natural landscape has increased, and the considerable changes in land cover have led to significant changes in the landscape pattern of the basin [4–6]. Therefore, studying the temporal and spatial evolution characteristics of land use types and controlling the overall ecological risk in the watershed is beneficial for prescribing the best methods to counteract landscape degradation based on the different risk levels, and give appropriate management and control suggestions. Sustainable development in the Yellow River Basin has important practical and theoretical significance [7–10].

A watershed is a comprehensive ecological regional system that connects natural ecosystems (land cover types and water cycles) with socio-economic systems (society and population) [11–13]. Owing to more complex, holistic, and special location characteristics, the excessive disturbance of a certain element in the basin ecosystem will inevitably threaten the overall stability. The excessive land use changes by humans leads to fragile regional ecosystems [14–17]. In this study, we performed landscape ecological risk assessments for landscape changes in land use and analyzed the threat of human activities to regional ecosystems. This new ecological management tool provides us with the theoretical support for policy-making, sustainable development, and ecological environment management in river basin risk management and control [18–20]. Therefore, this type of assessment has become the main method that is used by scholars to carry out regional ecosystem assessments and is currently a trending topic in landscape ecology [21–24].

The history of the development of ecological risk assessments has gone through four stages [25–30]: (1) the infancy stage (before the 1980s), based on qualitative analysis, mainly focusing on toxicological research on the impact of pollutants on the environment and humans [31–33]; (2) human health risk assessment stage (1980s), where the evaluation method was changed to quantitative and the evaluation process and framework were gradually systematic, focusing mainly on human health and their exposure to chemical pollution [34,35]; (3) ecological risk assessment stage (1990s), where the focus changed from environmental and human health risk assessments to ecological risk assessments, and the relevant standard documents of ecological risk assessment were promulgated by many countries and organizations, trying to transform the human health risk assessment framework to the ecological risk assessment framework. Thus, in 1998, the "Guidelines for Ecological Risk Assessment" were released, pioneering progress in the theory and technology of ecological risk assessments [36–39]. Finally, (4) the regional, landscape, and watershed ecological risk assessment stage (late 1990s to the present), where ecological risk assessment is combined with the theory of landscape ecology, and the research scale extends from population and ecosystem assessments to regional, landscape, and watershed assessments [40–42]. Sources and risk receptors both present a period of comprehensive risk assessment from a single development to a variety of risks, such as urbanization, human activities, meteorological changes, and land use changes and are all considered in ecological risk assessments, which uses spatial analysis tools to build models that are based on multi-scale and multi-factor analyses [43–46].

Shaanxi Province is in the middle reaches of the Yellow River and approximately 70% of the land area and 80% of the population belong to the Yellow River Basin [47]. It is an education, technology, energy, and equipment manufacturing base and the core area of economic development in China. Shaanxi is in an important regional location and has exceptional ecological functions; however, there are still practical problems such as a fragile ecological environment, shortage of water resources, insufficient carrying capacity, and uncoordinated regional social and economic development in the Yellow River Basin. The hilly and gully area of the Loess Plateau in northern Shaanxi is the main source of silt entering into the Yellow River [48–50]. A cumulative total of 2.69×10^4 km^2 of farmland has been restored to forest and grassland, and 15.7 million mu of desertified land has been treated. The Shaanxi section of the basin accounts for more than 83% of the industrial water and more than 78% of the domestic water in the province. The water resource of YRBS provides only 447 m^3 of water per capita [51], less than one-fifth of the national average. Therefore, in-depth development of the temporal and spatial evolution of land use and landscape patterns, and landscape ecological risk assessments that are based on the theory of landscape ecology in the YRBS is of great strategic significance to scientifically promote the ecological protection and sustainable development of the Yellow River Basin. This will help rationally allocate and utilize land resources and maintain a balanced state of economic and agricultural development [52–55].

This study was conducted in the Yellow River Basin in Shaanxi. We aim to provide a scientific basis for the ecological protection of the Yellow River Basin in Shaanxi. First, any law changes regarding the different land use types in the basin from 2000 to 2020 were determined; second, the spatial scale of the landscape pattern was studied, and the changes in the landscape pattern were analyzed from both the landscape and patch-type level. Based on the ecological risk assessment of the regional land use landscape pattern of the area, the temporal and spatial evolution and spatial correlation characteristics of the landscape ecological risk were revealed. Finally, considering the results of landscape ecological risk assessment, corresponding ecological risk management countermeasures are proposed.

2. Materials and Methods

2.1. Study Area

The Shaanxi section of the Yellow River Basin is in the center of the middle reaches of the Yellow River and passes through the north-central part of Shaanxi Province (Figure 1). It spans approximately 400 km and is connected to Gansu in the west, the Yellow River and Shanxi in the east, Inner Mongolia in the north, and the main beam of the Qinling Mountains in the south, 17.49% of the total area. This section has complex topography, such as undulating mountain ranges, vertical and horizontal rivers, high-lying areas in the north and south and low-lying in the middle, and slopes from west to east. The Shaanxi section of the Yellow River Basin is composed of 79 counties (districts and cities) in 8 cities, namely Yulin, Yan'an, Tongchuan, Baoji, Xianyang, Xi'an, Weinan, and Shangluo, with a population of about 29.15 million, accounting for 75.41% of the total population of the province [56]. It spans two climatic zones, roughly bounded by the Great Wall of China, the north is located in the middle temperate zone and the south is in the warm temperate zone. The natural vegetation is considerably varied, with grasslands and shrubs in the north and forests in the south.

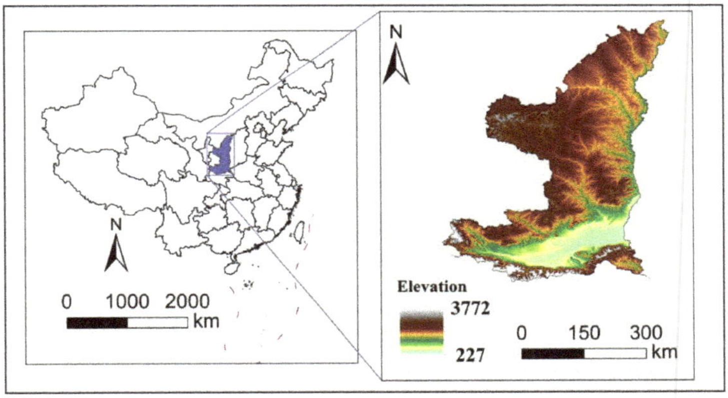

Figure 1. The location of the Yellow River Basin in Shaanxi Province, China.

2.2. Data Collection and Processing

There were three periods of land use data from Shaanxi Province (shp. Format), namely, 2000, 2010, and 2020, that were used in this study, derived from the GlobeLand 30 surface cover dataset (raster data, with a resolution of 30 m) (http://www.Globallandcover.com/ (accessed on 1 July 2022). The maps of the administrative division of the study area and the water system of the Yellow River Basin were obtained from the Resource and Environmental Science Data Center of the Chinese Academy of Sciences (http://www.resdc.cn (accessed on 1 July 2022)). The vector boundary of the YRBS used the hydrological model in ArcGIS software v10.2 (Esri, Redlands, CA, USA) The watershed delineation tool in ArcSWAT used the automatic watershed delineation command to generate watershed divisions. Based on the national land use classification system and according to the land use characteristics and research purposes of YRBS, the land use types in the study area were divided into eight categories: cultivated land, forest, grassland, shrubland, wetland, water body, artificial surface, and bare land.

2.3. Methods

In order to study the temporal and spatial variation characteristics of landscape ecological risk in the YRBS, the workflow is as follows (Figure 2). First, based on land use data in 2000, 2010, and 2020, Land use dynamics and the Land Use Transfer Matrix were used to explore the process of land use change in the YRBS. Then, we evaluated the landscape ecological risk and discuss its temporal and spatial variation characteristics. Finally, the spatial autocorrelation of ecological risk index was analyzed by Moran's I index and local spatial autocorrelation analysis method.

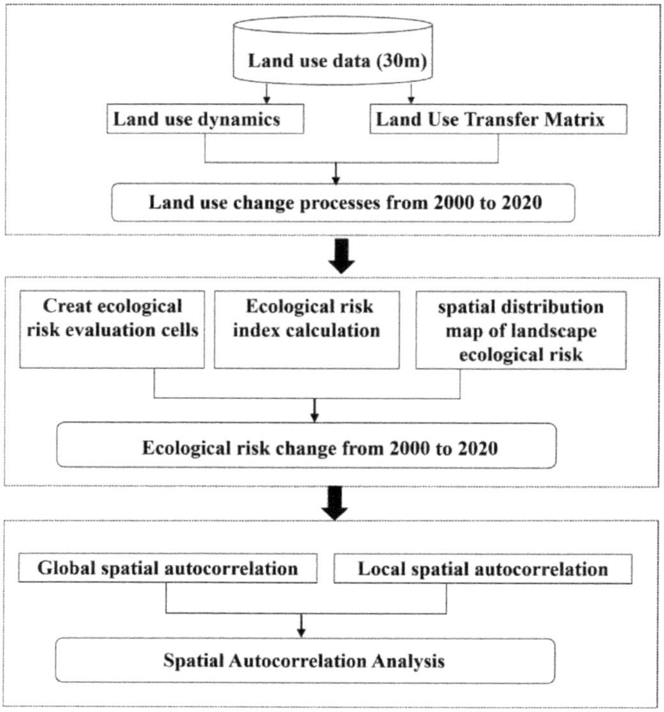

Figure 2. The workflow of the research.

2.3.1. Land Use Dynamics

The dynamic degree of land use quantitatively expresses the speed of land use change in certain periods, measures the difference in land use change between different regions, and predicts the future trend of land use change in the region. The two land-use dynamic degrees that were used in this study were the single land use K and the comprehensive land use S. The larger the absolute value of the single land use dynamic degree is, the faster the transformation speed of the land use type. The comprehensive land use dynamic degree indicates the degree of land use change in the study area from a macro perspective, and the larger the dynamic degree is the more severe the degree of change. The specific equations are as follows:

$$K = \frac{U_m - U_n}{U_n} \times \frac{1}{T} \times 100\%$$

$$S = \sum_{ij}^{n} \frac{\Delta S_{i-j}}{S_i} \times \frac{1}{T} \times 100\%$$

where K is the dynamic degree of a certain land use type in the research period; U_n and U_m are the area (km^2) of the land use types in the study area at the beginning and end of a certain period, respectively; and T is the research time (years). S represents the comprehensive land dynamic degree, $S \ldots$ is the total area (km^2) of the i-type land use that is converted to other land use types in the T period

2.3.2. Land Use Transfer Matrix

For the land use transition matrix, we used the Markon transition probability matrix. The Markon model can not only directly and specifically quantify the structural characteristics between the changes of land use types, but also show the number of transfers between different land types; thus, quantitatively showing the degree of system analysis on the system state and state transfer. The following equation was used:

$$S_{ij} = \begin{bmatrix} S_{11} & S_{12} & \cdots S_{1n} \\ S_{21} & S_{22} & \cdots S_{2n} \\ \vdots & \vdots & \vdots \quad \vdots \\ S_{n1} & S_{n2} & \cdots S_{nn} \end{bmatrix}$$

where S represents the transition matrix of land use change; n is the total number of different land types ($n = 8$); i and j represent the initial and final land types, respectively, in the study area ($i, j = 1, 2, \ldots, n$); and S is the area of the ith land type that is converted to the area of the jth land type, the larger the value, the more severe the change, and vice versa.

2.3.3. Landscape Ecological Risk Assessment

According to the characteristics of the area of the YRBS, the moderateness of the data sampling workload and the accuracy of the evaluation unit, the study area is divided into a 10 × 10 km grid size by using the equal-spaced systematic sampling method and the Create Fishnet tool of ArcGIS 10.2 software (Esri, Redlands, CA, USA). A total of 1500 risk cells (as shown in Figure 3). Then, ArcGIS 10.2 software was used to calculate the landscape ecological risk value of each risk area, and the ecological risk index was assigned to the center of each risk area, and then the ordinary Kring interpolation method of spatial interpolation was used to obtain the spatial distribution map of landscape ecological risks in the YRBS. In this way, the landscape ecological risk assessment of the entire watershed can be carried out.

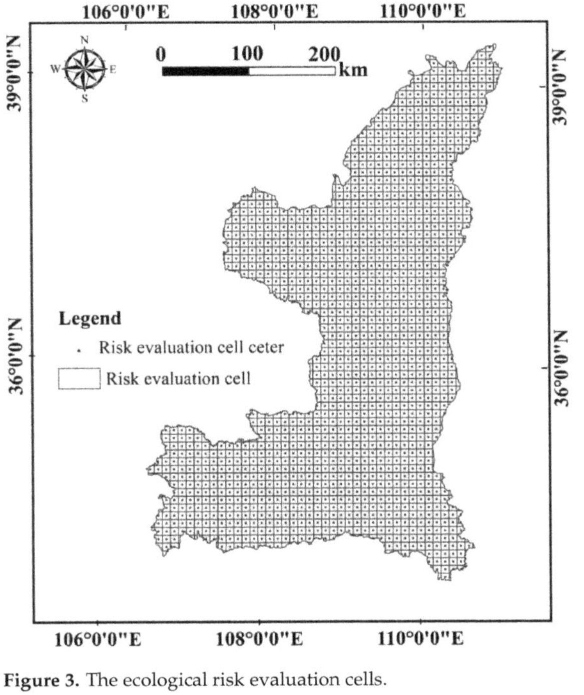

Figure 3. The ecological risk evaluation cells.

The landscape disturbance index (E_i) and landscape vulnerability index (F_i) were selected in this study for constructing a comprehensive Ecological Risk Index (*ERI*) model for the YRBS. The landscape ecological risk index was calculated as follows:

$$ERI_i = \sum_{i=1}^{n} \frac{A_{ki}}{A_k} \times R_i$$

where ERI_i is the landscape ecological risk index of the ith risk unit, n is the number of landscape types, A_{ki} is the area of the ith of landscape type in the kth risk unit, and A_k is the total area of the kth risk unit. R_i is the landscape loss index of the ith landscape type, which is calculated from the landscape disturbance degree and vulnerability index. The equations for each landscape pattern index in the model are shown in Table 1.

Table 1. Calculation formula of landscape index and ecological significance.

Index	Calculation Formula	Ecological Significance
Landscape loss degree index (R_i)	$R_i = E_i \times F_i$	R_i indicates the degree of loss of natural properties of ecosystems represented by different landscape types when they are subjected to natural and anthropogenic disturbances [54].
Landscape disturbance index (E_i)	$E_i = aC_i + bN_i + cD_i$	E_i describes the extent to which ecosystems located in different landscape types are disturbed by human activities and characterizes differences related to maintenance of ecological stability of different landscape types [57]; a, b, and c represent weights of the corresponding landscape indices; according to results of previous studies, values of a = 0.5, b = 0.3, and c = 0.2 are assigned.
Landscape fragmentation index (C_i)	$C_i = \frac{n_i}{A_i}$	Describes the degree of fragmentation of a landscape type in the region at a given time; such that, the higher its value, the lower the stability within the landscape unit and the greater the heterogeneity and discontinuity among patches [58]; n_i denotes the number of patches of landscape type *i* and A_i denotes the total area of landscape type *i*.
Landscape dominance index (D_i)	$D_i = \frac{Q_i + M_i}{4} + \frac{L_i}{2}$	The higher the value, the greater the influence of the landscape type on the overall landscape pattern [59]. Q_i = number of samples in which patch *i* occurs/total number of samples; M_i = number of patch *i*/total number of patches; and L_i = area of patch *i*/total area of samples.
landscape separateness index (N_i)	$N_i = \frac{A}{2A_i}\sqrt{\frac{n_i}{A}}$	The greater the degree of separation between different patches in a landscape type, the more discrete the distribution of the landscape type in the region for a correspondingly higher degree of fragmentation [60]; A is the total area of the landscape; N_i is the distance index of landscape type *i*.
Landscape vulnerability index (F_i)	Based on the previous studies	The higher the value, the more vulnerable and unstable the landscape type is and the more likely it will suffer ecological losses and physical changes due to external disturbances [61]. Based on the previous studies, in this study [62], vulnerability indices of six landscape types were assigned as follows: unused land 6, water 5, cultivated land 4, grassland 3, woodland 2, and residential land 1, with the landscape vulnerability index F_i obtained after normalization.

Finally, the natural breakpoint method is used to divide the ecological risk into five grades, as shown in Table 2.

Table 2. Landscape ecological risk classification in the Yellow River Basin, Shaanxi Province, China.

Ecological Risk	Risk Level				
	Low	Low-Medium	Medium	Medium-High	High
rank	I	II	III	IV	V
value	0.0135 < ERI	0.0135 ≤ ERI < 0.030	0.030 ≤ ERI < 0.060	0.060 ≤ ERI < 0.099	ERI ≥ 0.099

2.3.4. Spatial Autocorrelation Analysis

(1) Global spatial autocorrelation

Global spatial autocorrelation analysis was used to measure the agglomeration characteristics of the attribute values in the entire study area, reflecting the approximation of the attribute values of adjacent units. Generally, *Moran's I* index was used to characterize the degree and significance of spatial autocorrelation of ecological risks in the study area. The equation is as follows:

$$Moran's\ I = \frac{n\sum_i \sum_j W_{ij}(x_i - \overline{x})(x_j - \overline{x})}{\left(\sum_i \sum_j w_{ij}\right)\sum_i (x_j - \overline{x})^2}$$

where x_i and x_j are the attribute values of the variables in the adjacent units of the region, W_{ij} is the spatial weight matrix, and $\overline{x}$ is the average attribute value. The value range of *Moran's I* index is (1, 1) which means that the positive correlation of similar spatial proximity transitions to the negative correlation of spatial proximity dissimilarity. When *Moran's I* > 0, the spatial correlation is positive and the unit attribute value presents spatial clustering characteristics, and the closer the value is to 1, the higher the degree of agglomeration. When *Moran's I* < 0, the space is negatively correlated, and the unit attribute values show spatially discrete characteristics. When *Moran's I* = 0, there is no spatial correlation, and the unit attribute values are random with an independent distribution status.

(2) Local spatial autocorrelation

The global spatial autocorrelation represents the overall spatial state of the attribute value and cannot reflect the specific location of the agglomeration or abnormal attribute value in the spatial distribution. In this case, the local autocorrelation method needs to be used for further explanation. Local spatial autocorrelation mainly reveals the heterogeneity of the spatial distribution of local unit attribute values, including all spatial unit attribute values in the study area. Local *Moran's I* (I_i) is used to express the local spatial autocorrelation and it is calculated as follows:

$$I_i = \frac{(x_i - \overline{x})\left[(n-1) - \overline{x}^2\right]}{\sum_{j=1}^{n} x_{ij}^2 \sum_{i=1}^{n} \sum_{j=1}^{n} W_{ij}(x_j - \overline{x})}$$

where x_i and x_j are the attribute values of the variable in the adjacent units of the region, W_{ij} is the spatial weight matrix, and $\overline{x}$ is the average attribute value. When $I_i \geq 0$, the attribute value of the *i*th unit is similar to the attribute value of the adjacent unit, and the attribute value of the unit presents spatial discrete characteristics, which is a positive spatial correlation. When $I_i < 0$, the attribute value of the *i*th unit is related to the adjacent unit, the attribute values of the units are quite different, and the unit attribute values show spatial discrete characteristics, which is a negative spatial correlation.

3. Results

3.1. Land Use Change Processes from 2000 to 2020

3.1.1. Analysis of Land Use Dynamics Change

Using the dynamic degree formula to calculate the three-phase land use data in the Shaanxi Yellow River Basin from 2000 to 2020, we obtained the dynamic degree of each land use type in the study area from 2000 to 2010, 2010 to 2020, and 2000 to 2020 as shown in Table 3.

Table 3. Change in the area and single dynamic degree of land types in the Yellow River Basin, Shaanxi Province, China.

Land Type	2000–2010		2010–2020		2000–2020	
	Area Change (km^2)	Single Dynamics (%)	Area Change (km^2)	Single Dynamics (%)	Area Change (km^2)	Single Dynamics (%)
Cultivated Land	−1000.56	−0.17	−197.86	−0.03	−1198.41	−0.11
Forest Land	351.25	0.09	−118.57	−0.03	232.68	0.03
Grassland	−502.24	−0.11	−1232.16	−0.28	−1734.4	−0.2
Shrubland	159.93	3.14	−28.45	−0.41	131.48	1.36
Wetland	−54.8	−2.03	−10.77	−0.51	−65.56	−1.28
Water Body	−40.31	−0.69	30.08	0.56	−10.23	−0.09
Artificial Surface	1072.16	3.11	1665.78	3.59	2737.94	4.21
Bare Ground	14.47	0.09	−103.61	−0.64	−89.13	−0.29
Comprehensive Dynamics (%)	0.108		0.114		0.111	

As shown in Table 3, there are discernable differences in the single dynamic degree of each land use type during the study period, and the overall performance adheres to the following order: artificial surface > shrubland > wetland > bare land > grassland > cultivated land > water body > forest. The type of land cover with the greatest variation in single dynamics is always artificial surfaces. Between 2000 and 2010, the greatest change in the single dynamic degree was that of shrubland and artificial surfaces, which changed by 3.14 and 3.11%, respectively, and the area increased by 159.93 km^2 and 1072.16 km^2, respectively. Between 2010 and 2020, except for artificial surfaces and water bodies, the single dynamic degree of the utilization type is in a decreasing state, and the artificial surface is still the land use type with the largest change in single dynamic degree. The area has increased by 1665.78 km^2 in 10 years, and the dynamic degree is 3.59%. The single dynamic degree of the water bodies initially had a negative change but turned positive. The arable land, grassland, and wetland continued to decrease, and the change rate of grassland accelerated significantly compared to that in 2000–2010, the area decreased by 1232.16 km^2, and the change rate of arable land and wetland slowed down compared to the previous 10 years. Shrubland showed a significant slowdown with a single dynamic degree of 0.41% and went from a positive change at the beginning to a negative change. In general, artificial surfaces have grown positively throughout the research period from 2000 to 2020. The area increased by 2737.94 km^2 over 20 years, and the single dynamic degree is at most 4.21%. In 2010, there was rapid economic and social development in China, the pace of urbanization accelerated, anthropogenic interference increased, the change rate of land use types accelerated, and the change rate of man-made surfaces increased faster than that of other land types. In addition, all the remaining land use types showed fluctuating changes except for cultivated land, grassland, and wetlands, which showed a constant decreasing trend over the span of 20 years.

3.1.2. Analysis of Land Use Transfer Change

To further visualize the spatial evolution characteristics and mutual transformation rules of various land cover types in the Yellow River Basin of Shaanxi Province, we used GIS spatial analysis technology and the land use transfer matrix model to analyze the direction and quantity of changes between various land use types.

Figure 4A shows that during the conversion process of land use types in the Yellow River Basin of Shaanxi Province from 2000 to 2010, the conversion of cultivated land was the most severe, with a total transfer area of 3194.05 km^2. For artificial, grass, and forest land, the transfer area was 1559.61, 989.53, and 450.16 km^2, respectively; the conversion of grassland to other land types was more severe, with a transfer area of 1690.09 km^2. The top three areas with the largest conversion from grassland to other land types are, cultivated land, forest, and shrubland with transfer areas of 1060.96, 307.76, and 150.49 km^2, respectively. The transfer area of artificial surfaces is 615.90 km^2, the main transfer type is cultivated land with an area of 589.48 km^2, and the transfer area of forest is 433.85 km^2. The transfer areas were sorted as follows: Cultivated land > grassland and the transfer-out areas were 324.51 km^2 and 90.4 km^2, respectively. Wetlands and water bodies were relatively small with transfer areas of 168.92 km^2 and 252.12 km^2, respectively. The total area of shrubland and bare land was small, and the conversion was not significant. Figure 4B shows the conversion of land use types from 2010 to 2020, the conversion of grassland was most severe, with a total area of 5858.58 km^2, followed by cultivated land, forest, and artificial surfaces with transfer areas of 3155.60, 1405.16, and 565.91 km^2 respectively. The conversion of cultivated land is frequent, and the total transferred-out area is 4922.38 km^2. The top three areas with the largest area of cultivated land that were converted to other land types were grassland, artificial surfaces, and forest with transfer areas of 2450.65, 1583.78, and 688.29 km^2, respectively. The total forest transfer area is 2291.98 km^2, and the transferred land types are mainly grassland and cultivated land, with an area of 1465.67 and 719.80 km^2, respectively. Shrubland, artificial surfaces, and bare land have relatively small transfer-out areas of 422.106, 576.124, and 524.71 km^2, respectively. The total transfer area of wetlands was small, and the transformation was not significant.

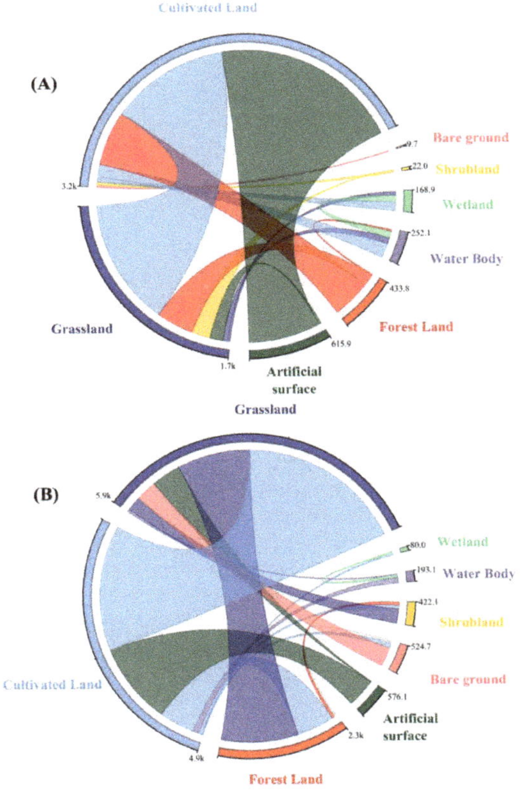

Figure 4. Land use type transfer area (km^2) matrix of the YRBS from 2000 to 2010 (**A**) and 2010 to 2020 (**B**).

3.2. Analysis of Spatial and Temporal Changes in Ecological Risks

According to the average values and change trend of landscape ecological risk from 2000 to 2020 (Figure 5), the overall landscape ecological risk of the Shaanxi Yellow River Basin is at a lower risk level, and the temporal change showed an increasing trend at first and then a decreasing trend. From 2000 to 2010, the increase rate was relatively clear, increasing from 0.0198 to 0.0203, and the decrease rate was significant from 2010 to 2020, decreasing from 0.203 to 0.0197. The landscape ecological risk value was in a state of decline over the span of 20 years (2000–2020). Since the 18th National Congress of the Communist Party of China proposed the concept of an ecological civilization to build a beautiful China, Shaanxi Province has actively responded by implementing a series of important ecological protection and restoration projects directly impacting the Yellow River Basin ecosystem. The landscape ecological risk showed a benign development trend; thus, it is necessary to coordinate the relationship between economic construction and environmental protection in the future.

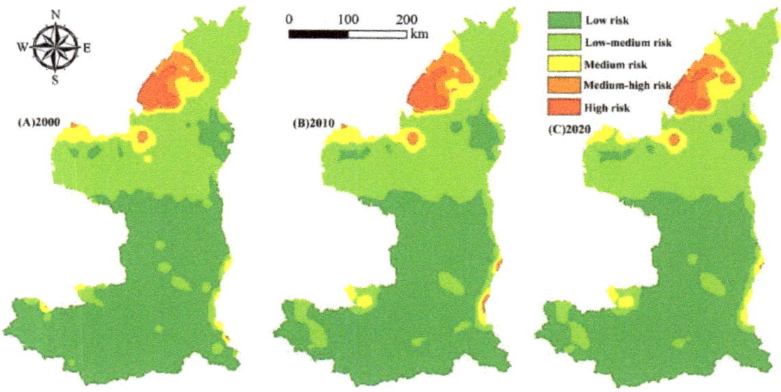

Figure 5. Spatial distribution changes in landscape ecological risk areas in the Yellow River Basin, Shaanxi Province from 2000 to 2020.

The proportion of different ecological risk areas in the Yellow River Basin from 2000 to 2020 and the temporal change characteristics of ecological risk levels in the basin over 20 years is shown in Figure 6. These data show the area distribution of each ecological risk area and indicate that the low-risk areas are always the largest and the high-risk areas are the smallest (low-risk > medium-risk > high-risk area). For example, in 2020, the low-risk areas accounted for 54.07% of the total basin area, the low-medium-risk areas accounted for 35.35%, and the medium-risk, medium-high-risk, and high-risk areas accounted for only 5.92%, 2.77%, and 1.89%, respectively.

Considering the temporal change characteristics of the ecological risk areas over the span of 20 years, the low-risk and low-medium-risk areas have always been dominant in terms of proportion, but there are still significant changes in the ecological risk areas. The size of the low-risk areas first decreased and then increased, decreasing from 74,208.75 km^2 in 2000 to 71,238.50 km^2 in 2010, and increasing to 72,031.50 km^2 in 2020. The low-risk areas showed first an increase and then a decrease. However, the increase was evident in the first 10 years, and only a slight decrease was observed from 2010 to 2020. The size of medium-risk areas increased continuously, but the increase in the first 10 years was substantially higher than that of the next 10 years. The area increased by 579.25 km^2 in 2010 and 175.5 km^2 in 2020. The size of the high-risk areas showed relatively clear growth from 2000 to 2010, increasing by 357.25 km^2, and then declined from 2010 to 2020, decreasing by 398.75 km^2. The size of the high-risk areas continued to decrease throughout the study period with a total decrease of 401.5 km^2 over the span of 20 years.

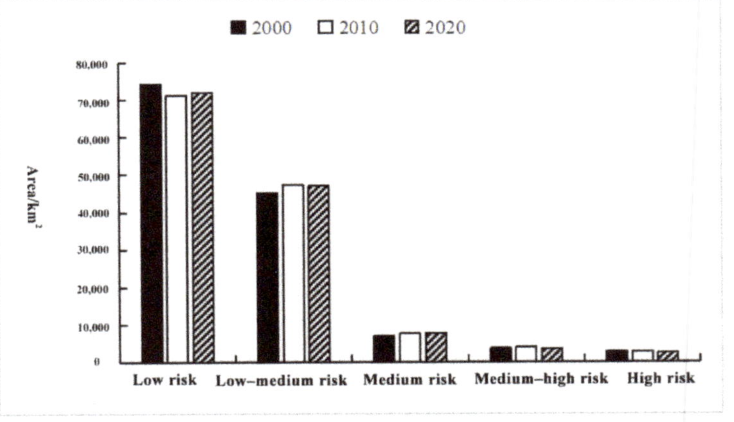

Figure 6. Area (km^2) change of landscape ecological risk area of the Yellow River Basin, Shaanxi Province from 2000 to 2020.

3.3. *Analysis of the Spatial Pattern of Ecological Risks*

3.3.1. Global Spatial Autocorrelation Analysis

According to Figure 7, the global Moran' I index of the landscape ecological risk values from 2000, 2010, and 2020 all exceeded 0.5, and were 0.698, 0.645, and 0.620, respectively, indicating that the landscape ecological risk values in the study area were positively correlated.

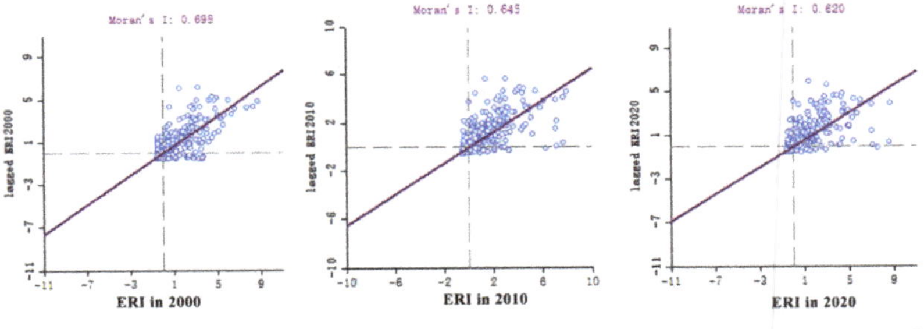

Figure 7. Moran's I index scatter distribution of the landscape ecological risk values in the Shaanxi Yellow River Basin from 2000 to 2020.

In addition, the distribution of the scattered points in Figure 4 is close to the regression line, indicating that the ecological risk values of the watershed landscape have characteristics of agglomeration in the spatial distribution. Meanwhile, the z-scores of the landscape ecological risk values from 2000, 2010, and 2020 were 36.281, 33.774, and 32.515, respectively, all exceeding 1.65, indicating that the elements in the spatial distribution are non-random processes, and the possibility of a random generation of clustering patterns is unlikely. Additionally, the calculated p-values are all equal to 0.001, indicating that the spatial autocorrelation is significant at the 99.9% confidence level.

3.3.2. Local Spatial Autocorrelation Analysis

According to Figure 8, the correlation changes of the landscape ecological risk index are consistent with the risk distribution map of ordinary kriging interpolation of the landscape ecological risk values. The ecological risk values of the watershed landscape are mainly distributed in high-high (H-H) and low-low(L-L) agglomeration. This area belongs to the ecological protection barrier area of the Mu Us Sandy Land, which has a high degree of land loss and a fragile natural ecosystem, reducing its ability to resist risks; thus, forming a high-risk area cluster that is consistent with the distribution of low-risk areas. This area is mainly dominated by cultivated land, forest, and artificial surface land types. It is relatively flat, landscape loss is low, landscape internal structure is stable, and anti-interference ability is strong, forming a cluster of low-risk areas.

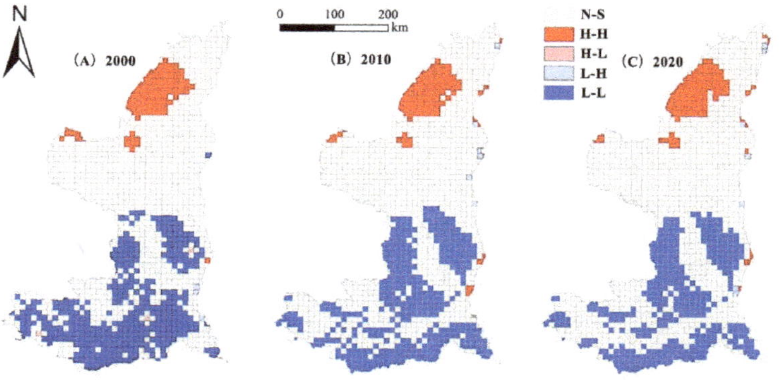

Figure 8. Local spatial autocorrelation distribution map of landscape ecological risks in the Yellow River Basin in Shaanxi Province from 2000 to 2020.

4. Discussion

4.1. Tempo-Spatial Changes of the Land Use and Landscape Ecological Risk in YRBS

Land use in YRBS experienced dramatic changes from 2000 to 2020. On the whole, the land use change was affected by human activities and natural factors. This is consistent with other research findings [63,64]. We found that cultivated land, forest land, and grassland had the largest changes and were the main land use types in the YRBS. It is closely related to the continuous promotion of ecological protection policies in Shaanxi Province. Shaanxi Province is one of the first provinces in China to pilot the policy project of returning farmland to forests and grasslands [65,66]. A series of policies of returning farmland to forests and grasslands have enabled the restoration of forest land and grassland areas, which relieved the pressure that was brought by the ecological footprint to a certain extent. In order to further improve the ecological carrying capacity of the YRBS and build a good ecological environment, it is necessary to continue to implement strict ecological environmental protection policies, and promoting the construction of ecological civilization is the key to promoting the sustainable development of the basin [67].

Our study found that high-risk areas in the YRBS are located in northern Shaanxi. This is related to the Loess Plateau region in the north, which has been studied by a large number of scholars [68–70]. However, the overall ecological risk management and control in northern Shaanxi has achieved preliminary results in the past 20 years. Ecological restoration projects such as forest (grass) and slope farmland improvement are closely related. A series of ecological restoration projects have adjusted the land use structure to a certain extent [71,72]. The area of cultivated land, forest land, and water area in northern Shaanxi has increased, which has improved the ecological conditions for agricultural production. However, there are still some problems of ecological degradation in the region. With the acceleration of urbanization, the demand for construction land has expanded rapidly, and human activities have exacerbated the division and occupation of cultivated land.

It is worth mentioning that we have studied the temporal and spatial evolution characteristics of ecological risks in the YRBS in the past 20 years, which can reflect the long-term trend of ecological quality in the basin to a certain extent. Overall, the distribution of risk areas is relatively stable. It can be judged that the vegetation of the YRBS has recovered significantly, and the ecological environment construction has achieved remarkable results, but the region is still a relatively fragile ecological environment [73,74]. In fragile areas, when the ecological service value is insufficient to maintain the self-circulation of the system, the ecological environment will deteriorate. In the context of global warming, extreme weather phenomena such as droughts, rainstorms, and floods have intensified, and it is difficult to maintain stable regional vegetation coverage. With economic development, the water demand for agriculture, industry, and urban domestic water increases, and the disparity between regional water supply and demand will become more prominent [75,76]. Therefore, it is necessary to strengthen the assessment of the impact of climate change on regional water resources and to improve the climate change response capabilities of key ecological function areas and ecological restoration and management areas, so as to better meet the needs of regional high-quality development.

4.2. Ecological Protection and High-Quality Development of the YRBS

Located in the middle reaches of the Yellow River, Shaanxi is an important national advanced manufacturing base, national defense science and technology industrial base, agricultural high-tech industrial base, energy and chemical base, and scientific, educational, and cultural base. It is the core area of ecological protection and economic and social development in Shaanxi Province [77]. The task of ecological protection is heavier. Compared with the goal of beautiful scenery in ecological space, the quality of forests in the Yellow River Basin is relatively low, the stock volume per unit area and the value of ecological service functions are lower than the national average level, the structure of forest and grass is unreasonable, the water conservation capacity is not high, the pressure of water ecological protection is high, and the area of some wetlands has shrunk [78,79]. Governance tasks remain daunting. Soil erosion control has a long way to go. The Baiyu Mountains, Weibei Dry Belt, and other key areas have a lot of historical debts, and the cost of control is high. The relationship between water and sediment is still inconsistent, and the water and sediment control system still needs to be improved. The construction of garbage and sewage treatment facilities in individual cities and towns is relatively lagging behind, and the water quality of some river sections such as the Yanhe River, Qingjian River, and Beiluo River cannot reach the standard stably. The progress of governance in coal mining subsidence areas is relatively slow. The problem of water shortage is prominent. The precipitation in the basin is low, the total water resources only account for one third of the province, and the per capita water resources are less than one fifth of the country. Xi'an, Xianyang, and other cities have serious over-exploitation of groundwater, and the ecological water volume of Yanhe, Wudinghe, Hongjiannao, and other rivers and lakes is insufficient, and some counties and districts are still seriously short of water resources, engineering, and water quality [80]. Development has caused certain constraints, and the way of water for production and living is relatively extensive. The quality of development needs to be improved urgently. The effect of implementing the new development concept is not obvious enough, the industrial structure adjustment and transformation and upgrading are relatively heavy tasks, the transformation of advantages in science, education, military industry, etc., is not sufficient, and the shortcomings of insufficient openness are still obvious. The central city's radiating and driving role is not strong, the regional urban and rural development is unbalanced, there are still shortcomings in the field of people's livelihood, and the modernization level of the governance system and governance capacity is not high [81,82].

According to the different regional characteristics and development orientations of the YRBS, we will coordinate the implementation of ecological governance measures that are differentiated, complete in governance elements, and reasonably and scientifically configured. Promoting the construction of check dams, changing slopes to ladders, and making land for ditch management according to local conditions. On the basis of following the laws of nature and resource endowments, implement suitable forests for forests, irrigation for irrigation, grasses for grasses, and famines for famines, and scientifically carry out land consolidation, high-standard farmland construction, and forest vegetation conservation and restoration. Coordinate the high-quality governance of mountains, rivers, forests, fields, lakes, grass, and sand and the construction of ecological projects, promote the restoration of ecosystem functions, comprehensively optimize the ecological environment of the Loess Plateau in northern Shaanxi, and consolidate the foundation for the healthy operation of the agricultural ecosystem [83]. To coordinate the relationship between ecological risk management and high-quality agricultural development, it is necessary to build a modern agricultural industrial cluster on the Loess Plateau that is based on regional ecological risks and resource endowments. Build an intensive, high-value green agricultural industry cluster that is focusing on grains (millet, barley, etc.), forest fruits (apples, red dates, etc.), edible fungi, and characteristic plants (medlar, sea buckthorn, hops, etc.), feed, meat (Beef cattle, sheep, etc.), dairy industry, cashmere-based agriculture, and animal husbandry combined with conservation agriculture and conservation grassland agriculture industry clusters. Strengthen the integration and matching of natural endowments and production factors; promote the development of regional agricultural industries to complement soil erosion control, farmland water conservancy construction, and circular agricultural projects; and support the development of eco-friendly new technologies, new formats, and new models [84].

4.3. Limitation

There are a few limitations and improvements that need further in-depth research and discussion: (1) Due to the difficulty of data collection, this study only started from the level of landscape structure changes that were caused by human activities as the source of risk. However, the basin is affected by natural factors such as soil erosion, meteorological changes, and geological disasters. Therefore, in the future, it is necessary to comprehensively evaluate the ecological risk from the perspective of multiple risk sources. (2) In this study, we only analyzed the temporal and spatial evolution of the existing land use data, therefore, there is no information on the future possibilities of land use in the river basin. Thus, the next step is to conduct prediction research. (3) Due to data availability and research time factors, this study only provides management suggestions that are based on the research results from the perspective of different risk levels. We did not build an indicator system for ecological risk influencing factors in detail but conducted a driving factor analysis. Generally, multi-dimensional analyses of the impact mechanism of the ecological risk in the basin are carried out.

5. Conclusions

The land use types in the YRBS have changed significantly over the span of 20 years. From 2000 to 2020, the overall landscape ecological risk value of the Yellow River Basin first increased then decreased, but the overall ecological risk level of the basin was lower, indicating that the ecological status of YRBS improved. The spatiotemporal evolution characteristics of five different levels of ecological risk areas in the YRBS that were discovered through this study provide scientific insights into the laws of ecological evolution in different regions in the river basin, which can further provide insights into regional ecological environment management, ecological security protection, risk early warning, and provide theoretical support for policy formulation in sustainable development and other aspects. The research conclusion can promote the sustainable development of land use and high-quality ecological protection in Changsha River beach.

In the future, further research can improve the evaluation index system of ecological risk, and not only carry out ecological risk evaluation on the basis of land use change, but also a more complex and richer index system and evaluation system should be considered. At the same time, it is also possible to analyze the driving mechanism of the evolution of ecological risks, especially considering the impact of socioeconomic factors and human activities on ecological risks.

Author Contributions: Conceptualization, Z.Z. and Y.F.; methodology, Z.Z.; software, Z.Z.; validation, Z.M. and Z.Z.; formal analysis, Z.Z.; investigation, Z.M. and X.X.; writing—original draft preparation, Z.Z.; writing—review and editing, G.R.; funding acquisition, Y.F. All authors have read and agreed to the published version of the manuscript.

Funding: This research was funded by the Shaanxi Provincial Forestry Science and Technology Innovation Program Special Project (grant number SXLK20200102) and the China Association for Science and Technology 2020 Postgraduate Science Popularization Ability Improvement Project (grant number kxyjs202034).

Institutional Review Board Statement: Not applicable.

Informed Consent Statement: Not applicable.

Data Availability Statement: The data that support the findings of this study are available from the corresponding author upon reasonable request.

Conflicts of Interest: The authors declare no conflict of interest.

References

1. Gao, J.J.; Liu, X.R.; Hao, H.; Yuan, H.; Zhou, H.D. Survey on typical organic pollutants and health risks of main water sources in yellow river basin. *Environ. Eng. Manag. J.* **2013**, *12*, 1751–1755.
2. Gao, P.; Wang, Y.M.; Li, P.F.; Zhao, G.J.; Sun, W.Y.; Mu, X.M. Land degradation changes in the Yellow River Delta and its response to the streamflow-sediment fluxes since 1976. *Land Degrad. Dev.* **2018**, *29*, 3212–3220. [CrossRef]
3. He, J.X.; Wang, Y.X. Study on shaanxi's leading role in the metropolitan area on ecological protection and high-quality development of the yellow river "ji" bay. *J. Environ. Prot. Ecol.* **2021**, *22*, 656–664.
4. Ma, L.J.; Liu, Z.; Zhao, B.F.; Lyu, J.W.; Zheng, F.M.; Xu, W.; Gan, X.B. Variations of runoff and sediment and their response to human activities in the source region of the Yellow River, China. *Environ. Earth Sci.* **2021**, *80*, 551. [CrossRef]
5. Mao, W.B.; Kang, S.Z.; Wan, Y.S.; Sun, Y.X.; Li, X.H.; Wang, Y.F. Yellow River Sediment as a Soil Amendment for Amelioration of Saline Land in the Yellow River Delta. *Land Degrad. Dev.* **2016**, *27*, 1595–1602. [CrossRef]
6. Wang, M.J.; Qi, S.Z.; Zhang, X.X. Wetland loss and degradation in the Yellow River Delta, Shandong Province of China. *Environ. Earth Sci.* **2012**, *67*, 185–188. [CrossRef]
7. Wohlfart, C.; Kuenzer, C.; Chen, C.; Liu, G.H. Social-ecological challenges in the Yellow River basin (China): A review. *Environ. Earth Sci.* **2016**, *75*, 1066. [CrossRef]
8. Xia, H.J.; Liu, L.S.; Bai, J.H.; Kong, W.J.; Lin, K.X.; Guo, F. Wetland Ecosystem Service Dynamics in the Yellow River Estuary under Natural and Anthropogenic Stress in the Past 35 Years. *Wetlands* **2020**, *40*, 2741–2754. [CrossRef]
9. Yang, X.; Zhang, H.Y.; Yang, J.; Li, H.W.; Zhu, X.B. Research on ecological compensation of water resources in the yellow river basin. *Fresenius Environ. Bull.* **2020**, *29*, 4426–4429.
10. Yin, J.; Lu, Y.; Ou, Z. Regional agricultural water footprint and crop water consumption study in yellow river basin, China. *Appl. Ecol. Environ. Res.* **2019**, *17*, 5539–5559. [CrossRef]
11. Bai, Y.; Chen, Y.Y.; Alatalod, J.M.; Yang, Z.Q.; Jiang, B. Scale effects on the relationships between land characteristics and ecosystem services—A case study in Taihu Lake Basin, China. *Sci. Total Environ.* **2020**, *716*, 137083. [CrossRef] [PubMed]
12. Cheng, G.D.; Li, X.; Zhao, W.Z.; Xu, Z.M.; Feng, Q.; Xiao, S.C.; Xiao, H.L. Integrated study of the water-ecosystem-economy in the Heihe River Basin. *Natl. Sci. Rev.* **2014**, *1*, 413–428. [CrossRef]
13. Flint, N.; Rolfe, J.; Jones, C.E.; Sellens, C.; Johnston, N.D.; Ukkola, L. An Ecosystem Health Index for a large and variable river basin: Methodology, challenges and continuous improvement in Queensland's Fitzroy Basin. *Ecol. Indic.* **2017**, *73*, 626–636. [CrossRef]
14. Ge, Q.Q.; Xu, W.J.; Fu, M.C.; Han, Y.X.; An, G.Q.; Xu, Y.T. Ecosystem service values of gardens in the Yellow River Basin, China. *J. Arid. Land* **2022**, *14*, 284–296. [CrossRef]
15. Huang, L.Y.; Wang, J.; Chen, X.J. Ecological infrastructure planning of large river basin to promote nature conservation and ecosystem functions. *J. Environ. Manag.* **2022**, *306*, 114482. [CrossRef] [PubMed]
16. Huang, X.; Ma, J.X. Changes in the ecosystem service values of typical river basins in arid regions of Northwest China. *Ecohydrology* **2013**, *6*, 1048–1056. [CrossRef]
17. Liu, H.; Zheng, L.; Wu, J.; Liao, Y.H. Past and future ecosystem service trade-offs in Poyang Lake Basin under different land use policy scenarios. *Arab. J. Geosci.* **2020**, *13*, 46. [CrossRef]
18. Liu, R.R.; Dong, X.B.; Zhang, P.; Zhang, Y.; Wang, X.W.; Gao, Y. Study on the Sustainable Development of an Arid Basin Based on the Coupling Process of Ecosystem Health and Human Wellbeing under Land Use Change—A Case Study in the Manas River Basin, Xinjiang, China. *Sustainability* **2020**, *12*, 1201. [CrossRef]
19. Sun, R.; Yao, P.P.; Wang, W.; Yue, B.; Liu, G. Assessment of Wetland Ecosystem Health in the Yangtze and Amazon River Basins. *Isprs Int. J. Geo Inf.* **2017**, *6*, 81. [CrossRef]

20. Sun, Y.J.; Li, J.; Liu, X.F.; Ren, Z.Y.; Zhou, Z.X.; Duan, Y.F. Spatially Explicit Analysis of Trade-Offs and Synergies among Multiple Ecosystem Services in Shaanxi Valley Basins. *Forests* **2020**, *11*, 209. [CrossRef]
21. Woldeyohannes, A.; Cotter, M.; Biru, W.D.; Kelboro, G. Assessing Changes in Ecosystem Service Values over 1985-2050 in Response to Land Use and Land Cover Dynamics in Abaya-Chamo Basin, Southern Ethiopia. *Land* **2020**, *9*, 37. [CrossRef]
22. Wu, C.S.; Ma, G.X.; Yang, W.S.; Zhou, Y.; Peng, F.; Wang, J.N.; Yu, F. Assessment of Ecosystem Service Value and Its Differences in the Yellow River Basin and Yangtze River Basin. *Sustainability* **2021**, *13*, 3822. [CrossRef]
23. Xu, Z.H.; Fan, W.G.; Wei, H.J.; Zhang, P.; Ren, J.H.; Gao, Z.C.; Ulgiati, S.; Kong, W.D.; Dong, X.B. Evaluation and simulation of the impact of land use change on ecosystem services based on a carbon flow model: A case study of the Manas River Basin of Xinjiang, China. *Sci. Total Environ.* **2019**, *652*, 117–133. [CrossRef]
24. Yang, J.; Xie, B.P.; Tao, W.Q.; Zhang, D.G. Ecosystem Services Assessment, Trade-Off, and Bundles in the Yellow River Basin, China. *Diversity* **2021**, *13*, 308. [CrossRef]
25. Abdullah, A.M.; Masrur, A.; Adnan, M.S.G.; Al Baky, M.A.; Hassan, Q.K.; Dewan, A. Spatio-Temporal Patterns of Land Use/Land Cover Change in the Heterogeneous Coastal Region of Bangladesh between 1990 and 2017. *Remote Sens.* **2019**, *11*, 790. [CrossRef]
26. Bai, Y.; Jiang, B.; Wang, M.; Li, H.; Alatalo, J.M.; Huang, S.F. New ecological redline policy (ERP) to secure ecosystem services in China. *Land Use Policy* **2016**, *55*, 348–351. [CrossRef]
27. Burkhard, B.; Kroll, F.; Nedkov, S.; Muller, F. Mapping ecosystem service supply, demand and budgets. *Ecol. Indic.* **2012**, *21*, 17–29. [CrossRef]
28. Connors, J.P.; Galletti, C.S.; Chow, W.T.L. Landscape configuration and urban heat island effects: Assessing the relationship between landscape characteristics and land surface temperature in Phoenix, Arizona. *Landsc. Ecol.* **2013**, *28*, 271–283. [CrossRef]
29. Cote, I.M.; Darling, E.S.; Brown, C.J. Interactions among ecosystem stressors and their importance in conservation. *Proc. R. Soc. B Biol. Sci.* **2016**, *283*, 20152592. [CrossRef] [PubMed]
30. Dadashpoor, H.; Azizi, P.; Moghadasi, M. Land use change, urbanization, and change in landscape pattern in a metropolitan area. *Sci. Total Environ.* **2019**, *655*, 707–719. [CrossRef]
31. Dai, L.; Liu, Y.B.; Luo, X.Y. Integrating the MCR and DOI models to construct an ecological security network for the urban agglomeration around Poyang Lake, China. *Sci. Total Environ.* **2021**, *754*, 141868. [CrossRef] [PubMed]
32. Fan, C.; Myint, S. A comparison of spatial autocorrelation indices and landscape metrics in measuring urban landscape fragmentation. *Landsc. Urban Plan.* **2014**, *121*, 117–128. [CrossRef]
33. Gibb, R.; Redding, D.W.; Chin, K.Q.; Donnelly, C.A.; Blackburn, T.M.; Newbold, T.; Jones, K.E. Zoonotic host diversity increases in human-dominated ecosystems. *Nature* **2020**, *584*, 398–402. [CrossRef]
34. Hale, R.; Swearer, S.E. Ecological traps: Current evidence and future directions. *Proc. R. Soc. B Biol. Sci.* **2016**, *283*, 20152647. [CrossRef] [PubMed]
35. He, L.; Shen, J.; Zhang, Y. Ecological vulnerability assessment for ecological conservation and environmental management. *J. Environ. Manag.* **2018**, *206*, 1115–1125. [CrossRef] [PubMed]
36. Hu, X.J.; Ma, C.M.; Huang, P.; Guo, X. Ecological vulnerability assessment based on AHP-PSR method and analysis of its single parameter sensitivity and spatial autocorrelation for ecological protection ? A case of Weifang City, China. *Ecol. Indic.* **2021**, *125*, 107464. [CrossRef]
37. Islam, A.M.T.; Talukdar, S.; Mahato, S.; Ziaul, S.; Eibek, K.U.; Akhter, S.; Pham, Q.B.; Mohammadi, B.; Karimi, F.; Linh, N.T.T. Machine learning algorithm-based risk assessment of riparian wetlands in Padma River Basin of Northwest Bangladesh. *Environ. Sci. Pollut. Res.* **2021**, *28*, 34450–34471. [CrossRef]
38. Jiang, B.; Bai, Y.; Wong, C.P.; Xu, X.B.; Alatalo, J.M. China's ecological civilization program-Implementing ecological redline policy. *Land Use Policy* **2019**, *81*, 111–114. [CrossRef]
39. Jones, N.A.; Shaw, S.; Ross, H.; Witt, K.; Pinner, B. The study of human values in understanding and managing social-ecological systems. *Ecol. Soc.* **2016**, *21*, 15. [CrossRef]
40. Jung, J.W.; Park, J.W.; Eo, S.; Choi, J.; Song, Y.K.; Cho, Y.; Hong, S.H.; Shim, W.J. Ecological risk assessment of microplastics in coastal, shelf, and deep sea waters with a consideration of environmentally relevant size and shape. *Environ. Pollut.* **2021**, *270*, 116217. [CrossRef]
41. Kroll, F.; Muller, F.; Haase, D.; Fohrer, N. Rural-urban gradient analysis of ecosystem services supply and demand dynamics. *Land Use Policy* **2012**, *29*, 521–535. [CrossRef]
42. Lasanta, T.; Arnaez, J.; Pascual, N.; Ruiz-Flano, P.; Errea, M.P.; Lana-Renault, N. Space-time process and drivers of land abandonment in Europe. *Catena* **2017**, *149*, 810–823. [CrossRef]
43. Lawler, J.J.; Lewis, D.J.; Nelson, E.; Plantinga, A.J.; Polasky, S.; Withey, J.C.; Helmers, D.P.; Martinuzzi, S.; Pennington, D.; Radeloff, V.C. Projected land-use change impacts on ecosystem services in the United States. *Proc. Natl. Acad. Sci. USA* **2014**, *111*, 7492–7497. [CrossRef]
44. Levers, C.; Muller, D.; Erb, K.; Haberl, H.; Jepsen, M.R.; Metzger, M.J.; Meyfroidt, P.; Plieninger, T.; Plutzar, C.; Sturck, J.; et al. Archetypical patterns and trajectories of land systems in Europe. *Reg. Environ. Chang.* **2018**, *18*, 715–732. [CrossRef]
45. Maanan, M.; Saddik, M.; Maanan, M.; Chaibi, M.; Assobhei, O.; Zourarah, B. Environmental and ecological risk assessment of heavy metals in sediments of Nador lagoon, Morocco. *Ecol. Indic.* **2015**, *48*, 616–626. [CrossRef]

46. More, S.J.; Hardy, A.; Bampidis, V.; Benford, D.; Bennekou, S.H.; Bragard, C.; Boesten, J.; Halldorsson, T.I.; Hernandez-Jerez, A.F.; Jeger, M.J.; et al. Guidance on harmonised methodologies for human health, animal health and ecological risk assessment of combined exposure to multiple chemicals. *Efsa J.* **2019**, *17*, e05634. [CrossRef]
47. Li, C.X.; Wu, J.Y. Land use transformation and eco-environmental effects based on production-living-ecological spatial synergy: Evidence from Shaanxi Province, China. *Environ. Sci. Pollut. Res.* **2022**, *29*, 41492–41504. [CrossRef]
48. Sayles, J.S.; Baggio, J.A. Social-ecological network analysis of scale mismatches in estuary watershed restoration. *Proc. Natl. Acad. Sci. USA* **2017**, *114*, E1776–E1785. [CrossRef]
49. Su, S.L.; Xiao, R.; Jiang, Z.L.; Zhang, Y. Characterizing landscape pattern and ecosystem service value changes for urbanization impacts at an eco-regional scale. *Appl. Geogr.* **2012**, *34*, 295–305. [CrossRef]
50. Tscharntke, T.; Tylianakis, J.M.; Rand, T.A.; Didham, R.K.; Fahrig, L.; Batary, P.; Bengtsson, J.; Clough, Y.; Crist, T.O.; Dormann, C.F.; et al. Landscape moderation of biodiversity patterns and processes—Eight hypotheses. *Biol. Rev.* **2012**, *87*, 661–685. [CrossRef]
51. Shaanxi Provincial Bureau of Statistics. *Shaanxi Statistical Yearbook*; China Statistics Press: Beijing, China, 2021.
52. Munteanu, C.; Kuemmerle, T.; Boltiziar, M.; Butsic, V.; Gimmi, U.; Halada, L.; Kaim, D.; Kiraly, G.; Konkoly-Gyuro, E.; Kozak, J.; et al. Forest and agricultural land change in the Carpathian region-A meta-analysis of long-term patterns and drivers of change. *Land Use Policy* **2014**, *38*, 685–697. [CrossRef]
53. Olsson, P.; Galaz, V.; Boonstra, W.J. Sustainability transformations: A resilience perspective. *Ecol. Soc.* **2014**, *19*, 4. [CrossRef]
54. Peng, J.; Zhao, M.Y.; Guo, X.N.; Pan, Y.J.; Liu, Y.X. Spatial-temporal dynamics and associated driving forces of urban ecological land: A case study in Shenzhen City, China. *Habitat Int.* **2017**, *60*, 81–90. [CrossRef]
55. Sarkodie, S.A. Environmental performance, biocapacity, carbon & ecological footprint of nations: Drivers, trends and mitigation options. *Sci. Total Environ.* **2021**, *751*, 141912. [CrossRef] [PubMed]
56. Shaanxi Provincial Bureau of Statistics. *Statistical Bulletin of Shaanxi Province*; China Statistics Press: Beijing, China, 2021.
57. Coops, N.C.; Wulder, M.A.; Iwanicka, D. Large area monitoring with a MODIS-based Disturbance Index (DI) sensitive to annual and seasonal variations. *Remote Sens. Environ.* **2009**, *113*, 1250–1261. [CrossRef]
58. Yao, Y.; Cheng, T.; Sun, Z.; Li, L.; Chen, D.; Chen, Z.; Wei, J.; Guan, Q. VecLI: A framework for calculating vector landscape indices considering landscape fragmentation. *Environ. Model. Softw.* **2022**, *149*, 105325. [CrossRef]
59. Cong, R.-G.; Ekroos, J.; Smith, H.G.; Brady, M.V. Optimizing intermediate ecosystem services in agriculture using rules based on landscape composition and configuration indices. *Ecol. Econ.* **2016**, *128*, 214–223. [CrossRef]
60. Wang, H.; Liu, X.; Zhao, C.; Chang, Y.; Liu, Y.; Zang, F. Spatial-temporal pattern analysis of landscape ecological risk assessment based on land use/land cover change in Baishuijiang National nature reserve in Gansu Province, China. *Ecol. Indic.* **2021**, *124*, 107454. [CrossRef]
61. Reddy, C.S.; Manaswini, G.; Satish, K.V.; Singh, S.; Jha, C.S.; Dadhwal, V.K. Conservation priorities of forest ecosystems: Evaluation of deforestation and degradation hotspots using geospatial techniques. *Ecol. Eng.* **2016**, *91*, 333–342. [CrossRef]
62. Mo, W.; Wang, Y.; Zhang, Y.; Zhuang, D. Impacts of road network expansion on landscape ecological risk in a megacity, China: A case study of Beijing. *Sci. Total Environ.* **2017**, *574*, 1000–1011. [CrossRef] [PubMed]
63. Yang, Y.Y.; Liu, Y.S.; Li, Y.R.; Du, G.M. Quantifying spatio-temporal patterns of urban expansion in Beijing during 1985–2013 with rural-urban development transformation. *Land Use Policy* **2018**, *74*, 220–230. [CrossRef]
64. Zhao, M.W.; Wei, G.W.; Wei, C.; Wu, J.; Wei, Y. Extended CPT-TODIM method for interval-valued intuitionistic fuzzy MAGDM and its application to urban ecological risk assessment. *J. Intell. Fuzzy Syst.* **2021**, *40*, 4091–4106. [CrossRef]
65. Wortley, L.; Hero, J.M.; Howes, M. Evaluating Ecological Restoration Success: A Review of the Literature. *Restor. Ecol.* **2013**, *21*, 537–543. [CrossRef]
66. Wu, J.G. Urban ecology and sustainability: The state-of-the-science and future directions. *Landsc. Urban Plan.* **2014**, *125*, 209–221. [CrossRef]
67. Ni, Q.; Zhao, M.J.; Li, C.Q.; Shi, Y.X.; Xu, T.; Khan, I. Multidimensional trust and its impact on the willingness to pay for ecological compensation in China's transboundary watersheds-taking the largest tributary of the Yellow River as an example. *J. Environ. Plan. Manag.* **2021**, *64*, 2257–2275. [CrossRef]
68. Li, C.; Yang, X.; Liu, H.; Zhu, H.; Wei, H.; Na, J.; Cui, X. Spatial distribution characteristics of discontinuous hillslope gullies on the Loess Plateau of China: A special focus on spoon gullies. *Catena* **2022**, *215*, 106327. [CrossRef]
69. Sun, J.; Li, G.; Zhang, Y.; Qin, W.; Wang, M. Identification of priority areas for afforestation in the Loess Plateau region of China. *Ecol. Indic.* **2022**, *140*, 108998. [CrossRef]
70. Sun, W.; Ding, X.; Su, J.; Mu, X.; Zhang, Y.; Gao, P.; Zhao, G. Land use and cover changes on the Loess Plateau: A comparison of six global or national land use and cover datasets. *Land Use Policy* **2022**, *119*, 106165. [CrossRef]
71. Xu, X.; Zhang, D.; Zhang, Y.; Yao, S.; Zhang, J. Evaluating the vegetation restoration potential achievement of ecological projects: A case study of Yan'an, China. *Land Use Policy* **2020**, *90*, 104293. [CrossRef]
72. Zhang, D.; Ge, W.; Zhang, Y. Evaluating the vegetation restoration sustainability of ecological projects: A case study of Wuqi County in China. *J. Clean. Prod.* **2020**, *264*, 121751. [CrossRef]
73. Wen, X.; Théau, J. Spatiotemporal analysis of water-related ecosystem services under ecological restoration scenarios: A case study in northern Shaanxi, China. *Sci. Total Environ.* **2020**, *720*, 137477. [CrossRef] [PubMed]

74. Zou, C.; Zhu, J.; Lou, K.; Yang, L. Coupling coordination and spatiotemporal heterogeneity between urbanization and ecological environment in Shaanxi Province, China. *Ecol. Indic.* **2022**, *141*, 109152. [CrossRef]
75. Du, H.; Liu, X.; Jia, X.; Li, S.; Fan, Y. Assessment of the effects of ecological restoration projects on soil wind erosion in northern China in the past two decades. *Catena* **2022**, *215*, 106360. [CrossRef]
76. Jiang, L.; Liu, Y.; Wu, S.; Yang, C. Analyzing ecological environment change and associated driving factors in China based on NDVI time series data. *Ecol. Indic.* **2021**, *129*, 107933. [CrossRef]
77. Wei, X.; Ye, Y.; Li, B.; Chen, T. Reconstructing cropland change since 1650 AD in Shaanxi province, central China. *Quat. Int.* **2022**, *130*, 105132. [CrossRef]
78. Sun, Y.; Wang, N. Development and correlations of the industrial ecology in China's Loess Plateau: A study based on the coupling coordination model and spatial network effect. *Ecol. Indic.* **2021**, *132*, 108332. [CrossRef]
79. Zhang, Q.; Li, F. Correlation between land use spatial and functional transition: A case study of Shaanxi Province, China. *Land Use Policy* **2022**, *119*, 106194. [CrossRef]
80. Zhang, Y.; Yuan, J.; You, C.; Cao, R.; Tan, B.; Li, H.; Yang, W. Contributions of National Key Forestry Ecology Projects to the forest vegetation carbon storage in China. *For. Ecol. Manag.* **2020**, *462*, 117981. [CrossRef]
81. Chen, Y.; Xu, F. The optimization of ecological service function and planning control of territorial space planning for ecological protection and restoration. *Sustain. Comput. Inform. Syst.* **2022**, *35*, 100748. [CrossRef]
82. Li, X.; Xiao, L.; Tian, C.; Zhu, B.; Chevallier, J. Impacts of the ecological footprint on sustainable development: Evidence from China. *J. Clean. Prod.* **2022**, *352*, 131472. [CrossRef]
83. Meng, Y.; Wang, M.; Xu, W.; Guan, X.; Yan, D. Structure construction, evolution analysis and sustainability evaluation of Water-Ecological-Economic system. *Sustain. Cities Soc.* **2022**, *83*, 103966. [CrossRef]
84. Sun, Y.; Wang, N. Sustainable urban development of the π-shaped Curve Area in the Yellow River basin under ecological constraints: A study based on the improved ecological footprint model. *J. Clean. Prod.* **2022**, *337*, 130452. [CrossRef]

International Journal of
Environmental Research and Public Health

MDPI

Article

Land Fragmentation, Technology Adoption and Chemical Fertilizer Application: Evidence from China

Liang Chi [1,†], Shuqing Han [1,†], Meili Huan [2,*], Yajuan Li [3] and Jifang Liu [1]

1 Agricultural Information Institute, Chinese Academy of Agricultural Sciences, Beijing 100081, China; chiliang@caas.cn (L.C.); hanshuqing@caas.cn (S.H.); liujifang@caas.cn (J.L.)
2 China Institute for Rural Studies, Tsinghua University, Beijing 100084, China
3 National Academy of Agriculture Green Development, Key Laboratory of Plant-Soil Interactions, Ministry of Education, College of Resources and Environmental Sciences, China Agricultural University, Beijing 100193, China; liyajuancau@cau.edu.cn
* Correspondence: huanmeili@mail.tsinghua.edu.cn
† These authors contributed equally to this work.

Abstract: Although it has been widely recognized that land fragmentation has increased chemical fertilizer application, little is known about the role of technology adoption in mitigating these adverse effects. To empirically examine the relationship between land fragmentation, technology adoption and chemical fertilizer application, we developed a mediation model. We applied our analysis to a survey data set encompassing 1388 farm-level samples collected in 14 Chinese provinces in 2019. Our study demonstrated that land fragmentation can not only directly increase chemical fertilizer application but also indirectly increase it by hindering the adoption of agricultural mechanization technologies (AMT's) and soil testing fertilization technologies (STFT's). Both are recognized as potent drivers of fertilizer use reductions. Moreover, the adoption of information and communications technologies (ICT's) can help mitigate the negative effects of land fragmentation on technology adoption, thus reducing chemical fertilizer application intensity (CFAI). However, the direct effects of land fragmentation on CAFI was unaffected by ICT's. Our findings suggest that ICT's have revolutionized farmer recognition, promotion and adoption of agricultural technologies by increasing awareness and diffusion of agricultural technology information.

Keywords: land fragmentation; agricultural mechanization; ICT's; soil testing fertilization; sustainable agricultural practices

Citation: Chi, L.; Han, S.; Huan, M.; Li, Y.; Liu, J. Land Fragmentation, Technology Adoption and Chemical Fertilizer Application: Evidence from China. *Int. J. Environ. Res. Public Health* **2022**, *19*, 8147. https://doi.org/10.3390/ijerph19138147

Academic Editors: Wei Song, Hualin Xie and Paul B. Tchounwou

Received: 17 May 2022
Accepted: 30 June 2022
Published: 2 July 2022

1. Introduction

Chemical fertilizers are widely adopted in agricultural production and play a significant role in increasing yields of agricultural products and ensuring food security [1]. However, the excessive use of chemical fertilizer has resulted in various problems such as food insecurity, soil degradation and greenhouse gas emissions in developing countries, especially in China [2]. More importantly, the overuse of chemical fertilizer in agricultural production has become a public concern, for social well-being and ecological balance are seriously threatened by massive chemical fertilizer use [3,4].

China's agricultural production features small-scale farming and severe land fragmentation. About 210 million rural households in China operate on cultivated land less than 10 mu (0.667 hectares) and the average farm size is only 7.46 mu (0.497 hectares) [5]. Compared with other Asian countries, the farm size in China is about one-third of that in South Korea and one-quarter of that in Japan [6]. Farm households have been the driving force of agricultural production since the implementation of the Household Contract Responsibility System in 1979.

Meanwhile, China is also the country with the largest amount of chemical fertilizer application in the world in terms of overall tonnage [7]. The agricultural growth in China depends heavily on the use of chemical fertilizer. The total agricultural output increased by 42.23% from 1978 to 1984, among which 45.79 percent of this output growth came from increases in inputs, including cultivated land, labor, fertilizer and capital, and fertilizer alone contributed 32.2% of the growth [8]. Some studies suggest that the extensive use of chemical fertilizers and other inputs is the fundamental reason for the rapid growth of Chinese agriculture [9]. Consequently, the development of sustainable agriculture in China is faced with severe challenges.

The unfavorable natural resource conditions have made it essential for China to develop intensive agriculture. However, the excessive and inefficient use of agricultural inputs were quite commonly seen at the early stages of agricultural production so as to ensure food security [10–14]. As a result, the extensive use of agricultural inputs has greatly damaged the environment [15–19].

In order to reduce the use of chemical fertilizers, the Chinese government has implemented a series of policies, such as the removal of subsidies for chemical fertilizers and the promotion of soil testing technologies [20,21]. Although early studies assume that these policies may not significantly decrease chemical fertilizer application [7], we believe that these policies have helped reduce the amount of chemical fertilizer application. According to the data from National Bureau of Statistics of China [22], the consumption of fertilizers in China has seen a steady increase since 1978, reached its peak at 60.33 million tons in 2015, and started to decrease thereafter, as is shown in Figure 1.

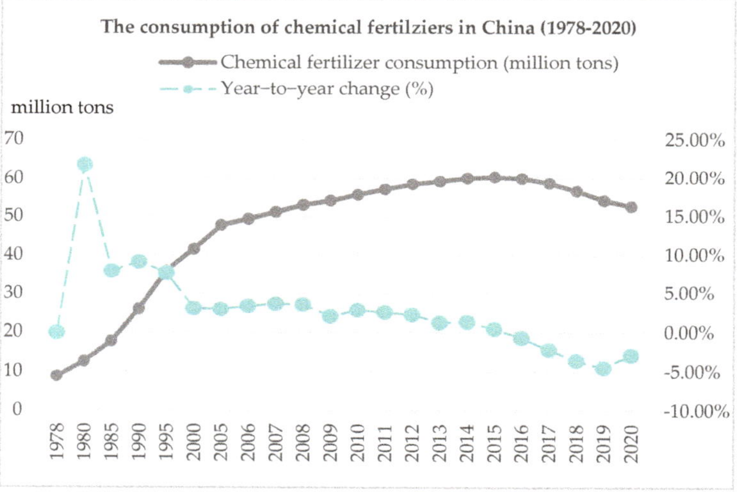

Figure 1. The consumption of chemical fertilizers in China from 1978 to 2020. Source of data: National Bureau of Statistics of China, 2021.

Although there has been a slight decrease in recent years due to the policies implemented, the consumption of chemical fertilizers remains large. In 2020, there were still 52.5 million tons of chemical fertilizers consumed. Moreover, the household-level survey data from the Research Center of Rural Economy (RCRE) of the Ministry of Agriculture and Rural Affairs of China shows a similar trend. The survey dataset with 17,000 farm-level observations in 31 provinces of China shows that the amount of chemical fertilizers applied by Chinese farmers basically remained at 464.18 kg/ha from 1995 to 2015 [6]. A question arises: what are the root causes of small farmers applying such an enormous amount of chemical fertilizer?

To answer this question, a growing body of literature has explored the drivers of chemical fertilizer application. The results, however, are unclear and even conflicting. While several studies suggested that farm household and farmer characteristics, including, farm size, cropping structure and resource endowment, have significant effects on the amount of chemical fertilizer application [23–27], others show that the effects of individual characteristics on fertilizer application tend to be weakened over time since smallholder farmers are likely to imitate each other and apply the same amount of fertilizers [28–30]. Therefore, land resource conditions are still regarded as one of the key drivers for chemical fertilizer application. In particular, China is faced with unfavorable land resource endowment, such as extremely small farm size and serious land fragmentation and whether it restricts the reduction of fertilizer application has raised a lot of concern.

Most of the existing studies investigating the impact of natural resource endowment on fertilizer use are mainly focused on farm size [31,32], and little is known about whether the characteristics of farmland affect chemical fertilizer use, and the influence mechanism remains unclear. Some studies in the literature have argued that land consolidation through land use rights circulation contributes to the reduction of chemical fertilizer use [33]. However, land use rights trading may increase the degree of land fragmentation, which increases the difficulty in agricultural production and farm management. It is unreasonable to discuss farm size only and ignore the role of land fragmentation. Furthermore, the results of the studies on the relationship between farm size and fertilizer use are unclear and even conflicting. While some studies show that increasing farm size can reduce chemical fertilizer application without decreasing or even increasing crop yield [7,31], others find that smaller farm size can lead to higher fertilizer use efficiency [32]. Moreover, precious few studies have explored the negative impact of land fragmentation on farmers' fertilizer use efficiency and discussed the heterogeneous effects of different contributing factors, such as farm size, crop structure and land quality [34]. However, the influence mechanism has not yet been fully understood.

More importantly, the existing literature has shown that land fragmentation may hinder the adoption of modern agricultural machinery [35], increase production costs [36] and cause the loss of technical efficiency [37] and land use efficiency [38]. Hence, land fragmentation may also have a direct impact on farmers' behavior regarding chemical fertilizer application. On one hand, instead of using machinery, smallholder farmers are likely to increase other inputs such as applying more chemical fertilizers and using more labor since land fragmentation increases the difficulty of mechanical operation, resulting in higher mechanical costs [39,40]. In particular, the low ratio of fixed inputs to total inputs is the key factor leading to over-fertilization on smallholder farms because smallholders lack fixed inputs and then compensate by over-applying fertilizer to attempt to achieve their yield goals [41]. On the other hand, land fragmentation also makes it possible for farmers to flexibly distribute labor and other inputs and thus improve efficiency [42,43].

Based on the above observations and previous studies, we hypothesize that land fragmentation has a significantly positive effect on chemical fertilizer application, and the adoption of agricultural technologies plays an important role in it. In other words, land fragmentation exerts a significant influence on farmers' chemical fertilizer application via its influence on the adoption of agricultural mechanization technologies (AMTs) and soil testing fertilization technologies (STFTs), and the information and communications technologies (ICTs) can help mitigate these negative effects. To fill in the literature gap, in this study, we provide a robust estimation of the effects of land fragmentation on farmers' chemical fertilizer application as well as the role of the adoption of three technologies in China's maize production.

The objectives of this study are two-fold. The first is to explore how land fragmentation and the adoption of two agricultural technologies, i.e., AMT and STFT, affect chemical fertilizer application intensity (CFAI) in maize production through a mediation model. The second is to investigate how ICT adoption mitigates the negative effects of land fragmentation on the adoption of two agricultural technologies and the reduction of chemical fertilizer application. Our analysis reveals the mechanism by which land fragmentation affects farmer's chemical fertilizer application via agricultural technology adoption. Specifically, land fragmentation changes the adoption of AMTs and STFTs, resulting in increasing farmers' chemical fertilizer application. Moreover, the adoption of ICTs can mitigate the process where land fragmentation negatively affects AMT and STFT adoption. To our knowledge, this study is among the first to investigate the effects of land fragmentation on chemical fertilizer application through the adoption of three technologies in rural China and therefore help shed light on the issue. Our study also has important implications for developing countries with agricultural characteristics similar to China.

The remainder of this paper is organized as follows. In Section 2 we provide the data and estimation strategy, followed by the estimation results in Section 3. Section 4 presents the discussion, and Section 5 concludes.

2. Methodology and Data

2.1. Empirical Model

2.1.1. Chemical Fertilizer Application Intensity

Chemical fertilizer application intensity (*CFAI*) is measured as the consumption of chemical fertilizers per hectare sown area. It is a general index to reflect chemical fertilizer use and corresponding ecological risks [44]. The *CFAI* can be calculated as:

$$CFAI = \frac{CCF}{SAC} \tag{1}$$

where *CCF* denotes the consumption of chemical fertilizers, which refers to the total amount of chemical fertilizers applied in maize production, including nitrogenous fertilizers, phosphate fertilizers, potash fertilizers and complex fertilizers. *SAC* denotes the total sown area of crops, which includes the land owned by the farmers themselves and that transferred from others.

2.1.2. Simpson's Index of Diversity

Before the assessment, we needed an indicator containing all important factors to measure the degree of land fragmentation. Three indicators are widely used in the existing literature to measure the degree of land fragmentation, i.e., number of plots, average plot size, average plot distance [31]. The Simpson Index of Diversity (*SI*), a general indicator to represent land fragmentation [45,46], is defined as:

$$SI = 1 - \frac{\sum_{i=1}^{n} a_i^2}{\left(\sum_{i=1}^{n} a_i\right)^2} \tag{2}$$

where $0 \leq SI \leq 1$, when $SI = 0$, which means that the household has only one piece of land, with a higher value of SI indicating a higher degree of land fragmentation. n is the number of plots that the household has. a_i is size of plot i.

2.1.3. Plot Distance Index

However, the Simpson index does not capture the distance of each plot [46]. Hence, we constructed a plot distance index (*PDI*) which captures the spatial distribution of plots of the farm household. The *PDI* is defined as:

$$PDI = \frac{d_1}{d_{max}} \times \frac{d_2}{d_{max}} \times \frac{d_3}{d_{max}} \times \cdots \times \frac{d_n}{d_{max}} \tag{3}$$

where d_i denotes the distance between the farmer's house and the plot i. d_{max} is the distance of the farthest plot to farmer's house, with a larger value of *PDI* indicating a higher degree of land fragmentation.

2.1.4. Mediation Model

In order to examine the mechanism of how land fragmentation affects *CFAI*, we employed a mediation model to explore if agricultural technology adoption mediates the effect of land fragmentation on *CFAI*. Here, we categorize agricultural technology adoption into two types, *AMT* and *STFT*. The mediating effect mainly tests the role of agricultural technology adoption in facilitating the process through which land fragmentation affects *CFAI*. The three-model system is widely used to examine the mediating effects of mediators [47], and we set up the three-model system as follows:

$$CFAI_i = \gamma_0 + \gamma_1 SI_i + \gamma_2 X_{ki} + \varepsilon_{1i} \quad (4)$$

$$M_i = a_0 + a_1 SI_i + a_2 X_{ki} + \varepsilon_{2i} \quad (5)$$

$$CFAI_i = \rho_0 + \rho_1 SI_i + \rho_2 M_{li} + \rho_3 X_{ki} + \varepsilon_{3i} \quad (6)$$

Here SI_i indicates the Simpson index of farm i; M_i is the mediator, namely, *AMT* adoption and *STFT* adoption of farm i; X_{ki} is a vector of other variables affecting agricultural technology adoption and *CFAI*, including factors such as farm household and farmer characteristics, farmland characteristics, region characteristics, policies, etc., following the existing studies [11,37,48]. ε_i is a random error term.

Specifically, we first test the direct effects of land fragmentation on *CFAI* without considering technology adoption in Equation (4). Then we explore the effects of land fragmentation on agricultural technology adoption in Equation (5). The last step is to investigate the effects of land fragmentation and technology adoption on *CFAI* in Equation (6). If we find a_1 equal to 0, ρ_2 equal to 0, or ρ_1 equal to γ_1, then we cannot reject the null hypothesis that there is not a mediating effect.

To better understand the role of *ICT* adoption in the relationship among land fragmentation, agricultural technology adoption and *CFAI*, we introduce a dummy variable (whether the farm household uses smart phone or personal computer (*PC*) to access information about agricultural production and selling via internet) to investigate whether *ICT* adoption mitigates the negative impacts of land fragmentation on agricultural technology adoption and *CFAI*. Both this dummy variable and its interaction with land fragmentation are incorporated into the regression so that:

$$CFAI_i = \gamma_0 + \gamma_1 SI_i + \gamma_2 X_{ki} + \gamma_3 ICT_i + \gamma_4 SI_i \times ICT_i + \varepsilon_{1i} \quad (7)$$

$$M_i = a_1 + a_2 SI_i + a_3 X_{ki} + a_4 ICT_i + a_5 SI_i \times ICT_i + \varepsilon_{2i} \quad (8)$$

$$CFAI_i = \rho_0 + \rho_1 SI_i + \rho_2 M_{li} + \rho_3 X_{ki} + \rho_4 ICT_i + \rho_5 SI_i \times ICT_i + \varepsilon_{3i} \quad (9)$$

where the dummy variable ICT_i takes a value of "1" if the farm uses smart phone or PC, and "0" otherwise. The internet use can help reduce chemical fertilizer use [49]. Hence, we hypothesize that *ICT* adoption has significant and negative coefficients, γ_3, and ρ_4, in Equations (7) and (9), respectively. Additionally, *ICT* adoption enables farmers to access more information about newly developed agricultural technologies and thus mitigate the negative effects of land fragmentation on both agricultural technology adoption and reduction of chemical fertilizer use. We therefore expect that $\gamma_4 < 0$ in Equation (7); $a_4 > 0$, $a_5 > 0$ in Equation (8); and $\rho_5 < 0$ in Equation (9).

2.2. Data

This study utilizes a dataset which was obtained by a face-to-face questionnaire survey administered by the National Agricultural and Rural Development Research Institute (NARI) of China Agricultural University (CAU) in 2019. The survey mainly focuses on grain production. Multistage sampling was employed for data collection. First, 14 provinces were chosen. Second, the towns were selected in each province based on the cultivated area of grains; that is, the sample towns should produce grains. Then 1–2 villages were randomly selected from each town. Next, 15–20 farm households were chosen from each village. As there might be farm households that are reluctant to participate in the survey, such a household would be replaced by another household.

From November to December of 2018, the NARI recruited students from CAU and trained them to guarantee that these students can collect appropriate data during the survey. The survey was conducted from January to February in 2019 when the university was on winter vacation. In the end, 2866 farm-level questionnaires from grain growers were obtained, covering a total of 14 provinces. The heads of the farm households were asked to answer the questionnaire based on their farm management in 2018. The survey data provide information on the inputs and outputs of crop production, land, income, expenditure and farm household characteristics.

Since our study is focused on smallholder farmers, we excluded the observations from farm sizes more than 2 hectares, according to classification by the World Bank. Additionally, inconsistent and incomplete questionnaires were dropped. The final dataset consists of 1388 farm households engaging in maize production, covering 144 villages from 119 counties across 14 provinces, namely Inner Mongolia, Jilin, Sichuan, Anhui, Shandong, Jiangsu, Jiangxi, Hebei, Henan, Hubei, Hunan, Gansu, Liaoning and Heilongjiang provinces, as shown in Figure 2.

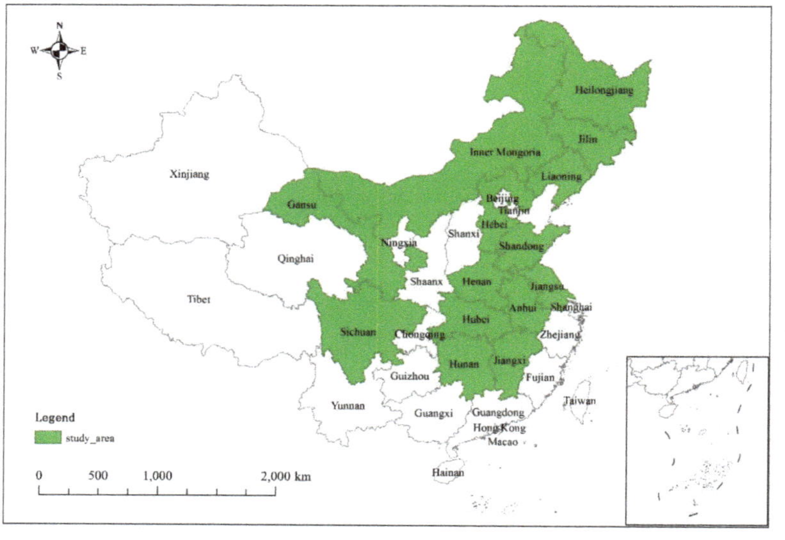

Figure 2. Profile of study areas.

The distribution of the 1388 observations from the 14 provinces is shown in Table 1. As one might want to know why there are only a few observations in Heilongjiang and Liaoning provinces, two major grain-producing provinces in China, the main reason is that most of the farms in these regions are larger than 2 hectares and were excluded from our analysis.

Table 1. Distribution of observations by province/autonomous region.

Province	N	Percentage (%)
Inner Mongolia	82	5.91%
Jilin	69	4.97%
Sichuan	166	11.96%
Anhui	38	2.74%
Shandong	257	18.52%
Jiangsu	110	7.93%
Jiangxi	74	5.33%
Hebei	139	10.01%
Henan	200	14.41%

Table 1. *Cont.*

Province	N	Percentage (%)
Hubei	111	8.00%
Hunan	58	4.18%
Gansu	15	1.08%
Liaoning	39	2.81%
Heilongjiang	30	2.16%
Total	1388	100.00%

2.3. Variables and Descriptive Statistics

The dependent variable is CFAI measured as kilograms per hectare. We calculated the CFAI using the consumption of chemical fertilizers per hectare sown area. Our core independent variable is land fragmentation, measured as Simpson's index. To provide robustness check, we used plot distance index (PDI), in the estimation.

Covariates in each equation are listed and explained in Table 2. For the two types of agricultural technology adoption, AMT is measured by a dummy variable, whether the household used agricultural machinery, and STFT is also measured by a dummy variable, whether the household adopted soil testing fertilization technologies. As one of our aims is to evaluate the moderating effects of ICT adoption on the relationship among land fragmentation, agricultural technology adoption and CFAI, we used a dummy variable, whether the household adopted ICTs, to measure ICT in our analysis.

Table 2. List of variables and definitions.

Variable	Definition and Descriptions	Mean	Std. Err.
Dependent Variable			
CFAI	Continuous variable, chemical fertilizer application intensity in maize production (kg/ha), measured using the CCF divided by SAC, expressed as a natural log (ln)	9.82	4.81
Variables of Interest			
SI	Continuous variable, land fragmentation, measured as Simpson's Index of Diversity	0.68	0.22
PDI	Continuous variable, plot distance index, proxy of land fragmentation, used for robustness test	0.18	0.36
AMT	Dummy variable, agricultural mechanization technology, "1" if the farm household used agricultural machinery during production, i.e., tillage, sowing, pest control, irrigation or harvesting, "0" otherwise	0.71	0.35
STFT	Dummy variable, soil testing fertilization technology, "1" if the farm household adopted the soil testing fertilization technology before the application of chemical fertilizer, "0" otherwise	0.23	0.38
ICT	Dummy variable, information and communication technology, "1" if the farmer used smart phone or personal computer, "0" otherwise	0.59	0.86
Control Variables			
Chemical fertilizer price	Continuous variable, the average price of chemical fertilizer purchased by farmers in 2018 (CNY/kg), expressed as a natural log (ln)	1.66	0.27
Herbicide	Continuous variable, the quantity of herbicide input in maize production per hectare in 2018 (kg/ha), expressed as a natural log (ln)	0.58	0.13
Farm size	Continuous variable, measured as the operated area of maize cropland (hectare), expressed as a natural log (ln)	2.01	1.36
Labor migration	Continuous variable, measured as the percentage of farm household members employed in non-agricultural sector	0.43	0.49
Agricultural investment	Continuous variable, measured as the depreciation expense of fixed assets used in maize production in 2018 (CNY), expressed as a natural log (ln)	10.65	15.38
Crop structure	Continuous variable, measured as the share of sales revenue of grains in agricultural income	0.75	0.51
Sell mode	Dummy variable, "1" if the sell mode is instant sale, "0" if the sale mode is contract sale	0.23	0.42
Village leader	Dummy variable, "1" if the farmer is village leader, "0" otherwise	0.16	0.37
Flat land ratio	Continuous variable, the percentage of flat land in the total operated land area (%)	0.64	0.33
Sloped land ratio	Continuous variable, the percentage of sloped land in the total operated land area (%)	0.21	0.12
Hilly land ratio	Continuous variable, the percentage of hilly land in the total operated land area (%)	0.15	0.11
Paddy land ratio	Continuous variable, the percentage of paddy fields in the total operated land area (%)	0.09	0.06
Dry land ratio	Continuous variable, the percentage of dry fields in the total operated land area (%)	0.91	0.77
Self-rated land quality	Ordered variable, indicating the self-rated quality of the operated land, "1" if the land is barren, "2" if low quality, "3" if medium, "4" if medium to high, and "5" if extremely fertile	3.03	0.88

Table 2. *Cont.*

Variable	Definition and Descriptions	Mean	Std. Err.
Land use rights	Dummy variable, "1" if the land use rights were registered and certificated, "0" otherwise	0.92	0.98
Age	Continuous variable, age of the household's head, expressed as a natural log (ln)	3.94	0.23
Education	Ordered variable, education level of the household's head (1–6), "1" illiterate, "2" elementary school, "3" middle school, "4" high school or vocational high school, "5" three-year college, and "6" college or post-graduate	2.76	0.95
Male	Dummy variable, "1" male, "0" female	0.77	0.42
Social capital	Continuous variable, measured as the frequency of the farms reach out to their friends, i.e., the number of friends or relatives the household says hi to via WeChat, phone calls or meetings during spring festival, expressed as a natural log (ln)	4.00	4.35
Technical guidance	Dummy variable, "1" if the farm household received technical guidance, "0" otherwise	0.16	0.27
Cooperative	Dummy variable, "1" if the farm household is member of cooperatives, "0" otherwise	0.25	0.41
Fixed assets investment	Continuous variable, measured as the depreciation expense of total fixed assets in 2018 (CNY), expressed as a natural log (ln)	9.27	11.32
Hired labor ratio	Continuous variable, measured as the number of work days of hired labor divided by the total number of work days devoted to maize production in 2018	0.12	0.54
Inward land transfer	Dummy variable, "1" if the farm household leased farmland from others, "0" otherwise	0.14	0.28
Outward land transfer	Dummy variable, "1" if the farm household transferred land use rights to others, "0" otherwise	0.02	0.12
Producer subsidy	Dummy variable, "1" if the farm household received a subsidy on maize production, "0" otherwise	0.15	0.36
Machinery subsidy	Dummy variable, "1" if the farm household received a subsidy on the purchase of agricultural machinery, "0" otherwise	0.05	0.22
East	Dummy variable, "1" if farm household is located in eastern region, "0" otherwise	0.34	0.47
Central	Dummy variable, "1" if farm household is located in central region, "0" otherwise	0.53	0.50
West	Dummy variable, "1" if farm household is located in western region, "0" otherwise	0.13	0.33

Notes: 1. The sum of flat land ratio, sloped land ratio and hilly land ratio equals 1. 2. Land use rights refers to the registration and certification of farmland. In particular, the rural land registration and certification program started since the No. 1 central document in 2013 was issued. It is the confirmation of land ownership, land tenure (land use rights) and other rights. The rights of each parcel must be subject to land registration procedures such as land registration application, cadastral investigation, verification of affiliation, registration and issuance of land certificates. 3. We categorize the 14 provinces into three regions according to the geographic location. Eastern region includes Hebei, Liaoning, Jiangsu, Shandong. Central region includes Inner Mongolia, Jilin, Heilongjiang, Anhui, Henan, Hubei, Hunan, Jiangxi. Western region includes Sichuan and Gansu.

In the mediation model, we control for farm household and farmer characteristics, such as age, gender, education, social capital, technical guidance, fixed assets investment and region characteristics, following the existing literature [6,32,50]. Characteristics of farmland, such as the terrain and structure of cropland (evaluated by flat land ratio, sloped land ratio, hilly land ratio, paddy land ratio, and dry land ratio), are also included as they are considered as crucial factors affecting household decisions regarding farming techniques [48,51–53]. Self-rated quality of cropland may also affect farmers' production decisions due to the endowment effects. Therefore, a variable for self-rated quality of cropland is included in the model. Moreover, since tenure security contributes to the reduction of chemical fertilizer use [54,55], a variable for land use rights certification is included in the CFAI equation. In addition, the rural–urban migration experience is conducive to reducing fertilizer use [56], so we control for the labor migration variable. Agricultural subsidies also reduced fertilizer use by promoting the adoption of agricultural techniques, a variable that indicates whether the farm received maize producer subsidy is included in the mediation model [57].

A statistical description of variables is presented in Table 2. The average CFAI is 335.89 kg/ha, which is very large compared with some developed countries such as the United States and Japan. Moreover, the average SI is 0.68 and the average PDI is 0.18, which means that land fragmentation is severe in China. In addition, more than half of the farms have adopted AMTs and ICTs, accounting for 71% and 58.8% of the total farms, respectively. However, only 23% of the farms have adopted STFTs, implying that the advantages of sustainable agricultural technologies have not yet been fully recognized.

3. Estimation Results

3.1. Baseline Regression

Based on our observation and previous studies, we establish a conceptual framework, considering the role of AMT, STFT and ICT adoption on the effects of land fragmentation on CFAI. A possible mechanism is shown in Figure 3, and we examine it using the survey data.

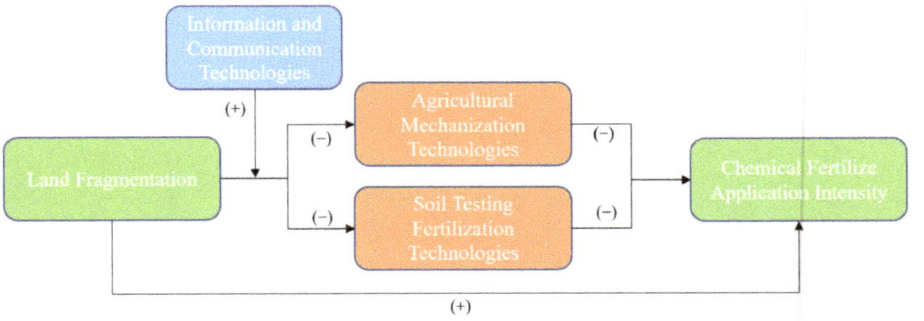

Figure 3. Influence mechanism of technology adoption on land fragmentation and CFAI.

Table 3 reports the mediating effect of STFT and AMT adoption on the relationship between land fragmentation and CFAI. It shows that the coefficient of SI on CFAI is significant and positive in columns (1), (3) and (5), implying that land fragmentation has a significantly positive effect on CFAI. Additionally, the coefficient of land fragmentation on STFT and AMT adoption is significant and negative, in columns (2) and (4), respectively, which means that land fragmentation has a negative impact on the adoption of these two agricultural technologies. Moreover, the coefficient of STFT and AMT adoption on CFAI is significantly negative in columns (3) and (5), respectively, meaning that the adoption of agricultural technologies has significantly decreased the CFAI. The results suggest the existence of the mediating effect of adopting two agricultural technologies, and the total effect mediated by the adoption of STFTs and AMTs are 11% and 29% respectively. As expected, land fragmentation can not only directly increase the CFAI, but it also indirectly increases the CFAI by decreasing the probability of farmers adopting agricultural technologies.

Table 3. Mediating effects of STFT and AMT adoption on the impact of SI on CFAI.

Variable	(1) CFAI	(2) STFT	(3) CFAI	(4) AMT	(5) CFAI
STFT			−0.121 ** (0.052)		
AMT					−0.238 *** (0.067)
SI	0.289 *** (0.051)	−0.268 * (0.162)	0.254 *** (0.056)	−0.352 *** (0.039)	0.205 *** (0.047)
Control	Yes	Yes	Yes	Yes	Yes
_cons	3.565 *** (1.100)	0.233 *** (0.089)	3.112 *** (0.655)	0.558 *** (0.110)	3.223 *** (0.724)
Obs.	1388	1388	1388	1388	1388
R-sqr	0.238	0.117	0.236	0.121	0.241
Sobel tests		0.0019 * (0.0010)		0.0022 *** (0.0006)	
Total effect mediated		11%		29%	

Notes: Robust errors are in parenthesis, *** $p < 0.01$, ** $p < 0.05$, * $p < 0.1$. CFAI refers to chemical fertilizer application intensity, SI refers to Simpson's Index of Diversity, AMT refers to agricultural mechanization technology adoption, STFT refers to soil testing fertilization technology adoption.

To examine the impact of ICT adoption on the relationship between land fragmentation and agricultural technology adoption, we apply OLS regression to Equations (5) and (7). The results are shown in Table 4.

Table 4. Effect of ICTs on the relationship between land fragmentation and STFT adoption.

Variable	(1)	(2)
ICT		0.644 *** (0.163)
SI × ICT		0.141 *** (0.043)
SI	−0.268 *** (0.162)	−0.286 *** (0.096)
Control	Yes	Yes
_cons	0.233 *** (0.089)	0.226 *** (0.077)
Obs.	1388	1388
R-sqr	0.117	0.285

Notes: Robust errors are in parenthesis, *** $p < 0.01$. The dependent variable is soil testing fertilization technology (STFT) adoption. SI refers to Simpson's Index of Diversity. ICT refers to information and communications technology adoption.

As previously analyzed, without considering the role of ICT adoption, land fragmentation has a significantly negative effect on STFT adoption, as shown in column (1). After ICT and the interaction term of ICT adoption and land fragmentation were introduced into the regression, we can see from column (2) that both ICT and the interaction term have significant and positive coefficients, implying that ICT adoption can significantly increase the probability of STFT adoption and mitigate the negative effects of land fragmentation on STFT adoption.

To sum up, the adoption of ICTs significantly affects the effect of land fragmentation on STFT adoption, which can be explained by the typical characteristics of farmers using the internet in rural areas. Based on our field research experience, for the vast majority of farmers who obtain agricultural service information through the internet, they use instant messaging software called WeChat (Shenzhen, China). WeChat has revolutionized farmers' technology adoption behaviors. On one hand, it pushes agricultural service information in real time. A large number of surveyed farmers subscribe to information services to receive the latest agricultural extension information and technical guidance. Internet use has significantly improved the availability of agricultural information. On the other hand, WeChat has significantly improved the intensity of farmers' social networks. For farmers, the impact of internet use on social network intensity is mutual. Farmers with strong social networks are more inclined to use the internet, and internet use further increases the social network intensity of farmers.

This interaction has a specific impact on the adoption of STFTs by farmers. They can share the obtained agricultural technology information through instant messaging tools such as WeChat and further exchange information, which significantly enhances the dissemination of technology adoption experience. Farmers who have adopted the STFTs can share the relevant experience and effectiveness of technology adoption with other farmers who use WeChat, which significantly affects other farmers' decisions regarding agricultural technology adoption. On the contrary, farmers who do not use WeChat have lower frequency and efficiency in agricultural technology information exchange. A field survey based on 1710 farmers in Hubei Province also confirmed that land fragmentation has a significant negative impact on the adoption of STFT by farmers with weak social networks [58].

Table 5 reports the effects of ICT adoption on the relationship between land fragmentation and AMT adoption. Without considering the role of ICT adoption, land fragmentation has a significantly negative effect on AMT adoption, as shown in column (1), which is consistent with the existing studies. After ICT and the interaction term of ICT adoption and land fragmentation were introduced into the regression, we can see from column (2) that both ICT and the interaction term have significant and positive coefficient, implying that ICT adoption can significantly increase the probability of AMT adoption and mitigate the negative effects of land fragmentation on AMT adoption.

Table 5. Effect of ICTs on the relationship between land fragmentation and AMT adoption.

Variable	(1)	(2)
ICT		0.552 ***
		(0.201)
SI × ICT		0.123 ***
		(0.031)
SI	−0.315 ***	−0.321 ***
	(0.072)	(0.058)
Control	Yes	Yes
_cons	0.245 ***	0.211 ***
	(0.051)	(0.063)
Obs.	1388	1388
R-squared	0.298	0.208

Notes: Robust errors are in parenthesis, *** $p < 0.01$. The dependent variable is agricultural mechanization technology (AMT) adoption. SI refers to Simpson's Index of Diversity. ICT refers to information and communications technology adoption.

3.2. Robustness Check

To provide a robustness check, we use the plot distance index (PDI) as an alternative variable of land fragmentation, following the existing studies [45,59]. As shown in Table 6, when PDI was used to replace SI as the independent variable, the results are completely consistent with the benchmark regression results, and land fragmentation significantly increases CFAI. In terms of the influence mechanism, land fragmentation has a significantly negative effect on the adoption of agricultural technologies and thus increases the CFAI.

Table 6. Effects of PDI on CFAI mediated by STFT and AMT adoption.

Variable	(1) CFAI	(2) STFT	(3) CFAI	(4) AMT	(5) CFAI
STFT			−0.163 ***		
			(0.031)		
AMT					−0.240 ***
					(0.076)
PDI	0.339 ***	−0.342 **	0.283 ***	−0.440 **	0.233 ***
	(0.128)	(0.144)	(0.050)	(0.191)	(0.068)
Control	Yes	Yes	Yes	Yes	Yes
_cons	3.441 ***	0.258 ***	3.625 ***	0.571 ***	3.145 ***
	(0.586)	(0.044)	(0.829)	(0.010)	(0.603)
Obs.	1388	1388	1388	1388	1388
R-sqr	0.366	0.185	0.268	0.167	0.308
Sobel Mediation Tests		0.0008 ***		0.0015 ***	
		(0.0002)		(0.0000)	
Total effect mediated		16%		32%	

Notes: Robust errors are in parenthesis, *** $p < 0.01$, ** $p < 0.05$. CFAI refers to chemical fertilizer application intensity, PDI refers to plot distance index, AMT refers to agricultural mechanization technology adoption, STFT refers to soil testing fertilization technology adoption.

3.3. Further Comparison

To further examine the role of technology adoption on the relationship between land fragmentation and CFAI, as shown in Figure 4, we introduced the interaction term of SI and the adoption of the three technologies into the regression. The estimation results are shown in Table 7.

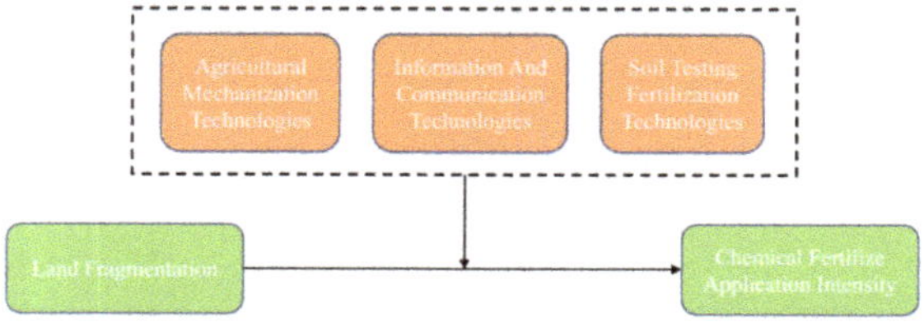

Figure 4. Heterogeneity in the moderating effects of the adoption of the three technologies.

Table 7. Effects of technology adoption.

Variable	(1)	(2)	(3)	(4)
SI	0.289 *** (0.096)	0.278 *** (0.086)	0.285 *** (0.070)	0.281 *** (0.091)
SI × ICT		−0.215 (0.187)		
SI × AMT			−0.176 *** (0.049)	
SI × STFT				−0.121 ** (0.056)
Control	Yes	Yes	Yes	Yes
cons	3.565 *** (1.100)	3.178 *** (0.739)	3.456 *** (1.197)	3.923 *** (0.713)
Obs.	1388	1388	1388	1388
R-sqr	0.238	0.222	0.233	0.235

Notes: Robust errors are in parenthesis, *** $p < 0.01$, ** $p < 0.05$. The dependent variable is chemical fertilizer application intensity (CFAI). SI refers to Simpson's Index of Diversity, AMT refers to agricultural mechanization technology adoption, STFT refers to soil testing fertilization technology adoption, STFT refers to soil testing fertilization technology adoption, ICT refers to information and communications technology adoption.

The results show that SI has significant and positive coefficients in column (1)–(4), implying that land fragmentation has increased CFAI. The coefficient of the interaction term of SI and AMT is significant and negative in column (3), which means that AMT adoption can mitigate the positive impact of land fragmentation on CFAI. Additionally, the results of column (4) suggest that STFT adoption can mitigate the positive impact of land fragmentation on CFAI. More importantly, the coefficient of the interaction term of SI and ICT is not statistically significant, and ICT has no significant effect on the direct effect of land fragmentation on CFAI, as shown in column (2).

Obviously, the use of agricultural machinery and the progress of agricultural technology are effective measures to reduce the intensity of chemical fertilizer use. When we examine the moderating effects of ICT on the direct effects of land fragmentation on CFAI alone, the role of information publicity and promotion brought with ICTs is limited. The unfavorable land resource endowment has restricted the adoption of advanced agricultural technologies and machinery. Therefore, even if farmers recognized the negative effects of excessive application of chemical fertilizer, they would increase the input of chemical fertilizer to ensure income and output. More importantly, the influence mechanism shown in Figure 3 confirmed that the rapid development of information technology has significantly increased the probability of farmers adopting AMTs and STFTs and thus leads to the reduction of CFAI.

4. Discussion

4.1. Role of AMT Adoption

According to the baseline regression results, the impact of agricultural mechanization exceeded our research expectations, which may be related to the stage of agricultural development in China. The issue of land fragmentation in China can be traced back to the implementation of the household contract responsibility system in the late 1980s. The arrangement of land property right system has led to the problem of land fragmentation to a certain extent [36]. However, there was no better choice for China back then. Moreover, land fragmentation did not have a negative impact on China's agricultural production and even dispersed agricultural risks and improved the utilization efficiency of the labor force [60–62]. However, with the rise of labor costs and popularization of agricultural machinery, land fragmentation has increased the commuting time between plots, which limits the application of large-scale agricultural machinery. It is rather difficult for small farms to adapt to the development of modern agriculture in China. Furthermore, compared with the adoption of STFTs, whether farmers use formula fertilizer and the proportion of formula fertilizer also have significant effects on CFAI. However, due to data availability, we cannot further verify the effect of actual use of formula fertilizer on CFAI in this study.

In fact, the application of chemical fertilizer by small farmers usually remains stable in the long run [6] since it is greatly affected by previous experience. It seems like another reasonable explanation for the effects of the adoption of agricultural technology. Before the promotion and popularization of chemical fertilizer, based on the fact that the soil nutrients of cultivated land were low, China carried out chemical fertilizer efficiency tests at the national level for several years and formulated the standards for the application amounts of chemical fertilizer to grain crops according to the test results. However, farmers' fertilization behavior is inertial and has a cumulative effect on fertilizer application. When there is no external change, such as the application of advanced technology, farmers are likely to overuse chemical fertilizer according to their experience and fertilization habits.

From the perspective of costs and benefits, from 1978 to 2014, the average annual growth rate of China's agricultural means of production prices was 5.4%, while it was 6.4% for agricultural producer prices. It is thus profitable for farmers to increase inputs [63]. However, China's unique urban and rural dual system and land system has diminished the advantages of China in competition with other countries with similar resource endowment conditions. The most predominant impact is the rise of agricultural labor costs and land rents. Moreover, with the continuous rise of the price of agricultural production inputs such as chemical fertilizer, the growth of marginal output and marginal income brought by increasing inputs has decreased significantly. Therefore, under the given cost constraints, it is feasible to reduce the input of other factors by increasing the use of agricultural machinery, which is reflected not only in the substitution of labor but also in the reduction of the input of means of production such as chemical fertilizer.

Based on the above analysis, taking into account the land management rights and other issues, land circulation and scale management aiming to improve the degree of mechanized operation are effective ways to reduce the intensity of chemical fertilizer application.

4.2. Role of STFT Adoption

The Chinese government began to implement policies to promote STFT adoption in 2005. After the implementation of the policy, the utilization rate of nitrogen fertilizer, phosphorus fertilizer and fertilizer addition in rice, wheat and corn was 33%, 24% and 42%, respectively, which increased by 5%, 12% and 10%, respectively. The effect of STFT adoption on improving the utilization rate of chemical fertilizer is obvious. However, our study shows that compared with AMT adoption, STFT adoption has a weaker influence on the effect of land fragmentation on CFAI. On one hand, the price of formula fertilizer is higher than that of chemical fertilizer. Although many farmers adopted STFTs because of policy incentives and financial subsidies, they did not use formula fertilizer. On the other hand, the effect of land fragmentation on AMT adoption is larger than that on STFT adoption.

Therefore, the popularization of STFTs can effectively promote the reduction of chemical fertilizer. Firstly, improve the market competitiveness of chemical fertilizers and pesticides use to produce agricultural products in accordance with scientific and reasonable methods. Secondly, increase the availability of professional production services for farmers, strengthen the promotion of soil testing fertilization and other technologies and increase the corresponding financial subsidies. Thirdly, reduce or even gradually abolish the preferential tax policies for chemical fertilizer production, strengthen the market supervision of excessive use of chemical fertilizer production products and improve the regulations on illegal production.

4.3. Role of ICT Adoption

In this study, we focus on internet use since smart phones and computers are quite powerful and play an increasingly important role in agricultural production. Based on our observation during field surveys, farmers can easily connect with the researchers from local agricultural research institutions by WeChat. For example, the National Green Manure Industry Technology System in China regularly records videos on the application of chemical fertilizer and green manure and provides corresponding information and technical services for farmers in rural areas through influential WeChat video subscription services.

Moreover, in recent years, Tik Tok is getting more and more popular and gaining wider influence. Considering that the education level of most of the farmers in rural China is relatively low, video is an efficient way to for them to access information. Tik Tok is quite significant because it has a large number of users in China. When we use agricultural planting technology as a keyword to search in Tik Tok, we can immediately get massive intuitive video information. It is noteworthy that these video providers do not rely on government support, and they are independent media practitioners. To our knowledge, there are millions of similar video providers who have created Tik Tok accounts China. At the same time, farmers can also communicate with these technical information providers through the comment function.

Our analysis shows that ICTs play the role of catalyst; that is, ICT adoption can slow down the positive effects of land fragmentation on CFAI by mitigating the negative impact of land fragmentation on the adoption of AMT and STFT. Apparently, ICT can also directly affect the fertilizer application intensity of farmers. According to our theoretical analysis and interviews with typical farmers, farmers are unlikely to change their decisions when they are faced with the pressure of crop yield and agricultural income, even if they fully understand the negative impact of excessive application of chemical fertilizer through ICT. Our results show the significant difference between ICT and typical agricultural technologies such as AMT and STFT. Most importantly, our study provides theoretical support for the Chinese government to formulate industrial policies to reduce the use of chemical fertilizer.

Based on the above analysis, we understand the mechanism of how ICTs affect the impact of land fragmentation on CFAI. It is noteworthy that it is not enough to rely merely on ICTs for policy encouragement and publicity without satisfying the demands of farmers through the application of agricultural technologies. The policies should be focused on supporting the development of agricultural technologies, give full consideration to the advantages of ICTs, and propagandize the important role of agricultural technology adoption in increasing productivity and efficiency so as to reduce the intensity of chemical fertilizer application.

5. Conclusions

In this study, we examined the relationship between land fragmentation and CFAI, and further explored the mediating effect of AMT and STFT adoption in China's maize production. We developed a mediation model to explore the influence mechanism of land fragmentation on CFAI through AMT and STFT adoption. Considering the important role of ICTs played in agricultural production, we explored the impact of the ICT adoption on the relationship between land fragmentation, agricultural technology adoption and CFAI and conducted an empirical analysis on a farm level survey data with 1388 observations.

Our results clearly indicate that land fragmentation has a positive impact on CFAI, and the adoption of both AMTs and STFTs has a significant negative effect on CFAI; land fragmentation reduced the probability of farmers adopting these technologies. Moreover, the adoption of ICTs can significantly reduce the negative effect of land fragmentation on the adoption of AMTs and STFTs, but it did not directly affect the process of land fragmentation decreasing the CFAI.

This study contributes to a better understanding of the relationship between land fragmentation and chemical fertilizer use in China's maize production. Moreover, the mediating effects of the adoption of AMTs and STFTs on the relationship between land fragmentation and chemical fertilizer use can provide insights on the influencing mechanism of land fragmentation, which is especially crucial to provinces suffering high chemical fertilizer application intensity. More importantly the adoption of ICTs can mitigate the negative impact of land fragmentation on technology adoption, which helps shed light on the issue of the low adoption rate of agricultural technologies in rural China. Therefore, policies should be carried out to continue to strengthen the extension, promotion and adoption of agricultural technologies such as AMTs and STFTs. In addition, it is of significance to give full consideration to the role of information technologies and to promote technology adoption in rural China.

The generalization of the findings of this study is subject to certain limitations. For example, the study was limited to maize production in 14 provinces in China. The results may not be able to be applied to other areas of grain production in the whole nation. China is a diverse country in terms of varying crop varieties and economic development across regions. Studies on other crops and other regions can be conducted to enrich the study in this field.

Author Contributions: Conceptualization, L.C. and M.H.; methodology, L.C.; software, L.C.; validation, L.C., S.H. and M.H.; formal analysis, L.C.; investigation, L.C.; resources, S.H.; data curation, L.C.; writing—original draft preparation, L.C.; writing—review and editing, L.C., M.H. and Y.L.; visualization, S.H.; supervision, S.H. and J.L.; project administration, S.H.; funding acquisition, J.L. All authors have read and agreed to the published version of the manuscript.

Funding: This research was funded by Youth Navigation Project, the Central Public-Interest Scientific Institution Basal Research Fund of China (Grant No. JBYW-AII-2022-40); the Central Public-Interest Scientific Institution Basal Research Fund of China (Grants No. JBYW-AII-2022-21, JBYW-AII-2022-13); the National Natural Science Foundation of China General Program (Grant No. 71973087); the National Natural Science Foundation of China Youth Fund Project (Grants No. 72003215, 71703159).

Institutional Review Board Statement: Not applicable.

Informed Consent Statement: Not applicable.

Data Availability Statement: The associated dataset of the study is available upon request from the corresponding author.

Acknowledgments: The authors would like to thank the anonymous reviewers for their comments and suggestions, which contributed to the further improvement of this paper.

Conflicts of Interest: The authors declare no conflict of interest.

References

1. Zheng, W.L.; Luo, B.L.; Hu, X.Y. The determinants of farmers' fertilizers and pesticides use behavior in China: An explanation based on label effect. *J. Clean. Prod.* **2020**, *272*, 123054. [CrossRef]
2. Huang, W.; Jiang, L. Efficiency performance of fertilizer use in arable agricultural production in China. *China Agric. Econ. Rev.* **2019**, *11*, 52–69. [CrossRef]
3. Fan, L.C.; Yuan, Y.M.; Ying, Z.C.; Lam, S.K.; Liu, L.; Zhang, X.C.; Liu, H.B.; Gu, B.J. Decreasing farm number benefits the mitigation of agricultural non-point source pollution in China. *Environ. Sci. Pollut. Res.* **2019**, *26*, 464–472. [CrossRef]
4. Yang, L.; Tang, K.; Wang, Z.H.; An, H.Z.; Fang, W. Regional eco-efficiency and pollutants' marginal abatement costs in China: A parametric approach. *J. Clean. Prod.* **2017**, *167*, 619–629. [CrossRef]
5. NBSC. *Bulletin of the Third Agricultural Census;* China Statitics Press: Beijing, China, 2018.
6. Gao, J.J.; Peng, C.; Shi, Q.H. Study on high amount of chemical fertilizer and fertilization behavior of small farmers in China. *Manag. World* **2019**, *35*, 12. [CrossRef]
7. Wu, Y.Y.; Xi, X.C.; Tang, X.; Luo, D.M.; Gu, B.J.; Lam, S.K.; Vitousek, P.M.; Chen, D.L. Policy distortions, farm size, and the overuse of agricultural chemicals in China. *Proc. Natl. Acad. Sci. USA* **2018**, *115*, 7010–7015. [CrossRef]
8. Lin, J.Y.F. Rural Reforms and Agricultural Growth in China. *Am. Econ. Rev.* **1992**, *82*, 34–51.

9. Deng, H.T.; Cui, B.M. The Nature of the Property Right of the Chinese Agricultural Land during Institution Change: The Angle of View of the History. *Nankai Econ. Stud.* **2007**, 118–141. [CrossRef]
10. Chen, X.P.; Cui, Z.L.; Fan, M.S.; Vitousek, P.; Zhao, M.; Ma, W.Q.; Wang, Z.L.; Zhang, W.J.; Yan, X.Y.; Yang, J.C.; et al. Producing more grain with lower environmental costs. *Nature* **2014**, *514*, 486–489. [CrossRef]
11. Gu, B.J.; Ju, X.T.; Chang, J.; Ge, Y.; Vitousek, P.M. Integrated reactive nitrogen budgets and future trends in China. *Proc. Natl. Acad. Sci. USA* **2015**, *112*, 8792–8797. [CrossRef]
12. Zhu, P.; Su, M.; Yan, J. Impact of farmland scale and stability on fertilizer input: Taking rice production of four counties of jiangsu province as example. *J. Nanjing Agric. Univ.* **2017**, *17*, 85–94.
13. Li, Y.X.; Zhang, W.F.; Ma, L.; Huang, G.Q.; Oenema, O.; Zhang, F.S.; Dou, Z.X. An Analysis of China's Fertilizer Policies: Impacts on the Industry, Food Security, and the Environment. *J. Environ. Qual.* **2013**, *42*, 972–981. [CrossRef] [PubMed]
14. Xiang, T.; Qi, Y. Food security and agricultural non-point source pollution: Taking the impact of agricultural land endowment on the intensity of fertilizer input as an example. *J. Financ. Econ.* **2015**, *41*, 132–144.
15. Carlson, K.M.; Gerber, J.S.; Mueller, N.D.; Herrero, M.; MacDonald, G.K.; Brauman, K.A.; Havlik, P.; O'Connell, C.S.; Johnson, J.A.; Saatchi, S.; et al. Greenhouse gas emissions intensity of global croplands. *Nat. Clim. Change* **2017**, *7*, 63–68. [CrossRef]
16. Foley, J.A.; Ramankutty, N.; Brauman, K.A.; Cassidy, E.S.; Gerber, J.S.; Johnston, M.; Mueller, N.D.; O'Connell, C.; Ray, D.K.; West, P.C.; et al. Solutions for a cultivated planet. *Nature* **2011**, *478*, 337–342. [CrossRef]
17. Fischer, G.; Winiwarter, W.; Ermolieva, T.; Cao, G.-Y.; Qui, H.; Klimont, Z.; Wiberg, D.; Wagner, F. Integrated modeling framework for assessment and mitigation of nitrogen pollution from agriculture: Concept and case study for China. *Agric. Ecosyst. Environ.* **2010**, *136*, 116–124. [CrossRef]
18. Wang, Y.; Zhu, Y.C.; Zhang, S.X.; Wang, Y.Q. What could promote farmers to replace chemical fertilizers with organic fertilizers? *J. Clean. Prod.* **2018**, *199*, 882–890. [CrossRef]
19. Vitousek, P.M.; Naylor, R.; Crews, T.; David, M.B.; Drinkwater, L.E.; Holland, E.; Johnes, P.J.; Katzenberger, J.; Martinelli, L.A.; Matson, P.A.; et al. Nutrient Imbalances in Agricultural Development. *Science* **2009**, *324*, 1519–1520. [CrossRef]
20. Zhang, W.F.; Cao, G.X.; Li, X.L.; Zhang, H.Y.; Wang, C.; Liu, Q.Q.; Chen, X.P.; Cui, Z.L.; Shen, J.B.; Jiang, R.F.; et al. Closing yield gaps in China by empowering smallholder farmers. *Nature* **2016**, *537*, 671–674. [CrossRef]
21. Cui, Z.L.; Zhang, H.Y.; Chen, X.P.; Zhang, C.C.; Ma, W.Q.; Huang, C.D.; Zhang, W.F.; Mi, G.H.; Miao, Y.X.; Li, X.L.; et al. Pursuing sustainable productivity with millions of smallholder farmers. *Nature* **2018**, *555*, 363–366. [CrossRef]
22. NBSC. *China Statistical Yearbook 2021*; China Statistics Press: Beijing, China, 2021.
23. Brunelle, T.; Dumas, P.; Souty, F.; Dorin, B.; Nadaud, F. Evaluating the impact of rising fertilizer prices on crop yields. *Agric. Econ. Blackwell* **2015**, *46*, 653–666. [CrossRef]
24. Takeshima, H.; Liverpool-Tasie, L.S.O. Fertilizer subsidies, political influence and local food prices in sub-Saharan Africa: Evidence from Nigeria. *Food Policy* **2015**, *54*, 11–24. [CrossRef]
25. Takeshima, H.; Adhikari, R.P.; Shivakoti, S.; Kaphle, B.D.; Kumar, A. Heterogeneous returns to chemical fertilizer at the intensive margins: Insights from Nepal. *Food Policy* **2017**, *69*, 97–109. [CrossRef]
26. Savari, M.; Gharechaee, H. Application of the extended theory of planned behavior to predict Iranian farmers' intention for safe use of chemical fertilizers. *J. Clean. Prod.* **2020**, *263*, 121512. [CrossRef]
27. Arriagada, R.A.; Sills, E.O.; Pattanayak, S.K.; Cubbage, F.W.; Gonzalez, E. Modeling fertilizer externalities around Palo Verde National Park, Costa Rica. *Agric. Econ. Blackwell* **2010**, *41*, 567–575. [CrossRef]
28. Shi, C.L.; Zhu, J.F.; Luan, J. Analysis of the efficiency and the influencing factors of wheat chemical fertilizer input in China: Positive study based on 15 major wheat-farming provinces in China. *J. Agrotech. Econ.* **2015**, *11*, 69–78.
29. Ji, Y.Q.; Zhang, H.; Lu, W.Y. Differentiation, incomplete information and farmer's excessive application of fertilizer. *J. Agrotech. Econ.* **2016**, *2*, 14–22.
30. Hu, H.; Yang, Y.B. Study on farmer's application of chemical fertilizer from the perspective of factor substitu tion: Based on the household data of national rural fixed observation point. *J. Agrotech. Econ.* **2015**, *3*, 84–91.
31. Ju, X.T.; Gu, B.J.; Wu, Y.Y.; Galloway, J.N. Reducing China's fertilizer use by increasing farm size. *Glob. Environ. Change* **2016**, *41*, 26–32. [CrossRef]
32. Hu, L.X.; Zhang, X.H.; Zhou, Y.H. Farm size and fertilizer sustainable use: An empirical study in Jiangsu, China. *J. Integr. Agric.* **2019**, *18*, 2898–2909. [CrossRef]
33. Yang, Q.; Zhu, Y.J.; Liu, L.; Wang, F. Land tenure stability and adoption intensity of sustainable agricultural practices in banana production in China. *J. Clean. Prod.* **2022**, *338*, 130553. [CrossRef]
34. Shi, C.; Zhang, Y.; Guo, Y.; Zhu, J. The impact of land fragmentation on farmer's chemical fertilizer use efficiency. *J. Nat. Resour.* **2019**, *34*, 2687–2700.
35. Wang, S.L.; Xin, X. Does land fragmentation hinder the implementation of sugarcane planting mechanization? *Chin. Rural Econ.* **2017**, *2*, 16–29.
36. Lu, H.; Hu, H. Does land fragmentation increase agricultural production costs? A microscopic investigation from Jiangsu province. *Econ. Rev.* **2015**, *5*, 129–140.
37. Tan, S.; Heerink, N.; Kuyvenhoven, A.; Qu, F. Impact of land fragmentation on rice producers' technical efficiency in South-East China. *NJAS-Wagening. J. Life Sci.* **2010**, *57*, 117–123. [CrossRef]

38. Li, X.; Ou, M.H.; Ma, X.L. Analysis on impact of fragmentation based on landscape index to cultivated land use efficiency: A case on Lixiahe district in Yangzhou city. *J. Nat. Resour.* **2011**, *27*, 1758–1767.
39. Yu, W.Y.; Qi, Y.B.; He, Y. The effect of rice irrigation efficiency and related factors on fertilizer non- point source pollution based on quantile regression. *J. Agro-Environ. Sci.* **2017**, *36*, 1274–1284.
40. Li, K.Q.; Ma, D.D. On the investment behavior and decision mechanism of farmers' fertilizer reduction in ecologically vulnerable regions. *J. Nanjing Agric. Univ.* **2018**, *18*, 138–145.
41. Ren, C.; Jin, S.; Wu, Y.; Zhang, B.; Gu, B. Fertilizer overuse in Chinese smallholders due to lack of fixed inputs. *J. Environ. Manag.* **2021**, *293*, 112913. [CrossRef]
42. Niroula, G.S.; Thapa, G.B. Impacts and causes of land fragmentation, and lessons learned from land consolidation in South Asia. *Land Use Policy* **2005**, *22*, 358–372. [CrossRef]
43. Tchale, H. The efficiency of smallholder agriculture in Malawi. *Afr. J. Agric. Resour. Econ.* **2009**, *3*, 101–121.
44. Zhang, Y.N.; Long, H.L.; Li, Y.R.; Ge, D.Z.; Tu, S.S. How does off-farm work affect chemical fertilizer application? Evidence from China's mountainous and plain areas. *Land Use Policy* **2020**, *99*, 104848. [CrossRef]
45. Wang, Y.H.; Li, X.B.; Xin, L.J. Characteristics of cropland fragmentation and its impact on agricultural production costs in mountainous areas. *J. Nat. Resour.* **2019**, *34*, 2658–2672. [CrossRef]
46. Tan, S.H.; Heerink, N.; Kruseman, G.; Qu, F.T. Do fragmented landholdings have higher production costs? Evidence from rice farmers in Northeastern Jiangxi province, PR China. *China Econ. Rev.* **2008**, *19*, 347–358. [CrossRef]
47. Wang, J.J.; Li, J.J.; Chang, J. Product co-development in an emerging market: The role of buyer supplier compatibility and institutional environment. *J. Oper. Manag.* **2016**, *46*, 69–83. [CrossRef]
48. Huan, M.L.; Dong, F.X.; Chi, L. Mechanization services, factor allocation, and farm efficiency: Evidence from China(dagger). *Rev. Dev. Econ.* **2022**, *99*, 104848. [CrossRef]
49. Yuan, F.; Tang, K.; Shi, Q.H. Does Internet use reduce chemical fertilizer use? Evidence from rural households in China. *Environ. Sci. Pollut. Res.* **2021**, *28*, 6005–6017. [CrossRef]
50. Zhang, W.S.; Qian, C.R.; Carlson, K.M.; Ge, X.L.; Wang, X.B.; Chen, X.P. Increasing farm size to improve energy use efficiency and sustainability in maize production. *Food Energy Secur.* **2021**, *10*, e271. [CrossRef]
51. Wang, X.B.; Yamauchi, F.; Huang, J.K. Rising wages, mechanization, and the substitution between capital and labor: Evidence from small scale farm system in China. *Agric. Econ. Blackwell* **2016**, *47*, 309–317. [CrossRef]
52. Wang, X.B.; Yamauchi, F.; Huang, J.K.; Rozelle, S. What constrains mechanization in Chinese agriculture? Role of farm size and fragmentation. *China Econ. Rev.* **2020**, *62*, 101221. [CrossRef]
53. Foster, A.D.; Rosenzweig, M.R. Are There Too Many Farms in the World? Labor Market Transaction Costs, Machine Capacities, and Optimal Farm Size. *J. Political Econ.* **2022**, *130*, 636–680. [CrossRef]
54. Liang, Z.H.; Zhang, L.; Zhang, J.B. Land Consolidation and Fertilizer Reduction: Quasi-natural Experimental Evidence from China's Well-facilitated Capital Farmland Construction. In *China Rural Economy*; Rural Development Institute: Beijing, China, 2021; pp. 123–144.
55. Zheng, L.Y.; Qian, W.R.; Liu, Q.; Guo, X.L. The Impact of the New Round of Farmland Certification on the Ecological Protection of Cultivated Land: Taking theApplication of Chemical Fertilizers and Pesticides as Examples. In *China Rural Economy*; Rural Development Institute: Beijing, China, 2021; pp. 76–93.
56. Zhang, C.; Sun, Y.D.; Hu, R.F.; Yang, F.; Shen, X. The impact of rural-urban migration experience on fertilizer use: Evidence from rice production in China. *J. Clean. Prod.* **2021**, *280*, 124429. [CrossRef]
57. Guo, L.L.; Li, H.J.; Cao, X.X.; Cao, A.D.; Huang, M.J. Effect of agricultural subsidies on the use of chemical fertilizer. *J. Environ. Manag.* **2021**, *299*, 113621. [CrossRef] [PubMed]
58. Yue, M.; Zhang, L.; Zhang, J.B. Land Fragmentation and Farmers' Environmental—Friendly Technology Adoption Decision: Taking Soil Measurement and Fertilization Technology as An Example. *Resour. Environ. Yangtze Basin* **2021**, *30*, 1957–1968. [CrossRef]
59. Xu, Q.; Yin, R.L.; Zhang, H. Economies of Scale—Returns to Scale and the Problem of Optimum-scale Farm Management: An Empirical Study Based on Grain Production in China. *Econ. Res. J.* **2011**, *46*, 59–71.
60. Xu, Q.; Tian, S.C.; XU, Z.; Shao, T. Rural land system, land fragmentation and farmer's income inequality. *Econ. Res. J.* **2008**, *2*, 83–92.
61. Wu, Z.P.; Liu, M.Q.; Davis, J. Land consolidation and productivity in Chinese household crop production. *China Econ. Rev.* **2005**, *16*, 28–49. [CrossRef]
62. Deininger, K.; Monchuk, D.; Nagarajan, H.K.; Singh, S.K. Does Land Fragmentation Increase the Cost of Cultivation? Evidence from India. *J. Dev. Stud.* **2017**, *53*, 82–98. [CrossRef]
63. Ye, X.Q. Evolution Path, Dilemma and Policy Choice of Transformation of Agricultural Development Mode in China. *Reform* **2016**, *6*, 22–39.

International Journal of
Environmental Research and Public Health

MDPI

Article

Comprehensive Risk Assessment of Typical High-Temperature Cities in Various Provinces in China

Xueru Zhang [1,2,*], Qiuyue Long [3], Dong Kun [1], Dazhi Yang [4,5] and Liu Lei [1]

1 School of Public Administration, Hebei University of Economics and Business, Shijiazhuang 050061, China; dongkun1113@heuet.edu.cn (D.K.); hbsjzliul@gmail.com (L.L.)
2 Collaborative Innovation Center of Urban-Rural Integration Development, Hebei University of Economics and Business, Shijiazhuang 050061, China
3 Chongqing Cybercity Sci-Tech Co., Ltd., Chongqing 401121, China; longqy0815@gmail.com
4 Key Laboratory of Land Surface Pattern and Simulation, Institute of Geographic Sciences and Natural Resources Research, Chinese Academy of Sciences, Beijing 100101, China; yangdazhi@igsnrr.ac.cn
5 College of Resources and Environment, University of Chinese Academy of Sciences, Beijing 100049, China
* Correspondence: zhangxueru5@pku.edu.cn

Abstract: Global climate change results in an increased risk of high urban temperatures, making it crucial to conduct a comprehensive assessment of the high-temperature risk of urban areas. Based on the data of 194 meteorological stations in China from 1986 to 2015 and statistical yearbooks and statistical bulletins from 2015, we used GIS technology and mathematical statistics to evaluate high-temperature spatial and temporal characteristics, high-temperature risk, and high-temperature vulnerability of 31 cities across China. Over the past 30 years, most Chinese cities experienced 5–8 significant oscillation cycles of high-temperature days. A 15-year interval analysis of high-temperature characteristics found that 87% of the cities had an average of 5.44 more high-temperature days in the 15-year period from 2001 to 2015 compared to the period from 1986 to 2000. We developed five high-temperature risk levels and six vulnerability levels. Against the background of a warming climate, we discuss risk mitigation strategies and the importance of early warning systems.

Keywords: GIS; high-temperature disaster risk; high-temperature disaster vulnerability; risk assessment

Citation: Zhang, X.; Long, Q.; Kun, D.; Yang, D.; Lei, L. Comprehensive Risk Assessment of Typical High-Temperature Cities in Various Provinces in China. *Int. J. Environ. Res. Public Health* **2022**, *19*, 4292. https://doi.org/10.3390/ijerph19074292

Academic Editor: Paul B. Tchounwou

Received: 19 March 2022
Accepted: 30 March 2022
Published: 3 April 2022

1. Introduction

High-temperature disaster refers to a meteorological disaster that causes discomfort to living and non-living things such as people, animals, plants, and inorganic environment due to extreme high-temperature weather and has adverse effects. Global climate change has resulted in more frequent extreme weather events, which have significant adverse effects on human health and the social economy [1]. One type of extreme climate event, extreme summer heat, is a global phenomenon: In 1993, the Southeastern United States was hit by heat waves, with most areas reaching their highest temperatures in history [2]. In 1994, continuous high summer temperatures in Northeast China, Japan, and other countries resulted in large-scale droughts [3]. In 1995, the heat wave across Europe seriously impacted ecological, socio-economic, and other aspects of society [4]. In the summer of 2003, temperatures in Europe hit a record high [5], with more than 10,000 people in France dying as a consequence of the heat [6]. In 2013, Southeast China suffered from abnormally high temperatures, and 167 excess deaths occurred in the Pudong New Area in Shanghai [7]; at the same time, 679 additional heat-related illnesses occurred in Ningbo [8]. Heat waves, which are occurring more frequently, affect not only human health [9,10] but also human wellbeing and productivity, resulting in urban water and power supply shortages [11] and, consequently, threatening food security [12]. Over the last few years, high-temperature events have become more frequent, and their frequency, range, and duration will continue to increase [13]. Analyzing the spatial and temporal characteristics of high-temperature events can facilitate our understanding of the intensity, frequency, and duration of such events and their causes. In this sense, high-temperature risk assessment is of great importance for avoiding the risks associated with high-temperature events.

Studies on high-temperature events mainly focused on the impacts and causes of these events, as well as early warning systems. For example, Rosenzweig et al. [13,14] pointed out that urban heat waves can have serious impacts on human health. According to Park et al. [15] and Guan et al. [3], the occurrence of high-temperature events is related to abnormal sea temperatures. While the US Hot-Weather Health Warning System (HHWS) can already assess the possible number of deaths caused by hot weather [16], Kalkstein et al. [17] evaluated the ability of HHWS to reduce the number of heat-related deaths based on the number of heat deaths and heat waves in major cities in the United States from 1975 to 2004. Yang et al. [18] pointed out that the urban heat island effect aggravated the scope and intensity of extremely high temperatures in cities, increased high-temperature health risks for urban residents, and also made an important contribution to the long-term upward trend of extremely high temperatures in cities.

High-temperature vulnerability research refers to the establishment of a vulnerability evaluation index system in terms of the sensitivity; exposure; and adaptability of natural resources, the environment, the population, and the social economy, with the aim of quantitatively expressing the regional high-temperature vulnerability and identifying its spatial distribution [19,20]. Against the background of global climate change, compared to vulnerability assessments of meteorological disasters such as floods, droughts, and typhoons, as well as geological disasters such as earthquakes and landslides, studies assessing the vulnerability of cities to high-temperature events are scarce [21–28]. Urban high-temperature vulnerability evaluation studies mainly used official statistics, obtained via remote sensing and GIS. Generally, indicators such as high-temperature stress, sensitivity, adaptability, and exposure were evaluated [29,30], and factors such as the number of high-temperature days, socioeconomic level, and education level were involved [19,20]. The determination of relevant evaluation indicators is highly subjective, requiring rigorous index demonstration, mostly using statistical data; remote sensing data and GIS data alone are insufficient [31].

Disaster risk assessment is a process of judging the nature and scope of risk by studying the disaster-causing factors and the vulnerability of disaster-bearing bodies that have potential impacts on life, property and environment [32]. Considering the combined effects of risk and vulnerability, the comprehensive risk of high temperature disaster pays more attention to the possible losses under high temperature stress; that is, it emphasizes the exposure of population, property, and ecosystem at high temperatures [20]. In the 1970s, several countries started to conduct risk assessments of meteorological disasters [33–38]. For example, Blaikei et al. [39] took each state as the research object and used natural disasters between 1957 and 1994 to conduct a natural disaster risk analysis, using disaster loss, population, and area data. The relationship between resource development and natural disasters was illustrated from the perspective of the comprehensive role of the disaster-prone environment, disaster-causing factors, and disaster-bearing factors, allowing the authors to obtain the disaster risk zoning of the United States. In 1982, Willam et al. [40] completed the book *Natural Disaster Risk Assessment and Disaster Reduction Policy*, in which the authors described natural disaster risk assessment. China's high-temperature risk assessment has mainly been carried out by considering hazards, exposure, vulnerability, disaster prevention, mitigation ability, etc., and high-temperature risk assessment is performed by considering disaster-causing factors, disaster-prone environments, disaster-bearing bodies, disaster resistance ability, etc. To date, although several studies have investigated the influences and causes of high-temperature events, including potential early warning signs [13–17], systematic studies on the risk assessment of high-temperature events are scarce.

The above-mentioned studies mainly focused on spatial-temporal characteristics, causes, impacts on human health, and a comprehensive assessment of high-temperature events. However, such research was largely carried out on the regional scale. On the national scale, based on different high-temperature risk assessment models and different evaluation index systems, this paper constructs a high-temperature risk assessment model from the comprehensive perspective of high-temperature spatiotemporal characteristics, risks, and vulnerability. We used statistical methods to analyze high-temperature spatiotemporal characteristics and risk assessment data from various cities within China based on provincial units. The identification of areas vulnerable to high temperatures and the assess of this vulnerability provide a scientific basis for the control of high-temperature risks in various cities with a high practical significance.

2. Data and Methods

2.1. Data

For this paper, we used meteorological and socioeconomic data. Meteorological data were obtained from the National Meteorological Science Data Sharing Center (http://data.cma.cn/site/index.html, accessed on 10 April 2020); we downloaded the daily maximum temperature data for 194 weather stations from the "China Ground International Exchange Station Climate Data Day Dataset" from 1986 to 2015 (Figure 1). Social and economic data were obtained from the statistical yearbooks of various provinces and cities, supplemented by departmental statistical yearbooks and statistical bulletins (Table 1).

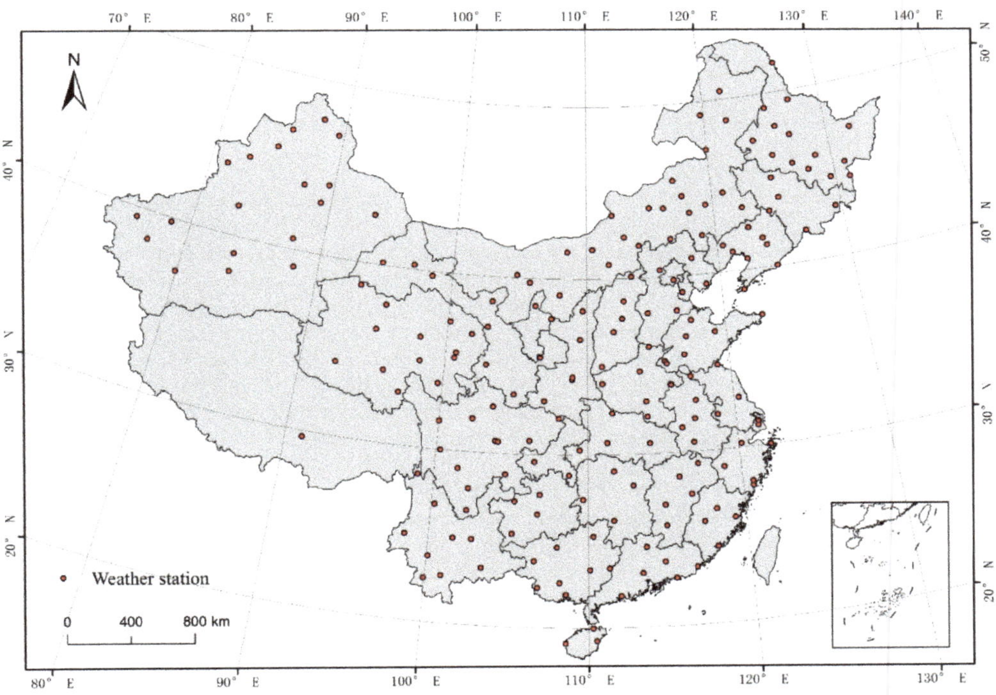

Figure 1. Map showing the spatial distribution of weather stations across China.

Table 1. Description of the data used for this paper.

Data Types	Data Description	Data Sources	Time Period
Meteorological data	Daily maximum temperatures from 194 meteorological stations	National Meteorological Science Data Sharing Center (http://data.cma.cn/site/index.html, accessed on 15 Apirl 2020)	1986–2015
Socio-economic data	Statistical data, including population, employment, income, finance, industry, education, healthcare, and other data from various administrative areas	Provincial statistical yearbooks from Anhui, Gansu, Guangdong, Guangxi, Hebei, Henan, Heilongjiang, Hubei, Hunan, Jilin, Jiangxi, Liaoning, Inner Mongolia, Shandong, Shanxi, Sichuan, Tianjin, Tibet, Xinjiang, Yunnan, Chongqing, Shanghai, Hainan, Beijing, Zhejiang, Guizhou, Qinghai, and Ningxia. Municipal statistical yearbooks from Nanjing, Nanping, Wuhan, Chaoyang, Shijiazhuang, Xi'an, Yuncheng, Kunming, and Zunyi. Statistical yearbooks are all from provincial and municipal statistical bureaus.	2016
	Statistics (As supplementary materials)	Heilongjiang Financial Yearbook and the national, economic, and social development statistical bulletins from the cities of Ganzhou and Pu 'er and other provinces and cities, provided by the Provincial and Municipal Statistics Bureau	2015

2.2. Methods

The fifth research report of the IPCC (2014) emphasizes the importance of risk assessment in global climate change research, describing a framework of natural disaster risk assessment based on "disaster stress-social vulnerability-exposure." Extremely high temperatures in summer can lead to high-temperature disasters, which pose serious threats to human health, the social economy, and ecosystems. Based on the natural disaster risk evaluation framework and "high-temperature risk-social vulnerability-population exposure," a high-temperature disaster risk assessment framework is constructed to comprehensively assess high-temperature risks. According to the above evaluation framework, the quantitative analysis of high-temperature characteristics, risks, and vulnerability is conducted to provide an assessment of urban high-temperature risks.

2.2.1. Analysis of High-Temperature Characteristics

Based on the daily maximum temperature data of 194 weather stations from the "China Ground International Exchange Station Climate Data Dataset" from 1986 to 2015, 31 cities were selected as typical cities with high temperatures, and the frequency characteristics of high-temperature events in these cities were determined. We used the SPSS software platform for statistical analysis and counted the days with high-temperature events from 1986 to 2015. In order to observe and compare these events, we used the equidistant grouping method that is frequently adopted [41]. The obtained dataset was divided into two groups, namely data from 1986 to 2000 and data from 2001 to 2015. For each group, which represents a 15-year period, we calculated the mean of the annual number of high-temperature days. Subsequently, we calculated the difference between the average number of high-temperature days from 2001 to 2015 and the average number of high-temperature days from 1986 to 2000 and compared the two 15-year periods.

2.2.2. Risk Analysis of High-Temperature Disasters

Using the temperature data of 31 typical stations, a high-temperature risk assessment model for 31 typical provinces and cities was constructed based on the following three aspects: the duration of high temperatures, high-temperature severity, and extreme high-temperature risk. The duration of high-temperature events was expressed by the number of high-temperature days (≥35 °C). The severity of a high-temperature event is expressed by the average difference between the daily maximum temperature (when it is ≥35 °C) and the temperature of 35 °C. The extreme high-temperature risk was expressed by the extreme high temperature ratio, which refers to the ratio of the number of days with a maximum daily temperature of ≥38 °C (China's Meteorological Administration defines weather with a daily maximum temperature ≥38 °C as hot summer weather) to the number of high-temperature days within a certain period of time. The model is constructed as follows:

$$R = d \times w_d + t \times w_t + p \times w_p \quad (1)$$

where R indicates the risk of urban high-temperature disasters; and d, t, and p represent the cumulative number of high-temperature days, the high-temperature severity, and the extreme high-temperature ratio of the standardized cities, respectively. w_d, w_t, and w_p are the standardized weights of the cumulative number of high-temperature days, high-temperature severity, and extreme high-temperature ratio for each city, respectively. The cumulative number of high-temperature days is the number of days with high-temperature weather; the high-temperature severity is the average of the difference between the daily maximum temperature (when it is ≥35 °C) and 35 °C; and the extreme high-temperature ratio refers to the ratio of the number of days with a daily maximum temperature ≥38 °C to the number of high-temperature days in a certain period of time.

In a previous study [19], w_d, w_t, and w_p were set to 0.6, 0.3, and 0.1, respectively. Based on the calculated risk index values of each typical city, the natural breakpoint method in ArcGIS was used to grade cities, and the spatial distribution maps of five high-temperature risk grades were obtained.

2.2.3. Vulnerability Analysis of High-Temperature Disaster and High-Temperature Risk Assessment

According to the high-temperature risk assessment framework, the risk can be determined using the following three indicators: high-temperature stress, social vulnerability, and population exposure. Social vulnerability refers to how vulnerable a specific group of people is to high temperatures and their ability to resist high-temperature hazards; it includes the sensitivity and adaptability of the population [42]. Based on previous studies, the multiplication and division of these indicators can reflect the synergistic relationships among the indicators more effectively than addition and subtraction [43–46]. Accordingly, the high-temperature vulnerability and the high-temperature risk models were obtained as follows:

$$DI = R \times F, \text{ and} \quad (2)$$

$$RI = R \times F \times E, \quad (3)$$

where DI is the high-temperature vulnerability index; RI is the high-temperature risk index; and R, F, and E are the high-temperature disaster risk, social vulnerability, and population exposure values, respectively. Parameter R is calculated using Equation (1); population exposure value E refers to the total population of each city in 2015. Social vulnerability, F, is determined via principal component analysis, which is used to analyze and calculate the selected social vulnerability indicators of sensitivity and adaptability; the specific indicators are shown in Appendix A. Principal component analysis (PCA) [47] is a multivariate dimensionality reduction technique for clustering and index reduction that uses the relationships between data points; it simplifies and organizes the relationships among a set of metrics and, thus, enables objective index confirmation. To calculate social vulnerability, the principal component score function of each typical site was first calculated based on the resulting component score coefficient matrix:

$$F_i = \sum Z_{ij} \times X \quad (4)$$

where X indicates the value of each index after standardization and Z_{ij} is the corresponding component score of the index.

Subsequently, we used the contribution rates of each principal component and calculated the social vulnerability value F of each city by applying the following equation:

$$F = \frac{e_i}{e} F_i \tag{5}$$

where e_i is the contribution rate of each principal component; e represents the total principal component contribution rate; and F_i represents each principal component score.

Based on the constructed high-temperature risk assessment framework, the high-temperature vulnerability and high-temperature risk values of the cities with the highest temperatures in various provinces were obtained. The natural breakpoint method in ArcGIS was used to generate the maps of high-temperature vulnerability and high-temperature risk zoning for each city (Table 2).

Table 2. Indicators used for the social vulnerability index calculation.

Primary Indicator	Sub-Indicator
Sensitivity	Proportion of the population that is female (%) Proportion of the population that works in the primary industry (%) Registered unemployment rate (%) Number of students in primary school (people)
Adaptability	Per capita disposable income of urban residents (CNY) Per capita disposable income of rural residents (CNY) Basic endowment insurance for urban workers (CNY) GDP per capita (CNY) Proportion of industrial output value in GDP (%) Local fiscal revenue (CNY 10,000) Number of health technicians (people) Local financial education expenditure (CNY 10,000) Social security and employment expenditure (CNY 10,000)

2.2.4. Jenks Natural Breaks

The natural breakpoint method is a standard method used to divide datasets into a certain number of classes; it is widely used in data analysis and map making [48]. By identifying the classification interval and dividing the elements into multiple classes, similar values can be appropriately grouped so that the difference between similar groups are small and the differences between less similar groups are large. Statistically, the variance can be used to perform the classification. The magnitude of the sum of the variance of various classifications can be used to classify the elements, and the lowest magnitude indicates the best classification result. The natural breaks method is the "best" method for finding an appropriate segmentation range. Most high-temperature risk studies divide the risk into five levels [47,49], which are not arbitrary decisions. Therefore, in this paper, five high-temperature risk levels are determined, and the thresholds of all levels are obtained using the natural breaks method.

3. Results

3.1. Analysis of the Spatial and Temporal Characteristics of Urban High-Temperature Events

Generally, 28 of the 31 typical cities showed cyclical fluctuations in the number of high-temperature days per year (Figures 2 and 3), and most cities experienced 5–8 significant oscillation cycles. Turpan, Xinjiang, had the largest number of high-temperature days per year, with a mean of 105.30 days. The number of high-temperature days per year in Baise, Guangxi; Nanping, Fujian; Ganzhou, Jiangxi; Chongqing; and Hangzhou, Zhejiang, showed obvious cyclical fluctuations, with a mean of greater than 30 days. Turpan had an average of 30.03 to 46.73 high-temperature days. Shaoguan, Guangdong; Haikou,

Hainan; Changsha, Hunan; Yuncheng, Shanxi; Wuhan, Hubei; Lu'an, Anhui; and Xi'an, Shaanxi, had 22 to 28.8 high-temperature days per year on average. Alxa League in Inner Mongolia; Nanchong, Sichuan; Jiuquan, Gansu; Zhengzhou, Henan; Shijiazhuang, Hebei; Nanjing, Jiangsu; Shanghai, Shandong; and Jinan, Shandong, averaged between 10 and 20 high-temperature days per year. Beijing; Tianjin; Chaoyang, Liaoning; Yinchuan, Ningxia; Zunyi, Guizhou; Pu 'er, Yunnan; Qiqihar, Heilongjiang; and Songyuan, Jilin, averaged between 1 and 10 high-temperature days per year. In every year, the number of high-temperature days was below 40 for these cities.

According to the number of high-temperature days, natural breakpoints were used in ArcGIS to create five intensity levels: very low, low, medium, high, and very high. From a spatial perspective, Turpan in Northwest China is a very high-intensity area. Most cities in South China, East China, and the Yangtze River basin are high-intensity areas, whereas cities in North China and Northeast China are medium- and low-intensity areas.

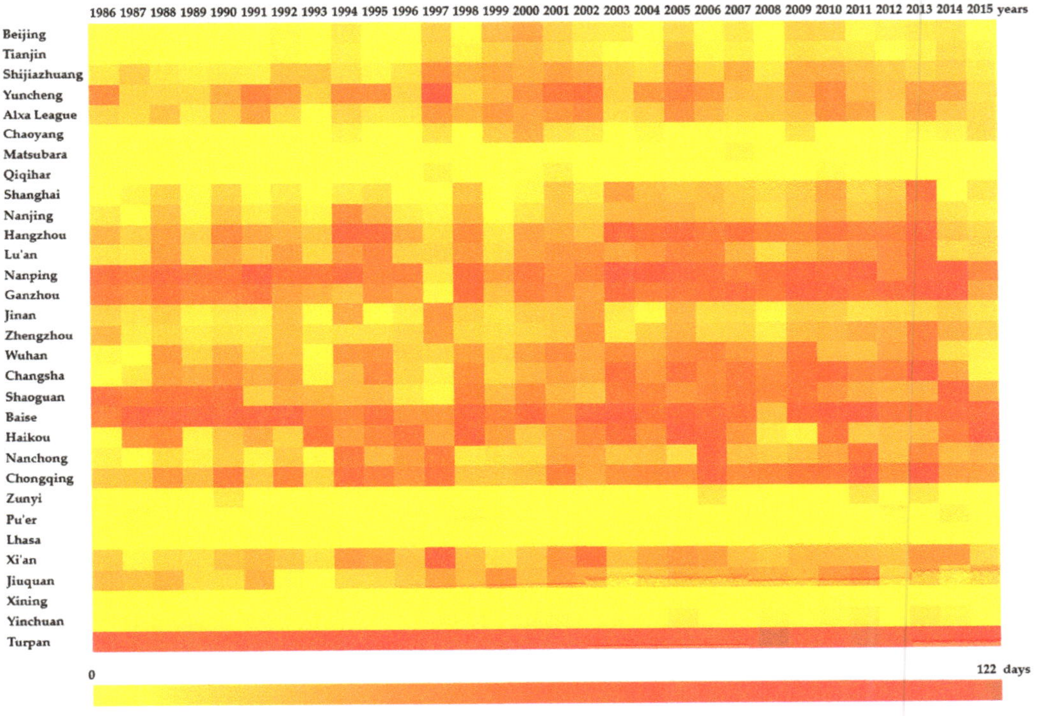

Figure 2. Number of high-temperature days (≥35 °C) in 31 typical Chinese cities from 1986 to 2015.

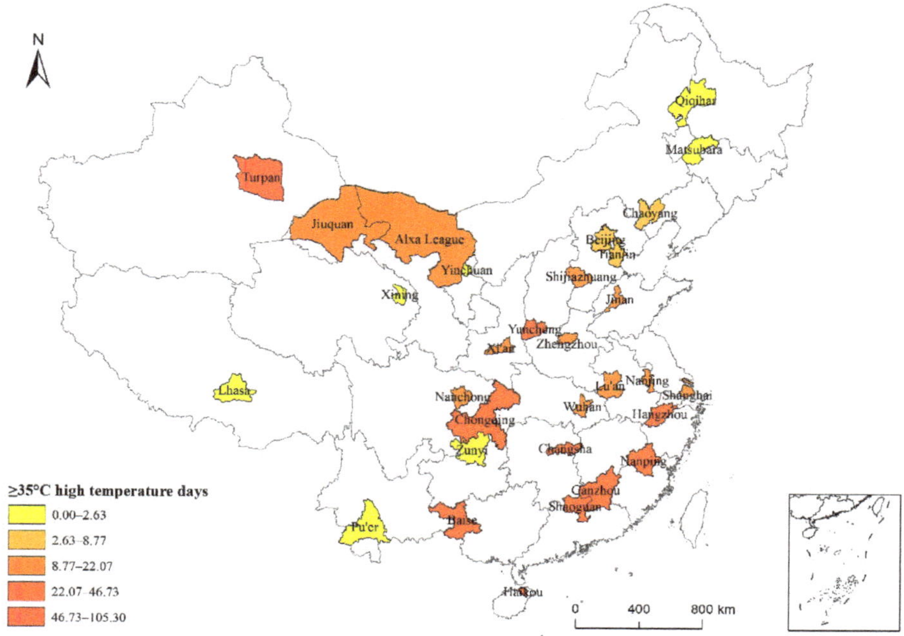

Figure 3. Spatial characteristics of high-temperature days in 31 typical cities in China from 1986 to 2015.

The results of the time series analysis of high-temperature events in the 31 typical cities are shown in Table 3. For Tianjin, Shanghai, Ganzhou, Zhengzhou, Changsha, Nanchong, Chongqing, Yinchuan, and Turpan, the number of high-temperature days gradually increased between 1986 and 2015. However, for 87.10% of the cities, we found an increase in the average number of high-temperature days from 2001 to 2015 compared to 1986–2000 (Figure 4); in two cities, the number of high-temperature days per year decreased. In Changsha, Ganzhou, Chongqing, Hangzhou, and Turpan, the number of high-temperature days increased significantly by more than 10 days; in Shanghai, Wuhan, Nanchong, Nanping, Lu'an, Haikou, Zhengzhou, Alxa League, and Nanjing, this number increased by 5.13–9.67 days. In contrast, in Beijing, Tianjin, Shijiazhuang, Yuncheng, Chaoyang, Songyuan, Qiqihar, Jinan, Shaoxing, Zunyi, Lhasa, Xi'an, Jiuquan, and Yinchuan, a significant increase of 0–5 days was found. The average number of high-temperature days in Lhasa and Xining was the same for the two 15-year periods. In Baise and Pu'er, the number of high-temperature days decreased by 2.80 and 0.33 days, respectively. The cities with a large increase in the number of high-temperature days were mainly located in East China, South China, and the Yangtze River basin.

Table 3. Time series analysis of high-temperature days from 1986 to 2015.

Statistical Metric	Tianjin	Shanghai	Ganzhou	Zhengzhou	Changsha	Nanchong	Chongqing	Yinchuan	Turpan
Pearson correlation	0.368 *	0.460 *	0.488 **	0.443 *	0.526 **	0.424 *	0.465 **	0.679 **	0.614 **
Sig. (2-tailed)	0.045	0.011	0.006	0.014	0.003	0.020	0.010	0.000	0.000
N	30	30	30	30	29	30	30	30	30

* Significant correlation at the 0.05 level (bilateral). ** Significant correlation at the 0.01 level (bilateral).

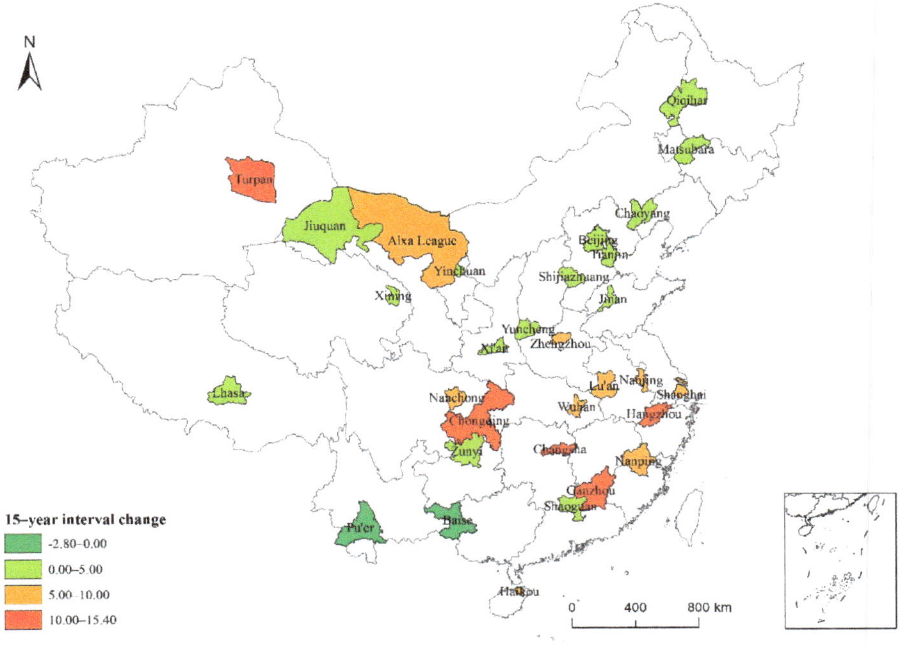

Figure 4. Map of China showing the 15-year interval variation of high-temperature weather in typical cities.

3.2. Risk and Vulnerability Analysis

Using ArcGIS combined with the obtained risk index value R, the natural breakpoint method was applied to divide the risk into five levels from high to low (Figure 5) as follows: Turpan—Level V (very high); Yuncheng, Hangzhou, Nanping, Ganzhou, Chongqing, and Baise—Level IV (high); Shijiazhuang, Alxa League, Lu'an, Zhengzhou, Wuhan, Changsha, Shaoguan, Haikou, Nanchong, Xi'an, and Jiuquan—Level III (medium); Beijing, Tianjin, Chaoyang, Qiqihar, Shanghai, Nanjing, and Jinan—Level II (low); Songyuan, Zunyi, Pu'er, Lhasa, Xining, and Yinchuan—Level I (very low).

Spatially, the high-risk areas were mainly located in East and Southwest China, whereas the areas with a medium-risk level were scattered throughout China (excluding Northeast China). In Central China, mostly medium-risk areas were found. The seven low-risk areas were mainly located in North, East, and Northeast China, and the six very-low-risk areas were distributed throughout high-elevation regions in the west, with a small number in the northeast.

The highest vulnerability level was found for Turpan, which is consistent with the high-temperature risk for this city (Figure 6). However, there were significantly fewer very-high risk-level (Level 5) areas than high-temperature risk areas, and they were located in Southeast China (Nanping, Ganzhou) and western South China (Baise). Yuncheng, Chongqing, Hangzhou, and other cities with high-level risk values did not show high vulnerability levels. Areas with vulnerability levels of 3 and 5 were scattered across various regions, including Shijiazhuang and Alxa League in North China; Chaoyang and Qiqihar in Northeast China; Jinan, Nanjing, Hangzhou, and Lu'an in East China; Zhengzhou, Wuhan, and Changsha in Central China; Shaoguan and Haikou in South China; Nanchong and Chongqing in Southwest China; and Xi'an and Jiuquan in Northwest China. The Level 2 cities were Tianjin, Shanghai, Yinchuan, and Songyuan, and the Level 1 cities were Beijing, Zunyi, Lhasa, Xining, and Pu'er.

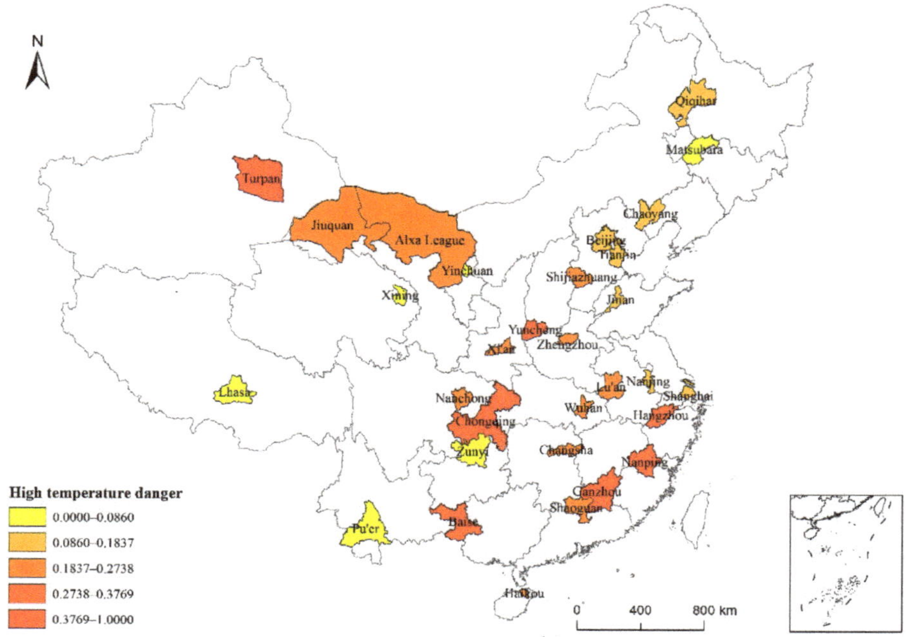

Figure 5. High-temperature risk levels for 31 typical cities across China based on data from 1986 to 2015.

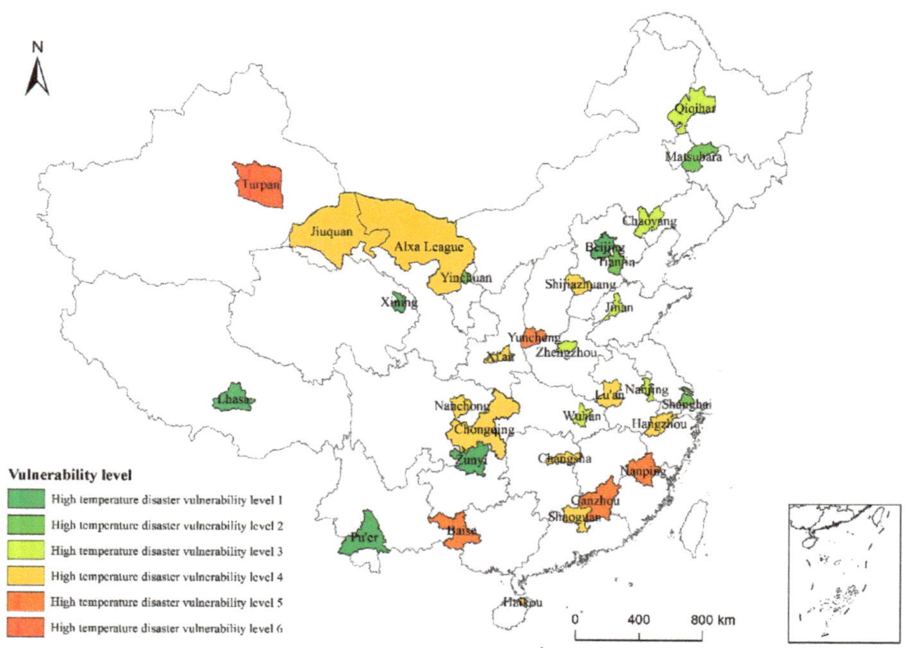

Figure 6. High-temperature vulnerability levels of 31 typical cities across China based on data from 1986 to 2015.

3.3. Comprehensive Assessment of High-Temperature Risk

According to the high-temperature risk assessment model, the high-temperature risk values for the 31 cities were calculated and rated using the natural breakpoint method to obtain the spatial distribution map of high-temperature risk areas (Figure 7).

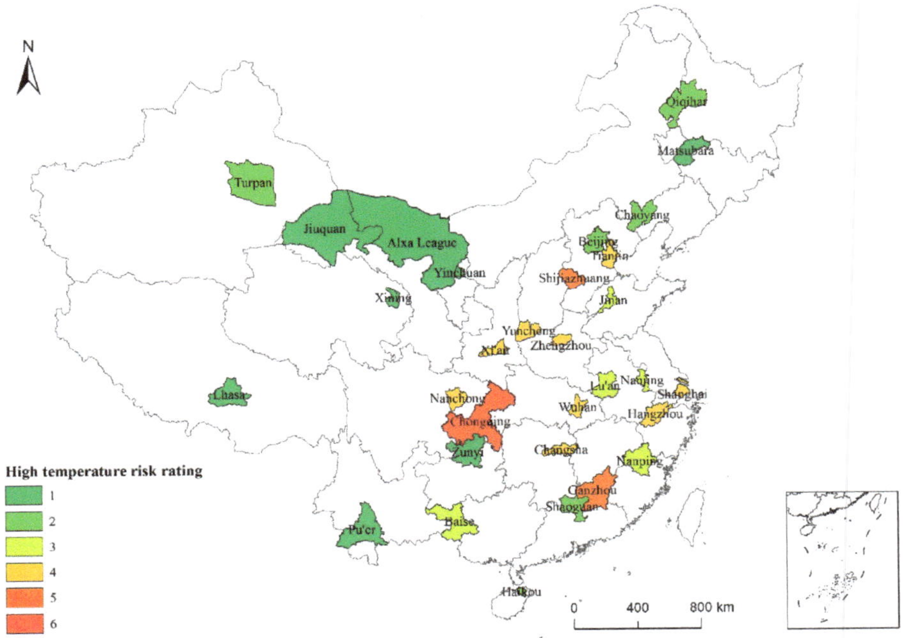

Figure 7. High-temperature risk distribution in 31 typical cities across China based on data from 1986 to 2015.

3.4. High-Temperature Risk Prevention Zoning

Risk prevention zoning is based on risk evaluation, which provides guidance for targeted risk prevention strategies [43]. In this paper, risk prevention and zoning were carried out for cities with higher high-temperature risk levels (Levels 4–6). First, the natural breakpoint method was used to divide the high-temperature risk and vulnerability factors into high and low levels. Subsequently, the cities with high risk and low vulnerability were called "high-temperature risk areas." Cities with low risk and high vulnerability were called "high-temperature vulnerability areas." Cities with both high risk and high vulnerability were called "high-temperature risk-vulnerability areas," and cities with both low risk and low vulnerability were called "population exposure areas." Among the 12 hotspot cities, there were 3, 0, 5, and 4 cities, respectively, in each of these four categories of risk factors (Figure 8); the highest risk values were found for Hangzhou, Xi'an, and Changsha. The cities with high risk and vulnerability values are Shijiazhuang, Yuncheng, Nanchong, Chongqing, and Guangzhou. In the densely populated cities of Tianjin, Zhengzhou, Shanghai, and Wuhan, there are no areas with high-temperature vulnerability. Based on risk prevention zoning, the areas subject to high-temperature risks are mainly located in plains, valleys, and river basins, whereas the areas with high vulnerability in terms of social aspects are mainly economically underdeveloped areas in densely populated regions.

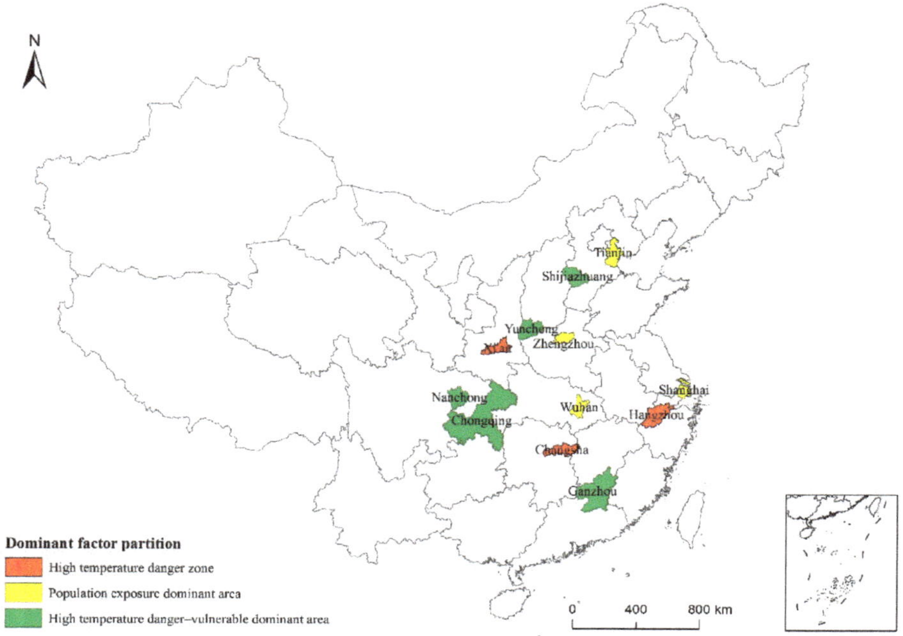

Figure 8. High-temperature risk dominant factor partition for 31 typical cities in China.

The zoning of high-temperature cities based on the leading factors can be helpful in determining the mechanisms underlying high-temperature risks. Areas with significantly high summer temperatures can be identified by examining the natural environment, whereas areas with high vulnerability can be determined based on the distribution of vulnerable groups and physiological or socioeconomic conditions. Regarding population exposure, vulnerable groups can be spatially separated. Risk prevention strategies can be formulated according to different high-temperature risk factors. For areas with a high-temperature risk, increasing the amount of vegetation and green space can reduce temperatures; potential actions could be to increase vegetation density along the streets or to create green roofs. Urban planning departments should also consider methods to reduce high-temperature risks, such as the ventilation of new buildings, and residents should be made more aware of high-temperature risks. Regarding vulnerability, risk mitigation can be achieved by relocating vulnerable groups or by installing air-conditioning equipment. As a reduction in population density is not realistic, early warning systems should be considered.

4. Discussion

4.1. Analysis of High-Temperature Spatiotemporal Characteristics and Comprehensive Risk Assessment

This paper used meteorological and socioeconomic data to analyze high-temperature spatiotemporal characteristics and constructed a high-temperature risk assessment model. We comprehensively evaluated urban high-temperature risks by considering high-temperature characteristics, high-temperature risk, high-temperature vulnerability, population exposure, and risk-prevention zoning, expanding urban high-temperature risk assessment research. The spatiotemporal distribution characteristics and comprehensive risk assessment results were obtained for 31 typical Chinese cities, yielding different results when compared to Xie et al. [47] and Dong et al. [49], mainly for the following two reasons: (1) There are differences in the models; for example, Dong et al. [49] constructed their assessment model based on the aspects of disaster-causing risk and the vulnerability (exposure) of the supporting body, without considering social vulnerability. (2) There are also differences in the evaluation index systems; according to the different models, the selected evaluation index factors were different, but all models included factors such as the number of high-temperature days, population exposure, and economic development status.

The selection of relevant evaluation indicators is highly subjective. In the analysis of high urban temperatures, due to the limited amount of data, only the hottest cities in each province were selected for analysis, instead of evaluating all cities in China. The lack of data concerning the proportion of the population working in the primary industries of Lhasa, Jiuquan, and Turpan, as well as data concerning the local financial education expenditure, social security, and the employment expenditure in Pu'er, may have had a certain impact on evaluation results.

4.2. Spatial and Temporal Distribution Characteristics of High Temperatures in Typical Chinese Cities

The analysis of the high-temperature characteristics of Chinese cities based on interprovincial units is one of the key contributions of this paper. The spatial distribution characteristics of 31 typical cities in China were explained according to data from between 1986 and 2015. Over the past 30 years, the number of days with high temperatures gradually increased in nine cities, and the average number of high-temperature days in most cities increased by 5.44 days within 15 years. The increasing number of high-temperature days is the most obvious manifestation of climate warming.

4.3. High-Temperature Risk Assessment and Risk-Prevention Zoning in Typical Chinese Cities

A comprehensive risk assessment of high temperatures in Chinese cities, based on interprovincial units, is another key part of this paper. On the basis of the natural disaster risk evaluation framework, a high-temperature risk assessment framework based on "high temperature risk–social vulnerability–population exposure" was constructed. Quantitative analysis was carried out based on the spatiotemporal characteristics of high temperatures, high-temperature risk, and high-temperature vulnerability. Finally, comprehensive evaluations and risk-prevention zoning were carried out for urban areas with high temperatures according to the following three aspects: the duration of high-temperature events, the severity of high-temperature events, and the extreme high-temperature risk based on weather characteristics. Vulnerability was estimated using the high-temperature risk and social vulnerability. Finally, by combining the high-temperature risk, high-temperature vulnerability, and population exposure, the high-temperature risk levels of the 31 typical cities were obtained. Generally, cities with high risk levels also showed high vulnerability levels; this was true for cities such as Turpan, Nanping, Ganzhou, and Baise.

Generally, areas with high risk levels, social vulnerability, and large populations being exposed to high temperatures are located in plains, valleys, and river basins; they are largely economically underdeveloped and densely populated areas. For high-temperature areas, increasing the vegetation density and creating green rooftops can be effective measures. Risk mitigation can be achieved by resettling vulnerable people. Regarding population exposure, early warning systems and evacuation strategies should be taken into consideration. These results can provide a realistic basis for decision making for meteorological departments and disaster prevention and mitigation departments; they have a certain guiding significance for understanding the regional high-temperature disaster risk and the vulnerability of disaster-bearing bodies, and they contribute to regional high-temperature risk management, high-temperature risk avoidance, and risk control.

5. Conclusions

Due to global climate change, extremely hot weather conditions are becoming frequent. Based on the provincial unit, the characteristics of high temperature in Chinese cities are analyzed. The results show that over the past 30 years, most cities have experienced

5–8 significant oscillation periods in terms of the number of high-temperature days, and the number of high-temperature days in nine cities, including Tianjin, Shanghai, and Chongqing, shows a significant positive time correlation. A comparative analysis of two 15-year intervals shows that in 87% of cities, from 2001 to 2015, the average number of high-temperature days per year increased by 5.44 days compared with 1986–2000.

By conducting a comprehensive assessment of the high-temperature risks, it was observed that the areas with the greatest high-temperature disaster risk in China are mainly concentrated in the central urban areas of plains, basins, and river basins. These areas have low social vulnerability due to the development of cities, but, similarly, the urban high-temperature risk and the exposure of the population to high temperatures are much higher than they are for cities in the western regions; thus, these cities have a significant high-temperature disaster risk.

The results of risk-factor zoning show that the areas with the greatest high-temperature risk are mainly plains, basins, and river basins, and the areas with the highest social vulnerability mainly include economically underdeveloped areas and areas where socially vulnerable people gather. The areas where a large proportion of the population is exposed to high temperatures consist mainly of densely populated areas.

Author Contributions: Conceptualization, X.Z.; methodology, X.Z. and Q.L.; formal analysis, Q.L. and D.K.; investigation, X.Z., Q.L. and D.K.; resources, X.Z. and Q.L.; writing–original draft preparation, Q.L. and D.Y.; writing–review and editing, Q.L., D.Y. and L.L.; supervision, X.Z. All authors have read and agreed to the published version of the manuscript.

Funding: This research was funded by the National Natural Science Foundation of China (Grant No. 41501202) and The Second Tibetan Plateau Scientific Expedition and Research (Grant No. 2019QZKK0603), as well as The Strategic Priority Research Program of Chinese Academy of Sciences (Grant No. XDA20040201).

Institutional Review Board Statement: Not applicable.

Informed Consent Statement: Not applicable.

Data Availability Statement: All relevant datasets in this study are described in the manuscript.

Conflicts of Interest: The authors have read the journal's guidelines and have the following competing interests: One of the co-authors (Qiuyue Long) is an employee of a company (Chongqing Cybercity Sci-tech Co., Ltd.). The other authors have no financial interests. The funders played no role in the design of the study; the collection, analysis, or interpretation of data; the writing of the manuscript; or the decision to publish the results. The authors declare that they have no non-financial conflict of interest. There are no personal relationships or competitive interests directly or indirectly related to this research, nor are there professional interests or personal beliefs that may have affected this study.

Appendix A PCA Analysis and SVI Calculation

We used the SPSS software platform to perform a principal component analysis of the standardized indicators and to calculate the social vulnerability of each city. The original index must pass the KMO test, which is a prerequisite of using the principal component analysis method. According to the standard provided by the statistician Kaiser, a KMO value of less than 0.6 means that the data are not suitable for factor analysis. KMO = 0.706 in this paper, so the data meet the requirements of principal component analysis. The orthogonal rotation method with maximum variance was used to make the coefficients in the factor load matrix more significant, and the initial factor load matrix can be rotated so that the relationships between the factors and the original variables can be redistributed and the correlation coefficient can be restricted to a range from 0 to 1.

The principal component analysis of the standardized data was carried out using the SPSS software platform, and four principal components—the characteristic values of each principal component, the contribution rate of each principal component, and the score coefficient matrix of each component—were obtained (Tables A1 and A2). Four principal components were obtained from the 13 variables in the social vulnerability index system, and the cumulative contribution rate was 83.66%.

Table A1. Time series analysis of high-temperature days.

Eigenvalue	Contribution Rate	Cumulative Contribution Rate
4.875	37.500	37.500
3.474	26.724	64.224
1.270	9.769	73.993
1.257	9.671	83.664

Table A2. Component function matrix.

	Components of Component Function Matrix			
X	Z1	Z2	Z3	Z4
X11	0.015	−0.073	0.623	−0.208
X12	−0.140	0.313	−0.146	0.049
X13	−0.083	−0.008	0.017	0.767
X14	0.199	0.010	−0.066	−0.107
X15	0.002	0.240	−0.040	−0.028
X16	−0.045	0.283	0.008	−0.067
X21	−0.109	0.302	0.200	0.043
X22	−0.076	0.068	0.576	0.262
X23	0.198	0.038	−0.056	−0.231
X24	0.188	−0.130	0.109	0.108
X25	0.212	−0.015	−0.014	−0.112
X26	−0.210	0.203	0.037	−0.286
X27	0.229	−0.066	−0.034	−0.045

To calculate social vulnerability, we first calculated the principal component score function of each typical city according to the obtained component score coefficient matrix:

$$F_i = \sum Z_{ij} \times X \quad \text{(A1)}$$

where X represents the standardized numerical value of each index, and Z_{ij} represents the corresponding component score of the index. Then, the social vulnerability value of each typical city was calculated using the contribution rate of each principal component as follows.

$$F = \frac{37.5}{83.664}F_1 + \frac{26.724}{83.664}F_1 + \frac{9.769}{83.664}F_1 + \frac{9.671}{83.664}F_1 \quad \text{(A2)}$$

Here, F represents social vulnerability, and F_1, F_2, F_3, and F_4 represent the scores of each principal component, respectively.

References

1. Easterling, D.R.; Meehl, G.A.; Parmesan, C.; Changnon, S.A.; Karl, T.R.; Mearns, L.O. Climate Extremes: Observations, Modeling, and Impacts. *Science* **2000**, *289*, 2068–2074. [CrossRef] [PubMed]
2. Lott, J.N. The US summer of 1993: A sharp contrast in weather extremes. *Weather* **1994**, *49*, 370–383. [CrossRef]
3. Guan, Z.; Yamagata, T. The unusual summer of 1994 in East A-sia: IOD teleconnections. *Geophys. Res. Lett.* **2003**, *30*, 1554. [CrossRef]
4. McCarthy, J.J.; Canziani, O.F.; Leary, N.; Dokken, D.J.; White, K.S. Climate Change 2001: Impacts, Adaptation, and Vulnerability. Contribution of Working Group II to the Third Assessment Report of the Intergovernmental Panel on Climate Change (IPCC). *Glob. Ecol. Biogeogr.* **2001**, *12*, 87–88.
5. Beniston, M.; Diaz, H.F. The 2003 heat wave as an example of summers in a greenhouse climate? Observations and climate model simulations for Basel, Switzerland. *Glob. Planet. Chang.* **2005**, *44*, 73–81. [CrossRef]
6. Pascal, M.; Laaidi, K.; Ledrans, M.; Baffert, E.; Caserio-Schnemann, C.; Tertre, A.L.; Manach, J.; Medina, S.; Rudant, J.; Empereur-Bissonnet, P. France's heat health watch warning system. *Int. J. Biometeorol.* **2006**, *50*, 144–153. [CrossRef] [PubMed]
7. Sun, X.; Sun, Q.; Zhou, X.; Li, X.; Yang, M.; Yu, A.; Geng, F. Heat wave impact on mortality in Pudong New Area, China in 2013. *Sci. Total Environ.* **2014**, *493*, 789–794. [CrossRef]

8. Bai, L.; Ding, G.; Gu, S.; Bi, P.; Su, B.; Qin, D.; Xu, G.; Liu, Q. The effects of summer temperature and heat waves on heat-related illness in a coastal city of China, 2011–2013. *Environ. Res.* **2014**, *132*, 212–219. [CrossRef]
9. Hansen, A.L.; Bi, P.; Ryan, P.; Nitschke, M.; Pisaniello, D.; Tucker, G. The effect of heat waves on hospital admissions for renal disease in a temperate city of Australia. *Int. J. Epidemiol.* **2008**, *37*, 1359–1365. [CrossRef]
10. Tian, Z.; Li, S.; Zhang, J.; Guo, Y.; Barengo, N.C. The Characteristic of Heat Wave Effects on Coronary Heart Disease Mortality in Beijing, China: A Time Series Study. *PLoS ONE* **2013**, *8*, e77321. [CrossRef]
11. Mazdiyasni, O.; AghaKouchak, A. Substantial increase in concurrent droughts and heatwaves in the United States. *Proc. Natl. Acad. Sci. USA* **2015**, *112*, 11484. [CrossRef] [PubMed]
12. Zaitchik, B.F.; Macalady, A.K.; Bonneau, L.R.; Smith, R.B. Europe's 2003 heat wave: A satellite view of impacts and land-atmosphere feedbacks. *Int. J. Climatol.* **2006**, *26*, 743–769. [CrossRef]
13. Parry, M. *Climate Change 2007, Working Group II Contribution to the Fourth Assessment Report of the IPCC Intergovernmental Panel on Climate Change*; Cambridge University Press: Cambridge, UK, 2007.
14. Rosenzweig, C.; Solecki, W.D.; Parshall, L.; Chopping, M.; Pope, G.; Goldberg, R. Characterizing the urban heat island in current and future climates in New Jersey. *Glob. Environ. Chang. Part B Environ. Hazards* **2005**, *6*, 51–62. [CrossRef]
15. Park, C.K.; Schubert, S.D. On the Nature of the 1994 East Asian Summer Drought. *J. Clim.* **2010**, *11*, 1056–1070. [CrossRef]
16. Kalkstein, L.S.; Jamason, P.F.; Greene, J.S.; Libby, J.; Robinson, L. The Philadelphia Hot Weather-Health Watch/Warning System: Development and Application, Summer 1995. *Bull. Am. Meteorol. Soc.* **1996**, *77*, 1519–1528. [CrossRef]
17. Kalkstein, L.S.; Greene, S.; Mills, D.M.; Samenow, J. An evaluation of the progress in reducing heat-related human mortality in major U.S. cities. *Nat. Hazards* **2011**, *56*, 113–129. [CrossRef]
18. Yang, X.; Chen, B.; Kejia, H.U.; University, Z. A review of impacts of urbanization on extreme heat events. *Prog. Geogr.* **2015**, *34*, 1219–1228.
19. Uejio, C.K.; Wilhelmi, O.V.; Golden, J.S.; Mills, D.M.; Gulino, S.P.; Samenow, J.P. Intra-urban societal vulnerability to extreme heat: The role of heat exposure and the built environment, socioeconomics, and neighborhood stability. *Health Place* **2011**, *17*, 498–507. [CrossRef]
20. Johnson, D.P.; Stanforth, A.; Lulla, V.; Luber, G. Developing an applied extreme heat vulnerability index utilizing socioeconomic and environmental data. *Appl. Geogr.* **2012**, *35*, 23–31. [CrossRef]
21. Bradford, K.; Abrahams, L.; Hegglin, M.; Klima, K. A Heat Vulnerability Index and Adaptation Solutions for Pittsburgh, Pennsylvania. *Environ. Sci. Technol.* **2015**, *49*, 11303–11311. [CrossRef]
22. Junzhe, B.; Xudong, L.; Chuanhua, Y. The Construction and Validation of the Heat Vulnerability Index, a Review. *Int. J. Environ. Res. Public Health* **2015**, *12*, 7220–7234.
23. Kates, R.W.; Colten, C.E.; Laska, S.; Leatherman, S.P. Reconstruction of New Orleans after Hurricane Katrina: A research perspective. *Proc. Natl. Acad. Sci. USA* **2006**, *103*, 14653–14660. [CrossRef] [PubMed]
24. Nandy, S.; Singh, C.; Das, K.K.; Kingma, N.C.; Kushwaha, S.P.S. Environmental vulnerability assessment of eco-development zone of Great Himalayan National Park, Himachal Pradesh, India. *Ecol. Indic.* **2015**, *57*, 182–195. [CrossRef]
25. Reid, C.E.; O'Neill, M.S.; Gronlund, C.J.; Brines, S.J.; Brown, D.G.; Schwartz, D.R. Mapping Community Determinants of Heat Vulnerability. *Environ. Health Perspect.* **2009**, *117*, 1730–1736. [CrossRef]
26. Weber, S.; Sadoff, N.; Zell, E.; de Sherbinin, A. Policy-relevant indicators for mapping the vulnerability of urban populations to extreme heat events: A case study of Philadelphia. *Appl. Geogr.* **2015**, *63*, 231–243. [CrossRef]
27. Wolf, T.; McGregor, G. The development of a heat wave vulnerability index for London, United Kingdom. *Weather. Clim. Extrem.* **2013**, *1*, 59–68. [CrossRef]
28. Masozera, M.; Bailey, M.; Kerchner, C. Distribution of impacts of natural disasters across income groups: A case study of New Orleans. *Ecolog. Econ.* **2007**, *63*, 299–306. [CrossRef]
29. Harlan, S.L.; Brazel, A.J.; Prashad, L.; Stefanov, W.L.; Larsen, L. Neighborhood microclimates and vulnerability to heat stress. *Soc. Sci. Med.* **2006**, *63*, 2847–2863. [CrossRef]
30. Alexander, L.V.; Allen, S.K.; Bindoff, N.L.; Bréon, F.-M.; Xie, S.P. Climate change 2013: The physical science basis, in contribution of Working Group I (WGI) to the Fifth Assessment Report (AR5) of the Intergovernmental Panel on Climate Change (IPCC). *Comput. Geom.* **2013**, *129*, 83–103.
31. CSDGSND. *Understanding the Changing Planet: Strategic Directions for the Geographical Sciences*; CSDGSND: La Prairie, QC, Canada, 2010.
32. Asian, D.R.C. *Living With Risk: A Global Review of Disaster Reduction Initiatives*; United Nations: Geneva, Switzerland, 2002.
33. Adrianto, L.; Matsuda, Y. Developing economic vulnerability indices of environmental disasters in small island regions. *Environ. Impact Assess. Rev.* **2002**, *22*, 393–414. [CrossRef]
34. Katz, R.W. Stochastic Modeling of Hurricane Damage. *J. Appl. Meteorol.* **2002**, *41*, 754. [CrossRef]
35. Lekes, V.; Dandul, I. Using airflow modelling and spatial analysis for defining wind damage risk classification (WINDARC). *For. Ecol. Manag.* **2000**, *135*, 331–344. [CrossRef]
36. Lynch, A.H.; Cassano, E.N.; Cassano, J.J.; Lestak, L.R. Case Studies of High Wind Events in Barrow, Alaska: Climatological Context and Development Processes. *Mon. Weather. Rev.* **2003**, *131*, 719. [CrossRef]
37. Whiteman, C.D.; Doran, J.C. The Relationship between Overlying Synoptic-Scale Flows and Winds within a Valley. *J. Appl. Meteor.* **2010**, *32*, 1669–1682. [CrossRef]

38. Moore, J.; Quine, C.P. A comparison of the relative risk of wind damage to planted forests in Border Forest Park, Great Britain, and the Central North Island, New Zealand. *For. Ecol. Manag.* **2000**, *135*, 345–353. [CrossRef]
39. Blaikie, M.P. *At Risk: Natural Hazards, People's Vulnerability and Disasters*; Routledge: London, UK, 1994; pp. 147–167.
40. Petak, W.J.; Atkisson, A.A. *Natural Hazard Risk Assessment and Public Policy*; Springer: New York, NY, USA, 1982.
41. Yao, R.; Hu, Y.; Sun, P.; Bian, Y.; Liu, R.; Zhang, S. Effects of urbanization on heat waves based on the wet-bulb temperature in the Yangtze River Delta urban agglomeration, China. *Urban Clim.* **2022**, *41*, 101067. [CrossRef]
42. Nelson, K.S.; Abkowitz, M.D.; Camp, J.V. A method for creating high resolution maps of social vulnerability in the context of environmental hazards. *Appl. Geogr.* **2015**, *63*, 89–100. [CrossRef]
43. Aubrecht, C.; Özceylan, D. Identification of heat risk patterns in the U.S. National Capital Region by integrating heat stress and related vulnerability. *Environ. Int.* **2013**, *56*, 65–77. [CrossRef]
44. El-Zein, A.; Tonmoy, F.N. Assessment of vulnerability to climate change using a multi-criteria outranking approach with application to heat stress in Sydney. *Ecol. Indic.* **2015**, *48*, 207–217. [CrossRef]
45. Vescovi, L.; Rebetez, M.; Rong, F. Assessing public health risk due to extremely high temperature events: Climate and social parameters. *Clim. Res.* **2005**, *30*, 71–78. [CrossRef]
46. Ebert, U.; Welsch, H. Meaningful environmental indices: A social choice approach. *J. Environ. Econ. Manag.* **2004**, *47*, 270–283. [CrossRef]
47. Xie, P.; Wang, Y.L.; Liu, Y.X.; Peng, J. Incorporating social vulnerability to assess population health risk due to heat stress in China. *Acta Geogr. Sin.* **2015**, *70*, 1041–1051.
48. North, M.A. A Method for Implementing a Statistically Significant Number of Data Classes in the Jenks Algorithm. In Proceedings of the 2009 Sixth International Conference on Fuzzy Systems and Knowledge Discovery, Tianjin, China, 14–16 August 2009; pp. 35–38.
49. Siyan, D.; Ying, X.; Botao, Z.; Meiting, H.; Yongxiang, Z. Projected Risk of Extreme Heat in China Based on CMIP5 Models. *Adv. Clim. Change Res.* **2014**, *10*, 365–369.

International Journal of
Environmental Research and Public Health

MDPI

Article

Cultivated Land Fragmentation and Its Influencing Factors Detection: A Case Study in Huaihe River Basin, China

Jiale Liang [1], Sipei Pan [1], Wanxu Chen [2,3,4,*], Jiangfeng Li [1,*] and Ting Zhou [5]

1 School of Public Administration, China University of Geosciences, Wuhan 430074, China; LJL0715@cug.edu.cn (J.L.); pamsp@cug.edu.cn (S.P.)
2 School of Geography and Information Engineering, China University of Geosciences, Wuhan 430074, China
3 Research Center for Spatial Planning and Human-Environmental System Simulation, China University of Geosciences, Wuhan 430078, China
4 State Key Laboratory of Earth Surface Processes and Resource Ecology, Beijing Normal University, Beijing 100875, China
5 School of Geographical Sciences, Fujian Normal University, Fuzhou 350007, China; zting1624@163.com
* Correspondence: cugcwx@cug.edu.cn (W.C.); jfli@cug.edu.cn (J.L.)

Abstract: The booming population and accelerating urbanization in the Huaihe River Basin have sped up the land use transformation and the cultivated land fragmentation (CLF), seriously impeded the advancement of agricultural modernization, and threatened regional stability and national food security as well. The analysis of CLF degree and its spatiotemporal distribution characteristics, along with the influencing factors in the Huaihe River Basin, is of great significance for promoting the intensive and efficient utilization of cultivated land resources and maintaining food security. Previous studies lack the measurement and cause analysis of CLF in Huaihe River Basin. To bridge the gap, this study introduces Fragstats4.2 and ArcGIS10.3 to analyze the spatiotemporal characteristics of CLF in county units in the Huaihe River Basin from 2000 to 2018 through the Lorentz curve, entropy method, and spatial auto-correlation method while the causes of the spatiotemporal differentiation of CLF in the basin were explored with the help of a geographic detector. The results show that the spatial distribution of cultivated land in the Huaihe River Basin is relatively balanced, and the Gini coefficients of cultivated land from 2000 to 2018 were 0.105, 0.108, and 0.113, respectively. More than 56% of the counties in the basin have a location entropy greater than 1. the percentage of landscape, area-weighted mean patch area, patch cohesion index, and aggregation index decrease year by year while the patch density and splitting index show an upward trend. The landscape pattern of cultivated land is highly complex, and the overall fragmentation degree is increasing. The county distribution pattern of the CLF degree with random and agglomeration is generally stable. The spatiotemporal differentiation of CLF in the Huaihe River Basin is affected by multiple factors, among which the influences of the normalized difference vegetation index, per capita cultivated land area, and intensity of human activity obviously stronger than other factors, and the contribution rate of the factors reached more than 0.4. The interaction effect among the factors is stronger than that of single factor, with dual-factor enhancement and nonlinear enhancement dominating. The results of this study have important implications for optimizing the agricultural structure in the Huaihe River Basin and alleviating the CLF in important grain production areas.

Keywords: cultivated land fragmentation; landscape pattern index; spatial autocorrelation; geographic detector; Huaihe River Basin; China

Citation: Liang, J.; Pan, S.; Chen, W.; Li, J.; Zhou, T. Cultivated Land Fragmentation and Its Influencing Factors Detection: A Case Study in Huaihe River Basin, China. *Int. J. Environ. Res. Public Health* **2022**, *19*, 138. https://doi.org/10.3390/ijerph19010138

Academic Editors: Wei Song and Hualin Xie

Received: 29 November 2021
Accepted: 21 December 2021
Published: 23 December 2021

1. Introduction

Serving as the fundamental resource for human survival and society development [1], cultivated land performs multiple functions, such as production and living, and plays an important role in maintaining ecological security and food security and promoting social and economic stability [2,3]. Since the 21st century, the disordered expansion of urban construction land [4,5], and the intensification of man-land contradiction brought about by the rapid development of social economy have forced land use changes and the intensification of cultivated land fragmentation (CLF) [6]. Cultivated land resources in China are facing serious threats, which, to a certain extent, hinder the development process of agricultural modernization and large-scale development [7], reduce the agricultural production efficiency [8], and threaten national food security and social stability. As an important grain production base in China, the Huaihe River Basin covers an area of only 2.9% of China's land area, but its cultivated land area accounts for 12% of the country's total [9]. In recent years, the degree of CLF in the basin has been deepened, and the food security has been threatened. Therefore, scientific measurement of the level and causes of CLF in the Huaihe River Basin has become a key to improve the efficiency of cultivated land use and to ensure regional food security.

CLF refers to the difficulty of concentrated and contiguous operation of cultivated land under the interference of human or natural factors, showing the state of interpolation, and scattered and disorderly utilization [1,2]. Corresponding to the scale operation of cultivated land, CLF not only features Chinese traditional agricultural production, but also exhibits one of the most distinctive characteristics of agricultural landscapes globally [10]. CLF can lead to an increase in the area of field cans and ditches, which directly results in the loss and waste of cultivated land resources [11]. A survey showed that cultivated land wasted by fragmentation in China accounts for about 19% of the net cultivated land areas [12], resulting in serious land waste. Simultaneously, some researchers have observed the increase of food production costs [13], and a reduction of productivity [14–16] and grain production as fragmentation increases. Moreover, ecological consequences invoked by fragmentation also include biodiversity loss, declined agricultural efficiency, and local micro-climate change, which will undoubtedly lessen the agricultural ecosystem provisioning service [17], further holding up China's agricultural modernization. Currently, research touching on CLF has achieved fruitful results owing to the hard work of numerous scholars [18–20]. Current research on CLF mainly focuses on three aspects: the evaluation of CLF, the causes of the spatial differentiation of CLF, and the impact of CLF.

Pieces of evidence from previous literature have proved that landscape metrics can effectively externalize agricultural fragmentation [21], and thus have been widely acknowledged and utilized by academia [22]. Currently, related research on the evaluation of CLF most adopt the landscape pattern index to characterize the degree of CLF. However, the processing of the landscape pattern index differs. Most scholars prefer methods, such as the moving window [23], principal component analysis [17], and multiple linear regression [24], to comprehensively deal with the landscape pattern index. Nevertheless, in previous studies on CLF evaluation, the multi-collinearity among them was seldom considered in the selection of the landscape pattern index, which had a certain impact on the accuracy of the evaluation results. In the study of the causes of CLF, the impacts of social and economic factors, such as the urbanization rate, population density, land use degree, road traffic, and land property rights, and natural factors, such as altitude, slope, annual precipitation, and water network density, on CLF were mainly explored. Scholars have used different research methods and focused on different directions. Most of them used geographically weighted regression models to identify the influencing factors of CLF, focusing on both natural and socioeconomic factors [24–29]. According to the results of scholars' studies, the influencing factors of CLF in different study areas are various, with some scholars considering soil quality diversity as the most important influencing factor of CLF [25], while others believe that socioeconomic factors play a more important role [26]. However, in the analysis of the causes of CLF, only the macro-qualitative analysis of the influencing factors of CLF was carried out, while the influence of each factor was not specifically quantified. Moreover, only the individual effects of each factor have been analyzed, but the interaction between the influencing factors has not been explored yet.

A review of previous studies found that a wide range of research areas of CLF tend to use qualitative analysis, and the quantitative analysis research mostly takes provinces or districts as the research object, but there is a lack of CLF measurement and influencing factors detection research on the watershed scale. It is necessary to analyze important grain-producing areas of the CLF mechanism from the perspective of the basin, which will provide a reference for ensuring food security and sustainable utilization of cultivated land resources. This study focuses on the following three aspects:

(1) The degree of CLF and the balance of the spatiotemporal distribution of cultivated land resources in the Huaihe River Basin from 2000 to 2018 were measured by the entropy method and Lorentz curve method.
(2) The spatial auto-correlation method was introduced to detect the spatial clustering characteristics of high and low values of CLF change in county units.
(3) In cooperation with the geographical detector, the influencing factors of spatiotemporal differentiation of CLF were explored.

The purpose of this study is to improve the agricultural modernization level and ensure food security of the Huaihe River Basin, and to provide a scientific reference for evaluating the CLF from the perspective of the basin and studying the driving mechanism of the CLF in important grain production areas.

2. Materials and Methods

2.1. Study Area

Covering a total watershed area of 270,000 km^2, the Huaihe River Basin is located in eastern China, between the Yangtze River and the Yellow River (Figure 1). It is divided into the Huaihe River and the Yishusi River with the abandoned Yellow River as the boundary. The Huaihe River enjoys a pleasant climate, with the south a subtropical zone and the north a warm temperate zone. The overall terrain of the basin is flat, dominated by plain, with extensive cultivated land. The total cultivated land area is 12.7 million hectares, accounting for 11.7% of China's total cultivated land area [30]. The grain output reaches 1/6 of the total grain output in China, making it an important grain production base in China. In recent years, with rapid development of the social economy and the acceleration of urbanization and industrialization in the Huaihe River Basin, the problem of CLF has become increasingly severe. The increase of CLF not only reduces the technical efficiency of cultivated land utilization [31], but also hinders the improvement of the scale efficiency of agricultural production and the process of agricultural modernization [14], which exerts a certain impact on the utilization of cultivated land resources and food production capacity in the Huaihe River Basin [32], threatening the food security and regional stability of the basin. According to the study of Zhou et al. [33], this study identified the Huaihe River Basin as the four provinces of Anhui, Henan, Shandong, and Jiangsu. A total of 218 counties in the four provinces were selected as the research objects, and the degree of CLF and its causes in the basin were analyzed from the county unit scale to provide ideas for improving the grain production efficiency in the Huaihe River Basin and solidifying its status as a grain production base.

2.2. Data Sources

The remote sensing monitoring data of land use in 2000, 2010, and 2018, annual average precipitation, road network, rivers, and normalized difference vegetation index (NDVI) of each county used in this study are all from the Resource and Environment Science and Data Center, Chinese Academy of Sciences (http://www.resdc.cn/ (accessed on 10 May 2021)). Among them, land use data is generated through manual visual interpretation, with a spatial resolution of 1 km and a comprehensive accuracy of over 90%. According to the national land use classification system, the land use types were divided into seven first-level types: cultivated land, forest land, grassland, water area, urban and rural construction land, unused land, and wetland [34]. Altitude data were obtained using 90 m resolution ASTERG DEM data from Geospatial Data Cloud (http://www.gscloud.cn/ (accessed on 15 May 2021)). The population density data is from the WorldPop (https://www.worldpop.org/ (accessed on 11 July 2021)), with a resolution of 100 m.

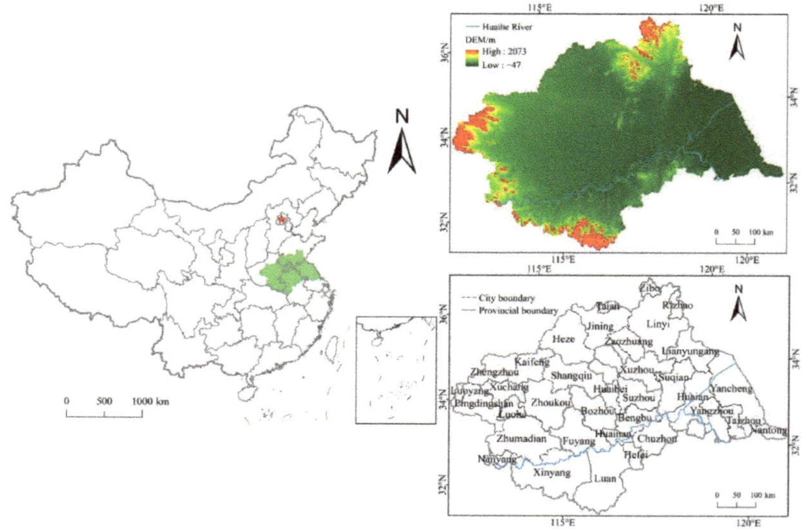

Figure 1. Location of the study area.

2.3. Method

To explore the CLF and its influencing factors in the Huaihe River Basin, this study first measured the spatiotemporal distribution characteristics of cultivated land in the Huaihe River Basin with the Lorentz curve, and then the degree of cultivated land fragmentation based on the entropy weight method was measured, and the spatial agglomeration characteristics of CLF with the spatial autocorrelation tool were analyzed. Eventually, the primary influencing factors of CLF in Huaihe River Basin were explored by the geographical detector (Figure 2).

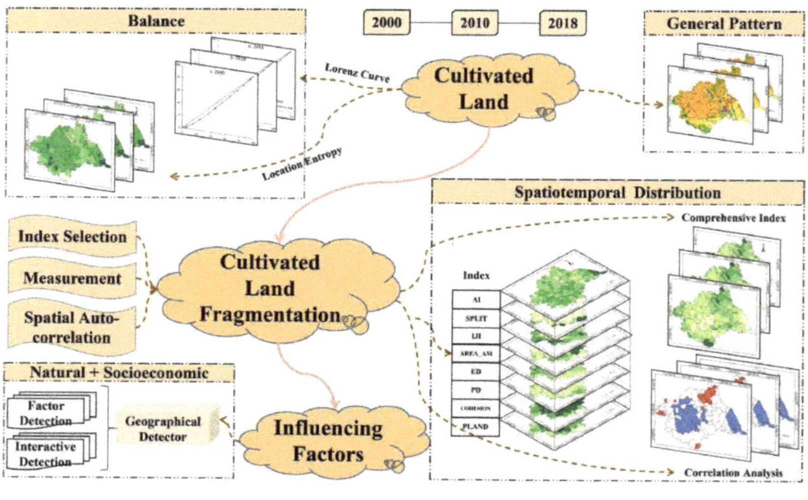

Figure 2. Conceptual framework for measuring cultivated land fragmentation and detecting its influencing factors.

2.3.1. Lorentz Curve

The Lorentz curve, proposed by Lorentz, an American statistician, is widely applied to reflect the fairness of income distribution in a country or region. In recent years, more and more studies have introduced the Lorentz curve into studies on cultivated land resource change, food security, land use structure analysis, and land use transition, etc. [35–37]. In this study, the Lorentz curve was introduced to identify the concentration degree of cultivated land spatial distribution in the Huaihe River Basin from 2000 to 2018, and on this basis, the difference degree of cultivated land distribution in different county units was further described quantitatively through the calculation of Gini coefficients, providing a basis for the measurement of CLF in the Basin. The calculation equation is as follows:

$$Q = \frac{P_1/P_2}{P_3/P_4}, \quad (1)$$

where Q is the location entropy of cultivated land in the study area, also known as the specialization rate; P_1 is the cultivated land area of a county in the study area; P_2 is the total area of cultivated land in the study area; P_3 is the total land area of a county; and P_4 is the total land area of the study area. The Q value can reflect the balance degree of cultivated land distribution in the basin. When $Q < 1$, it indicates that the proportion of cultivated land area in the county is less than that of the total land area in the study area, namely, the specialization degree is low, and the county is at a disadvantage. Otherwise, when $Q > 1$, it indicates that cultivated land has a high degree of specialization and is a regional advantage.

According to Equation (1), the location entropy Q of cultivated land is calculated, and the Q value is sorted from small to large. The cumulative percentage of cultivated land area and land area in each county is calculated, and then the cumulative percentage of cultivated land area in each county is used as the horizontal coordinate and the cumulative percentage of cultivated land area is plotted on the ordinate to draw the Lorentz curve.

Gini coefficient is used to describe the uniformity of the spatial distribution of research objects in the research area and to quantify the Lorentz curve. The calculation equation is as follows :

$$G = \sum_{i=1}^{m-1} (P_i Q_{i+1} - P_{i+1} Q_i), \quad (2)$$

where G is the Gini coefficient; P_i represents the cumulative percentage of a county's land area in the total land area of the study area; and Q_i represents the cumulative percentage of cultivated land area of a county in the total cultivated land area of the study area. With reference to related research [38], the larger G is, the more uneven the distribution of cultivated land in the study area is. When $G < 0.2$, the distribution is uniform; $0.2 < G < 0.3$, the distribution is relatively uniform; $0.3 < G < 0.4$, the distribution is basically reasonable; $0.4 < G < 0.6$, the distribution difference is large; and $G > 0.6$, the distribution difference is over the top.

2.3.2. Cultivated Land Fragmentation Index Selection

The degree of CLF is affected by different factors, such as the shape, size, and connectivity of cultivated land blocks, which cannot be simply described by a single dimension index. With reference to previous studies [1,24,39] and the actual situation of the Huaihe River Basin, 10 landscape indicators were selected from three aspects: size, edge-shape, and aggregation, to characterize the degree of CLF in the Huaihe River Basin. They are the percentage of landscape (PLAND), patch density (PD), edge density (ED), area-weighted mean patch area (AREA_AM), area-weighted mean shape index (SHAPE_AM), area-weighted mean patch fractal dimension (FRAC_AM), interspersion and juxtaposition index (IJI), patch cohesion index (COHESION), splitting index (SPLIT), and aggregation index (AI).

To reduce redundant indicators and improve the accuracy of the evaluation model, SPSS22 software (IBM, Armonk, NY, USA) was applied to test the multi-collinearity of the evaluation indicators to determine whether a variable should be excluded and the reciprocal of tolerance, namely the variance inflation factor (VIF), can be used for testing. When VIF > 10, it indicates that the multi-collinearity of this variable is very serious, which will affect the parameter estimation of the evaluation model and should be considered to remove this variable [40]. The multi-collinearity diagnosis results of various landscape metrics are shown in Table 1.

Table 1. Multi-collinearity diagnosis table of landscape metrics.

Metrics	PLAND	PD	ED	AREA_AM	SHAPE_AM
VIF	7.473	4.433	5.750	7.131	18.309
Metrics	FRAC_AM	IJI	COHESION	SPLIT	AI
VIF	11.263	1.223	5.409	1.823	9.625

According to the test results above, the area-weighted mean shape index (SHAPE_AM) and area-weighted fractal dimension (FRAC_AM) with VIF > 10 are finally removed to maintain the accuracy of the evaluation results. The calculation equation and ecological significance of each landscape metric are shown in Table 2 [41,42].

Table 2. Cultivated land fragmentation index and its description.

Aspects	Landscape Pattern Index	Calculation Formula	Description of Index
Size	Percentage of landscape (PLAND)	$PLAND = \frac{\sum_{j=1}^{n} a_{ij}}{LA}$ where a_{ij} is the area of patch ij; LA is the total landscape area of the study area.	PLAND indicates the area of a certain patch type accounts for the percentage of the total landscape area, and the value tends to be 0. The scarcer the patch type, the more fragmented the landscape pattern.
	Patch density (PD)	$PD = \sum_{i=1}^{m} N_i/LA$ where m is the total number of landscape types; LA is the total landscape area of the study area; N_i is the patch number of landscape i.	PD indicates the degree of influence of the patch boundary of the landscape type on the entire landscape. The larger the value, the more concentrated the patch type is distributed in the landscape.
Edge—shape	Edge density (ED)	$ED = \sum_{i=1}^{m} \sum_{j=1}^{m} P_{ij}/LA$ where P_{ij} is the boundary length between type i and type j landscape element patches.	ED refers to the degree of landscape type segmentation by element boundary, which is a direct reflection of landscape fragmentation. The larger the value, the more fragmented the landscape pattern of this type of element.
	Area-weighted mean patch area (AREA_AM)	$MPS = LA/NP$ where LA is the total landscape area of the study area; NP is the number of patches in the landscape.	AREA_AM reflects the degree of fragmentation of a certain type of landscape in landscape structure analysis. The higher the value, the lower the degree of fragmentation.
Aggregation	Interspersion and juxtaposition index (IJI)	$IJI = \frac{-\sum_{i=1}^{m} \sum_{j=i+1}^{m} \left[\left(\frac{E_{ij}}{E} \right) ln \left(\frac{E_{ij}}{E} \right) \right]}{ln[0.5m(m-1)]} \times 100$ where E_{ij} is the adjacent edge length between the type i and the type j of feature patch; m is the total number of landscape types; E is the edge length of the whole landscape.	IJI refers to the adjacent probability between certain patch element and other patches. The higher the value, the more adjacent other patch types, and the more fragmented the landscape of this type.
	Patch cohesion index (COHESION)	$COHESION = \left[1 - \frac{\sum_{i=1}^{m} \sum_{j=1}^{n} P_{ij}}{\sum_{i=1}^{m} \sum_{j=1}^{n} P_{ij} \sqrt{a_{ij}}} \right] \times \left[1 - \frac{1}{\sqrt{Z}} \right]^{-1} \times 100$ where P_{ij} represents the perimeter of patch i of landscape type j; a_{ij} represents the area of patch i of landscape type j; Z represents the number of patches in the landscape.	COHESION indicates the degree of agglomeration between different types of patches, the larger the value, the higher the degree of combination between dominant types of patches, and the lower the degree of fragmentation of this type of landscape.
	Splitting index (SPLIT)	$SPLIT = \frac{LA^2}{\sum_{i=1}^{m} \sum_{j=1}^{n} a_{ij}^2}$ where LA is the total landscape area of the study area; a_{ij} is the area of patch j of landscape type i; n is the number of patches of landscape type.	SPLIT refers to the degree of separation of landscape element. The larger the value is, the more dispersed among the same patch types and the higher the degree of fragmentation.
	Aggregation index (AI)	$AI = \left[\frac{g_{ij}}{max \to g_{ij}} \right]$ where g_{ij} is the number of similar adjacent patches of the landscape patch type	AI refers to the degree of agglomeration between patches of a certain type of landscape element. The larger the value, the more agglomerated patches of this type of element and the lower the degree of fragmentation.

2.3.3. Measurement of Cultivated Land Fragmentation

To reflect the degree of CLF in the basin from 2000 to 2018 more intuitively, the entropy weight method is introduced to determine the weight of the selected landscape indicators on the degree of CLF in the basin. The specific steps are as follows [43]:

(1) Data standardization

To avoid the inconsistency of the indicator units from affecting the calculation weight, the range method is adopted to standardize the data and the indicator data is converted to a range between 0 and 1. The larger the positive indicator value, the closer the evaluation target value is to the ideal value. The larger the negative indicator value, the more the evaluation target deviates from the ideal value.

Positive indicator:

$$y'_{ij} = \frac{y_{ij} - min\{y_{ij}\}}{max\{y_{ij}\} - min\{y_{ij}\}} \quad (3)$$

Negative indicator:

$$y'_{ij} = \frac{max\{y_{ij}\} - y_{ij}}{max\{y_{ij}\} - min\{y_{ij}\}} \quad (4)$$

where y'_{ij} represents the standardized value of indicator j in the year i; y_{ij} represents the actual value of indicator j in the year i; and $max\{y_{ij}\}$ and $min\{y_{ij}\}$ represent the maximum and minimum values of the indicator j, respectively.

(2) Calculation of the standardized value P_{ij} of indicator j in the year i:

$$P_{ij} = y'_{ij} / \sum_{i=1}^{n} y'_{ij} \quad (5)$$

(3) Calculation of the entropy value e_j of indicator j:

$$e_j = -k \sum_{i=1}^{n} P_{ij} = -\frac{1}{\ln m} \sum_{i=1}^{n} P_{ij} \ln P_{ij} \quad (6)$$

where k represents the proportionality coefficient, $k = 1/\ln m$, and m is the number of research samples.

(4) Calculation of weight W_j of indicator j:

$$W_j = (1 - e_j) / \sum_{j=1}^{n} (1 - e_j) \quad (7)$$

(5) Calculation of composite scores:

$$Z = \sum_{j=1}^{m} (W_j \cdot P_{ij}) \quad (8)$$

where Z is the composite score of the CLF index, W_j is the weight coefficient of indicator j, and P_{ij} is the standardized value of indicator j.

2.3.4. Spatial Autocorrelation

Spatial autocorrelation is often used to detect the potential interdependence between geographic data within a region [44]. With the help of Geoda1.12, this study uses the local spatial auto-correlation method to detect the random mode, discrete mode, and clustering characteristics of the spatial distribution of CLF in the Huaihe River Basin, and performs visual analysis. According to the spatial location of county units and the change of the CLF degree in the Huaihe River Basin, Local Moran's I statistic was introduced to measure the spatial autocorrelation of CLF degree change in the Huaihe River Basin from 2000 to 2018. The Local Moran's I can be calculated using Equation (9):

$$Local\ Moran's\ I = \frac{(x_i - \overline{x})}{S^2} \sum_{j=1}^{m} w_{ij} (x_j - \overline{x}) \quad (9)$$

$$S^2 = \frac{1}{m}\sum_{i=1}^{m}(x_i - \overline{x})^2 \tag{10}$$

where m is the number of county units; x_i and x_j are the measured values of spatial unit attributes, respectively; $\overline{x}$ is the mean value of the measured value; and w_{ij} is the spatial weight matrix, and S^2 is its variance.

2.3.5. Geographical Detector

Geographical detector is a statistical method to detect the spatial differentiation characteristics of geographical phenomena and reveal their influence [45], which has been widely applied in multiple disciplines (e.g., environmental science and resource utilization, regional economy, physical geography, tourism, agricultural basic science, etc.). It consists of factor detection, interactive detection, ecological detection, and risk detection. This study mainly uses factor detection and interactive detection methods to distinguish the influence of the driving factors of the spatial differentiation of CLF in the Huaihe River Basin, and reveals the correlation between the driving factors. The calculation is as follows:

$$q = 1 - \frac{1}{N\sigma^2}\sum_{i=1}^{L} N_i\sigma_i^2 \tag{11}$$

where q is the influence of the driving factor, $q\in[0,1]$; N is the number of samples in the study area, and i is the partition (i = 1,2, ... , L); and σ^2 and σ_i^2 are the variances of indicators in the study area and the variances of partition i, respectively. The size of q reflects the degree of spatial differentiation of the indicators. The larger the q value, the stronger the explanatory ability of each factor to the dependent variable, and vice versa.

Cultivated land is a complex of natural environment and socioeconomic conditions composed of topography, soil, climate, hydrology, vegetation, and human socioeconomic activities, which is affected by multi-dimensional factors. Considering previous studies [1,14,24] and data availability, this study selected the following indicators from two aspects of natural resource endowment and socioeconomic development as the influencing factors for the spatial distribution of CLF in the Huaihe River Basin. Among them, natural factors are the basis for the formation of the spatial pattern of cultivated land resources in the Huaihe River Basin. Altitude (X1) and slope (X2) directly affect the distribution and utilization of cultivated land. The distance to the river (X3), NDVI (X4), and average annual precipitation (X5) provide important conditions for the distribution and development and utilization of cultivated land resources. Socioeconomic factors mainly reflect the interference degree of different human activities on the cultivated land landscape pattern, including per capita cultivated land area (X6), intensity of human activities (X7), population density (X8), and distance from road network (X9). The selected influencing factor data is processed by the Arc Toolbox/Spatial Analyst Tools/Reclass and Arc Toolbox/Spatial Analyst Tools/Zonal/Zonal Statistics tools of ArcGIS10.3 (Esri, Redlands, CA, USA) for discretization and classification. The spatialization and quantification of the influencing factors of county units within the basin was realized, and the specific classification methods and descriptions are shown in Table 3.

Table 3. Description of the driving factors classification of the geographic detector.

Driving Factors	Classification Method	Level	Level Description
Altitude	NaturalBreaks	1–6	Calculation with Arc Toolbox/Spatial Analyst Tools/Reclass of ArcGIS10.3
Slope	NaturalBreaks	1–6	1. 0~5; 2. 6~10; 3. 11~15; 4. 16~20; 5. 21~25; 6. >25
Distance to river	NaturalBreaks	1–5	Calculation with Arc Toolbox/Spatial Analyst Tools/Reclass of ArcGIS10.3
NDVI	Zhang et al. [46]	1–5	1. ≤0.2; 2. 0.2~0.4; 3. 0.4~0.6; 4. 0.6~0.8; 5. 0.8~1
Average annual precipitation	NaturalBreaks	1–5	Calculation with Arc Toolbox/Spatial Analyst Tools/Reclass of ArcGIS10.3
Per capita cultivated land area	NaturalBreaks	1–5	Calculation with Arc Toolbox/Spatial Analyst Tools/Reclass of ArcGIS10.3
Intensity of human activities	Li et al. [47]	0–10	0. sparse woodland, shrub, sparse grass, barren land; 1. river, reservoir, ponds, tidal flat, natural and plantation forest land, Moderate grass; 2. other woodland, dense grass; 7. cultivated land; 8. rural settlements; 9. industrial land; 10. Urban built-up
Population density	Ge et al. [48]	1–5	1. 0~60; 2. 61~150; 3. 151~300; 4. 301~500; 5. >500
Distance from road network	NaturalBreaks	1–5	Calculation with Arc Toolbox/Spatial Analyst Tools/Reclass of ArcGIS10.3

3. Results

3.1. Spatiotemporal Distribution Characteristics of Cultivated Land

3.1.1. General Pattern of the Spatiotemporal Distribution

According to the statistics of land use data in the study area, cultivated land, forest land, grass land, and unused land decreased from 2000 to 2018. The most obvious change was in grass land, with a 22.08% reduction. Construction land and water area continued to increase, with increases of 27.96% and 12.17%, respectively. Specifically, there were 183,751 km^2 of cultivated land in the Huaihe River Basin in 2018, of which the dry land area was 134,591 km^2, accounting for 73.25% of the total cultivated land area, and the paddy land area was 49,160 km^2, accounting for 26.75% of the total cultivated land area. From the perspective of time scale, the cultivated land area decreased by 2363 km^2 from 2000 to 2010, and 6524 km^2 from 2010 to 2018, indicating that the cultivated land area in the Huaihe River Basin decreased year by year and the reduction amplitude increased. According to the spatial distribution of land use in the study area (Figure 3), the spatial distribution of cultivated land in the Huaihe River Basin is relatively balanced, and the spatial distribution boundary between dry land and paddy field is clear. Dry land is mainly distributed in the area north of the Huaihe River while paddy field is mostly distributed in the area south of the Huaihe River, and a small amount is also distributed in Xuzhou, Suqian, and other cities north of the Huaihe River.

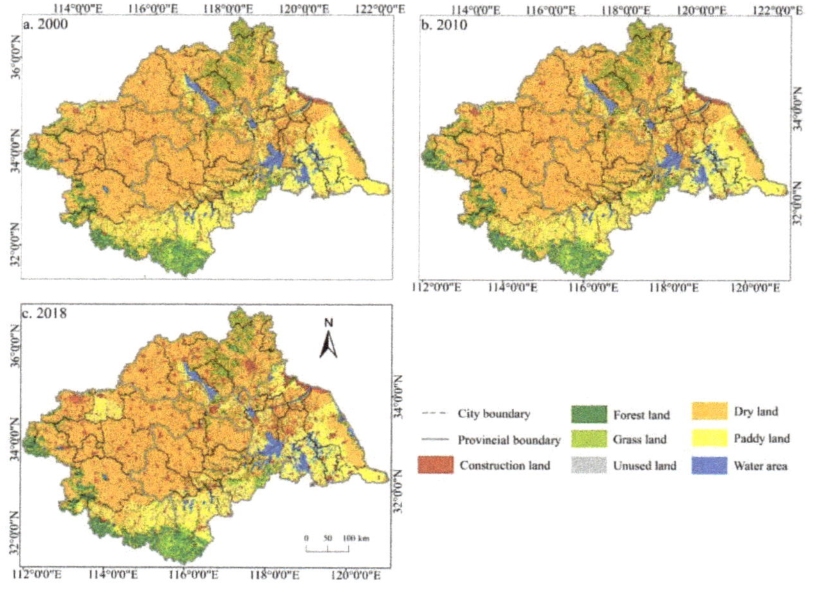

Figure 3. Spatial distribution of land use in the Huaihe River Basin from 2000 to 2018. Notes: (**a**) the spatial distribution of land use in 2000; (**b**) the spatial distribution of land use in 2010; (**c**) the spatial distribution of land use in 2018.

3.1.2. Balance of Spatiotemporal Distribution

To further explore the balance of the spatiotemporal distribution of cultivated land resources in the Huaihe River Basin during the study period, Equation (1) was used to calculate the location entropy Q of the cultivated area of each county in the basin, and then plot the cultivated land Lorentz curve in the Huaihe River Basin in 2000, 2010, and 2018, respectively (Figure 4), and the Gini coefficient of the cultivated land according to Equation (2) was calculated. The results showed that the Lorentz curve of cultivated land in the Huaihe River Basin was close to the absolute average line from 2000 to 2018, and the variation range was small, indicating that the spatial distribution of cultivated land resources in the basin was relatively scattered during the study period. The Gini coefficients of cultivated land in 2000, 2010, and 2018 were 0.105, 0.108, and 0.113, respectively, increasing year by year but less than 0.2, illustrating that the spatial distribution of cultivated land in the Huaihe River Basin was relatively uniform during the study period. According to the spatial distribution of the cultivated land locational entropy in the Huaihe River Basin from 2000 to 2018 (Figure 5), the locational entropy of cultivated land in more than 56% of the counties in the basin was greater than 1, indicating a high degree of specialization of the cultivated land in the basin. Simultaneously, the location entropy shows a coexistence of an increase and decrease, but the increasing trend dominated, indicating that the degree of cultivated land specialization in Huaihe River Basin increased. Specifically, the areas with a low location entropy of cultivated land were concentrated in Lu'an, Xinyang, and Zibo—Xuzhou—Huai'an, revealing that these areas are inferior areas of cultivated land with low specialization, which is possibly attributed to the rapid economic development and excessive occupation of cultivated land by urban expansion in some areas. From 2000 to 2018, the location entropy of some regions decreased significantly, most notably in Zhengzhou, Kaifeng, and Pingdingshan, indicating that the level of cultivated land agglomeration and specialization showed a conspicuous decrease. The location entropy of cultivated land in Zhumadian, Bozhou, Bengbu, and other cities increased significantly, and the agglomeration trend was constantly enhanced.

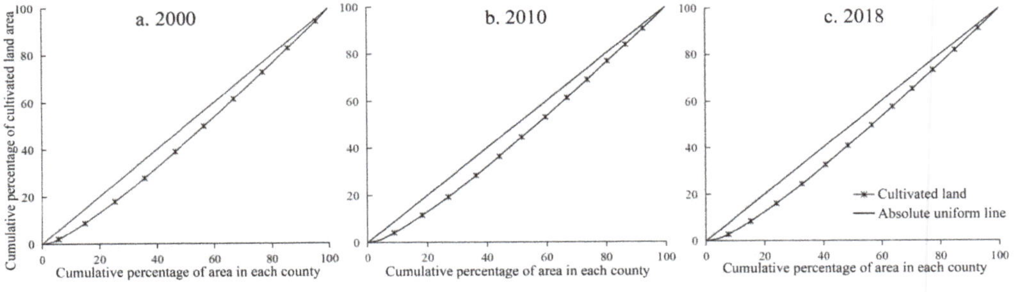

Figure 4. The Lorentz curve of cultivated land in the Huaihe River Basin from 2000 to 2018. Notes: (**a**) the Lorentz curve of cultivated land in 2000; (**b**) the Lorentz curve of cultivated land in 2010; (**c**) the Lorentz curve of cultivated land in 2018.

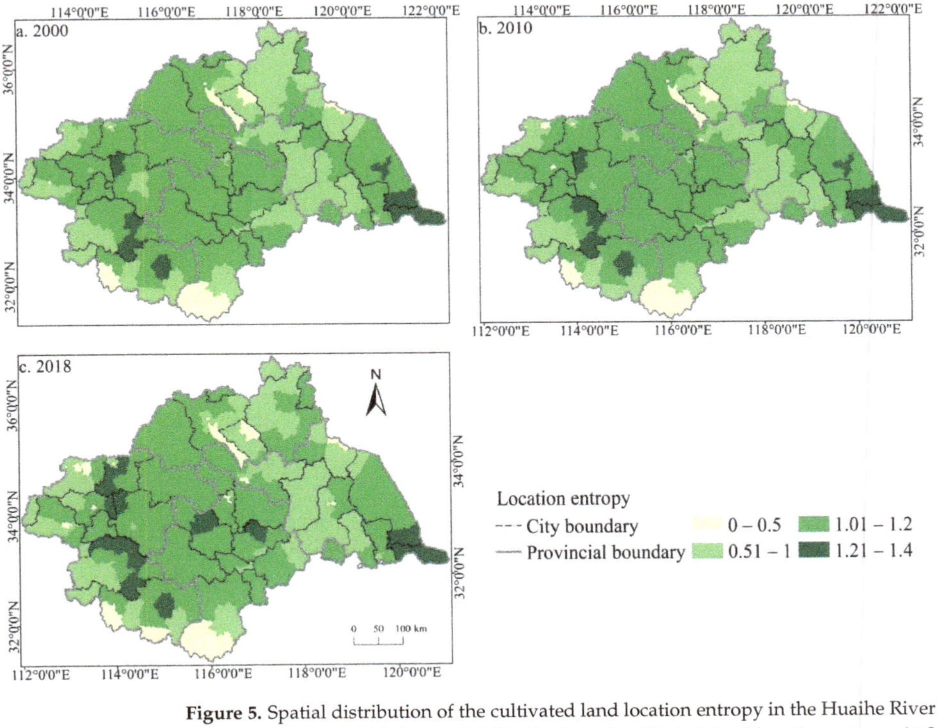

Figure 5. Spatial distribution of the cultivated land location entropy in the Huaihe River Basin from 2000 to 2018. Notes: (**a**) the spatial distribution of cultivated land location entropy in 2000; (**b**) the spatial distribution of cultivated land location entropy in 2010; (**c**) the spatial distribution of cultivated land location entropy in 2018.

3.2. Spatiotemporal Distribution of Cultivated Land Fragmentation

3.2.1. Spatiotemporal Distribution of the Cultivated Land Fragmentation Index

According to the measurement results of the CLF index in the Huaihe River Basin from 2000 to 2018 (Table 4), four negative indicators that characterize the CLF, namely PLAND, AREA_AM, COHESION, and AI, all show a decreasing trend year by year, which to some extent reflect that CLF increased in the Huaihe River Basin during the study period. Whereas PD and SPLIT of the positive indicators increased year by year, implying that the CLF in the Huaihe River Basin deteriorated. Among them, PLAND continued to decrease, and the decreasing amplitude grew, indicating that the composition of the cultivated land landscape in the basin increased and diversified. ED first increased and then decreased, indicating that the edge shape of the cultivated land landscape elements in the basin changed from irregular to regular, and the area of cultivated land patches also fluctuated significantly. COHESION decreased year by year, illustrating that the physical connection between cultivated land patches in the basin continued to decrease, and the phenomenon of cultivated land patch dispersion and fragmentation was severe. SPLIT kept on increasing and the increase rate gradually enlarged, demonstrating that cultivated land patches tended to be scattered, and the cultivated land landscape became more and more fragmented in the basin during the study period.

Table 4. Multi-collinearity diagnosis table of landscape metrics.

Landscape Index	PLAND (#/100 ha)	PD (m/km^2)	ED	AREA_AM (ha)	IJI (%)	COHESION	SPLIT	AI (%)
2000	69.236	0.009	5.350	83,991.047	49.444	97.346	10.220	76.025
2010	67.495	0.010	5.410	82,716.491	48.591	96.792	12.562	75.257
2018	63.603	0.011	5.357	79,706.491	46.155	93.374	17.511	73.439

The spatial distribution of the cultivated land landscape indexes PLAND and COHESION in the Huaihe River Basin differed significantly and had clear boundaries during 2000–2018 (Figures 6 and 7). The high value areas were concentrated in some counties of Henan and Anhui in the middle and west of the basin while the low value areas were only distributed in Zaozhuang, Xinyang, and Lu'an. During the study period, a large number of PLAND areas changed from high to low values, with significant changes in most counties in Huaibei, Zhengzhou, Heze, and other cities, which is closely related to the decreasing amount of cultivated land in these areas. The transition from a middle- and low-value area to a high-value area was rare, and the change was more obvious in some counties of Suzhou and Linyi, principally because of the remarkable achievement of land development and reclamation in these counties and the obvious increase of cultivated land patches. PD showed little difference in the spatial distribution (Figure 8). Low-value areas were widely distributed, and a small number of medium-value areas and high-value areas were scattered in some counties of Lu'an, Xinyang, Zibo, and other cities. During the study period, the spatial distribution did not change much. Only a very small number of counties in Zhengzhou, Huainan, Linyi, and other cities changed from low-value areas to medium- to high-value areas. The high-value areas of ED were concentrated in the northeast of the basin while the low-value areas were concentrated in Yancheng and Nantong in the east of the basin and some southern counties, such as Lu'an (Figure 9). The spatial distribution characteristics of AREA_AM and IJI were relatively similar (Figures 10 and 11). During the study period, the high-value areas of these two types were predominantly distributed in the south of the Huaihe River, whereas the low-value areas were concentrated in the middle of the basin. Some counties, such as Xuchang, Nanyang, and Lu'an, changed from high-value areas of AREA_AM to low-value areas, and only a few counties like Yancheng changed from low-value areas to high-value areas. Some counties, such as Xinyang, Huainan, and Huaibei City, changed from high-value areas of IJI to low-value areas, and only a few counties, such as Linyi City, changed from low-value areas to high-value areas. Low-value areas and high-value areas of SPLIT showed wide spatial distribution differences and clear boundaries (Figure 12). Low-value areas were concentrated in the central and eastern counties of the basin, whereas high-value areas were scattered, with Zibo, Huai'an, and Lu'an in the majority. During the study period, Luoyang, Xinyang, and Huai'an changed from low-value areas to high-value areas with significant changes. The spatial distribution of AI was balanced, with only a small number of low-value areas distributed in Lu'an and Zibo (Figure 13), whereas high-value areas were only distributed in some counties in Yancheng and Nantong.

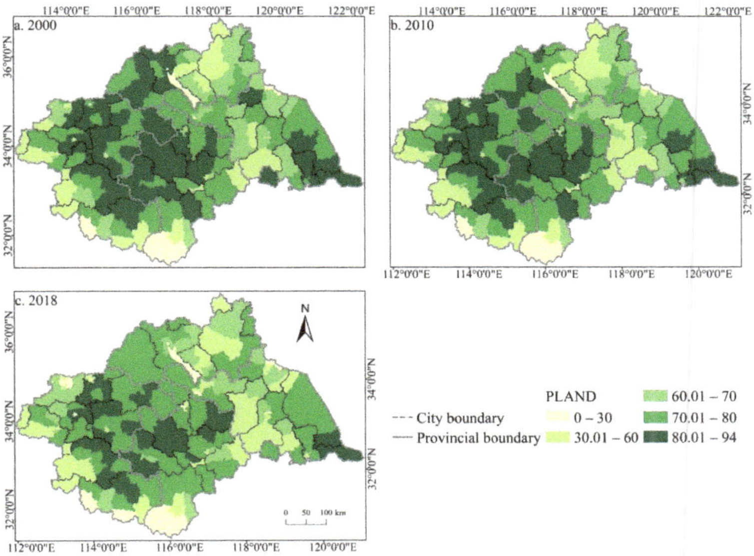

Figure 6. Spatial distribution of PLAND of cultivated land in the Huaihe River Basin from 2000 to 2018. Notes: (**a**) the spatial distribution of PLAND of cultivated land in 2000; (**b**) the spatial distribution of PLAND of cultivated land in 2010; (**c**) the spatial distribution of PLAND of cultivated land in 2018. PLAND: Percentage of landscape.

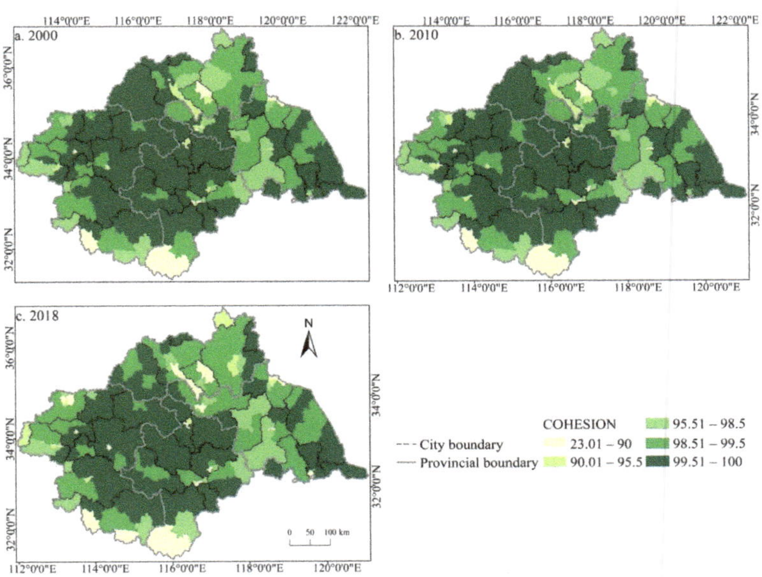

Figure 7. Spatial distribution of COHESION of cultivated land in the Huaihe River Basin from 2000 to 2018. Notes: (**a**) the spatial distribution of COHESION of cultivated land in 2000; (**b**) the spatial distribution of COHESION of cultivated land in 2010; (**c**) the spatial distribution of COHESION of cultivated land in 2018. COHESION: Patch cohesion index.

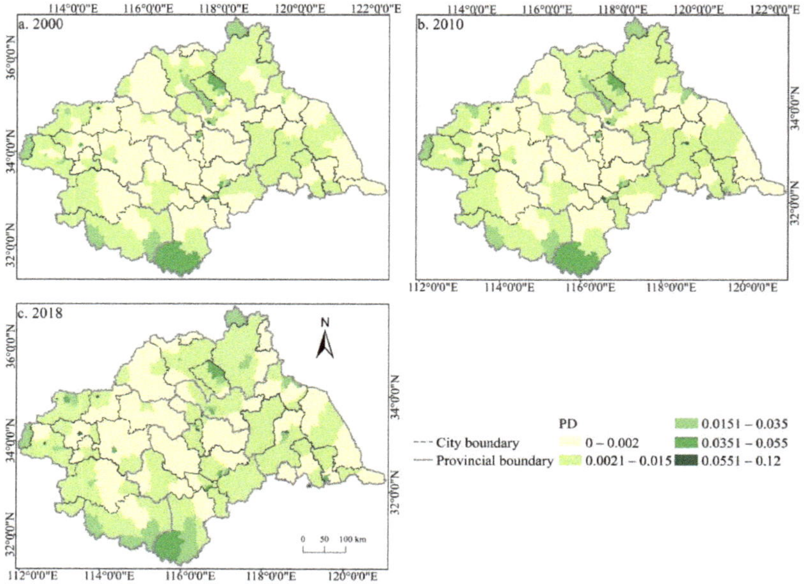

Figure 8. Spatial distribution of PD of cultivated land in the Huaihe River Basin from 2000 to 2018. Notes: (**a**) the spatial distribution of PD of cultivated land in 2000; (**b**) the spatial distribution of PD of cultivated land in 2010; (**c**) the spatial distribution of PD of cultivated land in 2018. PD: Patch density.

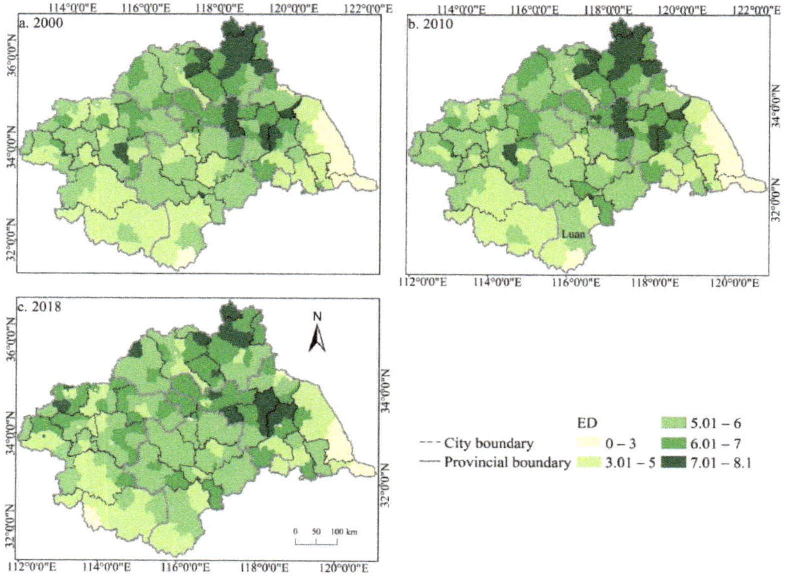

Figure 9. Spatial distribution of ED of cultivated land in the Huaihe River Basin from 2000 to 2018. Notes: (**a**) the spatial distribution of ED of cultivated land in 2000; (**b**) the spatial distribution of ED of cultivated land in 2010; (**c**) the spatial distribution of ED of cultivated land in 2018. ED: Edge density.

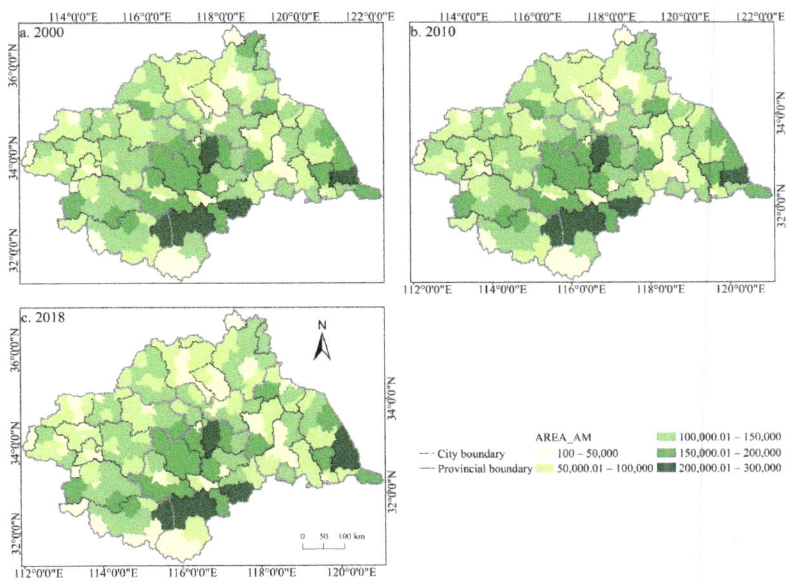

Figure 10. Spatial distribution of AREA_AM of cultivated land in the Huaihe River Basin from 2000 to 2018. Notes: (**a**) the spatial distribution of AREA_AM of cultivated land in 2000; (**b**) the spatial distribution of AREA_AM of cultivated land in 2010; (**c**) the spatial distribution of AREA_AM of cultivated land in 2018. AREA_AM: Area-weighted mean patch area.

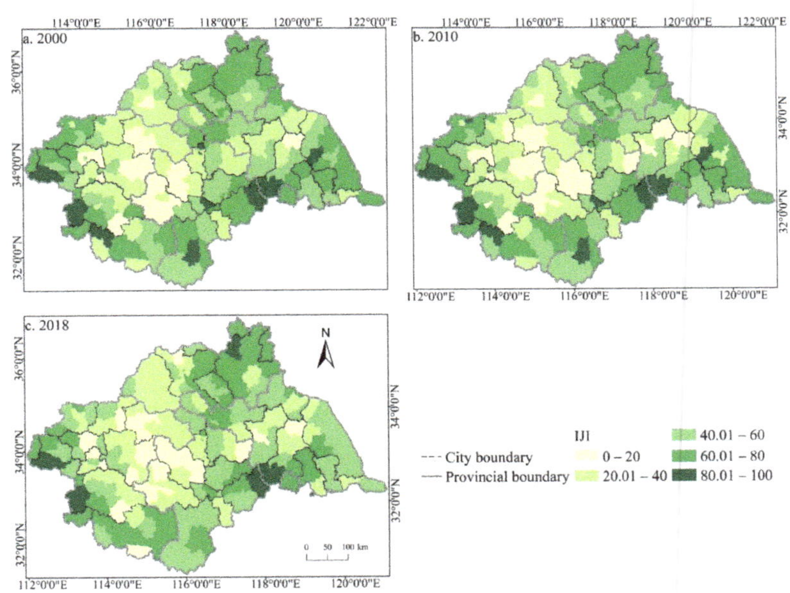

Figure 11. Spatial distribution of IJI of cultivated land in the Huaihe River Basin from 2000 to 2018. Notes: (**a**) the spatial distribution of IJI of cultivated land in 2000; (**b**) the spatial distribution of IJI of cultivated land in 2010; (**c**) the spatial distribution of IJI of cultivated land in 2018. IJI: Interspersion and juxtaposition index.

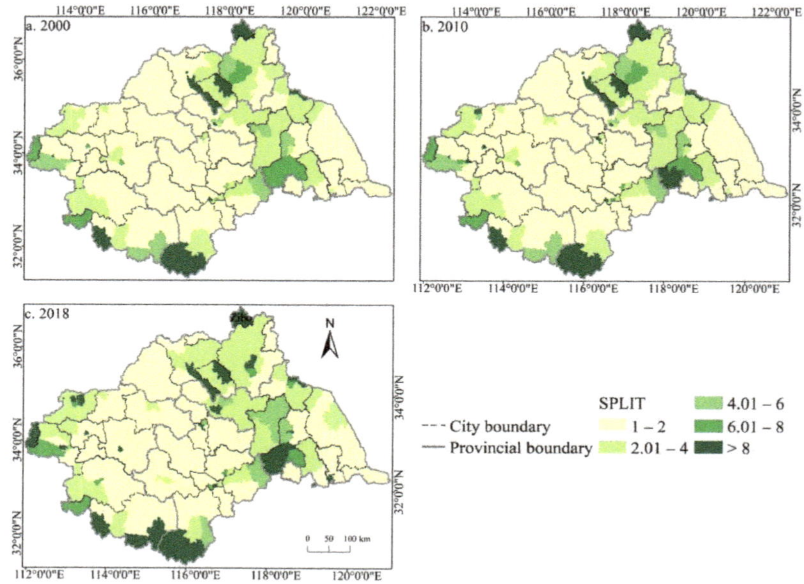

Figure 12. Spatial distribution of SPLIT of cultivated land in the Huaihe River Basin from 2000 to 2018. Notes: (**a**) the spatial distribution of SPLIT of cultivated land in 2000; (**b**) the spatial distribution of SPLIT of cultivated land in 2010; (**c**) the spatial distribution of SPLIT of cultivated land in 2018. SPLIT: Splitting index.

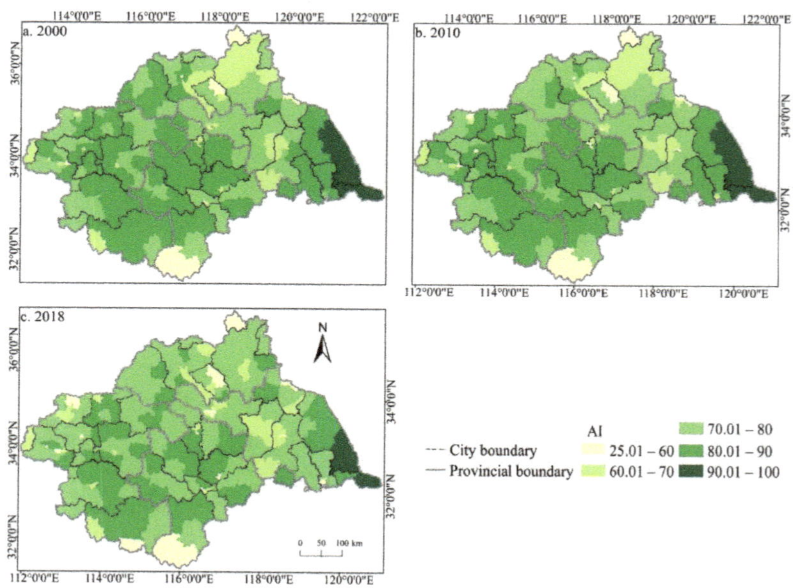

Figure 13. Spatial distribution of AI of cultivated land in the Huaihe River Basin from 2000 to 2018. Notes: (**a**) the spatial distribution of AI of cultivated land in 2000; (**b**) the spatial distribution of AI of cultivated land in 2010; (**c**) the spatial distribution of AI of cultivated land in 2018. AI: Aggregation index.

3.2.2. Spatiotemporal Distribution Characteristics of Comprehensive Index of Cultivated Land Fragmentation

According to the entropy weight method, the spatiotemporal distribution of the comprehensive fragmentation index of cultivated land from 2000 to 2018 was obtained (Figure 14). Overall, the spatial heterogeneity and complexity of the cultivated land landscape pattern in the Huaihe River Basin was comparatively high while the spatiotemporal differences of fragmentation varied significantly. The degree of fragmentation was high in the north and south of the basin but low in the middle. The overall degree of CLF deepened, but in some regions, the degree of fragmentation exhibited both increasing and decreasing trends. In 2000, Zibo, Zaozhuang, and Lu'an were the most seriously fragmented areas, whereas Bozhou, Huaibei, and Yancheng were the least fragmented. In 2010, areas with severe CLF in the basin had not yet been improved, and the degree of fragmentation in Zhengzhou, Jining, and other cities and the Bengbu-Huai'an-Taizhou line was significantly deepened. In 2018, the overall fragmentation of the basin further deteriorated, and the fragmentation degree of counties in Zhengzhou, Xuzhou, Taizhou, and other cities continued to deepen, while only some counties in Zhoukou, Shangqiu, Linyi, and other cities showed a decrease in the fragmentation degree. Additionally, it is worth noting that from 2000 to 2018, except for Hefei and Lu'an, the degree of CLF deepened in other counties along the Huaihe River, and the degree of CLF changed dramatically in the area of Huainan, Huai'an, and Yancheng, with a relatively high degree of fragmentation.

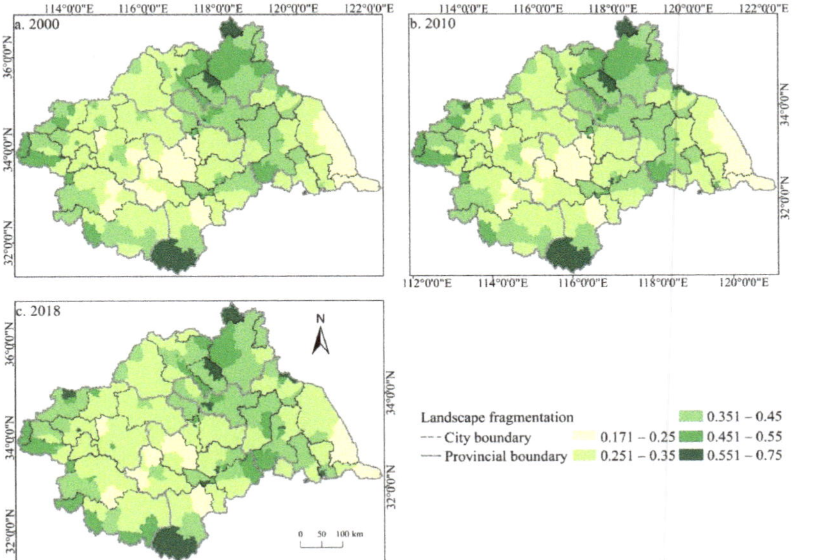

Figure 14. Spatial distribution of the comprehensive fragmentation degree of cultivated land in the Huaihe River Basin from 2000 to 2018. Notes: (**a**) the spatial distribution of comprehensive fragmentation degree of cultivated land in 2000; (**b**) the spatial distribution of comprehensive fragmentation degree of cultivated land in 2010; (**c**) the spatial distribution of comprehensive fragmentation degree of cultivated land in 2018.

3.3. *Correlation Analysis of Spatial Distribution of Cultivated Land Fragmentation*

To further explore the spatiotemporal distribution characteristics of the comprehensive index of CLF in the Huaihe River Basin from 2000 to 2018, Geoda 1.12 was applied in this study to conduct a univariate local spatial auto-correlation analysis of the comprehensive index of CLF in the basin during the study period, and a local indicators of spatial association (LISA) map was generated, namely, the aggregation distribution pattern of the comprehensive fragmentation index. In the LISA map, the comprehensive fragmentation index of the counties in the four quadrants of high-high, low-low, low-high, and high-low was significant at the level of 5%. To better display the comprehensive degree of fragmentation of each county in the basin, ArcGIS10.3 was used for visual analysis of changes in 2000, 2010, and 2018 (Figure 14). As shown in Figure 15, the distribution pattern of random and agglomeration counties in the Huaihe River Basin was generally stable during the study period. High-high areas represent the agglomeration areas of high value and high value, which were concentrated in some counties in Linyi, Zaozhuang, Pingdingshan, and other cities. The altitude of the counties in this region is higher than that of most counties in the basin, indicating that altitude exerted a certain impact on CLF. Low-low areas mean low-value and low-value agglomeration areas, which were concentrated in various cities at the junction of Anhui and Henan, counties mainly included Zhoukou, Shangqiu, Fuyang, Bozhou, and other cities, and the eastern part of some basin counties of Yancheng, Taizhou, and Nantong along the coast were also distributed. The low-high area reflects that the low value is surrounded by the high value. This type of area exhibits a small and scattered distribution, mainly scattered in a few counties in cities, such as Xuzhou and Pingdingshan. The high-low area indicates that the high value is surrounded by the low value. This type covers fewer counties, which did not appear during the study periods. Such counties are generally "outliers" and appear less frequently.

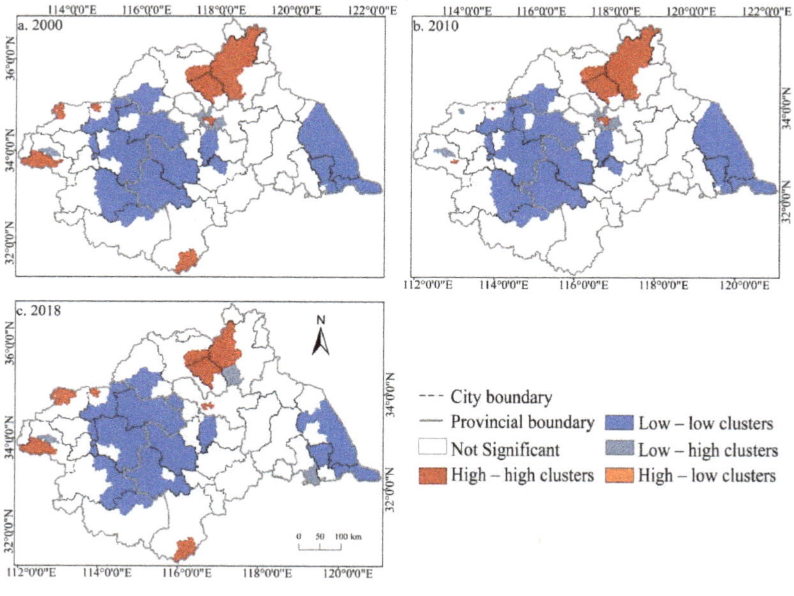

Figure 15. Local indicators of the spatial association (LISA) map of cultivated land fragmentation in the Huaihe River Basin from 2000 to 2018. Notes: (**a**) LISA map of cultivated land fragmentation in 2000; (**b**) LISA map of cultivated land fragmentation in 2010; (**c**) LISA map of cultivated land fragmentation in 2018.

3.4. *Influencing Factors of Cultivated Land Fragmentation*

Based on Equation (11), the spatiotemporal distribution mechanism of CLF in 218 counties of the Huaihe River Basin from 2000 to 2018 was explored, through which the spatiotemporal differentiation of CLF in the Huaihe River Basin was influenced by natural factors and socioeconomic factors can be observed. Additionally, the influence of different factors on CLF varied significantly. Overall, the comprehensive influence of socioeconomic factors on CLF in the Huaihe River Basin during the study period far exceeded that of natural factors. The comprehensive influence of NDVI (X4), per capita of cultivated land area (X6), and intensity of human activities (X7) reached more than 54% of the total factor contribution rate, which was significantly stronger than other factors (Table 5). Specifically, during the study period, the influence of the altitude (X1) and slope (X2) on CLF exhibited a fluctuating upward trend, revealing that the intensity of the impact of these natural background characteristics on CLF increased. This was mainly because of the ever-growing input of agricultural science and technology, which has enabled the natural geographical advantages of a lower elevation and lower slope in the Huaihe River Basin weigh heavily in accelerating the large-scale management of cultivated land in the basin. The influence of the distance to river (X3) on CLF first increased and then decreased because with the comprehensive promotion of irrigation projects in the Huaihe River Basin, the water consumption of cultivated land was effectively guaranteed, and the dependence on important water bodies and natural precipitation weakened. NDVI (X4) exerted a high impact on CLF, but it showed a decreasing trend year by year, with a decrease of 25.6%, which is closely related to the dramatic changes of vegetation cover types and land use structure in the process of rapid urbanization. During the study period, the per capita cultivated land area (X6) and the intensity of human activities (X7) exerted a high impact on the CLF, and seemed to gain momentum. The factor contribution rate reached 0.4. On the one hand, with the improvement of the socioeconomic development level in Huaihe River Basin, the rapid population growth burdened the load of cultivated land and decreased the per capita cultivated land area (X6) continuously, leading to widespread CLF management. On the other hand, the intensifying human activities, such as disordered urban sprawl and irrational land development and utilization, badly affected the cultivated land landscape pattern.

Table 5. Contribution rate of impact factors from 2000 to 2018.

Year	X1	X2	X3	X4	X5	X6	X7	X8	X9
2000	0.182	0.163	0.147	0.512	0.063	0.384	0.376	0.150	0.151
2010	0.169	0.159	0.165	0.406	0.096	0.424	0.378	0.259	0.141
2018	0.186	0.169	0.141	0.381	0.143	0.408	0.433	0.237	0.144

Notes: X1: altitude; X2: slope; X3: distance to river; X4: normalized difference vegetation index (NDVI); X5: average annual precipitation; X6: per capita cultivated land area; X7: intensity of human activities; X8: population density; X9: distance from road network.

To more clearly identify the main influencing factors of CLF and the changes in the intensity of interactions among the main influencing factors during the study period, we selected only the 10 interactions with high intensity. As shown in Table 6, the influences of each influencing factor on CLF in the Huaihe River Basin during the study period were not independent of each other, and the influence of the. interaction among factors was significantly stronger than that of a single factor, with dual-factor enhancement and nonlinear enhancement representing the majority. Specifically, in 2000, the interaction was dominated by dual-factor enhancement, and the main interaction factor intensity reached above 0.5, with the interaction intensity of human activity (X7) and NDVI (X4) the highest, reaching 0.677. In 2010, the interaction intensity of the main interaction factors increased significantly, and the dual-factor enhancement dominated. The interaction intensity between human activity intensity (X7) and per capita cultivated land area (X6) was the highest, reaching 0.729. In 2018, the dual-factor enhancement effect weakened, and the nonlinear enhancement gradually dominated. The interactive intensity of human activity (X7) and per capita cultivated land area (X6) was as high as 0.722. In general, the interaction between the intensity of human activity (X7) and other factors can better explain the CLF in the Huaihe River Basin than the interaction among other factors, indicating that the intensity of human activity (X7) played a major role in the process of CLF in the Huaihe River Basin. Mainly because the level of socioeconomic development in the Huaihe River Basin has improved, the man-land contradiction intensified, and the unreasonable human activities, such as urban sprawl and excessive reclamation, made the interaction with other factors more complex, further interfering with the landscape pattern of cultivated land.

Table 6. Main interaction factors and changes.

2000		2010		2018	
Interaction Factors	Interaction Intensity	Interaction Factors	Interaction Intensity	Interaction Factors	Interaction Intensity
X3∩X4	0.555 *	X1∩X4	0.532 *	X5∩X4	0.544 #
X5∩X4	0.567 *	X8∩X4	0.570 *	X5∩X7	0.558 #
X8∩X4	0.585 *	X6∩X4	0.580 *	X8∩X7	0.571 *
X8∩X6	0.588 #	X5∩X4	0.582 #	X8∩X6	0.579 *
X1∩X6	0.590 #	X8∩X7	0.594 *	X1∩X4	0.604 #
X1∩X4	0.594 *	X7∩X4	0.602 *	X1∩X6	0.608 #
X6∩X4	0.596 *	X2∩X6	0.609 #	X2∩X4	0.609 #
X2∩X6	0.598 #	X8∩X6	0.618 *	X2∩X6	0.612 #
X2∩X4	0.619 *	X1∩X6	0.627 #	X7∩X4	0.674 *
X7∩X4	0.677 *	X7∩X6	0.729 *	X7∩X6	0.722 *

Notes: * means dual-factor enhancement; # means nonlinear enhancement. X1: altitude; X2: slope; X3: distance to river; X4: normalized difference vegetation index (NDVI); X5: average annual precipitation; X6: per capita cultivated land area; X7: intensity of human activities; X8: population density.

4. Discussion

4.1. Comparing with Previous Studies

It can be observed in this study that the spatial distribution of cultivated land resources in the Huaihe River Basin was relatively uniform, the Gini coefficient of cultivated land increased year by year, and the degree of cultivated land specialization was high, which is similar to the results obtained by Liu et al. based on NDVI and the transfer matrix method [49]. Previous studies have shown that cultivated land in mountainous and hilly areas was more inclined to show the spatial separation characteristics of land structure than that in plain areas [20]. Profiting from the topographic characteristics of the wide-spread topography of the plain, the Huaihe River basin typically performs the above-mentioned cultivated land distribution characteristics. In terms of the indicators characterizing CLF, studies have constructed an evaluation index system from the aspects of the resource scale, spatial agglomeration, and convenience of utilization [1,24] but without considering the multi-collinearity of landscape indices. Principal component analysis has been applied in most relevant studies to reduce the redundancy among the selected metrics [50]. In this study, SPSS22 software was introduced for collinearity diagnostics testing of the fragmentation index selected, and the eight indicators filtered covered the three aspects of size, edge-shape, and aggregation, which can comprehensively reflect the complex process involved in fragmentation.

During the study period, the comprehensive index of CLF in the Huaihe River Basin increased by 3.8%, and the degree of fragmentation displayed an upward trend, which is consistent with the development trend of CLF across China, that is, an increasing trend of fragmentation in China was identified and southern China was characterized by more CLF than the other parts [17]. The analysis of influencing factors based on geographical detectors showed that the main factors leading to the increasing degree of fragmentation are the intensity of human activities (X7), per capita of cultivated land area (X6), and NDVI (X4), and their contribution rates were all above 0.4. Yet, based on the hierarchical linear model, Xu et al. found that the average patch area, gross domestic product (GDP), land use intensity, and urbanization rate at the county level were the main factors affecting the CLF in Jiangsu Province, and natural factors were considered to be the primary factor leading to CLF [1]. Sklenicka et al. drew the same conclusion on research concerning CLF in central and eastern Europe [25]. However, this study observed the opposite, namely the influence of socioeconomic factors on CLF weighed heavier than that of natural factors, and similar conclusions could be drawn from the study of Tan et al. [51]. This is mainly due to the rapid economic development in the Huaihe River Basin, where the impact of man-land contradiction on CLF has deepened. Meanwhile, the terrain of the basin is dominated by plain, and the impact exerted on CLF by natural factors, such as altitude and slope, was not prominent; thus, socioeconomic factors exceeded the natural ones.

Additionally, most scholars have taken socioeconomic factors [1,24], natural factors [1,20], and policy regulation factors [52] into account in the analysis touching on the. influencing factors of CLF. However, only a simple linear combination analysis of their forces weakened the chain reaction caused by the combination of multiple objectively existing factors, and lacks discussion on the intensity and mode of interaction between different influencing factors. This study found that the combination of different influencing factors of CLF produced multiple interactions, such as dual-factor enhancement and nonlinear enhancement, which increased the intensity of the impact on CLF. In addition, current research on the factors affecting CLF is mostly based on a single province or region [1,24,53], which is far-fetched to provide a reference for river basin units and important food production bases in China.

4.2. Policy Implications

CLF can exert a profound negative impact on the efficiency of cultivated land use, food production, and agricultural development [19,54]. In 2017, the Chinese government put forward the Rural Revitalization Strategy, which aims to accelerate the realization of agricultural and rural modernization [55]. However, the intensification of CLF has become a huge obstacle; hence, it is imperative to take measures to alleviate it. Acting as the important grain production base in China, the Huaihe River Basin covers an area of only 2.9% of China's land area, but its cultivated land area accounts for 12% of the country's land area [9]. However, because of the continuous urbanization occurring throughout the basin, population growth, excessive reclamation, and other high-intensity human activities in the Huaihe River Basin have resulted in frequent floods and droughts, soil erosion, ecological environment deterioration, and the associated aggravation of fragmentation, which has restricted sustainable socioeconomic development, increased the cost of food production, hindered the development of agricultural scale, and squeezed the space for future improvement of the total food production capacity in the basin, eventually seriously threatening national food security [51]. Therefore, it is of theoretical and practical significance to determine the characteristics of cultivated land resources among natural, spatial, and utilization attributes in the basin and adopt scientific and reasonable management measures to prevent and alleviate the exacerbation of the CLF phenomenon for formulating regional cultivated land utilization strategy, promoting agricultural modernization, and ensuring food security.

State-regulated consolidation is often perceived as a critical measure to tackle the CLF problem [56]. Based on the research results, the following suggestions are put forward:

(1) Strictly control the occupation of cultivated land by construction land in the basin, earnestly implement the policy of requisition-compensation balance [57], strictly observe the red line of cultivated land protection, ensure the efficient utilization of cultivated land resources in the Huaihe River Basin, guarantee food security, and stabilize the position of the grain production base in the basin.

(2) On the basis of protecting the ecological environment, implement differentiated conversion of cultivated land to avoid unreasonable conversion and excessive greening of cultivated land, and reduce the degree of dispersion of cultivated land patches.

(3) Formulate reasonable land use planning, realize the optimization of land use structure, take the Huaihe River as the boundary, adjust paddy field and dry land differently, implement a reasonable dry land conversion project, and avoid blindly pursuing high-income land use transformation.

(4) Leverage the advantages of extensive natural geography in the eastern and southern plain of the Huaihe River Basin, strengthen the support for large-scale agricultural industry management, reduce the phenomenon of land fragmentation management in the plain by support and encouragement, and improve the utilization efficiency of cultivated land resources.

4.3. Limitations and Future Directions

Certain limitations could be observed in this study. CLF is widely distributed globally [15,58]. CLF driven by the process of ownership fragmentation is a pervasive issue in various planning and managing activities in different countries [59]. Current research is more frequently carried out on characterizing CLF through the landscape pattern index, but the landscape scale cannot touch the cause of CLF caused by the division of ownership and the land use distribution system [60,61]. CLF is a complex process, which is formed under the combined effect of various factors, such as social economy, nature, policy, and culture. Therefore, it is not only necessary to pay attention to the impact of quantitative indicators, such as the level of socioeconomic development and natural geographical conditions on the CLF, but also to policy regulation [52], cultural customs [62], and institutional background [60,63] and other relevant indicators. Due to the limitations of data sources, it is beyond the scope of this study to indulge in a full-scale discussion on the above issues comprehensively, but these factors are crucial in exploring the study of CLF at the micro scale. Besides, the interaction between different driving force groups is complex [25], so it is extremely difficult to comprehensively study the influencing factors of CLF. Furthermore, although this study focused on the different intensities of the interactions of different influencing factors, due to the limitations of research methods, it failed to detect the driving mechanism of the interaction of different factors, which needs further improvement in follow-up study. Simultaneously, based on the understanding of the CLF in the county unit of the Huaihe River Basin, we will further explore it in township units and even villages in the future, with a view to providing support for the governance of CLF in the basin and ensuring food security.

5. Conclusions

Based on the land use remote sensing data from 2000 to 2018, Fragstats4.2 (Oregon State University, Corvallis, OR, USA) and ArcGIS10.3 were adopted, the spatiotemporal distribution characteristics of CLF index at county level in the Huaihe River Basin were measured by the entropy weight method, and multiple influencing factors of CLF were explored with the help of the geographical detector. The main conclusions are as follows:

(1) From 2000 to 2018, the Gini coefficients of cultivated land were 0.105, 0.108, and 0.113, respectively, increasing year by year but less than 0.15. More than 56% of the counties in the basin showed a cultivated land location entropy greater than 1, and the degree of specialization of cultivated land continued to increase. Whereas the location entropy of counties in Zhengzhou, Xinyang, Huaibei, and other cities showed a downward trend.

(2) From 2000 to 2018, PLAND, AREA_AM, COHESION, and AI, which are four negative indicators of cultivated land fragmentation in the Huaihe River Basin, showed a decreasing trend year by year while the positive indicators of PD and SPLIT increased year by year. Overall, the spatial and temporal differences of CLF in the Huaihe River Basin were distinguished, and the degree of CLF increased, but the degree of CLF in some areas showed a coexistence of an increase and decrease.

(3) The county distribution pattern of the CLF degree with random and agglomeration was generally stable. High-high areas were concentrated in some counties in Linyi, Xuzhou, Zhengzhou, and other cities; low-low areas were concentrated in cities at the junction of Anhui and Henan, with Zhoukou, Shangqiu, Fuyang, and Bozhou the main counties. The low-high areas had a small and scattered distribution, mainly distributed in a small number of counties in Xuzhou, Pingdingshan, Linyi, and other cities. High-low areas did not appear in the three time breakpoints, indicating that within the basin, there was no significant difference in the degree of CLF between adjacent counties, and no obvious polarization phenomenon was observed.

(4) During the study period, the spatiotemporal differentiation of CLF in the Huaihe River Basin was affected by multiple factors, such as nature, socioeconomic, etc., and the influence of different factors on CLF was significantly different. The comprehensive

influence of socioeconomic factors was significantly stronger than that of natural factors. The influence of NDVI (X4), per capita of cultivated land area (X6), and intensity of human activity (X7) was significantly stronger than that of other factors, with the factor contribution rate above 0.4. The intensity of human activity (X7) played a major role in the process of CLF. The influence of various factors on the CLF in the Huaihe River Basin was not independent of each other. The interaction effect among the factors was stronger than that of a single factor, with dual-factor enhancement dominant and nonlinear enhancement the supplement.

We hope that the results of this study and the proposed policy recommendations can provide references for identifying and alleviating CLF in important grain production areas to ensure regional food security.

Author Contributions: Conceptualization, W.C. and J.L. (Jiale Liang); methodology, J.L. (Jiale Liang) and S.P.; software, J.L. (Jiale Liang); validation, W.C. and T.Z.; formal analysis, S.P. and J.L. (Jiale Liang); resources, W.C. and J.L (Jiangfeng Li).; data curation, J.L. (Jiangfeng Li) and T.Z.; writing-original draft preparation, S.P. and T.Z.; writing-review and editing, S.P. and W.C. All authors have read and agreed to the published version of the manuscript.

Funding: This research was funded by the Natural Science Foundation of China, Grant number 42001187. The project was also supported by State Key Laboratory of Earth Surface Processes and Resource Ecology (2021-KF-03).

Institutional Review Board Statement: Not applicable.

Informed Consent Statement: Not applicable.

Data Availability Statement: The data that support the findings of this study are available from the corresponding author upon reasonable request.

Conflicts of Interest: The authors declare no conflict of interest.

References

1. Xu, W.; Jin, X.; Liu, J.; Zhou, Y. Analysis of influencing factors of cultivated land fragmentation based on hierarchical linear model: A case study of Jiangsu Province, China. *Land Use Policy* **2021**, *101*, 105119. [CrossRef]
2. Lu, H.; Xie, H.; Lv, T.; Yao, G. Determinants of cultivated land recuperation in ecologically damaged areas in China. *Land Use Policy* **2019**, *81*, 160–166. [CrossRef]
3. Lu, H.; Xie, H.; Yao, G. Impact of land fragmentation on marginal productivity of agricultural labor and non-agricultural labor supply: A case study of Jiangsu, China. *Habitat Int.* **2019**, *83*, 65–72. [CrossRef]
4. Jiang, Y.; Hou, L.; Shi, T.; Ning, Y. Spatial zoning strategy of urbanization based on urban climate co-movement: A case study in Shanghai mainland area. *Sustainability* **2018**, *10*, 2706. [CrossRef]
5. Surya, B.; Salim, A.; Hernita, H.; Suriani, S.; Menne, F.; Rasyidi, E.S. Land use change, urban agglomeration, and urban sprawl: A sustainable development perspective of Makassar City, Indonesia. *Land* **2021**, *10*, 556. [CrossRef]
6. Surya, B.; Ahmad, D.N.A.; Sakti, H.H.; Sahban, H. Land use change, spatial interaction, and sustainable development in the metropolitan urban areas, South Sulawesi Province, Indonesia. *Land* **2020**, *9*, 95. [CrossRef]
7. Han, H.; Lin, H. Patterns of agricultural diversification in China and its policy implications for agricultural modernization. *Int. Int. J. Environ. Res. Public Health* **2021**, *18*, 4978. [CrossRef] [PubMed]
8. Ma, L.; Long, H.; Tang, L.; Tu, S.; Zhang, Y.; Qu, Y. Analysis of the spatial variations of determinants of agricultural production efficiency in China. *Comput. Electron. Agr.* **2021**, *180*, 105890. [CrossRef]
9. Xu, L.; Ou, Z. Influences of agricultural drought on food production in Huaihe River Basin. *J. Econ. Water. Resour.* **2011**, *29*, 56–59.
10. Jürgenson, E. Land reform, land fragmentation and perspectives for future land consolidation in Estonia. *Land Use Policy* **2016**, *57*, 34–43. [CrossRef]
11. Li, J.; Chen, Y.; Jiang, Q.; Kuang, X. Study on the causes and countermeasures of cultivated land fragmentation in China. *Agric. Econ.* **2006**, *6*, 21–23.
12. Zhang, L.; Huang, J.; Rozelle, S. Land policy and land use in China. Agricultural Policies in China. *Organ. Econ. Co-Oper. Dev.* **1997**, 71–77.
13. Tan, S.; Qu, F.; Nick, H. Analysis of the causes and influencing factors of land fragmentation. *China Rural Surv.* **2003**, *6*, 24–30.
14. Wan, G.; Cheng, E. Effects of land fragmentation and returns to scale in the Chinese farming sector. *Appl. Econ.* **2001**, *33*, 183–194. [CrossRef]
15. Lu, H.; Xie, H.; He, Y.; Wu, Z.; Zhang, X. Assessing the impacts of land fragmentation and plot size on yields and costs: A translog production model and cost function approach. *Agr. Syst.* **2018**, *161*, 81–88. [CrossRef]

16. Orea, L.; Perez, J.A.; Roibas, D. Evaluating the double effect of land fragmentation on technology choice and dairy farm productivity: A latent class model approach. *Land Use Policy* **2015**, *45*, 189–198. [CrossRef]
17. Wei, L.; Luo, Y.; Wang, M.; Su, S.; Pi, J.; Li, G. Essential fragmentation metrics for agricultural policies: Linking landscape pattern, ecosystem service and land use management in urbanizing China. *Agr. Syst.* **2020**, *182*, 102833. [CrossRef]
18. Sklenicka, P.; Zouhar, J.; Trpáková, I.; Vlasák, J. Trends in land ownership fragmentation during the last 230 years in Czechia, and a projection of future developments. *Land Use Policy* **2017**, *67*, 640–651. [CrossRef]
19. Looga, J.; Jürgenson, E.; Sikk, K.; Matveev, E.; Maasikamäe, S. Land fragmentation and other determinants of agricultural farm productivity: The case of Estonia. *Land Use Policy* **2018**, *79*, 285–292. [CrossRef]
20. Yucer, A.A.; Kan, M.; Demirtas, M.; Kalanlar, S. The importance of creating new inheritance policies and laws that reduce agricultural land fragmentation and its negative impacts in Turkey. *Land Use Policy* **2016**, *56*, 1–7. [CrossRef]
21. Baker, W.L.; Cai, Y. The rle-programs for multiscale analysis of landscape structure using the GRASS geographical information system. *Landsc. Ecol.* **1992**, *7*, 291–302. [CrossRef]
22. Morelli, F.; Benedetti, Y.; Šímová, P. Landscape metrics as indicators of avian diversity and community measures. *Ecol. Indic.* **2018**, *90*, 132–141. [CrossRef]
23. Zhao, Z.; Zhang, B.; Jin, X.; Weng, B.; Yan, D.; Bao, S. Spatial gradients pattern of landscapes and their relations with environmental factors in Haihe River basin. *Acta. Ecol. Sin.* **2011**, *31*, 1925–1935.
24. Liu, J.; Jin, X.; Xu, W.; Sun, R.; Han, B.; Yang, X.; Gu, Z.; Xu, C.; Sui, X.; Zhou, Y. Influential factors and classification of cultivated land fragmentation, and implications for future land consolidation: A case study of Jiangsu Province in eastern China. *Land Use Policy* **2019**, *88*, 104185. [CrossRef]
25. Sklenicka, P.; Salek, M. Ownership and soil quality as sources of agricultural land fragmentation in highly fragmented ownership patterns. *Landsc. Ecol.* **2008**, *23*, 299–311. [CrossRef]
26. Gong, Z.; Zhang, Y.; Gong, H.; Zhao, W. Evolution of wetland landscape pattern and its driving factors in Beijing. *Acta. Geogr. Sin.* **2011**, *66*, 77–88.
27. Zang, L.; Liang, H.; Liang, W.; Zhang, C. Cultivated land fragmentation and affecting factors of Lulong county based on landscape pattern. *Res. Soil. Water Conserv.* **2018**, *25*, 265–269.
28. Zhang, X.; Cai, Z.; Li, G.; Sun, J.; Li, G. Assessment of determinants of cultivated land fragmentation and its impacts on rural income. *Sci. Surv. Mapp.* **2020**, *45*, 134–141.
29. Zhou, Z. Impact of the agricultural landscape change on ecosystem services in the process of rapid urbanization region: A case study of Xi'an metropolitan zone. *Arid. Land. Geogr.* **2015**, *38*, 1004–1013.
30. Gao, J.; Liu, Y.; Zhang, Y. Evolution of temporal-spatial pattern of grain production and its driving mechanism in Huaihe River Basin during 1990–2012. *Bull. Soli. Water Conserv.* **2016**, *36*, 179–185.
31. Manjunatha, A.; Anik, A.; Speelman, S.; Nuppenau, E. Impact of land fragmentation, farm size, land ownership and crop diversity on profit and efficiency of irrigated farms in India. *Land Use Policy* **2013**, *31*, 397–405. [CrossRef]
32. Yan, H.; Liu, J.; Huang, H.; Tao, B.; Cao, M. Assessing the consequence of land use change on agricultural productivity in China. *Global. Planet. Chang.* **2009**, *67*, 13–19. [CrossRef]
33. Zhou, L.; Xu, J.; Cai, B.; Jiang, J.; Sun, D. The spatio-temporal changes of grain production and fertilizer consumption and its impact on water environment in the Huaihe River Basin. *J. Nat. Resour.* **2014**, *29*, 1053–1064.
34. Liu, J.; Liu, M.; Tian, H.; Zhuang, D.; Zhang, Z.; Zhang, W.; Tang, X.; Deng, X. Spatial and temporal patterns of China's cropland during 1990–2000: An analysis based on Landsat TM data. *Remote Sens. Environ.* **2005**, *98*, 442–456. [CrossRef]
35. Yang, Y.; Zhang, S.; Wang, D.; Yang, J.; Xing, X. Spatiotemporal changes of farming-pastoral ecotone in northern China, 1954–2005: A case study in Zhenlai County, Jilin Province. *Sustainability* **2015**, *7*, 4978. [CrossRef]
36. Chen, W.; Chi, G.; Li, J. The spatial association of ecosystem services with land use and land cover change at the county level inChina, 1995–2015. *Sci. Total Environ.* **2019**, *669*, 459–470. [CrossRef] [PubMed]
37. Tang, J.; Wang, X. Analysis of the land use structure changes based on Lorenz curves. *Environ. Monit. Assess* **2009**, *151*, 175–180. [CrossRef]
38. Zheng, X.; Xia, T.; Yang, X.; Yuan, T.; Hu, Y. The land Gini coefficient and its application for land use structure analysis in China. *PLoS ONE* **2013**, *8*, e76165. [CrossRef] [PubMed]
39. Tian, Y.; Jim, C.; Tao, Y.; Shi, T. Landscape ecological assessment of green space fragmentation in Hong Kong. *Urban For. Urban Green.* **2011**, *10*, 79–86. [CrossRef]
40. Toebe, M.; Filho, A.C. Multicollinearity in path analysis of maize (*Zea mays* L.). *J. Cereal. Sci.* **2013**, *57*, 453–462. [CrossRef]
41. Cardille, J.; Turner, M.; Clayton, M.; Gergel, S.; Price, S. Metaland: Characterizing spatial patterns and statistical context of landscape metrics. *BioScience* **2005**, *55*, 983–988. [CrossRef]
42. Peng, J.; Wang, Y.; Zhang, Y.; Wu, J.; Li, W.; Li, Y. Evaluating the effectiveness of landscape metrics in quantifying spatial patterns. *Ecolo. Indic.* **2010**, *10*, 217–223. [CrossRef]
43. Chen, P. Effects of normalization on the entropy-based TOPSIS method. *Expert. Syst. Appl.* **2019**, *136*, 33–41. [CrossRef]
44. Anselin, L. Lagrange multiplier test diagnostics for spatial dependence and spatial heterogeneity. *Geogr. Anal.* **1988**, *20*, 1–17. [CrossRef]
45. Wang, J.; Xu, C. Geodetector: Principle and prospective. *Acta. Geogr. Sin.* **2017**, *72*, 116–134.

46. Zhang, L.; He, X.; Wei, M. Dynamic changes of NDVI-based vegetation coverage of Huaihe River Basin. *Resour. Environ. Yangtze Basin* **2012**, *21*, 51–56.
47. Li, S.; Bing, Z.; Jin, G. Spatially explicit mapping of soil conservation service in monetary units due to land use/cover change for the Three Gorges Reservoir Area, China. *Remote Sens.* **2019**, *11*, 468. [CrossRef]
48. Ge, M.; Feng, Z. Population distribution of China based on GIS: Classification of population densities and curve of population gravity centers. *Acta Geogr. Sin.* **2009**, *64*, 202–210.
49. Liu, F.; Qin, T.; Girma, A.; Wang, H.; Weng, B.; Yu, Z.; Wang, Z. Dynamics of land-use and vegetation change using NDVI and transfer matrix: A case study of the Huaihe River Basin. *Pol. J. Environ. Stud.* **2019**, *28*, 213–223. [CrossRef]
50. Morad, S.; Shacham, M.; Brenner, A. Utilization of collinearity in regressing modeling of activated sludge processes. *Chem. Eng. Process.* **2007**, *46*, 222–229. [CrossRef]
51. Tan, S.; Heerink, N.; Qu, F. Land fragmentation and its driving forces in China. *Land Use Policy* **2006**, *23*, 272–285. [CrossRef]
52. Hartvigsen, M. Land reform and land fragmentation in Central and Eastern Europe. *Land Use Policy* **2014**, *36*, 330–341. [CrossRef]
53. Qian, F.; Chi, Y.; Lal, R.; Lorenz, K. Spatio-temporal characteristics of cultivated land fragmentation in different landform areas with a case study in Northeast China. *Ecosyst. Health Sustain.* **2020**, *6*, 1800415. [CrossRef]
54. Latruffe, L.; Piet, L. Does land fragmentation affect farm performance? A case study from Brittany, France. *Agr. Syst.* **2014**, *129*, 68–80. [CrossRef]
55. Han, J. Prioritizing agricultural, rural development and implementing the rural revitalization strategy. *China Agr. Econ. Rev.* **2020**, *12*, 14–19. [CrossRef]
56. Ciaian, P.; Guri, F.; Rajcaniova, M.; Drabik, D.; Paloma, S.G. Land fragmentation and production diversification: A case study from rural Albania. *Land Use Policy* **2018**, *76*, 589–599. [CrossRef]
57. Liu, L.; Liu, Z.; Gong, J.; Wang, L.; Hu, Y. Quantifying the amount, heterogeneity, and pattern of farmland: Implications for China's requisition-compensation balance of farmland policy. *Land Use Policy* **2019**, *81*, 256–266. [CrossRef]
58. Sklenicka, P. Classification of farmland ownership fragmentation as a cause of land degradation: A review on typology, consequences, and remedies. *Land Use Policy* **2016**, *57*, 694–701. [CrossRef]
59. Carsjens, G.J.; van Lier, H.N. Fragmentation and land-use planning—An introduction. *Landsc. Urban Plan.* **2002**, *58*, 79–82. [CrossRef]
60. Ntihinyurwa, P.D.; de Vries, W.T.; Chigbu, U.E.; Dukwiyimpuhwe, P.A. The positive impacts of farm land fragmentation in Rwanda. *Land Use Policy* **2019**, *81*, 565–581. [CrossRef]
61. Sklenicka, P.; Janovska, V.; Salek, M.; Vlasak, J.; Molnarova, K. The farmland rental paradox: Extreme land ownership fragmentation as a new form of land degradation. *Land Use Policy* **2014**, *38*, 587–593. [CrossRef]
62. Niroula, G.S.; Thapa, G.B. Impacts and causes of land fragmentation, and lessons learned from land consolidation in South Asia. *Land Use Policy* **2005**, *22*, 358–372. [CrossRef]
63. Zang, L.; Araral, E.; Wang, Y. Effects of land fragmentation on the governance of the commons: Theory and evidence from 284 villages and 17 provinces in China. *Land Use Policy* **2019**, *82*, 518–527. [CrossRef]

International Journal of
Environmental Research and Public Health

MDPI

Article

Spatial-Temporal Pattern and Evolution Trend of the Cultivated Land Use Eco-Efficiency in the National Pilot Zone for Ecological Conservation in China

Zhenggen Fan [1], Chao Deng [1], Yuqi Fan [1,*], Puwei Zhang [1] and Hua Lu [2,*]

1 College of City Construction, Jiangxi Normal University, Nanchang 330022, China; fanzg@jxnu.edu.cn (Z.F.); 202040100793@jxnu.edu.cn (C.D.); zhang.p.w@jxnu.edu.cn (P.Z.)
2 Institute of Ecological Civilization, Jiangxi University of Finance and Economics, Nanchang 330013, China
* Correspondence: 005603@jxnu.edu.cn (Y.F.); luhua@jxufe.edu.cn (H.L.)

Abstract: The cultivated land use eco-efficiency (CLUE) is an important indicator to evaluate ecological civilization construction in China. Research on the spatial-temporal pattern and evolution trend of the CLUE can help to assess the level of ecological civilization construction and reveal associated demonstration and driving effects on surrounding areas. Based on the perspective of the CLUE, this paper obtains cultivated land use data pertaining to National Pilot Zones for Ecological Conservation in China and neighboring provinces from 2008 to 2018. In this study, the SBM-undesirable, Moran's I, and Markov chain models are adopted to quantitatively measure and analyze the CLUE and its temporal and spatial patterns and evolution trend. The research results indicate that the CLUE in the whole study area exhibited the characteristics of one growth, two stable, and two decline stages, with a positive spatial autocorrelation that increased year by year, and a spatial spillover effect was observed. Geographical spatial patterns and spatial spillover effects played a major role in the evolution of the CLUE, and there occurred a higher probability of improvement in the vicinity of cities with high CLUE values. In the future, practical construction experience should be disseminated at the provincial level, and policies and measures should be formulated according to local conditions. In addition, a linkage model between prefecture-level cities should be developed at the municipal level to fully manifest the positive spatial spillover effect. Moreover, we should thoroughly evaluate the risk associated with CLUE transition from high to low levels and establish a low-level early warning mechanism.

Keywords: land use; CLUE; temporal and spatial evolution; spatial spillover; national pilot zone for ecological conservation in China

Citation: Fan, Z.; Deng, C.; Fan, Y.; Zhang, P.; Lu, H. Spatial-Temporal Pattern and Evolution Trend of the Cultivated Land Use Eco-Efficiency in the National Pilot Zone for Ecological Conservation in China. *Int. J. Environ. Res. Public Health* **2022**, *19*, 111. https://doi.org/10.3390/ijerph19010111

Academic Editors: Wei Song and Hualin Xie

Received: 4 November 2021
Accepted: 21 December 2021
Published: 23 December 2021

1. Introduction

With increasing social and economic development levels, cultivated land notably functions as the basic means of agricultural production, provides ecological products, and plays a significant role in ensuring both national food and ecological security. However, with the rapid advancement of industrialization and urbanization in China, cultivated land also faces difficulties such as sharp reductions in quantity and quality, and idle abandonment. On the premise of ensuring national food security, this has led to changes in the cultivated land input and production structure, and an agricultural production mode dominated by petrol-agriculture has gradually been established. The accompanying changes in cultivated land use intensity have significantly undermined the integrity of biodiversity [1], and seriously threatened the quality of cultivated land habitat [2], accordingly resulting in food security problems in China [3]. According to the Second National Pollution Source Census released by China in 2020, in 2017, the ammonia nitrogen emissions of the planting industry in China reached 83,000 tons, the total phosphorus emissions reached 76,200 tons, the use of plastic films reached 1,419,300 tons, and the accumulated residue reached 1,184,800 tons over time. The anti-ecological effect of cultivated land utilization has gradually accumulated, thereby seriously restricting the green development of cultivated land utilization in China. Based on these aspects, the 18th Communist Party of China (CPC) National Congress report clearly proposed the five-in-one overall layout accounting for ecological civilization construction, which requires comprehensive consideration of regional economic development and ecological civilization construction to promote coordinated development. Moreover, to further promote the implementation of ecological civilization construction at the national level, the Central Committee of the CPC further selected and deployed Fujian, Jiangxi, and Guizhou provinces with a good ecological foundation and a high resource and environmental carrying capacity as the first batch of National Pilot Zones for Ecological Conservation in China to take the lead in exploration and provide model experience for ecological civilization construction in other regions. Agriculture constitutes an essential part of promoting regional ecological civilization construction, and cultivated land is the primary material carrier and production factor. Therefore, optimization of the input-output structure of cultivated land use and improvement of the cultivated land use efficiency have become critical paths to realize regional agricultural ecological civilization construction. Therefore, within this context, methods to enhance ecological use of cultivated land and ensure the coupling and coordination between cultivated land use and ecological environment have become critical aspects to promote ecological civilization construction.

The concept of eco-efficiency was proposed in 1990. It is an important indicator for measuring the construction of regional ecological civilization. It refers to the ratio of economic growth to environmental impact and emphasizes the coordinated development of economic growth and ecological environment [4]. In 1996, the World Business Council for Sustainable Development (WBCSD) further deepened and expanded the concept, and proposed that in the process of resource consumption, it is not only required to meet the basic needs of human society, but also to ensure that the ecological environmental impact is consistent with the environmental carrying capacity of the Earth [5]. This concept takes into account both social and economic development and resource environmental protection, which effectively solves the problem of how to quantify the two at the same level, and so it has been gradually expanded and applied by various institutions and scholars in different fields. The existing studies are mainly focused on the basic theory of eco-efficiency [6–8] and practical applications in different fields. The application fields cover agriculture [9,10], industry [11,12], manufacturing industry, etc. [13,14], and the application areas cover cities [15,16], regions [17,18], and countries [19]. Among them, the practical applications involved in the field of land use mainly focus on land management, cultivated land compensation, intensive land use, land use zoning, land use transformation, etc. [20–24].

The eco-efficiency of land use, as an important indicator to quantify the construction of ecological civilization in the field of land use, accurately reflects the degree of coordination between regional land use and the ecological environment, and accordingly has been widely applied and implemented by scholars. The existing studies mainly focus on the design of methods for measuring the eco-efficiency of land use, and the analysis of the spatial and temporal characteristics of regional eco-efficiency of land use and its influencing factors. The measurement methods mainly include ecological footprint [21,25], principal component analysis (PCA) [12,26] and data envelopment analysis (DEA), etc. [27–29]. Among these methods, the DEA-SBM model derived on the basis of DEA does not need to set the specific form and estimation parameters of the model in advance, and can effectively solve the slack problem of input and output variables, and thus has become the mainstream model for measuring the eco-efficiency of land use in the current academic circles. As regards the evolution of temporal and spatial characteristics, scholars mainly analyzed the temporal and spatial variation in eco-efficiency of land use [30,31], and summarized and elaborated the temporal and spatial evolution rules of eco-efficiency of land use in a specific study area [32,33]. In terms of the analysis of influencing factors, existing studies generally incorporate socio-economic development situations, the marketization level, the industrial development state, and the ecological input level into the analysis system of factors influencing the eco-efficiency of land use [33–37]. Specifically, the economic development level, industrial agglomeration and openness, and ecological input are considered to exert a positive effect on promoting the eco-efficiency of land use [33–36], while the land marketization level, urban-rural income gap, and ecological pressure are considered to play a negative hindering role in the eco-efficiency of land use [33,36,37].

As one of the most fundamental elements of agricultural production, cultivated land not only has the social and economic service functions of producing food to ensure the regional food security, but also has ecological service functions such as conserving water and soil resources, regulating climate, and protecting biodiversity [38,39]. Therefore, to strengthen the multi-functional value of cultivated land, quantify the environmental efficiency loss caused by cultivated land use, and realize the sustainable use of cultivated land, some scholars gradually shifted their research perspectives on the eco-efficiency of land use to the cultivated land use eco-efficiency (CLUE). At present, the research achievements of the CLUE are mainly concentrated on the measurement of application methods, including two categories: the ecological footprint method [25] and the DEA method [27]; at the same time, some scholars analyzed factors influencing the CLUE [34].

The above relevant research laid a solid theoretical foundation for this paper, but there is still room for further research in the following two aspects: (1) cultivated land use is an important activity in the process of agricultural production. Exploring the CLUE is highly important to promote agricultural ecological civilization construction. However, there exists relatively little research in this field in current academic circles, and it is difficult to establish a policy system, especially given the current measures aimed at vigorously promoting the experimental area of ecological civilization in China. The lack of relevant research on National Pilot Zones for Ecological Conservation in China makes it challenging to achieve a demonstration effect at the national level, which readily limits the effectiveness of these zones. (2) Existing research has only measured the CLUE within a certain region to analyze temporal and spatial evolution patterns, but the spillover effect has not been sufficiently explored. However, according to the first law of geography, each object in a geographical space exhibits a specific spatial autocorrelation. Therefore, in CLUE exploration research, we should consider the inherent spillover phenomenon. Based on this consideration, we can more accurately describe the temporal and spatial evolution rules of the CLUE.

Based on the above analysis of existing relevant research, the aims of this paper can be summarized as follows: (1) first, this paper aims to construct a CLUE evaluation index system from an ecological perspective. Taking the first batch of National Pilot Zones for Ecological Conservation in China and surrounding related prefecture-level cities as the research area, it measures the CLUE from 2008 to 2018 using the SBM-undesirable model, and analyzes the spatial-temporal evolution characteristics of the CLUE. (2) Then, traditional and spatial Markov chain models are constructed to empirically examine the evolution trend and the spatial spillover effect of the CLUE within the study area, and analyze the formation process of regional spatial agglomeration phenomena and the spatial spillover effect. (3) Finally, the fundamental strategies for low-carbon, low-pollution and efficient cultivated land use are put forward, and development ideas are assessed for agricultural ecological construction in National Pilot Zones for Ecological Conservation in China to provide policy recommendations for efficient cultivated land use and agricultural ecological construction in these zones.

2. Research Methods and Data Sources

2.1. Overview of the Study Area

Jiangxi, Fujian, and Guizhou comprise the first batch of National Pilot Zones for Ecological Conservation in China, with an excellent ecological environment, but the ecological advantage does not directly match the economic advantage. The economic development process of areas with a certain ecological advantage has occurred at a disadvantage for a long time [40], which is quite evident in the agricultural economy. Moreover, with increasing agricultural mechanization levels in China, regional carbon emissions have continuously risen. In addition, farmers use land intensively, but cultivated land is lightly maintained in the process of cultivated land utilization, while large amounts of chemical fertilizers and pesticides are applied, resulting in increasing pressure on regional resources and the environment and worsening environmental problems. In addition, ecological civilization construction is facing challenges. Therefore, in response to the call to reduce the application of chemical fertilizers and pesticides, decrease the discharge of harmless agricultural solid waste and improve green ecological agriculture, as proposed by the state in the Implementation Plan for National Ecological Civilization Pilot Zones, we must correctly understand the coupling and coordination of economic development and ecological protection in cultivated land utilization and strive to realize the unification and optimization of ecological, economic and social benefits, thereby promoting the construction of National Pilot Zones for Ecological Conservation in China. Based on these considerations, this paper selects 57 prefecture-level cities in the first batch of National Pilot Zones for Ecological Conservation in China (Jiangxi, Fujian, and Guizhou), in addition to Anhui and Hunan provinces (Figure 1), as the research object, studies the temporal and spatial patterns and evolution trend of the CLUE from the perspective of comparative analysis and provides theoretical support to improve the construction and leading role of National Pilot Zones for Ecological Conservation in China.

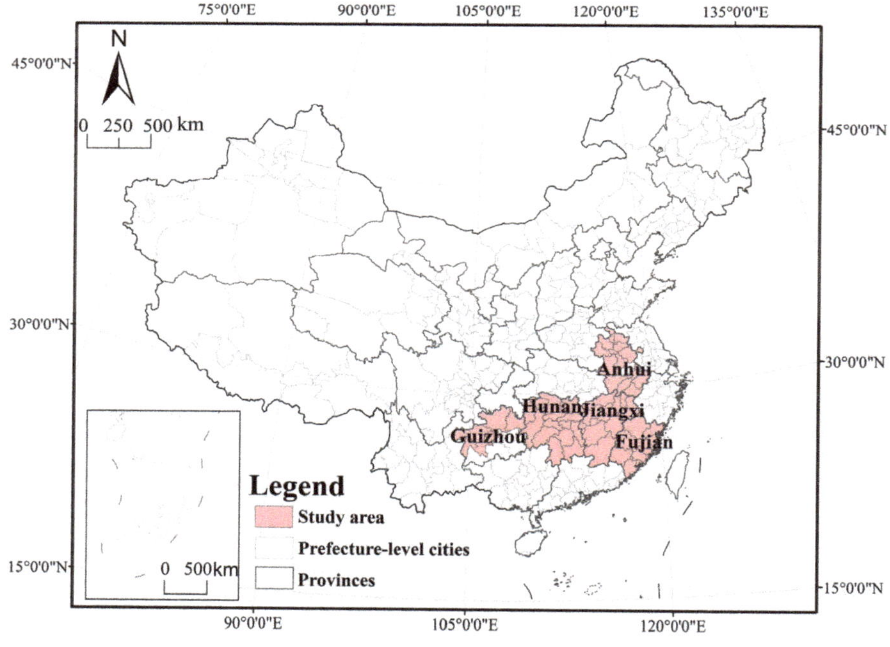

Figure 1. The geographical location of study area.

2.2. Research Methods

2.2.1. SBM-Undesirable Model

The SBM-undesirable model can solve the problem of efficiency measurement, including unexpected outputs, and can avoid result deviations due to radial and angular problems. This model is a scientific evaluation method established through the continuous improvement of the traditional DEA model according to practical experience. It has been widely used in efficiency evaluation in various research fields [41–43], and it is also a mainstream measurement method for land use efficiency evaluation in recent years [44–46]. Because CLUE calculation results can differ between variable return to scale (VRS) and constant return to scale (CRS), combined with the research of Zhou P. et al. [47–50], the SBM-undesirable model based on the VRS is adopted, which can better reflect the essence of the CLUE. Its basic principles are expressed as follows:

In the process of cultivated land use, it is assumed that there are n decision-making units, all corresponding to m input indicators x_{i0}, a expected output indicators y^n_{r0} and b unexpected output indicators y^n_{h0}. Then, matrices X, Y^e, and Y^n can be defined as $(x_1, x_2, \ldots, x_n) \in R^{m\times n}$, $(y^e_1, y^e_2, \ldots, y^e_n) \in R^{a\times n}$, $(y^n_1, y^n_2, \ldots, y^n_n) \in R^{b\times n}$, respectively. Moreover, assuming that X, Y^e, and Y^n are greater than zero, the production possibility set can be defined as $P_{ij}(N)$, and the SBM-undesirable model can be expressed as:

$$\rho^* = \min \frac{1 - \frac{1}{m}\sum\limits_{i=1}^{m} \frac{D^-_i}{x_{i0}}}{1 + \frac{1}{a+b}\left(\sum\limits_{r=1}^{a} \frac{D^e_r}{y^e_{r0}} + \sum\limits_{h=1}^{b} \frac{D^n_h}{y^n_{h0}}\right)} \tag{1}$$

$s.t. x_0 = X\lambda + D^-, y^e_0 = Y^e\lambda - D^e$, and $y^n_0 = Y^n\lambda + D^n$.
$D^- \geq 0, D^e \geq 0, D^n \geq 0$, and $\lambda \geq 0$.

In Equation (1), ρ^* denotes the CLUE value in each region within the study area, and the value ranges from 0 to 1. When the ρ^* value is 1, this indicates that the process is entirely effective. For $\rho^* < 1$, efficiency loss occurs, and there exists room for further optimization. The number of inputs and the expected and unexpected outputs are denoted as m, a, and b, respectively, and the corresponding slack variables are denoted as D^-, D^e, and D^n, respectively. Furthermore, the corresponding input-output values are denoted as x_{i0}, y^n_{r0}, and y^n_{h0}, respectively, and λ denotes the weight vector.

2.2.2. Spatial Autocorrelation Model

Spatial autocorrelation analysis is an effective method to describe spatial correlation and spatial heterogeneity by panel data. Spatial geographical relations are integrated into data analysis through global Moran's I [51–53]. The global Moran's index (Moran's I) reflects the spatial correlation characteristics of the CLUE from a global perspective, and the model is expressed in Equation (2):

$$\mathrm{I} = \frac{\sigma^2 \sum\limits_{i=1}^{n}\sum\limits_{j=1}^{n} w_{ij}(x_i - \overline{x})(x_j - \overline{x})}{\sigma^2 \sum\limits_{i=1}^{n}\sum\limits_{j=1}^{n} w_{ij}} \tag{2}$$

In Equation (2), I is Moran's I, n is the total number of evaluation units in the study area, x_i and x_j are the attribute values of evaluation units i and j, respectively ($i = j$), $\overline{x}$ is the average CLUE value of the evaluation unit, and σ^2 is the sample variance. For $\mathrm{I} > 0$, a positive spatial correlation exists in terms of the CLUE. For $\mathrm{I} < 0$, a negative spatial correlation exists. The magnitude of the positive or negative I values reflects the degree of spatial positive or negative correlation, respectively, and w_{ij} is the spatial weight matrix, which reflects the spatial adjacency relationship between the evaluation units. The data in this paper are based on panel data pertaining to the study area, and the Queen adjacency matrix based on GeoDa is adopted. Then, a certain criterion is applied to construct the spatial weight matrix.

The global spatial autocorrelation reflects the average correlation and the different degrees of the CLUE in the overall space but cannot reflect the specific characteristics of local spatial aggregation or differentiation. Therefore, to implement the local spatial autocorrelation method for analysis purposes, this paper adopts local Moran's I, and the calculation equation is as follows:

$$I_i = \frac{(x_i - \overline{x})}{\sigma^2} \sum_{j=1}^{n} w_{ij}(x_j - \overline{x}) \tag{3}$$

In Equation (3), I_i is the local Moran's I of evaluation unit i. Positive or negative I_i values correspond to adjacent areas with similar or different CLUE values, respectively. The absolute value of I_i reflects the degree of spatial proximity.

2.2.3. Markov Chain Model

The Markov chain model determines the change trend of each state of objects through the initial probability of different states and the transition probability between states. In Environmental Science, it is applied to analyze the spatial-temporal dynamic evolution characteristics of things [54,55]. In this paper, traditional and spatial Markov chain models are adopted for analysis.

According to the state type of the CLUE, the traditional Markov chain model can construct an $N \times N$-order Markov probability transfer matrix to analyze the temporal evolution characteristics of regional CLUE values. Assuming that P_{ij} is the transition probability of the CLUE of a given unit in the study area from state E_i to state E_j from year t to year t + 1, the value can be estimated with Equation (4), as follows:

$$P_{ij}(E_i \rightarrow E_j) = \frac{n_{ij}}{n_i} \tag{4}$$

In Equation (4), n_{ij} denotes the total number of regional units as the state type of the CLUE transitions from E_i to E_j, and n_i denotes the number of regional units with E_i occurring at the i level.

The spatial Markov chain model combines the traditional Markov chain with the concept of the spatial lag, which can explore the mechanism of the spatial spillover effect in the temporal and spatial transfer processes of the CLUE and can be applied to analyze the possibility of CLUE transfer against different geospatial backgrounds to explore the internal relationship between the evolution process of the CLUE and the regional background. Under the condition of spatial lag N_i, the traditional $N \times N$-order Markov probability transfer matrix is decomposed into an $N \times N \times N$-order probability transfer matrix. $P_{ij}(N)$ indicates that under the condition of spatial lag N_i, the possibility of CLUE transfer shifts from type E_i into type E_j.

2.3. *Index System Construction and Data Sources*

Referring to relevant research results [37,56], the evaluation index system of the CLUE constructed from an ecological perspective should cover three aspects, namely, the input, expected output, and unexpected output, of the four systems of resources, economy, nature, and society. The constructed index system of the CLUE is provided in Table 1.

In terms of the input index, the actual sowing area of crops (1000 hm^2), number of employees (10,000 persons), net amount of pesticide application (t), and net amount of chemical fertilizer application (t) were selected as representative indicators. In terms of the expected output index, the total grain output (t) and planting output value (10,000 yuan) were selected as representative indicators. In terms of the unexpected output index, the difference between the total carbon emissions (t) and total carbon absorption (t) was selected as a representative indicator.

The basic input and output data required to measure the CLUE in the study area were retrieved from the *China Statistical Yearbook 2009–2020*, *China Rural Statistical Yearbook 2009–2020*, provincial and municipal statistical yearbooks, and statistical bulletins. The acquired carbon emission data were related to chemical fertilizers, pesticides, and agricultural films. These data were obtained by multiplying and summarizing basic data, such as mechanized operation and cultivated land plowing data. The carbon emission coefficients of the various carbon sources were determined based on the carbon emission model and calculation coefficients of West and Marland et al. [57,58]. The carbon absorption coefficient of cultivated land was set to 0.0070 t/hm, as reported by He Yong et al. [59]. Correlation measurement coefficients were obtained from Liang Liu Tao and Feng Yonggang et al. [60,61].

Table 1. Evaluation index system of the CLUE.

Variable Type	Variable	Index Meaning
Input index	Cultivated land input	Actual sown area of crops/1000 hm^2
	Labor input	Number of employees in the primary industry × (agricultural output value/total output value of agriculture, forestry, animal husbandry, and fishery)/10,000
	Pesticide and fertilizer input	Net amount of pesticide and chemical fertilizer application/t
Expected output index	Agricultural output value	Output value of the planting industry/10,000 yuan
	Grain yield	Total grain output/t
Unexpected output index	Net carbon emissions	Difference between the total carbon emissions of mechanical operation and chemical fertilizer and pesticide application and the total carbon absorption of cultivated land/t

3. Analysis of the Empirical Results

3.1. Temporal Dynamic Evolution Characteristics of the CLUE

The CLUE was measured with DEA-SOLVER PRO13 software. According to the overall observations, the CLUE in the study area was notably different from 2008 to 2018, thereby exhibiting the characteristics of one growth, two stable, and two decline stages (Figure 2).

The growth stage suggests that the CLUE in Guizhou Province experiences a growth trend, with an average annual growth rate of 3.21%, which occupies the leading position in the region (Figure 2). This suggests that ecological civilization construction in Guizhou Province achieved remarkable results in cultivated land utilization. From 2008 to 2018, the average input of pesticides and chemical fertilizers in Guizhou Province reached 187.42 t/1000 hm^2, the lowest in the whole region, 27.3% lower than that in Jiangxi Province, which ranked second-lowest. In terms of the land average net carbon emissions, the land average net carbon emissions in Guizhou Province decreased from 238.87 T/1000 hm^2 in 2008 to 1869.18 T/1000 hm^2 in 2018. These land average net carbon emissions were the lowest in the study area, and Guizhou Province was the only province indicating a decline in emissions. In terms of the average annual growth rate of the agricultural output value, Guizhou Province attained a rate of 15.79%, while Jiangxi, Fujian, Anhui, and Hunan attained rates of 7.82%, 7.57%, 1.49%, and 0.24%, respectively, of which the annual average growth rate in Guizhou Province was much higher than that in the other provinces within the study area. Based on the above three groups of data, it is observed that although the initial average agricultural output value in Guizhou Province was the lowest, the expected output value growth rate was the highest, and the average pesticide and chemical fertilizer input and average net carbon emissions remained the lowest. Moreover, these findings are the main reasons why the CLUE in Guizhou Province has taken the lead in the study area in recent years.

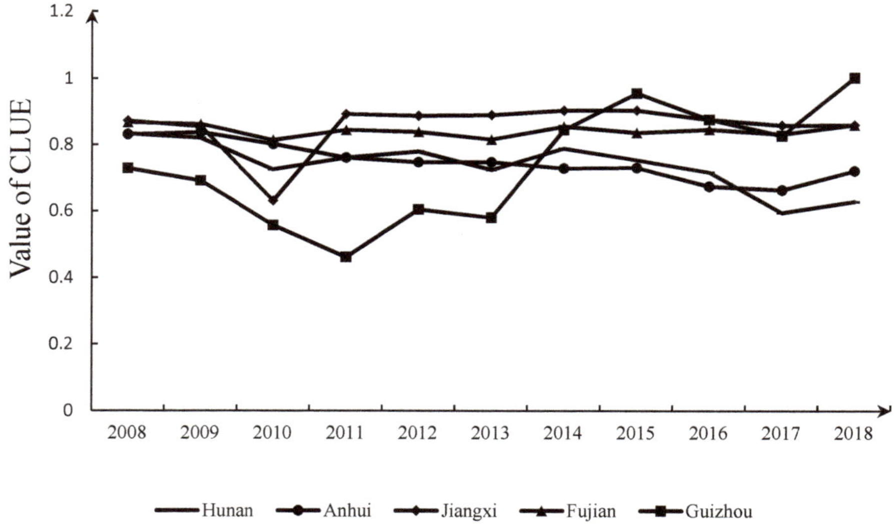

Figure 2. Change trend of the CLUE in the study region from 2008 to 2018.

The two stable stages and two decline stages suggest that the CLUE exhibited the characteristics of high and stable fluctuations in Jiangxi and Fujian and an overall downward trend in Hunan and Anhui, respectively (Figure 2). The average CLUE values in Jiangxi and Fujian were 0.856 and 0.842, respectively, ranking as the top two highest values in the study area, but the fluctuation range was smaller than 2%. The CLUE in Hunan and Anhui revealed a downward trend. In 2018, the CLUE in these two provinces reached 0.62 and 0.719, declines of 23.8% and 13.4%, respectively, compared to 2008. The reason why the CLUE in Jiangxi and Fujian remained high with stable fluctuations could be that the ecological basis of cultivated land use in these two provinces is good. The average grain production in Jiangxi and Fujian provinces from 2008 to 2018 reached 4113.6 and 2894.3 t/1000 hm^2, respectively, higher than the average grain production values of 2473.8 t/1000 hm^2 in Hunan and 2136.4 t/1000 hm^2 in Anhui Province. Moreover, the average annual net carbon emissions in these two provinces reached 1,955,700 and 1,642,300 tons, respectively, lower than those in Hunan (3,379,500 tons) and Anhui (3,942,800 tons). Compared to the data for Jiangxi and Fujian provinces within the ecological civilization construction experimental area in 2018, Anhui and Hunan provinces exhibited room for improvement by approximately 19% and 36%, respectively, in the CLUE.

3.2. *Spatial Evolution Characteristics of the CLUE*

3.2.1. Overall Spatial Evolution Characteristics

Table 2 indicates that the global spatial autocorrelation of the CLUE within the whole region is increasing, and the spatial correlation between adjacent regions is increasingly intensifying. From 2008 to 2018, global Moran's I value of the CLUE increased from 0.136 to 0.323, and the significance test result increased from 5% to 1%, indicating a fluctuating upward trend. This suggests that there occurs a significant positive spatial autocorrelation in regard to the CLUE in the study area.

Table 2. Global Moran's I of the CLUE in the study area from 2008 to 2018.

Year	Global Moran's I	Z-Value	*p-Value*
2008	0.136	1.759	0.039
2009	0.1869	2.1414	0.032
2010	0.1493	1.9436	0.029
2011	0.3403	3.9418	0.003
2012	0.2308	2.8181	0.004
2013	0.2362	2.7097	0.003
2014	0.1942	2.3543	0.009
2015	0.2561	3.0089	0.007
2016	0.191	2.3219	0.009
2017	0.2572	2.9803	0.002
2018	0.3234	3.7426	0.001

3.2.2. Local Evolution Characteristics of the CLUE

To further analyze the specific spatial agglomeration characteristics of the CLUE, according to the calculation results of local Moran's I at the 10% significance level, a Local Indicators of Spatial Association (LISA) cluster diagram of the CLUE in the study area for 2008, 2013, and 2018 was generated (Figure 3). As shown in the figure, the CLUE in the study area exhibited significant high-high (H-H) and low-low (L-L) agglomeration phenomena within the geographical space encompassing prefecture-level cities, which became increasingly significant over time.

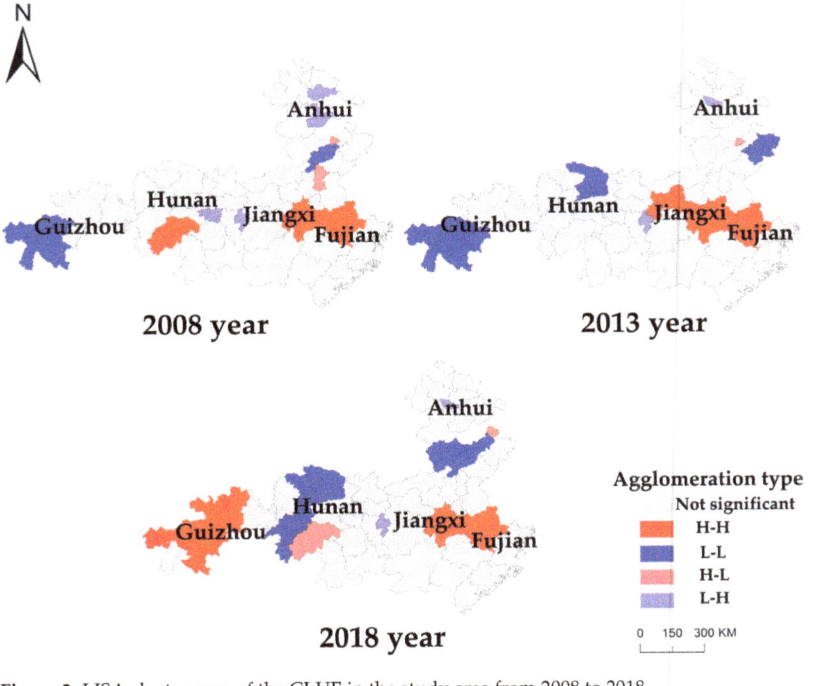

Figure 3. LISA cluster map of the CLUE in the study area from 2008 to 2018.

In 2008, the H-H and L-L agglomeration areas of the CLUE indicated the single-core agglomeration phenomenon. An H-H single-core aggregation area was located at the junction of Jiangxi and Fujian provinces in the National Pilot Zone for Ecological Conservation in China, comprising Fuzhou and Nanping. An L-L single-core aggregation area was located in western Guizhou Province, comprising Bijie, Liupanshui, and Anshun. The degree of single-core agglomeration was low in 2008.

In 2013, the H-H and L-L agglomeration areas of the CLUE revealed an increased scale of single-core agglomeration. An H-H single-core agglomeration area was still located at the junction of Jiangxi and Fujian provinces, comprising Yingtan, Yichun, Fuzhou, and Nanping. An L-L single-core agglomeration area was located in the west of Guizhou Province and comprised Guiyang, Bijie, Liupanshui, and Anshun. Compared to 2008, the geographical location of the H-H and L-L single-core agglomeration areas did not shift. According to the number of cities, the agglomeration scale expanded by 100% and 33.3%, respectively, and the positive spatial spillover effect was notable.

In 2018, the H-H and L-L agglomeration areas of the CLUE demonstrated the double-core agglomeration phenomenon. H-H dual-core agglomeration areas were located at the junction of Jiangxi and Fujian provinces and western Guizhou. The agglomeration area in western Guizhou Province represented a new agglomeration area. This agglomeration area comprised Guiyang, Zunyi, Bijie, and Anshun, exhibiting the characteristics of east–west H-H and dual-core agglomeration encompassing six cities. The L-L aggregation area shifted, and a new L-L dual-core aggregation area comprising seven cities was formed in Hunan and Anhui.

In summary, from 2008 to 2018, the spatial agglomeration of the CLUE in the study area exhibited the characteristics of agglomeration core-based expansion and transfer. In terms of agglomeration core-based expansion, a development trend was observed from two to four cores. In terms of agglomeration core-based transfer, the L-L agglomeration phenomenon in western Guizhou evolved into an H-H agglomeration phenomenon. The observed agglomeration phenomena indicated that there occurred a spatial spillover effect in the study area. In particular, when an adjacent area was observed with a high (low) level of the CLUE, the target area was more likely to become an area with a high (low) CLUE level. As of 2018, H-H aggregation areas of the CLUE were distributed among Fujian, Jiangxi, and Guizhou provinces, indicating that the positive spatial spillover effect of the National Pilot Zone for Ecological Conservation in China was notable. However, the effect was largely distributed within the National Pilot Zone for Ecological Conservation in China, and the driving effect on the surrounding provinces and cities of the National Pilot Zone for Ecological Conservation in China was not notable.

3.3. Markov Chain Analysis of the CLUE in the Study Area

3.3.1. Traditional Markov Chain Analysis

According to the quantile division method, thereby adopting the first, second, and third quantiles as boundaries, the 57 prefecture-level cities in the study area from 2008 to 2018 were divided into four adjacent but nonintersecting state spaces with low, medium-low, medium-high, and high efficiency values according to the difference in the CLUE, denoted as levels I, II, III and IV, respectively. The probability transition matrix of the traditional Markov chain analysis method was thus obtained (Table 3).

Table 3. Traditional Markov chain probability transition matrix of the CLUE in the study area from 2008 to 2018.

Local Status	Type I	Type II	Type III	Type IV
	<25%	25–50%	50–75%	>75%
Type I	0.832	0.117	0.007	0.044
Type II	0.190	0.647	0.085	0.085
Type III	0.058	0.385	0.365	0.192
Type IV	0.018	0.048	0.079	0.855

The CLUE in the whole region generally exhibited a consistent trend, and the convergence phenomenon was observed in regard to the extreme value. It was difficult to achieve a significant improvement over the short term, and there existed a certain transfer risk of to the medium-low state.

In terms of state maintenance, the probability along the diagonal at levels I, II and IV of the state space was significantly higher than that along the nondiagonal. Notably, the transformation probability of the CLUE at the same level was much higher than that between the different levels, in which the minimum value reached 0.647 and the maximum value reached 0.832. Under the above conditions, the CLUE attained a probability of at least 64.7% in the future development process and remained at the same level.

In terms of extreme value convergence, Table 3 demonstrates that the probability values of maintaining the current CLUE level along the diagonal followed the order of $P_{IV\text{-}IV}$ (0.855) > $P_{I\text{-}I}$ (0.832) > $P_{II\text{-}II}$ (0.674) > $P_{III\text{-}III}$ (0.365), and the probability values at both ends of the diagonal were significantly higher than the median value, indicating that the CLUE values were characterized by H-H and L-L agglomeration patterns, i.e., the core convergence phenomenon occurred.

In terms of efficiency improvement, except for $P_{III\text{-}II}$ (0.385), the state transition probability along the nondiagonal was significantly lower than that along the diagonal, of which the maximum value reached 0.192 and the minimum value reached 0.007, indicating that it is difficult to greatly and rapidly improve the CLUE within a short timeframe. Long-term and effective ecological civilization construction is thus needed.

In terms of risk prediction, the CLUE could indicate a certain transfer risk from medium-high to medium-low efficiency values over the short term. The probability of transferring from type III to type II was significantly higher than that of transferring from type III to types IV and I ($P_{III\text{-}II}$ (0.385) > $P_{III\text{-}III}$ (0.365) > $P_{III\text{-}IV}$ (0.192) > $P_{III\text{-}I}$ (0.058)). These areas are more likely to fall into the low eco-efficiency trap.

3.3.2. Spatial Markov Chain Analysis

Figure 3 shows that the spatial pattern of the CLUE in the study area exhibits significant spatial agglomeration characteristics. Therefore, a spatial lag was incorporated into the traditional Markov chain model, and a spatial Markov probability transfer matrix was constructed based on the spatial lag type of each regional unit in the first year. Similarly, according to the quantile division method, the spatial lag types in the study area were divided into four types, namely low, medium-low, medium-high, and high, denoted as types I, II, III and IV, respectively. The analysis results are listed in Table 4 below. Through comparison with the traditional Markov probability transfer matrix, the following spatial evolution characteristics of the CLUE could be obtained after considering the geospatial background:

Geospatial patterns play a significant role in the dynamic evolution process of the CLUE. Against the neighborhood background entailing different efficiency levels, the CLUE transfer probability varies, and the transfer probability further differs from that determined according to the corresponding traditional Markov probability transfer matrix. For example, in the traditional Markov probability transfer matrix, the transition probability of the CLUE level from type III to type II is the highest, at $P_{III\text{-}II}$ = 0.385, while in the spatial Markov probability transfer matrix, when a location is adjacent to a type-I area, the transition probability of the CLUE level from type III to type II is $P_{III\text{-}II\,(I)}$ = 0.333, which is lower than the probability that the CLUE level remains unchanged. When a given location is adjacent to a type II area, the transition probability is $P_{III\text{-}II\,(II)}$ = 0.400, higher than the probability that the CLUE level remains unchanged. When the adjacent area is a type-IV area, the transition probability is $P_{III\text{-}II\,(IV)}$ = 0.411 = $P_{III\text{-}III\,(IV)}$. The probability that the CLUE level remains unchanged is the same as the transition probability to a medium-low efficiency level. Therefore, geographical spatial patterns can exert a significant impact on CLUE evolution.

Table 4. Markov chain probability transition matrix of the CLUE in the study area from 2008 to 2018.

Spatial Lag	Local Status	Type I	Type II	Type III	Type IV
		<25%	25–50%	50–75%	>75%
Type I	I	0.770	0.148	0.000	0.082
	II	0.231	0.513	0.051	0.205
	III	0.000	0.333	0.444	0.222
	IV	0.077	0.077	0.077	0.769
Type II	I	0.854	0.122	0.000	0.024
	II	0.179	0.678	0.143	0.000
	III	0.100	0.400	0.300	0.200
	IV	0.000	0.048	0.065	0.887
Type III	I	0.926	0.074	0.000	0.000
	II	0.175	0.750	0.050	0.025
	III	0.063	0.375	0.313	0.25
	IV	0.016	0.049	0.098	0.836
Type IV	I	0.750	0.125	0.125	0.000
	II	0.139	0.639	0.111	0.111
	III	0.059	0.411	0.411	0.118
	IV	0.001	0.050	0.075	0.863

The spatial spillover effect plays a vital role in the dynamic transfer process of the CLUE. Generally, the transfer probability of the CLUE to reach a low level increases in areas adjacent to cities with low CLUE values, while the transfer probability of the CLUE to reach a high level increases at locations adjacent to areas with high CLUE values. For example, near areas with low CLUE values, $P_{II\text{-}I\,(I)}$ (0.231) > $P_{II\text{-}I}$ (0.190) and $P_{III\text{-}II\,(I)}$ (0.333) > $P_{III\text{-}II}$ (0.385), while at locations near areas with high CLUE values, $P_{I\text{-}II\,(IV)}$ (0.125) > $P_{I\text{-}II}$ (0.117) and $P_{II\text{-}III\,(IV)}$ (0.111) > $P_{II\text{-}III}$ (0.085). Areas with low CLUE values exert a negative impact on surrounding areas, while areas with high CLUE values impose a positive spillover effect on adjacent areas. Adopting Chizhou city, Anhui Province, and Ganzhou city, Jiangxi Province, as examples, from 2008 to 2018, the average CLUE value in the surrounding cities of Chizhou reached only 0.66, which affected the decline in CLUE in Chizhou city from 0.64 to 0.48 to a certain extent. The average CLUE value in the surrounding cities of Ganzhou reached as high as 0.82, which increased the CLUE in Ganzhou from 0.61 to 0.75.

4. Discussion

4.1. Policy Recommendations

Based on the above research results and analysis conclusions, the construction of National Pilot Zones for Ecological Conservation in China provides a suitable practical foundation and development trend in terms of cultivated land utilization. Through the spatial spillover effect at the municipal scale, CLUE improvement can be achieved in both cities within the National Pilot Zone for Ecological Conservation in China and surrounding areas, its demonstration and leading roles can be fully manifested, and national ecological civilization construction can be promoted. Based on these findings, the following policy recommendations are outlined:

Based on the determined unbalanced development of the CLUE at the provincial level, it is suggested to widely disseminate model construction experience, formulate provincial ecological measures, and realize CLUE enhancement in surrounding areas of National Pilot Zones for Ecological Conservation in China. Anhui and Hunan should learn from the construction experience of ecological civilization pilot areas similar to Jiangxi, Fujian, and Guizhou, including the formulation and implementation of safe utilization schemes of polluted cultivated land, the establishment of a classification list of cultivated land soil environmental quality categories, and the construction of a responsibility assessment system involving four-level cultivated land protection objectives at the provincial, municipal, county and township levels. Hence, policies and measures should be designed according to local conditions to reverse the downward trend of the CLUE to narrow the gap with areas exhibiting high CLUE values.

Based on the mechanism of the spatial adjacency spillover effect at the prefecture-level city scale, it is suggested to fully manifest the positive spillover effect of H-H CLUE agglomeration areas and establish a city linkage model. By improving the breadth and depth of opening-up policies, strengthening agricultural cooperation, resource flow, and personnel exchange processes, promoting coordinated and balanced development of the whole region, and learning from the improvement model of the CLUE in Ganzhou City, Jiangxi Province, under the influence of the positive spillover effect of neighboring cities, low-level areas should be encouraged to overcome their dilemmas.

Based on the downward transfer trend of the CLUE in the evolution process, it is recommended to thoroughly assess the transfer risk from high to low levels, establish a low-level early warning mechanism, and prevent downward transformation of the regional eco-efficiency. A cultivated land ecological utilization evaluation organization shall be established to promptly adjust and optimize the input-output structure of cultivated land, such as the allocation of low-carbon fertilizers and the optimization of the operation time of cultivated land machinery, based on the monitoring of cultivated land carbon emissions sources such as pesticides, chemical fertilizers and agricultural films and the prediction of cultivated land output, so as to ensure synchronous ecological construction in the process of cultivated land utilization.

4.2. Research Limitation and Future Research

In this paper, an evaluation index system of the CLUE is constructed, the SBM-undesirable model is adopted to measure the CLUE, and the spatial spillover effect of the CLUE is analyzed. The following two points should be further examined:

The undesired indicators in the eco-efficiency index system of cultivated land use constructed in this paper only consider carbon emissions, and do not consider nonpoint source pollution. In fact, the use of cultivated land can not only produce a large amount of carbon dioxide, but can also produce nonpoint source pollution, such as water and soil pollution attributed to pesticides and fertilizers. However, due to the regional characteristics of the correlation coefficient of nonpoint source pollution measurements, most current studies consider the same coefficient in China. The lack of regional characteristics may affect the accuracy of regional CLUE assessment. Based on this aspect, this paper only considers carbon emissions to ensure the accuracy of the measurements. The nonpoint source pollution emission coefficient based on regional characteristics should be measured in future research. Nonpoint source pollution and carbon emissions could be incorporated into the unexpected output index in the cultivated land use process to measure the CLUE more accurately.

Based on the prefecture-level city scale, this paper studies the temporal and spatial evolution characteristics and spatial spillover effect of the CLUE from the perspective of comparative analysis, thereby choosing 57 prefecture-level cities in Jiangxi, Fujian, Guizhou, Hunan, and Anhui as the research objects. In the future, microscale multilevel research can be carried out at the farm scale. In fact, as the main body of cultivated land use, based on the research scale, farms can better refine the input-output characteristics of cultivated land use, the cultivated land planting behavior of farmers can be elucidated, and the efficiency value can be determined more accurately. This could provide a greater reference value for follow-up optimization of cultivated land use and ecological civilization construction. Moreover, exploring the eco-efficiency of regional cultivated land use at the meso- and microscales could more comprehensively explain the construction effect of cultivated land use in National Pilot Zones for Ecological Conservation in China and provide a theoretical basis for other regions to learn from the model experience gained in these zones.

5. Conclusions

Based on the obtained panel data of cultivated land use pertaining to 57 cities in Jiangxi, Fujian, Guizhou, Hunan, and Anhui from 2008 to 2018, this paper adopts the SBM-undesirable model to measure the CLUE, and Moran's I and Markov chain models are employed to analyze the corresponding temporal and spatial evolution characteristics. The main conclusions are as follows:

In terms of temporal evolution, the CLUE in the whole region is significantly differentiated during the research period, exhibiting the characteristics of one growth, two stable, and two decline stages. Guizhou reveals a prominent growth trend, Jiangxi and Fujian exhibit high and stable fluctuation characteristics, and Hunan and Anhui demonstrate an overall downward trend. These results indicate that the effect of cultivated land use and ecological construction in the provinces within the National Pilot Zone for Ecological Conservation in China is better than that in Hunan and Anhui provinces.

In terms of the spatial pattern, during the research period, the CLUE in the whole region exhibits a positive spatial autocorrelation that increases year by year, and the spatial spillover effect is observed. In addition, local H-H and L-L agglomeration core areas exhibit expansion and transfer phenomena. Within the considered National Pilot Zone for Ecological Conservation in China, the positive spatial spillover effect is very pronounced. However, at present, the driving effect on the surrounding regions is not notable.

In terms of trend transfer, geospatial patterns and the spatial spillover effect play a significant role in CLUE evolution. The transfer probability of the CLUE against the different geographical backgrounds varies, and a high probability of improvement is attained near cities with high CLUE values. Proximity to cities with low CLUE values can inhibit enhancement, i.e., the core convergence phenomenon occurs.

Author Contributions: Conceptualization, Z.F., C.D. and Y.F.; writing—review and editing, C.D.; writing—original draft preparation, Y.F.; visualization, methodology, writing—review and editing, P.Z.; data curation, formal analysis, writing—review and editing, H.L. All authors drafted the manuscript and the approved the manuscript. All authors have read and agreed to the published version of the manuscript.

Funding: This research was funded by the National Natural Science Foundation of China (No. 72064020 and 71803071).

Institutional Review Board Statement: Not applicable.

Informed Consent Statement: Not applicable.

Data Availability Statement: The data presented in this study are available on request from the corresponding author.

Conflicts of Interest: The authors declare no conflict of interest.

References

1. Dullinger, I.; Essl, F.; Moser, D.; Erb, K.; Haberl, H.; Dullinger, S. Biodiversity models need to represent land-use intensity more comprehensively. *Glob. Ecol. Biogeogr.* **2021**, *30*, 924–932. [CrossRef]
2. Zhang, X.; Song, W.; Lang, Y.; Feng, X.; Yuan, Q.; Wang, J. Land use changes in the coastal zone of China's Hebei Province and the corresponding impacts on habitat quality. *Land Use Policy* **2020**, *99*, 104957. [CrossRef]
3. Chen, L.; Zhao, H.; Song, G.; Liu, Y. Optimization of cultivated land pattern for achieving cultivated land system security: A case study in Heilongjiang Province, China. *Land Use Policy* **2021**, *108*, 105589. [CrossRef]
4. Schaltegger, S.; Sturm, A. Okologische rationalitat:ansatzpunkte zur ausgestaltung von ökologieorientierten managementinstrumenten. *Die Unternehm.* **1990**, *44*, 273–290.
5. Development, W. *Eco-Efficient Leadership for Improved Economic and Environmental Performance;* World Business Council for Sustainable Development: Geneva, Switzerland, 1996; pp. 3–16.
6. Moll, S. Making sustainability accountable: Eco-efficiency, resource productivity and innovation. In Proceedings of the A Workshop on the Occasion of the Fifth Anniversary of the European Environment Agency (EEA), Copenhagen, Denmark, 28–30 October 1998; p. 45.
7. Vásquez, J.; Aguirre, S.; Fuquene-Retamoso, C.E.; Bruno, G.; Priarone, P.C.; Settineri, L. A conceptual framework for the eco-efficiency assessment of small-and medium-sized enterprises. *J. Clean. Prod.* **2019**, *237*, 117660. [CrossRef]

8. Desli, E.; Gkoulgkoutsika, A.; Sdrolia, E.; Zarotiadis, G. Eco-efficiency: A methodological framework and assessment. *Clean. Environ. Syst.* **2021**, *3*, 100049. [CrossRef]
9. Cortés, A.; Casillas-Hernández, R.; Cambeses-Franco, C.; Bórquez-López, R.; Magallón-Barajas, F.; Quadros-Seiffert, W.; Feijoo, G.; Moreira, M.T. Eco-efficiency assessment of shrimp aquaculture production in Mexico. *Aquaculture* **2021**, *544*, 737145. [CrossRef]
10. Grassauer, F.; Herndl, M.; Nemecek, T.; Guggenberger, T.; Fritz, C.; Steinwidder, A.; Zollitsch, W. Eco-efficiency of farms considering multiple functions of agriculture: Concept and results from Austrian farms. *J. Clean. Prod.* **2021**, *297*, 126662. [CrossRef]
11. Oliveira, R.; Camanho, A.S.; Zanella, A. Expanded eco-efficiency assessment of large mining firms. *J. Clean. Prod.* **2017**, *142*, 2364–2373. [CrossRef]
12. Davé, A.; Ball, P.; Salonitis, K. Factory eco-efficiency modelling: Data granularity and performance indicators. *Procedia Manuf.* **2017**, *8*, 479–486. [CrossRef]
13. Egilmez, G.; Park, Y.S. Transportation related carbon, energy and water footprint analysis of US manufacturing: An eco-efficiency assessment. *Transp. Res. Part D Transp. Environ.* **2014**, *32*, 143–159. [CrossRef]
14. Park, Y.S.; Egilmez, G.; Kucukvar, M. A novel life cycle-based principal component analysis framework for eco-efficiency analysis: Case of the United States manufacturing and transportation nexus. *J. Clean. Prod.* **2015**, *92*, 327–342. [CrossRef]
15. Yang, Z.; Zhou, X.; Xu, L. Eco-efficiency optimization for municipal solid waste management. *J. Clean. Prod.* **2015**, *104*, 242–249. [CrossRef]
16. Stanchev, P.; Ribarova, I. Complexity, assumptions and solutions for eco-efficiency assessment of urban water systems. *J. Clean. Prod.* **2016**, *138*, 229–236. [CrossRef]
17. Rybaczewska-Błażejowska, M.; Masternak-Janus, A. Eco-efficiency assessment of Polish regions: Joint application of life cycle assessment and data envelopment analysis. *J. Clean. Prod.* **2018**, *172*, 1180–1192. [CrossRef]
18. Basset-Mens, C.; Rhino, B.; Ndereyimana, A.; Kleih, U.; Biard, Y. Eco-efficiency of tomato from Rwamagana district in Rwanda: From field constraints to statistical significance. *J. Clean. Prod.* **2019**, *229*, 420–430. [CrossRef]
19. Maxime, D.; Marcotte, M.; Arcand, Y. Development of eco-efficiency indicators for the Canadian food and beverage industry. *J. Clean. Prod.* **2006**, *14*, 636–648. [CrossRef]
20. Sorvari, J.; Antikainen, R.; Kosola, M.L.; Hokkanen, P.; Haavisto, T. Eco-efficiency in contaminated land management in Finland–Barriers and development needs. *J. Environ. Manag.* **2009**, *90*, 1715–1727. [CrossRef]
21. You, H.Y.; Wu, C.F.; Yang, P.J.; Huang, L. Analysis on eco-efficiency and effectiveness of cultivated land quantity compensation. *Land Resour. Inf.* **2012**, *1*, 39–45.
22. Yang, H.R.; Wu, Q. Study on the Eco-efficiency of Land Use Transformation in Jiangsu Province from the Perspective of Carbon Emission—Based on the Mixed Directional Distance Function. *J. Nat. Resour.* **2017**, *32*, 1718–1730.
23. Zhu, Z.; Du, G. Coordinated Development of Urban Land Use and Ecological Economics in China. *J. Math.* **2021**, *2021*, 5599633. [CrossRef]
24. Zhang, C.; Su, Y.; Yang, G.; Chen, D.; Yang, R. Spatial-Temporal Characteristics of Cultivated Land Use Efficiency in Major Function-Oriented Zones: A Case Study of Zhejiang Province, China. *Land* **2020**, *9*, 114. [CrossRef]
25. Chen, B.; Wang, Q.; Liu, J. Comparative analysis on eco-efficiency of arable land ecological footprint in Hubei. *Wuhan Univ. J. Nat. Sci.* **2006**, *11*, 1052–1058.
26. Luo, J.X.; Cui, J.; Liu, Y.F. Coupling and Coordination Analysis of Urban Land Intensive Use and Ecological Efficiency—A Case Study of Northeast China. *Nat. Resour. Econ. China* **2021**, *34*, 81–89.
27. Bonfiglio, A.; Arzeni, A.; Bodini, A. Assessing eco-efficiency of arable farms in rural areas. *Agric. Syst.* **2017**, *151*, 114–125. [CrossRef]
28. Zhao, Z.; Bai, Y.; Wang, G.; Chen, J.; Yu, J.; Liu, W. Land eco-efficiency for new-type urbanization in the Beijing-Tianjin-Hebei Region. *Technol. Forecast. Soc. Chang.* **2018**, *137*, 19–26. [CrossRef]
29. Yang, H.; Wu, Q. Land use eco-efficiency and its convergence characteristics under the constraint of carbon emissions in China. *Int. J. Environ. Res. Public Health* **2019**, *16*, 3172. [CrossRef]
30. Liu, Y.; Sun, H.; Shi, L.; Wang, H.; Xiu, Z.; Qiu, X.; Chang, H.; Xie, Y.; Wang, Y.; Wang, C. Spatial-Temporal Changes and Driving Factors of Land-Use Eco-Efficiency Incorporating Ecosystem Services in China. *Sustainability.* **2021**, *13*, 728. [CrossRef]
31. Hong, K.R.; Li, B. Study on temporal-spatial differences and influencing factors of ecological efficiency of land resources. *Hubei Soc. Sci.* **2016**, *10*, 74–81.
32. Xie, M.M.; Li, X.X. Research on Spatiotemporal Evolution Regulation of Eco-efficiency of Land Utilization in Jilin Province Based on Data Envelopment Analysis. *Bull. Soil Water Conserv.* **2015**, *35*, 225–230.
33. Lu, Y.M.; Fang, S.M. Analysis of spatio-temporal evolution and influencing factors of eco-efficiency of urban construction land in Wuhan city circle based on SBM-DEA and Malmquist model. *Resour. Environ. Yangtze Basin* **2017**, *26*, 1575–1586.
34. Yang, B.; Wang, Z.; Zou, L.; Zou, L.; Zhang, H. Exploring the eco-efficiency of cultivated land utilization and its influencing factors in China's Yangtze River Economic Belt, 2001–2018. *J. Environ. Manag.* **2021**, *294*, 112939. [CrossRef] [PubMed]
35. Cui, W.; Miao, J.J.; Yang, J. Urban Non-agricultural Eco-efficiency and Affecting Factors Based on Carbon Emission. *China Popul. Resour. Environ.* **2013**, *23*, 63–69.
36. Huang, H.P.; Li, Y.L.; Wang, Z.P. Spatio-temporal changes of eco-efficiency and influencing factors of industrial land use at the provincial level of China. *Acta Ecol. Sin.* **2020**, *40*, 100–111.

37. Huang, H.P.; Wang, Z.P. Spatial-temporal Differences and Influencing Factors of Agricultural Land Eco-efficiency in Jiangxi Province: Based on the Dual Perspective of Non-point Source Pollution and Carbon Emission. *Resour. Environ. Yangtze Basin* **2020**, *2*, 412–423.
38. Costanza, R.; d'Arge, R.; De Groot, R.; Farber, S.; Grasso, M.; Hannon, B.; Limburg, K.; Naeem, S.; O'Neill, R.V.; Paruelo, J.; et al. The value of the world's ecosystem services and natural capital. *Nature* **1997**, *387*, 253–260. [CrossRef]
39. Jin, J.J.; Jiang, C.; Li, L.I. The economic valuation of cultivated land protection: A contingent valuation study in Wenling City, China. *Landsc. Urban Plan.* **2013**, *119*, 158–164.
40. Yu, H.; Zheng, J. The Research of Implementation Path on Transforming Ecological Advantages into Economic Advantages—A Case Study in National Ecological Civilization Experimental Zones. *For. Econ.* **2019**, *41*, 87–94.
41. Guo, Y.; Li, N.; Mu, H.; Li, L.; Duan, Y. Regional total-factor coal consumption efficiency in China: A meta-frontier SBM-undesirable approach. *Energy Procedia* **2017**, *142*, 2423–2428. [CrossRef]
42. Cecchini, L.; Venanzi, S.; Pierri, A.; Chiorri, M. Environmental efficiency analysis and estimation of CO_2 abatement costs in dairy cattle farms in Umbria (Italy): A SBM-DEA model with undesirable output. *J. Clean. Prod.* **2018**, *197*, 895–907. [CrossRef]
43. Zhang, R.L.; Liu, X.H. Evaluating ecological efficiency of Chinese industrial enterprise. *Renew. Energy* **2021**, *178*, 679–691. [CrossRef]
44. Jiang, H. Spatial–temporal differences of industrial land use efficiency and its influencing factors for China's central region: Analyzed by SBM model. *Environ. Technol. Innov.* **2021**, *22*, 101489. [CrossRef]
45. Kuang, B.; Lu, X.; Zhou, M.; Chen, D. Provincial cultivated land use efficiency in China: Empirical analysis based on the SBM-DEA model with carbon emissions considered. *Technol. Forecast. Soc. Chang.* **2020**, *151*, 119874. [CrossRef]
46. Chen, W.; Ning, S.; Chen, W.; Liu, E.N.; Wang, Y.; Zhao, M. Spatial-temporal characteristics of industrial land green efficiency in China: Evidence from prefecture-level cities. *Ecol. Indic.* **2020**, *113*, 106256. [CrossRef]
47. Zhou, P.A.B.W.; Ang, B.W.; Poh, K.L. Slacks-based efficiency measures for modeling environmental performance. *Ecol. Econ.* **2006**, *60*, 111–118. [CrossRef]
48. Li, L.B.; Hu, J.L. Ecological total-factor energy efficiency of regions in China. *Energy Policy* **2012**, *46*, 216–224. [CrossRef]
49. Tone, K.; Tsutsui, M. Network DEA: A slacks-based measure approach. *Eur. J. Oper. Res.* **2009**, *197*, 243–252. [CrossRef]
50. Paradi, J.C.; Rouatt, S.; Zhu, H. Two-stage evaluation of bank branch efficiency using data envelopment analysis. *Omega* **2011**, *39*, 99–109. [CrossRef]
51. Long, R.; Ouyang, H.; Guo, H. Super-slack-based measuring data envelopment analysis on the spatial–temporal patterns of logistics ecological efficiency using global Malmquist Index model. *Environ. Technol. Innov.* **2020**, *18*, 100770. [CrossRef]
52. Overmars, K.D.; De Koning, G.H.J.; Veldkamp, A. Spatial autocorrelation in multi-scale land use models. *Ecol. Model.* **2003**, *164*, 257–270. [CrossRef]
53. Kobayashi, A. *International Encyclopedia of Human Geography*; Elsevier: Amsterdam, The Netherlands, 2019.
54. Ahn, K.U.; Park, C.S. Temporal and spatial variation in the predictability of building occupancy. *Build. Environ.* **2019**, *149*, 477–489. [CrossRef]
55. Cui, Y.; Khan, S.U.; Deng, Y.; Zhao, M.; Hou, M. Environmental improvement value of agricultural carbon reduction and its spatiotemporal dynamic evolution: Evidence from China. *Sci. Total Environ.* **2021**, *754*, 142170. [CrossRef]
56. Hou, X.; Liu, J.; Zhang, D.; Zhao, M.; Xia, C. Impact of urbanization on the eco-efficiency of cultivated land utilization: A case study on the Yangtze River Economic Belt, China. *J. Clean. Prod.* **2019**, *238*, 117916. [CrossRef]
57. West, T.O.; Marland, G. A synthesis of carbon sequestration, carbon emissions, and net carbon flux in agriculture: Comparing tillage practices in the United States. *Agric. Ecosyst. Environ.* **2002**, *91*, 217–232. [CrossRef]
58. Dubey, A.; Lal, R. Carbon footprint and sustainability of agricultural production systems in Punjab, India, and Ohio, USA. *J. Crop Improv.* **2009**, *23*, 332–350. [CrossRef]
59. He, Y. *Study on Carbon Cycle of Climate and Terrestrial Ecosystem in China*; Meteorological Press: Beijing, China, 2006.
60. Liang, L.T. *Study on the Temporal and Spatial Evolution of Rural Ecological Enviroment*; Nanjing Agricultural University: Nanjing, China, 2009.
61. Feng, Y.G.; Peng, J.; Deng, Z.B.; Wang, J. Spatial-temporal Variation of Cultivated Land's Utilization Efficiency in China Based on the Dual Perspective of Non-point Source Pollution and Carbon Emission. *China Popul. Resour. Environ.* **2015**, *25*, 18–25.

International Journal of
Environmental Research and Public Health

MDPI

Article

Knowledge Mapping of Research on Land Use Change and Food Security: A Visual Analysis Using CiteSpace and VOSviewer

Peng Cheng [1], Houtian Tang [2], Yue Dong [1], Ke Liu [3], Ping Jiang [1,4,*] and Yaolin Liu [1,4]

1 College of Resource and Environmental Sciences, Wuhan University, Wuhan 430079, China; chengpeng1141@whu.edu.cn (P.C.); dongyue@whu.edu.cn (Y.D.); 00007101@whu.edu.cn (Y.L.)
2 School of Public Administration, Central South University, Changsha 410083, China; tanghoutian@csu.edu.cn
3 Graduate School of Huazhong Agricultural University, Wuhan 430070, China; liuke_1996@163.com
4 Key Laboratory of Geographic Information System, Ministry of Education, Wuhan University, Wuhan 430079, China
* Correspondence: longkangkang@whu.edu.cn; Tel.: +86-186-2793-1818

Abstract: Many scholars have conducted in-depth research on the theme of land use change and food security, and formed fruitful research results, but there is a lack of quantitative analysis and comprehensive evaluation of research achievements. Therefore, based on the relevant literature on the theme of land use change and food security in the core collection of the Web of Science (WOS) database, this paper takes the advantage of CiteSpace and VOSviewer bibliometric software to draw the cooperative network and keyword cooccurrence map to analyze the research progress and frontier. The results reveal that: (1) The research started in 1999 and can be divided into three stages: initial research, rapid development, and a stable in-depth stage. This topic has increasingly become a research hotspot in the academic community. (2) The distribution of research institutions is concentrated and forms a small cluster, and the research networks between developed and developing countries have been established, and developed countries are in the core position, but the cooperation network is not prominent. (3) The research content is becoming increasingly organized and systematic, and the research hot topics are divided into seven aspects. (4) The research area of the subject covers multiple levels, such as global, national, and specific natural geographical regions, and has formed a research system of geographic information technology and satellite remote sensing technology. It also presents the trend of cross integration with economics, land management and soil science. In the future, theoretical innovation still needs to be strengthened, and we should strengthen the research on the impact of agricultural chemical fertilizers on food security and study the impact of urban expansion on land use change.

Keywords: land use change; food security; visual analysis; CtieSpace; VOSviewer; progress and frontier

Citation: Cheng, P.; Tang, H.; Dong, Y.; Liu, K.; Jiang, P.; Liu, Y. Knowledge Mapping of Research on Land Use Change and Food Security: A Visual Analysis Using CiteSpace and VOSviewer. *Int. J. Environ. Res. Public Health* **2021**, *18*, 13065. https://doi.org/10.3390/ijerph182413065

Academic Editors: Wei Song and Hualin Xie

Received: 2 November 2021
Accepted: 9 December 2021
Published: 10 December 2021

1. Introduction

Land use change and its impact on food security have become one of the frontiers and hot topics studied by scholars worldwide [1–4]. With the development of human society, the structure, depth and intensity of land use are constantly changing, which not only affects biodiversity but also has a great impact on human food security [5]. The issue of "food security" was first put forward by the United Nations International Food and Agriculture Organization at the First World Food Summit in November 1974. The definition of food security is to ensure that anyone can obtain enough food for survival and health at any time, including food supply, food access, food stability and food utilization [6,7]. Over the years, food security has been a major issue related to the overall political and economic situation of a country or region. In particular, regional food production and food security have become hot issues of concern to governments and scholars [8–10]. According to the prediction of the United Nations [11], the global population will exceed

9.8 billion in 2050, food demand will increase by more than 50%, and food problems will be extremely serious [12]. However, there are many factors affecting grain production, including institutional and policy innovation, scientific and technological progress, material and labor investment, climate change, but cultivated land resources in land resources are the most important factor in grain production [13,14]. Cultivated land resources are the most basic natural condition of agricultural production. Food security is closely related to changes in cultivated land. The change in the quantity and quality of cultivated land directly affects grain output and then affects the effective supply of grain and the level of food security [15,16].

In recent years, research related to the theme of land use change and food security has emerged in the academic community, mainly combined with issues of climate change, carbon emissions, agricultural intensification [5,17,18]. For example, Galeana-Pizaña et al. [1] used a GIS-based food environmental efficiency (FEE) index to evaluate the trend of land use change and regional food security, and the FEE index proved useful assessment of land use policies. Moore et al. [19] used the regional climate model to compare the impacts of projected future greenhouse gases and future land use change on spatial variability of grain yields in East Africa. These show that this theme is an evolving knowledge field, but there is a lack of systematic review of research results. Accurately understanding the research progress and academic trends of land use change and food security is of great significance for carrying out follow-up research. Therefore, based on the core collection of the Web of Science (WOS) database, this paper comprehensively uses the advantages of CiteSpace and VOSviewer software to conduct bibliometric analysis, systematically and visually analyze and summarize the literature in the fields of land use change and food security, and explore the status of research. This research field objectively reveals the trends, accurately evaluates the research progress on land use change and food security and provides a reference for combining the research framework and expanding new ideas and methods in this field.

2. Data Collection and Research Methods

2.1. Data Collection

The data on the relevant literature used in this paper come from the core collection of the WOS database (http://apps.webofknowledge.com, accessed on 6 September 2021) and adopt the method of group retrieval. The WOS Citation database is an information retrieval platform developed by Thomson Reuters of the United States. With the Science Citation Index, Social Science Citation Index, and Arts and Humanities Citation Index as the core, it contains more than 9000 world authoritative and influential academic journals, and the documents in the database have high authority in the academic community [20,21]. The search prerequisites of this research are set as follows: (TS = "land use change" and "grain security") OR (TS = "land use change" and "food security"), TS is the theme, time spans are unlimited, the language is "English", and the literature types are "article" and "review". There were 628 literature records related to the subject that were retrieved. To avoid duplicate literature, CiteSpace's deduplication function was used for inspection, and no duplicate publications were found.

2.2. Research Methods

As an auxiliary procedure of bibliometrics, science mapping provides a spatial representation of network structures. Science mapping involves the interdisciplinary fields of applied mathematics, information science and computer science. It is a new development of scientometrics and information metrology. In recent years, with the rapid development of computer science, many scholars have used various science mapping tools to analyze the potential dynamic mechanism of discipline evolution [22–24]. CiteSpace and VOSviewer software are two powerful and complementary science mapping analysis tools. CiteSpace (https://sourceforge.net/projects/citespace, accessed on 6 September 2021) is a Java-based application software proposed by Professor Chen of Drexel University in 2004. It is based on the co-citation analysis theory and pathfinder, minimum spanning trees algorithm to make a quantitative analysis of the literature in specific fields, which is used to analyze and visualize the emerging trends and patterns in the knowledge field of scientific publications [25,26]. It has unique advantages in literature keyword analysis, cluster analysis, subject words, author information. VOSviewer (https://www.vosviewer.com, accessed on 6 September 2021) is also a literature analysis and knowledge visualization software tool developed by van Eck and Waltman of the Centre for Science and Technology Studies at Leiden University [27]. It can realize the construction and visualization of the keyword cooccurrence network in various fields. Compared with other visualization software, VOSviewer software has advantages in processing big data and drawing images, which can more clearly show the hot spots and topics in the research field [28].

3. Results

3.1. Trend Analysis of Literature Publication

The annual distribution of the number of published articles can reflect the research level and degree of development of a certain discipline [29]. The number of published articles on land use change and food security is shown in Figure 1. From 1999 to 6 September 2021, the number of published studies on land use change and food security showed a stable growth trend on the whole, which experienced three stages, i.e., initial research, rapid development, and a stable in-depth stage.

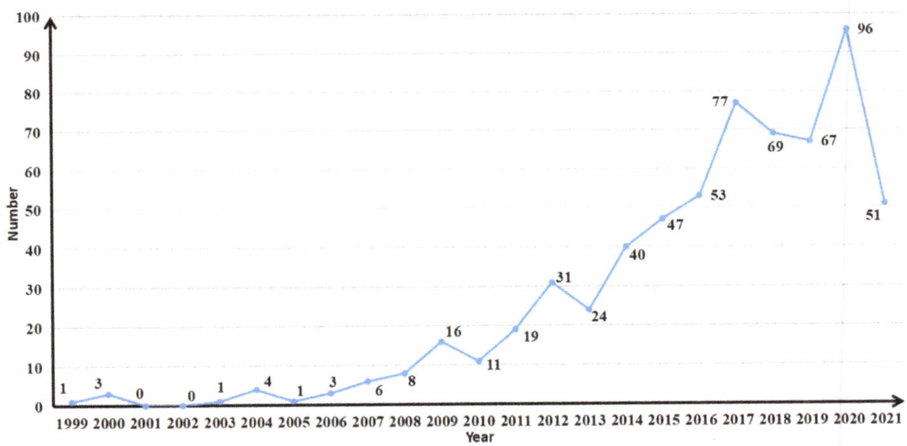

Figure 1. Number of articles published annually on the theme of land use change and food security.

The initial stage of the research (1999–2008): At this stage, the number of relevant research studies was relatively small, the research topic was relatively limited, and the number of research scholars in this field was also small, mainly because there was not much research on land use change and food security. Before this stage, scholars studied the theme of land use change or food security separately. In 1999, Sporton et al. first studied the theme of land use change and food security [30]. Subsequently, Murdiyarso and Verburg et al. paid more attention to this research topic [31,32].

The rapid development stage of the research (2009–2017): During this stage, scholars showed great interest in the research topic, the number of published articles continued to increase, the research questions and perspectives were further expanded, and several leading studies emerged, such as Alexander and Verburg et al. [33,34]. In 2009, famous scholars such as Khan, Garnett, Mertz and Yan et al. [35–38] published a series of high-level papers, meanwhile, the number of people suffering from hunger in the world will reach 1.02 billion in 2009 announced by the United Nations, reversing the continuous decline of hungry people, which made the research on land use change and food security widely concerned in the academic community. With the intensification of global land use change and food security, scholars researched climate change, biodiversity, policy development, agriculture, greenhouse gas emissions, and the research contents and methods were further enriched [19,39].

Stable in-depth stage of the research (2018-present): During this stage, the literature publication trend was relatively smooth, and the research content and perspective gradually increased in depth. The research perspective has focused on both macro and micro issues, including population growth, land systems, rural development, soil organic carbon, and life cycle assessment, in the research agenda of land use change and food security, and the overall research has continued to deepen [40,41].

3.2. Network Analysis of Author Cooperation, Institutional Cooperation and National Cooperation

3.2.1. Analysis of Author Cooperation Network

By analyzing the author's cooperation network [42], we can determine the strength of representative scholars and core research teams in the field of land use change and food security. VOSviewer software was used to overlay and visualize the author collaborative networks with more than 5 published articles. Through the color gradient, it can intuitively reflect the cooperation of various scholars in recent years (Figure 2) and present the author information with the number of published articles for the top 20 publications (Table 1). We found that the authors with a large number of published articles showed obvious network characteristics, mainly including the cooperative network of Verburg, Smith, Havlik and Popp. This indicates that these are core authors who have developed a high-yield author research team in the field of land use change and food security that has initially formed a scale. According to the Price Law [43], the formula for calculating the minimum number of published articles of core authors in a field is $m = 0.749 \times \sqrt{n_{max}} = 2.996$ (where n_{max} is the number of published articles of the top 1 author). Therefore, authors with more than three published articles are regarded as the core authors in this field. The top three scholars in the number of published articles are Verburg (16 articles), Smith (13 articles) and Havlik (10 articles). According to the data, there are 43 core authors and 206 articles, accounting for 32.8% of the total articles published in this field, which is less than the standard of 50% of the Price Law. This shows that after more than 20 years of development, the core author group in the field of land use change and food security has initially formed, but still needs further development.

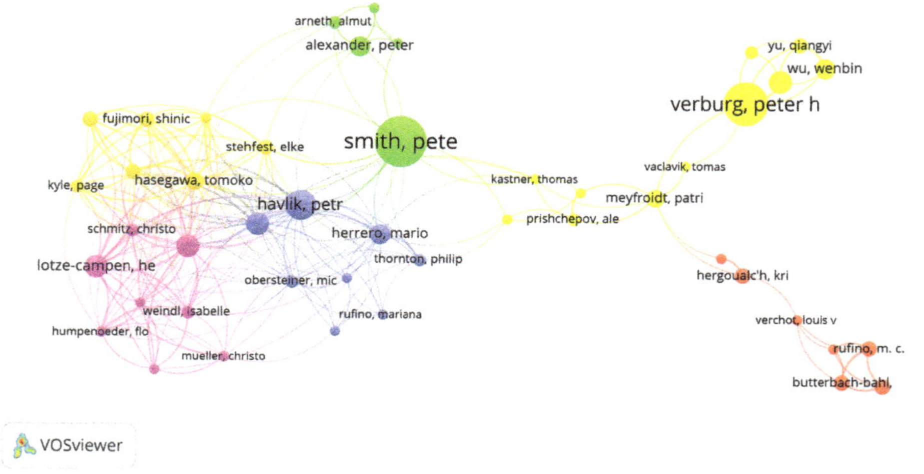

Figure 2. Author cooperation network map.

Table 1. Author information table of the top 20 published articles.

Ranker	Count	Centrality	Year	Authors	Ranker	Count	Centrality	Year	Authors
1	16	0	2016	Peter H Verburg	11	5	0	2018	Wenbin Wu
2	13	0.03	2008	Pete Smith	12	5	0	2015	Isabelle Weindl
3	10	0.01	2014	Petr Havlik	13	5	0	2013	Alexander V Prishchepov
4	9	0.03	2014	Alexander Popp	14	5	0	2017	K Butterbachbahl
5	7	0	2014	Hermann Lotzecampen	15	5	0	2014	Hans Van Meijl
6	6	0	2014	Tomoko Hasegawa	16	5	0	2017	M C Rufino
7	6	0	2009	Jiyuan Liu	17	5	0	2017	Jasper Van Vliet
8	6	0	2014	Hugo Valin	18	4	0	2016	Almut Arneth
9	5	0	2014	Shinichiro Fujimori	19	4	0	2017	Kamini Yadav
10	5	0	2014	Christoph Schmitz	20	4	0	2014	Andrzej Tabeau

Note: The centrality indicator measures the importance of network nodes [44]. The larger the value of centrality, the more articles published by the author in cooperation with other authors.

3.2.2. Analysis of Institutional Cooperation Network

Taking the research institution as the node for visual analysis, we can obtain the cooperation network map of the research institution (Figure 3) and show the network with connections. According to the information of the top 20 major research institutions (Table 2), the Chinese Academy of Sciences has the highest number of published articles (52), followed by Vrije University Amsterdam (26) and Wageningen University (22), and a research network has been formed of these three research institutions as the core. This shows that these institutions have strong scientific research and influence in the field, and there are cooperative relations and large-scale collaborations between the different institutions. It is worth noting that the reason for the highest number of documents issued by the Chinese Academy of Sciences may be related to China's national conditions. The main reasons include the following points: (1) In terms of policy, the Chinese Government has put forward the "red line of 1.8 billion mu of cultivated land" and other cultivated land protection policies to control land use changes and ensure food security. (2) In terms of economics, the Chinese Government has adjusted agricultural protection policies, increased investment in agricultural science and technology, and continuously improved the rate of grain self-sufficiency. (3) In terms of society, China is a populous country in the world, it is required to ensure food supply and firmly put its rice bowl in its own hands. (4) In terms of the environment, the deterioration of land and other production factors had a great impact on food security. To ensure food security, the Government has always taken measures to prevent land resource degradation and improve the ecological environment.

In Figure 3, the research institutions are in a local aggregation state, indicating that the distribution of research institutions is relatively concentrated and that a small aggregation cluster is formed; that is, there are some cooperative relations among institutions. Generally, there are a large number of research institutions related to the theme of land use change and food security, but a large cross-national institutional cooperation group has not yet formed.

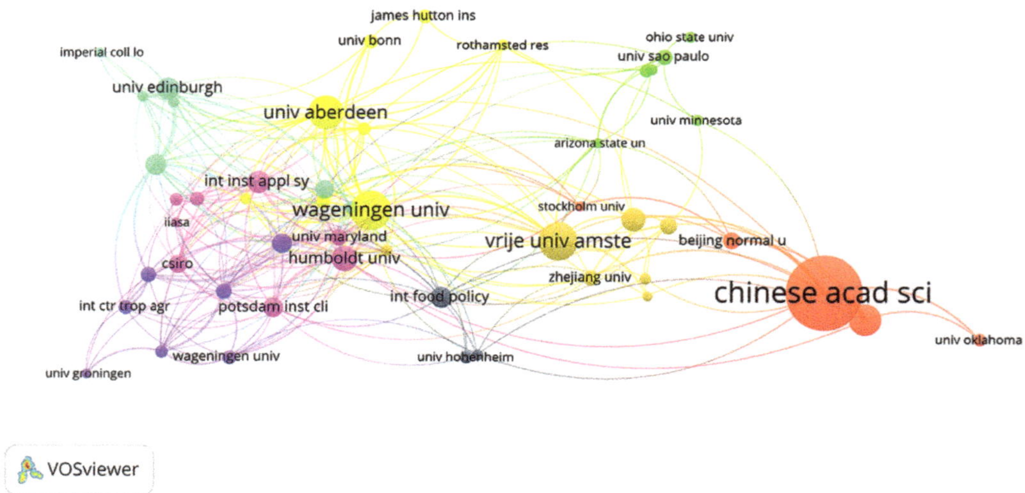

Figure 3. Institutional cooperation network map.

Table 2. Information table of the top 20 major research institutions with published articles.

Ranker	Count	Centrality	Year	Research Institutions	Ranker	Count	Centrality	Year	Research Institutions
1	52	0.18	2003	Chinese Academy Science	11	12	0.1	2006	Potsdam Institute for Climate Impact Research
2	26	0.12	2014	Vrije University Amsterdam	12	12	0.01	2009	Beijing Normal University
3	22	0.22	2010	Wageningen University	13	12	0.03	2010	University of Edinburgh
4	19	0.02	2013	University of Chinese Academy of Sciences	14	12	0.03	2016	Karlsruhe Institute of Technology
5	18	0.13	2008	University of Aberdeen	15	11	0.05	2008	University of Maryland
6	18	0.12	2011	Humboldt University	16	10	0.03	2006	University of Copenhagen
7	16	0.1	2009	Michigan State University	17	10	0.02	2009	Lancaster University
8	14	0.09	2011	International Food Policy Research Institute	18	10	0.06	2000	Chinese Academy of Agricultural Sciences
9	13	0.07	2014	Commonwealth Scientific and Industrial Research Organization	19	9	0.04	2004	Columbia University
10	12	0.05	2000	International Institute for Applied Systems Analysis	20	9	0.03	2017	PBL Netherlands Environmental Assessment Agency

3.2.3. Analysis of Country Cooperation Network

According to Figure 4 and Table 3, there are more than 100 articles published in the USA, China, and Germany, which is significantly higher than that in other countries. The number of articles published in these three countries accounted for 31.84%, 19.9% and 18.15% of the total number of articles published in this field, respectively. It can be seen from the connectivity in Figure 4 that the connections between nodes are dense and complex, indicating that there are many cooperative relations between different countries. In Figure 4, purple appears at the edge of some nodes, indicating that the centrality is $\geq$0.1, which also indicates that the node is in an important position within the network structure. Among them, the centrality values of the USA (0.58), Germany (0.13), England (0.11) and France (0.11) are higher than 0.1, indicating that these countries are in a relatively core area in the research field of land use change and food security and that the relevant research studies have a significant impact on this field.

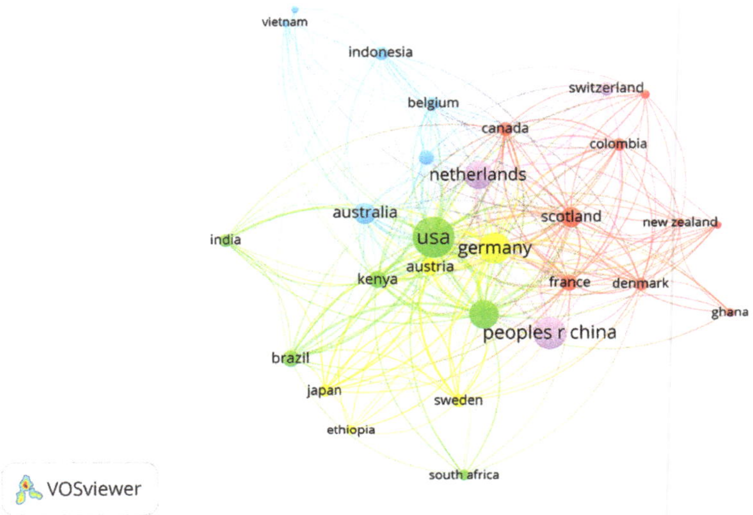

Figure 4. Country cooperation network map.

Table 3. Information table of the top 20 major research countries with published articles.

Ranker	Count	Centrality	Year	Countries	Ranker	Count	Centrality	Year	Countries
1	200	0.58	2003	USA	11	30	0.11	2010	France
2	125	0.1	2009	China	12	26	0.01	2007	Italy
3	114	0.13	2009	Germany	13	25	0.06	2013	Switzerland
4	93	0.11	2009	England	14	23	0.01	2011	Indonesia
5	92	0.1	2008	The Netherlands	15	22	0.02	2008	Belgium
6	56	0.04	2009	Australia	16	22	0.03	2008	Canada
7	47	0.04	2010	Scotland	17	20	0.09	2009	Denmark
8	36	0.02	2007	Austria	18	20	0.01	2009	Sweden
9	35	0.03	2008	Kenya	19	19	0.01	2015	Colombia
10	33	0	2009	Brazil	20	18	0.01	2012	India

3.3. Analysis of Hot Research Topics and Frontiers Trending

3.3.1. Analysis of Hot Research Topics

Keywords capture the core idea of the article. Through the research on keywords in a field, we can quickly grasp the hot topics in the field [44]. In this study, VOSviewer software was used to visualize keywords. Nodes in the knowledge map represent keywords. The larger the node is, the higher the frequency, and the lines between nodes represent the cooccurrence of particular keywords. In addition, in the VOSviewer knowledge map, different colors represent different clusters, and the same color represents the same cluster. By analyzing the keyword cooccurrence knowledge map (Figure 5), we find that the whole keyword knowledge map takes "food security", "land use change" and "climate change" as the core, producing a radial shape. Considering that high-frequency keywords can be clearly displayed, a total of 279 high-frequency keywords are obtained with the threshold of five of each keyword. The cooccurrence map of keywords is relatively clear, and the top 20 high-frequency keywords are shown in Table 4. As seen from Figure 5 and Table 4, "land use change" (185), "food security" (141), "climate change" (119), "impact" (102) and other high-frequency keywords constitute representative terms in this field. In terms of layout, these high-frequency keywords are also key hub nodes. Other nodes around them have together formed the hot cutting-edge research topics in this field in recent years.

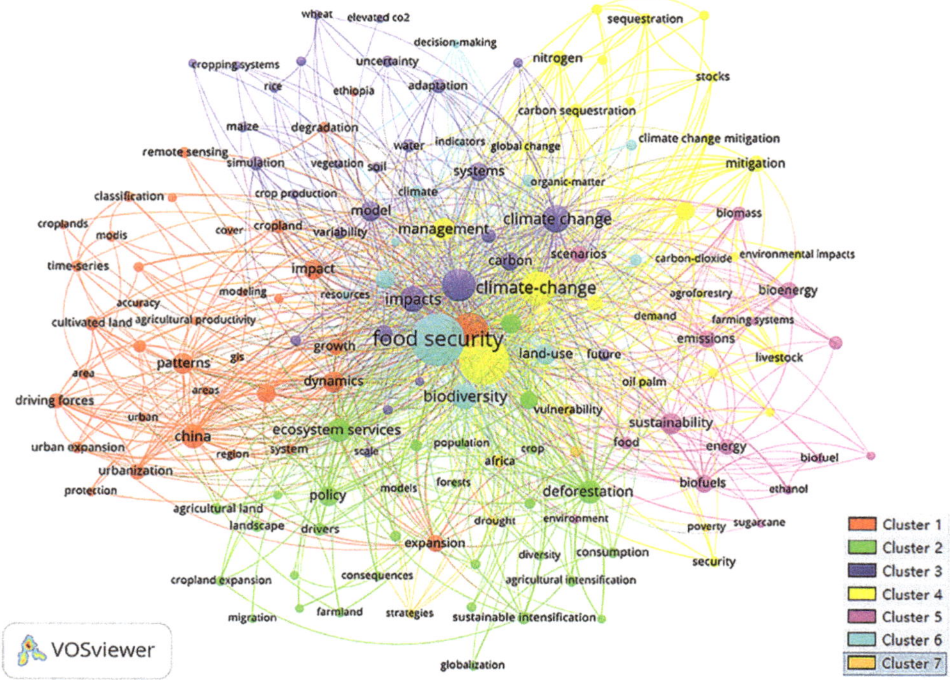

Figure 5. Keyword cooccurrence network map.

Table 4. Information table of top 20 keywords with cooccurrence.

Ranker	Count	Centrality	Year	Keywords	Ranker	Count	Centrality	Year	Keywords
1	185	0.29	2000	Land use change	11	42	0.08	2009	Biodiversity
2	141	0.21	2006	Food security	12	41	0.07	2006	Deforestation
3	119	0.13	2006	Climate change	13	37	0.05	2010	Land use
4	102	0.08	1999	Impact	14	34	0.05	2000	Policy
5	62	0.14	2003	Agriculture	15	33	0.06	2008	Conservation
6	45	0.08	2009	Management	16	32	0.02	2014	Ecosystem service
7	45	0.04	2011	System	17	31	0.04	2012	Greenhouse gas emission
8	45	0.02	2000	Model	18	30	0.02	2006	Cover change
9	43	0.05	2006	Dynamics	19	30	0.08	2005	Carbon
10	43	0.06	2000	Pattern	20	27	0.02	2012	Expansion

To refine the research topics more intuitively and effectively in this field, we use the unique clustering density map function of VOSviewer software to visualize the keyword cooccurrence clustering results (Figure 6). In the cluster density map, the density of an element depends on the number and weight of its surrounding elements. From the cold tone to the warm tone, the representative clustering density gradually increases; that is, the frequency of keyword cooccurrence increases, and the heat of related research topics increases [45]. According to the clustering results in Figure 6, combined with professional knowledge, we can extract seven frontier hot topics in the current research field of land use change and food security (Table 5) and further analyze and discuss the research contents and important achievements of each frontier hot topic.

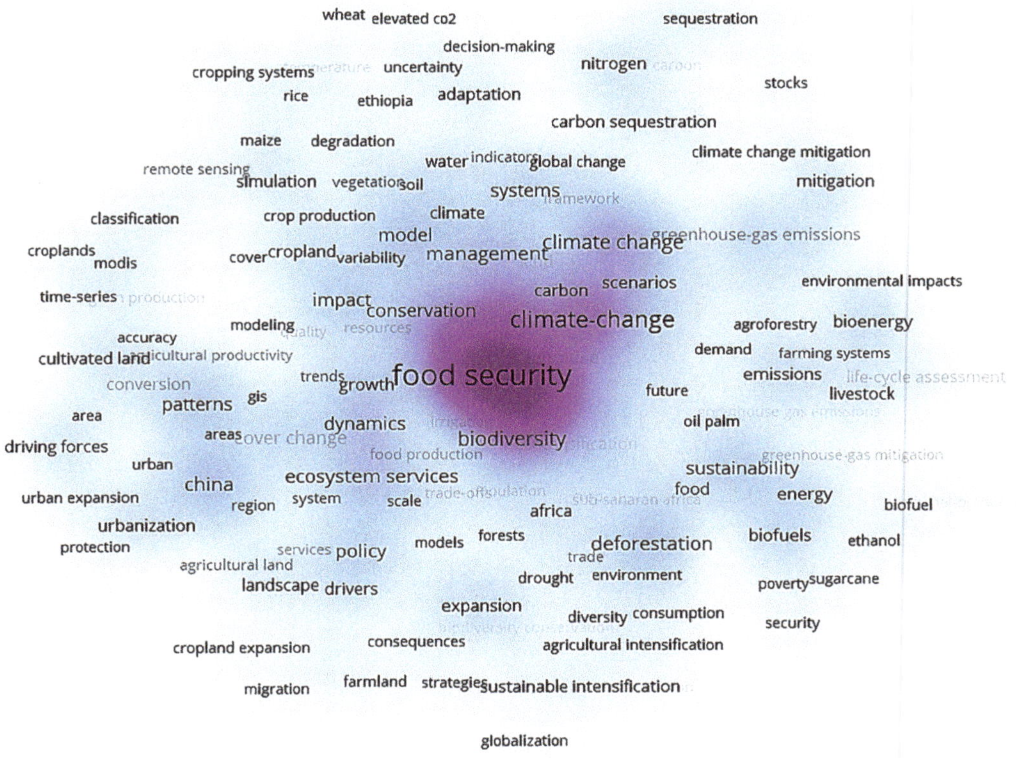

Figure 6. Keyword cooccurrence clustering density map produced by VOSviewer software.

Climate Change and Carbon Emissions

With excessive carbon emissions produced in the process of human production and consumption, global warming and abnormal climate events frequently occur, which have a direct impact on land use, especially on changes in cultivated land area, threatening global food security [46–48]. At present, with the continuous intensification of abnormal climate change, the geographical distribution of food-deficient areas will further expand. At the same time, grain production areas have been chronically affected by energy crops, feed crops, forestry and other economic crops, as well as the continuous expansion of vegetation areas caused by climate warming, which forces people to reduce the land allocated to grain production, thus, causing a global food supply crisis [49–51]. Hasegawa et al. built a comprehensive assessment model of the impact of climate change mitigation policies on food security [52]. The research found that if mitigation policies to address climate change are strictly implemented, they will have a huge negative impact on global food production and consumption, especially in low-income countries in Africa and South Asia. Moreover, Nobre and Beltrán-Tolosa et al. found that the development of traditional agriculture and animal husbandry will inevitably reduce the area of vegetation coverage, resulting in environmental problems such as soil and water loss and soil erosion, leading to drought with climate change, thus affecting the production of the main food crops [53–55]. Relevant studies show that by the end of this century, due to the impact of climate change, grain prices may rise by 110% or more over the prices in the baseline year. Similarly, Hasegawa and Popp et al. also confirmed that food prices in parts of Asia and Africa will be more affected, increasing the potential risk of a food crisis [49,56]. Therefore, the impact of climate change and carbon emissions on land use change and food security will still be one of the key topics that scholars continue to pay attention to in the future [57,58].

Table 5. Keyword cooccurrence clustering induction.

Cluster-ID	Research Topics	Main Keywords Included
1	Climate change and carbon emissions	Climate change, global change, climate change mitigation, change impacts, greenhouse gas emissions, carbon sequestration, carbon stocks, greenhouse gas emissions, soil carbon sequestration, soil organic carbon
2	Sustainable land management policy	Land management, policy, protection policies, cropland protection, farmland abandonment, rapid urbanization, transformation, urban expansion, urban sprawl, urbanization
3	Agricultural intensive development	Agricultural intensification, sustainable intensification, agricultural productivity, ecosystem, environmental change, food security, biodiversity conservation, impacts, risk
4	Land degradation	Cropping systems, land use change, degradation, desertification, land degradation, pollution, soil erosion, water resources, croplands, climate change impacts, rice, river basin
5	Renewable bioenergy	Carbon, carbon footprint, water footprint, bioenergy, biofuel, energy, environmental impact, farming systems, life cycle assessment, production systems, renewable energy, soil erosion, sustainable agriculture
6	Food production	Crop productivity, crop yield, efficiency, food production, human appropriation, impact assessment, yield gap, use efficiency, net primary production, irrigation, maize, wheat
7	Agricultural benefits	Agriculture, benefits, biodiversity, certification, costs, crop, food demand, integrated assessment, intensification, plantations, policies, scenarios, validation, yields

Sustainable Land Management Policy

A sustainable land management policy can help mitigate climate change, protect the land from soil erosion and ensure food security. Its core is to emphasize the resilience of management methods, that is, to seek maximum synergy through the combination of different land management policies [59,60]. Russo and Pavone believe that the mitigation potential of multiple land management policies that work together on the same land is generally greater than that of a single policy [61]. Moreover, Dax et al. also confirm that the combination of multiple land management policies can save resources, enhance social resilience and promote ecological restoration to better mitigate and adapt to climate change, prevent desertification and land degradation, and strengthen food security [62]. For example, (1) strengthening the combination of fire management and afforestation can increase land carbon sequestration, enhance the potential to mitigate climate change and land degradation, reduce management costs and ensure food production areas [63]. (2) Reducing food waste and a carnivorous diet will help to reduce carbon emissions, achieve sustainable land use management, and ensure food security and low carbon emissions [64]. (3) The construction of urban green infrastructure is also a solution to mitigate climate change. Through measures such as vertical greening, roof gardens, suburban agriculture and vertical agriculture, it can not only meet some food needs of urban residents but may also reduce the pressure of rural land food production and land degradation [65,66]. (4) In addition, improving land market management policies, ensuring land ownership and integrating environmental costs into food security and ecological compensation will help to achieve sustainable land management and eliminate poverty to achieve food security with stable food production [67,68]. A successful sustainable land management policy requires the participation of more stakeholders, especially local stakeholders such as local farmers and community residents who are easy to ignore, which can fully mobilize their enthusiasm to understand and practice land management policies [69,70]. However, there are great differences in the actual situation in diverse regions. Therefore, in future research, scholars should realistically build sustainable land management policies suitable for each region to achieve the stable production of food crops and ensure food security.

Agricultural Intensive Development

In the face of the food security crisis, although the development of marginal ecological land can improve food output in the short term, this process is mostly irreversible, and the opportunity for agricultural land expansion is limited [71]. Excessive exploitation of natural resources can easily lead to land degradation and reduction of ecological land area, resulting in greater social and ecological costs [12]. Therefore, the intensive development of land use is considered to be the fundamental approach to not only ensure the needs of human land products and functions but also to effectively reduce marginal land development and protect the ecological environment [72,73]. Compared with the traditional intensification realized by changing management practices and decisions, sustainable land use intensification has been widely discussed and explored as a necessary way to improve global food security and reduce ecological vulnerability and environmental pollution. Moreover, Charles and Struik et al. sustainable intensification can balance the competing demands for land use, improve ecosystem services and maintain biodiversity while increasing production to achieve common growth [74,75].

Research on agricultural intensification can be traced back to 1990. Vlek took sub-Saharan Africa as an example to explore the role of alternative soil fertility and other measures in agricultural production [76]. There are relatively many studies on sustainable intensification, focusing on the sustainable intensification of agricultural land, farms and agricultural production; the specific research contents include the conceptual connotation, empirical evaluation, impact mechanism, biodiversity, and improvement of soil organic matter [72,77]. For example, Wezel et al. [78] distinguished between the concepts of "ecological intensification", "sustainable intensification" and "agricultural ecological intensification" and analyzed the subtle differences of the three concepts. Mulwa et al. [79] used the dynamic random effect probit model and the control function method to evaluate the vitality of adopting sustainable agricultural inputs and the effect of large grain traders strengthening the adoption of these sustainable agricultural inputs at the farm level. However, with the in-depth development of agricultural intensification, ecological and environmental problems have gradually appeared. How to stabilize food production under the condition of coordinating land use types and protecting the ecological environment still needs further research.

Land Degradation

Climate change has changed the process of surface change and terrestrial ecosystems and their composition, structure and function [80], triggered changes in land use, accelerated the process of desertification and land degradation in many areas, reduced agricultural output and agricultural income, and deeply affected the security of world food production. According to the relevant data released by the Intergovernmental Science-Policy Platform on Biodiversity and Ecosystem Services (IPBES) of the United Nations, human intervention has degraded the ecological function of approximately 80% of the world's agricultural land, 10–20% of pasture land and 87% of wetlands, which brings economic losses ranging from 450 billion to 10.6 trillion US dollars to the global ecological service system every year; it also directly or indirectly affects the well-being of approximately 3.2 billion people around the world [81,82].

Land degradation seriously affects food production and distribution through soil erosion, the decline of land fertility and salinization [83]. Paoloni and Onorati found that it also directly threatens the well-being of the rural population, children and women and affects food security worldwide [84]. At present, a large number of studies have analyzed the factors of land degradation by exploring the driving force of land use change to analyze its impact on food security, especially from the aspects of geographical conditions, population characteristics, economic growth, road traffic, meteorological factors, Government policies, and technological evolution [75,85–87]. Among them, Prokop [86] analyzed the degree and type of land degradation of the Meghalaya Plateau through remote sensing data and found that the impact of different land degradation types and degrees on grain yield showed differentiated trends. In the face of land degradation, in response to the increase in food demand, Ranasinghe and Piyadasa [87] argue that we should integrate the main environmental, natural and socioeconomic factors in a region to build a productive land management system and explore an optimal mode of land production use to ensure the stable production of food. Land degradation is closely related to food production. Adopting sustainable land management policies not only effectively curbs the trend of land degradation and optimizes land use structure but also gives full play to the overall efficiency of different land types.

Renewable Bioenergy

With increasing attention given to energy security and ecological security, governments worldwide are pursuing multiple goals of energy security, reducing greenhouse gas emissions, and developing rural economies. Tian and Renzaho et al. found that the government have invested much money or established tax incentive mechanisms to develop renewable bioenergy represented by fuel ethanol and biodiesel to replace nonrenewable fossil fuels (coal, oil and natural gas) [88,89]. While the world vigorously advocates for the development of bioenergy to ensure energy security, the demand for land for bioenergy production is also increasing [90]. With the sharp rise of global food prices, whether bioenergy threatens food security is not only the focus of major international organizations and governments but also the main topic of debate within the academic community [46,91]. Although the use of bioenergy instead of fossil fuels can effectively reduce greenhouse gas emissions to a certain extent, the large-scale increase in bioenergy demand may also cause forest degradation and reduce food production [56,92]. Moreover, a large number of agricultural products are used to produce bioenergy, which greatly reduces the food supply in the international market and will inevitably lead to an increase in food prices [93], threatening global food security, especially the basic living needs of people in low-income countries with food shortages [94,95]. However, at present, there is no systematic research on how much-cultivated land is occupied by the development of bioenergy, what impact it has on land use change, how energy crops compete with other crop types at the household scale, and how to stabilize food production, which are worthy of further exploration by scholars in the future.

Food Production

Food security is a multidimensional security goal and is affected by many factors, among which food production is the most critical link in the food security system [96–98]. Land use change affects regional food production through changes in area and spatial location among different land use types, and temporal and spatial changes in cultivated land are one of the main forms of land use change [99], which affects the global food security supply [100]. At present, the research focus of most scholars is on quantifying the impact of cultivated land change on food security. However, due to the differences in research methods, regions and periods, the research results are also quite different [100]. For example, Wang et al. constructed the evaluation framework of "land food water" to quantify the impact of temporal and spatial changes in cultivated land on food production and water resource consumption and proposed the sustainable development policy of cultivated land and the optimal management policy of water resources [101].

In addition, the research results of some scholars show that grain production is affected by a variety of natural and socioeconomic factors, among which regional factors, family size, farming system, land use intensity, land tenure, climate change and environmental cost have a great influence on grain productivity. The actual grain yield is affected by the quantity and quality of cultivated land, climate, agricultural technology, and planting methods [13,102,103]. However, on the whole, although the current research helps to alleviate the contradiction between cultivated land change and food production, there are still some aspects to be optimized. For example, (1) the relevant research in this field is carried out at the national or a natural area level, which makes it difficult to guide practical work at the provincial level, and (2) weak supervision of newly reclaimed cultivated land and insufficient reserve resources of cultivated land easily leads to potential questions of food security production.

Agricultural Benefits

Global land use change is affected not only by climate change, land degradation and other factors but also by economic factors such as agricultural benefits [104]. The former is an irresistible natural factor, while the latter is the spontaneous change of land use types by farmers in pursuit of better comprehensive benefits [105]. In the environment of the market economy, farmers, as "rational economic people", their subjective will and choice of land production mode are the main influencing factors of cultivated land resource utilization and management and grain production capacity. Moreover, Wang and Tian et al. found that the price of agricultural products directly affects the type of cultivated land utilization and the result of grain production [106]. At present, there is a realistic situation that is not optimistic; that is, the economic benefits of food production are generally lower than those of other economic crops. Therefore, when there is no government subsidy or it is too low, farmers' willingness to plant food will continue to decrease and then switch to other economic crops [107]. In recent years, scholars have recognized that the change of land use types poses a greater threat to world food security than the small reduction of cultivated land area, and called on the Government to take effective measures to curb the drastic change of modes of man-made land use, which will help to stabilize the production area of cultivated food [108].

In addition, facing the problems of land fragmentation, higher agricultural production costs, lower agricultural productivity and lower grain output, most countries in the world have generally used effective measures, promoting moderately intensive land and large-scale production and management to transform and upgrade the agricultural system, to reduce agricultural production costs and to improve agricultural benefits [109]. Moreover, with the deepening of people's understanding of environmental pollution and biodiversity [110], scholars' attention to agricultural benefits has increased social and ecological benefits from a single economic benefit to emphasis more on the comprehensive benefits of agricultural production [111]. In this way, improving agricultural benefits not only protects the ecological environment but also stabilizes food production and ensures global food security.

3.3.2. Analysis of Frontier Trending Topics

Although the keyword clustering density map of the VOSviewer software can intuitively show the hot research topics in the field, the time factor is not considered. The time zone map of the CiteSpace software arranges keywords according to time series, which can more intuitively show the distribution of hot topics in each period [112]. Therefore, this paper combines the time zone map with the burst word detection function of CiteSpace software, which vividly shows the evolution of the research topic over time. We selected keywords with a frequency of more than five every one year (slice length = 1) from 1999 to

6 September 2021, to build a keyword cooccurrence network map (Figure 7). Furthermore, three burst words were detected (Table 6).

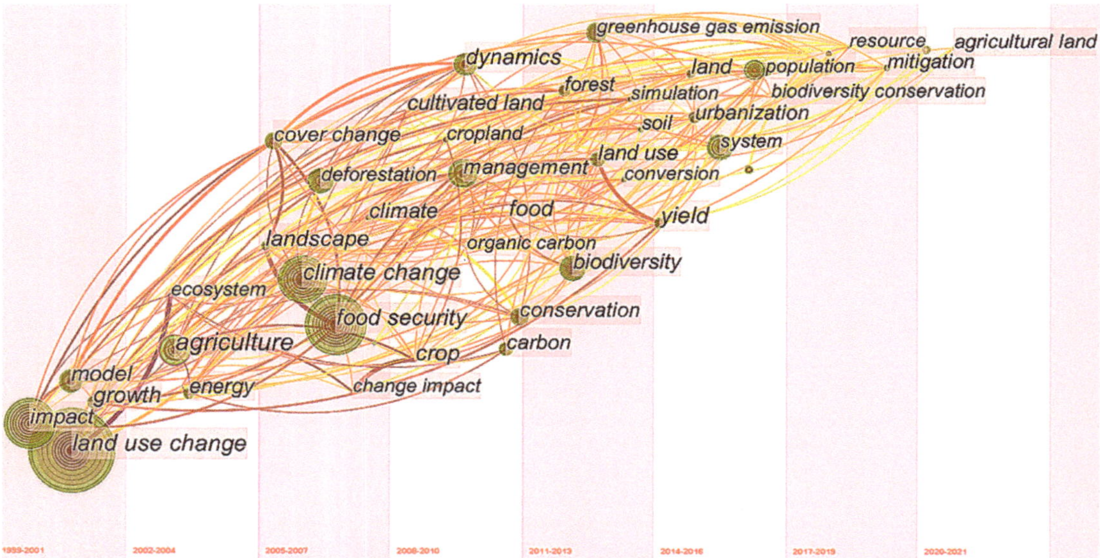

Figure 7. Time zone map for studying the evolution path produced by CiteSpace software.

Table 6. Top three burst keywords detection with the CiteSpace software.

Keywords	Year	Strength	Begin	End	1999–2021
Area	1999	3.56	2015	2016	
Consumption	1999	3.21	2015	2017	
Ecosystem service	1999	3.4	2019	2021	

Combined with Figure 7 and Table 6, according to the time distribution of key nodes, we summarize the development trend in the research field of land use change and food security as follows: (1) The research in the field of land use change and food security started from the field of land use change and then combined the research with food security. (2) From 2009 to 2017, there were a large number of key nodes related to the theme of land use change and food security, including agriculture, land use, forest, rice, food demand and yield. During this period, research on land use change and food security was in a stage of rapid development, which once again shows that research on this topic has attracted the continuous attention of scholars. (3) In addition, scholars generally pay attention to key nodes, including climate change, carbon, urban expansion, and environmental impact, which suggests that scholars prefer to further analyze the impact of land use change on food security by researching the current situation and influencing factors of land use change. (4) Since 2017, scholars have paid more attention to global research on land use change and food security, raised the issue of food security to the field of risk research, and advocated the formulation and implementation of agricultural protection policies to ensure food security.

The burst detection algorithm was proposed by Kleinberg to explore the research frontier trend in a field by studying the strength and duration of keyword bursts [113]. The research on land use change and food security includes three top burst keywords: "area", "consumption" and "ecosystem service" (Table 6). In 2015, the keywords "area" and "consumption" were the research hotspots, focusing on "differences in different research regions" and "energy and food consumption", but the duration was short. In 2019, the keyword "ecosystem service" has become a new research hotspot and continues until now. It focuses on agricultural intensive development, grain production and environmental protection, which shows that ecosystem service will become a hotspot and trend in future research. As a whole, there are few burst words in this research field, indicating that the research concentration in this field is poor, and further research is needed.

4. Discussion

To ensure food security, countries all over the world have generally adopted strict land use restriction measures, such as China's cultivated arable land minimum policy, Japan's land classification management system and the United States' land fallow policy. They hope to strictly restrict land use change through administrative control methods to stabilize food production and limit food risk within a controllable range [114–116]. At the same time, the academic research results on the theme of land use change and food security have increased significantly in the past two decades, and these results show a significant positive development trend.

4.1. Research Process

The results of this paper show that the first research article on the theme of land use change and food security was published in 1999. Since then, the number of published articles has shown a slow-growth trend. To deeply analyze the evolution of the research field of land use change and food security, this paper divides these research studies into three stages according to the number of articles published annually and the category and frequency of keywords. The first stage is the initial stage of research (1999–2008), during which the number of published articles was small, the research theme concentrated on a single topic, and overall research progress went slowly. Related research mainly focuses on conceptual and technical analysis, as well as direct analysis of the impact of land use change on food security. The second stage is the rapid development stage (2009–2017), during which basic research knowledge increased to a certain extent. The research focused on the influencing factors of land use change, including climate change, carbon emissions, and sustainable land management, climate change and sustainability would continue to be focused on in the future. During this stage, scholars also paid more attention to improving the ability to deal with land use change and food security risks, and to explore countermeasures and governance schemes at multiple levels, such as technology and policy. The number of articles published on the theme of land use change and food security did not increase significantly until 2009, which may attribute to the theme of World Food Day in that year, which was described as "coping with the crisis and achieving food security", and emphasized the serious plight of malnutrition of 1.02 billion people in the world, as well as the need to help solve the problem of hungry people under conditions of economic crisis. The third stage is the stable deepening stage (2018–present). In this stage, as the global food security problem becomes increasingly serious, scholars pay more attention to global land use change and food security, and they raise the issue of food security to a risk problem.

4.2. The Impact of Land Use Changes on Food Security

We found that the research in this field mainly explores the impact on food security from four land use change factors: environmental change, land quality, crop planting type and agricultural production mode. First, environmental change involves climate change and carbon emissions. Excessive carbon emissions will lead to global warming and extreme climate events, resulting in changes in production factors such as moisture, heat, humidity, and temperature, which will lead to changes in land use patterns to varying degrees, eventually affecting the cultivated land area of food production and endangering food security. Second, land quality is related to land degradation. Due to the unreasonable use of land and changes in the natural environment, part of land in the world has experienced serious degradation (soil erosion), which leads to a decline in land productivity and has a serious impact on food production. Third, crop planting types involve renewable bioenergy and food production. Compared with the economic benefits of food crops, the economic benefits of other cash crops are higher, especially with the rapid development of clean energy (bioenergy), which leads to the conversion of some cultivated land originally planted with food crops to other cash crops. The reduction in the planting area of food crops will inevitably lead to a decline in total food production, then threatening global food security. Fourth, the agricultural production model involves sustainable land management policy, agricultural intensive development, food production and agricultural benefits. Facing practical problems such as land degradation, reduction of ecological land, land pollution and decline of soil fertility, people urgently need to change the extensive agricultural production model to the production model with higher overall efficiency. For example, the intensive agricultural model pays more attention to the stability of grain production and the protection of the ecological environment, which can give full play to the economic, social and ecological effects of agricultural production [117].

4.3. Research Hotspots

At present, the research topic in this field is mainly aimed at the complex practical problems of global land use change and food security, and the research content is becoming increasingly organized and systematic. According to the clustering results, we can extract seven frontier hot topics in this field: climate change and carbon emissions, sustainable land management policy, agricultural intensive development, land degradation, renewable bioenergy, food production and agricultural benefits. These results present the trend of cross integration with economics, land management, soil science, public policy, politics, geography, and other disciplines, and indicate that the research in this field continues to expand. Meanwhile, it is worth noting that although the theme of land use change and food security has become very popular in recent years, scholars' research perspective is not limited to the direct analysis of the impact of land use change on food security but also considers climate change, carbon emissions, renewable bioenergy, agricultural intensive development models and other relevant aspects [7,8,54]. In terms of research methods, this field has formed a research system of geographic information technology, satellite remote sensing technology, theoretical models, investigations and interviews and other methods. Besides, through the analysis and summary of recent relevant literature, we found that scholars mainly focus on the hot issues including the utilization efficiency of chemical fertilizer (active nitrogen, etc.), urbanization expansion, greenhouse effects, and pay more attention to the combination of economic, social and ecological benefits of agricultural production to reduce the threat of land use change to food security [118–121].

4.4. Research Deficiency

This paper also has some research limitations that can be improved in the future. Firstly, because we only selected the WOS database as the data source for the bibliometric analysis of this study, and did not choose other databases (such as China National Knowledge Infrastructure, Scopus, et al.), the data used in our study cannot include all literature. However, as one of the most comprehensive databases in the world, the WOS database contains high-quality documents, it can represent the research hotspot and frontier in this field. Secondly, we can fully analyze the most influential publications in the database in the future, which will help to understand the real impact of the most important research in the scientific community.

5. Conclusions

This paper comprehensively used CiteSpace and VOSviewer software for bibliometric analysis and performed a visual analysis of the knowledge map of the literature with the theme of land use change and food security in the WOS database, and explored the research status, knowledge structure and evolution context. Since 1999, the number of annual published articles in the field of land use change and food security has shown an overall upward trend, which can be divided into three stages: initial research, rapid development, and a stable in-depth stage. Although a core author group was initially formed, the overall cooperation network is still relatively scattered. The distribution of research institutions is concentrated and forms a small cluster, which shows that there are only a few cooperative relationships among research institutions, and large institutional collaborative groups have not been formed across countries or regions. In national cooperation networks, developed countries are in the core position, but the cooperation network is not prominent. Meanwhile, keywords such as food security, land use change and climate change are taken as the core issues, they exhibit a radial shape and form seven frontier hot topics in this field. Moreover, due to the complexity of the research theme of land use change and food security, the research methods in this field are in-depth and diverse, and multidisciplinary development is constantly integrated. In addition, with the emergence of factors or problems such as climate change, carbon emission limitations, the application of new land use models and technologies, and the imbalance between food production and demand, there are new opportunities and challenges to the research on land use change and food security.

The research field of land use change and food security can be strengthened in the following aspects in the future. (1) Theoretical innovation still needs to be strengthened. At present, the research in this field is carried out through technology and mathematical models, which lack theoretical construction and innovation. The research perspective can also be innovated from the dimensions of land property rights systems and land management systems. (2) We should strengthen the research on the impact of agricultural chemical fertilizers on food security. Agricultural chemical fertilizer plays an important role in slowing down land use change and ensuring the sufficiency of food production, especially synthetic nitrogen fertilizer. However, the loss of active nitrogen will not only affect grain yields but also pollute the ecological environment. (3) Further research can focus on the impact of urban expansion on land use change. Due to the urbanization process of countries worldwide, a large amount of high-quality cultivated land around cities is occupied, which seriously threatens food security, especially in developing countries.

Author Contributions: Conceptualization, P.C. and P.J.; methodology, P.C., Y.D. and H.T.; software, P.C. and H.T.; formal analysis, P.C. and H.T.; data curation, P.C. and K.L.; writing—original draft preparation, P.C. and H.T.; writing—review and editing, P.C., H.T., Y.D., K.L. and P.J.; visualization, P.C.; supervision, P.J. and Y.L.; project administration, P.J. and Y.L.; funding acquisition, P.J. and Y.L. All authors have read and agreed to the published version of the manuscript.

Funding: This research was funded by the Major Projects of National Social Science Foundation (20ZDA086).

Institutional Review Board Statement: Not applicable.

Informed Consent Statement: Not applicable.

Data Availability Statement: The data presented in this study are available in the Web of Science core database.

Conflicts of Interest: The authors declare no conflict of interest.

References

1. Galeana-Pizaña, J.M.; Couturier, S.; Monsivais-Huertero, A. Assessing food security and environmental protection in Mexico with a GIS-based Food Environmental Efficiency index. *Land Use Policy* **2018**, *76*, 442–454. [CrossRef]
2. Daioglou, V.; Doelman, J.C.; Wicke, B.; Faaij, A.; van Vuuren, D.P. Integrated assessment of biomass supply and demand in climate change mitigation scenarios. *Glob. Environ. Chang.* **2019**, *54*, 88–101. [CrossRef]
3. Chigbu, U.E.; Ntihinyurwa, P.D.; de Vries, W.T.; Ngenzi, E.I. Why tenure responsive land-use planning matters: Insights for land use consolidation for food security in rwanda. *Int. J. Environ. Res. Public Health* **2019**, *16*, 1354. [CrossRef] [PubMed]
4. Long, H.; Ge, D.; Zhang, Y.; Tu, S.; Qu, Y.; Ma, L. Changing man-land interrelations in China's farming area under urbanization and its implications for food security. *J. Environ. Manag.* **2018**, *209*, 440–451. [CrossRef] [PubMed]

5. Chen, C.; Yu, L.; Choguill, C.L. "Dipiao", Chinese approach to transfer of land development rights: The experiences of Chongqing. *Land Use Policy* **2020**, *99*, 104870. [CrossRef]
6. Ericksen, P.J. Conceptualizing food systems for global environmental change research. *Glob. Environ. Chang.* **2008**, *18*, 234–245. [CrossRef]
7. Verburg, P.H.; Mertz, O.; Erb, K.H.; Haberl, H.; Wu, W. Land system change and food security: Towards multi-scale land system solutions. *Curr. Opin. Environ. Sustain.* **2013**, *5*, 494–502. [CrossRef] [PubMed]
8. Guo, Y.; Fu, Y.; Hao, F.; Zhang, X.; Wu, W.; Jin, X.; Robin Bryant, C.; Senthilnath, J. Integrated phenology and climate in rice yields prediction using machine learning methods. *Ecol. Indic.* **2021**, *120*, 106935. [CrossRef]
9. Ge, D.; Long, H.; Zhang, Y.; Ma, L.; Li, T. Farmland transition and its influences on grain production in China. *Land Use Policy* **2018**, *70*, 94–105. [CrossRef]
10. Wang, J.; Zhang, Z.; Liu, Y. Spatial shifts in grain production increases in China and implications for food security. *Land Use Policy* **2018**, *74*, 204–213. [CrossRef]
11. UNDESA. *World Family Planning;* United Nations, Department of Economic and Social Affairs: New York, NY, USA, 2020; ISBN 9789211483482.
12. Schiefer, J.; Lair, G.J.; Blum, W.E.H. Potential and limits of land and soil for sustainable intensification of European agriculture. *Agric. Ecosyst. Environ.* **2016**, *230*, 283–293. [CrossRef]
13. Singirankabo, U.A.; Ertsen, M.W. Relations between land tenure security and agricultural productivity: Exploring the effect of land registration. *Land* **2020**, *9*, 138. [CrossRef]
14. Lu, D.; Wang, Y.; Yang, Q.; He, H.; Su, K. Exploring a moderate fallow scale of cultivated land in China from the perspective of food security. *Int. J. Environ. Res. Public Health* **2019**, *16*, 4329. [CrossRef] [PubMed]
15. Zhou, Y.; Li, Y.; Xu, C. Land consolidation and rural revitalization in China: Mechanisms and paths. *Land Use Policy* **2020**, *91*, 104379. [CrossRef]
16. Li, Q.; Yan, J. Assessing the health of agricultural land with emergy analysis and fuzzy logic in the major grain-producing region. *Catena* **2012**, *99*, 9–17. [CrossRef]
17. Qi, X.; Si, Z.; Zhong, T.; Huang, X.; Crush, J. Spatial determinants of urban wet market vendor profit in Nanjing, China. *Habitat Int.* **2019**, *94*, 102064. [CrossRef]
18. Zhong, T.; Si, Z.; Scott, S.; Crush, J.; Yang, K.; Huang, X. Comprehensive Food System Planning for Urban Food Security in Nanjing, China. *Land* **2021**, *10*, 1090. [CrossRef]
19. Moore, N.; Alagarswamy, G.; Pijanowski, B.; Thornton, P.; Lofgren, B.; Olson, J.; Andresen, J.; Yanda, P.; Qi, J. East African food security as influenced by future climate change and land use change at local to regional scales. *Clim. Chang.* **2012**, *110*, 823–844. [CrossRef]
20. Liao, H.; Tang, M.; Luo, L.; Li, C.; Chiclana, F.; Zeng, X.J. A bibliometric analysis and visualization of medical big data research. *Sustainability* **2018**, *10*, 166. [CrossRef]
21. Abati, R.; Sampaio, A.R.; Maciel, R.M.A.; Colombo, F.C.; Libardoni, G.; Battisti, L.; Lozano, E.R.; de Castilhos Ghisi, N.; Costa-Maia, F.M.; Potrich, M. Bees and pesticides: The research impact and scientometrics relations. *Environ. Sci. Pollut. Res.* **2021**, *28*, 32282–32298. [CrossRef]
22. Chen, C.; Chen, Y.; Horowitz, M.; Hou, H.; Liu, Z.; Pellegrino, D. Towards an explanatory and computational theory of scientific discovery. *J. Informetr.* **2009**, *3*, 191–209. [CrossRef]
23. Ali, M.; Prakash, K.; Hossain, M.A.; Pota, H.R. Intelligent energy management: Evolving developments, current challenges, and research directions for sustainable future. *J. Clean. Prod.* **2021**, *314*, 127904. [CrossRef]
24. Ji, B.; Zhao, Y.; Vymazal, J.; Mander, Ü.; Lust, R.; Tang, C. Mapping the field of constructed wetland-microbial fuel cell: A review and bibliometric analysis. *Chemosphere* **2021**, *262*, 128366. [CrossRef] [PubMed]
25. Chen, C.; Song, I.Y.; Yuan, X.; Zhang, J. The thematic and citation landscape of Data and Knowledge Engineering (1985–2007). *Data Knowl. Eng.* **2008**, *67*, 234–259. [CrossRef]
26. Fan, J.; Gao, Y.; Zhao, N.; Dai, R.; Zhang, H.; Feng, X.; Shi, G.; Tian, J.; Chen, C.; Hambly, B.D.; et al. Bibliometric Analysis on COVID-19: A Comparison of Research Between English and Chinese Studies. *Front. Public Health* **2020**, *8*, 477. [CrossRef] [PubMed]
27. van Eck, N.J.; Waltman, L. Software survey: VOSviewer, a computer program for bibliometric mapping. *Scientometrics* **2010**, *84*, 523–538. [CrossRef]
28. Donthu, N.; Kumar, S.; Pattnaik, D. Forty-five years of Journal of Business Research: A bibliometric analysis. *J. Bus. Res.* **2020**, *109*, 39. [CrossRef]
29. Long, H.; Zhang, Y.; Ma, L.; Tu, S. Land use transitions: Progress, challenges and prospects. *Land* **2021**, *10*, 903. [CrossRef]
30. Sporton, D.; Thomas, D.S.G.; Morrison, J. Outcomes of social and environmental change in the Kalahari of Botswana: The role of migration. *J. S. Afr. Stud.* **1999**, *25*, 441–459. [CrossRef]
31. Murdiyarso, D. Adaptation to climatic variability and change: Asian perspectives on agriculture and food security. *Environ. Monit. Assess.* **2000**, *61*, 123–131. [CrossRef]
32. Verburg, P.H.; Chen, Y.; Veldkamp, T.A. Spatial explorations of land use change and grain production in China. *Agric. Ecosyst. Environ* **2000**, *82*, 333–354. [CrossRef]

33. Alexander, P.; Rounsevell, M.D.A.; Dislich, C.; Dodson, J.R.; Engström, K.; Moran, D. Drivers for global agricultural land use change: The nexus of diet, population, yield and bioenergy. *Glob. Environ. Chang.* **2015**, *35*, 138–147. [CrossRef]
34. Verburg, P.H.; Ritsema van Eck, J.R.; de Nijs, T.C.M.; Dijst, M.J.; Schot, P. Determinants of land-use change patterns in the Netherlands. *Environ. Plan. B Plan. Des.* **2004**, *31*, 125–150. [CrossRef]
35. Khan, S.; Hanjra, M.A.; Mu, J. Water management and crop production for food security in China: A review. *Agric. Water Manag.* **2009**, *96*, 349–360. [CrossRef]
36. Mertz, O.; Mbow, C.; Reenberg, A.; Diouf, A. Farmers' perceptions of climate change and agricultural adaptation strategies in rural sahel. *Environ. Manag.* **2009**, *43*, 804–816. [CrossRef]
37. Garnett, T. Livestock-related greenhouse gas emissions: Impacts and options for policy makers. *Environ. Sci. Policy* **2009**, *12*, 491–503. [CrossRef]
38. Yan, H.; Liu, J.; Huang, H.Q.; Tao, B.; Cao, M. Assessing the consequence of land use change on agricultural productivity in China. *Glob. Planet. Chang.* **2009**, *67*, 13–19. [CrossRef]
39. Song, X.; Ouyang, Z.; Li, Y.; Li, F. Cultivated land use change in China, 1999-2007: Policy development perspectives. *J. Geogr. Sci.* **2012**, *22*, 1061–1078. [CrossRef]
40. Ziem Bonye, S.; Yenglier Yiridomoh, G.; Derbile, E.K. Urban expansion and agricultural land use change in Ghana: Implications for peri-urban farmer household food security in Wa Municipality. *Int. J. Urban Sustain. Dev.* **2021**, *13*, 383–399. [CrossRef]
41. Wang, C.; Siriwardana, M.; Meng, S. Effects of the Chinese arable land fallow system and land-use change on agricultural production and on the economy. *Econ. Model.* **2019**, *79*, 186–197. [CrossRef]
42. Jiang, Y.; Ritchie, B.W.; Benckendorff, P. Bibliometric visualisation: An application in tourism crisis and disaster management research. *Curr. Issues Tour.* **2019**, *22*, 1925–1957. [CrossRef]
43. de Price, D.J.S. *Little Science, Big Science*; Columbia University Press: New York, NY, USA, 1963; ISBN 023188575X.
44. Chen, C.; Ibekwe-SanJuan, F.; Hou, J. The structure and dynamics of cocitation clusters: A multiple-perspective cocitation analysis. *J. Am. Soc. Inf. Sci. Technol.* **2010**, *61*, 1386–1409. [CrossRef]
45. Colares, G.S.; Dell'Osbel, N.; Wiesel, P.G.; Oliveira, G.A.; Lemos, P.H.Z.; da Silva, F.P.; Lutterbeck, C.A.; Kist, L.T.; Machado, Ê.L. Floating treatment wetlands: A review and bibliometric analysis. *Sci. Total Environ.* **2020**, *714*, 136776. [CrossRef] [PubMed]
46. Zimmermann, A.; Benda, J.; Webber, H.; Jafari, Y. *Trade, Food Security and Climate Change: Conceptual Linkages and Policy Implications*; Food and Agriculture Organization of the United Nations: Rome, Italy, 2018; ISBN 9789251311103.
47. Yawson, D.O. Estimating virtual land use under future conditions: Application of a food balance approach using the UK. *Land Use Policy* **2021**, *101*, 105132. [CrossRef]
48. Yawson, D.O.; Mulholland, B.J.; Ball, T.; Adu, M.O.; Mohan, S.; White, P.J. Effect of climate and agricultural land use changes on UK feed barley production and food security to the 2050s. *Land* **2017**, *6*, 74. [CrossRef]
49. Hasegawa, T.; Fujimori, S.; Shin, Y.; Tanaka, A.; Takahashi, K.; Masui, T. Consequence of Climate Mitigation on the Risk of Hunger. *Environ. Sci. Technol.* **2015**, *49*, 7245–7253. [CrossRef]
50. Krishnamurthy, P.K.; Lewis, K.; Choularton, R.J. A methodological framework for rapidly assessing the impacts of climate risk on national-level food security through a vulnerability index. *Glob. Environ. Chang.* **2014**, *25*, 121–132. [CrossRef]
51. Jabbar, A.; Wu, Q.; Peng, J.; Sher, A.; Imran, A.; Wang, K. Mitigating catastrophic risks and food security threats: Effects of land ownership in Southern Punjab, Pakistan. *Int. J. Environ. Res. Public Health* **2020**, *17*, 9258. [CrossRef]
52. Hasegawa, T.; Fujimori, S.; Havlík, P.; Valin, H.; Bodirsky, B.L.; Doelman, J.C.; Fellmann, T.; Kyle, P.; Koopman, J.F.L.; Lotze-Campen, H.; et al. Risk of increased food insecurity under stringent global climate change mitigation policy. *Nat. Clim. Chang.* **2018**, *8*, 699–703. [CrossRef]
53. Nobre, C.A.; Sampaio, G.; Borma, L.S.; Castilla-Rubio, J.C.; Silva, J.S.; Cardoso, M. Land-use and climate change risks in the amazon and the need of a novel sustainable development paradigm. *Proc. Natl. Acad. Sci. USA* **2016**, *113*, 10759–10768. [CrossRef] [PubMed]
54. Beltrán-Tolosa, L.M.; Navarro-Racines, C.; Pradhan, P.; Cruz-Garcia, G.S.; Solis, R.; Quintero, M. Action needed for staple crops in the Andean-Amazon foothills because of climate change. *Mitig. Adapt. Strateg. Glob. Chang.* **2020**, *25*, 1103–1127. [CrossRef]
55. Liu, X.; Liu, Y.; Liu, Z.; Chen, Z. Impacts of climatic warming on cropping system borders of China and potential adaptation strategies for regional agriculture development. *Sci. Total Environ.* **2021**, *755*, 142415. [CrossRef]
56. Popp, A.; Calvin, K.; Fujimori, S.; Havlik, P.; Humpenöder, F.; Stehfest, E.; Bodirsky, B.L.; Dietrich, J.P.; Doelmann, J.C.; Gusti, M.; et al. Land-use futures in the shared socio-economic pathways. *Glob. Environ. Chang.* **2017**, *42*, 331–345. [CrossRef]
57. Ranjan, R. Land use decisions under REDD+ incentives when warming temperatures affect crop productivity and forest biomass growth rates. *Land Use Policy* **2021**, *108*, 105595. [CrossRef]
58. Wilts, R.; Latka, C.; Britz, W. Who is most vulnerable to climate change induced yield changes? A dynamic long run household analysis in lower income countries. *Clim. Risk Manag.* **2021**, *33*, 100330. [CrossRef]
59. Dunning, R.J.; Moore, T.; Watkins, C. The use of public land for house building in England: Understanding the challenges and policy implications. *Land Use Policy* **2021**, *105*, 105434. [CrossRef]
60. Wang, L.; Zheng, W.; Tang, L.; Zhang, S.; Liu, Y.; Ke, X. Spatial optimization of urban land and cropland based on land production capacity to balance cropland protection and ecological conservation. *J. Environ. Manag.* **2021**, *285*, 112054. [CrossRef] [PubMed]
61. Russo, M.; Pavone, P. Evidence-based portfolios of innovation policy mixes: A cross-country analysis. *Technol. Forecast. Soc. Chang.* **2021**, *168*, 120708. [CrossRef]

62. Dax, T.; Schroll, K.; Machold, I.; Derszniak-Noirjean, M.; Schuh, B.; Gaupp-Berghausen, M. Land abandonment in mountain areas of the EU: An inevitable side effect of farming modernization and neglected threat to sustainable land use. *Land* **2021**, *10*, 591. [CrossRef]
63. Le Page, Y.; Hurtt, G.; Thomson, A.M.; Bond-Lamberty, B.; Patel, P.; Wise, M.; Calvin, K.; Kyle, P.; Clarke, L.; Edmonds, J.; et al. Sensitivity of climate mitigation strategies to natural disturbances. *Environ. Res. Lett.* **2013**, *8*, 15018. [CrossRef]
64. Shabanali Fami, H.; Aramyan, L.H.; Sijtsema, S.J.; Alambaigi, A. The relationship between household food waste and food security in Tehran city: The role of urban women in household management. *Ind. Mark. Manag.* **2021**, *97*, 71–83. [CrossRef]
65. Edmondson, J.L.; Davies, Z.G.; Gaston, K.J.; Leake, J.R. Urban cultivation in allotments maintains soil qualities adversely affected by conventional agriculture. *J. Appl. Ecol.* **2014**, *51*, 880–889. [CrossRef] [PubMed]
66. Wilhelm, J.A.; Smith, R.G. Ecosystem services and land sparing potential of urban and peri-urban agriculture: A review. *Renew. Agric. Food Syst.* **2018**, *33*, 481–494. [CrossRef]
67. Mwambo, F.M.; Fürst, C.; Nyarko, B.K.; Borgemeister, C.; Martius, C. Maize production and environmental costs: Resource evaluation and strategic land use planning for food security in northern Ghana by means of coupled emergy and data envelopment analysis. *Land Use Policy* **2020**, *95*, 104490. [CrossRef]
68. Zhou, Y.; Li, X.; Liu, Y. Rural land system reforms in China: History, issues, measures and prospects. *Land Use Policy* **2020**, *91*, 104330. [CrossRef]
69. Therville, C.; Antona, M.; de Foresta, H. The policyscape of agroforestry within Mediterranean protected landscapes in France. *Sustain. Sci.* **2020**, *15*, 1435–1448. [CrossRef]
70. Domingo, A.; Charles, K.A.; Jacobs, M.; Brooker, D.; Hanning, R.M. Indigenous community perspectives of food security, sustainable food systems and strategies to enhance access to local and traditional healthy food for partnering williams treaties first nations (Ontario, Canada). *Int. J. Environ. Res. Public Health* **2021**, *18*, 4404. [CrossRef]
71. Liu, Y.; Song, W.; Deng, X. Understanding the spatiotemporal variation of urban land expansion in oasis cities by integrating remote sensing and multi-dimensional DPSIR-based indicators. *Ecol. Indic.* **2019**, *96*, 23–37. [CrossRef]
72. Djurfeldt, A.A.; Hall, O.; Isinika, A.; Msuya, E.; Yengoh, G.T. Sustainable agricultural intensification in four Tanzanian villages—A view from the ground and the sky. *Sustainbility* **2020**, *12*, 8304. [CrossRef]
73. Kyalo Willy, D.; Muyanga, M.; Jayne, T. Can economic and environmental benefits associated with agricultural intensification be sustained at high population densities? A farm level empirical analysis. *Land Use Policy* **2019**, *81*, 100–110. [CrossRef] [PubMed]
74. Charles, H.; Godfray, H.; Garnett, T. Food security and sustainable intensification. *Philos. Trans. R. Soc. B Biol. Sci.* **2014**, *369*, 6–11. [CrossRef]
75. Struik, P.C.; Kuyper, T.W. Sustainable intensification in agriculture: The richer shade of green. A review. *Agron. Sustain. Dev.* **2017**, *37*, 39. [CrossRef]
76. Vlek, P.L.G. The role of fertilizers in sustaining agriculture in sub-Saharan Africa. *Fertil. Res.* **1990**, *26*, 327–339. [CrossRef]
77. Fischer, G.; Darkwah, A.; Kamoto, J.; Kampanje-Phiri, J.; Grabowski, P.; Djenontin, I. Sustainable agricultural intensification and gender-biased land tenure systems: An exploration and conceptualization of interactions. *Int. J. Agric. Sustain.* **2020**, *19*, 403–422. [CrossRef]
78. Wezel, A.; Soboksa, G.; McClelland, S.; Delespesse, F.; Boissau, A. The blurred boundaries of ecological, sustainable, and agroecological intensification: A review. *Agron. Sustain. Dev.* **2015**, *35*, 1283–1295. [CrossRef]
79. Mulwa, C.K.; Muyanga, M.; Visser, M. The role of large traders in driving sustainable agricultural intensification in smallholder farms: Evidence from Kenya. *Agric. Econ.* **2021**, *52*, 329–341. [CrossRef]
80. Seddon, A.W.R.; Macias-Fauria, M.; Long, P.R.; Benz, D.; Willis, K.J. Sensitivity of global terrestrial ecosystems to climate variability. *Nature* **2016**, *531*, 229–232. [CrossRef] [PubMed]
81. Zhang, B.; Pan, Y.; Xu, J.; Tian, Y. IPBES thematic assessment on land degradation and restoration and its potential impact. *Biodivers. Sci.* **2018**, *26*, 1243–1248. [CrossRef]
82. Gonzalez-Roglich, M.; Zvoleff, A.; Noon, M.; Liniger, H.; Fleiner, R.; Harari, N.; Garcia, C. Synergizing global tools to monitor progress towards land degradation neutrality: Trends.Earth and the World Overview of Conservation Approaches and Technologies sustainable land management database. *Environ. Sci. Policy* **2019**, *93*, 34–42. [CrossRef]
83. Prăvălie, R.; Patriche, C.; Borrelli, P.; Panagos, P.; Roșca, B.; Dumitraşcu, M.; Nita, I.A.; Săvulescu, I.; Birsan, M.V.; Bandoc, G. Arable lands under the pressure of multiple land degradation processes. A global perspective. *Environ. Res.* **2021**, *194*, 110697. [CrossRef]
84. Paoloni, L.; Onorati, A. Regulations of large-scale acquisitions of land: The case of the voluntary guidelines on the responsible governance of land, fisheries and forests. *Law Dev. Rev.* **2014**, *7*, 369–400. [CrossRef]
85. Ren, Y.; Lü, Y.; Fu, B.; Comber, A.; Li, T.; Hu, J. Driving factors of land change in china's loess plateau: Quantification using geographically weighted regression and management implications. *Remote Sens.* **2020**, *12*, 453. [CrossRef]
86. Prokop, P. Remote sensing of severely degraded land: Detection of long-term land-use changes using high-resolution satellite images on the Meghalaya Plateau, northeast India. *Remote Sens. Appl. Soc. Environ.* **2020**, *20*, 100432. [CrossRef]
87. Ranasinghe, T.K.G.P.; Piyadasa, R.U.K. Optimising usage of salinized lands in the lower part of the river basin for the coastal community in Bentota, Sri Lanka. *J. Natl. Sci. Found. Sri Lanka* **2020**, *48*, 379–396. [CrossRef]
88. Tian, Y.; Zhao, L.; Meng, H.; Sun, L.; Yan, J. Estimation of un-used land potential for biofuels development in China. *Appl. Energy* **2009**, *86*, S77–S85. [CrossRef]

89. Renzaho, A.M.N.; Kamara, J.K.; Toole, M. Biofuel production and its impact on food security in low and middle income countries: Implications for the post-2015 sustainable development goals. *Renew. Sustain. Energy Rev.* **2017**, *78*, 503–516. [CrossRef]
90. Wicker, R.J.; Kumar, G.; Khan, E.; Bhatnagar, A. Emergent green technologies for cost-effective valorization of microalgal biomass to renewable fuel products under a biorefinery scheme. *Chem. Eng. J.* **2021**, *415*, 128932. [CrossRef]
91. Benites Lazaro, L.L.; Giatti, L.L.; Puppim de Oliveira, J.A. Water-energy-food nexus approach at the core of businesses—How businesses in the bioenergy sector in Brazil are responding to integrated challenges? *J. Clean. Prod.* **2021**, *303*, 127102. [CrossRef]
92. Harris, E.; Ladreiter-Knauss, T.; Butterbach-Bahl, K.; Wolf, B.; Bahn, M. Land-use and abandonment alters methane and nitrous oxide fluxes in mountain grasslands. *Sci. Total Environ.* **2018**, *628–629*, 997–1008. [CrossRef]
93. Hasegawa, T.; Sands, R.D.; Brunelle, T.; Cui, Y.; Frank, S.; Fujimori, S.; Popp, A. Food security under high bioenergy demand toward long-term climate goals. *Clim. Chang.* **2020**, *163*, 1587–1601. [CrossRef]
94. Carrino, L.; Visconti, D.; Fiorentino, N.; Fagnano, M. Biofuel production with castor bean: A win-win strategy for marginal land. *Agronomy* **2020**, *10*, 1690. [CrossRef]
95. Brinkman, M.; Levin-Koopman, J.; Wicke, B.; Shutes, L.; Kuiper, M.; Faaij, A.; van der Hilst, F. The distribution of food security impacts of biofuels, a Ghana case study. *Biomass Bioenergy* **2020**, *141*, 105695. [CrossRef]
96. Jiang, L.; Wu, S.; Liu, Y.; Yang, C. Grain security assessment in Bangladesh based on supply-demand balance analysis. *PLoS ONE* **2021**, *16*, e0252187. [CrossRef]
97. Serra-Majem, L.; Tomaino, L.; Dernini, S.; Berry, E.M.; Lairon, D.; de la Cruz, J.N.; Bach-Faig, A.; Donini, L.M.; Medina, F.X.; Belahsen, R.; et al. Updating the mediterranean diet pyramid towards sustainability: Focus on environmental concerns. *Int. J. Environ. Res. Public Health* **2020**, *17*, 8758. [CrossRef]
98. He, J.; Liu, Y.; Yu, Y.; Tang, W.; Xiang, W.; Liu, D. A counterfactual scenario simulation approach for assessing the impact of farmland preservation policies on urban sprawl and food security in a major grain-producing area of China. *Appl. Geogr.* **2013**, *37*, 127–138. [CrossRef]
99. Cheng, C.; Liu, Y.; Liu, Y.; Yang, R.; Hong, Y.; Lu, Y.; Pan, J.; Chen, Y. Cropland use sustainability in Cheng–Yu Urban Agglomeration, China: Evaluation framework, driving factors and development paths. *J. Clean. Prod.* **2020**, *256*, 120692. [CrossRef]
100. Li, Y.; Li, X.; Tan, M.; Wang, X.; Xin, L. The impact of cultivated land spatial shift on food crop production in China, 1990–2010. *Land Degrad. Dev.* **2018**, *29*, 1652–1659. [CrossRef]
101. Wang, X.; Xin, L.; Tan, M.; Li, X.; Wang, J. Impact of spatiotemporal change of cultivated land on food-water relations in China during 1990–2015. *Sci. Total Environ.* **2020**, *716*, 137119. [CrossRef] [PubMed]
102. Pan, T.; Du, G.; Dong, J.; Kuang, W.; De Maeyer, P.; Kurban, A. Divergent changes in cropping patterns and their effects on grain production under different agro-ecosystems over high latitudes in China. *Sci. Total Environ.* **2019**, *659*, 314–325. [CrossRef] [PubMed]
103. Saddique, Q.; Liu, D.L.; Wang, B.; Feng, P.; He, J.; Ajaz, A.; Ji, J.; Xu, J.; Zhang, C.; Cai, H. Modelling future climate change impacts on winter wheat yield and water use: A case study in Guanzhong Plain, northwestern China. *Eur. J. Agron.* **2020**, *119*, 126113. [CrossRef]
104. Jalilov, S.M.; Keskinen, M.; Varis, O.; Amer, S.; Ward, F.A. Managing the water-energy-food nexus: Gains and losses from new water development in Amu Darya River Basin. *J. Hydrol.* **2016**, *539*, 648–661. [CrossRef]
105. Santisteban, J.I.; Celis, A.; Mediavilla, R.; Gil-García, M.J.; Ruiz-Zapata, B.; Castaño, S. The transition from climate-driven to human-driven agriculture during the Little Ice Age in Central Spain: Documentary and fluvial records evidence. *Palaeogeogr. Palaeoclimatol. Palaeoecol.* **2021**, *562*, 110153. [CrossRef]
106. Wang, S.; Tian, Y.; Liu, X.; Foley, M. How farmers make investment decisions: Evidence from a farmer survey in China. *Sustainbility* **2020**, *12*, 247. [CrossRef]
107. Qiu, T.; Boris Choy, S.T.; Li, S.; He, Q.; Luo, B. Does land renting-in reduce grain production? Evidence from rural China. *Land Use Policy* **2020**, *90*, 104311. [CrossRef]
108. Schlindwein, S.L.; Feitosa de Vasconcelos, A.C.; Bonatti, M.; Sieber, S.; Strapasson, A.; Lana, M. Agricultural land use dynamics in the Brazilian part of La Plata Basin: From driving forces to societal responses. *Land Use Policy* **2021**, *107*, 105519. [CrossRef]
109. Gao, X.; Li, B.; Jiang, S.; Nie, Y. Can increasing scale efficiency curb agricultural nonpoint source pollution? *Int. J. Environ. Res. Public Health* **2021**, *18*, 8798. [CrossRef] [PubMed]
110. Kong, X.; Zhou, Z.; Jiao, L. Hotspots of land-use change in global biodiversity hotspots. *Resour. Conserv. Recycl.* **2021**, *174*, 105770. [CrossRef]
111. Al Masud, M.M.; Gain, A.K.; Azad, A.K. Tidal river management for sustainable agriculture in the Ganges-Brahmaputra delta: Implication for land use policy. *Land Use Policy* **2020**, *92*, 104443. [CrossRef]
112. Chen, C.; Hu, Z.; Liu, S.; Tseng, H. Emerging trends in regenerative medicine: A scientometric analysis in CiteSpace. *Expert Opin. Biol. Ther.* **2012**, *12*, 593–608. [CrossRef] [PubMed]
113. Kleinberg, J. Bursty and hierarchical structure in streams. *Data Min. Knowl. Discov.* **2003**, *7*, 373–397. [CrossRef]
114. Liu, Y.; Zhou, Y. Reflections on China's food security and land use policy under rapid urbanization. *Land Use Policy* **2021**, *109*, 105699. [CrossRef]
115. Huang, W.; Hashimoto, S.; Yoshida, T.; Saito, O.; Taki, K. A nature-based approach to mitigate flood risk and improve ecosystem services in Shiga, Japan. *Ecosyst. Serv.* **2021**, *50*, 101309. [CrossRef]

116. Mutiibwa, D.; Fleisher, D.H.; Resop, J.P.; Timlin, D.; Reddy, V.R. Regional food production and land redistribution as adaptation to climate change in the U.S. Northeast Seaboard. *Comput. Electron. Agric.* **2018**, *154*, 54–70. [CrossRef]
117. Ickowitz, A.; Powell, B.; Rowland, D.; Jones, A.; Sunderland, T. Agricultural intensification, dietary diversity, and markets in the global food security narrative. *Glob. Food Sec.* **2019**, *20*, 9–16. [CrossRef]
118. Chang, J.; Havlík, P.; Leclère, D.; de Vries, W.; Valin, H.; Deppermann, A.; Hasegawa, T.; Obersteiner, M. Reconciling regional nitrogen boundaries with global food security. *Nat. Food* **2021**, *2*, 700–711. [CrossRef]
119. Wang, X.; Liu, F. Effects of elevated co2 and heat on wheat grain quality. *Plants* **2021**, *10*, 1027. [CrossRef]
120. Youssef, A.; Sewilam, H.; Khadr, Z. Impact of Urban Sprawl on Agriculture Lands in Greater Cairo. *J. Urban Plan. Dev.* **2020**, *146*, 05020027. [CrossRef]
121. Zhang, Y.; Wang, H.; Xie, P.; Rao, Y.; He, Q. Revisiting Spatiotemporal Changes in Global Urban Expansion during 1995 to 2015. *Complexity* **2020**, *2020*, 6139158. [CrossRef]

International Journal of
Environmental Research and Public Health

MDPI

Article

Spatio-Temporal Characteristics of Landscape Ecological Risks in the Ecological Functional Zone of the Upper Yellow River, China

Fuwei Qiao [1], Yongping Bai [2], Lixia Xie [2,*], Xuedi Yang [3] and Shuaishuai Sun [4]

[1] College of Economics, Northwest Normal University, Lanzhou 730070, China; qfw279@nwnu.edu.cn
[2] College of Geography and Environmental Science, Northwest Normal University, Lanzhou 730070, China; baiyp@nwnu.edu.cn
[3] College of Earth and Environmental Sciences, Lanzhou University, Lanzhou 730070, China; yxdlz01@163.com
[4] Northwest Branch of Shanghai Tongji Urban Planning & Design Institute Co., Xi'an 710000, China; sun62817@126.com
* Correspondence: xlxyuqiao@126.com

Abstract: The Ecological Functional Zone of the Upper Yellow River (EFZUYR) is a critical water-catching area in the Yellow River Basin, the ecological security of which affects the sound development of the ecosystem in the entire basin. Recently, significant land use changes have aggravated regional ecological risks and seriously affected the sustainable development of EFZUYR. In this context, this paper provides an in-depth study of the ecological risks caused by land use landscape changes. With the help of land use data and dynamic degree analysis, the land use transfer matrix, and the landscape pattern index, this paper quantifies the distribution trends of land use landscape patterns in EFZUYR from 1990 to 2018. In addition, this research explores the temporal and spatial dynamic distribution characteristics of landscape ecological risks in this functional zone. The research results show the following: (1) The transfer of land use in EFZUYR from 1990 to 2018 mainly occurred among cultivated land, grassland, and woodland, with the transferred area accounting for 87.16% of the total changed area. (2) The fragmentation degree of built-up areas is 0.1097, 0.1053, 0.0811 and 0.0762 in 1990, 2000, 2010 and 2018, respectively, with a decreasing trend. The dominance degree of grassland has been maintained at the highest level for a long time, with all values above 0.59. The separation degree and the interference degree of built-up areas were the highest and the values of the four periods were above 1.2 and 0.44, respectively. The loss degree of water was the highest, with a value above 0.67, while the value of other land use was mostly below 0.4. (3) The landscape ecological risk of EFZUYR presented a fluctuating rising, falling, and then rising trend. The spatial distribution characteristic of EFZUYR presented "high in the north and south, low in the middle.", which has been maintained for a long time. The proportion of low-risk areas is as high as 70%, and the overall ecological risk of the region was low. However, the ecological risk of some areas, such as Linxia City and Magu County, increased. These findings can provide theoretical support for land use planning and achieving sustainable development of EFZUYR.

Keywords: land use/land cover change; landscape pattern index; landscape ecological risk; EFZUYR

Citation: Qiao, F.; Bai, Y.; Xie, L.; Yang, X.; Sun, S. Spatio-Temporal Characteristics of Landscape Ecological Risks in the Ecological Functional Zone of the Upper Yellow River, China. *Int. J. Environ. Res. Public Health* **2021**, *18*, 12943. https://doi.org/10.3390/ijerph182412943

Academic Editors: Wei Song and Hualin Xie

Received: 21 October 2021
Accepted: 6 December 2021
Published: 8 December 2021

1. Introduction

Watershed is an important area in which humans engage in social production, and the population distributed in significant watersheds in the world is as high as 2.24 billion, accounting for about one-third of the world's population [1]. As the fifth-longest river in the world and the second-longest river in China, the Yellow River is famous for being among the birthplaces of the ancient Chinese civilization and the most extensive sand content. In 2019, China identified the ecological protection and high-quality development of the Yellow River Basin as the fifth national strategy. The policy of ecological environment management has reached an unprecedented high, and the construction of the ecological security pattern of the Yellow River Basin has continued to advance [2]. With these efforts, breakthroughs have been made in restoring the ecosystem and the environmental protection of the Yellow River Basin. However, due to the fragility of the ecological environment of the Yellow River Basin and the intensification of human interference, the ecological security problems of the entire basin have taken severe forms (such as soil erosion, land degradation, and weakening of ecosystem function). Ecological governance remains highly arduous [3]. Since 2001, to promote the implementation of ecological security protection project in the Yellow River Basin, the relevant departments of the State of China began to study the zoning of ecological functions and clarify the crucial areas for safeguarding national ecological security. The ecological functional zone is a comprehensive ecosystem that integrates conserving river sources, mediating the relationship between humans and nature, and promoting ecological protection. It plays a critical role in maintaining the safety of the region's ecological environment and the whole country. The protection of the stable development of the ecological functional zone depends on the rational use of land [4]. Located in the north-eastern part of the Tibetan Plateau, EFZUYR is known as the reservoir of the Yellow River and has a prominent ecological strategic position. Land use or land cover change (LUCC) in China has undergone a complex series of changes over the past three decades due to fast economic growth and the adoption of several land use policies [5], and EFUYR is no exception. However, assessing the ecological risk of EFZUYR based on LUCC is particularly important for ecological restoration and water conservation.

Scientific regional ecological risk assessment can provide an essential basis for the policy formulation, planning, and land management of natural resource sustainability [6]. The demand for environmental decision-making and planning management has promoted the continuous expansion of the scope and content of ecological risk research and, as a result, ecological risk assessment has received increasing attention from academic circles [7]. Moreover, ecological risk assessment has changed from a traditional ecological risk assessment to a regional ecological risk assessment and a landscape ecological risk assessment [8]. Compared with the traditional ecological risk assessment, the landscape ecological risk assessment can better express spatial heterogeneity, which has become the most popular method for assessing ecological risk [9]. Numerous scholars have made breakthroughs in landscape ecological risk. The typical research areas selection mainly covers research areas such as watershed [10], cities [11], mountain [12,13], wetland [14] and nature reserves [15,16]. The research content mainly focuses on the assessment of landscape ecological risk [17,18], spatial and temporal patterns [16], and the impact of different factors on landscape ecological risk [19]. With regard to research methods, the measurement and calculation of landscape ecological risk mainly include the risk "source-collection" method and landscape index method [11]. The evaluation method based on risk source collection is more suitable for evaluating specific ecological risks with apparent stress factors in certain areas. However, this method must be combined with the specific ecological processes or disaster risks to identify the landscape type that promotes or hinders the sound development of the ecosystem [20] and does not take landscape heterogeneity and ecosystem change patterns into consideration. However, the evaluation method based on the landscape index method focuses on assessing ecological risk from the spatial pattern of the landscape, which can comprehensively evaluate the ecological impact and cumulative effects of multiple risk sources in the landscape mosaic [21]. In this method, land use/cover change (LUCC) is the basis of ecological risk assessment [22]. Compared with the evaluation method of risk source collection, the landscape index method quantitatively evaluates the overall ecological quality of the region and focuses on analysing the spatial-temporal variation characteristics of risks and the risks of land use status to ecological functions and processes. This is the reason that, in recent years, landscape ecological risk assessment based on the landscape index method has witnessed its most comprehensive application.

EFZUYR is an important ecological barrier of the Yellow River, with important functions of water recharge, maintaining biodiversity and regulating regional climate, and has an irreplaceable role in maintaining the water resources and ecological security of the Yellow River basin. Therefore, this study quantitatively analyzed the spatial distribution characteristics of EFZUYR ecological risks using the land transfer matrix, landscape ecological risk model and cold-hot spot analysis, which fills the knowledge gap of landscape ecological risk assessment in the upper Yellow River Basin. The main goal of this study was to assess the landscape ecological risk of EFZUYR based on the changes in LUCC using the landscape index approach. The specific objectives of this study were (1) to analyze the area change of EFZUYR land use and the transfer characteristics between different land uses from 1900 to 2018, (2) to explore the change characteristics of the landscape index of EFZUYR land use, and (3) to evaluate the spatial and temporal evolution characteristics of EFZUYR landscape ecological risk.

2. Materials and Methods

2.1. Study Area

EFZUYR is located in the north-eastern extension of the Qinghai-Tibet Plateau and is the drainage divide between the Yellow River Basin and the Yangtze River Basin in China. Its unique geographical location and natural geographic characteristics have determined its fundamental ecological attributes, such as an ecological transition zone and a fragile zone, which is of great significance for maintaining social stability and ecological security [23]. The rapid socio-economic development and overuse of land have led to an increase in regional ecological risks. In addition, EFZUYR is not only the largest plateau wetland at the eastern end of the Qinghai-Tibet Plateau and a vital water replenishment region for the upper reaches of the Yellow River, but also an important conservation area for rare flora and fauna of the Tibetan Plateau. Its administrative region includes most counties in Linxia Hui Autonomous Prefecture (Linxia Prefecture) and Gannan Tibetan Autonomous Prefecture (Gannan Prefecture) in Gansu Province of China (Figure 1). The terrain of the study area is high in the southwest and low in the northeast, with altitudes ranging from 1500 to 4900m. The vegetation is mainly grassland, wetland, and mountain woodland, such as alpine meadow, which is mainly composed of Carex and Kobresia, and subalpine shrub, which is mainly composed of Rh. przewalskii and Rh. rufum. The climate is temperate continental monsoon with a large diurnal temperature difference. The annual average temperature in the southwest area is 4 °C. Additionally, precipitation in the southwest is unevenly distributed with large interannual variations. Linxia Prefecture and Gannan Prefecture are the gathering and living areas for Hui, Tibetan, Dongxiang, Salar, and Tu ethnic groups. Data from the seventh census of China shows that the resident population of the study area is 2,143,900.

2.2. Data

The land use data used in this paper are from 1990, 2000, 2010 and 2018, which were obtained from the Resources and Environmental Science and Data Center (http://www.resdc.cn/; accessed on 5 March 2021) and have a spatial resolution of 30 m × 30 m. These data have been widely used and their accuracy meets the criteria of the present study. Referring to the classification standard of China's land use status (GB/T21010—2007) and considering the characteristics of different land use in EFZUYR, this study utilized the reclassification function of ArcGIS 10.6 to classify the land use data into six categories: cultivated land, woodland, grassland, water, built-up areas and unused land (Figure 2). The Digital Elevation Model (DEM) data comes from the Geospatial Data Cloud (http://www.gscloud.cn/; accessed on 5 March 2021), with a spatial resolution of 90 m× 90 m.

2.3. Methods

2.3.1. Land Use Dynamic Degree

The dynamic degree of land use is mainly used to describe the area change in land use in a certain period, to express the intensity of land use in a region and the differences among different land uses, periods, or regions. In this study, the dynamic degree of various land uses of EFZUYR for 1990–2000, 2000–2010, 2010–2018 and 1990–2018 was calculated to analyze the land use changes in the study area. The formula for the dynamic degree of land use is detailed in the literature (Equation (A1)) [24].

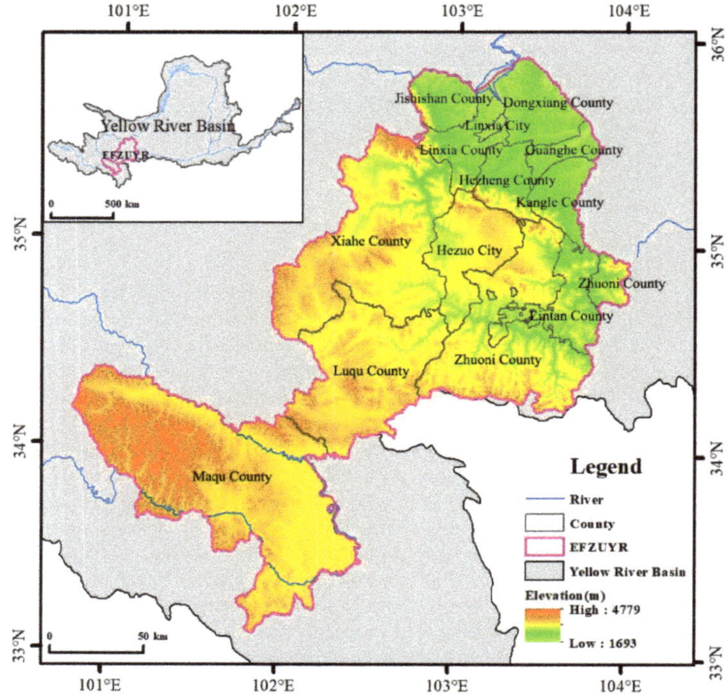

Figure 1. Overview of the study area in the background of Yellow River Basin.

2.3.2. Landscape Ecological Risk Index

Landscape ecological risk refers to the possible adverse consequences of the interaction between the landscape pattern and ecological process under the influence of natural or human factors [20]. The landscape ecological risk index (ERI) consists of the landscape disturbance index and the landscape vulnerability index, reflecting the relationship between landscape patterns of land use and ecological risk [25]. Based on fully considering the impact of land use and landscape variability on the ecological environment under human activity disturbance, this study used the landscape pattern index to establish the assessment method of the landscape ecological risk of EFZUYR. The formula for ERI is detailed in the literature (Equation (A2)) [16,26].

Combining the research results of Chen [26] and the actual condition of the research area, this study divided the ecological risk of the ecological function area in the upper Yellow River Basin into five grades with an equal interval division method: the lowest risk area (ERI $\leq$ 0.20), the lower risk area (0.20 < ERI $\leq$ 0.22), the medium risk area (0.22 < ERI $\leq$ 0.24), the higher risk area (0.24 < ERI $\leq$ 0.26), and the highest risk area (ERI > 0.26).

To present the EFZUYR landscape ecological risk index's spatial distribution characteristics, this study utilized landscape ecology theory to conduct equidistant sampling of land use data in 1990, 2000, 2010 and 2018. After many experiments and comparisons, it was found that 5 km $\times$ 5 km is the optimal scale for ecological risk research in the area, so the research area was divided into 5 km$\times$ 5 km grids (Figure 3). In data processing, grids were used as small research units for spatial sampling. Ecological risk values were calculated for all grids, and landscape ecological risk values were assigned to the center of each ecological risk unit.

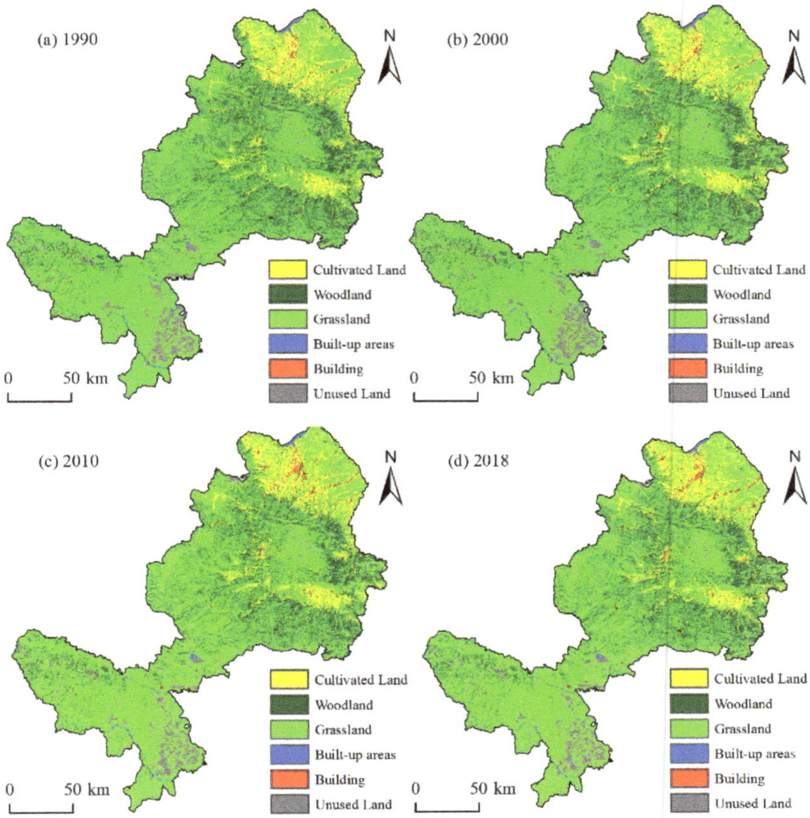

Figure 2. Land use of EFZUYR from 1990 to 2018. (**a**) 1990 (**b**) 2000, (**c**) 2010 and (**d**) 2018.

2.3.3. Cold-Hot Spot Analysis

Cold-hot spot analysis is often used to find the spatial distribution characteristics of landscape ecological risks in the study area. In this study, the Getis-Ord General G was used to explore the overall pattern and trend of landscape ecological risk in the study area, and the Getis-Ord Gi* index of landscape ecological risk was used to describe the spatial distribution of cold and hot spots of ecological units. Hot spots indicate areas where high values of landscape ecological risk are clustered, and cold spots represent areas where low values of landscape ecological risk are clustered. The formulas for the Getis-Ord General G and the Getis-Ord Gi* are detailed in the literature (Equations (A3)–(A5)) [27].

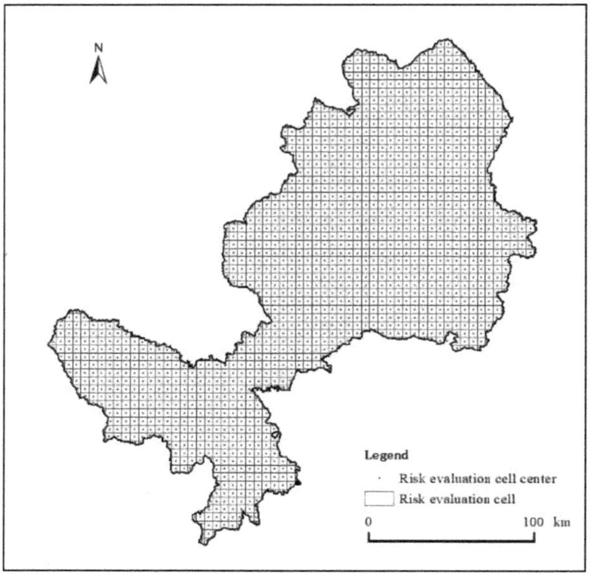

Figure 3. Grid division of EFZUYR ecological risk.

3. Results

3.1. Change Characteristics of Land Use

3.1.1. Area Changes in Land Use

This study performed a statistical analysis on the land use data of EFZUYR of four periods from 1990 to 2018 and obtained statistics on the area changes in the six land use types in functional areas (Table 1). During the research period, grassland and woodland were the main land use of EFZUYR, and the characteristics of the number changes in each land use type were widely different. In 1990, the ranking of the EFZUYR land use area ratio was as follows: grassland (61.07%) > woodland (21.82%) > cultivated land (8.72%) > unused land (6.67%) > water (0.92%) > built-up areas (0.72%); by 2018, the proportion transitioned into the following order: grassland (60.81%) > woodland (21.68%) > cultivated land (8.87%) > unused land (6.57%) > built-up areas (1.14%) > water (0.93%)). It can be seen that the sum of EFZUYR grassland and woodland maintained at more than 80%, and the increased speed of built-up areas exceeded water. Considering the number change degree of each land use type, unused land had the most significant reduction in area. From 1990 to 2018, unused land witnessed a reduction of 72.13 km^2 and a reduced rate of 3.03%, and the grassland area showed a fluctuant decreasing trend in area, with a decrease of 130.13 km^2. However, due to the large base of grassland area, the reduction area only accounted for 0.61% of the total area, and thus the change rate was rather low; woodland also showed a fluctuant decreasing trend in area, with a decrease of 64.94 km^2 compared with 1990 and a reduced rate of 0.85%. In EFZUYR, the land use with the largest increased area was built-up areas which increased by 148 km^2 during the research period, with an increased rate of 59.06%. From 1900 to 2018, the cultivated land area showed a fluctuating increasing trend of 47.07 km^2, and the rate of increase was 1.54%. Water also showed a trend of fluctuant increase in area and increased by 4.57 km^2, with an increased rate of 1.42%. The above statistics show that woodland and grassland cover a greater portion of ecological land use in EFZUYR, and the built-up areas showed a rapid increasing trend.

3.1.2. Dynamic Changes in Land Use

According to the dynamic degree of EFZUYR land use (Table 1), differences in the change rate of land use in different periods can be noticed. Cultivated land showed an increasing trend first and then decreased slowly from 1990 to 2018. The overall change showed the characteristics of growth, with a change rate of 0.04 %. The increase was the most obvious from 1990 to 2000, with a growth rate of 0.74 %. Woodland showed a decreasing trend first and then increased from 1990 to 2018. The overall change showed a decreasing characteristic, with a reduction rate of 0.02%. From 1990 to 2000, the rate of decrease was the largest, with a reduction rate of 0.16%. Grassland first showed a decreasing trend, then increased and decreased from 1990 to 2018. The overall change showed a characteristic of decreasing, with a reduction rate of 0.02%. The reduction rate in 1990–2000 and 2010–2018 was the same-both 0.06%. Water first showed a decreasing trend and then increased from 1990 to 2018. The overall change showed an increasing characteristic, with a growth rate of 0.04%. It increased rapidly from 2010 to 2018, with a growth rate of 6.21%. Built-up areas showed a continuous increasing trend from 1990 to 2018, with a growth rate of 1.51%, and was the fastest growing from 2000 to 2010, with a growth rate of 3.53%. Unused land showed a constant trend first and then decreased, and the overall change showed the decreasing characteristics. The decrease in 2000–2018 was the largest, with a decrease rate of 0.29%.

Table 1. Dynamics of different land use in EFZUYR.

Type of Land Use	Dynamic Degree of Land Use from 1990 to 2000 (%)	Dynamic Degree of Land Use from 2000 to 2010 (%)	Dynamic Degree of Land Use from 2010 to 2018 (%)	Dynamic Degree of Land Use from 1990 to 2018 (%)
Cultivated Land	0.74	−0.39	−0.18	0.04
Woodland	−0.16	0.05	0.02	−0.02
Grassland	−0.06	0.05	−0.06	−0.02
Water	−0.33	−3.27	6.21	0.04
Built-up areas	1.07	3.53	0.69	1.51
Unused land	0.00	−0.05	−0.29	−0.08

This study used the land use transfer matrix to reveal the detailed transfer status among each land use type (Figure 4, Table 2). According to the land use transfer of EFZUYR from 1990 to 2018, the total land use change was 5774.01 km^2, with a change rate of 16.45%. In the study area, grassland is 2530.28 km^2, accounting for 7.21% of the total transferred area, which is the largest land use transfer area. These areas were mainly transformed into woodland, cultivated land and unused land. Cultivated land was one of the major land-uses that transferred into other types, mainly grassland, woodland, and built-up areas. The transferring area of cultivated land was 818.61 km^2, accounting for 2.33% of the transferred land. Built-up areas were the main type of inflow, which was mainly derived from cultivated land and grassland. The area change reached 250.22 km^2, accounting for 76.80% of the total change area of built-up areas. Woodland mainly transferred into grassland and cultivated land, with a changing area of 1683.49 km^2, accounting for 22.14% of the total woodland area. Water was mainly transferred from grassland and cultivated land, with a changing area of 115.71 km^2, accounting for 35.52% of the total water area. Unused land mainly transferred into grassland and woodland, with the transferring area accounting for 14.28% and 1.07% of the total unused land. Overall, compared with built-up areas and cultivated land, the transfer area of other land use was smaller.

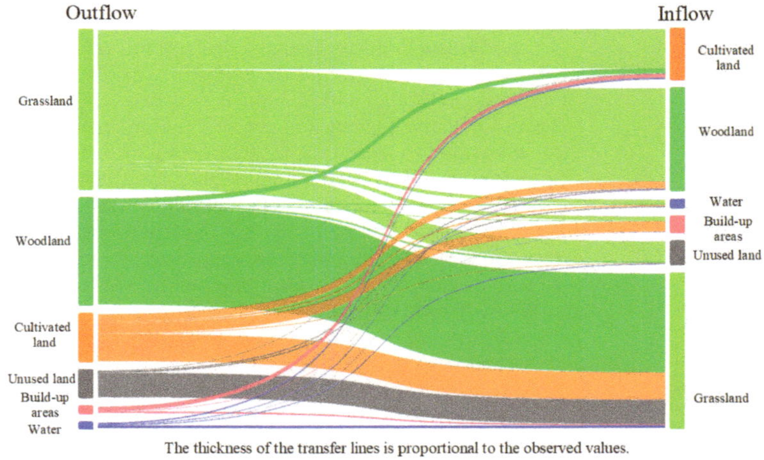

Figure 4. Transformation distribution of different land use from 1990 to 2018.

Table 2. Land use transfer matrix in ecological function zones of the Upper Yellow River from 1990 to 2018.

Type of Land Use		Area of Land Use in 2018 (km^2)					
		Cultivated Land	Woodland	Grassland	Water	Built-Up Areas	Unused Land
Area of land use in 1990 (km^2)	Cultivated Land	3064.99	8.72	3291.26	9.36	3162.44	8.99
	Woodland	7675.08	21.82	7553.57	21.48	7594.41	21.60
	Grassland	21,475.31	61.07	21,353.60	60.72	21,455.15	61.01
	Water	321.99	0.92	311.49	0.89	209.52	0.60
	Built-up areas	252.04	0.72	278.95	0.79	377.32	1.07
	Unused land	2377.96	6.76	2378.52	6.76	2367.47	6.73

In sum, the transfer of cultivated land, grassland, and woodland in EFZUYR was obvious, with a transfer area of 5023.37 km^2 from 1990 to 2018. The data showed that the expansion of urban built-up areas has taken up many cultivated land resources.

3.2. Analysis of Landscape Index Changes of Land Use

The landscape indexes of different land types in 1990, 2000, 2010, and 2018, including the fragmentation degree (C), the separation degree (N), the dominance degree (K), the interference degree (S) and the loss degree (R), were evaluated in this study. From Table 3, the following results can be drawn: (1) There was a relatively small change in the overall fragmentation degree of each land use type. The fragmentation degree of cultivated land, water area, and unused land increased during the research period. The fragmentation degree of woodland and grassland was unchanged. The fragmentation degree of built-up areas showed a decreasing trend, which indicates that built-up areas possess an obvious contiguous development trend, reducing their fragmentation degree. (2) The separation degree index of cultivated land, grassland, and unused land increased, while the index of woodland, grassland, and built-up areas are downward. Therefore, the separation degree of built-up areas was above 1.2, which is much higher than that of other land use type. (3) Grassland had the highest dominance degree, followed by woodland. The unique geographical environment and climatic conditions allowed the alpine grassland and forest ecosystem in EFZUYR to develop into a complete ecosystem, with the extensive distribution of grassland and woodland. (4) Built-up areas had the highest interference degree because people have the greatest interference degree regarding the environment. In addition, the interference degree of cultivated land and unused land showed a continuous increase during the research period. (5) The loss degree was affected by the interference degree and the vulnerability degree, and its development trend is the same as the interference degree. Water area had the highest loss degree among all land types, with the loss degree index value of the four phases exceeding 0.67. Built-up areas had the second-largest loss degree, with the loss degree index value exceeding 0.44. Due to its relatively large vulnerability degree index, water had the most extensive loss degree. Although the vulnerability of built-up areas was relatively low, the degree of loss was also influenced by the degree of disturbance. Therefore, the maximum disturbance degree of built-up areas led to an increased loss degree.

3.3. *Temporal and Spatial Evolution Characteristics of Landscape Ecological Risk*

3.3.1. Temporal Variations of Landscape Ecological Risk

To clearly express the temporal and spatial characteristics of ecological risks, this study calculated the EFZUYR ecological risk grade area and proportion (Table 4). We visualized ecological risks through the ordinary Kriging interpolation method (Figure 5).

Table 3. Changes in the fragmentation degree (C), the separation degree (N), the dominance degree (K), the interference degree (S), and the loss degree(R).

Type of Land Use	Year	C	N	K	S	R
Cultivated Land	1990	0.0047	0.1163	0.1584	0.0536	0.2142
	2000	0.0048	0.1138	0.1683	0.0539	0.2155
	2010	0.0050	0.1180	0.1655	0.0550	0.2198
	2018	0.0052	0.1206	0.1635	0.0556	0.2224
Woodland	1990	0.0083	0.0975	0.4212	0.0764	0.1527
	2000	0.0085	0.0992	0.4168	0.0765	0.1530
	2010	0.0085	0.0990	0.4189	0.0767	0.1533
	2018	0.0084	0.0985	0.4154	0.0762	0.1523
Grassland	1990	0.0011	0.0217	0.5977	0.0670	0.2009
	2000	0.0012	0.0220	0.5955	0.0668	0.2005
	2010	0.0011	0.0209	0.5935	0.0663	0.1988
	2018	0.0012	0.0219	0.5944	0.0667	0.2001
Water	1990	0.0064	0.4190	0.0560	0.1351	0.6757
	2000	0.0065	0.4289	0.0560	0.1382	0.6909
	2010	0.0113	0.6875	0.0416	0.2172	1.0858
	2018	0.0100	0.5179	0.0616	0.1675	0.8377
Built-up areas	1990	0.1097	1.9559	0.1348	0.6660	0.6660
	2000	0.1053	1.8213	0.1376	0.6233	0.6233
	2010	0.0811	1.3746	0.1460	0.4757	0.4757
	2018	0.0762	1.2911	0.1470	0.4477	0.4477
Unused land	1990	0.0039	0.1199	0.1480	0.0531	0.3186
	2000	0.0039	0.1207	0.1478	0.0534	0.3202
	2010	0.0042	0.1256	0.1556	0.0558	0.3346
	2018	0.0044	0.1287	0.1563	0.0569	0.3412

Table 4. The area and proportion of different risk levels from 1990 to 2018.

Risk Level	1990		2000		2010		2018	
	Area (km^2)	Proportion (%)	Area (km^2)	Proportion (%)	Area (km^2)	Proportion (%)	Area (km^2)	Proportion (%)
Lowest Risk	16,655.00	47.36%	16,509.25	46.94%	16,440.50	46.75%	16,331.00	46.44%
Lower Risk	10,113.50	28.76%	10,093.25	28.70%	10,660.25	30.31%	9341.75	26.56%
Middle Risk	4627.75	13.16%	4725.00	13.44%	4278.75	12.17%	5167.00	14.69%
Higher Risk	1514.25	4.31%	1571.75	4.47%	1644.75	4.68%	2019.00	5.74%
Highest Risk	2258.25	6.42%	2269.50	6.45%	2144.50	6.10%	2310.00	6.57%

From the characteristic of temporal variations evolution (Table 4), the change in landscape ecological risk in EFZUYR was characterized by an "N" type, with relatively minor changes. Therefore, the ecological risk was relatively stable. As the proportion of low ecological risk areas, including the lowest and lower ecological risk areas, remained above 70% for a long time, the whole area was in a low ecological risk state. Specifically, the average ecological risk values for the four periods of EFZUYR were 0.2122, 0.2126, 0.2115, and 0.2123, showing a fluctuant rising, falling, and then rising trend. However, the change rate was minimal. From the area change in different land use types, the proportion of low ecological risk type shows a decreasing trend from 1990 to 2018, and the area decreased by 1095.75 km^2, with a decreasing percentage of 3.11%. Among them, the area of the lowest ecological risk regions shows a continuous decline, with a total decrease of 324.00 km^2. The area of the lower ecological risk regions showed a trend of rising and then falling, with a decrease of 771.75 km^2. Contrary to the change characteristic in the low-risk regions, the area changes in the medium risk regions showed an "increasing-decreasing-increasing" trend, with an overall increase of 539.25 km^2. The highest risk regions showed an "increasing-decreasing-increasing" trend in the entire research area, with the total area increasing from 3772.50 km^2 in 1990 to 4329.00 km^2 in 2018. Among them, the higher risk regions had a relatively sizeable increasing range, with an increased area of 504.75 km^2 which accounted for 90.84% of the area change in the high ecological risk regions. The area changes in the highest risk regions increased, but the increased area was only 51.75 km^2, with an increasing percentage of only 0.15% compared with the increased area in 1990.

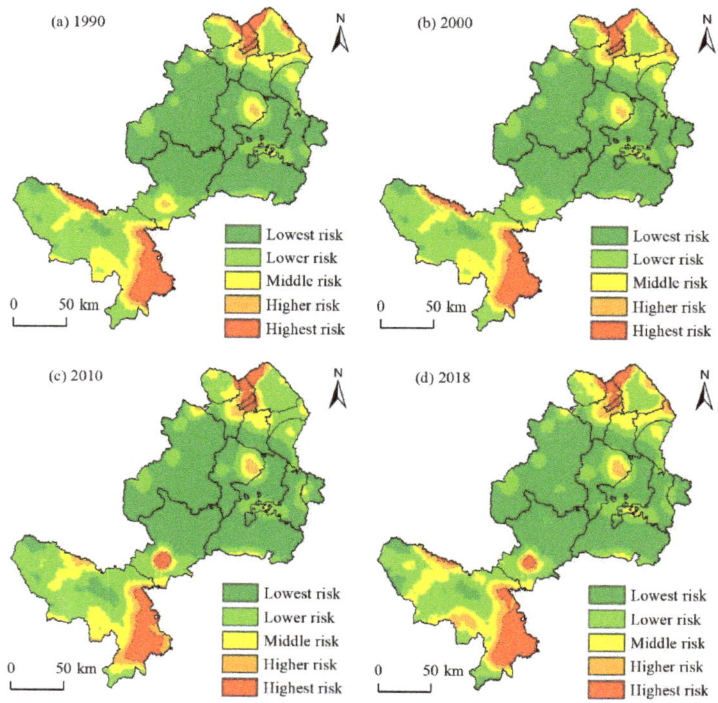

Figure 5. Spatial distribution map of landscape ecological risk. (**a**) 1990; (**b**) 2000; (**c**) 2010; (**d**) 2018.

3.3.2. Spatial Evolution of Landscape Ecological Risk

From the spatial distribution map of landscape ecological risk (Figure 5), we found that the characteristics of the spatial evolution of EFAUYR ecological risk were that the overall landscape ecological risk was relatively low. The ecological risk level increased in some areas, maintaining a long-term spatial pattern of "high at the north and south ends and low in the middle" from 1990 to 2018. Specifically, the highest-risk areas were mainly distributed in Linxia City, Linxia County and Dongxiang County in the north, and Maqu County and Luqu County in the south. Among them, Luqu County, which did not present the highest risk regions until 2010, has suffered a significant increase in highest risk regions, which has increased to 184.75 km^2, since 2010. Regions with higher risk were mainly concentrated in Maqu County in the south. In 2018, the area of the higher risk regions in this area reached 1059.75 km^2, accounting for 52.49% of the higher ecological risk regions in the entire region, followed by Linxia County and Hezuo City in the north. Medium risk regions were mainly distributed in Maqu County. From 1990 to 2018, the area of the medium risk regions in Maqu County accounted for more than 45% of the total area of the medium-risk regions in the research area, and this number exceeded 50% in 2010. The remaining medium ecological risk regions were mainly distributed in Dongxiang County, Jishishan County, Guanghe County, Luqu County and others. The lower risk regions were mainly distributed in Maqu County and Luqu County. The area of the lower risk regions in Maqu County, accounted for approximately 40% of the total area of the lower risk regions in the research area, and this proportion in Luqu County was approximately 10%. The remaining lower ecological risk regions were distributed in Zhuoni County, Xiahe County and other regions. The lowest ecological risk regions were mainly distributed in Xiahe County, Zhuoni County and Luqu County, which are in the central part of EFAUYR. Among them, the area of the lowest ecological risk regions in Xiahe County accounted for approximately 30–35% of the total area of the lowest risk regions in the research area, and the proportions in Zhuoni County and Luqu County were approximately 25% and 17%. The results showed that from 1990 to 2018, EFZUYR had two high risk agglomeration areas for a long time, which were distributed in the south and north. The low-risk area was mainly concentrated in the central area, and the degree of agglomeration showed a downward trend.

3.3.3. Spatial Clustering Characteristics of Landscape Ecological Risk

The spatial clustering of landscape ecological risks can better identify the spatial clustering characteristics of high and low ecological risk regions. According to the General G Index (Table 5), the Observed General G was greater than Expected General G, which indicated that high-value clusters were more obvious. Moreover, the Observed General G in the four periods has an increasing trend, which indicated that the spatial high-value clustering characteristic of the ecological risk in the research area was continuously increasing.

Table 5. Related parameters of the General G Index.

Year	Observed General G	Expected General G	z-Score	*p*-Value
1990	0.002394	0.00237	4.742	0.000
2000	0.002396	0.00237	5.123	0.000
2010	0.002455	0.00237	13.328	0.000
2018	0.002428	0.00237	10.156	0.000

In terms of spatial distribution characteristics, the spatial clustering pattern of EFAUYR ecological risk has long been characterized by "high in the north and south and low in the central". In this pattern, the ecological risks of various counties are relatively different. Among them, the ecological risks of Maqu County, Linxia City and Dongxiang County are the highest, and the ecological risks of Lintan County, Zhuoni County and other regions are lower, which forms higher ecological security (Figure 6). Specifically, EFZUYR has long been distributed with two hot spot clustering regions in the north and south. The hot spot regions in the north are mainly concentrated in the Daxia River valley area, and hot spot clustering regions cover the whole area of Linxia City. At the same time, the hot spot regions in the south are mainly concentrated in Maqu County. In addition, the cold spot of the landscape ecological risk is mainly in Zhuoni County and Lintan County, which are in the central part of EFZUYR.

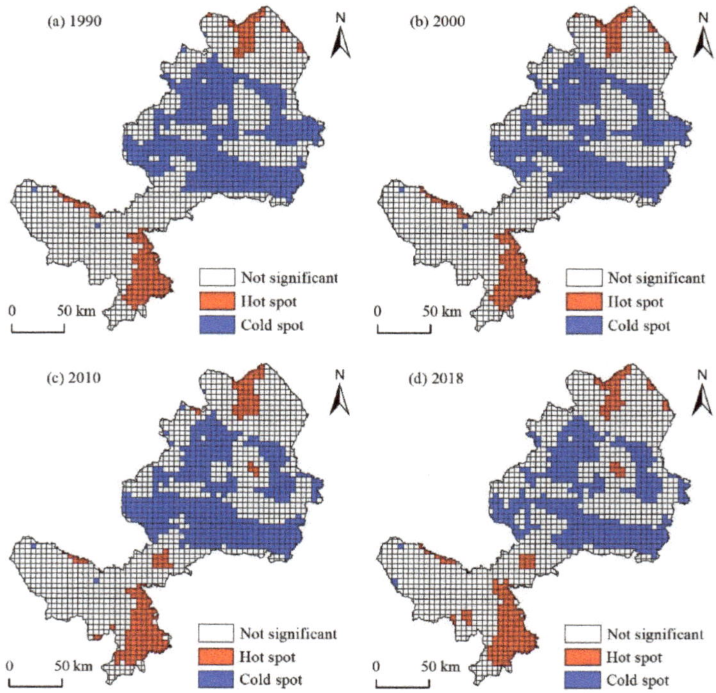

Figure 6. Analysis on Cold and Hot Spots of Landscape Ecological Risks in EFZUYR. (**a**) 1990; (**b**) 2000; (**c**) 2010 and (**d**) 2018.

4. Discussion

4.1. Reasons for Spatial Clustering Characteristics of Landscape Ecological Risks

According to the above research results, EFZUYR has long been distributed with two hot spot clustering regions in the north and south. Although there are two hot spot regions in the north and south of EFZUYR, the causes for the formation are different. The hot-spot regions in the north have a high level of economic development, which means that humans have greater interference with the ecological environment, and the destruction of the ecological environment is severe. In the south, Maqu County has become a hot-spot region. The causes are both natural and anthropogenic. Among them, natural causes are the most fundamental factor in the deterioration of Maqu County's ecological environment. Anthropogenic causes are an influencing factor but do not play the leading role. Climate warming has led to increased evaporation, surface drought, vegetation degradation, and lake retreat, reducing the stability of the originally fragile ecosystem and weakening its resilience, which has become the main driving force for the degradation of the ecological environment [28]. It is worth noting that the role of anthropogenic causes cannot be ignored [29]. The cold spot of the landscape ecological risk is mainly in Zhuoni County and Lintan County, which are in the central part of EFZUYR. The main reason for this is that there is a large area of forest and grassland in Zhuoni County and Lintan County. The attractive natural ecological environment facilitates the spatial clustering of low values of ecological risks, which constitutes the cold-spot clustering region. Due to temporal and spatial change, the clustering characteristic of the cold-spot region is gradually weakening. In addition, a small hot-spot area in Hezuo City, located in the central part of EFZUYR, was gradually formed and rapidly expanding, mainly due to the fact that Hezuo City is the political, economic and cultural center of Gannan Prefecture. In recent years, with the implementation of the Chinese government's specific poverty alleviation policy, the level of urbanization in poor areas such as Linxia Prefecture has rapidly increased [30]. At the same time, the continuous improvement of the urban infrastructure, the increasing concentration of the urban population and the increasing proportion of land urbanization means that human interference with the ecological environment is increasing, and ecological risks are rising.

4.2. Partition Management of EFZUYR

The study divided the ecological functional service zones in the upper Yellow River Basin into different levels and calculated the area proportion of the five risk levels in each county (city) in 2018 (Figure 7). It was found that Maqu County, Dongxiang County, and Linxia County are the main distribution areas of the highest, higher, and medium landscape ecological risk areas. The whole area of Linxia City is the highest and higher risk regions, and Xiahe County, Zhuoni County and Luqu County are the main distribution area of the lowest and lower risk regions. The reasons for the spatial difference of ecological risk distribution are very different. First, Maqu County has become a high-risk area mainly due to natural factors. From 1990 to 2018, the annual average temperature in Maqu County showed a significant upward trend. Climate warming led to a reduction in wetlands, grassland degradation, a reduction in biodiversity and the weakening of ecosystem functions [31,32], resulting in a larger proportion of landscape ecological risks in the region. Of course, human economic activities have also exacerbated the ecological risks in the area to a certain extent. Second, the entire area of Linxia City is a high-risk area, and a relatively high proportion of high-risk areas in Linxia Country and Dongxiang County are caused by human activities. Linxia City is the seat of the Linxia Hui Autonomous Prefecture and is the area with the highest degree of economic development in the study area. The population urbanization rate is as high as 88.69%. The land expansion is the most obvious in the study area [30]. This indicates that Linxia City is the most frequent human activity in the study area. In the process of rapid urbanization, the development of the land use landscape has changed greatly, resulting in a high concentration of landscape ecological risks. Dongxiang County and Linxia County are close to Linxia City, and their economic development is relatively fast. Regional development has shown that the ecological environment restricts economic development [33]. Ecological risks have also been aggravated due to excessive economic development. Finally, Xiahe County, Zhuoni County, and Luqu County are restricted by natural conditions, their economic development is relatively slow, the population is small, and human activities are relatively weak. At the same time, the vegetation in Gannan has generally improved since 2000. The increase in vegetation in Xiahe County, Luqu County and Zhuoni County is the most obvious [34], so the landscape ecological risk is relatively low.

The partition management of EFZUYR is of great significance for improving the ecological environment of the research area and promoting a virtuous cycle of the ecosystem. Conducting partitioning based on the different ecological environments means proposing a more targeted ecological environment management plan and future development plan [35,36], which is a suitable solution. The following suggestions are made for different ecological risk areas: (1) ecological restoration should be emphasized in areas with the highest, higher, and medium risks, such as Linxia City, Linxia County, Hezuo City and others with higher economic development levels. In these areas, the human demand for natural systems is accelerating, leading to imbalances in natural systems and a significant loss of biodiversity [37]. Therefore, these areas should strictly control the construction area and reduce land abuse. By rationally adjusting land use and increasing the green area in urban areas, the ecological benefits of urban areas will be improved, and financial and technical support for environmental protection will be strengthened. In Maqu County, where the natural environment is poor, excessive development must be prohibited to reduce human interference. As the recharge water volume of Maqu County accounts for 45% of the total flow of the Yellow River [38], nine provinces along the middle and lower Yellow River in China will be severely affected if the landscape ecological risks of Maqu County further rise. (2) The lowest and lower risk regions should emphasize protecting the original landscape. For example, areas such as Xiahe County and Zhuoni County should actively respond to national, provincial, and municipal ecological protection policies to continuously improve their ecological environment. Since ethnic clustering regions are endowed with unique cultural, historical, and natural landscape values, ecotourism can be carried out within the carrying capacity of their ecological environments to increase the fiscal revenue, which can be invested t back into the ecological construction projects. (3) The ecological security of the research area is highly dependent on natural climatic conditions. It is not easy to support the continued improvement of the ecology of the whole area with the measures taken by humans such as ecological construction and protection engineering. Therefore, the critical points for the development of EFZUYR are balanced between coordinating the conservation of the ecological environment and pursuing sustainable economic development, formulating corresponding partition control measures, and maximizing the role of the research area as a safety barrier for the entire river basin.

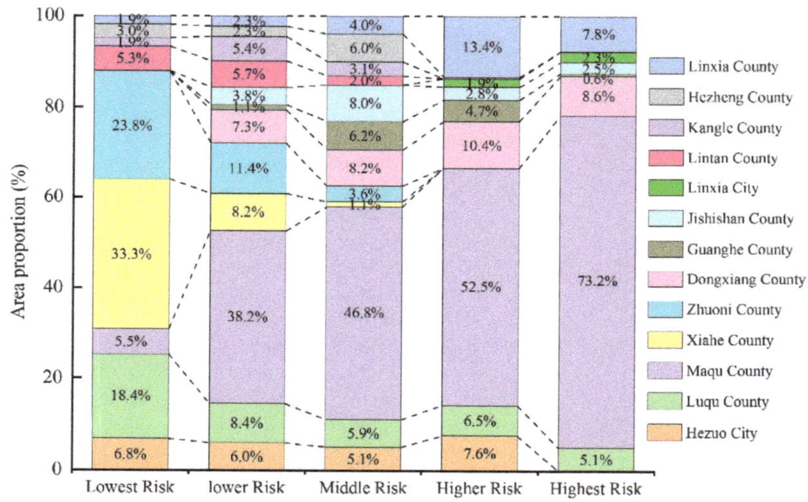

Figure 7. The proportion of the area of the five risk levels in each county (city) in 2018.

4.3. The Advantage and Limitations of This Research

The study explored the spatial and temporal evolution of land use change and landscape ecological risk in EFZUYR, supported by long-term data. The advantage of using long-term data is that they can eliminate the effects of the short-term data mutation on the research results, reducing uncertainty [39]. In addition, the ecological risk evaluation method based on landscape pattern in this study, to a certain extent, gets rid of the traditional ecological risk evaluation inherent model of "risk source identification-receptor analysis-exposure and hazard evaluation" and pays more attention to the spatial and temporal change characteristics of risk, which helps to better elucidate understand the current situation of regional ecological risk [20]. Furthermore, the optimization of landscape spatial pattern is closely related to land planning and design [40], this study fully analyzes the change of land use landscape pattern in EFZUYR, which can effectively provide support for the sustainable development of land use in the area. However, there are still some limitations. For example, the data underlying this study are land use data, and the accuracy of land use data interpretation significantly impacts the study. Errors in the characteristics of remote sensing data and technical methods, etc., may lead to uncertainty in the assessment results. Although the study adopted unified land use classification standards in the data processing to eliminate the errors of remote sensing data as much as possible and ensure the accuracy of the data, data errors still exist. Additionally, the spatial and temporal evolution of landscape ecological risk is an integrated and complex process influenced by various aspects of natural and human activities. Therefore, further research and analysis are needed and, thus, a more accurate assessment of regional landscape ecological risks is needed.

5. Conclusions

Using the theoretical knowledge of landscape ecology, this study constructed an ecological risk index based on the landscape ecological index and analyzed the spatial and temporal pattern evolution and spatial clustering of ecological risks in EFZUYR with the help of spatial statistical analysis methods. The main conclusions are:

(1) From 1990 to 2018, land use showed different changes. Here, built-up areas and grassland showed the largest increase and the largest decrease, with an increase of 148.84 km^2 and a decrease of 130.13 km^2, respectively. Among different types of land transfer, the transfer among cultivated land, grassland, and woodland had prominent advantages, with a transfer area of 5023.37 km^2, which accounted for 87.16% of the changing area. The order of change rate was as follows: built-up areas > unused land > cultivated land = water area > woodland = grassland. Among them, the dynamic degree of built-up areas is the only one that always showed a growth trend during the research period.

(2) During the study period, the landscape index changes in various land use types had obvious differences. As for the degree of fragmentation, built-up areas were the highest and decreased year by year, with the value dropping from 0.1053 to 0.07962, and the changes in other land types were relatively small. In terms of separation degree, the value of built-up areas dropped from 1.8213 to 1.2911, which was the most dramatic change. The separation degree of water fluctuated from 0.41 to 0.68 and then fell to 0.51. The value of other land use did not change significantly. Regarding the dominance degree, the values of grassland and woodland were maintained at 0.59 and 0.41, respectively, for a long period of time, occupying an absolute advantage. The dominance degree of other land use was below 0.2. For the interference degree, the value of built-up areas was the largest, and the value showed a downward trend, from 0.666 to 0.4477. The interference degree of water fluctuated, increasing from 0.1351 to 0.2172 and then decreasing to 0.1675. The value of cultivated land and unused land increased slightly. In terms of the loss degree, the value of water was the highest, with a value above 0.67, reaching the highest value of 1.0858 in 2010. The loss degree of built-up areas was above 0.44, in second place. The value of woodland was the smallest, which was maintained at around 0.15 for a long time.

(3) The ecological risks of EFZUYR presented a fluctuant rising, falling and then rising trend. The area of the low-risk regions accounted for more than 70% of the entire area, and the overall ecological risk of the area was relatively low. The ecological risks in the EFZUYR long maintained the spatial distribution characteristic of "high at the north and south ends and low in the middle." In addition, the spatial clustering characteristics of high-risk regions were more obvious: two large-scale hot-spot regions formed in the north and south, and cold spot regions mainly concentrated in the central area, with a decrease in the clustering degree.

Author Contributions: Conceptualization, F.Q., Y.B. and L.X.; writing—review and editing, F.Q.; writing—original draft preparation, L.X.; visualization, Methodology, writing—review and editing, X.Y.; Data curation, Formal analysis, writing—review and editing, S.S. All authors have read and agreed to the published version of the manuscript.

Funding: This research was funded by Higher Education Innovation Fund Projects in Gansu Province, grant number 2021B-087 and Project of Improving Young Teachers' Scientific Research Ability in Northwest Normal University, grant number NWNU-SKQN2021-22.

Institutional Review Board Statement: Not applicable.

Informed Consent Statement: Not applicable.

Data Availability Statement: Not applicable.

Acknowledgments: Thanks to the hard-working editors and valuable comments from reviewers.

Conflicts of Interest: The authors declare no conflict of interest.

Appendix A

Table 1. The area and proportion of different land use in EFZUYR from 1990 to 2018.

Type of Land Use	1990		2000		2010		2018	
	Area (km^2)	Proportion (%)	Area (km^2)	Proportion (%)	Area (km^2)	Proportion (%)	Area (km^2)	Proportion (%)
Cultivated Land	3064.99	8.72	3291.26	9.36	3162.44	8.99	3112.06	8.87
Woodland	7675.08	21.82	7553.57	21.48	7594.41	21.60	7610.14	21.68
Grassland	21,475.31	61.07	21,353.60	60.72	21,455.15	61.01	21,345.18	60.81
Water	321.99	0.92	311.49	0.89	209.52	0.60	326.56	0.93
Built-up areas	252.04	0.72	278.95	0.79	377.32	1.07	400.88	1.14
Unused land	2377.96	6.76	2378.52	6.76	2367.47	6.73	2305.83	6.57

Equation (A1)

The dynamic degree of land use can demonstrate the number of different land use changes in a certain period. It can be used to quantitatively describe the strength of the variance of different land use in the research area. The specific calculation formula is as follows:

$$K = \frac{U_t - U_0}{U_0} \times \frac{1}{T} \times 100\% \tag{A1}$$

In Equation (A1), K represents the dynamic degree of a certain land use in a certain period; U_t and U_0 are the amounts of different land use at the beginning and end of the study period; T is the research interval. The absolute value of K indicates the transferring speed of a certain land use. When K is greater than zero, the area of land use types shows an increasing trend, but if K is less than zero, the area of land use types shows a decreasing trend.

Equation (A2)

Landscape ecological risk index is composed of landscape disturbance index and landscape vulnerability index, reflecting the relationship between landscape pattern based on land use and ecological risk. The calculation formulas are as follows:

$$ERI_k = \sum_{i=1}^{m} \frac{A_{ki}R_i}{A_k} \tag{A2}$$

In Equation (A2), ERI is the landscape ecological risk index of ecological research unit k; i represents different land use landscapes; A_{ki} represents the type i landscape area of ecological unit k; A_k is the total area of ecological unit; R_i is the landscape loss degree index of the type i land use, which is calculated from the landscape pattern index and the specific required landscape index is shown in Table 2.

Equations (A3)–(A5)

Global statistics such as high/low clustering (Getis-Ord General G) are used to evaluate the overall pattern and trend of the data. The score of z has a greater impact on the research results. If the value of z score is positive, the observed General G index will be larger than the expected one, indicating that high-value attributes will cluster in the research area. If the value of z score is negative, the observed General G index will be smaller than the expected one, showing that low-value attributes will cluster in the research area.

The local spatial autocorrelation of landscape ecological risk describes the spatial correlation characteristics among the attribute values of ecological risk units with the Getis-Ord Gi* index, characterizing the spatial distribution of cold and hot spots. Hot spots indicate the areas where the high values of landscape ecological risks cluster and cold spots represent the areas where the low values of landscape ecological risks cluster. The calculation formula is as follows:

$$G_i^* = \frac{\sum_{j=1}^{n} w_{i,j}x_j - \overline{X}\sum_{j=1}^{n} w_{i,j}}{S\sqrt{\frac{\left[n\sum_{j=1}^{n} w_{i,j}^2 - \left(\sum_{j=1}^{n} w_{i,j}\right)^2\right]}{n-1}}} \tag{A3}$$

$$\overline{X} = \sum_{j=1}^{n} x_i \Big/ n \tag{A4}$$

$$S = \sqrt{\frac{\sum_{j=1}^{n} x_i}{n} - \left(\overline{X}\right)^2} \tag{A5}$$

In Equations (A3)–(A5), X_i and X_j represent the *ERI* of ecological risk cells i and j, respectively. W_{ij} is the binary matrix of the adjacent space; when ecological cell i is adjacent to ecological cell j, $W_{ij} = 1$; otherwise, $W_{ij} = 0$. $\overline{X}$ and S are the mean value and standard deviation.

Table 2. The construction method of landscape pattern index.

Landscape Index	Calculation Formula	Ecological Meaning
Landscape Fragmentation Degree Index (C_i)	$C_i = n_i / A_i$	In the formula, n_i is the patch number of the type i landscape; A_i is the distribution area of type i landscape. This index can reflect the degree of fragmentation of different landscape types and partly reflect the complexity of the spatial distribution of the landscape as well as the interference by human activities in the landscape. The smaller the landscape fragmentation, the higher the continuity of different landscape patch types distribution in the research unit. Additionally, the more significant the proportion of the area of different patch types, the more stable the corresponding landscape ecosystem.
Landscape Separation Degree Index (N_i)	$N_i = \frac{1}{2}\sqrt{\frac{n_i}{A}} \times A / A_i$	In the formula, A is the total area of all landscapes. Its value represents the separation degree between patches in the same landscape type. The greater the landscape separation degree, the longer the distance between the patch boundaries of the same landscape type, and the higher the heterogeneity of the landscape patch types in the research unit.
Landscape Dominance Degree Index (K_i)	$K_i = \frac{(B_i + L_i)}{4} + \frac{D_i}{2}$	In the formula, B_i equals the ratio between the quadrat numbers of patch i and the total quadrat number; L_i equals to the ratio between the numbers of patch i and the total patch number; D_i equals to the ratio between the distribution areas of patch i and the total area of the quadrat. Its value can reflect whether the distribution of a certain landscape type occupies the dominant position of the landscape. The landscape type with a greater dominance degree can directly affect the evolution of the landscape pattern.

Table 2. *Cont.*

Landscape Index	Calculation Formula	Ecological Meaning
Landscape Interference Degree Index (S_i)	$S_i = aC_i + bN_i + cK_i$	In the formula, a, b, c are the weights of their corresponding landscape indexes, and the sum of a, b and c equals one. Based on the analysis and combined with the research experience, a, b and c are assigned with the weights of 0.6, 0.3, and 0.1. These values are used to express the interference degree of human production activities on different types of landscapes.
Landscape Vulnerability Degree Index (F_i)	Obtained by expert scoring, assignment and normalization	It refers to the vulnerability of the landscape ecosystem when encountering different factors. This value has a greater relationship with the level of the landscape ecosystem. Generally, the lower the ecosystem level, the higher the internal vulnerability of the system. According to the research results, the vulnerability of unused land, water, cultivated land, grassland, woodland, and built-up areas is 6, 5, 4, 3, 2 and 1.
Landscape Loss Degree Index (R_i)	$R_i = S_i \times F_i$	The loss degree is related to each stage in the development of the landscape ecosystem. Typically, the lower the level of the landscape ecosystem, the higher its vulnerability. Conversely, the higher the level of the landscape ecosystem, the more stable its internal organizational structure and the lower the interference degree, and therefore the lower the vulnerability.

References

1. Yin, D.; Li, X.; Li, G.; Zhang, J.; Yu, H. Spatio-Temporal Evolution of Land Use Transition and Its Eco-Environmental Effects: A Case Study of the Yellow River Basin, China. *Land* **2020**, *9*, 514. [CrossRef]
2. Li, J.; Sun, W.; Li, M.; Meng, L. Coupling coordination degree of production, living and ecological spaces and its influencing factors in the Yellow River Basin. *J. Clean. Prod.* **2021**, *298*, 126803. [CrossRef]
3. Wang, S.Y.; Liu, J.S.; Ma, T.B. Dynamics and changes in spatial patterns of land use in Yellow River Basin, China. *Land Use Policy* **2010**, *27*, 313–323. [CrossRef]
4. Ning, J.; Liu, J.; Zhao, G. Spatio-temporal characteristics of disturbance of land use change on major ecosystem function zones in China. *Chin. Geogr. Sci.* **2015**, *25*, 523–536. [CrossRef]
5. Song, W.; Deng, X. Land-use/land-cover change and ecosystem service provision in China. *Sci. Total Environ.* **2017**, *576*, 705–719. [CrossRef]
6. Pomara, L.Y.; Lee, D.C. The Role of Regional Ecological Assessment in Quantifying Ecosystem Services for Forest Management. *Land* **2021**, *10*, 725. [CrossRef]
7. Xue, L.; Zhu, B.; Wu, Y.P.; Wei, G.; Liao, S.; Yang, C.; Wang, J.; Zhang, H.; Ren, L.; Han, Q. Dynamic projection of ecological risk in the Manas River basin based on terrain gradients. *Sci. Total Environ.* **2019**, *653*, 283–293. [CrossRef] [PubMed]
8. Cao, Q.W.; Zhang, X.W.; Ma, H.K.; Wu, J.S. Review of landscape ecological risk and an assessment framework based on ecological services: ESRISK. *Acta Geogr. Sin.* **2018**, *73*, 843–855.
9. Wang, D.; Ji, X.; Li, C.; Gong, Y. Spatiotemporal Variations of Landscape Ecological Risks in a Resource-Based City under Transformation. *Sustainability* **2021**, *13*, 5297. [CrossRef]
10. Li, Y.; Huang, S. Landscape ecological risk responses to land use change in the Luanhe River Basin, China. *Sustainability* **2015**, *7*, 16631–16652. [CrossRef]
11. Jin, X.; Jin, Y.; Mao, X. Ecological risk assessment of cities on the Tibetan Plateau based on land use/land cover changes–Case study of Delingha City. *Ecol. Indic.* **2019**, *101*, 185–191. [CrossRef]
12. Cui, L.; Zhao, Y.; Liu, J.; Han, L.; Ao, Y.; Yin, S. Landscape ecological risk assessment in Qinling Mountain. *Geol. J.* **2018**, *53*, 342–351. [CrossRef]
13. Zimmermann, P.; Tasser, E.; Leitinger, G.; Tappeiner, U. Effects of land-use and land-cover pattern on landscape-scale biodiversity in the European Alps. *Agric. Ecosyst. Environ.* **2010**, *139*, 13–22. [CrossRef]
14. Li, Z.; Jiang, W.; Wang, W.; Chen, Z.; Ling, Z.; Lv, J. Ecological risk assessment of the wetlands in Beijing-Tianjin-Hebei urban agglomeration. *Ecol. Indic.* **2020**, *117*, 106677. [CrossRef]
15. Gaines, K.F.; Porter, D.E.; Dyer, S.A.; Wein, G.R.; Pinder, J.E.; Brisbin, I.L. Using wildlife as receptor species: A landscape approach to ecological risk assessment. *Env. Manag.* **2004**, *34*, 528–545. [CrossRef] [PubMed]
16. Wang, H.; Liu, X.; Zhao, C.; Chang, Y.; Liu, Y.; Zang, F. Spatial-temporal pattern analysis of landscape ecological risk assessment based on land use/land cover change in Baishuijiang National nature reserve in Gansu Province, China. *Ecol. Indic.* **2021**, *124*, 107454. [CrossRef]
17. Liu, J.; Wang, M.; Yang, L. Assessing Landscape Ecological Risk Induced by Land-Use/Cover Change in a County in China: A GIS-and Landscape-Metric-Based Approach. *Sustainability* **2020**, *12*, 9037. [CrossRef]

18. Li, X.; Li, S.; Zhang, Y.; O'Connor, P.J.; Zhang, L.; Yan, J. Landscape Ecological Risk Assessment under Multiple Indicators. *Land* **2021**, *10*, 739. [CrossRef]
19. Mann, D.; Anees, M.M.; Rankavat, S.; Joshi, P.K. Spatio-temporal variations in landscape ecological risk related to road network in the Central Himalaya. *Hum. Ecol. Risk Assess.* **2021**, *27*, 289–306. [CrossRef]
20. Peng, J.; Dang, W.X.; Liu, Y.X.; Zong, M.L.; Hu, X.X. Review on landscape ecological risk assessment. *Acta Geogr. Sin.* **2015**, *70*, 664–677.
21. Cao, Q.; Zhang, X.; Lei, D.; Guo, L.; Sun, X.; Wu, J. Multi-scenario simulation of landscape ecological risk probability to facilitate different decision-making preferences. *J. Clean. Prod.* **2019**, *227*, 325–335. [CrossRef]
22. Xie, H.; Wen, J.; Chen, Q.; Wu, Q. Evaluating the landscape ecological risk based on GIS: A case-study in the Poyang Lake region of China. *Land Degrad. Dev.* **2021**, *32*, 2762–2774. [CrossRef]
23. Chen, Z.; Zhang, Q.; Li, F.; Shi, J. Comprehensive Evaluation of Land Use Benefit in the Yellow River Basin from 1995 to 2018. *Land* **2021**, *10*, 643. [CrossRef]
24. Liao, N.; Gu, X.; Wang, Y.; Xu, H.; Fan, Z. Analyzing Macro-Level Ecological Change and Micro-Level Farmer Behavior in Manas River Basin, China. *Land* **2020**, *9*, 250. [CrossRef]
25. Zhang, W.; Chang, W.J.; Zhu, Z.C.; Hui, Z. Landscape ecological risk assessment of Chinese coastal cities based on land use change. *Appl. Geogr.* **2020**, *117*, 102174. [CrossRef]
26. Ju, H.; Niu, C.; Zhang, S.; Jiang, W.; Zhang, Z.; Zhang, X.; Yang, Z.; Cui, Y. Spatiotemporal patterns and modifiable areal unit problems of the landscape ecological risk in coastal areas: A case study of the Shandong Peninsula, China. *J. Clean. Prod.* **2021**, *310*, 127522. [CrossRef]
27. Guo, W.; Wu, T.; Jiang, G.; Pu, L.; Zhang, J.; Xu, F.; Yu, H.; Xie, X. Spatial Distribution, Environmental Risk and Safe Utilization Zoning of Soil Heavy Metals in Farmland, Subtropical China. *Land* **2021**, *10*, 569. [CrossRef]
28. Wei, J.P.; Li, P. Analysis on ecological frangibility and forming causes of an important water nourising echological function region for the Yellow River in southern Gansu. *Chin. J. Agric. Resour. Reg. Plan.* **2009**, *30*, 56–61. (In Chinese)
29. Mu, H.; Li, X.; Du, X.; Huang, J.; Su, W.; Hu, T.; Wen, Y.; Yin, P.; Han, Y.; Xue, F. Evaluation of Light Pollution in Global Protected Areas from 1992 to 2018. *Remote Sens.* **2021**, *13*, 1849. [CrossRef]
30. Yang, X.D.; Bai, Y.P.; Che, L.; Qiao, F.W.; Xie, L.X. Incorporating ecological constraints into urban growth boundaries: A case study of ecologically fragile areas in the Upper Yellow River. *Ecol. Indic.* **2021**, *124*, 107436. [CrossRef]
31. Yao, Y.B.; Deng, Z.Y.; Dong, Y.; Wang, R.; Zhang, X.Y. Study on causes and strategy of grassland ecology and natural environment degeneration in the first meander of Yellow River. *Prataculturul Sci.* **2007**, *24*, 87–93. (In Chinese)
32. Xue, P.F.; Li, W.L.; Zhu, G.F.; Zhou, H.K.; Liu, C.L.; Yan, H.P. Changes in the pattern of an alpine wetland landscape in Maqu County in the first meander of the Yellow River. *Chin. J. Plant Ecol.* **2021**, *45*, 467–475. (In Chinese) [CrossRef]
33. Ma, W.H. Research on the coupling of ecological and economic construction in the Linxia Hui Autonomous Prefectures. Master's Thesis, Lanzhou University, Lanzhou, China, 2016. (In Chinese).
34. Yun, L.; Wang, D.W. Analysis of vegetation changing trends and causes in Gannan since 2000. *Prataculturul Sci.* **2016**, *33*, 1102–1111. (In Chinese)
35. Xie, L.X.; Bai, Y.P.; Che, L.; Qiao, F.W.; Sun, S.S.; Yang, X.D. Construction of ecological zone based on value-risk ecological function area in the Upper Yellow River. *J. Nat. Resour.* **2021**, *36*, 196–207. [CrossRef]
36. Shi, W.; Qiao, F.; Zhou, L. Identification of Ecological Risk Zoning on Qinghai-Tibet Plateau from the Perspective of Ecosystem Service Supply and Demand. *Sustainability* **2021**, *13*, 5366. [CrossRef]
37. Mu, H.; Li, X.; Ma, H.; Du, X.; Huang, J.; Su, W.; Yu, Z.; Xu, C.; Liu, H.; Yin, D.; et al. Evaluation of the policy-driven ecological network in the Three-North Shelterbelt region of China. *Landsc. Urban Plan.* **2022**, *218*, 104305. [CrossRef]
38. Yue, D.X.; Zeng, J.J.; Yang, C.; Zou, M.L.; Li, K.; Chen, G.G.; Guo, J.J.; Xu, X.F.; Meng, X.M. Ecological risk assessment of the Gannan Plateau, northeastern Tibetan Plateau. *J. Mt. Sci.* **2018**, *15*, 1254–1267. [CrossRef]
39. Zhang, X.; Song, W.; Lang, Y.; Feng, X.; Yuan, Q.; Wang, J. Land use changes in the coastal zone of China's Hebei Province and the corresponding impacts on habitat quality. *Land Use Policy* **2020**, *99*, 104957. [CrossRef]
40. Xie, H.; Zhang, Y.; Zeng, X.; He, Y. Sustainable land use and management research: A scientometric review. *Landsc. Ecol.* **2020**, *35*, 2381–2411. [CrossRef]

International Journal of
Environmental Research and Public Health

MDPI

Article

Assessing Efficiency of Urban Land Utilisation under Environmental Constraints in Yangtze River Delta, China

Yue Zhou, Yi Chen * and Yi Hu

School of Geography and Ocean Science, Nanjing University, Nanjing 210023, China; zhouyue092798@163.com (Y.Z.); huyi0234@163.com (Y.H.)

* Correspondence: yichen@nju.edu.cn

Abstract: Measuring the efficiency of construction land utilisation is important for optimising the allocation of regional resources and guiding the sustainable development of the regional society and economy. Based on municipal panel data on urban land use from 2009 to 2017 from a municipal perspective, this research built a slacks-based measure of a super-efficiency model (SE-SBM) to evaluate the temporal and spatial differentiation characteristics of the construction land-use efficiency of 41 cities in the Yangtze River Delta. Following this, the driving force of construction land efficiency was calculated using the Malmquist–Luenberger index. Finally, the entropy-weight TOPSIS (technique for order preference by similarity to ideal solution) model and the k-means clustering method were applied to evaluate an input–output model of the cities. The main conclusions are as follows: (1) The construction land efficiency of the Yangtze River Delta remains at a low level and presents a spatial differentiation pattern, with the efficiency being higher in the east and lower in the west. Due to undesired outputs, the mean value has dropped by 4.67%, and the regional imbalance has decreased. (2) The degree of efficiency loss is significantly positively correlated with the intensity of urban pollution emissions—the higher the pollution emissions, the greater the efficiency loss. (3) The total factor productivity of urban construction land is mainly driven by technological progress, while the promotion of technical efficiency is low and unstable. (4) The evaluation of construction land efficiency must include resource allocation or pollution emission factors to scientifically measure the input–output level. These research results will help to formulate reasonable land-use countermeasures.

Keywords: SE-SBM; construction land-use efficiency; environmental constraints; Yangtze River Delta; China

Citation: Zhou, Y.; Chen, Y.; Hu, Y. Assessing Efficiency of Urban Land Utilisation under Environmental Constraints in Yangtze River Delta, China. *Int. J. Environ. Res. Public Health* **2021**, *18*, 12634. https://doi.org/10.3390/ijerph182312634

Academic Editors: Wei Song and Hualin Xie

Received: 15 October 2021
Accepted: 27 November 2021
Published: 30 November 2021

1. Introduction

The efficient use of urban construction land, the site of economic and social activities [1], is essential for the sustainable development of the urban economy [2,3]. There are many ways to measure whether urban land has been used efficiently, among which measuring the land-use efficiency is the most direct, effective, and universal. More importantly, the efficiency of land use is not only able to reflect the allocation of land and space resources but is also related to the scientific exploration of human settlements and human well-being. There has been abundant research on land-use efficiency evaluations, both nationally and internationally [4,5].

Regarding research approaches, scholars mostly use the traditional data envelopment analysis (DEA) model, the Cobb–Douglas production function, and stochastic frontier analysis, among other methods, to measure the land inputs and outputs of an entire country [6,7], regions [8], provinces [9], cities [10–12], urban agglomerations [13,14], or counties [15]. In the past, scholars mostly began from an economic perspective and measured the land-use efficiency using the single economic output of the land [16]. Chen used the traditional DEA model to measure the efficiency of industrial land use in China's resource-based cities, revealing regional differences [17]. Liu used an extended Cobb–Douglas production function to determine that the efficiency of construction land allocation in China needed to be further improved, while the intensive use of land resources was also necessary [18]. Gui introduced stochastic frontier analysis (SFA) to determine that urban land-use efficiency in the Yangtze River Economic Zone showed a significant growth trend with a cumulative growth rate of 54.07% [19].

However, in recent years, as the green "people-oriented" development concept has become the social consensus, relevant research has shifted its perspective from the economic benefits of land use to the ecological benefits [20] and has begun to focus on the spatial-distribution characteristics of comprehensive benefits [21], such as economic, social, and ecological benefits, within the region. Generally speaking, scholars characterize the concept of greenness by considering undesired outputs (e.g., wastewater and carbon emissions), which not only reflects the ecological connotations of urban land use but also highlights the coordination relationship between humans and land [22]. Yu attempted to incorporate ecological factors into the urban-agglomeration land-use efficiency evaluation index system and found that the efficiency of urban-agglomeration construction land under ecological constraints was significantly reduced, which is consistent with the actual situation [23]. Liang believes that only a measurement factor framework containing undesired outputs can be used to obtain scientific results for urban land-use efficiency and contribute to coordinated urban development [24]. Liu adopted a one-stage SFA model to reveal the potential to improve urban land-use efficiency [25]. On the basis of explaining the connotations of land-use efficiency, Hu constructed an evaluation index system for comprehensive land-use efficiency. The analysis found that the comprehensive benefits of land use within the Jiangsu Province differed significantly, and the gradient structure was more obvious, with the efficiency generally decreasing from southern Jiangsu to northern Jiangsu [26]. In addition, many scholars usually use Malmquist index, Tobit model, panel threshold model, and other spatial measurement methods to study the change mechanism behind land use efficiency, as well as the corresponding optimal allocation and intensive use of urban land [27–29].

These studies provide a reference for formulating countermeasures for the efficient use of urban construction land resources and the optimization of the corresponding industrial layouts. Conversely, they provide reference for countermeasures to promote the healthy coupling of high-quality economic development and the ecological environment. Zhao applied an extended STIRPAT (Stochastic Impacts by Regression on Population, Affluence and Technology) model to explore the relationship between new-type urbanization and land eco-efficiency. The evidence revealed that the relationship follows an N-shaped curve [30]. According to a Finnish study, at the macroeconomic level, the domestic use of biomass per unit of value added decreased (−2.2%/a) as the amount of human appropriation of net primary productivity (HANPP) per unit of biomass decreased (−1.1%/a), reflecting increased economic efficiency in land use [31]. Some scholars have also found that adding a HSR (High-Speed Rail) route will increase urban land-use efficiency by 0.012 in the case that the city has opened HSR [32]. Overall, the existing results provide a useful reference for further research on the use efficiency of urban construction land. However, scholars have mainly discussed the temporal and spatial characteristics and evolutionary dynamics of efficiency itself, and insufficient attention has been paid to quantitative research on the relationship between efficiency and input–output factors. Therefore, additional efforts are needed in the future to make up for the lack of research in this field.

Since the reform and opening up, land use in the Yangtze River Delta has been characterized by a spatial expansion of construction land, a sharp decline of high-quality arable land resources, and an increase in environmental pollution, which has limited urban development to a certain extent. Especially after the 2008 financial crisis, the Yangtze River Delta is facing more unstable factors, and economic development has entered a new growth cycle, which puts forward higher requirements for the coupling coordination between economic development, social progress and ecological protection. A scientific evaluation of construction land-use efficiency is a key step for sustainable regional development including economy, society and environment. However, there are few relevant studies on this area, which is not conducive to promoting the long-term development of the region. In conclusion, quantitative studies to explain the utilization efficiency of land input in economic growth in this period is warranted, which is also of important reference significance for the formulation and implementation of the new round of *Three-year Action Plan for The Integrated Development of the Yangtze River Delta region (2021–2023)*.

The aim of this study was to use a slacks-based measure of super-efficiency model (SE-SBM) to quantify the current status of construction land-use efficiency in 41 cities in the Yangtze River Delta, use the Malmquist–Luenberger index (ML) to identify the key driving forces of the evolution of urban construction land-use efficiency; use the entropy-weight TOPSIS model to explore the specific correlation between the city's input, output, and pollution emissions; and use the k-means clustering method to judge the urban construction land-use mode. Through the above calculations, the study explored the unevenness of the construction land-utilisation efficiency, technological innovation level, and management systems and mechanism levels of cities in the Yangtze River Delta region; identified the problems existing in the construction land-utilisation process of each city; and tried to identify the improvement directions of different types of cities in the process of future development. In the context of global integration, the subsequent development of cities and creative cooperation between cities require new developmental increments. Construction land has always been an important engine for regional economic and social development and an essential place for high-quality development. By providing relevant suggestions for the optimization and management of regional construction land, this research can help the Yangtze River Delta to form a truly stronger, larger, and more concentrated world-class integrated urban-development area, and help it to stabilize its strategic position in the overall situation of national modernization and all-round opening to the outside world.

2. Data Sources and Research Methods

2.1. Overview of the Study Area

The Yangtze River Delta includes three provinces and one city, namely, Jiangsu, Zhejiang, Anhui and Shanghai, respectively, with a total of 41 prefecture-level cities (Figure 1). The *Outline of integrated development of the Yangtze River Delta* [33] pointed out that the Yangtze River Delta is experiencing strong and active growth in the new era and is a new focus of reform and opening. At the end of 2017, the total permanent population of the region was 224 million, the per capita gross domestic product (GDP) was CNY 88,600, and the urbanisation level reached 66.35%, together giving the region a leading position within the country. Further, the economic and social development have expanded rapidly, along with the construction land in the region, which increased from 5.3063 million hectares in 2009 to 6.1865 million hectares in 2017. The average annual growth of construction land during this period was 1.94%, which significantly exceeded the national average. Furthermore, the expansion of construction land led to insufficient arable land reserve resources. At the end of 2017, the per capita arable land area in the Yangtze River Delta was less than 0.072 hm^2, which was far lower than the national average of 0.097 hm^2. In addition, ecological and environmental problems, such as water resources, solid waste, and air pollution in the region have become increasingly severe. Regional construction mostly relies on natural resource endowments, and the development model is restricted by traditional thinking. Although the benefits are good, the cost and the emissions are high, and they no longer meet the strategic requirements of ecological civilisation construction. Therefore, promoting the green development of construction land and tapping into the potential of construction land utilisation have become important tasks that must be solved during the urbanisation process of the Yangtze River Delta.

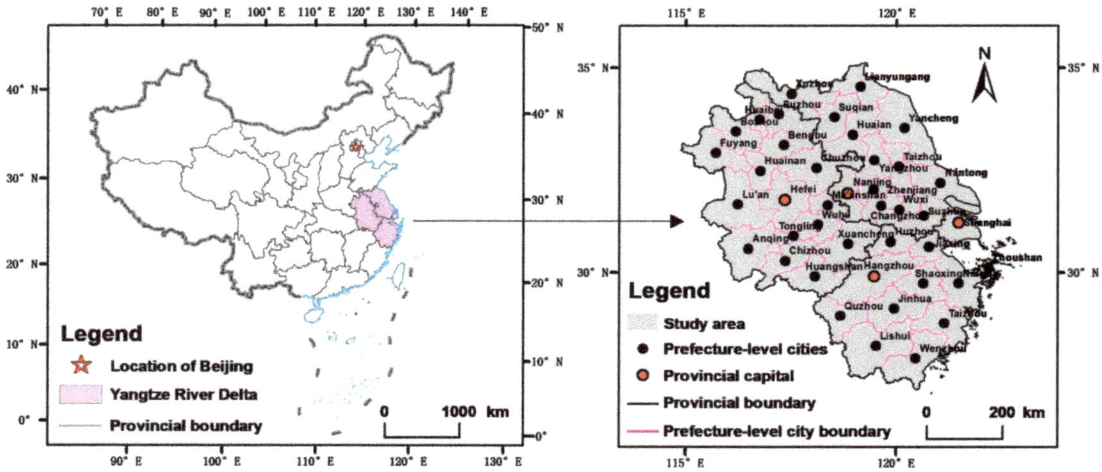

Figure 1. The geographical location and administrative divisions of study area.

2.2. Data Source

This study uses land use and socioeconomic panel data on 41 cities in the Yangtze River Delta region from 2009–2017 as a sample. In the data, construction land includes the following three types of land: urban-rural construction land, land for transportation and water conservancy, and other construction land. The data come from the Natural Resources Department's land-use change survey data, and the socioeconomic data come from the China City Statistical Yearbook (2010–2018) [34] and the provincial and municipal statistical yearbooks (2010–2018) [35–38]. In addition, during the study period, the administrative divisions of Chaohu, Tongling, Anqing, Lu'an and Huainan were adjusted. This study uses the new administrative division as the benchmark and decomposes and merges the corresponding indicators according to the adjustment of the administrative division.

2.3. Research Methods

2.3.1. The Research Design of the Paper

First, based on understanding the current status of construction land use in the Yangtze River Delta region, combined with literature reading, this study screened the input-output evaluation index system for construction land use in line with the actual development of the Yangtze River Delta from the perspectives of land, capital, labour, and output. Second, the article uses the SE-SBM model and the Malmquist–Luenberger model to calculate the static and dynamic efficiency of construction land use in cities in the Yangtze River Delta. Third, we used the entropy-weight TOPSIS method and K-means clustering method to determine the land use types of 41 cities based on the input, output, and pollution emission levels of each city. Finally, combining efficiency characteristics and input-output types, we put forward policy recommendations for optimizing construction land-utilization efficiency. The research design of this article is shown in Figure 2.

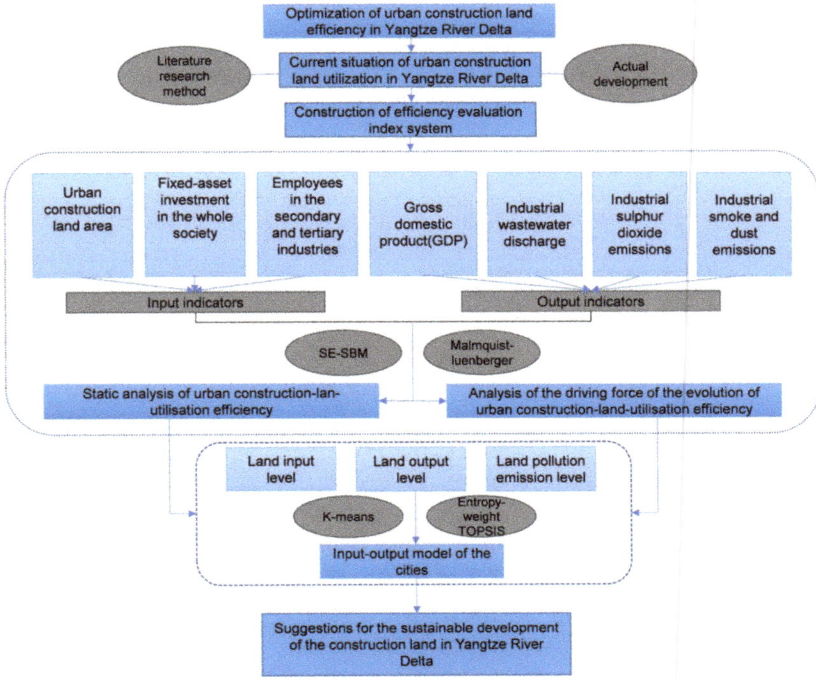

Figure 2. Flowchart of construction land-use optimization.

2.3.2. Construction of the Evaluation Index System

According to the current situation of construction land utilisation in the Yangtze River Delta region, by referring to existing studies [39–41] and based on the principles of a scientific, systematic, and representative selection of indicators, this work constructed an input–output evaluation index system for construction land in the Yangtze River Delta region (Table 1). Amongst the indicators, the input indicators include land, capital, and labour, which are characterised by the area of urban construction land, fixed-asset investments of the whole society, and employment numbers in the secondary and tertiary industries, respectively. The output indicators include the expected economic output and undesired environmental outputs, which are denoted by the gross regional product and the total discharge of industrial wastewater, exhaust gas, and dust waste. In addition, the entropy-weight TOPSIS method was used to synthesise the three types of waste data to characterise a comprehensive pollution-emission index before the calculations.

Table 1. Input–output indicator system for urban construction land-utilisation efficiency.

Indicator	Type	Index Content
Input indicators	Land	Urban construction land area/hectare
	Capital	Fixed-asset investment in the whole society/100 million yuan
	Labour force	Employees in the secondary and tertiary industries/ten thousand people
Output indicators	Expected output	GDP/100 million yuan
	Undesired output	Industrial wastewater discharge/ton
		Industrial sulphur dioxide emissions/ton
		Industrial smoke and dust emissions/ton

2.3.3. SE-SBM with Undesirable Output Model

Data envelopment analysis (DEA) [42] models are widely used to evaluate the efficiency status of decision-making units (DEA uses a decision-making unit (DMU) as its measurement object of efficiency). As it does not require there to be a lack of high correlation (collinearity) between the input indicators and output indicators, and there is no need to estimate parameters or weight assumptions in advance, DEA is especially suitable for systems with multiple inputs and multiple outputs. In 1978, Charnes, Cooper, and Rhodes created the first DEA model, called the CCR model, based on constant returns to scale (CRS). Its basic concept is to take one DMU as an evaluated unit and create an evaluation group with other DMUs, establish a mathematical model corresponding to the problem, and comprehensively analyse the relative efficiency (within the interval of (0, 1)) by solving the results of the model. Then, the production possibility set (PPS) and the production frontier (PF) are determined. According to the distance between the DMUs and the PF, we can determine whether the DMUs are DEA-effective or not. Then, we order the evaluation results. It should be noted that, in the DEA theory, the input and output vectors of the production activities of the decision-making unit are combined into the PPS. The PF is an "envelope surface" formed by the combination of input and output used to achieve maximum efficiency in the PPS. A DMU can be categorized into one of two states: effective or invalid. To judge whether a DMU is DEA effective is, essentially, to judge whether it falls into the PF of the PPS.

In the traditional DEA model, when multiple DMU are evaluated as effective, the maximum efficiency value obtained by the DEA model is 1, namely, the effective DMU efficiency value is the same. Therefore, The efficiency of these effective DMUs cannot be further distinguished.In addition, in the traditional DEA model, the weight coefficient used to calculate the efficiency value is set in a specific range that is most beneficial to the evaluated unit (maximizing its efficiency value), making it easy to exaggerate advantages and avoid shortcomings. As the traditional DEA model cannot evaluate the efficiency most reasonably and is unable to reorder multiple DMUs with an efficiency value of 1 [43]. Tone [44] further revised the DEA model and proposed the SE-SBM model to make up for the abovementioned defect. Based on the SE-SBM model with undesired outputs (SE-SBM-UN), this study measured the construction land-utilisation efficiency of cities in the Yangtze River Delta. The non-oriented CRS SE-SBM-UN model is expressed as follows:

$$\min p = \frac{1+\frac{1}{e}\sum_{i=1}^{e}\frac{s_i^-}{a_{ik}}}{1-\frac{1}{r_1+r_2}\left(\sum_{r=1}^{r_1}\frac{s_r^{g+}}{b_{rk}^{g}}+\sum_{t=1}^{r_2}\frac{s_t^{h-}}{b_{tk}^{h}}\right)} \quad (1)$$

$$s.t.\ \sum_{j=1,j\neq k}^{u} a_{ij}\lambda_j - s_i^- \leq a_{ik} \quad (2)$$

$$\sum_{j=1,j\neq k}^{u} b_{rj}^{h} - s_t^{h^-} \leq b_{tk}^{h} \quad (3)$$

$$\sum_{j=1,j\neq k}^{u} b_{tj}^{h}\lambda_j + s_r^{g^+} \geq b_{rk}^{g} \quad (4)$$

$$\lambda, s^-, s^g, s^h \geq 0 \quad (5)$$

$$i=1,\cdots,e; r=1,\cdots,q; j=1,\cdots,u;\ (j\neq k) \quad (6)$$

where *s.t.* denotes the set of constraints, and p indicates the efficiency value of the research unit. When $p < 1$, the DMU is invalid in the model, indicating that it has deviated from the PF, and the land-resource-use efficiency is low. Theoretically, improvement is based on reducing input and increasing output, making it possible to intensively use of land resources. When $p \geq 1$, the DMU is valid in the model, showing that it is at the PF, that is, the point of production efficiency. A larger p represents higher efficiency; λ represents the proportion of a DMU that is reassembled in a new effective DMU; a_{ij} represents the i-th input of the research unit j; b_{tj} is the t-th output of the research unit j; k is the DMU; e is the number of input indicators; r_1 and r_2 represent the numbers of expected and unexpected output factors, respectively; and s represents the slack variable of the input–output factors. Ferrier and Lovell believe that slack variables can ultimately be regarded as invalid resource allocation [45]. Among them, s^- is the input slack, which indicates the excess of input elements in the DMU, and s^g is the expected output slack, which indicates the output of the DMU is insufficient. Finally, s^h is the undesired output slack, which represents the surplus of the output factors of the DMU.

2.3.4. Malmquist–Luenberger Model

When the data of the evaluated DMU are panel data containing observations at multiple time points, the variation of productivity and decomposition factors of the variation can be analysed. The Malmquist model index is a non-parametric method commonly used to dynamically analyse changes in productivity. The Malmquist index model that includes undesired output is called the Malmquist–Luenberger (ML) model and is used to measure total factor productivity (TFP) with undesired output. It is also known as the ML index. It takes into account the intertemporal effect of dynamic factors and is strong in terms of practical applications [46]. In addition, compared to parametric methods, it has the following advantages: First, it does not need to provide the specific statistical distribution of the DMU. Second, it can deal with small amounts of data and classification variables. Third, it does not need to introduce a temporal trend into the data analysis, helping to avoid the phenomenon of smooth productivity change, which is an issue with most parametric methods [47]. The ML index is favoured by scholars based on these advantages.

Therefore, to study the evolution mechanism of dynamic efficiency, this study referred to the improved method of Fare [48,49] to calculate the ML index as follows:

$$ML\left(x^t, y^t, x^{t+1}, y^{t+1}\right) = \sqrt{\frac{E^t(x^{t+1}, y^{t+1})\, E^{t+1}(x^{t+1}, y^{t+1})}{E^t(x^t, y^t)\, E^{t+1}(x^t, y^t)}} \tag{7}$$

$$= \frac{E^{t+1}(x^{t+1}, y^{t+1})}{E^t(x^t, y^t)} \sqrt{\frac{E^t(x^t, y^t)\, E^t(x^{t+1}, y^{t+1})}{E^{t+1}(x^t, y^t)\, E^{t+1}(x^{t+1}, y^{t+1})}} \tag{8}$$

$$= EC * TC \tag{9}$$

where (x^t, y^t) is the input–output vector in the period t; (x^{t+1}, y^{t+1}) is the input–output vector in the period $t+1$; and E^t and E^{t+1} represent the distance function during t and t + 1. The ML index is used to measure the changes in construction land-use efficiency, and it expresses the productivity change of (x^{t+1}, y^{t+1}) relative to (x^t, y^t). If $ML > 1$, the productivity level increases; otherwise, it decreases. The ML can be split into two aspects to obtain two decomposition indices: the technological change (TC) and technical efficiency change (EC). The TC index reflects the contribution of technological progress, such as system-element optimisation and economic structural transformation, to improve construction land-utilisation efficiency [50]. If the value is greater than 1, the production technology has improved. The EC index reflects the distance of the evaluation unit relative to the PF in different periods and is called the "catch-up effect". When its value is greater than 1, this indicates progress in technical efficiency, meaning that the allocation of construction land input resources is reasonable, and the management level has improved [51].

2.3.5. Entropy-Weight TOPSIS Model

The entropy-weight TOPSIS model is derived from combining the entropy-weight and TOPSIS methods. Overall, it is a novel comprehensive evaluation method that scholars use to combine the advantages of the two methods and overcome the subjectivity of the index-weight setting [52]. The model has the advantages of less data loss in the calculation process, intuitive geometric meaning, and a lack of interference by the selection of reference sequences. It is also able to reflect the dynamic changes and laws of the evaluation indicators more scientifically, objectively, comprehensively, and reasonably, allowing a better explanation of the results [53]. Therefore, this study used this method to integrate the sub-indices that cover construction land-use inputs and outputs and pollution-emission levels. The calculation steps are as follows [54]:

First, the dimensional difference of measure index, S_{it}, was eliminated through standardisation as follows:

$$S_{it} = \frac{M_{it} - min(M_{it})}{max(M_{it}) - min(M_{it})}, M_{it} \text{ is a positive indicator} \quad (10)$$

Here, i represents the study area, M_{it} refers to the initial value of the indicator and S_{it} is the standardised value.

Second, j is a sub-index. By calculating the information entropy e_j and weight W_j of S_{it}, a weighting matrix R is constructed as follows:

$$e_j = \ln\frac{1}{x}\sum_{i=1}^{x}[(S_{it}/\sum_{i=1}^{x} S_{it})\ln(S_{it}/\sum_{i=1}^{x} S_{it})] \quad (11)$$

$$W_j = (1 - e_j)/\sum_{j=1}^{m}(1 - e_j) \quad (12)$$

$$R = (a_{ij})_{x\times m,}\ a_{ij} = W_j \times S_{ij} \quad (13)$$

Again, we determine the optimal scheme G_t^+, the worst scheme G_t^-, and the corresponding Euclidean distances h_i^+ and h_i^- as follows:

$$G_t^+ = (\text{max}a_{i1}, \text{max}a_{i2}, \cdots, \text{max}a_{im}) \quad (14)$$

$$G_t^- = (\text{min}a_{i1}, \text{min}a_{i2}, \cdots, \text{min}a_{im}) \quad (15)$$

$$h_i^+ = \sqrt{\sum_{j=1}^{m}\left(G_j^+ - a_{ij}\right)^2} \quad (16)$$

$$h_i^- = \sqrt{\sum_{j=1}^{m}\left(G_j^- - a_{ij}\right)^2} \quad (17)$$

Finally, the comprehensive index K of the evaluation object is calculated, and its value range is (0, 1):

$$K = h_i^- \Big/ h_i^+ + h_i^- \quad (18)$$

3. Results

3.1. Static Analysis of Urban Construction Land-Utilisation Efficiency

To obtain a more comprehensive understanding of the land-use situation in the Yangtze River Delta, this study used the MaxDEA(Beijing Rewomaidi Software Co., LTD, Beijing, China) professional software to calculate two types of construction land-utilisation efficiency in 41 prefecture-level cities in the Yangtze River Delta from 2009 to 2017. The results are shown below (Figure 3).

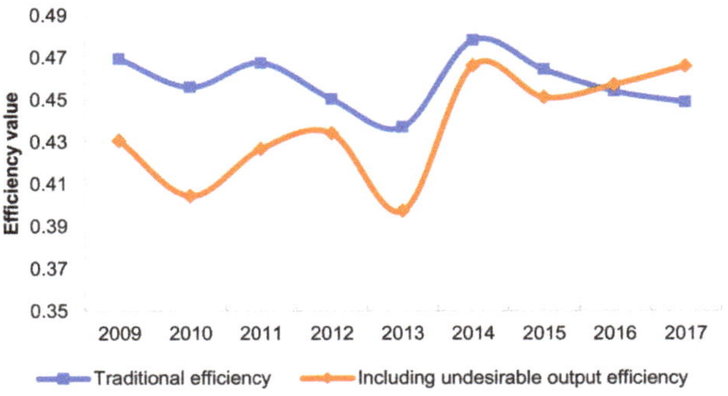

Figure 3. Average efficiency of construction land in the Yangtze River Delta from 2009 to 2017.

In general, the average construction land-use efficiency in the Yangtze River Delta from 2009 to 2017 was low, and the overall efficiency under the influence of pollution emissions (0.397–0.466) was slightly lower than the traditional efficiency (0.437–0.478). Although the efficiency dropped by 4.67% in total, over time, two types of efficiencies exhibited a fluctuating growth trend, and the gap between them gradually narrowed. Therefore, although the inclusion of undesired outputs will reduce the efficiency value, it is able to reflect the actual situation of regional construction land utilisation more scientifically. Recently, the Yangtze River Delta region has shown results based on the implementation of an ecological civilisation strategy, transforming the economic development mode, conserving energy, and reducing emissions, which has caused the construction land efficiency, including undesired outputs, to catch up with the traditional efficiency. However, overall, the low efficiency level reflects the current regional land-use pattern and is relatively extensive, and the land output is far from optimal. This is not conducive to promoting new urbanisation and high-quality city development, and there is room for the efficiency to be improved.

At the city level, traditional efficiency presents regional differentiation, being higher in the east and lower in the west (Figure 4a). Cities with high construction land efficiency are mostly located in the Jiangsu, Zhejiang, and Shanghai regions, which have significant geographical advantages, a high level of urbanisation, and developed economies. The Anhui region, which is located in a remote location in the Yangtze River Delta, has a relatively undeveloped economy. Therefore, coupled with insufficient radiation from the metropolitan area and core cities, the land-use efficiency of cities in this province is generally low. However, if environmental constraints are considered, this regional difference will be decreased (Figure 4b).

To better understand the constraints of undesired outputs, this study comprehensively analysed the impact of environmental factors on land-use efficiency based on the difference between the two efficiencies and the actual pollution-emission intensity of the city. Tests using the Stata16 software and an ordinary least squares (OLS) regression analysis method revealed a significant positive correlation between the efficiency differences and pollution emissions ($t = 2.29$; $p < 0.05$; where t is the regression coefficient and p is significance level), which indicates that the higher the pollution-emission intensity of a city, the greater the efficiency loss caused by undesired outputs.

(1) Southern Jiangsu, northern Zhejiang, and south-eastern Anhui have experienced greater efficiency losses (>0.1) due to a high pollution intensity. For example, Suzhou has been reduced from a high-efficiency city to a medium-high-efficiency city due to its pollution intensity of 0.192. Furthermore, six cities, namely, Jiaxing, Shaoxing, Lishui, Ma'anshan, Wuhu, and Tongling, are affected by high pollution emissions, and their land-use efficiency has changed from medium to low.

(2) The construction land utilisation efficiency of eastern Zhejiang, northern Jiangsu, and northern Anhui was moderately negatively affected by undesired outputs (0.05–0.1)

due to a higher pollution intensity. Amongst these regions, the land-use efficiency of Changzhou and Quzhou changed from medium-high to medium, and that of Yancheng changed from medium to medium-low. The efficiency of the four cities of Huai'an, Lianyungang, Huaibei, and Huainan were originally at the lower-middle level. Due to the impact of pollution emissions, the efficiency was further reduced.

(3) The difference between the two efficiencies is relatively small in most cities in southwestern Anhui due to the low total pollution emissions. For example, the pollution-emission indices of some cities, such as Wuhu and Anqing, are less than 0.008, which results in a small degree of efficiency loss. Therefore, the real efficiency of several high-efficiency cities in the region is greatly reduced due to exorbitant environmental pollution emissions, whereas some low-efficiency cities are less affected by relatively small emissions of pollution, and their efficiency only fluctuates within a small range. Accordingly, the efficiency differences within the region are reduced, and the spatial equilibrium of the efficiency is strengthened.

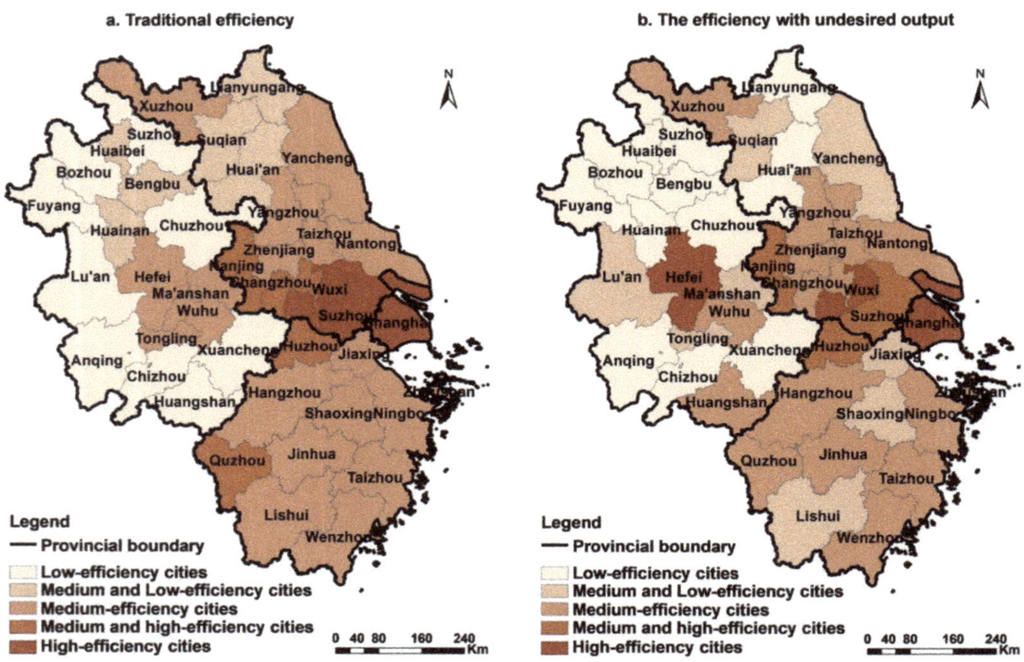

Figure 4. Mean value comparison between the traditional efficiency of urban construction land and the efficiency with undesired outputs in the Yangtze River Delta from 2009 to 2017. (**a**) Traditional efficiency; (**b**) efficiency with undesired outputs.

3.2. Analysis of the Driving Force of the Evolution of Urban Construction Land-Utilisation Efficiency

To further realise dynamic changes in urban construction land-utilisation efficiency and their driving factors, this study measured the ML and its decomposition index (TC and EC) and analysed these accordingly.

Using the ArcMap10.6 software and Natural Breaks (jenks) method, the annual average ML, TC, and EC indices of 41 prefecture-level cities in the Yangtze River Delta were divided into three levels, from small to large (Figure 5). Generally, the ML of the Yangtze River Delta had the highest spatial distribution characteristics in southern Jiangsu and southern Anhui, followed by northern Zhejiang and northern Jiangsu. Lower spatial distribution characteristics were present in Shanghai, southern Zhejiang, and northern Anhui.

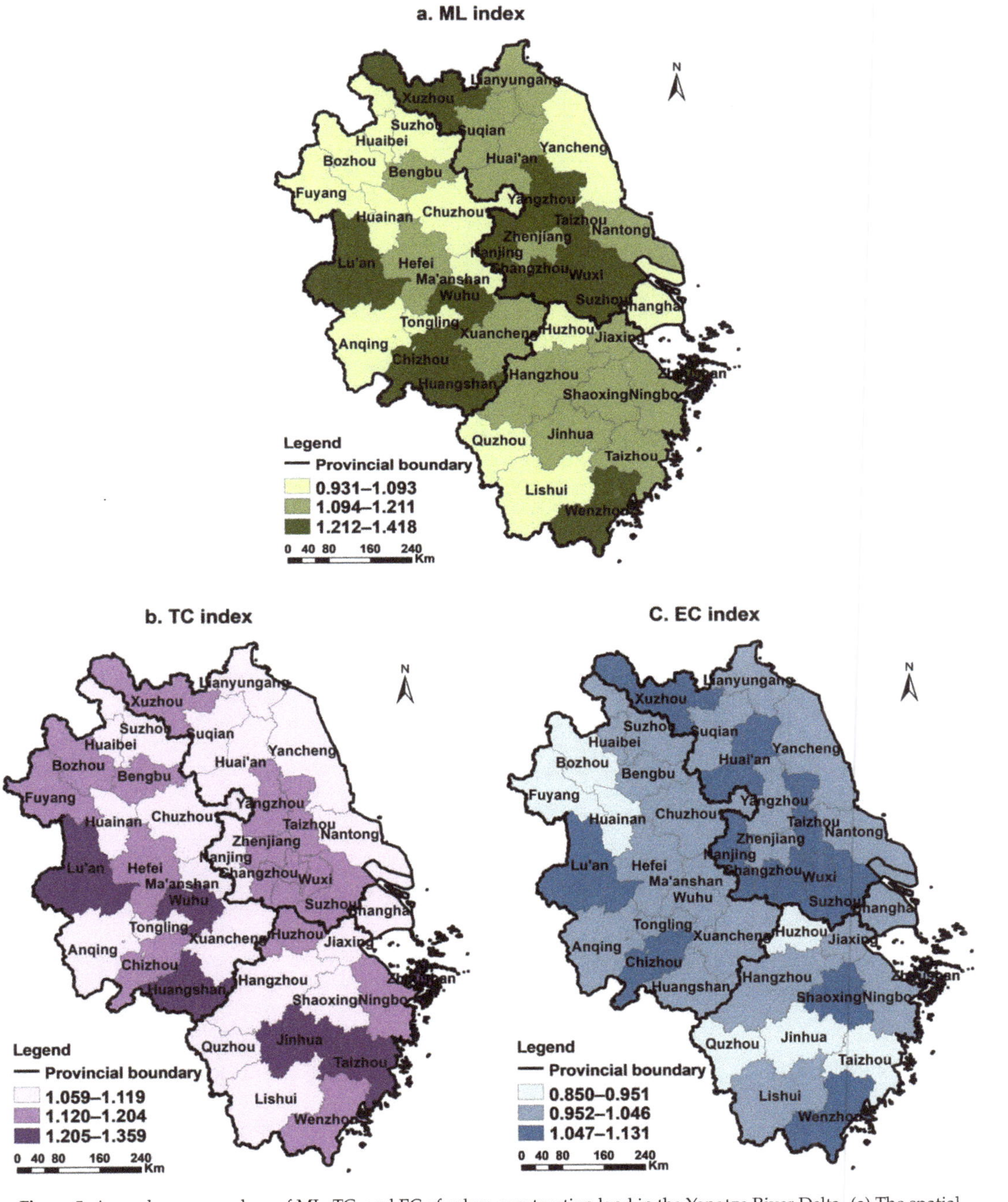

Figure 5. Annual average values of ML, TC, and EC of urban construction land in the Yangtze River Delta. (**a**) The spatial distribution of ML index; (**b**) spatial distribution of TC index; (**c**) spatial distribution of EC index.

(1) As the largest cities in the region, Suzhou, Wuxi, and Changzhou, among others, have gathered much high-level talent due to their advantages regarding their locations

for transportation, economic and technological development, and strong capital strength. These drive TC and EC towards improving regional industrial innovation and management system optimisation (TC > 1.12; EC > 1.05). This has resulted in a significant improvement in the ML (1.418 > ML > 1.212), and the efficiency of construction land increased significantly during the study period. Recently, Lu'an, Huangshan, and Chizhou have reorganised their input production factors, optimised resource allocation, and strengthened their technological updates and creations to promote the rapid growth of ML.

(2) The economic strengths of northern Zhejiang and northern Jiangsu are relatively lower than those of other regions. Although they have certain technological advancement capabilities, such as research and development (R&D) investment and technological output (1.12 < TC < 1.20), some cities are hindered regarding the improvement of ML due to a low EC (EC < 1), such as an imbalanced allocation of input resources or rigid management systems (1.094 < ML < 1.211). Additionally, though the efficiency of this type of urban construction land has improved, there is still plenty of room for improvement in the future. As a large city in the Yangtze River Delta, Shanghai has unique advantages in science, technology, policy, information, and capital. However, its land-use efficiency has reached the highest level on the common frontier, and there is limited room for its efficiency to improve. Therefore, its ML is low, but its efficiency is still in a slow growth status.

(3) The cities of Quzhou, Lishui, Huaibei, Suzhou, and Chuzhou in southern Zhejiang and northern Anhui have relatively little advantage regarding technology application and achievement conversion, and their technological progress is not obvious (1 < TC < 1.09). Coupled with their inadequacies regarding resource allocation and information circulation, they are restricted by technology efficiency (EC < 1), which results in little improvement in their ML and a lack of potential for the improvement of urban construction land efficiency.

In general, the average ML value for construction land in the Yangtze River Delta is 1.154, the TC index is 1.142, and the EC index is 1.010. The values of the three indicators are all greater than one, which indicates that the improvement of ML is affected by the combined effect of the TC and EC. However, the contribution rate for TC is 14 times that for EC. In addition, the average value of each index of the above case city shows that 98% of the cities' ML indices are greater than 1, and 41 cities are driven by TC, whereas only 63.41% of the cities are driven by EC. Moreover, the growth of the ML indices of the other 36.59% of the cities is restrained due to a low EC. Therefore, TC is the core driver for the improvement of ML, and the contribution of EC is relatively insignificant.

3.3. Correlation Analysis of Urban Construction Land-Utilisation Efficiency and Factor Inputs and Outputs

Using the method presented by Liu [55], the element input and output and pollution-emission levels during the construction land-utilisation process were analysed to further determine the internal causes of the differentiation in regional efficiency. As the factor input index, the unit land labour force and fixed-assets investment represent the land input level. Using the land-average production value as the expected output index for land use indicates the level of land output. Moreover, the land-average wastewater, sulphur dioxide, and dust emissions were used as undesired output indicators of land use to characterise the degree of land-use pollution emissions. In addition, using the entropy-weight TOPSIS method, three kinds of indicators of the input, expected output, and unexpected output were, respectively constructed into three kinds of comprehensive indices of input, expected output, and non-expected output, to scientifically and comprehensively measure the input–output situation of land. Finally, the K-means clustering method was used to divide the comprehensive indicators of construction land-use input and output and pollution emissions into the following three types: low, medium, and high (Table 2).

Table 2. Correlation types for input–output and pollution emissions of construction land use in cities of the Yangtze River Delta.

Input–Output Model of Construction Land	2009	2017
	City	City
High input, high output, low pollution	Shanghai (1.299)	/
High input, medium output, high pollution	Changzhou (0.369)	/
High input, medium output, medium pollution	Nanjing (0.445), Wuxi (0.592), Ma'anshan (0.366)	/
High input, medium output, low pollution	Hefei (1.234)	Wenzhou (1.004)
High input, low output, low pollution	Wuhu (0.301)	Zhoushan (0.462)
Medium input, high output, medium pollution	/	Shanghai (1.291)
Medium input, medium output, high pollution	Hangzhou (0.439)	/
Medium input, medium output, medium pollution	Suzhou (0.637)	Wuxi (1.04), Changzhou (0.721), Hangzhou (0.52)
Medium input, medium output, low pollution	Ningbo (0.549), Zhoushan (0.436)	Nanjing (1.006), Ningbo (0.525), Hefei (1.032)
Medium input, low output, high pollution	/	Shaoxing (0.366)
Medium input, low output, medium pollution	Shaoxing (0.206)	Jiaxing (0.323), Ma'anshan (0.28)
Medium input, low output, low pollution	Nantong (0.355), Zhenjiang (0.454) Taizhou (0.315), Jiaxing (0.226), Xuancheng (0.157), Tongling (0.265)	Yangzhou (0.494), Zhenjiang (0.524), Taizhou (0.62), Wuhu (0.399), Tongling (0.261)
Low input, medium output, medium pollution	Quzhou (1.044)	Suzhou (1.016)
Low input, medium output, low pollution	Huzhou (1.03), Jinhua (1.025), Taizhou (1.011)	Xuzhou (1.089)
Low input, low output, medium pollution	/	Quzhou (0.272)
Low input, low output, low pollution	Xuzhou (0.373), Lianyungang (0.229) Huaian (0.198), Yancheng (0.315), Yangzhou (0.379), Suqian (0.19), Wenzhou (0.46), Lishui (0.303), Huaibei (0.254), Bozhou (0.239), Suzhou (0.237), Bengbu (0.238), Fuyang (0.279), Huainan (0.263), Chuzhou (0.172), Lu'an (0.216), Chizhou (0.164), Anqing (0.203), Huangshan (0.189)	Nantong (0.445), Lianyungang (0.283), Huaian (0.381), Yancheng (0.289), Suqian (0.268), Huzhou (0.306), Jinhua (0.362), Taizhou (0.421), Lishui(0.288), Huaibei (0.229), Bozhou (0.16), Suzhou (0.177), Bengbu (0.276), Fuyang (0.135), Huainan (0.164), Chuzhou (0.188), Lu'an (0.422), Xuancheng (0.2), Chizhou (0.419), Anqing (0.192), Huangshan (0.265)

(1) Generally, a high input and output will result in high efficiency. Shanghai's land input and economic output are leading in the region, and its construction land utilisation maintained a high level of efficiency during the study period. Moreover, a low input and output often result in low efficiency. Cities with low input, output, and pollution comprise more than 46% of the Yangtze River Delta region, and most of them are located in marginal areas of northern Jiangsu, northern Zhejiang, northern Anhui, and southwest Anhui, such as Lianyungang, Huaian, Lishui, and Chizhou. Such cities are relatively deficient in economic and technological development and resource management and allocation, which results in insufficient input and a low output. Although the pollution degree is small, as the actual output of construction land at this time has become the fundamental constraint for efficiency evaluation, the land-use efficiency is low. With the transformation of the economic structure and the continuous progress of society, some areas, such as Yangzhou and Wenzhou, have gradually shifted from low input to medium-high input, while the land-use output has risen, and the efficiency of construction land has increased.

(2) However, during the process of urban development, the utilisation efficiency for construction land is not always consistent with the level of land input and output due to improper resource allocation or pollution discharge. Although Changzhou, Nanjing, Wuxi, Ma'anshan, and Wuhu, among other cities, have high input levels, they also have redundant or unreasonable input elements. Therefore, the land output failed to achieve synchronous growth. At the same time, these areas experience medium to high levels of

pollution emissions during the process of economic construction, which further reduces the efficiency of land use. In addition, low and medium inputs may result in a higher land output and land-use efficiency on the basis of controlling pollution emissions. In 2009, cities such as Huzhou, Jinhua, and Quzhou adopted low input, medium output, and low pollution land use model to achieve a relatively high level of construction land use efficiency. In 2017, some cities, such as Hefei and Nanjing, adopted a medium input and output and reduced the total pollution emissions to make more efficient use of construction land. However, although Hangzhou's land input and output levels are relatively high, its construction land efficiency is always in the lower position for the same level of cities due to negative impacts, such as high pollution emissions.

(3) In addition, medium inputs and low outputs will be the top priority of land-market-consolidation efforts in the future. The associated cities are mainly located in central Jiangsu and other areas with good economic development. They have strong comprehensive strength and development potential and can provide sufficient input for land use. However, due to their limitations, including technical conditions and inadequate means of resource allocation, the improvement of land efficiency has encountered bottlenecks, resulting in a clear deviation of land use from the optimal PF. In the future, we should improve the quality and efficiency of land use through comprehensive land consolidation and remould the land-use pattern.

4. Discussion

Urban construction land resources have become an important bottleneck restricting urban economic and social development. China's strict farmland-protection policies have significantly blocked the external supply of construction land resources [56]. In the context of new urbanization, each region should fully consider its own resource endowments and functional positioning, relying on the new era of territorial space planning, and comprehensively improve the use efficiency for urban construction land. The research of relevant scholars shows that the impact and rate of contribution of construction land expansion on economic growth gradually decreases with the evolution of economic development [57]. The development of construction land will play a more obvious driving economic role in underdeveloped areas [58]. The Yangtze River Delta region is one of the most economically developed regions in China, and the southern Jiangsu and northern Zhejiang and Shanghai areas have entered the later stages of urbanization. Therefore, these cities should pay attention to the connotative development of the city, supplement them with territorial space planning, and tap the potential of the urban stock land through the reconstruction of the old city, urban village renewal, and the development of idle and inefficient land. As northern Jiangsu, northern Anhui, and western Zhejiang are in the middle and later stages of urbanization, with the advancement of China's common prosperity, these areas need more indicators of construction land to drive economic development, which requires the overall coordination of construction land resource allocation in the compilation of territorial space planning. In view of the excellent ecological endowments in northern Jiangsu, northern Anhui, and western Zhejiang, the construction land index should be appropriately tilted to them under the conditions that the regional resource and environmental carrying capacity permit. Central Jiangsu, southern Anhui, and southern Zhejiang should give full play to their regional advantages, adjust the industrial structure to attract the transfer of industries in the core area of the Yangtze River Delta, and increase the level of land output.

Generally speaking, the economic development of certain regions is often accompanied by the excessive use of resources and greenhouse gas emissions. Given that the overall construction land-use efficiency in the Yangtze River Delta under environmental constraints is low, the regional imbalance is prominent, and the improvement of efficiency mainly comes from TC, we believe that further targeted efficiency-improvement measures should be taken in future. First, we should assume that the political responsibility and development mission of "putting ecology and green development first" strengthens the concept of ecological protection. Then, we should jointly build a strong environmental constraint mechanism and implement the negative list system for industrial access. In areas where the input and output levels are high, but the efficiency is reduced due to high pollution emissions, it is necessary to pay special attention to ecological protection and enhance the competitiveness of green development. Second, the Yangtze River Delta should prioritize the powerful driving effect of provincial capitals or large cities, promote overall regional efficiency, and strive to create high-quality integrated regional developmental growth. Third, the exchange culture of sharing resources and win–win cooperation should be promoted. By strengthening inter-regional communal work and rationally scheduling the spatial transfer and allocation of product factors between cities, the Yangtze River Delta region can overcome the contradiction between the production scale and technical structure that exists in economic, social, and environmental activities in various regions. With institutional innovation as the core, using policy innovation as the key factor and technological innovation as the driving force (based on continuous technological advancement), the Yangtze River Delta region can further improve EC and the technology catch-up effect by optimizing the scale allocation and management level of construction land input resources. Accordingly, the use efficiency of construction land will be evenly increased.

This paper has several shortcomings. First, due to the limited data acquisition regarding the measurement of urban construction land efficiency that includes undesired outputs, the total emissions made up of only three types of waste discharge were selected as environmental constraints to represent the level of urban pollution emissions. Follow-up studies could consider deepening and perfecting the evaluation indicators of undesired outputs in terms of environmental protection investment, carbon emissions, environmental governance, and air quality ($PM_{2.5}$). Second, this study used only 41 prefecture-level cities in the Yangtze River Delta as its research object. If county-level cities could be selected for research in the future, the efficiency of regional construction land use would be reflected more comprehensively and accurately. Finally, although this study performed a correlation analysis for the efficiency of urban construction land use, input factors, and output levels, it did not explore specific input and output types in depth, which must be expanded upon in future research.

5. Conclusions

Based on the land input–output data for 41 prefecture-level cities in the Yangtze River Delta, this study constructed an SE-SBM model to compare and evaluate the static efficiency of urban construction land use from 2009 to 2017 considering traditional and environmental constraints. Furthermore, using the ML index and its decomposition items, the dynamic evolution of the construction land-use efficiency was studied. In addition, the entropy-weight TOPSIS method was used to analyse the relationship between the urban construction land efficiency and input–output level. The research conclusions are as follows:

(1) Based on the results of the static efficiency measurement, it was found that the construction land efficiency of the Yangtze River Delta region remained at a low level, overall, during 2009–2017. Furthermore, due to the addition of undesired outputs, the efficiency dropped by 4.67%. However, with the transformation and upgrading of the economy structure, the land-use efficiency under environmental constraints has gradually caught up with the traditional efficiency, and the total efficiency has increased slightly. At the city level, the traditional construction land efficiency presents regional differentiation, being higher in the east and lower in the west. High-efficiency cities are concentrated in the economically developed Jiangsu, Zhejiang, and Shanghai regions, and most of the low-efficiency cities are located in the economically underdeveloped Anhui region. Due to the influence of pollution emissions, the regional imbalance in efficiency has been weakened. This change stems from the efficiency loss caused by pollution emissions, and the degree of efficiency loss is significantly positively correlated with the intensity of urban pollution emissions. For example, southern Jiangsu, northern Zhejiang, and south-eastern

Anhui have lost more than 0.1 efficiency due to their high pollution emission intensity, and eastern Zhejiang, northern Jiangsu, and northern Anhui suffered 0.05–0.1 efficiency losses due to their medium-to-high pollution emission intensity. Cities in southwest Anhui only lost small amounts of efficiency due to their lower pollution emissions. Compared to the traditional-efficiency values, the gap between high and low-efficiency cities in the region has been narrowed, decreasing the spatial differentiation of efficiency.

(2) Based on the results of the dynamic efficiency measurements, it was found that the ML in southern Jiangsu and southern Anhui has the fastest growth, followed by northern Zhejiang and northern Jiangsu, while the growth rates of Shanghai, southern Zhejiang, and northern Anhui are relatively slow. Amongst them, the ML productivity of construction land in 98% of cities is mainly improved by TC, whereas EC has a limited promoting effect and does not cover the whole area. Therefore, although, during the process of construction land use for production, the economic structure of most regions has been transformed and upgraded, the institutional system has been optimized, and scientific and technological innovations have been iterated, small and medium-sized cities still have outstanding problems. These include a lack of ecological protection, an unbalanced allocation of resource elements, unreasonable investment scales, and low management levels, restricting the further improvement of construction land-use efficiency and exacerbating regional differentiation.

(3) Using the entropy-weight TOPSIS model and K-means clustering method, the construction land input and output and pollution emission indicators of 41 cities in the Yangtze River Delta were divided into the following three types: low, medium, and high. The correlation between the efficiency and the input and output was then investigated. The results show that, under normal circumstances, there is a correlation between construction land-use efficiency and land input and output; namely, a high input and output lead to high efficiency, and a low input and output lead to low efficiency. However, the two are not always consistent. A medium input and output and a low input and medium output may result in higher land-use efficiency resulting from the control of pollution emissions. Medium-input and low-output types have sufficient room for efficiency improvement and should be the top priority in comprehensive land consolidation in the future.

Author Contributions: Conceptualization, Y.Z. and Y.C.; methodology, Y.Z. and Y.C.; validation, Y. H. and Y.C.; formal analysis, Y.Z.; data curation, Y.Z. and Y.C.; writing—original draft preparation, Y.Z.; writing—review and editing, Y.Z., Y.C., and Y.H. All authors have read and agreed to the published version of the manuscript.

Funding: This research was funded by the project from National Natural Science Foundation of China (Grant No. 41501185).

Institutional Review Board Statement: Not applicable.

Informed Consent Statement: Not applicable.

Data Availability Statement: All the socio-economic data can be found in the China statistical Yearbook, and the land data can be found in the Ministry of Natural Resources. All the socio-economic data can be found here: https://data.cnki.net/Yearbook/Navi?type=type&code=A. The land data are not public.

Conflicts of Interest: The authors declare no conflict of interest.

References

1. Chen, Y.; Chen, Z.; Xu, G.; Tian, Z. Built-up land efficiency in urban China: Insights from the General Land Use Plan (2006–2020). *Habitat Int.* **2016**, *51*, 31–38. [CrossRef]
2. Yao, Q.; Shu, B.; Yong, X. Comprehensive evaluation and influencing factors of urban land utilization total factor productivity in Huaihai economic zone. *Resour. Environ. Yangtze Basin* **2019**, *28*, 1823–1832.
3. Koroso, N.H.; Zevenbergen, J.A.; Lengoiboni, M. Urban land use efficiency in Ethiopia: An assessment of urban land use sustainability in Addis Ababa. *Land Use Policy* **2020**, *99*, 105081. [CrossRef]
4. Yan, S.; Peng, J.; Wu, Q. Exploring the non-linear effects of city size on urban industrial land use efficiency: A spatial econometric analysis of cities in eastern China. *Land Use Policy* **2020**, *99*, 104944. [CrossRef]

5. Du, J.; Thill, J.-C.; Peiser, R.B. Land pricing and its impact on land use efficiency in post-land-reform China: A case study of Beijing. *Cities* **2016**, *50*, 68–74. [CrossRef]
6. Han, H.; Zhang, X. Static and dynamic cultivated land use efficiency in China: A minimum distance to strong efficient frontier approach. *J. Clean. Prod.* **2020**, *246*, 119002. [CrossRef]
7. Cao, X.; Liu, Y.; Li, T.; Liao, W. Analysis of Spatial Pattern Evolution and Influencing Factors of Regional Land Use Efficiency in China Based on ESDA-GWR. *Sci. Rep.* **2019**, *9*, 520. [CrossRef]
8. Jiang, H. Spatial–temporal differences of industrial land use efficiency and its influencing factors for China's central region: Analyzed by SBM model. *Environ. Technol. Innov.* **2021**, *22*, 101489. [CrossRef]
9. Zhang, L.; Zhang, L.; Xu, Y.; Zhou, P.; Yeh, C.-H. Evaluating urban land use efficiency with interacting criteria: An empirical study of cities in Jiangsu China. *Land Use Policy* **2020**, *90*, 104292. [CrossRef]
10. Gao, X.; Zhang, A.; Sun, Z. How regional economic integration influence on urban land use efficiency? A case study of Wuhan metropolitan area, China. *Land Use Policy* **2020**, *90*, 104329. [CrossRef]
11. Zhu, X.; Zhang, P.; Wei, Y.; Li, Y.; Zhao, H. Measuring the efficiency and driving factors of urban land use based on the DEA method and the PLS-SEM model—A case study of 35 large and medium-sized cities in China. *Sustain. Cities Soc.* **2019**, *50*, 101646. [CrossRef]
12. Zhu, X.; Li, Y.; Zhang, P.; Wei, Y.; Zheng, X.; Xie, L. Temporal–spatial characteristics of urban land use efficiency of China's 35mega cities based on DEA: Decomposing technology and scale efficiency. *Land Use Policy* **2019**, *88*, 104083. [CrossRef]
13. Ding, T.; Yang, J.; Wu, H.; Liang, L. Land use efficiency and technology gaps of urban agglomerations in China: An extended non-radial meta-frontier approach. *Socio-Econ. Plan. Sci.* **2021**, 101090. [CrossRef]
14. Xie, H.; Chen, Q.; Lu, F.; Wu, Q.; Wang, W. Spatial-temporal disparities, saving potential and influential factors of industrial land use efficiency: A case study in urban agglomeration in the middle reaches of the Yangtze River. *Land Use Policy* **2018**, *75*, 518–529. [CrossRef]
15. Liu, J.; Jin, X.; Xu, W.; Gu, Z.; Yang, X.; Ren, J.; Fan, Y.; Zhou, Y. A new framework of land use efficiency for the coordination among food, economy and ecology in regional development. *Sci. Total Environ.* **2020**, *710*, 135670. [CrossRef]
16. Wu, C.; Wei, Y.D.; Huang, X.; Chen, B. Economic transition, spatial development and urban land use efficiency in the Yangtze River Delta, China. *Habitat Int.* **2017**, *63*, 67–78. [CrossRef]
17. Chen, W.; Chen, W.; Ning, S.; Liu, E.-n.; Zhou, X.; Wang, Y.; Zhao, M. Exploring the industrial land use efficiency of China's resource-based cities. *Cities* **2019**, *93*, 215–223. [CrossRef]
18. Liu, Y.; Zhang, Z.; Zhou, Y. Efficiency of construction land allocation in China: An econometric analysis of panel data. *Land Use Policy* **2018**, *74*, 261–272. [CrossRef]
19. Jin, G.; Deng, X.; Zhao, X.; Guo, B.; Yang, J. Spatio-temporal patterns of urban land use efficiency in the Yangtze River Economic Zone during 2005–2014. *Acta Geogr. Sin.* **2018**, *73*, 1242–1252.
20. Hu, Y.; Qiao, W.; He, T. Changes of the land use pattern and ecosystem service value in Yangtze—Hwai ecological economic zone. *Resour. Environ. Yangtze Basin* **2020**, *29*, 2450–2461.
21. Li, H.; Li, W.; Zheng, F. Optimal Land Use Structure for Sustainable Agricultural Development—A Case Study in Changsha County, South Central China. *J. Resour. Ecol.* **2021**, *12*, 203–213.
22. Liu, S.; Lin, Y.; Ye, Y.; Xiao, W. Spatial-temporal characteristics of industrial land use efficiency in provincial China based on a stochastic frontier production function approach. *J. Clean. Prod.* **2021**, *295*, 126432. [CrossRef]
23. Yu, J.; Zhou, K.; Yang, S. Land use efficiency and influencing factors of urban agglomerations in China. *Land Use Policy* **2019**, *88*, 104143. [CrossRef]
24. Liang, J.; Chen, S. Research on the land-use efficiency and driving factors of urban construction in Fujian province under environmental constraints. *J. Nat. Resour.* **2020**, *35*, 2862–2874. [CrossRef]
25. Liu, S.; Xiao, W.; Li, L.; Ye, Y.; Song, X. Urban land use efficiency and improvement potential in China: A stochastic frontier analysis. *Land Use Policy* **2020**, *99*, 105046. [CrossRef]
26. Hu, Y.; Qiao, W.; Wan, Y.; He, T.; Chai, Y.; Bi, Y. Comprehensive Evaluation and Spatial Distinction of Land Use Efficiency in County Area of Jiangsu Province. *Econ. Geogr.* **2020**, *40*, 186–195.
27. Yao, M.; Zhang, Y. Evaluation and Optimization of Urban Land-Use Efficiency: A Case Study in Sichuan Province of China. *Sustainability* **2021**, *13*, 1771. [CrossRef]
28. Song, J.; Chen, S. Impact of economic agglomeration on land use eco-efficiency of three major urban agglomerations in China. *J. Nat. Resour.* **2021**, *36*, 2865–2877. [CrossRef]
29. Guastella, G.; Pareglio, S.; Sckokai, P. A spatial econometric analysis of land use efficiency in large and small municipalities. *Land Use Policy* **2017**, *63*, 288–297. [CrossRef]
30. Zhao, Z.; Bai, Y.; Wang, G.; Chen, J.; Yu, J.; Liu, W. Land eco-efficiency for new-type urbanization in the Beijing-Tianjin-Hebei Region. *Technol. Forecast. Soc. Chang.* **2018**, *137*, 19–26. [CrossRef]
31. Saikku, L.; Mattila, T.J. Drivers of land use efficiency and trade embodied biomass use of Finland 2000–2010. *Ecol. Indic.* **2017**, *77*, 348–356. [CrossRef]
32. Qiao, W.; Huang, X. Change in Urban Land Use Efficiency in China: Does the High-Speed Rail Make a Difference? *Int. J. Environ. Res. Public Health* **2021**, *18*, 10043. [CrossRef] [PubMed]

33. The Central Committee of the Communist Party of China and the State Council. The Outline of the Regional Integration Development Plan of the Yangtze River Delta. Available online: https://www.ndrc.gov.cn/xwdt/ztzl/cjsjyth1/ghzc/202007/t20200728_1234708.html (accessed on 28 July 2020).
34. China Economic and Social Big Data Research Platform. China City Statistical Yearbook. Available online: https://www.tongjinianjian.com/111090.html (accessed on 20 April 2021).
35. China Economic and Social Big Data Research Platform. Jiangsu Statistical Yearbook. Available online: http://tj.jiangsu.gov.cn/col/col4009/index.html (accessed on 20 April 2021).
36. China Economic and Social Big Data Research Platform. Zhejiang Statistical Yearbook. Available online: http://tjj.zj.gov.cn/ (accessed on 20 April 2021).
37. China Economic and Social Big Data Research Platform. Anhui Statistical Yearbook. Available online: http://tjj.ah.gov.cn/ssah/qwfbjd/tjnj/index.html (accessed on 20 April 2021).
38. China Economic and Social Big Data Research Platform. Shanghai Statistical Yearbook. Available online: https://tjj.sh.gov.cn/tjnj/index.html (accessed on 20 April 2021).
39. Lu, X.; Kuang, B.; Li, J. Regional difference decomposition and policy implications of China's urban land use efficiency under the environmental restriction. *Habitat Int.* **2018**, *77*, 32–39. [CrossRef]
40. Lin, J.; Wang, Y.; Zhang, X.; Liu, X. Spatial and temporal characteristics and influencing factors of urban resources and environmental efficiency in the Yellow River Basin. *J. Nat. Resour.* **2021**, *36*, 208–222.
41. Xie, X.; Fang, B.; Xu, H.; He, S.; Li, X. Study on the coordinated relationship between Urban Land use efficiency and eco-system health in China. *Land Use Policy* **2021**, *102*, 105235. [CrossRef]
42. Cheng, G. *Data Envelopment Analysis Method and MaxDEA Software*, 1st ed.; Intellectual Property Publishing House: Beijing, China, 2014; pp. 11–181.
43. Tone, K. A slacks-based measure of efficiency in data envelopment analysis. *Eur. J. Oper. Res.* **2001**, *130*, 498–509. [CrossRef]
44. Tone, K. A slacks-based measure of super-efficiency in data envelopment analysis. *Eur. J. Oper. Res.* **2002**, *143*, 32–41. [CrossRef]
45. Ferrier, G.D.; Lovell, C.A.K. Measuring cost efficiency in banking. *J. Econom.* **1990**, *46*, 229–245. [CrossRef]
46. Bai, Y.; Zhang, X.; Hao, Y.; Song, X. Research on Regional Environmental Performance and Its Influential Factors Based on SBM-Malmquist-Tobit Model. *Areal Res. Dev.* **2013**, *32*, 90–95.
47. Qin, X.; Wang, J. Improvement and Application of Malmquist-luenberger Index Without Solution—A Case of Commercial Bank Productivity Evaluation. *Syst. Eng.* **2021**, 1–13, (Unpublished).
48. Malmquist, S. Index numbers and indifference surfaces. *Trab. Estad.* **1953**, *4*, 209–242. [CrossRef]
49. Chung, Y.; Färe, R.; Grosskopf, S. Productivity and undesirable outputs: A directional distance function approach. *J. Environ. Manag.* **1997**, *51*, 229–240. [CrossRef]
50. Su, X.; Gi, D.; He, H. Study on spatial and temporal differences and affecting factors of agricultural water resources green efficiency in Huang-Huai-Hai Plain. *Ecol. Econ.* **2021**, *37*, 106–111.
51. Zhang, X.; Gui, B. The analysis of total factor productivity in China: A review and application of Malmquist index approach. *J. Quant. Tech. Econ.* **2008**, *6*, 111–122.
52. Huang, C.; Wu, C. Research on the synergetic effect of industrial green transformation and ecological civilization construction in the Yangtze River economic belt. *Resour. Environ. Yangtze Basin* **2021**, *30*, 1287–1297.
53. Han, L.; Qi, X.; Hao, J. Natural resources development conditions and environmental carrying capacity in Mongolia. *J. Arid. Land Resour. Environ.* **2021**, *35*, 93–99.
54. Li, X. TOPSIS model with entropy weight for eco geological environmental carrying capacity assessment. *Microprocess. Microsyst.* **2021**, *82*, 103805. [CrossRef]
55. Lui, S.; Ye, Y.; Li, L.; Xiao, W. Research on urban land use efficiency in China based on the stochastic frontier analysis. *J. Nat. Resour.* **2021**, *36*, 1268–1281.
56. Nie, L.; Guo, Z.; Peng, C. Construction land utilization efficiency based on SBM-Undesirable and Meta-frontier model. *Resour. Sci.* **2017**, *39*, 836–845.
57. Zhang, W.; Li, G. Study on the Regional Economic Growth and Its Sustainability in China—Analyses Based on the Decoupling Index. *Econ. Geogr.* **2015**, *35*, 8–14.
58. Zhang, H.; Ye, Y.; Yang, L.; Wu, Q. Construction land expansion and its contribution to economic growth since reform and opening up in guangdong province. *Ecol. Econ.* **2008**, *28*, 904–908.

International Journal of *Environmental Research and Public Health*

MDPI

Article

Zoning of Ecological Restoration in the Qilian Mountain Area, China

Lin Liu [1], Wei Song [2,*], Yanjie Zhang [2], Ze Han [2], Han Li [2,3], Dazhi Yang [2,3], Zhanyun Wang [2,4] and Qiang Huang [3,5]

1 School of Land Science and Space Planning, Hebei GEO University, Shijiazhuang 050031, China; liulin7801@126.com
2 Key Laboratory of Land Surface Pattern and Simulation, Institute of Geographic Sciences and Natural Resources Research, Chinese Academy of Sciences, Beijing 100101, China; zhangyanjie_cdlg@yeah.net (Y.Z.); hanze1125@163.com (Z.H.); lih.19s@igsnrr.ac.cn (H.L.); yangdazhi@igsnrr.ac.cn (D.Y.); wangzhanyunwzy@163.com (Z.W.)
3 College of Resources and Environment, University of Chinese Academy of Sciences, Beijing 100049, China; qiang12031@foxmail.com
4 Institute of Mountain Hazards and Environment, Chinese Academy of Sciences and Ministry of Water Conservancy, Chengdu 610041, China
5 State Key Laboratory of Resources and Environmental Information System, Institute of Geographic Sciences and Natural Resources Research, Chinese Academy of Sciences, Beijing 100101, China
* Correspondence: songw@igsnrr.ac.cn

Citation: Liu, L.; Song, W.; Zhang, Y.; Han, Z.; Li, H.; Yang, D.; Wang, Z.; Huang, Q. Zoning of Ecological Restoration in the Qilian Mountain Area, China. *Int. J. Environ. Res. Public Health* **2021**, *18*, 12417. https://doi.org/10.3390/ijerph182312417

Academic Editor: Paul B. Tchounwou

Received: 13 October 2021
Accepted: 22 November 2021
Published: 25 November 2021

Abstract: Ecosystem restoration has been widely concerned with the damage and degradation of ecosystems worldwide. Scientific and reasonable formulations of ecological restoration zoning is the basis for the formulation of an ecological restoration plan. In this study, a restoration zoning index system was proposed to comprehensively consider the ecological problems of ecosystems. The linear weighted function method was used to construct the ecological restoration index (ERI) as an important index of zoning. The research showed that: (1) the ecological restoration zones of the Qilian Mountains can be divided into eight basins, namely the headwaters of the Datong River Basin, the Danghe-Dahaerteng River Basin, the northern confluence area of the Qinghai Lake, the upper Shule River to middle Heihe River, the Oasis Agricultural Area in the northern foothills of the Qilian Mountain, the Huangshui Basin Valley, Aksay (corridor region of the western Hexi Basin), and the northeastern Tsaidam Basin; (2) the restoration index of the eight ecological restoration zones of the Qilian Mountains was between 0.34–0.8, with an average of 0.61 (the smaller the index, the more prominent the comprehensive ecological problem representing the regional mountains, rivers, forests, cultivated lands, lakes, and grasslands, and thus the greater the need to implement comprehensive ecological protection and restoration projects); and (3) the ecological problems of different ecological zones are frequently numerous, and often show the phenomenon of multiple overlapping ecological problems in the same zone.

Keywords: land use zoning; ecosystem system; ecological restoration; Qilian Mountains; China

1. Introduction

Ecosystems as a whole connect biological organisms and the inorganic environment through a variety of ecological functions, which feature comprehensive, holistic, and systematic characteristics. The unreasonable use of natural resources has caused many ecological problems, such as soil erosion, land desertification, and the degradation of forest and grass vegetation. Many ecological problems have caused widespread concern for the restoration of ecosystems. For the formulation of ecological restoration plans in a scientific and reasonable manner, it is necessary to comprehensively analyze the problems existing in ecosystems and identify the key restoration areas. From this point of view, the demarcation of ecological restoration zoning is of great significance.

The earliest studies of ecosystem zoning can be traced back to the zoning studies of natural ecosystems. In the early 19th century, the isothermal graph of the German geographer A.V. Humboldt marked the beginning of the study of natural ecological zoning [1]. On this basis, the concept of ecological zones and their divisions, as proposed by Merriam, became the prototype of ecological divisions [2]. Since then, ecologists have continued to research the principles of ecological divisions. Ecological zoning during this period mainly considered the natural factors affecting the ecosystem, and the functional factors of the ecosystem itself were less prioritized. In the 1980s, ecosystem functional divisions were created, which were widely used for large-scale regional natural resource management by the Environmental Cooperation Commission [3], the World Wildlife Fund [4], and the Food and Agriculture Organization of the United Nations. In the initial stage of ecosystem functional zoning, the research areas were mainly considered on a large or medium scale [5,6], as in the case of the construction of large-scale ecosystem zoning systems in Europe, North America, and Africa [7,8]. Since then, researchers have focused more on the zoning of single ecosystems, especially in water environment ecosystems [9–11].

In recent years, the deterioration of ecosystems has become an important factor restricting socio-economic development [12,13], and the study of the restoration zoning of ecosystems has received increasing attention. For example, Li et al. [14] took China's National Ocean Park, in the Haizhou Bay National Ocean Park, as their research area. Based on the evaluation of the ecological environment of the Haizhou Bay Reserve, the ecosystem health of the island's terrestrial ecosystem, intertidal ecosystem, and shallow sea ecosystem were assessed, and the ecosystem restoration zones were divided according to a vulnerability assessment. Yang et al. [15], using the ecological compensation zoning index of cultivated lands and the financial payment model of cultivated lands' ecology, divided the cultivated lands' ecological compensation areas in Wuhan. Wang et al. [16], taking the Tarbagatay Basin in the Xinjiang province as an example, constructed an evaluation index system from three aspects of the suitability of ecological protection, urban development, and agricultural production, and divided the ecological, agricultural, and urban space regions of the Tarbagatay Basin.

For the zoning research method, the current commonly used methods mainly include main component analysis, index partition, empirical partition, graph cascade adding, and cluster partition. In recent years, due to the rapid development of computer and remote sensing technologies, the means of ecological zoning research has become more and more advanced. The quantity of regional research is increasing due to software and technologies such as ArcGIS, SPSS, Matlab, and neural networks. For example, Wang et al. [17], using the Sunan Yugu Autonomous County in the Qilian Mountain District as an example, discussed ecological space demarcation and zoning methods based on the ArcGIS platform. Some scholars [18,19] have also carried out zoning based on ArcGIS technologies based on the analysis of regional environmental statuses and spatial variations in ecological sensitivity. In another example, Jing et al. [20] proposed a regional-scale ecological protection zone division method based on an improved artificial bee colony.

Overall, the zoning of the ecosystem underwent three stages: natural-element-based zoning, functional-based zoning, and ecological-restoration-oriented zoning. Given the urgency for ecosystem restoration, research on ecosystem restoration zoning has received increasing attention in recent years. However, in this study, there were some problems both in the concept of zoning and its indicators. For example, how ecosystem zoning changes from the partitioning of a single ecosystem component into an integrated ecosystem restoration partition is worthy of attention [21].

In summary, research on ecosystem zoning has entered a new era in which more attention is being given to ecosystem restoration zoning. However, due to the inadequacy of zoning concepts and index system construction, there are still great challenges related to ecosystem restoration zoning. In light of this, taking the Qilian Mountains of China as an example, the ecological restoration zone of the Qilian Mountains was developed based on the concept of the integrated protection and restoration of mountains, rivers, forests, cultivated lands, lakes, and grasslands. Specifically, the research goals of this paper were to: (1) reveal the land use restoration of the Qilian Mountains in recent years; (2) put forth a new set of ecological restoration zone index systems based on the concept of the life communities of mountains, rivers, forests, cultivated lands, lakes, and grasslands; and (3) to delimit the ecological restoration zones of the Qilian Mountains scientifically and reasonably. The paper is structured as follows:

- In Section 2, we introduce the study area and data sources.
- In Section 3, we introduce the methods used for ecological restoration zoning.
- In Section 4, we document the key areas of ecological restoration and their ecological problems.
- The limitations and future research prospects are discussed in Section 5.
- The conclusions are then presented in Section 6.

2. Study Area and Data Sources

2.1. Study Area

The Qilian Mountain area is one of the main mountainous provinces in China (Figure 1), situated in the northeast of the Qinghai Province and the western border of the Gansu Province (northeast of the Qinghai–Tibet Plateau) between 94°20′–103° E, 36–40° N, with a total area of 237,000 km^2. The annual average temperature in the study area is 1.4 °C below zero to 9.6 °C, and the total amount of solar radiation is 5916–15,000 MJ/m^2. The annual average precipitation is between 0–700 mm. The Qilian Mountain area consists of a number of northwest–southeast parallel mountains and wide valleys. The mountains mainly include the Daxue Mountain, Tuolai Mountain, Tuolai South Mountain, Yema South Mountain, Shule South Mountain, Danghe South Mountain, Tuergen Daban Mountain, Chai Damu Mountain, and Zongwulong Mountain. The Qilian Mountains have an average elevation of 4000–4500 m, and many peaks are over 5000 m.

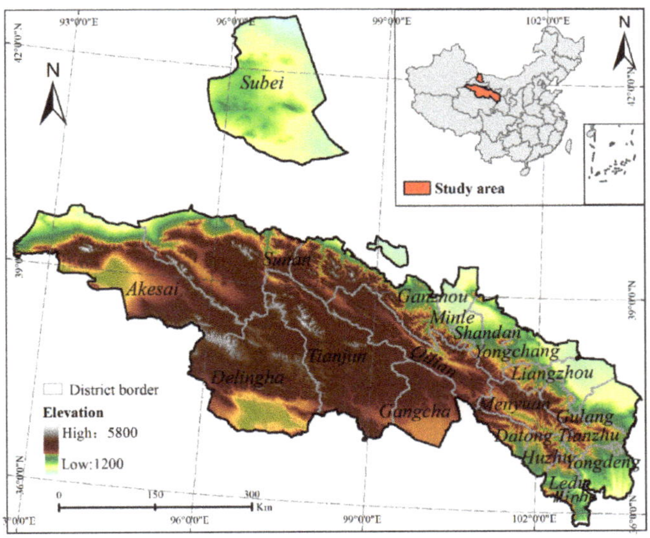

Figure 1. Geographic location of the Qilian Mountains in China.

2.2. Data Sources

The data used in this paper mainly include soil data, land use data, desertification data, cultivated land quality data, and statistical data. Among them, the soil organic matter content and soil texture data were derived from the Harmonised World Soil Database (HWSD) [22]. Land use data come from remote sensing monitoring data of land uses in

China in 2015 [23]. In this monitoring data, land use types include six primary types and twenty-four secondary types. The desertification data were obtained from the 1:100,000 Chinese Desert Gobi Distribution Map for 2000 [24]. The cultivated land quality data come from a 1:1 million land resource map [25]. The assessment of the change status of forest grassland vegetation is based on the 16-day synthetic NDVI products (MYD13A2 and MOD13A2) of MODIS from 2001–2015 with a spatial resolution of 1000 m [26]. The animal husbandry statistics at the county level come from the provincial statistical yearbooks, and the corresponding rural income per capita income come from the Gansu Development Yearbook for 2015 [27] and the Qinghai Statistical Yearbook for 2015 [28]. The mine distribution density was obtained from a large network crawler data search. Precipitation data were derived from the China Ground Climate Data Daily Value Dataset (V3.0) [29], and the spatial distribution was obtained by interpolation.

3. Research Methods

3.1. Index System Construction

The ecological development, overall protection, and comprehensive governance of all ecological and environmental elements are required for the ecological restoration of the "life community" of mountains, rivers, forests, cultivated lands, lakes, and grasslands. The demarcation of the ecological restoration zones of the Qilian Mountains aims to consider the comprehensive ecological problems related to the ecological factors of mountains, rivers, forests, cultivated lands, lakes, and grasslands in the Qilian Mountains. These comprehensive ecological problems include soil erosion, forest and grass quality degradation, cultivated land quality degradation, and water and soil erosion caused by mining, etc. We constructed the regional evaluation index system of ecological restoration in the Qilian Mountains according to the principles of scientific nature, systems, correlations, and operability; eight evaluation indicators were considered, including the amount of soil conservation, mine distribution densities, rainstorm days, annual precipitation, interannual change rates of forestry area vegetation, agricultural production potential, interannual change rates of grassland vegetation, and rural income per capita (Table 1).

Table 1. Diagnosis index system of mountains, forestry, fields, lakes, and grassland ecology systems in the Qilian Mountains.

Type	Ecological Issues	Indicators	Meaning of the Index
Mountain	Soil erosion	Amount of soil conservation	Application of land use and management methods to prevent soil erosion by human or natural factors to maintain the total amount of natural soil functions
	Mining	Mine distribution density	Reference to the number of mines within a certain geographical space range
Water	Extreme precipitation	Rainstorm days	Number of days with daily precipitation exceeding 50 mm
	Uneven precipitation distribution	Annual precipitation	The sum of the average monthly precipitation in the year represents the annual precipitation
Forestry areas	Forest is degraded	Interannual change rate of forest vegetation	Changes of forest vegetation within one year
	Poor forest score quality		
Cultivated land	Low farm quality	Agricultural production potential	Agricultural production potential is the maximum possible output to be achieved annually on lands per unit of land
Grassland	Meadows are degraded	Interannual change rate of grassland vegetation	Changes of grassland vegetation within one year
People	Poverty situation	Rural per capita income	Average income of rural individuals within one year

Among them, soil erosion and mining are the main ecological problems related to mountain elements. The diagnostic index of the ecological problems of soil erosion is soil conservation [30], and the diagnostic index of ecological problems in mining is the mine distribution density [31]. The imbalance of extreme precipitation and precipitation distribution is an ecological problem related to water [32]. The diagnostic index of extreme precipitation is represented by rainstorm days, and the diagnostic index of unbalanced precipitation distribution is the spatialized annual precipitation. The degradation of forested areas is an ecological problem related to forests, and its diagnostic index is the interannual change rate of forest land vegetation [33]. The quality of cultivated lands is an ecological problem related to cultivated lands, and its diagnostic index is its agricultural production potential [34]. Grassland degradation is an ecological problem related to grasslands, and its diagnosis index is the interannual change rate of grassland vegetation [33]. Poverty status is a human-related problem. Although poverty may not be an ecological problem, poverty is closely related to the emergence of ecological problems. The diagnostic index of poverty status selection is the rural per capita income [35]. The specific calculation process of indicators is shown in Appendix A.

3.2. Ecological Restoration Index (ERI) of the Qilian Mountains

We used the linear weighted function method to construct the regional ecological restoration index (ERI) and analyze the spatial differences by using the spatial clustering and grouping method. The specific steps are as follows:

First, with different units, meanings, and contents, there are differences in data dimensions and trend directions. Therefore, each index must be standardized. The formula was as follows [36]:

$$\begin{cases} Positive\ index: x'_{ij} = \dfrac{x_{ij} - x_j^{\min}}{x_j^{\max} - x_j^{\min}} \\ Negative\ index: x'_{ij} = \dfrac{x_j^{\max} - x_{ij}}{x_j^{\max} - x_j^{\min}} \end{cases} \quad (1)$$

Second, according to the number of ecological factors of mountains, rivers, forests, cultivated lands, lakes, grasslands, and people involved in various ecological indicators, the weight of various ecological indicators was determined. In other words, if this index involved only one ecological problem related to mountains, rivers, forests, cultivated lands, lakes, and grasslands, the value was 1; if two ecological problems are involved, the value was 2; and so on. Then, the score of each indicator was expressed as a percentage (Table 2) representing the indicator weight.

Table 2. Weight of the ecological restoration evaluation index in the Qilian Mountains.

Evaluation Indicators	Ecological Restoration Object	Indicator Weight
Amount of soil conservation	Mountains; rivers; and forests	0.3
Mine distribution density	Mountains	0.1
Extreme precipitation	Rivers	0.1
Annual precipitation	Rivers	0.1
Interannual change rate of forest vegetation	Forests	0.1
Interannual change rate of grassland vegetation	Grasslands	0.1
Cultivated land grade	Cultivated lands	0.1
Rural income per capita	People	0.1

Finally, based on the constructed ERI system and weight, the linear weighted function method was used to measure the ERI. The comprehensive situation of the ecological problems of regional mountains, rivers, forests, cultivated lands, lakes, and grasslands can be reflected through this indicator. The overall rule is that the smaller the ERI, the more prominent the comprehensive ecological problems representing mountains, rivers, forests, cultivated lands, lakes, and grasslands, and the more necessary it is to carry out comprehensive ecological restoration projects. The calculation formula was as follows [37]:

$$ERI = \sum_{i=1}^{n} w_i \times E_i \tag{2}$$

where "E_i" is the value of the *i*-th ecological indicator and "w_i" is the weight of the *i*-th ecological indicator.

3.3. Division of Basic Evaluation Units Based on River Basin Division

From the perspective of systematic restoration, we extracted small basins in the study area as the basic units for the evaluation of ecological restoration zoning. Basin extraction, or catchment extraction, is the joint determination of its spatial scope based on the river's flow direction and outlet. From a hydrology and geography perspective, its region must correspond to that of the river. Therefore, rivers must be designated before the watershed extraction. River data can be extracted from digital elevation model (DEM) data or converted from existing vector rivers. We used a slope runoff simulation algorithm to realize automatic water system extraction and river basin segmentation. The main steps were as follows:

First, we determined the direction of the water flow of the grid unit. After preprocessing the DEM data, the flow direction was calculated. We then extracted the depression, analyzed the threshold of the depression, and set the threshold to fill the DEM data. This step was repeated until all existing depressions in the DEM were eliminated to lay the basis for the hydrological analysis.

Second, the drainage basin was calculated, and the flow accumulation matrix was determined by the flow direction. Next, the upstream catchment area of each grid unit was obtained. Then, the appropriate confluence threshold was determined. The confluence threshold value needed to be measured repeatedly, and the threshold value was inversely proportional to the number of river basins. The water exchange area extraction was conducted based on the flow direction. After determining the basin outlet grid, all grids to the outlet could be searched according to the flow direction matrix to obtain the basin boundary and that of the sub-basin in order for basin segmentation to be realized.

Finally, the Qilian Mountain area was divided into 108 small river basins (Figure 2).

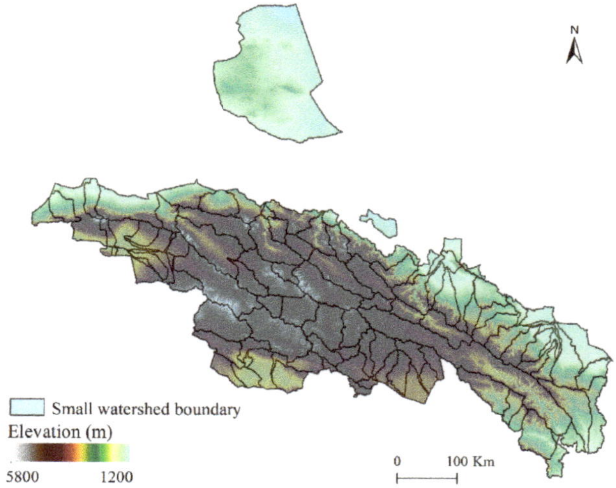

Figure 2. Small watershed division of the Qilian Mountains.

4. Results

4.1. Land Use Change in the Qilian Mountains from 1990–2015

The cultivated lands in the Qilian Mountains are mainly distributed between Ganzhou, Minle, Yongchang, Shandan, and Gulang in the northeast and Huzhu, Ledu, and Minle in the southeast, and most of them are dry lands (Figure 3). The forestry areas are mainly distributed in Qilian and Menyuan in the central region. Grasslands and unused lands collectively account for approximately 80% of the study area, mainly at high elevations. Among them, the grasslands are mainly distributed in the center of the study area, while the unused lands are distributed in the center and the northeast. The water areas comprise the smallest type in the research area, featuring blocks distributed in the middle and south of the study area. The built-up areas are scattered in the eastern and southeastern regions, which also correspond to the distribution areas of cultivated lands.

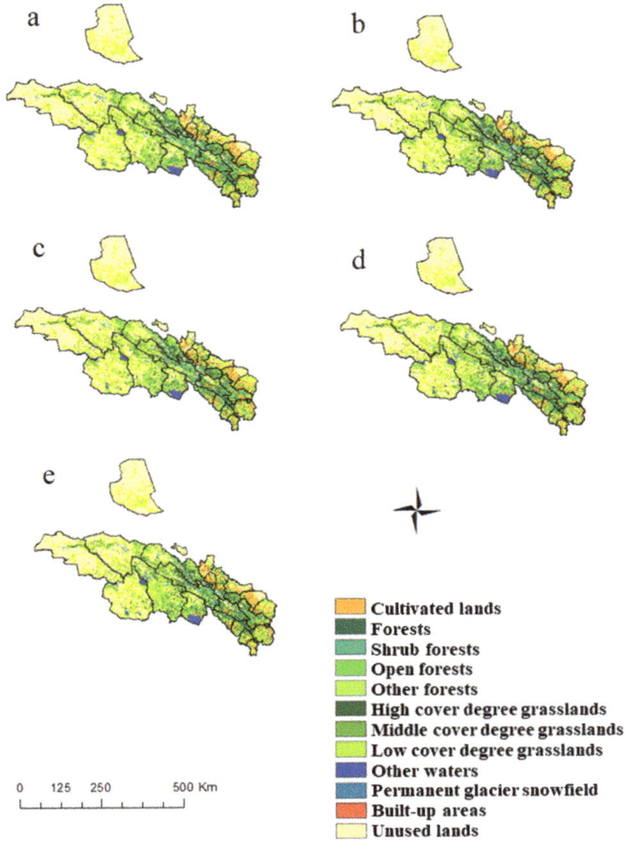

Figure 3. Types of land use in the Qilian Mountains in 1990 (**a**), 2000 (**b**), 2005 (**c**), 2010 (**d**), and 2015 (**e**).

From 1990–2015, the cultivated lands, grasslands, and built-up areas of the Qilian Mountains generally increased (Figure 4), while forestry areas, water bodies, and unused lands generally declined. The cultivated lands in the Qilian Mountains increased by 0.31% from 1990 (5.8%) to 2015 (6.11%). Of these cultivated lands, 0.79% were converted into other land use types, and the proportion of converted to cultivated lands was 1.11%. In terms of spatial change, the area of cultivated lands is mainly part of Ganzhou and

Sunan; meanwhile, in the southeast of Sunan, Menyuan, Tianzhu, and Yongdeng, a large proportion of the cultivated lands has been transformed into grasslands and built-up areas. As with cultivated lands, the area of grasslands has also shown an increasing trend over the years, increasing by 1.28% from 1990 (36.85%) to 2015 (38.13%). The Qilian Mountain area is mainly dominated by medium- and low-cover grasslands. From 1990–2015, the conversion of high-, medium-, and low-complexity grasslands to other land use types was 0.71%, 2.12%, and 2.72%, respectively. The proportions of conversion to high-, medium-, and low-complexity grasslands were 0.015%, 1.05%, and 1.84%, respectively. Spatially, the grasslands increased in the northeast and southeast of the Qilian Mountains, and the main reason for the decrease in grasslands was that the grasslands were transformed into unused lands; however, because the proportion was small, the spatial performance was not obvious.

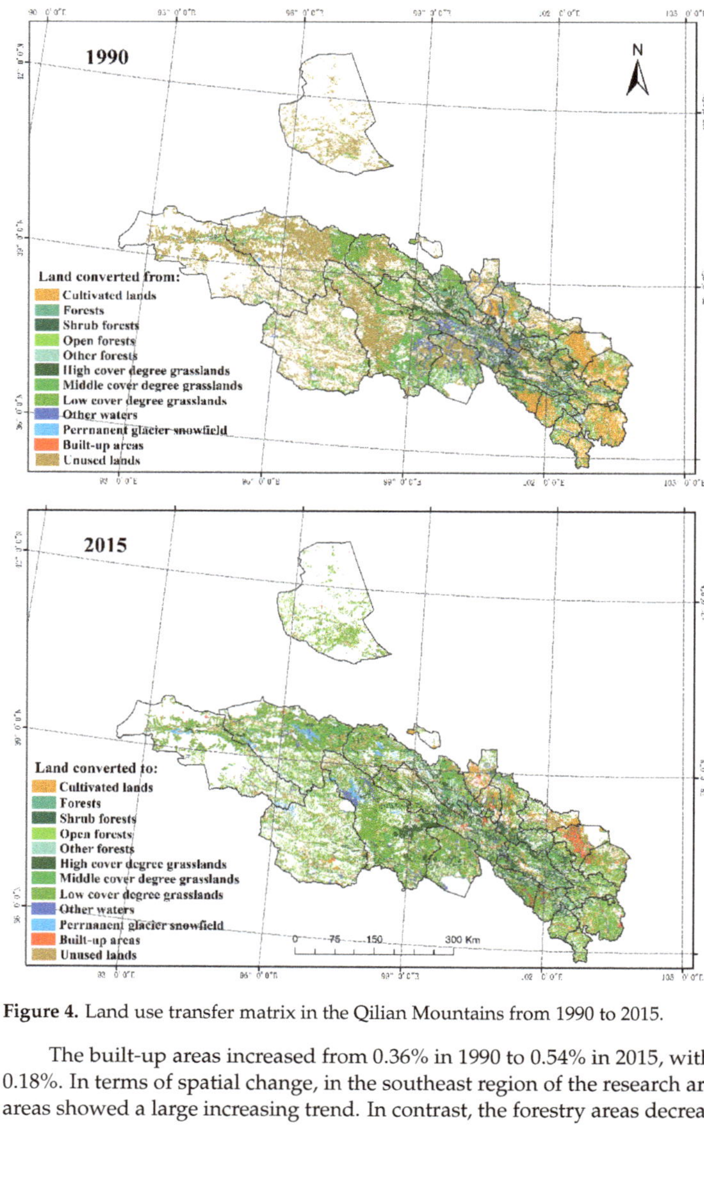

Figure 4. Land use transfer matrix in the Qilian Mountains from 1990 to 2015.

The built-up areas increased from 0.36% in 1990 to 0.54% in 2015, with an increase of 0.18%. In terms of spatial change, in the southeast region of the research area, the built-up areas showed a large increasing trend. In contrast, the forestry areas decreased from 6.12%

in 1990 to 6.10% in 2015, with a decrease of 0.02%. Of this latest percentage, the forestry areas, scrubs, sparse, and other land use types made up 0.18%, 0.48%, 0.22%, and 0.01% of the total, respectively. The proportion of lands converted from other lands into forested, shrub lands, open woodlands, and other wooded lands accounted for 0.19%, 0.46%, 0.21%, and 0.00% of the total, respectively. Spatially, the areas with forests were basically converted into grasslands, so the areas where grasslands increased just happened to be the areas of forest reduction. The water area increased by 0.20%, from 1.75% in 1990 to 1.95% in 2015. Within these areas, the conversion from water to other land types was 0.17%, and 3.94% from other land types into water bodies. Due to the impact of global warming, the Qilian Mountain glaciers have experienced a large-scale retreat. The glaciers are in a state of material loss, generally receding and thinning. Coupled with the reduction in wetlands, the waters generally show a downward trend. Spatially, the areas with water decreased mainly in Tianjun, Gangcha, and Qilian. The proportion of unused lands decreased from 48.6% in 1990 to 46.38% in 2015, with a total decrease of 2.22%. The proportion of unused lands converted to other land use types was 4.43%, and that of other land use types converted to unused lands was 2.21%. The unused reductions were concentrated in large areas in the northeastern part of the study area, including in places such as Aksai, Tianjun, Delingha, and Sunan.

4.2. Key Areas of Ecological Restoration in the Qilian Mountains

Considering the eight factors of soil conservation, mine distribution density, extreme precipitation, annual precipitation, the annual change rates of areas with forestry vegetation and poverty level, quality of cultivated lands, and annual change rates of grassland vegetation and poverty level, and using the comprehensive evaluation system of ecological restoration, we calculated the comprehensive ERI of mountains, forests, cultivated land, lakes, and grasslands in the planning area (Figure 5). The index reflects the degree necessary to implement the comprehensive ecological restoration of mountains, rivers, forests, cultivated lands, lakes, and grasslands. The average of the ecological restoration indicator in the Qilian Mountains was 0.57, and the average in each basin ranged between 0.34 and 0.80. It can be seen that the ERI of the Qilian Mountain area showed a law of decreasing from the edges to inland.

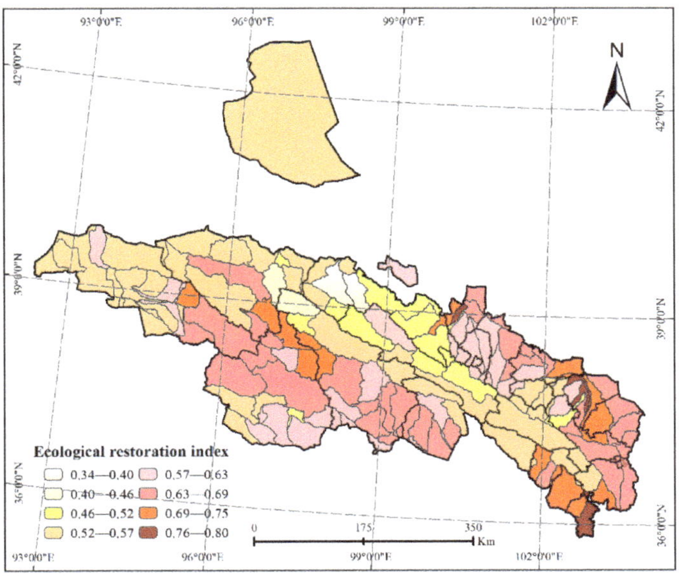

Figure 5. Ecological restoration index (ERI) of the Qilian Mountains.

The ERI is an index that measures the key areas of ecological restoration. The greater the value, the greater the need to carry out ecological restoration work in the ecological area. It can be seen that the ERI in the southeast and southwest of the study area is relatively large. This means that there is an urgent need for ecological restoration in these areas.

4.3. Ecological Restoration Zoning of the Qilian Mountains

Due to the different natural resource endowments of the Qilian Mountains and the strong spatial heterogeneity of various ecological restoration indicators, the use of ecological restoration zones divided by the comprehensive ERI cannot further reflect the differences in the ecological restoration directions in each district, and it is not conducive to the appropriate measures. Therefore, this paper used an ArcGIS cluster analysis tool based on the calculation results of eight ecological restoration indicators and ERIs in each district, and divided them into two level divisions in accordance with the principle of natural division order.

Finally, the Qilian Mountain area was divided into three primary districts and eight secondary districts. The first division included the forests and grassland water conservation areas, the ecological restoration areas of cultivated lands, and ecological control areas of deserted grassland (Table 3). The secondary ecological restoration zone included the headwaters of the Datong River Basin, the Danghe-Dahaerteng River Basin, the northern confluence area of the Qinghai Lake, the upper Shule River to middle Heihe River, the Oasis Agricultural Area at the northern foothills of the Qilian Mountains, the Huangshui Basin Valley, Aksay (corridor region of the western Hexi Basin), and the northeastern Tsaidam Basin (Table 3 and Figure 6).

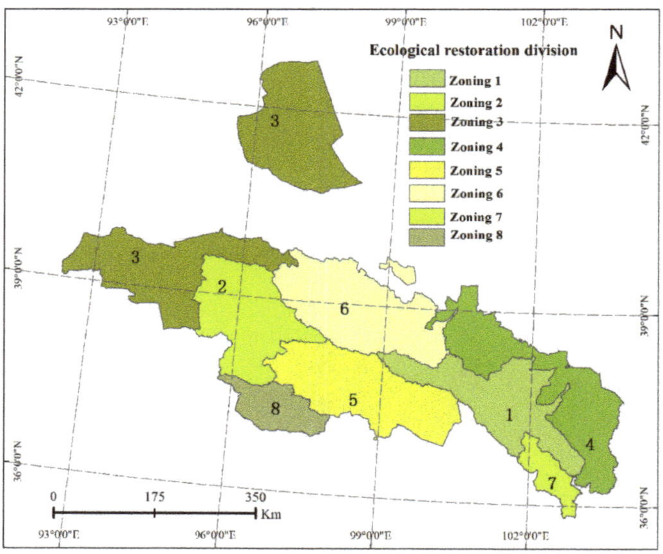

Figure 6. Ecological restoration zoning of the Qilian Mountains. Zoning 1 represents the headwaters of the Datong River Basin; Zoning 2 represents the Danghe-Dahaerteng River Basin; Zoning 3 represents Aksay (corridor region of the western Hexi Basin); Zoning 4 represents the Oasis Agricultural Area at the northern foothills of the Qilian Mountains; Zoning 5 represents the northern confluence area of the Qinghai Lake; Zoning 6 represents the upper Shule River to middle Heihe River; Zoning 7 represents the Huangshui Basin Valley; and Zoning 8 represents the northeastern Tsaidam Basin.

Table 3. Ecological restoration zones of the Qilian Mountains planning area.

First-Level Division	Secondary Division	Regional Area: Ten Thousand (km^2)	Area Ratio (%)	Ecological Restoration Index
I—forests' and grasslands' water conservation areas	1—headwaters of the Datong River Basin	2.75	11.61	0.54
	2—Danghe-Dahaerteng River Basin	2.80	11.83	0.65
	5—northern confluence area of the Qinghai Lake	3.28	13.84	0.62
	6—upper Shule River to middle Heihe River	3.62	15.26	0.50
II—cultivated lands' and grasslands ecological restoration areas	4—Oasis Agricultural Area at the northern foothills of the Qilian Mountains	3.43	14.45	0.68
	7—Huangshui Basin Valley	0.66	2.77	0.73
III—deserted grasslands' ecological control area	3—Aksay, corridor region of the western Hexi Basin	6.11	25.78	0.56
	8—northeastern Tsaidam Basin	1.06	4.46	0.57

4.4. Ecological Problems of Different Ecological Restoration Zones

Combined with the land use data, we further compared the differences of various ecological restoration indicators in different divisions (Figure 7) and discussed the main ecological problems of each ecological restoration division.

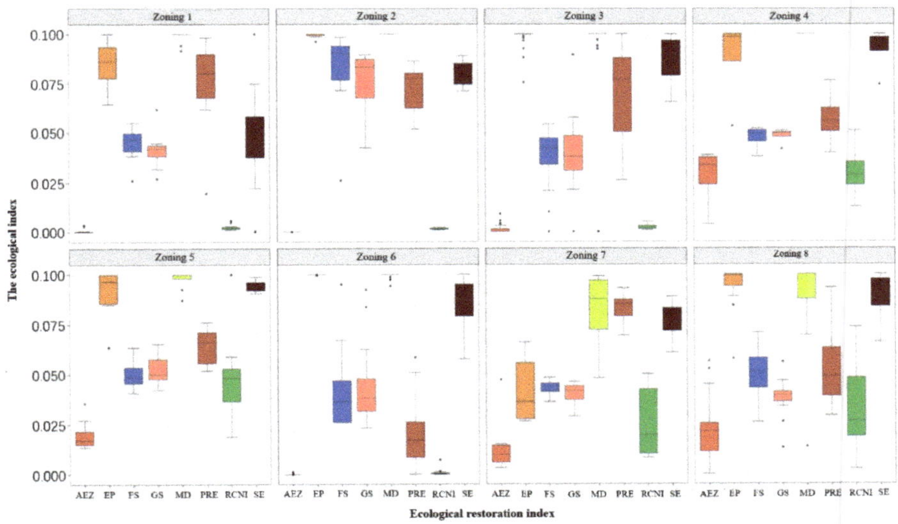

Figure 7. Ecological indicators of different ecological restoration zones of the Qilian Mountains. "AEZ" represents the agricultural production potential; "EP" represents extreme precipitation; "FS" represents the interannual change rate of the normalized difference vegetation index (NDVI) in forestry areas' vegetation; "Rural per capita income" represents regional GDP; "GS" represents the NDVI interannual change rate of grassland vegetation; "MD" represents the mine distribution density; "PRE" represents the total annual precipitation; "SE" represents the soil erosion volume; Zoning 1 represents the headwaters of the Datong River Basin; Zoning 2 represents the Danghe-Dahaerteng River Basin; Zoning 3 represents Aksay (corridor region of the western Hexi Basin); Zoning 4 represents the Oasis Agricultural Area at the northern foothills of Qilian Mountain; Zoning 5 represents the northern confluence area of the Qinghai Lake; Zoning 6 represents the upper Shule River to middle Heihe River; Zoning 7 represents the Huangshui Basin Valley; and Zoning 8 represents the northeastern Tsaidam Basin.

The headwaters of the Datong River Basin are located in the eastern section of the Qilian Mountains, and they are also the source of the Heihe, Shiyang River, and other river basins in the Qinghai and Gansu provinces. The main ecological functions of the area are water conservation and soil conservation. The district includes four counties: Qilian, Sunan, Menyuan, and Datong. The area covers 2.75 × 104 km^2, accounting for 11.61% of the total regional area. Compared with other zones, the area is seriously affected by extreme precipitation, with greater extreme precipitation days and a large mine density (Figure 7). The interannual growth of forest vegetation is weak, and the production potential of cultivated land resources is very low.

The Danghe-Dahaerteng River Basin is located in the high mountains in the western section of the Qilian Mountains, including the three sub-basins of the Danghe River, Dahaerteng River, and Yema River. The counties comprising this area are Delingha, Subei, and Aksay. The main ecological functions of the area are water conservation, soil conservation, windbreaks, and sand fixation. The basin covers an area of 2.80 × 104 km^2, accounting for 11.86% of the total regional area. The main ecological problem is the uneven distribution of water resources (Figure 7), showing a decline from east to west; such a low production potential further increases poverty.

The northern part of Aksay (corridor region of the western Hexi Basin) is the western section of the Hexi Corridor, with the Aksay Basin to its south. The two counties in this region Subei and Aksay. The main ecological functions of the area are windbreaks and sand fixation, as well as water–soil conservation. The area measures 6.11 × 10^4 km^2, accounting for 25.78% of the total regional area. The biggest natural problem of this zone compared to others is the sparse local precipitation and its uneven distribution (Figure 7). Droughts, less rain, and serious wind erosion have become the leading causes for its restricted ecological functions, and are also the source of other problems.

The Oasis Agricultural Area at the northern foothills of the Qilian Mountains is the core area of human activity and economic development in the Gansu Province. The main functions of the area are water and soil conservation, in addition to food production. The region covers an area of 3.43 × 104 km^2, accounting for 14.45% of the total regional area. Compared with other subdivisions, the development of agriculture, animal husbandry, and mineral resources is intense; ecological problems such as overgrazing, vegetation degradation, soil erosion, and desertification are thus prominent, and the relationship between humans and the natural environment has gradually become unbalanced (Figure 7).

The northern confluence area of the Qinghai Lake is located in the middle of the Qilian Mountains, centering around the Qinghai Lake Basin, and reaching the Shule Nanshan Mountain in the north and the Qinghai Nanshan one in the south. The main functions of the area are water and soil conservation. The area is 3.28 × 104 km^2, accounting for 13.84% of the total regional area. Compared with other divisions, the main ecological problems in the area include serious mountain soil erosion (Figure 7), as seen in the Buha River Basin and the southern foot of the Datong Mountain, thus often causing landslides and debris flows. Vegetation degradations in the northeast of the Qinghai Lake have also made the local microclimates unstable or even caused them to deteriorate. Finally, less cultivated land resources with a low production potential and poverty have destroyed the ecological environment.

The upper Shule River to middle Heihe River is located in the middle of the Qilian Mountains, south of the Shule Henan Mountain and north of the Hexi Corridor, including the upper reaches of the Shule River and the middle reaches of the Heihe River. The main functions of the area are water and soil conservation. The area covers 3.62 × 104 km^2, accounting for 15.26% of the total regional area. Compared with other divisions, the soil development is poor, both the quantity and quality of cultivated lands are rather low, and the poverty index is high (Figure 7).

The Huangshui Basin Valley is located in the river valley area of the Huangshui River Basin, and is its central area of human activity and economic development. The main functions of the area are soil conservation and food supply. The region covers an area of

0.66×104 km^2, accounting for 2.77% of the total regional area. The area is adjacent to the Oasis Agricultural Area at the northern foothills of the Qilian Mountains, but the current agricultural productivity and production potential of the area are significantly lower than those of the latter, and the problem of soil erosion is more prominent (Figure 7).

The northeast Tsaidam Basin is situated on the northeast edge of the Tsaidam Basin, with the Zongwu Mountain in the north. The main functions of the area are water and soil conservation. The area is 1.06×104 km^2, accounting for 4.46% of the total regional area. The subdivision has similar natural conditions to Aksay (corridor region of the western Hexi Basin) and is spatially divided into different ecological restoration zones due to discontinuity. The ecological problems in this zone mainly include less droughts and sparse vegetation, which indicate a fragile natural environment, a low agricultural production potential, and the fact that agricultural and animal husbandry production operations are extensive (Figure 7).

5. Discussion

In previous studies, many scholars have carried out ecological vulnerability [38] and ecological security assessments [39], as well as other related studies to provide decision-making guidelines based on ecological protection and restoration by identifying important and key areas of ecological restoration [40]. The commonly used assessment models for ecosystem vulnerability studies are the pressure-state-response (PSR) [41] and the exposure-sensitive-adaptive (ESA) models [42]. Ecological security pattern studies are based on ecological networks, especially those from Europe [43] and the United States [44]. The framework of ecological security patterns comprises the "ecological sources-ecological resistance surfaces-ecological corridors" [45]. Compared with these studies, China's ecological restoration projects emphasize the combination of important ecological areas (mountains, rivers, forests, farmlands, lakes, grasslands, etc.). Ecological restoration involves paying more attention to the systematic restoration of total regional elements to ensure the integrity of ecosystem structures and functions. China's 13th five-year plan puts forward the concept that "mountains, rivers, forests, fields, lakes, and grasslands are a community of life" in ecological restoration. Therefore, for ecological restoration zoning, the ecosystem as a community must also be comprehensively diagnosed. Based on this, this study proposed a comprehensive diagnosis method for identifying ecological problems, and a comprehensive restoration zoning method based on different ecosystem problems, which can provide better technical support for the implementation of ecological restoration projects.

In this study, some key indicators were selected to characterize each ecosystem element. For example, the assessment of soil conservation was used to describe "mountains, water, and forests" in the ecosystem elements of the Qilian Mountains. However, there were some limitations and uncertainties. Firstly, considering the natural environmental conditions and special ecological and environmental problems in different regions, the different needs for ecological services and ecological restoration will affect the selection of various ecosystem element indicators. Therefore, the scientific index selection method for evaluating ecological restoration needs to be further explored. Secondly, the index weight assignment involved the difficulty of constructing the ERI. The application of inappropriate methods may thus have directly affected the distribution characteristics of the evaluation results and significantly increased the uncertainty. Although the method of determining index weights in this study could better reflect the extent to which various indicators involve ecological problems, the weight value will be influenced by subjective selected indicators. Further ecological restoration zoning could be studied by combining different index weight determination methods. For example, research has shown that spatial principal component analysis has advantages in ecosystem vulnerability assessment [46]. It can objectively determine the weight of the evaluation indicators and avoid subjective arbitrariness, but there is a certain problem of information loss.

When planning ecological restoration projects, the allocation of restoration areas and the cost of restoration measures are two major problems faced by decision makers, which can be further studied by considering the construction of appropriate ecological restoration development frameworks. For example, Zhang et al. [47] planned wetland restoration projects based on the framework of interval fuzzy linear programming, which can deal with the trade-off between eco-environmental benefits and economic costs. Under the background of global climate change, ecosystem restoration should not only combine the characteristics of the ecosystem itself, but also consider the impact of climate change [48]. In addition, with the development and application of remote sensing (RS), geographic information systems (GIS), global positioning systems (GPS), and other technologies, ecological restoration zoning results have become more dynamic. In future research, it is also worth paying attention to carrying out the real-time diagnosis of ecological restoration problems and make timely policy adjustments by using remote sensing image data with high spatial and temporal resolutions.

6. Conclusions

Based on the land use data of the resources and environmental data centers of the Chinese Academy of Sciences, the cultivated land quality data of the 1:1 million land resource map, the NDVI data of MODIS, and the big data search of the network crawler, the ecological problems related to the mountains, water, forests, fields, lakes, and grasslands of the Qilian Mountains were systematically analyzed. The study found that mining exploration and hydropower projects have been the main reasons for the ecological damage of the Qilian Mountains for nearly half a century. From 1990–2015, the land use of the Qilian Mountains changed significantly: the forestry and wetland areas diminished while the proportion of grassland areas increased. From 1990–2015, cultivated lands and built-up areas expanded significantly. The high-coverage grasslands, low-coverage grasslands, and other woodlands showed increasing trends. The shrub forestry areas, medium-coverage grasslands, sparse forests, unused lands, and water areas decreased significantly.

Here, we have presented the index of the Qilian Mountains' ERI, from which we identified the key areas of the latter's ecological restoration and distinguished different restoration zones. The ERIs in the southeast and southwest of the research area were relatively large, implying that an urgent restoration of ecological protection is needed in these areas. Finally, we divided the study area into three primary ecological restoration zones and eight secondary ecological restoration zones. The average restoration index of the eight ecological restoration zones was 0.61, and the partition restoration index was between 0.34 and 0.8. Of these indices, the zone with the lowest values occurred in the upper Shule River to middle Heihe River, which had the most serious comprehensive ecological problems and the highest urgency of repair. In contrast, the ERI of the Huangshui River Basin, the area with the highest ecological quality in the Qilian Mountain area, was 0.73.

Our findings can serve as a scientific basis for policy implementations for the diagnosis and restoration of ecological problems. However, there were some limitations and areas of uncertainty, such as those involving the selection of indicators and the determination of indicator weights. The scientific index selection method for evaluating ecological restoration should be further explored. Ecological restoration zoning can be studied by combining different index weight determination methods. It is also worth paying attention to carrying out real-time diagnoses of ecological restoration problems and making timely policy adjustments by combining remote sensing image data with high spatial and temporal resolutions.

Author Contributions: Conceptualization, W.S.; methodology, L.L. and W.S.; formal analysis, L.L., W.S., Y.Z., Z.H., H.L., D.Y., Z.W. and Q.H.; investigation, L.L., W.S., Y.Z., Z.H., H.L., D.Y., Z.W. and Q.H.; resources, W.S.; writing—original draft preparation, L.L. and W.S.; writing—review and editing, W.S.; supervision, W.S. All authors have read and agreed to the published version of the manuscript.

Funding: This research was supported by the National Key R&D Program of China (grant number 2019YFC0507805), the Second Tibetan Plateau Scientific Expedition and Research (grant number 2019QZKK0603) and the Projects of National Natural Science Foundation of China (grant number 42071233).

Institutional Review Board Statement: Not applicable.

Informed Consent Statement: Not applicable.

Data Availability Statement: All relevant datasets in this study are described in the manuscript.

Conflicts of Interest: The authors declare no conflict of interest.

Appendix A

The Specific Calculation Process of Indicators

(1) Soil Conservation

The soil conservation determination method of the Qilian Mountains adopted the general equation of soil erosion, and its calculation was performed using MATLAB. First, the potential soil erosion was calculated according to rainfall erosivity, "*R*", soil erodibility, "*K*", and slope length slope factor, "*LS*". Second, under the conditions of the vegetation cover, management factor, "*C*", and engineering measure factor, "*P*", a value was assigned according to different land use types. Finally, the actual soil erosion was calculated by using the general soil erosion equation. The difference from the potential soil erosion was the soil conservation value. The specific formula was as follows:

First, the potential soil erosion amount was calculated. The formula was as follows [49]:

$$RKLS = R \times K \times LS \tag{A1}$$

Second, in the case of vegetation cover and engineering measures, the general soil loss equation was used to calculate the actual soil loss [50]. The formula was as follows:

$$USLE = R \times K \times LS \times C \times P \tag{A2}$$

Finally, the amount of soil conservation was defined as the difference between *RKLS* and *USLE*, expressed by the soil holding conservation (*SHC*), as follows:

$$SHC = RKLS - USLE \tag{A3}$$

where "*R*" is the rainfall erosion force, "*K*" is the soil erodibility, "*LS*" is the long slope factor, "*C*" is the vegetation cover and management factor, and "*P*" is a project measure factor.

Rainfall erosivity, "*R*", reflects the potential capacity of soil erosion caused by rainfall, in addition to Wischmeier's empirical formula [51,52], which was used in the calculation:

$$R = \sum_{1}^{12} 1.735 \times 10^{[(1.5 \times lg \frac{p_i^2}{p}) - 0.8188]} \tag{A4}$$

where "p_i" is the average monthly rainfall and "p" is the average annual rainfall.

Soil erodibility, "*K*", is an important index of the soil erosion equation, and the Erosion–Productivity Impact Calculator (EPIC) formula established by Williams et al. [53] was used for calculations, namely:

$$K = \left\{0.2 + 0.3\exp\left[-0.025san\left(1 - \frac{sil}{100}\right)\right]\right\} \times \left[\frac{sil}{cla + sil}\right]^{0.3} \times \left[1 - 0.025\frac{c}{c + \exp(3.72 - 2.95c)}\right] \times \left[1 - 0.7\frac{sn1}{sn1 + \exp(22.9sn1 - 5.51)}\right] \tag{A5}$$

where "*san*", "*sil*", "*cla*", and "*c*" represent the contents of sand, silt, clay particles, and organic carbon in the soil (%), respectively, and "*snl*" is equal to $1 - san/100$.

The topographic factor "*LS*" reflects the relationship between slope and surface conditions, and is the distance from which rain drops or sediments flow until the energy disappears, reflecting the impact of the topographic and geomorphological characteristics on soil erosion. For different slopes, the model can automatically calculate the value of the terrain factor and set the slope threshold to 25%, depending on the actual situation. The surface cover management, factor "C", refers to the ratio of the soil loss of a specific crop or natural vegetation under the same soil, terrain, and rainfall, mainly affected by soil water, early land use mode, and vegetation crops' planting sequence to reflect the impact of vegetation or crops and management measures on soil loss. This value is between 0–1. The soil conservation measure, factor "P", indicates the ratio of soil loss to soil loss after slope planting, reflecting the difference in soil loss caused by the differences in vegetation management measures. This is one of the most important factors to inhibit soil erosion, whose range is between 0–1. "P" and "C" can be obtained from the relevant literature [54].

(2) Mine Distribution Acquisition

The mining strength was quantitatively analyzed according to the mine distribution density (MD). Mine distribution data were obtained using mine information searching technologies. Mine information searching was mainly based on the Amap application programming interface (API), and was completed through secondary development. In other words, relevant means such as a point of interest (POI) search, geographic position inverse queries, and so on, were used to obtain the longitude and latitude coordinates of the mines, and then GIS software was used to generate the spatial distribution of the mines. This research used a Python programming crawler program to access the Amap WEB service, search for place names related to "mines", and determine the density of the Qilian Mountain mines based on their geographical location.

(3) Precipitation and Extreme Precipitation

The level of precipitation in China was mainly based on daily precipitation values. In Northern China, the standard of daily precipitation is 10 mm/d or less for light rain, 10–25 mm/d for moderate rain, 25–50 mm/d for heavy rain, and 50 mm/d and up for extreme precipitation. Therefore, this study selected days with daily precipitation values greater than 50 mm as the extreme precipitation index (rainstorms). In this paper, the daily precipitation and radiation data with a resolution of 250 m from 2001–2015 were obtained from the daily dataset of China's surface climate data (V3.0), and the spatial distribution was obtained by interpolation.

(4) Interannual Change Rate of Forests and Grassland Vegetation

We evaluated the 16-day synthetic normalized difference vegetation index (NDVI) products (MYD13A2 and MOD13A2) of grassland vegetation based on the 1000 MODIS for the years 2001–2015. To obtain the vegetation NDVI time series curve of the Qilian Mountains from 2001–2015, the noise time series curve of "cloud pollution" with quality control documents was eliminated, and the NDVI time series curve of each pixel was reconstructed with a Savitzky–Golay filter.

The change in vegetation in the Qilian Mountains was characterized by the interannual change rate of vegetation's NDVI, which was calculated using the slope of the trend line in linear regression analyses. The calculation formula was as follows [55–57]:

$$\theta_{slope} = \frac{n \times \sum_{i=1}^{n} i \times NDVI_i - \sum_{i=1}^{n} i \times \sum_{i=1}^{n} NDVI_i}{n \times \sum_{i=1}^{n} i^2 - \left(\sum_{i=1}^{n} i\right)^2} \tag{A6}$$

where "θ_{slope}" is the slope of the fitted trend line, "n" is the study period, and "$NDVI$" is the vegetation's $NDVI$ for the i-th year. When $\theta_{slope} > 0$, there was a downward trend in "$NDVI$" in the time series of the pixel; otherwise, "$NDVI$" had an upward trend.

References

1. Milan, R. The ecological and environmental education. *Ekol. Bratisl.* **1995**, *14*, 145–151.
2. Fischer, S.; Thatje, S. Temperature effects on life-history traits cause challenges to the management of brachyuran crab fisheries in the Humboldt Current: A review. *Fish. Res.* **2016**, *183*, 461–468. [CrossRef]
3. Omernik, J.M.; Griffith, G.E. Ecoregions of the Conterminous United States: Evolution of a Hierarchical Spatial Framework. *Environ. Manag.* **2014**, *54*, 1249–1266. [CrossRef]
4. Olson, D.M.; Dinerstein, E. The Global 200: A Representation Approach to Conserving the Earth's Most Biologically Valuable Ecoregions. *Conserv. Biol.* **1998**, *12*, 502–515. [CrossRef]
5. Omernik, J.M. Ecoregions: A framework for managing ecosystems. *Georg. Wright Forum* **1995**, *12*, 35–50.
6. Box, W.E. Global classification of natural terrestrial ecosystems. *Vegetatio* **1976**, *32*, 75–81. [CrossRef]
7. Bailey, R.G.; Zoltai, S.C.; Wiken, E.B. Ecological regionalization in Canada and the United States. *Geoforum* **1985**, *16*, 265–275. [CrossRef]
8. Zogaris, S.; Economou, A.N.; Dimopoulos, P. Ecoregions in the Southern Balkans: Should Their Boundaries Be Revised? *Environ. Manag.* **2009**, *43*, 682–697. [CrossRef]
9. Zhou, L.; Sun, D.; Xu, J. Zoning assessment of water environmental supporting capacity for socioeconomic development in the Huaihe River Basin, China. *J. Geogr. Sci.* **2015**, *25*, 1199–1217. [CrossRef]
10. Yang, S.Y.; Tang, T.; Cai, Q.H.; Xiao, W.; Wang, X.Z.; Li, F.Q.; Tang, J. Aquatic eco-regionalization of Erhai Lake Basin, Yunnan Province of Southwest China. *Chin. J. Ecol.* **2012**, *31*, 1798–1806. [CrossRef]
11. Liu, X.; Liu, L.; Peng, Y. Ecological zoning for regional sustainable development using an integrated modeling approach in the Bohai Rim, China. *Ecol. Modell.* **2017**, *353*, 158–166. [CrossRef]
12. Deboudt, P.; Dauvin, J.-C.; Lozachmeur, O. Recent developments in coastal zone management in France: The transition towards integrated coastal zone management (1973–2007). *Ocean Coast. Manag.* **2008**, *51*, 212–228. [CrossRef]
13. Pikitch, E.K.; Santora, C.; Babcock, E.A.; Bakun, A.; Bonfil, R.; Conover, D.O.; Dayton, P.; Doukakis, P.; Fluharty, D.; Heneman, B.; et al. Sainsbury. Ecosystem-Based Fishery Management. *Science* **2004**, *305*, 346–347. [CrossRef]
14. Li, F.; Xu, M.; Liu, Q.; Wang, Z.; Xu, W. Ecological restoration zoning for a marine protected area: A case study of Haizhouwan National Marine Park, China. *Ocean Coast. Manag.* **2014**, *98*, 158–166. [CrossRef]
15. Yang, X.; Zhang, F.; Luo, C.; Zhang, A. Farmland Ecological Compensation Zoning and Horizontal Fiscal Payment Mechanism in Wuhan Agglomeration, China, From the Perspective of Ecological Footprint. *Sustainability* **2019**, *11*, 2326. [CrossRef]
16. Wang, G.; Yang, D.; Xia, F.; Zhong, R.; Xiong, C. Three Types of Spatial Function Zoning in Key Ecological Function Areas Based on Ecological and Economic Coordinated Development: A Case Study of Tacheng Basin, China. *Chin. Geogr. Sci.* **2019**, *29*, 689–699. [CrossRef]
17. Wang, Y.; Li, H.; Ren, J. Delimitation and Zoning of Natural Ecological Spatial Boundary Based on GIS. *Asian J. Agric. Res.* **2019**, *11*, 40–45. [CrossRef]
18. Mamat, K.; Du, P.; Ding, J. Ecological function regionalization of cultural heritage sites in Turpan, China, based on GIS. *Arab. J. Geosci.* **2017**, *10*, 90. [CrossRef]
19. Wang, L.; Wang, W.; Yang, X. Eco-Environmental Zoning: A GIS-Based Approach. In Proceedings of the 2008 2nd International Conference on Bioinformatics and Biomedical Engineering, Shanghai, China, 16–18 May 2008. [CrossRef]
20. Shao, J.; Yang, L.; Peng, L.; Chi, T.; Wang, X. An Improved Artificial Bee Colony-Based Approach for Zoning Protected Ecological Areas. *PLoS ONE* **2015**, *10*, e0137880. [CrossRef]
21. Read, A.D.; West, R.J. Qualitative risk assessment of multiple-use marine park effectiveness–A case study from NSW, Australia. *Ocean Coast. Manag.* **2010**, *53*, 636–644. [CrossRef]
22. Guo, Y. Spatial Distribution and Simulation of Cropland Abandonment in Wushan County, Chongqing, China. *Sustainability* **2019**, *11*, 1367. [CrossRef]
23. Liu, Y.; Song, W.; Deng, X. Understanding the spatiotemporal variation of urban land expansion in oasis cities by integrating remote sensing and multi-dimensional DPSIR-based indicators. *Ecol. Indic.* **2019**, *96*, 23–37. [CrossRef]
24. Abutaleb, K.A.A.; Asmaa, M.; Hassan, E.; Ahmed, M. Climate Change Impacts, Vulnerabilities and Adaption Measures for Egypt's Nile Delta. *Earth Syst. Environ.* **2018**, *2*, 183–192. [CrossRef]
25. Yulin, S. *1:1 Million Land Resources Map of China*; Science Press: Beijing, China; Shanghai, China, 1990.
26. Alcantara, C.; Kuemmerle, T.; Prishchepov, A.V.; Radeloff, V.C. Mapping abandoned agriculture with multi-temporal MODIS satellite data. *Remote Sens. Environ.* **2012**, *124*, 334–347. [CrossRef]
27. Qian, X.; Zhu, H. A 1-km grid dataset of industrial output value in China (2010). *China Sci. Date* **2018**, *3*, 1–13. [CrossRef]
28. Xiaoming, W.; Junxing, Y.; Wei, S. Pollution status of agricultural land in China: Impact of land use and geographical position. *Soil Water Res.* **2018**, *13*, 234–242. [CrossRef]
29. Bliss, S. United Nations International Year of Forests 2011. *Geogr. Bull.* **2011**, *43*, 33.
30. Wang, P.; Zhang, L.; Li, Y.; Jiao, L.; Wang, H.; Yan, J.; Lü, Y.; Fu, B. Spatio-temporal characteristics of the trade-off and synergy relationships among multiple ecosystem services in the Upper Reaches of Hanjiang River Basin. *Acta Geogr. Sin.* **2017**, *72*, 2064–2078. [CrossRef]
31. Gao, Y.; Liu, H.; Liu, G. The spatial distribution and accumulation characteristics of heavy metals in steppe soils around three mining areas in Xilinhot in Inner Mongolia, China. *Environ. Sci. Pollut. Res.* **2017**, *24*, 25416–25430. [CrossRef]

32. Li, H.; Song, W. Characteristics of Climate Change in the Lancang-Mekong Sub-Region. *Climate* **2020**, *8*, 115. [CrossRef]
33. Han, Z.; Song, W.; Deng, X.; Xu, X. Grassland ecosystem responses to climate change and human activities within the Three-River Headwaters region of China. *Sci. Rep.* **2018**, *8*, 9079. [CrossRef] [PubMed]
34. Deng, X.; Huang, J.; Rozelle, S.; Uchida, E. Cultivated land conversion and potential agricultural productivity in China. *Land Use Policy* **2006**, *23*, 372–384. [CrossRef]
35. Jian, S.; Deng, G. The Chinese Urban and Rural per Capita Income and Trend Analysis. *Appl. Math.* **2014**, *5*, 106–109. [CrossRef]
36. Ding, L.; Zhao, W.; Huang, Y.; Cheng, S.; Liu, C. Research on the Coupling Coordination Relationship between Urbanization and the Air Environment: A Case Study of the Area of Wuhan. *Atmosphere* **2015**, *6*, 1539–1558. [CrossRef]
37. Liao, C.; Yue, Y.; Wang, K.; Fensholt, R.; Tong, X.; Brandt, M. Ecological restoration enhances ecosystem health in the karst regions of southwest China. *Ecol. Indic.* **2018**, *90*, 416–425. [CrossRef]
38. Li, H.; Song, W. Spatiotemporal Distribution and Influencing Factors of Ecosystem Vulnerability on Qinghai-Tibet Plateau. *Int. J. Environ. Res. Public Health* **2021**, *18*, 6508. [CrossRef] [PubMed]
39. Su, X.; Zhou, Y.; Li, Q. Designing Ecological Security Patterns Based on the Framework of Ecological Quality and Ecological Sensitivity: A Case Study of Jianghan Plain, China. *Int. J. Environ. Res. Public Health* **2021**, *18*, 8383. [CrossRef]
40. Kang, J.; Zhang, X.; Zhu, X.; Zhang, B. Ecological security pattern: A new idea for balancing regional development and ecological protection. A case study of the Jiaodong Peninsula, China. *Glob. Ecol. Conserv.* **2021**, *26*, e01472. [CrossRef]
41. Jin, X.; Jin, Y.; Mao, X. Ecological risk assessment of cities on the Tibetan Plateau based on land use/land cover changes—Case study of Delingha City. *Ecol. Indic.* **2019**, *101*, 185–191. [CrossRef]
42. Dieleman, H. Urban agriculture in Mexico City; balancing between ecological, economic, social and symbolic value. *J. Clean. Prod.* **2017**, *163*, S156–S163. [CrossRef]
43. Jongman, R.H.G. Nature conservation planning in Europe: Developing ecological networks. *Landsc. Urban Plan.* **1995**, *32*, 169–183. [CrossRef]
44. Fábos, J.G. Greenway planning in the United States: Its origins and recent case studies. *Landsc. Urban Plan.* **2004**, *68*, 321–342. [CrossRef]
45. Peng, J.; Zhao, H.; Liu, Y. Urban ecological corridors construction: A review. *Acta Ecol. Sin.* **2017**, *37*, 23–30. [CrossRef]
46. Xue, L.; Wang, J.; Zhang, L.; Wei, G.; Zhu, B. Spatiotemporal analysis of ecological vulnerability and management in the Tarim River Basin, China. *Sci. Total Environ.* **2018**, *649*, 876–888. [CrossRef] [PubMed]
47. Zhang, Y.; Shen, J. Wetland Restoration Planning Approach Based on Interval Fuzzy Linear Programming under Uncertainty. *Int. J. Environ. Res. Public Health* **2021**, *18*, 9549. [CrossRef]
48. Jiang, C.; Wang, F.; Zhang, H.; Dong, X. Quantifying changes in multiple ecosystem services during 2000–2012 on the Loess Plateau, China, as a result of climate variability and ecological restoration. *Ecol. Eng.* **2016**, *97*, 258–271. [CrossRef]
49. Huber, S.; Prokop, G.; Arrouays, D.; Banko, G.; Bispo, A.; Jones, R.J.A.; Kibblewhite, M.G.; Lexer, W.; Moller, A.; Rickson, R.J.; et al. *Environmental Assessment of Soil for Monitoring: Volume I Indicators & Criteria*; Office for the Official Publications of the European Communities Press: Ispra, Italy, 2008.
50. Wischmeier, W.H.; Smith, D.D. *Predicting Rainfall Erosion Losses—A Guide to Conservation Planning*; Agriculture Handbook No. 537; US Department of Agriculture Science and Education Administration: Washington, DC, USA, 1978; Volume 58.
51. Silva, A.M.D. Rainfall erosivity map for Brazil. *Catena* **2004**, *57*, 259. [CrossRef]
52. Renard, K.G.; Foster, G.R.; Weesies, G.A.; Mccool, D.K.; Yoder, D.C. *Predicting Soil Erosion by Water: A Guide to Conservation Planning with the Revised Universal Soil Loss Equation (RUSLE)*; Agricultural Handbook; United States Department of Agriculture Press: New York, NY, USA, 1997.
53. Williams, J.R.; Arnold, J.G. A system of erosion—Sediment yield models. *Soil Technol.* **1997**, *11*, 43–55. [CrossRef]
54. Li, S.; Wang, Z.; Zhang, Y. Crop cover reconstruction and its effects on sediment retention in the Tibetan Plateau for 1900–2000. *J. Geogr. Sci.* **2017**, *27*, 786–800. [CrossRef]
55. Ma, M.; Frank, V. Interannual variability of vegetation cover in the Chinese Heihe River Basin and its relation to meteorological parameters. *Int. J. Remote Sens.* **2006**, *27*, 3473–3486. [CrossRef]
56. Piao, S.; Wang, X.; Ciais, P.; Zhu, B.; Wang, T.; Liu, J. Changes in satellite-derived vegetation growth trend in temperate and boreal Eurasia from 1982 to 2006. *Glob. Chang. Biol.* **2011**, *17*, 3228–3239. [CrossRef]
57. Xia, H.; Li, A.; Feng, G.; Li, Y.; Qin, Y.; Lei, G.; Cui, Y. The Effects of Asymmetric Diurnal Warming on Vegetation Growth of the Tibetan Plateau over the Past Three Decades. *Sustainability* **2018**, *10*, 1103. [CrossRef]

International Journal of
Environmental Research and Public Health

MDPI

Article

Spatiotemporal Changes of Chemical Fertilizer Application and Its Environmental Risks in China from 2000 to 2019

Yuanzhi Guo [1,2] and Jieyong Wang [1,2,*]

1 Institute of Geographic Sciences and Natural Resources Research, Chinese Academy of Sciences, Beijing 100101, China; guoyz.16b@igsnrr.ac.cn
2 Key Laboratory of Regional Sustainable Development Modeling, Chinese Academy of Sciences, Beijing 100101, China
* Correspondence: wjy@igsnrr.ac.cn

Abstract: Chemical fertilizers are important inputs in agricultural production. They not only increase crop yield but also bring many negative effects, such as agricultural non-point source pollution. Therefore, a scientific understanding of the regional differences in chemical fertilizer application and its environmental risks is of significance to promote China's agricultural development. In this study, we analyzed the spatiotemporal pattern of chemical fertilizer application intensity (CFAI) in China since 2000, evaluated the environmental risks of provincial CFAI, and investigated the internal mechanism behind them. The results showed that the total amount and intensity of chemical fertilizer application in China from 2000 to 2019 presented a trend of increasing first and then decreasing. In 2000 and 2019, provincial CFAI in eastern China was generally higher than that in central and western China, and the environmental risks of provincial CFAI were spatially characterized by "high in the north and low in the south". Factors such as poor soil conditions, unreasonable farming structure and backward fertilization methods are the main reasons for the continuous increase in the total amount and intensity of chemical fertilizer application, while the construction of ecological civilization and the transformation of society and economy are the main reasons for their decline. Finally, measures such as targeted fertilization, adjusting the use structure of chemical fertilizers, improving fertilization methods and replacing chemical fertilizers with organic fertilizers are proposed to promote the quantity reduction and efficiency increase of chemical fertilizer application in China.

Keywords: chemical fertilizer application; environmental risks; agricultural production; food security; high-quality development; rural revitalization

Citation: Guo, Y.; Wang, J. Spatiotemporal Changes of Chemical Fertilizer Application and Its Environmental Risks in China from 2000 to 2019. *Int. J. Environ. Res. Public Health* **2021**, *18*, 11911. https://doi.org/10.3390/ijerph182211911

Academic Editors: Wei Song and Hualin Xie

Received: 24 September 2021
Accepted: 3 November 2021
Published: 12 November 2021

1. Introduction

Agriculture is not only the main source of human food and clothing but also the foundation supporting economic development and social progress [1,2]. Before the industrial revolution, agricultural development mainly depended on the self-recovery of soil fertility or the increase of soil organic matter through returning animal manure and straw to the field [3]. As a result, the development of agriculture was slow and the scale was small. By the middle of the 19th century, the emergence of chemical fertilizers opened a channel to provide nutrients for crop growth from outside the agricultural system, expanded the contents of material flow and energy cycle in agricultural system, and greatly promoted the development of agricultural production [4]. Through balancing nutrient composition and improving soil fertility, the application of chemical fertilizers effectively ensures the stable and high yield of grain, which makes great contributions to human survival and development. According to the existing research, the contribution rate of chemical fertilizers to the growth of food production in various countries is generally between 30% and 60% [5–7].

China is the most populous developing country in the world [8,9]. The long-standing problem of insufficient food and clothing makes the Chinese government attach great importance to agricultural development and continue to strengthen financial and policy supports to improve grain production capacity and ensure national food security [10–12]. After nearly 70 years of efforts, China's grain yield per unit area has increased from 1029 kg/ha in 1949 to 6272 kg/ha in 2019, and the per capita output of grain has correspondingly increased from 209 kg/person to 475 kg/person [13], feeding about 22% of the world's population with just 7% of the world's farmland [14–18]. In terms of the driving force behind it, technological progress, especially the development of chemical fertilizer technology, has played a vital role [19–21]. Since 2006, China has been the biggest producer and consumer of chemical fertilizers in the world, producing more than a quarter of the world chemical fertilizers and consuming more than 30% every year.

As one of the important components of the green revolution [22], chemical fertilizers are the "food" of crop growth and the necessary means of production for agriculture. Thus, they have played a critical role in aspects such as ensuring national food security and guaranteeing social stability [21]. However, because of the traditional agricultural production mode and unreasonable fertilization methods, the overuse of chemical fertilizers is widespread in China's agricultural production [23,24], and has been one of the main sources of agricultural non-point source pollution [25–27]. With the construction of ecological civilization and the enhancement of people's ecological awareness, more and more attention has been paid to resource and environmental problems such as water pollution, air pollution and land degradation, which are caused by the excessive use of chemical fertilizers [16,19,25,28–30]. As a result, how to deal with these problems has become an important target of the government policies and measures [31]. Due to the lack of scientific understanding of soil nutrients, the unreasonable chemical fertilizer application structure is also common in agricultural production [32], especially the prominent problems such as attaching importance to chemical fertilizers and neglecting organic fertilizers, attaching importance to nitrogenous fertilizers and neglecting phosphate and potash fertilizers and attaching importance to major elements and neglecting trace elements, all of which have seriously hindered crop growth [20,33]. In general, the unreasonable and unscientific problems in China's chemical fertilizer application not only damage basic soil fertility and increase the cost of grain production, but also affect the quantity and quality of agricultural products and threaten national food security.

Currently, China's socialist modernization has entered a new stage of high-quality development, and the focus of rural development has shifted to rural revitalization [9,34]. In this context, the Chinese government actively promotes the construction of ecological civilization to optimize human–earth relationship and achieve the sustainable development of agricultural system [35,36]. These new situations inherently require promoting the quantity reduction and efficiency improvement of agricultural inputs, especially the chemical fertilizers, to realize the green and high-quality development of agriculture. China is a country with vast territory and significant regional differences [37]. A scientific understanding of regional differences in chemical fertilizer application and its environmental risks is of great significance to guide the rational application of chemical fertilizers and the transformation and upgrading of agriculture in different areas. Employing a dataset of chemical fertilizer application from 2000 to 2019, this study analyzes the spatiotemporal pattern of chemical fertilizer application intensity (CFAI) in China, measures provincial CFAI safety thresholds according to local soil conditions, and then reveals the environmental risks of chemical fertilizer application in different provinces. Ultimately, the driving mechanism behind the spatial inequality of chemical fertilizer application as well as the measures for agricultural sustainable development are discussed. These findings will contribute to the implementation of new development concepts, promote the construction of resource-saving and environment-friendly industrial system in rural China, and finally realize the modernization of agriculture and rural areas.

2. Materials and Methods

2.1. Chemical Fertilizer Application Intensity

CFAI is a concept which reflects the consumption of chemical fertilizers per unit area of land. It has two statistic calibers of cultivated area and sown area [38]. Based on relevant studies [38,39], the latter is employed in this study to investigate the regional pattern of China's CFAI and its environmental risks, and the formula is given as follows:

$$\text{CFAI} = \text{CCF}/\text{SAC} \quad (1)$$

where CCF denotes the consumption of chemical fertilizers, which refers to the quantity of chemical fertilizers applied in agricultural production in the year, including nitrogenous fertilizers, phosphate fertilizers, potash fertilizers, and compound fertilizers. For the convenience of comparison, CCF is calculated in terms of the volume of effective components by converting the gross weight of respective fertilizers into weight containing effective components, e.g., nitrogen content in nitrogenous fertilizers, phosphorous pentoxide contents in phosphate fertilizers, and potassium oxide contents in potash fertilizers. SAC denotes the total sown area of crops, which refers to the area of all land sown or transplanted with crops that are harvested within the calendar year. All crops harvested within the year are counted as sown area, regardless of being sown in the current year or the previous year, and crops that are sown this year but harvested in the coming year are excluded.

2.2. Environmental Risk Assessment

The concept of risk assessment generally has a long history [40], but environmental risk assessment (ERA), as a scientific field, originated only in the early 1970s [41]. An ERA is a process used to evaluate the quantitative and qualitative characteristics of environment that may be impacted due to exposure to one or more environmental stressors, such as chemicals, disease, invasive species, and climate change [42,43]. The environmental risks of CFAI are the possibility of ecological damage and environmental pollution caused by chemical fertilizer application in the process of agricultural production, which has the characteristic of being non-sudden. Here, the ERA of CFAI is calculated as follows:

$$R_i = F_i/(F_i + T_i) \quad (2)$$

where R_i is the environmental risk index of region i, F_i is the CFAI of region i, and T_i is the environmental safety threshold of CFAI of region i. According to the calculation of environmental risks, the value of R_i varies from 0 to 1. When R_i is 0.50, it means that F_i and T_i are equal, which is the critical point of environmental safety; when R_i approaches 1, it means that F_i greatly exceeds T_i, i.e., there are extremely serious environmental risks in chemical fertilizer application; while R_i approaching 0 means that F_i is much lower than T_i. According to the multiple of F_i to T_i, the environmental risks are divided into five types: safe, low risk, moderate risk, high risk and serious risk (Table 1).

Table 1. Classification of the environmental risks of chemical fertilizer application.

Classification	Type	Threshold	Criteria
Level I	Safe	$0.00 < R_i \leq 0.50$	$F_i \leq T_i$
Level II	Low risk	$0.50 < R_i \leq 0.67$	$T_i < F_i \leq 2T_i$
Level III	Moderate risk	$0.67 < R_i \leq 0.75$	$2T_i < F_i \leq 3T_i$
Level IV	High risk	$0.75 < R_i \leq 0.80$	$3T_i < F_i \leq 4T_i$
Level V	Serious risk	$0.80 < R_i < 1.00$	$F_i > 4T_i$

2.3. Data Source and Processing

The basic geographic data used in this study come from the Resource and Environment Science and Data Center of Chinese Academy of Sciences (https://www.resdc.cn/) (accessed on 5 September 2021). Data on sown area of farm crops and chemical fertilizer application are obtained from China Statistical Yearbook. Data on the grade of cultivated land are collected from the Ministry of Natural Resources of the PRC (http://www.mnr.gov.cn/) (accessed on 5 September 2021). According to the research design, Hong Kong, Macao, and Taiwan are excluded in the analysis. As a result, a total of 31 provincial-level administrative units are obtained to investigate the regional differences of CFAI and their environmental risks in China.

3. Results

3.1. Evolution of Chemical Fertilizer Application in China

In 2000, the total amount of chemical fertilizers applied in China's agricultural production was 41.46 million tons, and increased to 54.04 million tons in 2019, with an average annual growth rate of 1.40%. Accordingly, China's CFAI increased from 265.28 kg/ha to 325.65 kg/ha, and the average annual growth rate was 1.08%. Specifically, the consumption of chemical fertilizers maintained a stable growth trend before 2015, and the average annual growth rate during this period reached 2.52%. Then, as the Ministry of Agriculture and Rural Affairs of the PRC issued the action plan for zero growth of chemical fertilizer application by 2020 in February 2015 to promote the green development of agriculture, the total amount of chemical fertilizer application in China decreased at an average annual rate of 2.67% in 2015–2019. The evolution of CFAI also showed a trend of increasing first and then decreasing, but the turning year was advanced to 2014, and the average annual growth rate of the two stages changed from 2.26% to −2.15% (Figure 1).

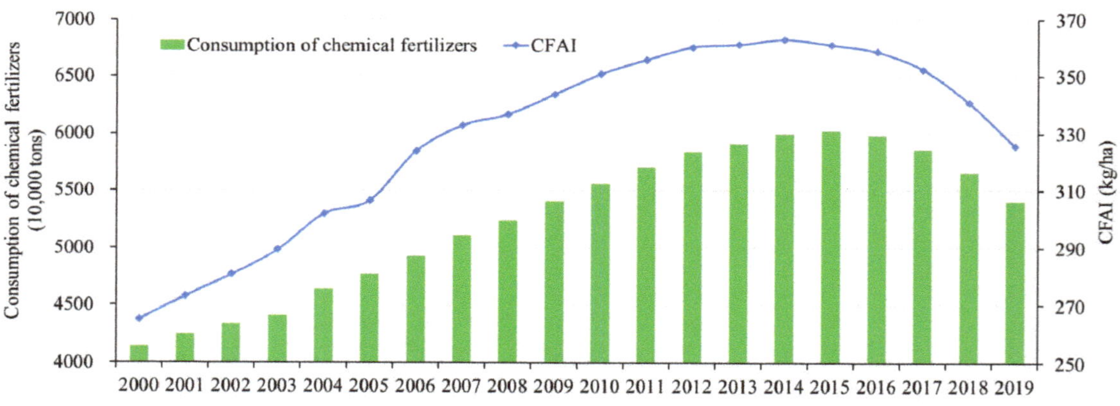

Figure 1. Evolution of chemical fertilizer application in China from 2000 to 2019.

In terms of the provincial CFAI, CFAI in Tibet was the lowest in 2000, only 108.18 kg/ha, followed by Inner Mongolia (126.47 kg/ha), Qinghai (130.03 kg/ha) and Heilongjiang (130.34 kg/ha). These provinces were the only four provinces with a CFAI lower than 150 kg/ha. The province with the highest CFAI was Fujian, followed by Jiangsu, Beijing and Shandong, where the value of CFAI was all greater than 375 kg/ha (Figure 2a). After nearly two decades of development, provincial CFAI has developed rapidly. In 2019, Qinghai was the province with the smallest CFAI, which was just 112.01 kg/ha and the only province with a CFAI lower than 150 kg/ha in the whole country. On the other end of the spectrum were three eastern provinces, Beijing, Hainan, and Fujian, where CFAI was 699.77 kg/ha, 684.71 kg/ha and 664.67 kg/ha, respectively; CFAIs of these three provinces were much greater than those of other provinces (Figure 2b). Spatially, provincial CFAI showed an obvious east-central-west gradient differentiation in 2000 and 2019, that is, provincial CFAI in the eastern coastal region was significantly greater than that in the central region, and provincial CFAI in the western region was generally small.

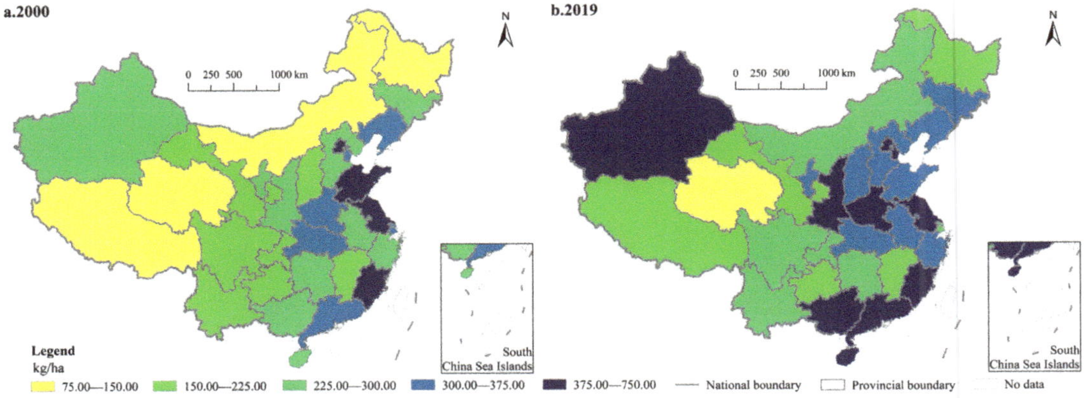

Figure 2. Spatial pattern of China's provincial CFAI in 2000 (**a**) and 2019 (**b**).

The analysis of the changes of provincial CFAI from 2000 to 2019 showed that CFAI decreased in only five provinces, namely, Shanghai, Qinghai, Jiangsu, Shandong and Guizhou, with a decrease of 22.59%, 13.86%, 8.94%, 4.76% and 0.02%, respectively. Among the provinces with an increase of CFAI, there were eleven provinces with an increase greater than 50%, and the greatest was Hainan (135.87%), followed by Inner Mongolia (94.36%), Xinjiang (78.93%) and Beijing (78.77%); only CFAI in Liaoning, Hubei and Sichuan increased by less than 10%, and the value was 9.43%, 7.56% and 3.89%, respectively. In general, the changes of provincial CFAI showed a characteristic of being small in traditional agricultural areas and large in non-traditional agricultural areas.

3.2. Environmental Risks of Chemical Fertilizer Application in China

To measure the environmental risks of chemical fertilizer application in different provinces, it is necessary to first determine the provincial safety upper limit of CFAI. According to existing research, the internationally recognized ceiling for safe CFAI is 225 kg/ha [44,45]. Here, we also set this value as the safety upper limit of CFAI in China, that is, the ceiling of CFAI corresponding to 9.96-grade cultivated land is 225 kg/ha. The calculation of China's CFAI over the years shows that CFAI in China has exceeded the safety upper limit since 1995, and has been running at a high level for a long time [13]. Because of the regional inequality in cultivated land grades derived from the spatial heterogeneity of natural and human conditions [37,46], the provincial safety upper limit of CFAI is also different. In line with the safety upper limit of CFAI, which is corresponding to the national average grade of cultivated land, the cultivated land grades of different provinces are employed to calculate their safety upper limits of CFAI. As shown in Figure 3, the environmental safety threshold of chemical fertilizer application in each province shows a significant gradient descent pattern from the southeast coastal areas to the northwest inland areas. Specifically, there are sixteen provinces with a safety upper limit of CFAI greater than 225 kg/ha, mainly distributed in the eastern coastal region and the central region with good natural conditions, such as abundant water and good soil. Among them, there are four provinces with a safety upper limit greater than 375 kg/ha, including Hubei (428.42 kg/ha), Guangdong (424.98 kg/ha), Shanghai (411.64 kg/ha) and Jiangsu (399.27 kg/ha). Jiangxi and Henan are also close to that level. It is noted that the provinces with a safety upper limit less than 225 kg/ha mainly distribute in the western and northeast regions, including Inner Mongolia, Gansu, Qinghai, Shanxi, and Tibet, where the upper safety limits are all less than 150 kg/ha.

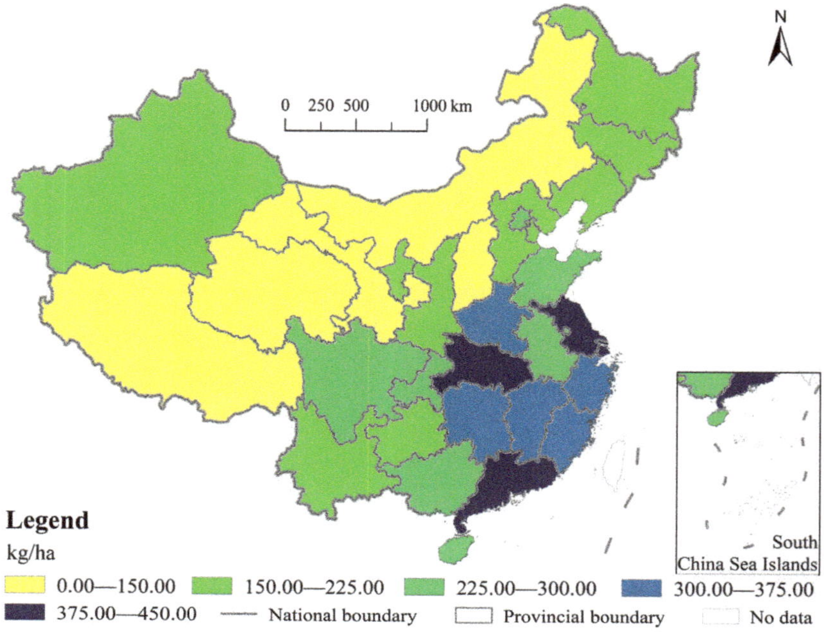

Figure 3. Spatial pattern of provincial environmental safety upper limits of CFAI in China.

Due to the relative stability of natural conditions [47], we assume that the cultivated land grade of each province is the same in 2000 and 2019. Based on provincial CFAI and their environmental safety upper limits, formula (2) is employed to calculate the environmental risks of chemical fertilizer application, and ArcGIS 10.4 is used for spatial visualization (Figure 4). In 2000, there were only two types of environmental risks for provincial CFAI, namely, safety and low risk. The former included sixteen provinces, mainly distributed in the south, of which the province with the lowest risk was Jiangxi (0.36), followed by Hubei (0.43) and Hunan (0.43). The other provinces belonged to the latter, and the province with the highest risk level was Gansu (0.66), with Shaanxi (0.64) and Liaoning (0.64) being close to that level. In 2019, there were only eight provinces whose environmental risk type was safety, mainly distributed in the middle and upper reaches of the Yangtze River. Among them, the province with the lowest environmental risk was Jiangxi (0.38), followed by Shanghai (0.41) and Hubei (0.45), and the other five provinces were on the edge of the safety line. The number of provinces with a risk type of moderate risk was seven; except for Hainan, Fujian and Beijing in the east, the rest were mainly distributed in the northwest. Only the risk type of Inner Mongolia was high risk.

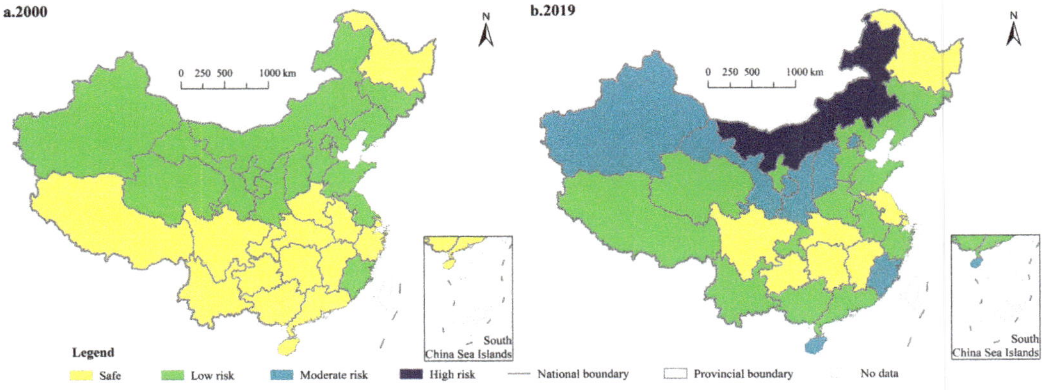

Figure 4. Spatial pattern of the environmental risks of China's CFAI in 2000 (**a**) and 2019 (**b**).

In terms of the changes of environmental risks of CFAI during the period of 2000–2019, there were five provinces with a decreased value of environmental risks, namely Shanghai, Qinghai, Jiangsu, Shandong and Guizhou, and the decrease was 13.31%, 6.36%, 4.55%, 2.15% and 0.01%, respectively. Among the provinces with an increased value of environmental risks, there were eight provinces with an increase greater than 20%, of which the highest was Hainan (41.06%), followed by Tibet (27.87%) and Guangxi (27.39%). The analysis of the changes of environmental risk types from 2000 to 2019 showed that there were eight provinces changing from safe to low risk and one province changing from safe to moderate risk; there was one province changing from low risk to safe, six provinces changing from low risk to moderate risk, and one province changing from low risk to high risk; there were seven provinces whose risk types remained safety, and the number of provinces whose risk type has always been low risk was also 7 (Table 2).

Table 2. Changes of the environmental risk type of provincial CFAI in China from 2000 to 2019.

		2019			
		Safe	**Low Risk**	**Moderate Risk**	**High Risk**
2000	Safe	Jiangxi, Shanghai, Hubei, Sichuan, Guizhou, Hunan, Heilongjiang	Anhui, Chongqing, Zhejiang, Guangdong, Tibet, Henan, Guangxi, Yunnan	Hainan	
	Low risk	Jiangsu	Qinghai, Shandong, Tianjin, Ningxia, Liaoning, Jilin, Hebei	Fujian, Shanxi, Gansu, Beijing, Xinjiang, Shaanxi	Inner Mongolia

4. Discussion

4.1. Understanding the Chemical Fertilizer Application in China

Through solving the problem that soil nutrients cannot meet the needs of crop growth, chemical fertilizers effectively promote agricultural development, thus ensuring national food security [19]. Wang et al. pointed out that 40% of the increase in China's agricultural production during the period of 1986–1990 came from the increase of chemical fertilizer application [48]. However, the growth rate of chemical fertilizer application in this process has far exceeded the growth rate of grain output, resulting in a serious problem of excessive and inefficient application of chemical fertilizers, which become an important challenge restricting the sustainable development of agriculture in China [16,49]. According to statistics from the Ministry of Agriculture and Rural Affairs of the PRC, the comprehensive utilization rate of chemical fertilizers for rice, wheat, and corn in 2020 was only 40.2%, which was far lower than that of 60%~70% in developed countries. As the products of technological progress, chemical fertilizers are essentially harmless. The problem related to chemical fertilizers in agricultural production is not the problem of chemical fertilizers itself but stems from the unreasonable and unscientific use of chemical fertilizers [50], including unreasonable farming structure and backward fertilization method, as well as insufficient scientific and technological guidance.

First, China's per capita cultivated land is small, less than 0.10 ha/person, and the basic fertility of cultivated land is low, with medium and low yield cultivated land accounting for nearly 65%. To ensure national food security, increasing the use of chemical fertilizers has become an important choice in agricultural production [19,51]. However, due to the imperfect fertilizer management system, the massive application of chemical fertilizers has caused many resources and environmental problems, such as soil degradation and the decline of land productivity, which further increases the use of chemical fertilizers, forming a vicious circle [52]. Meanwhile, the low-quality cultivated land is poor in water and fertilizer conservation, reducing the utilization efficiency of chemical fertilizers. Second, the adjustment of farming structure boosts the increase of chemical fertilizer application. Guided by a market economy and consumption upgrading, more and more cultivated land in China is used to grow non-grain crops [8], such as fruits and vegetables (Figure 5), which have a higher demand for chemical fertilizers [53–55]. Due to the traditional fertilization habits and the lack of understanding of fertilizer performance, the phenomenon of excessive chemical fertilizer application is widespread in the process of cash crops production, especially in the eastern developed region [56]. Third, the dominant position of small-scale peasant economy in rural China makes the agricultural mechanization develop slowly. In this context, the traditional artificial fertilization is still dominant, such as spreading and surface application, which leads to the volatilization and leaching of chemical fertilizers and reduces the utilization rate of chemical fertilizers. Fourth, due to the urban–rural dual structure and urban-biased development strategy, problems such as imperfect agricultural technology service system and few technicians are common in rural China, which make farmers unable to get guidance from technicians in chemical fertilizer application. Fifth, with the rapid development of industrialization and urbanization, many rural working people flow to urban areas. This results in the aging and weakening of agricultural producers, and causes the substitution of modern production factors such as chemical fertilizer for labor force, which further increases CFAI [57]. Additionally, in the process of circulation, fertilizer dealers tend to increase the recommended application number of chemical fertilizers to obtain more benefits, resulting in excessive chemical fertilizer application [58].

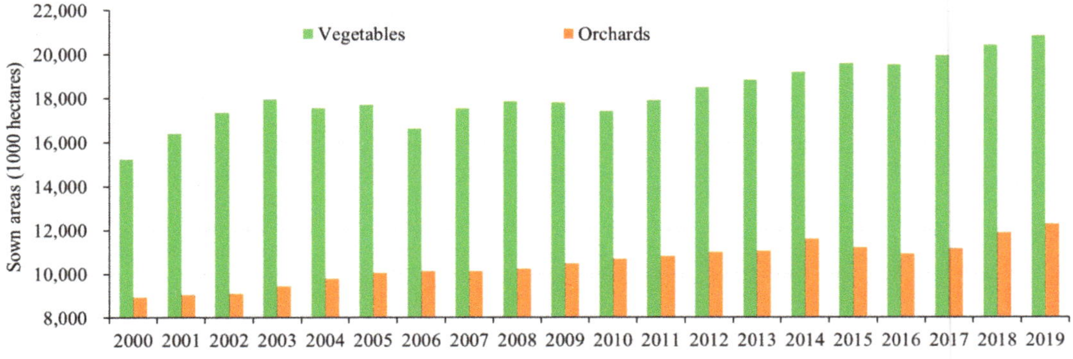

Figure 5. Changes of the sown areas of vegetables and orchards in China from 2000 to 2019.

After more than 30 years of reform and opening-up, China's economy developed to the new normal stage in 2014, which requires optimizing the economic structure and realizing the transformation and upgrading of economic growth from factor-driven to innovation-driven [59,60]. Against this background, China's agricultural development has gradually changed from extensive growth of scale–speed type to intensive growth of quality–efficiency type, constantly optimizing the structure of chemical fertilizer application and improving the efficiency of chemical fertilizer application. On the other hand, the report to the 18th National Congress of the Communist Party of China (CPC) proposed to vigorously promote the construction of ecological civilization and form resource-saving and environment-friendly spatial pattern, industrial structure, production and living mode to reverse the trend of ecological and environmental deterioration from the source [61]. To achieve this goal, the Chinese government has issued a series of policies and measures to promote the quantity reduction and efficiency improvement of agricultural inputs such as chemical fertilizers and pesticides, and paid attention to the comprehensive utilization of agricultural wastes, thus building a high-efficiency, low-carbon and green agriculture system. Moreover, the promotion of rural reform continues to help the rapid development of various new agricultural business entities [62], which promote the scale of agricultural production and improves the efficiency of chemical fertilizer application. Driven by the transformation of major national policies, the total amount and intensity of chemical fertilizer application in China has decreased steadily since the late 12th Five-Year Plan, and this trend will continue, thus boosting the high-quality development of agriculture and achieving the goal of agricultural and rural modernization.

4.2. Policy Implications for Agricultural Development

Since 2004, China's grain production has achieved bumper harvests for 16 consecutive years. However, the grain supply and demand is still in a tight balance, and the food-security situation remains grim [8,63]. As people's food demand changes from enough to high-quality and diversified, the grain consumed in the production of meat, eggs, milk and other food will continue to increase, which leads to rapid growth of grain consumption [64–66]. On the other hand, the restrictive factors such as fresh water resources and environmental carrying capacity have rigid constraints on grain production, and become increasingly prominent, which increases the difficulty of increasing grain production [67–69]. To ensure national food security and meet people's needs for a better life, chemical fertilizers must continue to be used. In the foreseeable future, chemical fertilizers will continue to play an important role in China's agricultural production, and more attention should be paid to scientific fertilization to maximize its socioeconomic benefits and avoid the adverse impacts of unreasonable and unscientific fertilization on resources and environment. Meanwhile, some targeted measures should be taken to establish and improve the chemical fertilizer application system to ensure production, save cost and increase efficiency according to local conditions.

First, it is necessary to promote targeted fertilization. According to soil conditions, crop yield potential and the requirements of comprehensive nutrient management, the fertilization quota standard per unit area of crop should be reasonably formulated in different regions to reduce blind fertilization. Second, more attention should be paid to adjust the use structure of chemical fertilizers. By optimizing the ratio of nitrogen, phosphorus and potassium fertilizers, the reasonable combination of major element and medium and trace element can be realized. Meanwhile, there is an urgent need to guide the optimization and upgrading of fertilizer products to meet the need of modern agriculture. Third, fertilization method needs to be improved. The technology of soil testing and formula fertilization (*Cetu Peifang Shifei*) should be popularized to improve farmers' skills of scientific fertilization. The government also should promote suitable fertilization equipment and guide the changes of fertilization methods from surface application and spreading application to mechanical deep application, water and fertilizer integration and foliar spraying. Fourth, chemical fertilizers should be replaced by organic fertilizers. Through rational utilization of organic nutrient resources, organic fertilizers are used to replace part of chemical fertilizers, realizing the rational combination of organic and inorganic fertilizers. In addition, some guarantee mechanisms need to be established and improved to ensure the implementation of these measures for quantity reduction and efficiency increase of chemical fertilizers, including a working mechanism of up–down linkage and multiparty cooperation, a national fertilizer efficiency monitoring network, publicity and training of new business entities, and supporting policies such as finance and taxation.

4.3. Limitations and Future Research Prospects

The objects of farming include not only grain crops, such as rice and corn, but also cash crops, such as vegetables and fruits [8], and there are significant differences in nutrient elements required for the growth of different crops. As a result, their demands for nitrogen fertilizers, phosphorus fertilizers, potassium fertilizers, compound fertilizers and other types of chemical fertilizers are different. Here, we only discuss the overall situation of regional chemical fertilizer application, but lack of understanding of the use structure of chemical fertilizers, the regional differences of different types of chemical fertilizers and their environmental risks. Therefore, the chemical fertilizer application of different crops is worthy of further investigation. When measuring the environmental risks of provincial CFAI, the environmental safety threshold used in this study is a value calculated by international general standards. In fact, the environmental safety threshold is closely related to the nutrient situations of cultivated land [70]. Thus, it is necessary to further promote soil testing in future research and comprehensively determine whether chemical fertilizer application is excessive according to regional soil fertility, thus scientifically guiding the chemical fertilizer application in agricultural production and avoiding various resource and environmental problems caused by the overuse of chemical fertilizers. In addition, with the rapid development of society and the economy, greenhouse gas emissions dominated by carbon dioxide are posing a serious threat to the global climate and ecology [71,72]. In this context, China is actively committed to the implementation of the Paris Agreement, striving to reach the peak of carbon dioxide emissions by 2030 and achieve carbon neutralization by 2060. The process of chemical fertilizer production is an important carbon source, and chemical fertilizer application also produces a large amount of carbon dioxide [73,74]. Therefore, scientific discussion on the relationship between chemical fertilizer production/application and carbon emission and its internal mechanism is of great significance to achieve the goal of carbon emission reduction in China.

5. Conclusions

Currently, China's agricultural development is in the critical stage of transformation and upgrading, and its driving force is changing from factor input to technological innovation [75]. Against this background, optimizing the structure and efficiency of agricultural inputs, such as chemical fertilizers, has become an important measure to promote agricultural development to a higher level in the new era. Affected by poor soil conditions, unreasonable farming structure and backward fertilization methods, the total amount and intensity of chemical fertilizer application in China have maintained an increasing trend for a long time. With the promotion of ecological civilization and rural reform, the total amount and intensity of chemical fertilizer application began to decline at the end of the 12th Five-Year Plan. Spatially, provincial CFAI in 2000 and 2019 showed a pattern of "high in the east and low in the west", and the changes of non-traditional agricultural areas were more obvious than those of traditional agricultural areas. The analysis of the CFAI safety thresholds showed that the provinces with a high safety threshold were mainly distributed in the third terrain ladder, and those with a low safety threshold were mainly distributed in northwest China and Qinghai–Tibet Plateau. As a result, the environmental risks of provincial CFAI in 2000 and 2019 were all characterized by "high in the north and low in the south". To solve excessive application of chemical fertilizers and its related problem in China, it is necessary to promote the quantity reduction and efficiency increase of chemical fertilizers through targeted fertilization, adjusting the use structure of chemical fertilizers, improving fertilization methods, and replacing chemical fertilizers with organic fertilizers, thus realizing agricultural green and high-quality development and supporting agricultural and rural modernization.

Author Contributions: Conceptualization, Y.G. and J.W.; methodology, Y.G. and J.W.; software, Y.G.; validation, Y.G. and J.W.; formal analysis, Y.G. and J.W.; investigation, Y.G.; resources, Y.G. and J.W.; data curation, Y.G.; writing—original draft preparation, Y.G. and J.W.; writing—review and editing, Y.G. and J.W.; visualization, Y.G.; supervision, Y.G. and J.W.; project administration, Y.G.; funding acquisition, Y.G. All authors have read and agreed to the published version of the manuscript.

Funding: This research was supported by the National Natural Science Foundation of China (Grant No. 42001203, 42171266) and the Strategic Priority Research Program of the Chinese Academy of Sciences (Grant No. XDA23070300, XDA28130400).

Institutional Review Board Statement: Not applicable.

Informed Consent Statement: Not applicable.

Data Availability Statement: The associated dataset of the study is available upon request to the corresponding author.

Acknowledgments: The authors would like to thank the anonymous reviewers for their comments and suggestions, which contributed to the further improvement of this paper.

Conflicts of Interest: The authors declare that they have no conflict of interest.

References

1. Gillespie, S.; Harris, J.; Kadiyala, S. The agriculture-nutrition disconnect in India: What do we know? *Gates Open Res.* **2019**, *3*, 1115.
2. Johnston, B.F.; Mellor, J.W. The role of agriculture in economic development. *Am. Econ. Rev.* **1961**, *4*, 566–593.
3. Zhang, W.; Shen, Z.; Shao, Y.; Shi, L.; Liu, S.; Shi, N.; Fu, L. Soil biota and sustainable agriculture: A review. *Acta Ecol. Sin.* **2020**, *10*, 3183–3206.
4. Food and Agriculture Organization of the United Nations (FAO). *The Future of Food and Agriculture: Trends and Challenges*; FAO: Rome, Italy, 2017.
5. Pradhan, P.; Fischer, G.; van Velthuizen, H.; Reusser, D.E.; Kropp, J.P. Closing yield gaps: How sustainable can we be? *PLoS ONE* **2015**, *6*, e0129487. [CrossRef] [PubMed]
6. Stewart, W.M.; Dibb, D.W.; Johnston, A.E.; Smyth, T.J. The contribution of commercial fertilizer nutrients to food production. *Agron. J.* **2005**, *1*, 1–6. [CrossRef]
7. Roberts, T.L. The role of fertilizer in growing the world's food. *Better Crops* **2009**, *2*, 12–15.

8. Guo, Y.; Wang, J. Identifying the determinants of nongrain farming in China and its implications for agricultural development. *Land* **2021**, *9*, 902. [CrossRef]
9. Guo, Y.; Liu, Y. Poverty alleviation through land assetization and its implications for rural revitalization in China. *Land Use Policy* **2021**, *105*, 105418. [CrossRef]
10. Yu, W.; Elleby, C.; Zobbe, H. Food security policies in India and China: Implications for national and global food security. *Food Secur.* **2015**, *2*, 405–414. [CrossRef]
11. Liu, Y.; Guo, Y.; Zhou, Y. Poverty alleviation in rural China: Policy changes, future challenges and policy implications. *China Agric. Econ. Rev.* **2018**, *2*, 241–259. [CrossRef]
12. Guo, Y.; Zhou, Y.; Liu, Y. Targeted poverty alleviation and its practices in rural China: A case study of Fuping County, Hebei Province. *J. Rural. Stud.* **2019**, in press. [CrossRef]
13. National Bureau of Statistics of China (NBS). *China Statistical Yearbook 2020*; China Statistics Press: Beijing, China, 2020.
14. Gong, P. China needs no foreign help to feed itself. *Nat. News* **2011**, *474*, 7. [CrossRef]
15. Zhang, F.; Cui, Z.; Fan, M.; Zhang, W.; Chen, X.; Jiang, R. Integrated soil–crop system management: Reducing environmental risk while increasing crop productivity and improving nutrient use efficiency in China. *J. Environ. Qual.* **2011**, *4*, 1051–1057. [CrossRef] [PubMed]
16. Cui, K.M.; Shoemaker, S.P. A look at food security in China. *NPJ Sci. Food* **2018**, *2*, 4. [CrossRef] [PubMed]
17. Chen, J. Rapid urbanization in China: A real challenge to soil protection and food security. *Catena* **2007**, *1*, 1–15. [CrossRef]
18. Chen, L.; Song, G.; Meadows, M.E.; Zou, C. Spatio-temporal evolution of the early-warning status of cultivated land and its driving factors: A case study of Heilongjiang Province, China. *Land Use Policy* **2018**, *72*, 280–292. [CrossRef]
19. Li, Y.; Zhang, W.; Ma, L.; Huang, G.; Oenema, O.; Zhang, F.; Dou, Z. An analysis of China's fertilizer policies: Impacts on the industry, food security, and the environment. *J. Environ. Qual.* **2013**, *4*, 972–981. [CrossRef]
20. Wang, Q.; Halbrendt, C.; Johnson, S.R. Grain production and environmental management in China's fertilizer economy. *J. Environ. Qual.* **1996**, *3*, 283–296. [CrossRef]
21. Cheng, X.; Han, C.; Taylor, D.C. Sustainable agricultural development in China. *World Dev.* **1992**, *8*, 1127–1144.
22. Rosset, P.; Collins, J.; Lappé, F.M. Lessons from the green revolution. *Third World Resurgence* **2000**, *2*, 11–14.
23. Wu, H.; Hao, H.; Lei, H.; Ge, Y.; Shi, H.; Song, Y. Farm size, risk aversion and overuse of fertilizer: The heterogeneity of large-scale and small-scale wheat farmers in Northern China. *Land* **2021**, *2*, 111. [CrossRef]
24. Liu, Y.; Zou, L.; Wang, Y. Spatial-temporal characteristics and influencing factors of agricultural eco-efficiency in China in recent 40 years. *Land Use Policy* **2020**, *97*, 104794. [CrossRef]
25. Ma, X.; Li, Y.; Zhang, M.; Zheng, F.; Du, S. Assessment and analysis of non-point source nitrogen and phosphorus loads in the Three Gorges Reservoir Area of Hubei Province, China. *Sci. Total. Environ.* **2011**, *412*, 154–161. [CrossRef]
26. Fischer, G.; Winiwarter, W.; Ermolieva, T.; Cao, G.Y.; Qui, H.; Klimont, Z.; Wiberg, D.; Wagner, F. Integrated modeling framework for assessment and mitigation of nitrogen pollution from agriculture: Concept and case study for China. *Agric. Ecosyst. Environ.* **2010**, *1–2*, 116–124. [CrossRef]
27. Guo, J.H.; Liu, X.J.; Zhang, Y.; Shen, J.L.; Han, W.X.; Zhang, W.F.; Christie, P.; Goulding, K.W.T.; Vitousek, P.M.; Zhang, F.S. Significant acidification in major Chinese croplands. *Science* **2010**, *5968*, 1008–1010. [CrossRef] [PubMed]
28. Zhang, Y.; Wang, J.; Dai, C. The adjustment of China's grain planting structure reduced the consumption of cropland and water resources. *Int. J. Environ. Res. Public Health* **2021**, *14*, 7352. [CrossRef] [PubMed]
29. Wu, Y. Chemical fertilizer use efficiency and its determinants in China's farming sector: Implications for environmental protection. *China Agric. Econ. Rev.* **2011**, *2*, 117–130. [CrossRef]
30. Sun, Y.; Hu, R.; Zhang, C. Does the adoption of complex fertilizers contribute to fertilizer overuse? Evidence from rice production in China. *J. Clean. Prod.* **2019**, *219*, 677–685. [CrossRef]
31. Xu, L.; Zhou, Z.; Du, J. An evolutionary game model for the multi-agent Co-governance of agricultural non-point source pollution control under intensive management pattern in China. *Int. J. Environ. Res. Public Health* **2020**, *7*, 2472. [CrossRef] [PubMed]
32. Huang, S.; Tang, J.; Li, C.; Zhang, H.; Yuan, S. Reducing potential of chemical fertilizers and scientific fertilization countermeasure in vegetable production in China. *J. Plant Nutr. Fertil.* **2017**, *6*, 1480–1493.
33. Huang, J.; Rozelle, S.; Zhu, X.; Zhao, S.; Sheng, Y. Agricultural and rural development in China during the past four decades: An introduction. *Aust. J. Agric. Resour. Econ.* **2020**, *1*, 1–13. [CrossRef]
34. Liu, Y. Modern human-earth relationship and human-earth system science. *Sci. Geogr. Sin.* **2020**, *8*, 1221–1234.
35. Zhang, M.; Liu, Y.; Wu, J.; Wang, T. Index system of urban resource and environment carrying capacity based on ecological civilization. *Environ. Impact Asses.* **2018**, *68*, 90–97. [CrossRef]
36. Calabi-Floody, M.; Medina, J.; Rumpel, C.; Condron, L.M.; Hernandez, M.; Dumont, M.; de la Luz Mora, M. Smart fertilizers as a strategy for sustainable agriculture. *Adv. Agron.* **2018**, *147*, 119–157.
37. Xu, B.; Lin, B. Differences in regional emissions in China's transport sector: Determinants and reduction strategies. *Energy* **2016**, *95*, 459–470. [CrossRef]
38. Liu, Q.; Sun, J.; Pu, L. Comparative study on fertilization intensity and integrated efficiency in China and Euro-American major countries. *Trans. Chin. Soc. Agric. Eng.* **2020**, *14*, 9–16.
39. Zhang, Y.; Long, H.; Li, Y.; Ge, D.; Tu, S. How does off-farm work affect chemical fertilizer application? Evidence from China's mountainous and plain areas. *Land Use Policy* **2020**, *99*, 104848. [CrossRef]

40. Bernstein, P.L. *Against the Gods: The Remarkable Story of Risk*; John Wiley & Sons: New York, NY, USA, 1996.
41. Aven, T. Risk assessment and risk management: Review of recent advances on their foundation. *Eur. J. Oper. Res.* **2016**, *1*, 1–13. [CrossRef]
42. Xu, L.; Liu, G. The study of a method of regional environmental risk assessment. *J. Environ. Manag.* **2009**, *11*, 3290–3296. [CrossRef] [PubMed]
43. McIntosh, A.; Pontius, J. *Science and the Global Environment: Case Studies for Integrating Science and the Global Environment*; Elsevier: Amsterdam, The Netherlands, 2016.
44. Cai, J.; Xia, X.; Chen, H.; Wang, T.; Zhang, H. Decomposition of fertilizer use intensity and its environmental risk in China's grain production process. *Sustainability* **2018**, *2*, 498. [CrossRef]
45. Fang, P.; Abler, D.; Lin, G.; Sher, A.; Quan, Q. Substituting organic fertilizer for chemical fertilizer: Evidence from apple growers in China. *Land* **2021**, *8*, 858. [CrossRef]
46. Liu, Y. Research on the urban-rural integration and rural revitalization in the new era in China. *Acta Geogr. Sin.* **2018**, *4*, 637–650.
47. Ricklefs, R.E. Community diversity: Relative roles of local and regional processes. *Science* **1987**, *4785*, 167–171. [CrossRef] [PubMed]
48. Wang, J.Y.; Wang, S.J.; Chen, Y. Leaching loss of nitrogen in double-rice-cropped paddy fields in China. *Acta Agric. Zhejiangensis* **1995**, *7*, 155–160.
49. Smith, L.E.D.; Siciliano, G. A comprehensive review of constraints to improved management of fertilizers in China and mitigation of diffuse water pollution from agriculture. *Agric. Ecosyst. Environ.* **2015**, *209*, 15–25. [CrossRef]
50. Arjjumend, H.; Koutouki, K.; Donets, O. Advantages of using the biofertilizers in Ukrainian agroecosystems. *Eur. J. Agric. Res.* **2020**, *2*, 92–123.
51. Chen, X. The core of China's rural revitalization: Exerting the functions of rural area. *China Agric. Econ. Rev.* **2019**, *1*, 1–13. [CrossRef]
52. Ma, W.; Abdulai, A.; Goetz, R. Agricultural cooperatives and investment in organic soil amendments and chemical fertilizer in China. *Am. J. Agric. Econ.* **2018**, *2*, 502–520. [CrossRef]
53. Luan, J.; Qiu, H.; Jing, Y.; Liao, S.; Han, W. Decomposition of factors contributed to the increase of China's chemical fertilizer use and projections for future fertilizer use in China. *J. Nat. Resour.* **2013**, *11*, 1869–1878.
54. Zhang, W.; Ji, Y.; Ma, J.; Wang, Y.; Ma, W.; Zhang, F. Driving forces of fertilizer consumption in China (Planting Structure). *Resour. Sci.* **2008**, *1*, 31–36.
55. Xin, L.; Li, X.; Tan, M. Temporal and regional variations of China's fertilizer consumption by crops during 1998–2008. *J. Geogr. Sci.* **2012**, *4*, 643–652. [CrossRef]
56. He, X.; Yin, Z. The small-scale peasant economy and path selection of agricultural modernization: A review of the agricultural modernization radicalism. *China Rev. Polit. Econ.* **2015**, *2*, 45–65.
57. Zhang, Y.; Long, H.; Wang, M.Y.; Li, Y.; Ma, L.; Chen, K.; Zheng, Y.; Jiang, T. The hidden mechanism of chemical fertilizer overuse in rural China. *Habitat Int.* **2020**, *102*, 102210. [CrossRef]
58. Lian, Y.; Liu, J.; Jin, S.; Liu, H.; Wu, S. Analysis of the reasons for overuse of chemical fertilizer from a seller's perspective. *Strateg. Stud. CAE* **2018**, *5*, 112–119. [CrossRef]
59. Guo, K. Theoretical rationales for China's economy entering the new normal: A political economic analysis of the socialism with Chinese characteristics. *Econ. Res. J.* **2016**, *9*, 5–16.
60. Jin, B. Study on the new normal of Chinese economic development. *China Ind. Econ.* **2015**, *1*, 5–18.
61. Hu, J. *Firmly March on the Path of Socialism with Chinese Characteristics and Strive to Complete the Building of a Moderately Prosperous Society in All Respects*; People's Publishing House: Beijing, China, 2012.
62. Zhao, F.; Yan, S.; Peng, H.; Yan, P.; Xiao, Z. Empirical research on the influence of Chinese rural financial reform on cultivation of new agricultural business entities. *J. Discret. Math. Sci. Cryptogr.* **2017**, *1*, 389–405. [CrossRef]
63. Wang, H.; Zhang, M.; Cai, Y. Problems, challenges, and strategic options of grain security in China. *Adv. Agric.* **2009**, *103*, 101–147.
64. Sheng, Y.; Song, L. Agricultural production and food consumption in China: A long-term projection. *China Econ. Rev.* **2019**, *53*, 15–29. [CrossRef]
65. Delgado, C.L. Rising consumption of meat and milk in developing countries has created a new food revolution. *J. Nutr.* **2003**, *11*, 3907–3910. [CrossRef] [PubMed]
66. Herrero, M.; Thornton, P.K.; Notenbaert, A.M.; Wood, S.; Msangi, S.; Freeman, H.A.; Bossio, D.; Dixon, J.; Peters, M.; van de Steeg, J.; et al. Smart investments in sustainable food production: Revisiting mixed crop-livestock systems. *Science* **2010**, *5967*, 822–825. [CrossRef] [PubMed]
67. Li, M.; Sicular, T. Aging of the labor force and technical efficiency in crop production: Evidence from Liaoning Province, China. *China Agric. Econ. Rev.* **2013**, *3*, 342–359. [CrossRef]
68. Heilig, G.K.; Fischer, G.; Van Velthuizen, H. Can China feed itself? An analysis of China's food prospects with special reference to water resources. *Int. J. Sustain. Dev. World* **2000**, *3*, 153–172. [CrossRef]
69. Xie, X.; Li, X.; He, W. A land space development zoning method based on resource–environmental carrying capacity: A case study of Henan, China. *Int. J. Environ. Res. Public Health* **2020**, *3*, 900. [CrossRef] [PubMed]
70. Tan, Y.; Chen, H.; Lian, K.; Yu, Z. Comprehensive evaluation of cultivated land quality at county scale: A case study of Shengzhou, Zhejiang Province, China. *Int. J. Environ. Res. Public Health* **2020**, *4*, 1169. [CrossRef] [PubMed]

71. Fearnside, P.M.; Pueyo, S. Greenhouse-gas emissions from tropical dams. *Nat. Clim. Chang.* **2012**, *6*, 382–384. [CrossRef]
72. Fajer, E.D.; Bazzaz, F.A. Is carbon dioxide a 'good'greenhouse gas?: Effects of increasing carbon dioxide on ecological systems. *Glob. Environ. Chang.* **1992**, *4*, 301–310. [CrossRef]
73. Rastogi, M.; Singh, S.; Pathak, H. Emission of carbon dioxide from soil. *Curr. Sci.* **2002**, *5*, 510–517.
74. Liu, H.; Li, J.; Li, X.; Zheng, Y.; Feng, S.; Jiang, G. Mitigating greenhouse gas emissions through replacement of chemical fertilizer with organic manure in a temperate farmland. *Sci. Bull.* **2015**, *6*, 598–606. [CrossRef]
75. Sheng, Y.; Tian, X.; Qiao, W.; Peng, C. Measuring agricultural total factor productivity in China: Pattern and drivers over the period of 1978-2016. *Aust. J. Agric. Resour. Econ.* **2020**, *1*, 82–103. [CrossRef]

International Journal of
Environmental Research and Public Health

Article

Does Coal Mining Have Effects on Land Use Changes in a Coal Resource-Based City? Evidence from Huaibei City on the North China Plain

Jing Guan [1] and Peng Yu [2,*]

[1] School of Geography and Tourism, Anhui Normal University, Wuhu 241002, China; guanjing1010@163.com
[2] School of Tourism and Cuisine, Yangzhou University, Yangzhou 225127, China
* Correspondence: pengyu@yzu.edu.cn

Abstract: Continuous coal mining results in dramatic regional land use change, and significantly influences the sustainable development of coal resource-based cities. Present studies pay little attention to the characteristics and regularities of land use change in coal resource-based cities, caused by underground coal mining in high groundwater areas. Based on the Landsat remote sensing images of 1999, 2000, 2010, and 2018 of Huaibei City, a typical coal resource-based city of a high ground water area on the North China Plain, this paper applies the dynamic degree and transition matrix of land use to analyze the land use change characteristics, and identify the regularity between land use type and coal mining production in this coal resource-based city. Results show that the land use change in the research area presents an overall characteristic of a constant increase in water area, urban construction land, and rural settlement land, and a continuous decrease in cultivated land. Cultivated land is converted into a water area, urban construction land, and rural settlement land, and rural settlement land and cultivated land are converted bidirectionally. The land use change in this coal resource-based city demonstrates significant reliance on coal resources, and coal mining is significantly related to the area of cultivated land, water area, and rural settlement land, which demonstrates that continuous large-scale coal mining results in damage to cultivated land, a decrease in rural settlement land, and an increase in water area. The research result contributes to the sustainable land use of coal resource-based cities.

Keywords: coal mining; high ground water area; land use change; resource-based city

Citation: Guan, J.; Yu, P. Does Coal Mining Have Effects on Land Use Changes in a Coal Resource-Based City? Evidence from Huaibei City on the North China Plain. *Int. J. Environ. Res. Public Health* **2021**, *18*, 11616. https://doi.org/10.3390/ijerph182111616

Academic Editors: Wei Song and Hualin Xie

Received: 18 September 2021
Accepted: 2 November 2021
Published: 4 November 2021

1. Introduction

Land has always been a crucial resource for human existence and development, and acts as the base and carrier that support human activities [1–3]. Land use change is a long-term activity that is based on certain social and economic aims. As an important research area of global change science and sustainable science, it is the most direct representation of terrestrial ecosystem change and the influence of human activities on the earth's surface system [4–6]. The measurement of land use change is a significant part of land resource management and observation. However, most studies often focus on the analysis of land use change on large scales, such as global, region, country, basin, island, or peninsula scales, while less attention is given to land use change on small scales in city or rural areas. Land use change is mainly caused by the complex interaction of human activities and ecological and social factors [7,8]. Present studies mainly probe the influence of macro human activities, such as urbanization and industrialization on land use change [9], while showing less interest in the micro driving force of regional land use change. Due to the difference between different types and degrees of human activities, as well as the heterogeneity of the earth's surface, land use changes in various areas also present different characteristics and regularities. In this regard, it is important to explore the dynamic change regularities of typical regional land use to promote sustainable regional land use, to further achieve the United Nations (UN) Sustainable Development Goals.

With the fast development of society and the economy, there is a growing demand for coal resources. Large-scale, high strength, and long-term coal mining inescapably affects the regional land use type change [10,11]. A coal resource-based city is one that develops based on exploring, processing, and producing coal resources, and mainly functions as the coal resource supplier. The activities of coal mining are critical driving factors of land use change in coal resource-based cities. However, the differentiation of coal resources determines various coal mining types, including open-cast coal mining and underground coal mining, which further contribute to different regional land use change characteristics and regularities [12–14]. The present research pays more attention to the land use change caused by the way of open-cast coal mining, instead of underground coal mining. Furthermore, there are few researchers studying the land use change characteristics in coal resource-based cities caused by coal mining [15–17]. Therefore, it is of crucial importance to carry out studies on land use spatial change paradigms in coal resource-based cities, and analyze the relationship between coal mining and land use type, further disclosing the interaction mechanism of regional land use change and coal mining.

However, there are still some limitations in the relative studies on land use change in coal resource-based cities. First, present studies tend to probe the transformation of the regional terrestrial ecological environment to a water ecological environment caused by underground coal mining in specific coal areas, instead of analyzing the land use change resulting from coal mining in populated, developed, high ground water areas on the plain from the perspective of the city. Second, land subsidence caused by coal mining is a long-term and complex evolution. Present studies fail to present the new characteristics of land use change after the transformation development in coal resource-based cities in today's era, which is facing structural transformation from energy resource use, obstacles of ecological environment, and constraints of traditional industry development. Therefore, the relevant remote sensing image data must also be updated. Lastly, the externality of coal mining greatly affects the land use change in coal resource-based cities. Moreover, the correlation analysis of land use type change and total coal production in coal resource-based cities is still unclear, and the relationship needs to be further illustrated.

Compared with the energy structures of other developed countries, the resource characteristics—coal-rich, oil- and gas-poor—contributes to China's long-term energy structure, which prioritizes coal. China is the largest coal-consuming country in the world, and approximately 90% of its total coal production comes from underground coal mining. Mining subsidence is the serious problem in the mining process [18]. There are great compositive areas of underground coal resources, cultivated land, and urban construction land and rural settlement land in high ground water areas of eastern China [19]. The subsidence caused by accumulated underground coal mining results in large-scale water-logged and inundated land in coal areas of the China Plains, which further leads to the decline in cultivated land and sharp conflicts between population, land, and agriculture [16]. In this regard, this paper selects Huaibei City—a coal resource-based city on the North China Plain—as an example to explore the characteristics and influencing factors of land use change, and attempts to answer the following questions: (1) is there any regularity in the mutual transformation of land use types in the coal resource-based city in high ground water areas on the plain with long-term scale? (2) What is the relationship between the amount of coal mining and land use types?

2. Research Method and Data Collection

2.1. Background of Research Area

The North China Plain is low and flat, with the majority of its land under 50 m above sea level, and is the most populous plain, making up approximately 24.2% of the total population in China. It is also the major grain-producing area, and an important high ground water area in China. As a typical mine–grain mixed zone, the North China Plain produces both coal and grain [20]. Huaibei City, a city of Anhui Province, is located in the North China Plain (Figure 1).

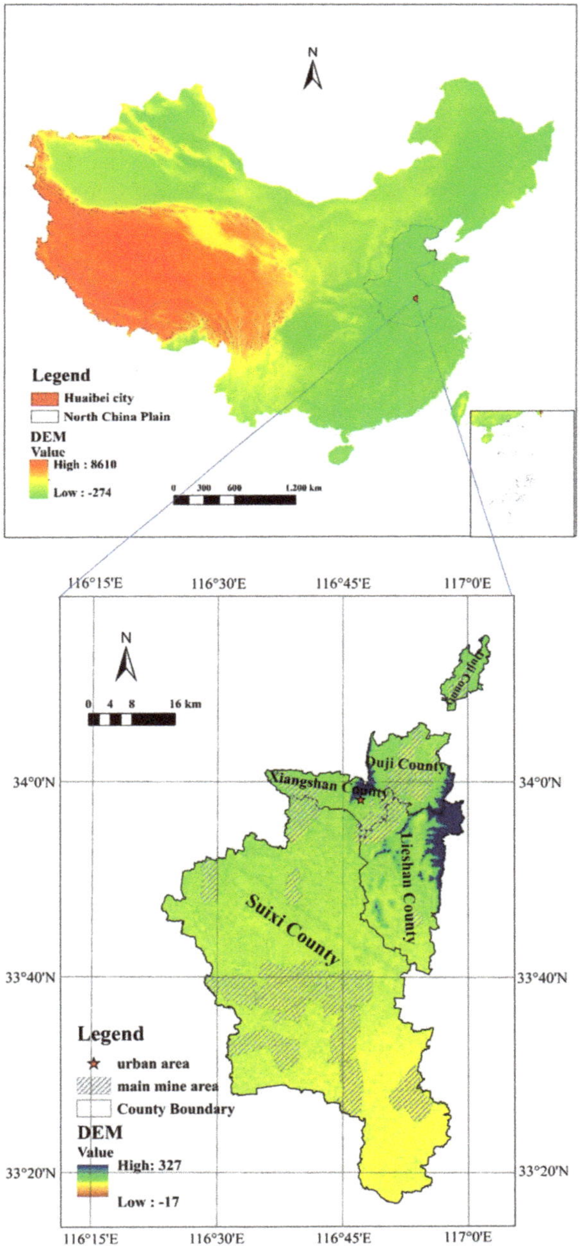

Figure 1. Huaibei's geographical location in China (note: the figure is drawn by the author).

Huaibei City, with a total land area of 2741 km^2, was founded in 1957 for coal mining. It is a typical coal resource-based city that governs three districts and one county, namely Xiangshan district, Duji district, Lieshan district, and Suixi county. It is a city with a dense population and a well-developed economy, whose permanent population density is 792 per square kilometers; the permanent population urbanization rate reached 65.9%, and the industrialization rate was 39% in 2019. At present, there are 23 large coal mines in Huaibei City, with an average annual total coal production of 33.97 million tons from 1990 to 2018 (Figure 2). Moreover, the largest annual coal production in the city occurred in 2012; since then, coal production has fallen markedly. On the one hand, China's economic growth has slowed since 2012, which has led to a decline in the demand for coal production. On the other hand, Huaibei City was listed as a resource-exhausted city by the state in 2009, and the amount of coal to be mined was insufficient (approximately 80% of the mines are in the coal-resource depletion phase). Under the influence of these two factors, the annual coal production in Huaibei City declined after 2012, and this trend will exist for a certain period of time.

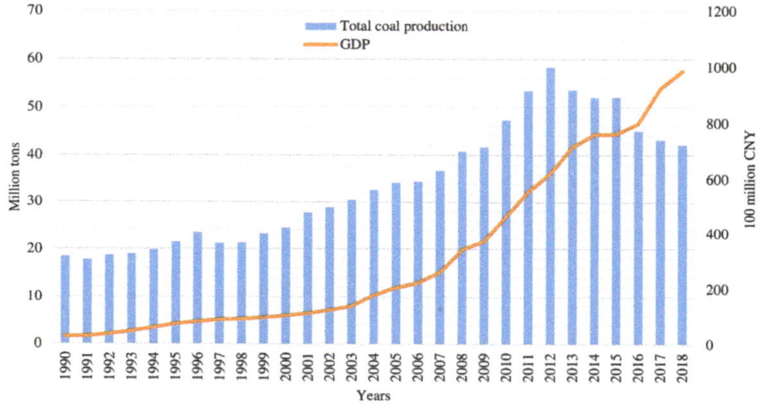

Figure 2. The total coal production and GDP in Huaibei City from 1990 to 2018.

Huaibei City's underground water level is 2–3 m, the annual average temperature is 14.5 °C, and the annual average amount of precipitation is 862.9 mm. As a coal–grain mixed zone in high ground water areas, land use in Huaibei prioritizes agriculture before coal mining development. With the development of coal mining, the coal industry has gradually become the pillar industry in the economic development of Huaibei since the 1990s. After 2010, the city entered a transition period, and the proportion of coal industry output value gradually decreased (Figure 3). With constant large-scale coal mining, the cultivated land is in the process of dynamic subsidence, which further results in waterlogging and becomes a water body, increasing the proportion of affected cultivated land. In the future, coal resource mining activities will continue, and their impact on land use will continue to exist.

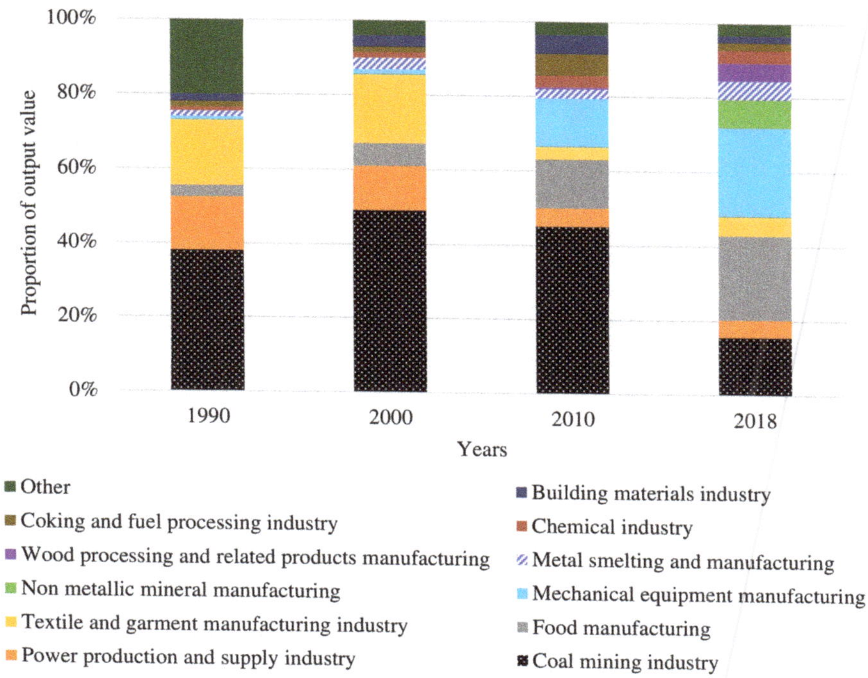

Figure 3. Structure of output value of major industrial sectors in Huaibei City.

2.2. Research Method

The data analysis methods in this article include land use dynamic degree (in order to describe the rate of change of a certain land use type in a region) and land use intensity (in order to describe the degree of change of a certain land use type during the study period). The specific formulas are as follows:

$$K = \left[\frac{U_b - U_a}{U_a}\right] \times \frac{1}{T} \times 100\% \quad (1)$$

where K indicates the dynamic degree of a certain land use type during the study period, U_a indicates the area of a certain land use type at the beginning of the study period, U_b indicates the area of a certain land use type at the end of the study period, and T indicates the length of study period.

$$S_i = \frac{U_i}{(U \times T)} \quad (2)$$

where S_i indicates the land use intensity of a certain land use type during the study period, U_i indicates the absolute change of the i-land type within the study period, U indicates the absolute change of all land types in the study area during the study period, and T indicates the length of study period.

2.3. Data Collection

The research data are mainly from the Landsat remote sensing image of Huaibei City in 1990, 2000, 2010, and 2018, with a spatial resolution of 30 × 30 m. The choice of research period mainly considers the availability of data, and the phased characteristics of Huaibei City's development. Although annual coal production began declining in 2012, the city has entered a transitional stage of development since 2010. Therefore, it is reasonable to select the year 2010 in this research. This paper applies the methods of geometric correction and image enhancement to process the remote sensing image, and adopts the human–machine interactive visual interpretation to extract the land use data of Huaibei. Referring to the classification of the International Geosphere–Biosphere Programme (IGBP), this paper divides the land use types in Huaibei City into six types, including cultivated land, forest land, grass land, water area, urban construction land, and rural settlement land, based on the actual situation of the case. In addition, the socio-economic data of this paper are primarily from the Huaibei City Statistical Yearbook [21].

3. Results

3.1. Stage Division of Land Use Change

Based on the analysis of total coal production, different land use type change and socio-economic development background in research area over the years (Figure 4), this research proposes that the land use change is generally divided into three stages, including low-speed dispersion development stage, medium-speed aggregation development stage, and high-speed equilibrium development stage. (1) During the low-speed dispersion development stage (from 1990 to 2000), the speed of land use change was low, and the development in coal resource-based cities mainly focused on the coal mining industry. Under the mainly influence of China's planned economic system, Huaibei City had a single industrial structure (Figure 3), and the extent of the city's economic development was low. The mining scope of this stage is small, and the influence of mining activities on the overall pattern of land use is not significant. (2) In the medium-speed aggregation development stage (from 2000 to 2010), the overall speed of land use change was faster. In the context of China's market economic system, the development of the coal industry and non-coal industry is accelerating in the city. Coal production increased from 24.59 million tons in 2000 to 47.35 million tons in 2010, and industrial output increased from CNY 3.84 billion in 2000 to CNY 27.37 billion [21]. The development of the industry has led to the expansion of the city's economic scope and population concentration. The urbanization process in Huaibei has accelerated, with the number of urban residents increasing from 0.73 million in 2000 to 1.161 million in 2010, which further promoted the rapid transformation of land types, such as cultivated land into urban construction land. (3) During the high-speed equilibrium development stage (from 2010 to 2018), the speed of land use change accelerated significantly. The total coal production first increased and then decreased, coal-related industries gradually declined, and the effect of coal resources on land use change was also weakened due to coal resource reserve shortages. However, under the influence of the central government support policy (by 2018, Huaibei City has received a total of approximately CNY 5.4 billion in financial transfer funds from the state), the city has carried out transformation and upgrading, the speed of urban construction has accelerated, and changes of the various types of land use in the city have been drastic.

3.2. Rate of Change of Different Land Use Types

According to dynamic degree and change intensity index models, this paper calculates the dynamic degree and change intensity index of each type of land use in all the stages in Huaibei City (Table 1). The results demonstrate that the overall land use structure in the city has dramatically changed since 1990, especially the decrease in cultivated land, and the increase in urban construction land and rural settlement land and water area.

The leading role of cultivated land was not significantly changed, although its area plummeted to 177.82 km^2 from 1990 to 2018. The changing intensity in different stages was significantly higher than other land use types. The dynamic degree of cultivated land improved from −0.07% at the low-speed dispersion development stage to −0.61% at the medium-speed aggregation development stage, and presented a constant accelerating decrease and decreasing pace characteristic. On the other hand, the urban construction land and rural settlement land continuously increased and showed an acceleration tendency, especially at the high-speed equilibrium development stage with a significant expansion area of 108.65 km^2—this was also observed for water land. Due to the promotion of coal mining development in Huaibei City, the changing intensity of water area has been continuously increasing since 2000. Both forest land and grass land presented the tendency of first increasing, and then declining, even though they were from comparatively minor bases.

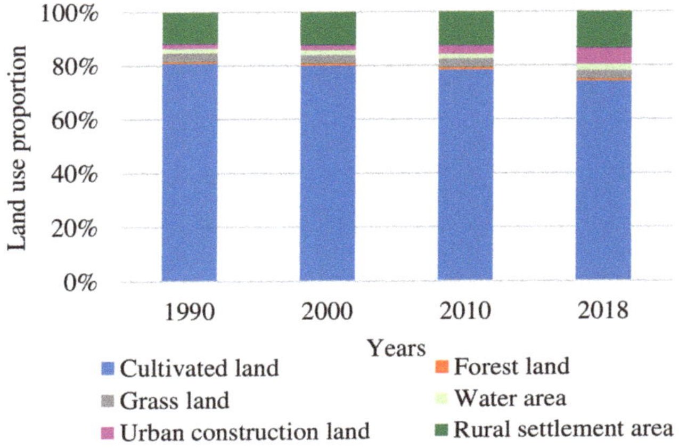

Figure 4. The total coal production and land use type in Huaibei City.

Table 1. Land use change rate and land use intensity at different stages.

Land Use Type	Low-Speed Dispersion Development Stage (1990–2000)			Medium-Speed Aggregation Development Stage (2000–2010)			High-Speed Equilibrium Development Stage (2010–2018)		
	Variation (km^2)	Land Use Dynamic Degree (%)	Land Use Intensity (%)	Variation (km^2)	Land Use Dynamic Degree (%)	Land Use Intensity (%)	Variation (km^2)	Land Use Dynamic Degree (%)	Land Use Intensity (%)
Cultivated land	−18.95	−0.07	4.94	−46.24	−0.20	4.99	−115.63	−0.61	6.05
Forest land	0.77	0.38	0.20	2.02	0.95	0.22	−1.51	−0.65	0.08
Grass land	0.07	0.01	0.02	0.01	-	0.00	−2.52	−0.29	0.13
Water area	1.60	0.38	0.41	3.77	0.7	0.42	10.47	2.14	0.55
Urban construction land	3.02	0.62	0.79	30.12	5.83	3.25	85.16	10.41	4.46
Rural settlement land	13.98	0.44	3.64	10.45	0.34	1.13	23.49	0.68	1.23

3.3. Dynamic Conversion between Different Land Use Types

To better reflect the land use types conversion, this paper probes the transition matrix (Table 2), and spatial distribution of land use change at different stages (Figure 5). Moreover, the table and figure only list the land use types that go through great change.

Based on the data analysis, this paper concludes that coal mining results in an increase in water land, which is mainly converted from cultivated land and rural settlement land. There was approximately 6.90 km^2, 9.57 km^2, and 28.86 km^2 of cultivated land and rural settlement land being converted into water area at the low-speed dispersion development stage, medium-speed aggregation development stage, and high-speed equilibrium development stage, respectively. This paper illustrates the stacking chart of water area change and major mine distribution to explore the relationship between the increase in water area and coal mining, and concludes that the distribution of increased water area is basically in line with major mine distribution, which further discloses that the increase in water area is closely related to the coal mining.

In addition, the increase in urban construction land and rural settlement land comes at the expense of a decrease in cultivated land. It is estimated that almost 199.04 km^2 of cultivated land was converted into construction and settlement land in the past 29 years, despite the fact that there were 15.00 km^2, 13.29 km^2, and 28.17 km^2 of other land use types being converted into cultivated land at the low-speed dispersion development stage, medium-speed aggregation development stage, and high-speed equilibrium development stage, respectively, which could have largely contributed to the land reclamation in the coal resource-based city. Huaibei City has implemented land reclamation since 1985, and has initiated many reclamation patterns such as deep digging and filling shallow.

Table 2. Land use transition matrix at different stages.

Research Period	Land Use Type	Cultivated Land (km^2)	Water Area (km^2)	Urban Construction Land (km^2)	Rural Settlement Land (km^2)
Low-speed dispersion development stage (1990–2000)	Cultivated land	2179.34	5.86	2.74	24.35
	Water area	4.22	44.97	0.04	1.08
	Urban construction land	0.38	0.04	48.17	0.01
	Rural settlement land	10.40	1.04	0.02	312.28
Medium-speed aggregation development stage (2000–2010)	Cultivated land	2153.52	6.57	28.97	21.99
	Water area	3.14	45.75	0.65	2.01
	Urban construction land	0.03	0.00	51.64	0.01
	Rural settlement land	10.12	3.00	0.43	336.68
High-speed equilibrium development stage (2010–2018)	Cultivated land	2020.84	22.81	62.61	58.38
	Water area	6.32	35.73	8.35	4.46
	Urban construction land	0.21	0.01	79.53	2.04
	Rural settlement land	21.64	6.05	9.24	332.80

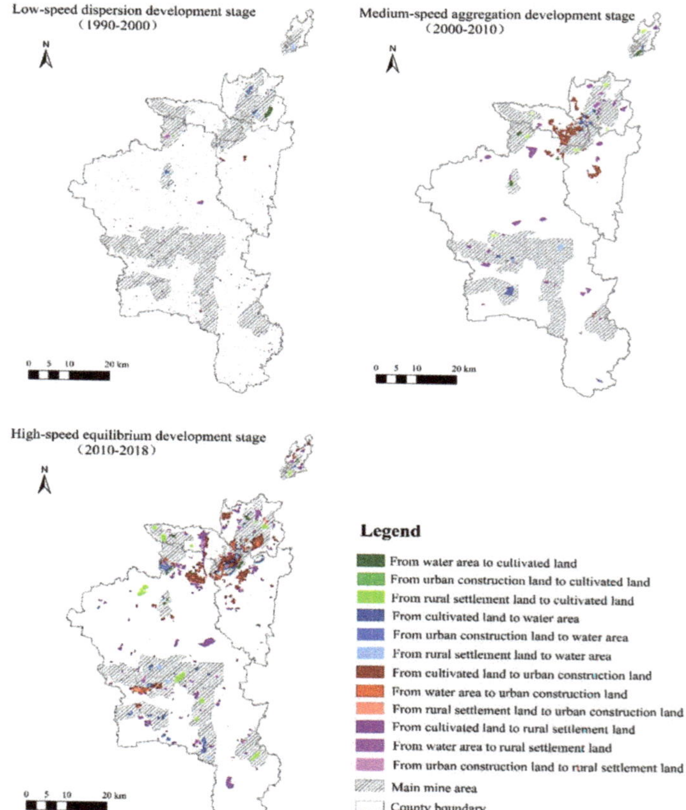

Figure 5. Land use type conversion spatial distribution map of Huaibei City (note: the figure is drawn by the author).

3.4. The Effect of Coal Mining on Land Use Types Change

Coal mining is the most prominent driving force of land use change in resource-based cities. Different from other city types, the development of coal resource-based cities presents significant resource-related reliance and orientation, and coal mining causes changes to the region of different land use types. Therefore, this paper analyzes the relationship between total coal production and cultivated land, forest land, grass land, water area, urban construction land, and rural settlement land in 1990, 1995, 2000, 2005, 2010, 2015, and 2018, respectively. This paper concludes that cultivated land, water area, and rural settlement land correlate significantly with total coal production (Figures 6–8). There is a significant positive correlation between water area and total coal production ($r = 0.791$, $p = 0.034$), rural settlement land, and total coal production ($r = 0.863$, $p = 0.012$), and a significant negative correlation between cultivated land and total coal production ($r = -0.773$, $p = 0.042$). However, there is no significant correlation between urban construction land and total coal production ($r = 0.712$, $p = 0.072$), forest land and total coal production ($r = 0.744$, $p = 0.055$), and grass land and total coal production ($r = -0.673$, $p = 0.097$). These results demonstrate that constant coal mining and the accumulation of total coal production contribute to the continuous expansion of water area and rural settlement land, as well as a constant decrease in cultivated land in the research area.

Subsidence is the most significant externality caused by underground coal mining in high ground water areas on a plain. Constant underground coal mining and eventual mining result in the failing of ground support, leading to surface land subsidence. Due to the difference between deep subsidence and land nature, subsidence causes different land changes from seasonal and perennial waterlogging to subtle land deformation. Seasonal and perennial water logging converts cultivated land, forest land, and construction or settlement land into wetland, which completely changes the land use type and increases the area of water area in the research area directly. Mining subsidence further influences the research area—water area, rural settlement land, and cultivated land. In this regard, the former three land use types are significantly correlated with coal mining. There is no significant correlation between urban construction land and coal mining, due to the following two aspects. First, the construction of urban construction land is generally based on the planning of the government, instead of the layout of a coal mine. Second, urban land use change is a result of multiple factors such as urbanization and government behavior, rather than the single effect of the coal mining industry. Furthermore, forest land and grass land are not significantly correlated with total coal production, mainly due to the scarcity of research area of these two land use types, thus there is no stable change regularity.

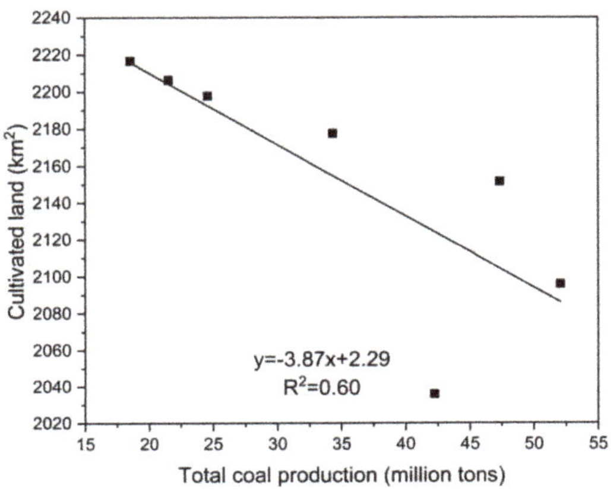

Figure 6. The relationship between total coal production and cultivated land.

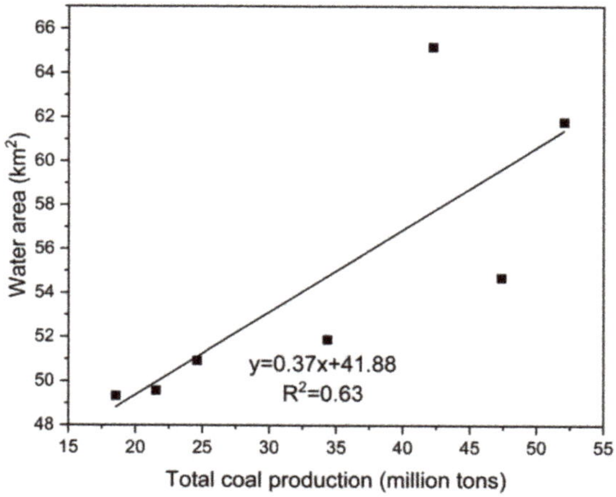

Figure 7. The relationship between total coal production and water land.

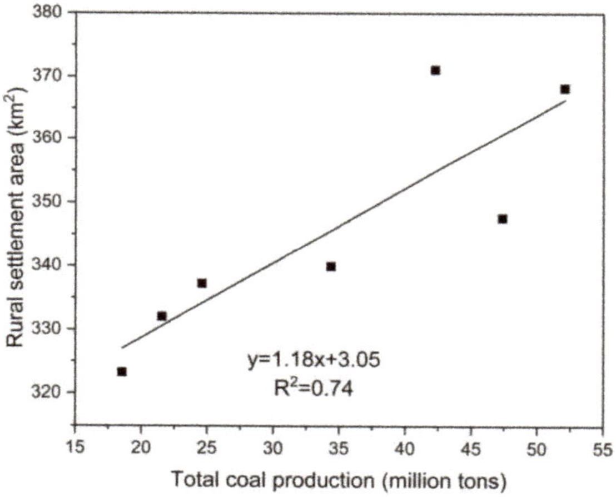

Figure 8. The relationship between total coal production and rural settlement land.

4. Discussion and Conclusion

4.1. Discussion

4.1.1. Land Use Changes in Coal Resource-Based Cities

Underground coal mining exerts a far-reaching influence on land use change in coal resource-based cities in high ground water areas on a plain. China's energy structure, dominated by coal, will not change in the short term. Moreover, the influence of coal mining on land use continues in many other coal-producing countries [22]. There are three major aspects of land use change that are affected by coal mining in high ground water areas of coal-producing countries.

The first is the worsening decline of cultivated land. Coal mining not only causes the subsidence of cultivated land and the loss of soil nutrients, which directly causes the reduction of cultivated land area and the decrease of soil quality in the research area, but also increases the content of heavy metal pollutants in water sources, which leads to the reduction of crop yield. The destruction of high-quality cultivated land caused by coal mining threatens regional, national, and even global, food security [23,24]. There were more than 0.268 million hectares of high-quality cultivated land damaged due to coal mining in the research area by the year 2018, with an urban per capita cultivated land area of 0.062 hectares, slightly above the minimum size of 0.053 hectares recommended by the United Nations Food and Agriculture Organization (UN-FAO).

The second is the expansion of water area with the development of coal mining. The formation of subsidence water land and its expansion alters the city water structure, and influences the city's ecosystem [25]. The terrain in the research area was previously dominated by plains and had few landscape types, such as wetland and lakes, before coal mining. However, the large-scale subsidence of water land—resulting from coal mining—was converted into a wetland landscape with collapsed lakes, which increased the city's water area.

The third is the destruction of rural settlement land. Mining subsidence forces a large number of villagers to move out. Since 1990, there have been 275 villages moving due to mining subsidence—more than 0.3 million people involved in the research area. Village moving in coal mining areas reconstructs the living space and converts the lifestyle of rural residents [26]. This new movement pattern leads to increased construction land that is smaller than the released land after moving, which contributes to the rural settlement land conservation, mitigates rural-urban land use conflict, and optimizes the urban land use structure. In 2009–2020, the relocation of villages in coal mining collapse areas could save 1275.55 hm^2 of land in the research area.

In addition, this paper discloses that there is no significant correlation between coal mining and urban construction land change, which is different from previous scholars, who hold that urban construction land in resource-based cities is significantly correlated with coal mining [27,28]. This is mainly due to the co-effectiveness of economic development and government regulation on urban construction land in coal resource-based cities in China. Influenced by the city's development planning, as well as the fiscal reliance on land, local government activates the real estate industry (real estate investment in Huaibei increased nearly 50-fold from 264 million in 2000 to 13.169 billion in 2018), which results in uncontrolled expansion and increased speed of construction land in the city [29]. This is highlighted in the fast expansion of urban construction land, even though the coal production decreased and economic growth slowed down (Huaibei's economic growth rate fell to 6.62% from 24.14%) in the city's transition development period.

4.1.2. Impact on Coal Resource-Based Cities Development Planning

The complex land use change in coal resource-based cities exerts a profound influence on the social economy, such as construction land destruction, cultivated land degeneration, and moving villages in coal mining areas. Coal resource-based cities need to coordinate the conflicts between mining, land reclamation, cultivated land protection, and city development before land exploitation and land use.

First, a plan for reclamation in advance and safeguard food security should be constructed. Despite the results that the present cultivated land protection policy has achieved, the degeneration of high-quality farmland caused by mining subsidence greatly threatens local food security [30]. Accelerating land reclamation is a comparatively effective way to recover the losses. Land reclamation in collapsed areas is of great practical significance for restoring cultivated land, repairing the ecological environment, and alleviating land use contradictions. Land reclamation in China was developed in practice in the early 1980s. The state has made compulsory provisions for land reclamation and ecological restoration in abandoned industrial and mining areas, and has promulgated relevant laws and regulations such as the Provisions on Land Reclamation (1988) and the Regulations on Land Reclamation (2011). However, traditional land reclamation carried out after land subsidence is lengthy, and has comparatively low efficiency. In this regard, it is important to consider production and land reclamation, integrate mining and reclamation, and develop overall planning of mining and reclamation, to achieve the goal of cultivated land protection before complete subsidence, reduce the reclamation time of disturbed land, and promote the land reclamation efficiency.

Second, reasonable city planning can improve land use efficiency. China is in the stage of rapid urbanization, and the conflicts between development, coal mining, ecological optimization, and cultivated land protection need to be highlighted in city planning. It is of crucial importance to predict the land use type after coal mining accurately, and grasp the rules of land use to formulate city planning in coal resource-based cities [31]. According to the statistics, more than one-third of underground land is coal-bearing land in the research area. The distribution of villages is generally in line with the distribution of mines, which results in the destruction of local residents' living environment. Moreover, forced evacuation causes land resource waste. Therefore, it is significant to develop coordinated planning of land reclamation and village moving in coal mining areas, and optimize the regional land use efficiency.

Third, water areas should be used rationally, and the ecological environment optimized. Mining subsidence enables single terrestrial landscapes to transform into terrestrial–wetland landscapes, which enhances the diversity of land use type, and enhances the anti-jamming capacity of the ecosystem [32]. In this regard, formulating reasonable planning and utilizing subsidence water land contributes to the city's ecological environment protection. In addition, the government should integrate ecological management concepts into all parts of land planning, including design, formulation, implementation, and feedback.

4.2. Limitations and Recommendations for Further Study

This paper has achieved certain progress in exploring the influence of coal mining in high ground water areas on land use change in a coal resource-based city, while failing to undertake further research on the following three topics. First, this study does not establish separate classification standards of land use type for coal mining and other uses, which might help to accurately measure the influence of coal mining on land use. Second, the present study does not probe the influence of land use change on ecosystem service in a coal resource-based city [33,34]. Third, this paper does not explore the coupling relationship between land use change and social economy in a coal resource-based city, especially the influence of coal mining on residents' livelihood transitions. It is also worth considering the costs related to industrial activity—mining damages and reclamation versus profits related to mining in future research.

4.3. Conclusions

Based on the adoption of RS and GIS, this paper takes Huaibei City on the North China Plain as a case to analyze the influence of coal mining in a high ground water area on land use change in a coal resource-based city, and examines the relationship between coal mining production and the dynamic change of land use type in Huaibei City, which provides support for decisions of regional sustainable land use and socio-economic development.

The significant expansion of urban construction land and water area, and constant decrease in cultivated land are major characteristics of land use change in coal resource-based cities. There are frequent conversions between different land use types, mainly dominated by cultivated land converted into construction and settlement land. Compared with other type cities, land use structure change is faster and more complex in coal resource-based cities, due to mining subsidence. The contradiction between land supply and demand becomes even more acute, and severely threatens the sustainable development of cities.

There is an apparent periodical characteristic in the land use of coal resource-based cities. The land use change is generally divided into three stages, including (1) the low-speed dispersion development stage, (2) the medium-speed aggregation development stage, and (3) the high-speed equilibrium development stage. The land use change in coal resource-based cities demonstrates significant reliance on resources, and coal mining is significantly related to the area of cultivated land, water area, and rural settlement land, which demonstrates that continuous large-scale coal mining results in damage to cultivated land, decrease in rural settlement land, and increase in water area.

Author Contributions: Conceptualization, J.G.; Formal analysis, J.G.; Funding acquisition, J.G.; Supervision, P.Y.; Writing—original draft, J.G.; Writing—review & editing, P.Y. All authors have read and agreed to the published version of the manuscript.

Funding: This research is supported by the National Natural Science Foundation of China (Grant No. 41171144).

Institutional Review Board Statement: Not applicable.

Informed Consent Statement: Not applicable.

Data Availability Statement: Not applicable.

Acknowledgments: We want to express our gratitude to Huafu Jiao for providing valuable comments and suggestions to improve the paper.

Conflicts of Interest: The authors declare no conflict of interest.

References

1. Foley, J.A.; DeFries, R.; Asner, G.P.; Barford, C.; Bonan, G.; Carpenter, S.R.; Chapin, F.S.; Coe, M.T.; Daily, G.C.; Gibbs, H.K.; et al. Global consequences of land use. *Science* **2005**, *309*, 570–574. [CrossRef] [PubMed]
2. Petkova, V.; Lockie, S.; Rolfe, J.; Ivanova, G. Mining developments and social impacts on communities: Bowen Basin case studies. *Rural. Soc.* **2009**, *19*, 211–228. [CrossRef]
3. Salazar, A.; Baldi, G.; Hirota, M.; Syktus, J.; McAlpine, C. Land use and land cover change impacts on the regional climate of non-Amazonian South America: A review. *Glob. Planet. Chang.* **2015**, *128*, 103–119. [CrossRef]
4. Lawler, J.J.; Lewis, D.J.; Nelson, E.; Plantinga, A.J.; Polasky, S.; Withey, J.C.; Helmers, D.P.; Martinuzzi, S.; Pennington, D.; Radeloff, V.C. Projected land-use change impacts on ecosystem services in the United States. *Proc. Natl. Acad. Sci. USA* **2014**, *111*, 7492–7497. [CrossRef] [PubMed]
5. Mooney, H.A.; Duraiappah, A.; Larigauderie, A. Evolution of natural and social science interactions in global change research programs. *Proc. Natl. Acad. Sci. USA* **2013**, *110*, 3665–3672. [CrossRef] [PubMed]
6. Wulder, M.A.; White, J.C.; Goward, S.N.; Masek, J.G.; Irons, J.R.; Herold, M.; Cohen, W.B.; Loveland, T.R.; Woodcock, C.E. Landsat continuity: Issues and opportunities for land cover monitoring. *Remote. Sens. Environ.* **2008**, *112*, 955–969. [CrossRef]
7. Robson, J.P.; Berkes, F. Exploring some of the myths of land use change: Can rural to urban migration drive declines in biodiversity? *Glob. Environ. Chang.* **2011**, *21*, 844–854. [CrossRef]
8. Turner, B.L.; Lambin, E.F.; Reenberg, A. The emergence of land change science for global environmental change and sustainability. *Proc. Natl. Acad. Sci. USA* **2007**, *104*, 20666–20671. [CrossRef]
9. Tiando, D.S.; Hu, S.-G.; Fan, X.; Ali, M.R. Tropical coastal land-use and land cover changes impact on ecosystem service value during rapid urbanization of Benin, West Africa. *Int. J. Environ. Res. Public Health* **2021**, *18*, 7416. [CrossRef] [PubMed]
10. Areendran, G.; Rao, P.; Raj, K.; Mazumdar, S.; Puri, K. Land use/land cover change dynamics analysis in mining areas of Singrauli district in Madhya Pradesh, India. *Trop. Ecol.* **2013**, *54*, 239–250.
11. Bian, Z.-F.; Lu, Q.-Q. Ecological effects analysis of land use change in coal mining area based on ecosystem service valuing: A case study in Jiawang. *Environ. Earth Sci.* **2013**, *68*, 1619–1630. [CrossRef]
12. Karan, S.K.; Samadder, S.R. Improving accuracy of long-term land-use change in coal mining areas using wavelets and Support Vector Machines. *Int. J. Remote Sens.* **2018**, *39*, 84–100. [CrossRef]
13. Larondelle, N.; Haase, D. Valuing post-mining landscapes using an ecosystem services approach—An example from Germany. *Ecol. Indic.* **2012**, *18*, 567–574. [CrossRef]
14. Mi, J.-X.; Yang, Y.-J.; Zhang, S.-L.; An, S.; Hou, H.-P.; Hua, Y.-F.; Chen, F.-Y. Tracking the land use/land cover change in an area with underground mining and reforestation via continuous Landsat classification. *Remote Sens.* **2019**, *11*, 1719. [CrossRef]
15. Miller, A.J.; Zégre, N. Landscape-scale disturbance: Insights into the complexity of catchment hydrology in the mountaintop removal mining region of the eastern United States. *Land* **2016**, *5*, 22. [CrossRef]
16. Hu, Z.-Q.; Xiao, W. Optimization of concurrent mining and reclamation plans for single coal seam: A case study in northern Anhui, China. *Environ. Earth Sci.* **2013**, *68*, 1247–1254. [CrossRef]
17. Damigos, D. An overview of environmental valuation methods for the mining industry. *J. Clean. Prod.* **2006**, *14*, 234–247. [CrossRef]
18. Xiao, W.; Hu, Z.-Q.; Chugh, Y.P.; Zhao, Y.-L. Dynamic subsidence simulation and topsoil removal strategy in high groundwater table and underground coal mining area: A case study in Shandong Province. *Int. J. Min. Reclam. Environ.* **2014**, *28*, 250–263. [CrossRef]
19. Hu, Z.-Q.; Yang, G.-H.; Xiao, W.; Li, J.; Yang, Y.-Q.; Yu, Y. Farmland damage and its impact on the overlapped areas of cropland and coal resources in the eastern plains of China. *Resour. Conserv. Recycl.* **2014**, *86*, 1–8. [CrossRef]

20. Xiao, W.; Hu, Z.-Q.; Fu, Y.-H. Zoning of land reclamation in coal mining area and new progresses for the past 10 years. *Int. J. Coal Sci. Technol.* **2014**, *1*, 177–183. [CrossRef]
21. Huaibei Bureau of Statistics. Huaibei Statistical Yearbook (1990–2018). Anhui Kuaima Printing Co.,Ltd.: Hefei, China. (In Chinese)
22. Wang, W.-D.; Li, S.-J.; Han, J.-X. Analysis of the main global coal resource countries' supply-demand structural trend and coal industry outlook. *China Min. Mag.* **2015**, *24*, 5–9.
23. Hu, Z.-Q.; Luo, Y.-M. Suggestions on environmental quality and food safety in overlapped areas of crop and mineral production. *Sci. Technol. Rev.* **2006**, *3*, 93–94.
24. Xiao, W.; Fu, Y.-H.; Wang, T.; Lv, X.-J. Effects of land use transitions due to underground coal mining on ecosystem services in high groundwater table areas: A case study in the Yanzhou coalfield. *Land Use Policy* **2018**, *71*, 213–221. [CrossRef]
25. Xiao, W.; Hu, Z.-Q.; Zhang, R.-Y.; Zhao, Y.-L. A simulation of mining subsidence and its impacts to land in high ground water area– An integrated approach based on subsidence prediction and GIS. *Disaster Adv.* **2013**, *6*, 142–148.
26. Bian, Z.-F.; Inyang, H.I.; Daniels, J.L.; Otto, F.; Struthers, S. Environmental issues from coal mining and their solutions. *Min. Sci. Technol.* **2010**, *20*, 215–223. [CrossRef]
27. Pei, W.-M.; Yao, S.-P.; Knight, J.F.; Dong, S.-C.; Pelletier, K.; Rampi, L.P.; Wang, Y.; Klassen, J. Mapping and detection of land use change in a coal mining area using object-based image analysis. *Environ. Earth Sci.* **2017**, *76*, 125. [CrossRef]
28. Wen, B.; Pan, Y.-H.; Zhang, Y.-Y.; Liu, J.-J.; Xia, M. Does the exhaustion of resources drive land use changes? Evidence from the influence of coal resources–exhaustion on coal resources-based industry land use changes. *Sustainability* **2018**, *10*, 2698. [CrossRef]
29. Yao, S.-M.; Zhang, P.-Y.; Yu, C.; Li, G.-Y.; Wang, C.-X. The theory and practice of new urbanization in China. *Sci. Geogr. Sin.* **2014**, *34*, 641–647.
30. Lichtenberg, E.; Ding, C. Assessing farmland protection policy in China. *Land Use Policy* **2008**, *25*, 59–68. [CrossRef]
31. Xiao, W.; Hu, Z.-Q.; Li, J.; Zhang, H.-Y.; Hu, J.-L. A study of land reclamation and ecological restoration in a resource-exhausted city– A case study of Huaibei in China. *Int. J. Min. Reclam. Environ.* **2011**, *25*, 332–341. [CrossRef]
32. Dolný, A.; Harabiš, F. Underground mining can contribute to freshwater biodiversity conservation: Allogenic succession forms suitable habitats for dragonflies. *Biol. Conserv.* **2012**, *145*, 109–117. [CrossRef]
33. Antwi, E.K.; Krawczynski, R.; Wiegleb, G. Detecting the effect of disturbance on habitat diversity and land cover change in a post-mining area using GIS. *Landsc. Urban Plan.* **2008**, *87*, 22–32. [CrossRef]
34. Yirsaw, E.; Wu, W.; Shi, X.-P.; Temesgen, H.; Bekele, B. Land use/land cover change modeling and the prediction of subsequent changes in ecosystem service values in a coastal area of China, the Su-Xi-Chang Region. *Sustainability* **2017**, *9*, 1204. [CrossRef]

International Journal of
Environmental Research and Public Health

MDPI

Article

Land Use Change and Its Impact on Landscape Ecological Risk in Typical Areas of the Yellow River Basin in China

Yanbo Qu [1,*], Haining Zong [1], Desheng Su [1], Zongli Ping [2] and Mei Guan [2]

1 School of Public Administration and Policy, Shandong University of Finance and Economics, Jinan 250014, China; zhn202109034@mail.sdufe.edu.cn (H.Z.); 20170927328@mail.sdufe.edu.cn (D.S.)
2 Shandong Institute of Territorial and Spatial Planning, Government-Affiliated Institution, Jinan 250014, China; zwy212109016@mail.sdufe.edu.cn (Z.P.); ll212109037@mail.sdufe.edu.cn (M.G.)
* Correspondence: qyb20126008@sdufe.edu.cn

Abstract: The basic premise of regional ecological construction would be to scientifically and effectively grasp the characteristics of land use change and its impact on landscape ecological risk. The research objects of this paper are the typical areas of the Yellow River Basin in China and "process-change-drive" as the logical main line. Moreover, this paper is based on multi-period land use remote sensing data from 2000 to 2020, the regional land use change process and influencing factors are identified, the temporal and spatial evolution and response process of landscape ecological risk are discussed, and the land use zoning control strategy to reduce ecological risk is put forward. The results indicated: (1) The scale and structure of land use show the characteristics of "many-to-one" and "one-to-many"; (2) the process of land use change is affected by the alternation of multiple factors. The natural environment and socio-economic factors dominate in the early stage and the location and policy factors have a significant impact in the later stage; (3) the overall landscape ecological risk level and conversion rate show a trend of "high in the southeast, low in the northwest", shift from low to high and landscape ecological risks gradually increase; and (4) in order to improve the regional ecological safety and according to the characteristics of landscape ecological risk and spatial heterogeneity, we should adopt the management and control zoning method and set different levels of control intensity (from key intensity to strict intensity to general intensity), and develop differentiated land use control strategies.

Keywords: land use; landscape ecological risk; ecological risk assessment; influencing factors; Yellow River Basin

Citation: Qu, Y.; Zong, H.; Su, D.; Ping, Z.; Guan, M. Land Use Change and Its Impact on Landscape Ecological Risk in Typical Areas of the Yellow River Basin in China. *Int. J. Environ. Res. Public Health* **2021**, *18*, 11301. https://doi.org/10.3390/ijerph182111301

Academic Editors: Wei Song and Hualin Xie

Received: 10 October 2021
Accepted: 26 October 2021
Published: 28 October 2021

1. Introduction

Ecological environmental risks have gradually become an important factor affecting national security and restricting the sustainable development of the economy and healthy society [1,2]. The continuous implementation of the "Five Development Concepts" and the "Two Mountains Theory" [3], as well as analyzing and resolving ecological risks in time in order to guarantee the environmental safety as a crucial part of achieving ecological civilization in China based on its construction goals, include harmonious symbiosis, virtuous circle, and comprehensive development. Ecological risk refers to the potential damage to the structure or function of the ecosystem, caused by accidents or hazards in the region [4,5]. The relationship between land use change and landscape ecological risk (LER) is complex. The process of land use type change affects a series of ecological processes, such as the atmosphere, soil, water bodies, and organisms. As a result, the ecosystem structure and function are changed by land cover changes, caused by land use changes and the extensive effects of ecological changes [6–8]. The landscape ecological risk assessment (LERA), based on the pattern of land use change, can measure the adverse effects of the combined landscape pattern and ecological process, and it is of great significance to analyze the global aspects, dynamic evolution, and optimization of prevention and control risk of regional ecological risks [9–11]. The rapid economic development has caused many ecological safety issues to be solved urgently. The protection of ecologically fragile areas has an irreplaceable role in promoting ecological safety and harmony between human and land. Existing studies have mostly carried out beneficial explorations on the LER in small-scale and intense human activities, while less attention has been given to large and medium scale ecologically sensitive and fragile natural areas. The study of land use change and ecological risk response in fragile areas provides a foundation for regional ecological construction, environmental restoration, and high-quality sustainable development [12].

The concept of landscape was first proposed by the German geographer, Carl Troll. As a complex of natural surface, the landscape is a collection of highly spatially heterogeneous regional ecosystems [13]. At present, LER research has gradually shifted from unilateral examination on risk sources and risk receptors to the overall impact of the ecosystem and the spatial correlation of LER, paying more attention to the macroscopicity and practicability [14–16]. Scholars have conducted in-depth research on land use and ecological risks, and LER patterns assessment based on the perspective of land use change, which has gradually become the mainstream of research [17]. At the beginning of the 1990s, scholars such as Heggem et al. [18], Kapustka et al. [19], and Estoque et al. [20] introduced landscape pattern analysis methods to assess the impact of human activities on ecological changes in the watershed. Most of the studies use high-risk communities as the evaluation unit, disregarding the spatial differences and natural geography relation, which is not helping in grasping the overall landscape pattern. However, the geographical significance of the evaluation cannot be ignored. As the evaluation unit, the administrative area should be ideal for studying the prevention and risk control in sensitive and fragile areas, in order for different policies to be defined for the ecological risk situation of different administrative regions [21,22]. The research on LER, brought by land use changes, is generally established on two evaluation models. One is based on the traditional "source-receptor analysis, exposure evaluation, and hazard evaluation" inherent mode. The ecological risk assessment index system is constructed from the risk source intensity, the receptor exposure, and the risk effect. The other model directly evaluates the LER from the landscape pattern and uses the landscape ecological index from the perspective of landscape ecology to reflect the ecological effect of land use and land cover change (LULCC) [23–26]. Comparing these two models, the conclusion is that the model based on the landscape perspective is more suitable for evaluating the ecological risks caused by human activities, since human activities will have an immense influence on the landscape pattern. The consequences of this influence will directly lead to changes in the ecological environment, and the research in this area becomes more reliable. Many academics have conducted research, analysis, and experiments on various landscape indices [10,27,28]. Therefore, based on the landscape pattern, the ecological risk assessment is relatively scientifically founded and feasible [29]. Existing studies mostly focused on the construction of LER models and spatial analysis [30,31]. The lack of attention to the LER and land use change process and the lack of time-period process research have led to a decline in the credibility and applicability of the risk assessment results. At the same time, according to the administrative division, the meso-scale study which takes into account the core field of economic development and the area with fragile ecological environment has gradually become a research hotspot [32]. In summary, scholars mainly explore landscape risks by constructing evaluation models, and have formed relatively mature evaluation methods and systems. They rarely involve a quantitative analysis of the influencing factors or driving forces of land use changes that lead to the differentiation of LER, and lack zoned explorations of local LER control measures [33].

The ecological protection and high-quality development of the Yellow River Basin are related to China's social development and ecological protection, and have risen to a major national strategy. As the only province along the Yellow River in the East, along the Yellow River and along the coast, Shandong Province is in a dominant position in the "Yellow River Strategy". The areas along the Yellow River in Shandong Province have the most developed economy and the largest permanent population among the provinces and cities along the Yellow River, with the highest urbanization rate and obvious geographical advantages of river sea intersection. At the same time, since the entire territory is downstream, the degradation of the natural ecological environment has become increasingly prominent, the pressure of ecological protection is great, and the problem of ecological security is prominent. The areas along the Yellow River in Shandong Province are a typical representative of the Yellow River Basin. Studying the landscape ecological risk response and countermeasures in typical areas of the Yellow River Basin is conducive to the healthy and sustainable development of the Yellow River Basin.

Therefore, this article selects the typical areas of the Yellow River Basin that are both ecologically fragile and economically developed as the study area. Taking the county area as the research scale, based on the land use data of the typical areas of the Yellow River Basin in 2000, 2010, and 2020, the landscape pattern index is calculated with the help of FRAGSTATS 4.2 (a program designed to compute a wide variety of landscape metrics for categorical map patterns) to construct the LERA model. It depicts the process of land use change, based on multiple dimensions of "process-change-drive" [34], explores and analyzes the influencing factors of land use change, fully reveals the temporal and spatial differentiation and LER transfer laws, explores the LER response of land use changes and conducts zoned prevention and control, proposing targeted management and control recommendations, in order to provide a useful reference for the realization of ecological protection and high-quality development strategy in the Yellow River Basin.

2. Materials and Methods

2.1. Study Area

The typical areas of the Yellow River Basin (34°26′–38°16′ N, 114°45′–119°19′ E) are located in the west of Shandong Province (Figure 1). The Yellow River enters from Dongming County of Shandong Province, flows north to east and through Heze, Jining, Taian, Liaocheng, Jinan, Dezhou, Binzhou, Zibo, and Dongying. The typical areas of the Yellow River Basin are a warm temperate humid and semi-humid monsoon climate type with four distinct seasons, significantly dry and wet, followed by rain and heat in the same season. The total land area in nine cities of seventy-seven counties along the Yellow River is 82,500 square kilometers (km^2), accounting for 53.4% of the provincial total land area. The terrain is complex, i.e., mountains and hills account for about 35% of the area, while plains, depressions, and beaches account for about 65% of the total area of the Yellow River Basin in Shandong. The mountains in central and southern Shandong are protruding. The northwest of Shandong is low-lying and flat, and the gentle hills in the southwest of Shandong are undulating, forming a general terrain with mountains and hills as the skeleton, while plains and basins are interlaced and ringed in between. In 2020, the regional gross domestic product (GDP) of the nine cities along the Yellow River is CNY 3891.7 billion, accounting for 50.9% of Shandong regional GDP, with a permanent population of 54.272 million. It is an important part of the economic circle for the capital of Shandong Province and southern Shandong. Ecological and environmental problems, such as soil salinization, desertification, and soil erosion in the region, are becoming more serious. Environmental pollution and degradation are prominent, and the LER prevention and control is imminent. Taking this as a case area to carry out LER research has an important significance.

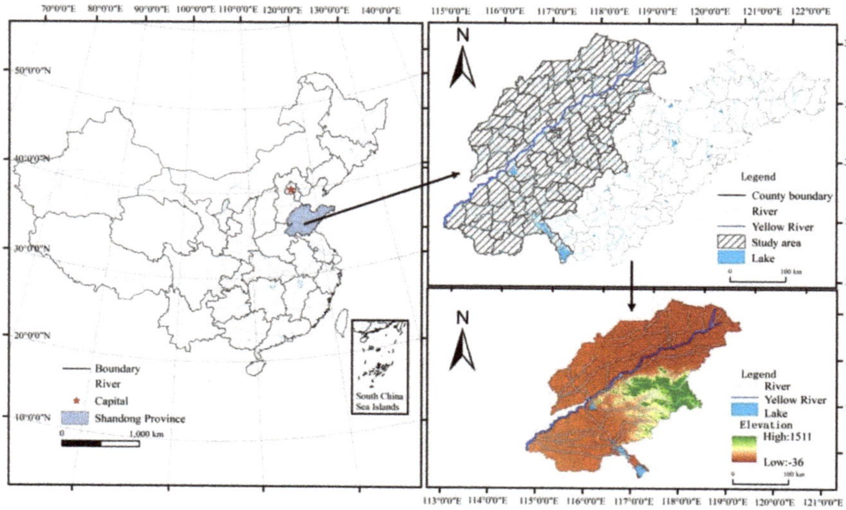

Figure 1. Location of the study area.

2.2. Data

The land use data are from the Resource and Environment Science and Data Center (RESDC) of Chinese Academy of Sciences (http://www.resdc.cn) (accessed on 3 February 2021), including the national land use remote sensing data in 2000, 2010, and 2020 (raster data, resolution of 1 km) and China's administrative division data (vector data). The land use data were cropped, according to the administrative boundaries of the study field, and the land use data of the typical areas of the Yellow River Basin in 2000, 2010, and 2020 were obtained. The test data with ENVI 5.3 (tool for processing remote sensing images), show the comprehensive accuracy of the three periods over 92%. On this basis, with the help of the ArcGIS operating platform, the relevant data of land use can be extracted from the administrative area of the study area, and construct a land use database of the typical area in the Yellow River Basin. Referring to the existing literature, it can reclassify the relevant data of land use in ArcGIS, and divide the land use types into six categories: Cultivated land, wetland, grass land, forest land, construction land, and bare land [17]. The data in the analysis of influencing factors include relevant data, such as natural environment foundation, social and economic conditions, traffic and location conditions, policy and institutional environment, which are from the geospatial data cloud (http://www.gscloud.cn) (accessed on 5 March 2021), the global nightlight remote sensing data (https://www.nature.com/sdata) (accessed on 5 March 2021), Chinese Soil Database of Nanjing Institute of Soil, Chinese Academy of Sciences (http://vdb3.soil.csdb.cn) (accessed on 5 March 2021), National Geomatics Center of China (NGCC, http://ngcc.sbsm.gov.cn/) (accessed on 5 March 2021), and statistical yearbooks of the nine cities along the Yellow River in 2000, 2021, and 2020. See Table 1 for details:

Table 1. Data sources of land use change influencing factors.

Influencing Factor	Variable	Data Sources
Natural environment foundation	Elevation N1	The geospatial data cloud (http://www.gscloud.cn) (accessed on 5 March 2021)
	Average annual precipitation N2 Average annual temperature N3	The Resource and Environmental Science Data Center of the Chinese Academy of Sciences (http://www.resdc.cn) (accessed on 5 March 2021)
	Soil organic matter content N4	Chinese Soil Database of Nanjing Institute of Soil, Chinese Academy of Sciences (http://vdb3.soil.csdb.cn) (accessed on 5 March 2021)
Social and economic conditions	Change rate of urbanization S1 Change rate of per capita social consumer goods sales S2 Change rate of ground-average agricultural machinery S3 Population change rate S4	Statistical yearbooks of the nine cities along the Yellow River in 2000, 2021, and 2020.
	Night light remote sensing S5	The global night light remote sensing data (https://www.nature.com/sdata) (accessed on 5 March 2021)
Traffic and location conditions	Change rate of road density T1 Distance from the town center T2	National Basic Geographic Information Center (http://ngcc.sbsm.gov.cn/) (accessed on 5 March 2021)
Policy and institutional environment	Change rate of ground-average investment in fixed assets P1 Change rate of public financial expenditure P2	Statistical yearbooks of the nine cities along the Yellow River in 2000, 2021, and 2020.

2.3. Methods

Changes in natural factors and human interference, directly or indirectly, affect the land use structure and function, and changes in land use types further lead to changes in landscape patterns [35,36]. Due to the significant landscape pattern heterogeneity and its relation to ecological process, under the stress and involvement of natural background and human geography elements, it will cause potential adverse effects or harms. Moreover, it indicates that LER and land use changes respond to each other [30,37]. LER are caused by land use changes and are driven by natural factors and human activities. To avoid, adapt, and comprehensively manage risks, it is necessary to improve the related factors of land use change and carry out the zoned management and control (Figure 2). Therefore, this paper first applies the chord diagram and geographical information system (GIS) map to identify the overall and process characteristics of regional land use change, and then analyzes the influencing factors and driving mechanism of land use change process with the help of factor detectors in geographic detectors. Second, the LERA model constructed by the landscape disturbance index and landscape vulnerability index is used to explore the temporal and spatial transfer and evolution of LER and the distribution of land types, and the LER response of land use change is obtained with the help of ecological risk contribution rate model. Finally, based on the above research results, it divides the LER management and control area of the typical area of the Yellow River Basin and proposes relevant measures to guide the high-quality sustainable development of the region [38].

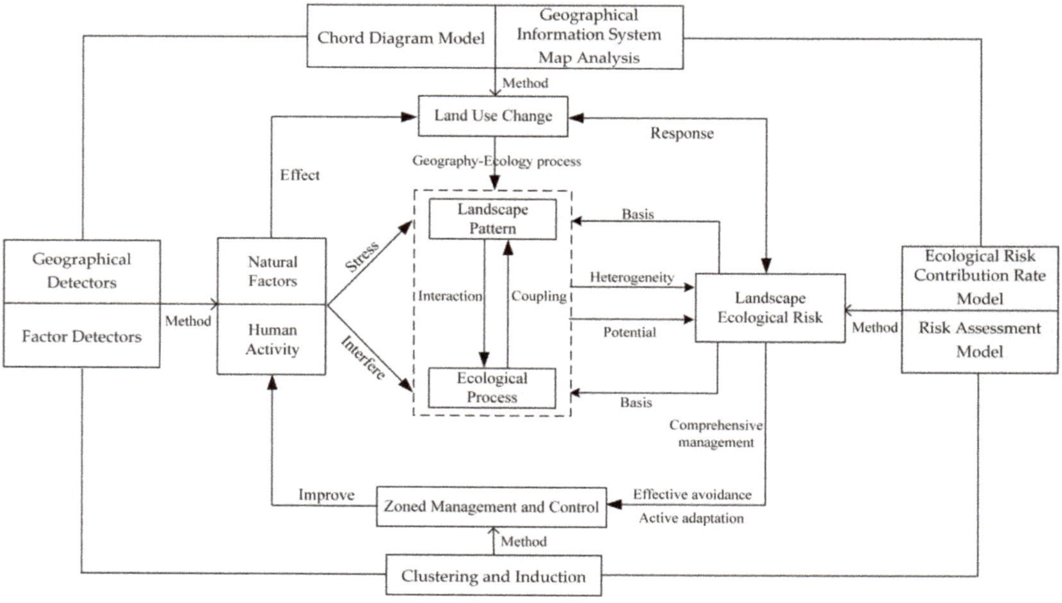

Figure 2. The framework of this study.

2.3.1. Chord Diagram of Land Use Changes

The chord diagram is a graphical method that shows the inter-relationship between data. The data points in the chord diagram are arranged radially in the form of circles, and lines are used to show the connections between the data. The chord diagram can reflect the number of conversions and relationship-flow between different land use types in the process of land use change and visualize it [39]. The wider the chord (connecting line), the higher the number of conversions between land use types. This paper uses the

Multi-Charts 1.8 software (https://jshare.com.cn/new) (accessed on 5 March 2021) to visualize the change process of different land use types.

2.3.2. GIS Map Analysis of Land Use Changes

The GIS map analysis is used to reflect the degree of quantitative changes in land use types, and is the basic manifestation of the impact of social and economic activities on land use [40]. The process of land use change reflects the relation of change from one land use type into another, including two conversion directions, transfer-in and transfer-out. The former pays attention to the increase in the transfer-in land use type, and the latter focuses on the reduction of the land use type. This paper uses the grid calculator of ArcGIS 10.3 to superimpose the land use types from 2000 to 2020 (Equation (1)), and obtain the land use change map of the typical area in the Yellow River Basin at different periods as follows:

$$W = A \times 10 + B \tag{1}$$

where W represents the newly generated graph coding; A represents the land use atlas unit coding at the beginning of the study; and B represents the land use atlas unit coding at the end of the study (secondary classification code). For example, $W = 12$ indicates a GIS map of land use types converted from forest land to wetland.

2.3.3. Land Use Change Influencing Factors Analysis

The geographical detectors can be used to analyze spatial differentiation characteristics and explore the interaction between factors. It is convenient to operate and is less affected by the sample size [41]. The main types of geographical detectors are risk, factor, ecological, and interaction detectors, among which factor detectors can disclose the explanatory power of independent variables to dependent variables [42,43]. This study selects factor detectors, takes typical counties (cities, districts) in the Yellow River Basin as the basic unit, and the six types of land use change rates from 2000 to 2020 as the geographical detector indicators, and carries out an analysis on the influencing factors of LER in typical areas of the Yellow River Basin by GeoDetector (http://www.geodetector.cn) (accessed on 8 March 2021). The method is shown in the following equation:

$$q = 1 - \frac{1}{N\delta^2}\sum_{h=1}^{H} N_h\delta_h^2 \tag{2}$$

where q represents the influencing factors index of land use change; N represents the number of global samples; N_h represents the number of samples in the secondary region; H represents the factors stratification; δ^2 represents the total variance of the whole region; and δ_h^2 represents the secondary region discrete variance. The value interval of q is [0, 1], and the greater the q value, the greater the influence force on land use change.

Land use change occurs within the three-fold framework of natural system, socio-economic system, and institutional system. The typical areas of the Yellow River Basin are greatly undulating, with rich landform types, strongly affected by the monsoon climate, and the significant change rate of average annual precipitation and average annual temperature. At the same time, with the advancement of the Yellow River Basin regional development strategy, the typical areas in the Yellow River Basin are committed to industrial structure adjustment and upgrading, infrastructure construction, and ecosystem restoration. Government departments provide continuous and strong financial support for the industrial development of the region. In addition, the level of urbanization, industrial structure, quality of life, and ecological environment in the region have been significantly improved. Therefore, based on the actual conditions of the typical areas in the Yellow River Basin, such as significant topographic fluctuations and rapid regional economic development, thirteen indicators were selected as the detection factors of land use changes, including natural environment foundation, social and economic conditions, traffic and location conditions, and policy and institutional environment (Table 2). The ArcGIS is used to rasterize all of the influence factors and unify the projection coordinate system. In addition, the natural break point method is used to process the spatial discretization of the influencing factors, to measure the degree of influence between land use changes, and the influencing factors in typical areas of the Yellow River Basin.

Table 2. Land use change influencing factors index.

Influencing Factor	Variable	Index Description
Natural environment foundation	Elevation N1	Terrain condition factors
	Average annual precipitation N2	Precipitation condition factors
	Average annual temperature N3	Weather condition factor
	Soil organic matter content N4	Soil condition factors
Social and economic conditions	Change rate of urbanization S1	Development level of urbanization
	Change rate of per capita social consumer goods sales S2	Resident consumption level
	Change rate of ground-average agricultural machinery S3	Level of technological progress
	Population change rate S4	Human-factor level
	Night light remote sensing S5	Level of economic development
Traffic and location conditions	Change rate of road density T1	Traffic accessibility
	Distance from the town center T2	Location advantage degree
Policy and institutional environment	Change rate of ground-average investment in fixed assets P1	Investment level
	Change rate of public financial expenditure P2	Fiscal expenditure level

2.3.4. LERA Method

The landscape pattern production is the result of differences of human impact on natural ecosystems. This ecological impact presents regional and cumulative characteristics. The intensity of external disturbance and the ability of internal resistance to the disturbance of the ecosystems, represented by different landscapes, determine the size of LER [44].

In view of the relation between landscape pattern and ecological risk, the landscape disturbance index and landscape vulnerability index are used to construct a LERA model [21–25]. The landscape disturbance index is composed of landscape fragmentation index, landscape separation index, and landscape subdimension index.

1. Landscape disturbance index (*Li*)

Li indicates the ability of different landscape ecosystems to resist interference from the outside world and self-recovery. The sensitivity of landscape ecosystems increases with the rise of the landscape disturbance pattern, which leads to a greater LER. By selecting the landscape fragmentation index (*Bi*), landscape separation index (*Si*), and landscape subdimension index (*Fi*), the Li is constructed. Bi represents the degree of fragmentation of the landscape space, from single continuous to complex discontinuous, reflecting the degree of natural or human disturbance to the landscape. It shows that the larger the value, the lower the stability of the corresponding landscape type. Si refers to the degree of separation in different patches of the landscape. The larger the value, the more scattered the corresponding landscape and the more complex the landscape distribution. Fi is an index used to determine the patch shape influence on the internal patch ecological process. The larger the value, the more complex the corresponding patch shape. The indices calculation equations are as follows:

$$L_i = aB_i + bS_i + cF_i \tag{3}$$

$$B_i = \frac{n_i}{A_i} \tag{4}$$

$$S_i = \frac{A}{2A_i}\sqrt{\frac{n_i}{A}} \tag{5}$$

$$F_i = \frac{2ln(P_i/4)}{lnA_i} \tag{6}$$

where a represents the weight value of Bi; b represents the weight value of Si; c represents the weight value of Fi; and $a + b + c = 1$. According to the relevant research results and the actual situation of the study area, the values are assigned to 0.5, 0.3, 0.2 [3,21]; n_i represents the number of patches in the i-th landscape; A_i represents the area of the i-th landscape, in km^2; A represents the total area of all the landscapes, in km^2; and P_i represents the perimeter of the i-th landscape, in km.

2. Landscape vulnerability index (Wi)

Wi represents the sensitivity of different landscape ecosystems to external disturbances. The larger the value, the lower the stability of the ecosystem and the higher the possibility of damage. Based on previous studies and the actual situation in typical areas of the Yellow River Basin, the six types of landscapes from high to low are assigned as follows: 6—construction land, 5—bare land, 4—cultivated land, 3—grass land, 2—wetland, and 1—forest land [24,25], which are normalized.

3. LERA index (ERI_k)

ERI_k is constructed by Li and Wi. It separates the spatial ecological risk from the landscape spatial structure and represents the degree of ecological loss within the assessment unit. The greater the value, the higher the corresponding LER, as shown in the following equation:

$$ERI_k = \sum_{i=1}^{z} \frac{A_{ki}}{A_k}(L_i W_i) \quad (7)$$

where ERI_k represents the LERA index unit k; z represents the number of landscape types; A_{ki} represents the area of the i-th landscape in the LERA unit k, in km^2; and A_k represents the area of landscape ecological risk assessment unit k, in km^2.

2.3.5. Land Use Change Ecological Risk Contribution Rate

The ecological risk contribution rate of land use changes refers to the degree of change in the regional LER, which is caused by a certain land use type change [45]. A positive value indicates that this type of change has aggravated the LER in the region, and a negative value indicates that this type of change improves the LER in the region. Isolating the main land use types that affect the LER changes is conducive to exploring the leading factors of changes in regional LER [46]. The calculation is as follows:

$$LEI = \frac{(ERI_1 - ERI_0)LA}{TA} \quad (8)$$

where LEI represents the ecological risk contribution rate of land use changes; ERI_0 represents a LER index at the early stage of change of land use type, and ERI_1 represents a LER index at the end of change of land use type; LA represents the area of the change type; and TA represents the total area of the region.

The positive and negative analysis of the LEI can be used to comprehensively determine the land use types that affect the LER change index in typical areas of the Yellow River Basin, which is helpful in distinguishing the leading factors of the improvement and degradation of LER in typical areas of the Yellow River Basin.

3. Results

3.1. Analysis of the Overall Characteristics of the Land Use Change

From 2000 to 2020, the land use is dominated by cultivated and construction land (Table 3), with the largest proportion of cultivated land in typical areas of the Yellow River Basin. This is consistent with the characteristics of Shandong Province as a major agricultural province and the rapid development of economic society and urbanization. From the perspective of time series characteristics, the area of forest land first increased and then decreased. The fluctuation range in the first 10 years was small, and the area of forest land decreased by 4.41 hectare (ha) in the next 10 years. The area of grass land, cultivated land, and bare land has shown a downward trend in the past two decades. The area of cultivated land has relatively stable changes and it decreased by 3.41%. This is inseparable from the Shandong Province emphasis on cultivated land safety and the effective implementation of basic farmland protection policy. The area of grass land has decreased by 42.11%, indicating that the implementation of the 2003 policy of returning cultivated land to forests and grass land needs to be strengthened. In addition, the area of bare land is significantly smaller, and a lot of bare land is fully utilized for project construction. The area of wetland and construction land is increasing gradually, and the increase rate of wetland, from 2010 to 2020, is as high as 51.86%. This is a remarkable result of paying attention to the protection of wetland system and vigorously investing in the construction of nature reserves and wetland parks in typical areas of the Yellow River Basin. The area of construction land has increased for 33.83%, reflecting the common needs of rapid urbanization and economic development.

Table 3. Different types of land use area and change rate in typical areas of the Yellow River Basin from 2000 to 2020.

Land Use Type	Area/km^2			Change Rate/%		
	2000	2010	2020	2000–2010	2010–2020	2000–2020
Forest land	0.3585	0.3592	0.3151	0.20	−12.28	−12.11
Wetland	0.3786	0.4034	0.6126	6.55	51.86	61.81
Grass land	0.4819	0.4135	0.2790	−14.19	−32.53	−42.11
Cultivated land	5.7116	5.6426	5.5167	−1.21	−2.23	−3.41
Bare land	0.2026	0.1534	0.0401	−24.28	−73.86	−80.21
Construction land	1.1369	1.2986	1.5215	14.22	17.16	33.83

3.2. Land Use Change Process Analysis

1. Scale feature analysis

From a staged perspective, the scale of land use change from 2000 to 2010 showed the characteristics of "many to one" and "one to many" in typical areas of the Yellow River Basin (Figure 3). Among them, the main construction land transfer sources are cultivated land, bare land, and wetland. The transferred areas are 1144, 314, and 125 km^2. The main wetland transfer sources are cultivated land, grass land, and bare land, and the transferred areas are 239, 128, and 62 km^2. The forest land transferred area is relatively small, and the main sources are cultivated land and grass land, and the transfer area is 21 and 17 km^2, respectively. The construction land, wetland, and forest land transferred area decreases successively, showing a "many-to-one" characteristic. Cultivated land is mainly converted into construction land and wetland, with the transferred area as 1144 and 239 km^2. Grass land is mainly converted into cultivated land, wetland, and construction land, with the transferred area as 460, 128, and 53 km^2, respectively. Bare land is mainly converted into construction land, cultivated land, and wetland, and the transferred area is 314, 174, and 62 km^2, respectively. The transferred area of cultivated land, grass land, and bare land successively decreased, showing the characteristics of "one-to-many".

Except for bare land, from 2010 to 2020, the dominant characteristics of "one-to-many" transfers continue, mainly for cultivated land, wetland, and construction land, with transfer areas as 828, 412, and 191 km^2, respectively. Other land use types show balanced conversion attributes, but their respective transfer-in and transfer-out dominant types show significant differences. Among them, the reciprocal conversion of cultivated land and construction land is crucial. The main source of cultivated land transfer is construction land, with an area of 7477 km^2, while the main source of construction land transfer is cultivated land with an area of 9922 km^2. Wetland, construction land, and cultivated land are mutually transformed, but the transfer-in scale is larger than the transfer-out scale. Grass land, wood land, and cultivated land are mutually transformed, and the transfer-out scale is larger than the transfer-in scale. Forest land, cultivated land, and construction land are mutually transformed, and the transfer-out scale is not much different from the transfer-in scale. From the perspective of the whole time period, the land use change trend scale in typical areas of the Yellow River Basin, from 2000 to 2020, follows the elements of transformation from 2010 to 2020, and the scale of conversion of some internal land use types slightly increases or decreases.

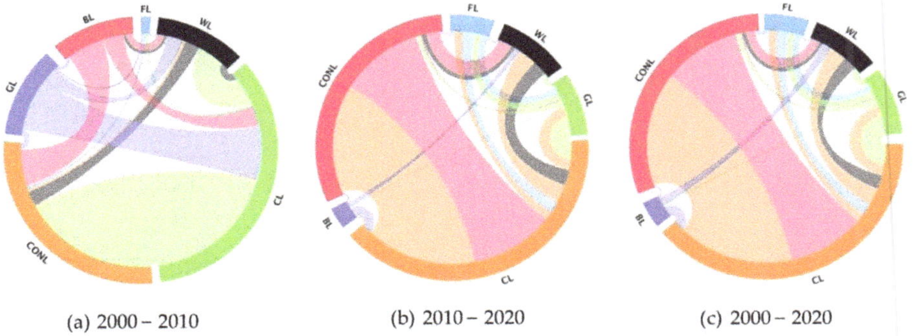

Figure 3. Chord diagram of land use changes in typical areas of the Yellow River Basin from: (**a**) 2000–2010, (**b**) 2010–2020, and (**c**) 2000–2020.

Here, FL, WL, GL, CL, BL, and CONL represent forest land, wetland, grass land, cultivated land, bare land, and construction land, respectively.

2. The spatial map analysis

The land use change map of the typical areas of the Yellow River Basin generates 30 types of map units from 2000 to 2010, and 24 types of map units are changed (Figure 4). Among them, the map type of "cultivated land → construction land" (code 46) is the most obvious, and widely distributed in urban agglomerations, mostly in urban fringe areas, which is in line with the features of rapid urbanization expansion in this region. At times when the government paid insufficient attention to the cultivated land safety, part of the cultivated land was transformed into urban and rural construction land. The "cultivated land → wetland" (code 42) and "grass land → wetland" (code 32) are mainly distributed in the coastal areas of Dongying City. To alleviate the ecosystem vulnerability, large amounts of cultivated land and grass land are converted into wetland. From 2010 to 2020, there are 36 types of land use change map units in typical areas of the Yellow River Basin, while 30 types of map units have changed. The spatial aggregation degree is particularly stronger than in the previous stage, and the coastal area is higher than in the inland area. Among them, the most important is the rapid expansion and wide distribution of "construction land → cultivated land" (code 64). The main reason is that the state gives a great authority to the protection of cultivated land and the implementation of the occupation and compensation balance policy. Substantial construction land is converted into cultivated land, which is conducive to the response to the slogan that the total area of Chinese farmland must remain above the red line of 120 million hectares. The "construction land → wetland" (code 62) and "bare land → wetland" (code 52) are becoming more dominant in the transformation, and they are distributed in the northern coastal areas. In the development theory of transformation, from focusing on speed growth to high-quality development and ecological protection, Shandong Province pays more attention to ecological environment protection, in order to gradually improve the ecological conditions of key protection areas, and the scale of wetland has been greatly expanded.

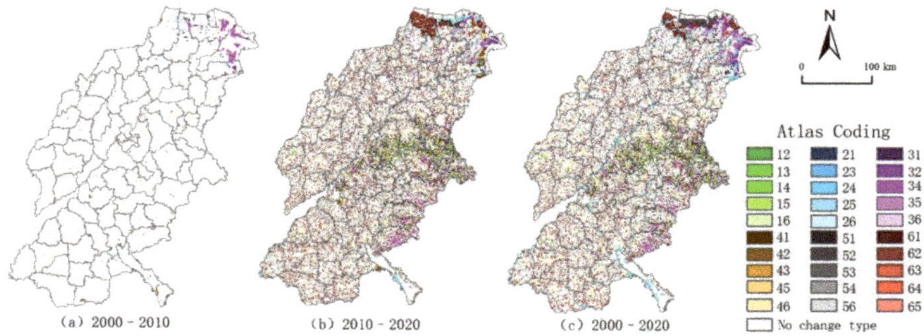

Figure 4. Map of land use changes in typical areas of the Yellow River Basin from: (**a**) 2000–2010, (**b**) 2010–2020, and (**c**) 2000–2020.

Regarding the whole period, the land use change map of the typical areas of the Yellow River Basin generated a total of 36 types of map units from 2000 to 2020, and a total of 30 types of map units have changed. The overall change is similar to the spatial distribution of the period from 2010 to 2020. Among them, "cultivated land → construction land" (code 46) and "construction land → cultivated land" (code 64) are more evenly distributed in typical areas of the Yellow River Basin, with obvious spatial dispersion. The "construction land → wetland" (code 62) and "bare land → wetland" (code 52) are mainly distributed in the northern coastal areas, while "cultivated land → wetland" (code 42) is mainly distributed near the Yellow River and its tributaries.

Here, codes 1–6 represent forest land, wetland, grass land, cultivated land, bare land, and construction land, respectively. The map unit represents land use type transformations, and the coding is a combination of secondary classification coding from two transformations, e.g., "forest land → wetland" (code 12).

From the fluctuation process point of view, from 2000 to 2020, the types of land use growth in typical areas of the Yellow River Basin are relatively obvious and complex. Among them, the land use was relatively stable from 2000 to 2010, with more than 95% of the area unchanged. The types of land use that have changed are mainly new construction land and cultivated land, which are mainly distributed in offshore areas. Urbanization expansion was the mainstream of the development of the typical areas in the Yellow River Basin during this period. Relying on marine resources, many ports and salt fields were built, while plenty of cultivated lands were supplemented through the saline-alkali land management. By 2010–2020, the overall land use changes are quite drastic, with the change area accounting for more than 35%, mainly cultivated land, construction land, and wetland. The newly added cultivated land is distributed more along the Yellow River banks and at the sea mouth. The newly-added construction land is mainly distributed around the existing urban and town areas, and the newly-added wetland is concentrated in the coastal areas. In this period, a national agricultural high-tech industry demonstration zone was established, and it provides support for the development of ecological and circular agriculture. Soil improvement has encouraged the planting and promotion of salt-tolerant crops such as cotton, vegetables, and forests, which has led to a significant increase in cultivated land. Driven by industrial transformation and development, the urbanization of the population has been brisk, and construction land has also increased. At the same time, ecological protection began to be carried out vigorously. The original overexploited and constructed industrial and mining land and the heavily polluted chemical enterprises gradually withdrew and changed from construction land to wetland.

3.3. Analysis of Land Use Change Influencing Factors

Altogether, the spatial differentiation characteristics of land use changes in typical areas of the Yellow River Basin are affected by natural and socio-economic factors, traffic, location, and policy, but the influencing factors and intensity of land use change rates of various types of land show some difference (Figure 5).

From the forest land perspective, the top three influencing factors, leading to the differentiation of forest land spatial change rates, from 2000 to 2010, are elevation (0.67), soil organic matter content (0.53), and change rate of public financial expenditure (0.37). Elevation affects the rate of conversion of other land types to forest land, and soil conditions affect the scale of conversion of forest land to grass land and cultivated land. Since the 2003 policy of farm land to forest return, the tendency of fiscal expenditure has prompted part of the cultivated land to be converted into forest land. During this time, the natural environment and the policy environment were the main influencing factors. Compared with the previous period, from 2010 to 2020, the top three influencing factors have been transformed into elevation (0.57), change rate of urbanization (0.43), and change rate of ground-average agricultural machinery (0.39). With the accelerated urbanization, due to the lax industrial orientation and policy, the construction land has been invaded and occupied by forest land, and the continuous improvement of agricultural technology has promoted the use of forest land that is not suitable for planting crops for food production, which are the leading factor at this stage.

Figure 5. Influencing factors of land use changes: (**a**) From 2000 to 2010 and (**b**) from 2010 to 2020.

From the perspective of wetlands from 2000 to 2010, the top three influencing factors that led to the spatial differentiation of wetland change rates were average annual precipitation (0.52), public financial expenditure change rate (0.35), and population change rate (0.33). The decreasing trend of precipitation, from coastal to inland territory, significantly affected the spatial distribution of wetland. Government policies, financial support, and human interference have led to the conversion of cultivated land and bare land into wetland. From 2010 to 2020, the top three influencing factors have been transformed into the public financial expenditure change rate (0.65), population change rate (0.43), and per capita social consumer goods sales change rate (0.37). Due to the establishment of the Yellow River Delta High-Efficiency Eco-Economic Zone, the impact of fiscal expenditures sharply increased. With the growth of the total population and per capita consumption level, due to the ecosystem integrity and ecotourism demand, the wetland area is quickly replenished. The level of social and economic development and policy guidance jointly drive the differentiation of wetland change rates.

From the perspective of grass land from 2000 to 2010, the main three factors leading to the spatial differentiation of grass land change rates are average annual precipitation (0.41), elevation (0.39), and soil organic matter content (0.32). The natural world plays a leading role in the transformation between grass land and forest land and between grass land and cultivated land. During 2010–2020, the three dominant factors are average annual precipitation (0.43), elevation (0.42), and average ground agricultural machinery change rate (0.30). In addition to natural factors, the expansion of agricultural technology promotes the full growth of animal husbandry, and part of grass land is transformed into cultivated land, and these are all key influencing factors.

From the cultivated land perspective from 2000 to 2010, the main three influencing factors leading to the spatial differentiation of cultivated land change rates are urbanization change rate (0.58), average annual precipitation (0.51), and population change rate (0.43). Driven by the interests of urbanization, a large number of cultivated lands are converted to construction land. Regional precipitation conditions are restraining the cultivated land expansion scale. The demand for homesteads, due to the population increase, will appropriately reduce the scale of cultivated land appropriately. Compared with the previous period, the change rates of average ground agricultural machinery (0.42), average ground investment in fixed assets (0.39), and the town center distance (0.36) have become the dominant factors. The improvement of the agricultural technology accelerates the agricultural production efficiency, but also creates favorable conditions for the development of bare land, which is conducive to timely cultivated land renewal. The "Development Plan for the Yellow River Delta High-Efficiency Ecological Economic Zone" initiates a strategic layout for establishing a high-efficiency eco-agricultural demonstration zone, and a further division of variety of production land zoning, differentiated policy support, and capital investment, based on the development characteristics and regional location advantages.

From the bare land point of view from 2000 to 2010, the leading three influencing factors that cause spatial differentiation of the bare land were the change rates of urbanization (0.65), average ground agricultural machinery (0.54), and road density (0.36). The urbanization expansion forces various types of land to be supplemented in time, and technology and transportation advantages provide excellent conditions for the effective development and utilization of bare land. Compared with the previous period, the top three influencing factors are transformed into the average ground agricultural machinery change rate (0.57), the town center distance (0.49), and the average ground fixed assets investment change rate (0.38). As the available land area is declining, location advantages and national investment support policies are conducive to the conversion of bare land to construction land and wetlands that are beneficial to ecological protection, which have become the leading factors.

From the construction land perspective from 2000 to 2010, the dominant three factors leading to the spatial differentiation of the construction land change rate are the urbanization change rate (0.61), the population change rate (0.53), and the road density change rate (0.46). The Yellow River Basin typical area is rich in oil and salt resources, and has an advantage of location transportation. With the level of urbanization and population agglomeration, the demand for industrial, mining, and residential land continues to increase. Generally speaking, the level of ecological environment protection lags behind the urbanization development, and the rapidness of urbanization has aggravated the regional LER. In comparison with the previous period, the main three factors have been transformed into the average ground fixed assets investment change rate (0.65), the town center distance (0.45), and the urbanization change rate (0.43). With the urbanization acceleration, relying on the advantages of industry and location, the metropolitan areas of Jinan, Dongbin, and Jihe formed gradually. At the same time, the Agricultural High-Tech Industry Demonstration Area of the Yellow River Delta was established in 2015, and the Yellow River Delta Industrial Investment Fund provides strong financial support for economic and technological development zones and typical industrial parks.

In view of the influencing factors of land use changes in the two periods, and based on the main three influencing factors, Figure 6 presents the comparative analysis of the driving mechanism in two periods (from 2000 to 2010 and from 2010 to 2020).

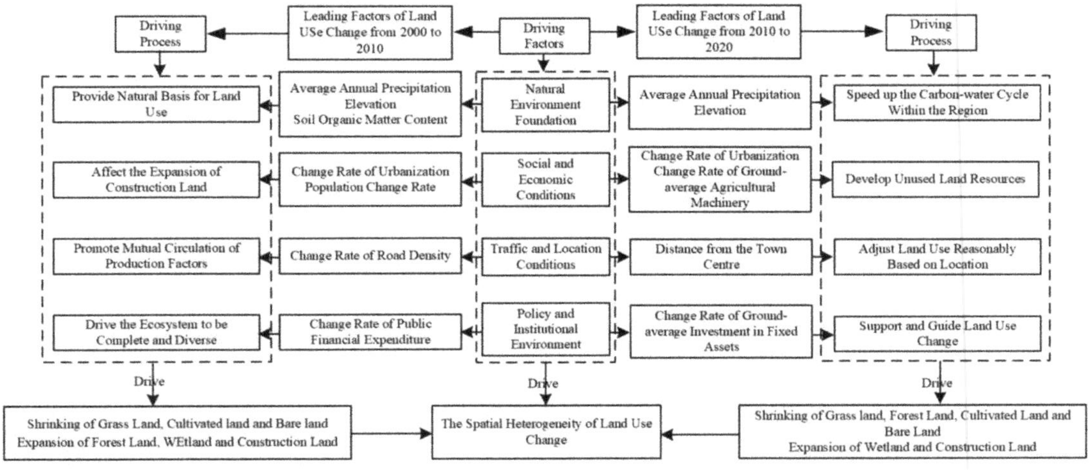

Figure 6. The land use change driving mechanism in typical areas of the Yellow River Basin.

During 2000–2010, a land use change pattern, affected by large-scale land development and high-intensity use, mainly resulted from the expansion of construction land, wetland, and forest land. On the one hand, the complex terrain and abundant rainfall provide a good natural basis for the changes in various land types in typical areas of the Yellow River Basin. At the same time, economic development and large-scale construction of transportation infrastructure, such as railways and highways, provide conditions for the mutual circulation of production factors [42]. During this period, the continuous industrialization, urbanization, and population expansion promoted the formation of more urban industrial, mining, and residential land. In addition, the country has introduced comprehensive and diverse policies for returning farm land to forest and grass land and other favorable ecosystems, prompting the expansion of forest land and wetland. From 2010 to 2020, the state began to pay attention to the importance of ecological protection in the Yellow River region, and successively established the Yellow River Delta High-Efficiency Eco-Economic Zone and the Agricultural High-Tech Industrial Demonstration Zone of the Yellow River Delta. The determination of the efficient eco-economic development promotes the corresponding changes of land use in this stage. In one way, affected by the increase in precipitation, the carbon and water cycle within the region has been accelerated. Additionally, with the continuous development of soil improvement technology, a large number of beaches have been developed and utilized, and the scale of new wetland and construction land is relatively large. In another way, the location advantages are prominent at this stage. The types of regional land use are constantly adjusted according to the location, and the industrial structure is more reasonable. Urbanization has shifted from incremental expansion to stock revitalization. At the same time, the investment of a large number of special funds and advances in technology provide economic support for land use changes. This has advanced the modern and efficient agricultural development and the replacement of old growth drivers with new ones, leading to a significant increase in the level of intensive land use in typical areas of the Yellow River Basin.

3.4. The LER Spatiotemporal Evolution Analysis

Through the LERA model, the value of each risk unit in 2000, 2010, and 2020 is calculated. Based on the natural break point method, the ecological risk is divided into five levels, corresponding to five levels of risk areas: Low ($ERI < 0.36$), relatively low ($0.36 < ERI < 0.50$), medium ($0.50 < ERI < 0.58$), relatively high ($0.58 < ERI < 0.60$), and high risk area ($ERI > 0.60$). Using GIS, the temporal and spatial evolution of risk conversion rate from 2000 to 2010, risk conversion rate from 2010 to 2020, and total conversion rate from 2000 to 2020 are plotted. It can be seen from Figure 7 that the temporal and spatial differentiation of LER in typical areas of the Yellow River Basin are significant. In 2000, high-risk areas accounted for 9.1%, mainly distributed in the central city of Jinan, the Dawen River basin, and Dongying area in northern Shandong. At the same time, taking the high-risk area as the center, the risk level shows a decreasing trend. In 2010, the high-risk area accounted for 7.8%, and the LER level of northern Shandong coastal area and the Dawen River basin decreased. In 2020, the proportion of high-risk areas increased to 14.3% and concentrated in Jinan metropolitan area. The overall LER of Northern Shandong coastal area continued to improve and the effect was significant, gradually showing the trend of "high in southeast, low in northwest" and "high in center, low around".

Based on the natural break point method, the LER conversion rate is divided into six levels, as shown in Figure 8. From 2000 to 2010, about 59.74% of the area had increased LER. Areas with excellent ecological transformation are distributed in the Dezhou Plain Ecological Zone in the northwest of Shandong, the Jihe Plain Ecological Zone in the southwest of Shandong, the Dongying Coastal Ecological Zone in the north of Shandong, and the Zibin Mountain and hills Ecological Zone in the middle east of Shandong. From 2010 to 2020, about 62.34% of the areas have increased LER, and the area of excellent ecological transformation areas has decreased. The overall landscape ecological transformation rate shows a decreasing trend from southeast to northwest. The landscape risks in the southeast have intensified, and the situation is concerning.

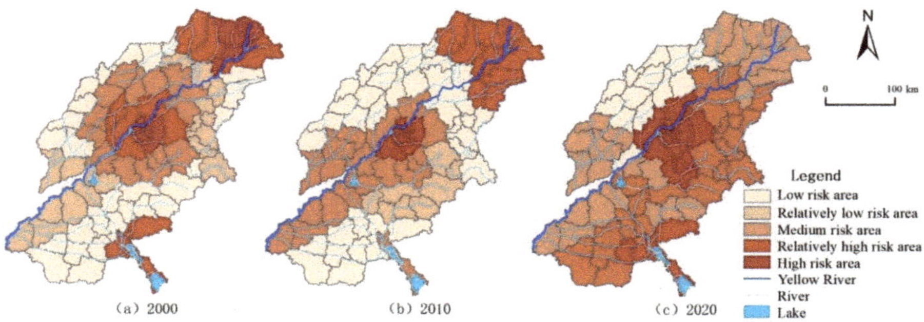

Figure 7. Spatiotemporal variation of LER grade in: (**a**) 2000, (**b**) 2010, and (**c**) 2020.

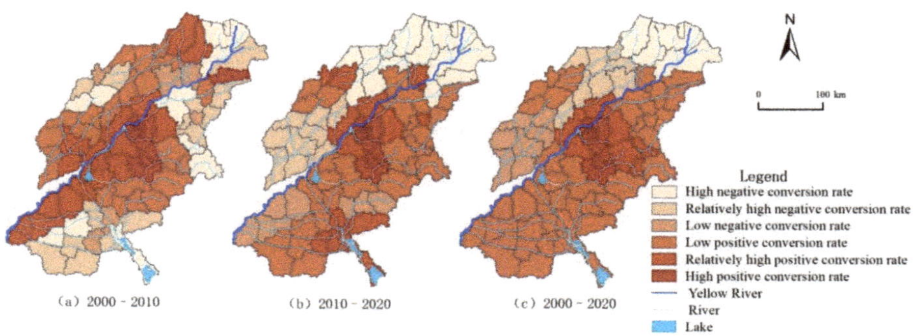

Figure 8. Spatiotemporal variation of LER conversion rate: (**a**) From 2000 to 2010, (**b**) from 2010 to 2020, and (**c**) from 2000 to 2020.

In summary, the conversion rate of LER in typical areas of the Yellow River Basin has gradually increased from 2000 to 2020. The highest value of the positive conversion rate of LER at the county scale has increased from 9.32% to 10.27%, and the negative conversion rate of LER has increased from 2.35% to 19.81%. In terms of spatial distribution, the high positive risk conversion areas are distributed stably and concentrated in the Jinan metropolitan area. The central urban area conversion rate is higher than the surrounding areas. The rapid urbanization and economic development of these areas have led to the continuous expansion of construction land, increasing the ecological pressure, and the LER index has crucially changed. During the study period, the LER in the typical areas of the Yellow River Basin continued to deteriorate. The overall landscape risk in the central urban area increased, the landscape ecology in the northwest region gradually improved, and the ecological risk index showed a decreasing trend, while the ecological risk index in the southeast showed an increasing trend.

By superimposing the LER distribution maps in 2000, 2010, and 2020, the LER transfer levels in typical areas of the Yellow River Basin are obtained (Figure 9). From 2000 to 2010, the risk transfer ratio was 59.74%, among which, the proportion of low risk to relatively low risk, low risk to medium risk, and medium risk to relatively low risk is larger in the number of administrative units. In the first 10 years, the LER levels in typical areas of the Yellow River Basin mostly shifted to relatively low and medium risk. Deterioration was accompanied by improvement. Altogether, the main development trend is a further increase in risk levels. From 2010 to 2020, the risk transfer ratio is 70.13%, among which, the proportion of relatively low risk to relatively high risk, relatively low risk to medium risk, and medium risk to low risk is larger in the number of transferred administrative units. In the next 10 years, the LER of the typical areas of the Yellow River Basin mostly turned to medium-high risk, and the overall risk increased. It can be seen that the LER mostly shifted from low-level to high-level, and the ecological risks are aggravated, which lead to consequential ecological challenges.

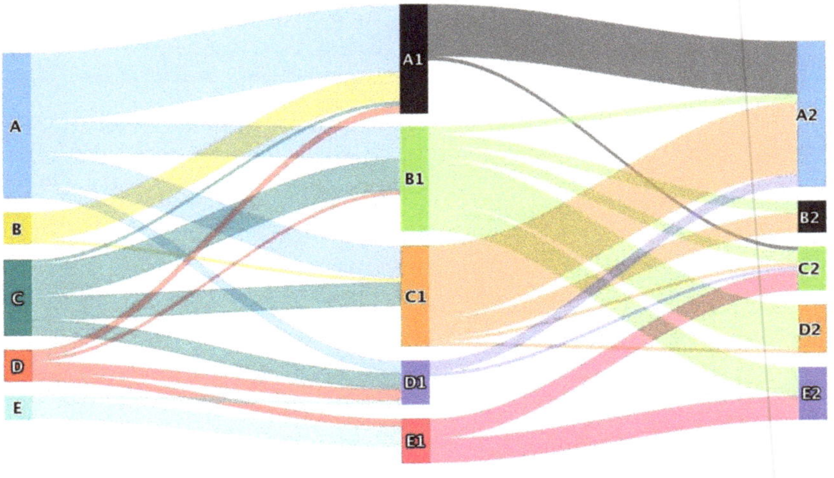

Figure 9. LER transfer situation in typical areas of the Yellow River Basin from 2000 to 2020.

Here, A, B, C, D, and E represent low risk area, relatively low risk area, medium risk area, relatively high risk area, and high risk area in 2000, respectively; A1, B1, C1, D1, and E1 represent low risk area, relatively low risk area, medium risk area, relatively high risk area, and high risk area in 2010, respectively; and A2, B2, C2, D2, and E1 represent low risk area, relatively low risk area, medium risk area, relatively high risk area, and high risk area in 2020, respectively.

3.5. Analysis of LER Response of Land Use Change

3.5.1. The Relationship between Land Use and LER Conversion

The ArcGIS statistical tools were used to obtain the area proportions of different types of land in the transfer of various levels of ecological risk from 2000 to 2020 (Table 4). In the first decade, forest land was mainly distributed in low positive risk conversion areas, accounting for 27.86% of the total forest area. In the second decade, forest land was mainly distributed in low negative risk conversion areas, accounting for 30.54%. The results show that the LER of the forest land is decreasing and the speed is relatively stable. The ecological improvement of forest land is closely related to the state policy on the cultivated land conversion to forest and forest land protection. The forest land ecological improvement is inseparable from the state policies of returning farm land to forest and forest land protection. From 2000 to 2010, wetlands were mainly distributed in low positive risk conversion areas, accounting for 30.05% of the total wetland area. From 2010 to 2020, wetlands areas in relatively high positive risk conversion regions increased significantly, with a total proportion of 33.96%, indicating that the LER of wetland increased sharply during the study period. In recent years, the excessive development of construction land has had a considerable impact on the wetland system. From 2000 to 2010, grass land was mainly distributed in low negative risk conversion areas, accounting for 34.13% of the total grass land area. From 2010 to 2020, the ratio of grass land in relatively high positive risk conversion areas increased significantly, accounting for 29.43%, indicating that the LER of grass land has increased drastically during the study period, but the proportion of the area distributed in the high risk conversion area is smaller. From 2000 to 2010, cultivated land was mainly distributed in low negative risk conversion areas, accounting for 36.35% of the total cultivated land areas. From 2010 to 2020, the ratio of cultivated land in negative risk conversion areas increased significantly, of which low negative risk accounted for 45.47%. This suggested that the LER of cultivated land has been continuously reduced during the research period, which is closely related to provincial emphasis on ensuring the safety of cultivated land and reducing the human interference.

Table 4. Proportion of LER transfer area of different types in typical areas of the Yellow River Basin from 2000 to 2020.

Period	Type	High Negative Risk Conversion Zone	Relatively High Negative Risk Conversion Zone	Low Negative Risk Conversion Zone	Low Positive Risk Conversion Zone	Relatively High Positive Risk Conversion Zone	High Positive Risk Conversion Zone
2000–2010	Forest land	10.35	15.59	20.04	27.86	18.24	7.92
	Wetland	12.85	11.90	16.32	30.05	20.66	8.22
	Grass land	13.56	15.89	34.13	16.34	18.59	1.49
	Cultivated land	10.23	12.45	36.35	15.75	16.34	8.88
	Bare land	7.32	6.34	19.73	20.23	28.96	17.42
	Construction land	8.47	7.40	12.93	21.56	26.45	23.19
2010–2020	Forest land	12.86	18.34	30.54	18.95	16.21	3.10
	Wetland	11.53	10.69	13.25	24.68	33.96	5.89
	Grass land	12.84	14.63	20.34	18.35	29.43	4.41
	Cultivated land	12.31	14.56	45.47	12.96	10.21	4.49
	Bare land	7.02	8.23	12.05	27.45	29.34	15.91
	Construction land	1.79	8.21	12.76	25.67	26.23	25.34

From 2000 to 2020, the distribution areas of construction land and bare land are mainly positive risk conversion areas, and the LER index crucially increased. Among them, the construction land is mainly distributed in high positive risk conversion areas, which is caused by the demand of rapid economic growth and accelerated urbanization process in typical areas of the Yellow River Basin, and the continuous expansion of the construction land is at the cost of the bare land prosperity and the existing land renewal.

3.5.2. The Impact of Land Use Type Conversion on LER

The regional LER often has two opposite trends of improvement and deterioration at the same time. In a certain area, the two trends can be offset to different degrees. On the one hand, the change of LER index reflects the overall LER development level. On the other hand, the LER stability does not mean that the internal ecological risk has not changed. The process of land use change affects the regional LER differentiation, and the mutual transformation between the different types may have a positive or negative effect on LER.

According to the actual situation of the study area, based on the related research results [47,48], the change in the regional LER index caused by the change of a certain land use type is obtained through the ecological risk contribution rate of land use change, and then the impact of land use type transformation on the LER of the typical areas of the Yellow River Basin was determined. Table 5 shows the change of LER index and its main land use change types contribution rate that promote the improvement and degradation of LER in typical areas of the Yellow River Basin during 2000–2020.

Table 5. The main types of land use changes that affect LER and their contribution rate to ecological risks.

Mode	2000–2010			2010–2020		
	The Main Types of Land Use Changes	**Index Change**	**Contribution Proportion (%)**	**The Main Types of Land Use Changes**	**Index Change**	**Contribution Proportion (%)**
Leading to deterioration of LER	Cultivated land -Construction land	0.01628	39.82	Bare land -Construction land	0.01821	37.98
	Wetland -Construction land	0.01069	15.24	Cultivated land -Construction land	0.00953	23.17
	Wetland -Cultivated land	0.00729	12.13	Wetland -Construction land	0.01083	10.86
	Bare land -Construction land	0.00217	11.65	Grass land-Cultivated land	0.00603	6.39
	Forest land -Construction land	0.00372	8.24	Wetland -Cultivated land	0.00527	2.38
	Grass land—Cultivated land	0.00598	5.67	Forest land-Cultivated land	0.00386	1.28
	Total	0.04613	92.75	Total	0.05373	82.06
Leading to the improvement of LER	Cultivated land -Wetland	−0.01023	39.11	Construction land -Cultivated land	−0.00289	36.10
	Bare land -Wetland	−0.00253	18.94	Bare land -Wetland	−0.00363	28.34
	Bare land -Cultivated land	−0.00162	14.83	Construction land -Wetland	−0.00928	10.04
	Grass land-Wetland	−0.00115	9.34	Bare land -Cultivated land	−0.00135	8.96
	Cultivated land -Grass land	−0.00102	5.23	Cultivated land -Wetland	−0.00296	6.48
	Cultivated land -Forest land	−0.00154	3.72	Cultivated land -Grass land	−0.00423	3.27
	Total	−0.01809	91.17	Total	−0.02434	93.19

From 2000 to 2010, land use changes generally promoted the increase of the regional LER index. The conversion of cultivated land to construction land was the leading factor in the deterioration of LER at this stage, accounting for 39.82% of the positive effect of LER. Due to the inadequacy of government policies and the demand for rapid economic development, a large amount of cultivated land has been converted to construction land, which has caused a rise in the LER index and substantial deterioration in ecological safety. At the same time, the conversion of wetland to construction land and the conversion of wetland to cultivated land also contributed to the deterioration of LER to a certain extent, and both accounted for nearly one-third of the LER contribution rate of positive effects. Second, the types of land use changes that worsen the LER are relatively concentrated. One is that other types of land use are converted to construction land and the other is that other types of land use, with ecological protection functions, are converted to cultivated land. This is mainly the construction and cultivated land transfer that accounted for more than 92% of the total contribution rate of the positive effects of LER. On the contrary, the transfer of wetland, the utilization of bare land, and the return of farm land to forests and grass land are important factors to improve the LER in the study area. Among them, the occupation of cultivated land and bare land by wetland is dominant, which accounts for more than 58% of the contribution rate of the LER positive effect.

From 2010 to 2020, the conversion of bare land to construction land is the leading factor in the deterioration of LER in typical areas of the Yellow River Basin at this stage, accounting for 37.98% of the LER positive effect. Under the background of urban expansion, the urgency of social and economic development makes a large number of bare lands exploited and utilized. However, due to the lack of institutional guidance and ecological considerations, the large-scale development of bare land leads to the increase of LER index. The rapid urbanization in this period was at the expense of the environment. At the same time, due to the increase in demand for construction land, some cultivated land and wetland have also been converted to construction land which, to a certain extent, has also contributed to the deterioration of the LER in the study area. The two accounted for more than one-third of the LER positive effect, and the ecological safety problem should not be underestimated. During this period, the type of land use change that worsened the LER was the transfer of construction land and cultivated land, and the contribution rate of forest land and grass land occupied by cultivated land slightly increased. The important factors for the LER improvement in the study area are the conversion of construction land to cultivated land and bare land, and the transfer of construction land to wetland, with a contribution rate of more than 70%. The transfers of cultivated land, wetland, and grass land are important reasons for the LER improvement.

Changes in land use types will change the landscape structure and vulnerability index, leading to original landscape fragmentation and increasing landscape ecological risks. In summary, LER improvement and deterioration coexist in typical areas of the Yellow River Basin, but the overall deterioration trend is greater than the improvement trend, and the degree of deterioration of LER continues to increase.

4. Discussion

4.1. The Relationship between Land Use Change and LER

Land is the carrier of the main social and economic activities, and an important part of the global environmental change and sustainable development research. Driven by economic and social changes and innovation, the types of regional land use have also changed. Land resources are the basis for the survival and development of human society [3]. In recent years, with the acceleration of urbanization, the loss and fragmentation of cultivated land have become more critical. Land itself is the macroscopic representation of the surface landscape, while frequent human activities and high-intensity development and construction make the landscape fragmented and complex, threatening the harmony of humans and land relationship [9]. Changes in land cover, caused by changes in land use, have created changes in the structure and function of ecosystems, and the pattern of surface landscapes has continued to change. Ecosystems may have adverse effects under the direct or indirect effects of land use, and their impacts involve a series of ecological processes such as the atmosphere, soil, water bodies, and organisms and have a wide range of ecological effects. This leads to a variety of real or potential LER, including land degradation [21]. Land use change is highly correlated with the temporal and spatial distribution and dynamics of LER. In fact, the land use change induces LER. To sum up, the LER based on land use change refers to the possibility of changes in landscape structure and reduction of corresponding ecological functions caused by land use and its changes.

LER management refers to the effective prevention and governing measures taken by risk managers for early warning, response, and restoration of LER according to the differences in risk levels and land use types changes in the process of LERA, in order to avoid and reduce LER [44]. The results of LERA of land use can enable managers to understand the spatial distribution of regional LER, identify high risk and medium-high risk areas, and put forward feasible risk control strategies for the temporal and spatial differentiation of different land use types. Land use management is an important approach to LER management. By optimizing the types of land use changes and spatial layout, the regional LER can be effectively reduced. At the same time, land use LERA and management can be continuously improved through mutual feedback. Land use LER management has stages and timeliness. In addition, it is connected with the dynamics of land use research and can promote a virtuous circle of the evaluation process. Land use LER management is a very important part of the ecological risk assessment process. The LERA results can be combined with complex factors, such as regional laws, politics, society, and economy, and its management results can be used for the next landscape risk. Since land use change is driven by nature, social economy, transportation, location, and policies, the LER management of land use can be carried out based on the results of LERA and the factors affecting land use change [48]. Therefore, through the LER differences based on the type of land use, according to the response of the regional LER to the type of land use change, the prevention and control strategies for the LER response and restoration can be proposed.

An important approach for ecological restoration is the construction of ecological projects, most of which involve changes in land use. Carrying out land use LERA provides a strong support and guarantee for ecological restoration. As a new field and important branch of LER research, the LERA of land use can provide a scientific basis and strong decision support for spatial planning and ecological restoration under the background of ecological civilization construction.

4.2. Land Use Control Strategy to Reduce Landscape Ecological Risk

The spatial difference in LER is large in typical areas of the Yellow River Basin. Comprehensively considering that the LER grade and conversion rate in the study area show the trend of "high in the center, low around", the idea of core-peripheral management and control zoning is proposed. Based on the results of county-scale ecological risk diagnosis, in accordance with the need for ecological risk prevention and control and for the convenience of regional management, the typical areas of the Yellow River Basin are divided into "two districts and six pieces" LER key control area, strict control area, and general control area according to the principle of not crossing the municipal administrative region and changing the LER index conversion rate. The "two districts and six pieces" include the core area along the Yellow River and the peripheral linkage area. The core area is divided into four control areas, and the peripheral linkage area is divided into two areas (Figure 10).

1. The control strategy of the core area

The core area covers the main stream of the Yellow River and counties (cities and districts) where Dawen River flows, covering an area of 42,800 km^2. Its development guideline is to build a core area and demonstration leading area for the ecological protection and high-quality progress of the Yellow River Basin. The focus should be on the latter two and on coordinating the inconsistencies among various types of land use, as well as the land use pattern optimization. The essence has to be on improving the central cities' influence and radiation, as well as strengthening the role of Jinan metropolitan area in ecological protection and high-quality development.

The LER of the strict control area in the upper section belongs to the relatively low risk level and medium risk level, and the conversion rate is first reduced and then increased, while the overall conversion rate is positive. Its development orientation is the Yellow River Ecological Corridor Construction Demonstration Zone. The main types of land use changes are "wetland-construction land" and "construction land-cultivated land", which lead to increasing LER in recent years. In the future, the region should pay attention to the implementation of the policy regarding the balance between farmland occupation and compensation. In order to effectively protect wetland, forest land, and other lands from occupation, the ecological security barrier system should be optimized and the radiation ability of Heze metropolitan area to the surrounding areas should be increased.

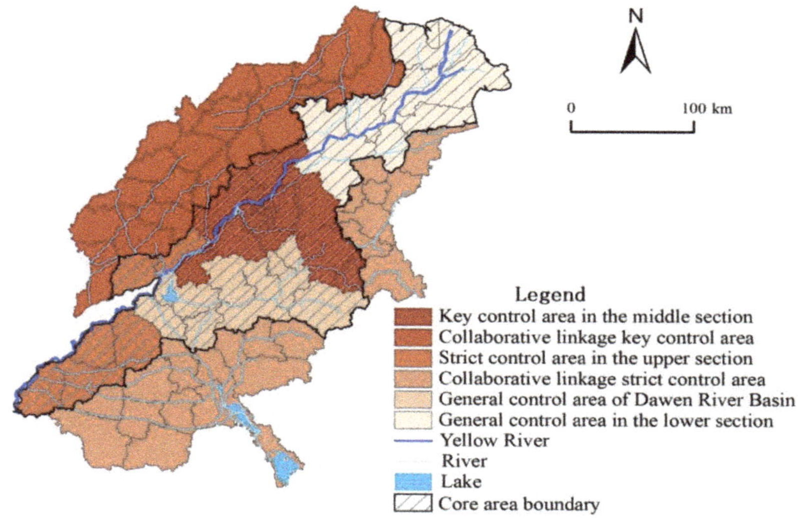

Figure 10. Spatial control distribution of LER in typical areas of the Yellow River Basin.

The LER of the general control area of the Dawen River Basin is dominated by medium and high risks. The risk conversion rate is first increased and then reduced, and the overall conversion rate is negative. Its development is positioned as a long-term unruffled demonstration area of the Yellow River. The main types of land use changes are grass land-wetland and grass land-forest land. Moreover, the intensive use of land reduces the ecological protection function. This region should make full use of mountain landscape and water resources. Relying on the existing nature reserves and tourist attractions, a certain protection zone should be set forth, and the region of adjacent forest land and wetland should be expanded to make them concentrated and contiguous to ensure the ecological environment. Then, the fragmentation of the landscape should be reduced and the anti-risk ability of the ecosystem should be improved.

The LER of the key control area in the middle section is mainly high risk, and the conversion rate showed the simultaneous increase and high positive conversion rate. The main types of land use changes are "cultivated land-construction land" and "bare land-construction land". The economic expansion leads to "the blind expansion" of construction land, resulting in the loss of cultivated land, the transition of bare land development, and the destruction of landscape ecosystem structure. This area is a key area leading the high-quality development of the Yellow River Basin. The maintenance of forest and grass land should be strengthened to reduce land loss and improve the stability of the ecosystem. In addition, it is necessary to control the regional population and reduce the occupation of cultivated land resources due to the expansion of construction land, and then rationally develop the unused land and adjust the land use structure based on location factors.

The LER of the general control area in the lower section belongs to relatively low and medium risk, and the conversion rate shows the simultaneous decrease, and the overall conversion rate is negative. The main land use change types are "cultivated land-wetland" and "construction land-cultivated land". The ecological risk has reduced significantly. The development is oriented towards collaborative protection and developing demonstration zones. We should establish wetland parks and other nature reserves, strengthen the maintenance of forest and grass land around rivers, reduce land loss and water fragmentation, and continue to provide policy and financial support, by exploring regional ecological protection and linkages to development mechanisms, while promoting the integrated development of the cities, such as Dongying, Binzhou and Lijin.

2. The control strategy of the peripheral linkage area

The peripheral linkage area includes 42 counties (cities, districts), except the core area. The LER level of the collaborative linkage of the key control area mainly belongs to the low risk, while the conversion rate shows a simultaneous increase and a low positive conversion rate. It is located in the ecological area of the northwest plain of Shandong Province. The types of land use change are relatively scattered, and the proportion of cultivated land to construction land is relatively large. With accelerated urbanization, the construction land continues to spread outward, the ecological load is severe, and the ecological risk continues to rise. As Shandong Province is largely agricultural and the protection and utilization of cultivated land is very important, it is recommended to divide the cultivated land based on quality and implement a policy of returning the farm land to forests on cultivated land of poor quality. In addition, the pressure of construction land expansion should be appropriately eased and shifted from incremental development to stock renewal with intensive and efficient utilization. The level of LER in the collaborative linkage strict control area is mainly medium and high risk, and the conversion rate is first increased and then reduced, while the overall conversion rate is positive. The main land use change types in this area are "wetland-construction land" and "forest land-construction land". We should adopt the leading role of land and space planning to reduce the human interference intensity, make a gradual development of construction land to go together with stock renewal, and protect forest land and wetland from intrusion. While mobilizing the enthusiasm of economic development, the negative impact on landscape ecology will be minimized, in order to maintain the current trend of ecological risk transformation and further improve the landscape ecology. The peripheral linkage area should focus on the joint protection and governance of environment, the integration of living spaces, and the coordinated development of urban and rural areas, while comprehensively reinforcing the collaborative linkage with the core area.

There are extensive distinctions in resource endowments and industrial development levels in typical areas of the Yellow River Basin. The changes in land use types should be promoted according to regional resource endowment conditions, ecological environment capacity, and positioning of main functions. Moreover, its ecological effects should be considered in the formulation of relevant policies and plans. In the process of land use change, it is necessary to actively adjust the structure and layout of land use to strengthen the self-healing function of the ecosystem.

5. Conclusions

Based on the land cover data in typical areas of the Yellow River Basin in 2000, 2010, and 2020, this study conducts a research on landscape ecological risk response and countermeasures of land use change. The main conclusions are as follows:

(1) The analysis of land use structure demonstrates that the main types of land use in typical areas of the Yellow River Basin are cultivated land and construction land. The change processes of various land use types are significantly different, showing the characteristics of "many-to-one", "one-to-many", and "balanced". Among them, the scale of forest land first increases and then decreases, the area of wetland and construction land increases sharply, and the areas of grass land, cultivated land, and bare land continue to shrink. In the conversion of different land use types, the exchange of cultivated land and construction land, the transfer of construction land to wetland, and the transfer of bare land to wetland are more prominent, as well as denser in coastal areas and more scattered in inland areas;

(2) The process of land use change is affected by the factors of nature, society, economy, location, and policy. Within the first decade, the natural environment, society, and economy played a leading role in land use changes. In the second decade, the

influence of natural factors declined, while the influence of location and policy factors increased significantly;

(3) The results show that the overall LER grades have the characteristics of "high in the southeast, low in the northwest" and "high in the center, low in the surroundings". The conversion rate of LER increased gradually, and the spatial distribution showed a decreasing trend from southeast to northwest. Most of the ecological risks have shifted from low level to high level. In recent years, the ecological risks of bare land and construction land have increased severely, which should cause concern;

(4) The change of land use type will change the landscape structure and vulnerability index, resulting in the original landscape fragmentation and the increase of LER. The landscape ecological improvement and deterioration coexist in typical areas of the Yellow River Basin, but the general landscape ecological deterioration trend is greater than the improvement trend, and the deterioration degree of the landscape ecological environment is increasing; and

(5) According to the results of the diagnosis of county-scale LER and the need of ecological risk prevention and control, the typical areas of the Yellow River Basin are divided into "two districts and six pieces" LER with the key control area, strict control area, and general control area. It is committed to transform the Yellow River Basin in Shandong Province into "Shandong model for ecological protection of the Yellow River Basin and a core growth pole for high-quality development".

This paper attempts to develop differentiated land use regulation strategies for research areas with different landscape ecological risk levels and regional characteristics from the influencing factors and driving mechanism of land use change for the first time. However, the following contents in the future should get further exploration. First, the LER temporal and spatial characteristics have obvious scale effects, and there may be a certain degree of difference in the research results with various exploration scales. This article is based on the remote sensing of land use (raster data, resolution of 1 km) from 2000 to 2020 obtained from the Resource and Environment Science and Data Center (RESDC) of Chinese Academy of Sciences. The results of this study reflect the LER distinctions and their change process at the macro-level in typical areas of the Yellow River Basin, but it is difficult to accurately describe the risk traits of some local areas or land types with a small area. In order to reflect the change of ecological risk more sensitively, it should be considered to study the scale characteristics of ecological risk change by setting grid cells of different sizes. Second, remote sensing data are not only an effective method for land use change research, especially dynamic monitoring, but also play an important role in analyzing the land use change pattern. Remote sensing is able to obtain a large range of data by virtue of its advantages, such as multiple means of obtaining information, fast speed, and not being blocked by terrain, but it may not be able to fully display the LUCC process. Therefore, the combination of traditional ground survey methods, such as land survey, topographic map query, field visit, and questionnaire survey with remote sensing technology will be the focus of the next step. Moreover, it is an important means to deepen the understanding of land use dynamics. Furthermore, this exploration will help in providing more reasonable suggestions for the high-quality development of the basin, and provide more effective regulation and control strategies for decision makers.

Author Contributions: Conceptualization, Y.Q.; methodology, Y.Q., H.Z., and D.S.; formal analysis, Y.Q., H.Z., and D.S.; data curation, Y.Q., H.Z., D.S., and M.G.; writing—original draft preparation, Y.Q., H.Z., and Z.P.; writing—review and editing, Y.Q., H.Z., and Z.P.; visualization, Y.Q., H.Z., and M.G.; supervision, Y.Q. and Z.P. All authors have read and agreed to the published version of the manuscript.

Funding: This research was funded by the National Natural Science Foundation of China, grant numbers 42077434, 41771560, and the Shandong Provincial Institutions of Higher Learning "Youth Innovation Team Development Plan" Project, grant number 2019RWG016.

Institutional Review Board Statement: Not applicable.

Informed Consent Statement: Not applicable.

Data Availability Statement: All of the relevant data sets in this study are described in the manuscript.

Acknowledgments: The authors sincerely thank the editors and the anonymous reviewers for their constructive feedback.

Conflicts of Interest: The authors declare no conflict of interest.

References

1. Peng, J.; Dang, W.X.; Liu, Y.X.; Zong, M.L.; Hu, X.X. Review on landscape ecological risk assessment. *Acta Geogr. Sin.* **2015**, *70*, 664–677.
2. Liu, Y.C.; Liu, Y.X.; Li, J.L.; Lu, W.Y.; Wei, X.L.; Sun, C. Evolution of Landscape Ecological Risk at the Optimal Scale: A Case Study of the Open Coastal Wetlands in Jiangsu, China. *Int. J. Environ. Res. Public Health* **2018**, *15*, 1691. [CrossRef] [PubMed]
3. Gao, B.P.; Li, C.; Wu, Y.M.; Zheng, K.J.; Wu, Y. Landscape ecological risk assessment and influencing factors in ecological conservation area in Sichuan-Yunnan provinces, China. *Chin. Appl. Ecol.* **2021**, *32*, 1603–1613.
4. Xu, W.X.; Wang, J.M.; Zhang, M.; Li, S.J. Construction of landscape ecological network based on landscape ecological risk assessment in a large-scale opencast coal mine area. *J. Clean Prod.* **2021**, *286*, 125523. [CrossRef]
5. Suter, G.W.; Norton, S.B.; Barnthouse, L.W. The evolution of frameworks for ecological risk assessment from the Red Book ancestor. *Hum. Ecol. Risk. Assess.* **2003**, *9*, 1349–1360. [CrossRef]
6. Wang, D.; Ji, X.; Li, C.; Gong, Y.X. Spatiotemporal variations of landscape ecological risks in a resource-based city under transformation. *Sustainability* **2021**, *13*, 5297. [CrossRef]
7. Wang, H.; Liu, X.M.; Zhao, C.Y.; Chang, Y.P.; Liu, Y.Y.; Zang, F. Spatial-temporal pattern analysis of landscape ecological risk assessment based on land use/land cover change in Baishuijiang National nature reserve in Gansu Province, China. *Ecol. Indic.* **2021**, *124*, 107454. [CrossRef]
8. Chen, C.; Park, T.; Wang, X.; Piao, S.; Xu, B.; Chaturvedi, R.K.; Fuchs, R.; Brovkin, V.; Ciais, P.; Fensholt, R.; et al. China and India lead in greening of the world through land-use management. *Nat. Sustain.* **2019**, *2*, 122–129. [CrossRef]
9. Wang, J.; Bai, W.Q.; Tian, G.H. Spatiotemporal characteristics of landscape ecological risks on the Tibetan Plateau. *Resour. Sci.* **2020**, *42*, 1739–1749. [CrossRef]
10. Xie, H.L.; Wang, P.; Huang, H. Ecological risk assessment of land use change in the Poyang Lake eco-economic zone, China. *Int. J. Environ. Res. Public Health* **2013**, *10*, 328–346. [CrossRef]
11. Turner, B.L. The sustainalility princilpe in global agendas: Implication understanding land use/cover change. *Geogr. J.* **1997**, *163*, 133–140.
12. Foley, J.A.; DeFries, R.; Asner, G.; Barford, C.; Bonan, G.; Carpenter, S.R.; Chapin, F.S.; Coe, M.; Daily, G.C.; Gibbs, H.K.; et al. Global consequences of land use. *Science* **2005**, *309*, 570–574. [CrossRef]
13. Turner, M.G. *Landscape Ecology in Theory and Practice: Pattern and Process*; Springer: New York, NY, USA, 2001.
14. Tang, L.; Ma, W. Assessment and management of urbanization-induced ecological risks. *Int. J. Sustain. Dev. World* **2018**, *25*, 383–386. [CrossRef]
15. Hua, L.Z.; Liao, J.F.; Chen, H.X.; Chen, D.K.; Shao, G.F. Assessment of ecological risks induced by land use and land cover changes in Xiamen City, China. *Int. J. Sustain. Dev. World* **2018**, *25*, 439–447. [CrossRef]
16. Ju, H.R.; Niu, C.Y.; Zhang, S.R.; Jiang, W.; Zhang, Z.H.; Zhang, X.L.; Yang, Z.Y.; Cui, Y.R. Spatiotemporal patterns and modifiable areal unit problems of the landscape ecological risk in coastal areas: A case study of the Shandong Peninsula, China. *J. Clean Prod.* **2021**, *310*, 127522. [CrossRef]
17. Liu, X.Z.; Li, X.S.; Jiang, D.M. Landscape pattern identification and ecological risk assessment using land-use change in the Yellow River Basin. *Trans. Chin. Soc. Agric. Eng.* **2021**, *37*, 265–274.
18. Daniel, T.; Heggem, C.M.; Edmonds, A.C.; Neale, L.; Bice, K.; Bruce, J. A landscape ecology assessment of the Tensas River Basin. *Environ. Monit. Assess.* **2000**, *64*, 41–54.
19. Lawrence, A.K. Using landscape ecology to focus ecological risk assessment and guide risk management decision-making. *Toxicol. Ind. Health* **2001**, *17*, 236–246.
20. Estoque, R.C.; Murayama, Y.; Lasco, R.D.; Myint, S.W.; Pulhin, F.B.; Wang, C.; Ooba, M.; Hijioka, Y. Changes in the landscape pattern of the La Mesa Watershed: The last ecological frontier of Metro Manila, Philippines. *For. Ecol. Manag.* **2018**, *430*, 280–290. [CrossRef]
21. Liu, Z.H.; Zhang, G.J.; Fu, F.J. Assessing landscape ecological risk based on landscape pattern and services in Guangzhou during 1990–2015. *Acta Ecol. Sin.* **2020**, *40*, 3295–3302.
22. Chen, X.Y.; Xie, G.Z.; Zhang, J.P. Landscape ecological risk assessment of land use changes in the coastal al'ea of Haikou City in the past 30 years. *Acta Ecol. Sin.* **2021**, *41*, 975–986.
23. Zhang, W.; Long, J.H.; Zhang, X.R.; Shen, W.N.; Wei, Z.Y. Pollution and ecological risk evaluation of heavy metals in the soil and sediment around the HTM tailings pond, Northeastern China. *Int. J. Environ. Res. Public Health* **2020**, *17*, 7072. [CrossRef]
24. Zhang, W.J.; Sun, X.Y.; Shan, R.F.; Liu, F. Spatio-temporal quantification of landscape ecological risk changes and its driving forces in the Nansihu Lake basin during 1975–2018. *Ecol. Sci.* **2020**, *39*, 172–181.

25. Li, Q.P.; Zhang, Z.D.; Wan, L.W.; Yang, C.X.; Yan, J.; Ye, C.; Chen, Y.C. Landscape pattern optimization in Ningjiang River Basin based on landscape ecological risk assessment. *Acta Geogr. Sin.* **2019**, *74*, 1420–1437.
26. Cui, L.; Zhao, Y.H.; Liu, J.C.; Han, L.; Ao, Y.; Yin, S. Landscape ecological risk assessment in Qinling Mountain. *Geol. J.* **2018**, *53*, 342–351. [CrossRef]
27. Gong, J.; Yang, J.X.; Tang, W.W. Spatially explicit landscape-level ecological risks induced by land use and land cover change in a national ecologically representative region in China. *Int. J. Environ. Res. Public Health* **2015**, *12*, 14192–14215. [CrossRef]
28. Chen, L.D.; Fu, B.J.; Zhao, W.W. Source-sink landscape theory and its ecological significance. *Acta Ecol. Sin.* **2006**, *26*, 1444–1449. [CrossRef]
29. Liu, D.; Chen, H.; Zhang, H.; Geng, T.W.; Shi, Q.Q. Spatiotemporal evolution of landscape ecological risk based on geomorphological regionalization during 1980–2017: A case study of Shaanxi Province, China. *Sustainability* **2020**, *12*, 941. [CrossRef]
30. Zhang, W.; Chang, W.J.; Zhu, Z.C.; Hui, Z. Landscape ecological risk assessment of Chinese coastal cities based on land use change. *Appl. Geogr.* **2020**, *117*, 102174. [CrossRef]
31. Cui, Y.L.; Gao, X.; Dong, B.; Wei, H.M. Landscape Ecological Risk Assessment of County. *J. Zhejiang A F Univ.* **2021**, *38*, 541–551.
32. Peng, J.; Li, H.L.; Liu, Y.X.; Hu, Y.N.; Yang, Y. Identification and optimization of ecological security pattern in Xiong'an New Area. *Acta Geogr. Sin.* **2018**, *73*, 701–710.
33. Zhao, W.W.; Fang, X.N. Landscape sustainability and landscape sustainability science. *Acta Ecol. Sin.* **2014**, *34*, 2453–2459.
34. Sun, H.B.; Yang, G.S.; Su, W.Z.; Zhu, T.M.; Wang, R.R. Ecological risk assessment of land use in the area along Changjiang River: A case study of Nanjing, China. *Acta Ecol. Sin.* **2010**, *30*, 5616–5625.
35. Wang, J.; Bai, W.Q.; Tian, G.H. A review on ecological risk assessment of land use. *J. Nat. Resour.* **2020**, *35*, 576–585.
36. Gong, J.; Cao, E.J.; Xie, Y.C.; Xu, C.X.; Li, H.Y.; Yan, L.L. Integrating ecosystem services and landscape ecological risk into adaptive management: Insights from a western mountain-basin area, China. *J. Environ. Manag.* **2021**, *281*, 111817. [CrossRef]
37. Ma, Y.Y.; Liu, Z.F. *Assessment of Landscape Ecological Risk and Its Dynamic Response with Urbanization Development of Jiangsu Province*; Journal of Nanjing Forestry University (Natural Sciences Edition): Nanjing, China, 2021; pp. 1–18.
38. Blaikie, N.; Priest, J. *Social Research: Paradigms in Action*, 3rd ed.; Polity Press: Cambridge, UK, 2017; p. 280.
39. Xiao, D.Y.; Niu, H.P.; Yan, H.X.; Fan, L.X.; Zhao, S.X. Spatiotemperal evolution of land use pattern in the Yellow River Basin(Henan section) from 1990 to 2018. Trans. *Chin. Soc. Agric. Eng.* **2020**, *36*, 271–281.
40. Zhang, W.H.; Lv, X.; Shi, Y.Y.; Sun, P.L.; Zhang, Y.S. Graphic characteristics of land use transition in the Yellow River Basin. China. *Land Sci.* **2020**, *34*, 80–88.
41. Qu, Y.B.; Wang, S.L.; Zhu, W.Y.; Ping, Z.L. Spatial-temporal differentiation characteristics and driving force of territorial space evolution in the Yellow River Delta. *Trans. Chin. Soc. Agric. Eng.* **2021**, *37*, 252–263.
42. Wang, J.F.; Xu, C.D. Geodetector: Principle and prospective. *Acta Geogr. Sin.* **2017**, *72*, 116–134.
43. Ma, S.C.; Xie, F.F.; Ding, C.; Zhang, H.B. Spatio-temporal change of landscape ecological quality and influencing factors based on four-quadrant model in overlapped area of cropland and coal production. *Trans. Chin. Soc. Agric. Eng.* **2020**, *36*, 259–268.
44. Zeng, Y.N.; Jin, W.P.; Wang, H.M.; Zhang, H.H. Simulation of land-use changes and landscape ecological assessment in eastern part of Qinghai Plateau. *Trans. Chin. Soc. Agric. Eng.* **2014**, *30*, 185–194.
45. Yang, Q.K.; Duan, X.J.; Wang, L.; Jing, Z.F. Land use transformation based on ecological-production-living spaces and associated eco-environment effects: A case study in the Yang-tze River Delta. *Sci. Geogr. Sin.* **2018**, *38*, 97–106.
46. Chakraborty, D.; Redd, M.; Tiwari, S.; Umapathy, G. Land use change increases wildlife parasite diversity in Anamalai hills, Western Ghats, India. *Sci. Rep.* **2019**, *9*, 11975. [CrossRef]
47. Gao, X.; Liu, Z.W.; Li, C.X.; Cha, L.S.; Song, Z.Y.; Zhang, X.R. Land use function transformation in the Xiong'an New Area based on ecological-production- living spaces and associated eco-environment effects. *Acta Ecol. Sin.* **2020**, *40*, 7113–7122.
48. Huang, T.N.; Zhang, Y.L. Transformation of land use function and response of eco-environment based on "production-life-ecology space": A case study of resource-rich area in western Guangxi. *Acta Ecol. Sin.* **2021**, *41*, 348–359.

International Journal of
Environmental Research and Public Health

MDPI

Article

Spatiotemporal Variation in Rainfall Erosivity and Correlation with the ENSO on the Tibetan Plateau since 1971

Bohao Cui [1,2], Yili Zhang [1,2,*], Linshan Liu [1], Zehua Xu [2,3], Zhaofeng Wang [1,2], Changjun Gu [1,2], Bo Wei [1,2] and Dianqing Gong [1,2]

1 Key Laboratory of Land Surface Pattern and Simulation, Institute of Geographic Sciences and Natural Resources Research, CAS, Beijing 100101, China; cuibh.19b@igsnrr.ac.cn (B.C.); liuls@igsnrr.ac.cn (L.L.); wangzf@igsnrr.ac.cn (Z.W.); gucj.18b@igsnrr.ac.cn (C.G.); weib.20b@igsnrr.ac.cn (B.W.); gongdq.18s@igsnrr.ac.cn (D.G.)
2 College of Resources and Environment, University of Chinese Academy of Sciences, Beijing 100049, China; zhxu_st@rcees.ac.cn
3 State Key Laboratory of Urban and Regional Ecology, Research Center for Eco-Environmental Sciences, Chinese Academy of Sciences, 18 Shuangqing Road, Beijing 100085, China
* Correspondence: zhangyl@igsnrr.ac.cn

Citation: Cui, B.; Zhang, Y.; Liu, L.; Xu, Z.; Wang, Z.; Gu, C.; Wei, B.; Gong, D. Spatiotemporal Variation in Rainfall Erosivity and Correlation with the ENSO on the Tibetan Plateau since 1971. *Int. J. Environ. Res. Public Health* **2021**, *18*, 11054. https://doi.org/10.3390/ijerph182111054

Academic Editor: Paul B. Tchounwou

Received: 17 August 2021
Accepted: 19 October 2021
Published: 21 October 2021

Abstract: Soil erosion is a serious ecological problem in the fragile ecological environment of the Tibetan Plateau (TP). Rainfall erosivity is one of the most important factors controlling soil erosion and is associated with the El Niño southern oscillation (ENSO). However, there is a lack of studies related to the spatial distribution and temporal trends of rainfall erosivity on the TP as a whole. Additionally, the understanding of the general influence of ENSO on rainfall erosivity across the TP remains to be developed. In this study, long-term (1971–2020) daily precipitation data from 91 meteorological stations were selected to calculate rainfall erosivity. The analysis combines co-kriging interpolation, Sen's slope estimator, and the Mann–Kendall trend test to investigate the spatiotemporal patten of rainfall erosivity across the TP. The Oceanic Niño Index (ONI) and multivariate ENSO Index (MEI) were chosen as ENSO phenomenon characterization indices, and the relationship between ENSO and rainfall erosivity was explored by employing a continuous wavelet transform. The results showed that an increasing trend in annual rainfall erosivity was detected on the TP from 1971 to 2020. The seasonal and monthly rainfall erosivity was highly uneven, with the summer erosivity accounting for 60.36%. The heterogeneous spatial distribution of rainfall erosivity was observed with an increasing trend from southeast to northwest. At the regional level, rainfall erosivity in the southeastern TP was mainly featured by a slow increase, while in the northwest was more destabilizing and mostly showed no significant trend. The rainfall erosivity on the whole TP was relatively high during non-ENSO periods and relatively low during El Niño/La Niña periods. It is worth noting that rainfall erosivity in the northwest TP appears to be more serious during the La Niña event. Furthermore, there were obvious resonance cycles between the rainfall erosivity and ENSO in different regions of the plateau, but the cycles had pronounced discrepancies in the occurrence time, direction of action and intensity. These findings contribute to providing references for soil erosion control on the TP and the formulation of future soil conservation strategies.

Keywords: rainfall erosivity; soil erosion; spatiotemporal variation; ENSO; Tibetan Plateau

1. Introduction

Soil erosion has already emerged as one of the most serious ecological and environmental problems globally, which not only threatens terrestrial ecosystems, but also severely restricts the security of human existence and the sustainable development of economy and society [1,2]. Soil erosion not only contributes to land degradation, but even interferes with the ability of the soil carbon cycle to mitigate the greenhouse effect [3,4]. Soil erosion by water is considered to be one of the most detrimental types of soil erosion, causing a loss of soil nutrients, which reduces crop yields, pollutes water quality, contributes to the sedimentation of rivers, and raises flooding [5–9]. Therefore, the accurate prediction of water erosion is of great significance for the comprehensive management of soil erosion and effective soil protection.

The causes of water erosion are related to a series of natural factors involving rainfall, soil, topography, vegetation, and other human factors such as land use and crop cultivation management [10,11]. In particular, rainfall is the principal climatic factor responsible for water erosion, which influences water erosion through the duration, amount, and intensity of rainfall events [12]. The principal predictive tools for water erosion are the Universal Soil Loss Equation (USLE) and the Revised Universal Soil Loss Equation (RUSLE), which have been applied worldwide, and rainfall erosivity (R-factor), one of the key input parameters in the model, is the potential capacity of rainfall to induce water erosion [12,13]. The R-factor is defined as the product of the rainfall energy and the maximum rainfall intensity in a 30 min period (EI30), and the calculation requires the use of consecutive rainfall data series with a temporal resolution of at least 15 min, which is however, hardly available in many countries and regions. Even if an adequate rain gauge data can be accessed, the complicated calculation process is time-consuming and laborious, which dramatically restricts model promotion and implementation [14].

In this context, as alternative algorithms based on the relationship between R-factor and available rainfall data were developed, including the calculation of rainfall erosivity based on annual [15,16], monthly [17,18], and daily [19,20] rainfall data from meteorological stations or satellite radar [21,22]. Among these, daily rainfall data are widely used due to their relative accessibility, which provides more characteristic information of rainfall and facilitates the precision and reliability of R-factor estimation [23]. In the daily rainfall data model, the rainfall erosivity algorithm was divided into linear exponential, logarithmic, and power functions to fit the relationship between rainfall and rainfall erosivity, and the models are mostly combined experimental and empirical based [24–26]. In general, these models perform an important role in the quantitative evaluation of rainfall erosivity, and provide scientific reference for the forming mechanism of water erosion, evolutionary process and even the mechanics of climate change.

It has become an indisputable fact that the global climate is changing remarkably, with extreme weather events growing stronger, more frequent, and lasting longer [27]. El Niño southern oscillation (ENSO) is the most intense sea-air interaction event affecting the global climate, and although it usually occurs in the eastern equatorial Pacific region, it can be responsible for rainfall anomalies spreading globally [28]. The ENSO cycle has a pronounced periodic character as a result of the interaction between the ocean and the atmosphere, with El Niño (warm phase) and La Niña (cold phase) as the two extreme phases of the ENSO cycle. Considerable work has been conducted on the relationship between ENSO and precipitation events, anomalous temperature, wet and dry variability, and atmospheric circulation [29–32]. The studies also pointed out that El Niño and La Niña showed diverse rainfall patterns, for example, compared with the La Niña period, northern China is more arid during El Niño in the northern hemisphere, while rainfall in the southeast of China appears to increase substantially, while the contrary phenomenon is present in the southern hemisphere [33,34]. Although these studies have enhanced our comprehension of atmospheric tele-correlation model (ENSO) effects on rainfall, currently the effect of ENSO on rainfall erosivity is still only shown in a few studies [35–39]. A significant dependence between rainfall erosivity and the ENSO indices has been observed in eastern China [35,37], northeastern Spain [40], and the southwestern United States [41], while studies on how ENSO affects rainfall erosivity on the TP are still unknown.

The Tibetan Plateau (TP) is the largest and highest geographical unit in the world, with an average altitude of over 4000 m, and is called the Earth's "third pole". It is of extreme importance to regional economic development and ecological security, as well as global climate, water resources, and ecosystem functioning [42]. Since the 21st century, however, drastic environmental changes have been remarkably observed on the TP [43]. These changes have become key drivers of increased soil erosion risk. Studies have demonstrated that grassland ecosystems on the TP are suffering from severe degradation due to the combined effects of climate change and human activities. This in turn has triggered a decline in biomass, biodiversity, and landscape complexity, fragmentation or complete loss of services such as soil and water conservation, and an increase in rainfall erosivity and sandstorms [44–47]. Permafrost degradation can reduce the stability of soil aggregates and the water content in the surface soil is abnormally high during the thawing stage, thus shortening the time of runoff generation and exacerbating erosion caused by rainfall [48,49]. The glaciers' retreat and the rise of the snow line in cold areas at high altitudes have changed the surface albedo and atmospheric heat circulation and thus have affected the local rainfall intensity, and the form of erosion caused by glacial meltwater and snowmelt runoff generated is one of the main reasons for increased erosion [50,51]. Additionally, according to observations and climatological models, the TP has suffered a faster rate of warming since the 1960s, which is three times the global average [52]. Notable changes in the plateau climate system, such as short periods of intense rainfall triggered by extreme precipitation events, may have led to an increasing trend in the rainfall erosivity on the TP [51]. Some studies have analyzed the variation in rainfall erosivity in the catchment and local scales of the TP, indicating an increasing trend of rainfall erosivity [53–55]. These studies provided useful information on the variation in rainfall erosivity, but a further analysis is necessary for the TP as a whole.

Soil erosion is serious on the TP, with 70% of the area suffering from varying degrees of soil erosion [56]. In both sides of the Yarlung Tsangpo River and the South Qiangtang area, gully erosion is widely distributed, while in the interior of the plateau scale erosion becomes the main type of erosion in grasslands [57]. Soil erosion on the TP has caused irreparable soil degradation and land area reduction, and is leading to the sedimentation of downstream rivers, landslides, mudslides and other disasters, posing a threat to transportation, agriculture, and animal husbandry. Moreover, soil conservation is particularly important in the TP due to its harsh physical environment, widespread permafrost and fragile alpine ecosystems making it the most sensitive and fragile region [58]. Once erosion happens, its rehabilitation process is prolonged and difficult.

Detection of long-term trends in rainfall erosivity can provide information regarding the potential impact of rainfall changes on soil erosion. It is particularly useful for the TP region, which is more sensitive to water erosion and climate change because of the fragile biophysical conditions [59]. However, these unique geographical features and complicated terrain have restricted soil erosion studies due to the scarce observational data on precipitation and soil erosion. Previous studies have focused on local watersheds or small areas of the TP, while the spatial and temporal characteristics of how rainfall erosivity vary over the entire TP have not been adequately studied [60,61]. Moreover, periodic factors lead to 'poverty years' and 'abundant years' of precipitation in the highlands in different years. The interannual variation in precipitation erosivity on the TP may be the result of ENSO action, but the general effect of ENSO on rainfall erosivity in the TP is not clear at present, and it is necessary to expand the related understanding.

In view of this, the TP as a whole was chosen as the study area, and daily rainfall data from 91 meteorological stations were selected to calculate the rainfall erosivity and to explore its relationship with ENSO. The objectives of the study are as follows: (1) to characterize the temporal trends of rainfall erosivity during 1971–2020 across the entire TP; (2) to present the spatial distribution of rainfall erosivity on the TP; (3) and to investigate the impacts of ENSO on rainfall erosivity in different regions of the TP.

2. Materials and Methods

This study was based on a single case study of the TP. In this section, the basic information about the study area, the required data handling process and the methods related to rainfall erosivity were described in detail.

2.1. *Study Area*

The Tibetan Plateau is located in the southwestern part of China, with an area of 2.74×10^6 km^2 and an average altitude of over 4000 m. It is known as the "roof of the world" [62]. It is included in the Tibet Autonomous Region and Qinghai Province, and the southern part of the Xinjiang Uygur Autonomous Region, the western part of Gansu Province, the western part of Sichuan Province and the northern part of Yunnan Province. The main mountain ranges are the Kunlun Mountains, Qilian Mountains, Karakorum Mountains, Himalayas, and Hengduan Mountains. The climate ranges from a humid monsoon climate in the southeast to an alpine arid plateau climate in the northwest, controlled by the Pacific monsoon, Indian monsoon, and prevailing westerly winds, and is influenced by the mountain terrain [63]. Diverse climate types form subtropical rainforests, shrubs, alpine meadows, alpine grasslands, and alpine desert vegetation types are present. TP precipitation exhibits a distinct gradient, gradually decreasing from more than 1000 mm in the southeast to less than 50 mm in the northwest [64]. The region has experienced soil erosion, desertification and landslide hazard [51,65]. Referring to [66], the criteria for the physical geographic zoning of the TP divided the plateau into Region I (arid zone) and Region II (humid zone) (Figure 1).

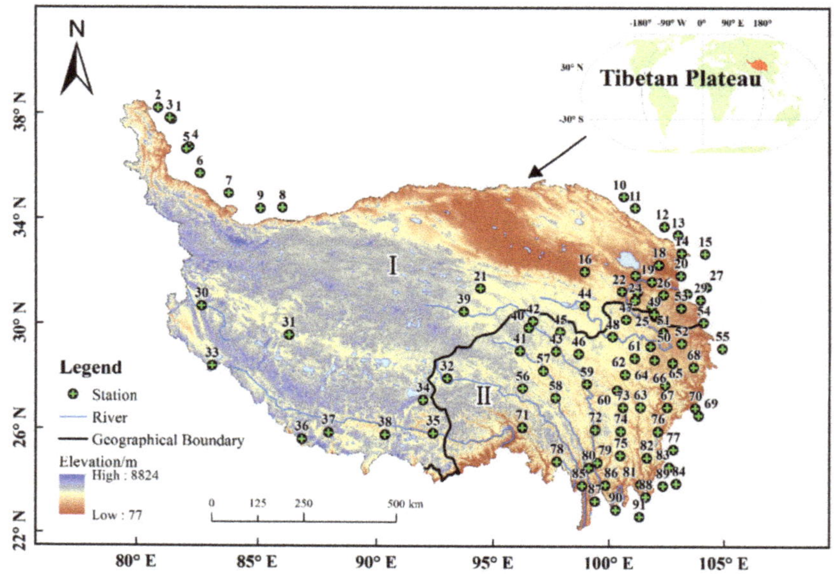

Figure 1. Study area of Tibetan Plateau (TP) and the distribution of meteorological stations.

2.2. *Database*

2.2.1. Daily Rainfall Data

The observed daily precipitation data used in this study was obtained from the Climatic Data Center, National Meteorological Information Center of the China Meteorological Administration (CMA) (http://data.cma.cn (accessed on 10 October 2021)). The data included a total of 91 meteorological stations (Figure 1), with complete data series, covering the time period 1971 to 2020 (Table S1). Moreover, considering the continuity of spatial interpolation and the stations spreading over the entire TP as much as possible, 27 meteorological stations around the study area were selected with the criterion that the shortest linear distance from a meteorological station to the TP boundary is not greater than 100 km. The observation records of all surrounding stations were recorded at the same time as the study period. In order to ensure data reliability and continuity, each meteorological data record was evaluated by the National Meteorological Center [67].

2.2.2. ENSO Indices

ENSO is a phenomenon of irregular periodic changes in sea surface temperature and wind occurring in the equatorial eastern Pacific Ocean, one of the strongest natural signals of interannual climate change worldwide. The typical characteristics of ENSO events are commonly known to be anomalous SSTs (±0.5 °C) in the eastern Pacific Ocean for more than 5 months, where warm episodes are El Niño events and cold episodes are La Niña events [68,69]. The multivariate ENSO Index (MEI) was obtained as the first non-rotating principal component (PC) of the six variables (sea-level pressure, zonal and meridional components of the surface wind, sea surface temperature, surface air temperature, total cloudiness fraction of the sky) over the tropical Pacific [70,71]. it is considered as a better index for detecting the ENSO phenomena with respect to other indices because it takes into account more information and fewer data failures [37]. Therefore, in this study, the occurrence and duration of the El Niño event and La Niña event were determined based on the Oceanic Niño Index (ONI), and MEI was selected as the ENSO proxy to probe the relationship between rainfall erosivity and ENSO during the time period of 1971–2020. These indexes are obtained from the National Oceanic and Atmospheric Administration (NOAA). Specifically, ONI was acquired from NOAA Climate Prediction Center [72], and MEI was acquired from NOAA Earth System Research Laboratory.

2.3. Methods

2.3.1. Technical Route

The study was divided into four steps (Figure 2):

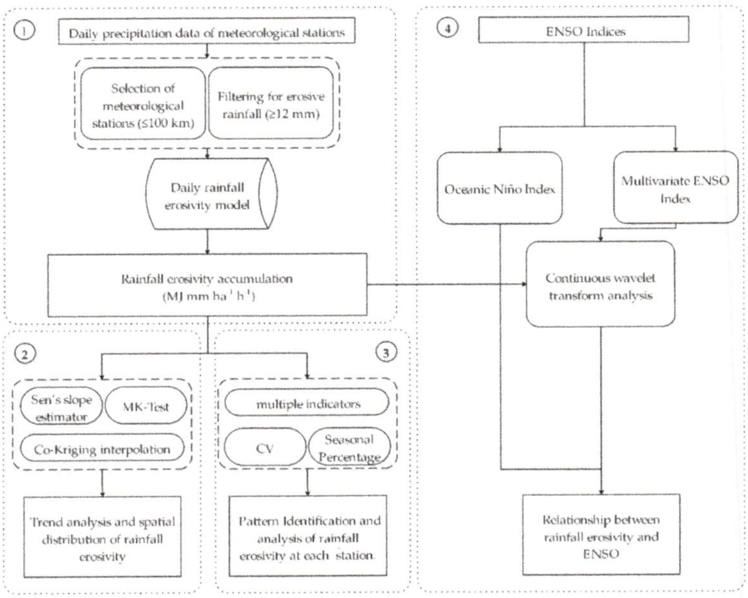

Figure 2. Technical route. Note: MK-test refers to Mann–Kendall test, CV refers to the coefficient of variation.

Step 1: Estimation of rainfall erosivity at different time scales. Based on the daily rainfall data of 91 stations, the annual, seasonal, and monthly average rainfall erosivity of TP from 1971 to 2020 were calculated using the daily rainfall erosivity model.

Step 2: Trend analysis and spatial distribution of rainfall erosivity in TP. Firstly, the temporal changes trend of rainfall erosivity was evaluated by using Sen's slope estimation and the MK trend test, and then the co-kriging method was used for the spatial mapping of rainfall erosivity for the period of 1971–2020.

Step 3: Pattern identification and analysis of rainfall erosivity change at each meteorological station. Firstly, the change trend of rainfall erosivity at each station was classified by integrating multiple indicators. Then, the coefficient of variation (CV) and the seasonal spatial distribution of rainfall erosivity of each site was analyzed.

Step 4: Relationship between rainfall erosivity and ENSO. Based on ONI, the variation in monthly mean rainfall erosivity during El Niño and La Niña events from 1971 to 2020 was analyzed; based on the MEI index, a continuous wavelet transform analysis method was used to examine the influence of ENSO on rainfall erosivity, and to clarify the response of the resonance period in different regions of the TP.

2.3.2. Calculation of Rainfall Erosivity

The half-monthly rainfall erosivity was estimated for each of the 91 meteorological stations from 1971 to 2020 using the daily rainfall erosivity model. The agent model was originally built from Richardson's equation [73], and later improved by Zhang [20]. Previous studies have demonstrated that this method is reliable and has been widely used on the national and regional scales in China [36,74–76]. This method is based on daily rainfall data to obtain monthly, seasonal, and annual rainfall erosivity. The calculation procedures are as follows:

$$R_i = \alpha \sum_{j=1}^{k} (P_j)^{\beta} \tag{1}$$

$$\alpha = 21.586\beta^{-7.1981} \tag{2}$$

$$\beta = 0.8363 + \frac{18.144}{P_{(d12)}} + \frac{24.455}{P_{(y12)}} \tag{3}$$

where R_i is the rainfall erosivity in the i-th half-month period (MJ mm ha^{-1} h^{-1}), k is the number of days in the half-month period, and P_j is the daily erosive rainfall amount (mm) on the j-th day during the half-month period. The half-month interval method is as follows: with the fifteenth day of each month as the dividing point, the whole year is divided into 24 half months. The half-month period as a basic statistical unit is used to calculate the corresponding half-month rainfall erosivity.

According to the national rainfall and runoff analysis, $\geq$12 mm is defined as erosive rainfall [77]. Therefore, the daily rainfall $\geq$12 mm is applied to Formula (1), otherwise, it is regarded as value of 0 in the calculations.

The terms α and β are two parameters to be determined in the model. $P_{(d12)}$ is the average daily erosive rainfall amount (mm) and $P_{(y12)}$ is the average annual erosive rainfall amount (mm). In this study, Formulas (1)–(3) are used to calculate the half-month rainfall erosivity of each meteorological station. The annual and seasonal rainfall erosivity of is the cumulative value of rainfall erosivity in every half-month period.

2.3.3. Sen's Slope Estimator and Mann–Kendall Test

In this study, the trends magnitude of annual rainfall erosivity was estimated with the non-parametric Sen's method. The trends and significance of annual and seasonal (monthly) rainfall erosivity were detected with the non-parametric Mann–Kendall test.

Sen's slope estimation is a non-parametric method of slope calculation, which is commonly used in the trend analysis due to its high robustness and computational efficiency [78]. The determination for the slope of annual rainfall erosivity is as follows: first, the values of Q_i calculated by the Formula (4) are ranked in order of magnitude, and then determines the overall estimator ($SLOPE_{med}$) as the median of these Q_i by Formula (5).

The slope in the N pairs of samples is calculated as follows:

$$Q_i = \frac{x_j - x_k}{j - k} \ (i = 1, \ldots, N) \tag{4}$$

where x_j and x_k are values of the rainfall erosivity corresponding to periods j and k, respectively ($j > k$). $SLOPE_{med}$ is calculated according to the following formula:

$$SLOPE_{med} = \begin{cases} Q_{\left[\frac{N+1}{2}\right]} < 0 & \text{if } N \text{ is odd} \\ \frac{Q_{\left[\frac{N+1}{2}\right]} + Q_{\left[\frac{N+1}{2}\right]}}{2} & \text{if } N \text{ is even} \end{cases} \tag{5}$$

where $SLOPE_{med} > 0$ indicates an upward trend, and vice versa. Its value indicates the magnitude of the trend change.

The non-parametric Mann–Kendall test is a widely used technique for to assess the significance of trends in long time series [79,80]. It is distribution free and not affected by missing values and outliers, and is highly recommended by the World Meteorological Organization [81]. This method is primarily based on two parameters, S and Z, to determine whether a time series has a significant trend. The intermediate variable S is computed as:

$$S = \sum_{k=1}^{n-1} \sum_{j=k+1}^{n} sgn(x_j - x_k) \tag{6}$$

where n is the length of the time series, x_j and x_k are values of the rainfall erosivity corresponding to periods j and k, respectively ($j > k$). S is the summation of $sgn(x_j - x_k)$, which takes the value of -1, 0, or 1 when $(x_j - x_k)$ is less than, equal to, or greater than 0, respectively. The variance of S can be acquired as follows:

$$var(S) = \frac{n(n-1)(2n+5)}{18} \tag{7}$$

Then the normalized statistical value Z is denoted as follows:

$$Z = \begin{cases} \frac{S-1}{\sqrt{var(S)}} & \text{if } S > 0 \\ 0 & \text{if } S = 0 \\ \frac{S+1}{\sqrt{var(S)}} & \text{if } S < 0 \end{cases} \tag{8}$$

where a positive (negative) value of Z indicates an upward (downward) trend. In bilateral trend detection, a time series with a significant trend is indicated if $|Z| \geq Z_{1-\alpha/2}$ at a certain significance level α, where $Z_{1-\alpha/2}$ is obtained from the standard normal cumulative distribution tables. The trend is statistically significant at the 0.1, 0.05, and 0.01 significance level when $|Z| > 1.645$, 1.96 and 2.576, respectively. Besides, the Mann–Kendall test can also be used to detect the abrupt points. The abrupt points and the approximate time of occurrence can be located according to the intersection of the progressive and retrograde sequences within the sequence. More details of the abrupt points calculation on the Mann–Kendall Test are available from the network resources.

2.3.4. Spatial and Statistical Analysis

The mean annual rainfall erosivity at each of the 91 stations was calculated by a long-term (1971–2020) average value of annual rainfall erosivity. Based on these station's values, the co-kriging interpolation method was used to interpolate the spatial distribution of the average annual erosivity of the TP, using the geostatistical analysis tool ArcGIS 10.4.

Different from the inverse distance weighting (IDW) method which only considered one assumption: nearby points should be closer to the value of the interpolation position than distant points, the co-kriging interpolation method allowed the addition of covariates to improve the accuracy of estimation or prediction [74,82]. Considering the complex terrain of the TP, the elevation factor was defined as a co-variable in the co-kriging interpolation method [36]. The elevation data of each meteorological station was provided by the China Meteorological Administration (CMA). Based on the values of rainfall erosivity for 91 meteorological stations on the TP, the co-kriging interpolation method was performed and generated the spatial distribution map of rainfall erosivity for the period of 1971–2020.

In this study, the rainfall erosivity anomalies of the TP is expressed as the difference between the annual erosivity value of the observation year and the 50-year average value. The 5-year moving average anomaly can smooth fluctuations and reduce potential errors, and was used to analyze the temporal changes of the rainfall erosivity across the TP. The seasonal (monthly) average rainfall erosivity of the TP was calculated by averaging the seasonal (monthly) erosivity of 91 meteorological stations during the same time span, from 1971 to 2020. The annual variation from the meteorological site is represented by the coefficient of variation (CV), which is expressed as a percentage of the standard deviation of the annual erosivity to the average of the observation year. The map of seasonal spatial distribution of rainfall erosivity is expressed as a percentage of the seasonal rainfall erosivity in the annual total erosivity at each weather station.

2.3.5. Identification of Rainfall Erosivity Trend Patterns

With reference to the time series trend identification method of Ray [83], rainfall erosivity change patterns were identified for each meteorological station from 1971 to 2020. Multiple indicators were integrated: the Z values calculated by Mann–Kendall, $SLOPE\%$ and R_{ST}.

The $SLOPE\%$ value represents as a percentage of $SLOPE$ for the average rainfall erosivity for each meteorological station during 1971–2020. The calculation formula is as follows:

$$SLOPE\% = \frac{SLOPE}{(\sum_{i=1}^{n} x_i)/n} \times 100 \tag{9}$$

where $SLOPE$ is the Sen's slope value of rainfall erosivity changes, x_i is the value of the rainfall erosivity corresponding to period i, and n is the length of the time series. The R_{ST} is defined as the ratio of the average rainfall erosivity for the last 3 years to the maximum 3 year moving average. This is used to identify whether the increasing trend of annual rainfall erosivity is interrupted, shifting to a decline at later stages. The equation is expressed below:

$$R_{ST} = \frac{\text{ave}(x_{n-2}, x_{n-1}, x_n)}{max(\text{AVE}(x_1, x_2, x_3), \text{AVE}(x_2, x_3, x_4), \ldots AVE(x_{n-2}, x_{n-1}, x_n))} \tag{10}$$

where x_i is the value of the rainfall erosivity corresponding to period i, and n is the length of the time series.

The trend of rainfall erosivity was classified by the above-mentioned three indicators into four patterns of decreasing, stagnant, increasing-stagnant and increasing (Table 1), abbreviated as DE, ST, IN-ST and IN, respectively. The Z value indicates whether there is a significant trend of rainfall erosivity ($|Z| > 1.96$ at the 0.05 significance level), a non-significant change trend ($0.675 < |Z| \leq 1.96$ at the 0.05–0.5 significance level), and no change trend ($|Z| \leq 0.675$ at the below 0.5 significance level); $SLOPE\%$ indicates whether the magnitude of the rainfall erosivity trend change is significant. References [84,85] used 0.25% as a criterion, i.e., a change greater than 0.25% is assumed to be significant. R_{ST} is used to determine whether the annual rainfall erosivity growth trend is interrupted, or turns down and mitigates at a later stage.

erosivity for the year. The monthly rainfall erosivity was 168.73 MJ·mm·ha^{-1}·h^{-1}, 264.75, and 229.10 MJ·mm·ha^{-1}·h^{-1}, respectively. Rainfall erosivity was highest on the Tibetan Plateau in July, the proportion of erosivity reached 24.19% for the year. November was the lowest with 1.12%, but a higher variability was found in this period, with a high extreme ratio of 18.71 (Figure 4). The trends in the monthly average rainfall erosivity from 1971 to 2020 were further examined, showing an increasing trend in eight months and a decreasing trend in four months. Specifically, the greatest increasing trend in mean rainfall erosivity was in August, but the increasing trend was insignificant in all months. Three months showed a significant decreasing trend, including March and April at 90% and 95% confidence levels, respectively, and the most significant decreasing trend was in November, which passed 99% confidence level (Figure S3).

3.3. Spatial Patten of Rainfall Erosivity in the Tibetan Plateau

In order to reduce the boundary effect on the annual rainfall erosivity spatial pattern, 91 meteorological stations were selected for the interpolation better to reveal the spatial variation in rainfall erosivity. In general, the rainfall erosivity on the TP from 1971 to 2020 had obvious spatial differences, roughly exhibiting a spatial pattern of decreasing distribution from southeast to northwest (Figure 5). The high value zone of rainfall erosivity was approximately in the southeastern part of the TP, mainly distributed in the lower altitude regions such as the Hengduan Mountains and the Yarlung Tsangpo Valley. There were three stations with an average annual rainfall erosivity greater than 6000 MJ·mm·ha^{-1}·h^{-1}, of which the highest value occurs at Dujiangyan station in Sichuan Province, with an average annual rainfall erosivity of 6605.21 MJ·mm·ha^{-1}·h^{-1}. The other two stations are Gongshan station and Huaping station in Yunnan Province, with an average annual rainfall erosivity of 6152.44 MJ·mm·ha^{-1}·h^{-1} and 6193.75 MJ·mm·ha^{-1}·h^{-1}, respectively. The zones with low average annual rainfall erosivity were mainly found in the northern and western parts of the TP, including concentrations in the Qiangtang Plateau and the Qaidam Basin. For example, the lowest value was at Shiquanhe station in the Tibet Autonomous Region, where the annual rainfall erosivity was only 103.46 MJ·mm·ha^{-1}·h^{-1}. In summary, the average annual rainfall erosivity was less than 500 MJ·mm·ha^{-1}·h^{-1} which accounted for 48.35% of all stations, 500–1000 MJ·mm·ha^{-1}·h^{-1} for 24.16% and that of more than 1000 MJ·mm·ha^{-1}·h^{-1} accounted for 27.47%.

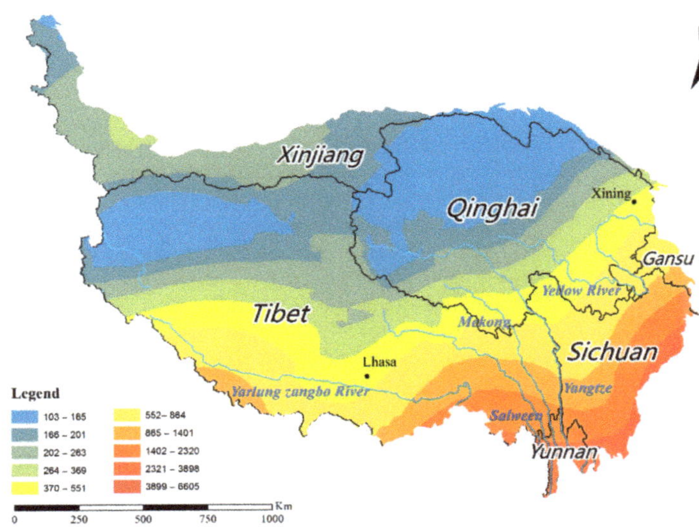

Figure 5. Spatial distribution of rainfall erosivity on the TP during the period from 1971 to 2020.

The long time series trend analysis of each meteorological station on the TP from 1971 to 2020 showed that the annual rainfall erosivity exhibited an increasing trend at 49 meteorological stations, accounting for 54% of all stations. Thereby this demonstrated the reason for the increasing rainfall erosivity across the entire TP since 1971 (Figure S4). Of these, 13 meteorological stations increased consistently, mainly in the eastern Hengduan Mountains of the TP, accounting for 27% of the total increases. The meteorological stations that showed a pattern of rainfall erosivity increased first and then gradually stabilized were mainly located in the Hehuang valley in the northeastern part of the Tibetan Plateau and the southern Tibetan valley in the southeast, accounting for 27% of the total increase in stations. The annual rainfall erosivity showed a long-term stable (no significant trend) pattern at 37 meteorological stations, accounting for 41% of all meteorological stations. Only five stations showed a gradual decrease in annual rainfall erosivity, accounting for 5% of all meteorological stations, but both patterns of change were not significant. In addition, as shown in Table S2, further analysis based on the statistics results of the multiple indicators at each station revealed that the types of trends at each site differ significantly in terms of significance levels and magnitude of change. Of all the increasing pattern stations, the significant increases were seen at Min station ($p = 0.05$) and Derong station ($p = 0.01$). The largest and smallest increases were at Min station and Pishan station, respectively. Of all the increasing-stagnating pattern stations, there are five meteorological stations at more than 95% confidence level, namely Zekog station ($p = 0.01$), Gerze station ($p = 0.01$), Dulan station ($p = 0.01$), Wuwei station ($p = 0.01$), and Guinan station ($p = 0.05$). Of all the decreasing pattern stations, the largest and smallest increases were at Gongshan station and Artux station, respectively.

As shown in Figure 6, the mean coefficient of variation (CV, the ratio of the standard deviation to the mean) in the interannual rainfall erosivity for 91 stations since 1971 was 0.61, indicating moderately high rainfall erosivity variability across the plateau. The spatial distribution pattern of CV had high consistency with annual rainfall erosivity. In other words, the CV increased from south-east to north-west. Specifically, 11 meteorological stations were in regions of intense variation (CV > 1), mostly in the north-western flank of the Kunlun Mountains on the TP, the Ali Mountains in the Tibet Autonomous Region, and the northern Qilian Mountains in eastern Qinghai Province. while 54 meteorological stations were in regions of lesser variation (CV < 0.5), accounting for 59% of the total meteorological stations, mainly in the south-eastern part of the TP. Furthermore, the meteorological station with the smallest CV in the interannual rainfall erosivity was Jiulong Station, located in Sichuan Province in the southeastern part of the Tibetan Plateau, while the largest CV was Pishan Station, located in Xinjiang Autonomous Region in the northwestern part of the Tibetan Plateau. In summary, over the past half century, rainfall erosivity exhibited clear spatial disparities on the TP, specifically, annual rainfall erosivity in the southeast were mainly characterized by a slowly and steadily increase, while annual rainfall erosivity in the northwestern part of the plateau showed greater fluctuations and instability, with no significant trends.

The seasonal spatial distribution pattern of rainfall erosivity varied widely across the TP (Figure 7). Five meteorological stations (5.5% of the total) with the highest percentage of spring rainfall erosivity were concentrated in the Kunlun Mountains on the Tibetan Plateau near the Pamir Plateau and in the Nu River basin in the Eastern Himalaya. In particular, the erosivity of spring rainfall accounted for more than 50% of Pishan station, Kashgar station, and Zayu station, and Pishan station was as high as 79%. (Figure 7a). Nearly 92% of the meteorological stations (total 84) had the highest percentage of summer rainfall erosivity. The largest was Shiquanhe station, which surprisingly had 94.28% of the annual rainfall erosivity (Figure 7b). There was only one meteorological station with the highest percentage of fall rainfall erosivity, with three stations accounting for more than 30%, namely Nyalam, Burang, and Keriya station (Figure 7c). Winter was the season with the lowest percentage of rainfall erosivity, all stations had less than 30% of rainfall erosivity (Figure 7d). Summer and autumn were the most erosive seasons. It is worth noting that rainfall erosivity was generally higher across the plateau in summer, particularly in the southeastern part of the Tibetan Plateau, whereas rainfall erosivity in autumn and winter was still higher proportion in the south-western part of the plateau near the Himalayas. Thus, extra caution will be needed to prevent aggravation of soil erosion in this region.

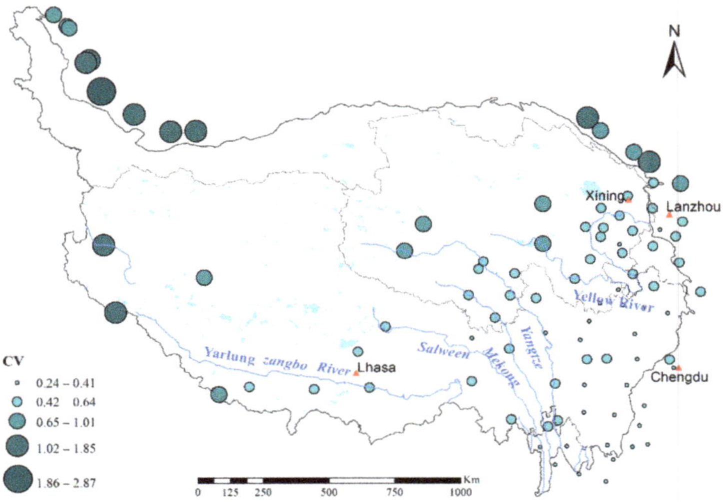

Figure 6. Spatial distribution of the coefficient of variation (CV) in rainfall erosivity during 1971–2020.

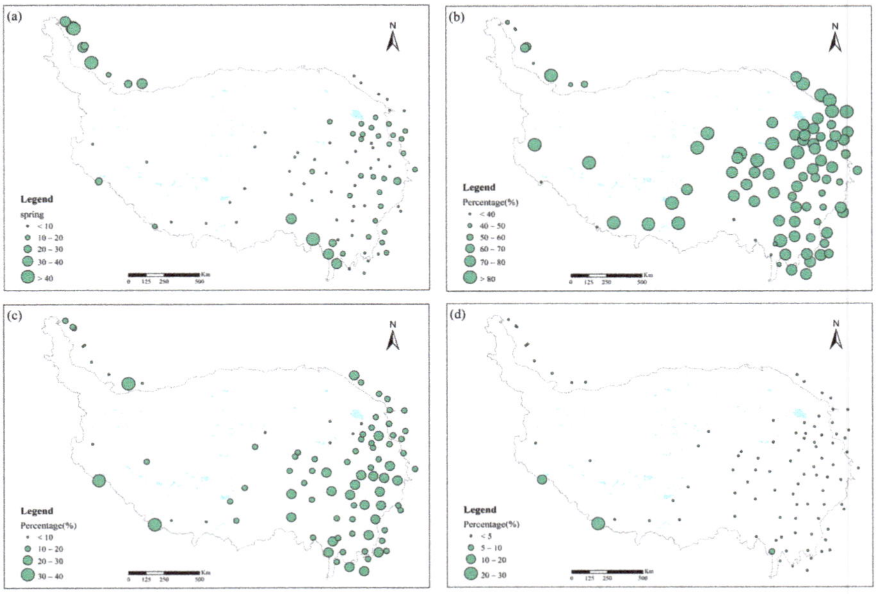

Figure 7. Spatial pattern of rainfall erosivity from 1971 to 2020 in spring (**a**), summer (**b**), autumn (**c**), and winter (**d**). Note: Percentage refers to the proportion of seasonal rainfall erosivity to total annual erosivity.

3.4. Relationship between Rainfall Erosivity and ENSO

3.4.1. Influence of ENSO on Rainfall Erosivity in the Different Regions of TP

The El Niño and La Niña events are the ENSO cycles of the warm and cold periods, respectively. Statistics on the rainfall erosivity in Region I (arid zone) and Region II (humid zone) of the TP and whole plateau during the El Niño (ENSO warm event) and La Niña (ENSO cold event) periods are presented in Table 2.The occurrence and duration of El Niño event and La Niña event were determined based on ONI.

Table 2. Average monthly rainfall erosivity in different regions (Region I and Region II) and TP for El Niño and La Niña events from 1971 to 2020. The bottom table provides a summary of the average monthly rainfall erosivity.

No.	Time Internal	Duration Time	Region I Erosivity	Region II Erosivity	TP Erosivity
		El Niño events			
1	1972.05–1973.03	11	282.09	233.00	201.92
2	1976.09–1977.02	6	90.57	216.53	198.37
3	1977.09–1978.01	5	114.40	132.34	120.58
4	1979.10–1980.02	5	104.33	209.91	175.40
5	1982.04–1983.06	15	83.32	214.03	174.49
6	1986.09–1988.02	18	315.97	221.63	225.18
7	1991.05–1992.06	14	106.09	256.33	210.08
8	1994.09–1995.03	7	74.89	167.78	142.50
9	1997.05–1998.05	13	144.15	235.53	193.00
10	2002.06–2003.02	9	141.13	251.03	208.63
11	2004.07–2005.02	8	155.69	247.67	133.83
12	2006.09–2007.01	5	84.40	167.47	188.51
13	2009.07–2010.03	9	128.03	217.52	170.04
14	2014.10–2016.04	19	63.75	193.68	204.68
15	2018.09–2019.06	10	318.80	169.55	201.92
		La Niña events			
1	1971.01–1972.01	12	105.87	162.81	142.12
2	1973.05–1974.07	15	101.82	327.52	197.69
3	1974.10–1976.04	19	136.34	246.86	234.16
4	1983.09–1984.01	5	170.46	216.68	191.36
5	1984.10–1985.08	11	125.46	251.72	216.24
6	1988.05–1989.05	13	408.05	261.23	486.03
7	1995.08–1996.03	8	117.67	249.63	190.13
8	1998.07–2001.02	32	153.16	264.05	209.01
9	2005.11–2006.03	5	474.19	124.57	142.75
10	2007.06–2008.06	13	155.75	201.97	178.10
11	2008.11–2009.03	5	283.13	123.77	138.83
12	2010.06–2011.05	12	102.80	214.74	185.26
13	2011.07–2012.04	10	195.09	192.46	187.99
14	2016.08–2016.12	5	109.84	251.42	223.47
15	2017.10–2018.04	7	48.87	116.56	101.67
16	2020.08–2020.12	5	93.12	156.05	229.36
	Average monthly erosivity El Niño		147.17	208.93	181.94
	Average monthly erosivity La Niña		173.85	210.13	203.39
	Average monthly erosivity ENSO		160.51	209.53	192.67
	Average monthly erosivity non-ENSO		140.48	253.09	213.77
	Average monthly erosivity 1971–2020		146.25	233.68	208.12

In terms of the degree of influence of cold and warm events on the rainfall erosivity, the average monthly rainfall erosivity for the El Niño event was slightly lower than the La Niña event across the TP, but the average of both events was less than the monthly rainfall erosivity for the period 1971–2020. During the El Niño event, the maximum monthly average rainfall erosivity was 225.18 MJ·mm·ha^{-1}·h^{-1} and the minimum value was 120.58 MJ·mm·ha^{-1}·h^{-1}, with an extreme value ratio of 1.87; during the La Niña event, the maximum monthly average rainfall erosivity was 486.03 MJ·mm·ha^{-1}·h^{-1} and the minimum value was 101.67 MJ·mm·ha^{-1}·h^{-1}, with an extreme value ratio of 4.78; thus, higher variability occurred during the La Niña event. In terms of the presence or absence of ENSO events, the average rainfall erosivity during the ENSO and Non-ENSO periods were 192.67 MJ·mm·ha^{-1}·h^{-1} and 213.77 MJ·mm·ha^{-1}·h^{-1}, respectively. It was evident that the average monthly rainfall erosivity during the non-ENSO period was not only greater than that during the ENSO period, but also greater than the total average monthly rainfall erosivity for the whole study period.

The impact of ENSO events on the monthly mean rainfall erosivity in different regions was notably dissimilar. For Region I, the average monthly rainfall erosivity for the El Niño and La Niña events were 147.17 MJ·mm·ha^{-1}·h^{-1} and 173.85 MJ·mm·ha^{-1}·h^{-1}, respectively. This was higher than the average monthly rainfall erosivity for this region for 1971–2020. Moreover, we found that when El Niño events or La Niña events occurred, there was a significant increase in rainfall erosivity in Region I relative to the Non-ENSO period, but La Niña events had a greater impact on the monthly average rainfall erosivity, compared with El Niño events; For Region II, the average monthly rainfall erosivity for the El Niño and La Niña events were 208.93 MJ·mm·ha^{-1}·h^{-1} and 210.13 MJ·mm·ha^{-1}·h^{-1}, respectively, with the El Niño event slightly lower than the La Niña event. There was a modest gap between the two events. However, compared with Region I, the direction of influence of ENSO in Region II was in the opposite direction. In other words, when ENSO occurs, the average monthly rainfall erosivity in this region decreased more significantly than the average for the study period. Due to the difference in the magnitude of rainfall erosivity on the Tibetan Plateau during the ENSO period and the non-ENSO period, under the premise that other contributing factors was fixed, rainfall erosivity was stronger during the non-ENSO period and soil erosion concerns and soil conservation measures should be strengthened during this period. Considering the obvious spatial heterogeneity of the impact of ENSO on the Tibetan Plateau, the emphasis should be on erosion in the north-west during El Niño or La Niña events, especially during the La Niña event when control measures should be enhanced.

3.4.2. Correlation between Rainfall Erosivity and Multivariate ENSO Index

To examine the extent and impact of ENSO on rainfall erosivity, an XWT and WTC analysis were conducted on the time series of rainfall erosivity and MEI index in different regions of the Tibetan Plateau from 1971 to 2020, revealing the periodicity characteristics of both. As shown in Figure 8, in the time-frequency space domain of Region I, it is obvious that there was 3–5 years of high-energy resonance cycle between rainfall erosivity and the MEI index for the period of 1981–1988, during which there was a negative correlation between both time series. In the Region II power spectrum, there were two significant high-energy domains, specifically a 3–5 years resonance cycle from 1981 to 1988 was similar to that of Region I, indicating a consistent ENSO effect across the plateau during this period, but the intensity of the Region II resonance cycle was higher. The other was that there was a 2–5 years resonance cycle of rainfall erosivity and the MEI index from 1995 to 1999, and the mean phase angle was nearly 90° vertically upwards, indicating that the rainfall erosivity change was later than the MEI index. In other words, rainfall erosivity had a lag compared with ENSO over the same period. As shown in Figure 9, in the Region I WTC power spectrum, there were negative phase cycles of 3–5 years and 2–3 years in 1985–1992 and 2006–2009, respectively, indicating a negative correlation between rainfall erosivity and the MEI index during this period. Regarding Region II, there were negative phase cycles of 3–7 years from 1977 to 1988, 1–3 years from 1994 to 1998 and 2–4 years from 2007 to 2013, indicating a negative correlation between rainfall erosivity and the MEI index during these periods. While positive phase cycles of 1–2 years from 1987 to 1989 indicate a positive correlation between rainfall erosivity and the MEI index during this period.

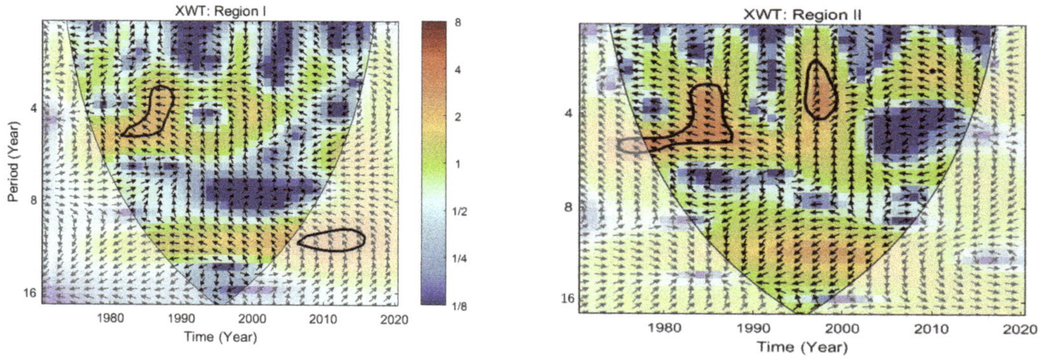

Figure 8. Cross wavelet transforms (XWTs) for annual rainfall erosivity and multivariate ENSO Index (MEI) in Region I (**a**) and Region II (**b**). NOTE: The thick black outline indicates the 95% significance level against red noise, the white translucent area indicates the cone of influence, and the color bar indicates the magnitude of the XWT cross spectral power. That is, red is strong and blue is weak. The arrows (vectors) designate the phase difference between rainfall erosivity and MEI. Where the left arrow indicates the opposite phase relationship between the rainfall erosivity and MEI and vice versa. The north-pointing arrow indicates that the peak rainfall erosivity are lower than the peak MEI.

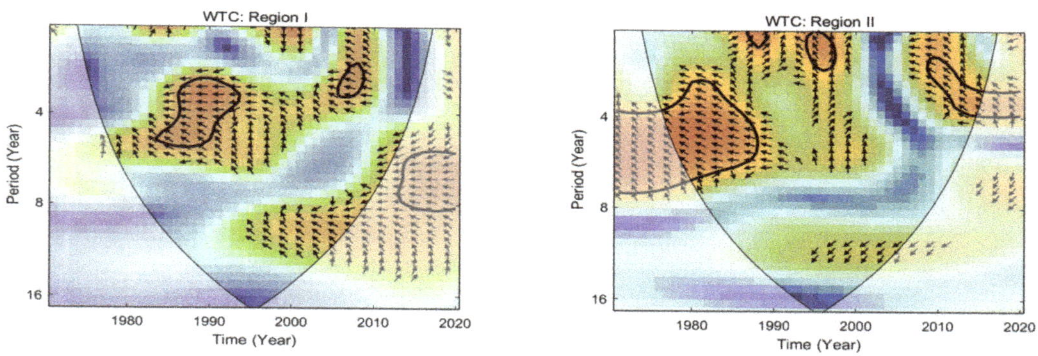

Figure 9. Wavelet transform coherence (WTC) for annual rainfall erosivity and multivariate ENSO Index (MEI) in Region I (**a**) and Region II (**b**). NOTE: The thick black outline indicates the 95% significance level, the white translucent area indicates the cone of influence, and the color bar indicates the significance level of the Monte Carlo test. That is, red means strong correction and blue is weak. where the left arrow indicates the opposite phase relationship between the two-time series, vice versa.

4. Discussion

This study revealed the spatial and temporal characteristics of rainfall erosivity on the TP from 1971 to 2020 and their relationship with the ENSO Index. The results showed that the average annual rainfall erosivity on the TP since 1971 was 1071.42 MJ·mm·ha^{-1}·h^{-1}. According to previous studies, this value is higher than in northwestern China but lower than in southeastern China [74], and the overall degree of erosion is light, approximately 0.5 times the global average [91]. Upward trends are shown for rainfall erosivity during 1971 to 2020. Gu et al. also found an increasing trend in rainfall erosivity from 1981 to 2015 in the Tibet Autonomous Region (TAR) [60], and Wang et al. found a same uptrend in rainfall-runoff erosivity from 1961 to 2012 in Sanjiangyuan region, Qinghai Province [53], which is consistent with this study. Fan et al. [61] used TRMM 3B42 data to assess the spatial and temporal variability of rainfall erosivity in the TAR from 2000 to 2010 and found that the average rainfall erosivity was 768 MJ·mm·ha^{-1}·h^{-1}, which is lower than the results of this study, probably due to the different extent of the study area and the accuracy of the data. Moreover, the reasons for the trend of increasing annual rainfall erosivity but significant decreasing rainfall erosivity in spring and autumn may be related to the variation in rainfall on the TP [92]. Previous studies have shown that since 1961, the Tibetan Plateau is gradually warming and humidifying [93], which may contribute to an increase in rainfall erosivity. Meanwhile, while changes in the westerly circulation lead to a reduction in rainfall in spring and autumn which in turn affects rainfall erosivity [94].

In the previous section, this work indicated rainfall erosivity on the TP varied greatly not only seasonally but also monthly. This may be caused by its complicated geography and dominant atmospheric circulation conditions [95]. The plateau spans a wide range of latitudes and longitudes and has a variety of climate types. It is also at the crossroads of monsoon and non-monsoon zones, and is influenced by the prevailing westerly winds, the South Asian monsoon and the East Asian monsoon circulation, resulting in an uneven spatial and temporal distribution of rainfall. Besides, according to previous reports, the amount and intensity of rainfall are the main factors affecting the rainfall erosivity [96]. The spatial and temporal variability of rainfall at various magnitudes will certainly contribute to soil erosion by water at different times and regions to different degrees [97,98]. The average heavy rainfall and average heavy rainfall erosivity for each station from 1971 to 2020 are presented in Table S3. It can be seen from Table S3 that the average heavy rainfall ($\geq$25 mm) of 91 meteorological stations in the plateau was 33.05 mm, and the average rainfall erosivity was 243.35 MJ·mm·ha^{-1}·h^{-1}. The absolute amount of erosivity caused by heavy rainfall on the TP is low compared with the eastern coastal areas of China, mainly because the total rainfall erosivity in the plateau is much lower than those in the east [35,99,100]. Furthermore, the comparison of the proportion of heavy rain and rainfall erosivity found that the average heavy rainfall on the TP accounted for only 27.02% of the total rainfall, but the rainfall erosivity caused by heavy rainfall occupied by 43.3% of the total rainfall erosivity (Figure 10). In addition, the change in the percentage of heavy rainfall and heavy rainfall erosivity has a high consistency on the TP. In case of heavy rainfall, it can be easily seen that rain erosivity becomes more intense. The maximum values of heavy rainfall and heavy rainfall erosivity were 71.26% and 81.46%, respectively, both of which occurred at Dujiangyan station, and the minimum values of 6.77% and 8.15%, respectively, which occurred at Zhidoi station. Studies have revealed that rainfall is the most important climatic factor contributing to soil erosion, and in particular, heavy rainfall ($\geq$25 mm) is one of the main factors affecting the rainfall erosivity [101]. In general, the higher the intensity of the heavy rainfall, the greater the amount of soil erosion. The frequency of extreme rainfall events is growing as a result of global climate change [102,103], Future research should pay more attention to high intensity rainfall and soil erosion.

The spatial distribution of rainfall erosivity on the TP decreased roughly from south-east to north-west, with significant spatial heterogeneity. Previous studies have shown that mountain tectonics and topographic gradients are essential factors influencing precipitation [104]. The high rainfall erosivity in the south-east is largely attributed to the roughly north-south alignment of the Hengduan Mountains and the gradual rise in elevation from south to north, which is a natural water vapor corridor and facilitates the deeper uplift of monsoon air masses and the formation of rainfall. The north-west of the plateau, on the other hand, is located in an inland region, with high altitude and low temperatures. It is extremely hard for the monsoon to reach it, and rainfall is minimal throughout the year, resulting in lower rainfall erosivity. Besides, in the southern part of the plateau, although it is on the leeward slopes of the Himalayas, where water vapor is not easily accessible, the Yarlung Tsangpo valley is at a relatively low altitude and has locally better hydrothermal conditions, resulting in a higher rainfall erosivity in the region. In addition, the role of anthropogenic activities should not be neglected. Nearly 38.8% of the grasslands on the TP have been degraded [105], and the degradation of meadows caused by overgrazing is a serious environmental and ecological problem. Consequently, soil erosion caused by the contradiction between people and land may exacerbate the rainfall erosivity.

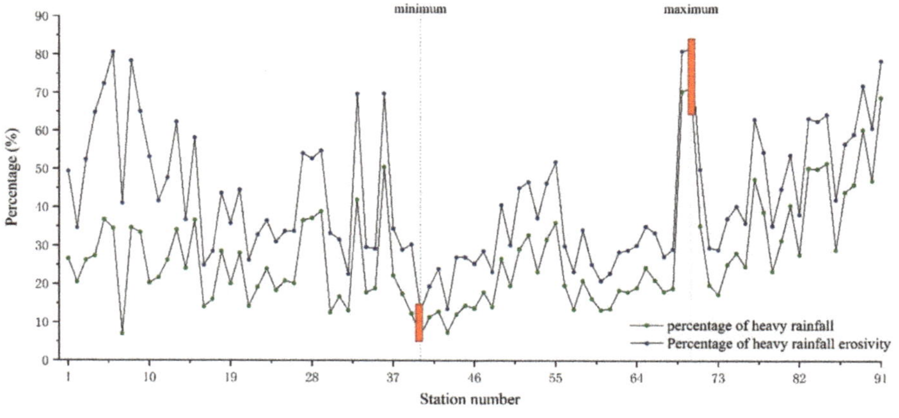

Figure 10. Comparison of the proportion of heavy rain (daily rainfall $\geq$ 25 mm) and rainfall erosivity for 91 meteorological stations on the TP from 1971 to 2020. Note: the designation of station numbers is shown in Table S3; and the percentage of heavy rainfall represents the proportion of heavy rain amount to total annual precipitation; percentage of heavy rainfall erosivity represents the proportion of heavy rainfall erosivity to annual rainfall erosivity.

Clarifying the temporal and spatial patterns of rainfall erosivity on the TP over the past 50 years is of great significance for soil conservation and future land use planning. This paper indicates that the tendency of increasing rainfall erosivity was identified in the southeastern part of the TP. Published studies have shown that the Hengduan Mountains region has the highest soil erosion modulus of the TP [106]. Furthermore, the alternation of steep slopes and deep ravines, the superimposed effect of complex topography and intensified rainfall erosivity may further magnify soil erosion in this region. The Yarlung Tsangpo Valley and the Hehuang Valley are areas of intensive human activity on the TP, and are also major wheat and barley cultivation areas, with crops mostly grown on the slopes of the valleys [107]. Although rainfall erosivity has not shown a sharp rise over the past 50 years in these origins, the region's originally high rainfall erosivity may still result in frequent natural hazards and elevated ecological risks. The deployment of measures against landslides and debris flows should be considered as a priority. In the western and northern parts of the Tibetan Plateau, there is higher variability, although with lower rainfall erosivity. On the one hand, in the context of increased rainfall erosivity across the plateau, it is still possible to damage the low-cover turf. In particular, the interactive effects of human activities such as grazing and soil erosion processes can also exacerbate the 'black beach' degradation of plateau grasslands. On the other hand, the north-west has a higher altitude and fragile natural environment where seasonal differences in rainfall erosivity is more likely to cause damage to alpine ecosystems. Therefore, soil erosion management strategies in this region should not be neglected either.

The impact of global-scale climate oscillation regimes on climate change has received widespread attention. This paper used MEI to characterize the global-scale climate oscillation model ENSO in an attempt to explore the relationship between ENSO and rainfall erosivity on the Tibetan Plateau, with a view to providing guidance for collaborative work on climate change and soil conservation. This study found that rainfall erosivity was lower during ENSO than during non-ENSO, and other studies have found the same results [35,36]. ENSO is a major factor influencing temperature and precipitation in China. Some studies have shown that precipitation anomalies can reach up to 30% of the average precipitation during ENSO periods [108]. It is worth noting that rainfall erosivity was higher during the La Niña period than during the El Niño period on the TP. This is not in agreement with previous studies, with results in places such as Fujian in southeastern China [37], but is consistent with studies in Guizhou in southwestern China [36]. The other research indicated that El Niño occurs with a delayed arrival of the southwest monsoon, while the opposite occurs at La Niña, so this may be related to a weakening of the Indian monsoon [109,110]. These findings would explain the difference of rainfall erosivity between the La Niña period and El Niño period. According to the CWT results, a noticeable resonance cycle between rainfall erosivity and MEI was found in different regions of the Tibetan Plateau, but there were also significant differences in cycle duration, direction of action, and intensity. This may be due to the fact that ENSO events themselves present diversities of climatic features at each stage of occurrence, development, maturation and decline [108]. Additionally, it also suggested that global climate anomalies are an important driver of changes in the rainfall erosivity on the TP.

Due to the vast expanses of land and sparse populations as well as the harsh natural conditions, the distribution of weather stations on the Tibetan Plateau is extremely irregular, which may affect the accuracy of the interpolation. Although this study improves the comprehension of the impact of ENSO on rainfall erosivity, it still lacks further explanation from a mechanistic perspective. Furthermore, in light of the known results, it is clear that not only ENSO but also topography, altitude, and microclimate are associated with rainfall erosivity, and detailed knowledge will be necessary for future studies.

5. Conclusions

This study carried out an insightful analysis of the spatial, interannual, and seasonal variability of rainfall erosivity on the TP from 1971 to 2020 and its relationship with ENSO. Daily rainfall data from 91 meteorological stations were collected, and the change trend of rainfall erosivity calculated based on a daily rainfall erosivity model were detected at a regional and site-scale using methods such as Mann–Kendall test and Sen's slope. The potential influence of ENSO on rainfall erosivity was revealed using the continuous wavelet transform method. The main findings were summarized below:

Rainfall erosivity has shown a fluctuating trend of increasing over the past half century. Seasonal and monthly rainfall erosivity showed high heterogeneity, which was greatly related to heavy rainfall. The rainfall erosivity in order was summer > autumn > spring > winter. July was the most erosive month, accounting for 24.19% of the year, while November was the lowest, accounting for only 1.12%. The rainfall erosivity in spring and autumn showed a significant decreasing trend ($p < 0.05$), and in summer it showed an increasing trend but not significant. There was generally an obvious spatial variation in rainfall erosivity on the TP from 1971 to 2020, presenting a roughly spatial pattern of decreasing distribution from southeast to northwest. Annual rainfall erosivity in the south-eastern part of the plateau was mainly characterized by a slow increase, while in the north-western part annual rainfall erosivity was more unstable with mostly no significant trends.

ENSO events had a significant impact on rainfall erosivity on the TP. The rainfall erosivity in the non-ENSO period was higher than that in the ENSO period, and the La Niña event was higher than the El Niño event. It was also found that there was a clear resonance cycle between rainfall erosivity and ENSO in different regions of the plateau, with an average cycle of about 3–5 years in the high energy region, but there were differences in the timing of occurrence, direction of action, and intensity of the cycle. The rainfall erosivity on the TP was relatively large during non-ENSO periods and relatively small during El Niño/La Niña periods. In addition, the response of rainfall erosivity to ENSO was spatially heterogeneous. Rainfall erosivity in the northwest of TP appears to be more serious during the La Niña event and less severe during the El Niño event. It can be concluded that soil erosion may become more intense during the La Niña event in the northwest TP. Therefore, during the La Niña event, soil protection should be enhanced to diminish soil spattering and disturbance.

This study contributes to the understanding of the spatial and temporal variability of annual rainfall erosivity across the entire TP over the last half century and extends the cognition of the possible impact of changes in ENSO characteristics associated with climate change. Uncertainties may be involved due to limited data availability and interpolation bias errors. Future studies should integrate the effects of multiple factors on rainfall erosivity, more carefully relate the effects of climate extremes, and improve the insights from the mechanistic aspects of change.

Supplementary Materials: The following are available online at https://www.mdpi.com/article/10.3390/ijerph182111054/s1, Figure S1: Rainfall erosivity anomalies from 1971 to 2017, the blue line shows the 5–year moving average curve, Figure S2: Sequential MK test for 50-year rainfall erosivity of TP. Note: UF and UB refer to progressive and retrograde sequences within the sequence, respectively. UF > 0 indicates an increasing trend, UF < 0 indicates a decreasing trend. The mutation year exists at the intersection of UF and UB. The dashed line represents the 95% confidence interval, Figure S3: Seasonal and monthly average rainfall erosivity Mann–Kendall trends of plateau from 1971 to 2020. ** $p < 0.01$; * $p < 0.05$; + $p < 0.1$, Figure S4: Long–term trend pattern of rainfall erosivity at each station from 1971 to 2020. Note: DE: decreasing, ST: stagnating, IN-ST: increasing-stagnating, IN: increasing, Table S1: The basic information of 91 meteorological stations in this study. 'No' refers to station number, which is the same as the station number in Figure 1, Table S2: Statistical results of *Z*, *SLOPE*, and R_{st} associated with the rainfall erosivity trend classification at each meteorological station on the TP from 1971 to 2020, Table S3: The average heavy rainfall (AHR) and its proportion (HR) and the proportion of average heavy rain erosivity (HRE) of each station from 1971 to 2020.

Author Contributions: Y.Z. and B.C. had the original idea and designed the study. B.C. processed and analyzed the data and wrote the manuscript; Y.Z., L.L. and Z.W. had insights on the revision of the manuscript and suggestions for improvement. Z.X. provided help and guidance for data processing and model adjustment. C.G., B.W. and D.G. revised the paper and polished the language. All authors have read and agreed to the published version of the manuscript.

Funding: This research was funded by the Strategic Priority Research Program of the Chinese Academy of Sciences (Grant No. XDA20040201), the Second Tibetan Plateau Scientific Expedition and Research (Grant No. 2019QZKK0603), and the National Natural Science Foundation of China (Grant No. 41671104).

Institutional Review Board Statement: Not applicable.

Informed Consent Statement: Not applicable.

Data Availability Statement: All relevant data sets in this study are described in the manuscript.

Conflicts of Interest: The authors declare no conflict of interest.

Abbreviations

ENSO	El Niño southern oscillation
TP	Tibetan Plateau
ONI	Oceanic Niño Index
MEI	Multivariate ENSO Index
CWT	Continuous wavelet transforms
XWT	Cross Wavelet Transform
WTC	Wavelet Transform Coherence
CV	Coefficient of variation

References

1. Amundson, R.; Berhe, A.A.; Hopmans, J.W.; Olson, C.; Sztein, A.E.; Sparks, D.L. Soil and human security in the 21st century. *Science* **2015**, 348. [CrossRef] [PubMed]
2. Pimentel, D.; Kounang, N. Ecology of soil erosion in ecosystems. *Ecosystems* **1998**, *1*, 416–426. [CrossRef]
3. Lal, R.; Bruce, J.P. The potential of world cropland soils to sequester C and mitigate the greenhouse effect. *Environ. Sci. Policy* **1999**, *2*, 177–185. [CrossRef]

4. Pimentel, D.; Harvey, C.; Resosudarmo, P.; Sinclair, K.; Kurz, D.; McNair, M.; Crist, S.; Shpritz, L.; Fitton, L.; Saffouri, R.; et al. Environmental and economic costs of soil erosion and conservation benefits. *Science* **1995**, *267*, 1117–1123. [CrossRef] [PubMed]
5. Borrelli, P.; Robinson, D.A.; Panagos, P.; Lugato, E.; Yang, J.E.; Alewell, C.; Wuepper, D.; Montanarella, L.; Ballabio, C. Land use and climate change impacts on global soil erosion by water (2015–2070). *Proc. Natl. Acad. Sci. USA* **2020**, *117*, 21994–22001. [CrossRef] [PubMed]
6. Panagos, P.; Borrelli, P.; Poesen, J.; Ballabio, C.; Lugato, E.; Meusburger, K.; Montanarella, L.; Alewell, C. The new assessment of soil loss by water erosion in Europe. *Environ. Sci. Policy* **2015**, *54*, 438–447. [CrossRef]
7. Blanco, H.; Lal, R. *Principles of Soil Conservation and Management*; Springer: New York, NY, USA, 2008; p. 167169.
8. Montgomery, D.R. Soil erosion and agricultural sustainability. *Proc. Natl. Acad. Sci. USA* **2007**, *104*, 13268–13272. [CrossRef]
9. Acquaotta, F.; Faccini, F.; Fratianni, S.; Paliaga, G.; Sacchini, A.; Vil I Mek, V.I.T. Increased flash flooding in Genoa Metropolitan Area: A combination of climate changes and soil consumption? *Meteorol. Atmos. Phys.* **2019**, *131*, 1099–1110. [CrossRef]
10. Pennock, D.J. *Soil Erosion: The Greatest Challenge for Sustainable Soil Management*; Food and Agriculture Organization of the United Nations: Rome, Italy, 2019.
11. Li, Z.; Fang, H. Impacts of climate change on water erosion: A review. *Earth-Sci. Rev.* **2016**, *163*, 94–117. [CrossRef]
12. Renard, K.G. *Predicting Soil Erosion by Water: A Guide to Conservation Planning with the Revised Universal Soil Loss Equation (RUSLE)*; United States Government Printing: Washington DC, USA, 1997.
13. Wischmeier, W.H.; Smith, D.D. *Predicting rainfall-erosion losses from cropland east of the Rocky Mountains*; No. 282; US Department of Agriculture, Government Printing Office: Washington, DC, USA, 1965.
14. Diodato, N. Estimating RUSLE's rainfall factor in the part of Italy with a Mediterranean rainfall regime. *Hydrol. Earth Syst. Sci.* **2004**, *8*, 103–107. [CrossRef]
15. Bonilla, C.A.; Vidal, K.L. Rainfall erosivity in central Chile. *J. Hydrol.* **2011**, *410*, 126–133. [CrossRef]
16. Lee, J.; Heo, J. Evaluation of estimation methods for rainfall erosivity based on annual precipitation in Korea. *J. Hydrol.* **2011**, *409*, 30–48. [CrossRef]
17. De Santos Loureiro, N.; de Azevedo Coutinho, M. A new procedure to estimate the RUSLE EI30 index, based on monthly rainfall data and applied to the Algarve region, Portugal. *J. Hydrol.* **2001**, *250*, 12–18. [CrossRef]
18. Renard, K.G.; Freimund, J.R. Using monthly precipitation data to estimate the R-factor in the revised USLE. *J. Hydrol.* **1994**, *157*, 287–306. [CrossRef]
19. Angulo-Martínez, M.; Beguería, S. Estimating rainfall erosivity from daily precipitation records: A comparison among methods using data from the Ebro Basin (NE Spain). *J. Hydrol.* **2009**, *379*, 111–121. [CrossRef]
20. Zhang, W.; Xie, Y.; Liu, B. Rainfall erosivity estimation using daily rainfall amounts. *Sci. Geogr. Sin.* **2002**, *22*, 705–711.
21. Chen, Y.; Xu, M.; Wang, Z.; Gao, P.; Lai, C. Applicability of two satellite-based precipitation products for assessing rainfall erosivity in China. *Sci. Total Environ.* **2021**, *757*, 143975. [CrossRef] [PubMed]
22. Teng, H.; Ma, Z.; Chappell, A.; Shi, Z.; Liang, Z.; Yu, W. Improving rainfall erosivity estimates using merged TRMM and gauge data. *Remote Sens.* **2017**, *9*, 1134. [CrossRef]
23. Liu, B.T.; Tao, H.P.; Song, C.F.; Guo, B.; Shi, Z.; Zhang, C.; Kong, B.; He, B. Temporal and spatial variations of rainfall erosivity in China during 1960 to 2009. *Geogr. Res.* **2013**, *32*, 245–256.
24. Mello, C.D.; Viola, M.R.; Beskow, S.; Norton, L.D. Multivariate models for annual rainfall erosivity in Brazil. *Geoderma* **2013**, *202*, 88–102. [CrossRef]
25. De Luis, M.; Gonz A Lez-Hidalgo, J.C.; Longares, L.A. Is rainfall erosivity increasing in the Mediterranean Iberian Peninsula? *Land Degrad. Dev.* **2010**, *21*, 139–144. [CrossRef]
26. Yu, B.; Hashim, G.M.; Eusof, Z. Estimating the R-factor with limited rainfall data: A case study from peninsular Malaysia. *J. Soil Water Conserv.* **2001**, *56*, 101–105.
27. Easterling, D.R.; Evans, J.L.; Groisman, P.Y.; Karl, T.R.; Kunkel, K.E.; Ambenje, P. Observed variability and trends in extreme climate events: A brief review. *Bull. Am. Meteorol. Soc.* **2000**, *81*, 417–426. [CrossRef]
28. Diaz, H.F. *El Niño and the Southern Oscillation: Multiscale Variability and Global and Regional Impacts*; Cambridge University Press: Cambridge, UK, 2000.
29. McGregor, G.R.; Ebi, K. El Niño Southern Oscillation (ENSO) and health: An overview for climate and health researchers. *Atmosphere* **2018**, *9*, 282. [CrossRef]
30. Dong, B.; Sutton, R.T.; Scaife, A.A. Multidecadal modulation of El Niño-Southern Oscillation (ENSO) variance by Atlantic Ocean sea surface temperatures. *Geophys. Res. Lett.* **2006**, 33. [CrossRef]
31. Adams, R.M.; Houston, L.L.; McCarl, B.A.; Tiscare N O., M.; Matus, J.; Weiher, R.F. The benefits to Mexican agriculture of an El Niño-southern oscillation (ENSO) early warning system. *Agr. For. Meteorol.* **2003**, *115*, 183–194. [CrossRef]
32. Ropelewski, C.F.; Halpert, M.S. North American precipitation and temperature patterns associated with the El Niño/Southern Oscillation (ENSO). *Mon. Weather Rev.* **1986**, *114*, 2352–2362. [CrossRef]
33. Lu, B.; Li, H.; Wu, J.; Zhang, T.; Liu, J.; Liu, B.; Chen, Y.; Baishan, J. Impact of El Niño and Southern Oscillation on the summer precipitation over Northwest China. *Atmos. Sci. Lett.* **2019**, *20*, e928. [CrossRef]
34. Xu, Z.X.; Takeuchi, K.; Ishidaira, H. Correlation between El Niño-Southern Oscillation (ENSO) and precipitation in South-east Asia and the Pacific region. *Hydrol. Process.* **2004**, *18*, 107–123. [CrossRef]

35. Xu, Z.; Pan, B.; Han, M.; Zhu, J.; Tian, L. Spatial-temporal distribution of rainfall erosivity, erosivity density and correlation with El Niño-Southern Oscillation in the Huaihe River Basin, China. *Ecol. Inf.* **2019**, *52*, 14–25. [CrossRef]
36. Zhu, D.; Xiong, K.; Xiao, H.; Gu, X. Variation characteristics of rainfall erosivity in Guizhou Province and the correlation with the El Niño Southern Oscillation. *Sci. Total Environ.* **2019**, *691*, 835–847. [CrossRef] [PubMed]
37. Chen, S.; Zha, X. Effects of the ENSO on rainfall erosivity in the Fujian Province of southeast China. *Sci. Total Environ.* **2018**, *621*, 1378–1388. [CrossRef] [PubMed]
38. Romero, C.C.; Baigorria, G.A.; Stroosnijder, L. Changes of erosive rainfall for El Niño and La Niña years in the northern Andean highlands of Peru. *Clim. Chang.* **2007**, *85*, 343–356. [CrossRef]
39. Lee, J.H.; Lee, J.; Julien, P.Y. Global climate teleconnection with rainfall erosivity in South Korea. *Catena* **2018**, *167*, 28–43. [CrossRef]
40. Angulo-Mart I Nez, M.; Beguer I A., S. Do atmospheric teleconnection patterns influence rainfall erosivity? A study of NAO, MO and WeMO in NE Spain, 1955–2006. *J. Hydrol.* **2012**, *450*, 168–179. [CrossRef]
41. D'Odorico, P.; Yoo, J.C.; Over, T.M. An assessment of ENSO-induced patterns of rainfall erosivity in the southwestern United States. *J. Clim.* **2001**, *14*, 4230–4242. [CrossRef]
42. Yao, T.; Thompson, L.G.; Mosbrugger, V.; Zhang, F.; Ma, Y.; Luo, T.; Xu, B.; Yang, X.; Joswiak, D.R.; Wang, W.; et al. Third pole environment (TPE). *Environ. Dev.* **2012**, *3*, 52–64. [CrossRef]
43. Chen, D.; Xu, B.; Yao, T.; Guo, Z.; Cui, P.; Chen, F.; Zhang, R.; Zhang, X.; Zhang, Y.; Fan, J.; et al. Assessment of past, present and future environmental changes on the Tibetan Plateau. *Chin. Sci. Bull.* **2015**, *60*, 3025–3035.
44. Yuan, Q.; Yuan, Q.; Ren, P. Coupled effect of climate change and human activities on the restoration/degradation of the Qinghai-Tibet Plateau grassland. *J. Geogr. Sci.* **2021**, *31*, 1299–1327. [CrossRef]
45. An, R.; Zhang, C.; Sun, M.; Wang, H.; Shen, X.; Wang, B.; Xing, F.; Huang, X.; Fan, M. Monitoring grassland degradation and restoration using a novel climate use efficiency (NCUE) index in the Tibetan Plateau, China. *Ecol. Indic.* **2021**, *131*, 108208. [CrossRef]
46. Li, L.; Zhang, Y.; Liu, L.; Wang, Z.; Zhang, H.; Li, S.; Ding, M. Mapping Changing Population Distribution on the Qinghai–Tibet Plateau since 2000 with Multi-Temporal Remote Sensing and Point-of-Interest Data. *Remote Sens.* **2020**, *12*, 4059. [CrossRef]
47. Harris, R.B. Rangeland degradation on the Qinghai-Tibetan plateau: A review of the evidence of its magnitude and causes. *J. Arid. Env.* **2010**, *74*, 1–12. [CrossRef]
48. Gao, X.; Shi, X.; Lei, T. Influence of thawed soil depth on rainfall erosion of frozen bare meadow soil in the Qinghai–Tibet Plateau. *Earth Surf. Proc. Land* **2021**, *46*, 1953–1963. [CrossRef]
49. Cheng, G.; Wu, T. Responses of permafrost to climate change and their environmental significance, Qinghai-Tibet Plateau. *J. Geophys. Res. Earth Surf.* **2007**, *112*. [CrossRef]
50. Yao, T.; Thompson, L.; Yang, W.; Yu, W.; Gao, Y.; Guo, X.; Yang, X.; Duan, K.; Zhao, H.; Xu, B.; et al. Different glacier status with atmospheric circulations in Tibetan Plateau and surroundings. *Nat. Clim. Chang.* **2012**, *2*, 663–667. [CrossRef]
51. Yu, C.; Zhang, Y.; Claus, H.; Zeng, R.; Zhang, X.; Wang, J. Ecological and Environmental Issues Faced by a Developing Tibet. *Environ. Sci. Technol.* **2012**, *46*, 1979–1980. [CrossRef]
52. Liu, X.; Chen, B. Climatic warming in the Tibetan Plateau during recent decades. *Int. J. Climatol. A J. R. Meteorol. Soc.* **2000**, *20*, 1729–1742. [CrossRef]
53. Wang, Y.; Cheng, C.; Xie, Y.; Liu, B.; Yin, S.; Liu, Y.; Hao, Y. Increasing trends in rainfall-runoff erosivity in the Source Region of the Three Rivers, 1961–2012. *Sci. Total Environ.* **2017**, *592*, 639–648. [CrossRef]
54. Liu, B.; Tao, H.; Song, C.; Guo, B.; Shi, Z. Temporal and Spatial Variations of Rainfall Erosivity in Southwest China from 1960 to 2009. *Adv. Earth Sci.* **2012**, *27*, 499–509.
55. Hren, M.T.; Chamberlain, C.P.; Hilley, G.E.; Blisniuk, P.M.; Bookhagen, B. Major ion chemistry of the Yarlung Tsangpo-Brahmaputra river: Chemical weathering, erosion, and CO_2 consumption in the southern Tibetan plateau and eastern syntaxis of the Himalaya. *Geochim. Cosmochim. Acta* **2007**, *71*, 2907–2935. [CrossRef]
56. He, Q.; Chen, S.; Dai, X. Assessing the effects of vegetation and precipitation on soil erosion in the Three-River Headwaters Region of the Qinghai-Tibet Plateau, China. *J. Arid. Land* **2020**, *12*, 865–886. [CrossRef]
57. Mengmei, W.; Suhua, F.U.; Baoyuan, L. Quantitative Research of Water Erosion on the Qinghai-Tibet Plateau. *Adv. Earth Sci.* **2021**, *36*, 740–752.
58. Piao, S.; Cui, M.; Chen, A.; Wang, X.; Ciais, P.; Liu, J.; Tang, Y. Altitude and temperature dependence of change in the spring vegetation green-up date from 1982 to 2006 in the Qinghai-Xizang Plateau. *Agr. For. Meteorol.* **2011**, *151*, 1599–1608. [CrossRef]
59. Zhu, D.; Xiong, K.; Xiao, H. Multi-time scale variability of rainfall erosivity and erosivity density in the karst region of southern China, 1960–2017. *Catena* **2021**, *197*, 104977. [CrossRef]
60. Gu, Z.; Feng, D.; Duan, X.; Gong, K.; Li, Y.; Yue, T. Spatial and temporal patterns of rainfall erosivity in the Tibetan Plateau. *Water* **2020**, *12*, 200. [CrossRef]
61. Fan, J.; Chen, Y.; Yan, D.; Guo, F. Characteristics of rainfall erosivity based on tropical rainfall measuring mission data in Tibet, China. *J. Mt. Sci.* **2013**, *10*, 1008–1017. [CrossRef]
62. Zhang, Y.; Li, B.; Zheng, D. Datasets of the boundary and area of the Tibetan Plateau. *Acta Geogr. Sin.* **2014**, *69*, 164–168.
63. Hu, Y.; Xu, J.; Huang, Y.; Zhou, Y.; Pang, Y.; Shi, Z.; Chen, X. Spatial and Temporal Variations in the Rainy Season Onset over the Qinghai—Tibet Plateau. *Water* **2019**, *11*, 1960. [CrossRef]

64. Zheng, J.; Yin, Y.; Li, B. A new scheme for climate regionalization in China. *Acta Geogr. Sin.* **2010**, *65*, 3–12.
65. Gao, J.; Yao, T.; Masson-Delmotte, V.E.R.; Steen-Larsen, H.C.; Wang, W. Collapsing Glaciers Threaten Asia's Water Supplies. *Nature* **2019**, *565*, 19–21. [CrossRef]
66. Zheng, D. The system of physico-geographical regions of the Qinghai-Xizang (Tibet) Plateau. *Sci. China Ser. D-Earth Sci.* **1996**, *39*, 410–417.
67. Li, X.; Jiang, F.; Li, L.; Wang, G. Spatial and temporal variability of precipitation concentration index, concentration degree and concentration period in Xinjiang, China. *Int. J. Clim.* **2011**, *31*, 1679–1693. [CrossRef]
68. Kaplan, A.; Cane, M.A.; Kushnir, Y.; Clement, A.C.; Blumenthal, M.B.; Rajagopalan, B. Analyses of global sea surface temperature 1856–1991. *J. Geophys. Res. Ocean.* **1998**, *103*, 18567–18589. [CrossRef]
69. Trenberth, K.E. The definition of el nino. *Bull. Am. Meteorol Soc.* **1997**, *78*, 2771–2778. [CrossRef]
70. Wolter, K.; Timlin, M.S. El Niño/Southern Oscillation behaviour since 1871 as diagnosed in an extended multivariate ENSO index (MEI. ext). *Int. J. Clim.* **2011**, *31*, 1074–1087. [CrossRef]
71. Wolter, K.; Timlin, M.S. Monitoring ENSO in COADS with a Seasonally Adjusted Principal component index. In Proceedings of the 17th Climate Diagnostics Workshop, Noman, OK, USA, 18–23 October 1992; Volume 52, pp. 52–57.
72. Huang, B.; Thorne, P.W.; Banzon, V.F.; Boyer, T.; Chepurin, G.; Lawrimore, J.H.; Menne, M.J.; Smith, T.M.; Vose, R.S.; Zhang, H. Extended reconstructed sea surface temperature, version 5 (ERSSTv5): Upgrades, validations, and intercomparisons. *J. Clim.* **2017**, *30*, 8179–8205. [CrossRef]
73. Richardson, C.W.; Foster, G.R.; Wright, D.A. Estimation of Erosion Index from Daily Rainfall Amount. *Trans. Am. Soc. Agric. Eng.* **1983**, *26*, 153–160. [CrossRef]
74. Qin, W.; Guo, Q.; Zuo, C.; Shan, Z.; Ma, L.; Sun, G. Spatial distribution and temporal trends of rainfall erosivity in mainland China for 1951–2010. *Catena* **2016**, *147*, 177–186. [CrossRef]
75. Zhang, W.B.; Fu, J.S. Rainfall erosivity estimation under different rainfall amount. *Resour. Sci.* **2003**, *25*, 35–41.
76. Liu, S.; Huang, S.; Xie, Y.; Leng, G.; Huang, Q.; Wang, L.; Xue, Q. Spatial-temporal changes of rainfall erosivity in the loess plateau, China: Changing patterns, causes and implications. *Catena* **2018**, *166*, 279–289. [CrossRef]
77. Xin, Z.; Yu, X.; Li, Q.; Lu, X.X. Spatiotemporal variation in rainfall erosivity on the Chinese Loess Plateau during the period 1956–2008. *Reg. Env. Chang.* **2011**, *11*, 149–159. [CrossRef]
78. Sen, P.K. Estimates of the regression coefficient based on Kendall's tau. *J. Am. Stat. Assoc.* **1968**, *63*, 1379–1389. [CrossRef]
79. Kendall, M.G. *Rank Correlation Methods*; Griffin: London, UK, 1975.
80. Mann, H.B. Nonparametric tests against trend. *Econom. J. Econom. Soc.* **1945**, *13*, 245–259. [CrossRef]
81. Mitchell, J.M.; Dzerdzeevskii, B.; Flohn, H.; Hofmeyr, W.L.; Lamb, H.H.; Rao, K.N.; Wallen, C.C. *Climate Change, WMO Technical Note No.79*; World Meteorological Orgazination: Geneva, Switzerland, 1966; p. 79.
82. Beguería, S.; Serrano-Notivoli, R.; Tomas-Burguera, M. Computation of rainfall erosivity from daily precipitation amounts. *Sci. Total Environ.* **2018**, *637*, 359–373. [CrossRef]
83. Ray, D.K.; Ramankutty, N.; Mueller, N.D.; West, P.C.; Foley, J.A. Recent patterns of crop yield growth and stagnation. *Nat. Commun.* **2012**, *3*, 1293. [CrossRef]
84. Zhang, Z.; Lu, C. Identification of Maize Yield Trend Patterns in the North China Plain. *Int. J. Plant. Prod.* **2021**, *15*, 125–137. [CrossRef]
85. Schauberger, B.; Ben-Ari, T.; Makowski, D.; Kato, T.; Kato, H.; Ciais, P. Yield trends, variability and stagnation analysis of major crops in France over more than a century. *Sci. Rep.* **2018**, *8*, 16865. [CrossRef]
86. Zhang, Q.; Xu, C.; Jiang, T.; Wu, Y. Possible influence of ENSO on annual maximum streamflow of the Yangtze River, China. *J. Hydrol.* **2007**, *333*, 265–274. [CrossRef]
87. Jevrejeva, S.; Moore, J.C.; Grinsted, A. Influence of the Arctic Oscillation and El Niño-Southern Oscillation (ENSO) on ice conditions in the Baltic Sea: The wavelet approach. *J. Geophys. Res. Atmos.* **2003**, *108*. [CrossRef]
88. Torrence, C.; Compo, G.P. A practical guide to wavelet analysis. *Bull. Am. Meteorol. Soc.* **1998**, *79*, 61–78. [CrossRef]
89. Grinsted, A.; Moore, J.C.; Jevrejeva, S. Application of the cross wavelet transform and wavelet coherence to geophysical time series. *Nonlinear Proc. Geoph.* **2004**, *11*, 561–566. [CrossRef]
90. Schiemann, R.; Lüthi, D.; Schär, C. Seasonality and interannual variability of the westerly jet in the Tibetan Plateau region. *J. Clim.* **2009**, *22*, 2940–2957. [CrossRef]
91. Panagos, P.; Borrelli, P.; Meusburger, K.; Yu, B.; Klik, A.; Lim, K.J.; Yang, J.E.; Ni, J.; Miao, C.; Chattopadhyay, N.; et al. Global rainfall erosivity assessment based on high-temporal resolution rainfall records. *Sci. Rep.* **2017**, *7*, 4175. [CrossRef]
92. Shin, J.; Kim, T.; Heo, J.; Lee, J. Spatial and temporal variations in rainfall erosivity and erosivity density in South Korea. *Catena* **2019**, *176*, 125–144. [CrossRef]
93. Li, L.; Yang, S.; Wang, Z.; Zhu, X.; Tang, H. Evidence of warming and wetting climate over the Qinghai-Tibet Plateau. *Arct. Antarct. Alp. Res.* **2010**, *42*, 449–457. [CrossRef]
94. Sato, T.; Kimura, F. How does the Tibetan Plateau affect the transition of Indian monsoon rainfall? *Mon. Weather Rev.* **2007**, *135*, 2006–2015. [CrossRef]
95. Zhou, T.; Gao, J.; Zhao, Y.; Zhang, L.; Zhang, W. Water Vapor Transport Processes on Asian Water Tower. *Bull. Chin. Acad. Sci.* **2019**, *34*, 1210–1219.
96. Stocking, M.A.; Elwell, H.A. Rainfall erosivity over Rhodesia. *Trans. Inst. Br. Geogr.* **1976**, *1*, 231–245. [CrossRef]

97. Paliaga, G.; Donadio, C.; Bernardi, M.; Faccini, F. High-resolution lightning detection and possible relationship with rainfall events over the Central Mediterranean Area. *Remote Sens.* **2019**, *11*, 1601. [CrossRef]
98. Nearing, M.A.; Yin, S.; Borrelli, P.; Polyakov, V.O. Rainfall erosivity: An historical review. *Catena* **2017**, *157*, 357–362. [CrossRef]
99. Lai, C.; Chen, X.; Wang, Z.; Wu, X.; Zhao, S.; Wu, X.; Bai, W. Spatio-temporal variation in rainfall erosivity during 1960–2012 in the Pearl River Basin, China. *Catena* **2016**, *137*, 382–391. [CrossRef]
100. Huang, J.; Zhang, J.; Zhang, Z.; Xu, C. Spatial and temporal variations in rainfall erosivity during 1960–2005 in the Yangtze River basin. *Stoch. Environ. Res. Risk Assess.* **2013**, *27*, 337–351. [CrossRef]
101. Panagos, P.; Ballabio, C.; Borrelli, P.; Meusburger, K.; Klik, A.; Rousseva, S.; Tadić, M.P.; Michaelides, S.; Hrabalíková, M.; Olsen, P.; et al. Rainfall erosivity in Europe. *Sci. Total Environ.* **2015**, *511*, 801–814. [CrossRef] [PubMed]
102. Chen, H. Projected change in extreme rainfall events in China by the end of the 21st century using CMIP5 models. *Chin. Sci. Bull.* **2013**, *58*, 1462–1472. [CrossRef]
103. Immerzeel, W.W.; Van Beek, L.P.; Bierkens, M.F. Climate change will affect the Asian water towers. *Science* **2010**, *328*, 1382–1385. [CrossRef]
104. Zhang, J.; Li, D.; Wang, W. Influence of terrain on precipitation in Qinghai-Tibet Plateau during summer monsoon. *Sci. Geogr. Sin.* **2008**, *28*, 235.
105. Wang, Z.; Zhang, Y.; Yang, Y.; Zhou, W.; Gang, C.; Zhang, Y.; Li, J.; An, R.; Wang, K.; Odeh, I.; et al. Quantitative assess the driving forces on the grassland degradation in the Qinghai–Tibet Plateau, in China. *Ecol. Inf.* **2016**, *33*, 32–44. [CrossRef]
106. Chen, T.; Jiao, J.; Wang, H.; Zhao, C.; Lin, H. Progress in Research on Soil Erosion in Qinghai-Tibet Plateau. *Acta Pedol. Sin.* **2019**, *57*, 547–564.
107. Zhang, Y.; Liu, L.; Wang, Z.; Bai, W.; Ding, M.; Wang, X.; Yan, J.; Xu, E.; Wu, X.; Zhang, B.; et al. Spatial and temporal characteristics of land use and cover changes in the Tibetan Plateau. *Chin. Sci. Bull.* **2019**, *64*, 2865–2875.
108. Cao, Q.; Hao, Z.; Yuan, F.; Su, Z.; Berndtsson, R.; Hao, J.; Nyima, T. Impact of ENSO regimes on developing-and decaying-phase precipitation during rainy season in China. *Hydrol. Earth Syst. Sci.* **2017**, *21*, 5415–5426. [CrossRef]
109. Cai, Z.; Tian, L.; Bowen, G.J. ENSO variability reflected in precipitation oxygen isotopes across the Asian Summer Monsoon region. *Earth Planet. Sci. Lett.* **2017**, *475*, 25–33. [CrossRef]
110. Kumar, K.K.; Rajagopalan, B.; Hoerling, M.; Bates, G.; Cane, M. Unraveling the mystery of Indian monsoon failure during El Niño. *Science* **2006**, *314*, 115–119. [CrossRef] [PubMed]

International Journal of
Environmental Research and Public Health

MDPI

Article

Ecological Effect of Ecological Engineering Projects on Low-Temperature Forest Cover in Great Khingan Mountain, China

Shuqing Wang [1], Run Zhong [1], Lin Liu [2,*] and Jianjun Zhang [1]

[1] School of Land Science and Technology, China University of Geosciences (Beijing), Beijing 100083, China; wangsq@cugb.edu.cn (S.W.); z_xb_2003@163.com (R.Z.); zhangjianjun_bj@126.com (J.Z.)
[2] School of Land Science and Space Planning, Hebei GEO University, Shijiazhuang 050031, China
* Correspondence: liulin7801@126.com; Tel.: +86-135-2108-8014

Abstract: The evaluation of ecological restoration projects can provide support for further strengthening the efforts of ecological restoration work and implementing the strategic objectives of the ecological region. Considering the current problem of the single evaluation index, this study evaluated the implementation effect of ecological projects from different temporal and spatial dimensions. Based on the MODIS vegetation index time series data, this study first computed the Sustainable Development Goal (SDG) indicator 15.3.1 of Great Khingan Mountain (GKM) to evaluate the impact of ecological engineering on land use change and land productivity. As a common indicator, the Normalized Difference Vegetation Index (NDVI) values showed a trend of a decrease and then gradual increase after the start of the Natural Forest Protection Project (NFPP) II, which was related to the land use changes from the forest to the grassland during the implementation of the NFPP. However, land productivity maintained a steady trend because of the transition between the forest and grassland. Meanwhile, to detect changes in vegetation at a smaller scale, the LandTrendr algorithm was used to identify the magnitude of forest disturbance, the years when it occurred, and the year of restoration. After implementing the ecological project, the forests in the GKM region were only partially disturbed, and most of the forests in most areas maintained a stable trend. Our study highlighted the varying effectiveness of different indexes for NFPP and evaluated the ecological impact of ecological projects from multiple perspectives.

Keywords: SDG indicator; forest degradation; land productivity; land degradation

Citation: Wang, S.; Zhong, R.; Liu, L.; Zhang, J. Ecological Effect of Ecological Engineering Projects on Low-Temperature Forest Cover in Great Khingan Mountain, China. *Int. J. Environ. Res. Public Health* **2021**, *18*, 10625. https://doi.org/10.3390/ijerph182010625

Academic Editors: Wei Song and Hualin Xie

Received: 1 September 2021
Accepted: 8 October 2021
Published: 11 October 2021

1. Introduction

Ecosystems are mainly influenced by human activities and climate variations [1]. With the development of the global modernization trend, a larger scale of environmental pollution and destruction is accompanied by a new wave of worldwide environmental protection. Many large-scale ecological restoration projects, especially forest restoration projects were underway for more than 20 years to reverse ecological degradation and achieve environmental sustainability [2]. Of these ecological projects, forest conservation and restoration projects are the most important, as the effects of forest loss and degradation are felt on all scales, from global climate change to the decline in the economic value of forest resources and biodiversity, and threats to local livelihoods [3,4]. The New Zealand government announced the One Billion Trees (1 BT) program which aimed to plant one billion trees in the country by 2028 [3,5]. Reducing Emissions from Deforestation and Forest Degradation plus (REDD+), as the best-known international forestry-based policy for carbon dioxide removal, was created by the United Nations Framework Convention on Climate Change (UNFCCC) Conference and aimed to implement schemes by national governments to reduce human impact on forests, an activity which results in greenhouse gas emissions at the national level [6].

China has a long history of forest degradation and restoration. To minimize the human impact on the environment and restore vegetation coverage, a series of vegetation-related policies were implemented by China [7]. The largest project is the Three-North Shelter Forest Program (TNSFP), focused on increasing forest cover from 5% to 15% from 1978 to 2050 [8]. In 2000, the Chinese government fully implemented the Natural Forest Protection Project (NFPP), which instilled logging bans and harvesting reductions in 68.2 million ha of forest land. The main goal of this project was to protect the existing natural forests, increase vegetation cover and mitigate soil erosion and desertification through the revegetation and conversion of agricultural land to forest land [9]. In 2014, the State Forestry Administration (SFA) expanded the scope of NFPP and tried to ban commercial logging in the state-owned natural forests of Heilongjiang Province. The log supply in Heilongjiang Province historically accounted for more than 30% of China's domestic log supply.

A large number of ecological projects were implemented and estimating the effects of these ecological projects became an important research direction. Several studies investigated the spatial and temporal effects of climatic factors and human activities on changes in vegetation productivity from 1985 to 2015, where actual net primary productivity (ANPP) and net primary productivity (NPP) were often used to quantify vegetation dynamics [3,7]. Several studies used multiple satellite images to assess the changes of the leaf area index (LAI), gross primary productivity (GPP), Normalized Difference Vegetation Index (NDVI), and aboveground biomass in different regions in China [10,11]. Previous studies focused on long-term change analysis, often failing to capture such subtle changes which occurred over long periods, but could have significant impacts on forest structure, composition, and function, and thus ultimately limited the successful implementation of sustainable development goals (SDG) [12].

Trend and change analysis using different time series data is the most common method when performing calculations using the different indexes. To find the trend of NDVI or GPP time series, the linear regression technique is a viable and robust method [13]. As vegetation change is influenced not only by human activities but also by the climate, a residual trend analysis is now commonly used to separate human-induced and climate-driven vegetation changes [14,15]. The Breaks for Additive Season and Trend (BFAST) algorithm and LandTrendr algorithm are used to identify long-term trends and abrupt changes (breaks) in the time series [16–18]. These methods use multivariate (parametric) time series analysis and effective algorithms to comprehensively assess the effects of vegetation restoration, which provide us with the data, information, and knowledge necessary to better implement and manage large-scale ecological projects. Although all of the above methods are used in many current studies, there is relatively little research on applying them to analyze the impact of ecological engineering on the environment.

To integrate the application of different methods to evaluate the impact of ecological engineering, this study uses the alpine region of Great Khingan Mountain (GKM) as the research object and integrates the application of various methods. In this study design, the three sub-indicators used for SDG 15.3.1 are first used to measure different aspects of land cover which relate to vegetation. Primary productivity, the first sub-indicator, can directly measure changes in the biomass present in an area, but it is not able to determine whether this change is positive, as not all increases in plant biomass are interpreted as improvements. As the second sub-indicator, land cover fills this gap. It explains the landscape from a thematic point of view, looking at the features that were there previously and those that are there now. It includes changes in vegetation, as well as bare land, cities, and water. The last sub-indicator, the Soil Organic Carbon Index, uses a land cover map to inform changes in the organic carbon of the soil over time. Although the results of this method are not ideal, considering the status of global soil science and measurement, researchers agree that this method is the best method to implement on a global scale. Vegetation is a key component of most ecosystems and a good representation of its overall function and health. For example, within a particular land cover type, the land productivity or soil carbon levels may change over time. However, changes in land cover types usually lead to changes in the level and dynamics of land productivity, which in turn affect the carbon storage of a particular area. Therefore, when evaluating the degradation status against these three sub-indices, it is convenient to summarize the fine-scale results (such as the pixel-scale evaluation in this method) into the spatial characteristics determined in the land cover and land cover change sub-indices. In addition to the SDG factor, in order to find changes, disturbances, and causes of vegetation at a smaller scale, the LandTrendr algorithm is used in this study to identify disturbances in the vegetation of the study area.

Forest change can be influenced by many factors. In this study, the influence of other climatic factors are removed from the calculations to study the impact of ecological projects. This means that the study assumes that external factors such as the climate are consistent over the period of forest change in the study area. To assess the interest and potential of ecological engineering in forests from different perspectives, this study uses images at different scales to identify the forest degradation and improvement in order to explore the possible impacts of different contexts and policies.

2. Study Area and Data

2.1. Study Area

As the northernmost border area in China, the Greater Khingan Mountains are located at 121°12′–127°00′ East longitude and 50°10′–53°33′ North latitude. With an average altitude of 573 meters, it has a cold temperate and a continental monsoon climate with short summers and long winters. The annual average frost-free period is only 80–110 days. The annual average temperature is −2.1 °C, and the historical minimum temperature is −52.3 °C. In the study area, the mountains are mainly covered by boreal coniferous forest and grass (Figure 1). The huge mountains and forests effectively block the Siberian cold current and the cold wind of the Mongolian plateau and become a natural barrier for the Songnen Plain in northeast China and the Hulunbuir grassland in inner Mongolia. This area is also an important water conservation area for the Heilongjiang and Nenjiang rivers and other water sources, guaranteeing the water supply for residents in the middle and lower reaches of the two rivers.

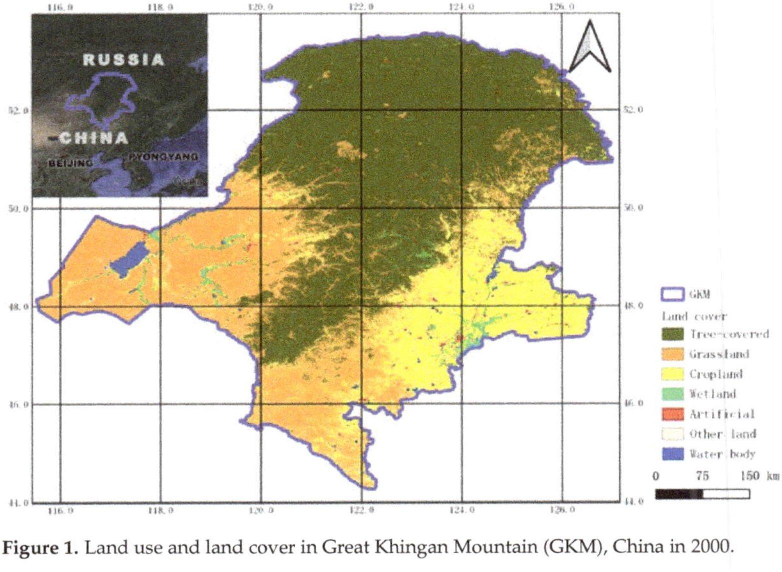

Figure 1. Land use and land cover in Great Khingan Mountain (GKM), China in 2000.

2.2. Data

2.2.1. NDVI Datasets

Normalized Difference Vegetation Index (NDVI) (Exelis Visual Information Solutions, Boulder, CO, USA) is often used to measure vegetation cover and health. Moderate Resolution Imaging Spectroradiometer (MODIS) and Landsat provide two-scale NDVI data sets. MODIS products provide vegetation indices with 16-day intervals and multiple spatial resolutions, providing a consistent spatial and temporal comparison of vegetation canopy greenness, which is a comprehensive attribute of leaf area, chlorophyll, and canopy structure. The MODIS Vegetation Indices product, MOD13Q1, at 250 m spatial resolution, was selected to compute the annual value and land productivity, as well as degradation. Landsat data sets provided the Surface Reflectance-derived NDVI with a 30-meter resolution, derived from Landsat 4–5 Thematic Mapper (TM), Landsat 7 Enhanced Thematic Mapper Plus (ETM+), and Landsat 8 Operational Land Image (OLI)/Thermal Infrared Sensor (TIRS). The finer resolution NDVI dataset was used to quantify vegetation greenness and was more helpful to assess changes in plant health. The data used in this study are shown in Table 1.

Table 1. Remote sensing image data source for Great Khingan Mountain (GKM), China.

Sensor	Satellite	Frequency	Data Source	Data Record	Spatial Resolution	Time Step
MODIS	Terra/Aqua	1–2 days	MOD13 vegetation index	2000–present	250 m, 500 m, 1 km	8-day, 16-day
TM	Landsat 4–5	16 days	USGS/EROS	1982–2011	30 m	Distributed by scene
ETM+	Landsat 7	16 days	USGS/EROS	1999–presnet	30 m	Distributed by scene
OLI	Landsat 8	16 days	USGS/EROS	2013–present	30 m	Distributed by scene

2.2.2. Land Cover and Climate Data

To assess changes in land cover, we used land cover maps downloaded from ESA CCI (European Space Agency Climate Change Initiative) datasets, which covering the study area for the baseline and target years. The land use classes included forest, grassland, cropland, wetland, artificial areas, bare areas and water, with a 300 m resolution.

Land productivity was affected by several factors, such as temperature, the availability of light, nutrients, and water. Water availability had a significant influence on the amount of plant tissue produced every year. It was important to interpret the results in the context of the information on historical precipitation. When the land trends were driven by regional patterns of changes in water availability, the declining productivity of these trends could be identified as human-caused land degradation. The precipitation product CHIRPS (Climate Hazards Group InfraRed Precipitation) was used to eliminate the effects of the climate. From 1981 to the present, CHIRPS combined our internal climatology, CHPclim, 0.05° resolution satellite imagery, and in situ station data to create a gridded rainfall time series for trend analysis and seasonal drought monitoring.

3. Method

The effect of ecological engineering projects policies on forests is a long-term process, so time series analyses with multivariable (parameters) and effective algorithms are necessary to comprehensively assess the effectiveness of vegetation restoration which provide essential data, information, and knowledge. Due to the long duration of ecological projects and the number of influencing factors, a multi-parameter time series analysis was used to assess the impact of ecological projects in providing the necessary data and information for forest restoration. The temporal dynamics of terrestrial ecosystems commonly consisted of continuous changes, discontinuous changes, and no changes [10]. The ability to identify the trend or dynamic change in policies, not the effect of natural factors (precipitation, fire, temperature, diseases) is important for better assessing the effects of policies. In this study design, we conducted a time series analysis using multi-scale satellite images, climate data, and statistical data (Figure 2).

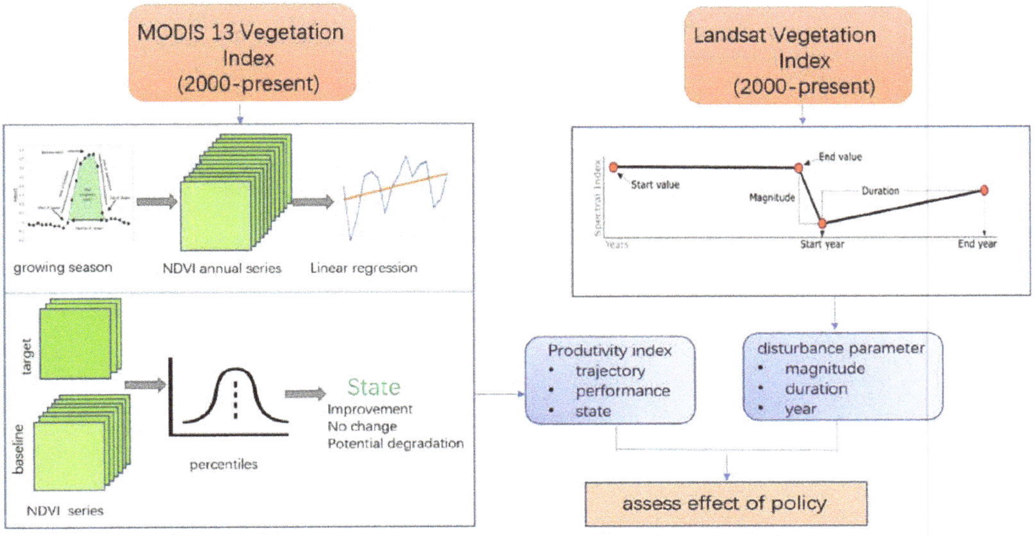

Figure 2. Scheme for calculating land productivity and disturbance for assessment effect of policy in Great Khingan Mountain (GKM), China.

3.1. Land Productivity Index

The land productivity is the maximum productivity of the land at the current level of farming technology and the measures relevant to it. It is the source of all food, fiber, and fuel that sustains human life. As land productivity can reflect long-term changes that indicate the health and productivity of the land and the net effect of changes in the ecosystem functions of plants and biomass growth, land productivity can be used as an indicator in the evaluation of ecological projects. Land productivity can be measured over large areas by the satellite Earth observations of NPP. The productivity index is the algorithm used to measure land productivity levels from image data. This study used *NDVI* [19] as the surrogates of NPP. To separate degradation effects from other sources of variation in productivity observations, a calibration was performed by calculating the ratio of *iNDVI* to *iET* (Evapotranspiration) to minimize the influence of climatic or seasonal factors [20]. The method for calculating water use efficiency (WUE) with corrected *iNDVI* (*iNDVIw*) per year is:

$$iNDVIw = \frac{iNDVI}{iET} \tag{1}$$

where $iNDVI$ is the $NDVI$ integrated over the growing season or relevant period each year, and iET is ET integrated over the same period.

Productivity was assessed in terms of trajectory, performance, and status. Trajectories could be used to identify degradation in areas with increasing productivity trends, and performance could identify low productivity compared to other areas with similar land cover types and similar climatic conditions. Status was used to compare the historical range of productivity levels at the site over time.

Productivity trajectory was calculated using the Thiel-Sen median, a robust, non-parametric linear regression method. The trend significance was determined by Mann–Kendall [21–23]. Positive and negative z-scores indicated trends of increasing productivity or decreasing productivity. The significance of the slope of the trajectory, calculated at the $p = 0.05$ level for more than eight data points, should be reported on three scales: improved, degraded, and insignificant.

Productivity state represented the level of productivity in a given spatial unit compared to the observed productivity levels for that spatial unit over time. Productivity state could be interpreted as an indicator of the relative standing biomass [24]. Existing degradation based on productivity performance could be identified by a mean productivity performance in the baseline period of less than 50% of the potential maximum. The $iNDVIw$ values were classified during the baseline period into ten decile classes using the unsupervised ISODATA classification. These become the baseline $iNDVIw$ classes. Productivity state change was assessed by comparing $iNDVIw$ in the assessment year to the baseline $iNDVIw$ classes.

As the NFPP was a long-term project, this study used 2000 as a cut-off point to analyze the different land use changes and land degradation before and after the NFPP. The year 2010, the start of the second phase of the NFPP, was chosen as the interval. The average productivity of the early epoch (2001 to 2010) and late epoch (2011 to 2018) could be calculated. Then, pixels in which the productivity level decresed between early and late epochs were identified for this metric.

3.2. Land Trendr

To detect trends in forest disturbance and recovery at multi-scale using yearly Landsat time series, we used LandTrendr as temporal segmentation algorithms. LandTrendr reduced image data to a single band or spectral index and then divided it into a series of straight-line segments by breakpoint (vertex) identification [3,18]. In practice, LandTrendr looks at the spectral history of a pixel to identify the breakpoints which separate periods of persistent change or stability in the spectral trajectory and records the year in which the change occurred. In this study, the breakpoints, durable changes, and the year could be used to detect trends in forest disturbance and recovery. Meanwhile, the typical area could be selected and compared with the Landsat remote sensing images selected on Google Earth to explore the possible reason for disturbance.

4. Results

4.1. Trend and Changes of Land Productivity Index

Land cover addresses the state and changes in the structure and composition of the landscape, from natural events and human activities. As the NFPP was a long-term project this study used 2000 as a cut-off point to analyze the different land use changes and land degradation before and after the NFPP. The forest was the dominant land use type in the study area throughout the project period. The western side of GKM was mainly grassland, while the eastern side was cropland (Figure 3a,b). The land cover transition map in Figure 3c showed that most land use types maintained the original state for a long time before the NFPP. After project implementation, the type of land use conversion in the area was dominated by the loss of forest (Figure 4a,b), and the distribution of forest loss was more concentrated (Figure 4c). According to the definition of land cover degradation in SDG, Figures 3d and 4d showed the results of the land cover degradation in the study area. After the project implementation, there was a higher concentration of land degradation appearing in the north and northeast area and some areas of land improvement (Figure 4d). Additionally, prior to the start of the project, the study area was less prone to land degradation and land improvement (Figure 3d).

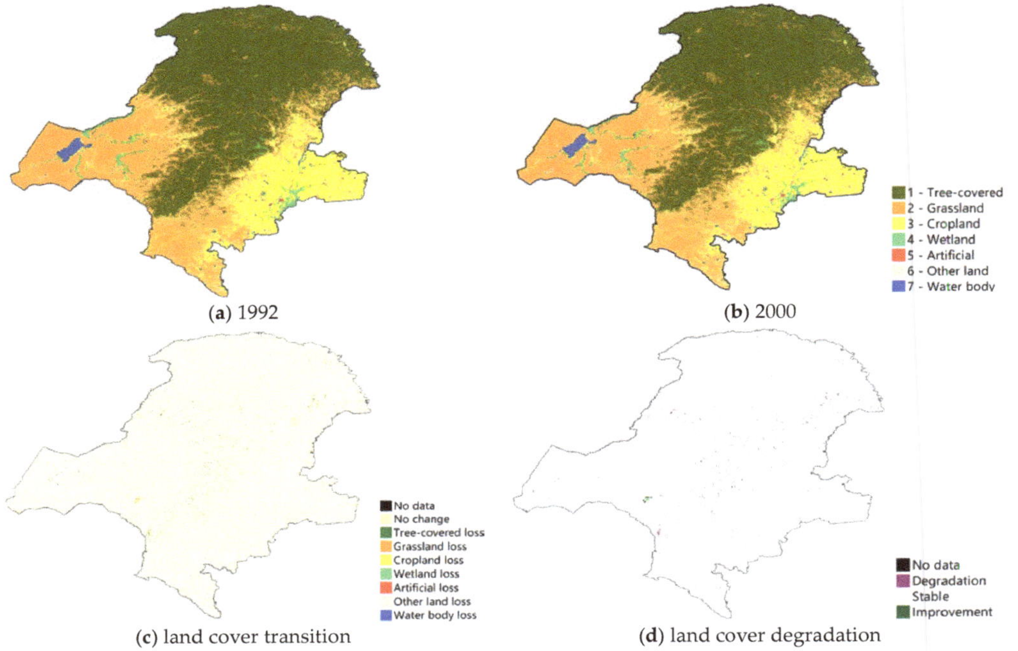

Figure 3. Land cover change and degradation in Great Khingan Mountain (GKM), China (1992–2000).

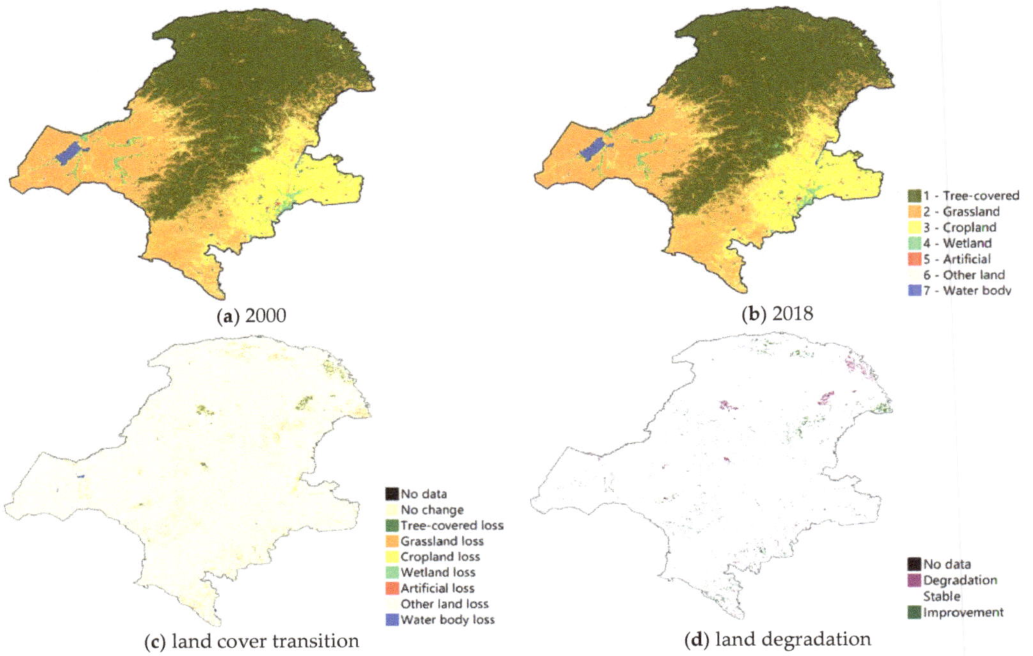

Figure 4. Land cover change and degradation in Great Khingan Mountain (GKM), China (2000–2018).

Figure 5 compared the different changes in the quantity of land cover before and after the implementation of the project. Before 2000, land use losses were dominated by forest losses. This was inseparable from the deforestation of that period. After 2000, the natural forest protection project was implemented. After the implementation of the policy, land use losses became dominated by grassland losses.

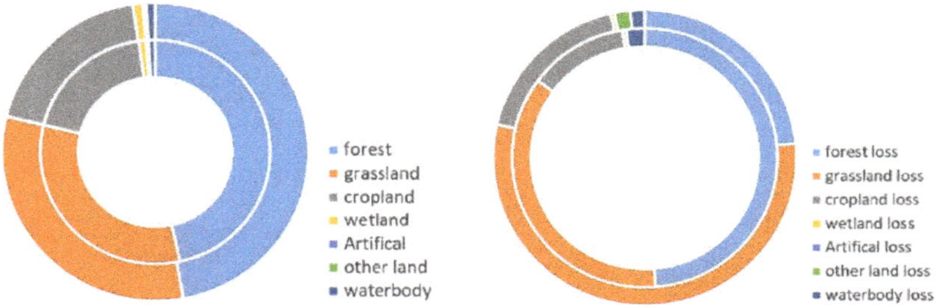

Figure 5. Land cover change and loss from 1992 to 2018 in Great Khingan Mountain (GKM), China (inner circle: 1992–2000; outer circle: 2000–2018).

The ecological engineering projects had an effect not only on the land use change but also on forest degradation or deforestation. Figure 6 showed the trend of NDVI using the annual NDVI value. The NDVI showed a stable increase trend while experiencing a dramatic decrease from 2002 to 2003. The annual NDVI values maintained a decreasing trend under the trend line until 2010. At that time, all large-scale ecological engineering projects, especially NFPP, were launched by the Chinese Government to speed up the restoration of the forest ecosystems.

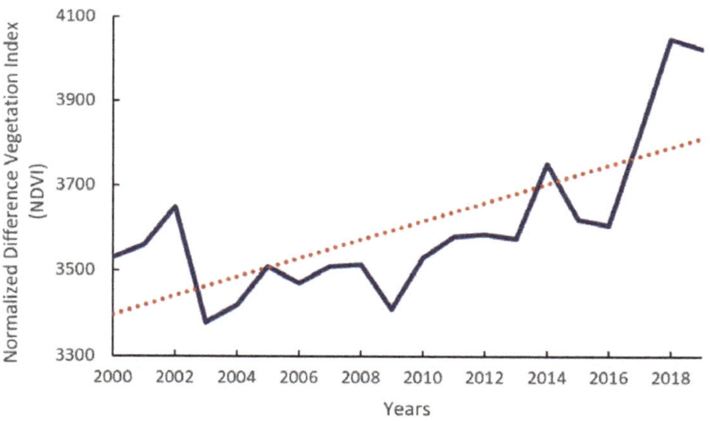

Figure 6. Normalized Difference Vegetation Index (NDVI) trend from 2000 to 2019, in Great Khingan Mountain (GKM), China.

Land productivity reflected the net effects of changes in the ecosystem functioning on plant and biomass growth. Figure 7 interpretated the trend and its significance regarding the variability of annual NPP in the time series. Throughout the project period, most of the study area showed an increase in land productivity, a small amount of stability and a very small trend of degradation: 67.3%, 31.8%, and 0.9%, respectively.

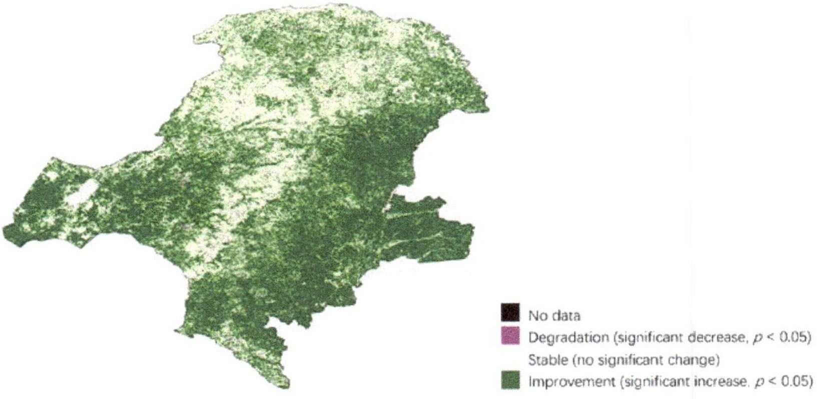

Figure 7. Productivity trajectory degradation in Great Khingan Mountain (GKM), China (2000–2018).

Productivity state change can be reported in terms of three classes relative to the baseline productivity state: (1) Improvement: observed productivity in baseline iNDVI classes 9 or 10. (2) Stable: observed productivity in baseline iNDVI classes 6, 7 or 8. (3) Degradation: observed productivity in baseline iNDVI classes 1, 2, 3, 4 or 5. The forest experienced a degradation between the two periods, whereas the other land showed an improvement (Figure 8). The degradation areas, especially in forest area, consisted of productivity trajectory degradation. Figure 9 showed that most of the land use types had no degradation, except 0.6% grassland and cropland located in the eastern and southeastern areas.

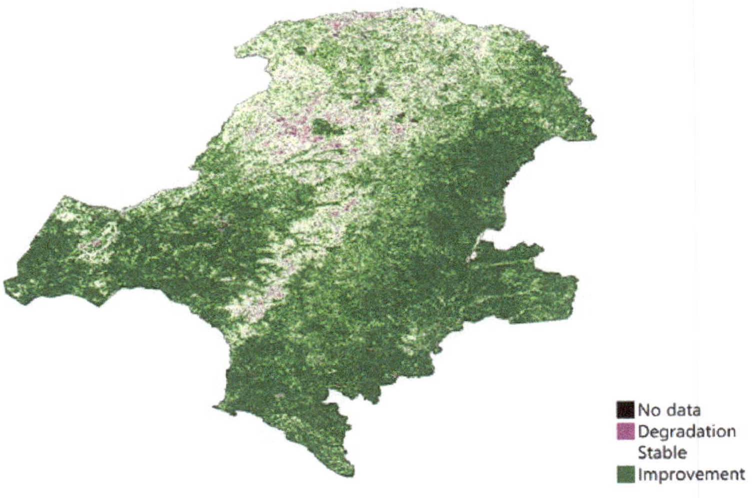

Figure 8. Productivity state degradation in Great Khingan Mountain (GKM), China (2001–2010 vs. 2011–2018).

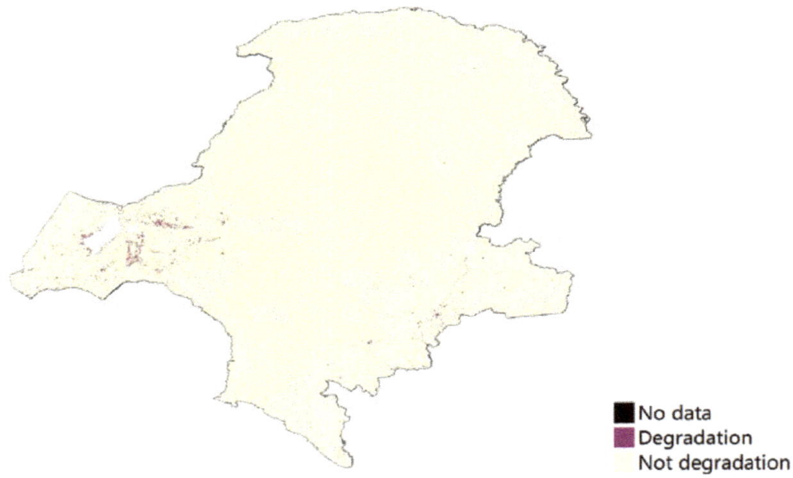

Figure 9. Land productivity performance degradation (2000–2018) in Great Khingan Mountain (GKM), China.

4.2. Forest Change Monitoring

4.2.1. Magnitude, Duration, and Years of Land Use Changes

Corse-scale image analyses revealed that the land cover/use and land productivity of the study area only had a little degradation (Figure 9). The magnitude, duration, and years of land use changes were computed by the LandTrendr algorithm shown in Figure 10. The higher disturbances of magnitude occurred only in some parts of the central area. The relatively high disturbances occurred in the northeastern area near the border. Most of the remaining areas were largely free of disturbances. The larger disturbances occurred in 2003 and 2006. The vast majority of vegetation disturbance across the study region was mostly short-term disturbance with rapid recovery, lasting up to three years (e.g., yellow areas in the northeast of the map).

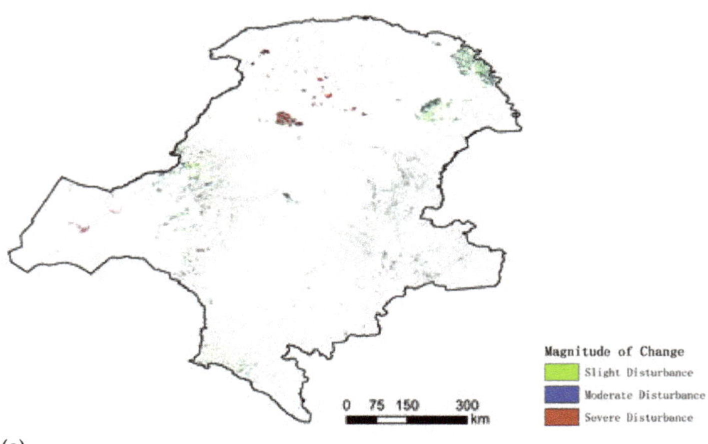

(a)

Figure 10. *Cont.*

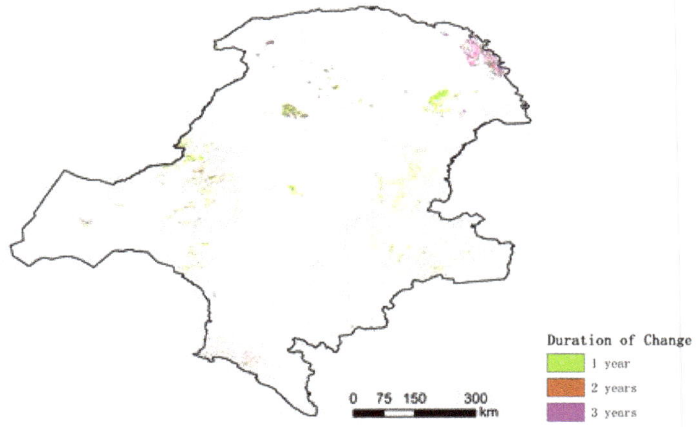

(b)

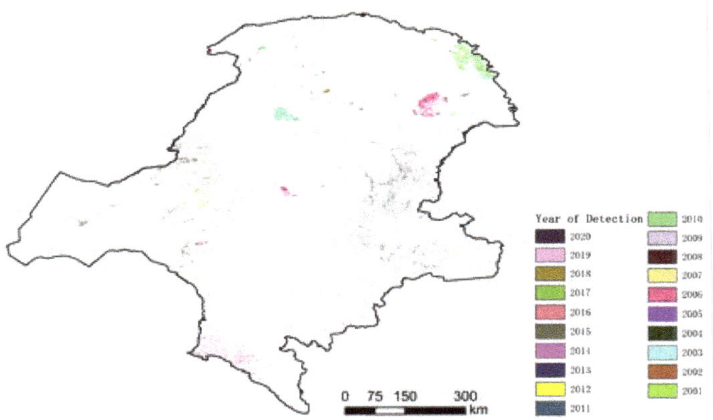

(c)

Figure 10. Forest disturbance and recovery at multi-scale and time series in Great Khingan Mountain (GKM), China: (**a**) magnitude of change, (**b**) duration of change, and (**c**) year of detection.

Taking 2000 as the starting point of the disturbance, the obtained annual time series of the disturbed areas from 2001 to 2020 is shown in Figure 11a. It can be seen that the disturbed area in 2001 reached 330,000 ha, accounting for 35.8%, followed by larger disturbed areas in 2003, 2006, 2015, and 2020.

The study area was mainly disturbed for a short period and the disturbance durations were all within three years. This indicated that the vegetation in the Daxinganling area recovered within a relatively short time after the disturbance occurred. In this area, the disturbance duration was approximately 1–2 years, and the disturbance duration area ratio reached 72%, as shown in Figure 11b.

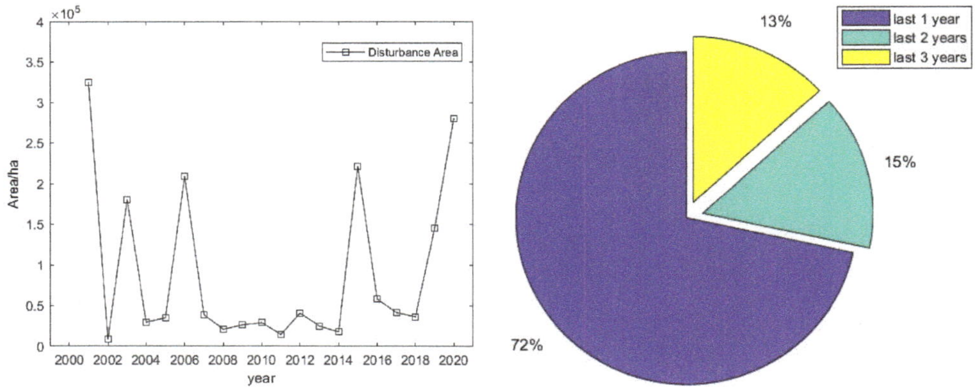

(a) (b)

Figure 11. (**a**) Disturbance area and (**b**) duration year, in Great Khingan Mountain (GKM), China.

4.2.2. Detection of Typical Deforest Area

The typical sites were selected to explore the fine-scale changes and reasons for changes using the LandTrendr algorithms. Figure 12 shows the distribution of typical area and the high-resolution images at different years. From the typical area selected in Figure 12a, it can be seen that the main disturbance in this area occurred in the year 2003, and the Landsat remote sensing images of 2004, 2005, and 2006 were selected from Google Earth for comparison in Figure 12b. Additionally, it was found that there was a certain amount of disturbance in all years in this area, but none of these disturbances were the main disturbance, and the disturbance in 2003 was the main interference, consistent with the main interference obtained by LandTrendr algorithm.

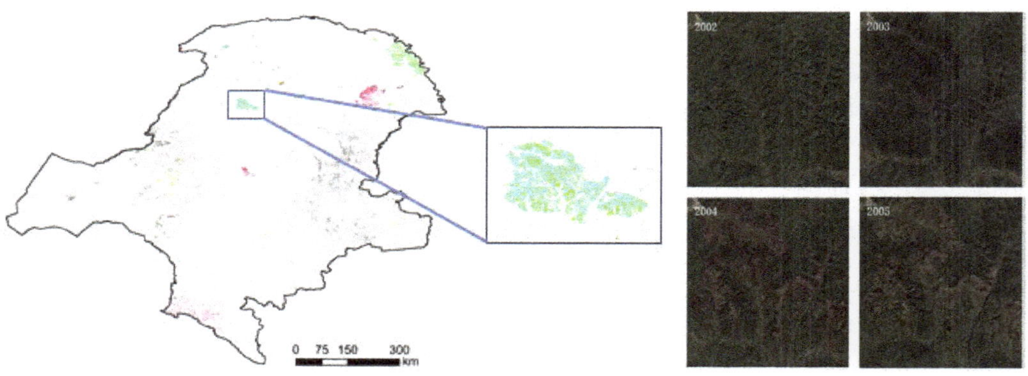

(a) (b)

Figure 12. (**a**) Typical disturbance area and (**b**) remote sensing images, in Great Khingan Mountain (GKM), China.

In the typical region, the sampling point (123°15′53.70″ E, 51°26′5.26″ N) was selected for the LandTrendr algorithm pixel time series trajectory extraction, and the time series was 2000–2020. The Normalized Burn Ratio (NBR) index (multiplied by 1000) was fitted to make the change information more intuitive, shown in Figure 13. From the index trajectory of the sampling point, the NBR value of this point ranged from 0.355 to 0.487.

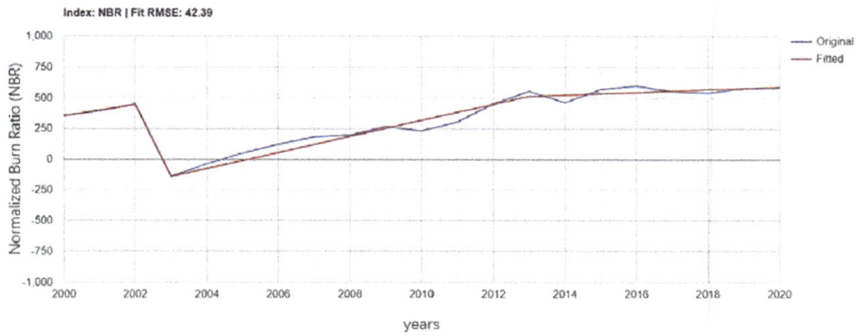

Figure 13. Normalized Burn Ratio (NBR) index trend from 2000 to 2020, in typical disturbance area.

In 2003 after a serious disturbance, the NBR value was −0.118, demonstrating bare soil. The area was continuously disturbed for a short period of 1 year and started to recover gradually after 1 to 2 years. After 8 to 10 years, it returned to the initial vegetation cover level after the disturbance.

5. Discussion

5.1. Land Use Change Trend

This research compared the similarities and differences in land use changes before and after NFPP implementation, as shown in Figure 5. Until the late 1990s, forests were reported to show a trend of shrinkage [25,26]. After the implementation of the policy in 2000, the loss of forests decelerated, while the loss of grasslands became the most significant loss. The trends in forest changes were consistent with those previously reported [27]. The results of land use change were related to policies in the natural forest protection project: reducing the amount of natural forest resources harvested in state-owned forest areas in northeastern and inner Mongolia, strictly controlling timber consumption and eliminating over-limit harvesting. However, after the implementation of the natural forest protection policy, while deforestation was halted under the constraints of a strict policy, the loss of grasslands increased. The dramatic increase in population led to the occupation of grassland and the clearing of arable land and other land use types [22,28]. Meanwhile, the decline in grassland trends found in this study were consistent with those found in the other literature. The local peasants generally prefer to convert their cropland into forest rather than grassland to receive a higher compensation [29].

This study showed the trend of the annual mean NDVI; that the NDVI began to decline sharply from 2003 and maintained a downward trend until 2013, when an upward trend was observed. The trend in NDVI change is consistent with the trend in land use change: after the implementation of the 2000 policy, there was also a loss of forests, especially grasslands, which led to a decrease in the annual average NDVI value [30]. In addition, the forest gain after 2004 was in line with investments at the early stages of the ecological programs [31].

5.2. Land Productivity Trend

In addition to the first indication of land cover change, representing to some extent the underlying use, as well as land conversion and the resulting habitat fragmentation, land productivity provided an indication of ecosystem function and health and brought increased attention to ecosystem services. On average, land productivity in GKM maintained a stable state throughout the implementation of the NFPP. The improvement in trajectory was seen in grasslands and croplands (on the sides of the forest in Figure 9). Most of the forest remained intact, with a small proportion degraded to some extent. Land productivity trends showed that this research was consistent with the results of other methods and field surveys [32].

From 2010, the government began implementing the second phase of the NFPP in key state-owned forest areas in northeast China and inner Mongolia [31]. The aim of the second phase of the project was to continue to protect the natural forest and, at the same time, to treat the previously cultivated tree species on a merit basis [22,33,34]. In this study, land productivity degradation was calculated using 2010 as the time transition point (shown in Figure 8). After 2010, there was a certain degree of productivity degradation in the forested part of GKM, which was related to the purpose of the NFPP II. In Phase II, there were many low-yielding and inefficient forests that needed to be renovated and nurtured. Additionally, natural forests only recently entered the stage of recovery and development, and the quality of forest resources remains insufficient, with a large proportion of medium and young forests. Moreover, within the NFPP provinces, Heilongjiang and Inner Mongolia in north-eastern China, were designated in the NFPP as "key state-owned forest regions." Logging was not as strictly banned in this region as it was in the watersheds of the Yangtze and Yellow Rivers [35]. So, in the forest core of GKM, there was a land productivity degradation, due to the felling of weak and ecologically worthless tree species and deforestation. In 2014, the State Forestry Administration (SFA) expanded the NFPP with the launch of a trial ban on commercial logging in state-owned natural forests in Heilongjiang Province [33]. The NDVI trends shown in Figure 6 were consistent with the new commercial logging ban.

This study used SDG 15.3.1 as an important indicator to evaluate the effectiveness of large-scale ecological projects, with three sub-objectives in land productivity trajectory degradation, state degradation and performance degradation. The results were consistent with the situation of actual land use change and policy direction. The sub-indicator based on SDG 15.3.1 provided a reliable and practical method to assess the effect of ecological projects.

5.3. LandTrendr Analyses

The LandTrendr algorithm detected the magnitude of vegetation change, the year of change, and the timing of recovery at a more granular level. In terms of the magnitude of disturbance occurrence, there was a concentration of disturbances in the forest area, mainly in three areas: the Genhe area in the central part of the forest area and the Mohe area in the northeast. The disturbances occurred at different times in the three areas, in 2003, 2006, and 2010. The magnitude of disturbance was moderate and light, except in the Genhe area. To further identify the causes of disturbance in the Genhe area, the high-resolution images of the surrounding years in which the disturbances occurred are given in Figure 11. It can be seen from the images that a dramatic change occurred in the Genhe region in 2003. The identification of the images and the comparison of the information shows that a forest fire occurred in the area in 2003 [36]. Following the forest fires, the forest gradually recovered. Compared to other disturbed areas, the recovery period in this area was longer: around 3 years. Since 2008 there was a decreasing trend in forest disturbance. The implementation of the "NFPP II" campaign for nature reserves, in 2011, with the Ministry of Environmental Protection as the main enforcement agency, served as a catalyst for the protection and supervision of nature reserves at the provincial and municipal levels.

The distribution of disturbance outside the forest area was more dispersed, mainly on grassland and cultivated land to the east and west of the forest area. Disturbances occurred to a lesser extent and largely occurred around 2004. In terms of the impact of policy, the implementation of strict natural forest protection projects after 2000 did effectively protect the forests from further deforestation. Most of the disturbances which occurred during the implementation of the policy were dominated by natural disturbances (e.g., forest fires), and man-made deforestation did not pose a threat.

6. Conclusions

Based on the long-term Modis and Landsat time-series, this study explored the impact of major ecological projects on the ecological environment using the SDG index and land disturbance index in GKM. Our study found that the browning trends of vegetation coexisted with tree transition to grass. Only in some areas was there some degree of land productivity degradation, while most other areas showed improvement. Multiple factors such as precipitation, temperature, and the inappropriate afforestation led to a contrasting pattern of vegetation restoration. To remove the effect of precipitation, the effect of precipitation is considered in the calculation of land productivity and its effect is removed in the algorithm. Meanwhile, to identify the causes of the forest disturbance, the LandTrendr algorithm explored several places with high disturbance and analyzed the causes. The land degradation index of SDG 15.3.1 could be used to identify the land type change and changes in land productivity. As discussed in the previous section, land use change as the first indicator of ecological effects provided a direct indication of land change under the influence of the policy.

The study emphasizes that the continuous monitoring and effective management are the keys to the successful implementation of large-scale ecological projects. In addition, in the goal of the ecological restoration project, strengthening the functions of integrated ecosystems should be prioritized over increasing vegetation coverage. We suggest that future research should pay more attention to the scientific planning of large-scale ecological restoration projects, whilst fully considering local conditions and the goal of promoting ecosystem functions.

Author Contributions: Data curation, R.Z.; Methodology, L.L.; Validation, J.Z.; Writing—original draft, S.W. All authors have read and agreed to the published version of the manuscript.

Funding: This research was funded by National Key R&D Program of China (Grant No: 2019YFC0507804).

Institutional Review Board Statement: Not applicable.

Informed Consent Statement: Not applicable.

Data Availability Statement: Not applicable.

Conflicts of Interest: The authors declare no conflict of interest.

References

1. Haberl, H.; Erb, K.H.; Krausmann, F.; Gaube, V.; Bondeau, A.; Plutzar, C.; Gingrich, S.; Lucht, W.; Fischer-Kowalski, M. Quantifying and mapping the human appropriation of net primary production in earth's terrestrial ecosystems. *Proc. Natl. Acad. Sci. USA* **2007**, *104*, 12942–12947. [CrossRef] [PubMed]
2. Ren, Y.; Lü, Y.; Fu, B.; Zhang, K. Biodiversity and Ecosystem Functional Enhancement by Forest Restoration: A Meta-analysis in China. *Land Degrad. Dev.* **2017**, *28*, 2062–2073. [CrossRef]
3. White, J.; Wulder, M.; Hermosilla, T.; Coops, N.C.; Hobart, G.W. A nationwide annual characterization of 25 years of forest disturbance and recovery for Canada using Landsat time series. *Remote Sens. Environ.* **2017**, *194*, 303–321. [CrossRef]
4. Mitchell, A.L.; Rosenqvist, A.; Mora, B. Current remote sensing approaches to monitoring forest degradation in support of countries measurement, reporting and verification (MRV) systems for REDD+. *Carbon Balance Manag.* **2017**, *12*, 1–22. [CrossRef]
5. West, T.A.; Monge, J.J.; Dowling, L.J.; Wakelin, S.J.; Yao, R.T.; Dunningham, A.G.; Payn, T. Comparison of spatial modelling frameworks for the identification of future afforestation in New Zealand. *Landsc. Urban Plan.* **2020**, *198*, 103780. [CrossRef]
6. Vonhedemann, N.; Wurtzebach, Z.; Timberlake, T.J.; Sinkular, E.; Schultz, C.A. Forest policy and management approaches for carbon dioxide removal. *Interface Focus* **2020**, *10*, 20200001. [CrossRef]
7. Naeem, S.; Zhang, Y.; Tian, J.; Qamer, F.M.; Latif, A.; Paul, P.K. Quantifying the Impacts of Anthropogenic Activities and Climate Variations on Vegetation Produc-tivity Changes in China from 1985 to 2015. *Remote Sens.* **2020**, *12*, 1113. [CrossRef]
8. Wang, X.; Zhang, C.; Hasi, E.; Dong, Z. Has the Three Norths Forest Shelterbelt Program solved the desertification and dust storm problems in arid and semiarid China? *J. Arid. Environ.* **2010**, *74*, 13–22. [CrossRef]
9. Bryan, B.A.; Gao, L.; Ye, Y.; Sun, X.; Connor, J.; Crossman, N.D.; Smith, M.S.; Wu, J.; He, C.; Yu, D.; et al. China's response to a national land-system sustainability emergency. *Nat. Cell Biol.* **2018**, *559*, 193–204. [CrossRef]
10. Niu, Q.; Xiao, X.; Zhang, Y.; Qin, Y.; Dang, X.; Wang, J.; Zou, Z.; Doughty, R.B.; Brandt, M.; Tong, X.; et al. Ecological engineering projects increased vegetation cover, production, and biomass in semiarid and subhumid Northern China. *Land Degrad. Dev.* **2019**, *30*, 1620–1631. [CrossRef]
11. Zhao, H.; Wu, R.; Hu, J.; Yang, F.; Wang, J.; Guo, Y.; Zhou, J.; Wang, Y.; Zhang, C.; Feng, Z. The contrasting east–west pattern of vegetation restoration under the large-scale ecological restoration programmes in southwest China. *Land Degrad. Dev.* **2020**, *31*, 1688–1698. [CrossRef]
12. Mondal, P.; McDermid, S.S.; Qadir, A. A reporting framework for Sustainable Development Goal 15: Multi-scale monitoring of forest degradation using MODIS, Landsat and Sentinel data. *Remote Sens. Environ.* **2020**, *237*, 111592. [CrossRef]

13. Mao, D.; Wang, Z.; Wu, B.; Zeng, Y.; Luo, L.; Zhang, B. Land degradation and restoration in the arid and semiarid zones of China: Quantified evidence and implications from satellites. *Land Degrad. Dev.* **2018**, *29*, 3841–3851. [CrossRef]
14. Evans, J.; Geerken, R. Discrimination between climate and human-induced dryland degradation. *J. Arid. Environ.* **2004**, *57*, 535–554. [CrossRef]
15. Wu, Z.; Wu, J.; He, B.; Liu, J.; Wang, Q.; Zhang, H.; Liu, Y. Drought Offset Ecological Restoration Program-Induced Increase in Vegetation Activity in the Beijing-Tianjin Sand Source Region, China. *Environ. Sci. Technol.* **2014**, *48*, 12108–12117. [CrossRef]
16. De Jong, R.; Verbesselt, J.; Zeileis, A.; Schaepman, M.E. Shifts in Global Vegetation Activity Trends. *Remote Sens.* **2013**, *5*, 1117–1133. [CrossRef]
17. Tong, X.; Brandt, M.; Yue, Y.; Horion, S.; Wang, K.; De Keersmaecker, W.; Tian, F.; Schurgers, G.; Xiao, X.; Luo, Y.; et al. Increased vegetation growth and carbon stock in China karst via ecological engineering. *Nat. Sustain.* **2018**, *1*, 44–50. [CrossRef]
18. Kennedy, R.E.; Yang, Z.G.; Cohen, W.B. Detecting trends in forest disturbance and recovery using yearly Landsat time series: 1. LandTrendr—Temporal segmentation algorithms. *Remote Sens. Environ.* **2010**, *114*, 2897–2910. [CrossRef]
19. Tucker, C.J. Red and photographic infrared linear combinations for monitoring vegetation. *Remote Sens. Environ.* **1979**, *8*, 127–150. [CrossRef]
20. Ponce-Campos, G.E.; Moran, M.S.; Huete, A.; Zhang, Y.; Bresloff, C.; Huxman, T.E.; Eamus, D.; Bosch, D.D.; Buda, A.R.; Gunter, S.; et al. Ecosystem resilience despite large-scale altered hydroclimatic conditions. *Nat. Cell Biol.* **2013**, *494*, 349–352. [CrossRef]
21. Onyutha, C.; Tabari, H.; Taye, M.T.; Nyandwaro, G.N.; Willems, P. Analyses of rainfall trends in the Nile River Basin. *J. Hydro-Environ. Res.* **2016**, *13*, 36–51. [CrossRef]
22. Sannigrahi, S.; Zhang, Q.; Joshi, P.; Sutton, P.; Keesstra, S.; Roy, P.; Pilla, F.; Basu, B.; Wang, Y.; Jha, S.; et al. Examining effects of climate change and land use dynamic on biophysical and economic values of ecosystem services of a natural reserve region. *J. Clean. Prod.* **2020**, *257*, 120424. [CrossRef]
23. Costa, R.L.; Baptista, G.M.D.M.; Gomes, H.B.; Silva, F.D.D.S.; Júnior, R.L.D.R.; Salvador, M.D.A.; Herdies, D.L. Analysis of climate extremes indices over northeast Brazil from 1961 to 2014. *Weather. Clim. Extremes* **2020**, *28*, 100254. [CrossRef]
24. Ivits, E.; Cherlet, M.; Tóth, T.; Lewińska, K.E.; Tóth, G. Characterisation of productivity limitation of salt-affected lands in different climatic regions of europe using remote sensing derived productivity indicators. *Land Degrad. Dev.* **2013**, *24*, 438–452. [CrossRef]
25. Zhan, J.Y.; Deng, X.Z.; Yue, T.X.; Bao, Y.H.; Zhao, T.; Ma, S.N. Land use change and its environmental effects in the farming-pasturing interlocked areas of Inner Mongolia. *Resour. Sci.* **2004**, *26*, 80–88.
26. Zhang, Y.; Song, C. Impacts of Afforestation, Deforestation, and Reforestation on Forest Cover in China from 1949 to 2003. *J. For.* **2006**, *104*, 383–387. [CrossRef]
27. Altan, Z. Analysis on the Current Status and the Changes of Land Use in Hulunbuir, Inner Mongolia. *China Land Sci.* **2011**, *25*, 43–48.
28. Trac, C.J.; Harrell, S.; Hinckley, T.M.; Henck, A.C. Reforestation programs in Southwest China: Reported success, observed failure, and the reasons why. *J. Mt. Sci.* **2007**, *4*, 275–292. [CrossRef]
29. Tong, X.; Wang, K.; Yue, Y.; Brandt, M.; Liu, B.; Zhang, C.; Liao, C.; Fensholt, R. Quantifying the effectiveness of ecological restoration projects on long-term vegetation dynamics in the karst regions of Southwest China. *Int. J. Appl. Earth Obs. Geoinformation* **2017**, *54*, 105–113. [CrossRef]
30. Yin, H.; Pflugmacher, D.; Li, A.; Li, Z.; Hostert, P. Land use and land cover change in Inner Mongolia—Understanding the effects of China's re-vegetation programs. *Remote Sens. Environ.* **2018**, *204*, 918–930. [CrossRef]
31. Liu, J.; Li, S.; Ouyang, Z.; Tam, C.; Chen, X. Ecological and socioeconomic effects of China's policies for ecosystem services. *Proc. Natl. Acad. Sci. USA* **2008**, *105*, 9477–9482. [CrossRef]
32. Zhang, S.; Zhang, G.; Hui, G. Analysis of spatial temporal pattern of forest Net Primary Productivity of the Great Khingan in Inner Mongola. *For. Res.* **2019**, *32*, 74–82.
33. Sun, X.; Canby, K.; Liu, L. China's Logging Ban in Natural Forests:Impacts of Extended Policy at Home and Abroad. Forest Trends Information Brief. 2016. Available online: https://www.forest-trends.org/publications/china%c2%92s-logging-ban-in-natural-forests/ (accessed on 1 September 2021).
34. Zhang, Y.; Liu, X.; Gao, W.; Li, H. Dynamic changes of forest vegetation carbon storage and the characteristics of carbon sink (source) in the Natural Forest Protection Project region for the past 20 years. *Acta Ecol. Sin.* **2021**, *41*, 5093–5105.
35. Ren, G.; Young, S.S.; Wang, L.; Wang, W.; Long, Y.; Wu, R.; Li, J.; Zhu, J.; Yu, D.W. Effectiveness of China's National Forest Protection Program and nature reserves. *Conserv. Biol.* **2015**, *29*, 1368–1377. [CrossRef] [PubMed]
36. Chang, Y.; He, H.S.; Bishop, I.; Hu, Y.; Bu, R.; Xu, C.; Li, X. Long-term forest landscape responses to fire exclusion in the Great Xing'an Mountains, China. *Int. J. Wildland Fire* **2007**, *16*, 34–44. [CrossRef]

International Journal of
Environmental Research and Public Health

MDPI

Article

Simulation and Analysis of Urban Production–Living–Ecological Space Evolution Based on a Macro–Micro Joint Decision Model

Yuanyuan Tao [1], Qianxin Wang [1,*] and Yan Zou [1,2]

1 School of Environment and Spatial Informatics, China University of Mining and Technology, Xuzhou 221116, China; tyy@cumt.edu.cn (Y.T.); zouyan@bucea.edu.cn (Y.Z.)
2 School of Humanity and Law, Beijing University of Civil Engineering and Architecture, Beijing 102616, China
* Correspondence: wqx@cumt.edu.cn

Abstract: The precise simulation of urban space evolution and grasping of the leading factors are the most important basis for urban space planning. However, the simulation ability of current models is lacking when it comes to complicated/unpredictable urban space changes, resulting in flawed government decision-making and wasting of urban resources. In this study, a macro–micro joint decision model was proposed to improve the ability of urban space evolution simulation. The simulation objects were unified into production, living and ecological space to realize "multiple planning in one". For validation of the proposed model and method, remote sensing images, geographic information and socio-economic data of Xuzhou, China from 2000 to 2020 were collected and tested. The results showed that the simulation precision of the cellular automata (CA) model was about 87% (Kappa coefficient), which improved to 89% if using a CA and multi-agent system (MAS) joint model. The simulation precision could be better than 92% using the prosed model. The result of factor weight determination indicated that the micro factors affected the evolution of production and living space more than the macro factors, while the macro factors had more influence on the evolution of ecological space than the micro factors. Therefore, active policies should be formulated to strengthen the ideological guidance towards micro individuals (e.g., a resident, farmer, or entrepreneur), and avoid disordered development of living and production space. In addition, ecological space planning should closely link with the local environment and natural conditions, to improve urban ecological carrying capacity and realize urban sustainable development.

Keywords: urban space evolution simulation; cellular automata; multi-agent system; leading factors analysis; urban sustainable development

Citation: Tao, Y.; Wang, Q.; Zou, Y. Simulation and Analysis of Urban Production–Living–Ecological Space Evolution Based on a Macro–Micro Joint Decision Model. *Int. J. Environ. Res. Public Health* **2021**, *18*, 9832. https://doi.org/10.3390/ijerph18189832

Academic Editors: Wei Song and Hualin Xie

Received: 14 July 2021
Accepted: 9 September 2021
Published: 18 September 2021

1. Introduction

Land use/cover change (LUCC) has been generally considered a main driving force of global ecosystem and climate change [1]. Urbanization is the most typical form of LUCC, and has a significant impact on biological diversity and ecosystem services [2]. Nowadays, about 55% of the world's population lives in cities with this rate expected to reach 68% by 2050 [3]. Therefore, the study of urban LUCC is of great importance to understand and grasp global LUCC. The process of urbanization influences the flow of material, energy, and information, and affects the structure and function of ecosystems [4]. Therefore, on the one hand, the socio-economic level may be significantly improved; on the other hand, it leads to loss of farmland, fragmentation of habitats, and increases in the heat island effect [5–7]. However, these problems can be alleviated through reasonable urban space planning and efficient utilization of urban resources [8]. Therefore, urban space planning is considered an effective tool/means to improve urban sustainable development.

The precise simulation of urban space evolution and analysis of the leading factors are the most important bases and essential prerequisites for urban space planning. Therefore, many studies of urban space evolution simulation and leading factors analysis have been carried out, and some typical simulation models have been constructed, such as the econometric statistics model (ES) [9], the system dynamics model (SD) [10], the cellular automata model (CA) [11], the multi-agent system model (MAS) [12], etc. Each of these models has its own advantages and disadvantages. In ES models, mathematical statistics methods are used to simulate the variations in scale of urban space, such as the logistic regression model [13], Kuznets curve model [14], and panel econometric model [15]. These ES models are easy to construct and use, but they cannot simulate the dynamic process and variation in urban spatial distribution. In SD models, the process of urban space evolution is expressed by simulating the interactions of urban elements, e.g., MEPLAN model [16], Dortmund model [17] and LILT model [18]. However, these SD models also cannot describe variation in urban spatial distribution. The ES and SD models are considered "top-down" models.

To improve the ability to simulate urban spatial variation, the CA model was first introduced by Chapin and Weiss in 1968 [19]. From the late 1990s to the early 21st century, the CA model entered a high-speed development period, and saw widespread use in urban space planning [20,21]. However, it cannot well represent macro-scale political, economic and cultural driving forces that influence urban spatial variations [22]. Therefore, some improved methods have been proposed, for example taking the macro factors as constraint conditions in the CA model [23]. With the development of artificial intelligence (AI) technology, the study and application of the MAS model became popular over the past two decades. MAS defines a set of agents living in a common environment, with all agents coming to a joint decision on urban space use type within this system. The MAS model can provide a more powerful tool for simulating the multi-level decision-making processing in urban space evolution than the other existing methods [24,25]. However, the MAS model has to be used in conjunction with CA for considering the effects of neighborhood space use type on urban space evolution [26]. Both CA and MAS are considered "bottom-up" models. Table 1 shows the comparison of the characteristics of the four kinds of models.

Table 1. Performance comparison of current models for urban spatial evolution simulation.

Performance	ES [(a)]	SD [(b)]	CA [(c)]	MAS [(d)]
Scale changes simulation	Strong	Strong	Strong	Strong
Spatial distribution simulation	Weak	Weak	Strong	Strong
Time varying simulation	Weak	Normal	Strong	Strong
Macro factors simulation	Strong	Strong	Weak	Strong
Micro factors simulation	Weak	Weak	Normal	Strong
Model operation mechanism	Top-down		Bottom-up	

[(a)] econometric statistics model; [(b)] system dynamics model; [(c)] the cellular automata model; [(d)] the multi-agent system model.

From Table 1, it is apparent that the simulation ability of the MAS model is stronger than that of other models. However, there are still some defects in the current MAS model. For example, the interaction between macro and micro factors is rarely considered, despite having a significant effect on the simulation precision of the model. Macro factors (e.g., the natural condition, government policy) often restrict the decision-making behaviors of micro factors (a resident, farmer, entrepreneur or environmentalist), particularly in developing countries [27]. The choice preferences of micro factors may also have an important influence on the decision-making behaviors of macro factors [28]. If the interaction and mutual influence between macro and micro factors are neglected, the simulation precision of the model will inevitably decrease. Moreover, the simulation objects are usually unit plots of different types of land use (e.g., urban construction land, cultivated land, woodland, grassland, water bodies, etc.) in current models. If the types of land use are different from the categories of urban planning (e.g., traffic planning, garden planning, land planning) for the same unit plot at the same time, it will lead to land use conflicts.

To solve these problems, a macro–micro joint decision model is proposed in this study for improving the simulation ability of urban space evolution. In this model, all simulation objects are unified into production, living and ecological space for realizing "multiple planning in one". Compared with current models, the proposed model has two main characteristics. One is that the interaction and mutual influence between the macro and micro factors are fully considered; the other is that "Production–Living–Ecological" (PLE) space is taken as the simulation object, replacing exclusive types of land use. For validation of the proposed model, remote sensing images, geographic information and socio-economic data from Xuzhou, China between 2000–2020 were collected and tested. The results show that the simulation precision of CA model was 87.14% (Kappa coefficient), with an increase to 89.63% when using CA + MAS model. An additional 2.68% improvement of simulation precision was achieved by using the CA + MAS + Correlation model. Moreover, the result of factor weight determination indicated that the micro factors affected the evolution of production and living space more than the macro factors. However, the macro factors had more influence on the evolution of ecological space evolution than the micro factors. Therefore, we should pay more attention to the micro factors to realize the orderly development of living and production spaces. For example, some active policies should be formulated to strengthen the ideological guidance for micro individuals (e.g., residents, farmers, entrepreneurs), helping them establish scientific views of urban space utilization. Meanwhile, macro factors should be paid more attention to ensure the sustainable development of ecological space. For example, ecological space planning should closely link with the local environment and natural conditions to improve the urban ecological carrying capacity.

2. Study Area and Data Sources

2.1. Study Area

To analyze the leading factors and construct a precise simulation model of urban space evolution, Xuzhou, an eastern city of China, was selected as the study area. It is located in the northwest of Jiangsu Province, China (between 33°43′–34°58′ N and 116°22′–118°40′ E). It includes five districts, three counties and two county-level cities, with a total area of 11,765 km^2. Figure 1 shows the geographic location and administrative divisions of Xuzhou. The majority of the Xuzhou region consists of plains, which account for 90% of the total area. This area has a temperate continental monsoon climate and receives 44–54% possible sunshine. The annual average temperature is 14 °C, and the annual average rainfall is 900 mm. In addition, Xuzhou is rich in mineral resources and well placed for easy access to the other Chinese cities. Therefore, it is an important coal production base and a transportation hub in China [29]. In the past two decades (2000–2019), the population of Xuzhou increased from 8,964,400 to 10,417,300, and the Gross National Product (GDP) improved from 61.630 billion CNY to 715.135 billion CNY; the per capita green area increased from 10.5 m^2 to 15.4 m^2, and the urbanization rate increased from 25.8% to 66.7% [30].

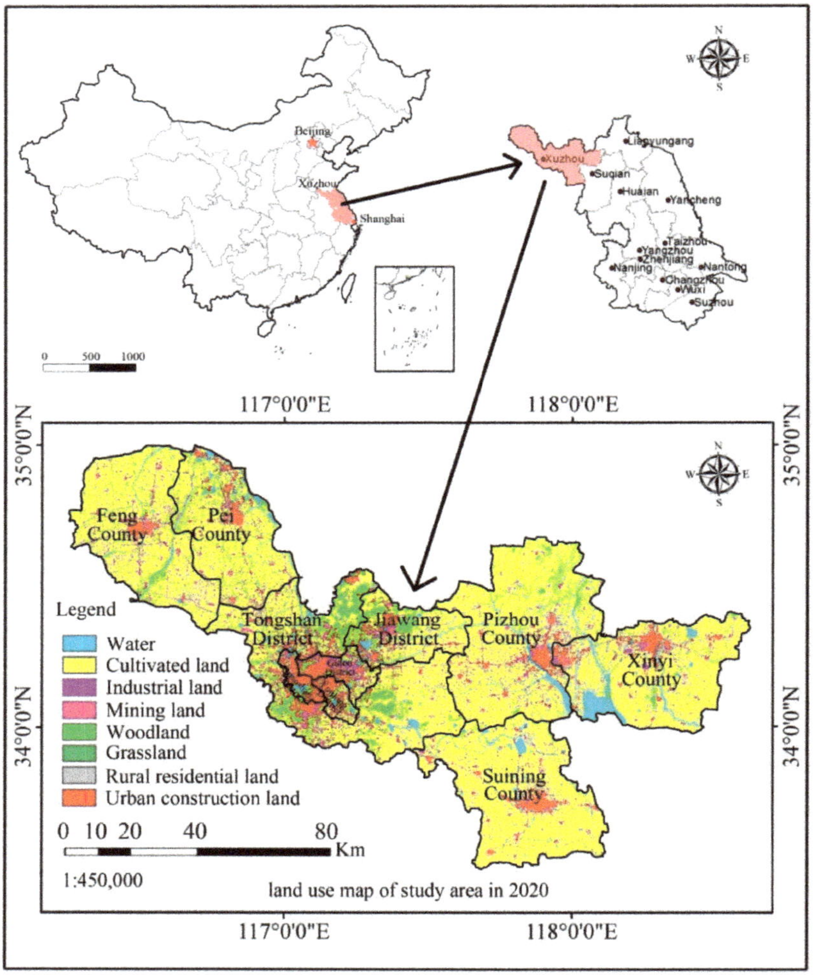

Figure 1. The geographical location and administrative divisions of Xuzhou in China.

From the above statistics, it is evident that the socio-economic level and living standards of Xuzhou have improved significantly in the past two decades. The types of land use/coverage have changed significantly in Xuzhou, which can provide a good basic dataset to investigate the process of urban space evolution and leading factors. However, it should be noted that the level of economic development is still low and the cost of economic development is high in Xuzhou compared with other cities in Jiangsu Province. In 2018, the per capita GDP of Xuzhou was 76,915 CNY, which was 66.88% of the average level of Jiangsu Province, but the comprehensive energy consumption of Xuzhou was 0.37 tons (consumed standard coal for obtaining 10,000 CNY of industrial output), which was 3.12 times the average level of Jiangsu Province [31]. Therefore, more energy needs to be consumed in order to realize the same amount of economic growth in Xuzhou as in other cities. Therefore, it is important and urgent to strengthen the urban space planning and optimize the industrial structure as soon as possible, to realize the sustainable development of social economy and the ecological environment in Xuzhou.

2.2. Data Sources

In this study, three kinds of data were collected and used: the first was remote sensing (RS) images of Xuzhou from 2000 to 2020, which provided the basic data for Production-Living-Ecological (PLE) space recognition; the second was geographic information system (GIS) data, which provided vector files of the administrative divisions, traffic networks and distribution of public facilities in Xuzhou, such as schools, hospitals, and shopping malls, as downloaded from a geographic national conditions monitoring cloud platform; the third was socio-economic statistical data from the literature and statistical yearbooks from 2000 to 2020, including population, industrial economy, natural resources, etc. Details on the three types of data are listed in Table 2.

Table 2. Study data sources and contents.

Data Type	Content	Time	Source
RS image data	GF [(a)]-2 1 m × 1 m RS [(b)] image	2000–2020	Natural Resources Satellite RS Cloud Service Platform
Urban GIS [(c)] data	Vector files of administrative division Road	2018	Geographical Information Monitoring Cloud
Socioeconomic statistics	Population, industrial economy, natural resources	2000–2020	Literature, statistical yearbooks

[(a)] GaoFen-2 remote sensing images; [(b)] Remote sensing; [(c)] Geographic information system.

It should be noted that the RS images needed to be chosen and processed carefully in order to obtain the precise urban space evolution information. Therefore, RS image data was collected during summer (from mid-July to mid-August), because identifying RS images of vegetation in the growth season peak is easier than during other seasons [31]. Environment for Visualizing Images (ENVI) software (Research System Comp., Boulder, CO, USA) was used for data processing. The main procedures included radiative correction, atmospheric correction, geometric correction, contrast stretching, graphics clipping, etc. Finally, the land was classified into urban construction land, rural residential land, cultivated land, woodland, grassland, water area, industrial land, and mining land by the supervised classification and visual interpreted method. To improve the accuracy of land use classification, convolutional neural network (CNN) technology was adopted [32]. Through land function evaluation, the above eight types of land use could be amalgamated into production, living and ecological space [33]. The recognition precision of PLE space was better than 96% (Kappa coefficient) based on the RS image data. Therefore, it could meet the requirements of PLE space evolution simulation and leading factor analysis.

3. Research Methods

3.1. Production-Living-Ecological Space Evolution Simulation Model Based on Macro-Micro Joint Decision

According to Section 1, many models have been constructed to simulate urban space evolution, and the simulation ability of CA + MAS is the strongest among the current models. However, there are still some defects in the CA + MAS model; for example, the correlations between macro and micro factors are neglected. To improve the performance of the CA + MAS model, a macro–micro joint decision model is proposed in this study. Compared with the current CA + MAS model, the proposed model has two main characteristics. One is that the interaction between and mutual influence of the macro and micro factors are considered to improve the simulation precision over the CA + MAS model; the other is that PLE space is taken as the planning object to facilitate the unified implementation of multiple planning. The specific steps of the proposed model are as follows: (i) Areas of increased living space are obtained by predicting the growth of the urban population in the future. (ii) Locations of increased living space are determined by the CA + MAS model. It should be noted that the probability of the ith unit space to be transformed into living space is calculated by the joint decision of macro and micro factors; the highest-probability unit space is transformed into living space. (iii) After completing the transformation of the ith unit space, the model checks if the total of transformed area is equal to the area of increased living space. If not, the above work is done iteratively until the total transformed area is equal to the area of increased living space. (iv) After completing the simulation of living space evolution, production and ecological space evolution are simulated using the same method. However, the areas of increased production and ecological space are instead determined by historical data and a Markov model. (v) After completing the nth simulation of PLE space evolution, the simulation precision is calculated ($Kappa_n$) by comparing with the results of RS image recognition. If the value of ($Kappa_n - Kappa_{n-}$) is smaller than the threshold, the solution has converged and the nth simulation result is outputted. If not, the weights of macro and micro factors aree adjusted and the above steps are done iteratively. The method of factor weight adjustment is introduced below in Section 3.2. Figure 2 shows the data processing flow of PLE space evolution simulation based on the macro and micro factor joint decision model.

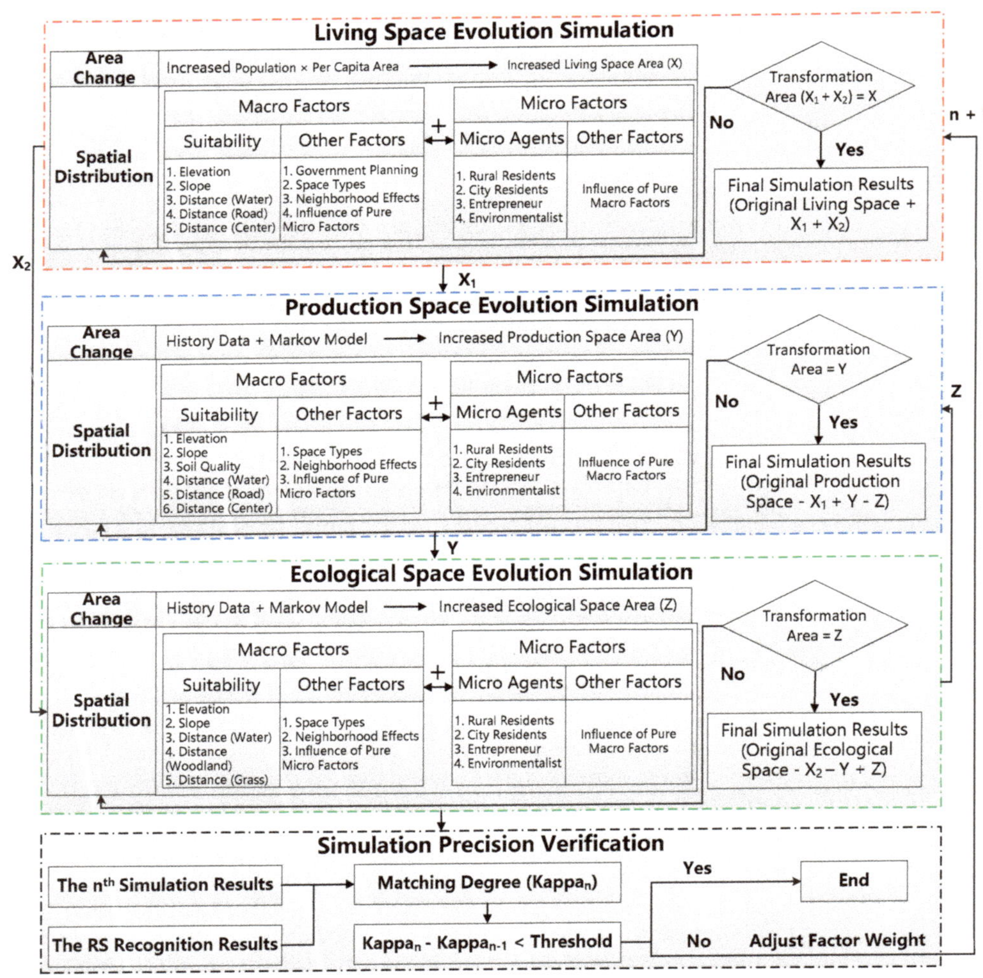

RS-remote sensing.

Figure 2. The data processing flow for simulating Production–Living–Ecological space evolution based on the macro and micro factor joint decision model.

3.1.1. Simulation Method of Living Space Evolution

According to Figure 2, the areas of increased living space should be determined first in order to simulate the evolution of urban space. Two methods are often used to realize this. One is based on the historical data (areas of increased living space in previous years) and a Markov model. The other is based on predicting the growth of the urban population and the per capita area of living space in the region. Existing studies indicate that the estimation precision of areas of increased living space from the latter method is better than that from the former method, if the urban population is predicted accurately [34]. Therefore, the latter method is adopted in this study and the calculation formula is:

$$A^k = P^k \times S \tag{1}$$

A^k and P^k are the areas of increased living space and urban population in the kth year, respectively. S is the per capita area of living space, which refers to the historical data of this city or its urban planning and design standards. In China, the per capita area of living space is divided into seven types, according to the urban population size and climatic conditions. The per capita area of living space ranges from 65 m^2 to 115 m^2 based on the different type of city. Xuzhou belongs to the second type of city, where the per capita area of living space is 110 m^2.

After obtaining the area of increased living space, one of most important problems is determining the locations of these increases. In this model, locations of increased living space are determined by the joint decision of macro and micro factors. The greater the transformation probability of unit space, the higher its priority for transformation into living space. The calculation formula is:

$$F^k_{i,j} = W^{macro} F'^k_{i,j} + W^{micro} F'''^k_{i,j} \tag{2}$$

$F^k{}_{i,j}$ is the probability of unit space (i, j) to be transformed into living space in the kth year. For convenience, probability is replaced with a score, where 0–100 scores denote 0–100%. $F'^k{}_{i,j}$ and $F''^k{}_{i,j}$ are the scores from the macro and micro factors, respectively. W^{macro} and W^{micro} are the weights of macro and micro factors, respectively, where $W^{macro} + W^{micro} = 1$. The initial values of W^{macro} and W^{micro} can be determined by statistical analysis. The specific implementation steps are introduced below in Section 3.2. Final weights of the macro and micro factors are adjusted by comparing the simulation results with RS image recognition results. The scores from macro factors include two components: suitability and all other factors. The calculation formula is:

$$F'^k_{i,j} = W^E E_{i,j} + W^Q Q^k_{i,j} \tag{3}$$

$E_{i,j}$ and $Q^k{}_{i,j}$ are the scores of unit space (i, j) from the suitability evaluation and the other factors, respectively. W^E and W^Q are their weights and $W^{macro} + W^{micro} = 1$. The calculation formula is:

$$E_{i,j} = \sum\nolimits_{m=1}^{n} W^m F^m_{i,j} \tag{4}$$

$F^m{}_{i,j}$ and W^m are the score and weight of the mth factor, respectively. In this model, the factors of suitability evaluation include two parts: one is denoted the natural factor (e.g., elevation, slope); the other is denoted the location factor (e.g., the distances to water, main road and city center). These can be obtained with ArcGIS (ESRI, Redlands, CA, USA). The specific marking standards are as follows: the smaller the difference between the elevation of the unit space (i, j) and the average elevation of the region, the higher the likelihood of its transformation into living space. As the height difference increases from 0 to 10 km, the suitability score decreases from 100 to 0. The steeper the slope, the lower the score; as the slope increases from 0 to 90 degrees, the score decreases from 100 to 0. Likewise, the shorter the distance to a water area (0–5 km), the lower the score (0–100); the shorter the distance to a main road (0–5 km), the higher the score (100–0); and the shorter the distance to the city center (0–50 km), the higher the score (100–0). The calculation formula of scores from the other factors is:

$$Q_{i,j}^{k} = W^{1}F_{g} + W^{2}F_{l}^{k-1} + W^{3}F_{N}(\sum_{m_2=1}^{n_2} N_{i,j}^{m_2}) + W^{4}F_{q}(\sum_{m_3=1}^{n_3} w^{m_3}q_{i,j}^{m_3}) \quad (5)$$

F_g, F_l^{k-1}, F_n, F_q and W^1, W^2, W^3, W^4 are the scores and weights from government planning, land type, neighborhood influence, and influence of micro factors, respectively. $W^1 + W^2 + W^3 + W^4 = 1$. The specific marking standards are as follows: If the unit space belongs to the first, second, or third class of urban construction land in government planning, the score of F_g is 100, 80, or 50 respectively. If the probability of lth type of land to be transformed into urban construction land is between 0 and 1 in the $(k-1)^{th}$ year, the score of F_l^{k-1} is correspondingly between 0 and 100. If the number of living spaces permitted is between 0 and 8 (in a 3 × 3 neighborhood), the score of F_n is correspondingly between 0 and 100. The score of F_q is a weighted average of the scores of $q_{i,j}^{m_3}$, where $m^3 = 4$ (rural resident, city resident, entrepreneur and environmentalist), and $w^1 + w^2 + w^3 + w^4 = 1$. The calculation formula of $F_{i,j}^{'''k}$ in Equation (2) and the marking standards of $q_{i,j}^{m_3}$ are as follows:

$$F_{i,j}^{'''k} = W^{F''}F_{i,j}^{''k} + W^{F'}F_{i,j}^{'k} \quad (6)$$

$F_{i,j}^{''k}$ denotes the scores of the pure micro factors, which are calculated by Equation (7). $F_{i,j}^{'k}$ is the scores from the macro factors in Equation (3). $W^{F''}$ and $W^{F'}$ are their weights, and $W^{F''} + W^{F'} = 1$.

$$F_{i,j}^{''k} = \sum_{m_3=1}^{n_3} w^{m_3}q_{i,j}^{m_3} \quad (7)$$

The meanings of the symbols in Equation (7) are the same as those in Equation (5). The marking standards of $q_{i,j}^{m_3}$ can be introduced in detail: The shorter the distance to cultivated land (0–10 km), the lower the score (0–100) from the rural resident agent. The shorter the distance to public facilities (e.g., hospital, school, mall, 0–10 km), the higher the score (100–0) from the city resident agent. The higher the housing price per unit of space (0–20,000 CNY), the higher the score (0–100) from the enterprise agent. The shorter the distance to ecological space (0–10 km), the lower the score (0–100) from the environmentalist agent. Through Equations (1)–(7), the scores of all unit spaces transformed into living space can be obtained. The highest scoring unit space is then transformed into living space. If the accumulated transformed area is smaller than the area of increased living space (X), the above work is done iteratively.

3.1.2. Simulation Method of Production and Ecological Space Evolution

After completing the simulation of living space evolution, production and ecological space evolution is simulated by the above method. Similarly, the areas of increased production and ecological space need to be determined first. In general, they are determined by historical data and a Markov model. If the areas of increased production and ecological space are Y and Z, and the areas of production and ecological space transformed into living space are X_1 and X_2 ($X_1 + X_2 = X$), the following equation can be employed:

$$(S_L + X_1 + X_2) + (S_P - X_1 - Z + Y) + (S_E - X_2 + Z - Y) = S_L + S_P + S_E \quad (8)$$

S_L, S_P and S_E are the areas of original living, production and ecological space, respectively. Equation (8) requires that the total area before and after transformation be equal, and the method of determining the locations of increased production and ecological space is the same as that for increased living space. However, their factors and marking standards are different. In the simulation of production space evolution, the macro factors include the natural factor (elevation, slope, soil quality), location factor (distance to water area, main road, city center), and the other factors. The marking standards are: the larger the height difference between the elevation of unit space (i, j) and the average elevation of the region (0–10 km), the lower the score for its transformation into production space (0–100); the steeper the slope (0–90 degrees), the lower the score (0–100); the higher the soil quality (level 1–10), the lower the score (100–0); the shorter the distance to a water area (0–5 km), the higher the score (100–0); the shorter the distance to a main road (0–5 km), the higher the score (100–0); and the shorter the distance to the city center (0–50 km), the higher the score (100–0). The other factors include the land type, the neighborhood and the influence of micro factors. The marking standards of land type and neighborhood influence are similar to those of increased living space. However, the influence rules of micro factors are different to those of increased living space.

In the simulation of production space evolution, the shorter the distance to cultivated land (0–10 km), the higher the score (100–0) from the rural resident agent. The shorter the distance to living space (0–10 km), the lower the score (0–100) from the city resident agent. The higher the cost of land (0–10,000 CNY), the lower the score (100–0) from the enterprise agent. The shorter the distance to ecological space (0–10 km), the lower the score (0–100) from the environmentalist agent. Following the above rules, the final score of each unit space can be obtained from the weighted average of the scores of all factors. The highest-scoring unit space is transformed into production space. If the accumulated transformed area is smaller than the area of increased production space (Y), the above work is done iteratively.

In the simulation of ecological space evolution, the macro factors also include the natural factor (elevation, slope), location factor (distance to water area, woodland, grassland), and the other factors. The marking standards are: the larger the height difference between the elevation of unit space (i, j) and the average elevation of the region (0–10 km), the higher the score for its transformation into ecological space (100–0); the steeper the slope (0–90 degrees), the higher the score (100–0); the shorter the distance to a water area (0–5 km), the higher the score (100–0); the shorter the distance to woodland (0–5 km), the higher the score (100–0); and the shorter the distance to grassland (0–5 km), the higher the score (100–0). The other factors include the land type, the neighborhood and the influence of micro factors. The marking standards of land type and neighborhood influence are similar to those of increased living space. However, the influence rules of micro factors are different.

The shorter the distance to cultivated land (0–10 km), the lower the score (0–100) from the rural resident agent for the simulation of ecological space evolution. The shorter the distance to living space (0–10 km), the higher the score (100–0) from the city resident agent. The shorter the distance to production space (0–10 km), the lower the score (0–100) from the enterprise agent. The shorter the distance to ecological space (0–10 km), the higher the score (100–0) from the environmentalist agent. The highest-scoring unit space is transformed into ecological space, and the above work is done iteratively until the accumulated transformed area is equal to the area of increased ecological space (Z).

3.2. Method of Factor Weight Determination

In general, factor weight determination includes two main steps: the first step is the initial weight determination and the second step is the final weight adjustment. The initial weights of all factors can be determined through the Delphi method (expert scoring) or through statistical analysis. Although the Delphi method is easy to use, it has strong subjectivity. Therefore, the statistical analysis method is adopted in this study. The implementation steps of this method are introduced as follows, taking living space as an example: (i) One hundred experimental units are selected as the samples and they are evenly distributed across the study region. The area of each sample unit is 3 km × 3 km, which include 10,000 grids (30 m × 30 m). (ii) Those grids are picked out which belonged to production or ecological spaces in 2000. If the number of grids is n, n equations can be formed in one sample unit; see Equation (9).

$$\begin{cases} W_1F_1^1 + \ldots W_jF_j^1 + \ldots W_mF_m^1 = V^1 \\ \vdots \\ W_1F_1^i + \ldots W_jF_j^i + \ldots W_mF_m^i = V^i \\ \vdots \\ W_1F_1^n + \ldots W_jF_j^n + \ldots W_mF_m^n = V^n \end{cases} \tag{9}$$

W_j is the weight of the jth factor, which is a known parameter; F_j^i is the score of the jth factor in the ith grid, which can be obtained by the methods in Section 3.1; V^i is the total score of the ith grid, which can be determined based on the time order of the transformation into living space. The earlier the grid is transformed into living space (from 2001 to 2010), the higher the score (from 100 to 0), which can be obtained by comparing the results of RS image recognition in different years. For example, if the ith grid was transformed into living space in 2001, V^i is 100; if it was transformed into living space in 2002, V^i is 90; if it had not been transformed into living space in 2010, V^i is 0. m is the number of factors for living space evolution simulation, which is 16 in this study (see Table 5). n is the number of grids which belonged to production or ecological spaces in one sample unit in 2000. In general, n is larger than m. In this example, n is approximately 6000. Therefore, in order to strengthen the stability of solutions, n equations of one sample unit are divided into k groups, with each group including approximately 200 equations. The unknown parameters (factor weight W_j) of each group can then be estimated by the adjustment method (e.g., least square adjustment, LSQ), and solution precision (root mean square error, RMS) can be obtained. If the RMS of this group is more than three times larger than the minimal RMS of all groups, it is treated as an outlier and the estimated results of this group are removed. Then, the weighted average values of the remaining groups are taken as the estimated results of this sample unit. (iii) Finally, the estimated results of one hundred units are analyzed by statistical methods (e.g., Shapiro–Wilk Test [35]). If the estimated result of the factor shows a normal distribution, the weighted average value of all units is taken as the initial weight of this factor. Otherwise, sample units are deleted from the samples where the estimated values obviously deviate from the mean value of all samples, and the estimated results are tested for a normal distribution again.

After the factor initial weights are obtained, they must be further adjusted to determine the final weights, because the estimated precision and reliability of factor initial weight is strongly related to the sample selection. Therefore, they need to be adjusted by comparing the overall consistency of simulation and recognition results from the study region. In this model, the best-fit method is adopted to determine the final weights of all factors. The basic idea is that the results of simulation are always compared with those of RS image recognition, in order to test the rationality of factor weight allocation. For example, if the initial weights of macro and micro factors are 0.6 and 0.4, they will be reset to 0.59 and 0.41 on the first try. If the matching rate (Kappa) improves, they will be further set to 0.58 and 0.42 in the second try and similar attempts will be carried out until the matching rate starts to decline. Otherwise, if the matching rate in the first try is declined, the weights of macro and micro factors will be adjusted in the opposite direction (e.g., be set to 0.61 and 0.39) and similar attempts will be carried out until the matching rate starts to decline. The purpose of this method is to hunt for the optimal weight allocation of factors by a continuous adjustment. However, some rules of weight adjustment are defined to improve operating efficiency: (i) the overall weights between macro and micro factors must be adjusted first, followed by the local weights of internal elements in macro and micro factors; (ii) the lower the matching rate of space simulations, the more likely the factor weight of this type of space is adjusted first. Based on the above rules, the implementation steps of this method are as follows: (i) The matching rates of production, living and ecological spaces are calculated, based on the initial weights. (ii) The weights of macro and micro factors are adjusted first in the type of space with the lowest matching rate. If the matching rate improves, the adjustment is increased. If not, the adjustment is repeated in the opposite direction and increased until the matching rate starts to decline. (iii) Then, the weights of internal elements in the macro and micro factors are adjusted, using a method similar to that of (ii). (iv) Finally, the weights of macro factors, micro factors and their internal elements are adjusted one by one in the spaces with the second highest and highest matching rates. After the weights of all factors are determined, the effect of each factor on PLE space evolution can be analyzed by its weight. Figure 3 shows the data process flow of factor weight determination by the statistical analysis and best-fit method.

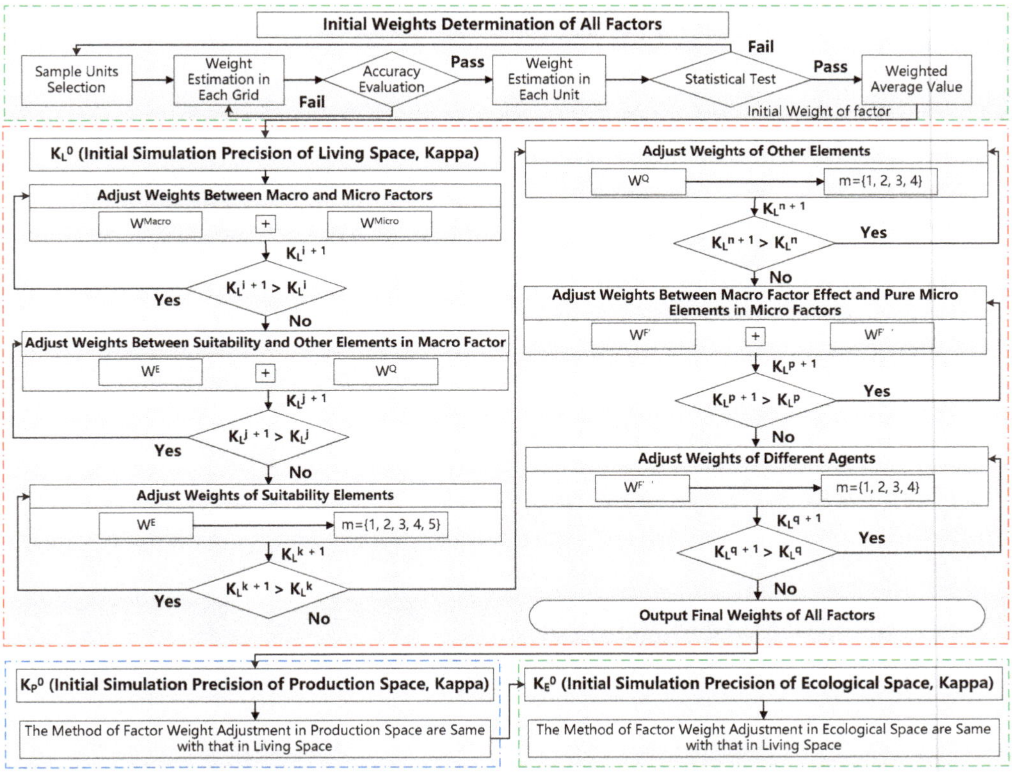

Figure 3. The data processing flow of factor weight determination using statistical analysis and best-fit method. The meanings of symbols are same with those in Equations (1)–(9).

4. Experimental Results and Analysis

4.1. Simulation and Prediction Results of Production-Living-Ecological Space Evolution

For validation of the proposed model and method in Section 3, GaoFen-2 (GF) remote sensing images, geographic information and socio-economic data from Xuzhou, China between 2000 and 2020 were collected. Three experimental schemes were designed. In scheme 1, the results of PLE space recognition from RS imaging in 2000 were taken as the basic data, and used to simulate the PLE space evolution in 2010 based on the CA model. The simulation precision was obtained by comparing its results with the result of RS image recognition in 2010. In schemes 2 and 3, the basic data were the same as in scheme 1. However, the CA + MAS model and the macro–micro joint decision model (CA + MAS

+ Correlation) were adopted in schemes 2 and 3, respectively. Figure 4 shows the spatial distribution of patches where the simulation results of the three schemes were different from the results of RS imaging. Table 3 shows the statistical results of the simulation precisions (Kappa coefficient) of the three schemes.

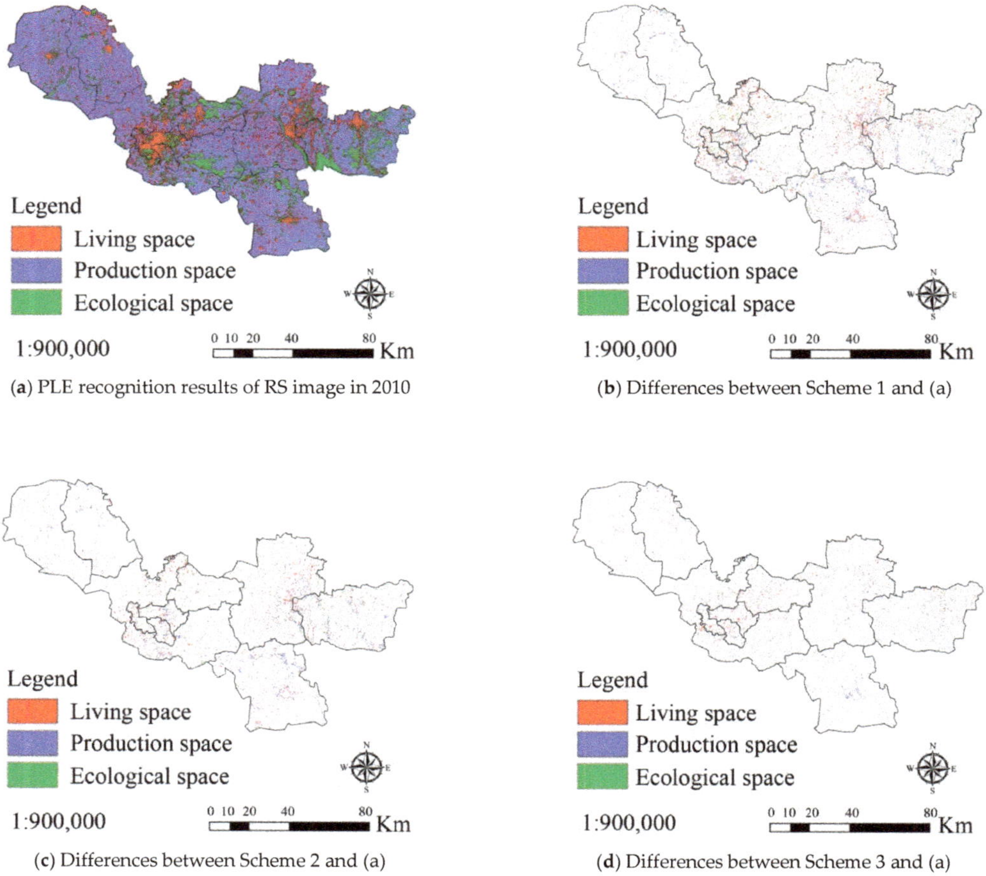

(a) PLE recognition results of RS image in 2010
(b) Differences between Scheme 1 and (a)
(c) Differences between Scheme 2 and (a)
(d) Differences between Scheme 3 and (a)

Figure 4. Differences between production-living-ecological (PLE) space evolution results from three experimental simulation schemes and remote sensing image recognition in Xuzhou, China in 2010. (**a**) is PLE recognition results of RS image in 2010; (**b–d**) are the differences between the simulation results of scheme 1, 2, 3 and the recognition results of (**a**), respectively.

Table 3. Statistical precisions of PLES evolution simulation from three schemes Unit: %.

Evaluated Object	Scheme 1	Scheme 2	Scheme 3
Model	CA	CA + MAS	CA + MAS + Correlation
Kappa	87.14	89.63	92.31

From Figure 4 and Table 3, it can be seen that the simulation precision of the CA model was poor (87.14%) and that most of errors resulted from the simulation of living space evolution. The simulation precision of production space evolution was better, and that of ecological space evolution was best. In scheme 2, the simulation precision of PLE space evolution was improved to 89.63% using the CA + MAS model. The precision improvement in living space evolution simulation was the maximum. In scheme 3, the simulation precision of PLE space evolution increased to 92.31% by using the CA + MAS + Correlation model. These experimental results prove that the proposed model has a stronger ability to simulate PLE space evolution than the current models (e.g., CA and CA + MAS).

To further verify the prediction ability of PLE space evolution of our proposed model, three further experimental schemes were designed. In scheme 1, the recognition result of PLE space from RS imaging in 2010 was taken as the basic data source. It was used to predict the PLE space evolution in 2020 based on the CA model, and the prediction precision was calculated by comparison with the results of RS image recognition in 2020. It should be noted that the weights of all factors remained unchanged in the prediction, and were derived from the simulation results comparing 2000 to 2010. The main reason was that the actual distribution of PLE space in 2020 could not be known in advance during the prediction. Therefore, the weights of all factors could not be adjusted over time in the simulation. This led to a decrease in prediction precision. In schemes 2 and 3, the data and method were the same as those used in scheme 1. However, the CA + MAS model and the CA + MAS + Correlation were adopted to predict PLE space evolution in 2020 in schemes 2 and 3 respectively. Figure 5 shows the spatial distribution of patches where the prediction results of three schemes were different from the recognition results from RS imaging in 2020. Table 4 is the statistical result of the prediction precisions of the three schemes.

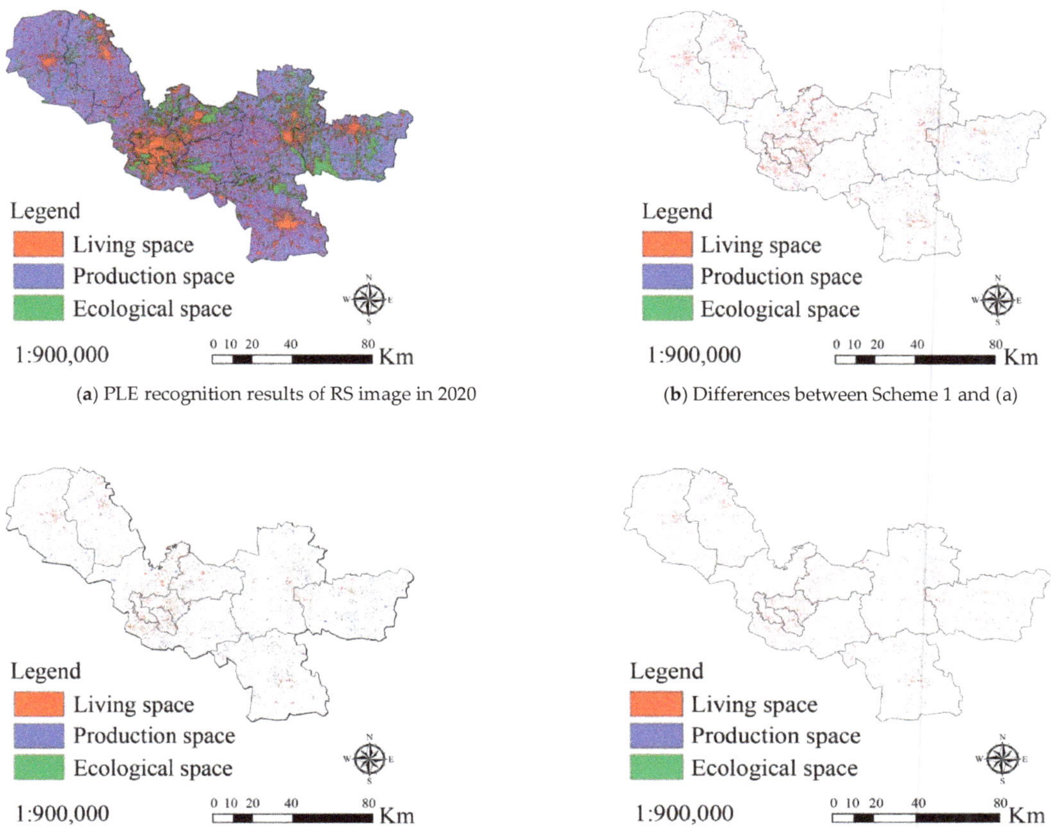

(**a**) PLE recognition results of RS image in 2020 (**b**) Differences between Scheme 1 and (a)

(**c**) Differences between Scheme 2 and (a) (**d**) Differences between Scheme 3 and (a)

Figure 5. Differences between PLE space evolution predictions from three experimental schemes and that determined from remote sensing image recognition in Xuzhou, China in 2020.

Table 4. Statistical precisions of PLES evolution prediction from three schemes Unit: %.

Evaluated Object	Scheme 1	Scheme 2	Scheme 3
Model	CA	CA + MAS	CA + MAS + Correlation
Kappa	84.59	86.19	89.33

From Figure 5 and Table 4, it can be seen that the prediction precisions of PLE space evolution were lower than their simulation precisions in the earlier schemes, as the factor weights from the prediction models were replaced with those from the simulation models. This inevitably leads to errors. However, the prediction precision of the CA + MAS + Correlation model remained better than that of the CA and CA + MAS models, and the improvement in prediction precision was larger than that in simulation precision, after considering the interactions of macro and micro factors. This proves that the proposed model had a stronger ability to predict PLE space evolution than the CA and CA + MAS models. Thus, the future distribution of urban PLE space may best be predicted by the CA + MAS + Correlation model. Policy makers and city administrators can determine the problems of urban space development in advance based on prediction results, and some positive policies can be formulated to avoid these problems and realize the sustainable development of urban PLE space.

4.2. Results of Factor Weight Determination

The initial weights of all factors can be obtained by statistical analysis methods and the final weights are determined by the best-fit method, as introduced in Section 3.2. The effect of each factor on PLE space evolution can be obtained based on the weight determination results. It is helpful for policy makers to grasp the leading factors and formulate scientific planning of urban PLE space development. Tables 5–7 show the initial weight determination results of factors using the CA + MAS + Correlation model to simulate the evolution of production, living and ecological space as described in Section 4.1, respectively. Figure 6 shows the error distribution of the estimated initial weights of some factors (e.g., elevation, slope, distance to water and distance to road) in the simulation of living space evolution. Figure 7 shows the final weight determination results of all factors in the simulation of production, living, and ecological space evolution and their differences with the initial weights, respectively.

Table 5. Initial weight determination results of factors in living space evolution simulation.

Elevation	**Slope**	**Distance to Water**	**Distance to Road**	**Distance to Center**	**Government Planning**	**Type of Land**	**Neighborhood Influence**
0.028	0.104	0.035	0.064	0.043	0.076	0.047	0.056
Influence of Micro factor	Protected land	House Price	Hospital	Mall	School	Distance to Ecological Space	Influence of Macro factor
0.014	0.076	0.095	0.109	0.058	0.089	0.064	0.042

Table 6. Initial weight determination results of factors in production space evolution simulation.

Elevation	**Slope**	**Soil Quality**	**Distance to Water**	**Distance to Road**	**Distance to Center**	**Type of Land**
0.042	0.046	0.052	0.059	0.106	0.043	0.109
Neighborhood Influence	Influence of Micro factor	Protected land	Land Price	Distance to living space	Distance to Ecological Space	Influence of Macro factor
0.065	0.032	0.184	0.091	0.102	0.035	0.034

Table 7. Initial weight determination results of factors in ecological space evolution simulation.

Elevation	**Slope**	**Distance to Water**	**Distance to Woodland**	**Distance to Grassland**	**Type of Land**	**Neighborhood Influence**
0.069	0.076	0.072	0.148	0.076	0.064	0.091
Influence of Micro factor	Protected land	Distance to production space	Distance to living space	Distance to Ecological Space	Influence of Macro factor	/
0.081	0.037	0.056	0.057	0.149	0.024	/

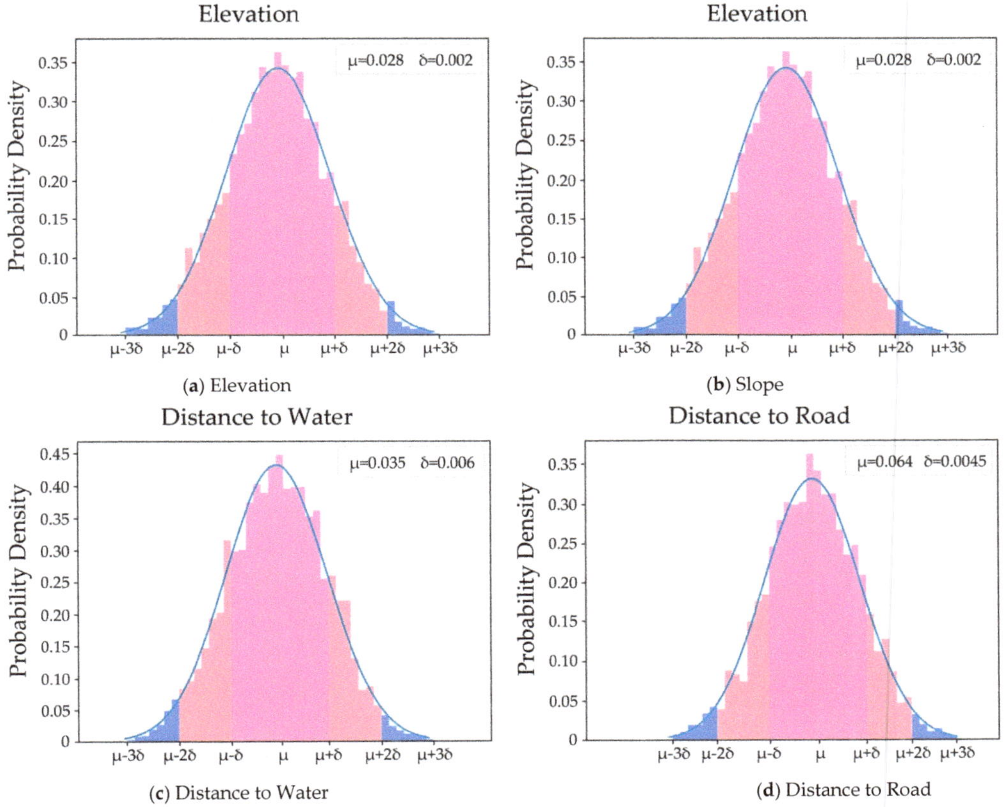

Figure 6. Error distributions of initial weight estimation of Elevation (**a**), Slope (**b**), Distance to Water (**c**) and Distance to Road (**d**) in living space evolution simulation.

(a) Living space (b) Production space

(c) Ecological space (d) Differences between the initial weight and final weight

Figure 7. Final weight determination results of living space (**a**), production space (**b**) and ecological space (**c**) evolution simulation factors and their differences with the initial weights (**d**).

From Figures 5 and 6, it is known that the initial weights of all factors could be obtained by statistical analysis methods and that the estimated results were reliable. In general, the weights obeyed normal distributions, although the estimated precisions were different. Some were high (e.g., elevation and distance to water), and some were low (e.g., slope and distance to road), because the estimated precision of factor initial weight was strongly related to the sample selection. Therefore, it is a key to selecting representative and diverse samples.

From Figure 7, it can be seen that the final weights of all factors had some changes from the initial weights in Tables 5–7. The reason was that the local consistency of simulation and recognition results from the samples was exclusively considered in the initial weight determination, while the overall consistency from the study region was emphasized in the final weight determination. It is noted that the differences between the initial and final weight determination were not obvious, proving that the initial weight determination and sample selection were suitable for this study. Based on the results of factor weight determination, the effect of each factor on PLE space evolution could be obtained. The micro factors affected the evolution of production and living space more than the macro factors. By contrast, the macro factors influenced the evolution of ecological space more than the micro factors. For living space evolution, the most important factors were the city resident (0.215), the entrepreneur (0.127) and the environmentalist (0.123). This indicates that the decision-making behaviors of city residents play the most important role in urban living space development. For production space evolution, the most important factors were the rural resident (0.202), the entrepreneur (0.147), and the city resident (0.098). This means that production space development depends more on human wishes than on natural resources and conditions or the development of science and technology. For ecological space evolution, the most important factors were the location condition (0.230), natural condition (0.150), and neighborhood influence (0.142). This indicates that the location and natural condition have an important influence on the development of ecological space, and with the increase in awareness of environmental protection, human intervention in the evolution of ecological space is decreasing. It has instead developed according to local conditions in Xuzhou in the past decades.

5. Discussions

The scientific planning of urban space is an important way to realize urban sustainable development. The precise simulation and understanding of the leading factors in urban space evolution are essential prerequisites for urban space planning. However, there are still some defects in existing models. For example, the econometric statistics model and the system dynamics model have a strong ability to predict the size of urban space evolution, but they are unable to simulate the variation in urban spatial distribution [36]. For another example, the cellular automata model is a key milestone in the development of urban spatial simulation technology, which can simulate not only the variation in urban space size but also the variation in urban spatial distribution, but it is unable to simulate the effects of macro-scale factors (e.g., policy, economy, culture) on urban space evolution, and nor can it simulate the decision-making behaviors of different urban agents (e.g., resident, entrepreneur, environmentalist) [37]. With the development of artificial intelligence (AI) technology, the study of a joint model of MAS and CA has become a hot spot for urban space evolution simulation. This type of model can simulate the influences of macro factors (e.g., natural environment, geographical location, policy) and micro factors (e.g., different urban agents) on urban space evolution [38]. Therefore, the joint model of MAS and CA has a stronger ability to simulate urban space than other models.

However, the current CA + MAS model does not consider the correlations between macro and micro factors. It is assumed that the effects of macro and micro factors are completely independent, which is inconsistent with the facts, as macro factors (e.g., the natural condition, government policy) often restrict the decision-making behaviors of micro factors (e.g., residents, entrepreneurs, environmentalists). Meanwhile, the preferences of micro factors also have an important influence on the decision-making behaviors of macro factors. Therefore, if the interaction between and mutual influence of macro and micro factors are neglected, the simulation precision of CA + MAS model will decrease. In addition, the simulation objects of current models are usually unit plots with different types of land use (e.g., woodland, grassland, water body). Therefore, the simulation result ascribes a specific type of land use to each unit plot. However, the type of land use may be different from the urban planning for a given unit plot (e.g., traffic planning, garden planning, land planning) [39]. Therefore, if multiple plans are implemented in the same place at the same time, it will lead to difficulties in decision-making and implementation (e.g., the conflict areas) [40].

To alleviate the above problems, a macro–micro joint decision model is proposed based on the CA + MAS model in this study to improve the simulation of urban space evolution. The simulation objects are unified into production, living and ecological space for the convenience of unified implementation of multiple plans. For validation of the proposed model, remote sensing images, geographic information and socio-economic data from Xuzhou, China between 2000 and 2020 were collected and tested. The results proved that the simulation and prediction precisions of the proposed model were better than those of current models (e.g., CA, CA + MAS) for urban space evolution simulation, particularly for the simulation of living space evolution.

It is very important for urban planning and sustainable development strategies that the precision of PLE evolution simulation models be improved, because urban space evolution simulation models are often used to compare the implementation effects of different urban planning schemes. Therefore, it is beneficial to select the optimal planning scheme for realizing sustainable urban development, if the model simulation precision is high. Otherwise, this can result in faulty government decision-making and the waste of urban resources if the model simulation precision is poor. A high-precision simulation model helps urban planners discover the problems associated with different plans in advance, and is also an important tool to analyze the effects of different factors on the evolution of urban space. Therefore, it is very useful to grasp the rules governing urban space evolution and operation mechanisms.

However, there are still some limitations of the proposed model in this study. (i) The urban space is expressed as a regular grid (30 m × 30 m) in the proposed model, but actual urban space is an irregular polygon. This led to inconsistencies between the simulation result and the RS imaging-observed results, as well as a decrease in simulation precision [41]. (ii) In theory, the smaller the area of unit space, the higher the precision of urban space simulation. However, the computational burden increases at exponential levels with decreasing areas of unit space. Therefore, the selection of optimal geographic unit scale for simulation of urban space evolution remains an unsolved problem. For larger areas (e.g., an urban agglomeration or economic zone), an adaptive theory should be applied to define the grid scale. Regions with rapid land use change could be defined using a small grid (e.g., 30 m × 30 m), while regions with slow land use change could be simulated using a large grid (e.g., 300 m × 300 m). On the one hand, the differentiating grids could improve simulation accuracy; on the other hand, they could ensure operational efficiency. (iii) The estimated precision of the factor initial weight is strongly related to the sample selection when using a statistical method. Therefore, the selection of representative and diverse samples is a key problem. In this study, an even sampling strategy was adopted to select the experimental samples. However, this was not an optimal solution. In addition, the variation in urban population size and spatial distribution is an important factor that affects urban space evolution [42]. In this study, the information on urban population size was used to predict the variation in living space size, but the data of population spatial distribution is not used. Therefore, urban space evolution simulation models should be further investigated considering the variations in population spatial distribution in the future.

6. Conclusions

The accurate simulation of urban space evolution and understanding of the leading factors are key issues to improve the sustainability of urban development. In this study, a macro–micro joint decision model was constructed based on the CA + MAS model in order to improve the ability to simulate urban space. A method of factor weight determination was proposed to analyze the effects of different factors on urban space evolution. For validation of the proposed model and method, experimental data (e.g., RS data, GIS data and socio-economic data) were collected and tested. The results proved that the proposed model and method were valid and reliable, and could improve the simulation and prediction of urban space evolution. The main conclusions of this study are as follows:

(1) The simulation precision of urban space evolution from the CA + MAS model was better than that from the CA model, because the decision-making behaviors of different urban agents (e.g., resident, entrepreneur, environmentalist) were considered in the CA + MAS model and not the CA model. Moreover, if the interactions and influences of macro and micro factors were considered (e.g., CA + MAS + Correlation model), the simulation precision of the CA + MAS model could be further improved. In this study, the simulation precisions (Kappa coefficient) of urban space evolution from the CA, CA + MAS and CA + MAS + Correlation models were 87.14%, 89.63% and 92.31%, respectively. The improvement in living space simulation precision was the most significant when using the CA + MAS+ Correlation model, compared with simulation of other types of spaces.

(2) The prediction precisions of CA, CA + MAS and CA + MAS + Correlation model were worse than their simulation precisions, as the factor weights in prediction models were replaced with those from the simulation models, disregarding changes that occurred during the prediction period. Therefore, errors were inevitably introduced. However, the prediction precision of the CA + MAS + Correlation model remained better than that of the CA and CA + MAS models, and the improvement in prediction precision was larger than that in simulation precision using the CA + MAS + Correlation model, compared to the CA and CA + MAS models. It was proved that the CA + MAS + Correlation model had a stronger ability to predict PLE space evolution than the CA and CA + MAS models.

(3) According to the results of factor weight determination, it was determined that the effects of micro factors on the evolution of living and production space were greater than those of macro factors. This indicated that the influences of desires and behaviors of human beings on the evolution of living and production space are increasing, correlating with the development of science and technology. The decision-making behaviors of city residents played the most important role in urban living space development. In the evolution of production space, rural residents and entrepreneurs had more influence than the other factors. By comparison, the effects of macro factors on the evolution of ecological space were more significant than those of micro factors, where the three most important factors were the location condition, neighborhood influence and natural condition. This means that the location and natural condition have more influence on the evolution of ecological space than the other factors.

According to the above analysis and conclusions, some policy implications are proposed to improve the sustainability of urban development. At present, the micro factors (e.g., city and rural residents and entrepreneurs) are the leading factors in the evolution of living and production space. Therefore, active policies should be formulated to strengthen the ideological guidance for these micro individuals, help them establish scientific views of urban space utilization, and realize the ordered development of living and production space. In addition, macro factors (e.g., location condition, natural environment) have the most important influence on the evolution of ecological space. Therefore, urban ecological space planning should closely link with the local environmental and natural conditions, to improve urban ecological carrying capacity and realize urban sustainable development.

Author Contributions: Conceptualization, funding acquisition and writing—review & editing, Q.W.; Data curation, image processing, software and writing—original draft preparation, Y.T. and Y.Z.; Statistical analysis and methodology, Y.Z. and Y.T. All authors have read and agreed to the published version of the manuscript.

Funding: This work was supported by the "Research on Optimal Allocation of Urban Public Resources Based on Population Size and Spatial Distribution Information" from MOE (Ministry of Education in China) Liberal arts and Social Sciences Foundation (No. 19YJC840067).

Acknowledgments: The authors would like to thank Yuanxin Qi and Shan Han for their help in data collection, image processing and preliminary analysis. The authors would like to acknowledge the Natural Resources Jiangsu Satellite Application Technology Center and Geographical Information Monitoring Cloud Platform (GIM) for the provision of relevant data and products.

Conflicts of Interest: The authors declare no conflict of interest. The funders had no role in the design of the study, in the collection, analyses, or interpretation of data; in the writing of the manuscript, or in the decision to publish the results.

References

1. Eric, K.; Nikolas, C.; Brain, S. Global warming and the carbon balance of boreal forests. *Ecol. Appl.* **1995**, *5*, 437–451.
2. Bryan, B.A.; Gao, L.; Ye, Y.Q.; Sun, X.F.; Connor, J.D.; Crossman, N.D.; Stafford-Smith, M.; Wu, J.G.; He, C.Y.; Yu, D.Y.; et al. China's response to a national land-system sustainability emergency. *Nature* **2018**, *559*, 193–204. [CrossRef]
3. United Nations, Department of Economic and Social Affairs, Population Division. *World Urbanization Prospects: The 2018 Revision*; United Nation: New York, NY, USA, 2019; pp. 1–126.
4. Liu, J.; Xu, X.; Zhuang, D.; Gao, Z. Impacts of LUCC processes on potential land productivity in China in the 1990s. *Sci. China D* **2005**, *48*, 1259–1269. [CrossRef]
5. Allen, C.; Nejdawi, R.; El-Baba, J.; Hamati, K.; Metternicht, G.; Wiedmann, T. Indicator-based assessments of progress towards the sustainable development goals (sdgs): A case study from the arab region (special feature): Sustainability science and implementing the sustainable development goals). *Sustain. Sci.* **2017**, *12*, 975–989. [CrossRef]
6. Chiara, C.; Dagmae, H.; Bruno, Z.; Davide, G. Is urban spatial development on the right track? Comparing strategies and trends in the European Union. *Landsc. Urban Plan.* **2019**, *181*, 22–37.
7. Fullman, N.; Barber, R.M.; Abajobir, A.A. Measuring progress and projecting attainment on the basis of past trends of the health-related sustainable development goals in 188 countries: An analysis from the global burden of dis-ease study 2016. *Lancet* **2017**, *390*, 1423–1459. [CrossRef]
8. Arbabi, H.; Punzo, G.; Meyers, G.; Tan, L.; Li, Q.; Tingley, D.; Mayfield, M. On the use of random graphs in analyzing resource utilization in urban systems. *R. Soc. Open Sci.* **2020**, *7*, 200087. [CrossRef] [PubMed]
9. Hu, G.; Li, X.; Zhou, B.B.; Ma, Q.; Liu, X. How to minimize the impacts of urban expansion on farmland loss: Developing a few large or many small cities? *Landsc. Ecol.* **2020**, *35*, 2487–2499. [CrossRef]
10. Berger, T. Agent-based spatial models applied to agriculture: A simulation tool for technology diffusion, resource use changes and policy analysis. *Agric. Econ.* **2015**, *25*, 245–260. [CrossRef]
11. Hu, Y.; Wu, X.; Ma, S.; Guo, X.; Liu, X. An empirical analysis of the system dynamics model of spatial coupling of wetland's "Production-Living-Ecological Space"—Taking Xixi national wetland park as an example. *Econ. Geogr.* **2018**, *38*, 173–180.
12. He, C.; Okada, N.; Zhang, Q.; Shi, P.; Li, J. Modelling dynamic urban expansion processes incorporating a potential model with cellular automata. *Landsc. Urban Plan.* **2008**, *86*, 79–91. [CrossRef]
13. Hu, Z.; Lo, C. Modeling urban growth in Atlanta using logistic regression. *Compu. Environ. Urban Syst.* **2007**, *31*, 667–688. [CrossRef]
14. Yang, D.; Yin, C.; Long, Y. Urbanization and sustainability in China: An analysis based on the urbanization Kuznets-curve. *Plan. Theory* **2013**, *12*, 391–405.
15. Jiang, L.; Zhang, Y. Modeling urban expansion and agricultural land conversion in Henan province, China: An integration of land use and socioeconomic data. *Sustainability* **2016**, *8*, 920. [CrossRef]
16. Hunt, J.D.; Simmonds, D. Theory and application of an integrated land-use and transport modeling framework. *Environ. Plan. B Plan. Des.* **1993**, *20*, 221–224. [CrossRef]
17. Wegener, M. Urban/regional models and planning cultures: Lessons from cross-national modelling projects. *Environ. Plan. B Plan. Des.* **1994**, *21*, 629–641. [CrossRef]
18. Roger, L.M. LILT and MEPLAN: A comparative analysis of land-use and transport policies for Leeds. *Trans. Rev.* **1991**, *11*, 131–154.
19. Chapin, F.S.; Weiss, S.F. A probabilistic model for residential growth. *Trans. Res.* **1968**, *2*, 375–390. [CrossRef]
20. Clark, K.C.; Hoppen, S.; Gaydos, L. A self-modifying cellular automaton model of historical urbanization in the San Francisco bay area. *Comput. Environ. Urban Syst.* **1997**, *24*, 247–261. [CrossRef]
21. He, C.Y.; Okada, N.; Zhang, Q.F.; Shi, P.J.; Zhang, J.S. Modeling urban expansion scenarios by coupling cellular automata model and system dynamic model in Beijing, China. *Appl. Geogr.* **2006**, *26*, 323–345. [CrossRef]
22. Ward, D.; Murray, A.; Phinn, S. A stochastically constrained cellular model of urban growth. *Comput. Environ. Urban Syst.* **2000**, *24*, 538–559. [CrossRef]
23. Wu, F. Polycentric urban development and land-use change in a transitional economy: The case of Guangzhou. *Environ. Plan. A* **1998**, *30*, 1077–1100. [CrossRef]
24. Ligmann-Zielinska, A.; Wachowicz, M.; Bregt, A.K.; Beulens, A.; Kettenis, D.L. A design and application of a multi-agent system for simulation of multi-actor spatial planning. *J. Environ. Manag.* **2004**, *72*, 43–55.
25. Torrens, P.M. Simulating sprawl. *Ann. Assoc. Am. Geogr.* **2006**, *96*, 248–275. [CrossRef]
26. Zeng, Y.; Yu, M.; Li, S. Urban expansion modeling approach based on multi-agent system and cellular automata. *ISPRS* **2018**, *42*, 2213–2229. [CrossRef]
27. Tan, M.; Li, X.; Xie, H.; Lu, C. Urban land expansion and arable land loss in Chinas a case study of Beijing-Tianjin-Hebei region. *Land Use Policy* **2005**, *22*, 187–196. [CrossRef]

28. Bone, C.; Dragicevic, S.; White, R. Modeling-in-the-middle: Bridging the garp between agent-based modeling and multi-objective decision-making for land use change. *Int. J. Geogr. Inf. Sci.* **2011**, *25*, 717–737. [CrossRef]
29. Li, H.; Li, L.; Chen, L.; Zhou, X.; Cui, Y.; Liu, Y. Mapping and characterizing spatiotemporal dynamics of impervious surfaces using Landsat images: A case study of Xuzhou, east China from 1995 to 2018. *Sustainability* **2019**, *11*, 1224. [CrossRef]
30. Liang, X.; Ji, X.; Guo, N.; Meng, L. Assessment of urban heat islands for land use based on urban planning: A case study in the main urban area of Xuzhou city, China. *Environ. Earth Sci.* **2021**, *80*, 1–22. [CrossRef]
31. Liu, F.; Zhang, Z.; Wang, X. Forms of urban expansion of Chinese municipalities and provincial capitals, 1970s–2013. *Remote Sens.* **2016**, *8*, 930. [CrossRef]
32. Akodad, S.; Bombrun, L.; Xia, J.; Berthoumieu, Y.; Germain, C. Ensemble learning approaches Based on covariance pooling of CNN features for high resolution remote sensing scene classification. *Remote Sens.* **2020**, *12*, 3292. [CrossRef]
33. Tao, Y.; Wang, Q. Quantitative Recognition and Characteristic Analysis of Production-Living-Ecological Space Evolution for Five Resource-Based Cities: Zululand, Xuzhou, Lota, Surf Coast and Ruhr. *Remote Sens.* **2021**, *13*, 1563. [CrossRef]
34. Xu, X.; Min, X.; Tian, Z. Simulation and prediction of urban expansion in Shanghai city based on GIS and LTM model. China Population. *Resour. Environ.* **2010**, *20*, 136–139.
35. Alva, J.A.V.; Estrada, E.G. A generalization of Shapiro-Wilk's Test for multivariate normality. *Commun. Stat.* **2009**, *38*, 1870–1883. [CrossRef]
36. He, C.; Shi, P.; Chen, J.; Pan, Y.; Li, X.; Li, J.; Li, Y.; Li, J. Research on land use scenario model based on system dynamics model and cellular automata model. *Sci. China D* **2005**, *35*, 463–474. (In Chinese)
37. Ligtenberg, A.; Bregt, A.K.; Van Lammeren, R. Multi-actor-based land use modelling: Spatial planning using agents. *Landsc. Urban Plan.* **2001**, *56*, 21–33. [CrossRef]
38. Zhu, J.; Tian, S. A dynamic urban lake area evolution model based on multilevel grid, cellular automata, and multiagent system. *Complexity* **2020**, *2020*, 1845090. [CrossRef]
39. Wang, D.; Jiang, D.; Fu, J.; Lin, G.; Zhang, J. Comprehensive assessment of Production–Living–Ecological space based on the coupling coordination degree model. *Sustainability* **2020**, *12*, 2009. [CrossRef]
40. Huang, J.; Lin, H.; Qi, X. Research progress of production–living–ecological space oriented to the optimization of national space. *Adv. Geogr. Sci.* **2017**, *36*, 378–391.
41. Zhou, S.; Tao, H.; Zhou, L. Vector-based multi-agent simulation of urban expansion: A case study in Panyu District in Guangzhou City. *Prog. Geogr.* **2014**, *33*, 202–210.
42. Zou, Y.; Zhang, S.; Min, Y. Exploring urban population forecasting and spatial distribution modeling with artificial intelligence technology. *CMES* **2019**, *119*, 295–310. [CrossRef]

International Journal of
*Environmental Research
and Public Health*

MDPI

Article

The Adjustment of China's Grain Planting Structure Reduced the Consumption of Cropland and Water Resources

Yu Zhang [1,2,3], Jieyong Wang [1,2,*] and Chun Dai [1,2,3]

[1] Institute of Geographic Sciences and Natural Resources Research, Chinese Academy of Sciences, Beijing 100101, China; zhangy1.19b@igsnrr.ac.cn (Y.Z.); daic.19s@igsnrr.ac.cn (C.D.)
[2] Key Laboratory of Regional Sustainable Development Modeling, Chinese Academy of Sciences, Beijing 100101, China
[3] College of Resources and Environment, University of Chinese Academy of Sciences, Beijing 100049, China
* Correspondence: wjy@igsnrr.ac.cn

Abstract: Driven by technological progress and market demand, the optimization and adjustment of grain planting structure played an important role in increasing grain output. Due to the great difference between the yield per unit area of different types of food crops, the consumption of cropland and water resources has a significant change during the grain growth. From the perspective of structural adjustment, rather than the usual productive factor input, we analyze the process of adjustment for grain planting structure in China and its effect on the consumption of cropland and water resources by using the scenario comparative analysis method. The results show that: (1) From 2003 to 2019, China's grain output has increased steadily and the planting structure has changed greatly. Rice was replaced by corn to become the grain crop with the maximum proportion of planting area since 2007. The increase of corn planting structure proportion is concentrated in the northern regions. (2) At the national level, according to the adjustment of grain planting structure, the saving of cropland and water resources consumption showed a "cumulative effect" as time went on. (3) The saving effects of structural adjustment in the northern regions on cropland and water resources consumption are better than that in the southern regions, such as Northeast China Plain, Northern arid and semiarid region and Loess Plateau. (4) In reality, although the adjustment of grain planting structure saved lots of cropland and water resources, the continuous growth of grain output has increased the pressure on the ecological environment in the northern regions according to theirs water limits. Therefore, it is necessary to continuously optimize the grain planting structure and restrict land reclamation in northern China. In addition, to ensure food security, it is feasible to encourage the southern regions with abundant water and heat resources to increase the grain planting area and meet its self-sufficiency in grain demand.

Keywords: grain production; structure adjustment; cropland and water resources; food security

Citation: Zhang, Y.; Wang, J.; Dai, C. The Adjustment of China's Grain Planting Structure Reduced the Consumption of Cropland and Water Resources. *Int. J. Environ. Res. Public Health* **2021**, *18*, 7352. https://doi.org/10.3390/ijerph18147352

Academic Editor: Paul B. Tchounwou

Received: 9 June 2021
Accepted: 6 July 2021
Published: 9 July 2021

1. Introduction

Grain production is the main source of cropland and water resource consumption in the world [1]. Statistics also indicate that agricultural water use accounts for about 80% of the global total water consumption [2]. China is an important country for food production and consumption in the world [3]. However, the supply of cropland and water resources is insufficient in China, with per capita occupancy equal to 28% and 40% of the world average, respectively [4,5]. The evolution of China's grain production pattern is not only reflected in the movement of grain production gravity [6], but also accompanied by the adjustment of grain planting structure [7]. Further, these will cause the problem of spatial allocation of cropland and water resources and ultimately affect the sustainable development of the whole agriculture. Therefore, this is crucial to ensuring China's food security and promoting agricultural sustainable development, through the analysis of grain planting structure adjustment and its impact on the consumption of cropland and water resources.

With the rapid development of the regional economy and society, great changes have taken place in terms of grain production and supply-demand patterns in China. It has been found the barycenter of grain crops presents a trend of goes to north and center movement in China [6], which is driven by factors such as economic benefits, per capita farmland acreage and grain yield per unit area [8]. In addition, the gravity of China's grain output is also moving northward due to urbanization, land-use policies and climate change. The pattern of grain supply-demand has changed, from the traditional pattern of "grain in the south being transported to the north" to the present pattern of "grain in the north being transported to the south" [9,10]. Due to 0 °C isotherm moved northward, the annual accumulated temperature (AAT) increased and the rice planting area continued to expand in Northeast China [7,11,12]. The shortage and spatial imbalance of cropland and water resources restrict the sustainable development of grain production. Other studies analyze the utilization of cropland and water resources and their spatial allocation and evaluate the relationship between grain production and the spatial allocation of cropland and water resources [13–15].

As one of the most populous countries in the world, China has always paid high attention to food security. Many studies have identified several ways to help increase grain production. We summarize as follows: (1) Expand arable land to increase the area of grain crop planting [16]. (2) Increase the input of productive factors, such as chemical fertilizers, pesticides, etc. [17,18]. (3) Increase the intensity of grain crop planting [19,20]. There are some problems with this approach to increasing food production. On the one hand, the abuse of pesticides and fertilizers causes serious ecological and environmental problems in grain production, such as agricultural non-point source pollution [21], which affects the food quality and further threatens human health. On the other hand, factor input has a marginal effect. That is to say, the increase of production factor input has less and less effect on improving grain output [22]. Moreover, overuse of cultivated land resources leads to the decline of soil fertility, which is not conducive to agricultural sustainable development. In this case, we need to explore more effective strategies to increase grain production [23]. As such, the previous studies focused on the input amount of grain production and crop yield, as well as the driving factors of crop structure adjustment. Few studies have analyzed the effects of grain production on cropland and water resources from the perspective of planting structure. Different grain crops not only have different yields per unit area, but also have great differences in cropland and water consumption. The adjustment of the grain planting structure has become a "bridge" to observe the relationship between grain demand and consumption of cropland and water resources. Therefore, while the total grain yield increases, the adjustment of the internal grain planting structure has an impact on the consumption of cropland and water resources in grain production. In other words, it is very feasible and meaningful to analyze the consumption of cropland and water resources in grain production through adjustment of planting structure rather than the input of production factors. Therefore, this paper abandons the traditional research idea of analyzing grain production based on productive factor input. From the perspective of the adjustment of grain planting structure, to calculate its effects on cropland and water resources consumption while the grain yield continues to increase.

Considering the condition of the shortage and spatial imbalance of cropland and water resources in China, we want to answer two questions: (1) Does the adjustment of the grain planting structure contribute to the economic utilization of cropland and water resources? (2) What are the differences in the consumption of cropland and water resources between different agricultural areas? These questions are posed based on the scenario comparison method, taking the grain planting structure without adjustment as the base scenario in 2003. This paper aims to explore the impact of cropland and water resources by adjusting the grain planting structure from 2003 to 2019. It provides a reference for the rational

adjustment of grain production structure and relevant departments to make decisions in the future.

2. Materials and Methods

2.1. Data Sources and Processing

The agricultural division is an important approach to guide agricultural production [17]. China is divided into nine agricultural regions based on geographical zoning and regional characteristics of the grain planting system, which are Northeast China Plain, Northern arid and semiarid region, Huang-Huai-Hai Plain, Loess Plateau, Qinghai Tibet Plateau, Sichuan Basin and surrounding regions, Middle-lower Yangtze Plain, Yunnan-Guizhou Plateau and Southern China. The study region includes 31 provinces (except Hong Kong, Macao and Taiwan) that participated in the complete information (Figure 1). The basic map of the nine agricultural regions is taken from the Resources and Environmental Sciences and Data Center, Chinese Academy of Science (http://www.resdc.cn/Default.aspx (accessed on 5 March 2021)).

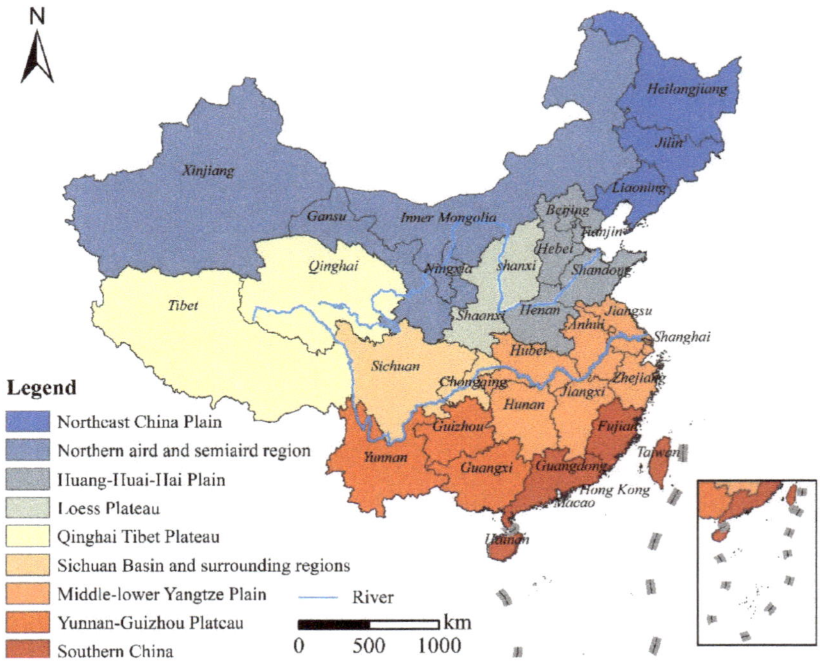

Figure 1. Map of agricultural regions in China.

All of the study data are public data obtained from different sources. These data were derived from the following sources. Data sets on grain planting area and yield of rice, wheat, corn, beans and tubers were obtained from the China Statistical Yearbook (2004–2020) (http://www.yearbookchina.com/index.aspx (accessed on 20 March 2021)). Agricultural water consumption data were obtained from the China Water Resources Bulletin of the corresponding years. The basic data of the nine agricultural regions were collected by each province, such as grain output, agricultural water consumption, cropland resources and so on.

Due to the different climatic conditions and soil environment in nine agricultural regions, water resource consumption varies greatly among different crops in the same

agricultural region or the same crop in different agricultural regions [24]. Before the water consumption analysis, virtual water content per unit mass of grain crops needs to be considered. Firstly, the weight mean of virtual water content per unit mass of grain crops was calculated in each province by referring to relevant research results. Secondly, the proportions of different grain production were measured in each province. Finally, the virtual water content per unit mass of different grain crops was obtained in nine agricultural regions of China.

2.2. Model Design

2.2.1. Calculation Formula of Cropland Resource Consumption

To obtain the same grain yield, the gap between the cropland resources consumption with structural adjustment and without structural adjustment is the number of cropland resources saved by the adjustment of the grain structure. The specific calculation process is shown as follows:

$$Q_t = A_t \times Y_t = A_t \times \sum(s_{it} \times y_{it}) \tag{1}$$

$$Q_{t+j} = A_{t+j} \times \sum(s_{i,t+j} \times y_{i,t+j});\ j = 1,2,3\ldots16 \tag{2}$$

where Q_t, Q_{t+j} represents the grain yield in t and $t + j$ period, j is the number of the year (in this study, j = 16). A_t is the sown area of grain, S_{it} represents the sown area of crop i accounted for the proportion of the sown total area and y_i is the yield per unit area of crop i. Similarly, A_{t+j}, $S_{i,t+j}$, $y_{i,t+j}$ represent sown area of grain, sown area of crop i accounted for the proportion of the sown total area and the yield per unit area of crop i in $t + j$ period, respectively.

$$Q_{t+j} = A'_{t+j} \times \sum(s_{it} \times y_{i,t+j}) \tag{3}$$

$$A'_{t+j} = A_{t+j} \times \frac{\Sigma(s_{i,t+j} \times y_{i,t+j})}{\Sigma(s_{it} \times y_{i,t+j})} \tag{4}$$

A'_{t+j} represents sown area of grain when without structural adjustment. Meanwhile, $S_{it} = S_{i,t+j}$. Therefore, the gap between the cropland resources consumption with structural adjustment and without structural adjustment can be calculated using Equation (5):

$$A'_{t+j} - A_{t+j} = \left[\frac{\Sigma(s_{i,t+j} \times y_{i,t+j})}{\Sigma(s_{it} \times y_{i,t+j})} - 1\right] \times A_{t+j} \tag{5}$$

If $A'_{t+j} - A_{t+j} > 0$, it means that the adjustment of grain planting structure can help save cropland resources, while if $A'_{t+j} - A_{t+j} < 0$, it means that the adjustment of grain planting structure increases the consumption of cropland resources and if $A'_{t+j} - A_{t+j} = 0$, it means there is no effect.

2.2.2. Calculation Formula of Water Resource Consumption

Following the same train of thought, the amount of water resources saving can be calculated by the adjustment of the grain structure. The formula is:

$$\beta_{i,t+j} = \frac{\Sigma((s_{i,t+j} \times y_{i,t+j}))}{\Sigma(s_{it} \times y_{i,t+j})} \tag{6}$$

$$W'_{t+j} - W_{t+j} = A_{t+j} \times \left[\sum(s_{it} \times \beta_{i,t+j} \times y_{i,t+j} \times m_{i,t+j}) - \sum(s_{i,t+j} \times y_{i,t+j} \times m_{i,t+j})\right] \tag{7}$$

where $\beta_{i,t+j}$ is the increase or decrease coefficient, it indicates that the changing intensity is caused by structural adjustment compared with no structural adjustment. In Equation (7), W'_{t+j}, W_{t+j} represent the amount of water resources consumption with structural adjustment and without structural adjustment, respectively. $m_{i,t+j}$ represents water consumption per unit mass of crop i.

If $W'_{t+j} - W_{t+j} > 0$, it means that the adjustment of grain planting structure contributes to the economic utilization of water resources, while if $W'_{t+j} - W_{t+j} < 0$, it means that the adjustment of grain planting structure increases the consumption of water resources and if $W'_{t+j} - W_{t+j} = 0$, it means there is no effect.

2.2.3. Consumption Reduction Contribution (*CRC*) of Cropland and Water Resources

The consumption reduction contribution is used to measure the saving effect of cropland and water resources in an agricultural region. It can provide a concise and intuitive result regardless of the total amount of cropland and water resources among agricultural regions. The formula is

$$CRC_l = \frac{A'_{t+j} - A_{t+j}}{A'_{t+j}} \tag{8}$$

$$CRC_w = \frac{W'_{t+j} - W_{t+j}}{W'_{t+j}} \tag{9}$$

In Equations (8) and (9), CRC_l and CRC_w represent consumption reduction contribution of cropland resources and consumption reduction contribution of water resources, respectively. The values of CRC_l and CRC_w range from −1 to 1. The higher the value, the higher the consumption reduction contribution (*CRC*) of cropland and water resources. On the contrary, the smaller the value, the lower the consumption reduction contribution (*CRC*) of cropland and water resources.

The last required a bit of explanation that the study results were calculated by Excel and visualized by ArcGIS and Origin software.

3. Results

3.1. Changes of Grain Yield and Consumption of Cropland and Water Resources in China

From 2003 to 2019, China's output of major farm products has shown a steady increase (Figure 2), which can be divided into two stages: the rapid growth stage (2003–2015) and the fluctuating growth stage (2016–2019). During the rapid growth stage (2003–2015), China's grain output increased from 431 million tons to 621 million tons, with an average annual growth rate of 3.10%. In the fluctuating growth stage (2016–2019), although grain production has declined in some years, it has maintained a rapid growth overall, reaching 664 million tons in 2019.

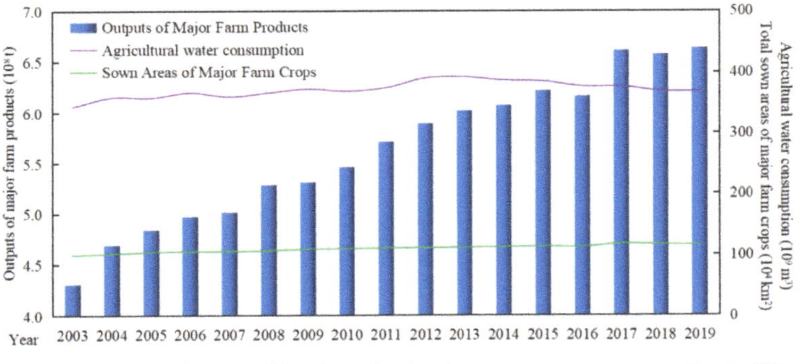

Figure 2. Changes of grain yield and cropland and water resources consumption in China from 2003 to 2019. Sown area of major farm crops (rice, wheat, corn, beans and tubers) and its outputs data were obtained from the China Statistical Yearbook (2004–2020). Agricultural water consumption data were obtained from the China Water Resources Bulletin (2004–2020).

However, with the rapid increasing of the total grain yield, the consumption of cropland and water resources for grain production remained stable on the whole. The water resource consumption showed a trend of small fluctuation, which remain between 350 × 10^9 m^3 and 390 × 10^9 m^3. The cropland resource consumption increased from 99.41 × 10^4 km^2 in 2003 to 116.06 × 10^4 km^2 in 2019, which was only 1.17 times higher. However, grain production increased 1.54 times in 17 years. Therefore, the rapid increase of grain output is not only related to the use of chemical fertilizers and pesticides, technological progress [25] and other input elements, but also closely related to the adjustment of grain planting structure.

3.2. Spatio-Temporal Changes of Grain Planting Structure in China

3.2.1. Temporal Changes of Grain Planting Structure in China

We selected five types of grain crops, including rice, wheat, corn, beans and tubers, to analyze the change of grain planting structure in China. Figure 3 shows that significant changes have taken place in grain planting structure during the steady growth stage of grain outputs. Concretely, the proportion of rice, wheat and corn changed from 27.85%, 23.11% and 25.29% in 2003 to 26.30%, 21.01% and 36.56% in 2019, respectively. Since 2007, corn, the maximum proportion of planting structure, has replaced rice as China's largest grain crop, which has continued to increase.

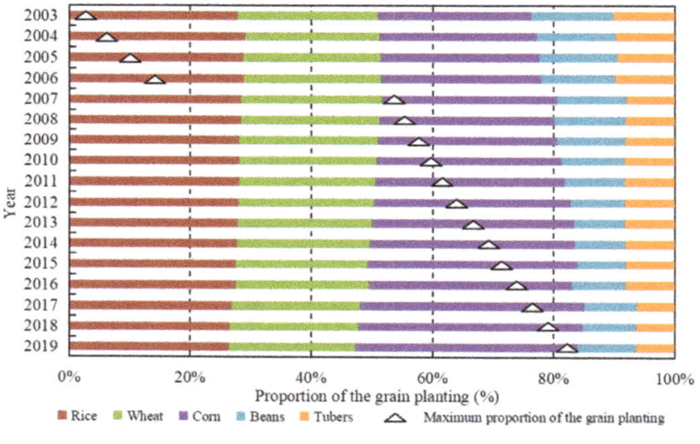

Figure 3. Changes of grain planting structure in China from 2003 to 2019.

After China acceded to the World Trade Organization (WTO), a large number of international agricultural products entered the Chinese market, especially the import of soybeans increased sharply. The soybean imports have reached 88.51 million tons in 2019, accounting for 83.10% of China's total soybean consumption. As a result, China's grain planting structure has also been strategically adjusted, with the proportion of soybeans planted gradually declining from 13.55% in 2003 to 9.81% in 2019.

3.2.2. Spatial Characteristics of Grain Planting Structure in China

Figure 3 reflects the dynamic change trend of China's grain planting structure from 2003 to 2019, but does not reflect the spatial distribution characteristics among the nine agricultural regions. Therefore, Using the agricultural division in Figure 1, we show the grain planting structure of different agricultural regions in China at two time points in 2003 and 2019 (Figure 4).

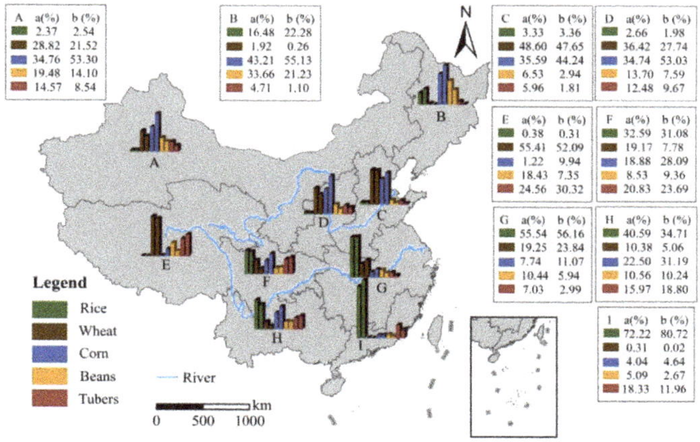

Figure 4. Changes of grain planting structure of China's nine agricultural regions in 2003 and 2019. (**A–I**) represents each agricultural regions, (**A**,**B**) refer to 2003 and 2019, respectively. The values are the planting structure of different food crops.

Significant differences in grain planting structure among the nine agricultural regions in China and the proportion of corn planted in the northern region has increased rapidly, which can be presented by Figure 4. For example, the proportion of corn planted in the Northeast China Plain, the Loess Plateau and the Northern arid and semiarid region has increased from 43.21%, 34.74% and 34.76% in 2003 to 55.13%, 55.03% and 55.30% in 2019, respectively. Compared with 2003, in 2019, the planting area of rice continued to expand and the planting proportion increased by 5.8% in the Northeast China Plain, while the planting proportion of beans decreased by 12.43%. Wheat is the dominant crop in Huang-Huai-Hai Plain and the proportion of planting area is stable at about 48%. Rice is the main grain crop in southern China, but its planting proportion decreased in some agricultural regions and was gradually replaced by corn, such as Yunnan-Guizhou, Plateau Sichuan Basin and surrounding regions.

3.3. The Impact Analysis of Grain Planting Structure Adjustment on Cropland Resource Consumption

We used Equation (5) to calculate the cropland resource consumption with or without the adjustment of grain planting structure from 2003 to 2019 (Figure 5) to analyze the evolution trend in both scenarios and then used Equation (8) to calculate the consumption reduction contribution (Table 1) to measure the saving effect of cropland resource.

It can be seen from Figure 5 that the cropland resource consumption with structural adjustment is less than that without structural adjustment, indicating that the adjustment of the grain planting structure contributes to saving cropland resources under certain grain output. Note the difference value of the slope of the trend line in both scenarios, it represents the extent to which the structural adjustment affects the cropland resource consumption in this region. From 2003 to 2019, due to the adjustment of grain planting structure, the saving of cropland resources consumption showed a "cumulative effect" at the national total. Consistent with the overall trend of the country, the northern region also showed a significant "cumulative effect", such as Northeast China Plain and the Northern arid and semiarid region. For the southern region where arable cropland resources are relatively scarce, such as Middle-lower Yangtze Plain, Yunnan-Guizhou Plateau and Southern China, the trend lines of cropland resource consumption under the two scenarios are nearly parallel, indicating that grain planting structural adjustment plays a certain role in the reduction of cropland resources, but the reduction scope is almost to the limit.

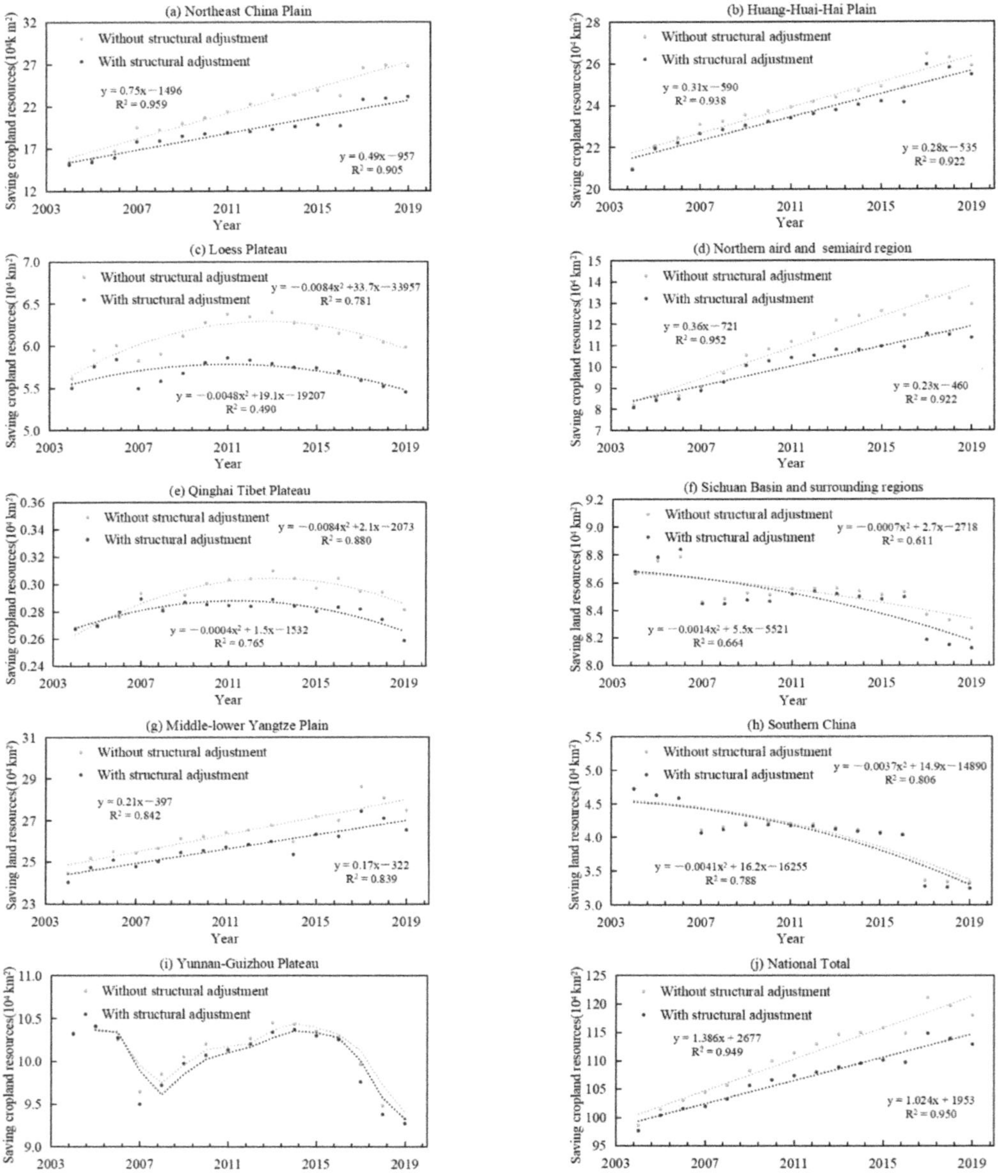

Figure 5. The evolution trend of cropland resources consumption with or without the adjustment of grain planting structure. (**a**,**b**,**d**,**g**,**j**) fit the linear approximation, (**c**,**e**,**f**,**h**) are described by polynomials, (**i**) uses the moving average method to express the evolution trend. y represents the cropland resources consumption with or without the adjustment of grain planting structure, x represents the year, R^2 expresses the explanatory degree of the relationship between the x and y variables. The values of R^2 range from 0 to 1. The higher the value, the closer relationship between the x and y variables. Green and blue line are fitted curve, which means the evolution trend of cropland resources consumption without structural adjustment or with the structural adjustment, respectively.

Table 1. Effect of grain planting structure adjustment on cropland resources consumption during 2003–2019.

Regions	Agricultural Division	Actual Area Changes (10^4 km^2)	Area Saved (10^4 km^2)	CRC_l (%)
The Northern Regions	Northeast China Plain	9.01	3.62	15.61
	Huang-Huai-Hai Plain	4.45	0.43	1.70
	Loess Plateau	0.09	0.54	9.83
	Northern arid and semiarid region	3.53	1.58	13.93
The Southern Regions	Qinghai Tibet Plateau	−0.01	0.02	8.97
	Sichuan Basin and surrounding regions	−0.44	0.15	1.79
	Middle-lower Yangtze Plain	3.51	0.95	3.56
	Southern China	−1.51	0.08	2.57
	Yunnan-Guizhou Plateau	−0.88	0.06	0.62
National Total	-	17.75	5.21	4.62

CRC_l means consumption reduction contribution of cropland resources.

At the national total, 5.21×10^4 km^2 of cropland resources saved by the adjustment of grain planting structure and consumption reduction contribution of cropland resources (CRC_l) was 4.62%. From the perspective of regions, both the area saved and the CRC_l of the northern region were the highest. More concretely, because high-yielding rice and corn replaced the relatively low-yielding wheat and soybean, area saved of cropland resources and CRC_l in Northeast China Plain reached 3.62×10^4 km^2 and 15.61%, respectively, which were higher than other agricultural regions. Followed by the Northern arid and semiarid region, area saved of cropland resources and CRC_l were 1.58×10^4 km^2 and 13.93% respectively, which were mainly caused by the substitution of planting corn for wheat. However, compared with the substantial expansion of the actual grain planting area in northern China, the cropland resources saved by grain planting structural adjustment were still quite limited, accounting for only 29.35% of the actual increase of the grain sown area in the same period.

Different from the northern region, the proportion of high-yielding rice planting is relatively high in the southern region, such as Middle-lower Yangtze Plain and Southern China, so the scope for the adjustment of the grain planting structure becomes very limited. Therefore, the cropland resource area saved by the adjustment of the grain planting structure is less than 1×10^4 km^2 and the CRC_l value is also quite low in each agricultural region. In other words, the adjustment of the grain planting structure has a small effect on saving cropland resources in China's southern region. However, in reality, except for the Middle-lower Yangtze Plain, the grain planting area of other agricultural regions has shrunk to some extent.

3.4. The Impact Analysis of Grain Planting Structure Adjustment on Water Resource Consumption

Figure 6 shows the evolution trend of water resource consumption with or without the adjustment of grain planting structure in nine agricultural regions and national total from 2003 to 2019. At the national level, the adjustment of the grain planting structure has a crucial effect on saving water resources. The gap of water resource consumption under the two scenarios became larger over time in the process of grain production, which also showed a significant "cumulative effect".

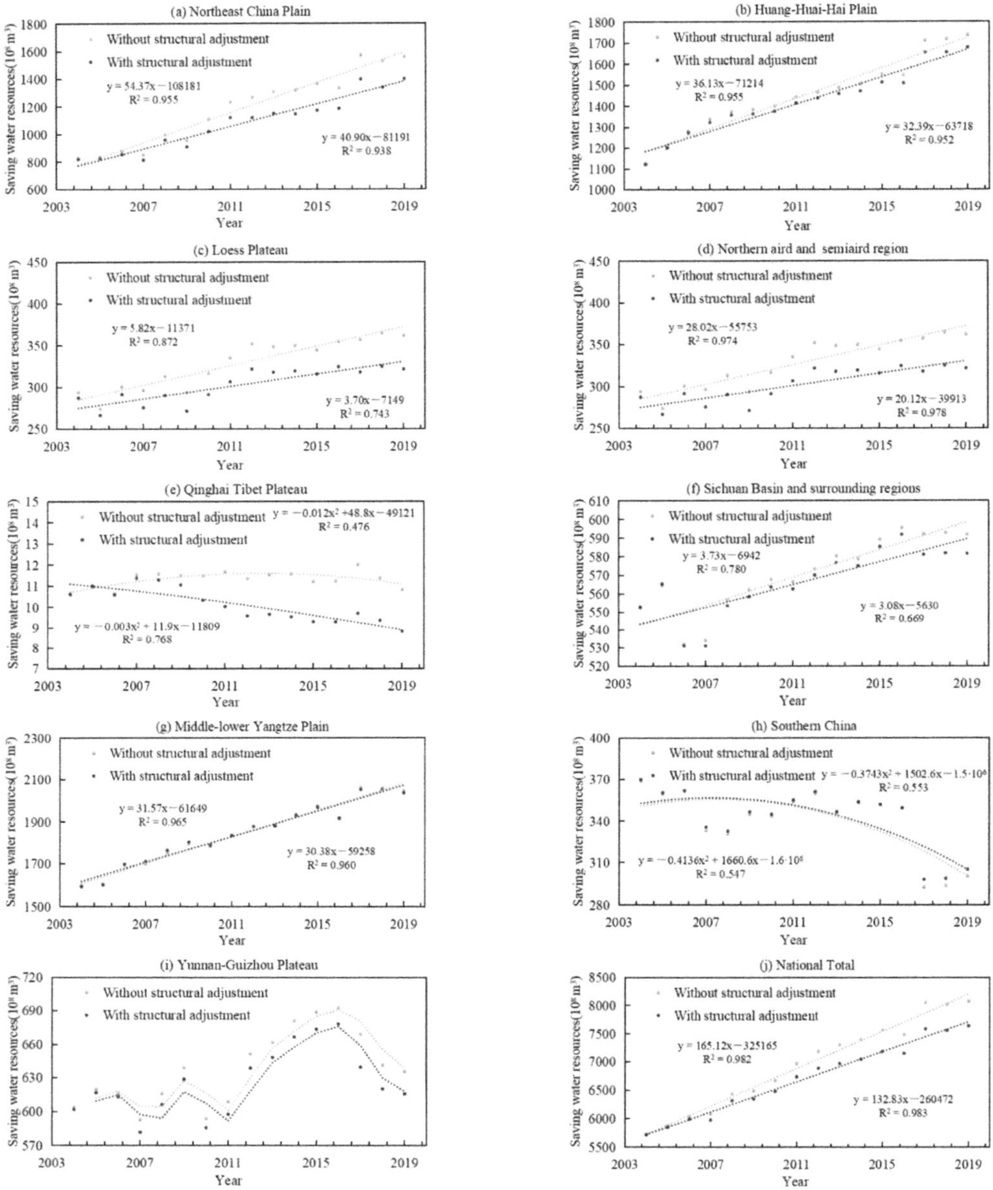

Figure 6. The evolution trend of water resource consumption with or without the adjustment of grain planting structure. (**a–d**,**f**,**g**,**j**) fit the linear approximation, (**e**,**h**) are described by polynomials, (**i**) uses the moving average method to express the evolution trend. y represents the water resources consumption with or without the adjustment of grain planting structure, x represents the year, R^2 expresses the explanatory degree of the relationship between the x and y variables. The values of R^2 range from 0 to 1. The higher the value, the closer relationship between the x and y variables. Green and blue line are fitted curve, which means the evolution trend of water resources consumption without structural adjustment or with the structural adjustment, respectively.

Compared with the southern regions and northern regions, we found that in the northern regions where water resources supply is insufficient, the adjustment of grain planting structure has a more obvious effect on saving water consumption. For example, compared with the non-adjustment scenario, water resource consumption continues to decrease in the Northeast China Plain, Loess Plateau and Northern arid and semiarid region. From the perspective of development trends, it is still possible to continue to save more water resources. However, in water-rich southern regions, such as Middle-lower Yangtze Plain and Southern China, the trend lines of water consumption almost coincide with or without the adjustment of grain planting structure, indicating that the adjustment of grain planting structure has a small effect on water resource reduction and reaches the upper limit.

It can be seen from Table 2 that the grain planting structural adjustment has helped to save an appropriate 437.09×10^8 m^3 water resource for the country, compared with the scenarios without structure adjustment, accounting for 38.58% of the increased water resource consumption in actual grain production. At the regional level, Qinghai Tibet Plateau has the largest value of consumption reduction contribution of water resources (CRCw)under both scenarios, followed by the Northern arid and semiarid region, Loess Plateau and Northeast China Plain. However, the Qinghai-Tibet Plateau has a low grain planting area and limited water resource consumption, so its reference value is insufficient. Thus, the Northern Region has a greater effect on water resource reduction. In particular, because of the rare rainfall all the year-round, the water resource consumption of grain production mainly depends on irrigation water in the Northern arid and semiarid region [26], which is part of the temperate continental climate region. CRCw is 14.68%, higher than in other agricultural regions. Through grain planting structural adjustment, 102.27×10^8 m^3 of water resources were saved, accounting for 47.98% of the actual increase in the total water demand for grain production in this region.

Table 2. Effect of grain planting structure adjustment on water resources consumption during 2003–2019.

Regions	Agricultural Division	Actual Water Changes (10^8 m^3)	Water Saved (10^8 m^3)	CRCw (%)
The Northern Regions	Northeast China Plain	576.42	159.33	11.37
	Huang-Huai-Hai Plain	268.52	59.21	3.53
	Loess Plateau	−2.66	40.44	12.57
	Northern arid and semiarid region	213.14	102.27	14.68
The Southern Regions	Qinghai Tibet Plateau	−1.53	2.02	22.86
	Sichuan Basin and surrounding regions	−31.37	10.33	1.78
	Middle-lower Yangtze Plain	320.05	11.53	0.57
	Southern China	−123.96	−4.62	-1.51
	Yunnan-Guizhou Plateau	−75.65	19.82	3.22
National Total	-	1133.05	437.09	5.72

CRCw—means consumption reduction contribution of water resources.

The southern region is located in subtropical and tropical climates and has abundant water resources. Statistics also show that the water resources per unit arable area in southern regions is 8 times that in northern regions [27]. The value of water saved and CRCw are relatively low in the southern regions. Middle-lower Yangtze Plain, which is the most important grain planting area in the southern regions, the water resources saved accounted for about 1/30 of the actual increase in water resources consumption due to the expansion of grain planting area. Even, the water resources consumption has further increased by grain planting structural adjustment in Southern China and CRCw is −1.51%.

4. Discussion

As the population grows and wealth increases, people can buy more varied and resource-intensive diets. For now, the human demand for food is still increasing. It has been found that global agricultural production may need to increase by 70–110% to meet the growing demands associated with human uses and livestock feed by 2050 [28]. There is increased competition for land, water, energy and other inputs into food production [29]. The expansion of built-up areas occupies a lot of arable lands in the process of rapid urbanization, crowding out the space for grain production [30]. Recent studies also show that future urbanization will result in a 1.8–2.4% loss of global cultivated lands by 2030 [31,32]. The sudden outbreak of COVID-19 has alerted the world to the importance and seriousness of food security [33,34]. Global food security faces even more severe challenges in the future. However, China has achieved a sustained and steady increase in grain output in the process of rapid urbanization, which has ensured the country's food security. The study results prove that the proportion of grain planting increased, such as corn, which alleviated the pressure of grain production on the consumption of cropland and water resources to a certain extent.

Remarkably, with the improvement of technology and living standard, people pay more attention to the nutritional content in food consumption, which further enhances the substitutability among different food crops demands. For the past few years, the consumption quantity of animal food increases year by year, especially the consumption of milk and dairy product. At the same time, grain consumption decreases significantly. This reflects the replacement of basic rations with feed and industrial grains. In China, the food structure of residents is transforming from vegetable fiber orientation to animal fat and high protein orientation. FAO statistics on the food supply of countries and regions indicate that per capita nutritional level of the Chinese Mainland is close to Japan, Taiwan and Korea, which is 3 kcal each day [35,36]. Consumer demand tends to be more about the nutrient content of a grain crop than its type. Therefore, the structure of grain production weakens the constraints on food supply. In that sense, the adjustment of the grain planting structure becomes an inevitable trend. It is possible to increase grain production and reduce the pressure of cropland and water resource consumption by increasing the planting area of high-yielding crops in the context of marketization.

However, the study also found that the adjustment of grain planting structure of the corn and other crops were concentrated in the northern regions, such as the Northern arid and semiarid region, the Northeast China Plain and the Loess Plateau, which are also areas with severe water shortage. Although the structural adjustment has improved the utilization efficiency of cropland and water resources, the growing acreage of food crops has further aggravated the contradiction of water shortage in northern regions. It has brought more pressure to the ecologically fragile northern regions. By contrast, the southern regions with superior water and heat conditions are suitable for grain production. The loss of the agricultural labor force has resulted in the abandonment of arable land [37,38], which threatens agricultural production and food security. In a word, the natural conditions for grain production in the northern regions, such as heat and water resources, are inferior to those in the southern regions. The spatial movement of grain production barycenter and the spatial dynamic change of grain planting structure aggravated the spatial imbalance of cropland and water resources to some extent. What should be paid attention to is the spatial dislocation between the expansion of grain production and the natural resources, such as cropland, heat and water resources.

Therefore, future research should focus on the following points: First, the regional types of grain production in China are classified, such as core area, potential area and buffer area. The purpose is to clarify the ability to ensure self-sufficiency in food demands. Second, the trend of grain planting structure adjustment was predicted from the perspective of the nutrient content of crops. In the end, it is also of great significance to further explore the effects of the adjustment of grain planting structure on the consumption of cropland and water resources at the microscale, such as prefecture-level city or county units. In addition, in this paper, the scenario analysis was used to analyze the process of adjustment for grain planting structure in China and its effect on the consumption of cropland and water resources in the past ten years, but it lacks the prediction of grain planting structure

Table 1. Definition and indicators of rainfall erosivity trend patten.

Trend Patten	Definition	Identification Indicators
DE	Rainfall erosivity decreased significantly during the study period	$Z < -1.96$, or $1.96 \leq Z < -0.675$ and $SLOPE\% < -0.25\%$
ST	Rainfall erosivity showed no significant change during the study period	$\lvert Z \rvert < 0.675$, or $0.675 \leq \lvert Z \rvert < 1.96$ and $\lvert SLOPE\% \rvert < 0.25\%$
IN-ST	Rainfall erosivity showed an increasing trend in the early stage, but showed a stable trend in the later stage	$0.675 \leq Z < 1.96$ and $SLOPE\% > 0.25\%$ and $R_{ST} < 1$, or $Z \geq 1.96$ and $R_{ST} < 1$
IN	Rainfall erosivity showed a gradual increase trend during the study period	$0.675 < Z < 1.96$ and $SLOPE\% > 0.25\%$ and $R_{ST} = 1$, or $Z \geq 1.96$ and $R_{ST} = 1$

Note: The Z value indicates the MK trend detection value, the *SLOPE%* is the trend change rate percentage, and the R_{ST} refers to the ratio of the rainfall erosivity in the past 3 years to the maximum 3-year moving average. DE: decreasing, ST: stagnating, IN-ST: increasing-stagnating, IN: increasing.

2.3.6. Continuous Wavelet Transform Analysis

The continuous wavelet transform (CWT) is a method to decompose a time series into a two-dimensional phase plane of the time-frequency simultaneously. It is commonly applied to the analysis of various hydrological and meteorological processes with high variability to detect non-stationary trends, periodicities, and durations as it better characterizes oscillatory behaviors of signals than discrete wavelet transforms [86–88]. Specifically, two CWTs, cross wavelet transform (XWT), and wavelet transform coherence (WTC), were constructed to investigate whether there is any periodicities or correlations between rainfall erosivity and ENSO. The XWT reveals regions of high common power in the time-frequency spectrum, and calculates the phase relationships between signals. The WTC identifies two time series variation correlations in both time and frequency space, even in the absence of high-power regions. In this study, the wavelet power spectrum of CWT was employed to analyze the relationship and the possible periodicity between rainfall erosivity in different regions and changing patterns of MEI. XWT revealed high common power regions and phase relationships between the two variables, and WTC was used to determine the correlation position of the two variables at local scales. The CWT toolbox package for MATLAB was used to perform all wavelet analyses. For further details about CWT, refer to [89].

3. Results

In the following section, the results are represented according to the technical approach mentioned in Section 2. This section analyzed the variability characteristics of rainfall erosivity at different time scales and the spatial distribution pattern of rainfall erosivity at each station, while identifying the relationship between rainfall erosivity and ENSO on the TP.

3.1. Variation Characteristics of Annual Rainfall Erosivity

Sen's slope estimation analysis showed an increasing trend of rainfall erosivity on the Tibetan Plateau from 1971 to 2020, with a Sen's slope value of 2.69 for annual rainfall erosivity (Figure 3). The average annual erosivity range from 713.50 to 1495.41 MJ·mm·ha^{-1}·h^{-1}, with a multi-year average erosivity of 1071.42 MJ·mm·ha^{-1}·h^{-1}. An anomaly analysis indicated obvious inter-annual fluctuation in rainfall erosivity. The magnitude of rainfall erosivity undulation was relatively small until 1996 and increased significantly after 1996, with longer fluctuation periods. For the entire study period, the annual rainfall erosivity was above the mean for the same duration as the periods below the mean, with the highest value of 1495.41 MJ·mm·ha^{-1}·h^{-1} in 2020; the lowest value of 713.50 MJ·mm·ha^{-1}·h^{-1} in 2009; the extreme value ratio was 2.1 (Figure S1). Meanwhile, the Mann–Kendall trend analysis had a Z value of 1.67, indicating that this trend passed the significance test at the 90% confidence level, with the mutation point occurring in approximately 2017 (Figure S2).

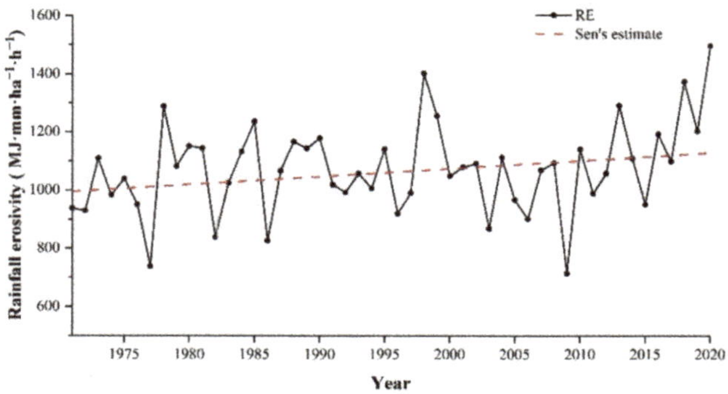

Figure 3. Annual variation in rainfall erosivity on the TP from 1971 to 2020, the red dotted line represents Sen's estimate.

3.2. Changes in Seasonal and Monthly Rainfall Erosivity

The seasonal mean rainfall erosivity showed significant discrepancies at 91 meteorological stations on the TP. The rainfall erosivity in order was summer > autumn > spring > winter, with a range of 89.33–662.58 MJ·mm·ha^{-1}·h^{-1}. In particular, the average rainfall erosivity in summer was the highest, accounting for 60.36%, while winter was the lowest, accounting for only 8.14% of the total annual erosivity (Figure 4). This phenomenon was mainly influenced by the heterogeneity of the seasonal distribution of precipitation. With the transport of water vapor from the North Indian and Western Pacific monsoons, the summer monsoon brought 58.5% of the year's rainfall, while late spring and early autumn accounted for 90% of the year's rainfall [90]. As shown in Figure S3, the summer rainfall erosivity showed a non-significant increasing trend, with the MK statistical value of 1.54. In contrast, there was a decreasing trend in spring, autumn, and winter rainfall erosion; the spring and autumn MK statistic passed the significance test (p = 0.05), which were −2.19 and −2.09, respectively.

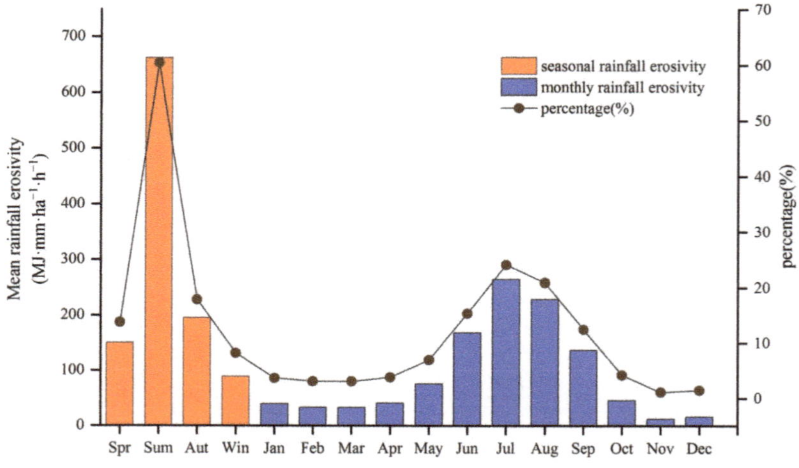

Figure 4. Statistics of seasonal and monthly average rainfall erosivity and its percentage.

Although the monthly average rainfall erosivity was highly variable, there was a clear temporal consistency with the seasonal rainfall erosivity. June, July, and August, corresponding to summer, were the three months with the highest percentage of rainfall and its impact in the future. Next, different scenarios can be set to predict the adjustment of grain planting structure, such as policy intervention, natural disasters and international market environment. Based on this, it is of great significance to explore the impact on the consumption of cropland and water resources for agricultural sustainable development.

Ensuring agricultural production and food security is a systematic project that involves not only natural factors such as water, heat and land, but also social and economic factors such as economic development, urbanization and the international market environment. Therefore, in other words, a sustainable agricultural system is one which is environmentally sound, nonexploitative and which contributes to economic development and social progress [39]. For the stable operation of the agricultural system, it is necessary to encourage the southern regions with rich-water resources to increase the planting area of grain crops and meet its self-sufficiency in grain demand and guide the agricultural areas with water shortage to plant crops in the same season of rain and heat, to improve the spatial matching relationship between the planting of grain crops and cropland and water resources consumption. It is also very important to balance the supply and demand of domestic agricultural products for ensuring food security through international grain market cooperation.

5. Conclusions

In this paper, a mathematical econometric model was established by using a scenario comparison analysis method to analyze the process of grain planting structural adjustment and its impact on the consumption of cropland and water resources from 2003 to 2019. The main conclusions were drawn as follows:

During the study period, China's outputs of major farm products exhibited a marked upward trend, which can be divided into the rapid growth stage (2003–2015) and the fluctuating growth stage (2016–2019). Meanwhile, the grain planting structure has changed greatly and rice was replaced by corn to become the grain crop with the maximum proportion of planting area since 2007. The increase of corn planting structure proportion was concentrated in the northern regions, such as Northeast China Plain, Northern arid and semiarid region and Loess Plateau.

At the national level, the saving of cropland and water resources consumption showed a "cumulative effect" by the adjustment of the grain planting structure. CRC_l and CRCw were 4.62% and 5.72%, respectively. The structural adjustment of grain planting still has a certain effect on saving the consumption of cropland and water resources. From the regional perspective, the impacts of structural adjustment in the northern regions, where cropland resources are relatively abundant but water resources are scarce, are "saving cropland and water", such as Northeast China Plain and Northern arid and semiarid region. Meanwhile, that in the southern regions, where water resources are relatively abundant, but cropland resources are scarce, are "saving cropland but not the water", such as Southern China and Middle-lower Yangtze Plain.

Author Contributions: Conceptualization, Y.Z. and J.W.; methodology, Y.Z. and J.W.; formal analysis, Y.Z.; data curation, Y.Z. and C.D.; writing–original draft preparation, Y.Z.; writing–review and editing, Y.Z. and J.W.; visualization, Y.Z. and C.D.; supervision, J.W. All authors have read and agreed to the published version of the manuscript.

Funding: This research was supported by the Strategic Priority Research Program of the Chinese Academy of Sciences, Grant No. XDA19040402.

Institutional Review Board Statement: Not applicable.

Informed Consent Statement: Not applicable.

Data Availability Statement: Data available in a publicly accessible repository.

Acknowledgments: We would like to thank the Resources and Environmental Sciences and Data Center, Chinese Academy of Science (http://www.resdc.cn/Default.aspx (accessed on 5 March

2021)). We would like to thank anonymous reviewers for their valuable comments and suggestions for improving this paper.

Conflicts of Interest: The authors declare no conflict of interest.

References

1. Tang, Y.; Lu, X.; Yi, J.; Wang, H.; Zhang, X.; Zheng, W. Evaluating the spatial spillover effect of farmland use transition on grain production—An empirical study in Hubei Province, China. *Ecol. Indic.* **2021**, *125*, 107478. [CrossRef]
2. Wei, S.; Liu, Y.; Arowolo, A.; Ying, Z.; Qian, X. Optimal Water Allocation Scheme in Integrated Water-Ecosystem-Economy System. *River Basin Manag.* **2018**, 1–28. [CrossRef]
3. Zhou, C.; Zhang, R.; Ning, X.; Zheng, Z. Spatial-Temporal Characteristics in Grain Production and Its Influencing Factors in the Huang-Huai-Hai Plain from 1995 to 2018. *Int. J. Environ. Res. Public Health* **2020**, *17*, 9193. [CrossRef] [PubMed]
4. Carter, C.A.; Zhong, F.; Zhu, J. Advances in Chinese Agriculture and its Global Implications. *Appl. Econ. Perspect. Policy* **2012**, *34*, 1–36. [CrossRef]
5. Liu, J.; Yang, W.; Chu, J.-H.; Kuo, H.-H.; Analytis, J.G.; Fisher, I.R. Water Sustainability for China and Beyond. *Science* **2012**, *337*, 649–650. [CrossRef] [PubMed]
6. Liu, Y.; Li, J.T.; Yang, Y. Strategic adjustment of land use policy under the economic transformation. *Land Use Policy* **2018**, *74*, 5–14. [CrossRef]
7. Liu, Y.; Wang, D.; Gao, J.; Deng, W. Land Use/Cover Changes, the Environment and Water Resources in Northeast China. *Environ. Manag.* **2005**, *36*, 691–701. [CrossRef]
8. Xu, Y.; McNamara, P.; Wu, Y.; Dong, Y. An econometric analysis of changes in arable land utilization using multinomial logit model in Pinggu district, Beijing, China. *J. Environ. Manag.* **2013**, *128*, 324–334. [CrossRef]
9. Su, W.; Yu, D.; Sun, Z.; Zhan, J.; Liu, X.; Luo, X. Spatial-Temporal Changes in Grain Production, Consumption and Driving Mechanism in China. *J. Integr. Agric.* **2016**, *15*, 1145–1156. [CrossRef]
10. Liu, Y.S.; Wang, J.Y.; Guo, L.Y. The Spatial-Temporal Changes of Grain Production and Arable Land in China. *Sci. Agric. Sin.* **2009**, *42*, 4269–4274.
11. Gao, J.; Liu, Y. Climate warming and land use change in Heilongjiang Province, Northeast China. *Appl. Geogr.* **2011**, *31*, 476–482. [CrossRef]
12. Liu, X.; Liu, Y.; Liu, Z.; Chen, Z. Impacts of climatic warming on cropping system borders of China and potential adaptation strategies for regional agriculture development. *Sci. Total Environ.* **2021**, *755*, 142415. [CrossRef]
13. Zhang, Y.; Yan, Z.; Song, J.; Wei, A.; Sun, H.; Cheng, D. Analysis for spatial-temporal matching pattern between water and land resources in Central Asia. *Hydrol. Res.* **2020**, *51*, 994–1008. [CrossRef]
14. Kuang, W.; Hu, Y.; Dai, X.; Song, X. Investigation of changes in water resources and grain production in China: Changing patterns and uncertainties. *Theor. Appl. Clim.* **2015**, *122*, 557–565. [CrossRef]
15. Tao, J.; Xiaoyu, Q.; Liyan, H. Changes in Grain Production and the Optimal Spatial Allocation of Water Resources in China. *J. Resour. Ecol.* **2016**, *7*, 28–35. [CrossRef]
16. Levers, C.; Butsic, V.; Verburg, P.; Müller, D.; Kuemmerle, T. Drivers of changes in agricultural intensity in Europe. *Land Use Policy* **2016**, *58*, 380–393. [CrossRef]
17. Pan, J.; Chen, Y.; Zhang, Y.; Chen, M.; Shailaja, F.; Bo, L.; Wang, F.; Meng, D.; Liu, Y.; Jiao, L.; et al. Spatial-temporal dynamics of grain yield and the potential driving factors at the county level in China. *J. Clean. Prod.* **2020**, *255*, 120312. [CrossRef]
18. Kravchenko, A.N.; Snapp, S.; Robertson, G.P. Field-scale experiments reveal persistent yield gaps in low-input and organic cropping systems. *Proc. Natl. Acad. Sci. USA* **2017**, *114*, 926–931. [CrossRef] [PubMed]
19. Meng, Q.; Wang, H.; Yan, P.; Pan, J.; Lu, D.; Cui, Z.; Zhang, F.; Chen, X. Designing a new cropping system for high productivity and sustainable water usage under climate change. *Sci. Rep.* **2017**, *7*, srep41587. [CrossRef] [PubMed]
20. Wu, W.; Yu, Q.; You, L.; Chen, K.; Tang, H.; Liu, J. Global cropping intensity gaps: Increasing food production without cropland expansion. *Land Use Policy* **2018**, *76*, 515–525. [CrossRef]
21. Smith, L.E.D.; Siciliano, G. A comprehensive review of constraints to improved management of fertilizers in China and mitigation of diffuse water pollution from agriculture. *Agric. Ecosyst. Environ.* **2015**, *209*, 15–25. [CrossRef]
22. Zhang, Q.; Zhang, F.; Wu, G.; Mai, Q. Spatial spillover effects of grain production efficiency in China: Measurement and scope. *J. Clean. Prod.* **2021**, *278*, 121062. [CrossRef]
23. Erb, K.-H.; Lauk, C.; Kastner, T.; Mayer, A.; Theurl, M.C.; Haberl, H. Exploring the biophysical option space for feeding the world without deforestation. *Nat. Commun.* **2016**, *7*, 11382. [CrossRef] [PubMed]
24. Sun, C.; Zhang, L. Changes in spatial and temporal differences of agricultural product virtual water versus cultivated land in China. *Resour. Sci.* **2009**, *31*, 84–93.
25. Zheng, W.; Luo, B.; Hu, X. The determinants of farmers' fertilizers and pesticides use behavior in China: An explanation based on label effect. *J. Clean. Prod.* **2020**, *272*, 123054. [CrossRef]
26. Tan, Q.; Zhang, S.; Li, R. Optimal Use of Agricultural Water and Land Resources through Reconfiguring Crop Planting Structure under Socioeconomic and Ecological Objectives. *Water* **2017**, *9*, 488. [CrossRef]
27. Li, T.; Long, H.; Zhang, Y.; Tu, S.; Ge, D.; Li, Y.; Hu, B. Analysis of the spatial mismatch of grain production and farmland resources in China based on the potential crop rotation system. *Land Use Policy* **2017**, *60*, 26–36. [CrossRef]

28. Tilman, D.; Balzer, C.; Hill, J.; Befort, B.L. Global food demand and the sustainable intensification of agriculture. *Proc. Natl. Acad. Sci. USA* **2011**, *108*, 20260–20264. [CrossRef]
29. Garnett, T.; Appleby, M.C.; Balmford, A.; Bateman, I.; Benton, T.G.; Bloomer, P.; Burlingame, B.; Dawkins, M.; Dolan, L.; Fraser, D.; et al. Sustainable Intensification in Agriculture: Premises and Policies. *Science* **2013**, *341*, 33–34. [CrossRef]
30. Shi, W.; Tao, F.; Liu, J. Changes in quantity and quality of cropland and the implications for grain production in the Huang-Huai-Hai Plain of China. *Food Secur.* **2012**, *5*, 69–82. [CrossRef]
31. D'Amour, C.B.; Reitsma, F.; Baiocchi, G.; Barthel, S.; Güneralp, B.; Erb, K.-H.; Haberl, H.; Creutzig, F.; Seto, K.C. Future urban land expansion and implications for global croplands. *Proc. Natl. Acad. Sci. USA* **2017**, *114*, 8939–8944. [CrossRef] [PubMed]
32. Jiang, L.; Deng, X.; Seto, K.C. The impact of urban expansion on agricultural land use intensity in China. *Land Use Policy* **2013**, *35*, 33–39. [CrossRef]
33. Clapp, J.; Moseley, W.G. This food crisis is different: COVID-19 and the fragility of the neoliberal food security order. *J. Peasant. Stud.* **2020**, *47*, 1393–1417. [CrossRef]
34. Seleiman, M.F.; Selim, S.; Alhammad, B.A.; Alharbi, B.M.; Juliatti, F.C. Will novel coronavirus (Covid-19) pandemic impact agriculture, food security and animal sectors? *Biosci. J.* **2020**, *36*, 1315–1326. [CrossRef]
35. Li, Y.; Wang, L.-E.; Liu, G.; Cheng, S. Rural household food waste characteristics and driving factors in China. *Resour. Conserv. Recycl.* **2021**, *164*, 105209. [CrossRef]
36. Qun, Y. Influence of Food Consumption Structure Change on the Development of Agricultural Economical Development in the Accelerated Urbanization of China. *Adv. J. Food Sci. Technol.* **2015**, *7*, 455–458.
37. Li, J.; Liu, Y.; Hai, X.; Shangguan, Z.; Deng, L. Dynamics of soil microbial C:N:P stoichiometry and its driving mechanisms following natural vegetation restoration after farmland abandonment. *Sci. Total Environ.* **2019**, *693*, 133613. [CrossRef]
38. Zhou, T.; Koomen, E.; Ke, X. Determinants of Farmland Abandonment on the Urban–Rural Fringe. *Environ. Manag.* **2020**, *65*, 369–384. [CrossRef]
39. Sarah, V.; Julia, L.; Nicolas, J.; Jens, N. What Is Sustainable Agriculture? A Systematic Review. *Sustainability* **2015**, *7*, 7833–7865.

International Journal of
Environmental Research and Public Health

MDPI

Article

The Impact of Urban Development Intensity on Ecological Carrying Capacity: A Case Study of Ecologically Fragile Areas

Jinjing Hu [1], Yong Huang [1,2,*] and Jie Du [3]

[1] School of Architecture and Urban Planning, Chongqing University, Chongqing 400045, China; joyameteor@163.com
[2] Key Laboratory of New Technology for Construction of Cities in Mountain Area (Chongqing University), Ministry of Education, Chongqing 400045, China
[3] School of Microelectronics and Communication Engineering, Chongqing University, Chongqing 400045, China; dujie@cqu.edu.cn
* Correspondence: cqhy2001@126.com

Abstract: In ecologically fragile areas, an uncontrolled increase in urban development intensity (UDI) will erode the ecological carrying capacity (ECC). This study aimed to explore the relationship between UDI and ECC and quantify the impacts of UDI on ECC. The Three Gorges Reservoir Area (Chongqing section) was chosen for the case study. Firstly, the UDI and ECC were comprehensively evaluated. Then, the coupling coordination relationship between the two was analyzed by a coupling coordination degree model. Finally, the influences of UDI on the coordinated development of the two were analyzed by a geographically weighted regression model. The results show that the distributions of UDI and ECC are opposite; UDI and ECC are mutually restricted to some extent. UDI and ECC are moderately coupled and poorly coordinated, and a higher UDI is mostly correlated to a higher coordination degree of UDI and ECC. In areas with higher UDI, an appropriate control on population and economy may benefit the coordinated development. Meanwhile, in areas with lower UDI, the promotion of population aggregation and economic investment would enhance the coordinated development between UDI and ECC. This study could optimize the dimensional control of UDI, which contributes to the long-term sustainability of ecologically fragile areas.

Keywords: ecologically fragile area; ecological carrying capacity; urban development intensity; the Three Gorges Reservoir Area (Chongqing section); coupling coordination degree model; geographically weighted regression model; ordinary least squares model; pressure-state-response model

Citation: Hu, J.; Huang, Y.; Du, J. The Impact of Urban Development Intensity on Ecological Carrying Capacity: A Case Study of Ecologically Fragile Areas. *Int. J. Environ. Res. Public Health* **2021**, *18*, 7094. https://doi.org/10.3390/ijerph18137094

Academic Editors: Nir Krakauer and Tom Cole-Hunter

Received: 1 April 2021
Accepted: 25 June 2021
Published: 2 July 2021

1. Introduction

According to the annual report of the China Council for international cooperation in environment and development (CCICED) (2012), more than 360 million people are living in ecologically fragile areas, accounting for about one-quarter of the total population of China [1]. Meanwhile, there is a certain overlap between ecologically fragile areas and poverty-stricken areas [2–4]. Therefore, equal attention to urban development in ecologically fragile areas is necessary for the sustainable development of the whole region.

Ecological fragile areas are located in the transition zone of two different types of ecosystems [5]. They are more sensitive to climate change, weak in anti-interference, and prone to various ecological problems [6,7]. Lots of cities in ecologically fragile areas are facing challenges [7–9]. For example, various human activities (e.g., urban construction) tend to have a huge impact on the environment, even posing a threat to ecological security [7,9]. Cities in ecologically fragile areas are in urgent need of development, which could alleviate poverty. Therefore, for cities in ecologically fragile areas, how to achieve reasonable urban development yet prevent environmental problems and ensure ecological security is an important practical question.

To address this issue, we need to investigate the impacts of urban development on the environment. Recently, many scholars have described the process of urban development from the perspectives of urbanization [10–12], urban expansion [13,14], land-use change [7,15], and urban forms [16], and explored the various impacts of urban development on the environment. However, these descriptions, focused on the physical changes of urban development, do not apply to the real situation of cities in ecologically fragile areas. In particular, due to the constraints of a complex terrain, the change in physical space in ecologically fragile areas is not as significant an influencing factor as the socioeconomic and demographic level [7,17]. Therefore, physical and socioeconomic explanations should be suitable for revealing the realistic changes of urban development in ecologically fragile areas.

Urban development intensity (UDI) can be defined as the multiple impacts of various human activities on urban areas [18–20], including land-use intensity, population density, economic intensity, etc. This term reflects the compound state of urban land use, population, and economic development in a specific period [19,20]. It can well reflect the compound changes of urban physical space and socioeconomic level, making it applicable to cities in ecologically fragile areas. Furthermore, UDI has been widely used in urban planning, land use, and other fields [18,20,21], and the control and guidance of UDI are a direct reflection of relevant planning and policy [22]. Therefore, considering UDI as the basis to reveal the relationship between urban development and the ecological environment would be beneficial for urban managers and policymakers in ecologically fragile areas.

At present, there are some studies on the impact of UDI on the environment. It is pointed out that excessive land-use intensity will reduce biodiversity [15,23,24]. Similarly, an increase in land-use intensity would reduce the ecosystem regulation services [7,14]. In addition, an increase in UDI leads to an increase in carbon dioxide emissions [16,19,25], which may aggravate the urban heat island effect. Additionally, land-use intensity is positively correlated with the $PM_{2.5}$ concentration [26,27], which means a negative impact on air quality. These studies give some suggestions for the control and guidance of UDI. However, most of the evaluation factors for the impact selected in the above works are relatively one-sided, and attention to ecological security issues is missing. Therefore, they are not suitable for revealing the practical problems of ecologically fragile areas.

Ecological carrying capacity (ECC) [28], i.e., the environmental carrying capacity, is an important index to reflect the state of the environment. According to the theory of urban complex ecosystem [29], it is also the ability of the ecosystem to provide services, prevent ecological problems, and protect regional ecological security. Those are exactly what cities in ecologically fragile areas are concerned about. Specifically, the ECC is the comprehensive "social–economic–natural" capacity of the urban complex ecosystem [30–32], including support capacity, supply capacity, and coordination capacity. Therefore, it can reflect the sustainable development ability of a region and has been widely developed in urban planning, resource and environmental management, regional development, and other fields.

The evaluation methods of ECC are rich, some of which can be applied to a variety of different research scenarios [28,33,34]. In particular, the state space method, such as the pressure-state-response (PSR) model [35–37], can better reflect the dynamic changes of regional ECC in a certain period. This is conducive to the prevention of environmental problems caused by the weakening of ECC in advance, which is very useful for ecologically fragile areas. Thus, it is appropriate to consider the ECC assessed by the PSR model as the basis for revealing the impact of UDI on the environment in ecologically fragile areas.

The Three Gorges Reservoir Area (Chongqing section) [38,39] is located in the upper reaches of the Yangtze River, accounting for more than 85% of the whole Three Gorges Reservoir area. As an ecotone of the karst landscape ecosystem and the karst forest ecosystem in Southwest China, the Three Gorges Reservoir Area (Chongqing Section) is a typical ecologically fragile area [6,40]. Due to the complex geological conditions, the area is easily affected by various natural disasters, e.g., landslides, bank collapses, debris flows, etc. [39]. At present, research on the Three Gorges Reservoir Area (Chongqing Section) mainly focuses on the evaluation of environmental sensitivity [6], landscape change [41], and ecosystem services in the area [42], with less attention paid to an analysis of the relationship between urban development and the environment. However, urban development in the Three Gorges Reservoir Area (Chongqing section) also encounters challenges faced by other cities in ecologically fragile areas. As a case study, it has typicality and can reveal the impact of UDI on ECC.

To demonstrate the impact of UDI on ECC, it is also necessary to analyze the relationship between UDI and ECC. However, little research has been conducted on the relationship between UDI and ECC. Most research has been conducted on the relationship between urban development and the environment, which can also indirectly illustrate the relationship between UDI and ECC [10,12,43]. Meanwhile, the coupling coordination degree model has been widely used to quantify the relationship between urbanization and the environment (hereafter called relationships). For instance, relationships in the whole country [12,44], river basin [10,45], and city [44,46] have been analyzed from the macro, middle, and micro perspectives. It was found that the relationships of most regions in China was in a state of moderately coupled and weakly coordinated [12]. Subsequently, we found that there were obvious spatial differentiations in the coordination degree between urban development and the environment [12,45]. Specifically, the coordination degree in the eastern region of China with a mature urban development level was higher than that in the middle and western regions, and this trend had been constantly strengthening from 2005 to 2016 [12]. Similarly, in research from 2021 [45], it was found that the relationships of most areas in the Pearl River Delta were in a middle coupling and coordination state from 2000 to 2015. Some cities located in the relatively central parts of the Pearl River Delta, i.e., developed cities, had a higher coordination degree than the surrounding areas, forming a core-periphery spatial distribution pattern. Furthermore, we found that this phenomenon was not only affected by the spatial distribution of the cities, but also seemed to be relevant to the urban development trajectories. In 2017, a case conducted in Shanghai [46], a developed city in China, found that the coordination degree of urbanization and the environment was growing gradually from a barely coordinated stage to a highly coordinated stage from 1980 to 2013. Therefore, we believe that there is a linkage between the urban development trajectory and coordination degree growth: the higher the level of urban development, the higher the coordination degree.

These previous studies have enriched our understanding of the interaction between UDI and the ECC based on the relationship between urban development and the environment. It is mostly defined as a coupling and coordination relationship [44–47]. However, most of the studies focus on urban developed areas, while less attention is paid to developing, ecologically fragile areas. Thus, there is a lack of targeted research that can better reveal the practical problems in ecologically fragile areas. Meanwhile, it has become a common phenomenon that in areas with higher urban development, such as developed cities, the coordination degree of urbanization and the environment is higher than in other areas. However, they have, as a result, failed to provide clear and adequate evidence of what caused that common phenomenon, which is critical for further explanation of the specific impact of urban development on the environment, as well as the impact of UDI on the ECC.

To further explore the relationship between UDI and ECC in ecologically fragile areas, this paper takes the Three Gorges Reservoir Area (Chongqing section) as a typical case to explore the relationship between UDI and ECC in 2010, 2014, and 2019. First, we establish a comprehensive index system to evaluate UDI and ECC. Then, we reveal the coupling coordinated relationship between UDI and ECC by a coupling coordinated degree model. Finally, we analyze the impact of UDI on the coordination degree of UDI and ECC by a geographic weighted regression model. It is expected that the specific impact of UDI on ECC in ecologically fragile areas can be revealed to provide valuable reference suggestions for urban development strategies and management measures in ecologically fragile areas. Based on this, the urban development practice can be optimized, so that a win–win situation between high-quality urban development and environmental protection can be achieved in ecologically fragile areas.

The aims of this paper can be summarized as follows: (1) We use UDI and ECC as crucial indicators to evaluate the realistic situation of urban development and the environment in ecologically fragile areas from a dynamic perspective. (2) Motivated by this, we analyze the spatiotemporal distribution characteristics of UDI and ECC, and explore the reasons behind this phenomenon, as well as the potential relationship between UDI and ECC. (3) Then, we analyze the spatiotemporal distribution characteristics of the coupling coordination relationship between UDI and ECC, as well as the reasons behind this phenomenon. (4) Based on this, we present the in-depth influences of different realistic factors of UDI (e.g., economy, population) on the coordinated development of UDI and ECC in the ecologically fragile areas in the long term. (5) Finally, we propose some effective ways to achieve coordinated development for cities in ecologically fragile areas.

2. Materials and Methods

2.1. Study Area

The Three Gorges Reservoir Area (Chongqing section) is mainly located at the end of the upper reaches of the Yangtze River [40,48]. Geographical coordinates range from 28°28′–31°44′ N and 105′N°49′–110°12′ E. It involves 22 districts and counties in Chongqing, starting from Jiangjin District in the west, Wushan County in the East, Wulong District in the south, Wuxi County, and Kaizhou District in the north, as shown in Figure 1. The whole Chongqing section covers an area of 46,158.53 km^2 [38], accounting for 80% of the Three Gorges Reservoir Area.

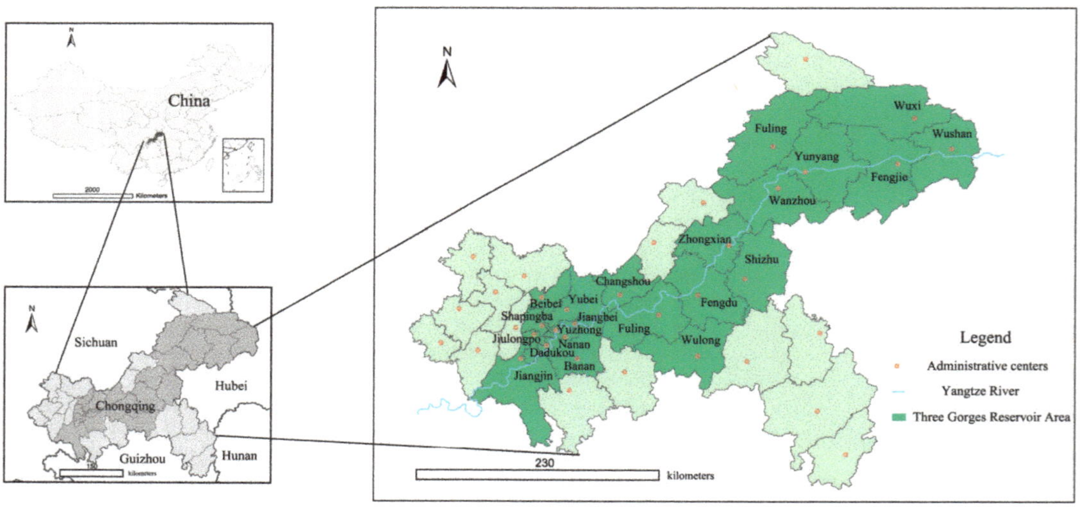

Figure 1. Study area. Sources: Standard Chinese Maps issued by the Ministry of Natural Resources (China) in 2020.

According to the National Plan for the Protection of Ecologically Fragile Areas issued in 2008 [42], the Three Gorges Reservoir area (Chongqing section) is located in the southwest karst rocky desertification ecologically fragile area. It is often affected by various natural disasters, such as debris flow, soil erosion, etc. Meanwhile, the Three Gorges Reservoir area plays an important role in national ecological security. As an ecological barrier area with important ecological functions, the ecological security of this area is related to the overall security of the entire upper reaches of the Yangtze River [38].

By the end of 2009, the Three Gorges Project was completed, and the Resettlement Project in the Three Gorges Reservoir Area (Chongqing section) smoothly came to an end [42]. The Resettlement Project led to rapid urbanization in this area and caused some problems for the environment. Therefore, at the end of 2013, the government positioned the Three Gorges Reservoir Area (Chongqing section) as an Ecological Conservation and Development Zone [49]. With more emphasis on environmental protection and ecological development, this area entered the new era of urbanization transformation.

To authentically illustrate the development track of the Three Gorges Reservoir Area (Chongqing section), the years 2010, 2014, and 2019 are selected as three key time points of this research.

2.2. The Interaction between UDI and ECC

Based on previous studies [10,12,43], urban development and the environment are mutually constrained and coordinated in a long time series. Therefore, the relationship between UDI and ECC can be defined as a coupling coordination relationship. That is to say, there is a dynamic equilibrium between the UDI and ECC. Under the impact of UDI, the ECC shows a law of resilience and constantly converts from a balance to a new balance for a certain period of time. We summarize the coupling mechanism between UDI and ECC and represent it in Figure 2. The specific explanation is as follows.

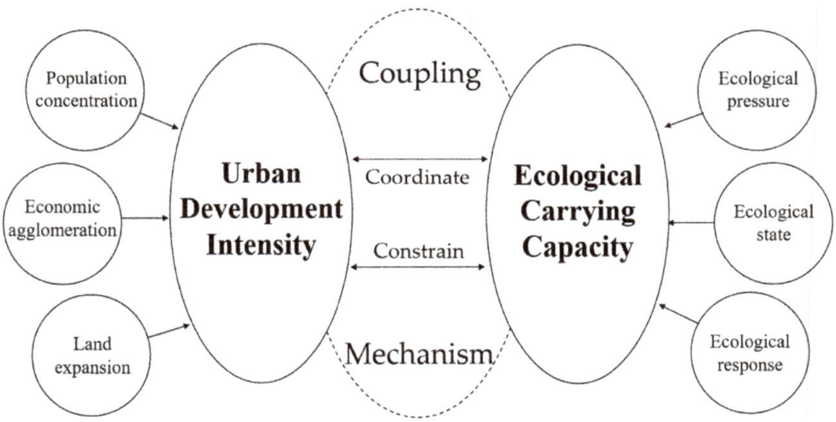

Figure 2. Coupling mechanism between UDI and ECC. Source: Summarized according to the literature [10,12,43].

For instance, if the UDI continues to increase without regulation, the environmental pressure would be exacerbated, especially for ecologically fragile areas [50,51]. With the development of human society, human residences have moved from the countryside to the city, the population has grown excessively, and the urban space has expanded. Following this, the construction area of the city gradually expanded, and more and more skyscrapers have sprung up. These activities resulted in the erosion of ecological spaces and the continuous reduction of ecological resources such as green space and cultivated land. The ecological security of the whole area will be under serious threat. As a result, the ECC would inevitably decrease, and the initial equilibrium state would become imbalanced.

On the other hand, urban development is inseparable from the rigid constraints of the environment [12,43]. The construction and expansion of a city, the growth of the population, and the thriving of the economy inevitably consume many types of environmental resources: land resources, water resources, forest resources, etc. When the consumption of various resources exceeds the carrying capacity, it will affect the development of agriculture, industry, and even tertiary industry. What is worse, in extreme cases, it can lead to terrible natural disasters, such as soil erosion and drought. In that situation, the process of urban development has to slow down, or even stop. Then, the urban development intensity decreases.

In addition, a coupling coordination degree (CCD) model could well reflect the mutual influence and coordinated development of the two systems [52,53], which has been widely used to analyze the coupling coordination relationship between urban development and the environment [44–47]. It is also suitable for analyzing the relationship between UDI and ECC.

2.3. Research Framework

We took 22 districts and counties in the Three Gorges Reservoir Area (Chongqing section) as representatives, using the data from 2010, 2014, and 2019 to explore the interaction between UDI and ECC in ecologically fragile areas.

First, we comprehensively evaluated UDI from the aspects of population, economy, and land use in 2010, 2014, and 2019. We used a PSR model to assess the ECC in 2010, 2014, and 2019. In addition, the space–time characteristics of UDI and ECC were represented, which is useful for discussing the potential relationship between them.

Secondly, based on the understanding of the relationship of UDI and ECC, we used a coupling coordination degree (CCD) model to analyze the coupling coordination relationship between UDI and ECC in ecologically fragile areas, and the coupling degree of UDI and ECC (UDI–ECC coupling degree) and the coordination degree of UDI and ECC (UDI–ECC coordination degree) were obtained.

Then, we further explored the specific impact of the UDI on the coordinated development of UDI and ECC. The geographic weighted regression (GWR) model was adopted, which could embed its geographical coordinates into the model during a regression analysis and form a regression coefficient related to the location [54,55]. Thus, it could well reveal the impact of UDI on the UDI–ECC coordination in the three dimensions of population, economy, and land use.

Finally, according to the analysis results, we provide a discussion and give our conclusions. The research framework is shown in Figure 3.

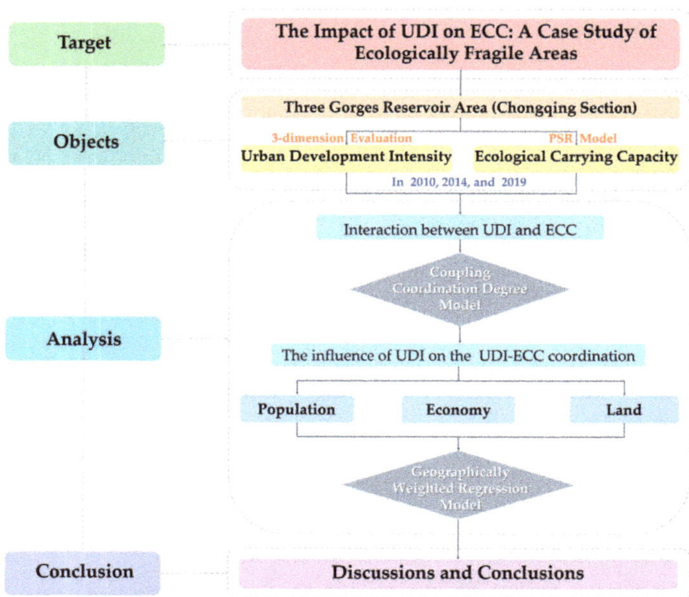

Figure 3. Research framework. Sources: Established by the authors.

2.4. Methods

2.4.1. Evaluations of UDI and ECC

Evaluation of UDI

UDI reflects the compound state of urban land use, population, and economic development in a specific period [18–20]. Considering previous studies on the measurements of UDI [18–20] and the realistic situation in this paper, the comprehensive evaluation index system of UDI constructed included three first-grade indicators (population concentration intensity, economic agglomeration intensity, and land-use intensity). In addition, two second-grade indicators were selected under the first-grade indicators. The final index system is illustrated in Table 1, including six indicators.

Table 1. Evaluation index system of UDI and ECC.

Target Layer	Element Layer	Index Layer	Unit	Weight	Direction
Urban development intensity (UDI)	Population concentration intensity	Population density	Person/km^2	0.2455	+
		Urbanization rate	%	0.1116	+
	Economic agglomeration intensity	Economic density	10,000 yuan/km^2	0.1759	+
		Per capita GDP	Ten thousand yuan	0.1129	+
	Land-use intensity	Per capita construction land area	100 m^2/person	0.1324	+
		Proportion of urban built-up area in total area	%	0.2217	+
Ecological carrying capacity (ECC)	Ecological pressure	Per capita industrial waste water discharge	T/person	0.1036	–
		Per capita industrial solid waste discharge	10,000 m^3/person	0.0917	–
		Per capita industrial emission	T/person	0.0937	–
		Energy consumption per 10,000 yuan output value	T standard coal/person	0.1262	–
	Ecological state	Green coverage rate of built-up area	%	0.0656	+
		Forest coverage	%	0.0832	+
		Per capita cultivated land area	Hm^2/person	0.0889	+
		Per capita water resources	10,000 m^3/person	0.1666	+
	Ecological response	Standard rate of industrial waste water discharge	%	0.0581	+
		Synthesis utilization rate of industrial waste	%	0.0558	+
		Proportion of environmental protection investment in GDP	%	0.0666	+

Source: Chongqing Statistical Yearbook (the districts and counties data section) from 2011, 2015, and 2020 (the Statistical Yearbook for a certain year only contains the data up to the previous year, which was decided by the Statistics Department in China); Historical Remote Sensing Data of 22 districts and counties for 2010, 2014, and 2019; Environmental Statistics Bulletin and Water Resources Statistics Bulletin of Chongqing from 2010, 2014, and 2019. The missing data were obtained by interpolation method.

Based on the multifactor comprehensive evaluation method, we also constructed a measurement model of UDI as follows:

$$f(x_i) = \sum_{i=1}^{n} x_i \cdot W_j \quad (1)$$

where n is the number of research units; $f(x_i)$ is the UDI value of the i-th unit; x_i is the standardized value of the i-th unit on the j-th index; and W_j is the index weight of the j-th index.

Referring to previous studies [18–20] and the real situation of the case, the calculation results of UDI were classified with the Jenk's Natural Breaks Classification method, as shown in Table 2.

Table 2. Classification standard of UDI.

Low Level	Middle Level	High Level	Over High Level
<0.22	0.22~0.96	0.96~1.93	>1.93

Sources: Literature [18–20].

Evaluation of ECC

The ECC is the comprehensive capacity of the social–economic–natural complex ecosystem [30–32], including support capacity, supply capacity, and coordination capacity. The pressure-state-response (PSR) model can well reflect the dynamic changes of this comprehensive capacity in a certain period of time, which is helpful for assessing ecologically fragile areas. Therefore, we used the PSR model (Figure 4) to establish a comprehensive evaluation method for ECC.

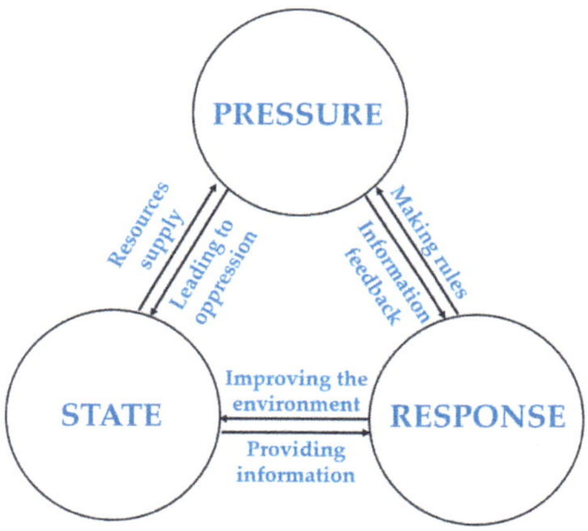

Figure 4. The framework of the PSR model. Source: Summarized by the authors according to the literature [35–37].

The PSR model is composed of three parts [35–37]. The first part is the pressure level, i.e., the impacts on the environment caused by the activities that exert some pressure on the system. The second part is the state level, i.e., the description of the current situation of the system. The last one is the response level, which refers to the response to the system in the form of laws, regulations, and standards or environmental management behavior.

Based on the framework of the PSR model and previous studies [35–37] on the measurements of ECC, four indicators were selected to represent the pressure level, four to represent the state level, and three to represent the response level. The final index system is shown in Table 1, including 11 indicators.

At the same time, based on the multifactor comprehensive evaluation method, a measurement model of ECC was constructed. The details are as follows:

$$g(y_i) = \sum_{i=1}^{n} y_i \cdot W_j \tag{2}$$

where $g(y_i)$ is the ECC value of the i-th unit; y_i is the standardized value of the i-th unit on the j-th index; and W_j is the index weight of the j-th index.

Referring to the previous studies [35–37] and the realistic situation of the case, the calculation results of ECC are classified with Jenk's Natural Breaks Classification method, as in the following Table 3.

Table 3. Classification standard of ECC.

Low Level	Middle Level	Higher Level	High Level
<0.34	0.34~0.85	0.85~1.19	>1.19

Sources: Literature [35–37].

Index Weight and Direction

The entropy method and the analytic hierarchy process (AHP) were used to calculate the weight of each index, which could achieve a good balance between subjective and objective [36,52]. The calculation formula is as follows:

$$W_j = aW_m + (1 - a)W_n, \tag{3}$$

where n is the number of research units; W_j is the combination weight; W_m is the weight given by the AHP; W_n is the weight given by the entropy method; a is the proportion of the AHP weight in combination weight, with a value of 0.5; and $1 - a$ is the proportion of the entropy method in combination weight.

The positives and negatives of all the indicators were determined according to whether their roles are beneficial or harmful in the system in reality, as shown in Table 3.

2.4.2. Coupling Coordination Degree Model

The coupling coordination degree model can well analyze the coupling coordination relationship between UDI and ECC [44–47]. It consists of a coupling degree model, a comprehensive evaluation index, and a coordination degree model.

Coupling Degree

Coupling degree can measure the interaction degree between the system or elements. To measure the coupling degree of UDI and ECC quantitatively, according to the coupling coordination model, the measurement model of the UDI–ECC coupling degree is constructed as follows:

$$C = 2\left\{f(x) \times g(y)/(f(x) + g(y))^2\right\}^{1/2}, \tag{4}$$

where C is the coupling degree, f(x) and g(y) are the index system function of UDI and the index system function of ECC, respectively.

Comprehensive Evaluation Index

The comprehensive evaluation index indicates the comprehensive development level of UDI and ECC. The function is as follows,

$$T = af(x) + bg(y) \tag{5}$$

where a and b respectively represent the weight of UDI and ECC. According to previous studies [44–47] and comprehensive consideration, this paper considers that they have the same contribution, so a = b = 0.5 in the formula.

Coordination Degree

Coordination degree was constructed from the comprehensive evaluation index T and the coupling degree C, which can well measure the coordinated development level of the whole system [44–47]. The function is as follows:

$$D = \sqrt{C \bullet T}, \tag{6}$$

where D is the coordination degree. Moreover, $C \in [0,1]$, $D \in [0,1]$, the coupling degree is directly proportional to the C value, and the coordination degree is directly proportional to the D value. When C = 1, the UDI and ECC reach the optimal coupling state. When D = 1, UDI and ECC reach the optimal coordination state.

At the same time, referring to the existing research results, the calculation results of the coupling degree and coordination degree were classified by the Equal-interval Classification method [44–47], as in Table 4.

Table 4. Classification standard of coupling coordination degree.

Coupling Degree C Value	Coupling Degree	Coordination Degree D Value	Coordination Level
$0.8 < C < 1$	Highly coupled	$0.8 < D \leq 1$	Highly coordinated
$0.5 < C \leq 0.8$	Moderately coupled	$0.6 < D \leq 0.8$	Moderately coordinated
$0.3 < C \leq 0.5$	Weakly coupled	$0.4 < D \leq 0.6$	Weakly coordinated
$0 < C \leq 0.3$	Barely coupled	$0 < D \leq 0.4$	Barely coordinated

Sources: Literature [44–47].

2.4.3. Moran's I

Testing the spatial autocorrelation of dependent variables is a prerequisite for the application of the geographically weighted regression model [54,55]. We used the global Moran's I to test whether there was a significant spatial correlation between the coordination degree of UDI and ECC. The specific calculation formula is as follows:

$$I = \frac{n \sum_{i=1}^{n} \sum_{j=1}^{n} W_{ij}(x_i - \overline{x})(x_j - \overline{x})}{(\sum_{i=1}^{n} \sum_{j=1}^{n} W_{ij}) \sum_{i=1}^{n} (x_i - \overline{x})^2} \tag{7}$$

where n is the number of research units; x_i and x_j are the observed value of the marked i and j of research units; and W_{ij} is the spatial weight matrix. If the research units are adjacent, the result is 1; if they are not adjacent, the result is 0.

When $I > 0$, it means that the results are positively correlated, which represents that the research units belong to the aggregation spatial layout, and there is an autocorrelation. When $I < 0$, it means that the results are negatively correlated, which indicates that the research units and the research units belong to a decentralized spatial layout.

The global Moran's I exponential statistics are generally tested by constructing value statistics (The Z value is the deviation between the attribute of element i and its average value); the calculation formula is as follows:

$$Z = \frac{I - E(I)}{\sqrt{var(I)}}, \tag{8}$$

where E(I) is the expected value, var(I) is the variance, and the evaluation standard of the Z value is as in Table 5.

Table 5. Value evaluation criteria.

Z Value (Standard Deviation)	*p* Value (Probability)	Confidence
<−1.65 or >+1.65	<0.10	90.0000%
<−1.96 or >+1.96	<0.05	95.0000%
<−2.58 or >+2.58	<0.01	99.0000%

Sources: Literature [54,55].

2.4.4. Geographically Weighted Regression Model

The geographically weighted regression (GWR) model is a spatial econometric regression analysis method [54,55]. It can reflect different influences of a variable in the region based on different geographical coordinates.

Therefore, this paper analyzes the spatiotemporal differentiation characteristics of the impact of UDI on the UDI–ECC coordination by the GWR model. The function of the model is presented as follows:

$$y_i = a_{io}(u_i, v_i) + \sum_{i=1}^{n} a_{ik}(u_i, v_i) \cdot x_{ik} + \varepsilon_i \tag{9}$$

where the coordinate of the i-th point is (u_i, v_i); x_{ik} is the independent variable of the ith point; $a_{io}(u_i, v_i)$ is the estimated value of the constant term of the i-th point, $a_{ik}(u_i, v_i)$ is the estimated value of the regression parameter of the k-th independent variable at the i-th point; n is the number of regression terms, and ε_i is the residual correction term.

To verify the fitting effect of the GWR model in this paper, the fitting parameters of the GWR model will be compared with the traditional least square linear regression model. The ordinary least squares (OLS) model is a global linear regression model, which uses the best fitting line method to analyze the relationship between explained variables and explanatory variables [55]. The calculation formula is as follows:

$$y_i = \beta_o + \sum_{k=1}^{n} \beta_k x_{ik} + \varepsilon_i, \tag{10}$$

where y_i is the value of the dependent variable at i-th point; x_{ik} is the value of the k-th independent variable at the i-th point; k is the number of independent variables; β_o is the constant term; β_k is the regression coefficient of the k-th independent variable; n is the number of regression terms, ε_i is the residual.

In this paper, the weight function of geographically weighted regression is calibrated by the adaptive method, and the minimum Akaike information criterion (AIC) method [54,55] is used to determine the bandwidth. The Akaike information criterion (AIC) is a technique that measures the goodness of an estimated statistical model [56]. As for the calculation results, the corrected Akaike information criterion (*AICc*) can help to compare different regression models. R^2 (the goodness of fit) [57] indicates the degree of explanation of the regression equation to the changes of dependent variables. Adjusted R^2 is the calculation result of variable compensation based on R^2, which could minimize the calculation error [57].

Both geographically weighted regression analysis and least square linear regression analysis were completed in ArcGIS 10.2 (ESRI, Redlands, CA, USA).

2.5. Data

2.5.1. Data Resources

The social and economic data mainly came from the Statistical Yearbook published by the Chongqing Municipality or the governments of each district and county (in the Three Gorges Reservoir) from 2010 to 2020 (the statistical yearbook for a certain year only contains data up to the previous year, which was decided by the Statistics Department in

China). The environmental data mainly came from the Environmental Statistical Bulletin of Chongqing and the Water Resources Statistical Bulletin of Chongqing from 2010 to 2019. The spatial data mainly came from the geospatial cloud data of Chongqing from 2010 to 2019. The missing data were obtained by interpolation using the nearest year's data, to ensure the authenticity and integrity of the data.

2.5.2. Standardization Treatment

Due to the differences in dimensions, meanings, and attributes of the original data, it was necessary to eliminate the influence caused by the differences. Thus, we used the Z-score Equation (10) to standardize the original data, as follows:

$$x_i' = \frac{x_i - \overline{x}}{\sqrt{\sum_{i-1}^{n} (x_i - \overline{x})^2/(n-1)}} \tag{11}$$

where x_i is the original index value and x_i' is the standardized value of x_i.

3. Results

3.1. Spatiotemporal Variations of UDI

Based on the evaluation index system of UDI, we obtained the UDI values of the 22 districts and counties in the Three Gorges Reservoir Area (Chongqing section) for the years 2010, 2014, and 2019. Then, we analyzed their spatial distributions and temporal changes by ArcGIS 10.2 (ESRI, Redlands, CA, USA) and SPSS 25 (IBM, Armonk, NY, USA).

As shown in Figure 5, in 2010, 2014, and 2019, the cities with relatively high UDI were mostly distributed in the upper reaches of the Three Gorges Reservoir Area, close to the main urban area of Chongqing. On the contrary, the cities with relatively low UDI were mostly located in the lower reaches of the Three Gorges Reservoir Area. Yuzhong District had the highest UDI value, and the average of the UDI over the three years reached 7.82, which means it stayed at an over high level. Wuxi County had the lowest UDI value, and the UDI average in the three years was as low as 0.147, which means it stayed at a low level. This distribution phenomenon may be affected by the urban development pattern of Chongqing. In particular, most of the central urban areas of Chongqing were located in the upper reaches of the Three Gorges Reservoir Area, the UDI values of which were relatively high.

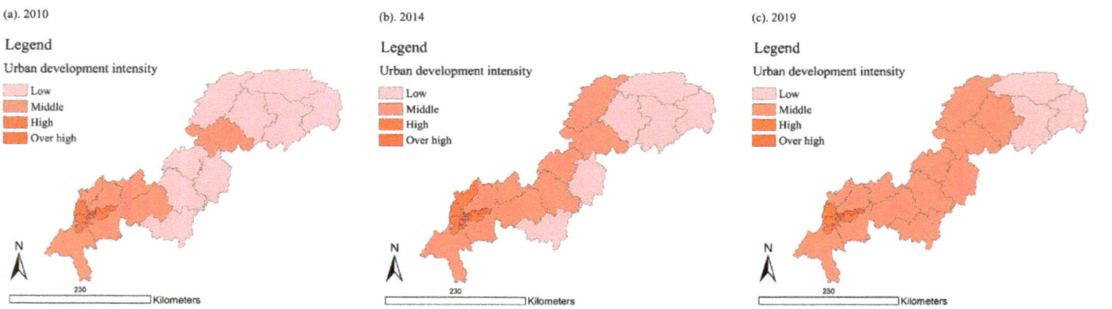

Figure 5. Spatial distribution of UDI in the Three Gorges Reservoir Area (Chongqing section) in year 2010, 2014, and 2019 (presented in subfigure **a–c**, respectively). Sources: Calculation results of UDI in the Three Gorges Reservoir Area (Chongqing section).

As shown in Figure 6, the variations of UDI average values of the counties were not significant in 2010, 2014, and 2019, which indicates that the UDI in the Three Gorges Reservoir Area was nearly constant on the whole. However, the UDI variance of districts and counties in the Three Gorges Reservoir Area (Chongqing section) was about 3.4 in 2010 and 2.92 in 2019, a decline of 0.48. That is, the UDI differences between districts and counties narrowed over time. This indicates that the western development projects in the western region of China have indeed achieved corresponding results for the past 10 years.

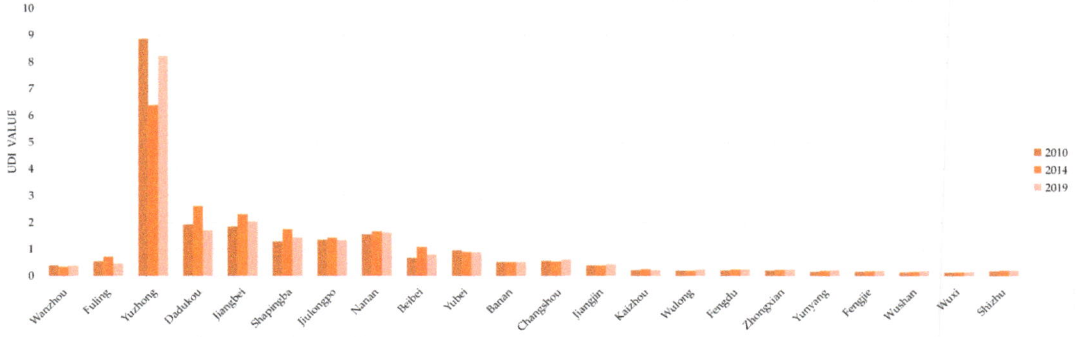

Figure 6. Changes of UDI in the Three Gorges Reservoir Area (Chongqing section). Sources: Calculation results of UDI in the Three Gorges Reservoir Area (Chongqing section).

On the whole, the spatial distribution of UDI in the Three Gorges Reservoir Area (Chongqing section) is high upstream and low downstream. In the terms of time, the differences between districts and counties are shrinking, and a balanced development trend between districts and counties is appearing.

3.2. Spatiotemporal Variations of ECC

After calculation, the ECC values in 2010, 2014, and 2019 were obtained. Then, the spatial distributions and temporal variations were analyzed by ArcGIS 10.2 and SPSS 25.

As shown in Figure 7, the results show that the areas with relatively high ECC in the Three Gorges Reservoir Area (Chongqing section) were mostly distributed in the middle–lower reaches. In 2010, 2014, and 2019, Wuxi County had the highest ECC value with an average value of 1.77, which means it stayed at a high level. Yuzhong District had the lowest ECC value with an average value of 0.39, which means it stayed at a low level. This situation was opposite to the distribution of UDI.

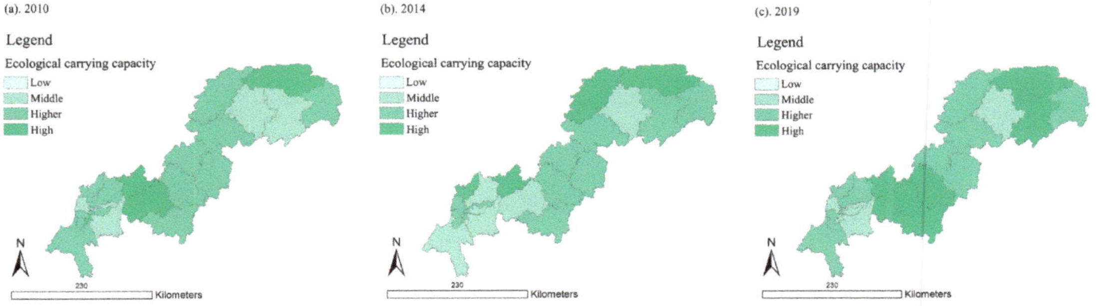

Figure 7. Spatial distribution of ECC in the Three Gorges Reservoir Area (Chongqing section) in year 2010, 2014, and 2019 (presented in subfigure **a–c**, respectively). Sources: Calculation results of ECC in the Three Gorges Reservoir Area (Chongqing section).

As shown in Figure 8, the average ECC values of the Three Gorges Reservoir Area (Chongqing section) in 2010, 2014, and 2019 were all about 1, which means almost the whole area was in the higher level for ECC. The variance values in 2010, 2014, and 2019 all stayed at about 0.1, indicating that the distributions of ECC between districts and counties were relatively concentrated and stable.

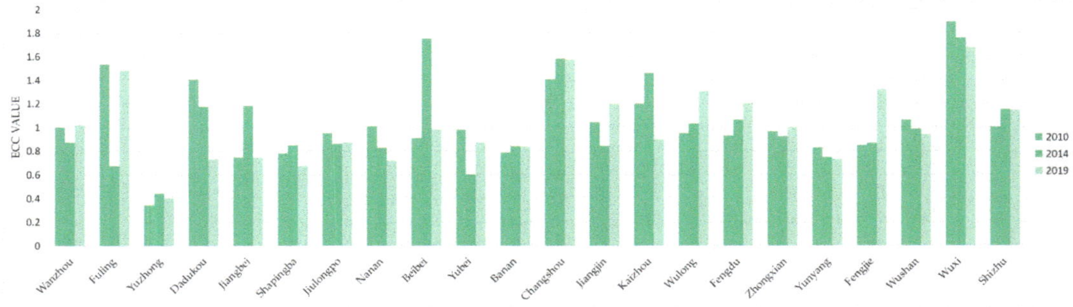

Figure 8. Changes of ECC in Three Gorges Reservoir Area (Chongqing section). Sources: Calculation results of ECC in Three Gorges Reservoir Area (Chongqing section).

In summary, the ECC of the Three Gorges Reservoir Area (Chongqing section) formed a spatial pattern of lower upstream and higher downstream. In terms of time, the ECC stayed at a high level, and the changes of ECC were not significant. At the same time, the high UDI areas had low ECC, while the low UDI areas had high ECC.

3.3. Coupling Coordination Degree between UDI and ECC

In 2010, 2014, and 2019, the UDI–ECC coupling degree in the Three Gorges Reservoir Area (Chongqing section) remained at a medium or low level.

As shown in Figure 9, Wuxi County and Yuzhong District had the lowest UDI–ECC coupling degree, showing that they were barely coupled. The districts in the high coupling stage were Nan'an District, Dadukou District, Shapingba District, Jiangbei District, and Jiulongpo District. Their UDI–ECC coupling degree values were all over 0.8 in 2010, 2014, and 2019. Meanwhile, most of them were distributed in the upper reaches, close to the main area of Chongqing.

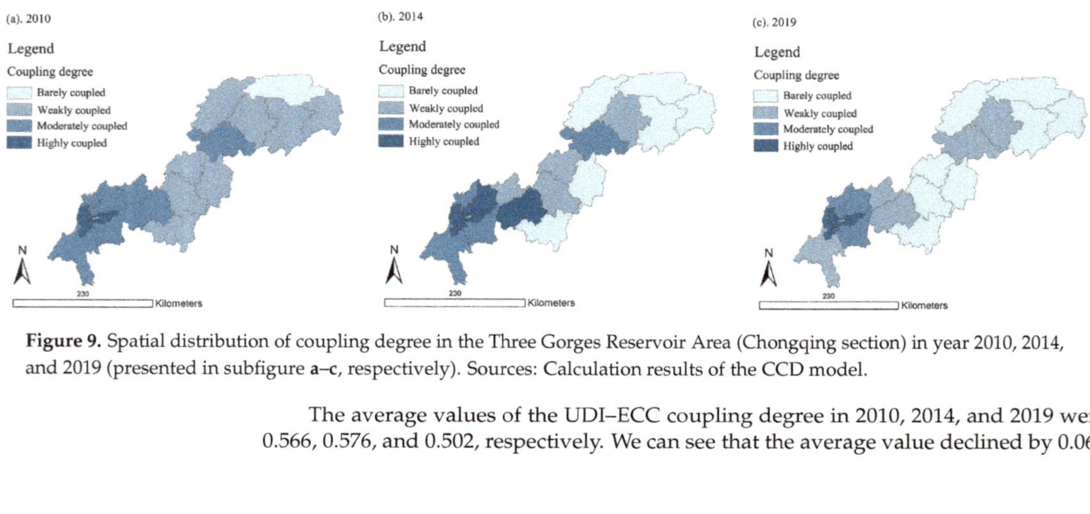

Figure 9. Spatial distribution of coupling degree in the Three Gorges Reservoir Area (Chongqing section) in year 2010, 2014, and 2019 (presented in subfigure **a–c**, respectively). Sources: Calculation results of the CCD model.

The average values of the UDI–ECC coupling degree in 2010, 2014, and 2019 were 0.566, 0.576, and 0.502, respectively. We can see that the average value declined by 0.064

from 2010 to 2019. This may indicate that the overall UDI–ECC coupling degree of the Three Gorges Reservoir Area (Chongqing section) had a slight downward trend.

As shown in Figure 10, in 2010, 2014, and 2019, the UDI–ECC coordination degrees of the Three Gorges Reservoir Area (Chongqing section) were mostly at a low level. Specifically, there were 13, 15, and 13 districts and counties that had a UDI–ECC coordination degree values lower than 0.4, in 2010, 2014, and 2019, respectively. Meanwhile, most of the districts and counties with relatively low coordination were distributed in the lower reaches, and parts of them were distributed in the middle and upper reaches.

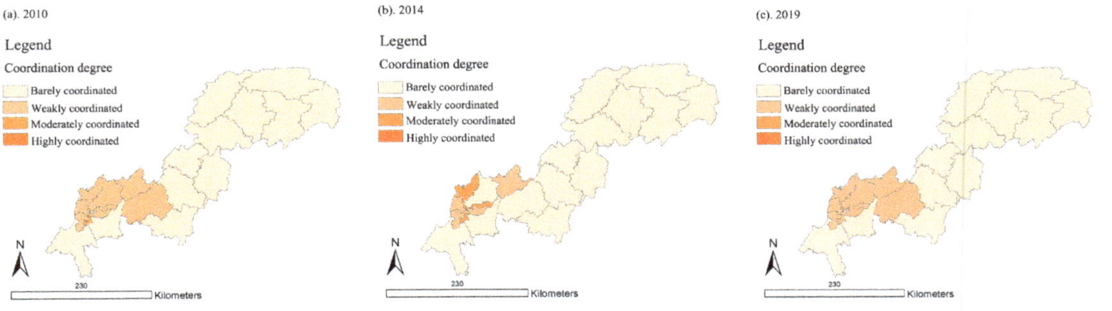

Figure 10. Spatial distribution of coordination degree in the Three Gorges Reservoir Area (Chongqing section) in year 2010, 2014, and 2019 (presented in subfigure **a**–**c**, respectively). Sources: Calculation results of the CCD model.

On the whole, the average values of UDI–ECC coordination were 0.377, 0.367, and 0.338 in 2010, 2014, and 2019, respectively, showing a decreasing trend over time. This demonstrates that the UDI–ECC coordination degree value declined over the past 10 years, and most of the districts and counties in the Three Gorges Reservoir Area (Chongqing section) were in a barely coordinated state.

As we can see from Figure 11, it is worth noting that the UDI–ECC coupling degrees and UDI–ECC coordination degrees of Yuzhong District (high UDI, low ECC), Wuxi County (low ECC, high ECC), and Wushan County (low ECC, high ECC) were all at a relatively low level in 2010, 2014, and 2019. This indicates that the UDI–ECC coupling degrees and UDI–ECC coordination degrees of these areas were barely coupled and weakly coordinated for a long time.

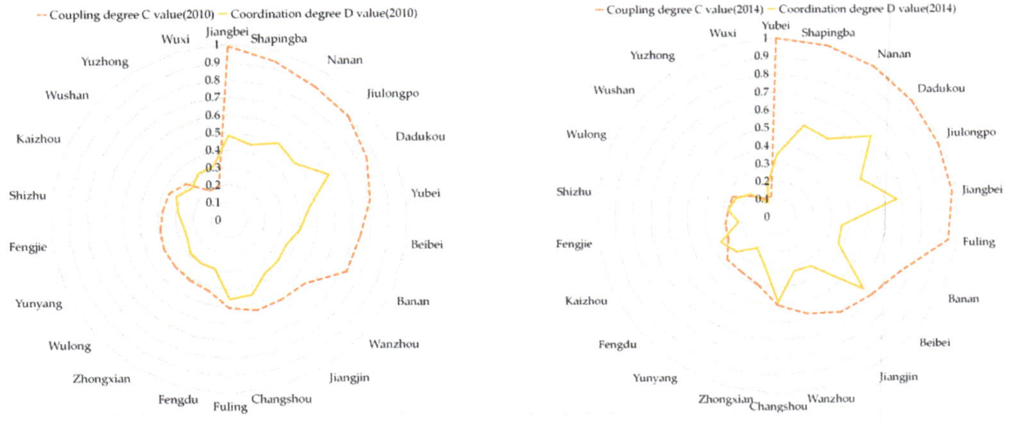

(a) (b)

Figure 11. *Cont.*

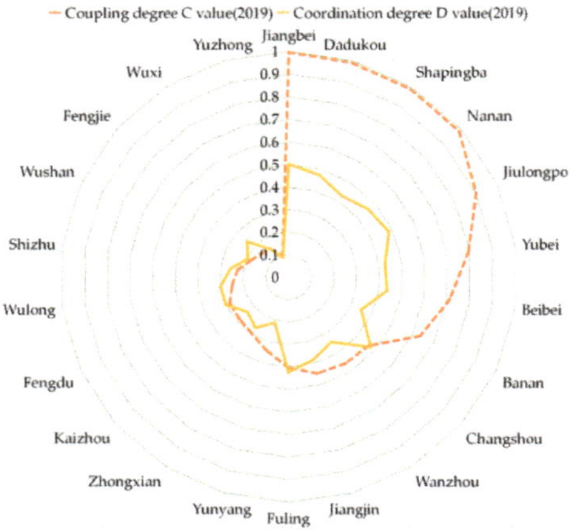

(c)

Figure 11. Comparison of UDI-ECC coupling degree and UDI–ECC coordination degree in 2010 (**a**), 2014 (**b**), and 2019 (**c**). Sources: Calculation results of the CCD model.

Therefore, the values of UDI and ECC should be kept in a relatively balanced range, which is more helpful for the coordination of UDI and ECC.

3.4. The Impacts of UDI on UDI–ECC Coordination Degree

According to the relevant research [11,18,20], the coordination degree value in the CCD model can well measure the coordinated development level of the whole system. Therefore, the UDI–ECC coordination degree can be used to represent the coordinated development levels between UDI and ECC.

After a comparison with the previous results in Sections 3.1 and 3.3, we found that the coordination degree of the regions with higher UDI is also higher. It is worth exploring what kind of impact UDI has on the UDI–ECC coordination degree.

To figure out the answers, this section will use a GWR model to analyze the internal influence of the UDI on the UDI–ECC coordination degree in different dimensions.

3.4.1. Spatial Autocorrelation Test

Based on the evaluation criteria in Table 5, the global Moran's I values of the UDI–ECC coordination degree were greater than 0 in 2010, 2014, and 2019, indicating that the spatial distribution of the UDI–ECC coordination had a positive spatial correlation. As shown in Table 6, in 2010, 2014, and 2019, the Z values were 5.225, 3.282, and 3.708, respectively, and higher than the test value of 2.58. In other words, they were all in the 99% confidence interval of normal distribution. There was indeed a significant spatial correlation between the coordination degree of UDI and ECC.

Table 6. Moran's I value of coordination in Three Gorges Reservoir Area (Chongqing section).

Year	Moran's I	Z Value (Standard Deviation)	p Value (Probability)	Confidence
2010	0.506	5.225	0	99%
2014	0.304	3.282	0.001	99%
2019	0.347	3.708	0.0002	99%

Sources: Calculation results of the global Moran's I.

3.4.2. Selection of Independent Variables and Model

For the independent variables, we first selected six indicators from the three dimensions of population, economy, and land: urbanization rate, population density, economic density, per capita GDP, construction land proportion, and per capita construction land area. After the significance test and multicollinearity test, only the indicators population density, per capita GDP, and per capita construction land area were chosen to represent the UDI from the dimensions of population, economy, and land, respectively.

Motivated by this, we used the GWR model and OLS model to analyze the impacts of UDI on UDI–ECC coordination in 22 districts and counties in 2010, 2014, and 2019, respectively. The calculation results of the GWR model and the OLS model are shown in Table 7. By comparison, we found that R^2 (the goodness of fit) of the GWR model were higher than that of the OLS model, and *AICc* were lower than that of the OLS model.

Table 7. The fitting parameters of the GWR model and OLS model (2010, 2014, and 2019).

Year	Index	GWR Model			OLS Model		
		AICc	R^2	Adjusted R^2	*AICc*	R^2	Adjusted R^2
2010	Population density	−48.4281	0.5502	0.4978	23.4375	0.4638	0.4370
	Per capita GDP	−46.8069	0.5175	0.4594	33.9026	0.4481	0.4205
	Per capita construction land area	−82.6492	0.8951	0.8898	3.5383	0.8951	0.8898
2014	Population density	−21.2195	0.4136	0.3581	30.4726	0.3539	0.3216
	Per capita GDP	−21.5387	0.4222	0.3669	32.4790	0.3776	0.3465
	Per capita construction land area	−42.8567	0.7662	0.7544	18.4766	0.7662	0.7544
2019	Population density	−52.3502	0.81	0.7568	109.4140	0.1081	0.0636
	Per capita GDP	−40.9364	0.6403	0.5625	34.7369	0.0588	0.0117
	Per capita construction land area	−58.5241	0.7879	0.7773	11.8778	0.7879	0.7773

Sources: Calculation results of the GWR model. Note: A lower *AICc* value means the model better fits the observed data. The R^2 value varies from 0 to 1. The larger the value, the better. The adjusted R^2 is usually lower than the R^2 value. Its evaluation standard is the same as for R^2.

Furthermore, we found that the fitting parameters of the OLS model and GWR model were almost unchanged with per capita construction land area as the independent variable. However, the fitting parameters of the two models changed significantly when population density or per capita GDP was an independent variable. This indicates that the impacts of UDI on UDI–ECC coordination under the land-use dimension may not have spatial differences, and the spatial differences were more reflected in the impacts of population and economy.

Therefore, in this study, the GWR model has a better fitting effect and a stronger explanation of geographic differences than the OLS model. It is appropriate to choose the GWR model to explain the impacts of UDI on UDI–ECC coordination.

3.4.3. Impacts of UDI on UDI–ECC Coordination Degree

The impacts of UDI on the UDI–ECC coordination degree in different dimensions are analyzed by the GWR model in this section.

For the UDI in population dimension (P-UDI), the average regression coefficients in 2010, 2014, and 2019 were 0.1214, 0.1667, and 0.1644, respectively, which indicates that the impact of P-UDI on the UDI-ECC coordination was gradually strengthened in the Three Gorges Reservoir Area (Chongqing section).

As shown in Figure 12, in 2010 and 2014, the regression coefficients in 22 districts and counties were all positive, and the values in the upstream area were lower than in the downstream area. However, in 2019, the impact of P-UDI on the UDI-ECC coordination appeared an obvious spatial differentiation. In the middle and upper reaches of the Three

Gorges Reservoir Area (Chongqing section), there were weak/negative values, while in the lower reaches, there were strong/positive values.

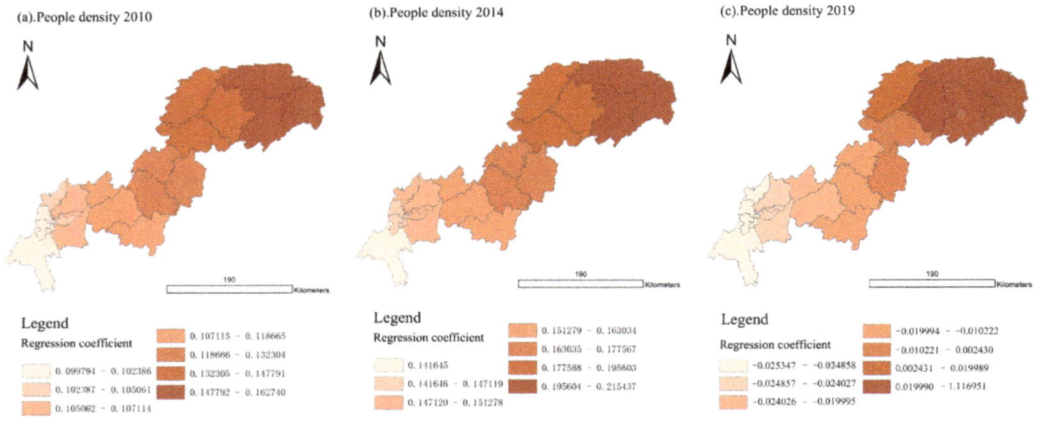

Figure 12. Spatial distribution of regression coefficient of P-UDI in year 2010, 2014, and 2019 (presented in subfigure **a**–**c**, respectively). Sources: Calculation results of the GWR model from the population dimension.

Specifically, as Figure 13 shows, in 2019, the P-UDI regression coefficients in sixteen districts and counties changed from positive to negative, and only six remained positive. In particular, the values of four districts and counties in Northeast Chongqing, Yunyang, Fengjie, Wushan, and Wuxi, increased significantly from 2014 to 2019. This shows that the impact of P-UDI on the UDI-ECC coordination has changed from a positive effect to a negative effect in most areas.

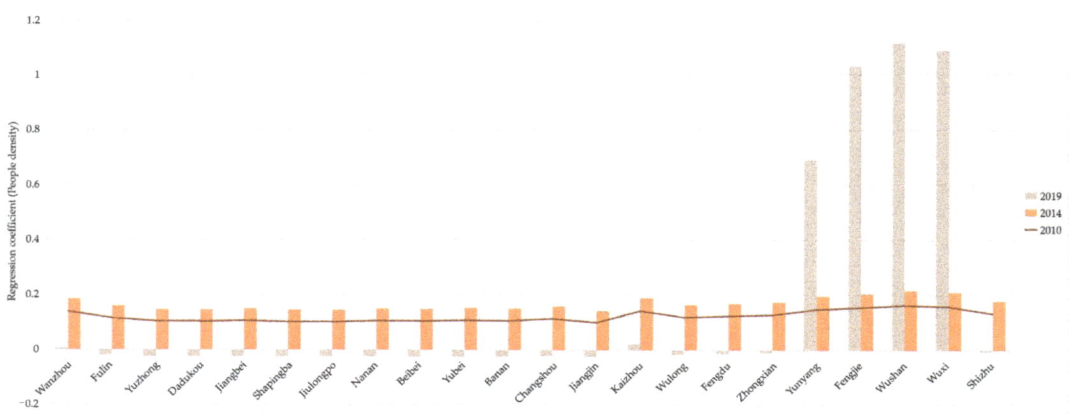

Figure 13. Changes of regression coefficient of P-UDI in the Three Gorges Reservoir Area (Chongqing section). Sources: Calculation results of the GWR model from the population dimension.

Therefore, for good coordination in the Three Gorges Reservoir Area (Chongqing section), those areas with negative values should appropriately control the excessive population aggregation, while those with positive values should appropriately guide the population growth.

For UDI in the economic dimension (E-UDI), the average regression coefficients of 2010, 2014, and 2019 were 0.0943, 0.1536, and 0.045, respectively, indicating that the E-UDI on the UDI–ECC coordination first strengthened and later weakened over the past 10 years.

As shown in Figure 14, similar to the population dimension, in 2010 and 2014, the regression coefficients of 22 districts and counties in the Three Gorges Reservoir Area (Chongqing section) were positive, and the upper reaches were lower than the lower reaches. When it came to 2019, the impact of E-UDI on the UDI–ECC coordination showed an obvious spatial differentiation, i.e., the values were negative in the upper reaches and positive in the middle and lower reaches.

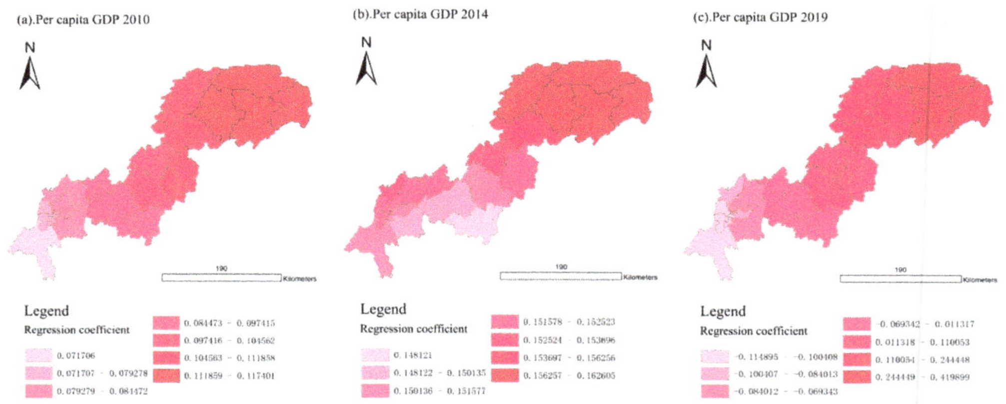

Figure 14. Spatial distribution of regression coefficient of E-UDI in year 2010, 2014, and 2019 (presented in subfigure **a**–**c**, respectively). Sources: Calculation results of the GWR model from the economic dimension.

Specifically, as Figure 15 shows, in 2019, the E-UDI regression coefficients of twelve districts and counties changed from positive to negative, and only ten remained positive. In particular, the values of five districts and counties located in the middle and lower reaches, including Kaizhou, Yunyang, Fengjie, Wushan, and Wuxi, have increased significantly during 2014–2019. This shows that the impact of E-UDI on the UDI–ECC coordination has changed from positive effect to negative effect in more than half of the regions in 2019.

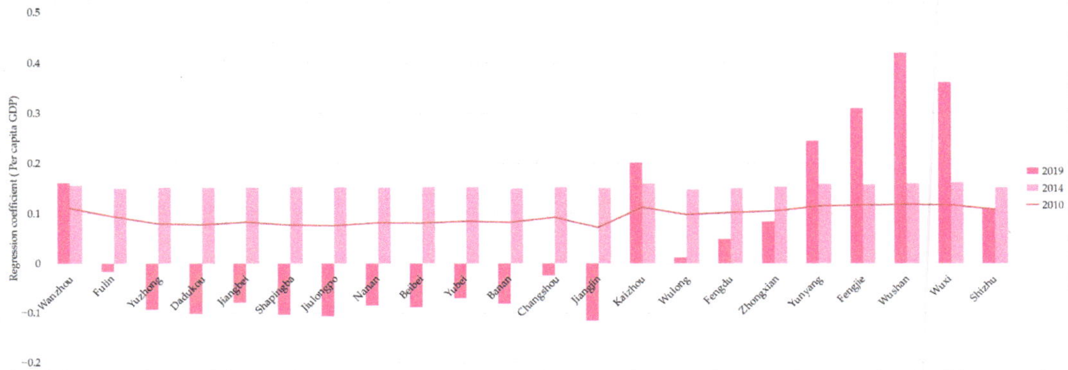

Figure 15. Changes of regression coefficient of E-UDI in the Three Gorges Reservoir Area (Chongqing section). Sources: Calculation results of the GWR model from the economic dimension.

Therefore, we should pay attention to ensuring a balanced economic development both in the upstream and downstream areas of the Three Gorges Reservoir Area (Chongqing section), and increasing investment in underdeveloped areas, so as to improve the coordination development for the whole region.

For UDI in the land-use dimension (L-UDI), the average regression coefficients of L-UDI in 2010, 2014, and 2019 were 0.1255, 0.2042, and 0.1685, respectively, indicating that the impact of L-UDI on UDI–ECC coordination degree was significantly enhanced from 2010 to 2014. From 2014 to 2019, urbanization in the Three Gorges Reservoir Area (Chongqing section) entered the transition period, and the growth of UDI slowed down. Therefore, the regression coefficients of L-UDI in 2019 were lower than those in 2014. To summarize, L-UDI has a low positive correlation with UDI–ECC coordination.

As shown in Figure 16, in 2010, 2014, and 2019, the regression coefficients of 22 districts and counties in the Three Gorges Reservoir area were all positive, and the values of regression coefficients were nearly close. Furthermore, as shown in Figure 17, in 2010, 2014, and 2019, the variance of the L-UDI regression coefficient of the whole region remained around 0. Therefore, there was no spatial differentiation. These were consistent with our inference in Section 3.4.2.

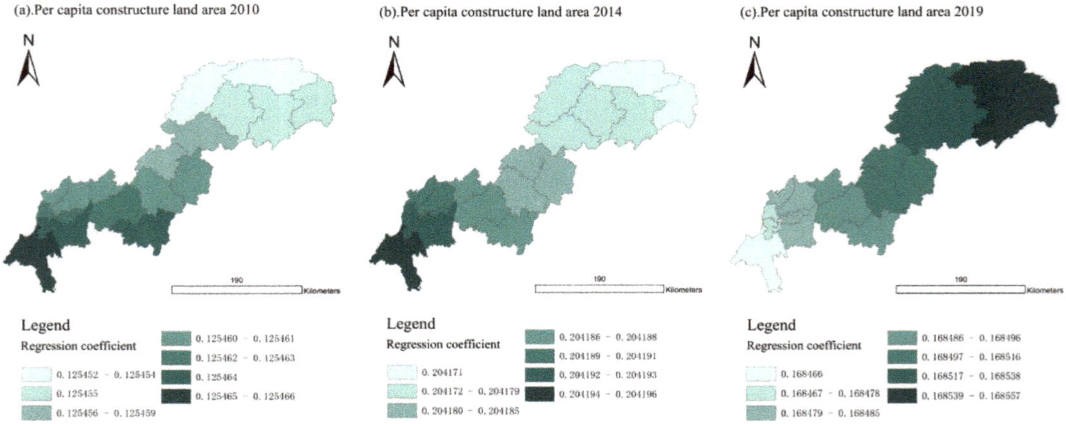

Figure 16. Spatial distribution of regression coefficient of L-UDI in year 2010, 2014, and 2019 (presented in subfigure **a**–**c**, respectively). Sources: Calculation results of the GWR model from the land-use dimension.

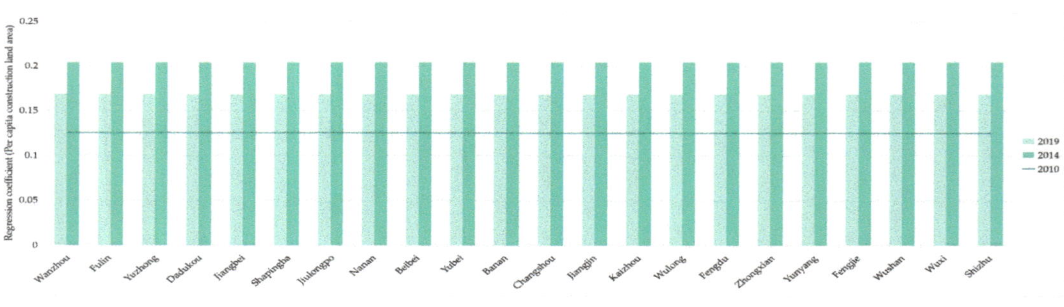

Figure 17. Changes of regression coefficient of L-UDI in the Three Gorges Reservoir Area (Chongqing section). Sources: Calculation results of the GWR model from the land-use dimension.

In addition, the regression coefficient of L-UDI has been kept at a low positive value in the past 10 years, which indicates that the appropriate improvement of L-UDI will be conducive to UDI–ECC coordination.

By comparing the mean values of the regression coefficient of UDI in different dimensions, we can find that the mean value of the L-UDI regression coefficient (0.166) > the mean value of the P-UDI regression coefficient (0.1508) > the mean value of E-UDI regression coefficient (0.0977). This shows that the UDI–ECC coordination degree was affected more by L-UDI and P-UDI than by E-UDI. What is more, the impacts of the dimensions on the UDI–ECC coordination degree all showed a low degree of positive correlation, which means that an appropriate increase in UDI may be helpful for the improvement of the UDI–ECC coordination degree.

4. Discussion

The fundamental spatial distribution pattern of ECC is determined by the natural environment [28,58]. The terrain is high in the east and low in the west in the Three Gorges Reservoir Area (Chongqing section) [38]. The area with steep terrain has more abundant natural resources, such as forest, water, etc. In this study, a stable spatial distribution with relatively low ECC upstream and relatively high ECC downstream reflects this point. The spatial distribution of ECC is also affected by social and economic factors [30,32]. In 2008, the National Council released the National Ecological Function Zoning, identifying the Three Gorges Reservoir Area as one of 25 key ecological function areas [42]. Then, in 2013, the Chongqing municipal government issued the Ecological Conservation Function Zone Planning, mentioning the important role of the Three Gorges Reservoir area (Chongqing section) in maintaining the overall ecological quality [42]. The varying of ECC in the Three Gorges Reservoir Area (Chongqing section) over the past 10 years is not significant, and the whole ECC stays at a high level. Therefore, we can see that these policies indeed played a positive role, and we should pay more attention to the implementation of relevant policies in the future.

Previous studies suggested that effective land use is of positive significance in terms of ecosystem services and ECC [7]. The results of this study demonstrate the same points. In the Three Gorges Reservoir Area (Chongqing section), the UDI values were relatively high upstream and relatively low downstream in 2010, 2014, and 2019. With the release of a series of policies [42], such as the Chengdu Chongqing Urban Agglomeration Planning in 2016, etc., the districts and countries in the upper reaches have had more economic investment and opportunities, and the UDI has developed faster than in the lower reaches. Meanwhile, the ECC has maintained a high level over the past 10 years, which indicates that reasonable guidance of UDI is conducive to maintaining a high ECC. However, it is worth noting that areas with extremely high UDI are linked to extremely low ECC. Therefore, we speculate that the UDI and ECC may be mutually restricted within a certain range, but this situation is not obvious outside the range.

Based on a further analysis of the relationship between UDI and ECC, we found that the UDI–ECC coupling degree and UDI–ECC coordination degree of the Three Gorges Reservoir Area (Chongqing section) were mostly in the state of moderate coupling and weak coordination in 2010, 2014, and 2019. Previous studies [12] have found that the coupling coordination relationship in most regions of China was at a low level. Therefore, the results of this study also verify this point, indicating that there are some contradictions between urban development and the environment in many parts of China, especially in ecologically fragile areas.

Previous studies [10,45,46] have shown that a higher urban development level, such as in the Yangtze River Delta and Pearl River Delta, is linked to a higher coordination degree between urban development and the environment. However, the specific impact of UDI on the coordination degree between urban development and the environment requires further analysis. In this study, we found that the areas with relatively high UDI correlate to the higher UDI–ECC coordination degree—that is, also the area close to the main urban area of Chongqing. At the same time, the UDI–ECC coordination degree often stayed at a low level in areas where the values of UDI and ECC were extreme (too high or too low). This verifies our initial assumption that too high a UDI value will restrict the ECC. Therefore, we believe that the UDI level can promote the coordinated development between UDI and ECC to a certain extent. The values of UDI and ECC in the region should be kept to a relatively narrow range, not too high or too low, for the coordinated development of urban spaces and the environment.

Moreover, we have analyzed the specific impact of UDI on UDI–ECC coordination by the GWR model. The UDI of different dimensions and UDI–ECC coordination degree all had a weakly positive correlation in 2010, 2014, and 2019. In particular, the UDI changes in population and land had a greater positive impact on UDI–ECC coordination than changes in the economy. This means that the reasonable improvement of UDI was, to a certain extent, conducive to the UDI–ECC coordination. This confirms our hypothesis about the impact of UDI on UDI–ECC coordination, and applies to most areas in the Three Gorges Reservoir Area (Chongqing section). Motivated by this, we also found that the impacts of P-UDI and E-UDI on the UDI–ECC coordination degree had obvious spatial differentiations, while the impacts of L-UDI on the UDI–ECC coordination degree had no spatial differentiation yet showed a slightly positive correlation in the whole region. Therefore, we put forward some corresponding suggestions for improvement based on the specific research results, as follows.

Specifically, in the regions with relatively high UDI, i.e., the upper reaches areas of the river, the impacts of P-UDI on the UDI–ECC coordination degrees went from positive to negative over time. This shows that the improvement in P-UDI used to have a positive effect on the degree of coordination in a certain range, but now has a moderately negative effect. The impacts of E-UDI on UDI–ECC coordination showed a similar development trajectory, and were relatively lower than those of P-UDI. Therefore, in areas with high UDI, more attention should be paid to controlling overpopulation and encouraging a balanced economic distribution, which will be useful for the coordinated development of urban spaces and the environment.

For regions with relatively low UDI, i.e., the lower reaches, the impacts of P-UDI and E-UDI on UDI–ECC coordination were positive and had a tendency to strengthen. In particular, the impacts of P-UDI were greater than those of E-UDI. This indicates that appropriate improvements in UDI in the population and economy will be conducive to UDI–ECC coordination in areas with low UDI. Therefore, we should improve the UDI in the region appropriately, such as by leading to population aggregation and enhancing the economic investments, to facilitate the coordinated development of urban spaces and the environment.

5. Conclusions

In this paper, taking the Three Gorges Reservoir Area (Chongqing section) as a typical ecologically fragile area, the UDI and ECC in this area in 2010, 2014, and 2019 were evaluated, and the internal relationship of UDI and ECC was analyzed using the CCD and GWR models. On this basis, the impact of UDI on the environment was discussed in depth.

Specifically, we found that: (1) The distributions of UDI and ECC are different. The UDI and ECC may be mutually restricted to some extent. (2) UDI and ECC are mostly moderately coupled and lowly coordinated. Extreme UDI and ECC values are linked to extreme coordination degrees. (3) The UDI–ECC coordination degree tends to be higher in areas with higher UDI. However, it is not suitable for a case with extreme values. (4) UDI can promote the coordinated development of UDI and ECC to some extent. (5) The UDI changes in population and land have a greater impact on UDI–ECC coordination than those in the economy. (6) For areas with different UDI, the measurements to promote the coordinated development of UDI and ECC should be different. Specifically, in areas with higher UDI, i.e., the upper reaches, more attention should be paid to controlling overpopulation and encouraging a balanced economy distribution, which will be conducive to the coordinated development of urban spaces and the environment. Meanwhile, in areas with lower UDI, i.e., the lower reaches, promotion of population aggregation and economic investment will encourage the coordinated development of urban spaces and the environment.

The results of this study have revealed the relationship between UDI and ECC and the specific impact of UDI on the coordinated development of the two, and enriched our theoretical understanding of the impact of urban development on the environment. This is helpful for city managers and policymakers in ecologically fragile areas to formulate different control measures for UDI under different dimensions and promote the realization of long-term sustainable development and a virtuous circle.

This paper has some limitations. The impacts of UDI in different dimensions on the coordinated development of UDI and ECC have been revealed. We also found that some areas in the Three Gorges Reservoir Area (Chongqing section) are facing extremely imbalanced coordinated degree values, if let be, which will be detrimental to the ecological security for the whole region. However, due to the limitations of the methods and the short time selected, the specific desirable range of UDI remains to be further studied, especially for the imbalanced areas. Therefore, we expect to conduct targeted research in imbalanced areas in the future, using system models such as neural network algorithms and system dynamics models to calculate the numerical rational range of UDI from a long time series. This will be useful for the dynamic monitoring of UDI, and potential ecological security problems can be prevented in time.

Author Contributions: J.H. participated in all phases; Y.H. helped in conceiving the research, project administration; J.D. helped in software, technical assistance, reviewing, and editing the paper. All authors have read and agreed to the published version of the manuscript.

Funding: This research is fund by National Key R&D Program of Science and Technology (Grant No.: 2018yfd1100804); Postgraduate Research and Innovation Foundation of Chongqing (Grant No.: cys20030).

Data Availability Statement: Data available on request due to restrictions eg privacy or ethical. The data presented in this study are available on request from the corresponding author. The data are not publicly available due to [Part of the data is secondary processing data, involving a large amount of original data and so many types of data, which are not suitable for forming a unified database.].

Acknowledgments: The authors sincerely thank the editors and the anonymous reviewers for their constructive.

Conflicts of Interest: The authors declare no conflict of interest. The funders had no role in the design of the study; in the collection, analyses, or interpretation of data; in the writing of the manuscript, or in the decision to publish the results.

References

1. Liu, Y.S.; Guo, Y.Z.; Zhou, Y. Poverty alleviation in rural China: Policy changes, future challenges and policy implications. *China Agric. Econ. Rev.* **2018**, *10*, 241–259. [CrossRef]
2. Zhou, Y.; Guo, Y.; Liu, Y.; Wu, W.; Li, Y. Targeted poverty alleviation and land policy innovation: Some practice and policy implications from China. *Land Use Policy* **2018**, *74*, 53–65. [CrossRef]
3. Baloch, M.A.; Danish; Khan, S.U.D.; Ulucak, Z.S. Poverty and vulnerability of environmental degradation in Sub-Saharan African countries: What causes what? *Struct. Chang. Econ. Dynam.* **2020**, *54*, 143–149. [CrossRef]
4. Cheng, X.; Shuai, C.M.; Liu, J.L.; Wang, J.; Liu, Y.; Li, W.J.; Shuai, J. Topic modelling of ecology, environment and poverty nexus: An integrated framework. *Agric. Ecosyst. Environ.* **2018**, *267*, 1–14. [CrossRef]
5. Niu, W. The discriminatory index with regard to the weakness, overlapness, and breadth of ecotone. *Acta Ecol. Sin.* **1989**, *9*, 97–105.
6. Hong, H.; Liao, H.; Wei, C.; Li, T.; Xie, D. Health assessment of a land use system used in the ecologically sensitive area of the Three Gorges reservoir area, based on the improved TOPSIS Method. *Acta Ecol. Sin.* **2015**, *35*, 8016–8027.
7. Huang, A.; Xu, Y.Q.; Sun, P.L.; Zhou, G.Y.; Liu, C.; Lu, L.H.; Xiang, Y.; Wang, H. Land use/land cover changes and its impact on ecosystem services in ecologically fragile zone: A case study of Zhangjiakou City, Hebei Province, China. *Ecol. Indic.* **2019**, *104*, 604–614. [CrossRef]

8. Deng, X.; Wang, Z.; Zhao, C. Economic Evolution in China Ecologically Fragile Regions. *J. Econ. Surv.* **2016**, *30*, 552–576. [CrossRef]
9. Yang, X.; Bai, Y.; Che, L.; Qiao, F.; Xie, L. Incorporating ecological constraints into urban growth boundaries: A case study of ecologically fragile areas in the Upper Yellow River. *Ecol. Indic.* **2021**, *124*, 107436. [CrossRef]
10. Zhao, Y.; Wang, S.; Zhou, C. Understanding the relation between urbanization and the eco-environment in China's Yangtze River Delta using an improved EKC model and coupling analysis. *Sci. Total Environ.* **2016**, *571*, 862–875. [CrossRef]
11. Wu, J.S.; Zheng, H.Q.; Zhe, F.; Xie, W.D.; Song, J. Study on the relationship between urbanization and fine particulate matter (PM2.5) concentration and its implication in China. *J. Clean. Prod.* **2018**, *182*, 872–882. [CrossRef]
12. Yao, L.; Li, X.; Li, Q.; Wang, J. Temporal and Spatial Changes in Coupling and Coordinating Degree of New Urbanization and Ecological-Environmental Stress in China. *Sustainability* **2019**, *11*, 1171. [CrossRef]
13. Wu, R.; Li, Z.; Wang, S. The varying driving forces of urban land expansion in China: Insights from a spatial-temporal analysis. *Sci. Total Environ.* **2021**, *766*, 142591. [CrossRef] [PubMed]
14. Zank, B.; Bagstad, K.J.; Voigt, B.; Villa, F. Modeling the effects of urban expansion on natural capital stocks and ecosystem service flows: A case study in the Puget Sound, Washington, USA. *Landsc. Urban. Plan.* **2016**, *149*, 31–42. [CrossRef]
15. Van Meerbeek, K.; Helsen, K.; Hermy, M. Impact of land-use intensity on the conservation of functional and phylogenetic diversity in temperate semi-natural plant communities. *Biodivers. Conserv.* **2014**, *23*, 2259–2272. [CrossRef]
16. Ou, J.P.; Liu, X.P.; Li, X.; Chen, Y.M. Quantifying the relationship between urban forms and carbon emissions using panel data analysis. *Landsc. Ecol.* **2013**, *28*, 1889–1907. [CrossRef]
17. Ding, Y.; Peng, J. Impacts of Urbanization of Mountainous Areas on Resources and Environment: Based on Ecological Footprint Model. *Sustainability* **2018**, *10*, 765. [CrossRef]
18. Xia, C.; Yeh, A.G.-O.; Zhang, A. Analyzing spatial relationships between urban land use intensity and urban vitality at street block level: A case study of five Chinese megacities. *Landsc. Urban. Plann.* **2020**, *193*, 103669. [CrossRef]
19. Wang, S.J.; Fang, C.L.; Wang, Y.; Huang, Y.B.; Ma, H.T. Quantifying the relationship between urban development intensity and carbon dioxide emissions using a panel data analysis. *Ecol. Indic.* **2015**, *49*, 121–131. [CrossRef]
20. Zhang, P.; Yang, D.; Qin, M.; Jing, W. Spatial heterogeneity analysis and driving forces exploring of built-up land development intensity in Chinese prefecture-level cities and implications for future Urban Land intensive use. *Land Use Policy* **2020**, *99*, 104958. [CrossRef]
21. Antoniucci, V.; Marella, G. Small town resilience: Housing market crisis and urban density in Italy. *Land Use Policy* **2016**, *59*, 580–588. [CrossRef]
22. Caparros-Midwood, D.; Dawson, R.; Barr, S. Low Carbon, Low Risk, Low Density: Resolving choices about sustainable development in cities. *Cities* **2019**, *89*, 252–267. [CrossRef]
23. Gan, J.; WU, Z.; Guo, G. The Influence of Built Environment on Urban Biodiversity in High-density Urban Areas: Case Study in Blocks along Century Avenue, Pudong New District, Shanghai. *Urban. Dev. Studies* **2018**, 97–106.
24. Wen, Z.; Zheng, H.; Zhao, H.; Xie, S.; Liu, L.; Ouyang, Z. Land-use intensity indirectly affects soil multifunctionality via a cascade effect of plant diversity on soil bacterial diversity. *Glob. Ecol. Conserv.* **2020**, *23*, e01061. [CrossRef]
25. Pu, Y.; Wang, Y.; Wang, P. Driving effects of urbanization on city-level carbon dioxide emissions: From multiple perspectives of urbanization. *Int. J. Urban. Sci.* **2020**, 1–21. [CrossRef]
26. Zhou, L.; Yuan, B.; Mu, H.; Dang, X.; Wang, S. Coupling relationship between construction land expansion and PM2.5 in China. *Environ. Sci. Pollut Res. Int.* **2021**, 1–13. [CrossRef]
27. Shao, J.; Ge, J.; Feng, X.; Zhao, C. Study on the relationship between PM2.5 concentration and intensive land use in Hebei Province based on a spatial regression model. *PLoS ONE* **2020**, *15*, e0238547. [CrossRef]
28. Peng, B.; Li, Y.; Elahi, E.; Wei, G. Dynamic evolution of ecological carrying capacity based on the ecological footprint theory: A case study of Jiangsu province. *Ecol. Indic.* **2019**, *99*, 19–26. [CrossRef]
29. Wang, R.S.; Li, F.; Hu, D.; Li, B.L. Understanding eco-complexity: Social-Economic-Natural Complex Ecosystem approach. *Ecol. Complex.* **2011**, *8*, 15–29. [CrossRef]
30. Wang, R.S.; Zhou, T.; Hu, D.; Li, F.; Liu, J.R. Cultivating eco-sustainability: Social-economic-natural complex ecosystem case studies in China. *Ecol. Complex.* **2011**, *8*, 273–283. [CrossRef]
31. Rusong, M.S.W. The social-economic-natural complex ecosystem. *Acta Ecol. Sin.* **1984**, *1*, 2006.
32. Yao, L.; Liu, J.R.; Wang, R.S.; Yin, K.; Han, B.L. A qualitative network model for understanding regional metabolism in the context of Social-Economic-Natural Complex Ecosystem theory. *Ecol. Inform.* **2015**, *26*, 29–34. [CrossRef]
33. Wei, X.; Shen, L.; Liu, Z.; Luo, L.; Wang, J.; Chen, Y. Comparative analysis on the evolution of ecological carrying capacity between provinces during urbanization process in China. *Ecol. Indic.* **2020**, *112*, 106179. [CrossRef]
34. Fang, Y.-p.; Zhu, F.-b.; Yi, S.-h.; Qiu, X.-p.; Ding, Y.-j. Ecological carrying capacity of alpine grassland in the Qinghai–Tibet Plateau based on the structural dynamics method. *Environ. Dev. Sustain.* **2021**, 12550–12578. [CrossRef]
35. Zhao, Y.W.; Zhou, L.Q.; Dong, B.Q.; Dai, C. Health assessment for urban rivers based on the pressure, state and response framework—A case study of the Shiwuli River. *Ecol. Indic.* **2019**, *99*, 324–331. [CrossRef]
36. Wu, M.; Wu, J.; Zang, C. A comprehensive evaluation of the eco-carrying capacity and green economy in the Guangdong-Hong Kong-Macao Greater Bay Area, China. *J. Clean. Prod.* **2021**, *281*, 124945. [CrossRef]

37. Shen, W.; Zheng, Z.; Qin, Y.; Li, Y. Spatiotemporal Characteristics and Driving Force of Ecosystem Health in an Important Ecological Function Region in China. *Int. J. Environ. Res. Public Health.* **2020**, *17*, 5075. [CrossRef] [PubMed]
38. Zhao, X.; Zhou, W.; Tian, L.; He, W.; Zhang, J.; Liu, D.; Yang, F. Effects of land-use changes on vegetation net primary productivity in the Three Gorges Reservoir Area of Chongqing. *Acta Ecol. Sin.* **2018**, *38*, 7658–7668.
39. Li, S.C.; Bing, Z.L.; Jin, G. Spatially Explicit Mapping of Soil Conservation Service in Monetary Units Due to Land Use/Cover Change for the Three Gorges Reservoir Area, China. *Remote Sens.* **2019**, *11*, 468. [CrossRef]
40. Liao, Q.; Wang, Z.; Huang, C. Green Infrastructure Offset of the Negative Ecological Effects of Urbanization and Storing Water in the Three Gorges Reservoir Area, China. *Int. J. Environ. Res. Public Health* **2020**, *17*, 8077. [CrossRef]
41. Han, J.; Meng, X.; Zhou, X.; Yi, B.; Liu, M.; Xiang, W.-N. A long-term analysis of urbanization process, landscape change, and carbon sources and sinks: A case study in China's Yangtze River Delta region. *J. Clean. Prod.* **2017**, *141*, 1040–1050. [CrossRef]
42. Xiao, Y.; Xiong, Q.L.; Pan, K.W. What Is Left for Our Next Generation? Integrating Ecosystem Services into Regional Policy Planning in the Three Gorges Reservoir Area of China. *Sustainability* **2019**, *11*, 3. [CrossRef]
43. Li, Y.; Li, Y.; Zhou, Y.; Shi, Y.; Zhu, X. Investigation of a coupling model of coordination between urbanization and the environment. *J. Environ. Manag.* **2012**, *98*, 127–133. [CrossRef]
44. Liu, N.N.; Liu, C.Z.; Xia, Y.F.; Da, B.W. Examining the coordination between urbanization and eco-environment using coupling and spatial analyses: A case study in China. *Ecol. Indic.* **2018**, *93*, 1163–1175. [CrossRef]
45. Sun, C.G.; Zhang, S.Y.; Song, C.C.; Xu, J.H.; Fan, F.L. Investigation of Dynamic Coupling Coordination between Urbanization and the Eco-Environment-A Case Study in the Pearl River Delta Area. *Land* **2021**, *10*, 190. [CrossRef]
46. He, J.Q.; Wang, S.J.; Liu, Y.Y.; Ma, H.T.; Liu, Q.Q. Examining the relationship between urbanization and the eco-environment using a coupling analysis: Case study of Shanghai, China. *Ecol. Indic.* **2017**, *77*, 185–193. [CrossRef]
47. Ai, J.Y.; Feng, L.; Dong, X.W.; Zhu, X.D.; Li, Y.F. Exploring coupling coordination between urbanization and ecosystem quality (1985-2010): A case study from Lianyungang City, China. *Front. Earth Sci.* **2016**, *10*, 527–545. [CrossRef]
48. Cheng, J.; Xu, L.; Fan, H.; Jiang, J. Changes in the flow regimes associated with climate change and human activities in the Yangtze River. *RRA* **2019**, *35*, 1415–1427. [CrossRef]
49. Huang, C.B.; Huang, X.; Peng, C.H.; Zhou, Z.X.; Teng, M.J.; Wang, P.C. Land use/cover change in the Three Gorges Reservoir area, China: Reconciling the land use conflicts between development and protection. *Catena* **2019**, *175*, 388–399. [CrossRef]
50. Xu, Y.; Tang, H.; Wang, B.; Chen, J. Effects of land-use intensity on ecosystem services and human well-being: A case study in Huailai County, China. *Environ. Earth Sci.* **2016**, *75*, 416. [CrossRef]
51. Bai, Y.; Deng, X.; Jiang, S.; Zhang, Q.; Wang, Z. Exploring the relationship between urbanization and urban eco-efficiency: Evidence from prefecture-level cities in China. *J. Clean. Prod.* **2018**, *195*, 1487–1496. [CrossRef]
52. Shen, L.; Huang, Y.; Huang, Z.; Lou, Y.; Ye, G.; Wong, S.-W. Improved coupling analysis on the coordination between socio-economy and carbon emission. *Ecol. Indic.* **2018**, *94*, 357–366. [CrossRef]
53. Dong, G.; Ge, Y.; Zhu, W.; Qu, Y.; Zhang, W. Coupling Coordination and Spatiotemporal Dynamic Evolution Between Green Urbanization and Green Finance: A Case Study in China. *Front. Environ. Sci.* **2021**, *8*, 621846. [CrossRef]
54. Wei, Q.; Zhang, L.; Duan, W.; Zhen, Z. Global and Geographically and Temporally Weighted Regression Models for Modeling PM2.5 in Heilongjiang, China from 2015 to 2018. *Int. J. Environ. Res. Public Health* **2019**, *16*, 5107. [CrossRef] [PubMed]
55. Yan, J.W.; Tao, F.; Zhang, S.Q.; Lin, S.; Zhou, T. Spatiotemporal Distribution Characteristics and Driving Forces of PM2.5 in Three Urban Agglomerations of the Yangtze River Economic Belt. *Int. J. Environ. Res. Public Health* **2021**, *18*, 2222. [CrossRef] [PubMed]
56. Burnham, K.; Anderson, D. Multimodel Inference: Understanding AIC and BIC in Model Selection. *Sociol. Methods Res.* **2004**, *33*, 261–304. [CrossRef]
57. Ohtani, K. Bootstrapping R2 and adjusted R2 in regression analysis. *Econ. Model.* **2000**, *17*, 473–483. [CrossRef]
58. Peng, J.; Du, Y.Y.; Liu, Y.X.; Hu, X.X. How to assess urban development potential in mountain areas? An approach of ecological carrying capacity in the view of coupled human and natural systems. *Ecol. Indic.* **2016**, *60*, 1017–1030. [CrossRef]

International Journal of
Environmental Research and Public Health

MDPI

Article

Spatiotemporal Distribution and Influencing Factors of Ecosystem Vulnerability on Qinghai-Tibet Plateau

Han Li [1,2] and Wei Song [1,*]

[1] Key Laboratory of Land Surface Pattern and Simulation, Institute of Geographic Sciences and Natural Resources Research, Chinese Academy of Sciences, Beijing 100101, China; lih.19s@igsnrr.ac.cn
[2] College of Resources and Environment, University of Chinese Academy of Sciences, Beijing 100049, China
* Correspondence: songw@igsnrr.ac.cn

Abstract: As the "Third Pole", the Qinghai-Tibet Plateau is threatened by environmental changes. Ecosystem vulnerability refers to the sensitivity and resilience of ecosystems to external disturbances. However, there is a lack of relevant studies on the driving factors of ecosystem vulnerability. Therefore, based on spatial principal components analysis and geographic detectors methods, this paper evaluates the ecosystem vulnerability and its driving factors on the Qinghai-Tibet Plateau from the years 2005 to 2015. The results were as follows: (1) The ecosystem vulnerability index (EVI) of the Qinghai-Tibet Plateau is mainly heavy and extreme, showing a gradually increasing trend from southeast to northwest. (2) The spatial heterogeneity of the EVI is significant in the southeast and northwest, but not in the southwest and central parts. (3) Analysis of influencing factors shows that environmental factors have more significant effects on EVI than socioeconomic variables, facilitating the proposal of adequate policy implications. More efforts should be devoted to ecological protection and restoration to prevent grassland degradation and desertification in the high-EVI areas in northwest. The government is also urged to improve the ecological compensation mechanisms and balance ecological protection and residents' development needs in the southeast.

Keywords: ecosystem vulnerability; spatiotemporal distribution; influencing factors; Qinghai-Tibet Plateau; principal components analysis

Citation: Li, H.; Song, W. Spatiotemporal Distribution and Influencing Factors of Ecosystem Vulnerability on Qinghai-Tibet Plateau. *Int. J. Environ. Res. Public Health* **2021**, *18*, 6508. https://doi.org/10.3390/ijerph18126508

Academic Editor: Paul B. Tchounwou

Received: 30 April 2021
Accepted: 15 June 2021
Published: 16 June 2021

1. Introduction

Ecosystem is the general term for all organisms and environments within a particular space. Ecosystems are complex open systems, mainly including social systems, natural systems and social-natural coupled systems [1]. According to Adger [2], vulnerability is the sensitivity of ecosystem under the stress of natural and social changes due to the lack of adaptability. In recent years, as a result of increased human activities and global climate changes, ecosystems have been under increasing pressure, aggravating their vulnerability towards a series of stressors [3]. Ecosystem vulnerability assessments are therefore critical in global environmental change research [4], providing a decision-making basis and technical support for ecological protection and environmental restoration and governance [5]. Ecosystem vulnerability has become a hot spot of global environmental change and sustainable development research [6–9]. Understanding the driving mechanisms of regional ecological vulnerability evolution can facilitate the establishment of guidelines for the use and protection of the regional ecological environment [10].

Several studies have considered the impacts of climate change and natural disasters on ecosystem vulnerability. For example, based on the prediction results of temperature and precipitation under low (B1), medium (A1B) and high (A2) emission scenarios, released in the fourth assessment report of the International Panel of Climate Change (IPCC), Gonzalez et al. [11] studied the changes in vegetation vulnerability patterns in global ecosystems in the 21st century. Based on their results, one-tenth to one-half of the global vegetation area may be highly (confidence level 0.80–0.95) to very highly (confidence ≥ 0.95)

vulnerable to climatic changes. Similarly, Alexander et al. [12] assessed the vulnerability of tropical ecosystems in southern Ecuador and found differences in ecosystem vulnerability under different climate scenarios. Patrick et al. [13] investigated the vulnerability of 52 major vegetation types in the western United States exposed to changes in temperature and precipitation under RCP 4.5 scenarios (RCP4.5, Representative Concentration Pathway 4.5, a moderate emission scenario proposed by the Coupled Model Intercomparison Project Phase 5). Their results showed that by the middle of the 21st century, 33 vegetation types will be faced with high or very high vulnerability, of which more than 50% will have higher regional vulnerability levels.

In recent years, the vulnerability of different types of ecosystems has gradually been studied, with a higher number of studies on the vulnerability of certain ecosystem systems, such as mining areas, economically developed areas and oceans. For example, Sarah et al. [14] assessed the vulnerability of marine ecosystems in California and found that tidal flats, beaches, salt marshes and intertidal rocky ecosystems were most vulnerable to human activities. Similarly, Zhang et al. [15] investigated the effects of extreme rainfall on ecosystem vulnerability in the middle and lower reaches of the Yangtze River in China and showed that both human-dominated ecosystems (e.g., agro-ecosystems) and natural ecosystems are vulnerable to extreme climate events. The current vulnerability studies of typical ecosystems that are particularly sensitive to global climate change mainly focus on coastal zones [16–18] and wetland regions [19,20]. However, there are no studies on the vulnerability of high-elevation ecosystems, such as the Qinghai-Tibet Plateau, the largest and highest plateau in the world. Due to its unique environment, it is highly sensitive to climate change and human activities, with a fragile ecosystem [21].

China is one of the countries with the most vulnerable ecosystem types in the world, and the research on its ecosystem vulnerability began in the 1980s. For example, Niu [22] conducted a study from the perspective of the ecotone. Early studies mainly focused on the impacts of climate change, extreme weather and natural disasters on ecosystems, such as the analysis of the vulnerability of China's forest ecosystems under global climate change [23]. In the 1990s, socio-economic factors started to become increasingly considered in the assessment of ecosystem vulnerability, such as the relationship between the fragile zone of the ecosystem and the population [24] or the relationship between ecosystem vulnerability and agricultural development [25]. Since the 21st century, natural and socio-economic factors have been regarded as important factors that play a crucial role assessing ecosystem vulnerability, and numerous related studies have been conducted in typical regions, such as the Three Rivers Source [26]. In addition, the vulnerability of different components of an ecosystem, such as grassland ecosystems [27], was further studied.

In recent years, as the government has started to increasingly consider the importance of environmental integrity, substantial investments have been made in the field of ecological protection. For example, in 2020, the Chinese government put forward the concept of building a "beautiful China" and promoted the construction of an ecological civilization. As a consequence, researchers are paying more attention to the evaluation of ecosystem vulnerability in typical regions with serious ecological and environmental problems. Even though in some areas of the Qinghai-Tibet Plateau, studies on ecosystem vulnerability have been performed, there is a lack of consideration of anthropogenic factors [28]. Due to the construction of infrastructure, such as the Qinghai-Tibet Railway, and the development of tourism, the intensity of human activities in the Qinghai-Tibet Plateau has increased sharply. Against the background of the implementation of China's ecological protection policy, it is now necessary to gain insights into the overall ecosystem vulnerability and driving forces of the Qinghai-Tibet Plateau. Such studies can provide theoretical references for the sustainable development of the Qinghai-Tibet Plateau and put forward feasible suggestions for the protection of this area.

The objective of this study is to explore the temporal and spatial changes of ecosystem vulnerability and the impacts of natural and socio-economic factors on the Qinghai Tibet Plateau. Specifically, we tested two main hypotheses: (1) the spatial distribution of ecosystem vulnerability has significant spatial patterns; (2) the impacts of natural factors on ecosystem vulnerability are greater than those of socio-economic factors.

2. Literature Review

2.1. The Concept of Ecosystem Vulnerability

Since the concept of ecological vulnerability has evolved from vulnerability, we start with a brief review of the development of the vulnerability concept. There are numerous statements about the concept of vulnerability. In 1945, White et al. [29] put forward the "adaptation and adjustment view" for the first time when studying flood disasters, which marked the beginning of vulnerability research. After that, White [30] defined vulnerability as a system, subsystem or system component due to its exposure and sensitivity, making it susceptible to external disturbance and pressure. Timmerman [31] defined vulnerability as the degree to which a system is adversely affected or damaged. After that, Dow [32], Cutter [33] and the IPCC [34] defined ecosystem vulnerability from different perspectives. In the 21st century, the concept of vulnerability has been widely used in many fields, including sustainable development [35], climate change [3] and ecology [12].

Ecosystem vulnerability was initially introduced into ecology by Clements, with the concept of the "ecological transition zone" [36], and a unified definition of ecosystem vulnerability as not yet been provided (Table 1). At present, the IPCC's definition of vulnerability has been widely accepted and adopted in the field of climate change research. Based on relevant literature, ecosystem vulnerability can be summarized as the sensitivity and resilience of ecosystems in response to external interference including human disturbance, climate change, etc.

Table 1. Some definitions of ecosystem vulnerability.

Organization/Author	Definition of Ecosystem Vulnerability
Williams et al. [37]	The potential of an ecosystem to modulate its response to stressors over time and space, where that potential is determined by the characteristics of an ecosystem with many levels of organization. It is an estimate of the inability of an ecosystem to tolerate stressors over time and space.
Birkmann [38]	The sensitive response and self-restoring ability of an ecosystem when it is subjected to external interference. It usually occurs within a specific time and space and is an inherent attribute of the ecosystem.
IPCC [39]	The degree of sensitivity and self-regulation of an ecosystem to disturbances caused by climate change, including extreme weather events.

2.2. Assessment of Ecosystem Vulnerability

Ecosystem vulnerability studies rely on building assessment models. At present, there is no unified model for ecosystem vulnerability assessment; the common evaluation models include the Pressure-State-Response model (PSR) [21] and the Exposure-Sensitive-Adaptive model (ESA) [40]. Based on the PSR model framework, some scholars have developed a series of models by adding factors, such as the Driving force-Pressure-State-Impact-Response (DPSIR) [41] and the Pressure-Support-State-Response (PSSR) [42]. Similar to the ESA models, there is the Vulnerability-scoping-Diagram (VSD) model [43]. Based on PSR and ESA models, some scholars have also proposed Pressure-Sensitivity-Elasticity (PSE) [44] and Sensitivity-Resilience-Pressure (SRP) models [45]. The ecosystem vulnerability assessment model is developing in the direction of integrating multiple systems and multiple factors.

On the basis of the indicator system, ecological vulnerability assessment needs to be carried out in conjunction with the assessment methodology, such as the hierarchical analysis method [46], the fuzzy evaluation method [47], the artificial neural network method [48], the entropy weight analysis method [49] and the expert scoring method [50]. With the development and application of RS (Remote Sensing), GIS (Geographic Information System), GPS (Global Positioning System) and other technologies, vulnerability assessment results have become more dynamic. For example, Yaw et al. [51] used GIS and RS to analyze the vulnerability of the Niger River Basin and its influencing factors. The spatial principal components analysis method (SPCA), based on principal components analysis and spatial feature extraction, has advantages in ecosystem vulnerability assessment [52]. For example, it not only adds spatial constraints to the traditional PCA but also considers the spatial dependence in data sets.

Since the purposes and regional characteristics of the studies, along with their emphasize, can largely differ, there is no unified index system. In recent years, ecosystem vulnerability assessment indicators for different regions have been selected (Table 2). In this study, the Sensitivity-Resilience-Pressure (SRP) model was used to construct the index system. This model is constructed based on the connotation of ecosystem stability and has been widely used in the Karst Mountains [53], the Yimeng Mountain area [45] and the Shiyang River region [54], among others. Here, sensitivity reveals the resistance of the ecosystem to various disturbances and is usually expressed by topographical and meteorological factors. In contrast, restoration refers to the ability of an ecosystem to be restored to the original state after damage by internal and external interference factors; it is mainly characterized by vegetation factors. Pressure refers to the variety of pressures from anthropogenic interference, often expressed by population pressure and intensity of economic activities. Since ecosystem vulnerability is generally the result of a combination of natural and human activities, the driving factors that affect changes in ecological vulnerability can be divided into two categories: natural and socio-economic factors [10,55].

Table 2. Different ecosystem vulnerability assessment indicators.

Year	Study Area	Level Indicators	Secondary Indicators
2017	Yellow River Delta, China [20]	Pressure, support, state, response	Land reclamation rate, population density, human disturbance index, normalized difference vegetation index (NDVI), afforestation area percentage, Shannon's evenness index, ecological water percentage, pollution load, elastic degree of wetland evaluation; wetland area of change, gross domestic product
2018	Southern Shaanxi, China [10]	Environmental topography and socio-economic level	Cultivation ratio, land use rate, natural growth rate, population density, gross domestic product (GDP) per capita, agricultural output, industrial output, NDVI, average precipitation, average annual temperature, hours of sunshine, average elevation
2018	Jiangsu, China [56]	Pressure, state, response	Soil erosion sensitivity, soil desertification sensitivity, landscape patch density, landscape evenness, land resource use degree
2020	Ningxia Hui Autonomous Region, China [57]	Natural and social factors	Digital elevation model, hours of sunshine, average annual precipitation, average annual temperature, NDVI soil erosion and degree of land use, GDP, agricultural output, industrial output, population density, grassland area
2020	Karst Mountains, China [53]	Sensitivity, resiliency, pressure	Climate, soil, terrain, water, geology, vegetation, land use, social development, economic development

3. Study Area and Data Sources

3.1. Study Area

The Qinghai-Tibet Plateau in southwest China is the highest plateau in the world, also known as the "Third Pole" (Figure 1). Its average elevation is more than 4000 m above sea level. The administrative regions include Tibet Autonomous Region, Qinghai Province and parts of Xinjiang Uygur Autonomous Region, Gansu, Sichuan and Yunnan Province. It is the birthplace of the Yangtze River, the Yellow River and the Lancang River, among others. The terrain is diverse, containing valleys and basins and the climate is highly complex and largely affected by the terrain. The spatial and temporal distribution patterns of air and heat on the Qinghai-Tibet Plateau are significant. The southeastern area is warm and humid, whereas the northwestern area is dry and cold. The annual average temperature of the entire region ranges between 5.6 and 17.6 °C. Annual precipitation is unevenly distributed, gradually decreasing from 2000 mm to less than 50 mm from southeast to northwest. Under the influence of temperature and precipitation, the surface cover type changes from southeast to northwest, gradually transitioning from forest and shrub areas to grassland, meadow and desert. As a result of overgrazing, the alpine grassland on the Qinghai-Tibet Plateau is subjected to serious desertification. The major ecological issues faced include freeze-thaw erosion, hydraulic erosion, desertification, salinization and water scarcity [58].

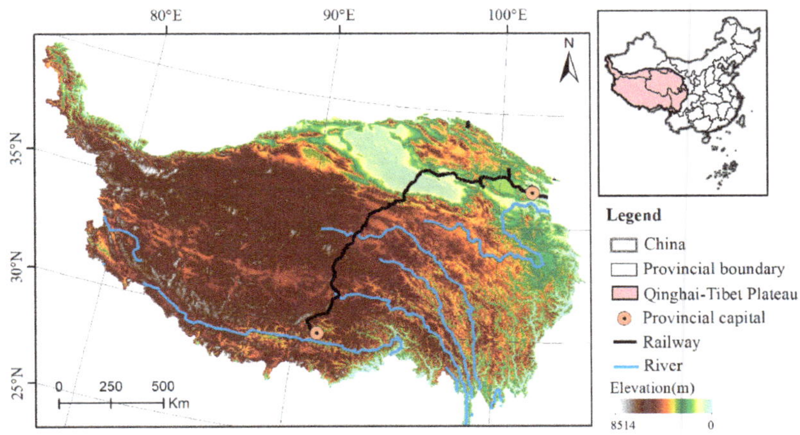

Figure 1. Geographical location of the Qinghai-Tibet Plateau.

3.2. Data Sources

For this study, the data used include socio-economic, remote sensing, topographic, meteorological and land use (Table 3) from 2010, 2010 and 2015. They were mostly obtained from the Resource and Environmental Science Data Center of the Chinese Academy of Sciences (RESDC) and include socio-economic (population and GDP (gross domestic product)), topographic (DEM (digital elevation model)), meteorological (annual precipitation and average annual temperature) and land use data with a spatial resolution of 1 km. Remote sensing data were obtained from MODIS (Moderate resolution imaging spectroradiometer) and include NDVI (Normalized difference vegetation index), NP (Net Primary Productivity) and ET (Evapotranspiration). The spatial resolution of NDVI and NP is 1 km ant that of ET 500 m.

Table 3. Basic data and sources of ecological vulnerability assessment for the Qinghai-Tibet Plateau.

Type	Source	Spatial Resolution	Temporal Resolution
NDVI	MODIS/MOD13A3 [59]	1 km	Monthly
Land use	RESDC [60]	1 km	Yearly
DEM	RESDC	1 km	Yearly
Annual average temperature	RESDC	1 km	Yearly
Annual precipitation	RESDC	1 km	Yearly
NPP	MODIS/MOD17A3	1 km	Yearly
ET	MODIS/MOD16A3	500 m	Yearly
Population	RESDC	1 km	Yearly
GDP	RESDC	1 km	Yearly

Notes: NDVI is Normalized difference vegetation index; DEM is digital elevation model; NPP is Net Primary Productivity; ET is Evapotranspiration; GDP is gross domestic product; RESDC is Resource and Environment Science and Data Center, Chinese Academy of Sciences.

All data were preprocessing using the ARCGIS 10.2 software. First, all data were projected into the same coordinate system (WGS_1984_UTM_45N) and then cut into the same spatial boundary according to the study area. Finally, the spatial resolution of data was unified to 1 km by bilinear interpolation. The NDVI represents the monthly data with 12 periods per year, and the annual NDVI was generated by selecting the annual maximum.

4. Research Method

4.1. Technical Route

The study was divided into the following four steps (Figure 2):

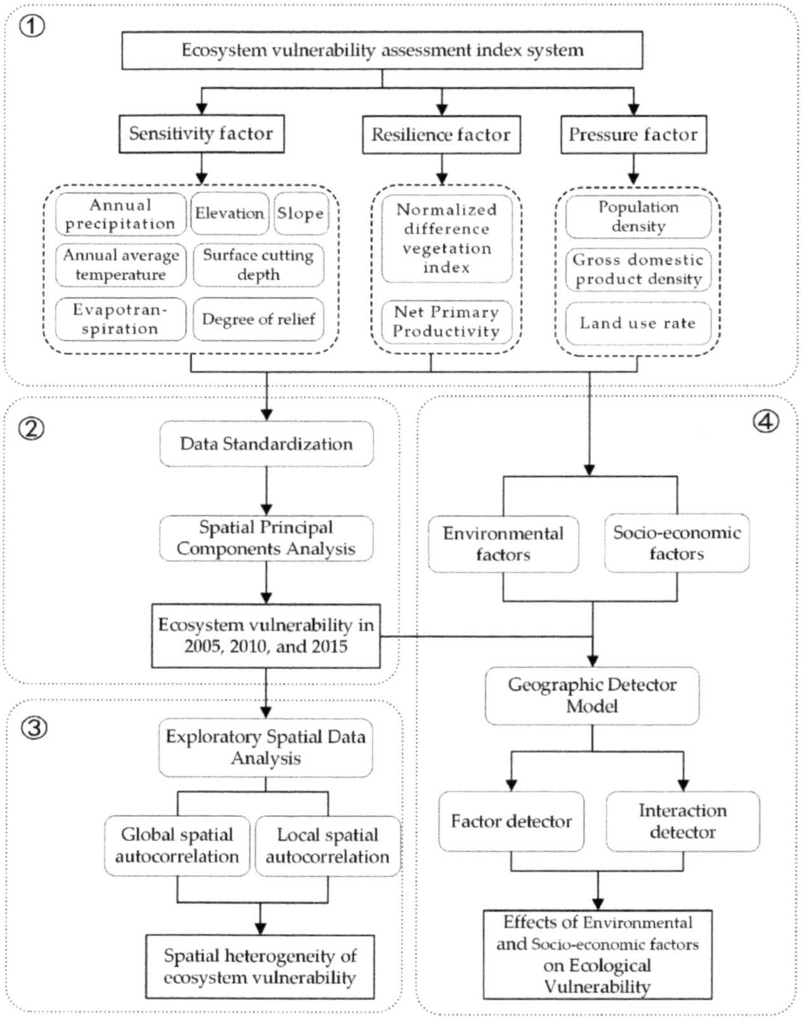

Figure 2. Flowchart showing the process followed in this analysis for assessing ecosystem vulnerability of the Qinghai-Tibet Plateau.

Step 1: Establishing the ecosystem vulnerability assessment index system. According to the Sensitivity-Resilience-Pressure (SRP) model, the indicators were selected from three aspects: ecological sensitivity, resilience and pressure.

Step 2: Mapping the distribution of ecosystem vulnerability. First, the indicators were standardized and uniformly mapped to the same value range to solve the problem of inconsistent original data units. Subsequently, the spatial scale of ecological vulnerability was determined using the ARCGIS 10.2 software and the SPCA method.

Step 3: Spatial heterogeneity analysis. The spatial and temporal distribution characteristics of ecosystem vulnerability were analyzed via exploratory spatial data analysis.

Step 4: Driving force analysis. Using the factor and interaction detector in the geodetector model, the effects of natural and socio-economic factors on ecosystem vulnerability were analyzed.

4.2. Establishing an Ecosystem Vulnerability Assessment Indicator System

In this study, the Sensitivity-Resilience-Pressure (SRP) model was used to construct the index system. The selected 12 indicators were divided into three categories, namely sensitivity, resilience and pressure (Table 4). In previous studies of ecosystem vulnerability in Shiyang River Basin [54], Karst [53] and Yimeng [45] mountainous areas using SRP model, sensitivity is considered to be the product of the interaction between the topographic factors and the distribution of meteorological factors. In this study, annual average temperature, annual precipitation and ET were just selected to reflect the hydrothermal conditions of the ecosystem [10,28]. Here they are not affected as external hazards. Elevation, slope, surface cutting depth and degree of relief were used to characterize the regional topography [61]. Resilience is usually characterized by vegetation factors [62], and NDVI and NPP were selected to reflect vegetation growth. The NDVI can detect the vegetation growth status and accurately reflect the surface vegetation coverage [63]. The NPP not only reflects the productive capacity of vegetation communities, but also represents ecosystem quality [64]. Pressure factors include population density, gross domestic product density and land use rate. Population density and GDP density represent the degree of population and economic concentration, reflecting the interference intensity of human activities. When the disturbance intensity exceeds the carrying capacity of the ecosystem, the ecological environment will be degraded, resulting in increased ecosystem vulnerability [26]. The land use rate (proportion of cultivated land) was selected to reflect the influence of human activities on land use.

Table 4. Ecosystem vulnerability assessment indicators for the Qinghai-Tibet Plateau.

Factor Category	Indicator	Type
Sensitivity	Annual precipitation (PRE)	−
	Annual average temperature (TEM)	−
	Evapotranspiration (ET)	−
	Elevation (ELE)	+
	Slope	+
	Surface cutting depth (SCD)	+
	Degree of relief (DR)	+
Resilience	Normalized difference vegetation index (NDVI)	−
	Net Primary Productivity (NPP)	−
Pressure	Population density (PD)	+
	Gross domestic product density (GDPD)	+
	Land use rate (LUR)	+

Note: "+" means positive action; the greater the value, the lower the quality of the ecological environment, the greater the probability of a fragile ecological environment; "−" means reverse action.

4.3. Mapping Ecosystem Vulnerability

4.3.1. Data Standardization

Standardization is generally carried out to solve the issue of inconsistent original data units [53]. There are two relationships between ecosystem vulnerability and evaluation factors [65]. The lower the index value, the lower the ecosystem vulnerability, representing a positive correlation. Conversely, there is a negative correlation, that is, the lower the index value, the stronger the ecosystem vulnerability. The maximum-difference normalization method was used to standardize the positive and negative indicators. For the positive indicators in the ecosystem vulnerability assessment index system, the standardized methods are as follows [10]:

$$M_i = \frac{X_i - X_{min}}{X_{max} - X_{min}}, \quad (1)$$

The negative indicators are treated as follows:

$$M_i = \frac{X_{max} - X_i}{X_{max} - X_{min}}, \quad (2)$$

where "M_i" is the standardized value of index i; "X_i" is the initial value of index i; "X_{min}" is the minimum value of index i; "X_{max}" is the maximum value of index i.

4.3.2. Spatial Principal Components Analysis

Spatial principal components analysis (SPCA) is a statistical analysis method that converts initial multiple indicators into irrelevant comprehensive indicators by dimension reduction [66,67]. At the same time, the correlation between the original evaluation indexes is reduced, and the information reflected by the original variables is kept to the maximum extent with less comprehensive indices to avoid the repetition of the indicators affecting the accuracy of the evaluation. In this study, we analyzed 12 standardized indices by principal components analysis to generate a new comprehensive index. By solving the correlation coefficient matrix of the index, the feature vector was obtained, and 12 principal component results are acquired. The principal component with a cumulative contribution rate of more than 85% was selected to replace the original index, and the principal factor was determined [67]. On this basis, the comprehensive index of the principal component was calculated as follows [66]:

$$PC_i = a_{1i}X_1 + a_{2i}X_2 + a_{3i}X_3 + \ldots a_{ni}X_n, \quad (3)$$

where "PC_i" is the i-th principal component; "$a_{1i}, a_{2i} \ldots a_{ni}$" are the feature vectors corresponding to the respective index factors of the i-th principal component; "$X_1, X_2 \ldots X_n$" are the respective index factors.

The ecosystem vulnerability index (EVI) was calculated based on the principal components analysis, using the following equation [68]:

$$\mathrm{EVI} = b_1PC_1 + b_2PC_2 + b_3PC_3 + \ldots b_nPC_n, \quad (4)$$

where "EVI" is the ecosystem vulnerability index; "bi" is the contribution rate corresponding to the i-th principal component; "PC_i" is the i-th principal component; "n" is the first n principal component whose cumulative contribution rate exceeds 85%. The SPCA in this study was calculated by the ArcGIS 10.2 software. The SPCA results for the years 2005, 2010 and 2015 are shown in Table 5.

To compare the EVI results of several years, the EVI was standardized as follows:

$$K_i = \frac{EVI_i - EVI_{min}}{EVI_{max} - EVI_{min}}, \quad (5)$$

where "K_i" is the standardized value of ecosystem vulnerability in the i-th year, with a value range of 0–1; "EVI_i" is the actual value of the ecosystem vulnerability index in the i-th year; "EVI_{max}" is the maximum value of the i-th ecosystem vulnerability index; "EVI_{min}" is the minimum value of the i-th ecosystem vulnerability index.

Table 5. Results of the SPCA (spatial principal components analysis) of ecosystem vulnerability on the Qinghai-Tibet Plateau.

PC	Eigenvalues			Contribution Ratio of Eigenvalues/%			Cumulative Contribution of Eigenvalues/%		
	2005	2010	2015	2005	2010	2015	2005	2010	2015
1	0.0669	0.0729	0.0763	48.7327	48.5814	51.9332	48.7327	48.5814	51.9332
2	0.0391	0.0429	0.0392	28.5202	28.5649	26.6587	77.2529	77.1463	78.5920
3	0.0101	0.0105	0.0092	7.3796	7.0087	6.2786	84.6325	84.1550	84.8706
4	0.0070	0.0079	0.0072	5.0880	5.2544	4.8803	89.7204	89.4094	89.7509

4.3.3. EVI Classification

We used natural breakage classification (NBC) to classify the EVI to reflect different degrees of ecosystem vulnerability. This method is generally used to analyze the statistical distribution of attribute, maximizing the difference between classes [56]. In this study, according to the results of the NBC for 2005, the EVI was divided into five grades, namely, slight, light, medium, heavy and extreme vulnerability (Table 6). Subsequently, the ArcGIS 10.2 software was used to visualize the spatial distribution of EVI.

Table 6. Classification of the ecosystem vulnerability index (EVI).

EVI	Slight	Light	Medium	Heavy	Extreme
Grading standard	<0.35	0.35–0.5	0.5–0.64	0.64–0.77	>0.77

4.4. Spatial Heterogeneity Analysis

The exploratory spatial data analysis method (ESDA) can be used to reveal the spatial interaction mechanism by describing and visualizing the spatial distribution pattern [69]. According to the different scales of analysis, global and local spatial autocorrelation are often used to study the spatial feature of the observation [56]. Here, this was performed using the OpenGeoda 1.16.0.16 software at a spatial resolution of 1 km. Global spatial autocorrelation analysis is mainly used to reflect the cluster degree of similar attributes in a study area [70]. The degree of spatial autocorrelation is usually measured by the Global Moran's I proposed by Moran [71]. Local spatial autocorrelation is mainly used to measure the spatial correlation and difference between the region of the research target and its surrounding areas [72].

The global Moran index is calculated as follows:

$$I_i = \frac{N}{\sum_{i=1}^{N}(x_i - \overline{x})^2} \times \frac{\sum_{i=1}^{N}\sum_{j=1}^{N}(x_i - \overline{x})(x_j - \overline{x})}{\sum_{i=1}^{N}\sum_{j=1}^{N} w_{ij}}, \tag{6}$$

where "I" is the Moran index, "N" is the number of research objects, "x_i" and "x_j" are the spatial attribute values of the research objects, and "w_{ij}" is the spatial weight matrix. The value range of "I" is [−1, 1]. If the index is greater than 0, the space is positively correlated; if it is smaller than 1, it is negatively correlated. At a value equal to 0, there is no correlation.

The specific equation to calculate local spatial autocorrelation is as follows:

$$I_i = \frac{N}{\sum_{i=1}^{N}(x_i - \overline{x})} \times (x_i - \overline{x}) \times \sum_{j=1}^{N} w_{ij}(x_i - \overline{x}), \tag{7}$$

where when $I_i > 0$, the local space of the research target is positively correlated, and the surrounding area presents a similar attribute value cluster. When the attribute values of the research target area and the surrounding research area are both high, they are hotspot clusters, generally represented by high-high (HH); when the attribute values of the research target area and its surrounding research area are low, they are coldspot clusters, generally represented by low-low (LL). When $I_i < 0$, the research target's local space is negatively correlated, and the surrounding area of the research target shows the opposite phenomenon of attribute value cluster. When the attribute value of the research target area itself is high, but that of the surrounding area is low, it is a high-low cluster, generally represented by high-low (HL). When the attribute value of the research target area itself is low, but that of the surrounding area is high, it is a low-high cluster, generally represented by low-high (LH).

4.5. Driving Force Analysis

The Geographic Detector Model (GDM) is a set of statistical methods to identify spatial differentiation among the geographical elements. This method can quantitatively analyze the driving mechanisms of geographical phenomena and is widely used to determine the explanatory power of driving factors and the interaction between factors without too many hypothetical conditions [73–75].

The GDM includes four detectors, namely risk detector, factor detector, ecological detector and interaction detector. In this study, factor detector and interaction detector were used to analyze the driving factors of ecosystem vulnerability on the Qinghai-Tibet Plateau, with the aim to explore the main driving mechanism of ecosystem vulnerability and to compare the spatial consistency between EVI and evaluation indices. If a factor dominates the cause of vulnerability, vulnerability will exhibit a spatial distribution similar to the evaluation index and the intra-layer variance is lower than the inter-layer variance. Using q-statistics to measure the decisive effect of each evaluation index on EVI, the calculation method is as follows [76]:

$$q = 1 - \frac{\sum_{h=1}^{L} N_h \delta_h^2}{N\delta^2}, \tag{8}$$

where "q" is the explanatory power of the influencing factors to the vulnerability of the ecosystem, "N" is the sample size, "L" is the classification number of the index factors and "N_h" and "δ_h^2" represent the variance of h-layer sample size and ecosystem vulnerability, respectively. The value of the q-statistic is in the range of [0, 1]; the larger the value, the stronger the explanatory power of the influence factor to the ecosystem vulnerability, and its spatial distribution is consistent with the EVI. When the q-statistic is equal to 0, there is no significant relationship between the given influence factor and the EVI distribution. When the value is 1, the impact factor can fully explain the spatial variation of the EVI.

The interaction detector was adopted to reveal the factor explanatory power to the results after multi-factor interaction, that is, whether the interaction of impact factors X1 and X2 will strengthen or weaken the impact on ecosystem vulnerability. The main types are shown in Table 7.

Table 7. Interaction Detector Model.

Description	Interaction Type
q (X1∩X2) < Min (q (X1), q (X2))	Non-linear-weaken
Min(q (X1), q (X2)) < q (X1∩X2) < Max(q (X1)), q (X2))	Uni-weaken
q (X1∩X2) > Max (q (X1), q (X2))	Bi-enhance
q (X1∩X2) = q (X1) + q (X2)	Independent
q (X1∩X2) > q (X1) + q (X2)	Non-linear-enhance

Note: q (X1∩X2) represents the interaction effect of influencing factors X1 and X2, and q (X1) and q (X2) represent the respective effects of X1 and X2, respectively.

5. Results

5.1. Spatiotemporal Variations in Ecosystem Vulnerability

5.1.1. Temporal Variations in Ecosystem Vulnerability

The ecosystem vulnerability levels in most areas of the Qinghai-Tibet Plateau were dominated by heavy and extreme vulnerability (Figure 3). In 2005, 2010 and 2015, heavily and extremely vulnerable areas accounted for 51.37, 51.64 and 53.08% of the total area, respectively. Heavily vulnerable areas accounted for the largest proportions, namely 28.10, 28.62 and 29.07%, respectively. From 2005 to 2015, the proportions of slightly and medium vulnerable areas decreased by 0.72 and 0.99%, respectively. The proportion of slightly vulnerable areas did not change, whereas those of heavily and extremely vulnerable areas increased by 0.97 and 0.77%, respectively.

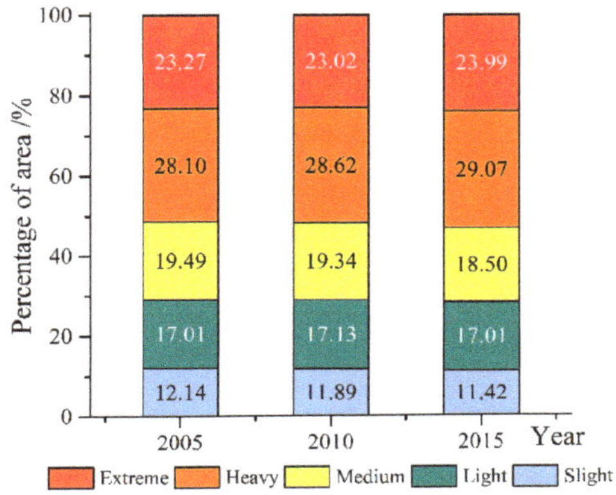

Figure 3. Area proportions of different ecosystem vulnerability levels on the Qinghai-Tibet Plateau in 2005, 2010 and 2015.

The transition areas of ecosystem vulnerability level were calculated for 2005, 2010 and 2015 (Figure 4). Area conversion mainly occurred between adjacent levels. For example, the increasing areas of heavily and extremely vulnerable areas were former medium and heavily vulnerable areas. From 2005 to 2015, highly vulnerable areas were mainly a result of the transformation of medium and extremely vulnerable areas, accounting for 76.32 and 23.64%, respectively. The extremely vulnerable areas are almost entirely transformed into heavily vulnerable ones. The main types of ecosystem vulnerability scale conversion include medium to high, light to medium, slight to light and high to extreme vulnerability.

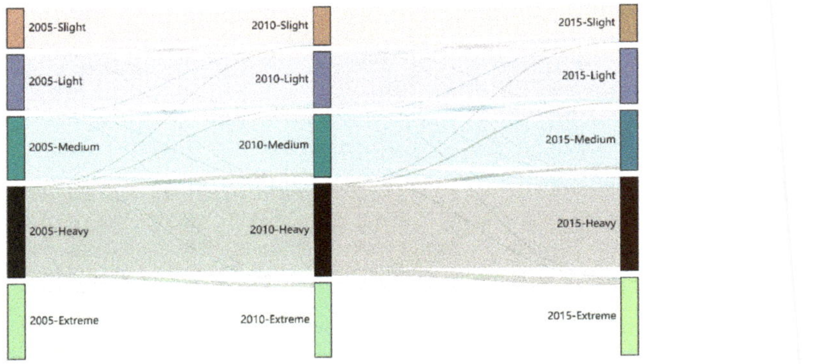

Figure 4. Area conversion of ecosystem vulnerability grades on the Qinghai-Tibet Plateau in 2005, 2010 and 2015.

5.1.2. Spatial Variations in Ecosystem Vulnerability

According to the spatial distribution pattern of EVI classification (Figure 5), the Qinghai-Tibet Plateau as a whole is mainly extremely vulnerable. The overall distribution of ecosystem vulnerability grades was higher in the northwest than in the southeast and gradually increased from southeast to northwest. The ecosystem vulnerability level in the northwest in 2005–2015 was mainly extreme; extremely vulnerable areas first decreased and then increased, whereas for lightly vulnerable areas, the opposite pattern was observed. Ecosystem vulnerability in the southeast was mainly slight and light, with a decrease in slightly vulnerable areas. From southeast to northwest, the vulnerability index increased, and the degree of vulnerability intensified. The middle area mainly showed a medium vulnerability, and the area with medium vulnerability decreased over time.

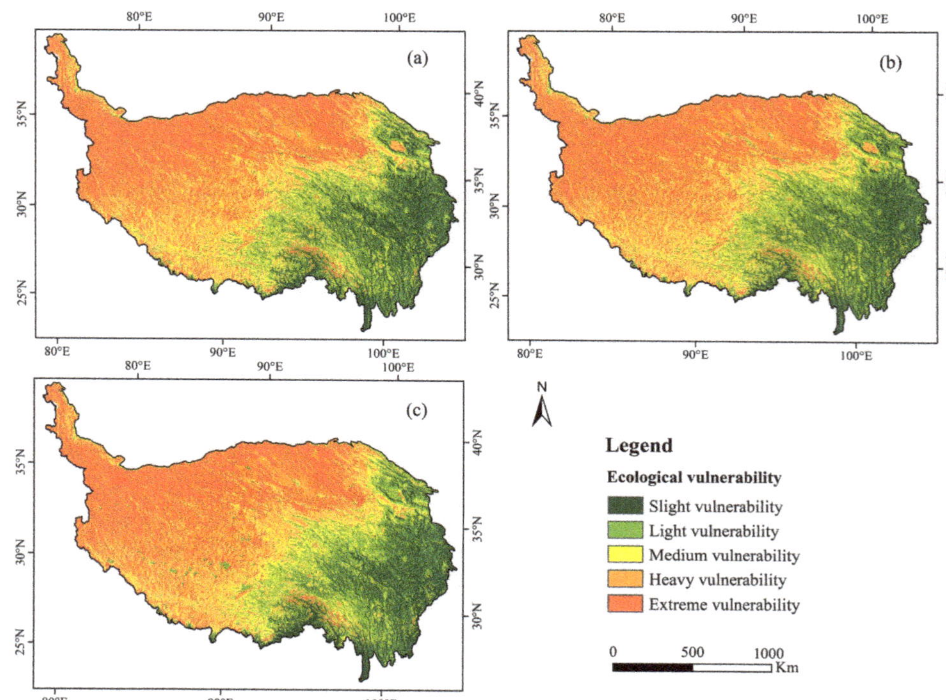

Figure 5. Spatial distribution of ecosystem vulnerability on the Qinghai-Tibet Plateau in (**a**) 2005, (**b**) 2010 and (**c**) 2015.

To analyze the transition between different levels of ecosystem vulnerability on each patch, we visualized the change in vulnerability grade from 2005 to 2015 (Figure 6). The main changes in ecosystem vulnerability levels consisted of the reduction of slightly and medium vulnerable areas and the increase in heavily and extremely vulnerable areas. From 2005 to 2015, changes in ecosystem vulnerability occurred in 14.80% of the study area, with 18 transformation types. The transition from medium vulnerability to heavy vulnerability accounted for 3.30% of the study area and mainly occurred in the northwest of the Qinghai-Tibet Plateau. The conversion of light vulnerability to medium vulnerability accounted for 2.52% of the study area, mainly in the central region.

5.2. Spatial Heterogeneity of Ecosystem Vulnerability

The EVI Global Moran Index for 2005, 2010 and 2015 passed the significance test, with values of 0.916, 0.915 and 0.929 (Figure 7). Ecosystem vulnerability showed positive spatial autocorrelation and high clustering. The overall cluster trend decreased first (from 2005–2010) and then slightly increased (from 2010–2015).

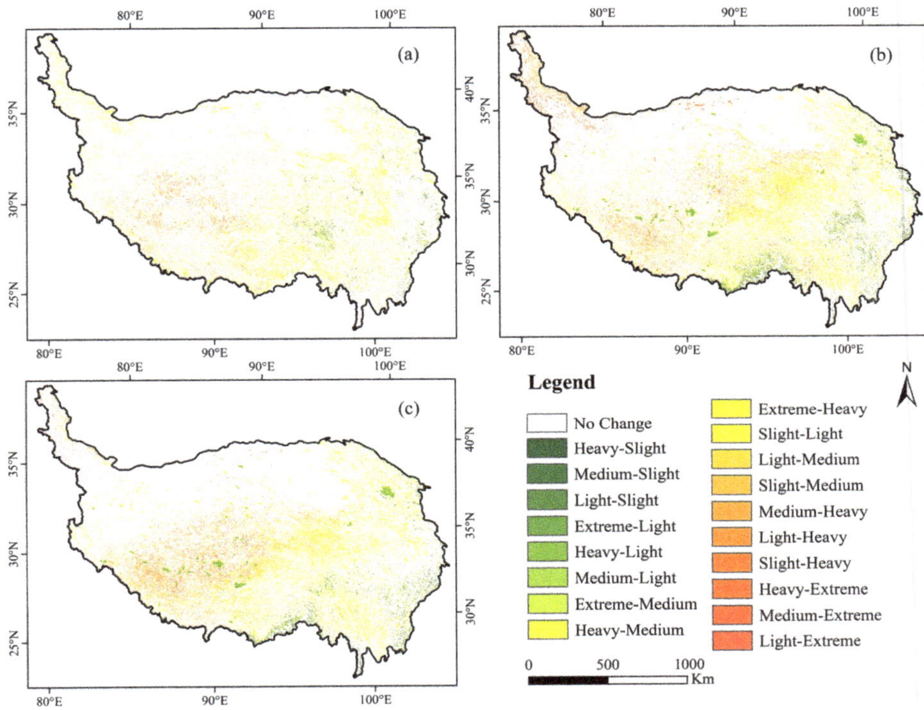

Figure 6. Temporal variations in ecosystem vulnerability on the Qinghai-Tibet Plateau in (**a**) 2005–2010, (**b**) 2010–2015, (**c**) 2005–2010.

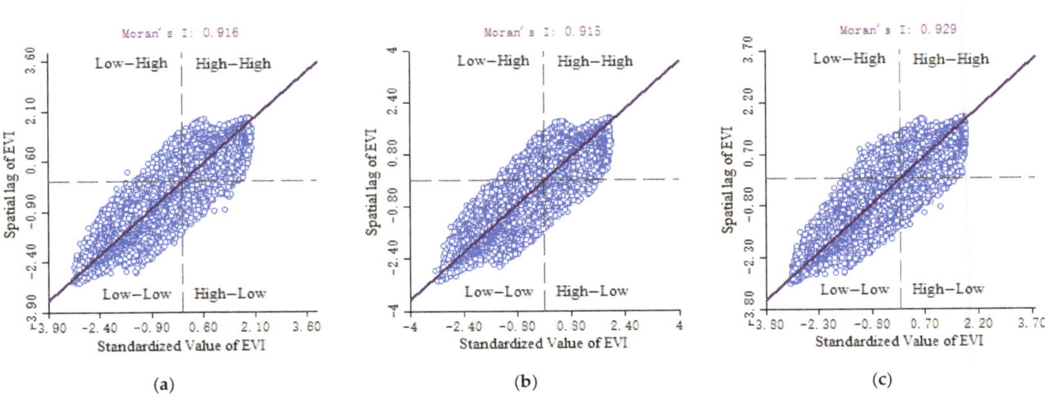

Figure 7. Moran scatterplot of EVI on the Qinghai-Tibet Plateau in (**a**) 2005, (**b**) 2010, (**c**) 2015.

Using the local spatial autocorrelation index, the distribution of EVI spatial clustering characteristics and the spatial variation difference on the time scale can be seen intuitively from 2005 to 2015 (Figure 8). In 2005, 2010 and 2015, the distribution of spatial clustering characteristics was similar. The spatial clustering characteristics of EVI on the Qinghai-Tibet

Plateau were mainly high-high and low-low. The high-high area was mainly distributed on in the Kunlun Alpine Plateau and in the Qaidam Basin in the northwest of the Qinghai-Tibet Plateau, with heavy and extreme vulnerability. The low-low are was mainly distributed in the southeast, with slight and light vulnerability. The southern part showed insignificant spatial clustering distribution, mainly with medium vulnerability. Compared with other cluster types, the distribution range of the low-high cluster type was lower. The distribution range of the high-low agglomeration was the smallest; such areas were scattered in the transition area from low-low to high-high clusters.

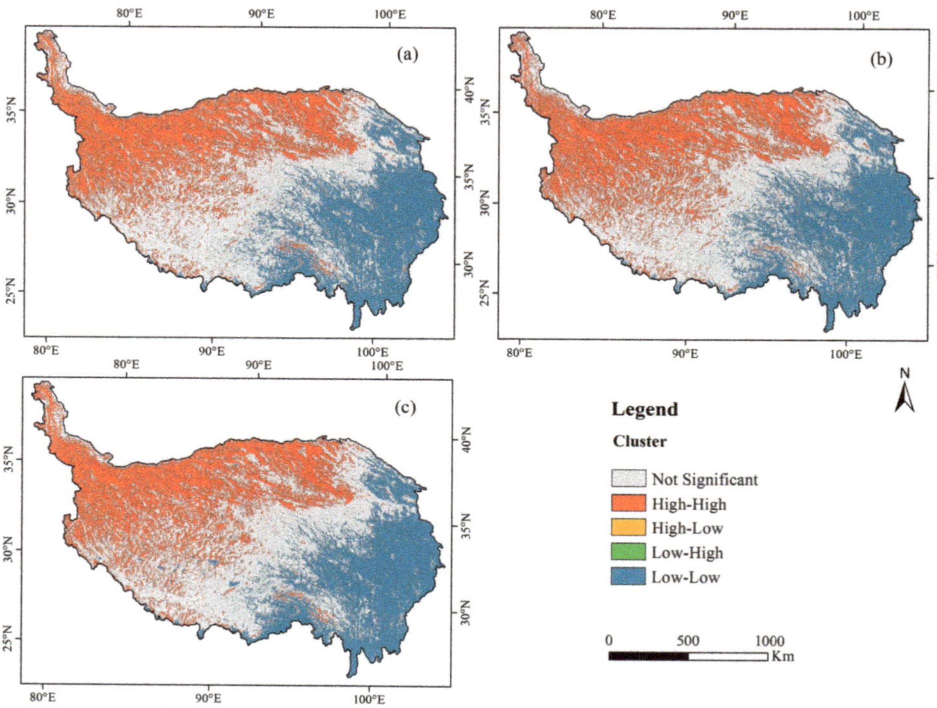

Figure 8. Local spatial autocorrelation diagram for the Qinghai-Tibet Plateau in (**a**) 2005, (**b**) 2010, (**c**) 2015.

5.3. Determinants and Interactions of EVI

In this study, we used the geographic detector method to determine the importance and mutual influence of potential determinants of ecosystem vulnerability. The EVI mean values for the years 2005–2015 were selected as dependent variables, and the corresponding assessment indicators included socio-economic and natural factors. One of the most important findings of this analysis is that natural factors contribute more significantly to EVI than socio-economic factors.

By using factor detectors in geographical detectors, the q-statistics of the explanatory power of each influencing factor to ecosystem vulnerability could be obtained (Table 8). The q-statistics for natural factors ranged from 0.036 to 0.918, with an average value of 0.449. All factors were statistically significant. The determinants of these factors (in descending order) were the normalized difference vegetation index (NDVI), net primary productivity (NPP), evapotranspiration (ET), annual precipitation (PRE), annual mean temperature (TEM), elevation (ELE), Degree of Relief (DR), slope and surface cutting degree (SCD). These results indicate that vegetation types and climatic characteristics are important determinants of the spatial distribution of ecosystem vulnerability, whereas the effect of topography is relatively weak.

Table 8. Results for different factors of EVI.

Factors	NDVI	NPP	ET	PRE	TEM	ELE	DR	Slope	LUR	PD	SCD	GDPD
q statistic	0.918	0.868	0.746	0.600	0.334	0.239	0.152	0.150	0.067	0.063	0.036	0.022
p Value	0.000	0.000	0.000	0.000	0.000	0.000	0.000	0.000	0.000	0.000	0.000	0.000

The q-statistical values of socio-economic factors ranged between 0.022 and 0.067, with an average value of 0.051 (Table 8). The determinants of the socio-economic factors obtained here can be ranked in descending order of land use ratio (LUR), population density (PD) and GDP density (GDPD). Overall, q-statistics show that LUR, PD and GDPD (in descending order) can significantly explain the spatial changes of EVI for the entire Qinghai-Tibet Plateau. The total value ranged between 2.21 and 6.71%.

In addition to exploring the effects of single factors on ecosystem vulnerability, we also used the interactive detection module in geographical detectors to analyze the effects of two factors on ecosystem vulnerability. The results show that the interaction between the two factors exceeded that only of a single factor (Figure 9). The effects of NDVI and ET interaction on ecosystem vulnerability were the most significant, indicating that vegetation and surface evapotranspiration were the main factors affecting ecosystem vulnerability on the Qinghai-Tibet Plateau. The q-statistics between socio-economic factors were small, but the interaction between the socio-economic and natural factors also strongly affected EVI.

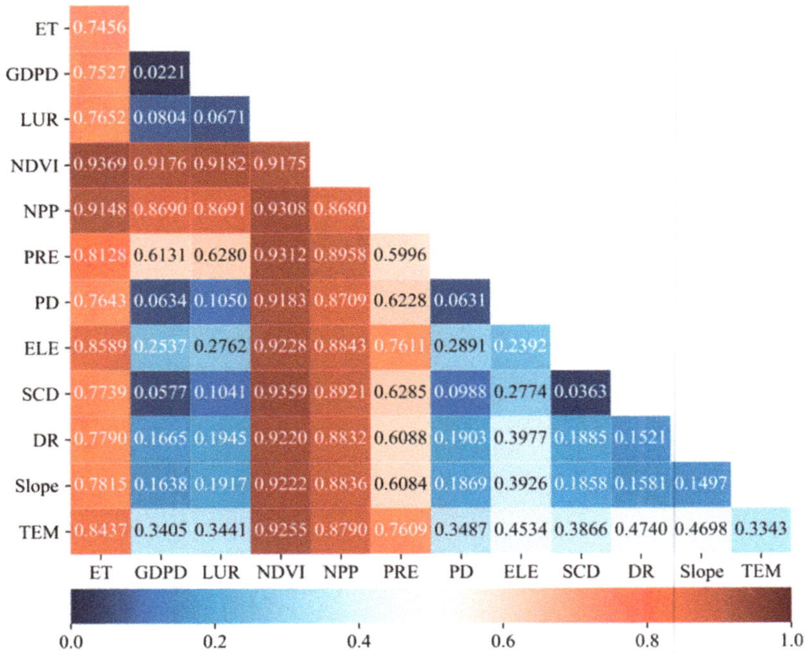

Figure 9. Interactions between pairs of forces influencing EVI. (Notes: the q-statistic on the diagonal line in each case denotes the separate effects of each variable (Table 2), whereas the lower periodic matrix includes values for interactive effects between private sources.).

There were two types of interactive detection results, namely bi-enhanced and nonlinear enhanced effects (Figure 10). Most of the interaction of the two factors showed bi-enhanced effects, and a few showed nonlinear enhanced effects. In fact, the bi-enhanced effects were most often observed for topographic factors (e.g., between elevation and slope and between elevation and slope), which means that the interaction effect was more significant than that produced by a single factor. The interaction effects exhibited nonlinear enhanced effects (such as NDVI, NPP, climate and other factors), indicating that they exceeded the effects of the sum of their individual factors.

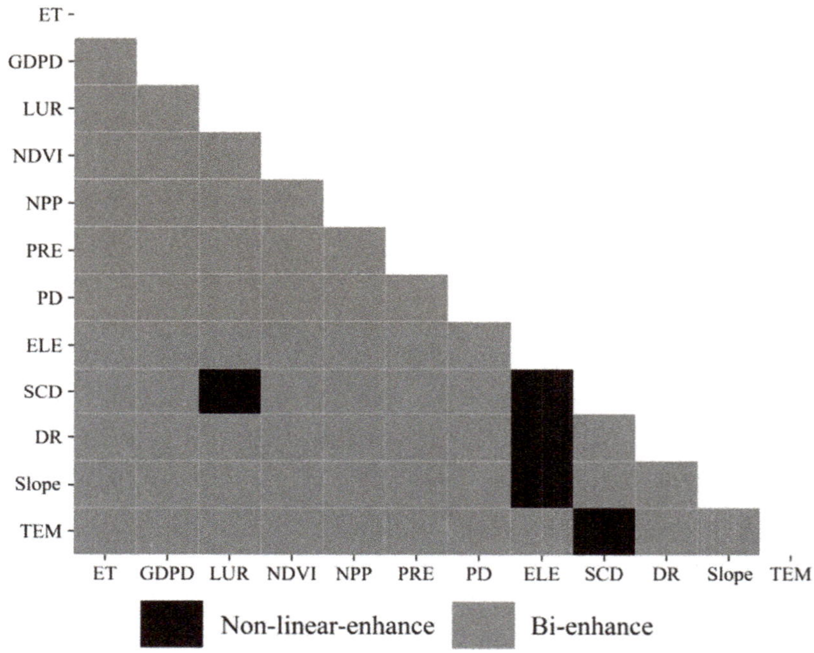

Figure 10. Interaction type between pairs of forces influencing EVI.

6. Discussion

6.1. Spatial Distribution of Ecosystem Vulnerability

Based on remote sensing data, we used spatial principal components analysis to evaluate the ecosystem vulnerability of the Qinghai-Tibet Plateau for the years 2005, 2010 and 2015 at a spatial resolution of 1 km. The distribution of ecosystem vulnerability showed significant spatial differences, and the overall distribution trend gradually increased from southeast to northwest. The spatial distribution characteristics were similar to those of previous studies on Tibetan Plateau vulnerability [61]. The ecosystem vulnerability of the Qinghai-Tibet Plateau is mainly heavy and extreme, whereas previous studies found medium or heavy vulnerability; these differences might be related to the boundary of the study area and the criteria of vulnerability classification. Previous studies have focused on some areas of the Qinghai-Tibet Plateau, such as the Tibet Plateau [61], the Three-River-Source Area [26], of Delhi City [21]. Compared with previous studies, we expanded the research area to cover the entire Qinghai-Tibet Plateau. However, there is a large desert area in the northwest, resulting in mainly heavy and extreme overall vulnerability.

6.2. Effects of Natural and Socio-Economic Factors on Ecological Vulnerability

The q-statistical values for natural factors based on GDM ranged from 0.036 to 0.918, with an average of 0.449, whereas those of socio-economic factors were between 0.022 and 0.067, with an average value of 0.051. Therefore, the spatiotemporal variation of EVI mainly depended on natural factors and their changes than on socio-economic factors. Based on analyzing the effects of single factors on vulnerability, we discuss the influences of two factors on vulnerability. The results indicate that NDVI and ET interaction showed the greatest explanatory power to ecosystem vulnerability, instead of NDVI and NPP with the highest single-factor explanatory power. The NPP of the Qinghai-Tibet Plateau decreased gradually from southeast to northwest, showing significant spatial correlation with NDVI, giving it a certain consistency in explaining ecosystem vulnerability. The parameters NDVI and ET can more accurately reflect ecosystem conditions in terms of vegetation and climate than NDVI and NPP interactions. Interaction detection can supplement the analysis results of single-factor detection. A previous study has shown that ecosystem quality is highly positively correlated with NDVI and NPP [26]. Therefore, a decrease in vegetation coverage will inevitably lead to an increase in ecosystem vulnerability. Affected by global warming, rising temperatures with result in increased ET and, subsequently, a loss in soil moisture.

6.3. Policy Implications

The Qinghai-Tibet Plateau is an important ecological security barrier for China and even Asia, and the Chinese government attaches great importance to the construction of an ecological civilization on the plateau. Based on the mapping of the vulnerability of the Qinghai-Tibet Plateau ecosystem, the spatial distribution of high- and low-vulnerability areas can be seen. This provides clear evidence for the selection of pilot projects for ecological protection and restoration of the Qinghai-Tibet Plateau. The vegetation types were mainly grassland and desert in the high-vulnerability area in the northwest. Ecological protection and restoration should be therefore be emphasized in this area to prevent grassland degradation and desertification. The vegetation coverage in the low-vulnerability areas in the southeast was high and there were significant human activity impacts. The regional government should therefore improve the ecological compensation mechanism and balance the needs of ecological protection and residential development. In the analysis of factor detection, the q-statistics for evapotranspiration and precipitation reached 0.746 and 0.6, respectively, indicating significant effects on ecosystem vulnerability. Therefore, when carrying out ecosystem restoration, it is not only necessary to combine the characteristics of the ecosystem itself, but also to consider the impacts of climate change. For example, Jiang et al. [77] studied the changes in ecosystem services on the Loess Plateau and stated that ecosystem protection needs to consider climate change. In addition, human activities, such as excessive livestock production, which leads to overgrazing, will also have a great impact on the ecological environment. Chen et al. [78], studying the ecosystem of the Mongolian Plateau, showed that the impact of human activities exceeds that of natural environmental changes. Therefore, the future protection of ecosystems should not ignore human interference, and sustainable human activity is a factor to be considered in ecological restoration. For example, a moderate grazing intensity can improve grassland adaptability and reduce grassland vulnerability [27].

6.4. Limitations and Future Research Perspectives

In this study, we investigated the influences of natural and socioeconomic factors on the spatial distribution of ecosystem vulnerability. However, there were some limitations and areas of uncertainty. First, ecosystem fragility covers many factors such as nature, economy, society and policies. Due to limitations, such as the inaccessibility of data sources or the difficulty of spatial expression, some indicators compared to other ecosystem vulnerability studies are not included in the indicator system. There is no uniform standard for the selection of sensitivity, resilience and pressure indices. In this paper, climatic conditions are classified as sensitivity index, but some scholars classify them as exposure index (exposure usually refers to the interference degree of environmental and socioeconomic pressure on the ecosystem) [79]. Even if the same ecosystem vulnerability assessment model is selected, climate factors are also divided into different index categories. For example, based on same Exposure-Sensitive-Adaptive model, Jiang et al. [79] takes meteorological factors as exposure, whereas Zheng et al. [80] divides them into sensitivity indicator. Therefore, the scientific index selection method to assess ecosystem vulnerability remains to be explored in depth. Second, the ecosystem vulnerability of the Qinghai-Tibet Plateau was divided into five levels, with only relative differences. For example, the slight vulnerability in this article may be medium or heavy in other areas. Therefore, the classification standard of ecosystem vulnerability is not applicable to areas outside the study area.

7. Conclusions

We explored the spatial and temporal differentiation characteristics of the Qinghai-Tibet Plateau ecosystem vulnerability and its driving factors. The Qinghai-Tibet Plateau was mainly in a heavy and extreme vulnerability state from the years 2005 to 2015. The ecosystem vulnerability in the northwest was greater than that in the southeast. The vulnerability grade gradually increased from southeast to northwest. Overall, ecosystem vulnerability deteriorated slightly in 2005–2015. The spatial distribution of EVI showed significant clustering. The high-value area was mainly concentrated in the northwest and the low-values are in the southeast.

The EVI spatial distribution was mainly affected by natural factors. The intensity of these effects followed the order NDVI, NPP, ET, PRE, TEM, ELE, DR, Slope and SCD. Vegetation growth and hydrothermal conditions had significant effects on changes in ecosystem vulnerability. We could also show that socio-economic factors exerted a less significant effect on EVI, on average, than natural factors. The q-statistics for these variables followed the order LUR, PD and GDPD. The types of factor interactions were mainly bi-enhanced, with some showing nonlinear enhanced effects. The explanatory power of factor interaction for EVI was greater than that of single factors. The interaction of NDVI and ET had the greatest explanatory power on ecological vulnerability.

Our findings can serve as a scientific base for the establishment of policy implications. Larger efforts are needed to ensure ecological protection and restoration and to prevent grassland degradation and desertification in the high-EVI areas in the northwest. The government should also improve the ecological compensation mechanism and balance ecological protection and residents' development needs in the southeast. In addition, in the process of ecosystem restoration, it is not only necessary to combine the characteristics of the ecosystem itself, but also to consider the impacts of a changing climate.

Author Contributions: Conceptualization, W.S.; methodology, H.L. and W.S.; formal analysis, H.L. and W.S.; investigation, W.S. and H.L.; resources, W.S.; writing—original draft preparation, H.L. and W.S.; writing—review and editing, W.S.; supervision, W.S. Both authors have read and agreed to the published version of the manuscript.

Funding: This research was supported by the Second Tibetan Plateau Scientific Expedition and Research (grant number 2019QZKK0603), the Strategic Priority Research of the Chinese Academy of Sciences (grant number XDA20040201) and the Projects of National Natural Science Foundation of China (grant number 42071233).

Institutional Review Board Statement: Not applicable.

Informed Consent Statement: Not applicable.

Data Availability Statement: All relevant data sets in this study are described in the manuscript.

Conflicts of Interest: The authors declare no conflict of interest.

References

1. Gallopín, G.C. Linkages between vulnerability, resilience, and adaptive capacity. *Glob. Environ. Chang.* **2006**, *16*, 293–303. [CrossRef]
2. Adger, W.N. Vulnerability. *Glob. Environ. Chang.* **2006**, *16*, 268–281. [CrossRef]
3. Abutaleb, K.A.A.; Asmaa, M.; Hassan, E.; Ahmed, M. Climate change impacts, vulnerabilities and adaption measures for Egypt's Nile Delta. *Earth Syst. Environ.* **2018**, *2*, 183–192. [CrossRef]
4. Wang, X.; Zhong, X.; Gao, P. A GIS-based decision support system for regional eco-security assessment and its application on the Tibetan Plateau. *J. Environ. Manag.* **2010**, *91*, 1981–1990. [CrossRef]
5. Turner, B.L.; Kasperson, R.E.; Matson, P.A.; McCarthy, J.J.; Corell, R.W.; Christensen, L.; Eckley, N.; Kasperson, J.X.; Luers, A.; Martello, M.L.; et al. A framework for vulnerability analysis in sustainability science. *Proc. Natl. Acad. Sci. USA* **2003**, *100*, 8074–8079. [CrossRef] [PubMed]
6. Polsky, C.; Neff, R.; Yarnal, B. Building comparable global change vulnerability assessments: The vulnerability scoping diagram. *Glob. Environ. Chang.* **2007**, *17*, 472–485. [CrossRef]
7. De Lange, H.J.; Sala, S.; Vighi, M.; Faber, J.H. Ecological vulnerability in risk assessment—A review and perspectives. *Sci. Total Environ.* **2010**, *408*, 3871–3879. [CrossRef] [PubMed]
8. Wu, G.; Li, L.; Ahmad, S.; Chen, X.; Pan, X. A dynamic model for vulnerability assessment of regional water resources in arid areas: A case study of Bayingolin, China. *Water Resour. Manag.* **2013**, *27*, 3085–3101. [CrossRef]
9. Ippolito, A.; Sala, S.; Faber, J.H.; Vighi, M. Ecological vulnerability analysis: A river basin case study. *Sci. Total Environ.* **2010**, *408*, 3880–3890. [CrossRef]
10. Kang, H.; Tao, W.; Chang, Y.; Zhang, Y.; Xuxiang, L.; Chen, P. A feasible method for the division of ecological vulnerability and its driving forces in Southern Shaanxi. *J. Clean. Prod.* **2018**, *205*. [CrossRef]
11. Gonzalez, P.; Neilson, R.; Lenihan, J.; Drapek, R. Global patterns in the vulnerability of ecosystems to vegetation shifts due to climate change. *Glob. Ecol. Biogeogr.* **2010**, *19*, 755–768. [CrossRef]
12. Velepucha, P.A.E.; Chamba, J.A.M.; Mendoza, N.A.A.; Luna, T.L.O.; Rojas, N.S.S.; Furniss, M.J.; Howe, C.; Mendoza, Z.H.A. Tropical ecosystems vulnerability to climate change in southern Ecuador. *Trop. Conserv. Sci.* **2016**, *9*. [CrossRef]
13. Comer, P.; Hak, J.; Reid, M.; Auer, S.; Schulz, K.; Hamilton, H.; Smyth, R.; Kling, M. Habitat climate change vulnerability index applied to major vegetation types of the western interior United States. *Land* **2019**, *8*, 108. [CrossRef]
14. Teck, S.; Halpern, B.; Kappel, C.; Micheli, F.; Selkoe, K.; Crain, C.; Martone, R.; Shearer, C.; Arvai, J.; Fischhoff, B.; et al. Using expert judgment to estimate marine ecosystem vulnerability in the California Current. *Ecol. Appl.* **2010**, *20*, 1402–1416. [CrossRef]
15. Zhang, Y.; Tao, B.; Li, Y. Assessment on the vulnerability of different ecosystems to extreme rainfalls in the middle and lower reaches of Yangtze River. *Theor. Appl. Climatol.* **2015**, *121*, 157–166. [CrossRef]
16. Xin, M.; Martin, D.J.; Baiqing, S.; Xin, B. Nouveauté or Cliché? Assessment on island ecological vulnerability to Tourism: Application to Zhoushan, China. *Ecol. Indic.* **2019**, *113*. [CrossRef]
17. Gao, S.; Sun, H.; Zhao, L.; Wang, R.; Xu, M.; Cao, G. Dynamic assessment of island ecological environment sustainability under urbanization based on rough set, synthetic index and catastrophe progression analysis theories. *Ocean Coast. Manag.* **2019**, *178*, 104790. [CrossRef]
18. Sun, B.; Ma, X.; de Jong, M.; Bao, X. Assessment on Island Ecological Vulnerability to Urbanization: A Tale of Chongming Island, China. *Sustainability* **2019**, *11*, 2536. [CrossRef]
19. Malekmohammadi, B.; Jahanishakib, F. Vulnerability assessment of wetland landscape ecosystem services using driver-pressure-state-impact-response (DPSIR) model. *Ecol. Indic.* **2017**, *82*, 293–303. [CrossRef]
20. Zhang, X.; Fu, X.; Zhang, L. Ecological vulnerability assessment of estuarine wetland of the Yellow River Delta. *J. Interdiscip. Math.* **2016**, *19*. [CrossRef]
21. Jin, X.; Jin, Y.; Mao, X. Ecological risk assessment of cities on the Tibetan Plateau based on land use/land cover changes—Case study of Delingha City. *Ecol. Indic.* **2019**, *101*, 185–191. [CrossRef]
22. Niu, W. The discriminatory index with regard to the weakness, overlapness, and breadth of ecotone. *Acta Ecol. Sin.* **1989**, *9*, 97–105.
23. Kerang, L.; Yufeng, C. Analysis on vulnerability of forest in China responsed to global climatie change. *Acta Geogr. Sin.* **1996**, *51*, 40–49.
24. Qang, G.Z.; Yan, L.J.; da Fang, Z. The relations analysis between ecological environmental quality of Chinese land resources and population. *J. Remote Sens.* **1999**, *1*, 66–70. [CrossRef]
25. Erda, L. Agricultural vulnerability and adaptation to global warming in China. *Water Air Soil Pollut.* **1996**, *92*, 63–73. [CrossRef]
26. Guo, B.; Zhou, Y.; Zhu, J.; Liu, W.; Wang, F.; Wang, L.; Yan, F.; Wang, F.; Yang, G.; Luo, W.; et al. Spatial patterns of ecosystem vulnerability changes during 2001–2011 in the three-river source region of the Qinghai-Tibetan Plateau, China. *J. Arid Land* **2016**, *8*, 23–35. [CrossRef]
27. Meng, L.; Zhang, X.; Yongtao, H.; Niu, B.; Wu, J. Assessment of the vulnerability of alpine grasslands on the Qinghai-Tibetan Plateau. *PeerJ* **2020**, *8*, e8513. [CrossRef]
28. Bhatt, R.; Hossain, A. Concept and consequence of evapotranspiration for sustainable crop production in the era of climate change. *Adv. Evapotranspiration Methods Appl.* **2019**, *1*, 1–13. [CrossRef]
29. Burton, I.; Kates, R.W.; White, G.F. *The Environment as Hazard*, 2nd ed.; Guilford Press: New York, NY, USA, 1993.

30. Zobler, L.; White, G. Natural hazards: Local, national, global. *Geogr. Rev.* **1976**, *66*, 247. [CrossRef]
31. Timmerman, P. Vulnerability. Resilience and the collapse of socieiy: A review of models and possible climatic applications. *Ecol. Monogr.* **1981**, 1–46. [CrossRef]
32. Dow, K. Exploring differences in our common future(s): The meaning of vulnerability to global environmental change. *Geoforum* **1992**, *23*, 417–436. [CrossRef]
33. Cutter, S.L. Vulnerability to environmental hazards. *Prog. Hum. Geogr.* **1996**, *20*, 529–539. [CrossRef]
34. Parry, M.; Parry, M.L.; Canziani, O.; Palutikof, J.; van der Linden, P.; Hanson, C. *Climate Change 2007-Impacts, Adaptation and Vulnerability: Working Group II Contribution to the Fourth Assessment Report of the IPCC*; Cambridge University Press: Cambridge, UK, 2007; Volume 4.
35. Lee, Y.-J. Social vulnerability indicators as a sustainable planning tool. *Environ. Impact Assess. Rev.* **2014**, *44*, 31–42. [CrossRef]
36. Kormondy, E.J. A brief introduction to the history of ecology. *Am. Biol. Teach.* **2012**, *74*, 441–443. [CrossRef]
37. Williams, L.R.R.; Kapustka, L.A. Ecosystem vulnerability: A complex interface with technical components. *Environ. Toxicol. Chem.* **2000**, *19*, 1055–1058. [CrossRef]
38. Birkmann, J. Risk and vulnerability indicators at different scales: Applicability, usefulness and policy implications. *Environ. Hazards* **2007**, *7*, 20–31. [CrossRef]
39. McCarthy, J.J.; Canziani, O.F.; Leary, N.; Dokken, D.J.; White, K.S. Climate change 2001: Impacts, adaptation, and vulnerability. Contribution of working group II to the third assessment report of the Intergovernmental Panel on Climate Change (IPCC). *Glob. Ecol. Biogeogr.* **2001**, *12*, 87–88. [CrossRef]
40. Dieleman, H. Urban agriculture in Mexico City; balancing between ecological, economic, social and symbolic value. *J. Clean. Prod.* **2017**, *163*, S156–S163. [CrossRef]
41. Zhang, F.; Liu, X.; Zhang, J.; Wu, R.; Ma, Q.; Chen, Y. Ecological vulnerability assessment based on multi-sources data and SD model in Yinma River Basin, China. *Ecol. Modell.* **2017**, *349*. [CrossRef]
42. Zhang, X.; Wang, L.; Fu, X.; Li, H.; Xu, C. Ecological vulnerability assessment based on PSSR in Yellow River Delta. *J. Clean. Prod.* **2017**, *167*. [CrossRef]
43. Zhou, Y.; Yuyang, Z.; Guofu, Z.; Qifen, H.; Li, Y. Evaluation of ecological vulnerability of county areas in Karst mountain area based on VSD model—A case study of Duyun City. In *IOP Conference Series: Earth and Environmental Science*; IOP Publishing: Bristol, UK, 2020.
44. Sun, P.J.; Xiu, C.L. Study on the vulnerability of economic development in mining cities based on the PSE Model. *Geogr. Res.* **2011**, *30*, 301–310. [CrossRef]
45. Liu, Z.; Yu, X.; Li, L.; Huang, M. Vulnerability assessment of eco-environment in Yimeng mountainous area of Shandong Province based on SRP conceptual model. *J. Appl. Ecol.* **2011**, *22*, 2084–2090. [CrossRef]
46. Thirumalaivasan, D.; Karmegam, M.; Venugopal, K. AHP-DRASTIC: Software for specific aquifer vulnerability assessment using DRASTIC model and GIS. *Environ. Model. Softw.* **2003**, *18*, 645–656. [CrossRef]
47. Dixon, B. Groundwater vulnerability mapping: A GIS and fuzzy rule based integrated tool. *Appl. Geogr.* **2005**, *25*, 327–347. [CrossRef]
48. Park, Y.S.; Chon, T.S.; Kwak, I.S.; Lek, S. Hierarchical community classification and assessment of aquatic ecosystems using artificial neural networks. *Sci. Total Environ.* **2004**, *327*, 105–122. [CrossRef]
49. Gang, H.; Keyu, B.; Wenwen, W.; Yanna, Z.; Shuzhou, L.; Lan, J. Assessment of ecological vulnerability of resource-based cities based on entropy-set pair analysis. *Environ. Technol.* **2019**, *42*, 874–1884. [CrossRef]
50. Wang, Y.; Ding, Q.; Zhuang, D. An eco-city evaluation method based on spatial analysis technology: A case study of Jiangsu Province, China. *Ecol. Indic.* **2015**, *58*, 37–46. [CrossRef]
51. Twumasi, Y.A.; Merem, E.C. Using remote sensing and GIS in the analysis of ecosystem decline along the River Niger Basin: The case of Mali and Niger. *Int. J. Environ. Res. Public Health* **2007**, *4*, 173–184. [CrossRef]
52. Xue, L.; Wang, J.; Zhang, L.; Wei, G.; Zhu, B. Spatiotemporal analysis of ecological vulnerability and management in the Tarim River Basin, China. *Sci. Total Environ.* **2018**, *649*. [CrossRef]
53. Guo, B.; Zang, W.; Luo, W. Spatial-temporal shifts of ecological vulnerability of Karst Mountain ecosystem-impacts of global change and anthropogenic interference. *Sci. Total Environ.* **2020**, *741*, 140256. [CrossRef]
54. Jia, J.J.; Zhao, J.; Wang, J.B.; Gao, C.; Chang, R.Y. Ecological vulnerability assessment of Shiyang River basin based on SRP model. *J. Arid Land Resour. Environ.* **2020**, *34*, 34–41.
55. Chuvieco, E.; Martínez, S.; Román, M.V.; Hantson, S.; Pettinari, M.L. Integration of ecological and socio-economic factors to assess global vulnerability to wildfire. *Glob. Ecol. Biogeogr.* **2014**, *23*, 245–258. [CrossRef]
56. Ding, Q.; Shi, X.; Zhuang, D.; Wang, Y. Temporal and spatial distributions of ecological vulnerability under the influence of natural and anthropogenic factors in an eco-province under construction in China. *Sustainability* **2018**, *10*, 87. [CrossRef]
57. Li, R.; Han, R.; Yu, Q.; Qi, S.; Guo, L. Spatial heterogeneous of ecological vulnerability in arid and semi-arid area: A case of the Ningxia Hui autonomous region, China. *Sustainability* **2020**, *12*, 4401. [CrossRef]
58. Parson, E.A.; Corell, R.W.; Barron, E.J.; Burkett, V.; Janetos, A.; Joyce, L.; Karl, T.R.; MacCracken, M.C.; Melillo, J.; Morgan, M.G.; et al. Understanding climatic impacts, vulnerabilities, and adaptation in the United States: Building a capacity for assessment. *Clim. Chang.* **2003**, *57*, 9–42. [CrossRef]
59. National Aeronautics and Space Administration. Available online: https://search.earthdata.nasa.gov (accessed on 10 March 2021).

60. Resource Environmental Science and Data Center, Chinese Academy of Sciences. Available online: http://www.resdc.cn/Default.aspx (accessed on 5 March 2021).
61. Gao, J.; Hou, W.; Zhao, D.; Wu, S. Comprehensive assessment of natural ecosystem vulnerability in Tibetan Plateau based on satellite-derived datasets. *Sci. Agric. Sin.* **2016**, *36*, 580–587. [CrossRef]
62. Zhao, M.; Running, S. Drought-induced reduction in global terrestrial net primary production from 2000 through 2009. *Science* **2010**, *329*, 940–943. [CrossRef]
63. Luo, H.; Wang, L.; Fang, J.; Li, Y.; Li, H.; Dai, S. *NDVI, Temperature and Precipitation Variables and Their Relationships in Hainan Island from 2001 to 2014 Based on MODIS NDVI*; Springer: Berlin/Heidelberg, Germany, 2016.
64. Imhoff, M.L.; Bounoua, L.; de Fries, R.; Lawrence, W.T.; Stutzer, D.; Tucker, C.J.; Ricketts, T. The consequences of urban land transformation on net primary productivity in the United States. *Remote Sens. Environ.* **2004**, *89*, 434–443. [CrossRef]
65. Zhao, J.; Ji, G.; Tian, Y.; Chen, Y.; Wang, Z. Environmental vulnerability assessment for mainland China based on entropy method. *Ecol. Indic.* **2018**, *91*, 410–422. [CrossRef]
66. Jolliffe, I.T.; Morgan, B.J. Principal component analysis and exploratory factor analysis. *Stat. Methods Med. Res.* **1992**, *1*, 69–95. [CrossRef]
67. Zou, T.; Yoshino, K. Environmental vulnerability evaluation using a spatial principal components approach in the Daxing'anling region, China. *Ecol. Indic.* **2017**, *78*, 405–415. [CrossRef]
68. Liu, Q.; Shi, T. Spatiotemporal differentiation and the factors of ecological vulnerability in the Toutun River Basin based on remote sensing data. *Sustainability* **2019**, *11*, 4160. [CrossRef]
69. Haining, R. *Spatial Data Analysis in the Social and Environmental Sciences*; Cambridge University Press: Cambridge, UK, 1993.
70. Anselin, L. *Spatial Econometrics: Methods and Models*; Springer Science & Business Media: Boston, MA, USA, 1988.
71. Moran, P.A. The interpretation of statistical maps. *J. R. Stat. Soc.* **1948**, *10*, 243–251. [CrossRef]
72. Anselin, L. Local Indicators of Spatial Association—LISA. *Geogr. Anal.* **1995**, *27*, 93–115. [CrossRef]
73. Shi, T.; Hu, Z.; Shi, Z.; Guo, L.; Chen, Y.; Li, Q.; Wu, G. Geo-detection of factors controlling spatial patterns of heavy metals in urban topsoil using multi-source data. *Sci. Total Environ.* **2018**, *643*, 451–459. [CrossRef]
74. Cao, W.; Yuan, X. Region-county characteristic of spatial-temporal evolution and influencing factor on land use-related CO_2 emissions in Chongqing of China, 1997–2015. *J. Clean. Prod.* **2019**, *231*, 619–632. [CrossRef]
75. Chen, S.; Liu, X.; Wu, Y.; Xu, G.; Zhang, X.; Mei, S.; Zhang, Z.; O'Meara, M.; O'Gara, M.C.; Tan, X.; et al. The application of meteorological data and search index data in improving the prediction of HFMD: A study of two cities in Guangdong Province, China. *Sci. Total Environ.* **2019**, *652*, 1013–1021. [CrossRef]
76. Wang, J.; Xu, C. Geodetector: Principle and prospective. *Acta Geogr. Sin.* **2017**, *72*, 116–134. [CrossRef]
77. Jiang, C.; Wang, F.; Zhang, H.; Dong, X. Quantifying changes in multiple ecosystem services during 2000–2012 on the Loess Plateau, China, as a result of climate variability and ecological restoration. *Ecol. Eng.* **2016**, *97*, 258–271. [CrossRef]
78. Chen, J.; John, R.; Shao, C.; Fan, Y.; Zhang, Y.; Amarjargal, A.; Brown, D.G.; Qi, J.; Han, J.; Lafortezza, R. Policy shifts influence the functional changes of the CNH systems on the Mongolian plateau. *Environ. Res. Lett.* **2015**, *10*. [CrossRef]
79. Jiang, L.; Huang, X.; Wang, F.; Liu, Y.; An, P. Method for evaluating ecological vulnerability under climate change based on remote sensing: A case study. *Ecol. Indic.* **2018**, *85*. [CrossRef]
80. Zang, Z.; Zou, X.; Zuo, P.; Song, Q.; Wang, C.; Wang, J. Impact of landscape patterns on ecological vulnerability and ecosystem service values: An empirical analysis of Yancheng Nature Reserve in China. *Ecol. Indic.* **2017**, *72*. [CrossRef]

International Journal of
Environmental Research and Public Health

MDPI

Article

Effects of Vegetation Restoration on Soil Erosion on the Loess Plateau: A Case Study in the Ansai Watershed

Hui Wei [1,2], Wenwu Zhao [1,2,*] and Han Wang [1,2]

1 State Key Laboratory of Earth Surface Processes and Resource Ecology, Faculty of Geographical Science, Beijing Normal University, Beijing 100875, China; irene1993weihui@163.com (H.W.); hanwang_226@163.com (H.W.)
2 Institute of Land Surface System and Sustainable Development, Faculty of Geographical Science, Beijing Normal University, Beijing 100875, China
* Correspondence: zhaoww@bnu.edu.cn

Abstract: Large-scale vegetation restoration greatly changed the soil erosion environment in the Loess Plateau since the implementation of the "Grain for Green Project" (GGP) in 1999. Evaluating the effects of vegetation restoration on soil erosion is significant to local soil and water conservation and vegetation construction. Taking the Ansai Watershed as the case area, this study calculated the soil erosion modulus from 2000 to 2015 under the initial and current scenarios of vegetation restoration, using the Chinese Soil Loess Equation (CSLE), based on rainfall and soil data, remote sensing images and socio-economic data. The effect of vegetation restoration on soil erosion was evaluated by comparing the average annual soil erosion modulus under two scenarios among 16 years. The results showed: (1) vegetation restoration significantly changed the local land use, characterized by the conversion of farmland to grassland, arboreal land, and shrub land. From 2000 to 2015, the area of arboreal land, shrub land, and grassland increased from 19.46 km^2, 19.43 km^2, and 719.49 km^2 to 99.26 km^2, 75.97 km^2, and 1084.24 km^2; while the farmland area decreased from 547.90 km^2 to 34.35 km^2; (2) the average annual soil erosion modulus from 2000 to 2015 under the initial and current scenarios of vegetation restoration was 114.44 t/(hm^2·a) and 78.42 t/(hm^2·a), respectively, with an average annual reduction of 4.81×10^6 t of soil erosion amount thanks to the vegetation restoration; (3) the dominant soil erosion intensity changed from "severe and light erosion" to "moderate and light erosion", vegetation restoration greatly improved the soil erosion environment in the study area; (4) areas with increased erosion and decreased erosion were alternately distributed, accounting for 48% and 52% of the total land area, and mainly distributed in the northwest and southeast of the watershed, respectively. Irrational land use changes in local areas (such as the conversion of farmland and grassland into construction land, etc.) and the ineffective implementation of vegetation restoration are the main reasons leading to the existence of areas with increased erosion.

Keywords: vegetation restoration; soil erosion; land use; Loess Plateau; Ansai Watershed

Citation: Wei, H.; Zhao, W.; Wang, H. Effects of Vegetation Restoration on Soil Erosion on the Loess Plateau: A Case Study in the Ansai Watershed. *Int. J. Environ. Res. Public Health* **2021**, *18*, 6266. https://doi.org/10.3390/ijerph18126266

Academic Editors: Wei Song, Hualin Xie and Andrew Hursthouse

Received: 9 April 2021
Accepted: 4 June 2021
Published: 10 June 2021

1. Introduction

Approximately 20% of the land area is currently experiencing a decline in productivity linked to erosion, wastage, and pollution in the world [1]. Among these factors, soil erosion not only causes problems such as soil quality decline, land degradation, and loss of farmland resources, but also leads to a series of ecological and environmental problems such as water environment deterioration, river siltation, debris flows, and even flood disasters [2–6]. The global soil erosion area has reached 25 million km^2, accounting for 16.8% of the total land area and threatening the security of 27% of the total farmland area [7]. To this end, soil erosion has become a global ecological and environmental problem [8–12]. The land that has undergone water erosion or wind erosion is up to 3 million km^2 in China, accounting for approximately 32% of the total land area [13]. The Loess Plateau is the region with the most severe soil erosion in China, where the area of soil and water loss is as high as 4.5×10^5 km^2, mainly dominated by intensive erosion (>5000 t/(hm^2·a)), and the average sediment transport over years is 1.6×10^9 t [14]. The area, intensity, and amounts of the soil erosion in the Loess Plateau are the largest in the world [14–16].

To effectively control soil erosion and ecological degradation, the Chinese government implemented the "Grain for Green Project" (GGP) since 1999 to return farmland with slopes of 25° or more to perennial vegetation [2,17,18]. Vegetation restoration triggered by the "GGP" is an effective approach to ecological construction and soil erosion control in the western region of China [19–21]. Since the implementation of the "GGP", the soil erosion environment in the Loess Plateau has been greatly changed by large-scale vegetation restoration [2,9,17–19]. The United Nations General Assembly announced the "United Nations Decade on Ecosystem Restoration 2021–2030 (UNDER)" on 1 March 2019, a movement aimed to expand the restoration of degraded and damaged ecosystems as an effective measure to address the climate crisis and enhance food security, water resources, and biodiversity [1,22]. Under the global background of the UNDER, assessing the effects of vegetation restoration on soil erosion over the past 20 years is significant to sustaining the water and soil conservation benefits of vegetation restoration in the Loess Plateau.

Selecting the Ansai Watershed as the case study area of the Loess Plateau, this study identified the effects of vegetation restoration on soil erosion by comparing the differences between the soil erosion modulus from 2000 to 2015 under two land use scenarios (the initial and current scenarios of vegetation restoration). The research results have important theoretical and practical significance for regional soil and water conservation and vegetation construction.

2. Materials and Methods

2.1. Study Area

The Ansai watershed (108°5′44″–109°26′18″ E, 36°30′45″–37°19′3″ N) is located in the upper reaches of the Yanhe River basin, in the inland hinterland of the northwestern Loess Plateau. This watershed lies in the northern part of Shaanxi Province and borders the Ordos basin (Figure 1). It belongs to the typical loess hilly and gully region, and covers a total area of 1334.00 km^2 [23]. The soil type in the study area is loess soil, with low fertility and high vulnerability to erosion [24,25]. The topography is complex and varied, and the land surface is fragmented into different land uses, dominated by rain-fed farmland, grassland, shrubland, and forest land [26]. The elevations within the watershed are low in the southeast and high in the northwest, ranging between 997 m and 1731 m above sea level [23]. The climate is a continental semi-arid monsoon climate in the middle temperate zone, and the average annual precipitation is 505.3 mm, and 74% of the rainfall occurs from June to September [26].

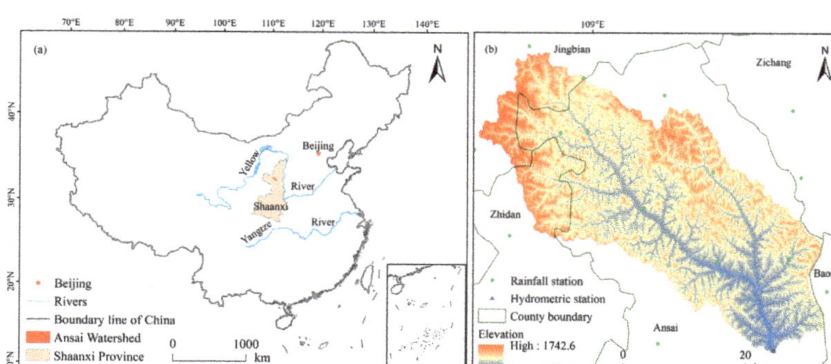

Figure 1. The study area: (**a**) The geographical location of the Ansai Watershed; (**b**) The elevation, rainfall and hydrometric stations distribution in the Ansai Watershed.

2.2. Data Sources

We used 25 m resolution DEM data, obtained from the 1:50,000 database of the National Center for Basic Geographic Information of China [27]. Vector land cover data in 2000 and 2015 was obtained from the Data Center for Resources and Environmental Sciences at the Chinese Academy of Sciences [28]. Daily rainfall data at 20 rainfall stations in and around the Ansai Watershed from 2000 to 2015 was collected from the *Hydrological Yearbook of the People's Republic of China* [29]. Remote sensing images from 2000 to 2015 were obtained from the Geospatial Data Cloud [30]. Terrace and silting dam data from 2000 to 2015 were collected from the *Statistical Yearbook of Ansai County*. Soil data, derived from a dataset of 151 sample points was obtained from a soil survey in the Ansai Watershed conducted in July to August of 2014. In Figure 2, 151 soil sample points are evenly distributed in the Ansai watershed, which can well represent the soil attribute conditions in the study area; the location of sample points is accurately located by handheld GPS.

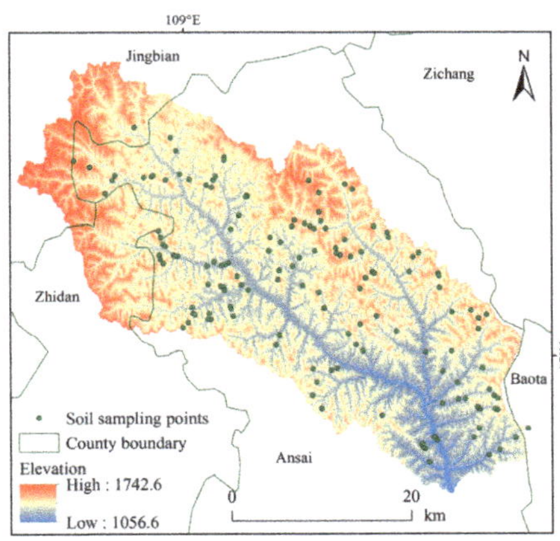

Figure 2. Soil sampling points in the Ansai Watershed.

2.3. Research Methods

Since the 1980s, Chinese scholars proposed some regional models for soil erosion estimation based on the Universal Soil Loss Equation (USLE) and combined with local topographical features [31]. Among these models, the Chinese Soil Loss Equation (CSLE) fully considers the impact of biological, engineering, and tillage measures on the process and results of soil erosion, making it more suitable and widely used in the soil erosion estimation in China [32]. The CSLE model expression is as follows:

$$A = R \cdot K \cdot L \cdot S \cdot B \cdot E \cdot T \quad (1)$$

where A is the average annual soil erosion modulus in t/(hm^2·a); R is the rainfall erosivity factor in MJ·mm/(hm^2·a); K is the soil erodibility factor in t·h/(MJ·mm); L and S are dimensionless factors of slope length and slope steepness, respectively; and B, E, T are dimensionless factors of biological-control, engineering-control, and tillage practices, respectively. The dimensionless factors of slope and soil conservation measures were defined as the ratio of soil erosion amounts from unit plot to actual plot with the aimed factor changed but the same sizes of other factors as the unit plot [32].

Based on the CSLE model and the control variable method, this study calculated the soil erosion modulus in the Ansai Watershed from 2000 to 2015 under two land use scenarios (the initial and current scenarios of vegetation restoration). The effect of vegetation restoration on soil erosion during the study period was identified by comparing the differences of average soil erosion modulus under two scenarios among 16 years. It should be noted that for the soil erosion modulus calculation under the two scenarios in the same year, the R, K, L, S, E, and T factors remained unchanged, while the B factor related to vegetation restoration was calculated based on the land use maps and remote sensing images in 2000 and 2015, respectively. Furthermore, the calculation method of each factor is as follows.

2.3.1. Rainfall Erosivity (R) Factor

Rainfall erosivity (R) factor reflects the influence of rainfall on soil erosion [33]. In this study, we calculated the R factor according to the method proposed by Zhang et al. (2002) [34], a method that has been widely used in China [34,35]. The R factor, based on aggradations of half-month rainfall erosivity, was estimated using daily rainfall data obtained from the *Hydrological Yearbook of the People's Republic of China* from 2000 to 2015. The calculation method is as follows:

$$M_i = \alpha \sum_{j=1}^{k} (D_j)^{\beta} \quad (2)$$

where M_i is the half-month rainfall erosivity in MJ·mm/(hm^2·h·a), k refers to the number of days in a half-month, and D_j represents the effective rainfall for day j in one half-month. D_j is equal to the actual rainfall if the actual rainfall is greater than the threshold value of 12 mm, which is the standard for China's erosive rainfall. Otherwise, D_j is equal to zero [34]. The terms α and β are the undetermined parameters of the model and are calculated as follows:

$$\beta = 0.8363 + \frac{18.177}{\overline{P}_{d12}} + \frac{24.455}{\overline{P}_{y12}} \quad (3)$$

$$\alpha = 21.586\beta^{-7.1891} \quad (4)$$

where $\overline{P}_{d12}$ is the daily average rainfall that is greater than 12 mm, and $\overline{P}_{y12}$ is the yearly average rainfall for days with rainfall more than 12 mm.

2.3.2. Soil Erodibility (K) Factor

Soil erodibility (K) factor indicates both the susceptibility of soil to erosion and the amount and rate of runoff, as measured under standard plot conditions [36]. Previous studies found that the existing foreign K factor estimation models cannot be directly applied to the K factor calculation in China, and their estimated values are far greater than the actual measured values, while there is a certain linear relationship among them [37]. To this end, based on soil data obtained from the soil survey conducted in the Ansai Watershed, the K factor was calculated according to the Equations (5) and (6) [37,38].

$$K_{shirazi} = 7.594\left\{0.0017 + 0.0494e^{-\frac{1}{2}\left[\frac{\log(D_g)+1.675}{0.6986}\right]^2}\right\} \quad (5)$$

$$K = -0.00911 + 0.55066K_{shirazi} \quad (6)$$

where D_g is the geometric mean diameter of soil grains, and $K_{shirazi}$ is the K value estimated by the Equation (5) proposed by Shirazi et al. (1988) [38].

2.3.3. Slope Length (L) and Steepness (S) Factor

Topography is an important factor that directly affects soil erosion. The slope length factor (L) and slope steepness factor (S) represent the effects of slope length and slope gradient on soil erosion, respectively [39]. The L factor and S factor can be calculated using the following equations:

$$L = (\lambda/22.13)^m \quad (7)$$

$$m = \beta/(1+\beta) \quad (8)$$

$$\beta = \left(\frac{\sin\theta}{0.0896}\right)/\left[3.0 \times (\sin\theta)^{0.8} + 0.56\right] \quad (9)$$

$$S = \begin{cases} 10.8 \times \sin\theta + 0.03, & \theta < 5.14^\circ \\ 16.8 \times \sin\theta - 0.50, & \theta \geq 5.14^\circ \end{cases} \quad (10)$$

where λ is the length of the slope, m is the variable length-slope exponent, β is a factor that varies with slope gradient, and θ is slope gradient calculated based on DEM.

2.3.4. Biological-Control (*B*) Factor

Biological-control (*B*) factor refers to the ratio of the soil erosion amounts of land with vegetation cover or field management, and that of continuously fallowed land under certain conditions [40,41]. In this study, we extracted NDVI values and calculated the vegetation coverage by using Equation (11) according to Li et al. (2020) [42] based on remote sensing images captured from June to September during 2000 to 2015; *B* factor was obtained according to the relationship between *B* factor and the land use types, and vegetation coverage (Table 1) [43]. The vegetation coverage was calculated as follows:

$$f = \frac{NDVI - NDVI_{\min}}{NDVI_{\max} - NDVI_{\min}} \quad (11)$$

where f is the vegetation coverage, and $NDVI_{\min}$ and $NDVI_{\max}$ are the minimum and maximum $NDVI$ values.

Table 1. *B* factor under different land use types and different vegetation coverage.

Land Use Type	Vegetation Coverage (%)	*B* Factor	Land Use Type	Vegetation Coverage (%)	*B* Factor
Arboreal and shrub land	0~20	0.100	Grassland	0~20	0.450
	20~40	0.080		20~40	0.240
	40~60	0.060		40~60	0.150
	60~80	0.020		60~80	0.090
	80~100	0.004		80~100	0.043
Water	–	0.000	Farmland	–	0.476
Construction land	–	0.353	Desert land	–	1.000

2.3.5. Engineering-Control (*E*) Factor

Engineering-control (*E*) factor refers to the ratio of the soil erosion amounts occurring under certain engineering measures to that occurring without engineering measures under the same conditions [32]. The engineering-control practices in the Ansai Watershed mainly include silting dams and terraces. Considering the difficulty of collecting data on engineering measures, this study obtained terrace and silting dam data based on the *Statistical Yearbook of Ansai County* and calculated the *E* factor by referring to Equation (12) proposed by Xie et al. (2009) [44]:

$$E = (1 - \frac{S_t}{S} \times \alpha)(1 - \frac{S_d}{S} \times \beta) \quad (12)$$

where S_t is the terrace area, S_d is the area controlled by silting dams, S is the total land area, and α and β refer to the sediment reduction coefficients of terrace and silting dam and are 0.836 and 1, respectively.

2.3.6. Tillage (*T*) Factor

Tillage (*T*) factor refers to the ratio of the soil erosion amounts occurring under a specific tillage measure to that occurring under consistent flat cropping or slope tillage [45]. In this study, the slope gradient was extracted based on the DEM, and *T* factor was calculated according to the relationship between the slope gradient and the *T* factor (Table 2).

Table 2. *T* factor under different slope gradient.

Slope Gradient	≤5°	5–10°	10–15°	15–20°	20–25°	>25°
T factor	0.100	0.221	0.305	0.575	0.735	0.800

3. Results

3.1. Dynamic Land Use Changes Since Vegetation Restoration

Large-scale vegetation restoration led to significant land use changes in the Ansai Watershed (Figure 3). Land use was dominated by grassland and farmland, while arboreal land and shrub land were scattered and did not form contiguous patterns in 2000. With the progress of vegetation restoration, grassland became the main land use type, and arboreal land and shrub land increased significantly and formed a distribution pattern which decreased gradually from southeast to northwest in 2015. Thanks to the relatively superior natural conditions and the location conditions closer to the urban area, compared with the upstream areas, the implementation of the "GGP" is more active and the benefits of vegetation restoration is more obvious in the downstream areas.

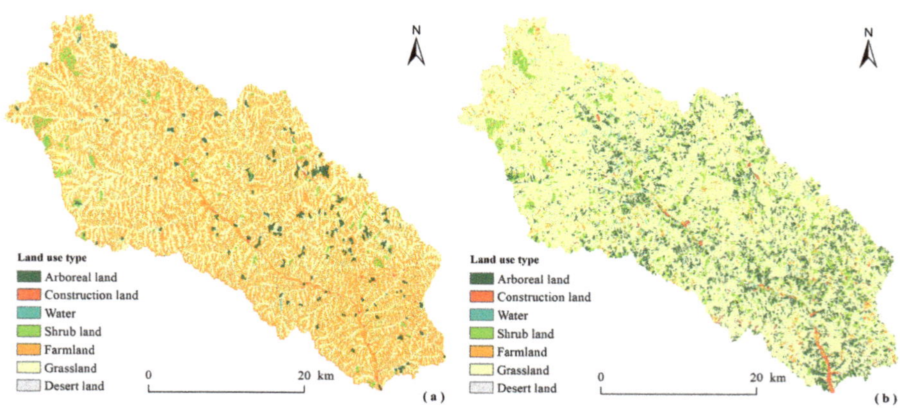

Figure 3. Land use map of the Ansai Watershed in 2000 (**a**) and in 2015 (**b**).

From 2000 to 2015, the area of arboreal land, shrub land, and grassland increased significantly in the Ansai Watershed, while the farmland decreased drastically, and the construction land, water land, and desert land increased slightly (Table 3). Furthermore, the farmland was mainly converted to grassland, followed by arboreal land, and shrub land; while a small part was converted to construction land, water land, and desert land. The primary driving factor of the changes was the implementation of "GGP" since 1999 [2,18,44].

3.2. Estimation of Soil Erosion under the Initial Scenario of Vegetation Restoration

The soil erosion modulus calculated based on the initial scenario of vegetation restoration was 31.18, 116.45, 170.88, 99.92, 147.21, 167.17, 88.56, 91.38, 55.03, 162.21, 80.11, 68.66, 115.97, 291.11, 115.96, and 31.19 t/(hm^2·a) from 2000 to 2015, respectively, and the average soil erosion modulus among the 16 years was 114.56 t/(hm^2·a) (Table 4). The light erosion accounted for the largest proportion, with 22.61%; the severe erosion followed, with an area of 300.67 km^2; the areas of moderate erosion, extreme erosion, and serious erosion all exceeded 150 km^2, accounting for 17.09%, 14.66%, and 12.29%, respectively, and the proportion of slight erosion was the smallest, with an area of 144.25 km^2. It can be seen that the soil erosion under the initial scenario of vegetation restoration in the Ansai Watershed was dominated by severe erosion and light erosion, and the soil erosion situation was relatively severe.

Table 3. Transfer matrix of land use changes in the Ansai Watershed from 2000 to 2015 (km^2).

Land Use Type		2015							Total
		Arboreal Land	Shrub Land	Grassland	Farmland	Construction Land	Water	Desert Land	
2000	Arboreal land	2.41	1.20	16.01	0.04	0.14	0.16	0.01	19.96
	Shrub land	1.22	16.07	2.00	0.07	0.02	0.04	0.01	19.43
	Grassland	44.83	23.43	643.08	1.66	2.13	3.13	1.24	719.49
	Farmland	50.72	35.25	422.78	32.57	3.84	1.86	0.89	547.90
	Construction land	0.03	0.02	0.18	0.02	1.89	0.01	0.00	2.14
	Water	0.04	0.01	0.20	0.00	0.00	0.00	0.00	0.25
	Desert land	0.00	0.00	0.00	0.00	0.00	0.00	0.00	0.00
Total		99.26	75.97	1084.24	34.35	8.02	5.20	2.14	–
Change from 2000 to 2015		79.30	56.54	364.75	−513.55	5.88	4.95	2.14	–

Table 4. Soil erosion in the Ansai Watershed from 2000 to 2015 under the initial scenario of vegetation restoration.

Year	Soil Erosion Modulus (t/(hm^2·a))	Area of Different Soil Erosion Intensity (%)					
		Slight	Light	Moderate	Serious	Extreme	Severe
2000	31.18	22.88	41.67	15.47	8.51	8.86	2.60
2001	116.45	8.18	18.44	18.64	13.75	16.19	24.80
2002	170.88	6.12	13.65	14.53	13.15	18.39	34.16
2003	99.92	9.55	21.30	19.14	13.77	14.60	21.63
2004	147.21	7.21	15.29	16.66	13.20	17.45	30.19
2005	167.17	6.59	13.96	15.02	13.09	17.97	33.36
2006	88.56	10.69	23.77	19.44	13.11	13.90	19.09
2007	91.38	10.25	22.74	19.37	13.52	14.23	19.88
2008	55.03	14.68	33.43	19.72	10.53	11.42	10.23
2009	162.21	6.74	14.46	15.26	13.31	17.67	32.55
2010	80.11	11.15	25.45	19.80	12.87	13.49	17.24
2011	68.66	15.07	29.23	18.47	11.36	12.02	13.85
2012	115.97	8.52	18.78	18.40	13.71	15.92	24.67
2013	291.11	4.45	9.45	8.81	10.59	17.53	49.16
2014	115.96	8.61	18.41	18.64	13.74	15.75	24.85
2015	31.19	22.34	41.79	15.98	8.41	9.10	2.38
Average	114.56	10.81	22.61	17.09	12.29	14.66	22.54

Note: Slight erosion (≤5 t/(hm^2·a)), light erosion (5–25 t/(hm^2·a)), moderate erosion (25–50 t/(hm^2·a)), serious erosion (50–80 t/(hm^2·a)), extreme erosion (80–150 t/(hm^2·a)), and severe erosion (>150 t/(hm^2·a)).

3.3. Estimation of Soil Erosion under the Current Scenario of Vegetation Restoration

The soil erosion modulus calculated based on the current scenario of vegetation restoration was 20.88, 80.67, 119.51, 68.41, 101.07, 114.77, 60.17, 62.51, 36.16, 111.59, 54.80, 46.05, 80.01, 97.60, 79.13, and 21.44 t/(hm^2·a) from 2000 to 2015, respectively, and the average soil erosion modulus among the 16 years was 78.42 t/(hm^2·a) (Table 5). The light erosion accounted for 23.71% of the total area, covering the largest area of 316.30 km^2; the moderate erosion and extreme erosion followed by 19.35% and 17.57%; the proportions of serious erosion and severe erosion all exceed 14%, with the area of 199.53 km^2 and 188.68 km^2, respectively, and the proportion of slight erosion was the smallest, with an area of 137.10 km^2. Therefore, in contrast from the soil erosion dominated by severe and light erosion under the initial stage of vegetation restoration, soil erosion under the current scenario of vegetation restoration was dominated by light erosion and moderate erosion in the Ansai Watershed. Furthermore, the proportion of severe erosion decreased from 22.54% to 14.14%, indicating that the soil erosion situation had been greatly improved.

Table 5. Soil erosion in the Ansai Watershed from 2000 to 2015 under the current scenario of vegetation restoration.

Year	Soil Erosion Modulus (t/(hm^2·a))	Area of Different Soil Erosion Intensity (%)					
		Slight	Light	Moderate	Serious	Extreme	Severe
2000	20.88	22.16	49.11	20.38	5.63	2.28	0.44
2001	80.67	7.82	18.43	19.00	18.51	22.07	14.16
2002	119.51	5.85	13.72	13.52	15.16	24.09	27.67
2003	68.41	8.93	20.98	21.86	18.03	20.51	9.69
2004	101.07	6.86	15.22	15.58	17.19	23.53	21.62
2005	114.77	6.19	13.98	13.84	15.71	24.07	26.22
2006	60.17	9.91	23.61	23.74	17.38	18.12	7.23
2007	62.51	9.42	22.58	23.50	17.54	19.26	7.70
2008	36.16	14.47	34.98	26.34	14.77	7.32	2.12
2009	111.59	6.16	14.84	14.87	15.96	23.49	24.67
2010	54.80	10.38	25.14	25.15	17.16	16.68	5.50
2011	46.05	14.21	31.27	23.00	14.28	12.81	4.44
2012	80.01	7.87	19.09	19.71	17.85	21.54	13.93
2013	197.60	4.22	9.32	8.86	9.85	21.00	46.75
2014	79.13	7.94	18.72	19.33	18.50	21.91	13.60
2015	21.44	22.05	48.37	20.87	5.78	2.37	0.56
Average	78.42	10.28	23.71	19.35	14.96	17.57	14.14

Note: Slight erosion (≤5 t/(hm^2·a)), light erosion (5–25 t/(hm^2·a)), moderate erosion (25–50 t/(hm^2·a)), serious erosion (50–80 t/(hm^2·a)), extreme erosion (80–150 t/(hm^2·a)), and severe erosion (>150 t/(hm^2·a)).

3.4. Changes in Soil Erosion before and after Vegetation Restoration

The average soil erosion modulus from 2000 to 2015 under the initial and the current scenarios of vegetation restoration was 114.44 t/(hm^2·a) and 78.42 t/(hm^2·a), respectively, with an average annual reduction of 4.81 × 10^6 t of soil erosion amount. The soil erosion condition was improved by vegetation restoration, and the soil erosion modulus decreased annually by 10.30, 35.78, 51.37, 31.51, 46.14, 52.40, 28.39, 28.87, 16.87, 50.62, 25.31, 22.61, 35.96, 93.51, 36.83, and 9.75 t/hm^2 from 2000 to 2015, respectively (Figure 4). In addition, the effects of vegetation restoration on soil erosion were different in different years, mainly because of the large differences of rainfall in each year. The water and soil conservation benefits of vegetation restoration was more obvious in the years with heavy rainfall such as 2002, 2005, 2009, and 2013.

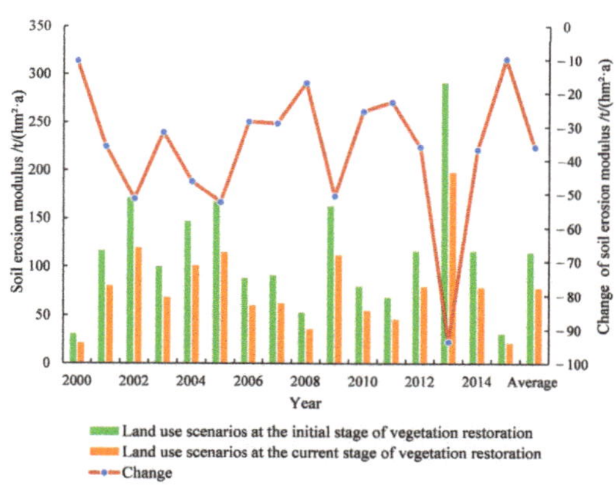

Figure 4. Effects of vegetation restoration on soil erosion from 2000 to 2015.

The average soil erosion modulus changes from 2000 to 2015 were divided into two categories, increased erosion (>0) and decreased erosion (<0), based on the reclassification function of ArcGIS 10.6 (Figure 5). During the study period, the areas with increased erosion and decreased erosion were alternately distributed in the Ansai Watershed. The south and southeast of the Ansai Watershed had obvious improvement effects on soil erosion and were the main areas with decreased soil erosion, while the northwest of the study area was the main region experiencing increased soil erosion. Although the areas with increased and decreased soil erosion distributed alternatively, the former was lower than the latter (Table 6). The area of decreased and increased soil erosion from 2000 to 2015 was 696.92 km^2 and 637.12 km^2, respectively, accounting for 52% and 48% of the total land area of the Ansai Watershed.

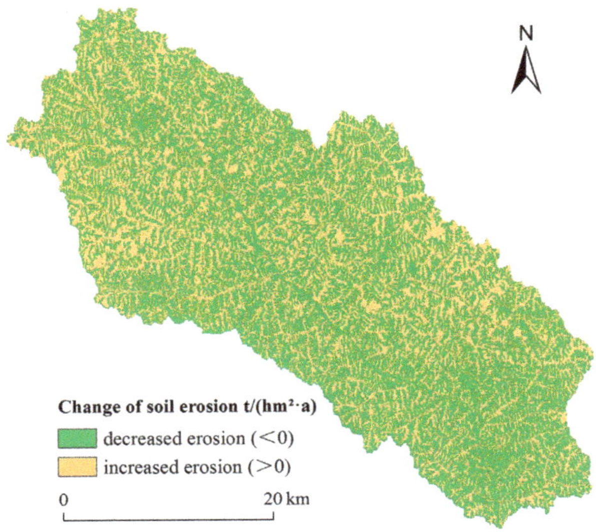

Figure 5. Spatial differentiation of the effect of vegetation restoration on soil erosion in 2000–2015.

Table 6. Changes in soil erosion areas and proportions from 2000 to 2015.

Year	Increased Erosion Area (km^2)	Proportion (%)	Decreased Erosion Area (km^2)	Proportion (%)
2000	630.34	47.25	703.66	52.75
2001	673.74	50.51	661.73	49.60
2002	625.30	46.87	707.83	53.06
2003	634.18	47.54	699.67	52.45
2004	614.26	46.05	719.81	53.96
2005	614.26	46.05	719.81	53.96
2006	640.90	48.04	693.10	51.96
2007	642.82	48.19	691.18	51.81
2008	640.46	48.01	693.54	51.99
2009	640.81	48.04	693.19	51.96
2010	638.50	47.86	695.50	52.14
2011	640.93	48.05	693.06	51.95
2012	638.06	47.83	695.94	52.17
2013	653.53	48.99	680.47	51.01
2014	637.94	47.82	696.06	52.18
2015	627.85	47.07	706.14	52.93
Average	637.12	47.76	696.92	52.24

4. Discussion

4.1. Effects of Vegetation Restoration on Soil Erosion

Land use types not only affect the properties of the underlying soil surface, but also influence the redistribution of rainfall and the transport of runoff and sediment [46]. According to previous research in Yanan city, the soil conservation modulus varied with the land use types; furthermore, forest land and grassland had the best soil conservation effects [47]. Since the implementation of the "GGP" in 1999, the land use structure in the Ansai Watershed has undergone significant changes, mainly characterized by the conversion of sloping farmland to grassland, arboreal land, and shrub land. The effective implementation of this project significantly improved the soil erosion environment in the study area, in accordance with previous research results [40,48,49]. Wang et al. (2016) [49] found that compared with the sloping farmland, the conversion of sloping farmland to grassland or woodland can reduce gully erosion by more than 90%. Results of this study indicated that the average annual soil erosion modulus dropped from 114.56 t/(hm^2·a) to 78.42 t/(hm^2·a), and the dominant soil erosion intensity changed from severe erosion and light erosion to moderate erosion and light erosion in the Ansai Watershed during 2000–2015. In addition, according to the data released by the China National Forestry and Grassland Administration (http://www.forestry.gov.cn/ accessed on 25 April 2021), the average annual soil erosion modulus in the Ansai County dropped from 140.00 t/(hm^2·a) in 1998 to 54.00 t/(hm^2·a) in 2018 since the implementation of the "GGP", which further confirmed the accuracy and credibility of our research results. By the end of 2018, a total of 94,920 hm^2 forest land was increased, of which 56,520 hm^2 was transferred from sloping farmland, and 36,470 hm^2 was transferred from desert land and grassland in Ansai County [50]. Thanks to massive vegetation restoration, with increasing vegetation coverage and biomass, the dense vegetation canopy reduced the effective precipitation in forest land, prolonged the precipitation and runoff duration, and cut off the kinetic energy of raindrops; surface mulch dispersed the kinetic energy of runoff, and the complex vegetation root system increased the resistance of the soil runoff erosion, effectively strengthened the regional soil and water conservation benefits, and improved soil erosion conditions [51]. Furthermore, previous studies showed that the soil profile structure destroyed by erosion became more and more complete, and soil properties were restored in loess hilly and gully regions after the implementation of "GGP" [16]. For example, soil bulk density and PH value decreased, while soil organic matter content, C, and N content increased. The conversion of sloping farmland to forest land with relatively little human interference was conducive to the accumulation of soil nutrients and the maintenance of porosity, and effectively enhanced the water and fertilizer retention performance of soil.

The spatial differentiation of soil erosion changes also further indicated the positive effect of vegetation restoration on soil erosion. Vegetation restoration was actively carried out in the southeast and south of the Ansai Watershed, where the land use change was relatively drastic, and mainly included transformations from farmland to grassland, shrub land and arboreal land, and from grassland to shrub land and arboreal land. The massive vegetation restoration in this area effectively strengthened the water and soil conservation benefits, and greatly improved the soil erosion condition. However, in the northwest of the Ansai Watershed, the implementation effect of vegetation restoration was poor, which led to more serious soil erosion in local areas.

4.2. Policy Implications

The main reason for the poor soil erosion control effect in the northwest of the Ansai Watershed was the ineffective implementation of the "GGP" and the unreasonable land use changes, such as the conversion of farmland and grassland into construction land. In view of this, local governments should actively carry out this project, strictly implement land use planning, control the occupation of farmland for non-agricultural construction, and prevent unreasonable land use changes.

The field survey in the Ansai Watershed found that although the "GGP" was also carried out in the northwest of the watershed and a certain number of sea-buckthorn and Caragana korshinskii plants were planted in this region where soil erosion was increasingly serious, the majority of these trees did not survive due to lack of supervision and management. Previous studies showed that the phenomenon of "seeing only the saplings but not the forest" was common in the process of returning farmland to forest in Ansai County [48]. The main reasons include the following two aspects: first, farmers lacked the initiative in the management of forest seedlings, and only focused on the subsidies for returning farmland to forest, but ignored the follow-up management of forest land; second, the lack of support for pest control and forest fire prevention directly affected the quality and subsequent benefits of this project. The implementation of the "GGP" is a long-term process, and the local government should strengthen the supervision and management of ecological restoration, follow the principle of "whoever builds, manages and benefits", strengthen the management and protection of vegetation seedlings, and ensure the normal growth of young forests. Considering the fragile ecological environment in the Loess Plateau region, excellent tree and grass species with strong adaptability and good quality should be selected to improve the survival rate of seedlings. Local governments should carry out inspections of vegetation restoration occasionally. To ensure the effects of vegetation restoration, measures such as supplementary planting, tending, pruning, watering, weeding, and pest control should be taken for forest land converted from farmland with substandard numbers of living plants and low survival rates.

4.3. Research Limitation and Future Research

The CSLE model proposed by the Chinese scholar is widely used to calculate the soil erosion amount in China. The applicability of this model in Ansai county, the loess hilly and gully regions, and even in China has been verified by previous studies [32,33,52–56], and can well reflect the soil erosion situation in China. Among them, by comparing the simulation results of the CSLE model with the soil erosion data measured by the Ministry of Water Resources of the People's Republic of China, the predecessors proved that the model has good applicability and credibility in the soil erosion evaluation in Shaanxi Province [52]. In view of this, this study assessed the soil erosion amount using the CSLE model in the Ansai Watershed belonging to Shaanxi Province. Furthermore, all parameters of the CSLE model in this study were calibrated based on previous research. Although model uncertainties are unavoidable, the calculation results of the soil erosion modulus in the Ansai watershed using the CSLE model can reflect the actual situation.

Unlike the soil erosion amount which can be measured on the field spot, the effect of vegetation restoration on soil erosion cannot be directly measured through field experiments. Through field control experiments, comparing the changes of soil erosion amount in the two watersheds with the same conditions only, except for implementing or not implementing the "GGP", can verify our research results to a certain extent. However, considering the effect of vegetation restoration on soil erosion is correlated with the length of time for reforestation [20], short-term field monitoring results through field control experiments cannot verify the model simulation results from 2000 to 2015. Furthermore, it takes a lot of manpower and time to conduct continuous monitoring of soil erosion for more than 10 years at the watershed scale, and the gap in existing data sources also makes it difficult to currently verify our research results on the field spot. Under the above constraints, we validated our research results based on the official data on the effect of vegetation restoration on soil erosion in Ansai County released by the National Forestry and Grass Administration of the People's Republic of China, and the previous research results in loess hilly and gully regions. It is worth mentioning that the data released by the National Forestry and Grass Administration was obtained through on-site monitoring of the Ansai Hydrological Station, which is not the model simulation result, and its consistency with our research results provided a good proof of the credibility of our research.

In follow-up studies, we will increase long-term field monitoring experiments to more accurately assess the effect of vegetation restoration on soil erosion.

5. Conclusions

Large-scale vegetation restoration triggered by the "Grain for Green Project" (GGP) since 1999 led to significant land use changes in the Loess Plateau. Using the CSLE model, this study calculated and compared the differences between the soil erosion modulus from 2000 to 2015 under two land use scenarios (the initial and current scenarios of vegetation restoration), and identified the effect of vegetation restoration on soil erosion in the Ansai Watershed. The results showed that the soil erosion conditions have greatly improved in the Ansai Watershed since vegetation restoration. The average soil erosion modulus under the initial scenario of vegetation restoration among the 16 years was 114.56 t/(hm^2·a), dominated by severe erosion and light erosion; while the average soil erosion modulus under the current scenario of vegetation restoration among the 16 years was 78.42 t/(hm^2·a), with light and moderate erosion as the dominant soil erosion intensity. However, due to the unreasonable land use changes (farmland, grassland was converted into construction land, etc.) and the ineffective implementation of vegetation restoration, soil erosion became more serious in some areas. Therefore, it is necessary to strengthen the supervision and management of the "GGP", control unreasonable land use changes by land use planning, and prevent decreases in vegetation coverage to control soil erosion in the Ansai Watershed.

Author Contributions: H.W. (Hui Wei) mainly conducted the article structure design, data processing, and article writing; W.Z. mainly engaged in the modification and polishing of the article, and H.W. (Han Wang) engaged in article polishing. All authors have read and agreed to the published version of the manuscript.

Funding: This research was funded by the National Natural Science Foundation of China (grant number 41771197) and the Fundamental Research Funds for the Central Universities.

Institutional Review Board Statement: Not applicable.

Informed Consent Statement: Not applicable.

Data Availability Statement: All relevant data sets in this study are described in the manuscript.

Acknowledgments: We would like to thank Jing Wang, Qiang Feng, Xuening Fang, Xiao Zhang, and Yuanxin Liu for their support and contributions during the fieldwork.

Conflicts of Interest: The authors declare no conflict of interest.

References

1. United Nations Environment Agency. United Nations Decade on Ecosystem Restoration (2021–2030) 2019, 73/284. Available online: https://undocs.org/A/RES/73/284 (accessed on 20 March 2021).
2. Zhou, J.; Fu, B.J.; Gao, G.Y.; Lü, Y.H.; Liu, Y.; Lü, N.; Wang, S. Effects of precipitation and restoration vegetation on soil erosion in a semi-arid environment in the Loess Plateau, China. *Catena* **2016**, *137*, 1–11. [CrossRef]
3. Huang, C.B.; Zhou, Z.X.; Teng, M.J.; Wu, C.G.; Wang, P.C. Effects of climate and land use/cover changes on soil loss in the Three Gorges Reservoir area, China. *Geogr. Sustain.* **2020**, *1*, 200–208. [CrossRef]
4. Piacentini, T.; Galli, A.; Marsala, V.; Miccadei, E. Analysis of soil erosion induced by heavy rainfall: A case study from the NE Abruzzo Hills area in central Italy. *Water* **2018**, *10*, 1314. [CrossRef]
5. Brandolini, P.; Pepe, G.; Capolongo, D.; Cappadonia, C. Hillslope degradation in representative Italian areas: Just soil erosion risk or opportunity for development? *Land Degrad. Dev.* **2017**, *29*, 3050–3068. [CrossRef]
6. Borrelli, P.; Robinson, D.; Panagos, P.; Lugato, E.; Yang, J.E.; Alewell, C.; Wuepper, D.; Montanarella, L.; Ballabio, C. Land use and climate change impacts on global soil erosion by water (2015–2070). *Proc. Natl. Acad. Sci. USA* **2020**, *117*, 1–8. [CrossRef] [PubMed]
7. Zhai, Z.N.; Wang, K.Q.; Su, B.; Zhang, X.Q.; Hua, J.X.; Zhu, X.T. Study on soil erodibility of different land utilization styles in the Songhuba water source area. *J. Southwest For. Univ.* **2016**, *5*, 118–124. (In Chinese)
8. Meliho, M.; Khattabi, A.; Mhammdi, N. Spatial assessment of soil erosion risk by integrating remote sensing and GIS techniques: A case of Tensift watershed in Morocco. *Environ. Earth Sci.* **2020**, *79*, 1–19. [CrossRef]
9. Fu, B.J.; Liu, Y.; Lü, Y.H.; He, C.S.; Zeng, Y.; Wu, B.F. Assessing the soil control service of ecosystems change in the Loess Plateau of China. *Ecol. Complex.* **2011**, *8*, 284–293. [CrossRef]

10. Yang, X.; Guo, B.; Lu, Y.F.; Zhang, R.; Zhang, D.F.; Zhen, X.Y.; Chen, S.T.; Wu, H.W.; Wei, C.X.; Yang, L.A.; et al. Spatial-temporal evolution patterns of soil erosion in the Yellow River Basin from 1990 to 2015: Impacts of natural factors and land use change. *Geomat. Nat. Hazards Risk* **2021**, *12*, 103–122.
11. Tuo, D.F.; Xu, M.X.; Gao, G.Y. Relative contributions of wind and water erosion to total soil loss and its effect on soil properties in sloping croplands of the Chinese Loess Plateau-ScienceDirect. *Sci. Total Environ.* **2018**, *633*, 1032–1040. [CrossRef] [PubMed]
12. Ryken, N.; Nest, T.V.; Al-Barri, B.; Blake, W.; Taylor, A.; Bode, S.; Ruysschaert, G.; Boeckx, P.; Verdoodt, A. Soil erosion rates under different tillage practices in central Belgium: New perspectives from a combined approach of rainfall simulations and measurements. *Soil Tillage Res.* **2018**, *179*, 29–37. [CrossRef]
13. Ministry of Water Resources of China; National Bureau of Statistics China. *Bulletin of First National Census for Water*; China Water Power Press: Beijing, China, 2013.
14. Xi, J.; Zhao, X.; Wang, X.; Zhang, Z. Assessing the impact of land use change on soil erosion on the Loess Plateau of China from the end of the 1980s to 2010. *J. Soil Water Conserv.* **2017**, *72*, 452–462. [CrossRef]
15. Zhao, G.J.; Kondolf, G.M.; Mu, X.M.; Han, M.W.; He, Z.; Rubin, Z.; Wang, F.; Gao, P.; Sun, W.Y. Sediment yield reduction associated with land use changes and check dams in a catchment of the Loess Plateau, China. *Catena* **2017**, *148*, 126–137. [CrossRef]
16. Li, J.; Li, Z.B.; Guo, M.J.; Li, P.; Cheng, S.D.; Yuan, B. Effects of vegetation restoration on soil physical properties of abandoned farmland on the Loess Plateau, China. *Environ. Earth Sci.* **2018**, *77*, 205. [CrossRef]
17. Guo, X.J.; Shao, Q.Q. Spatial pattern of soil erosion drivers and the contribution rate of human activities on the Loess Plateau from 2000 to 2015: A boundary line from northeast to southwest. *Remote Sens.* **2019**, *11*, 2429. [CrossRef]
18. Wang, X.F.; Xiao, F.Y.; Feng, X.M.; Fu, B.J. Soil conservation on the Loess Plateau and the regional effect: Impact of the 'Grain for Green' Project. *Earth Environ. Sci. Trans. R. Soc. Edinb.* **2019**, *109*, 461–471. [CrossRef]
19. Zhao, J.; Feng, X.M.; Deng, L.; Yang, Y.Z.; Zhao, Z.; Zhao, P.X.; Peng, C.H.; Fu, B.J. Vegetation quantifying the effects of vegetation restorations on the soil erosion export and nutrient loss on the Loess Plateau. *Front. Plant Sci.* **2020**, *11*, 573126. [CrossRef]
20. Ran, L.S.; Lu, X.X.; Xu, J.C. Effects of Vegetation Restoration on Soil Conservation and Sediment Loads in China: A Critical Review. *Crit. Rev. Environ. Sci. Technol.* **2013**, *43*, 1384–1415. [CrossRef]
21. Zheng, F.L. Effect of Vegetation Changes on Soil Erosion on the Loess Plateau. *Pedosphere* **2006**, *16*, 420–427. [CrossRef]
22. Zhang, X.; Wang, J.; Gao, Y.; Wang, L.X. Variations and controlling factors of vegetation dynamics on the Qingzang Plateau of China over the recent 20 years. *Geogr. Sustain.* **2021**, *23*, 74–85.
23. Zhao, W.W.; Wei, H.; Jia, L.Z.; Stefani, D.; Zhang, X.; Liu, Y.X. Soil erodibility and its influencing factors on the Loess Plateau of China: A case study in the Ansai watershed. *Solid Earth* **2018**, *9*, 1507–1516. [CrossRef]
24. Zhao, W.W.; Fu, B.J.; Chen, L.D. A comparison between soil loss evaluation index and the C-factor of RUSLE: A case study in the Loess Plateau of China. *Hydrol. Earth Syst. Sci.* **2012**, *16*, 2739–2748. [CrossRef]
25. Yu, Y.; Wei, W.; Chen, L.D.; Jia, F.Y.; Yang, L. Responses of vertical soil moisture to rainfall pulses and land uses in a typical loess hilly area, China. *Solid Earth* **2015**, *6*, 595–608. [CrossRef]
26. Feng, Q.; Zhao, W.W.; Qiu, Y.; Zhao, M.Y.; Zhong, L.N. Spatial heterogeneity of soil moisture and the scale variability of its influencing factors: A case study in the Loess Plateau of China. *Water* **2013**, *5*, 1226–1242. [CrossRef]
27. National Center for Basic Geographic Information of China. Available online: http://www.ngcc.cn (accessed on 10 December 2020).
28. Data Center for Resources and Environmental Sciences at the Chinese Academy of Sciences. Available online: https://www.resdc.cn/ (accessed on 12 December 2020).
29. Ministry of Water Resources. *Hydrological Yearbook of the People's Republic of China*; Yellow River Conservancy Commission of the Ministry of Water Resources: Beijing, China, 2000–2015.
30. Geospatial Data Cloud Website. Available online: http://www.gscloud.cn/ (accessed on 15 December 2020).
31. Chen, G.K.; Zhang, Z.X.; Guo, Q.K.; Wang, X.; Wen, Q.K. Quantitative assessment of soil erosion based on CSLE and the 2010 national soil erosion survey at regional scale in Yunnan Province of China. *Sustainability* **2019**, *11*, 3252. [CrossRef]
32. Liu, B.Y.; Zhang, K.L.; Xie, Y. An empirical soil loss equation proceedings-process of soil erosion and its environment effect. In Proceedings of the 12th International Soil Conservation Organization Conference, Beijing, China, 26–31 May 2002; Tsinghua University Press: Beijing, China, 2002; pp. 21–25.
33. Liu, S.Y.; Huang, S.Z.; Xie, Y.Y.; Leng, G.Y.; Huang, Q.; Wang, L.; Xue, Q. Spatial-temporal changes of rainfall erosivity in the loess plateau, China: Changing patterns, causes and implications. *Catena* **2018**, *166*, 279–289. [CrossRef]
34. Zhang, W.B.; Xie, Y.; Liu, B.Y. Rainfall erosivity estimation using daily rainfall amounts. *Sci. Geogr. Sin.* **2002**, *22*, 705–711. (In Chinese)
35. Cui, Y.S.; Pan, C.Z.; Liu, C.L.; Luo, M.J.; Guo, Y.H. Spatiotemporal variation and tendency analysis on rainfall erosivity in the Loess Plateau of China. *Hydrol. Res.* **2020**, *51*, 1048–1062. [CrossRef]
36. Sun, W.Y.; Shao, Q.Q.; Liu, J.Y.; Zhai, J. Assessing the effects of land use and topography on soil erosion on the Loess Plateau in China. *Catena* **2014**, *121*, 151–163. [CrossRef]
37. Zhang, K.L.; Peng, W.Y.; Yang, H.L. Soil erodibility and its estimation for agricultural soil in China. *Acta Pedol. Sin.* **2007**, *44*, 7–13. (In Chinese) [CrossRef]
38. Shirazi, M.A.; Hart, J.W.; Boersma, L. A unifying quantitative analysis of soil texture: Improvement of precision and extension of scale. *Soil Sci. Soc. Am. J.* **1988**, *52*, 181–190. [CrossRef]

39. Wischmeier, W.H.; Smith, D.D. *Predicting Rainfall Erosion Losses—A Guide to Conservation Planning*; Agriculture Handbook No. 537; U.S. Department of Agriculture: Washington, DC, USA, 1978; pp. 12–17.
40. Wen, X. Temporal and spatial relationships between soil erosion and ecological restoration in semi-arid regions: A case study in northern Shaanxi, China. *GISci. Remote Sens.* **2020**, *57*, 572–590. [CrossRef]
41. Yan, R.; Zhang, X.P.; Yan, S.J.; Chen, H. Estimating soil erosion response to land use/cover change in a catchment of the Loess Plateau, China. *Int. Soil Water Conserv. Res.* **2018**, *6*, 13–22. [CrossRef]
42. Li, P.F.; Zang, Y.Z.; Ma, D.Y.; Yao, W.Q.; Holden, J.; Irvine, B.; Zhao, G.J. Soil erosion rates assessed by RUSLE and PESERA for a Chinese Loess Plateau catchment under land-cover changes. *Earth Surf. Process. Landf.* **2020**, *45*, 707–722. [CrossRef]
43. Zhao, W.W.; Fu, B.J.; Chen, L.D.; Lü, Y.H.; Liu, Y.Q. Effects of land use pattern change on soil and water loss at the catchment scale in the hilly and gully area of the loess plateau of China. *Acta Ecol. Sin.* **2004**, *24*, 1358–1364. (In Chinese)
44. Xie, H.X.; Li, R.; Yang, Q.K.; Li, J.; Liang, W. Effect of returning farmland to forest (Pasture) and changes of precipitation on soil erosion in the Yanhe Basin. *Sci. Agric. Sin.* **2009**, *42*, 569–576. (In Chinese)
45. Xu, L.F.; Xu, X.G.; Meng, X.W. Risk assessment of soil erosion in different rainfall scenarios by RUSLE model coupled with information diffusion model: A case study of Bohai Rim, China. *Catena* **2012**, *100*, 74–82. [CrossRef]
46. Hao, H.X.; Wei, Y.J.; Cao, D.N.; Guo, Z.L.; Shi, Z.H. Vegetation restoration and fine roots promote soil infiltrability in heavy-textured soils. *Soil Tillage Res.* **2020**, *198*, 104542. [CrossRef]
47. Wang, S.; Wang, H.Y.; Xie, Y.S.; Luo, H. Evaluation of ecological service function of soil conservation before and after Grain for Green Project in Yanan City. *Res. Soil Water Conserv.* **2019**, *26*, 280–286. (In Chinese)
48. Jiang, C.; Zhang, H.Y.; Zhang, Z.D.; Wang, D.W. Model-based assessment soil loss by wind and water erosion in China's Loess Plateau: Dynamic change, conservation effectiveness, and strategies for sustainable restoration. *Glob. Planet. Chang.* **2019**, *172*, 396–413. [CrossRef]
49. Wang, Z.J.; Jiao, J.Y.; Rayburg, S.; Wang, Q.L.; Su, Y. Soil erosion resistance of "Grain for Green" vegetation types under extreme rainfall conditions on the Loess Plateau, China. *Catena* **2016**, *141*, 109–116. [CrossRef]
50. Wang, G.Q.; Bai, S.B.; Hu, P. Discussion on the work of returning farmland to forest in Ansai District, Yan'an City. *Mod. Agric. Sci. Technol.* **2020**, *17*, 130–131. (In Chinese)
51. Zhang, W.; Ren, C.J.; Deng, J.; Zhao, F.Z.; Yang, G.H.; Han, X.H.; Tong, X.G.; Feng, Y.Z. Plant functional composition and species diversity affect soil C, N, and P during secondary succession of abandoned farmland on the Loess Plateau. *Ecol. Eng.* **2018**, *122*, 91–99. [CrossRef]
52. Cheng, L.; Yang, Q.K.; Xie, H.X.; Wang, C.M.; Guo, M.L. GIS and CSLE based quantitative assessment of soil erosion in Shaanxi, China. *J. Soil Water Conserv.* **2009**, *23*, 61–66. (In Chinese)
53. Ma, Y.Y.; Wang, J.; Zhang, C.; Liu, G.B.; Pang, G.W. Evaluation of soil erosion based on CSLE model in Zhifanggou Watershed of Northern Shaanxi Province. *Bull. Soil Water Conserv.* **2018**, *38*, 95–102. (In Chinese)
54. Zhang, Y.; Liu, X.C.; Li, G.Z.; Zhu, Q.K. Surveying soil erosion condition in Loess Plateau using soil erosion model. *Trans. Chin. Soc. Agric. Eng.* **2012**, *28*, 165–171. (In Chinese)
55. Liang, Y.; Jiao, J.Y.; Tang, B.Z.; Cao, B.T.; Li, H. Response of runoff and soil erosion to erosive rainstorm events and vegetation restoration on abandoned slope farmland in the Loess Plateau region, China. *J. Hydrol.* **2020**, *584*, 124694.
56. Soil Conservation Monitoring Center. *Soil and Water Loss Equation for the Loess Plateau-User's Guide[Z]*; Ministry of Water Resources and Beijing Normal University: Beijing, China, 2005.

International Journal of
Environmental Research and Public Health

MDPI

Article

Control Models and Spatiotemporal Characteristics of Air Pollution in the Rapidly Developing Urban Agglomerations

Longwu Liang [1,2,3] and Zhenbo Wang [1,2,3,*]

1 Institute of Geographic Sciences and Natural Resources Research, Chinese Academy of Sciences, Beijing 100101, China; lianglw.17s@igsnrr.ac.cn
2 Key Laboratory of Regional Sustainable Development Modeling, Chinese Academy of Sciences, Beijing 100101, China
3 College of Resource and Environment, University of Chinese Academy of Sciences, Beijing 100049, China
* Correspondence: wangzb@igsnrr.ac.cn

Abstract: This paper systematically summarizes the hierarchical cross-regional multi-directional linkage in terms of air pollution control models implemented in the Beijing-Tianjin-Hebei urban agglomeration, including the hierarchical linkage structure of national-urban agglomeration-city, the cross-regional linkage governance of multiple provinces and municipalities, the multi-directional linkage mechanism mainly involving industry access, energy structure, green transportation, cross-regional assistance, monitoring and warning, consultation, and accountability. The concentration data of six air pollutants were used to analyze spatiotemporal characteristics. The concentrations of SO_2, NO_2, PM_{10}, $PM_{2.5}$, CO decreased, and the concentration of O_3 increased from 2014 to 2017; the air pollution control has achieved good effect. The concentration of O_3 was the highest in summer and lowest in winter, while those of other pollutants were the highest in winter and lowest in summer. The high pollution ranges of O_3 diffused from south to north, and those of other pollutants decreased significantly from north to south. Finally, we suggest strengthening the traceability and process research of heavy pollution, increasing the traceability and process research of O_3 pollution, promoting the joint legislation of different regions in urban agglomeration, create innovative pollution discharge supervision mechanisms, in order to provide significant reference for the joint prevention and control of air pollution in urban agglomerations.

Keywords: control models; spatiotemporal characteristics; air pollution; urban agglomeration; China

Citation: Liang, L.; Wang, Z. Control Models and Spatiotemporal Characteristics of Air Pollution in the Rapidly Developing Urban Agglomerations. *Int. J. Environ. Res. Public Health* **2021**, *18*, 6177. https://doi.org/10.3390/ijerph18116177

Academic Editor: Paul B. Tchounwou

Received: 17 April 2021
Accepted: 29 May 2021
Published: 7 June 2021

1. Introduction

Rapid economic development and urbanization have concurrently boosted energy consumption and pollutant emissions while promoting nations' overall power and social progress [1], which has led to increasingly serious environmental pollution [2] and health problems [3] in urban areas. As Samet [4] suggested, urban air pollution could become a public health and environmental problem of crisis proportions in the near future, if it is not so already. Since 2012, haze pollution has become a severe environmental problem that has impacted individuals' health and daily lives in China [5,6]. On the one hand, the continuous spread of air pollution on a large scale endangers the health of local residents, especially the elderly [7]. On the other hand, pollution has huge economic consequences. China will lose 2% of its total GDP and bear healthcare costs of USD 25.2 billion until 2030 [8]. As one of the five national-level urban agglomerations in China, the Beijing–Tianjin–Hebei urban agglomeration (BTHUA) is the implementation of the regional coordinated development strategy, and it is also the region with the most serious air pollution [9,10] and the most acute contradiction between resources, environment, and development [11] in China. Eight of the ten cities with the highest average annual concentration of $PM_{2.5}$ were located in BTHUA in 2015 [9]. Therefore, the status of high strategic status, high population density, high development speed, high environmental pollution, high energy emissions, and low environmental carrying capacity has led to the great conflict between regional economic development and environmental protection in BTHUA [12]. The central and regional governments have put forward many policies, models, and strategies for the air pollution control in BTHUA from the scale of nation, urban agglomeration, city, and county. It is urgent to systematically sort out and summarize them, so as to provide significant reference for the prevention and control of air pollution in other regions.

Air pollution is the result of the comprehensive effects of natural climate, topography, ecological environment, economic, and social factors in urban areas, determining the long-term, arduous, and complex prevention and control of air pollution. In recent years, scholars from all countries have created rich academic literature on air pollution chemical composition [13], spatiotemporal characteristics [14], source apportionment [15], impact mechanism [16], prevention and control model [17], regional division of joint control [18], and other related topics. Specifically, the influencing factors of air pollution include urbanization level [19], energy consumption structure [20], transportation infrastructure construction [21], foreign direct investment [22], climate conditions [23], policy elements [24], etc. In the international literature of urban air pollution, Liu et al. [25] found that the atmospheric pollutants revealed a stable trend from 2008 to 2018 in Germany. Hossain et al. [26] found that the emission control measures resulted in NO_2, SO_2, and PM reductions from 2000 to 2018. Chakraborty and Basu [27] considered that $PM_{2.5}$ concentration levels were significantly higher in more urbanized districts located predominantly in northern India. Practical experience shows that developing strict air quality standards and pollution control policies through cross-regional governance agencies, industrial restructuring, using clean energy and renewable energy, implementing and improving public transport, planning urban infrastructure, and pricing electricity for more efficient usage can effectively reduce air pollution [28]. In the short term, realistic approaches include energy conservation, promoting mass transit, reducing open burning, instituting motor vehicle inspection programs, and phasing out lead in petrol. In the longer term, air quality management strategies, such as urban and transportation planning, institutional and technological limitations should be incorporated in industrial and urban development [29].

In order to improve the air quality, the Chinese government has made great efforts in the control of air pollution, the ultra-low emission policy of China's thermal power industry has led to a significant reduction in the emissions of various air pollutants [30]. As the leading area of joint prevention and control models of air pollution in China, the air pollution control in the BTHUA has become a benchmark for measuring the effectiveness of urban environmental governance in China. However, there are few studies systematically sorting out and summarizing the air pollution control models from the regional perspective, and few studies have explored the spatiotemporal changes characteristics of the air pollution in BTHUA from the perspective of the air pollution prevention and control effectiveness.

Therefore, based on the air pollution control models of different cities in BTHUA, this paper forms the hierarchical cross-regional multi-directional linkage (HCML) in terms of air pollution prevention and control models, in order to reduce air pollutant concentration and adverse health effects. This paper introduces the structural characteristics and specific strategies of the models in detail from the model architecture, model management, and model guarantee. In order to explore the effectiveness of this model in air pollution control, this paper uses statistical method and spatial pattern analysis method to quantitatively analyze the annual, quarterly, and monthly and compliance rate characteristic of six air pollutants ($PM_{2.5}$, PM_{10}, SO_2, O_3, NO_2, and CO) in BTHUA from 2014 to 2017. Finally, we further discussed the problems of air pollution control and the recommendations of research and policy about the efficient control.

2. Hierarchical Cross-Regional Multi-Directional Linkage in Terms of Air Pollution Control Models

The atmosphere is a public resource that exhibits fluidity and infinity. Its state and composition are directly affected by natural geographical factors, such as solar radiation, temperature, humidity, air pressure, wind direction, precipitation, and other meteorological and topographical elements. The state and composition of the atmosphere are also indirectly affected by interactions between atmospheric particulate and pollution emissions resulting from human activity, such as the use of motor vehicles, industrial production, and coal burning. Based on the spatial and temporal attributes of pollution sources and pollution spread, the government also coordinates multiple levels of countries, urban agglomerations, cities, and enterprises and continuously adopts actions for the prevention and control of air pollution via direct and indirect governance. We summarize this action as the hierarchical cross-regional multi-directional linkage in terms of air pollution control models (Figure 1), which comprises the following framework and operational mechanism [17].

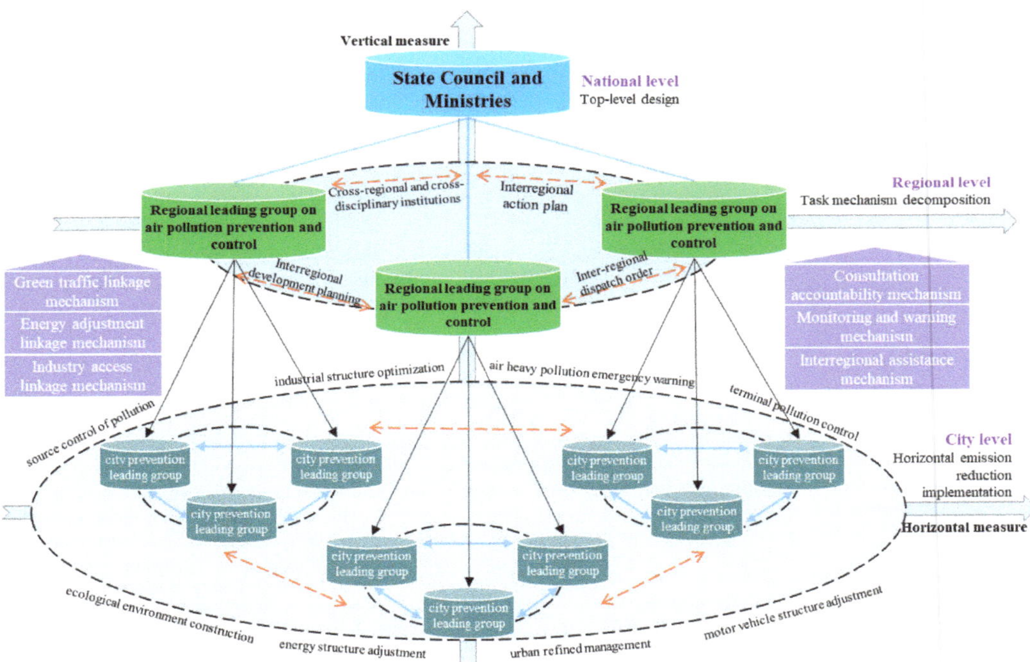

Figure 1. Hierarchical cross-regional multi-directional linkage model in terms of air pollution in BTHUA.

2.1. Linkage Architecture of Vertical Layer

First, a top-level design plan was created at the national level. In 2013, the governance issued the "Air Pollution Prevention and Control Action Plan" (hereinafter referred to as the "Plan"), proposing overall requirements, goals, and policy measures for the prevention and control of air pollution nationwide. After 2014, the relevant departments successively issued control documents, such as the "Air Pollution Prevention and Control Law", the new "Environmental Protection Law of the People's Republic of China", "Urban Environmental Air Quality Change Degree Ranking Scheme", and the "Fire Power Plant Pollution Prevention and Control Technology Policy", to further refine the air pollution control objectives. Second, task decomposition and accountability mechanisms were implemented in the urban agglomerations. The state council authorized the Ministry of Environmental Protection and the people's governments of all provinces (autonomous regions and municipalities) within the urban agglomeration to sign the responsibility letter for air pollution prevention and control; breakdown the emission reduction tasks to all levels of government, departments, and enterprises; and establish the air pollution control responsibility assessment system both vertically and laterally. Third, specific plans were implemented for lateral emission reductions at the city level. A city project management system of energy conservation and emission reduction was established, including source control, end pollution control, industrial structure optimization, energy structure upgraded, motor vehicle structure adjustment, ecological environment construction, and air heavy pollution emergency prevention and control projects. Finally, a "special responsibility list" was developed, the territorial management responsibilities of the governments at all levels determined, and the responsibilities divided between various departments in the city.

2.2. Linkage Governance of Horizontal Cross-Region

First, the air pollution control agency was established cross-regionally. In 2013, air pollution prevention and control cooperation groups were established in BTHUA and the surrounding area. In response to the pollution problem, regular meetings were organized to engage the government official for the provinces and cities, such as Beijing, Tianjin, Hebei, Shanxi, Shandong, Neimenggu, Henan, and other relevant ministries and commissions. Second, short-term prevention targets were identified in the interregional action plans. The "Beijing–Tianjin–Hebei Air Pollution Prevention and Control Action Plan Implementation Rules" (hereinafter referred to as the "Rules"), jointly issued by the six ministries and commissions, such as the Ministry of Environmental Protection, clearly defined the annual average concentration control targets for particulate matter and environmental standards for industrial access and elimination tasks of the high-pollution industries in provinces and cities. Third, cross-regional collaborative planning was used to develop long-term prevention and control routes. The "BTHUA Collaborative Development Plan" and the "BTHUA National Economic and Social Development Plan during the 13th Five-Year Plan" proposed the construction of an ecological restoration environment improvement demonstration zone, and defined the framework and models for pollution prevention and green development in BTHUA. Fourth, the dispatch order system was created cross-regionally. In 2016, the system was initiated and implemented in Hebei Province. The dispatching order is authorized by the provincial government to improve atmospheric environmental quality and address regionally polluted areas and outstanding environmental issues, using the results from research on pollution sources. The order can also schedule key tasks, such as the reduction of industrial enterprises, use of clean energy, motor vehicle restrictions, urban dust control, and non-point-source pollution control in specific regions.

2.3. Linkage Mechanism of Synergistic Multi-Direction

The industrial structure, energy structure, and transportation network pattern are the main influencing factors of air pollution in BTHUA. Based on this observation, the governments at all levels of the urban agglomeration formulated the following linkage measures.

The first is the heavy-industry access linkage mechanism. Excess and new capacity projects, such as steel, cement, electrolytic aluminum, flat glass, ship, coking, colored, calcium carbide, and ferroalloy, cannot be approved in BTHUA and its surrounding areas. General manufacturing and new capacity labor-intensive projects cannot be approved, and existing projects are gradually shifting outward from Beijing. The six major industries, thermal power, iron and steel, petrochemical, cement, nonferrous metals and chemical industries, and coal-fired boiler projects, must strictly impose special emission limits for atmospheric pollutants in the BTHUA. Concurrently, highly polluting companies' elimination lists and capacity reduction targets were formulated in various provinces.

The second is the energy structure adjustment linkage mechanism. The government implemented a control plan for total coal consumption, reducing total consumption by 83 million tons in 2017; implemented a clean energy replacement plan; increased the proportion of clean energy use; promoted a coal clean utilization plan; expanded the scope of high-pollution fuel-free zones; promoted efficient and clean heating and solar water heating systems; optimized industrial space layout; and completed the relocation and transformation of heavy-polluting enterprises, such as steel, petrochemical, and chemical industries, in the main urban area. As a result of these measures, coal accounted for less than 10% of energy production in Beijing.

The third is the green transportation linkage mechanism. The government optimized the intercity integrated transportation system for the urban agglomerations; implemented a bus priority strategy; increased the proportion of green transportation; increased the proportion of public transportation by more than 60% in Beijing and Tianjin in 2017; controlled the number of motor vehicles, simultaneously implemented total motor vehicle control; limited travel time according to license plate tail number in Beijing, Tianjin, and Langfang; eliminated all yellow label cars (those heavy-polluting vehicles) at the end of 2017; upgraded the fuel quality to meet the national fifth stage standard; promoted new energy vehicles; and increased the proportion of new energy and clean fuel vehicles to 60% in Beijing, Tianjin, and the provincial capital cities.

The fourth is the inter-regional assistance linkage mechanism. Simultaneously with the implementation of the regional internal linkage plan, the joint pollution control programs of Beijing and Tianjin assisting the key polluted cities in Hebei Province were executed. Beijing provided 460 million yuan to compensate Langfang and Baoding for joint pollution control activities, with reducing the coal-fired emissions by 0.770 million tons and the annual sulfur dioxide emissions by more than 6 thousand tons in 2014. Tianjin invested 400 million yuan in special funds to support air pollution control projects in Zhangzhou and Tangshan in 2016.

The fifth is the monitoring and early warning linkage mechanism. The government established a unified national air quality monitoring network in prefecture-level cities, improved the online monitoring system for pollution sources, built a monitoring platform for motor vehicle sewage discharge, established a monitoring and early warning system for heavy-pollution weather, prepared heavy-pollution emergency plans, and implemented a regional emergency response mechanism for heavy-pollution weather.

The sixth is the consultation and accountability mechanism. The provincial, municipality, and autonomous region governments and relevant ministries and commissions jointly coordinated efforts to address major environmental issues as well as to organize and implement environmental assessments, information sharing, joint law enforcement, early warning, and emergency measures. The state Council, provinces, and cities signed the responsibility book for air pollution prevention and control objectives, with each organization respectively implementing their pollution control task. The assessment system was established by focusing on government assessment whilst considering third-party input. Pollution control status reports from previous years are assessed in a timely manner, and the results reported to the state council and announced to the public.

3. Data Methods and Framework

3.1. Data Source and Initial Processing

The research scopes are the 13 cities above the prefecture level in the BTHUA, including Beijing, Tianjin, Shijiazhuang, Baoding, Zhangzhou, Chengde, Handan, Hengshui, Langfang, Qinhuangdao, Tangshan, Xingtai, and Zhangjiakou; these cities cover an area of 0.22 million km^2 (Figure 2). In 2017, the BTHUA carried 8% of the country's population and contributed 10% of the country's total economic output with a national land area of 2.3%. Data for SO_2, NO_2, PM_{10}, $PM_{2.5}$, O_3, and CO concentration were obtained from the urban air quality real-time release platform of the China Environmental Monitoring Station. According to the effectiveness requirements of air pollutant concentration data in the GB3095-2012 [31], with reference to the research results [6], the quality of monitoring data was controlled. Because the number of urban inspection points and daily data time points are different, this paper averaged the data to obtain mean daily concentration values for the six pollutants. Daily average refers to the arithmetic mean of the 24-h average concentration of a natural day, monthly average refers to the arithmetic mean of the average daily concentration in a calendar month, quarter average refers to the average concentration for each day in a calendar season, and annual average refers to the arithmetic mean of the average concentration of each day in a calendar year. In addition, spring season refers to the months of March–May, summer to June–August, autumn to September–November, and winter to December–February.

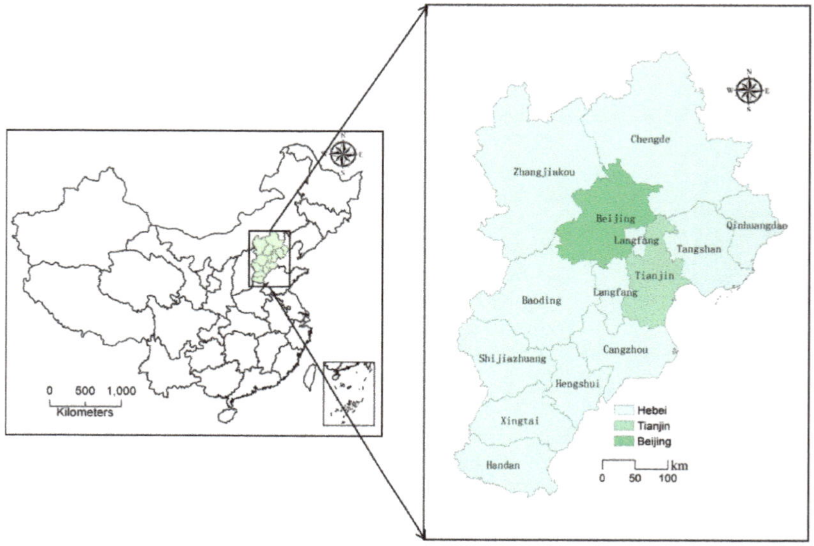

Figure 2. Location of the study area in China.

3.2. Spatial Pattern Analysis Method

Spatial interpolation is a method of estimating unknown points by using locally known points according to the correlation of adjacent regions in geospatial space. The surface is created by a highly accurate interpolation method with known sample selection. The spatial analysis of air pollution mainly uses kriging interpolation method, inverse distance weight interpolation, geographic weighted regression model [32], and geographic semivariogram method [33] to estimate spatial differentiation. This study uses the geostatistical wizard of geostatistical analyst in ArcGIS software to perform spatial interpolation and to accurately analyze the spatial distribution of six pollutants in the BTHUA. The commonly used methods mainly include inverse distance weighted, spline method, and kriging method. The geostatistical kriging interpolation method is the most flexible, and the result is the highest precision [34]. The formula is as follows:

$$h(x_0) = \sum_{i=1}^{n} \lambda_i h(x_i) \tag{1}$$

Among them, $h(x_0)$ is the monitoring value of the point x_0; $h(x_i)$ is the monitoring value of the point x_i; λ_i is the kriging weight coefficient; n is the total number of monitored cities. For the selection of different semivariograms (exponential model, triangular model, spherical model, Gaussian model, linear model) in kriging method, this study used the cross-validation of spatial interpolation results to compare the average absolute error and the root mean square error, and finally selected the higher precision function model as the method for analyzing the spatial pattern of atmospheric pollution in the BTHUA.

3.3. Research Framework

Air pollution has significant spatial and temporal differentiation in BTHUA. The fundamental reason is that its topography, industry, meteorology, and urban scale have significant spatial differentiation. The differentiation of time is reflected in the difference between heating season and non-heating season. The phenomenon of "scattering pollution" in Baoding and other cities is serious, and the amount of loose coal is large and the emission level is high. The differentiation of space is mainly determined by the spatial distribution of high-emission and high-pollution heavy industry. The total emissions and intensity are large with the steel, glass, petrochemical, chemical industry, and the supporting heavy transport vehicles in Tangshan, Tianjin, Shijiazhuang, Handan, and Xingtai. The main goal of the model governance is to prevent and control air pollution by adjusting the industrial structure, energy structure, transportation structure, land use structure, and emergency measures in BTHUA. However, topographical and meteorological elements are uncontrollable, and the government cannot regulate the flow of pollutants from outside the BTHUA into the area. Therefore, this model needs to be promoted in a larger area, even nationally and internationally. Based on the above analysis, this paper made the following logical framework diagram (Figure 3).

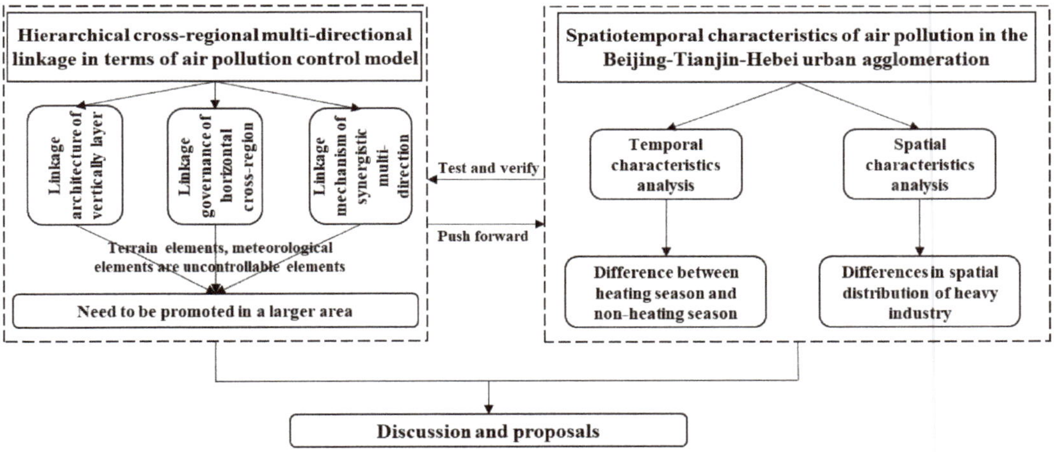

Figure 3. The logical framework of this study.

4. Spatiotemporal Characteristics of Air Pollution in the BTHUA

This section uses the kriging interpolation method to analyze the timing evolution and the spatial pattern characteristics of air pollution from 2014 to 2017 in BTHUA, respectively. The thorough comparison of 112 ground monitoring stations, primarily across the BTHUA, was conducted to evaluate the concentrations of the main atmospheric pollutants from 2014 to 2017.

4.1. Air Pollution Concentration Analysis

Air quality significantly improved from 2014 to 2017 in the BTHUA. Except for O_3, the concentration of other air pollutants declined, the high-pollution spatial scope decreased from north to south. The mean annual concentrations of $PM_{2.5}$, PM_{10}, SO_2, NO_2, and CO decreased by 30%, 23%, 52%, 3%, and 17%, respectively. Therefore, a large overall improvement of air quality was achieved, confirming the successful implementation of air pollution controls (Figure 4); however, the overall concentration value is still high, and the decline range is far from reaching the environmental capacity, so the task of air pollution prevention and control is still arduous. From a year-on-year comparison, the concentrations of $PM_{2.5}$, PM_{10}, SO_2, and CO continued to decline from 2014 to 2017, but the downward trend slowed down, indicating that the difficulty of air pollution control increased with the overall improvement of air quality [35]. The concentration of NO_2 showed a downward trend as a whole but increased in 2016, indicating that the effect of air pollution control has certain volatility [36]. The concentration of O_3 continues to rise, and the rising trend is more significant, which indicates that the problem of O_3 pollution has become more and more serious in recent years, which needs to arouse the attention of relevant government departments. At the same time, it also needs the academic community to carry out in-depth research on the chemical composition, impact mechanism, and prevention and control models of O_3 in urban agglomerations [37].

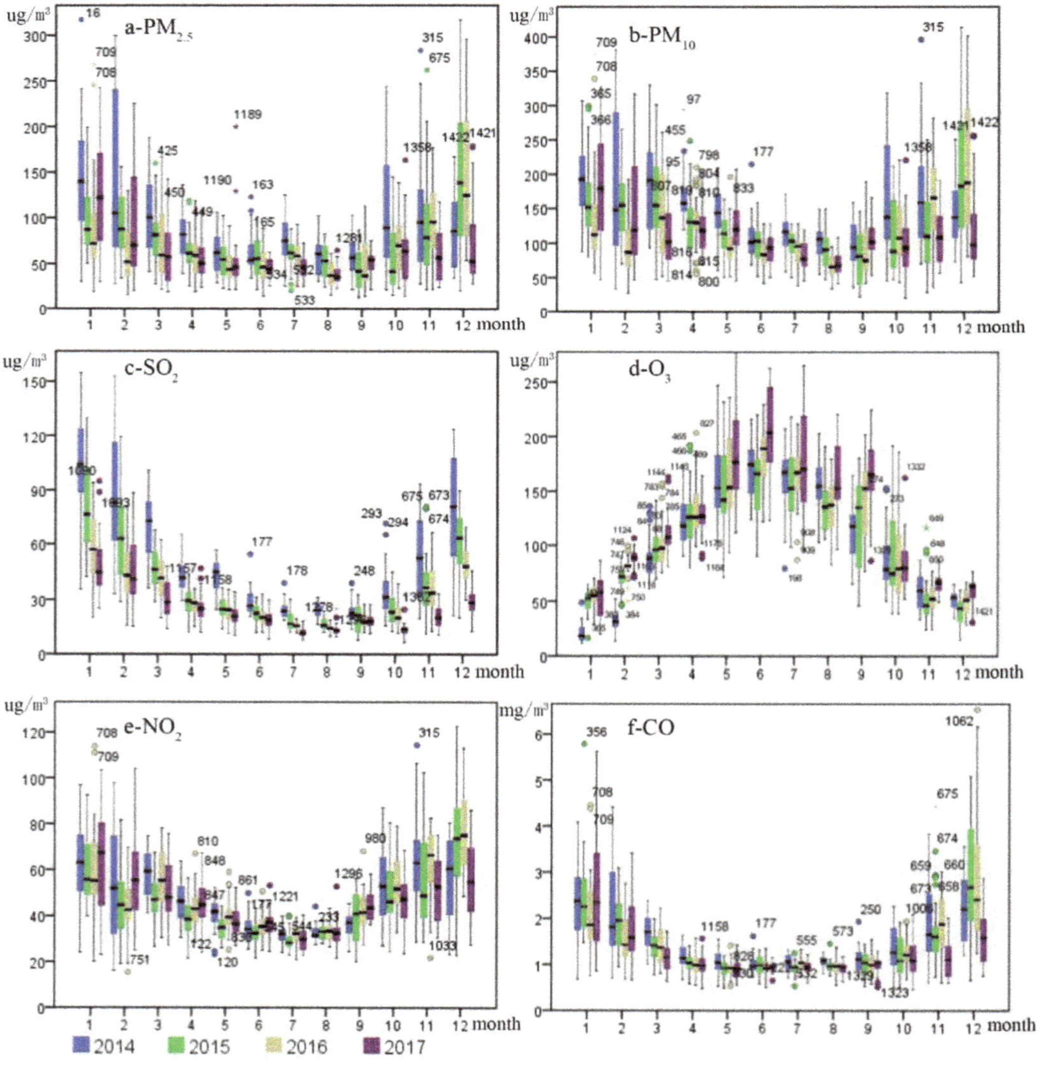

Figure 4. Trends in the seasonal variations of six air pollutant concentrations in BTHUA from 2014 to 2017.

In terms of the monthly change trend, the concentrations of $PM_{2.5}$, PM_{10}, and SO_2 kept a downward trend year by year from 2014 to 2017, and the prevention and control measures were significantly effective. The concentrations of NO_2 and CO showed a trend of first rising and then decreasing, it is still necessary to further strengthen pollution prevention and control; the concentration of O_3 kept increasing year by year on the whole. Therefore, it is necessary to attach great importance to O_3 pollution and strictly implement an ultra-low emission policy [38]. Specifically, compared with the monthly pollutant concentration in 2014, the concentrations of $PM_{2.5}$ and SO_2 decreased significantly in all months in 2017, the concentrations of PM_{10} and CO increased only in January, and decreased significantly in other months; the concentration of NO_2 increased in January, February, June, September, December, and decreased in other months; the concentration of O_3 only decreased in October and increased significantly in other months.

In terms of the seasonal average concentration, the concentrations of $PM_{2.5}$, NO_2, CO, PM_{10}, and SO_2 were the highest in winter, the lowest in summer, the middle in spring and autumn, the concentrations of the first three were higher in autumn than in spring, and the concentrations of the last two were higher in spring than in autumn. The coal burning and heavy industrial production increased the emission of the above pollutants [39], the adverse climate conditions led to the increase of $PM_{2.5}$ concentrations [40]. The concentration of O_3 was the highest in summer, the lowest in winter, and medium in spring and autumn. Under strong sunlight, nitrogen oxides and volatile organic compounds (VOCs) can produce O_3 by photochemical reaction [41]. In terms of the seasonal change trend, compared with that in 2014, the concentrations of $PM_{2.5}$, PM_{10}, SO_2, and CO showed a downward trend in all quarters in 2017, the concentration of PM_{10} decreased significantly, followed by $PM_{2.5}$. The concentration of NO_2 only increased by 1.5% in winter, the other quarters showed a downward trend [42]. The concentration of O_3 showed an upward trend in all quarters. The reduction degrees of pollutant concentration in autumn and winter were higher than those in spring and autumn, and the government effects were more significant in autumn and winter.

4.2. Analysis of Pollutant Compliance Rate

According to the requirements of GB 3095-2012, the 24 h average concentration limits of a second-class region for $PM_{2.5}$, PM_{10}, SO_2, O_3, NO_2, and CO are 75 $\mu g/m^3$, 150 $\mu g/m^3$, 150 $\mu g/m^3$, 160 $\mu g/m^3$, 80 $\mu g/m^3$, and 4 mg/m^3, respectively. Excluding O_3 and CO, the concentrations of the other air pollutants showed a significant growth in the rate of compliance in the BTHUA from 2014 to 2017 (Figure 5). The compliance rates of $PM_{2.5}$, PM_{10}, and SO_2 continued to increase by 25.6%, 21.50%, 0.85%, respectively; $PM_{2.5}$ increased from 49.58% to 75.98%; PM_{10} increased from 60.34% to 81.84%; SO_2 increased from 99.15% to 100%. The compliance rate of NO_2 first decreased and then increased, with an overall increase of 0.9%, 95.47% in 2014; 93.48% in 2015; and 96.37% in 2017. The compliance rate of CO also first declined and then rose, with an overall decline of 0.83%, 99.43% in 2014; 97.17% in 2015; and 98.60% in 2017. The compliance rate of O_3 first increased and then decreased, with an overall decrease of 7.54%, 80.45% in 2014; 83.29% in 2015; and 72.91% in 2017.

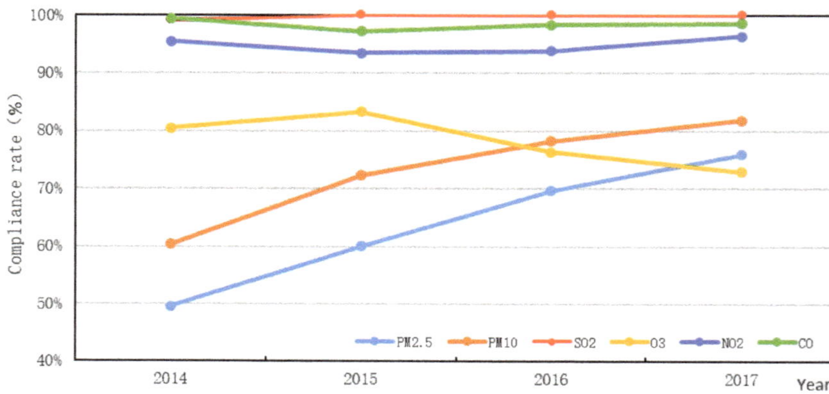

Figure 5. Trends in compliance rates for six air pollutants from 2014 to 2017.

4.3. Spatial Characteristics of Air Pollution

The paper uses kriging interpolation method in the ArcGIS software to perform spatial interpolation analysis on the site data of six atmospheric pollutants in the BTHUA. The air quality high-pollution range decreased significantly in the BTHUA from 2014 to 2017, excluding that the O_3 high-pollution range spread from south to north, the high pollution ranges for other pollutant decreased significantly from north to south in the BTHUA (Figure 6). Concentrations of $PM_{2.5}$ and PM_{10} both showed a pattern of high values in the south and low values in the north, and the concentrations decreased in the high-polluted southern areas more than those in the low-polluted northern areas. Pollution in excess of the 75 $\mu g/m^3$ standard line for $PM_{2.5}$ concentration retreated to the south of Beijing and Tianjin, but the PM_{10} concentration in the entire urban agglomeration was higher than the national annual limit of 70 $\mu g/m^3$. Areas with large fluctuations in $PM_{2.5}$ concentrations are mainly concentrated in the central of urban agglomeration, such as Beijing, Tianjin, Hengshui, Langfang, and Baoding, indicating that it is more difficult to control the $PM_{2.5}$ pollution in the central area of the urban agglomeration with Beijing–Tianjin as the core. SO_2 concentrations show a spatial pattern of high values in the south and low values in the north, these values decreased significantly in the four quarters, the decline in the north was more significant than that in the south, with the concentrations decreasing significantly in Xingtai, Baoding, Shijiazhuang, and Tangshan. The only pollutant with an increased concentration was O_3 in the urban agglomeration, the average annual concentration increased by 21 $\mu g/m^3$ (up 20.59%) in four years [43]. The O_3 concentrations decreased in the south of Beijing, such as Zhangzhou, Xingtai, Handan, Hengshui, Tianjin, and Shijiazhuang, increased in the north, with those increasing by 75% in Zhangjiakou. NO_2 concentrations declined first, then increased in 2016, and declined again in 2017, with an overall decline of 3% over four years. Cangzhou and Hengshui have become the pollution centers in the south, while other regions have shown a downward trend. Although the concentrations in Tangshan have decreased, it is still a northern pollution center due to the high concentration. The concentration of CO decreased by 17% in four years, with decreases in the central cities of Beijing, Tianjin, Zhangjiakou, and Baoding driving the overall concentration in the urban agglomeration, which decreased by 33% and 18% in Beijing and Tianjin, respectively, the concentration increased in the south, i.e., in Hengshui, Chengde, Langfang, and Xingtai.

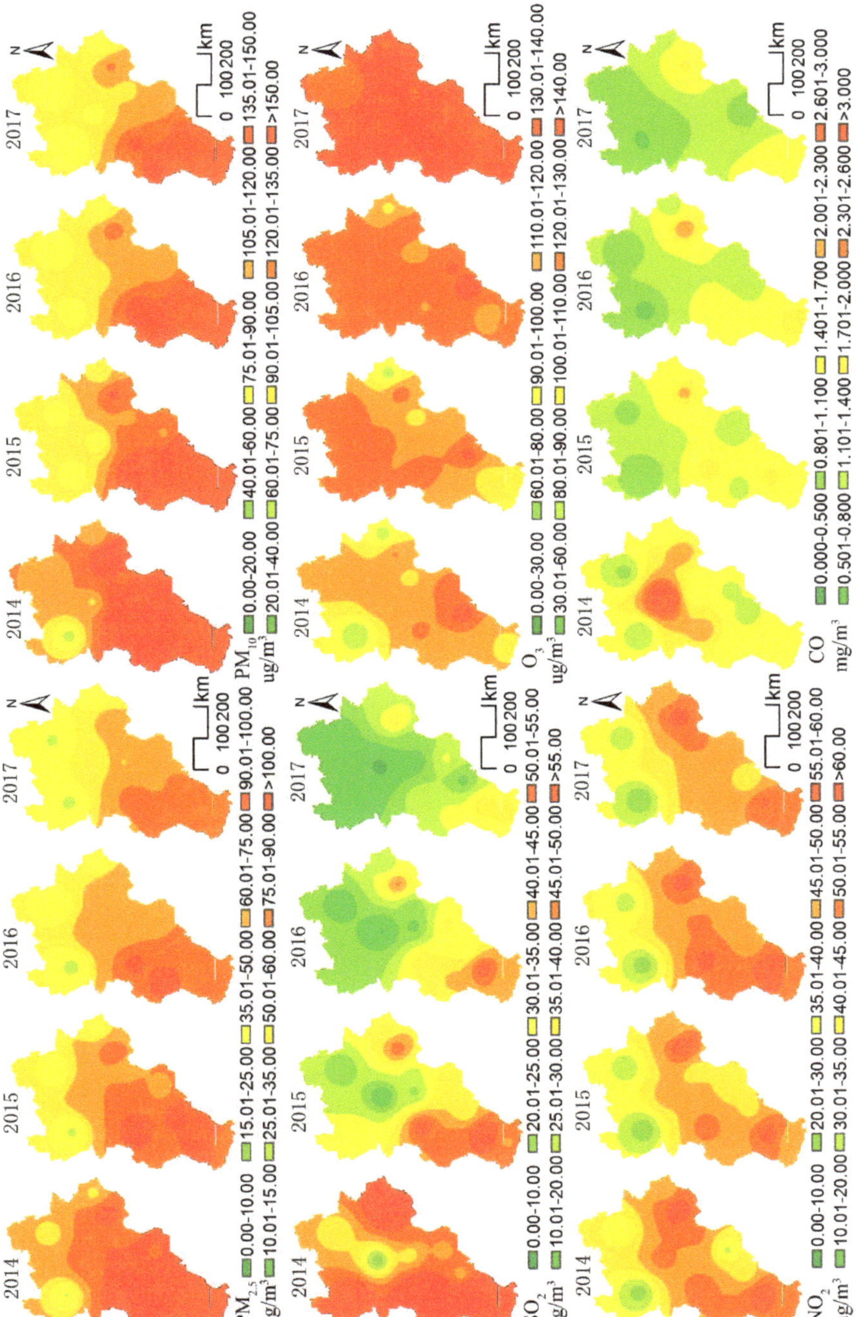

Figure 6. Spatial pattern of annual mean values of six air pollutants in BTHUA from 2014 to 2017.

5. Discussion and Proposals

5.1. Discussion on the Air Pollution Control Models in China

The implementation of the HCML model has effectively improved the air quality in heavily polluted areas over a short time period in China. On the whole, China's prevention and control strategies have the following unique characteristics. First, it has the most stringent laws in China, such as the Environmental Protection Law of the People's Republic of China (Revised in 2014) and Atmospheric Pollution Prevention and Control Law of the People's Republic of China (2015 Revision). Companies are subject to more stringent standards, with harsher penalties. Illegal sewage discharge from illegal enterprises is punished on a daily basis, and penalties are not capped. Furthermore, the four types of non-crime, such as the implementation of environmental impact assessment, are subject to administrative detainment. The local government is responsible for the quality of the environment in its jurisdiction, and the environmental protection objectives are linked to the administrative personnel evaluations. The government controls the total amount of pollution discharge and limits the projects that exceed the total amount in the region. Regulatory authorities have also been endowed more responsibility: the environmental protection department has greater power to directly shut down illegal enterprises, and relevant unregulated officials have strict administrative accountability.

Second, the model has the toughest plan to date for combating air pollution, the Air Pollution Prevention and Control Action Plan. This plan is the largest governance effort, with the tightest protection measures and most stringent assessment. For the first time, the action plan incorporates fine particulate matter into binding indicators and incorporates environmental quality improvement into the official assessment system. The state council and provincial (district, municipal) people's governments sign the responsibility book for air pollution prevention and control targets and breakdown the target tasks for distribution to the local people's government. Enterprises adopt administrative means to force the completion of capacity-reduction plans for 21 key industries, including steel, cement, electrolytic aluminum, and flat glass enterprises, according to the timetable.

Third, the model provides a fully covered weather monitoring and early warning system and the national–provincial–city–enterprise linkage action mechanism. In 2015, 31 provinces (autonomous regions and municipalities) implemented provincial air quality monitoring and early warning systems in China, and 32 cities and provincial capital cities launched municipal air quality monitoring and early warning systems. Aiming at the spatiotemporal characteristics of air pollution, the national–provincial–municipal–department–enterprise management structure was constructed, including the target decomposition models and the fast dispatching and execution models.

5.2. Discussion of Research Proposals

5.2.1. Traceability and Process Research of Heavy Pollution in Autumn and Winter

According to the trend analysis, air pollution is reduced significantly from the spring to the summer of the subsequent year and increased in the autumn and winter in BTHUA. The volatility results showed that the daily average concentration volatility of pollutants showed a weakening trend and tended to be stable, but the time performance was lower in the first half of the year and higher in the second half. Therefore, air pollution in autumn and winter is the focus of future prevention and control, especially in winter, which is the season in which the most serious air pollution occurs in the BTHUA. Using $PM_{2.5}$ as an example, in 2014 and 2015, the proportion of winter pollution days was more than 70% (60 days), accounting for 35% and 43% of the total number of those days in the year, respectively, and the number of days with heavy pollution accounted for 54% and 83% of the total number of those days in the year, respectively. Therefore, it is critical to strengthen the traceability and process research of heavy pollution in the winter coal burning period [44].

5.2.2. Traceability and Process Research of O_3 Pollution

Ozone is the main component of smog in Los Angeles; its formation is directly related to automobile exhaust and gas particle conversion due to photochemical reactions. It is also one of the most prominent air pollution problems in the Beijing–Tianjin–Hebei urban agglomeration [45]. Among the six pollutants in the BTHUA, only the levels of O_3 pollution are increasing, and the increase is more significant in the summer in the southern region with Hengshui as the general center. Pan et al. [46] proposed that the increase in the relative contribution of nitrate to PM_1 observed during the early stages of haze pollution was due to new particle formation, whereas the nitrate formed in $PM_{1-2.5}$ during the latter stages was due to heterogeneous formation and hygroscopic growth. Therefore, controlling NOx emissions should be a priority for improving air quality in mega cities of the BTHUA [47]. The BTHUA is located in the VOC control area, and VOC emissions have not been effectively controlled, resulting in an upward trend of O_3 and $PM_{2.5}$ pollution. The VOC emission reduction project should be implemented to control O_3 and $PM_{2.5}$ emissions [48]. Given the health hazards, the explanations, processes, and solutions for aggravated O_3 pollution in the BTHUA urgently warrant further research.

5.3. Policy Recommendations

5.3.1. Multi-Regionally Joint Legislation and Implementing Policies

There are significant spatial differences in the high-pollution range in BTHUA, the air pollution was serious with dense population and rapid industrial development in the central and southern region. However, the intensity of fighting against pollution is weak, and the legal treatment awareness of pollution is not strong in those areas; it is urgent to carry out coordinated legislation and multi-regional linkage measures and strengthen market-oriented policies and incentive measures in BTHUA [49]. Taking cues from the implementation of the new "Air Pollution Prevention and Control Law" and the "Environmental Protection Law", a coordinated legislative mechanism that is compatible with the coordinated development of the BTHUA may be developed. This mechanism will be mutually supportive and reinforcing and enable the development of coordinated legislation in key areas by integrating and matching relevant regulations from different locations, continuing to refine legal constraints, eliminating administrative barriers and local protection through legislation, and ensuring that all localities can conduct concerted and precise attacks on air-polluting behavior.

5.3.2. Innovative Pollution Discharge Supervision Mechanisms and Public Supervision

In recent years, the air quality has been improved, and the pollution control has certain effect, but it has obvious volatility; the air pollution problem is still relatively serious in BTHUA. Industrial pollution, motor vehicle pollution, and heating coal in winter are the main reasons for the air pollution of urban agglomeration. With people's desire for a better life, public participation is very important in air pollution control. Therefore, it is necessary to further innovate the emission supervision mode and improve the public supervision mechanism [50]. Both the "plan" and the "details" clarify the objectives for the closure and renovation of industrial enterprises and key industries in the BTHUA. However, the number of industrial enterprises within the urban agglomeration is huge. The proportion of small- and medium-size enterprises is over 96%, and their locations are scattered. There are many issues, such as undocumented smuggling, night smuggling, online data fraud, and no using pollution control facilities. Concurrently, the number of management supervisors cannot meet the need of enterprises, which, in turn, impacts the regulation of atmospheric pollutant discharge. Therefore, the implementation of a unified supervision mechanism for multi-sectors and multi-subjects and a joint supervision model for the implementation of responsibilities and improved public supervision are inevitable directions for pollution supervision work.

6. Conclusions

Air pollution caused by the rapid development of Chinese cities in the past few decades and its impact on resident health cannot be ignored [51]. The governments of the BTHUA have achieved certain progress in implementing the "stratified cross-region multi-directional linkage" air pollution prevention and control model. However, the industrial structure, energy structure, and traffic pattern in the current industrialization stage continues to pose long-term challenges to air pollution prevention and control measures in urban agglomerations.

Based on air pollution control in BTHUA, this paper constructs the hierarchical cross-regional multi-directional linkage in terms of air pollution control models to provide significant reference and theoretical support for air pollution control in other regions of China and other countries in the world. We hope to arouse the discussion and application of this models and the practical details and problems in the academic societies. However, we do not further reveal the issue of policy transferability by using the quantitative methods, which is the focus of our future research. In addition, each region and each country have different difficulties for air pollution control. Therefore, it is necessary to consider the specific characteristics of the city when controlling the air pollution.

From the perspective of the effectiveness of air pollution control, this paper explores the change characteristics of six pollutants, namely, SO_2, NO_2, PM_{10}, $PM_{2.5}$, O_3, and CO in BTHUA from 2014 to 2017. Due to the length of the article, we do not further reveal the impact mechanism of different air pollutants. Impact mechanism and traceability analysis is important for the radical cure of air pollution, which is a meaningful direction of our future research. In the spatial analysis section of this paper, we use the statistical cokriging interpolation method of statistics, which is the spatial analysis method used in geography. We can also use other advanced methods to explore the spatial characteristics of air pollutants.

Author Contributions: L.L.: data curation, formal analysis, investigation, methodology, visualization, writing—original draft, writing—review and editing. Z.W.: funding acquisition, project administration, resources, supervision, writing—review and editing. All authors have read and agreed to the published version of the manuscript.

Funding: This work was supported by National Natural Science Foundation of China (41771181).

Institutional Review Board Statement: Not applicable.

Informed Consent Statement: Not applicable.

Data Availability Statement: Data are available upon request.

Acknowledgments: Many thanks go to the editor and the anonymous reviewers for their careful reading of the manuscript with many insightful comments and suggestions on improving the quality of this paper.

Conflicts of Interest: No potential conflict of interest was reported by the authors.

References

1. Armeanu, D.S.; Joldes, C.C.; Gherghina, S.C.; Andrei, J.V. Understanding the multidimensional linkages among renewable energy, pollution, economic growth and urbanization in contemporary economies: Quantitative assessments across different income countries' groups. *Renew. Sustain. Energy Rev.* **2021**, *142*, 110818. [CrossRef]
2. Haque, M.; Fang, C.; Schnelle-Kreis, J.; Abbaszade, G.; Liu, X.; Bao, M.; Zhang, W.; Zhang, Y.-L. Regional haze formation enhanced the atmospheric pollution levels in the Yangtze River Delta region, China: Implications for anthropogenic sources and secondary aerosol formation. *Sci. Total Environ.* **2020**, *728*, 138013. [CrossRef] [PubMed]
3. Xue, T.; Liu, J.; Zhang, Q.; Geng, G.; Zheng, Y.; Tong, D.; Liu, Z.; Guan, D.; Bo, Y.; Zhu, T.; et al. Rapid improvement of $PM_{2.5}$ pollution and associated health benefits in China during 2013–2017. *Sci. China Earth Sci.* **2019**, *62*, 1847–1856. [CrossRef]
4. Samet, J.M. London Fog-The Biography. *Am. J. Public Health* **2016**, *106*, 1352. [CrossRef]
5. Wang, Z.B.; Fang, C.L. Spatial-temporal characteristics and determinants of $PM_{2.5}$ in the Bohai Rim Urban Agglomeration. *Chemosphere* **2016**, *148*, 148–162. [CrossRef]

6. Fang, C.L.; Wang, Z.B.; Xu, G. Spatial-temporal characteristics of $PM_{2.5}$ in China: A city-level perspective analysis. *J. Geogr. Sci.* **2016**, *26*, 1519–1532. [CrossRef]
7. Bretón, R.M.C.; Bretón, J.G.C.; Kahl, J.W.D.; Fuentes, M.D.L.L.E.; Lara, E.R.; Marrón, M.R.; Severino, R.D.C.L.; Chi, M.P.U. Short-term effects of atmospheric pollution on daily mortality and their modification by increased temperatures associated with a climatic change scenario in Northern Mexico. *Int. J. Environ. Res. Public Health* **2020**, *17*, 9219. [CrossRef] [PubMed]
8. Xie, Y.; Dai, H.; Dong, H.; Hanaoka, T.; Masui, T. Economic impacts from $PM_{2.5}$ pollution-related health effects in China: A provincial-level analysis. *Environ. Sci. Technol.* **2016**, *50*, 4836–4843. [CrossRef] [PubMed]
9. Xie, Z.; Li, Y.; Qin, Y. Value assessment of health losses caused by $PM_{2.5}$ pollution in cities of atmospheric pollution transmission channel in the Beijing–Tianjin–Hebei region, China. *Int. J. Environ. Res. Public Health* **2019**, *16*, 1012. [CrossRef]
10. Wang, Z.B.; Fang, C.L.; Xu, G. Spatio-temporal variation of $PM_{2.5}$ concentration in Chinese cities in 2014. *Acta Geogr. Sin.* **2015**, *70*, 1720–1734. (In Chinese)
11. Liang, L.W.; Wang, Z.B.; Li, J.X. The effect of urbanization on environmental pollution in rapidly developing urban agglomerations. *J. Clean. Prod.* **2019**, *237*, 117649. [CrossRef]
12. Wang, Z.B.; Liang, L.W.; Sun, Z.; Wang, X.M. Spatiotemporal differentiation and the factors influencing urbanization and ecological environment synergistic effects within the Beijing-Tianjin-Hebei urban agglomeration. *J. Environ. Manag.* **2019**, *243*, 227–239. [CrossRef]
13. Zhang, Y.; Lang, J.; Cheng, S.; Li, S.; Zhou, Y.; Chen, D.; Zhang, H.; Wang, H. Chemical composition and sources of PM_1 and $PM_{2.5}$ in Beijing in autumn. *Sci. Total Environ.* **2018**, *630*, 72–82. [CrossRef] [PubMed]
14. Xiong, H.H.; Liang, L.W.; Zeng, Z. Dynamic analysis of $PM_{2.5}$ spatial-temporal characteristics in China. *Resour. Sci.* **2017**, *39*, 136–146.
15. Yang, W.; Li, J.; Wang, Z.; Wang, L.; Dao, X.; Zhu, L.; Pan, X.; Li, Y.; Sun, Y.; Ma, S.; et al. Source apportionment of $PM_{2.5}$ in the most polluted Central Plains Economic Region in China: Implications for joint prevention and control of atmospheric pollution. *J. Clean. Prod.* **2021**, *283*, 124557. [CrossRef]
16. Liu, H.; Fang, C.; Zhang, X.; Wang, Z.; Bao, C.; Li, F. The effect of natural and anthropogenic factors on haze pollution in Chinese cities: A spatial econometrics approach. *J. Clean. Prod.* **2017**, *165*, 323–333. [CrossRef]
17. Wang, Z.B.; Liang, L.W.; Lin, X.B.; Liu, H.M. Mode summary of air pollution in Beijing-Tianjin-Hebei urban agglomeration and evaluation of treatment effects. *Environ. Sci.* **2017**, *38*, 4005–4014. (In Chinese)
18. Zhang, N.-N.; Ma, F.; Qin, C.-B.; Li, Y.-F. Spatiotemporal trends in $PM_{2.5}$ levels from 2013 to 2017 and regional demarcations for joint prevention and control of atmospheric pollution in China. *Chemosphere* **2018**, *210*, 1176–1184. [CrossRef]
19. Wang, Z.B.; Liang, L.W.; Wang, X.J. Spatio-temporal evolution patterns and influencing factors of $PM_{2.5}$ in Chinese urban agglomerations. *Acta Geogr. Sin.* **2019**, *74*, 2614–2630. (In Chinese)
20. Li, M.C.; Mao, C.M. Spatial effect of industrial energy consumption structure and transportation on haze pollution in Beijing-Tianjin-Hebei region. *Int. J. Environ. Res. Public Health* **2020**, *17*, 5610. [CrossRef] [PubMed]
21. Zhang, M.; Liu, X.X.; Ding, Y.T. Assessing the influence of urban transportation infrastructure construction on haze pollution in China: A case study of Beijing-Tianjin-Hebei region. *Environ. Impact Assess. Rev.* **2021**, *87*, 106547. [CrossRef]
22. Xu, H.L.; Deng, Y.P. Does foreign direct investment lead to environmental pollution in China? Spatial measurement based on Chinese provincial panel data. *Manag. World* **2012**, *2*, 30–43.
23. Wang, Z.B.; Li, J.X.; Liang, L.W. Spatio-temporal evolution of ozone pollution and its influencing factors in the Beijing-Tianjin-Hebei Urban Agglomeration. *Environ. Pollut.* **2020**, *256*, 113419. [CrossRef] [PubMed]
24. Fuller, G.W.; Font, A. Keeping air pollution policies on track. *Science* **2019**, *365*, 322–323. [CrossRef] [PubMed]
25. Liu, X.; Hadiatullah, H.; Tai, P.; Xu, Y.; Zhang, X.; Schnelle-Kreis, J.; Schloter-Hai, B.; Zimmermann, R. Air pollution in Germany: Spatio-temporal variations and their driving factors based on continuous data from 2008 to 2018. *Environ. Pollut.* **2021**, *276*, 116732. [CrossRef]
26. Hossain, S.; Frey, H.C.; Louie, P.K.; Lau, A.K. Combined effects of increased O_3 and reduced NO_2 concentrations on short-term air pollution health risks in Hong Kong. *Environ. Pollut.* **2021**, *270*, 116280. [CrossRef]
27. Chakraborty, J.; Basu, P. Air quality and environmental injustice in India: Connecting particulate pollution to social disadvantages. *Int. J. Environ. Res. Public Health* **2021**, *18*, 304. [CrossRef]
28. Novan, K. Valuing the wind: Renewable energy policies and air pollution avoided. *Am. Econ. J. Econ. Policy* **2015**, *7*, 291–326. [CrossRef]
29. Rao, S.; Klimont, Z.; Smith, S.J.; Van Dingenen, R.; Dentener, F.; Bouwman, L.; Riahi, K.; Amann, M.; Bodirsky, B.L.; van Vuuren, D.P.; et al. Future air pollution in the Shared Socio-economic Pathways. *Glob. Environ. Chang.* **2017**, *42*, 346–358. [CrossRef]
30. Tang, L.; Qu, J.; Mi, Z.; Bo, X.; Chang, X.; Anadon, L.D.; Wang, S.; Xue, X.; Li, S.; Wang, X.; et al. Substantial emission reductions from Chinese power plants after the introduction of ultra-low emissions standards. *Nat. Energy* **2019**, *4*, 929–938. [CrossRef]
31. National Environmental Protection Department. Environmental Air Quality Standards (for Trial Implementation). In *People's Republic of China National Environmental Protection Standards. (GB3095-2012)*; National Environmental Protection Department: Beijing, China, 2012.
32. Song, W.; Jia, H.; Huang, J.; Zhang, Y. A satellite-based geographically weighted regression model for regional $PM_{2.5}$ estimation over the Pearl River Delta region in China. *Remote Sens. Environ.* **2014**, *154*, 1–7. [CrossRef]

33. Song, W.; Jia, H.; Li, Z.; Tang, D. Using geographical semi-variogram method to quantify the difference between NO_2 and $PM_{2.5}$ spatial distribution characteristics in urban areas. *Sci. Total Environ.* **2018**, *631–632*, 688–694. [CrossRef] [PubMed]
34. Wu, C.G.; Lin, D.S.; Zhou, Z.X. Spatial interpolation method for precipitation in the three gorges reservoir area and time-space distribution. *Resour. Environ. Yangtze Basin* **2010**, *19*, 752–758.
35. Yang, W.X.; Yuan, G.H.; Han, J.T. Is China's air pollution control policy effective? Evidence from Yangtze River Delta cities. *J. Clean. Prod.* **2019**, *220*, 110–133. [CrossRef]
36. Wang, C.J.; Wang, T.; Wang, P.C. The spatial–temporal variation of tropospheric NO_2 over China during 2005 to 2018. *Atmosphere* **2019**, *10*, 444. [CrossRef]
37. Montzka, S.A.; Dutton, G.; Yu, P.; Ray, E.; Portmann, R.; Daniel, J.S.; Kuijpers, L.; Hall, B.D.; Mondeel, D.; Siso, C.; et al. An unexpected and persistent increase in global emissions of ozone depleting CFC-11. *Nature* **2018**, *557*, 413–417. [CrossRef] [PubMed]
38. Cheng, L.J.; Wang, S.; Gong, Z.Y. Pollution trends of ozone and its characteristics of temporal and spatial distribution in Beijing-Tianjin-Hebei region. *Environ. Monit. China* **2017**, *33*, 14–21.
39. Mei, M.; Xu, D.H.; Zhu, R. Quantitative estimation of air pollutant emission rate based on urban atmospheric load index. *China Environ. Sci.* **2020**, *40*, 465–474.
40. Zhang, X.; Xu, X.; Ding, Y.; Liu, Y.; Zhang, H.; Wang, Y.; Zhong, J. The impact of meteorological changes from 2013 to 2017 on $PM_{2.5}$ mass reduction in key regions in China. *Sci. China Earth Sci.* **2019**, *62*, 1885–1902. [CrossRef]
41. Li, K.; Jacob, D.J.; Liao, H.; Shen, L.; Zhang, Q.; Bates, K.H. Anthropogenic drivers of 2013-2017 trends in summer surface ozone in China. *Proc. Natl. Acad. Sci. USA* **2019**, *116*, 422–427. [CrossRef]
42. Chen, Z.; Liu, C.; Liu, W.; Zhang, T.; Xu, J. A synchronous observation of enhanced aerosol and NO_2 over Beijing, China, in winter 2015. *Sci. Total Environ.* **2017**, *575*, 429–436. [CrossRef]
43. Wang, L.; Zhang, F.; Pilot, E.; Yu, J.; Nie, C.; Holdaway, J.; Yang, L.; Li, Y.; Wang, W.; Vardoulakis, S.; et al. Taking action on air pollution control in the Beijing-Tianjin-Hebei (BTH) region: Progress, challenges and opportunities. *Int. J. Environ. Res. Public Health* **2018**, *15*, 306. [CrossRef]
44. Jiang, L.; He, S.X.; Zhou, H.F. Spatio-temporal characteristics and convergence trends of $PM_{2.5}$ pollution: A case study of cities of air pollution transmission channel in Beijing-Tianjin-Hebei region, China. *J. Clean. Prod.* **2020**, *256*, 120631. [CrossRef]
45. Ma, R.; Ban, J.; Wang, Q.; Zhang, Y.; Yang, Y.; He, M.Z.; Li, S.; Shi, W.; Li, T. Random forest model based fine scale spatiotemporal O_3 trends in the Beijing-Tianjin-Hebei region in China, 2010 to 2017. *Environ. Pollut.* **2021**, *276*, 116635. [CrossRef] [PubMed]
46. Pan, Y.; Wang, Y.; Zhang, J.; Liu, Z.; Wang, L.; Tian, S.; Tang, G.; Gao, W.; Ji, D.; Song, T.; et al. Redefining the importance of nitrate during haze pollution to help optimize an emission control strategy. *Atmos. Environ.* **2016**, *141*, 197–202. [CrossRef]
47. Xing, J.; Mathur, R.; Pleim, J.; Hogrefe, C.; Gan, C.-M.; Wong, D.C.; Wei, C.; Gilliam, R.; Pouliot, G. Observations and modeling of air quality trends over 1990–2010 across the Northern Hemisphere: China, the United States and Europe. *Atmos. Chem. Phys.* **2015**, *15*, 2723–2747. [CrossRef]
48. Li, K.; Jacob, D.J.; Liao, H.; Zhu, J.; Shah, V.; Shen, L.; Bates, K.H.; Zhang, Q.; Zhai, S. A two-pollutant strategy for improving ozone and particulate air quality in China. *Nat. Geosci.* **2019**, *12*, 906–910. [CrossRef]
49. Li, N.; Zhang, X.L.; Shi, M.J. Does China's air pollution abatement policy matter? An assessment of the Beijing-Tianjin-Hebei region based on a multi-regional CGE model. *Energy Policy* **2019**, *127*, 213–227. [CrossRef]
50. Wang, B.; Hong, G.; Qin, T.; Fan, W.-R.; Yuan, X.-C. Factors governing the willingness to pay for air pollution treatment: A case study in the Beijing-Tianjin-Hebei region. *J. Clean. Prod.* **2019**, *235*, 1304–1314. [CrossRef]
51. Ebenstein, A.; Lavy, V.; Roth, S. The long run economic consequences of high-stakes examinations: Evidence from transitory variation in pollution. *Am. Econ. J. Appl. Econ.* **2016**, *8*, 36–65. [CrossRef]

International Journal of
Environmental Research and Public Health

MDPI

Article

Spatiotemporal Changes of Ecosystem Service Value Determined by National Land Space Pattern Change: A Case Study of Fengdu County in The Three Gorges Reservoir Area, China

Haozhe Zhang [1,2,3], Qingyuan Yang [1,2,3,*], Zhongxun Zhang [1,2,3], Dan Lu [1,2,3] and Huiming Zhang [1,2,3]

1 School of Geographical Sciences, Southwest University, Chongqing 400045, China; ssaijj@email.swu.edu.cn (H.Z.); zhongxun2021@swu.edu.cn (Z.Z.); ludswu@email.swu.edu.cn (D.L.); zhm19981900@email.swu.edu.cn (H.Z.)

2 Chongqing Jinfo Mountain Kaster Ecosystem National Observation and Research Station, Chongqing 400715, China

3 The State Cultivation Base of Eco-Agriculture for Southwest Mountainous Land, Southwest University, Chongqing 400045, China

* Correspondence: yizyang@swu.edu.cn; Tel.: +86-023-6825-3911

Abstract: Exploring the spatiotemporal change characteristics of ecosystem service value (ESV) under the influence of national land space pattern (NLSP) changes is of great significance for promoting the rational use of land resources and the optimization of ecosystems. In this study, Fengdu County in the Three Gorges Reservoir Area was selected as a case study. We analyzed the changes in NLSP using land use data from 1990, 2000, 2010 and 2018. Then, we used the equivalent factor method and exploratory spatial data analysis method to explore the spatiotemporal change characteristics of the ESV of Fengdu County. The results show that: (1) From 1990 to 2018, the changes in NLSP in Fengdu County generally manifested in the transformation of agricultural space into urban space and ecological space; (2) The spatiotemporal change of ESV is a process that positively responds to the increase in ecological space and negatively responds to the expansion of urban space. From 1990 to 2018, the total ESV of Fengdu County showed a trend of continuous growth, with a total increase of CNY 11.10×10^8, and the change rate was 9.33%. The ESV gain area is mainly located along the Yangtze River and the southernmost part of the county, and the loss area is mainly located near the south bank of the Yangtze River; (3) ESV and its changes in Fengdu County have a significant positive spatial autocorrelation. The cold and hot spots of ESV change are mainly distributed along the Yangtze River and to the south of the Yangtze River. Therefore, it is suggested to integrate ESV as an important indicator into the decision-making of national land space planning. At the same time, it is necessary to strengthen the intensive use of urban space and protect the important ecological space from decreasing. Our study results provide useful insights for the development of regional NLS management and environmental protection policies. However, it is worth noting that the results of this paper are more applicable to areas where the terrain is dominated by mountains.

Keywords: ecosystem services value; national land space pattern; spatiotemporal changes; Fengdu County; the Three Gorges Reservoir Area

Citation: Zhang, H.; Yang, Q.; Zhang, Z.; Lu, D.; Zhang, H. Spatiotemporal Changes of Ecosystem Service Value Determined by National Land Space Pattern Change: A Case Study of Fengdu County in The Three Gorges Reservoir Area, China. *Int. J. Environ. Res. Public Health* **2021**, *18*, 5007. https://doi.org/10.3390/ijerph18095007

Academic Editor: Giulia Maisto

Received: 1 March 2021
Accepted: 5 May 2021
Published: 9 May 2021

1. Introduction

Ecosystem services are the sum of life-sustaining products and services that humans obtain from ecosystems, which are closely related to human well-being and sustainable development [1]. The ecosystem supports and maintains the balance of the human living environment by regulating the climate and maintaining biodiversity [2]. It also provides the food and raw materials needed in life and production and brings entertainment and aesthetic enjoyment to human beings [3]. Ecosystem service value (ESV) is an important indicator to measure ecosystem service functions [4]. National land space (NLS) is the territorial space under the jurisdiction of national sovereignty and sovereign rights [5]. According to the categories of products provided, the NLS can be divided into urban space, with the main function of providing industrial products and service products, agricultural space, with the main function of providing agricultural products, and ecological space, with the main function of providing ecological products [6]. NLS includes resources and industrial elements such as land, labor, and minerals. Land is the supporting carrier of NLS, and the type, scale, and intensity of land use determine the national land space pattern (NLSP). At the same time, land is also a material provider of ecosystem services. Land use change is considered to be one of the main driving forces of changes in ecosystem services at regional and global levels because it reflects the coupling relationship between natural systems and human systems and profoundly affects the structure, function, and process of ecosystems [7]. Therefore, the NLSP change is an important factor leading to changes in ESV.

At present, with the increasing economic downturn and structural adjustment pressure, China is in a period of economic and social transformation. They face the challenges of tightening resource constraints, insufficient ecological environment carrying capacity, and incomplete institutional systems. It has become a top priority to coordinate the scientific protection and rational use of natural resources and to improve regional development quality. In this context, the Chinese government has changed the value orientation and measures of land resource allocation. The main functional area plan, land-use plan, and urban and rural plan are integrated into a unified national land spatial plan, and "multiple planning integration" is realized [8]. The emphasis of the national land spatial plan is to optimize the structure and layout of urban space, agricultural space, and ecological space. At the same time, the Chinese government has also proposed building a spatial planning system based on the core value orientation of pushing the development of an ecological civilization and practicing the "Two Mountains" Theory (Appendix A). Ecosystem services are related to environmental quality and human well-being and can be used as carriers for spatial plans to shape the value of natural resources [9,10]. ESV change can reflect the spatiotemporal effects and causality of planning decisions, which is conducive to deepening the understanding of the priority of space uses under the influence of multiple factors [11,12]. Therefore, the ESV is an important basis for promoting NLSP to conform to the concept of ecological civilization. Exploring the spatiotemporal change characteristics of ESV under the influence of NLSP change can provide a decision-making basis for national land spatial planning in the new era.

ESV is a hot topic of study because of the quick degradation of ecosystem services under intensive human disturbance [13]. Thus far, scholars have conducted significant work, primarily regarding the definition [1,7,14–16], classification [1,7,14,16,17], and valuation [1,18–22] of ecosystem services. Scientifically assessing the ESV can provide an important reference for ecosystem management [23,24], biological conservation [25], ecosystem service trade-offs [26,27], and ecological restoration [28,29]. At present, three main approaches have been widely applied to assess ecosystem services, including equivalent factors, productivity, and biomass [30]. Among them, the result of the equivalent factor method is presented in the form of monetary quantity, which can reflect the willingness of humans to pay for ecological services, with strong operability and wider applicability [31]. In 1997, Costanza et al. [1] divided the global ecosystem service functions, estimated the service value of each ecosystem item by item, and proposed the principles and methods of ESV estimation. Subsequently, scholars have studied the evaluation methods proposed by Costanza et al. and have explored and improved the theoretical methods of valuing ecosystem services [18–20]. Based on the research of Costanza et al. [1], Xie et al. [21,22] conducted a questionnaire survey among 200 professionals with ecological backgrounds and developed and improved the equivalent coefficient of ESV per unit area according to the actual situation in China, which has been used to evaluate the service value of land ecosystems. This method has the advantages of simple use, low data demand, high comparability of results, and comprehensive evaluation. It has been widely used in the study of ESV spatiotemporal change characteristics of administrative regions [32–35], natural regions [36–40], and economic regions [41–43] at different scales in China.

At present, the studies of NLS mainly focus on the optimization and regulation of NLSP based on the relationship between land systems and the internal environment of natural systems [44,45]. The partitioning and adjustment of regional NLS are studied from the aspect of multifunctional land characteristics [10,46]. China's territorial imbalance has intensified [47,48] and there has been a coexistence of insufficient space for the development and over-development of space; therefore, the optimization and control methods of NLSP based on regional functional suitability and resource and environmental carrying capacity have gradually become a hot topic [49,50].

In summary, we find that the existing research methods are gradually mature and the research results are constantly enriched, but still need to be improved in the following three aspects: (1) Most scholars believe that land use/cover change (LUCC) is the main reason for the change in ESV, mainly studying the spatiotemporal change characteristics of ESV based on the perspective of LUCC [32–35,37,38,41,51]. However, attention has rarely been paid to changes in the urban–agricultural–ecological functions of NLSs. No scholar has explored the spatiotemporal change characteristics of ESV based on the perspectives of ecology, agriculture, and urban spatial changes. Therefore, the relevant studies have limited guiding value for national land spatial planning; (2) The exploratory spatial statistical analysis method based on grid cells can effectively express the spatial change characteristics of ESV and can also more accurately describe the local impact of changes in NLSP on ESV, but it has only been applied by a few studies [52,53]; (3) In terms of case area selection, most of the current studies have selected provincial or city-level administrative areas, while there are few studies on county-level administrative areas. Although these studies are typical, their results are insufficient for the practical significance of county-level administrative units.

The Three Gorges Reservoir Area (TGRA) is an important ecological barrier in the Yangtze River Basin. Its ecological environment is not only directly related to the long-term safe operation of the Three Gorges Dam (TGD) and the stable enrichment of millions of immigrants, but is also related to the ecological security and sustainable development of the whole Yangtze River Basin. However, the terrain of the TGRA is dominated by mountains and hills, with serious problems such as a fragile ecological environment, nonpoint source pollution, landscape fragmentation and soil erosion, and the contradiction between humans and land is prominent [54]. Located in the hinterland of the TGRA, Fengdu County is faced with multiple problems, such as immigration relocation, hollowing out of the countryside, rapid urban expansion, and deterioration of the ecological environment. The rational development of NLSs and ecological environmental protection are facing huge challenges. In view of this, we selected Fengdu County as the study area, using land use data from 1990, 2000, 2010, and 2018 to analyze the NLSP change based on the perspectives of urban, agricultural, and ecological spaces. Then, we used the equivalent factor method and exploratory spatial data analysis method to explore the spatiotemporal change characteristics of ESV in Fengdu County. The specific objectives of this study are described as follows: (1) to characterize the change characteristics of NLSP in Fengdu County from 1990 to 2018; (2) to analyze the spatiotemporal change characteristics of ESV in Fengdu County; and (3) to explore the spatial autocorrelation characteristics of ESV and its change.

2. Materials and Methods

2.1. Study Area

Fengdu County of Chongqing city is located in the upper reaches of the Yangtze River and in the hinterland of the TGRA. The geographical coordinates are 107°28′–108°12′ E and 29°33′–30°16′ N (Figure 1). The terrain is dominated by mountains, followed by hills and only a few flat dams in mountains and river valleys, showing a general pattern of high south and low north. The land uses are mainly cropland and forest, with abundant forest resources and diverse ecosystems. Fengdu County has a total area of 2900.86 km^2, with 2 subdistricts, 23 towns, and 5 townships. In 2018, the per capita GDP of Fengdu County was CNY 40,400, the per capita disposable income was CNY 21,300, and the population urbanization rate was 46.48%. Since the 21st century, the immigrant population in Fengdu County has been larger than the non-immigrant population, and the number of total permanent population dropped from 671,100 in 2000 to 585,200 in 2018. The outflow population mainly comes from rural areas. Between 2000 and 2018, the number of rural population decreased by 210,600.

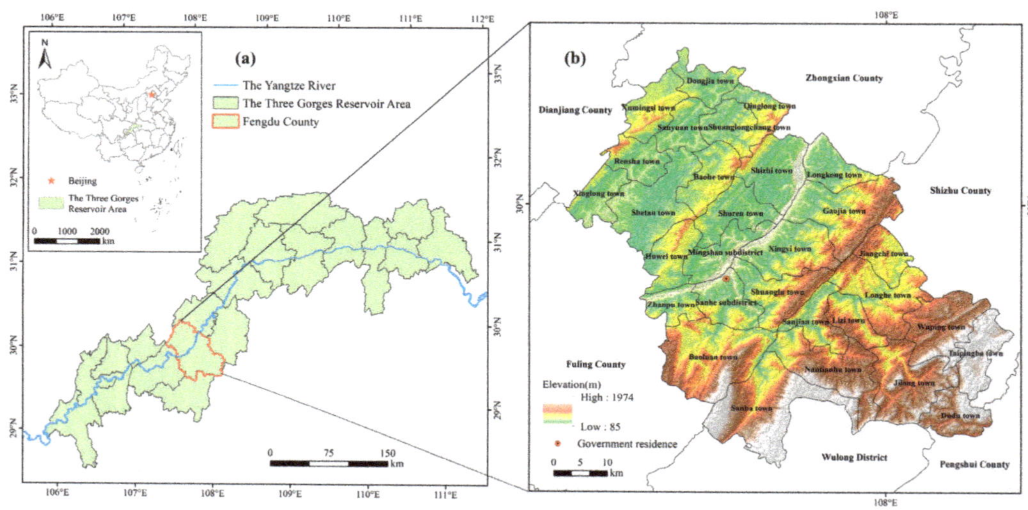

Figure 1. The study area: (**a**) the location of Fengdu County in the Three Gorges Reservoir Area; (**b**) the administrative divisions and elevation distribution of Fengdu County.

2.2. Data Sources

The data used in this paper mainly include: (1) Land use data obtained from the Chinese Academy of Sciences Resource and Environmental Science Data Center (https://www.resdc.cn/ (accessed on 1 September 2020)), including four periods of land use data in 1990, 2000, 2010 and 2018, with a spatial resolution of 30 m. The types of land use in the study area include 6 primary categories (cropland, forest, grassland, water, built-up and unused land) and 17 secondary categories (omitted here); (2) Net Primary Productivity (NPP) data were obtained from the National Aeronautics and Space Administration (https://www.nasa.gov/ (accessed on 1 September 2020)), and soil conservation simulation and precipitation data were obtained from the National Earth System Science Data Sharing Service Platform (https://www.geodata.cn/ (accessed on 1 September 2020)). These data were used to modify the ecosystem service value equivalent factors; (3) The data on the yield per unit area and average price of rice, wheat and corn are from the "China Agricultural Product Price Survey Yearbook", which is used for the calculation of the ESV of a standard equivalent factor. The other socioeconomic data required for this article come from the "China Statistical Yearbook", "Chongqing Statistical Yearbook", and "Fengdu Yearbook" of the corresponding years.

2.3. Methods

2.3.1. Quantitative Analysis of National Land Space Pattern Change

(1). National Land Space Classification

NLS classification is the process of considering the various functions of land in the natural environment and the development of human society, and dividing it into several functional areas [55]. Among the multiple functions of NLS, ecological function is the premise and foundation of other functions [56]. Based on the "Main Functional Area Planning" issued by the State Council of China, we took the strengthening of the basic status of ecological functions as the goal and established the NLS classification system of Fengdu County [57]. The classification system included 3 primary categories and 6 secondary categories (Table 1, Figure 2). As the spaces with different functions in the urban space are closely connected and the boundaries are blurred, the urban space is not divided into secondary categories. It is worth noting that the Chinese government added Fengdu County as an important area for soil conservation in the "National Ecological Function Zoning of China" program, and made the protection and restoration of vegetation as the focus of the regional ecosystem management. Therefore, we classified forest and grassland as vegetation ecological space, which not only emphasizes its ecological attributes, but is also conducive to the management of local soil erosion and rocky desertification. In addition, we classified the bare land as other ecological spaces, to prevent the blind adjustment of these lands into cropland and built-up in the process of subsequent NLS development, which is conducive to promoting the natural ecological restoration of degraded land.

Table 1. The classification of NLS in Fengdu County.

Primary Categories	Secondary Categories	Corresponding Land Use Type (Second-Level Categories)
Urban space (US)	—	Built-up (urban land, industrial and mining land)
Agricultural space (AS)	Agricultural production space (APS)	Cropland (paddy field, dry land)
	Agricultural living space (ALS)	Built-up (rural residential land)
Ecological space (ES)	Vegetation ecological space (VES)	Forest (closed forest land, open forest land, and other forest land) and Grassland (high coverage grassland, medium coverage grass, and low coverage grass)
	Water ecological space (WES)	Water (river, lake, reservoir, pond, and beach)
	Other ecological space (OES)	Unused land (bare land)

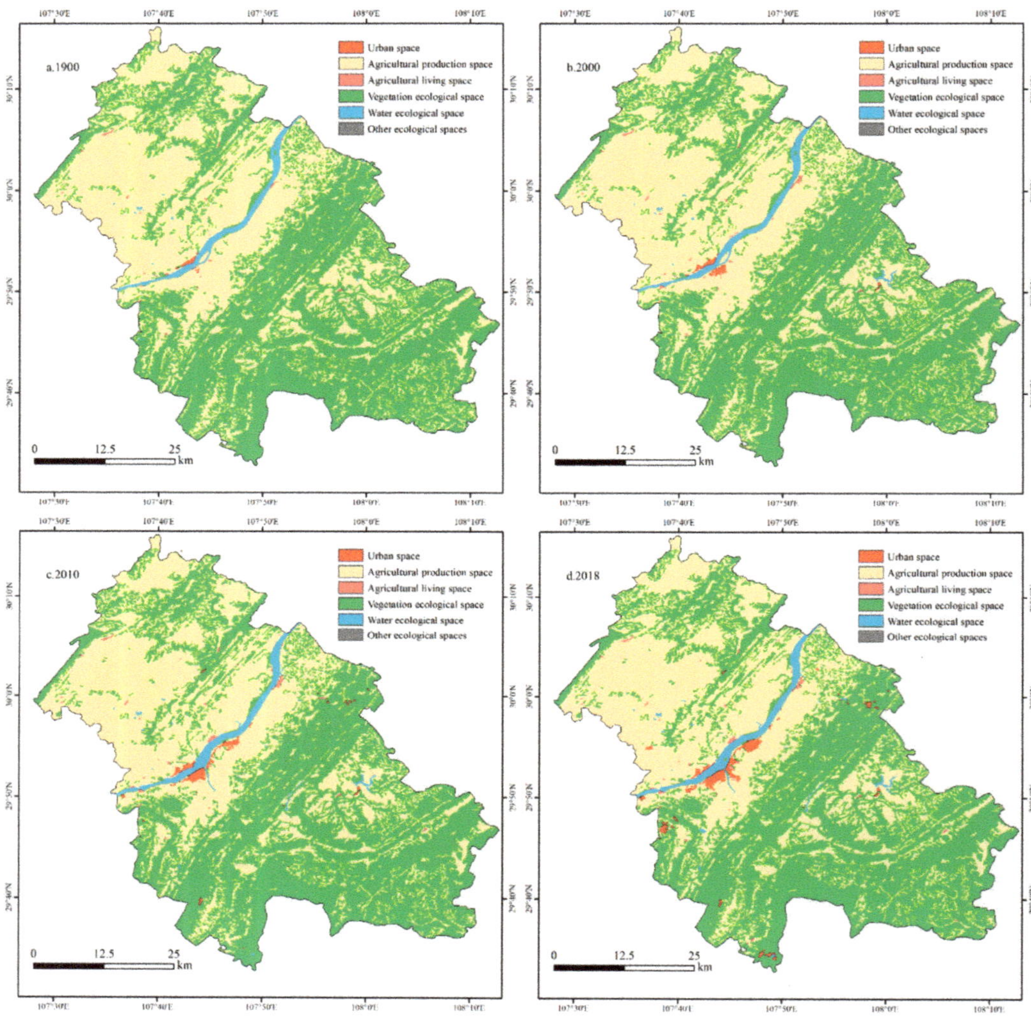

Figure 2. Spatial distribution of NLS in Fengdu County from 1990 to 2018.

(2). National land Space Transition Matrix

The land-use transition matrix can not only reflect the initial and final land type structure of the study area, but also describes the detailed changes of land use in the study area, including the source, composition, and direction of the changes [58]. Based on the theory of the land use transition matrix, we established the NLS transition matrix to analyze the change structure and direction of NLS during the study period; its mathematical expression is:

$$S_{ij} = \begin{vmatrix} S_{11} & S_{12} & \cdots & S_{1n} \\ S_{21} & S_{22} & \cdots & S_{2n} \\ \vdots & \vdots & \vdots & \vdots \\ S_{n1} & S_{n2} & \cdots & S_{nn} \end{vmatrix}, \tag{1}$$

In the formula, i and j are the NLS types in the previous stage and later stage, n is the number of NLS types, and S_{ij} is the area of the ith NLS type transferred to the jth NLS type during the research period. We used the raster calculator tool in ArcGIS 10.7 to calculate the NLS transition matrix, and to realize the visualization of the results. The formula is as follows:

$$G = 10G_a + G_b, \tag{2}$$

In Equation (2), G is the new NLS unit code formed by the change in NLS during the study stage, G_a is the NLS unit code in the previous stage, G_b is the NLS unit code in the later stage, and the codes for US, APS, ALS, VES, WES, and OES are 1, 2, 3, 4, 5, and 6, respectively.

2.3.2. Calculation of Ecosystem Services Value

This study used the equivalent factor method proposed by Xie et al. [59] to measure ESV. In this method, ecosystem services are divided into 4 primary categories and 11 secondary categories. Specifically, provisioning services include food supply, raw material supply, and water supply. Regulating services include air quality regulation, climate regulation, waste treatment, and regulation of water flows. Supporting services include erosion prevention, maintenance of soil fertility, and habitat services. Cultural services include cultural and amenity services. The equivalent value per unit area of food production of cropland was set to 1, and the equivalent value per unit area of other ecosystem services could be quantified by comparison with the standard value of 1. Generally, the economic value provided by natural ecosystems without human input is approximately 1/7 of the economic value of food provided by existing farmland per unit area. From 1990 to 2018, the annual average yield of the three main food crops (rice, corn, and soybean) in Fengdu County was 3589.90 kg/hm^2, and the average price of the three main food crops in 2018 was 2.98 CNY/kg. Therefore, the unit value of the equivalent factor (E_n) was 1528.27 CNY/hm^2, obtained using the following equation:

$$E_n = 1/7\ PQ, \tag{3}$$

In the formula, P is the annual average grain yield, and Q is the average grain price.

Considering the regional heterogeneity of the internal structure and external form of the ecosystem, there are obvious differences in ecosystem service functions and value in different regions [60]. The table of the ESV equivalent factors proposed by Xie et al. is applicable at the national scale. If applied directly to regional ESV research, major errors may occur. Therefore, we adjusted the ESV coefficients to suit the ecological characteristics of Fengdu County. The ecosystems were divided into 6 primary categories and 14 secondary categories in the study of Xie et al. [59]. Land use types cannot correspond to ecosystem types one by one; therefore, we selected the closest land use type for equivalent evaluation according to the actual situation of the study area. Specifically, the equivalent factors of the corresponding ecosystem types were used for paddy field, dry land, water, and bare land. Broadleaf forests are the main types of forest, and shrub grass is the main type of grassland in Fengdu County [61]; we used the equivalent factors of broadleaf forest and shrub grass as the representatives of forest and grassland, respectively. Simultaneously, we adjusted the ESV equivalent factors based on the spatial distribution raster data of national net primary productivity (NPP), precipitation per unit area, and soil conservation [22]. Among these raster data, the NPP, precipitation per unit area, and simulated soil conservation of each grid are the average values of 1990, 2000, 2010 and 2015, calculated by the raster calculator tool in ArcGIS 10.7. Thus, the ESV coefficients per unit area were calculated according to Equation (4).

$$VC_f = \begin{cases} D_{f1} \times E_n \times \overline{B}/\overline{B_t} \text{ or} \\ D_{f2} \times E_n \times \overline{W}/\overline{W_t} \text{ or} \\ D_{f3} \times E_n \times \overline{E}/\overline{E_t} \end{cases}, \tag{4}$$

In the formula, VC_f is the ESV coefficient per unit area of the fth ecological service type of a certain ecosystem in the study area, D_{f1}, D_{f2}, and D_{f3} are the equivalent factors of the ecosystem service function related to NPP, precipitation, and soil conservation, respectively [61], $\overline{B}$, $\overline{W}$, and $\overline{E}$ are the average NPP, precipitation per unit area, and simulated soil conservation in the study area, respectively, and $\overline{B_t}$, $\overline{W_t}$, and $\overline{E_t}$ are the national average NPP, precipitation per unit area, and soil conservation simulation quantity, respectively. In addition, several researchers believe that "built-up" does not belong to natural ecosystems and assign an ESV value of zero to built-up areas [62,63]. However, due to the strong disturbance of human activities, built-up spaces have a huge impact on regional ecosystem services. Human activities in built-up areas consume food and water, and emit exhaust gas, wastewater, and solid waste at the same time, which have a negative impact on provisioning services and regulating services. However, built-up zones have the function of maintaining soil, and also add the values of appreciation and entertainment; thus, it has a positive impact on supporting services and cultural services [51]. In this study, we determined the ESV coefficient per unit of built-up area based on the research of the Chengdu–Chongqing Economic Zone by Yuan et al. [64]. Finally, we obtained the ESV per unit area for Fengdu County (Table 2).

Table 2. Ecosystem service value per hectare of different land use categories in the study area. (Unit: CNY/hm^2/year).

Land Use	Paddy Field	Dry Land	Forest	Grassland	Water	Built-Up	Unused Land
Ecosystem	Paddy field	Dry land	Broad-leaved	Shrub grass	Weter	—	Barren
Provisioning service	−3257.20	3094.90	3285.32	3174.83	314,672.47	−406.34	0
Regulating service	12,310.37	3521.59	39,376.18	32,601.97	7200.90	−10,653.50	377.79
Supporting service	988.79	2339.02	10,797.07	9769.31	4592.60	4138.55	82.22
Cultural service	218.70	145.80	2575.75	2332.75	352,787.22	9.47	24.30
Total	10,260.66	9101.31	56,034.32	47,878.86	679,253.19	−6911.82	484.31

The formulae for estimating the ESV are as follows:

$$ESV_f = \sum A_k \times VC_{fk}, \tag{5}$$

$$ESV = \sum A_k \times VC_k, \tag{6}$$

where ESV_f is the value of the fth ecological service type, ESV is the total value of ecosystem services in the study area, A_k is the area of the kth land use type, VC_{fk} is the ESV per unit area of the fth ecological service type of the kth land use type, and VC_k is the ESV per unit area of the kth land use type.

Based on this, the formula for calculating the ESV of different NLS types is as follows:

$$ESV_g = \sum S_k \times VC_k, \tag{7}$$

In Equation (7), ESV_g is the ESV of a certain type of NLS, and S_k is the area of the kth land-use type included in this type of NLS.

2.3.3. Exploratory Spatial Data Analysis

Exploratory spatial data analysis is a collection of techniques for describing and visualizing spatial distributions, determining atypical locations or spatial outliers, discovering spatial associations, clusters, or hot spots, and inferring spatial characteristics or other forms of space heterogeneity [65]. In this study, the global spatial autocorrelation analysis method and hot spot analysis method are adopted to explore the spatial autocorrelation characteristics of ESV. We calculated Moran's I value, which is used to describe the global spatial autocorrelation characteristics of ESV. At the same time, we calculated Getis–Ord Gi^* statistics to describe the spatial locations of "cold spots" and "hot spots" of ESV changes. These statistics are used to reveal the spatial clustering pattern of the high and low values of ESV change. The Spatial Autocorrelation (Moran's *I*) and Hot Spot Analysis (Getis–Ord *Gi**) tools in ArcGIS 10.7 software were used for analysis.

3. Results

3.1. National Land Space Pattern Change

3.1.1. The Area Change of National Land Space

According to the land use data for the four periods of 1990, 2000, 2010 and 2018, the area change of NLS was calculated (Table 3), and the results showed that Fengdu County is dominated by ES, which accounted for approximately 53%, followed by AS, which accounted for more than 45%, while US is the smallest, accounting for less than 1%. From 1990 to 2018, the US of Fengdu County expanded significantly (1165.64%). There was a slight decrease in AS (−4.15%) and a small increase in ES (2.18%). From the changes in the secondary types, it can be seen that US, ALS, VES and WES have increased, while the rest of the space has decreased. Among them, the US has the fastest growth rate (41.69%/a), which reflects the intense urbanization process that Fengdu County has experienced in the past 28 years. The growth rate of ALS is 5.47%/a, which is second only to US. From the changes of different stages, it can be seen that US, ALS and WES continue to increase. US is gradually increasing, and the increase in ALS at each stage is relatively stable. The WES increased explosively from 2000 to 2010, mainly because the Three Gorges Reservoir began to store water at this stage, which led to the expansion of the water area of the Yangtze River and its tributaries. APS and OES continued to decrease, and both had the largest decrease from 2000 to 2010. The VES increased sharply from 2000 to 2010, which may be related to the effective implementation of policies such as the Grain for Green Project, the Natural Forest Protection Program, and the policy of rocky desertification control at this stage.

Table 3. Changes in NLS structure in Fengdu County from 1990 to 2018.

Type		US	AS			ES			
			APS	ALS	Total	VES	WES	OES	Total
1990	Area/km^2	1.95	1379.68	3.26	1382.94	1474.90	39.40	0.36	1514.66
	Proportion/%	0.07	47.58	0.11	47.69	50.87	1.36	0.01	52.24
2000	Area/km^2	5.76	1373.02	5.03	1378.05	1474.10	41.27	0.35	1515.72
	Proportion/%	0.20	47.35	0.17	47.52	50.84	1.42	0.01	52.27
2010	Area/km^2	13.71	1327.67	7.02	1334.69	1498.70	50.89	0.05	1549.64
	Proportion/%	0.47	45.81	0.24	46.05	51.71	1.76	1.83×10^{-3}	53.47
2018	Area/km^2	24.68	1317.37	8.24	1325.61	1495.37	52.33	0.05	1547.75
	Proportion/%	0.85	45.46	0.28	45.74	51.60	1.81	1.74×10^{-3}	53.41
Rate of change/%		1165.64	−4.52	152.76	−4.15	1.39	32.82	−86.11	2.18
Annual rate of change/%		41.63	−0.16	5.46	−0.15	0.05	1.17	−3.08	0.08

3.1.2. National Land Space Transition

Table 4 shows the NLS transition matrix of Fengdu County. It can be seen that VES has the largest transfer-in area, which is mainly from APS (100.45 km^2). Next is WES, which mainly comes from APS (8.35 km^2) and VES (5.04 km^2). US is mainly transferred from APS (18.24 km^2) and VES (5.24 km^2), and ALS is mainly transferred from APS (5.39 km^2). We found that the hot spot of transfer in Fengdu County is the transfer of APS to VES and US.

Table 4. NLS transition matrix of Fengdu County from 1990 to 2018. (Unit: km^2).

1990	2018					
	US	**APS**	**ALS**	**VES**	**WES**	**OES**
US	0.76	0.02	0	0.01	1.17	0
APS	18.24	1246.66	5.39	100.45	8.35	0
ALS	0.21	0.49	2.45	0.10	0	0
VES	5.24	68.44	0.33	1393.57	5.04	0.01
WES	0.23	1.18	0.06	0.36	37.56	0
OES	0	0.12	0	0.01	0.19	0.04

From the spatial distribution of NLS in Fengdu County (Figure 2), it can be seen that, initially, the APS was roughly distributed in mountain valleys and gentle slope areas. These areas have a flat terrain and good hydraulic conditions, which are suitable for agricultural production. VES were scattered in mountains and valleys. The complex natural environment in these areas restricts human activities and provides a good condition for the restoration of vegetation. The WES mainly includes the Yangtze River and Long River. The US and ALS are mainly distributed along the Yangtze River. According to the NLS transition map in Fengdu County from 1990 to 2018 (Figure 3), the most obvious NLS change was the expansion of US, which is mainly distributed in the north of the Sanhe subdistrict, northwest of Shuanglu town, and northwest of Xingyi town. This is mainly because Fengdu County started the construction of a new county town on the south bank of the Yangtze River after the old county town was flooded. Next, transitions between APS and VES occurred in Fengdu County at a large scale, but the locations were relatively scattered. Among them, the transfer from APS to VES in some areas with steep slopes in townships, such as Shuren Town, Dudu Town, Jilong Town, Wuping Town, and Taipingba Township is obvious. These areas were unsuitable for farming, therefore they were the key areas for the Grain for Green Project. The expansion of WES was distributed along the Yangtze River, mainly due to the operation of the TGD. The construction of many small reservoirs in Gaojia town, Longhe town and Baoluan town also resulted in the transfer of a small amount of APS into WES. The transfer-in of ALS was mainly distributed in northern Zhanpu town, eastern Mingshan town, and northeastern Gaojia town.

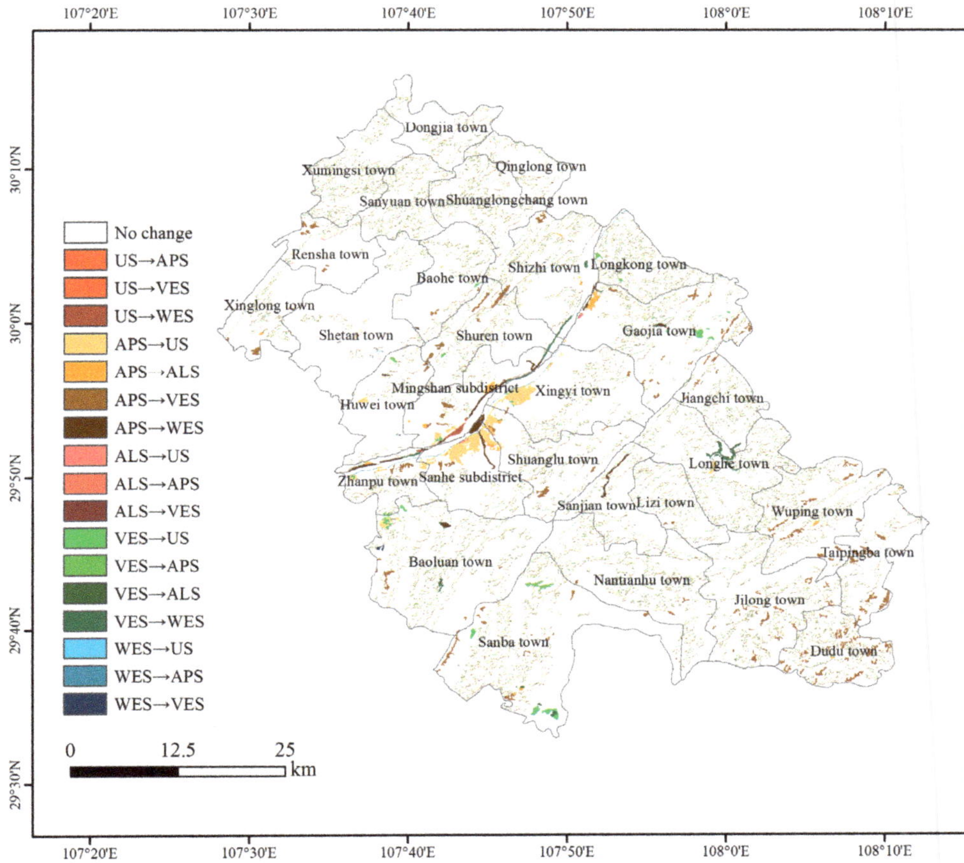

Figure 3. NLS transition map in Fengdu County from 1990 to 2018.

3.2. Spatiotemporal Changes of Ecosystem Service Value

3.2.1. The Changes of Ecosystem Service Value

The ESVs of Fengdu County in 1990, 2000, 2010, and 2018 were calculated in combination with the ESV per unit area (Table 5). The results show that the total ESV of Fengdu County showed a trend of continuous growth, with a total increase of CNY 11.10×10^8, and the change rate was 9.33%, during the 1990–2018 period. From the perspective of different stages, the total ESV growth rate of Fengdu County was the largest from 2000 to 2010, which was mainly due to the rapid expansion of the scale of VES and WES during this period. Among the four primary types of ecosystems services, the value of regulating services had the largest contribution to the total ESV. In the past 28 years, the value of the four ecosystem services has increased. Among them, cultural services (CNY 4.66×10^8) have the most value growth, followed by provisioning services (CNY 3.97×10^8) and regulating services (CNY 1.97×10^8); supporting services (CNY 0.49×10^8) have the least value growth.

Table 5. ESV of Fengdu County from 1990 to 2018 (unit: 10^8 CNY).

Ecosystem Service	1990	2000	2010	2018
Provisioning service	19.13	19.70	22.67	23.09
Regulating service	63.57	63.41	65.85	65.54
Supporting service	18.43	18.43	18.92	18.92
Cultural service	17.83	18.48	21.99	22.49
Total	118.95	120.03	129.43	130.05

The largest ESV per unit area is for the water and the lowest is for the unused land, whereas the ESV per unit area of the built-up is negative (Table 2). According to Equation (7), the ESV of different NLS types was calculated (Table 6), and it was found that the VES is the main contributor to the ESV of Fengdu County and has contributed more than 60% during the four periods; the next highest contributor is the WES, at approximately 25%. The contributions of the other NLS types, which have smaller areas, were relatively low; especially for OES, which was close to zero. From the perspective of ESV changes, we observed that the ESV of VES and WES increased by CNY 3.07×10^8 and CNY 8.78×10^8, respectively. The ESV of the remaining space types all declined. Among them, the ESV of APS decreased the most (CNY 0.57×10^8), followed by US (CNY 0.16×10^8) and ALS (CNY 0.03×10^8). In summary, the ESV loss caused by the expansion of the construction space (US and ALS) and the reduction in APS is offset by the ESV gain caused by the increase in ES. The significant increase in the ESV of the WES is the main reason for the increase in the total ESV of Fengdu County.

Table 6. The ESV of different NLS types in 1990–2018 (unit: 10^8 CNY).

Type	US	AS			ES			
		APS	ALS	Total	VES	WES	OES	Total
1990	−0.01	12.98	−0.02	12.96	79.24	26.77	1.72×10^{-4}	102.61
2000	−0.04	12.92	−0.03	12.89	79.15	28.03	1.72×10^{-4}	103.61
2010	−0.09	12.51	−0.05	12.46	82.50	34.57	2.44×10^{-5}	111.86
2018	−0.17	12.42	−0.06	12.36	82.31	35.55	2.44×10^{-5}	112.82
change	−0.16	−0.57	−0.03	−0.60	3.07	8.78	-1.48×10^{-4}	10.21

3.2.2. The Spatial Change of Ecosystem Service Value

To further analyze the spatial variation characteristics of ESV, the Fish net tool in ArcGIS 10.7 was used to divide the administrative area of Fengdu County into 12,052 square cells with a side length of 500 m. We calculated the ESV of each grid and used natural breaks to divide it into six levels from high to low. Among them, an ESV of level one was the lowest, and an ESV of level six was the highest (Figure 4). The results showed that the spatial difference in ESV in Fengdu County was obvious. Due to the vast water area, the Yangtze River is a concentrated area with the highest value of ecosystem services. In addition, the ESV in Fengdu County has obvious spatial distribution characteristics, with the Yangtze River as the boundary, high in the south and low in the north.

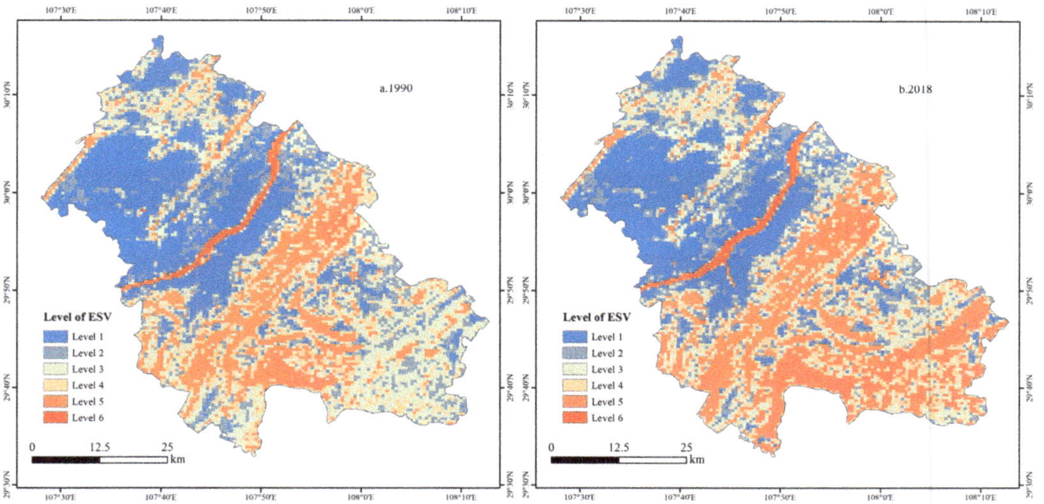

Figure 4. Spatial distribution of ESV in Fengdu County, 1990 and 2018.

Figure 5 is the spatial distribution map of ESV changes in Fengdu County from 1990 to 2018. The changes of ESV in Fengdu County were mainly distributed along the Yangtze River and the southernmost part of the county, while the changes of ESV in other areas were not obvious. The ESV gain area along the Yangtze River was distributed in a zonal pattern, mainly located to the north of Zhanpu town and the south of Mingshan subdistrict, Shuren town and Shizhi town. The loss areas were distributed in blocks, mainly located in the northern Sanhe subdistrict, Shuanglu town, Xingzhi town and Gaojia town. Among them, the gain in ESV was mainly due to expansion of the water area of the Yangtze River, and the loss of ESV was mainly due to the continuous encroachment of other spaces by US and ALS. The areas with high ESV values in Sanba Township, Jilong Township, Dudu Township, Taipingba Township and Wuping Town in the southernmost part of Fengdu County increased, which was mainly due to the transition of a large amount of grassland in the ES into forest. There were also some scattered ESV gain and loss areas. Among them, the loss of ESV was mainly due to the transformation of APS and VES into US and ALS. The main reasons for the gain of ESV on both sides of the Yangtze River are different. In the north of the Yangtze River, it was mainly due to the transformation of APS into VES. However, in the south of the Yangtze River, it was mainly due to the transformation of VES into WES.

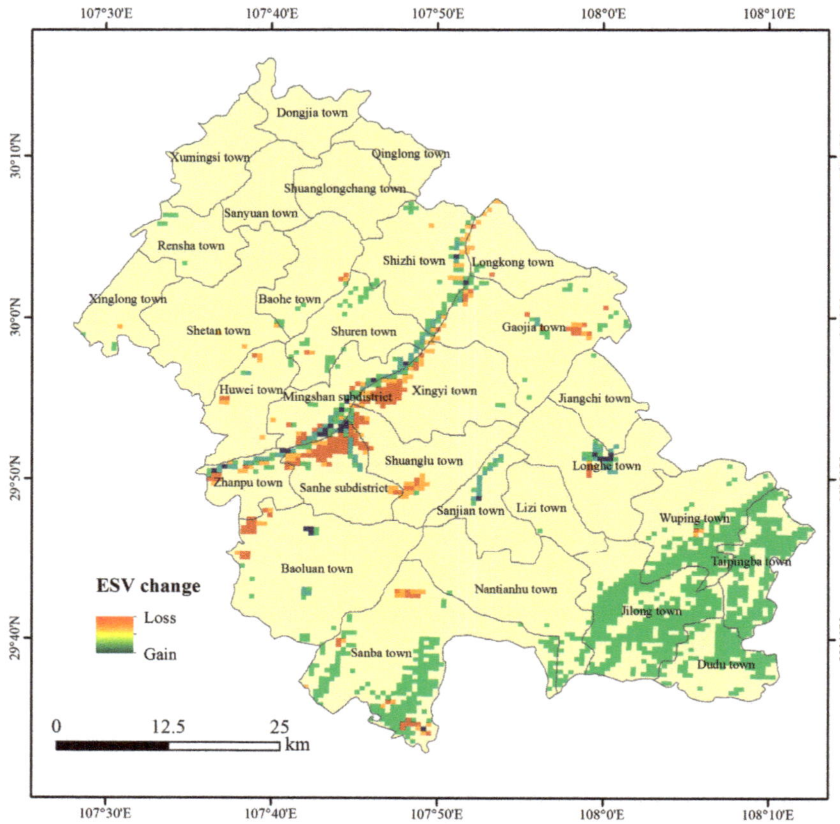

Figure 5. Spatial distribution of ESV change in Fengdu County from 1990 to 2018.

3.2.3. Spatial Autocorrelation Characteristics of Ecosystem Service Value

(1). Global Spatial Autocorrelation Analysis

The spatial correlation characteristics of ESV are shown in Table 7. It can be seen that the global Moran's I of ESV and ESV changes in each stage were all greater than 0 and significant at the threshold level of 1%, indicating that ESV and its changes in Fengdu County were not randomly distributed but positively correlated. This indicates that the spatial distribution of ESV and its changes showed strong spatial clustering characteristics. The Moran's I of ESV first decreased from 0.6324 to 0.6214, and then continually rose to 0.6437. This indicates that the clustering characteristics of ESV distribution were weakened from 1990 to 2000, but the clustering characteristics of ESV distribution were enhanced from 2000 to 2018 under the influence of the construction of the Three Gorges Project, the implementation of macro policies, and the rapid development of urbanization; the Moran's I of ESV changes at different stages showed "up–down" fluctuation characteristics. From 2000 to 2010, the Moran's I value of ESV changes was 0.4724, and the spatial distribution of ESV changes had the strongest clustering characteristics. From 2010 to 2018, the Moran's I value of ESV changes was only 0.2394, and the spatial distribution of ESV changes had the weakest clustering characteristics.

Table 7. Moran's I value of ESV in Fengdu County.

Index	1990	2000	2010	2018	1990–2000	2000–2010	2010–2018	1990–2018
Moran's I	0.6324	0.6214	0.6398	0.6437	0.34674	0.4724	0.2394	0.4542
p-value	<0.001	<0.001	<0.001	<0.001	<0.001	<0.001	<0.001	<0.001
z-scores	132.5123	134.4064	128.2431	147.6582	115.3241	123.1716	89.9226	96.9496

(2). Hot Spot Analysis

Figure 6 shows the hot spot spatial distribution pattern of ESV changes in Fengdu County. From 1990 to 2018, most of the cold spots and hot spots of ESV changes in Fengdu County were distributed along the Yangtze River and to the south of the Yangtze River, which was consistent with the spatial distribution of ESV changes.

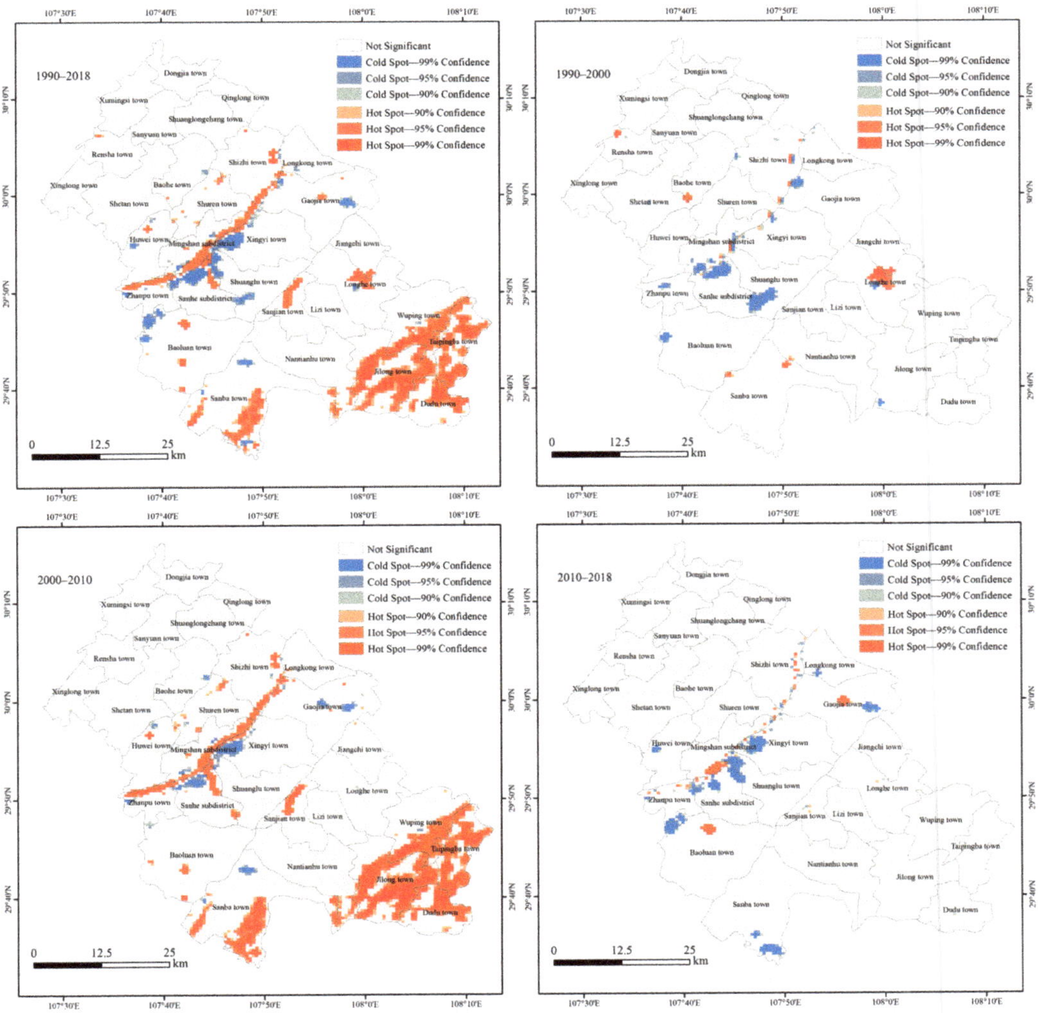

Figure 6. Hot spot spatial distribution pattern of ESV change in Fengdu County from 1990 to 2018.

From 1990 to 2000, cold spots and hot spots of ESV changes were few and scattered. Among them, the cold spots were mainly distributed in the Mingshan subdistrict, Sanhe subdistrict, Zhanpu town, Baoluan town, Shuanglu town, and Gaojia town. This was mainly due to the expansion of US and ALS. The hot spots were mainly distributed in Longhe town. The reason is that the construction of Shiban Reservoir (since 1997) led to the rapid expansion of local WES; from 2000 to 2010, the cold spots and hot spots in Fengdu County increased significantly. As the TGD began to store water, hot spots were concentrated along the Yangtze River. At the same time, the increase in the water area of the Long River also led to the emergence of hot spots within the Sanhe subdistrict and Sanjian township. Due to the transition of grassland to forest with higher ESV per unit area, large areas of hot spots appeared in Sanba township, Jilong town, Wuping town, Taipingba township and Dudu township. In addition, due to the construction of the Danzitai reservoir (since 2003), hot spots also appeared in Baoluan town. The cold spots were mainly distributed in the Sanhe subdistrict, Baohe township, Xingyi town, Gaojia town and Sanba township. From 2010 to 2018, the cold spots and hot spots decreased compared with the previous period. The further expansion of the water area of the Yangtze River triggered the agglomeration of hot spots. The hot spots were mainly distributed in the Mingshan subdistrict, Baoluan town, and Gaojia town. The main reasons were the further expansion of the water area of the Yangtze River, and Jiangjiagou Reservoir (since 2015), Guantiangou Reservoir (since 2017), and Liziping Reservoir (since 2017), which have been built one after another. The cold spots were mainly distributed in Huwei town, Baoluan town, Sanhe subdistrict, Shuanglu town, Xingyi town, Gaojia town and Sanba township. The main reasons are the construction of the Fengdu Railway Station (since 2013) and the Fuling–Fengdu–Shizhu Expressway (since 2013), and the continuous growth of urban space and agricultural living space in each township. Overall, the cold spots and hot spots were the most widely distributed from 2000 to 2010, and this was the period with the most dramatic changes in ESV in Fengdu County.

4. Discussion

The reform of China's ecological civilization system is the cornerstone of the reform of national land spatial planning. In the context of increasing emphasis on the development of an ecological civilization, ecosystem services have important reference significance for national land spatial planning. However, there is an objective contradiction between the complexity of the theory and methods of ecosystem services and the feasibility of national land spatial planning practices. How to integrate ecosystem services into the new spatial planning system has become an important task of China's NLS governance [66,67]. This paper starts with ESV, an important measure of ecosystem services, and attempts to analyze the spatiotemporal change characteristics of ESV under the influence of NLSP change in Fengdu County. The conclusions can provide a scientific basis for promoting the multifaceted supporting role of ecosystem services in the formulation of national land spatial plans.

4.1. Effects of NLSP Change on ESV

The ESV of Fengdu County increased by CNY 11.10 $\times$ 10^8 from 1990 to 2018. The construction and operation of the Three Gorges Reservoir and some small reservoirs has greatly increased the water ecological space in Fengdu County, which has made the most contribution to the total growth of ESV. This was similar to the research results in the upstream Xiong'an New Area [41]. In the ES, the large increase in forest has had a huge contribution to the total growth of ESV. Particularly, since 2000, with the implementation of the Grain for Green Project and Natural Forest Protection Program, the forest coverage rate has greatly increased, resulting in a significant increase in ESV. This was similar to the research results in northern Shaanxi [68]. The ESV per unit area of the construction space was negative. Over the past 28 years, a large number of other spaces have been transformed into construction space, which has had a large negative impact on the ESV of

Fengdu County. This also reflects the problems existing in the development and utilization of NLSs in Fengdu County. Firstly, the development and utilization model of US pays too much attention to scale and speed and neglects the intensive use of space. Secondly, with the massive outflow of the rural population, the ALS has not decreased but has increased. Relevant studies have shown that this pattern of dysfunctional development of rural human–land relationships is widespread in China, and it is one of the main problems in China's space governance [69–71].

In general, the spatiotemporal changes of ESV are processes that positively respond to the increase in ecological space but negatively respond to the expansion of urban space. With the completion and operation of the Three Gorges Reservoir and the long-term implementation of the Grain for Green Project, the water area and forest area of Fengdu County have basically stabilized. In the future, it will be difficult to significantly increase the ESV by increasing the water and forest. However, the urbanization process of Fengdu County will continue to advance, and it is foreseeable that ESV will face downward pressure in the future. Therefore, the NLSP should be guided to develop in the direction of ESV appreciation while ensuring that ESV does not depreciate.

We also found that ESV change shows obvious positive spatial autocorrelation characteristics, which indicates that NLSP change may have a certain spatial spillover effect on ESV. Moreover, NLSP changes may cause ESV gains and losses in the region. At the same time, they may also cause ESV increases and decreases in surrounding areas. The study of Lu et al. [72] reached a similar conclusion. The reason may be that NLSP changes have affected the material, energy, and information interactions between organisms and environmental components in the local ecosystem. The theory of landscape ecology shows that when the ES changes to AS and US, in which human activities are more intense, it will increase resistance to the migration and flow of species and energy between heterogeneous landscapes, which is not conducive to the progress of regional ecological processes [73], in turn leading to the weakening of ESV in surrounding areas. Therefore, preventing the expansion and penetration of spaces with higher ESV per unit area to spaces with lower ESV per unit area is an effective way to maintain the continuity of the ecosystem pattern and increase regional ESV.

4.2. Discussion on the Impact of the Three Gorges Dam on the Ecological Environment

This study mainly analyzed the impact of the construction of the TGD on ESV from the perspective of NLSP changes. From the research results, the construction of the TGD has increased the scale of water ecological space and has had positive significance for the regional ecosystem. However, it is worth discussing that the construction and operation of large dams also has a huge adverse impact on the ecological environment [74], which is confirmed by relevant studies in the Amazon Basin [75], Tennessee Valley [76] and the Mekong River Basin [77]. The TGD is the world's largest hydro project, and is of great significance in flood control, power generation, and shipping. However, the TGD has also caused ecosystem degradation, water pollution, biodiversity reduction, downstream river erosion, geology disasters, and many other ecological hazards [78,79]. The Chinese government has implemented a series of policy interventions to mitigate adverse eco-environmental impacts of the TGD. Initially, during the construction of the TGD (1993–2002), many ecological programs were planed and enforced. The Transforming Sloping Cropland to Terraced Land (since 1993), Grain for Green Program (since 1998), Natural Forest Protection Program (since 1998), and the Comprehensive Plan on Prevention and Control of Geological Hazards in the Three Gorges Reservoir Area (since 2001) are a few examples. Subsequently, since the start of operation of the TGD in 2003, projects such as the Water Pollution Prevention and Control Plan in the Three Gorges Reservoir Area and the Upstream (2001–2010), and the Outline of the Water Pollution Prevention and Water Pollution Prevention and Control Plan in the Key Basins (2011–2015) have gradually been implemented. More than ten ecological operation trials were carried out simultaneously to rehabilitate "four domestic fish species (herring, grass carp, silver carp, bighead carp)". These ecological programs have played a significant role in mitigating the negative ecological impacts of the TGD. From 1996 to 2016, 2118.47 km^2 of sloping cropland were returned to forest or grassland, 2196 km^2 of soil under erosion were curbed, and the forest coverage rate increased from 22% to 49% in the Three Gorges Reservoir Area. At the same time, water quality in the tributaries in the reservoir area improved, with the proportion of eutrophication being reduced from 39.4% in 2011 to 29.8% in 2016. Annual average spawning stocks of four domestic fish species increased by 137.4% in 2011–2016 on the basis of 2003–2010 levels [80]. In summary, we suggest that the pros and cons of the dams should be fully traded-off before construction. For dams that have already been built, it is necessary to carry out systematic ecological restoration measures, especially in some developing countries with increasing demand for water and energy but not enough awareness of ecological protection [81].

4.3. Policy Implications

Combined with our research, the following policy recommendations are suggested.

(1) This study concludes that NLSP is an important influencing factor of ESV, and ESV is an important basis for promoting NLSP to conform to the concept of ecological civilization. It is necessary to integrate ESV into the decision-making of national land space planning. Therefore, we suggest: (a) providing special training on ESV for spatial planners so that they can firmly grasp the relevant theories and evaluation methods of ESV; (b) based on relevant academic research, ESV should be included as a quantitative indicator in the work of delineating the "three zones and three lines" (three zones—ecological zone, agricultural zone, and urban zone; three lines—permanent basic farmland red line, urban development boundary, and ecological red line) and identifying key areas for ecological restoration [82]; and (c) monitoring and assessing the impact of the implementation of the national land spatial plan on the ESV to provide a basis for the revision of plans.

(2) Promoting the coordinated development of urban, agricultural, and ecological spaces is an important way to simultaneously achieve stable economic and social development and sustainable improvement of ecosystem services. Thus, we suggest the following: (a) It is necessary to strengthen the intensive use of urban space, fully tap the potential of existing urban land use, increase spatial compactness, and strictly enforce the control of urban development boundaries; (b) The protection of basic farmland should be strengthened, and sloping farmland and abandoned farmland should be gradually returned to forests. At the same time, rural residential land consolidation should be combined with the flow of urban and rural construction land indicators to maintain a balance between urban and rural construction land; (c) Land use control measures must be strictly implemented to protect the important ecological space from decreasing. At the same time, the integrated land consolidation and ecological restoration project of mountain–river–forest–field–lake–grassland should be implemented to improve the quality of the ecological environment.

(3) The spatial spillover effects of NLSP change on ESV should be fully considered. We suggest that a population withdrawal policy should be implemented in the ecological space, and the regional population should be encouraged to congregate in the urban spaces. At the same time, the development and construction activities in high ESV areas should be minimized. In addition, the urban space and agricultural living space should focus on the development of eco-friendly space uses on their natural edges, such as country parks and Linpan [83] (a kind of natural settlement of forests, water, houses, and fields widely distributed in southwestern China), so that they can be integrated with the surrounding ecosystem. Especially along the Yangtze River, it is even more necessary to plant a certain scale of ecological forests or build green parks to create ecological coastal zones to avoid pollution of the water environment of the Yangtze River from urban construction.

(4) The Three Gorges Reservoir area undertakes the important task of ecological environmental protection and restoration and has lost some opportunities for economic and social development to improve the service functions of the ecosystem. This has widened the gap in regional development and caused an imbalance between fairness and efficiency.

For the sake of achieving regional fairness and sustainable development, we suggest taking ESV as the foundation for defining the regional ecological compensation relationship, determining the ecological compensation standard, and dividing the ecological compensation zones. At the same time, we have explored the establishment of a market-oriented and diversified ecological compensation mechanism for different regions and different principal parts to improve the enthusiasm and sustainability of ecological environmental protection.

4.4. Limitations

This study also has limitations. The ESV assessment method adopted in this study does not consider the impact of different use methods and use conditions of built-up areas on the ESV coefficient [84]. The negative impact of built-up spaces on ecosystem service functions mainly comes from human disturbance. The population density on urban land is relatively high, and the interaction between humans and land is strong. However, rural residential land carries a smaller population per unit area and causes less damage to the natural ecosystem of the land. In addition, idle rural residential land is subject to little human disturbance, and its negative impact on ecosystem service functions is almost negligible. Therefore, if the types of built-up land are subdivided and assigned value coefficients consistent with their ecological functions, the evaluation result of ESV will be more accurate. For other land use types such as forest, grassland, and water, the internal differences between them has less influence on the ESV coefficient. This is because these land use types belong to natural ecosystems, which are weakly disturbed by human activities, and the interaction between man and land is simple; the ESV coefficient mainly depends on the regional natural endowment [59]. At the county scale, regional natural endowments have a certain degree of homogeneity [85], and the difference in the spatial distribution of the internal structure and external form of the ecosystem is not significant.

The scope of application of the research results in this paper also has certain limitations. A significant negative relationship existed between topographic gradients and human disturbance. With the increase in altitude, the disturbance of human activities to the land ecosystem continued to weaken, and the value of ecosystem services showed an upward trend [86]. The terrain of Fengdu County is dominated by mountains, and the population is mainly distributed in mountain troughs with a flat terrain. In addition, in recent years, a large number of people have moved out of mountainous areas, which has further strengthened the differences in population distribution on topographical gradients [87]. Therefore, the spatial distribution of ESV in Fengdu County showed obvious imbalance. However, the topographical gradients of plain areas have little influence on human activities, and the difference in the spatial distribution of ESV may not be obvious. Therefore, the results of this research are not necessarily applicable to plain areas, such as the Amazon Plain, the North American Prairie, the Gangetic Plain, and the Northeast China Plain. However, the analysis conclusions of the ESV change mechanisms in this research are universal and can provide references for the management of ecosystem services in plain areas.

5. Conclusions

(1) From 1990 to 2018, the changes of NLSP in Fengdu County generally manifested in the transformation of AS into US and ES. US, ALS, VES, and WES increased, while APS and OES decreased. The newly added US and WES were mainly located along the Yangtze River, and the newly added ALS and VES were scattered.

(2) The spatiotemporal changes of ESV are processes that positively respond to the increase in ES but negatively respond to the expansion of US. From 1990 to 2018, the total ESV of Fengdu County showed a trend of continuous growth, with a total increase of CNY 11.10×10^8, and the change rate was 9.33%. The significant increase in the ESV of the WES was the main reason for the increase in the total ESV of Fengdu County. The ESV gain area was mainly located along the Yangtze River and south of Fengdu County, and it has benefited from the implementation of ecological protection policies and the

construction of the Three Gorges Reservoir and some small reservoirs. The ESV loss area was mainly located in Sanhe subdistrict, Shuanglu town, and Xingyi town on the south bank of the Yangtze River. The reason is that the construction of the new city invaded many other spaces.

(3) ESV and its change have a significant positive spatial autocorrelation. From 1990 to 2018, Moran's I of ESV and its change in Fengdu County were all greater than 0, indicating that the spatial distribution of ESV and its changes showed strong spatial clustering characteristics. The spatial distribution of cold spots and hot spots of ESV changes at different stages was consistent with ESV changes, and these cold and hot spots were mainly located along the Yangtze River and to the south of the Yangtze River.

Author Contributions: The co-authors jointly contributed to the completion of this article. Specifically, their individual contributions are as follows: conceptualization, H.Z. (Haozhe Zhang) and Q.Y.; validation, Q.Y. and Z.Z.; data curation, H.Z. (Haozhe Zhang), H.Z. (Huiming Zhang) and D.L.; formal analysis, Q.Y.; methodology, H.Z. (Haozhe Zhang) and Z.Z.; supervision, project administration, Q.Y.; writing—original draft preparation, H.Z. (Haozhe Zhang); writing—review and editing, H.Z. (Haozhe Zhang) and Q.Y.; visualization, H.Z. (Haozhe Zhang). All authors have read and agreed to the published version of the manuscript.

Funding: This research was funded by the National Natural Science Foundation of China, grant number 42071234, and Chongqing Social Science Planning Project, grant number 2020YBZX15.

Institutional Review Board Statement: Not applicable.

Informed Consent Statement: Not applicable.

Data Availability Statement: Not applicable.

Conflicts of Interest: The authors declare no conflict of interest.

Abbreviations

The abbreviations in this article:

ESV	Ecosystem Service Value
NLS	National Land Space
NLSP	National Land Space Pattern
TGRA	Three Gorges Reservoir Area
TGD	Three Gorges Dam
US	Urban space
AS	Agricultural space
ES	Ecological space
APS	Agricultural production space
ALS	Agricultural living space
VES	Vegetation ecological space
WES	Water ecological space
OES	Other ecological space
NPP	Net Primary Productivity

Appendix A

The "Two Mountains" Theory is a scientific conclusion put forward in 2005 by Jinping Xi, the President of China. Its core idea is that "lucid waters and lush mountains are invaluable assets", which means that a good ecological environment is the most inclusive factor promoting human well-being, and maintaining the ecological environment means maintaining productivity.

References

1. Costanza, R.; d'Arge, R.; de Groot, R.; Farber, S.; Grasso, M.; Hannon, B.; Limburg, K.; Naeem, S.; Oneill, R.V.; Paruelo, J.; et al. The value of the world's ecosystem services and natural capital. *Nature* **1997**, *387*, 253–260. [CrossRef]
2. Tolessa, T.; Senbeta, F.; Kidane, M. The impact of land use/land cover change on ecosystem services in the central highlands of Ethiopia. *Ecosyst. Serv.* **2017**, *23*, 47–54. [CrossRef]
3. Baral, H.; Keenan, R.J.; Sharma, S.K.; Stork, N.E.; Kasel, S. Economic evaluation of ecosystem goods and services under different landscape management scenarios. *Land Use Policy* **2014**, *39*, 54–64. [CrossRef]
4. Redford, K.H.; Adams, W.M. Payment for ecosystem services and the challenge of saving nature. *Conserv. Biol.* **2009**, *23*, 785–787.
5. Gui, J. *Research on Comprehensive Functional Zoning of National Land Space*; China University of Geosciences: Beijing, China, 2014.
6. Li, S.; Zhao, X.; Pu, J.; Miao, P.; Wang, Q.; Tan, K. Optimize and control territorial spatial functional areas to improve the ecological stability and total environment in karst areas of Southwest China. *Land Use Pol.* **2021**, *100*, 104940. [CrossRef]
7. Millennium Ecosystem Assessment Board. Ecosystems and Human Well-Being: Desertification Synthesis. Millennium Ecosystem Assessment Board. 2005. Available online: http://hdl.handle.net/20.500.11822/8719 (accessed on 20 September 2020).
8. Yan, J.M.; Chen, H.; Xia, F.Z. Cognition, direction and path of future spatial planning based on the background of multiple planning integration. *China Land Sci.* **2017**, *21*, 21–27.
9. Li, R.Q.; Li, Y.F.; Hu, H. Support of ecosystem services for spatial planning theories and practices. *Acta Geogr. Sin.* **2020**, *75*, 2417–2430.
10. Scolozzi, R.; Morri, E.; Santolini, R. Delphi-based change assessment in ecosystem service values to support strategic spatial planning in Italian landscapes. *Ecol. Indic.* **2012**, *21*, 134–144. [CrossRef]
11. Ouyang, Z.; Zheng, H.; Xiao, Y.; Polasky, S.; Liu, J.; Xu, W.; Wang, Q.; Zhang, L.; Xiao, Y.; Rao, E.M.; et al. Improvements in ecosystem services from investments in natural capital. *Science* **2016**, *352*, 1455–1459. [CrossRef] [PubMed]
12. Rodriguez, J.P.; Beard, T.D.; Bennett, E.M.; Cumming, G.S.; Cork, S.J.; Agard, J.; Dobson, A.P.; Peterson, G.D. Trade-offs across space, time, and ecosystem services. *Ecol. Soc.* **2006**, *11*, 14. [CrossRef]
13. Harris, J.M. Global environmental challenges of the twenty-first century: Resources, consumption, and sustainable solutions. *Ecol. Econ.* **2004**, *50*, 315–316. [CrossRef]
14. Daily, G. *Nature's Services: Societal Dependence on Natural Ecosystems*; Island Press: Washington, DC, USA, 1997.
15. Boyd, J.; Banzhaf, S. What are ecosystem services? The need for standardized environmental accounting units. *Ecol. Econ.* **2007**, *63*, 616–626. [CrossRef]
16. De Groot, R.S.; Wilson, M.A.; Boumans, R.M.J. A typology for the classification, description and valuation of ecosystem functions, goods and services. *Ecol. Econ.* **2002**, *41*, 393–408. [CrossRef]
17. Wallace, K.J. Classification of ecosystem services: Problems and solutions. *Biol. Conserv.* **2007**, *139*, 235–246. [CrossRef]
18. Bjorklund, J.; Limburg, K.E.; Rydberg, T. Impact of production intensity on the ability of the agricultural landscape to generate ecosystem services: An example from Sweden. *Ecol. Econ.* **1999**, *29*, 269–291. [CrossRef]
19. Bolund, P.; Hunhammar, S. Ecosystem services in urban areas. *Ecol. Econ.* **1999**, *29*, 293–301. [CrossRef]
20. Ronnback, P. The ecological basis for economic value of seafood production supported by mangrove ecosystems. *Ecol. Econ.* **1999**, *29*, 235–252. [CrossRef]
21. Xie, G.; Zhen, L.; Lu, C.; Xiao, Y.; Chen, C. Expert knowledge based valuation method of ecosystem services in China. *J. Nat. Resour.* **2008**, *23*, 911–919.
22. Xie, G.; Zhang, C.; Zhang, L.; Chen, W.; Li, S. Improvement of the evaluation method for ecosystem service value based on per unit area. *J. Nat. Resour.* **2015**, *30*, 1243–1254.
23. Hummel, C.; Poursanidis, D.; Orenstein, D.; Elliott, M.; Adamescu, M.C.; Cazacu, C.; Ziv, G.; Chrysoulakis, N.; van der Meer, J.; Hummel, H. Protected area management: Fusion and confusion with the ecosystem services approach. *Sci. Total Environ.* **2019**, *651*, 2432–2443. [CrossRef] [PubMed]
24. Carrilho, C.D.; de Almeida Sinisgalli, P.A. Contribution to Araçá Bay management: The identification and valuation of ecosystem services. *Ocean Coast. Manag.* **2018**, *164*, 128–135. [CrossRef]
25. Huang, L.; Cao, W.; Xu, X.; Fan, J.; Wang, J. Linking the benefits of ecosystem services to sustainable spatial planning of ecological conservation strategies. *J. Environ. Manag.* **2018**, *222*, 385–395. [CrossRef]
26. Li, B.; Wang, W. Trade-offs and synergies in ecosystem services for the Yinchuan Basin in China. *Ecol. Indic.* **2018**, *84*, 837–846. [CrossRef]
27. Bennett, E.M.; Peterson, G.D.; Gordon, L.J. Understanding relationships among multiple ecosystem services. *Ecol. Lett.* **2009**, *12*, 1394–1404. [CrossRef]
28. Qi, W.; Li, H.; Zhang, Q.; Zhang, K. Forest restoration efforts drive changes in land-use/land-cover and water-related ecosystem services in China's Han River basin. *Ecol. Eng.* **2019**, *126*, 64–73. [CrossRef]
29. Bullock, J.M.; Aronson, J.; Newton, A.C.; Pywell, R.F.; Rey-Benayas, J.M. Restoration of ecosystem services and biodiversity: Conflicts and opportunities. *Trends Ecol. Evol.* **2011**, *26*, 541–549. [CrossRef]
30. Yin, D.; Li, X.; Li, G.; Zhang, J.; Yu, H. Spatio-temporal evolution of land use transition and its eco-environmental effects: A Case study of the Yellow River basin, China. *Land* **2020**, *9*, 514. [CrossRef]

31. Gashaw, T.; Tulu, T.; Argaw, M.; Worqlul, A.W.; Tolessa, T.; Kindu, M. Estimating the impacts of Land Use/Land Cover Changes on Ecosystem Service Values: The case of the Andassa watershed in the Upper Blue Nile basin of Ethiopia. *Ecosyst. Serv.* **2018**, *31*, 219–228. [CrossRef]
32. Zhang, X.M.; Xie, H.L.; Shi, J.Y.; Lv, T.G.; Zhou, C.H.; Liu, W.D. Assessing changes in ecosystem service values in response to land cover dynamics in Jiangxi Province, China. *Int. J. Environ. Res. Public Health* **2020**, *17*, 3018. [CrossRef]
33. Wu, C.Y.; Chen, B.W.; Huang, X.J.; Wei, Y.H.D. Effect of land-use change and optimization on the ecosystem service values of Jiangsu province, China. *Ecol. Indic.* **2020**, *117*, 14. [CrossRef]
34. Yuan, K.Y.; Li, F.; Yang, H.J.; Wang, Y.M. The Influence of Land Use Change on Ecosystem Service Value in Shangzhou District. *Int. J. Environ. Res. Public Health* **2019**, *16*, 1321. [CrossRef] [PubMed]
35. Hu, S.; Chen, L.Q.; Li, L.; Wang, B.Y.; Yuan, L.N.; Cheng, L.; Yu, Z.Q.; Zhang, T. Spatiotemporal dynamics of Ecosystem Service Value determined by Land-Use Changes in the urbanization of Anhui Province, China. *Int. J. Environ. Res. Public Health* **2019**, *16*, 5104. [CrossRef]
36. Guo, A.; Zhang, Y.; Zhong, F.; Jiang, D. Spatiotemporal patterns of Ecosystem Service Value Changes and their coordination with economic development: A case study of the Yellow River Basin, China. *Int. J. Environ. Res. Public Health* **2020**, *17*, 8474. [CrossRef] [PubMed]
37. Jiang, W.; Fu, B.; Lu, Y. Assessing impacts of Land Use/Land Cover Conversion on changes in Ecosystem Services Value on the Loess Plateau, China. *Sustainability* **2020**, *12*, 7128. [CrossRef]
38. Wang, Y.; Zhang, S.; Zhen, H.; Chang, X.; Shataer, R.; Li, Z. Spatiotemporal evolution characteristics in Ecosystem Service Values based on Land Use/Cover Change in the Tarim River Basin, China. *Sustainability* **2020**, *12*, 7759. [CrossRef]
39. Tan, Z.; Guan, Q.; Lin, J.; Yang, L.; Luo, H.; Ma, Y.; Tian, J.; Wang, Q.; Wang, N. The response and simulation of ecosystem services value to land use/land cover in an oasis, Northwest China. *Ecol. Indic.* **2020**, 118. [CrossRef]
40. Jiang, W.; Lu, Y.H.; Liu, Y.X.; Gao, W.W. Ecosystem service value of the Qinghai-Tibet Plateau significantly increased during 25 years. *Ecosyst. Serv.* **2020**, *44*, 10. [CrossRef]
41. Wang, Z.; Cao, J.; Zhu, C.; Yang, H. The Impact of Land Use Change on Ecosystem Service Value in the upstream of Xiong'an New Area. *Sustainability* **2020**, *12*, 704.
42. Yirsaw, E.; Wu, W.; Shi, X.P.; Temesgen, H.; Bekele, B. Land Use/Land Cover Change modeling and the prediction of Subsequent changes in Ecosystem Service Values in a Coastal Area of China, the Su-Xi-Chang Region. *Sustainability* **2017**, *9*, 1204. [CrossRef]
43. Xiao, R.; Lin, M.; Fei, X.F.; Li, Y.S.; Zhang, Z.H.; Meng, Q.X. Exploring the interactive coercing relationship between urbanization and ecosystem service value in the Shanghai-Hangzhou Bay Metropolitan Region. *J. Clean. Prod.* **2020**, *253*, 13. [CrossRef]
44. Hersperger, A.; Langhamer, D.; Dalang, T. Inventorying human-made objects: A step towards better understanding land use for multifunctional planning in a periurban Swiss landscape. *Landsc. Urban Plan.* **2012**, *105*, 307–314. [CrossRef]
45. Zhou, D.; Xu, J.; Lin, Z. Conflict or coordination? Assessing land use multi-functionalization using production-living-ecology analysis. *Sci. Total Environ.* **2016**, 577. [CrossRef] [PubMed]
46. Wende, W.; Huelsmann, W.; Marty, M.; Penn-Bressel, G.; Bobylev, N. Climate protection and compact urban structures in spatial planning and local construction plans in Germany. *Land Use Policy* **2010**, *27*, 864–868. [CrossRef]
47. Arbolino, R. Development policies in China: An analysis of the territorial imbalances. In *Multicriteria and Multiagent Decision Making with Applications to Economics and Social Sciences*; Ventre, A.G.S., Maturo, A., Hošková-Mayerová, Š., Kacprzyk, J., Eds.; Springer: Berlin, Germany, 2013; pp. 1–14.
48. Oppido, S.; Ragozino, S.; De Vita, G.E. Exploring territorial imbalances: A systematic literature review of meanings and terms. In *New Metropolitan Perspectives, Proceedings of International Symposium: New Metropolitan Perspectives, Italy (Online), 26–28 May 2020*; Bevilacqua, C., Calabrò, F., Della Spina, L., Eds.; Springer: Cham, Switzerland, 2020; pp. 90–100.
49. Liu, J.L.; Liu, Y.S.; Li, Y.R. Classification evaluation and spatial-temporal analysis of "production-living-ecological" spaces in China. *Acta Geogr. Sin.* **2017**, *72*, 1290–1304.
50. Ma, L.B.; Niu, S.W.; Shi, P.J.; Guo, X.D. The functional zoning of territorial space and the developmental pattern of future space—Based on the framework of the major function oriented zoning. *Econ. Geogr.* **2015**, *35*, 68–77.
51. Bryan, B.A.; Ye, Y.Q.; Zhang, J.E.; Connor, J.D. Land-use change impacts on ecosystem services value: Incorporating the scarcity effects of supply and demand dynamics. *Ecosyst. Serv.* **2018**, *32*, 144–157. [CrossRef]
52. Fu, B.J.; Wang, S.; Su, C.H.; Forsius, M. Linking ecosystem processes and ecosystem services. *Curr. Opin. Environ. Sustain.* **2013**, *5*, 4–10. [CrossRef]
53. Qiu, S.S.; Yue, W.Z.; Zhang, H.; Qi, J.G. Island ecosystem services value, land-use change, and the National New Area Policy in Zhoushan Archipelago, China. *Isl. Stud. J.* **2017**, *12*, 177–197. [CrossRef]
54. Meng, Q.H.; Fu, B.J.; Yang, L.Z. Effects of land use on soil erosion and nutrient loss in the Three Gorges Reservoir Area, China. *Soil Use Manag.* **2001**, *17*, 288–291. [CrossRef]
55. Shi, Z.; Deng, W.; Zhang, S. Spatio-temporal pattern changes of land space in Hengduan Mountains during 1990–2015. *J. Geogr. Sci.* **2018**, *28*, 529–542. [CrossRef]
56. Kates, R.; Clark, W.; Corell, R.; Hall, J.; Jaeger, C.; Lowe, I.; McCarthy, J.; Schellnhuber, H.; Bolin, B.; Dickson, N.; et al. Environment and development. Sustainability science. *Science* **2001**, *292*, 641–642. [CrossRef] [PubMed]
57. Zhao, X.Q.; Li, S.N.; Pu, J.W.; Miao, P.P.; Wang, Q.; Tan, K. Optimization of the national land space based on the coordination of urban-agricultural-ecological functions in the Karst Areas of Southwest China. *Sustainability* **2019**, *11*, 6752. [CrossRef]

58. Liao, N.; Gu, X.; Wang, Y.; Xu, H.; Fan, Z. Analyzing macro-level ecological change and micro-level farmer behavior in Manas River Basin, China. *Land* **2020**, *9*, 250. [CrossRef]
59. Xie, G.; Zhang, C.; Zhen, L.; Zhang, L. Dynamic changes in the value of China's ecosystem services. *Ecosyst. Serv.* **2017**, *26*, 146–154. [CrossRef]
60. Poudyal, N.C.; Elkins, D.; Nibbelink, N.; Cordell, H.K.; Gyawali, B. An exploratory spatial analysis of projected hotspots of population growth, natural land loss, and climate change in the conterminous United States. *Land Use Policy* **2016**, *51*, 325–334. [CrossRef]
61. Wen, Z.; Wu, S.; Chen, J.; Lü, M. NDVI indicated long-term interannual changes in vegetation activities and their responses to climatic and anthropogenic factors in the Three Gorges Reservoir Region, China. *Sci. Total Environ.* **2017**, *574*, 947–959. [CrossRef]
62. Cao, L.D.; Li, J.L.; Ye, M.Y.; Pu, R.L.; Liu, Y.C.; Guo, Q.D.; Feng, B.X.; Song, X.Y. Changes of ecosystem service value in a Coastal Zone of Zhejiang Province, China, during rapid urbanization. *Int. J. Environ. Res. Public Health* **2018**, *15*, 1301. [CrossRef]
63. Song, W.; Deng, X.Z. Land-use/land-cover change and ecosystem service provision in China. *Sci. Total Environ.* **2017**, *576*, 705–719. [CrossRef]
64. Yuan, X.Z.; Xiao, H.Y.; Yan, W.T.; Li, B. Dynamic analysis of land use and ecosystem services value in Cheng-Yu Economic Zone, Southwest China. *Chin. J. Ecol.* **2012**, *21*, 182–188.
65. Haining, R. *Spatial Data Analysis in the Social and Environmental Sciences*; Cambridge University Press: Cambridge, UK, 1997.
66. Liu, J.; Jin, X.B.; Xu, W.Y.; Fan, Y.T.; Ren, J.; Zhang, X.L.; Zhou, Y.K. Spatial coupling differentiation and development zoning trade-off of land space utilization efficiency in Eastern China. *Land Use Policy* **2019**, *85*, 310–327. [CrossRef]
67. Liu, W.; Zhan, J.Y.; Zhao, F.; Yan, H.M.; Zhang, F.; Wei, X.Q. Impacts of urbanization-induced land-use changes on ecosystem services: A case study of the Pearl River Delta Metropolitan Region, China. *Ecol. Indic.* **2019**, *98*, 228–238. [CrossRef]
68. Deng, Y.J.; Hou, M.Y.; Xie, Y.F.; Gao, Q.; Yao, S.B.; Gong, Z.W.; Lu, Y.N.; Jia, L.; Li, Y.Y. Impact of the Grain for Green Project on the temporal and spatial evolution of ecosystem service value in northern Shaanxi. *Acta Ecol. Sinica* **2020**, *40*, 6597–6612.
69. Zhu, S.Y.; Kong, X.S.; Jiang, P. Identification of the human-land relationship involved in the urbanization of rural settlements in Wuhan city circle, China. *J. Rural Stud.* **2020**, *77*, 75–83. [CrossRef]
70. Su, K.C.; Hu, B.Q.; Shi, K.F.; Zhang, Z.X.; Yang, Q.Y. The structural and functional evolution of rural homesteads in mountainous areas: A case study of Sujiaying village in Yunnan Province, China. *Land Use Policy* **2019**, *88*, 12. [CrossRef]
71. Cai, E.X.; Chen, W.Q.; Wei, H.J.; Li, J.W.; Wang, H.; Guo, Y.L.; Feng, X.W. The coupling characteristics of population and residential land in rural areas of China and its implications for sustainable land use. *Sustain. Dev.* **2020**, *28*, 646–656. [CrossRef]
72. Xing, L.; Zhu, Y.; Wang, J. Spatial spillover effects of urbanization on ecosystem services value in Chinese cities. *Ecol. Indic.* **2021**, *121*, 107028. [CrossRef]
73. Fu, Y.J.; Shi, X.Y.; He, J.; Yuan, Y.; Qu, L.L. Identification and optimization strategy of county ecological security pattern: A case study in the Loess Plateau, China. *Ecol. Indic.* **2020**, *112*, 10. [CrossRef]
74. Yalcin, E.; Tigrek, S. Hydropower production without sacrificing environment: A case study of Ilisu Dam and Hasankeyf. *Int. J. Water Resour. Dev.* **2016**, *32*, 247–266. [CrossRef]
75. Latrubesse, E.; Arima, E.; Dunne, T.; Park, E.; Baker, V.; d'Horta, F.; Wight, C.; Wittmann, F.; Zuanon, J.; Baker, P.; et al. Damming the rivers of the Amazon basin. *Nature* **2017**, *546*, 363–369. [CrossRef] [PubMed]
76. Bednarek, A.; Hart, D. Modifying dam operations to restore rivers: Ecological responses to Tennessee River Dam mitigation. *Ecol. Appl.* **2005**, *15*, 997–1008. [CrossRef]
77. Winemiller, K.O.; McIntyre, P.B.; Castello, L.; Fluet-Chouinard, E.; Giarrizzo, T.; Nam, S.; Baird, I.G.; Darwall, W.; Lujan, N.K.; Harrison, I.; et al. Balancing hydropower and biodiversity in the Amazon, Congo, and Mekong. *Science* **2016**, *351*, 128–129. [CrossRef]
78. Wu, J.; Huang, J.; Han, X.-G.; Xie, Z. Three-Gorges Dam—Experiment in habitat fragmentation? *Science* **2003**, 300. [CrossRef] [PubMed]
79. Stone, R. The Legacy of the Three Gorges Dam. *Science* **2011**, *333*, 817. [CrossRef]
80. Xu, X.; Yang, G.; Tan, Y.; Liu, J.; Zhang, S.; Bryan, B. Unravelling the effects of large-scale ecological programs on ecological rehabilitation of China's Three Gorges Dam. *J. Clean. Prod.* **2020**, *256*, 120446. [CrossRef]
81. Moran, E.; Lopez, M.C.; Moore, N.; Müller, N.; Hyndman, D. Sustainable hydropower in the 21st century. *Proc. Natl. Acad. Sci. USA* **2018**, *115*, 201809426. [CrossRef]
82. Wang, Y.; Pan, J.H. Building ecological security patterns based on ecosystem services value reconstruction in an arid inland basin: A case study in Ganzhou District, NW China. *J. Clean. Prod.* **2019**, *241*, 12. [CrossRef]
83. Wan, A.; Liu, Y.; Xie, X.; Tu, R.; Qi, X. Study on spatial layout optimization of Linpan settlements based on point pattern analysis. *Environ. Dev. Sustain.* **2020**, 1–19. [CrossRef]
84. Mamat, A.; Halik, U.; Rouzi, A. Variations of Ecosystem Service Value in Response to Land-Use Change in the Kashgar Region, Northwest China. *Sustainability* **2018**, *10*, 200. [CrossRef]
85. Zhan, C.; Zhao, R.; Hu, S. Emergy-based sustainability assessment of forest ecosystem with the aid of mountain eco-hydrological model in Huanjiang County, China. *J. Clean. Prod.* **2019**, *251*, 119638. [CrossRef]

86. Brentrup, F.; Küsters, J.; Lammel, J.; Kuhlmann, H. Life Cycle Impact assessment of land use based on the Hemeroby concept. *Int. J. Life Cycle Assess.* **2002**, *7*, 339–348. [CrossRef]
87. Lu, D.; Wang, Y.; Yang, Q.; Su, K.; Zhang, H.; Li, Y. Modeling spatiotemporal population changes by integrating DMSP-OLS and NPP-VIIRS nighttime light data in Chongqing, China. *Remote Sens.* **2021**, *13*, 284. [CrossRef]

International Journal of
Environmental Research and Public Health

MDPI

Article

Identifying Ecological Corridors and Networks in Mountainous Areas

Di Zhou [1,2] and Wei Song [1,*]

[1] Key Laboratory of Land Surface Pattern and Simulation, Institute of Geographic Sciences and Natural Resources Research, Chinese Academy of Sciences, Beijing 100101, China; zd15071032867@163.com

[2] School of Geosciences, Yangtze University, Wuhan 430100, China

* Correspondence: songw@igsnrr.ac.cn

Abstract: Since the 1950s, human activities have been driving economic development and land changes, hindering the conservation of biological habitats and landscape connectivity. Constructing ecological networks is an effective means to avoid habitat destruction and fragmentation. Mountain areas are hotspots of biological habitats and biodiversity; however, the pace of urbanization in mountain areas is also accelerating. To protect an ecosystem more effectively, it is necessary to identify ecological corridors and ecological networks. Therefore, based on the Minimal Cumulative Resistance model and taking Chongqing in China as an example, the identification of potential ecological corridors and the construction of an ecological network in Chongqing were realized using the Linkage Mapper software. The results were as follows: (1) From 2005 to 2015, the patch area of cultivated land and grassland in Chongqing decreased by 0.08% and 1.46%, respectively, while that of built-up areas increased by 1.5%. The fragmentation degree of cultivated land was higher, and the internal connectivity of forestry areas was worse. (2) In total, 24 ecological sources were selected, and 87 potential ecological corridors and 35 ecological nodes were generated using the Morphological Spatial Pattern Analysis and the Conefor2.6 software. The total length of the ecological network in Chongqing is 2524.34 km, with an average corridor length of 29.02 km. (3) The overall complexity and network efficiency are high, but the spatial distribution of ecological corridors is uneven, especially in the southwest of Chongqing.

Keywords: ecological corridor; ecological network; landscape pattern; Linkage Mapper software; Chongqing; China

Citation: Zhou, D.; Song, W. Identifying Ecological Corridors and Networks in Mountainous Areas. *Int. J. Environ. Res. Public Health* **2021**, *18*, 4797. https://doi.org/10.3390/ijerph18094797

Academic Editor: Paul B. Tchounwou

Received: 2 March 2021
Accepted: 24 April 2021
Published: 30 April 2021

1. Introduction

Currently, the global scale of cities continues to grow [1,2], with an expanding scope of human activities [3], putting the global biodiversity at risk of further degradation [4,5]. Although considerable efforts have been made by governments around the world to protect biodiversity [6–10], the coordination of economic development and biodiversity conservation is still challenging [11,12]. Among the factors affecting biodiversity, fragmentation and loss of biological habitat are considered the most important ones [13]. As an ecological network can adequately depict the ecological processes [14] and can protect habitats and maintain landscape connectivity [15,16], the construction of ecological networks has become a research hotspot in ecosystem protection [17,18].

An ecological network is a complex network having ecological corridors and ecological nodes as its constituent elements that connect core areas, nature reserves, and other landscape elements [19–22]. Ecological networks have first been used in the study of biological protection [23]. Along with the expansion and improvement of ecological network functions, they are widely used in biodiversity conservation [21]. For example, from the 1970s to the 1990s, some Eastern European countries, the Netherlands, Canada, and the United States developed network plans for nature reserves [24] and began to emphasize the role of ecological interconnection. In the early 21st century, Europe established an ecological network framework that emphasized green protection [23]. In the Convention on Biological Diversity 2010, the Aichi Target 11 (Aichi refers to Japan's Aichi Prefecture) defined the goal of ecosystem protection by 2020, which is to protect at least 17% of the land and inland water areas and 10% of the marine area by linking ecological areas [10,25]. It further highlights the significance of using ecological networks for biodiversity conservation.

An ecological corridor is an important tool connecting ecological networks [26]. Ecological corridors were originally designed to connect natural habitats for wildlife protection [23]. However, due to the interference from human activities [27], habitat fragmentation around the world has become increasingly prominent [28], and against this background, in 1975, Wilson and Willis put forward the view of connecting broken patches with corridors to weaken the impact of habitat fragmentation [23]. The concept of ecological corridors was put forward for the first time in the early 21st century by Jordan, based on reliability theory [29]. Subsequently, scholars have engaged in extensive discussions and studies on ecological corridors. For example, Bowers and McKnight put forward the necessity of constructing ecological corridors in North America [30], and Chettri et al. discussed the importance of constructing ecological corridors in the Kangchenjunga landscape from the view of maintaining forest ecosystems and protecting biodiversity [31].

Although the models and methods of describing ecological corridors and ecological networks are slightly different, the construction of an ecological network generally includes three steps [25,32]: (1) selecting the ecological source; (2) determining the resistance surface; (3) extracting ecological corridors and nodes. Based on these steps, the ecological network is constructed. The early selection of ecological sources was mostly limited to large landscape patches such as nature reserves and scenic spots [33,34], lending this method a certain subjectivity. With the development of relevant research methods, MSPA (Morphological Spatial Pattern Analysis) and other methods are gradually used to quantitatively evaluate the importance of ecological sources and landscape connectivity [4,35], thereby improving the scientific nature of ecological source selection. The resistance surface reflects the resistance of species passing through ecological patches during migration, which is one of the important factors to measure the risk of species migration. Currently, most scholars prefer to choose terrain conditions, human activity intensity [36], and land use type [37], among others, as the factors impacting the resistance surface, and some prefer to build the resistance surface based on land use types [35]. An ecological corridor is an important tool for the formation of an ecological network, and the research methods are constantly being improved. For example, Pomianowski and Solon used the GraphScape software to generate an ecological corridor [38], whereas Guo et al. constructed an ecological corridor based on the least-cost distance (LCD) and the least-cost path (LCP) [39]. Peng et al. combined the circuit theory and the Linkage Mapper software to identify the ecological corridor [40].

Ecological nodes are generally located at the convergence between the least-cost paths of the ecological corridors, at the site of the functional weakness area to connect scattered and isolated patches, which is the key to enhance the connectivity of ecological sources and promote the operation of ecological flows among ecological networks. The ecological resistance surface model is widely used in the extraction of ecological nodes [20]. Considering the resistance factors, the *MCR* (Minimal Cumulative Resistance) model is widely used in ecological corridor identification and ecological network construction [36]. For example, it has been used by Yang et al. [26] to identify the ecological corridor of Wuhan City, China, with the aim to construct an ecological network of urban agglomeration. Similarly, Dai et al. [41] combined the *MCR* model and the DOI (Duranton and Overman Index) to construct the ecological network of the Poyang Lake urban agglomeration in China.

Overall, current studies on ecological networks mainly focus on the construction of an urban agglomeration ecological network [42], a wetland ecological network, and a desert oasis ecological network [20], whereas studies in mountain areas are scarce [31,43]. In China, mountain areas account for 70% of the country's terrestrial area [44], and ecosystem protection in mountain areas plays an important role for ecosystem protection in China [45]. In this context, this paper takes Chongqing, China's "Mountain City", as the research area, discussing ecological corridor identification and ecological network construction in mountain areas. The specific objectives are as follows: (1) to reveal the landscape pattern evolution of Chongqing from 2005 to 2015; (2) to use the Linkage Mapper software (The Nature Conservancy, Seattle, WA, USA) to identify potential ecological corridors in Chongqing; (3) in combination with GIS (Geographic Information System), to construct the potential ecological network of Chongqing.

2. Study Area and Data Sources

2.1. Study Area

The city of Chongqing (28°10′ N~32°13′ N, 105°11′ E~110°11′ E) is located in the southwest of inland China (Figure 1) and is the sole municipality in central and western China. It stretches over 470 km from east to west and 450 km from north to south, covering an area of 82,400 square kilometers. The terrain of Chongqing decreases from north and south in the direction of the Yangtze River valley; the area is mainly covered by mountains and hills, also known as the "Mountain City". Chongqing is located in the Subtropical Zone, with abundant precipitation, four distinct seasons, and a high annual relative humidity. Driven by China's reform and opening up, urbanization in Chongqing is rapid [46], with a rate above 81.04% in 2017 [47]. The gross domestic product in 2019 amounted to 236.06 billion RMB. The surge of the urban population and the rapid economic development result in an increased land development [48], with conflicts between humans and the ecological environment becoming increasingly prominent [49], resulting in a continuous deterioration of ecosystems [50].

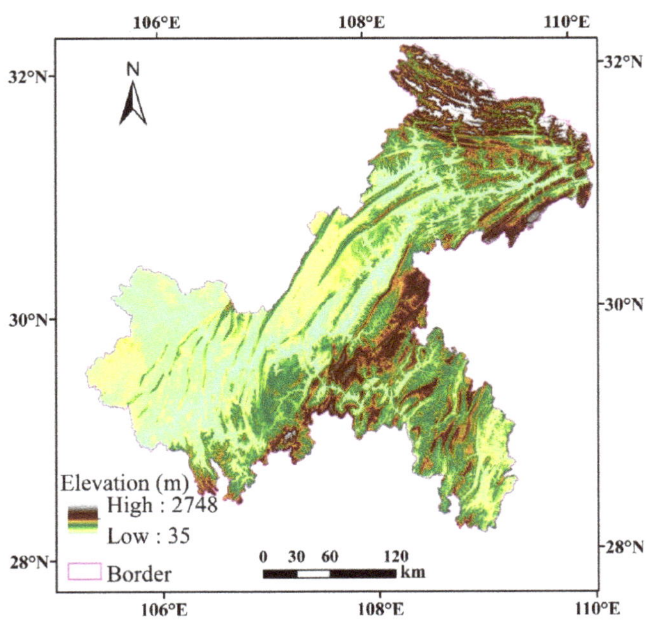

Figure 1. Geographical location of Chongqing, China.

2.2. Data Sources

We used four data types, namely land use data, DEM (Digital Elevation Model data), road data, and river data. Land use data were derived from the Resources and Environment Science and Data Center of the Chinese Academy of Sciences [51]. We selected data from the years 2005 and 2015, with a spatial resolution of 1 km; the data type was raster data. The main types of land use data included 6 first-class categories (cultivated land, forest land, grassland, water area, built-up areas, and unused land) and 17 s-class categories (paddy fields, dry land, forested land, shrub land, sparse forest land, other forestry areas, high-coverage grassland, medium-coverage grassland, low-coverage grassland, river channel, lake, reservoir pit, beach land, urban land, rural residential area, other built-up areas, and other unused land). The DEM data were derived from the geospatial data cloud [52]; the data type was raster data with a resolution of 250 m. Data for roads and rivers were derived from the Resources and Environment Science and Data Center of the Chinese Academy of Sciences [51]; the data type was vector data. Road data included data from national highways, railways, and highways. River data had a spatial distribution of 1–5 rivers.

3. Research Method

3.1. Technical Route

The study was divided into four steps (Figure 2). First, based on the landscape pattern analysis method, we used the Fragstats4.2 software (Oregon State University, Corvallis, OR, USA) to calculate the index values of PLAND (Percentage of Landscape), PD (Patch Density), COHESION (Patch Cohesion Index), DIVISION (Landscape Division Index), and AI (Aggregation Index) in Chongqing to analyze the evolution process of land use in Chongqing from 2005 to 2015. Second, using land use data from 2015, applying the MSPA method and the Conefor2.6 software (Jenness Enterprises, Flagstaff, AZ, USA), the ecological sources of Chongqing were selected. Then, based on available land use data, DEM data, road data, and river data, the appropriate GIS method was applied to obtain the resistance value, determine the landscape resistance surface, and generate the comprehensive resistance surface with reference to previous studies and specific conditions in Chongqing. Finally, we used the Linkage Mapper software to input the obtained ecological source data and the comprehensive resistance surface data to generate the ecological corridor and the accumulated resistance surface. By overlaying ecological sources, ecological corridors, and ecological nodes, Chongqing's ecological network was finally constructed. According to the number of ecological corridors and nodes, the ecological network was evaluated and analyzed using the α index, β index, and γ index.

3.2. Landscape Pattern Index Analysis

The landscape pattern index is a quantitative method that provides different landscape indices for streamlining the structure of the landscape spatial pattern and the concise landscape spatial morphological characteristics [53]. Taking into account the specific ecological situation of Chongqing and other relevant data [54,55], we selected five landscape indices, namely PLAND, PD, COHESION, DIVISION, and AI, to depict the landscape characteristics of Chongqing. Among them, PLAND quantifies the proportion of each patch type area in the landscape; the larger the proportion, the larger the corresponding patch type area. The Patch Density represents the degree of landscape fragmentation; the greater the value, the higher the fragmentation degree. The Patch Cohesion Index indicates the degree of physical connection between patch types; greater values indicate better landscape connectivity. The Landscape Division Index reflects cuts of the landscape by urban roads and other factors, with greater values indicating that cutting is more obvious and landscape fragmentation is high. The Aggregation Index reflects the connectivity inside the patch; the larger the value, the better the connectivity. Related work has been completed using the Fragstats4.2 software [56].

3.3. Selection of Ecological Sources

3.3.1. Identification of Core Ecological Patches by the MSPA Method

The MSPA method originates from mathematical morphology and was initially applied in studies on forest fragmentation [57,58]. When Peter Vogt developed the Guidos software (European Commission Joint Research Centre, Ispra, Italy), MSPA was formally applied to the research of landscape connectivity [59]. By using the ArcGIS software (Environmental Systems Research Institute, Redlands, CA, USA), the land use type data in raster data format were divided into raster binary maps of foreground and background. Forestland, grassland, water area, among others, are generally taken as the foreground and other land use types as the background. Using the Guidos Toolbox software, the foreground was divided into the seven categories core area, bridge area, edge area, branch line, isolated island, ring road, and pore [60]. Among them, core areas generally refer to ecological sources, and the bridging zone is a corridor connecting ecological source patches; both have significant impacts on the connectivity of regional ecological landscapes [61].

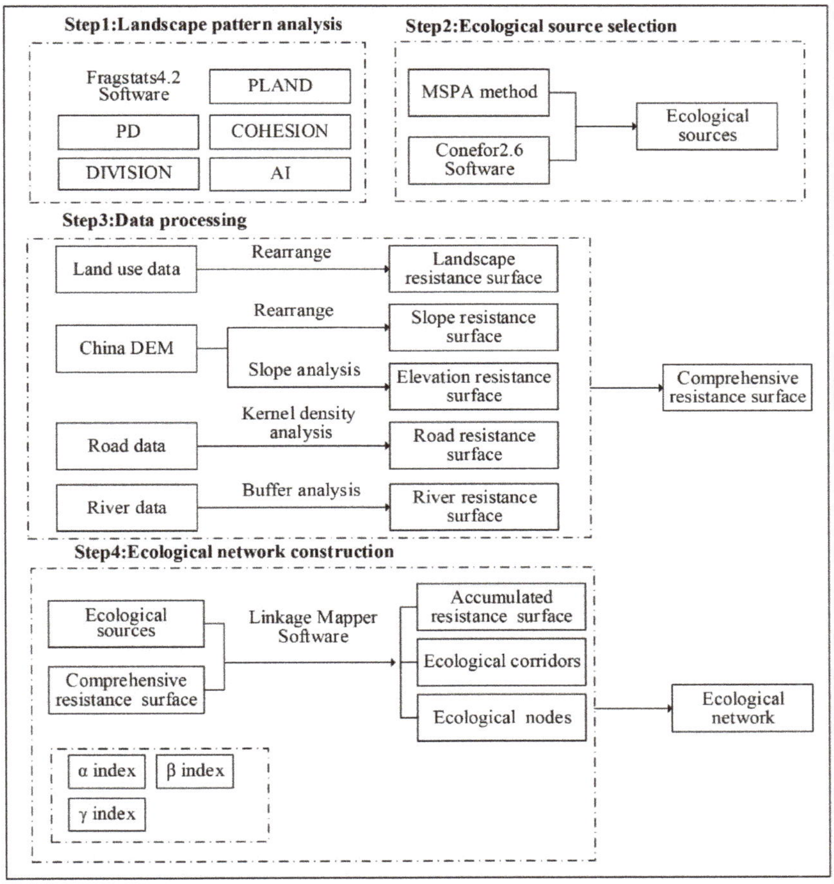

Figure 2. Technical route. Notes: PLAND (Percentage of Landscape) refers to landscape percentage, PD (Patch Density) refers to patch density, COHESION (Patch Cohesion Index) refers to the cohesion index, DIVISION (Landscape Division Index) refers to the landscape separation index, AI (Aggregation Index) refers to the aggregation index. α index refers to the network closure index, β index refers to the network connectivity index, γ index refers to the network connectivity rate.

We conducted MSPA analysis on land use data of Chongqing using the Guidos Toolbox software platform to identify the core ecological patches of Chongqing, with the following steps:

1. Standardized Data Processing

The land use data of Chongqing for 2015 were loaded into the ArcMap, and the forest land, grassland, and water areas were set as the foreground with a value of 2. The other land classes were assigned as the background with a value of 1 and converted into a binary grid map. Finally, the land use data were exported to the TIFF format.

2. Neighborhood Rule Setting

After loading the binary grid map of Chongqing in the Guidos Toolbox software, the neighborhood structure rules of foreground pixels were set. The software provides two kinds of neighborhood structure: the eight-neighborhood structure and the four-neighborhood structure. Structural elements refer to the units that deal with the target landscape map. The selection of structural elements affects the movement law and the osmotic critical threshold of species in habitat patches. Here, we selected the eight-neighborhood structure for MSPA analysis.

3. Edge Width Setting

The width of the ecological corridor has a direct impact on the analysis results for an ecological network. The terrain and other features analyzed in previous studies [62,63] are similar compared to our study. In addition, because the width increases, a small core area becomes an island, and a narrow core area becomes a bridge. At the same time, combined with the actual situation in Chongqing, we selected a 30 m edge width. In the MSPA method, corridor width is equal to the edge width multiplied by the pixel resolution, that is, the edge width can be determined according to the value of the corridor width and the pixel resolution. The raster data pixel size was 1000 × 1000 m, the width of the ecological corridor was set to 30 m, and the corresponding edge width parameter was set to 0.03.

4. Generation of Seven Landscape Types

After importing the binary grid diagram in the Guidos Toolbox software, we set the required parameters. We used MSPA to generate seven landscape types, non-overlapping and independent of each other: core area, island, pore, edge area, bridge area, ring road, and branch line.

3.3.2. Identification of Ecological Sources by the Landscape Connectivity Index

The MSPA method can identify regional core ecological patches, but it cannot distinguish their importance. This paper uses the landscape connectivity index to classify the importance of the identified ecological patches in the core area [64], and the larger area and higher connectivity in the patches in the core area were selected as ecological sources. Currently, the commonly used landscape connectivity index includes overall connectivity (*IIC*, Equation (1)), possible connectivity (*PC*, Equation (2)), and plaque importance (*dPC*, Equation (3)) [65]. We therefore selected the overall connectivity index, *IIC*, and the possible connectivity index, *PC*, to calculate the patch importance value, *dPC* and, consequently, to evaluate the relative importance of ecological connectivity in the core area. We used equations described elsewhere [66]. The overall connectivity index was calculated as

$$IIC = \frac{\sum_{i=1}^{n}\sum_{j=1}^{n}\frac{a_i \times a_j}{1+nl_{ij}}}{{A_l}^2} \quad (1)$$

where $0 < IIC < 1$; when *IIC* equals 1, all landscapes are occupied by habitats. Factors. a_i and a_j refer to the area of patch "*i*" and patch "*j*", respectively, and nl_{ij} is the number of connections between patch "*i*" and patch "*j*". A_l represents the total landscape area of the study area. The *PC* was calculated as

$$PC = \frac{\sum_{i=1}^{n}\sum_{j=1}^{n} a_i \times a_j \times {P^*}_{ij}}{{A_l}^2} \quad (2)$$

where p_{ij}^* refers to the maximum connection probability of two patches. The calculation result of *PC* ranges between 0 and 1, and the *PC* value represents the high or low possibility of landscape connection. p_{ij}^* represents the maximum possible connectivity probability between patch "i" and patch "j". In general, the smaller the distance between patches, the higher the probability of maximum possible connectivity will be and vice versa. The dPC was calculated as follows:

$$dPC = \frac{PC - PC'}{PC} \times 100\% \quad (3)$$

where the dPC refers to the important value of the possible connectivity index of the patch, PC refers to the possible connectivity index of a patch in the landscape, and PC' refers to the possible connectivity index of the landscape after removing the patch. For example, when PC is 80 and PC' is 20, the calculated dPC is 75.

We calculated dPC in the Conefor2.6 software. First, the Conefor Toolbox plug-in in ArcGIS was used to extract the distance information between the core patches generated by MSPA, and subsequently, we entered the generated file into the Conefor2.6 software to solve the important value of connectivity. According to the geography of Chongqing, and based on relevant literature [25], the threshold of connectivity distance was set as 500 m, and the connectivity probability was 0.5. According to relevant studies [61], when $dPC > 1$, the patch's connectivity is better and the patch is more important. Therefore, the core patch of the case where $dPC > 1$ was selected as the ecological source in this paper.

3.4. Resistance Surface Construction

Different types of patches will cause different resistances to species migration. The magnitude of resistance can reflect the difficulty of species migration. For example, the resistance value of forest land and green land is small, which is conducive to species migration, whereas that of cultivated land, built-up areas, and unused land is larger, impeding species migration. In addition, resistance factors such as elevation, slope, roads, and rivers also affect species migration. Based on the actual situation in Chongqing, we comprehensively considered the influencing factors of land use types, elevation, slope, roads, and rivers and finally determined the resistance value of various landscape patches in Chongqing (Table 1), based on relevant studies [25,37].

Table 1. Resistance values and weights of resistance factors in Chongqing.

Resistance Factors	Weight	Classification Indicators	Resistance Value
Land use types	0.30	Forests	1
		Shrubs, sparse forests	3
		Paddy fields, dry lands	50
		Other forestry areas	300
		High-coverage grassland	10
		Medium-cover grassland	15
		Low-coverage grassland	20
		Rivers	600
		Lakes	300
		Reservoir Pit Tong	100
		Beach	1
		Urban land, rural settlements	900
		Other construction sites	1000
		Others	700
Elevation	0.10	Elevation range	Resistance value
		<450 m	150
		450–700 m	300
		700–1000 m	500
		1000–1400 m	800
		1400–1800 m	1000
		>1800 m	1500

Table 1. *Cont.*

Resistance Factors	Weight	Classification Indicators	Resistance Value
Slope	0.10	Slope range	Resistance value
		<3°	1
		3–6°	20
		6–10°	100
		10–16°	200
		>16°	600
Roads	0.25	Road types	Resistance value
		Railways	700
		National Highway	2000
		Other roads	500
Rivers	0.25	Buffer	Resistance value
Level I rivers		<50 m	5
		50–120 m	25
		120–300 m	50
Level II rivers		<50 m	10
		50–120 m	50
		120–300 m	100
Level III rivers		<50 m	20
		50–120 m	100
		120–300 m	500

Note: Resistance value data in Table 1 are derived from literature [39,42,67].

3.4.1. Construction of Landscape Resistance Surfaces

Based on the GIS software platform and the characteristics of Chongqing, we converted the land use raster data for 2015 into vector data and assigned corresponding resistance values according to different types, which were finally converted into landscape type raster data to generate the landscape resistance surface. The DEM data of Chongqing were reclassified and assigned to generate the elevation resistance surface. We then performed slope analysis and reclassification of Chongqing DEM data and generated the slope resistance surface after assignment. The core density of road element data in Chongqing was calculated, and the corresponding search radius parameters were set to generate the road resistance surface. The river data of Chongqing were graded and assigned, and the river resistance surface was generated. All landscape resistance values are shown in Table 1.

3.4.2. Construction of the Integrated Resistance Surface

Each resistance surface contributes to the comprehensive resistance surface according to a certain weight, making it necessary to adequately allocate the weight of each resistance surface before constructing the comprehensive resistance. In this paper, we adopted the analytic hierarchy process [12] to determine the weight of each resistance surface, in combination with the opinions of relevant experts. The greater the weight given, the higher the importance. Based on this, the weights of a total of five resistance surfaces of the landscape type resistance surface, elevation resistance surface, slope resistance surface, road resistance surface, and river resistance surface were set to 0.30, 0.10, 0.10, 0.25, and 0.25, respectively (Table 1). The integrated resistance surface was obtained by weighted superposition of the grid calculator tool of the ArcMap software as the cost data of the minimum cost distance model.

3.4.3. Construction of the Cumulative Resistance Surface

The cumulative resistance surface was generated using the Linkage Mapper software tool in ArcGIS. This tool facilitates the analysis of the connectivity of regional animal habitat [68] and identifies and maps the lowest-cost relationship among ecological sources through ecological source vector data and integrated resistance surface raster data. Each pixel of the integrated resistance surface has a value that reflects the difficulty of the species passing through that pixel. It is usually the pixel feature that determines the resistance value, such as land use type or elevation, slope, road, river, among others. The cost-weighted distance will produce a cumulative resistance surface when the species leaves a specific ecological source area.

3.5. Minimal Cumulative Resistance Model

The *MCR* model, first proposed by Knaapen et al. in 1992, is used to simulate the minimum path of species passing through different types of spatial resistance from ecological sources [69]. The model has been improved to identify the ecological corridor [70]. The equation is as follows [71]:

$$MCR = f_{\min} \sum_{j=n}^{i=m} (D_{ij} \times R_i) \tag{4}$$

where *MCR* is the minimum cumulative resistance, "f" is the positive function relationship between the minimum cumulative resistance and the ecological process, "min" is the minimum cumulative resistance of the evaluated patches to different sources, "$\sum$" is the cumulative value of the distance and resistance between landscape unit "i" and ecological source patch "j" across all units, "D_{ij}" is the spatial distance of species from landscape unit "i" to ecological source patch "j", and "R_i" represents the resistance coefficient of the landscape unit "i" to the movement of a certain species.

3.6. Evaluation of the Ecological Network Index

An ecological network is formed by connecting ecological sources with corridors and various ecological function nodes, and its quality can be evaluated via landscape connectivity. In this paper, we selected the network closure index (α index), the network connectivity index (β index), and the network connectivity rate (γ index) to analyze the ecological corridor network structure and to analyze and evaluate the network closure and connectivity of the Chongqing ecological network [72], with the aim of quantifying the landscape connectivity and complexity of Chongqing. It can not only provide a scientific basis for the construction of the Chongqing ecological network but also represents a reference for the further optimization of this network. The equations for the above three indices are as follows [73]:

$$\alpha = \frac{L - V + 1}{2V - 5} \tag{5}$$

$$\beta = \frac{L}{V} \tag{6}$$

$$\gamma = \frac{L}{3\,(V - 2)} \tag{7}$$

where "L" represents the number of corridors, "V" is the number of nodes, "$L - V + 1$" is the actual number of loops, "$2V - 5$" is the maximum number of possible loops, and "$3\,(V - 2)$" represents the maximum number of possible corridors in the network.

4. Results and Discussion

4.1. Landscape Pattern Index

According to the land use data of Chongqing for 2005 and 2015 (Figure 3a,b), the values of each landscape pattern index in Chongqing were obtained (Table 2). Furthermore, the landscape pattern index was evaluated and compared, and the evolution trend of land use types in Chongqing from 2005 to 2015 was obtained.

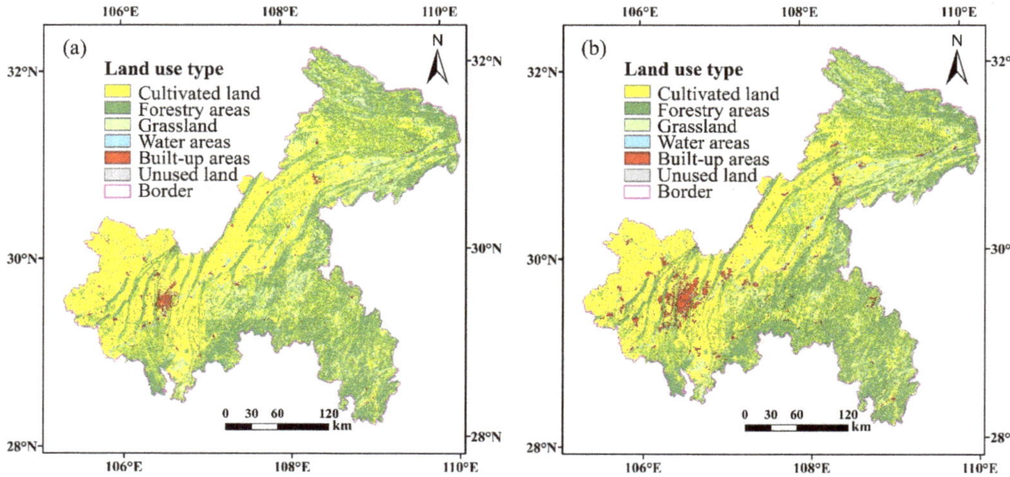

Figure 3. Land use types in Chongqing in 2005 (**a**) and 2015 (**b**).

Table 2. Calculation results of the landscape pattern index in Chongqing for 2005 and 2015.

Landscape Index	Year	Cultivated Land	Forestland	Grassland	Water Areas	Built-Up Areas	Unused Land
PLAND (100%)	2005	46.49	40.45	11.00	1.16	0.88	0.02
	2015	45.61	41.01	9.54	1.45	2.38	0.01
PD (n/ha)	2005	0.31	0.15	0.08	0.01	0.02	0.00
	2015	0.30	0.15	0.08	0.01	0.04	0.00
COHESION	2005	99.83	99.95	99.47	99.77	98.32	93.23
	2015	99.84	99.95	99.10	99.81	98.84	93.01
DIVISION	2005	0.98	0.98	1.00	1.00	1.00	1.00
	2015	0.98	0.97	1.00	1.00	1.00	1.00
AI	2005	94.84	95.08	92.66	93.73	93.34	87.19
	2015	94.87	95.15	92.59	94.09	95.21	86.64

Notes: "PLAND" represents landscape percentage; "PD" represents patch density; "COHESION" represents the cohesion index; "DIVISION" represents the landscape separation index; "AI" represents the aggregation index; "n/ha" represents the number of landscape patches per 100 hectares.

Among the land use types of Chongqing in 2005 (Table 2), the proportions of cultivated land and forested land both exceeded 40%, and the proportion of unused land was lowest with only 0.02%. This indicates that the cultivated land area and the forested area in Chongqing were larger in 2005. The patch density of cultivated land and forestry areas in Chongqing was higher, and the patch density of the water area and unused land was lower, indicating a high degree of fragmentation for cultivated and forest areas and a low degree for water areas and unused land. The patch cohesion index of all landscape types was above 90, and landscape connectivity was high. For cultivated and forested land, the landscape separation degree was 0.98, and that of other landscape types was 1, indicating serious landscape segmentation. The aggregation index of unused land was 87.19 and that of other landscape types was above 90, indicating a high connectivity, except for unused land.

Compared to 2005, in 2015, the proportions of forest land area, water area, and built-up areas were higher. This increase was most significant for built-up areas with 1.5 percentage points, reflecting the large-scale expansion of built-up areas in the process of urbanization. The proportions of cultivated land, grassland, and unused land were lower, with the greatest decline for grassland (1.46%); the cultivated land area only decreased by 0.88%. The cultivated land area in Chongqing was 0.01 n/ha lower in 2015 than in 2005, whereas the built-up area was 0.02 n/ha higher. The patch density of other landscape types did not change considerably, reflecting the occupation and destruction of cultivated land in the process of urbanization in Chongqing over 10 years. The patch cohesion index values of cultivated land, water area, and built-up areas increased slightly, indicating that the connectivity of cultivated land, water area, and built-up areas in Chongqing was high during this period. Affected by human activities, the connectivity of grassland and unused land reduced. When comparing the landscape separation index of each landscape type, the division of landscape types in Chongqing was not obvious during the 10 years. Except for grassland and unused land, the aggregation index values of the other landscape types increased, which was most significant for the built-up areas.

When comparing the land use types in 2005 and 2015, cultivated land was the dominant landscape type in both years. Cultivated land fragmentation increased over time, and the connectivity between forested areas decreased. As the ecological network has the function of repairing the ecosystem, these changes show that the urgency of constructing the ecological network in Chongqing is more apparent.

4.2. Ecological Source Selection

4.2.1. Identification of Core Ecological Patches

Based on the generated binary grid, the foreground area was 42,973 km^2, accounting for 52.03% of the total landscape area. The MSPA showed that, for the seven landscape types generated (Figure 4), the proportions of all kinds of landscapes followed the order bridge (50.69%), core area (16.70%), edge (14.35%), branch (11.36%), isolated island (5.68%), loop (1.02%), and perforation (0.20%). Among them, the bridge area is the current corridor in the region, which is conducive to the spread of species and to energy flows. The area of the bridge area was 21,781 km^2, accounting for 50.69% of the total foreground area. The number of ecological corridors in Chongqing is larger, facilitating the construction of ecological corridors in Chongqing. The core area, which plays a key role in the level of ecological network connectivity, has a large scale, with an area of 7177 km^2, accounting for 16.70% of the total area of the foreground land category. Compared with other related studies [65,74], the core area is relatively small, most likely because of the specific natural geographical conditions of Chongqing, with a more fragmented landscape and a smaller core area. However, the bridge area of Chongqing is larger than that in other cities, which is more conducive to the construction of ecological corridors.

In addition, the patches of core area are unevenly distributed, with most of its area being in the northeastern and the southeastern parts of Chongqing. Most likely, this is because of the large nature reserves such as Daba Mountain, Yintiaoling Mountain, and Wushan in the northeast and the large landscapes such as Jinfo Mountain, Black Valley, and Qingxi Gou Reservoir in the southeast. The edge area is the transition area between the core area and the background landscape and can reduce the interference of external factors, with an area of 6167 km^2, accounting for 14.35% of the total foreground area. The branch also has a certain connectivity, with an area of 4881 km^2, accounting for 11.36% of the total foreground area. The isolated islands are mainly small patches distributed inside the building land, with an area of 2441 km^2, accounting for 5.68% of the total foreground area. The loop is a shortcut for species movement within the same core area, covering an area of 438 km^2 and accounting for 1.02% of the total foreground area. The perforation with the same edge effect is the inner edge of the core area, with an area of only 88 km^2, accounting for 0.20% of the prospects. In general, Chongqing fulfills the requirements for the construction of an ecological network, and further analyses of the core area patches are crucial to screen out ecological sources.

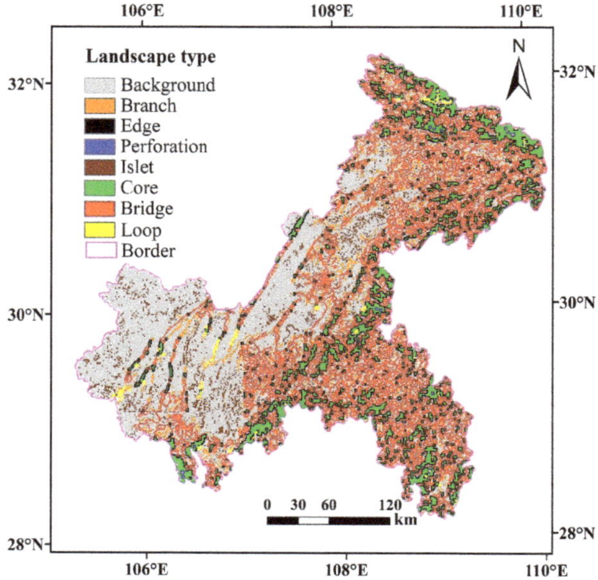

Figure 4. Landscape type map of Chongqing generated via Morphological Spatial Pattern Analysis.

4.2.2. Identification of Ecological Sources

The important value of connectivity (*dPC*) of the core patch generated by MSPA shows that the greater the *dPC* values of the core patch, the greater the contribution of the patch to the overall landscape connectivity. The ecological network constructed in this paper should not only play a macro-role in maintaining regional ecological security, but also guide the layout of land use in ecological construction. Overall, we obtained 24 source plots (Figure 5).

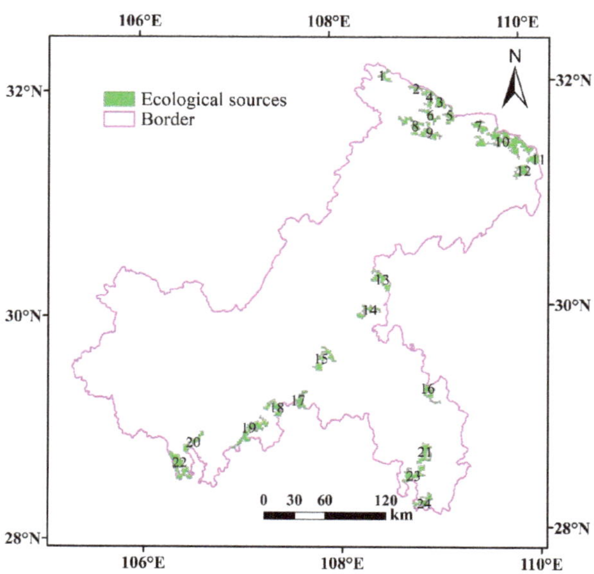

Figure 5. Distribution of ecological sources in Chongqing.

We selected 24 core patches with high connectivity as ecological sources. These patches not only have large area but can also protect the biodiversity and strengthen the landscape connectivity. The *IIC* value, the *PC* value, and *dPC* value of the 24 ecological sources were calculated according to Equations (1)–(3); the total area was 2961 km^2, accounting for 68.90% of the total foreground area. The largest ecological source was source No. 10, with an area of 447 km^2. The *IIC* and *PC* values were also the largest, indicating that the connectivity between them and other ecological sources is the strongest and plays a key role in the ecological process.

4.3. Resistance Surface Analysis

Based on all data types for Chongqing, by the construction method of each front resistance surface, the landscape resistance surface map, the comprehensive resistance surface map, and the accumulative resistance surface map were obtained (Figure 6), and analysis and evaluation were carried out.

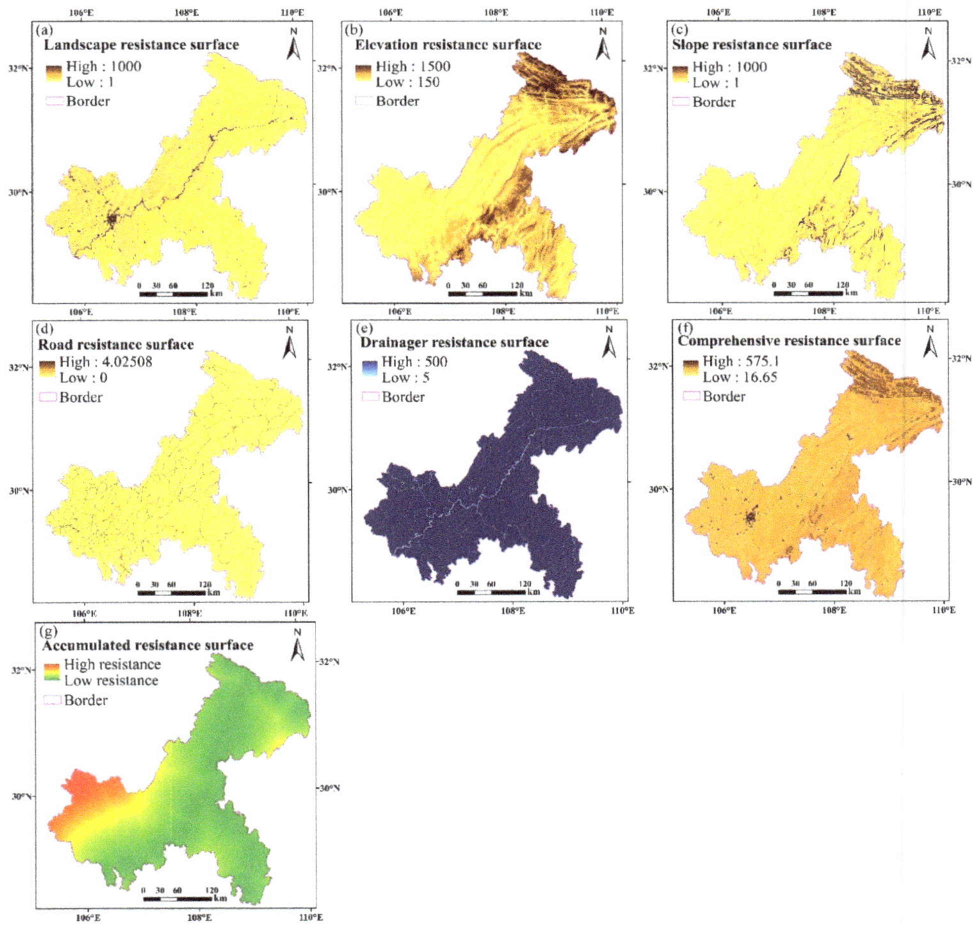

Figure 6. Landscape resistance surface (**a**), elevation resistance surface (**b**), slope resistance surface (**c**), road resistance surface (**d**), river resistance surface (**e**), comprehensive resistance surface (**f**), and cumulative resistance surface (**g**) obtained for Chongqing.

Based on Figure 6a, the landscape types of Chongqing are rich and diverse. The resistance value of the western main urban area is high, and the resistance value of species movement in this area is high, followed by the central part of the area and the northeast of Chongqing. The resistance values of other areas with rich forest and grassland resources are low, and the resistance value of species movement or migration in these areas is also low. The spatial distribution of elevation and slope resistance values in Chongqing is similar. The resistance values in the northeast and southeast of Chongqing are higher, along with those of species migration, whereas the values for other areas are relatively low, indicating easier species migration. Based on the road resistance surface (Figure 6d), the resistance value is higher in the southwest of Chongqing, where the road network is dense and it is difficult for species to move or migrate, whereas the resistance value in the southeast is smaller and the road network is less distributed, which is conducive to species movement or migration. According to the river resistance surface, in the southeast, species movement and migration are easier. Finally, according to the comprehensive resistance surface (Figure 6f), the resistance values of the southwestern and northern parts of Chongqing are larger, impeding species movement.

In general, the accumulative resistance surface of Chongqing is lower in the south and east, higher in the southwest, and unevenly distributed in the north. Considering that, generally, species choose paths with lower resistance for movement and migration, the cumulative resistance surface provides a data basis for the construction of an ecological networks.

4.4. Construction of Potential Ecological Corridors in Chongqing

4.4.1. Identification of Potential Ecological Corridors in Chongqing

Based on the ArcGIS software, the corresponding parameters were set up by the Linkage Mapper software to construct the potential ecological corridors of Chongqing (Figure 7a). Overall, there are 87 potential corridors in Chongqing, with a total length of 2524.34 km and an average length of 29.02 km. Spatially, the ecological corridors are mainly concentrated in the northern and southeastern areas, connecting 24 ecological patches.

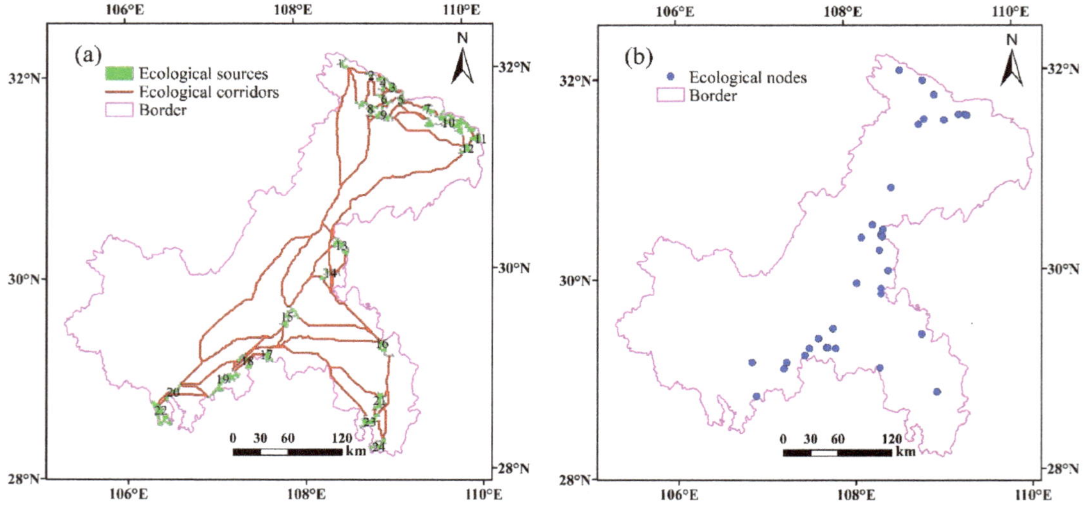

Figure 7. Potential ecological corridor (**a**) and ecological node distribution (**b**) in Chongqing.

4.4.2. Ecological Node Identification

Ecological node quantity, quality, and spatial distribution affect the efficiency of species activity or migration. The results of ecological node extraction in Chongqing are shown in Figure 7b above. There are 35 ecological nodes in Chongqing, mainly distributed in the northeast and south, with an uneven distribution. Therefore, based on the particular geographical location of Chongqing, it is necessary to improve the stability of regional ecological nodes and promote their coordination, ensuring that each node can adequately be protected and restored.

4.5. Construction of Chongqing's Ecological Network

4.5.1. Identification of Chongqing's Ecological Network

The ecological network of Chongqing was constructed by overlaying ecological sources, ecological corridors, and ecological nodes (Figure 8). The distribution of ecological corridors and ecological nodes in the north and southeast of Chongqing is relatively dense, and the large core areas in the southwest consist of only a few patches. The rapid urbanization of Chongqing has brought great pressure to the ecological environment, and the landscape is more fragmented. The ecological network provides the best path choice for species movement or migration, avoiding risks and maintaining biodiversity to a large extent.

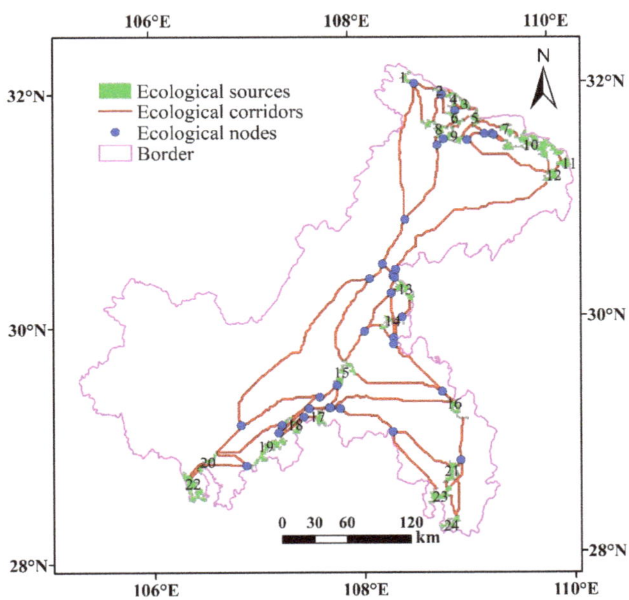

Figure 8. Chongqing ecological network.

The unique geographical conditions of mountainous areas challenge species survival and migration. However, with the development of urbanization worldwide, the destruction of mountainous ecosystems is unstoppable [73]. As a country with a significant amount of mountainous terrain, China needs to pay more attention to the protection of mountain environments. In recent years, the deterioration of the ecological environment in Chongqing has also attracted the attention of the Chinese government and scholars [75]. The establishment of ecological networks as an effective means to maintain biodiversity should also be considered. In addition, the construction of ecological networks requires financial and material support. On the basis of existing nature reserves, the Chinese government should increase investments in the construction of ecological networks in Chongqing, especially in the southwest, and the following measures should be taken: generating urban green spaces and other green landscapes, formulating relevant policies, strictly managing urban industrial development, and scientifically planning built-up areas.

4.5.2. Ecological Network Index Analysis

The ecological network in Chongqing contains 87 ecological corridors and 24 ecological sources. By calculating the α index, the β index, and the γ index, the structural characteristic index of the ecological network in the study area was obtained (Table 3), and the constructed ecological network was evaluated and analyzed.

Table 3. Characteristics of the Chongqing ecological network.

	Index	L	V	L − V + 1	2V − 5	3 (V − 2)	Results
Ecological Network	α index	87	35	53	65		0.82
	β index	87	35				2.49
	γ index	87	35			99	0.89

The value of the α index ranges from 0 to 1, with larger values indicating a higher number of loops in the network, a higher accessibility of the ecological flows in the network, and a higher number of paths that can be selected for migration. For the ecological network constructed in this paper, the actual number of loops is 53, the maximum possible number of loops is 65, and the value of the α index is 0.82, indicating a high connectivity of the network. Species flow or migration and other ecological processes are extremely smooth. However, the distribution of network circuits in Chongqing is uneven, and the network circuits composed of ecological sources and ecological corridors in the central and southwestern regions are less. The construction of ecological sources needs to be increased in this part of the region to strengthen the network circuits, providing more path choices for species migration.

The value of the β index ranges between 0 and 3; the larger the value, the higher the complexity of the network. Here, the β index is 2.49, which indicates that the ecological network is more complex as a whole, and the interaction among ecological sources in the network is facilitated.

The value of the γ index ranges between 0 and 1, and the larger the value, the higher the connection degree of the ecological sources. In this paper, this value is 0.89, indicating that the ecological corridor in the ecological network connects more ecological sources and that the network efficiency is better. However, due to the small number of ecological sources and ecological corridors in the southwest of Chongqing, species movement and migration are impeded. Therefore, to promote the diffusion of ecological flows in Chongqing and improve biodiversity protection, corridor connection in the southwest should be strengthened.

5. Conclusions

Based on land use data for Chongqing, China, from 2005 and 2015, we used MSPA and the Conefor2.6 software to identify 24 important ecological sources. Based on the *MCR* models and the Linkage Mapper software, 87 potential ecological corridors and 35 ecological nodes were constructed. Finally, the ecological sources, corridors, and nodes were superimposed, and Chongqing ecological network was generated. The total length of the potential ecological corridor is 2524.34 km, with an average corridor length off 29.02 km. The α index, β index, and γ index values indicate a good connectivity between the selected ecological sources, with a high network efficiency.

Against the background of topography and urbanization, the number of ecological sources in Chongqing is small and the spatial distribution is uneven. They are mainly distributed in the northeast and southeast, with fewer sources in the southwest. The spatial distribution of ecological nodes is similar to that of ecological sources. The spatial distribution of potential ecological corridors is not balanced, as they are concentrated in the northeast and southeast, indicating a poor efficiency of the ecological network.

Our results provide a scientific basis for the protection of Chongqing's ecological environment. Regarding the southwestern part of Chongqing, it is necessary to strengthen the connectivity between the sources, thereby further improving the ecological network. In addition, the parameter setting of the resistance value is relatively subjective. Based on the current land use data for Chongqing, economic factors should be considered, and the effective resistance surface data for landscape, elevation, slope, roads, and rivers should be further combined.

Author Contributions: Conceptualization, W.S.; methodology, D.Z. and W.S.; formal analysis, D.Z. and W.S.; investigation, W.S. and D.Z.; resources, W.S.; writing—original draft preparation, D.Z. and W.S.; writing—review and editing, W.S.; supervision, W.S. All authors have read and agreed to the published version of the manuscript.

Funding: This research was supported by the Projects of National Natural Science Foundation of China (grant number 42071233), the Second Tibetan Plateau Scientific Expedition and Research (grant number 2019QZKK0603), and the Strategic Priority Research of the Chinese Academy of Sciences (grant number XDA20040201).

Institutional Review Board Statement: Not applicable.

Informed Consent Statement: Not applicable.

Data Availability Statement: All relevant data sets in this study are described in the manuscript.

Conflicts of Interest: The authors declare no conflict of interest. The funders played no role in the design of the study; in the collection, analyses, or interpretation of data; in the writing of the manuscript; or in the decision to publish the results.

References

1. Song, W.; Pijanowski, B.C. The effects of China's cultivated land balance program on potential land productivity at a national scale. *Appl. Geogr.* **2014**, *46*, 158–170. [CrossRef]
2. Li, F.; Ye, Y.; Song, B.; Wang, R. Evaluation of urban suitable ecological land based on the minimum cumulative resistance model: A case study from Changzhou, China. *Ecol. Model.* **2015**, *318*, 194–203. [CrossRef]
3. Su, Y.; Chen, X.; Liao, J.; Zhang, H.; Wang, C.; Ye, Y.; Wang, Y. Modeling the optimal ecological security pattern for guiding the urban constructed land expansions. *Urban For. Urban Green.* **2016**, *19*, 35–46. [CrossRef]
4. Carlier, J.; Moran, J. Landscape typology and ecological connectivity assessment to inform Greenway design. *Sci. Total Environ.* **2019**, *651*, 3241–3252. [CrossRef]
5. Song, W.; Deng, X. Land-use/land-cover change and ecosystem service provision in China. *Sci. Total Environ.* **2017**, *576*, 705–719. [CrossRef]
6. Troumbis, A.Y.; Zevgolis, Y. Biodiversity crime and economic crisis: Hidden mechanisms of misuse of ecosystem goods in Greece. *Land Use Policy* **2020**, *99*. [CrossRef]
7. Hugé, J.; de Bisthoven, L.J.; Mushiete, M.; Rochette, A.-J.; Candido, S.; Keunen, H.; Dahdouh-Guebas, F.; Koedam, N.; Vanhove, M.P.M. EIA-driven biodiversity mainstreaming in development cooperation: Confronting expectations and practice in the DR Congo. *Environ. Sci. Policy* **2020**, *104*, 107–120. [CrossRef]
8. Carranza, D.M.; Varas-Belemmi, K.; De Veer, D.; Iglesias-Müller, C.; Coral-Santacruz, D.; Méndez, F.A.; Torres-Lagos, E.; Squeo, F.A.; Gaymer, C.F. Socio-environmental conflicts: An underestimated threat to biodiversity conservation in Chile. *Environ. Sci. Policy* **2020**, *110*, 46–59. [CrossRef]
9. Wang, W.; Feng, C.; Liu, F.; Li, J. Biodiversity conservation in China: A review of recent studies and practices. *Environ. Sci. Ecotechnology* **2020**, *2*, 100025. [CrossRef]
10. McNeely, J.A. The Convention on Biological Diversity. In *Encyclopedia of the Anthropocene*; Elsevier Inc.: London, UK, 2018; pp. 321–326. [CrossRef]
11. Chu, E.W.; Karr, J.R. Environmental Impact: Concept, Consequences, Measurement. In *Reference Module in Life Sciences*; Elsevier BV: Amsterdam, The Netherlands, 2017. [CrossRef]
12. Wu, X.; Zhang, J.; Geng, X.; Wang, T.; Wang, K.; Liu, S. Increasing green infrastructure-based ecological resilience in urban systems: A perspective from locating ecological and disturbance sources in a resource-based city. *Sustain. Cities Soc.* **2020**, *61*. [CrossRef]
13. Jongman, R. Ecological Networks are an Issue for All of US. *J. Landsc. Ecol.* **2008**. [CrossRef]
14. Linehan, J.; Gross, M.; Finn, J. Greenway planning: Developing a landscape ecological network approach. *Landsc. Urban Plan.* **1995**, *33*. [CrossRef]

15. An, Y.; Liu, S.; Sun, Y.; Shi, F.; Beazley, R. Construction and optimization of an ecological network based on morphological spatial pattern analysis and circuit theory. *Landsc. Ecol.* **2020**, 1–18. [CrossRef]
16. De Montis, A.; Ganciu, A.; Cabras, M.; Bardi, A.; Peddio, V.; Caschili, S.; Massa, P.; Cocco, C.; Mulas, M. Resilient ecological networks: A comparative approach. *Land Use Policy.* **2019**, *89*, 104207. [CrossRef]
17. Jongman, R.H.G.; Külvik, M.; Kristiansen, I. European ecological networks and greenways. *Landsc. Urban Plan.* **2004**, *68*, 305–319. [CrossRef]
18. Takemoto, K.; Iida, M. Ecological Networks. In *Encyclopedia of Bioinformatics and Computational Biology*; Elsevier BV: Amsterdam, The Netherlands, 2019; pp. 1131–1141.
19. Nor, A.N.M.; Corstanje, R.; Harris, J.A.; Grafius, D.R.; Siriwardena, G.M. Ecological connectivity networks in rapidly expanding cities. *Heliyon* **2017**. [CrossRef]
20. Yu, Q.; Yue, D.; Wang, Y.; Kai, S.; Fang, M.; Ma, H.; Zhang, Q.; Huang, Y. Optimization of ecological node layout and stability analysis of ecological network in desert oasis: A typical case study of ecological fragile zone located at Deng Kou County(Inner Mongolia). *Ecol. Indic.* **2018**, *84*, 304–318. [CrossRef]
21. Hepcan, Ş.; Hepcan, Ç.C.; Bouwma, I.M.; Jongman, R.H.G.; Özkan, M.B. Ecological networks as a new approach for nature conservation in Turkey: A case study of İzmir Province. *Urban Plan.* **2009**, *90*, 143–154. [CrossRef]
22. Closset-Kopp, D.; Wasof, S.; Decocq, G. Using process-based indicator species to evaluate ecological corridors in fragmented landscapes. *Biol. Conserv.* **2016**, *201*, 152–159. [CrossRef]
23. Peng, J.; Zhao, H.; Liu, Y. Urban ecological corridors construction: A review. *Acta Ecol. Sin.* **2017**, *37*, 23–30. [CrossRef]
24. Jongman, R.H.G. Connectivity and Ecological Networks. In *Encyclopedia of Ecology*; Elsevier BV: Amsterdam, The Netherlands, 2019; pp. 366–376. [CrossRef]
25. Shi, F.; Liu, S.; Sun, Y.; An, Y.; Zhao, S.; Liu, Y.; Li, M. Ecological network construction of the heterogeneous agro-pastoral areas in the upper Yellow River basin. *Agric. Ecosyst. Environ.* **2020**, *302*, 107069. [CrossRef]
26. Yang, J.; Zeng, C.; Cheng, Y. Spatial influence of ecological networks on land use intensity. *Sci. Total Environ.* **2020**, *717*, 137151. [CrossRef] [PubMed]
27. Zhang, Y.; Li, X.; Song, W. Determinants of cropland abandonment at the parcel, household and village levels in mountain areas of China: A multi-level analysis. *Land Use Policy* **2014**, *41*, 186–192. [CrossRef]
28. Pili, S.; Serra, P.; Salvati, L. Landscape and the city: Agro-forest systems, land fragmentation and the ecological network in Rome, Italy. *Urban For. Urban Green.* **2019**, *41*, 230–237. [CrossRef]
29. Jordán, F. A reliability-theory approach to corridor design. *Ecol. Model.* **2000**, *128*, 211–220. [CrossRef]
30. Bowers, K.; McKnight, M. Reestablishing a Healthy and Resilient North America—Linking Ecological Restoration with Continental Habitat Connectivity. *Ecol. Restor.* **2012**, *30*, 267–270. [CrossRef]
31. Chettri, N.; Sharma, E.; Shakya, B.; Bajracharya, B. Developing Forested Conservation Corridors in the Kangchenjunga Landscape, Eastern Himalaya. *Mt. Res. Dev.* **2007**, *27*, 211–214. [CrossRef]
32. Su, K.; Yu, Q.; Yue, D.; Zhang, Q.; Yang, L.; Liu, Z.; Niu, T.; Sun, X. Simulation of a forest-grass ecological network in a typical desert oasis based on multiple scenes. *Ecol. Model.* **2019**, *413*. [CrossRef]
33. Aminzadeh, B.; Khansefid, M. A case study of urban ecological networks and a sustainable city: Tehran's metropolitan area. *Urban Ecosyst.* **2009**, *13*, 23–36. [CrossRef]
34. Zhao, X.Q.; Xu, X.H. Research on landscape ecological security pattern in a Eucalyptus introduced region based on biodiversity conservation. *Russ. J. Ecol.* **2015**, *46*, 59–70. [CrossRef]
35. Wang, C.; Yu, C.; Chen, T.; Feng, Z.; Hu, Y.; Wu, K. Can the establishment of ecological security patterns improve ecological protection? An example of Nanchang, China. *Sci. Total Environ.* **2020**, *740*, 140051. [CrossRef] [PubMed]
36. Dong, J.; Peng, J.; Liu, Y.; Qiu, S.; Han, Y. Integrating spatial continuous wavelet transform and kernel density estimation to identify ecological corridors in megacities. *Landsc. Urban Plan.* **2020**, *199*, 103815. [CrossRef]
37. Liu, S.; Dong, Y.; Cheng, F.; Zhang, Y.; Hou, X.; Dong, S.; Coxixo, A. Effects of road network on Asian elephant habitat and connectivity between the nature reserves in Xishuangbanna, Southwest China. *J. Nat. Conserv.* **2017**, *38*, 11–20. [CrossRef]
38. Pomianowski, W.; Solon, J. Modelling patch mosaic connectivity and ecological corridors with GraphScape. *Environ. Model. Softw.* **2020**, *134*, 104757. [CrossRef]
39. Guo, X.; Zhang, X.; Du, S.; Li, C.; Siu, Y.L.; Rong, Y.; Yang, H. The impact of onshore wind power projects on ecological corridors and landscape connectivity in Shanxi, China. *J. Clean. Prod.* **2020**, *254*, 120075. [CrossRef]
40. Peng, J.; Yang, Y.; Liu, Y.; Hu, Y.; Du, Y.; Meersmans, J.; Qiu, S. Linking ecosystem services and circuit theory to identify ecological security patterns. *Sci. Total Environ.* **2018**, *644*, 781–790. [CrossRef]
41. Dai, L.; Liu, Y.; Luo, X. Integrating the MCR and DOI models to construct an ecological security network for the urban agglomeration around Poyang Lake, China. *Sci. Total Environ.* **2021**, *754*, 141868. [CrossRef]
42. Zhang, Y.; Song, W. Identify Ecological Corridors and Build Potential Ecological Networks in Response to Recent Land Cover Changes in Xinjiang, China. *Sustainability* **2020**, *12*, 8960. [CrossRef]
43. Liu, G.; Yang, Z.; Chen, B.; Zhang, L.; Zhang, Y.; Su, M. An Ecological Network Perspective in Improving Reserve Design and Connectivity: A Case Study of Wuyishan Nature Reserve in China. *Ecol. Model.* **2015**, *306*, 185–194. [CrossRef]
44. Peng, L.; Chen, T.; Liu, S. Spatiotemporal Dynamics and Drivers of Farmland Changes in Panxi Mountainous Region, China. *Sustainability* **2016**, *8*, 1209. [CrossRef]

45. Liang, J.; He, X.; Zeng, G.; Zhong, M.; Gao, X.; Li, X.; Li, X.; Wu, H.; Feng, C.; Xing, W.; et al. Integrating priority areas and ecological corridors into national network for conservation planning in China. *Sci. Total Environ.* **2018**, *626*, 22–29. [CrossRef]
46. Liu, Y.; Yue, W.; Fan, P.; Zhang, Z.; Huang, J. Assessing the urban environmental quality of mountainous cities: A case study in Chongqing, China. *Ecol. Indic.* **2017**, *81*, 132–145. [CrossRef]
47. Huang, L.; Qian, S.; Li, T.; Jim, C.Y.; Jin, C.; Zhao, L.; Lin, D.; Shang, K.; Yang, Y. Masonry walls as sieve of urban plant assemblages and refugia of native species in Chongqing, China. *Landsc. Urban Plan.* **2019**, *191*. [CrossRef]
48. Long, H.; Wu, X.; Wang, W.; Dong, G. Analysis of Urban-Rural Land-Use Change during 1995–2006. *Sensors* **2008**, *8*, 681–699. [CrossRef]
49. Lin, Y.; Li, Y.; Ma, Z. Exploring the Interactive Development between Population Urbanization and Land Urbanization: Evidence from Chongqing, China (1998–2016). *Sustainability* **2018**, *10*, 1741. [CrossRef]
50. Zhang, J.F.; Deng, W. Industrial Structure Change and Its Eco-environmental Influence since the Establishment of Municipality in Chongqing, China. *Procedia Environ. Sci.* **2010**, *2*, 517–526. [CrossRef]
51. Resource Environmental Science and Data Center, Chinese Academy of Sciences. Available online: http://www.resdc.cn/Default.aspx (accessed on 22 November 2020).
52. Geospatial Data Cloud Homepage. Availabe online: http://www.gscloud.cn/sources/ (accessed on 22 November 2020).
53. Zhang, Q.; Chen, C.; Wang, J.; Yang, D.; Zhang, Y.; Wang, Z.; Gao, M. The spatial granularity effect, changing landscape patterns, and suitable landscape metrics in the Three Gorges Reservoir Area, 1995–2015. *Ecol. Indic.* **2020**, *114*, 106259. [CrossRef]
54. Li, H.; Peng, J.; Yanxu, L.; Yi'na, H. Urbanization impact on landscape patterns in Beijing City, China: A spatial heterogeneity perspective. *Ecol. Indic.* **2017**, *82*, 50–60. [CrossRef]
55. Availabe online: http://www.umass.edu/landeco/research/fragstats/documents/fragstats.help.4.2.pdf (accessed on 22 November 2020).
56. Zhang, J.; Qu, M.; Wang, C.; Zhao, J.; Cao, Y. Quantifying landscape pattern and ecosystem service value changes: A case study at the county level in the Chinese Loess Plateau. *Glob. Ecol. Conserv.* **2020**, *23*, e01110. [CrossRef]
57. Vogt, P.; Riitters, K.H.; Iwanowski, M.; Estreguil, C.; Kozak, J.; Soille, P. Mapping landscape corridors. *Ecol. Indic.* **2007**, *7*, 481–488. [CrossRef]
58. Vogt, P.; Riitters, K.H.; Estreguil, C.; Kozak, J.; Wade, T.G.; Wickham, J.D. Mapping Spatial Patterns with Morphological Image Processing. *Landsc. Ecol.* **2006**, *22*, 171–177. [CrossRef]
59. Wang, J.; Xu, C.; Pauleit, S.; Kindler, A.; Banzhaf, E. Spatial patterns of urban green infrastructure for equity: A novel exploration. *J. Clean. Prod.* **2019**, *238*, 238. [CrossRef]
60. Zhang, Y.; Shen, W.; Li, M.; Lv, Y. Assessing spatio-temporal changes in forest cover and fragmentation under urban expansion in Nanjing, eastern China, from long-term Landsat observations (1987–2017). *Appl. Geogr.* **2020**, *117*, 102190. [CrossRef]
61. Cui, L.; Wang, J.; Sun, L.; Lv, C. Construction and optimization of green space ecological networks in urban fringe areas: A case study with the urban fringe area of Tongzhou district in Beijing. *J. Clean. Prod.* **2020**, *276*, 124266. [CrossRef]
62. Rosot, M.A.D.; Maran, J.C.; de Luz, N.B.; Garrastazú, M.C.; de Oliveira, Y.M.M.; Franciscon, L.; Clerici, N.; Vogt, P.; de Freitas, J.V. Riparian forest corridors: A prioritization analysis to the Landscape Sample Units of the Brazilian National Forest Inventory. *Ecol. Indic.* **2018**, *93*, 501–511. [CrossRef]
63. Hernando, A.; Velázquez, J.; Valbuena, R.; Legrand, M.; García-Abril, A. Influence of the resolution of forest cover maps in evaluating fragmentation and connectivity to assess habitat conservation status. *Ecol. Indic.* **2017**, *79*, 295–302. [CrossRef]
64. Saura, S.; Pascual-Hortal, L. A new habitat availability index to integrate connectivity in landscape conservation planning: Comparison with existing indices and application to a case study. *Landsc. Urban Plan.* **2007**, *83*, 91–103. [CrossRef]
65. Ye, H.; Yang, Z.; Xu, X. Ecological Corridors Analysis Based on MSPA and MCR Model—A Case Study of the Tomur World Natural Heritage Region. *Sustainability* **2020**, *12*, 959. [CrossRef]
66. Pascual-Hortal, L.; Saura, S. Comparison and development of new graph-based landscape connectivity indices: Towards the priorization of habitat patches and corridors for conservation. *Landsc. Ecol.* **2006**, *21*, 959–967. [CrossRef]
67. Zeller, K.A.; McGarigal, K.; Whiteley, A.R. Estimating landscape resistance to movement: A review. *Landsc. Ecol.* **2012**, *27*, 777–797. [CrossRef]
68. Jewitt, D.; Goodman, P.S.; Erasmus, B.F.; O'Connor, T.G.; Witkowski, E.T. Planning for the Maintenance of Floristic Diversity in the Face of Land Cover and Climate Change. *Environ. Manag.* **2017**, *59*, 792–806. [CrossRef] [PubMed]
69. Jan, P.K.; Marten, S.; Bert, H. Estimating habitat isolation in landscape planning. *Landsc. Urban Plan.* **1992**, *23*, 1–16.
70. Yu, K.J. Security patterns and surface model in landscape ecological planning. *Landsc. Urban Plan.* **1996**, *36*, 1–17. [CrossRef]
71. Shi, H.; Shi, T.; Yang, Z.; Wang, Z.; Han, F.; Wang, C. Effect of Roads on Ecological Corridors Used for Wildlife Movement in a Natural Heritage Site. *Sustainability* **2018**, *10*, 2725. [CrossRef]
72. Li, H.; Chen, W.; He, W. Planning of Green Space Ecological Network in Urban Areas: An Example of Nanchang, China. *Int. J. Environ. Res. Public Health* **2015**, *12*, 12889–12904. [CrossRef]
73. Grêt-Regamey, A.; Weibel, B. Global assessment of mountain ecosystem services using earth observation data. *Ecosyst. Serv.* **2020**, *46*, 101213. [CrossRef]
74. Miao, Z.; Pan, L.; Wang, Q.; Chen, P.; Yan, C.; Liu, L. Research on Urban Ecological Network Under the Threat of Road Networks—A Case Study of Wuhan. *ISPRS Int. Geo-Inf.* **2019**, *8*, 342. [CrossRef]
75. Yang, C.; Zeng, W.; Yang, X. Coupling coordination evaluation and sustainable development pattern of geo-ecological environment and urbanization in Chongqing municipality, China. *Sustain. Cities Soc.* **2020**, *61*, 102271. [CrossRef]

International Journal of
Environmental Research and Public Health

MDPI

Review

Status, Trend, and Prospect of Global Farmland Abandonment Research: A Bibliometric Analysis

Bo Liu [1,2], Wei Song [1,3,*] and Qian Sun [4]

1 Key Laboratory of Land Surface Pattern and Simulation, Institute of Geographic Sciences and Natural Resources Research, Chinese Academy of Sciences, Beijing 100101, China
2 School of Geomatics, Liaoning Technical University, Fuxin 123000, China
3 Hebei Collaborative Innovation Center for Urban-Rural Integration Development, Shijiazhuang 050061, China
4 National Research Center for Geoanalysis, China Geological Survey, Beijing 100037, China
* Correspondence: songw@igsnrr.ac.cn

Abstract: Farmland abandonment is one of the most important land use changes in the world today and crucial to the sustainable development of the global environment. The authors carried out extensive research on farmland abandonment from many perspectives, but, due to the variety of the research contents, rich research perspectives, and complex research objects, the current research in this field lacks comprehensiveness, objectivity, and systematization. In this study, the bibliometric R software packages bibliometrix and biblioshiny (K-Synth Srl, Naples, Italy) were used to analyze the development history and current situation of 896 articles on farmland abandonment in the Web of Science core collection database from 1980 to 2021, revealing their research hotspots and predicting the future development trends. Over the past 40 years, the number of published papers on abandoned farmland has continuously increased. Research mainly focused on the ecological environment, with natural succession, biodiversity, and vegetation restoration being high-frequency keywords in this field. Research on the social aspects of farmland abandonment has developed rapidly in the past 6 years. Based on these findings, this paper put forward four future research directions: the data source for the extraction of abandoned farmland should transform to high spatial-temporal resolution and hyperspectral remote sensing images; the method should pay more attention to the time series change detection and the application of the model; future research should focus on the economic costs of the reclamation of abandoned farmland and the ecological consequences of such reclamation; and the global ecological impact of vegetation succession after the abandonment of farmland should be further discussed from a broader perspective.

Keywords: farmland abandonment; research progress; theme evolution; biodiversity; reclamation; bibliometric analysis

Citation: Liu, B.; Song, W.; Sun, Q. Status, Trend, and Prospect of Global Farmland Abandonment Research: A Bibliometric Analysis. *Int. J. Environ. Res. Public Health* **2022**, *19*, 16007. https://doi.org/10.3390/ijerph192316007

Academic Editor: Lingxin Chen

Received: 8 November 2022
Accepted: 28 November 2022
Published: 30 November 2022

1. Introduction

Land use change seriously affects the sustainable development of the global environment [1,2], especially in terms of ecosystem services and biodiversity [3]. Globally, farmland abandonment is one of the major land use changes today [4,5] and a result of the marginalization of farmland driven by social, economic, and environmental factors [6,7]. The main consequences of farmland abandonment are vegetation succession, environmental problems, as well as landscape and socio-economic impacts, with global impacts. Extensive farmland abandonment in one area may lead to large-scale grain imports, often resulting in deforestation [8] in other countries, with effects on ecology, goods, production, and services [9,10]. Recently, farmland abandonment has attracted wide attention from scholars, organizations, research institutions, and the public.

The term "farmland abandonment" has different definitions, depending on the method and content of the study [7]. The Joint Research Centre (JRC) defined farmland abandonment as changes in ecosystem services due to a significant reduction in farmland management. In some studies, farmland abandonment was defined as the cessation of agricultural activity on a given land surface and the subsequent natural restoration of vegetation to grassland [11], shrub vegetation [12], or forest [13], rather than its occupation by other activities (e.g., urbanization or afforestation) [14]. Some studies classified the conversion to forest into a special category of abandonment [15]. At a temporal scale, some studies defined abandoned farmland as farmland left idle for more than 1 year [7]. However, according to the Food and Agriculture Organization (FAO), an idle period of 2 to 5 years must be guaranteed to define farmland as abandoned [16]. Fallows have been assessed as part of a crop rotation cycle to determine whether a plot of land was abandoned or awaiting future use [17]. Some studies distinguished abandoned farmland as a long-term fallow from temporary fallow (defined as farmland that was not cultivated within 1 year) [18].

Previous studies reviewed numerous aspects of research in the field of farmland abandonment. For example, the impact of farmland abandonment on the ecosystem and biodiversity [19] could be regarded as positive or negative [20]. Regarding the positive aspects, farmland abandonment is conducive to the regional biodiversity [21–23], increasing the soil organic carbon and vegetation water retention capacity [24,25], and promoting reforestation [26]. However, early succession vegetation after farmland abandonment provides fuel for wildfires [27] and increases the reproductive fitness of weeds, pests, and pathogens on the remaining farmland [28]. Farmland abandonment may also lead to the marginalization of the agricultural cultural landscape, causing the loss of cultural and aesthetic value [29,30]. These effects vary with the geographic area distribution and abandonment period. For example, in Spain, farmland abandonment led to the loss of bird habitat, reduced biodiversity [31], decreased the soil erosion potential within 50 years of farmland abandonment, and increased the incidence of soil erosion within 80–100 years of abandonment [32]. Against the background of food security and the local demand for agricultural products, in some areas, abandoned farmland is being reclaimed. Some examples are the reclamation of some areas after the disintegration of the former Soviet Union [24,33] and the cultivation of formerly abandoned farmland in mountainous areas of China [34,35]. To better understand the ecological consequences and reclamation potential of farmland abandonment, the spatial and temporal patterns and drivers of farmland abandonment distribution have been investigated [36–38]. Numerous factors can affect the distribution of abandoned farmland, including natural conditions (such as climate [39], terrain [40], soil [41]), market demand, traffic location conditions, urbanization, and policy [42–44].

Generally, literature reviews on farmland abandonment only cover a certain country or region [45,46], thereby lacking a global perspective. Moreover, they largely focus on one aspect of farmland abandonment, failing to outline the current situation or to predict the future development direction. Such an approach limits the ability of researchers to investigate farmland abandonment from different viewpoints and perspectives.

Bibliometrics is an alternative approach to analyze the distribution structure, research topics, and changing trends of a large number of academic publications via mathematical and statistical methods and visualizes the results through visual and intuitive charts [47,48]. Compared with classical review writing, researchers do not need to screen large amounts of literature one by one, which is more convenient, objective, and reliable. The commonly used bibliometric analysis tools include HistCite [49], SATI [50], and CiteSpace [51]. Bibliometrix is an open-source tool developed by Massimo Aria in the R language in 2017. Compared to other bibliometric software packages, such as CITAN [52] and ScientoText, bibliometrix can import and transform data from various database sources, such as the Web of Science, Scopus, Dimensions, and Lens.org, and has more literature information analysis and result visualization functions [53]. Based on the secondary development of bibliometrix, the entire bibliometric steps were assembled into an automated online workflow of biblioshiny.

The difference between the two packages is that the bibliometrix's mode of operation consisted of code commands, whereas biblioshiny uses the Shiny package to package the bibliometrix's core code and create a web-based online data analysis framework, making it fully available for researchers without any programming skills. It can provide multiple types of statistical methods and a wide variety of visual charts [54].

Bibliometrix has been widely used to quantitatively analyze the literature in geography-related fields. For example, Soler et al. [55] retrieved 1150 publications on rural population reduction, published in the Web of Science and Scopus databases from 1979 to 2018, and used co-word analysis and the co-citation network to conclude that important themes of rural depopulation were relevant to specific geographic areas. Xie et al. [54] conducted data mining and quantitative analysis by retrieving publications in the field of land degradation from 1990 to 2019 from the core collection database of the Web of Science, revealing the research status of global land degradation and evaluating the direction of future research on land degradation. However, publications in the field of farmland abandonment have rarely been reviewed via bibliometrics.

Here, we used the bibliometrix and biblioshiny R language packages to comprehensively analyze the literature in the field of farmland abandonment, considering studies published in the Web of Science core database between 1980 and 2021. First, the development process, current situation, and trend of the research period in the field were identified through the statistical analysis of the annual output of related publications, citations, journal sources, and the main research countries and regions. In addition, collaborative network analysis and clustering of authors, keywords, and topics determined the research hotspots and future frontiers in this field. Based on the above analysis, the problems to be addressed are as follows: (1) What is the trend of article production and citation in the field of farmland abandonment? (2) What are the fluctuation trends regarding the authors, journals, and countries in the field of farmland abandonment? (3) What are the main research directions in the field of farmland abandonment? (4) What is the future research focus in the field of farmland abandonment?

2. Data Sources and Methods

2.1. Data Sources

To improve the representativeness and usability of the data, the data were collected from the Web of Science core collection database [56], which is the largest and one of the most comprehensive bibliometric analysis databases in the world, including data from natural sciences, computer sciences, biology, arts, humanities, and other fields from more than 12,000 high-impact academic journals. By referring to relevant articles in the field of farmland abandonment, we identified the most commonly used keyword "farmland abandonment" [35,57–60], and the search formula was as follows: T1 = (cropland abandonment) or T2 = (abandoned cropland) or T3 = (farmland abandonment) or T4 = (abandoned farmland) or T5 = (agricultural land abandonment) or T6 = (abandoned agricultural land) or T7 = (cultivated land abandonment) or T8 = (abandoned cultivated land) or T9 = (land abandonment) or T10 = (abandoned land). We limited the publication years to 1980–2021 as some articles will still be published in 2022, whereas the number of articles published before 1980 was extremely small. Overall, 977 publications were found. We focused on journal articles, conference transcript articles, and reviews, excluding comments, news, and letters. After filtering, 926 criteria-compliant publications were downloaded, exported as complete records, and referenced in the format ".txt". To ensure the completeness and accuracy of the retrieved data, duplicates were manually removed. Finally, a total of 896 publications on farmland abandonment were obtained (Figure 1, Supplementary Table S1). We searched by title rather than by theme, author, abstract, and keywords. In many bibliometric studies, search by title is considered as the most accurate search method [61–63], whereas searching by theme/abstract/keywords will produce many false-positive results (the main concern is not the abandonment of farmland itself). Although searching by title can result in the loss of some documents (false-negatives), there will be more false-positive results when searching by theme. In this study, we searched by title and theme, respectively, resulting in 8603 documents being different between the two scenarios. Sorting according to the global citation rate of articles, manually reviewing the titles, abstracts, and keywords of the top 100 articles with the highest citation rates (Supplementary Tables S2 and S3), we found that only 12% of the articles appeared by a theme search focused on farmland abandonment, and most of them also appeared in the articles searched by title. Overall, 42% of the articles on farmland abandonment appeared as marginal research, and 46% of the articles did not investigate farmland abandonment. Among the articles found by the title search, 72% focused on farmland abandonment, 19% only marginally studied farmland abandonment, and 9% did not carry out research on farmland abandonment. Based on this, it was decided to search by title.

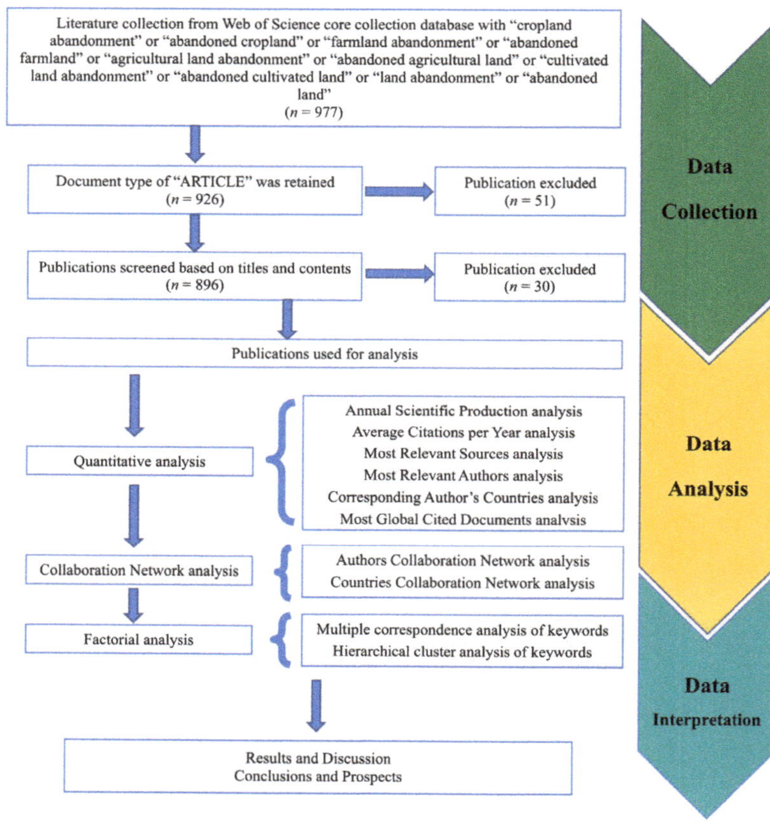

Figure 1. Technical flow of the applied bibliometrics approach.

2.2. Methods

Using the bibliometrix and biblioshiny software packages, the research status and trends in the field of farmland abandonment were reviewed and analyzed. Data analysis included a quantitative analysis of annual scientific production, citations, and journal sources, a collaborative network analysis of authors, countries, and regions, a clustering analysis of document keywords, and the evolution analysis of thematic trends and monitoring methods in abandoned farmland areas during the research period (Figure 1). The present situation and law of research in abandoned farmland were expounded through multiple perspectives to provide a reference and suggestions for future research in this field. The workflow is as follows:

(1) Problems and spans of design research: This paper identified farmland abandonment as the research topic and searched the core database of the Web of Science for articles from 1980 to 2021 by determining search terms, search time, and document types. (2) Data collection: An article database was built [64] by exporting the retrieved documents into the appropriate format of bibliometric tools. (3) Data analysis and visualization: This included co-word analysis [65], collaborator analysis [66], citation analysis [67–69], coupling analysis [70,71], and co-citation analysis [72]. For example, two-dimensional maps, dendrograms, and collaborative network diagrams were generated. This step was implemented in the bibliometric software packages bibliometrix and biblioshiny. (4) Data interpretation: the results were analyzed and described, and future research directions in the field of farmland abandonment were proposed.

3. Results and Discussion

3.1. Analysis of Annual Scientific Results Output and Citation Number

To analyze the general situation and development trend of the research field, the number of published articles (Figure 2) and the mean number of citations (Figure 3) were summarized. From 1980 to 2000, fewer than 11 articles were published in any given year, representing only 10.9% of the total articles published. The mean number of citations can reflect the quality and influence of the article, with four peaks during this period, i.e., in 1987, 1994, 1997, and 2000. The research content was focused on the ecological impacts of farmland abandonment, such as species recovery [73], carbon fixation [74], water retention potential [75], and soil erosion [76]. Immediately after abandonment, because of the lack of vegetation, soil erosion and runoff were serious. In sites that had been abandoned for several years, with a dense vegetation cover, soil erosion and runoff were negligible. There were also studies evaluating the possibility of abandonment reclamation from the ecological perspective [77]. Although in this period studies were scarce, investigating a series of influences after the succession of abandoned farmland from an ecological perspective laid the foundation for subsequent research.

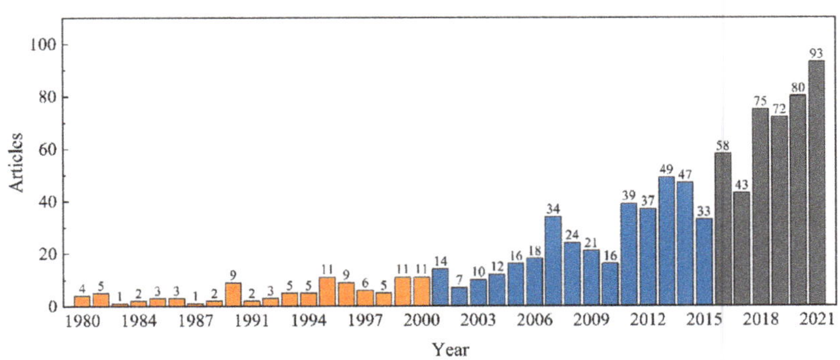

Figure 2. Number of articles on farmland abandonment published from 1980 to 2021 (Note: The different colors in the figure represent the three stages of the number of articles published).

From 2001 to 2015, the number of published articles on farmland abandonment showed an upward trend, albeit with fluctuations, with an average of 25.13 articles per year, accounting for 42.1% of the total published volume. Mean citations peaked in 2003, 2008, and 2015. During this period, sustainability issues, such as climate change and food security, became major concerns. With increasing population size, urbanization, and farmland intensification, the migration of the rural population to cities has led to the large-scale abandonment of farmland. In this period, the contribution of the rational use of abandoned farmland to sustainable development was recognized, and the administration of abandoned farmland has become a hotspot [78,79]. For example, Calcerrada et al. predicted landscape dynamics after farmland abandonment by developing a landscape change transfer matrix model. The authors assumed that, in addition to the ecological value, abandoned farmland can also be used for a series of human activities, such as agriculture, animal husbandry, and tourism [80]. Vegetation succession on abandoned sites and the ecological impacts are some of the main research directions [81–84]. At the same time, with the emergence of machine learning and different resolution sensors, the use of remote sensing means combined with mathematical models to extract abandoned farmland also attracted wide attention [41,85].

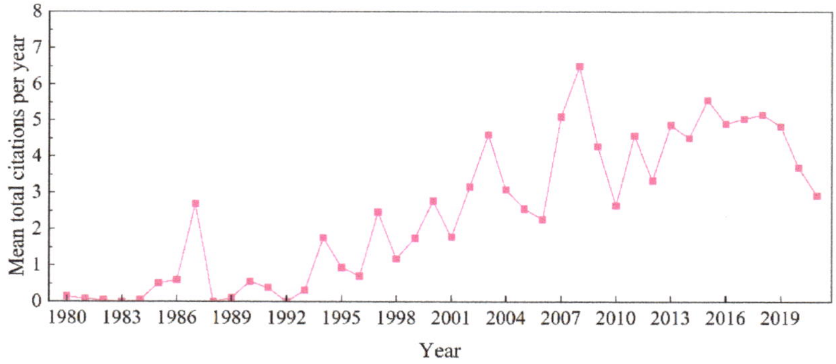

Figure 3. Mean total citations per year on farmland abandonment from 1980 to 2021.

From 2016 to 2021, the number of articles on farmland abandonment increased dramatically, especially in 2020 and 2021, with 80 and 93 articles, respectively. The papers published during this period accounted for 46.99% of the total articles published. In this period, in addition to a change from the traditional ecological perspective to a more modern approach, the research perspective was more diversified, involving the driving factors of farmland abandonment [86–88], social, economic, and environmental consequences [6,89], spatial patterns [90], and policy enlightenment [91]. For example, Xu et al. used survey data of rural households in 27 provinces of China to analyze the spatial patterns of abandoned farmland and studied the impact of farmers' employment on farmland abandonment. Based on the results, labor migration and non-agricultural employment were the main factors driving the abandonment of rural farmland in China [92]. The methods and means of extracting abandoned farmland were also enriched, resulting in a higher accuracy. For example, Yin et al. developed a new method for the extraction of abandoned farmland using a 30-m resolution Landsat satellite time series, with an extraction accuracy of 97% [59]. Although the research content was richer and more diverse, the mean number of citations per project declined after 2018, and the influence of respective articles decreased in 2020 and 2021.

3.2. Journal Sources

To study the development of the research area in different journals, the sources of journals were ranked considering the number of publications, the H index, and TC (total citations) (Table 1). The H index was used to measure the importance and impact of the authors' cumulative research contributions, and the TC represented the total citations for all articles from a journal source. The number of articles published in these journals ranged from 13 to 32. Articles published by *Land Use Policy* over the entire period accounted for 3.6% of the total published articles, followed by those published in *Sustainability*, with 2.9%. In terms of total citations, *Land Use Policy* was the most-cited source, followed by *Catena* with 1642 citations; the ninth-ranked journal, *Remote Sensing*, had the lowest number of citations (165). Taking into account the H index, *Remote Sensing* was again at the bottom of the list, whereas the seventh-ranked journal, *Forest Ecology and Management*, had an H index of 17, which was the same as the first-ranked journal *Land Use Policy*.

Table 1. Journal sources with the most published articles on farmland abandonment from 1980 to 2021 (Note: TC = total citations).

Rank	Total	Articles	H Index	TC
1	*Land Use Policy*	32	17	1856
2	*Sustainability*	26	12	423
3	*Catena*	25	14	1462
4	*Land Degradation & Development*	24	14	690
5	*Land*	24	9	190
6	*Agriculture Ecosystems & Environment*	23	14	1170
7	*Forest Ecology and Management*	23	17	970
8	*Science of the Total Environment*	23	13	527
9	*Remote Sensing*	14	7	165
10	*Ecological Engineering*	13	10	357

According to the journal source dynamics, the evolution curve of the cumulative number of published articles for the 10 most productive journals was plotted over time (Figure 4), obtaining the following information: (1) *Land Use Policy* had the largest production volume and the fastest growth rate during the research period. (2) *Forest Ecology and Management* had a significant production volume before 2014, with a decrease thereafter. (3) *Catena* published the first papers on farmland abandonment in 1999, with stagnation in the middle part of the research period and a significant increase after 2016; *Land Degradation & Development* showed a similar trend. In general, over time, farmland abandonment become the focus of several journals.

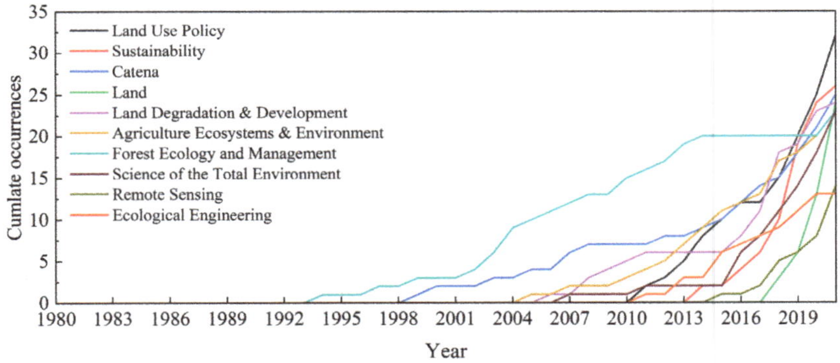

Figure 4. Growth curves of journal sources on farmland abandonment from 1980 to 2021.

3.3. Author Analysis

We first ranked the most relevant authors in the field of farmland abandonment, along with several bibliometric metrics (Table 2), and subsequently generated the author collaboration network (Figure 5), which reflects the connection of different authors' knowledge or social levels. This network can be used to identify previously unknown research groups and represents the similarity of research themes [93]. Finally, we plotted the authors' changing results over time (Figure 6). Articles fractionalized (AF) indicated the contribution rate of the author to the article in the article jointly published by the author and other authors. Overall, 2747 authors contributed to publications in the field of farmland abandonment. Among them, the top three authors were Guobin Liu, Teodoro Lasanta, and Tobias Kuemmerle, with 19, 18, and 16 published papers, respectively (Table 2). Using the H index to measure the importance and influence of the cumulative research contribution, the author with the highest H index was the German scholar Tobias Kuemmerle, with a H index of 15 and a total citation number of 1813, indicating the high quality and influence of his published papers.

Table 2. Top 10 most relevant authors in the field of farmland abandonment from 1980 to 2021 (Note: AF = articles fractionalized; TC = total citations).

Rank	Authors	Articles	AF	H Index	TC
1	Guobin Liu	19	4.12	12	789
2	Teodoro Lasanta	18	4.28	12	949
3	Tobias Kuemmerle	16	2.8	15	1813
4	Estela Nadal-Romero	15	3.43	9	359
5	Alexander V Prishchepov	15	2.6	12	1505
6	Daniel Muller	12	2.92	12	1197
7	Sha Xue	12	2.63	8	527
8	Artemi Cerda	11	3.16	10	612
9	Volker C Radeloff	11	1.78	11	1453
10	Ju-Ying Jiao	9	1.83	7	229

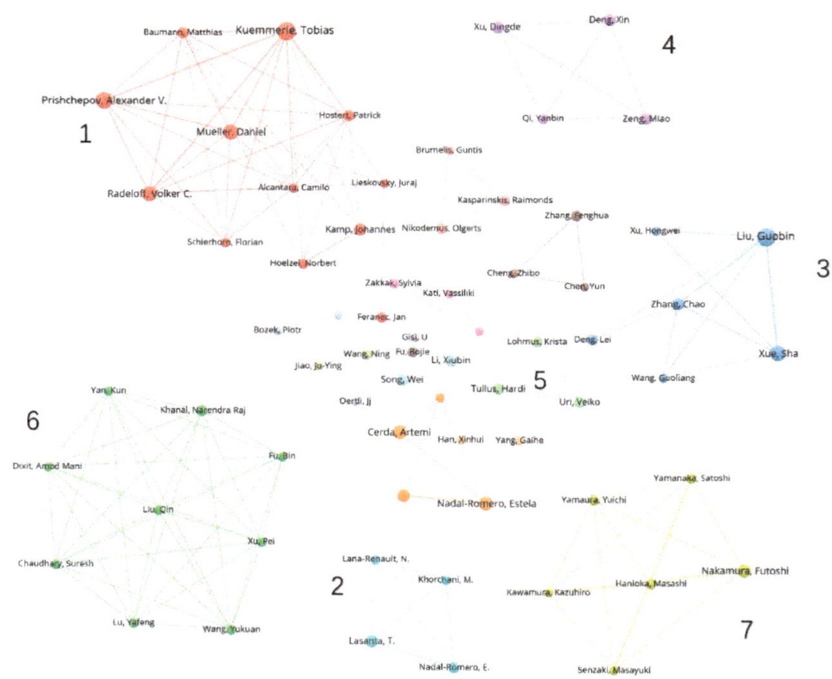

Figure 5. Collaborative network of authors publishing papers on farmland abandonment from 1980 to 2021.

The results in the Author Collaborative Network (Figure 5) were divided into 13 clusters, with each color representing a group of collaborative authors. The larger the circle, the more productive the author. The most representative seven groups of authors were selected, and their studies were analyzed.

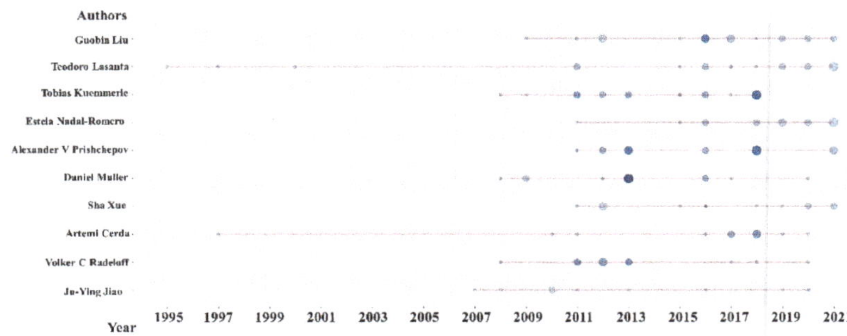

Figure 6. Works of different authors in the field of farmland abandonment over time (Note: N. Articles = number of articles; TC per Year = total citations per year).

The first group examined farmland abandonment in eastern Europe and the former Soviet Union after the disintegration of socialism. Different land policies, economic development, and reform strategies led to large differences in the abandonment rates among different countries [94,95]. During the transition period of 1990–2000, the agricultural sector in Russia, Latvia, and Lithuania declined sharply, while the socio-economic and agricultural infrastructure in Lithuania were more suitable for agriculture, leading to a low abandonment rate. In this period, the Russian government almost completely abolished agricultural subsidies, which led to a high abandonment rate [96]. In contrast, the Belarusian government provided more support for agriculture, leading to a low abandonment rate. Poland and Slovakia, located near the edge of the Carpathian Mountains, are largely affected by terrain and market access, which resulted in a high abandonment rate. However, due to the adoption of private agriculture in the socialist period in Poland's plain area, the agricultural sector in this area quickly adapted to the post-socialist framework, which resulted in a low abandonment rate [94].

The second author group mainly studied the effects of farmland abandonment on soil and hydrological conditions in the Mediterranean region [97,98]. For example, in the southeast of Spain, soil degradation after farmland abandonment is serious. Droughts led to the generation of a surface crust, which reduces the permeability and increases surface runoff and soil erosion [76]. In the Pyrenees Mountains, farmland abandonment has greatly increased the contents of suspended sediment in the water, which has affected the water supply and agricultural activities in the downstream area. However, the local environment is humid, which is conducive to afforestation. Vegetation has increased the soil stability in this area, resulting in reduced soil erosion, surface runoff, and flood probability [5,97,98].

The third group of authors focused on the effects of farmland abandonment on the Loess Plateau of China, such as the effects on soil carbon sequestration [99] and the microbial community [100]. For example, Zhang et al. reported that the vegetation coverage rate decreased in the 10 years before abandonment and returned to normal levels after 15–20 years of abandonment [101]. Deng et al. studied the organic carbon reserves of soil after abandonment on the Loess Plateau, which decreased significantly in the first 20 years and recovered to pre-abandoned levels 30 years later [102]. Similar to the second group, this author group studied the impacts of farmland abandonment on the land. However, whilst the second group focused on soil properties and runoff, this group mainly investigated the succession of vegetation communities and organic carbon reserves over time.

The fourth group of authors mainly analyzed the factors driving and influencing the abandonment of rural farmland in China. In this country, more areas are abandoned in mountainous and hilly areas than in the plains [101]. Xu et al. revealed a significant inverted U-shaped relationship between farmers' employment outside the agricultural sector and farmland abandonment; based on their results, non-agricultural labor does not necessarily lead to farmland abandonment. On the one hand, when rural laborers are employed in non-agricultural sectors, this leads to a shortage of the agricultural labor force, a lack of the land transfer market in rural areas, and insufficient labor resources for families to maintain the production of their existing land, often resulting in extensive farmland abandonment. On the other hand, most farmers are unable to integrate into urban areas and are unwilling to sacrifice their land; they may, therefore, choose outsourcing to avoid giving up their farmland [92,103]. Another study shows that the internet can help farmers significantly reduce farmland abandonment [104].

Similar to the second and third groups, the fifth group of authors also discussed the relationships between farmland abandonment and vegetation and species diversity but paying more attention to the afforestation on abandoned farmland in eastern and northern Europe. For example, hybrid aspen plantations planted on abandoned farmland in Estonia were monitored, and their impacts on understory vegetation and species numbers were discussed [105], as well as the effect of alder on soil nitrogen [106]. In another study, the aboveground biomass, belowground biomass, and nutrients of natural birch forests on abandoned farmland in Estonia were studied [107].

The sixth group of authors focused on abandoned farmland in the mountainous areas of Nepal, such as the Himalayan mountains [108,109] and the Dordi River basin [110]. For example, Chaudhary et al. evaluated the temporal and spatial degradation of abandoned farmland in the Dordi River basin and analyzed its causes and the resulting ecological environment risks. Of the total farmland surveyed, 92% had been completely destroyed, leading to landslides, debris flow, rock fall, gully formation, soil erosion, and limestone pits, thus increasing the negative impact on land resources and vegetation succession [109]. The authors also evaluated the temporal and spatial distribution and driving factors of farmland abandonment in the mountainous areas of Nepal and discussed the ecological and social impacts of abandonment. Based on their results, farmland abandonment is widespread in the hills and mountains of Nepal, with one fifth of the total farmland area being abandoned from 2001 to 2010. Population growth, migration, urbanization, inconvenient transportation, scattered living, poverty, and lack of land policies are the main factors driving farmland abandonment in Nepal. In turn, abandoned farmland restricts the supporting structure of terraced fields in this area, causing natural disasters, such as landslides and mud rock flows, and impeding natural succession. Overall, the results on the ecological environment and the rural society are negative [108,111].

The seventh group of authors studied the changes in species distributions after farmland abandonment, focusing on birds. The main study area was Hokkaido, Japan, where abandoned farmland and wetlands coexist. The authors assessed the landscape patterns of the abandoned farmland [112] and compared the bird biodiversity of the abandoned farmland with that of other local land use types [113,114], indicating that the bird abundance in abandoned areas is equivalent to that of wetlands. Different from the European region, the positive impact of farmland abandonment in this region is greater than the negative impact. Abandoned farmland may evolve into wetlands and forests, which can provide habitat and recovery opportunities for local species. In this context, the management of abandoned farmland in this area is crucial to the protection of bird species in the agricultural landscape.

At least five of these clusters were interested in ecological issues, whereas the remaining two clusters reported social concerns. Based on different natural conditions, policies, and systems, studies on farmland abandonment showed obvious regional differences.

Considering the authors' production over time (Figure 6), Teodoro Lasanta and Artemi Cerda were the two authors with the longest-standing involvement in research on farmland abandonment. Daniel Muller published four articles with other authors in 2013, with the highest annual total citation number of 68.6. Abandoned farmland in Europe was extracted mainly by means of remote sensing methods. For example, the abandoned farmland [115] in 30 countries, including central and eastern Europe, was quantified by support vector machine classification in an article in cooperation with Alcantara et al., using MODIS NDVI time series satellite images. A spatial allocation model that obtained a time series of abandoned farmland [24] in Russia, Ukraine, and Belarus from 1990 to

2009 was developed in cooperation with Schierhorn et al. Four articles published by Tobias Kuemmerle in collaboration with other authors received 58.2 annual total citations, also studying farmland abandonment in Europe [116,117]. Two authors, Daniel Muller and Tobias Kuemmerle, belong to the first cluster of the author collaboration network (Figure 5), with most of the total citations.

3.4. Analysis of the Distribution Characteristics of the Main Research Countries/Regions

To identify the key countries and regions in terms of publication number and impact, the cooperation network between countries and regions was analyzed (Figure 7). Between 1980 and 2021, a total of 62 countries or regions published papers. Among the top 20 countries regarding publication number (Table 3), there were three Asian countries (China, Japan, Korea), two American countries (the United States, Canada), 14 European countries (Spain, Italy, Germany, Poland, the Netherlands, the United Kingdom, France, Estonia, Sweden, Switzerland, Russia, Greece, Slovakia, Portugal), and one Oceanian country (Australia). Among the top five countries with the largest numbers of publications, three were European countries. Based on our findings, European authors were more concerned about the issue of farmland abandonment. China was the only developing country among the top five countries regarding publication number, ranking first and having published 193 papers. However, the mean number of citations per article was low, with only 20.71, indicating that Chinese authors need to pay more attention to the quality of their articles. The countries with the highest mean citation numbers were the Netherlands, followed by Switzerland, with 96.88 and 95.07, respectively, but with fewer publications, ranking ninth and 15th, respectively.

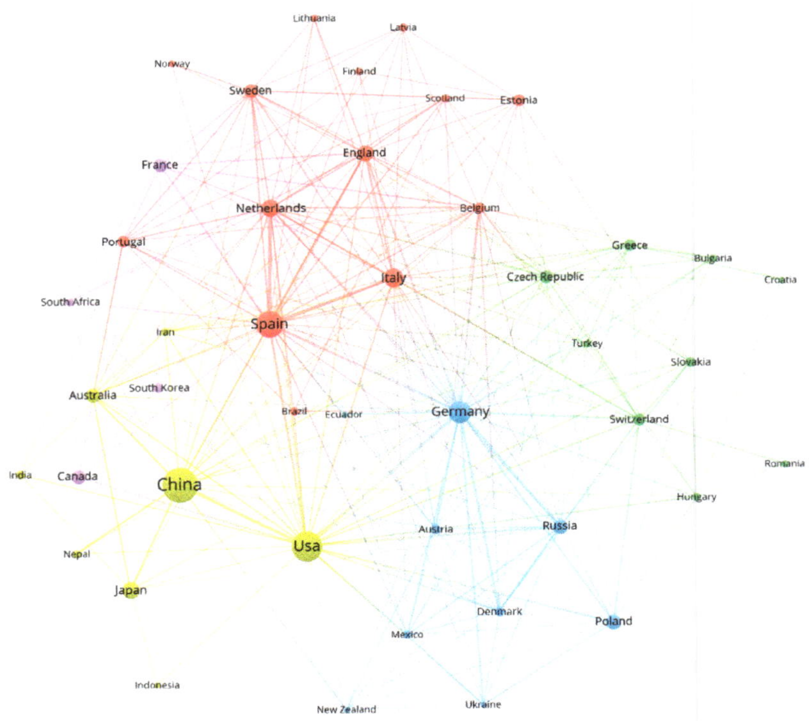

Figure 7. National collaborative network for research on farmland abandonment from 1980 to 2021.

Table 3. Numbers of publications and citations on farmland abandonment in different countries from 1980 to 2021.

Country	Articles	Total Citations	Average Citations
China	193	3997	20.71
USA	105	3936	37.49
Spain	83	2856	34.41
Italy	42	1223	29.12
Germany	39	2325	59.62
Japan	34	484	14.24
Poland	30	423	14.1
Canada	26	990	38.08
Netherlands	25	2422	96.88
Australia	20	1164	58.2
United Kingdom	18	1021	56.72
France	17	723	42.53
Estonia	16	418	26.12
Sweden	15	833	55.53
Switzerland	14	1331	95.07
Russia	13	61	4.69
Greece	12	376	31.33
Slovakia	12	146	12.17
Korea	11	81	7.36
Portugal	11	310	28.18

Based on the national collaboration network (Figure 7), the most closely cooperating countries were Germany and the United States, Russia and Germany, and Spain and the Netherlands, with 17, 15, and 19 cooperation articles, respectively (the thicker the lines, the closer the connection between countries). The node size represents the number of articles published in cooperation with other countries. The countries with the most co-published articles were China, Spain, and Germany, with 42, 32, and 27 articles, respectively (Figure A1), accounting for 16%, 12%, and 10% of the total output of co-published articles, respectively. China and Spain published a larger number of articles independently, with Germany, the UK, and the Netherlands focusing on collaborative research with other countries.

3.5. Document Analysis

3.5.1. Analysis of Highly Cited Papers

The top 10 cited articles (Table 4) reflected the most influential content in the field of farmland abandonment from the perspective of citations, including seven research papers and three reviews. The article with the highest overall number of citations proposed that determining the ecological restoration goal of abandoned farmland is a challenge that people will face in the future. It was also proposed that the future research direction should be combined with land use policy [78]. Gellrich et al. [118] developed an economic model to analyze the driving factors of natural forest regeneration in the Swiss mountains, showing that natural forest regeneration is more likely to occur in areas where farmland has been abandoned. Similar findings were found in other mountainous areas, such as the Alps as well as in Sweden, Poland, Denmark, the Baltic Sea, and parts of Slovenia. Other studies focused on Europe; for example, the 10th article [119] used the revised CAPRI model to assess the consequences of farmland abandonment in the EU under the Common Agricultural Policy (CAP). In the whole of the EU, the distribution of abandoned farmland in different countries, regions, and farm types is uneven. Abandoned farmland caused by agricultural and trade policy reforms may have a significant impact on rural livelihoods and the environment, such as a reduced biodiversity.

In addition, the paper ranked No. 2 [120] used the Dyna CLUE model to conduct a clear simulation and spatial evaluation on the natural succession track of abandoned farmland and vegetation in Europe from 2000 to 2030. The simulation results show that abandoned farmland in Europe mainly occurs in the mountainous regions. The natural succession after abandonment mainly depends on local conditions, such as climate, population density, and terrain [81,121–123]. The seventh-ranked article assessed the hydrological impacts of land abandonment in the Mediterranean region of Europe, focusing on water resource availability and soil erosion [45]. This study shows that vegetation succession after abandonment is related to many factors, including soil depth and fertility, slope aspect, and climate (annual average precipitation, evapotranspiration), among others [124]. In the Mediterranean region, abandoned farmland is mainly concentrated in mountain terraces and semi-arid areas, such as southern France, southeast Spain, south central Portugal, and Italy [81,125,126]. In terraced areas, such as the Iberian Mountains and the Pyrenees Mountains in Spain [84,127,128], farmland abandonment has resulted in landscape degradation and soil erosion, sediment deposition in river channels, and increased surface runoff. In semi-arid areas, higher temperatures, less precipitation, and long-term drought impede vegetation succession on abandoned farmland. The resulting surface crust reduces the infiltration rate, increases surface runoff, and leads to surface erosion [129]. On the contrary, in humid areas, the succession of vegetation dominated by herbaceous and shrub species is rapid [120].

After the collapse of the socialist system in eastern Europe and the former Soviet Union, under the dual impact of the system and the economy, farmland abandonment differed among different European countries. The article ranked No. 8 [42] studied the spatial distribution and driving factors of farmland abandonment in western Ukraine after the collapse of the socialist system, identifying differences with other parts of Europe. In western Europe, abandonment is mainly driven by industrialization, market orientation, and urbanization, such as in areas around the Alps and Romania [130]. The abandonment rate in the plains of western Ukraine is high, whereas that in the marginal areas is low. In the transitional period, the agriculture in this area was not yet industrialized and dominated by traditional self-supporting agriculture. The plain soil is relatively barren and not suitable for farming, and the reduction of the agricultural labor force due to large-scale migration and regular remittances from family members seeking non-agricultural employment in other regions can maintain the livelihood of the farmers. In addition, compared with other eastern European countries, such as Latvia, Poland, Russia, Lithuania, and Estonia [91,131,132], the abandonment mode in western Ukraine is different, most likely because of differences in socialist land ownership patterns, post-socialist land reform strategies, and rural population densities [94,115]. With this in mind, the factors driving farmland abandonment cannot be generalized among countries that are also impacted by the collapse of socialism.

The third-ranked article mainly studied the driving mechanism of grassland abandonment on forest expansion. Although the total citation number was high, the local citation rate (LCR) was 0, indicating that the article was not correlated to the field of farmland abandonment. The fourth-ranked article estimated the abandoned global farmland area via land use data from the global Environment 3.0 historical database and MODIS satellite images, demonstrating the bioenergy potential of the global abandoned farmland [79] using the CASA ecosystem model. Another article reviewed farmland abandonment and its impact on biodiversity from a global perspective by investigating studies from different countries, ranking No. 9 [19]. Reviewing 276 papers, this paper found that the impact of abandoned farmland on biodiversity differs regionally. Whilst most studies in Europe reported a negative impact of farmland abandonment on biodiversity, studies in central and southern America largely found positive impacts on biodiversity. In the latter region, agricultural expansion is the primary issue in biodiversity conservation. Most studies in North America discussed the processes and mechanisms that lead to or accompany farmland abandonment, largely neglecting its impact on biodiversity. Through further analysis, the authors found that the differences were mainly related to the environment, the methods used by researchers, and the concerns about the landscape before and after abandonment.

Table 4. Highly cited articles on farmland abandonment from 1980 to 2021 (Note: TC per Year = total citations per year; LCR = local citation rate).

Paper	Year	Total Citations	TC per Year	LCR	Source
What's new about old fields? Land abandonment and ecosystem assembly [78]	2008	531	35.4	12.81	*Trends in Ecology and Evolution*
Combining top-down and bottom-up dynamics in land use modeling: exploring the future of abandoned farmlands in Europe with the Dyna-CLUE model [120]	2009	449	32.07	8.02	*Landscape Ecology*
Tree line shifts in the Swiss Alps: Climate change or land abandonment? [133]	2007	433	27.06	0	*Journal of Vegetation Science*
The global potential of bioenergy on abandoned agriculture lands [79]	2008	418	27.87	9.09	*Environmental Science & Technology*
The potential for carbon sequestration through reforestation of abandoned tropical agricultural and pasture lands [74]	2000	347	15.09	2.88	*Restoration Ecology*
Agricultural land abandonment and natural forest re-growth in the Swiss mountains: A spatially explicit economic analysis [118]	2007	342	21.38	23.98	*Agriculture Ecosystems & Environment*
Hydrological and erosive consequences of farmland abandonment in Europe, with special reference to the Mediterranean region—A review [45]	2011	288	24	17.71	*Agriculture Ecosystems & Environment*
Patterns and drivers of post-socialist farmland abandonment in Western Ukraine [42]	2011	287	23.92	27.87	*Land Use Policy*
Farmland abandonment: threat or opportunity for biodiversity conservation? A global review [19]	2014	262	29.11	21.76	*Frontiers in Ecology and the Environment*
Policy reform and agricultural land abandonment in the EU [119]	2013	259	25.9	21.24	*Land Use Policy*

3.5.2. Keyword Analysis

By comparing the top 20 author keywords (DE) and keywords-plus (ID) most related to the field of farmland abandonment (Table A1), farmland abandonment/agricultural land abandonment, succession, soil erosion, biodiversity, restoration, and Europe were the six keywords shared by the two. Different author keywords clearly defined the content, approach, or method of the author's research, and the common ones were farmland abandonment, land use change, land use, secondary succession, and afforestation. The keywords-plus in the Web of Science database could objectively describe the content of the article from a macro-perspective, such as farmland abandonment, vegetation, dynamics (mostly referred to soil dynamics), forest, and management. This indicates that, currently, studies focus on the natural succession of abandoned farmland and the impacts of human activities on the ecological environment.

3.5.3. High-Frequency Keyword Cluster Analysis and Multiple Correspondence Analysis

Using the functions of cluster analysis and multiple correspondence analysis, biblioshiny could reflect the hotspots, themes, and writing directions in the field of farmland abandonment. Cluster analysis was used to cluster complex keyword–network relationships into several relatively simple groups [134]. The keywords with the highest similarity were merged into one cluster, and the cluster with the highest similarity was merged into a large cluster, and so on, until all categories were finally merged into one class, forming a dendrogram representing the close or alienation relationship of the keywords in the field of farmland abandonment (Figure 8). Multiple correspondence analysis (MCA) resulted in the formation of an intuitive two-dimensional graph by reducing multi-dimensional data to low dimensions, using a plane distance to reflect the similarity among keywords. A close proximity to the center indicates that this type of topic had received higher attention, whereas a greater distance of keywords from the center indicates less attention or a greater degree of deviation from the topic [135] (Figure A2).

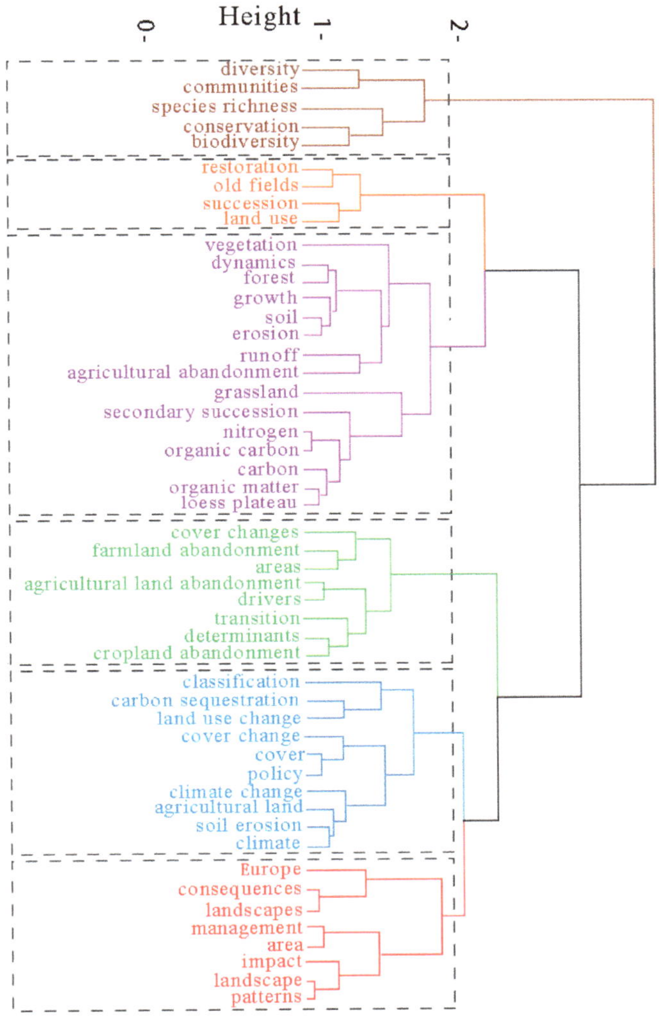

Figure 8. Hierarchical clustering analysis of keywords in the field of farmland abandonment.

The first cluster category focuses on the study of biodiversity, species richness, and plant communities after farmland abandonment. There are obvious regional differences in the ecological and environmental consequences of farmland abandonment. For example, most studies in European regions show that the encroachment of vegetation on bare land and original farmland after abandonment will have a negative impact on lizards, birds, and other animal species [136]. One study in the Mediterranean Sea shows that vegetation encroachment after farmland abandonment will cause the loss of bird habitats. Regular burning, grazing, and other means may help to protect bird diversity. However, a study on Hokkaido, Japan, obtained opposite results. Most wetlands in this area have been transformed into farmland, which is now abandoned. The authors assume that the abandoned farmland can be used as a habitat for wetland birds through management, helping to promote biodiversity [137,138]. In southeast Europe, such as in Bulgaria and Croatia, plant species diversity is greatly affected by elevation and temperature. In areas with low elevation and cold temperatures, species are more abundant [139]. However, the richness of bird species declines after farmland abandonment, highlighting the importance of the management and maintenance of traditional rural landscapes in this area.

The second cluster category mainly contains studies on vegetation and species restoration over time after farmland abandonment. Zhang et al. [140] studied the vegetation community and soil characteristics of abandoned farmland in the Qinling Mountains of China after a period of natural succession, indicating that natural succession is conducive to the ecological recovery of abandoned farmland, which is characterized by high biodiversity, high soil fertility, and high species richness. The species richness of the community increased in the first 15 years after abandonment [100,102,141]. However, Yannelli et al. [142] reported that, in arid regions, natural succession may not be the best method for ecological restoration, and soil erosion may occur during succession, preventing the establishment of a vegetation cover. Restoration should be adapted to local conditions, taking into account environmental factors such as climate and topography in different regions. Standish et al. [143] also considered the impact of alien species invasion on the restoration of abandoned farmland vegetation, indicating that, during natural succession, alien species are more competitive than native ones. The authors also highlighted that restoration should be more focused on maintaining ecosystem services at the site, rather than attempting to restore historic ecosystem states.

The third category contains publications on the impacts of natural succession on forests, soils, and microbial communities, mostly in the mountainous areas of Europe and on the Loess Plateau of China. Cojzer et al. [144], in Slovenia, compared the species structure, quantity, and tending time of young forests formed by secondary succession and artificial reforestation after farmland abandonment. The results showed that the young forests formed by secondary succession had great structural complexity and biodiversity, and the management of these stands was more conducive to local ecological restoration than reforestation. Treml et al. [145] evaluated the changes in tree coverage and number in the Sudeten Mountains of Central Europe through aerial images; based on their results, farmland abandonment is the main cause of forest intensification. Chinese authors studied the vegetation succession and soil carbon reserves after farmland abandonment on the Loess Plateau, reporting that vegetation coverage, species richness, and soil carbon reserves significantly increased over time, recovering to the original levels [146] within about 15 years, increasing significantly within 15 to 25 years; however, after 50 years, they were still lower than those of natural grassland [147,148]. In one study, the introduction of alien species reduced the time to restore vegetation to its original level to 11 years [149]. Another study showed that the reforestation of farmland after abandonment can significantly increase soil fertility [150].

The fourth category focused on methods to extract abandoned farmland, studying the mode, driving force, and spatial determinants of abandoned farmland. The commonly used method identified abandoned farmland by analyzing the inter-annual land use change in a given study area. Subsequently, the driving factors and spatial determinants of farmland abandonment [151,152] were determined by mathematical models. For example, Baumann et al. [42] plotted the abandoned farmland in the post-socialist period of western Ukraine, using the support vector machine classification method and Landsat images from 1986 to 2008, and determined the spatial determinants of abandonment using a combination of the optimal subset linear regression model and hierarchical division. Based on the results, 30% of the farmland was abandoned in the post-socialist period, mainly as the result of the combined impact of the system and the economy. A study in southern Chile [153] extracted the abandoned farmland by comparing land use in 1985 and 2007, using a spatial explicit statistical model. The authors showed that soil quality, distance to secondary roads, and agricultural subsidies were important drivers of local farmland abandonment. In eastern Poland, Zgłobicki et al. [154] showed that natural conditions (topography, soil), socioeconomic characteristics (farmland area, forest cover changes, farm size), and agricultural policies were the largest drivers of local farmland abandonment.

The fifth category was similar to the third and fourth categories. However, this category not only constructed the extraction method of farmland abandonment but also focused on the bioenergy potential abandoned farmland [155], estimating that the potential is 8% of the global demand for primary energy [79]. Schierhorn et al. developed a model for mapping abandoned farmland and calculating carbon sinks in such farmland, based on acreage statistics. The authors provided the means to map abandoned farmland in areas in which remote sensing is difficult to perform and stated that the rich carbon sinks of abandoned farmland can mitigate global climate change; reclamation of such areas is likely to lead to large amounts of carbon emission [24].

In Europe, with the highest number of publications on farmland abandonment, studies on the distribution, patterns, consequences, and impacts of farmland abandonment formed a separate cluster. The causes of farmland abandonment in Europe are various, depending on the region and period under consideration. Due to natural conditions, historical development, and economic, social, and demographic backgrounds, agricultural conditions vary from region to region. For example, Terres et al. established a unified indicator of the driving factors of abandonment through a comprehensive analysis of farmland abandonment in different regions of Europe, including the aging population, the low population density, the low farm income, the small scale, and the poor implementation of agricultural plans. This indicator covers the possible drivers of abandonment in all EU countries and can be flexibly applied according to the background of different regions [156–158]. Prishchepov et al. [132] discussed the impacts of institutional changes in different countries in eastern Europe and the former Soviet Union on farmland abandonment after the collapse of socialism. The authors stated that the reform of systems and policies is an important factor indirectly driving farmland abandonment in this region. In places where the institutional change of agricultural land management is relatively small, and where the new institutions after the institutional change are relatively strong, the abandonment rate is the lowest. On the contrary, countries that delay the formulation of new agricultural production systems and regulations have a higher abandonment rate. In addition, the European region also focuses on the impact of farmland abandonment on ecology. For example, in the Alps, farmland abandonment leads to spontaneous reforestation, reducing grassland species and negatively impacting plant diversity [159]. Another study assessed the carbon storage dynamics of abandoned farmland in the former Soviet Union and showed that the abandoned farmland is transformed into an ecosystem dominated by grassland, resulting in an increase in carbon sequestration and a significant impact on the carbon sink in the region [160].

3.5.4. Analysis of the Research Theme Evolution

Theme evolution analysis can explore the changing laws, relationships, paths, and trends of the content, and intensity and structure of the research topic over time. Here, the study period was divided into three stages, with strategic coordinates being drawn (Figure A3). Overall, the research topics in the field of farmland abandonment experienced a process from increase to decrease to increase. In the initial stage of 1980–2000 (Figure A3), the core themes with high maturity were vegetation, landscape, dynamic degree, and management (the first quadrant in Figure A3). For example, a Swiss study provided advice on the local agricultural management policy [161] by interviewing people regarding their preferences for reforestation after farmland abandonment. However, Puerto Rico was the area of concern for this phase, mainly because it was funded by the US National Fund at the time [74] and because of the support of the University of Puerto Rico in this field [162]. The themes competition and basin disappeared in the following stage. In addition, the natural succession of abandoned farmland resulted in the formation of new forests after the abandonment, and forest communities, soil, and vegetation became research hotspots.

During the development stage from 2001 to 2015 (Figure A3), the theme type with high maturity was relatively single, and the dynamic degree (forests dynamic, land cover dynamics, vegetation dynamics) was still the core theme of this period. The new core themes, such as land cover and agricultural abandonment, were consistent with the study of land cover changes [81] following agricultural abandonment in the Mediterranean mountains and the role [80] of farmland abandonment in landscape dynamics in central Spain, using three Landsat land cover maps. Runoff and soil erosion were emerging themes in this period. However, the occurrence in the third quadrant showed that the research on this theme was not mature enough, and the topics such as vegetation, forest, and diversity are still hot topics for future research, getting closer to the first quadrant. Notably, the top 10 most-cited articles (Table 4) were all published in the second half of this period.

During the active phase of 2016–2021 (Figure A3), studies on forest, vegetation, dynamic degree, and biodiversity formed the core of the field of farmland abandonment research. Europe, as an important area in respective research, is distributed in the fourth quadrant as a keyword with a high number of occurrences. Along with the analysis of the driving factors and the determinants of abandonment, this field may become a research hotspot in the future. Whereas previous studies on farmland abandonment in Europe were mainly conducted in typical European countries and regions, with a small scale, with the increase in achievements and the improvement of technical means, large-scale studies covering the entire European continent appeared in this stage. For example, Walter et al. evaluated the causes and consequences of farmland abandonment in Europe by summarizing previous studies [6], whereas Ustaoglu et al. evaluated the drivers of farmland abandonment and their sustainability impact on society and the environment [86]. Lasanta studied the spatial and temporal processes and drivers of farmland abandonment in Europe [87]. Research on abandoned farmland in typical regions such as Ukraine [163], Russia, and Kazakhstan [37] has also been revitalized. Compared with previous stages, this stage of research pays more attention to the determinants of the reclamation of abandoned farmland, with the aim to avoid the ecological impacts of farmland expansion.

3.5.5. Analysis of the Monitoring Method Evolution

This study reviewed the abandoned farmland monitoring methods used in the database. The representative methods in this field changed over time (Table 5) and can be divided into three types: (1) Field survey; (2) remote sensing image classification methods, such as visual interpretation, supervised classification, object-oriented classification; and (3) change detection methods, including vegetation index change detection and multi-temporal remote sensing image time series change detection. At present, the monitoring of abandoned farmland mainly considers multi-source remote sensing images and the integration of multiple monitoring methods. Over time, the monitoring methods changed from field survey and the visual interpretation of aerial photos to remote sensing. Remote sensing data also changed from small-scale- and medium low-spatial-resolution data to large-scale- and high-spatial-resolution data [94,116,131,164,165].

Table 5. Evolution of methods for the extraction of abandoned farmland.

Countries and Regions	Data Source	Research Methods	Notes
Cal Rodo catchment (southern margin of the Pyrenees)	Two aerial photos from 1957 and 1996 (20 m)	Visual interpretation	Poyatos et al., 2003
Swiss mountains	Land use survey data during the 1980s and 1990s	Field survey	Gellrich and Zimmermann, 2006
Peyne in France	Aerial photos (1 m) from 1946, 1954, 1970, 1971, 1983, and 1988, field geographic data	Field survey, visual interpretation	Sluiter and Jong, 2006
Ijuez River Valley (Central Spanish Pyrenees)	Aerial photos from 1957, 1977, and 2002 (1 m)	Visual interpretation	Pueyo and Beguería, 2007
Poland, Slovakia, and Ukraine	Landsat TM/ETM+ images from 1986, 1988, and 2000 (30 m)	Support Vector Machines	Kuemmerle et al., 2008
Galicia (Spain)	Aerial photographs from 1956 and 1957, and land use of the plot size specified in SIGPAC (2004)	Visual interpretation	Corbelle et al., 2011
Western Ukraine	Landsat TM images from 1986, 1989, 2006, and 2008 (30 m)	Support Vector Machines	Baumann et al., 2011
Smolensk, Kaluga, Tula, Rjazan, and Vladimir in European Russia	Landsat TM/ETM+ satellite images from 1990 and 2000 (30 m)	Support Vector Machines	Prishchepov et al., 2012

Table 5. *Cont.*

Countries and Regions	Data Source	Research Methods	Notes
Baltic countries, Belarus, and Poland	MODIS NDVI time series from 2003 to 2008 (250 m)	Vegetation index change detection, Support Vector Machines	Alcantara et al., 2012
Belarus, Lithuania, and Poland	Landsat TM/ETM+ satellite images from 1989 and 1999 (30 m)	Support Vector Machines	Prishchepov et al., 2012
Poland, Belarus, Latvia, Lithuania, and European Russia	Landsat TM/ETM+ satellite images from 1990 and 2000 (30 m)	Support Vector Machines	Prishchepov et al., 2012
Covering 6.4 Mkm^2 across central and eastern Europe and the Balkan Peninsula, including 30 countries fully or partly	MODIS NDVI time series from 2004 to 2006 (250 m)	Vegetation index change detection, Support Vector Machines	Alcantara et al., 2013
European Russia, Ukraine, and Belarus	GLC2000, MODIS, national sown area statistics (1 km)	A spatial allocation model was developed to allocate national area statistics to remote sensing image data	Schierhorn et al., 2013
Throughout Europe	MODIS NDVI time series from 2000 to 2012 (250 m)	Vegetation index change detection, Random Forest	Estel et al., 2015
Mountainous areas in China	Household survey data of 262 counties from 2011 to 2012	Field survey	Li et al., 2017
Parts of Georgia and the North Caucasian Federal District of Russia	Landsat images from 1985 to 2015 (30 m)	LandTrendr time series change detection, object-oriented classification	Yin et al., 2018
Northern Kazakhstan	Landsat images from 1988 to 2013 (30 m)	LandTrendr time series change detection, Random Forest	Dara et al., 2018
14 regions in the world (Iraq, Nebraska, Shaanxi, Orenburg, Uganda, Belarus, Bosnia and Herzegovina, Sardinia, Volgograd, Wisconsin, Chongqing, Goias, Mato Grosso, Nepal)	Landsat images from 1987 to 2017 (30 m)	Time series change detection, Random Forest	Yin et al., 2020
Global	CCI-LC data from 1992 and 2015 (300 m)	Change detection after classification	Næss et al., 2021

(1) Field survey. The largest advantage of this method is that it can explain the mechanism behind farmland abandonment, although it is difficult to obtain a complete view of all abandoned farmland areas in a given region due to the lack of spatial details. Therefore, the extraction of abandoned farmland often uses field survey data coupled with other models. For example, Florian et al. distributed the survey data of European Russia, Ukraine, and Belarus to the remote sensing image data through the spatial allocation model, obtaining an accuracy of 65% [117]. Raymond et al. combined geographic data obtained from field surveys with aerial photos to extract abandoned farmland in southern France, which accounted for 5% of the study area [166].

(2) Classification of remote sensing images. Visual interpretation methods are mostly applied to Mediterranean mountains [41,42,81], and the data sources are mostly aerial photos. Visual interpretation has the advantages of simple operation and high accuracy, although it is greatly influenced by the interpreter, and the degree of automation is low. It requires considerable manpower and time to process data, and the data are difficult to obtain. The application of supervised classification originated in eastern Europe and the former Soviet Union [96,121,132,167], generally applying Landsat TM/ETM+ data. Supervised classification can use prior knowledge to improve classification accuracy by learning sample features. This method can be applied to high-, medium-, and low-spatial-resolution remote sensing data at the same time, allowing its wide use in abandoned farmland monitoring. The disadvantage is that the sample selection is highly subjective, and it is difficult to distinguish land types with similar spectral characteristics.

(3) Change detection method. Some authors took more than 30 countries in central and eastern Europe as the research area and used MODIS NDVI time series products to extract abandoned farmland. This product has a high time resolution, and the characteristics of abandoned farmland can be compared with the growth cycle characteristics of vegetation types. However, the spatial resolution is low (250 m). This study also combines SVM classification, obtaining an accuracy of 50.7% [115]. Estel et al. used MODIS NDVI time series products combined with random forest classification to monitor abandoned farmland throughout Europe and mapped the reclamation range of farmland, with an accuracy of 90.1%. Farmland was mainly abandoned in eastern Europe, southern Scandinavia, and the European mountains. Reclamation is also common, mainly in eastern Europe (such as

Russia in Europe, Poland, Belarus, Ukraine, and Lithuania) and the Balkans [85]. Vegetation index change detection can identify the growth difference between crops and natural vegetation, with low data redundancy and high method fault tolerance, but, due to the influence of image spatial resolution, the recognition accuracy of abandoned farmland is low [79,115]. In recent years, the most commonly used method was Landsat remote sensing image time series change detection. Its key is the initial classification accuracy of the image, and the change detection accuracy is accumulated by different time phase classification accuracies. Yin et al. used Landsat images from 1985 to 2015 and the LandTrendr model to detect changes in the time series of multi-temporal remote sensing images, applying the object-oriented classification method. This method is not only based on the spectral information and texture information of abandoned farmland but also considers the geometric information. By formulating a variety of rules to constrain the target land class, it has significant advantages in improving the initial classification accuracy of remote sensing images. The overall accuracy of the study on the classification of abandoned farmland is 97%, which is better than that obtained using the pixel level change detection (82%) [59].

4. Conclusions and Prospects

We conducted a quantitative analysis and visualization of articles published in the field of farmland abandonment from 1980–2021 at the macro-scale by using the bibliometrix and biblioshiny software packages, based on the Web of Science core collection database. From this, we determined the current status and future development trends of scientific production in this field, overcoming the shortcomings of previous literature reviews that rarely comprehensively analyzed the research results from multiple perspectives and revealed the potential changes.

The annual scientific output results showed that the number of articles in the field of farmland abandonment fluctuated and increased rapidly, especially after 2016. According to the number of articles issued, the research period can be divided into three stages: budding stage, development stage, and active stage. Judging from the average number of citations of papers, the years with the strongest development in the field of farmland abandonment were 1987, 2003, 2008, and 2015. Over time, more attention was paid to abandoned farmland as such, with most papers being published from 2007 to 2017; however, the overall quality of the publications declined after 2018.

Journal source analysis showed that *Land Use Policy* was the most productive journal in the field of farmland abandonment from 1980 to 2021, with the highest number of citations and the highest H index. In terms of research strength, China was the country that published the most articles, although countries in Europe were more influential. Cooperation between countries suggested that some countries in Europe, such as Germany, the UK, and the Netherlands, focused more on collaborative research. When considering both the number of papers published by independent authors and international cooperations, papers published by countries such as China and the United States through independent research were more common.

Studies on farmland abandonment were two-branched: one branch mainly analyzed the social factors, such as policies and trade, which may promote or inhibit the abandonment of farmland; the other focused on the ecological impacts of farmland abandonment, such as impacts on soil quality, carbon sequestration, and biodiversity. The same trend was found in the authors' collaborative network, with five of the seven most productive research groups focusing on ecological aspects and the other two on social factors.

The most common keywords were farmland abandonment, land use, vegetation, dynamics, forest, and management. Analysis of the theme evolution of abandoned farmland showed that vegetation, landscape, dynamic degree, and management were the most commonly used keywords in the first stage. The influence of the most used keywords in the second stage was weakened; in this stage, the focus was on the improvement of the extraction technology. The last period was related to the driving forces and determinants of farmland abandonment, in addition to vegetation succession and biodiversity. The theme that remained constant throughout the study period was the succession of vegetation after farmland abandonment.

This study has two limitations. First, it relies more on dynamic databases that are constantly updated as the number of indexed journals increases or decreases. Therefore, the bibliometric analysis of farmland abandonment may change in the next few years. Another limitation is that this study largely relies on the results obtained from visual charts. However, the display of results lacks a richer global perspective. The global perspective can, however, be improved by incorporating other databases, such as Scopus.

Overall, the results obtained can help researchers determine the specific research areas and contents of farmland abandonment. The analysis of the cooperation between the author and the country may facilitate the investigation of the authors in the field of farmland abandonment or the formulation of policies at the national level. In addition, this study provides a database and analyzes the different topics and methods for future researchers, assisting in the selection of research areas and topics.

According to the results, farmland abandonment research needs to be further expanded in the following aspects:

(1) In terms of data sources, farmland abandonment experienced a process from the traditional family survey data to the remote sensing image data; the latter underwent a transformation from low to high spatial resolution. However, the high-resolution images used for farmland abandonment generally have a low temporal resolution. In the future, the use of multi-source remote sensing data to combine high temporal resolution and high spatial resolution for farmland abandonment extraction may become the mainstream of research. Nowadays, there are more and more applications of hyperspectral images, and improvements in spectrum number and spectral resolution can significantly enrich spectral information, revealing more details of ground objects. Such an approach would greatly improve the accuracy of remote sensing in the identification of ground objects. In the future, this direction can be explored with the use of hyperspectral images to extract abandoned farmland.

(2) To extract abandoned farmland, change detection methods are mostly used. The extraction accuracy largely depends on the classification accuracy, and the classification error of each temporary phase will be accumulated, affecting the final result. Moreover, the classification of heterogeneous images is independent and does not consider the correlation of time, making it easy to ignore vegetation changes in the growth cycle, resulting in unreasonable change detection results. Classification time series data can effectively avoid this problem, and models such as the Landsat-based detection of trends in disturbance and recovery (LandTrendr), breaks for additive season and trend (BFAST), and continuous change detection and classification (CCDC) can demonstrate this. Today, only the LandTrendr model is used for the extraction of abandoned farmland. With the rise of deep learning, the remote sensing time series change detection research based on deep learning has attracted the interest of many researchers. Deep learning is an automated learning based on the end-to-end mechanism. It does not rely on prior knowledge, but it is still difficult to automatically mine data and acquire the spatiotemporal features of images. In the future, how to use these models to effectively extract abandoned farmland may be an important development trend.

(3) Since the 21st century, the impact of global climate change on ecology and the social economy has been an important topic. In recent years, the global outbreak of COVID-19 and regional armed conflict have caused unprecedented crises in the world, seriously affecting agriculture, the economy, human health, and food security. It is, therefore, of great significance to meet the global food crisis by rationally reclaiming abandoned farmland to rapidly increase food production and the global grain supply. However, there is still controversy about the economic cost of abandoned farmland reclamation and the ecological consequences of such reclamation. Should the abandoned farmland be reclaimed, continue to expand, or be subjected to intensive farming? Research around this area may be a future focus.

(4) According to the results of this study, future studies will still focus on the ecological impacts of abandoned fields, such as vegetation succession, carbon sequestration, and impacts on biodiversity. However, most of the current studies focus on a certain country and region, with conflicting ecological impacts along with different geographical locations and environmental conditions. At present, the ecological impact of abandoned farmland is still controversial among scholars. In the future, further discussions on this impact are: Is this a threat or an opportunity? Is this negative or positive? The formulation of policies according to local conditions when facing farmland abandonment may be an important research topic in the near future.

Supplementary Materials: The following supporting information can be downloaded at: https://www.mdpi.com/article/10.3390/ijerph192316007/s1, Table S1: Database Article; Table S2: Most Global Cited Documents 100 (theme); Table S3: Most Global Cited Documents 100 (title).

Author Contributions: Conceptualization, W.S.; methodology, B.L. and W.S.; formal analysis, B.L. and W.S.; investigation, Q.S.; resources, W.S.; writing—original draft preparation, B.L. and W.S.; writing—review and editing, W.S.; supervision, W.S. All authors have read and agreed to the published version of the manuscript.

Funding: This research was supported by the Project of National Natural Science Foundation of China (Grant No. 42071233), the Second Tibetan Plateau Scientific Expedition and Research (Grant No. 2019QZKK0603), and the Strategic Priority Research Program of Chinese Academy of Sciences (Grant No. XDA20040201).

Institutional Review Board Statement: Not applicable.

Informed Consent Statement: Not applicable.

Data Availability Statement: All relevant data sets are mentioned in the manuscript.

Conflicts of Interest: The authors declare no conflict of interest.

Appendix A

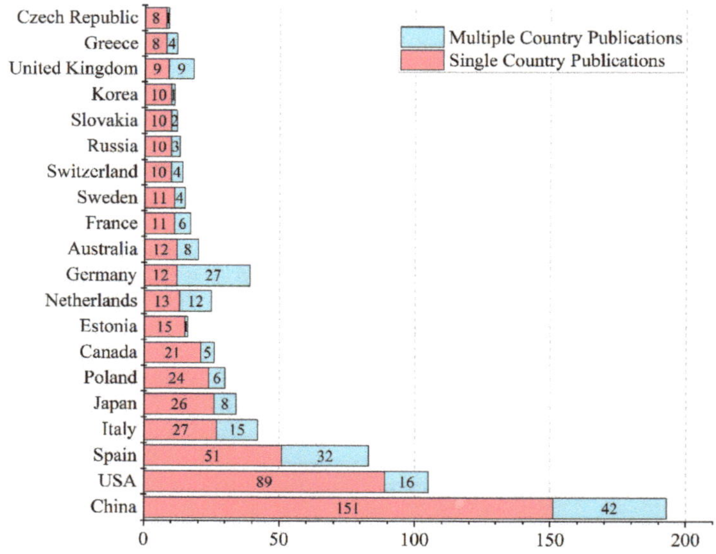

Figure A1. Number of independent and cooperative publications in the 20 countries with the most published articles from 1980 to 2021.

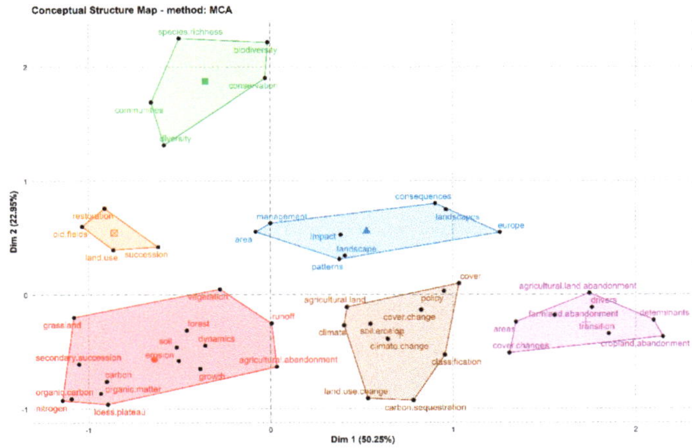

Figure A2. Multiple correspondence analysis of keywords in the field of farmland abandonment.

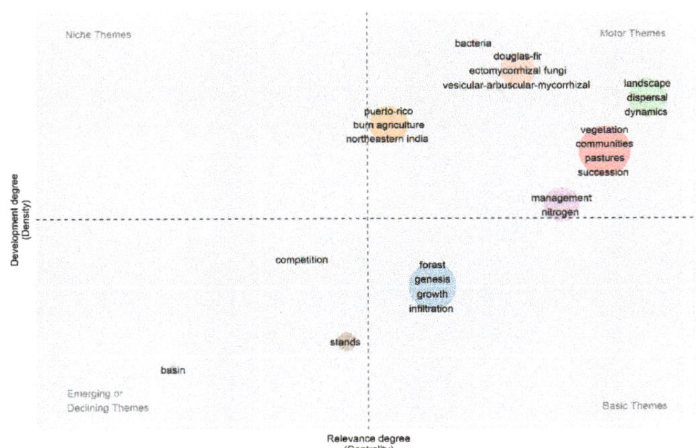

(A)

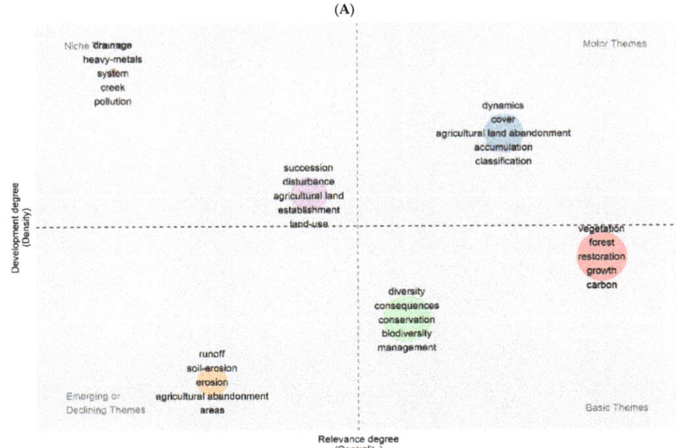

(B)

Figure A3. *Cont.*

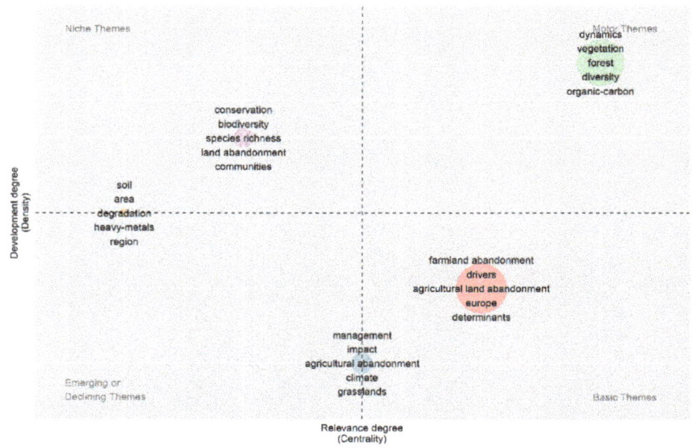

(C)

Figure A3. Strategic coordinates of the farmland abandonment period 1980–2000 (**A**), 2001–2015 (**B**), and 2016–2021 (**C**).

Appendix B

Table A1. Most relevant keywords when searching for publications in farmland abandonment from 1980 to 2021.

Rank	Author Keywords (DE)	Articles	Keywords-Plus (ID)	Articles
1	Abandonment/land abandonment/farmland Abandonment/abandoned farmland/cropland abandonment/abandoned land/abandoned cropland/agricultural abandonment/agricultural land abandonment/abandoned agricultural land	284	Farmland abandonment/agricultural land abandonment	107
2	Land use change	81	Vegetation	90
3	Land use	42	Dynamics	86
4	Secondary succession	32	Forest	81
5	Afforestation	25	Management	79
6	Soil erosion	24	Diversity	74
7	Remote sensing	22	Europe	62
8	Agriculture	21	Consequences	61
9	China	21	Conservation	61
10	Loess plateau	21	Biodiversity	56
11	Succession	20	Landscape	54
12	Mediterranean	15	Patterns	54
13	Soil organic carbon	15	Drivers	46
14	Biodiversity	14	Restoration	46
15	Reforestation	14	Impact	44
16	Restoration	14	Cover	43
17	Spain	14	Nitrogen	43
18	Europe	11	Carbon	42
19	GIS	11	Succession	40
20	Natural regeneration	11	Soil erosion	39

References

1. Pereira, H.M.; Leadley, P.W.; Proença, V.; Alkemade, R.; Scharlemann, J.P.; Fernandez-Manjarrés, J.F.; Araújo, M.B.; Balvanera, P.; Biggs, R.; Cheung, W.W. Scenarios for global biodiversity in the 21st century. *Science* **2010**, *330*, 1496–1501. [CrossRef] [PubMed]
2. Lambin, E.F.; Meyfroidt, P. Global land use change, economic globalization, and the looming land scarcity. *Proc. Natl. Acad. Sci. USA* **2011**, *108*, 3465–3472. [CrossRef] [PubMed]

3. Foley, J.A.; DeFries, R.; Asner, G.P.; Barford, C.; Bonan, G.; Carpenter, S.R.; Chapin, F.S.; Coe, M.T.; Daily, G.C.; Gibbs, H.K. Global consequences of land use. *Science* **2005**, *309*, 570–574. [CrossRef] [PubMed]
4. Van der Sluis, T.; Kizos, T.; Pedroli, B. Landscape change in Mediterranean farmlands: Impacts of land abandonment on cultivation terraces in Portofino (Italy) and Lesvos (Greece). *J. Landsc. Ecol.* **2014**, *7*, 23–44. [CrossRef]
5. Nadal-Romero, E.; Cammeraat, E.; Pérez-Cardiel, E.; Lasanta, T. Effects of secondary succession and afforestation practices on soil properties after cropland abandonment in humid Mediterranean mountain areas. *Agric. Ecosyst. Environ.* **2016**, *228*, 91–100. [CrossRef]
6. Leal Filho, W.; Mandel, M.; Al-Amin, A.Q.; Feher, A.; Chiappetta Jabbour, C.J. An assessment of the causes and consequences of agricultural land abandonment in Europe. *Int. J. Sustain. Dev. World Ecol.* **2017**, *24*, 554–560. [CrossRef]
7. Keenleyside, C.; Tucker, G.; McConville, A. *Farmland Abandonment in the EU: An Assessment of Trends and Prospects*; Institute for European Environmental Policy: London, UK, 2010.
8. Zhang, X.; Brandt, M.; Tong, X.; Ciais, P.; Yue, Y.; Xiao, X.; Zhang, W.; Wang, K.; Fensholt, R. A large but transient carbon sink from urbanization and rural depopulation in China. *Nat. Sustain.* **2022**, *5*, 321–328. [CrossRef]
9. Mottet, A.; Ladet, S.; Coqué, N.; Gibon, A. Agricultural land-use change and its drivers in mountain landscapes: A case study in the Pyrenees. *Agric. Ecosyst. Environ.* **2006**, *114*, 296–310. [CrossRef]
10. Viviroli, D.; Dürr, H.H.; Messerli, B.; Meybeck, M.; Weingartner, R. Mountains of the world, water towers for humanity: Typology, mapping, and global significance. *Water Resour. Res.* **2007**, *43*, W07447. [CrossRef]
11. Brinkert, A.; Hölzel, N.; Sidorova, T.V.; Kamp, J. Spontaneous steppe restoration on abandoned cropland in Kazakhstan: Grazing affects successional pathways. *Biodivers. Conserv.* **2016**, *25*, 2543–2561. [CrossRef]
12. Rey-Benayas, J.M.; Galván, I.; Carrascal, L.M. Differential effects of vegetation restoration in Mediterranean abandoned cropland by secondary succession and pine plantations on bird assemblages. *For. Ecol. Manag.* **2010**, *260*, 87–95. [CrossRef]
13. Stanturf, J.A.; Madsen, P. Restoration concepts for temperate and boreal forests of North America and Western Europe. *Plant Biosyst.-Int. J. Deal. All Asp. Plant Biol.* **2002**, *136*, 143–158. [CrossRef]
14. Song, W.; Deng, X. Land-use/land-cover change and ecosystem service provision in China. *Sci. Total Environ.* **2017**, *576*, 705–719. [CrossRef] [PubMed]
15. Song, W. Mapping cropland abandonment in mountainous areas using an annual land-use trajectory approach. *Sustainability* **2019**, *11*, 5951. [CrossRef]
16. Pointereau, P.; Coulon, F.; Girard, P.; Lambotte, M.; Stuczynski, T.; Sánchez Ortega, V.; Del Rio, A. *Analysis of the Driving Forces Behind Farmland Abandonment and the Extent and Location of Agricultural Areas that are Actually Abandoned or are in Risk to be Abandoned*; EUR-OP: Luxembourg, 2008.
17. Löw, F.; Fliemann, E.; Abdullaev, I.; Conrad, C.; Lamers, J.P. Mapping abandoned agricultural land in Kyzyl-Orda, Kazakhstan using satellite remote sensing. *Appl. Geogr.* **2015**, *62*, 377–390. [CrossRef]
18. Loboda, T.; Krankina, O.; Savin, I.; Kurbanov, E.; Hall, J. Land management and the impact of the 2010 extreme drought event on the agricultural and ecological systems of European Russia. In *Land-Cover and Land-Use Changes in Eastern Europe after the Collapse of the Soviet Union in 1991*; Springer: Cham, Switzerland, 2017; pp. 173–192. [CrossRef]
19. Queiroz, C.; Beilin, R.; Folke, C.; Lindborg, R. Farmland abandonment: Threat or opportunity for biodiversity conservation? A global review. *Front. Ecol. Environ.* **2014**, *12*, 288–296. [CrossRef]
20. Li, S.; Li, X. Global understanding of farmland abandonment: A review and prospects. *J. Geogr. Sci.* **2017**, *27*, 1123–1150. [CrossRef]
21. Woodhouse, S.; Good, J.; Lovett, A.; Fuller, R.; Dolman, P. Effects of land-use and agricultural management on birds of marginal farmland: A case study in the Llŷn peninsula, Wales. *Agric. Ecosyst. Environ.* **2005**, *107*, 331–340. [CrossRef]
22. Lark, T.J.; Spawn, S.A.; Bougie, M.; Gibbs, H.K. Cropland expansion in the United States produces marginal yields at high costs to wildlife. *Nat. Commun.* **2020**, *11*, 4295. [CrossRef] [PubMed]
23. Dantas de Miranda, M.; Pereira, H.M.; Corley, M.F.; Merckx, T. Beta diversity patterns reveal positive effects of farmland abandonment on moth communities. *Sci. Rep.* **2019**, *9*, 1549. [CrossRef] [PubMed]
24. Schierhorn, F.; Müller, D.; Beringer, T.; Prishchepov, A.V.; Kuemmerle, T.; Balmann, A. Post-Soviet cropland abandonment and carbon sequestration in European Russia, Ukraine, and Belarus. *Glob. Biogeochem. Cycles* **2013**, *27*, 1175–1185. [CrossRef]
25. Romero-Díaz, A.; Ruiz-Sinoga, J.D.; Robledano-Aymerich, F.; Brevik, E.C.; Cerdà, A. Ecosystem responses to land abandonment in Western Mediterranean Mountains. *Catena* **2017**, *149*, 824–835. [CrossRef]
26. Benayas, J.R.; Martins, A.; Nicolau, J.M.; Schulz, J.J. Abandonment of agricultural land: An overview of drivers and consequences. *CAB Rev. Perspect. Agric. Vet. Sci. Nutr. Nat. Resour.* **2007**, *2*, 1–14. [CrossRef]
27. Pereira, P.; Cerdà, A.; Úbeda, X.; Mataix-Solera, J.; Arcenegui, V.; Zavala, L. Modelling the impacts of wildfire on ash thickness in a short-term period. *Land Degrad. Dev.* **2015**, *26*, 180–192. [CrossRef]
28. Smelansky, I.E. Biodiversity of Agricultural Lands in Russia: Current State and Trends. Available online: https://www.researchgate.net/publication/283504120_Biodiversity_of_Agricultural_Lands_in_Russia_Current_State_and_Trends (accessed on 14 October 2022).
29. Wei, S.; Ying, Z. Farmland abandonment research progress: Influencing factors and simulation model. *J. Resour. Ecol.* **2019**, *10*, 345–352. [CrossRef]

30. Tieskens, K.F.; Schulp, C.J.; Levers, C.; Lieskovský, J.; Kuemmerle, T.; Plieninger, T.; Verburg, P.H. Characterizing European cultural landscapes: Accounting for structure, management intensity and value of agricultural and forest landscapes. *Land Use Policy* **2017**, *62*, 29–39. [CrossRef]
31. Plieninger, T. Habitat loss, Fragmentation, and Alteration–Quantifying the Impactof Land-use Changes on a Spanish Dehesa Landscape by Use of Aerial Photography and GIS. *Landsc. Ecol.* **2006**, *21*, 91–105. [CrossRef]
32. Ruiz-Flan, P.; Garci, J.; Ortigosa, L. Geomorphological evolution of abandoned fields. A case study in the Central Pyrenees. *Catena* **1992**, *19*, 301–308. [CrossRef]
33. Saraykin, V.; Yanbykh, R.; Uzun, V. Assessing the Potential for Russian Grain Exports: A Special Focus on the Prospective Cultivation of Abandoned Land. In *The Eurasian Wheat Belt and Food Security*; Springer: Cham, Switzerland, 2017. [CrossRef]
34. Shi, T.; Li, X.; Xin, L.; Xu, X. The spatial distribution of farmland abandonment and its influential factors at the township level: A case study in the mountainous area of China. *Land Use Policy* **2018**, *70*, 510–520. [CrossRef]
35. Li, S.; Li, X.; Sun, L.; Cao, G.; Fischer, G.; Tramberend, S. An estimation of the extent of cropland abandonment in mountainous regions of China. *Land Degrad. Dev.* **2018**, *29*, 1327–1342. [CrossRef]
36. Ioffe, G.; Nefedova, T.; Kirsten, D.B. Land abandonment in Russia. *Eurasian Geogr. Econ.* **2012**, *53*, 527–549. [CrossRef]
37. Meyfroidt, P.; Schierhorn, F.; Prishchepov, A.V.; Müller, D.; Kuemmerle, T. Drivers, constraints and trade-offs associated with recultivating abandoned cropland in Russia, Ukraine and Kazakhstan. *Glob. Environ. Chang.* **2016**, *37*, 1–15. [CrossRef]
38. Khamzina, A.; Lamers, J.; Vlek, P.L. Conversion of degraded cropland to tree plantations for ecosystem and livelihood benefits. In *Cotton, Water, Salts and Soums*; Springer: Dordrecht, The Netherlands, 2012; pp. 235–248. [CrossRef]
39. Hatna, E.; Bakker, M.M. Abandonment and expansion of arable land in Europe. *Ecosystems* **2011**, *14*, 720–731. [CrossRef]
40. Müller, D.; Leitão, P.J.; Sikor, T. Comparing the determinants of cropland abandonment in Albania and Romania using boosted regression trees. *Agric. Syst.* **2013**, *117*, 66–77. [CrossRef]
41. Gellrich, M.; Zimmermann, N.E. Investigating the regional-scale pattern of agricultural land abandonment in the Swiss mountains: A spatial statistical modelling approach. *Landsc. Urban Plan.* **2007**, *79*, 65–76. [CrossRef]
42. Baumann, M.; Kuemmerle, T.; Elbakidze, M.; Ozdogan, M.; Radeloff, V.C.; Keuler, N.S.; Prishchepov, A.V.; Kruhlov, I.; Hostert, P. Patterns and drivers of post-socialist farmland abandonment in Western Ukraine. *Land Use Policy* **2011**, *28*, 552–562. [CrossRef]
43. Lambin, E.F.; Meyfroidt, P. Land use transitions: Socio-ecological feedback versus socio-economic change. *Land Use Policy* **2010**, *27*, 108–118. [CrossRef]
44. Stellmes, M.; Röder, A.; Udelhoven, T.; Hill, J. Mapping syndromes of land change in Spain with remote sensing time series, demographic and climatic data. *Land Use Policy* **2013**, *30*, 685–702. [CrossRef]
45. García-Ruiz, J.M.; Lana-Renault, N. Hydrological and erosive consequences of farmland abandonment in Europe, with special reference to the Mediterranean region–A review. *Agric. Ecosyst. Environ.* **2011**, *140*, 317–338. [CrossRef]
46. Tarolli, P.; Preti, F.; Romano, N. Terraced landscapes: From an old best practice to a potential hazard for soil degradation due to land abandonment. *Anthropocene* **2014**, *6*, 10–25. [CrossRef]
47. Rousseau, D.M. *The Oxford Handbook of Evidence-Based Management*; Oxford University Press: Oxford, UK, 2012. [CrossRef]
48. Bosserman, P. Invisible Colleges: Diffusion of Knowledge in Scientific Communities. *Am. J. Sociol.* **1973**, *79*, 180–182. [CrossRef]
49. Garfield, E. From the science of science to Scientometrics visualizing the history of science with HistCite software. *J. Informetr.* **2009**, *3*, 173–179. [CrossRef]
50. Liu, Q.; Ye, Y. A Study on Mining Bibliographic Records by Designed Software SATI: Case Study on Library and Information Science. *J. Inf. Resour. Manag.* **2012**, *2*, 50–58.
51. Wang, W.; Lu, C. Visualization analysis of big data research based on Citespace. *Soft Comput.* **2020**, *24*, 8173–8186. [CrossRef]
52. Gagolewski, M. Bibliometric impact assessment with R and the CITAN package. *J. Informetr.* **2011**, *5*, 678–692. [CrossRef]
53. Aria, M.; Cuccurullo, C. bibliometrix: An R-tool for comprehensive science mapping analysis. *J. Informetr.* **2017**, *11*, 959–975. [CrossRef]
54. Xie, H.; Zhang, Y.; Wu, Z.; Lv, T. A bibliometric analysis on land degradation: Current status, development, and future directions. *Land* **2020**, *9*, 28. [CrossRef]
55. Rodríguez-Soler, R.; Uribe-Toril, J.; Valenciano, J.D.P. Worldwide trends in the scientific production on rural depopulation, a bibliometric analysis using bibliometrix R-tool. *Land Use Policy* **2020**, *97*, 104787. [CrossRef]
56. Zhang, X.; Zhang, Y.; Wang, Y.; Fath, B.D. Research progress and hotspot analysis for reactive nitrogen flows in macroscopic systems based on a CiteSpace analysis. *Ecol. Model.* **2021**, *443*, 109456. [CrossRef]
57. Zhang, Y.; Li, X.; Song, W. Determinants of cropland abandonment at the parcel, household and village levels in mountain areas of China: A multi-level analysis. *Land Use Policy* **2014**, *41*, 186–192. [CrossRef]
58. Song, W.; Pijanowski, B.C. The effects of China's cultivated land balance program on potential land productivity at a national scale. *Appl. Geogr.* **2014**, *46*, 158–170. [CrossRef]
59. Yin, H.; Prishchepov, A.V.; Kuemmerle, T.; Bleyhl, B.; Buchner, J.; Radeloff, V.C. Mapping agricultural land abandonment from spatial and temporal segmentation of Landsat time series. *Remote Sens. Environ. Interdiscip. J.* **2018**, *210*, 12–24. [CrossRef]
60. Quintas-Soriano, C.; Buerkert, A.; Plieninger, T. Effects of land abandonment on nature contributions to people and good quality of life components in the Mediterranean region: A review. *Land Use Policy* **2022**, *116*, 106053. [CrossRef]
61. Ekundayo, T.C.; Okoh, A.I. A global bibliometric analysis of Plesiomonas-related research (1990–2017). *PLoS ONE* **2018**, *13*, e0207655. [CrossRef]

62. Zyoud, S.H.; Waring, W.S.; Al-Jabi, S.W.; Sweileh, W.M. Global cocaine intoxication research trends during 1975-2015: A bibliometric analysis of Web of Science publications. *Subst Abus. Treat Prev Policy* **2017**, *12*, 6. [CrossRef] [PubMed]
63. Okolie, C.C.; Ogundeji, A.A. Effect of COVID-19 on agricultural production and food security: A scientometric analysis. *Humanit. Soc. Sci. Commun.* **2022**, *9*, 64. [CrossRef]
64. Waltman, L. A review of the literature on citation impact indicators. *J. Informetr.* **2016**, *10*, 365–391. [CrossRef]
65. Callon, M.; Courtial, J.-P.; Turner, W.A.; Bauin, S. From translations to problematic networks: An introduction to co-word analysis. *Soc. Sci. Inf.* **1983**, *22*, 191–235. [CrossRef]
66. Glänzel, W. National characteristics in international scientific co-authorship relations. *Scientometrics* **2001**, *51*, 69–115. [CrossRef]
67. Zhao, D.; Strotmann, A. Evolution of research activities and intellectual influences in information science 1996–2005: Introducing author bibliographic-coupling analysis. *J. Am. Soc. Inf. Sci. Technol.* **2008**, *59*, 2070–2086. [CrossRef]
68. White, H.D.; Griffith, B.C. Author cocitation: A literature measure of intellectual structure. *J. Am. Soc. Inf. Sci.* **1981**, *32*, 163–171. [CrossRef]
69. Yan, E.; Ding, Y. Scholarly network similarities: How bibliographic coupling networks, citation networks, cocitation networks, topical networks, coauthorship networks, and coword networks relate to each other. *J. Am. Soc. Inf. Sci. Technol.* **2012**, *63*, 1313–1326. [CrossRef]
70. Kessler, M.M. Bibliographic coupling between scientific papers. *Am. Doc.* **1963**, *14*, 10–25. [CrossRef]
71. Yang, S.; Han, R.; Wolfram, D.; Zhao, Y. Visualizing the intellectual structure of information science (2006–2015): Introducing author keyword coupling analysis. *J. Informetr.* **2016**, *10*, 132–150. [CrossRef]
72. Small, H. Co-citation in the scientific literature: A new measure of the relationship between two documents. *J. Am. Soc. Inf. Sci.* **1973**, *24*, 265–269. [CrossRef]
73. Gibson, C.; Watt, T.; Brown, V. The use of sheep grazing to recreate species-rich grassland from abandoned arable land. *Biol. Conserv.* **1987**, *42*, 165–183. [CrossRef]
74. Silver, W.L.; Ostertag, R.; Lugo, A.E. The potential for carbon sequestration through reforestation of abandoned tropical agricultural and pasture lands. *Restor. Ecol.* **2000**, *8*, 394–407. [CrossRef]
75. Lasanta, T.; Garcıa-Ruiz, J.; Pérez-Rontomé, C.; Sancho-Marcén, C. Runoff and sediment yield in a semi-arid environment: The effect of land management after farmland abandonment. *Catena* **2000**, *38*, 265–278. [CrossRef]
76. Cerdà, A. Soil erosion after land abandonment in a semiarid environment of southeastern Spain. *Arid Land Res. Manag.* **1997**, *11*, 163–176. [CrossRef]
77. Molinillo, M.; Lasanta, T.; García-Ruiz, J.M. Managing mountainous degraded landscapes after farmland abandonment in the Central Spanish Pyrenees. *Environ. Manag.* **1997**, *21*, 587. [CrossRef]
78. Cramer, V.A.; Hobbs, R.J.; Standish, R.J. What's new about old fields? Land abandonment and ecosystem assembly. *Trends Ecol. Evol.* **2008**, *23*, 104–112. [CrossRef] [PubMed]
79. Campbell, J.E.; Lobell, D.B.; Genova, R.C.; Field, C.B. The global potential of bioenergy on abandoned agriculture lands. *Environ. Sci. Technol.* **2008**, *42*, 5791–5794. [CrossRef]
80. Romero-Calcerrada, R.; Perry, G.L. The role of land abandonment in landscape dynamics in the SPA 'Encinares del río Alberche y Cofio, Central Spain, 1984–1999. *Landsc. Urban Plan.* **2004**, *66*, 217–232. [CrossRef]
81. Poyatos, R.; Latron, J.; Llorens, P. Land use and land cover change after agricultural abandonment. *Mt. Res. Dev.* **2003**, *23*, 362–368. [CrossRef]
82. Bowen, M.E.; McAlpine, C.A.; House, A.P.; Smith, G.C. Regrowth forests on abandoned agricultural land: A review of their habitat values for recovering forest fauna. *Biol. Conserv.* **2007**, *140*, 273–296. [CrossRef]
83. Lugo, A.E.; Helmer, E. Emerging forests on abandoned land: Puerto Rico's new forests. *For. Ecol. Manag.* **2004**, *190*, 145–161. [CrossRef]
84. Koulouri, M.; Giourga, C. Land abandonment and slope gradient as key factors of soil erosion in Mediterranean terraced lands. *Catena* **2007**, *69*, 274–281. [CrossRef]
85. Estel, S.; Kuemmerle, T.; Alcántara, C.; Levers, C.; Prishchepov, A.; Hostert, P. Mapping farmland abandonment and recultivation across Europe using MODIS NDVI time series. *Remote Sens. Environ.* **2015**, *163*, 312–325. [CrossRef]
86. Ustaoglu, E.; Collier, M.J. Farmland abandonment in Europe: An overview of drivers, consequences, and assessment of the sustainability implications. *Environ. Rev.* **2018**, *26*, 396–416. [CrossRef]
87. Lasanta, T.; Arnáez, J.; Pascual, N.; Ruiz-Flaño, P.; Errea, M.; Lana-Renault, N. Space–time process and drivers of land abandonment in Europe. *Catena* **2017**, *149*, 810–823. [CrossRef]
88. Yan, J.; Yang, Z.; Li, Z.; Li, X.; Xin, L.; Sun, L. Drivers of cropland abandonment in mountainous areas: A household decision model on farming scale in Southwest China. *Land Use Policy* **2016**, *57*, 459–469. [CrossRef]
89. Kolecka, N.; Kozak, J.; Kaim, D.; Dobosz, M.; Ostafin, K.; Ostapowicz, K.; Wężyk, P.; Price, B. Understanding farmland abandonment in the Polish Carpathians. *Appl. Geogr.* **2017**, *88*, 62–72. [CrossRef]
90. Su, G.; Okahashi, H.; Chen, L. Spatial pattern of farmland abandonment in Japan: Identification and determinants. *Sustainability* **2018**, *10*, 3676. [CrossRef]
91. Han, Z.; Song, W. Abandoned cropland: Patterns and determinants within the Guangxi karst mountainous area, China. *Appl. Geogr.* **2020**, *122*, 102245. [CrossRef]

92. Xu, D.; Deng, X.; Guo, S.; Liu, S. Labor migration and farmland abandonment in rural China: Empirical results and policy implications. *J. Environ. Manag.* **2019**, *232*, 738–750. [CrossRef]
93. Peters, H.; Van Raan, A. Structuring scientific activities by co-author analysis: An expercise on a university faculty level. *Scientometrics* **1991**, *20*, 235–255. [CrossRef]
94. Kuemmerle, T.; Hostert, P.; Radeloff, V.C.; van der Linden, S.; Perzanowski, K.; Kruhlov, I. Cross-border Comparison of Post-socialist Farmland Abandonment in the Carpathians. *Ecosystems* **2008**, *11*, 614–628. [CrossRef]
95. Müller, D.; Munroe, D.K. Changing Rural Landscapes in Albania: Cropland Abandonment and Forest Clearing in the Postsocialist Transition. *Ann. Assoc. Am. Geogr.* **2008**, *98*, 855–876. [CrossRef]
96. Prishchepov, A.V.; Müller, D.; Dubinin, M.; Baumann, M.; Radeloff, V.C. Determinants of agricultural land abandonment in post-Soviet European Russia. *Land Use Policy* **2013**, *30*, 873–884. [CrossRef]
97. Lana-Renault, N.; Regüés, D. Seasonal patterns of suspended sediment transport in an abandoned farmland catchment in the Central Spanish Pyrenees. *Earth Surf. Process. Landf.* **2009**, *34*, 1291–1301. [CrossRef]
98. Khorchani, M.; Nadal-Romero, E.; Tague, C.; Lasanta, T.; Zabalza, J.; Lana-Renault, N.; Dominguez-Castro, F.; Choate, J. Effects of active and passive land use management after cropland abandonment on water and vegetation dynamics in the Central Spanish Pyrenees. *Sci. Total Environ.* **2020**, *717*, 137160. [CrossRef] [PubMed]
99. Deng, L.; Wang, G.-l.; Liu, G.-b.; Shangguan, Z.-p. Effects of age and land-use changes on soil carbon and nitrogen sequestrations following cropland abandonment on the Loess Plateau, China. *Ecol. Eng.* **2016**, *90*, 105–112. [CrossRef]
100. Zhang, C.; Liu, G.; Xue, S.; Zhang, C. Rhizosphere soil microbial properties on abandoned croplands in the Loess Plateau, China during vegetation succession. *Eur. J. Soil Biol.* **2012**, *50*, 127–136. [CrossRef]
101. Zhang, C.; Liu, G.; Xue, S.; Wang, G. Soil bacterial community dynamics reflect changes in plant community and soil properties during the secondary succession of abandoned farmland in the Loess Plateau. *Soil Biol. Biochem.* **2016**, *97*, 40–49. [CrossRef]
102. Deng, L.; Shangguan, Z.P.; Sweeney, S. Changes in soil carbon and nitrogen following land abandonment of farmland on the Loess Plateau, China. *PLoS ONE* **2013**, *8*, e71923. [CrossRef]
103. Deng, X.; Xu, D.; Qi, Y.; Zeng, M. Labor Off-Farm Employment and Cropland Abandonment in Rural China: Spatial Distribution and Empirical Analysis. *Int. J. Env. Res. Public Health* **2018**, *15*, 1808. [CrossRef]
104. Deng, X.; Xu, D.; Zeng, M.; Qi, Y. Does Internet use help reduce rural cropland abandonment? Evidence from China. *Land Use Policy* **2019**, *89*, 104243. [CrossRef]
105. Tullus, T.; Tullus, A.; Roosaluste, E.; Lutter, R.; Tullus, H. Vascular plant and bryophyte flora in midterm hybrid aspen plantations on abandoned agricultural land. *Can. J. For. Res.* **2015**, *45*, 1183–1191. [CrossRef]
106. Uri, V.; Lõhmus, K.; Mander, Ü.; Ostonen, I.; Aosaar, J.; Maddison, M.; Helmisaari, H.-S.; Augustin, J. Long-term effects on the nitrogen budget of a short-rotation grey alder (*Alnus incana* (L.) Moench) forest on abandoned agricultural land. *Ecol. Eng.* **2011**, *37*, 920–930. [CrossRef]
107. Uri, V.; Lõhmus, K.; Ostonen, I.; Tullus, H.; Lastik, R.; Vildo, M. Biomass production, foliar and root characteristics and nutrient accumulation in young silver birch (Betula pendula Roth.) stand growing on abandoned agricultural land. *Eur. J. For. Res.* **2007**, *126*, 495–506. [CrossRef]
108. Chaudhary, S.; Wang, Y.; Dixit, A.M.; Khanal, N.R.; Xu, P.; Yan, K.; Liu, Q.; Lu, Y.; Li, M. Eco-environmental risk evaluation for land use planning in areas of potential farmland abandonment in the high mountains of Nepal Himalayas. *Sustainability* **2019**, *11*, 6931. [CrossRef]
109. Chaudhary, S.; Wang, Y.; Dixit, A.M.; Khanal, N.R.; Xu, P.; Fu, B.; Yan, K.; Liu, Q.; Lu, Y.; Li, M. Spatiotemporal degradation of abandoned farmland and associated eco-environmental risks in the high mountains of the Nepalese Himalayas. *Land* **2019**, *9*, 1. [CrossRef]
110. Chaudhary, S.; Wang, Y.; Khanal, N.R.; Xu, P.; Fu, B.; Dixit, A.M.; Yan, K.; Liu, Q.; Lu, Y. Social impact of farmland abandonment and its eco-environmental vulnerability in the high mountain region of Nepal: A case study of Dordi River Basin. *Sustainability* **2018**, *10*, 2331. [CrossRef]
111. Chaudhary, S.; Wang, Y.; Dixit, A.M.; Khanal, N.R.; Xu, P.; Fu, B.; Yan, K.; Liu, Q.; Lu, Y.; Li, M. A Synopsis of Farmland Abandonment and Its Driving Factors in Nepal. *Land* **2020**, *9*, 84. [CrossRef]
112. Yamanaka, S.; Hanioka, M.; Nakamura, F. Changes in Ground Beetle and Bird Species After Farmland Abandonment. In *Biodiversity Conservation Using Umbrella Species*; Springer: Singapore, 2018. [CrossRef]
113. Hanioka, M.; Yamaura, Y.; Senzaki, M.; Yamanaka, S.; Kawamura, K.; Nakamura, F. Assessing the landscape-dependent restoration potential of abandoned farmland using a hierarchical model of bird communities. *Agric. Ecosyst. Environ.* **2018**, *265*, 217–225. [CrossRef]
114. Kitazawa, M.; Yamaura, Y.; Senzaki, M.; Kawamura, K.; Hanioka, M.; Nakamura, F. An evaluation of five agricultural habitat types for openland birds: Abandoned farmland can have comparative values to undisturbed wetland. *Ornithol. Sci.* **2019**, *18*, 3–16. [CrossRef]
115. Alcantara, C.; Kuemmerle, T.; Baumann, M.; Bragina, E.V.; Griffiths, P.; Hostert, P.; Knorn, J.; Müller, D.; Prishchepov, A.V.; Schierhorn, F. Mapping the extent of abandoned farmland in Central and Eastern Europe using MODIS time series satellite data. *Environ. Res. Lett.* **2013**, *8*, 035035. [CrossRef]

116. Dara, A.; Baumann, M.; Kuemmerle, T.; Pflugmacher, D.; Rabe, A.; Griffiths, P.; Hölzel, N.; Kamp, J.; Freitag, M.; Hostert, P. Mapping the timing of cropland abandonment and recultivation in northern Kazakhstan using annual Landsat time series. *Remote Sens. Environ.* **2018**, *213*, 49–60. [CrossRef]
117. Levers, C.; Schneider, M.; Prishchepov, A.V.; Estel, S.; Kuemmerle, T. Spatial variation in determinants of agricultural land abandonment in Europe. *Sci Total Env.* **2018**, *644*, 95–111. [CrossRef]
118. Gellrich, M.; Baur, P.; Koch, B.; Zimmermann, N.E. Agricultural land abandonment and natural forest re-growth in the Swiss mountains: A spatially explicit economic analysis. *Agric. Ecosyst. Environ.* **2007**, *118*, 93–108. [CrossRef]
119. Renwick, A.; Jansson, T.; Verburg, P.H.; Revoredo-Giha, C.; Britz, W.; Gocht, A.; McCracken, D. Policy reform and agricultural land abandonment in the EU. *Land Use Policy* **2013**, *30*, 446–457. [CrossRef]
120. Verburg, P.H.; Overmars, K.P. Combining top-down and bottom-up dynamics in land use modeling: Exploring the future of abandoned farmlands in Europe with the Dyna-CLUE model. *Landsc. Ecol.* **2009**, *24*, 1167–1181. [CrossRef]
121. Pueyo, Y.; Beguería, S. Modelling the rate of secondary succession after farmland abandonment in a Mediterranean mountain area. *Landsc. Urban Plan.* **2007**, *83*, 245–254. [CrossRef]
122. Lesschen, J.; Cammeraat, L.; Kooijman, A.; van Wesemael, B. Development of spatial heterogeneity in vegetation and soil properties after land abandonment in a semi-arid ecosystem. *J. Arid Environ.* **2008**, *72*, 2082–2092. [CrossRef]
123. MacDonald, D.; Crabtree, J.R.; Wiesinger, G.; Dax, T.; Stamou, N.; Fleury, P.; Lazpita, J.G.; Gibon, A. Agricultural abandonment in mountain areas of Europe: Environmental consequences and policy response. *J. Environ. Manag.* **2000**, *59*, 47–69. [CrossRef]
124. García-Ruiz, J.; Ruiz Flaño, P.; Lasanta, T. Soil erosion after farmland abandonment in submediterranean mountains: A general outlook. In *Soil Degradation and Desertification in Mediterranean Environments*; Geoforma Ediciones: Logroño, Spain, 1996; pp. 165–183.
125. Nunes, A.; Coelho, C.d.O.A.; De Almeida, A.; Figueiredo, A. Soil erosion and hydrological response to land abandonment in a central inland area of Portugal. *Land Degrad. Dev.* **2010**, *21*, 260–273. [CrossRef]
126. Duarte, F.; Jones, N.; Fleskens, L. Traditional olive orchards on sloping land: Sustainability or abandonment? *J. Environ. Manag.* **2008**, *89*, 86–98. [CrossRef] [PubMed]
127. Lesschen, J.; Cammeraat, L.; Nieman, T. Erosion and terrace failure due to agricultural land abandonment in a semi-arid environment. *Earth Surf. Process. Landf. J. Br. Geomorphol. Res. Group* **2008**, *33*, 1574–1584. [CrossRef]
128. Llorens, P.; Latron, J.; Gallart, F. Analysis of the role of agricultural abandoned terraces on the hydrology and sediment dynamics in a small mountainous basin (High Llobregat, Eastern Pyrenees). *Pirineos* **1992**, *139*, 27–46. [CrossRef]
129. Cammeraat, L.; Imeson, A. The evolution and significance of soil–vegetation patterns following land abandonment and fire in Spain. *Catena* **1999**, *37*, 107–127. [CrossRef]
130. Müller, D.; Kuemmerle, T.; Rusu, M.; Griffiths, P. Lost in transition: Determinants of post-socialist cropland abandonment in Romania. *J. Land Use Sci.* **2009**, *4*, 109–129. [CrossRef]
131. Alcantara, C.; Kuemmerle, T.; Prishchepov, A.V.; Radeloff, V.C. Mapping abandoned agriculture with multi-temporal MODIS satellite data. *Remote Sens. Environ.* **2012**, *124*, 334–347. [CrossRef]
132. Prishchepov, A.V.; Radeloff, V.C.; Baumann, M.; Kuemmerle, T.; Müller, D. Effects of institutional changes on land use: Agricultural land abandonment during the transition from state-command to market-driven economies in post-Soviet Eastern Europe. *Environ. Res. Lett.* **2012**, *7*, 024021. [CrossRef]
133. Gehrig-Fasel, J.; Guisan, A.; Zimmermann, N.E. Tree line shifts in the Swiss Alps: Climate change or land abandonment? *J. Veg. Sci.* **2007**, *18*, 571–582. [CrossRef]
134. Ding, Y. Scientific collaboration and endorsement: Network analysis of coauthorship and citation networks. *J. Informetr.* **2011**, *5*, 187–203. [CrossRef]
135. Abdi, H.; Valentin, D. Multiple correspondence analysis. *Encycl. Meas. Stat.* **2007**, *2*, 651–657.
136. Zakkak, S.; Halley, J.M.; Akriotis, T.; Kati, V. Lizards along an agricultural land abandonment gradient in Pindos Mountains, Greece. *Amphib. -Reptil.* **2015**, *36*, 253–264. [CrossRef]
137. Katayama, N.; Mashiko, M.; Koshida, C.; Yamaura, Y. Effects of rice-field abandonment rates on bird communities in mixed farmland–woodland landscapes in Japan. *Agric. Ecosyst. Environ.* **2021**, *319*, 107539. [CrossRef]
138. Hanioka, M.; Yamaura, Y.; Yamanaka, S.; Senzaki, M.; Kawamura, K.; Terui, A.; Nakamura, F. How much abandoned farmland is required to harbor comparable species richness and abundance of bird communities in wetland? Hierarchical community model suggests the importance of habitat structure and landscape context. *Biodivers. Conserv.* **2018**, *27*, 1831–1848. [CrossRef]
139. Zakkak, S.; Radovic, A.; Panitsa, M.; Vassilev, K.; Shuka, L.; Kuttner, M.; Schindler, S.; Kati, V.; Woods, K. Vegetation patterns along agricultural land abandonment in the Balkans. *J. Veg. Sci.* **2018**, *29*, 877–886. [CrossRef]
140. Zhang, K.; Dang, H.; Tan, S.; Wang, Z.; Zhang, Q. Vegetation community and soil characteristics of abandoned agricultural land and pine plantation in the Qinling Mountains, China. *For. Ecol. Manag.* **2010**, *259*, 2036–2047. [CrossRef]
141. Wang, G.; Liu, G.; Xu, M. Above- and belowground dynamics of plant community succession following abandonment of farmland on the Loess Plateau, China. *Plant Soil* **2008**, *316*, 227–239. [CrossRef]
142. Yannelli, F.A.; Tabeni, S.; Mastrantonio, L.E.; Vezzani, N. Assessing degradation of abandoned farmlands for conservation of the Monte Desert biome in Argentina. *Environ. Manag.* **2014**, *53*, 231–239. [CrossRef] [PubMed]
143. Standish, R.J.; Cramer, V.A.; Hobbs, R.J. Land-use legacy and the persistence of invasive Avena barbata on abandoned farmland. *J. Appl. Ecol.* **2008**, *45*, 1576–1583. [CrossRef]

144. Cojzer, M.; Diaci, J.; Brus, R. Tending of young forests in secondary succession on abandoned agricultural lands: An experimental study. *Forests* **2014**, *5*, 2658–2678. [CrossRef]
145. Treml, V.; Šenfeldr, M.; Chuman, T.; Ponocná, T.; Demková, K. Twentieth century treeline ecotone advance in the Sudetes Mountains (Central Europe) was induced by agricultural land abandonment rather than climate change. *J. Veg. Sci.* **2016**, *27*, 1209–1221. [CrossRef]
146. Zhang, C.; Liu, G.; Song, Z.; Qu, D.; Fang, L.; Deng, L. Natural succession on abandoned cropland effectively decreases the soil erodibility and improves the fungal diversity. *Ecol. Appl.* **2017**, *27*, 2142–2154. [CrossRef]
147. Chang, X.; Chai, Q.; Wu, G.; Zhu, Y.; Li, Z.; Yang, Y.; Wang, G. Soil organic carbon accumulation in abandoned croplands on the Loess Plateau. *Land Degrad. Dev.* **2017**, *28*, 1519–1527. [CrossRef]
148. Jiang, J.-P.; Xiong, Y.-C.; Jiang, H.-M.; De-You, Y.; Ya-Jie, S.; Feng-Min, L. Soil microbial activity during secondary vegetation succession in semiarid abandoned lands of Loess Plateau. *Pedosphere* **2009**, *19*, 735–747. [CrossRef]
149. Yuan, Z.-Q.; Yu, K.-L.; Epstein, H.; Fang, C.; Li, J.-T.; Liu, Q.-Q.; Liu, X.-W.; Gao, W.-J.; Li, F.-M. Effects of legume species introduction on vegetation and soil nutrient development on abandoned croplands in a semi-arid environment on the Loess Plateau, China. *Sci. Total Environ.* **2016**, *541*, 692–700. [CrossRef]
150. Zhang, Y.; Wei, L.; Wei, X.; Liu, X.; Shao, M. Long-term afforestation significantly improves the fertility of abandoned farmland along a soil clay gradient on the Chinese Loess Plateau. *Land Degrad. Dev.* **2018**, *29*, 3521–3534. [CrossRef]
151. Pazúr, R.; Lieskovský, J.; Feranec, J.; Ot'ahel', J. Spatial determinants of abandonment of large-scale arable lands and managed grasslands in Slovakia during the periods of post-socialist transition and European Union accession. *Appl. Geogr.* **2014**, *54*, 118–128. [CrossRef]
152. Janus, J.; Bozek, P. Aerial laser scanning reveals the dynamics of cropland abandonment in Poland. *J. Land Use Sci.* **2019**, *14*, 378–396. [CrossRef]
153. Díaz, G.I.; Nahuelhual, L.; Echeverría, C.; Marín, S. Drivers of land abandonment in Southern Chile and implications for landscape planning. *Landsc. Urban Plan.* **2011**, *99*, 207–217. [CrossRef]
154. Zgłobicki, W.; Karczmarczuk, K.; Baran-Zgłobicka, B. Intensity and driving forces of land abandonment in Eastern Poland. *Appl. Sci.* **2020**, *10*, 3500. [CrossRef]
155. Tremblay, S.; Ouimet, R. White spruce plantations on abandoned agricultural land: Are they more effective as C sinks than natural succession? *Forests* **2013**, *4*, 1141–1157. [CrossRef]
156. Terres, J.-M.; Scacchiafichi, L.N.; Wania, A.; Ambar, M.; Anguiano, E.; Buckwell, A.; Coppola, A.; Gocht, A.; Källström, H.N.; Pointereau, P. Farmland abandonment in Europe: Identification of drivers and indicators, and development of a composite indicator of risk. *Land Use Policy* **2015**, *49*, 20–34. [CrossRef]
157. Rajpar, H.; Zhang, A.; Razzaq, A.; Mehmood, K.; Pirzado, M.B.; Hu, W. Agricultural land abandonment and farmers' perceptions of land use change in the indus plains of Pakistan: A case study of Sindh province. *Sustainability* **2019**, *11*, 4663. [CrossRef]
158. Dax, T.; Schroll, K.; Machold, I.; Derszniak-Noirjean, M.; Schuh, B.; Gaupp-Berghausen, M. Land abandonment in mountain areas of the EU: An inevitable side effect of farming modernization and neglected threat to sustainable land use. *Land* **2021**, *10*, 591. [CrossRef]
159. Orlandi, S.; Probo, M.; Sitzia, T.; Trentanovi, G.; Garbarino, M.; Lombardi, G.; Lonati, M. Environmental and land use determinants of grassland patch diversity in the western and eastern Alps under agro-pastoral abandonment. *Biodivers. Conserv.* **2016**, *25*, 275–293. [CrossRef]
160. Vuichard, N.; Ciais, P.; Belelli, L.; Smith, P.; Valentini, R. Carbon sequestration due to the abandonment of agriculture in the former USSR since 1990. *Glob. Biogeochem. Cycles* **2008**, *22*, GB4018. [CrossRef]
161. Hunziker, M. The spontaneous reafforestation in abandoned agricultural lands: Perception and aesthetic assessment by locals and tourists. *Landsc. Urban Plan.* **1995**, *31*, 399–410. [CrossRef]
162. Rivera, L.; Zimmerman, J.; Aide, T. Forest recovery in abandoned agricultural lands in a karst region of the Dominican Republic. *Plant Ecol.* **2000**, *148*, 115–125. [CrossRef]
163. Smaliychuk, A.; Müller, D.; Prishchepov, A.V.; Levers, C.; Kruhlov, I.; Kuemmerle, T. Recultivation of abandoned agricultural lands in Ukraine: Patterns and drivers. *Glob. Environ. Chang.* **2016**, *38*, 70–81. [CrossRef]
164. Yin, H.; Brandão Jr, A.; Buchner, J.; Helmers, D.; Iuliano, B.G.; Kimambo, N.E.; Lewińska, K.E.; Razenkova, E.; Rizayeva, A.; Rogova, N. Monitoring cropland abandonment with Landsat time series. *Remote Sens. Environ.* **2020**, *246*, 111873. [CrossRef]
165. Næss, J.S.; Cavalett, O.; Cherubini, F. The land–energy–water nexus of global bioenergy potentials from abandoned cropland. *Nat. Sustain.* **2021**, *4*, 525–536. [CrossRef]
166. Sluiter, R.; de Jong, S.M. Spatial patterns of Mediterranean land abandonment and related land cover transitions. *Landsc. Ecol.* **2006**, *22*, 559–576. [CrossRef]
167. Prishchepov, A.V.; Radeloff, V.C.; Dubinin, M.; Alcantara, C. The effect of Landsat ETM/ETM + image acquisition dates on the detection of agricultural land abandonment in Eastern Europe. *Remote Sens. Environ.* **2012**, *126*, 195–209. [CrossRef]

International Journal of
Environmental Research and Public Health

MDPI

Review

Evolutionary Overview of Terrace Research Based on Bibliometric Analysis in Web of Science from 1991 to 2020

Qianru Chen *, Yuyang Wen, Xinmin Zhang and Zhenhong Zhu

Institute of Ecological Civilization, Jiangxi University of Finance and Economics, Nanchang 330013, China; 2201920708@stu.jxufe.edu.cn (Y.W.); 1201800118@jxufe.edu.cn (X.Z.); 2202010105@stu.jxufe.edu.cn (Z.Z.)
* Correspondence: 1202000027@jxufe.edu.cn; Tel.: +86-791-8382-0732

Abstract: Based on the Web of Science core collection database, this paper retrieves 349 research papers on terraced fields published during 1991–2020. Keyword co-occurrence analysis, cluster analysis, and thematic evolutionary analysis were used to identify the evolutionary path of terrace research. The findings were as follows: (1) In the past 20 years, the study of terraced fields has shown an upward trend. The number of annual published papers during 2012–2020 was much more than that during 1991–2011, but papers during 1991–2011 were more academically influential than those during 2012–2020. (2) Regional analysis showed that terrace research in China is the most abundant currently, and is mainly focused on agricultural production, agricultural engineering, cultural tourism, and ecological environment. (3) Keyword co-occurrence analysis showed that terrace landscape, terrace agriculture, terrace abandonment, land use change, soil and water conservation, and sustainable utilization of typical terraces are the main modules of current terrace studies. (4) In a temporal dynamic perspective, terrace research presented 10 main evolutionary paths during 1991–2020, reflecting the trend of terrace research towards sustainable terrace development of ecological agriculture and ecosystem service. (5) Finally, this paper suggests that here is a need to deepen studies on terrace ecosystem services and human well-being based on their structure and processes, to analyze the interaction and comprehensive effect of natural process and humanistic driving forces on terrace abandonment, and to explore the multi-functional benefits and sustainable management of high quality terraced landscape.

Keywords: agricultural terraces; land use; bibliometrics; co-word analysis; thematic evolution

Citation: Chen, Q.; Wen, Y.; Zhang, X.; Zhu, Z. Evolutionary Overview of Terrace Research Based on Bibliometric Analysis in Web of Science from 1991 to 2020. *Int. J. Environ. Res. Public Health* **2022**, *19*, 7796. https://doi.org/10.3390/ijerph19137796

Academic Editors: Wei Song and Hualin Xie

Received: 16 May 2022
Accepted: 23 June 2022
Published: 25 June 2022

1. Introduction

Terraces refer to cultivated land with terraced distributions, horizontal surfaces, and ridges [1]. Terraced fields are recognized as living fossils created by human beings in nature, and they are great symbols of human utilization and transformation of nature. Terraces built at different times are scattered all over the world, especially in Asia and Europe. Terraced fields first appeared in prehistoric times. Terraced fields were initially used to build fortifications on hillsides and to cultivate land. Later, they evolved into an intensive agricultural production mode on mountains all over the world [2]. Nowadays, various new terraces have been built in mountainous and hilly areas, and they can be divided into different types according to different classification standards. For example, terraces can be divided into table terraces and wave terraces by their cross-section structure [3]. They can be divided into horizontal terraces, slope terraces, reverse slope terraces, and separated slope terraces by the field surface structure, among which the horizontal terrace is the most common [4]. Terraces can be divided into earth terraces and stone terraces by construction material. In modern times, cement terraces have become more and more common due to their role in landslide prevention and control. In addition, terraces can be divided into rice terraces and dry farming terraces, where cash crops such as grapes and kiwifruit are usually planted. According to literature review, current research on terraces mainly focuses on their history and distribution, and their value and utilization. The world-famous terraces listed on the World Cultural Heritage list by UNESCO, including the Honghe Hani rice terraces and Yuanyang terraces in China, the Ifugao rice terraces in the Philippines, and the Lavaux vineyard terraces in Switzerland, have attracted most scholars' attention in fields such as agricultural planting benefits, soil and water conservation, and landscape aesthetics.

Terraced fields are of great value in broadening agricultural production sources, alleviating soil erosion and degradation [5,6], saving agricultural water resources, reducing runoff, improving biodiversity [7], and increasing ecological and cultural values. Terrace archaeology has also shown that terraced fields in different periods were built for different reasons, such as increasing per capita cultivated land area and grain output, soil and water conservation, and cultural landscape promotion. Due to the limitations of population and terrain, people in areas with low per capita cultivated land area have to develop terraces in hilly areas to increase food output. Modern scientific research in areas with severe soil erosion has proven that terraced fields effectively slow surface runoff [8–10]. Terraced fields help reduce water loss in soil [11] and prevent the deposition of soil organic matter from terraces on the bottom of sloping land or prevent losses [12,13]. Some studies have also found that terraced fields play an important role in microclimate and flood relief [14,15]. More than material practicability, the value of terraced fields is also reflected in intangible cultural value [16], such as providing landscape aesthetics [17–19] or cultural heritage [20,21].

Although terraced fields are of great value, their utilization is not optimal. Studies show that influenced by agricultural input cost, grain yield, agricultural facilities, labor quantity, and labor opportunity cost [22–26], the abandoned terrace area has reached 77.4% in the past 60 years [27]. Compared with ordinary farmland, terraced fields are more likely to be abandoned due to poor farming, irrigation, or transportation conditions, which will result in low grain yield. Especially in recent years, with the rapid development of secondary and tertiary industries, the proportion of farmers' agricultural income continues to decrease, while the cost of farming continues to rise. In this context, terraced fields will be abandoned preferentially among farmers' contracted land.

The influence of terraced land abandonment is multifaceted. Although terrace abandonment may benefit biodiversity improvement and soil habitat restoration, its adverse effects cannot be ignored. Research show that terrace abandonment will cause resource waste, ecosystem service value reduction, land cover change, and an overall decline in water resource availability [28], resulting in wildfire risk and land degradation [29,30]. However, the adverse influence of abandonment will diminish over time [31].

More than typical ecological agriculture modes in mountainous areas, terraced fields are also outstanding ecological and cultural landscapes and precious agricultural cultural heritage, which have important enlightenment and reference significance to the sustainable development of agriculture. The second section of this paper introduces data sources and research methods. The third and fourth sections describe quantitative and qualitative scientometric analysis on terraced fields by bibliometrix and VOSviewer software, respectively. The fifth section draws conclusions and describes prospects for future research. This review provides a broader vision of terrace study, which has a positive effect on terrace landscape maintenance, ecological protection, and sustainable development in hilly and mountainous areas.

2. Data Sources and Research Methods

2.1. Data Sources

The data in this study were sourced from the world's largest comprehensive information resource, the Web of Science core collection database of the Institute for Science Information (ISI). This database includes over 8700 core academic journals in various fields of natural sciences, engineering, biomedicine, social sciences, arts, and humanities. In this paper, the core collection of the Web of Science was used as the data source, the SCI-EXPANDED and SSCI databases were indexed, articles were selected as the literature type, and English was set as the language. After screening and removing the duplicated data,

349 papers on terraced fields during 1991–2020 were obtained, and the downloaded data was saved in a text format.

2.2. Research Methods

Bibliometric analysis provides a comprehensive overview of a large body of research literature and further develops previously unevaluated insights by allowing quantitative and objective identification of past and present research topics [32]. The Bibliometrix *R* package provides a set of tools for quantitative research in scientific metrology. R-language-based bibliometric software developed by Dr. Aria and others provides support for importing and processing literature information from SCOPUS and WoS, statistically analyzing relevant scientific literature indices, constructing a data matrix, and conducting research and visualization processing on co-citation, coupling, cooperative analysis, and co-word analysis [33]. It has been widely used in scientific research in various fields in recent years [34–36]. VOS viewer, developed by Jan van Eck and Ludo Waltman, is widely used in different kinds of co-occurrence analysis software, especially in keyword co-occurrence analysis [37–41]. To accurately and comprehensively analyze the research status and popular topics on terraced fields, this study used VOS viewer and bibliometrix software packages to analyze and visualize current research on terraced fields. This paper analyzes the annual number of published papers, research power (country and author), and popular research topics quantitatively and qualitatively. At the same time, this paper analyzes the trends in terrace research through historical citation analysis, theme evolution analysis, and coupling analysis. The specific research steps are shown in Figure 1.

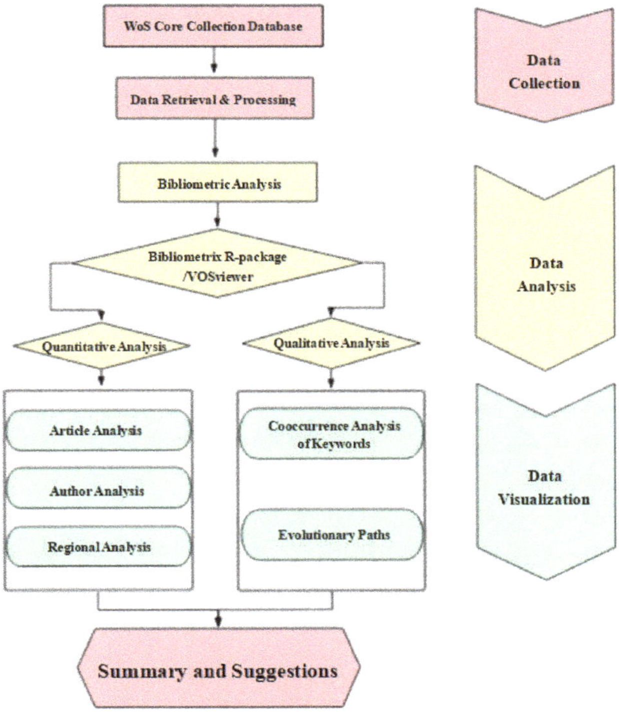

Figure 1. Bibliometrix and science-mapping workflow.

3. Summary of Quantitative Research

3.1. Article Publication and Citation Analysis

Based on the statistical analysis of 349 papers on terraces during 1991–2020 from WoS, the research was divided into two stages. The first stage was the exploration of the period of low yield from 1991 to 2011, which was characterized by the small number of published papers on terraces. In this period, the average number of annual published papers was 7.1 and the average annual number of citations was 23.1. The second stage was the period of increasing production during 2012–2020. The average number of annual published papers had increased to 21.7 and the average annual citations reached 10.9. While the number of published papers in the second stage was approximately three times that of the first stage, the number of citations in the second stage was only half that of the first stage, indicating a greater academic influence in the first stage and a higher research enthusiasm in the second stage. It also indicates that the early pioneering studies from varied perspectives laid a foundation for subsequent terrace research. Based on current research trend, it is expected that terrace research will be further enriched.

In the first stage during 1991–2011, there were four small peaks of citations, namely, 1994, 1998, 2003, and 2007 (line chart of average citations in Figure 2). The paper *"Studying the role of old agricultural terraces on runoff generation in small Mediterranean mountainous basin"* published in 1994 had been cited 122 times by the end of the study period. It was proven in this paper that surface runoff in the Mediterranean mountainous area was related to the spontaneous reorganization of the artificial drainage network in abandoned terraced fields [42]. By the end of 2020, the article *"Land use change effects on abandoned terraced soils in a Mediterranean catchment, NE Spain"* published in 2003 had been cited 146 times. In this paper, the different vegetation stages were divided into four main land-use types according to the age of abandonment: cultivated fields (vineyard and olive tree, 0 years), recent abandonment (dense and cleared shrubs, 5 years), moderate abandonment (cleared cork trees and dense olive trees, 25 years) and early abandonment (dense cork trees and pine trees reforestation, 50 years), and variance analysis indicated significant differences in the main soil quality parameters such as soil organic matter (SOM), total nitrogen (N), water holding capacity (WHC), and pH, among the selected environments under different land-use conditions [43]. The paper *"Land environment and slope gradient as key factors of soil erosion in Mediterranean terraced lands"* published in 2007 had been cited 211 times by the end of 2020. The research results showed that the influence of traditional extensive cultivation abandonment on soil sediment losses varied with slope gradient. When slope gradient was steep (25%), soil erosion increased significantly. When slope gradient was very steep (40%), soil sediment losses remained at the same high levels after cultivation abandonment as slope gradient was the main factor controlling soil erosion, although soil and vegetation properties were changing [44]. The highly cited papers in the first period solved key basic problems such as classification standards and environmental impacts in the initial stage of terrace research through exploratory attempts, so their influence was greater. The average number of annual published papers and citations are shown in Figure 2 below.

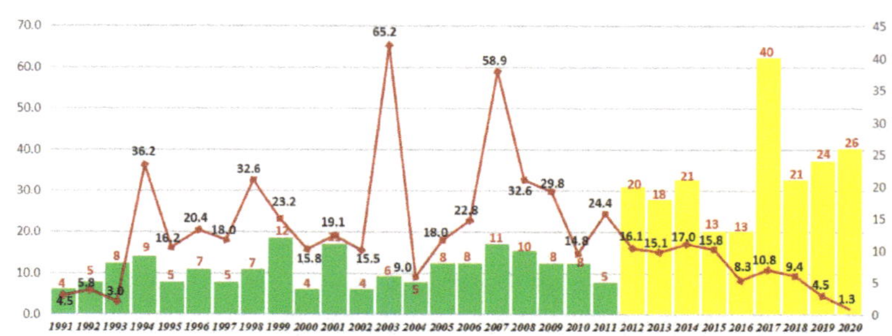

Figure 2. Agricultural-terrace research articles and citations during 1991–2020.

3.2. Analysis of Authors' Papers

As shown in Figure 3, the bubble chart of authors' papers directly reflects the published number of articles and citations of each author in different years. The circle size represents authors' published article numbers, the color depth represents the number of the paper's citations. The top ten authors in terrace research field are shown in Figure 3. H-indexes, total numbers of published articles, total citations, and publication dates of the ten authors are shown in Table 1. An author's H-index means that at least H of his/her papers had been cited at least H times in a given period [45]. It can be seen from Table 1 that the H-index of Min Q. W. and Li Y. was 5, indicating that their studies were the most influential. Taking rice-fish system in Hani Terrace as an example, Min Q.W. studied the standards of ecological compensation for traditional eco-agriculture [46]. Li Y. studied crops planted in terraced fields from a biological viewpoint for 19 years. He proposed that the accumulation of H_2O_2 and MDA helped induce the improvement of antioxidant enzyme activity, thus improving the tolerance of plants to UV-B radiation. The enhancement of UV-B radiation directly affected rice growth, and changed the system of *Magnaporthe grisea* indirectly, which was conducive to terraced-field planting [47,48]. Li Y.'s team quantified the soil quality parameters of terraced and steep slopes on the Loess Plateau of China, and cleared the influence of tillage erosion and water erosion on soil quality parameters. It was concluded that water erosion was the main reason for the overall decline in the soil quality of steeply sloping farmland, and that tillage had a controlling effect on the spatial pattern of organic matter, nitrogen, and phosphorus in terraced and steep-slope soil [49]. These studies on terraced fields from different angles, such as economy, ecology, and landscape, have laid a foundation for multidisciplinary comprehensive study of terraced fields.

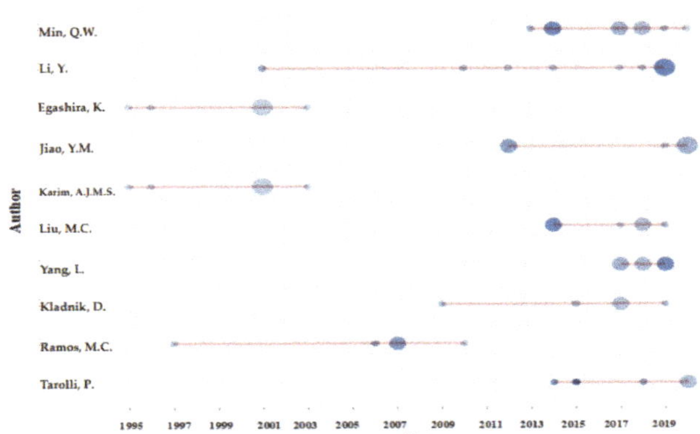

Figure 3. Authors' publications on agricultural terraced fields over time.

Table 1. Top 10 authors in the field of agricultural-terrace research.

Author	H_Index	TC	NP	PY_Start
Min Q.W.	5	96	9	2013
Li Y.	5	121	7	2001
Egashira K.	3	20	6	1995
Jiao Y.M.	4	53	6	2012
Karim A.J.M.S.	3	20	6	1995
Liu M.C.	4	69	6	2014
Yang L.	4	41	6	2017
Kladnik D.	4	49	5	2009
Ramos M.C.	4	159	5	1997
Tarolli P.	3	137	5	2014

Notes: TC: total article citations; NP: number of papers; PY_start: publication year.

3.3. Analysis of National Documents

3.3.1. Global Research Analysis

In Figure 4, the shades of blue indicate the number of papers published in the country, the gray indicates that there is no relevant literature on terraces in the country, the red dots indicate the location of the 20 countries with the largest number of papers, and the yellow triangle indicates the famous terraces that have been listed as world heritage sites by FAO or UNESCO. The distribution map of numbers of papers published by different countries shows that there is little research on terraces in Africa and Western Asia. As can be seen from Table 2, related articles published in China are more abundant, indicating a strong research force and phased research focus in China. However, in terms of research influence, articles from developed countries in Europe and America seem to be more influential, which can be seen from the higher citations of articles from developed countries. On the whole, research on terraces is consistent with the distribution of terraced fields, that is, terrace research is more abundant in regions with more terraced fields, such as East Asia, Western Europe, and North America. Asian terraces are mainly planted with rice, and these large-scale terraced landscapes support a large population and form a common and distinct "terraced cultural circle" [50], while European terraces were mainly built to facilitate the mechanized production of vineyards during their long history. They are represented by the Lavaux vineyard terraces in Switzerland, and terraces in Hungary, Italy, Portugal, and Spain. However, agricultural industrialization and abandonment in Europe led to the degradation and disappearance of traditional terraced landscapes after the 1960s.

Table 2. Top 10 countries in terms of article number.

Region	Articles	Total Citations	Average Number of Article Citations
China	203	732	3.61
USA	95	784	8.25
Japan	62	224	3.61
Spain	60	1136	18.93
Italy	38	401	10.55
United Kingdom	35	510	14.57
India	30	40	1.33
Germany	28	234	8.36
Philippines	26	85	3.27
France	18	74	4.11

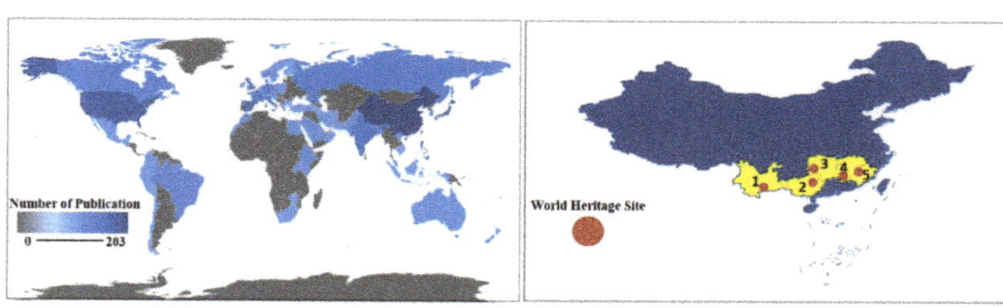

Figure 4. Number of publications on terraced fields in each region of the world.

3.3.2. Analysis of Major Research Countries

China is the country with the largest number of publications. There are 203 articles written by Chinese authors, accounting for 58.2% of the total publications. Due to the large population and limited per capita cultivated land area in China, the Chinese government attaches great attention to terrace development and protection to ensure food security. In recent years, owing to the Transforming Slope into Terrace Project on the Loess Plateau, terraced area has increased significantly in China [51–54]. Data from WoS show that the five Chinese world terraced fields, namely Honghe Hani Terraced Fields, Ziquejie Terrace, Longji Terrace, Youxi United Terrace, and Chongyi Hakka Terraces, are the main research areas of Chinese publications (Table 3, Figure 4).

Table 3. Main terraces in China.

Name	Location	Area (hm)	Crop	World Heritage Site
Honghe Hani Terraced Fields	Yunnan	16,603	Rice	UNESCO(2013), GIAHS(2010)
Longji Terrace	Guangxi	7010	Rice	GIAHS(2018)
Ziquejie Terrace	Hunan	1334	Rice	GIAHS(2018)
Chongyi Hakka Terraces	Jiangxi	2000	Rice	GIAHS(2018)
Youxi United Terrace	Fujian	713	Rice	GIAHS(2018)

Discipline analysis shows that terrace research in China is mainly focused on the following four aspects (Figure 5). ① Agricultural production. Rice and local rice varieties are the main varieties, among which, red rice in Hani Terrace is a traditional high-quality variety. ② Agricultural engineering. To solve soil and water loss in the Loess Plateau and control Yellow River bed aggradation, some scholars suggest transforming sloping land into terraced farmland, so as to reduce nutrient loss on the surface of cultivated land. In addition, the identification and extraction of terrace spatial information has developed rapidly in recent years, but its application on a large scale still needs to be improved. ③ Cultural tourism. Represented by Hani Terrace and Longji Terrace, terraced field tourism combines visual enjoyment brought by terraced landscapes and ethnic culture, thus bringing positive travel opportunities and economic benefits. In this way, a new terrace-tourism mode combining terrace protection with landscape, local culture, and tourism revenue, can be formed [55–57]. ④ Ecological environment. Biodiversity conservation, ecological compensation, and ecosystem service value, especially terraced cultural service value which involves cultural, aesthetic, spiritual, and religious aspects, has attracted more and more attention in terrace field [58,59].

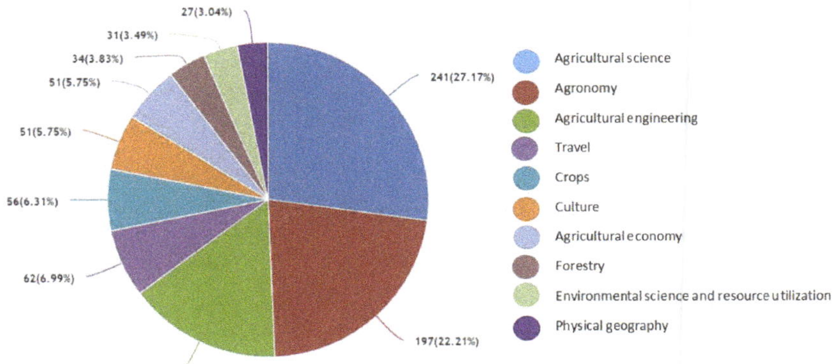

Figure 5. The main terrace research subjects in China.

4. Summary of Qualitative Research

4.1. Analysis of Keyword Co-Occurrence Network

The 20 most frequent keywords in terrace studies were obtained by eliminating meaningless or repetitive words through statistics, as shown in Figure 6. "Agricultural terraces" was the most frequent keyword, followed by "soil erosion", "surface runoff", "terraced landscape", and "land environment". A single keyword can identify the core information of terrace research, but it cannot identify the necessary relationship between the core information. Therefore, the co-occurrence analysis in VOS viewer was introduced to set the author keywords at least five times to obtain keyword co-occurrence (Figure 7). In Figure 7, the lines represent the usage of keywords. The larger the node, the higher the frequency of keywords. The five main clusters of keywords, labeled #1–#5, in terrace studies are highlighted by rectangular boxes of different colors in Figure 7. They also reflect the main modules of current terrace studies.

Figure 6. High frequency keywords and their occurrence in agricultural terrace publications.

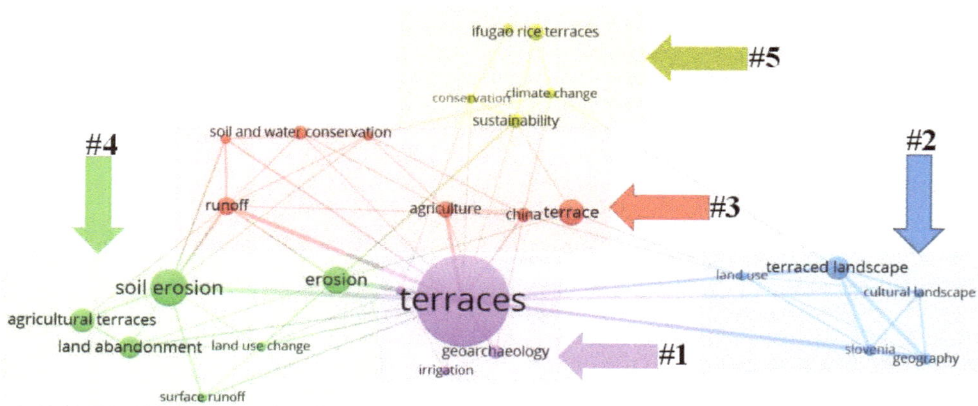

Figure 7. Co-occurrence network of agricultural terrace keywords.

Cluster 1 mainly focuses on the geological and archaeological characteristics of terraced fields. Geoarchaeology not only helps understand terrace history, function, and sustainability, but also provides evidence for vegetation change and erosion over time [60]. Terrace construction is a long process. Terraces are rarely constructed in a single stage but gradually develop or even evolve, and new terraces are usually built gradually based on maintaining ancient terraces [2,61,62]. Archaeological study of isotopic carbon or nitrogen on terraced soil indicates that the Longji terrace in China was constructed in the late Chinese Yuan Dynasty. In the late Yuan Dynasty, the strong survival pressure brought by social upheaval motivated minority populations to migrate to the Longji Mountain area to open up new living space. As a result, agricultural terraces and gravity irrigation networks were built on hillside land [63]. Archaeological research in Bali proved that the spatial pattern of

Balinese terraces built hundreds of years ago was more about farmers' decision-making and rice field ecology [64].

Cluster 2 mainly involves studies on terrace landscape and its cultural value. Cultural landscape is formed in the long-term interaction between humans and nature [65]. It also represents humans' living landscape formed by long-term intensive and continuous cultivation [66]. As the cultural value of terraced landscape is derived from the combination of terrace maintenance and value creation [67], Japan attaches great importance to the protection and inheritance of terraced landscape culture. Local people believe that terraced field is the manifestation of hard work, which represents people's belief in natural spirit [68]. In addition, beautiful terrace landscape brings visual enjoyment to tourists and even affects people's world outlook, which tourists are willing to pay for. In many places, owing to their cultural value, terraces are mainly retained as decorative elements of the landscape rather than for their original agricultural function although farmers actively cultivate their fields [69]. Especially in some tourist attractions, terrace cultivation is maintained mainly to protect its cultural landscape value, which is much higher than its food production value.

Cluster 3 mainly involves studies on soil and water conservation and agricultural utilization of terraced fields. Traditional slope tillage causes severe soil erosion, while terrace construction helps prevent soil erosion effectively [70]. However, the effectiveness is limited by factors including climate, soil properties, topography, land use, culture, population, and socioeconomic status [71]. To maximize the function of terraces in soil and water conservation, numerous studies have been carried out. It is proposed that terraces should be well maintained to retain more water, and vegetation restoration is essential in areas prone to water flow [72,73]. Studies in abandoned terraced fields, which are more prone to soil erosion, show that terraces planted with vegetation produce less runoff than those planted with crops [74]. Therefore, adjusting planting types and avoiding terrace abandonment helps reduce soil erosion.

Cluster 4 mainly involves studies on terrace soil erosion, land abandonment, and land-use change. In recent years, with the development of the social economy, farmland abandonment has become more and more serious [75]. In comparison to flat land, abandonment is more likely to occur in terraced fields due to their inferior topography and irrigation conditions. Currently, studies on terrace abandonment are mainly focused on the identification [76], driving forces, impacts, and recovery. Some reasons for cultivated land abandonment and terrace abandonment might be similar, but the degree of influence is different. For example, the influence of slope on abandonment is greater for terraces than for cultivated land [44]. Terrace abandonment may cause soil erosion and wildfires [30], and further reduce biodiversity and landscape cultural value [77]. It is noteworthy that soil erosion and terrace abandonment are mutually causal. Therefore, to alleviate the adverse effects of terrace abandonment, it is proposed to restore and maintain terraced fields. Local endemic species are encouraged to be replanted to promote the restoration of local ecosystems, so as to enable the ecology and land use in terraced fields to evolve in a better way [78].

Cluster 5 mainly involves studies on the protection and sustainable utilization of typical terraces, which are represented by the Ifugao rice terraces. Although some ancient terraced fields function fully, it cannot be ignored that inefficient utilization still widely exists. Combining historical and cultural landscape perspectives, scholars have studied well-maintained and typical terraced fields to seek the sustainable utilization of terraced fields. Studies have found that both natural and social-economic factors affect the sustainable use of terraced fields. For example, the El Niño phenomenon has caused drought and insufficient irrigation infrastructure in the Ifugao rice terraces, resulting in decreased ecosystem value, and obvious changes in extensive planting and ripening [79]. At the same time, the influence of organisms is striking too. In the Ifugao rice terraces, golden apple snails seriously harm crops, and earthworms and mice disintegrate terraces [80,81]. Similar problems exist in the Hani terraces. The introduction of new rice varieties significantly increases the plant diseases and insect pests [82]. In addition, extensive management and abandonment begin to spread as farmers' willingness to plant decreases. To realize the sustainable utilization of terraced fields, dynamic protection approaches are proposed. First, it is necessary to enhance farmers' cultural cognitive abilities, especially their consciousness to traditional agricultural knowledge. Second, it is necessary encourage young people to participate in the management and maintenance of terraced fields [83,84]. Third, there is a need to improve the ecological compensation mechanism to promote ecological agriculture and ecological services [85]. For terraced areas with cultural value, tourism development and community monitoring can be combined to explore an endogenous development strategy [86].

The above analysis shows that the five clusters are closely related and mutually supported. Despite their own focus, the contents of the five clusters are both dependent and intersecting, reflecting the interdisciplinary development trend in terrace research. As a whole, the five clusters reflect the general research situation of terraced fields involving utilization, functions, values, and development.

4.2. Analysis of Thematic Evolution

To further analyze the research context reflected by keywords, Figure 8 is charted to present the temporal information of keywords and research hotspots, thus providing support to study the relationship between different research fields from temporal and causal dimensions. In Figure 8, the darker the color, the hotter the research. To visualize the theme evolution of terrace research over time, a Sankey diagram (Figure 9) is charted to present the qualitative information and flow status of different themes [87], such as theme flow and its direction and transformation relationships.

Three evolutionary time nodes of 2004, 2013, and 2017 were given through the software. Figure 9 shows that both single research paths and extended research paths exist in terrace research over time. The single research path means that the research theme remained unchanged during the study period, and continues to be the research focus for a period of time in the future. Single research paths are divided into traditional paths and emerging paths based on when they formed, among which, emerging paths are more likely to become popular. For extended research paths, the former and latter research themes are closed related but different; to be specific, the former enlightens the latter while the latter deepens the former.

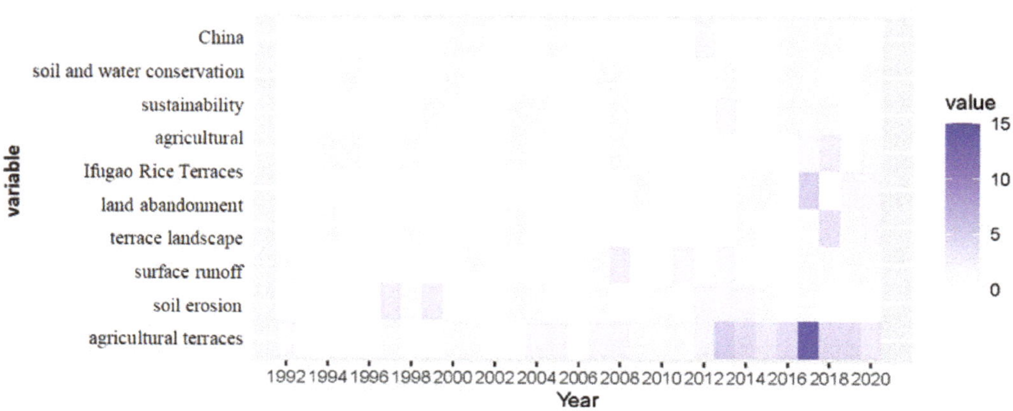

Figure 8. Thermal map of terrace research.

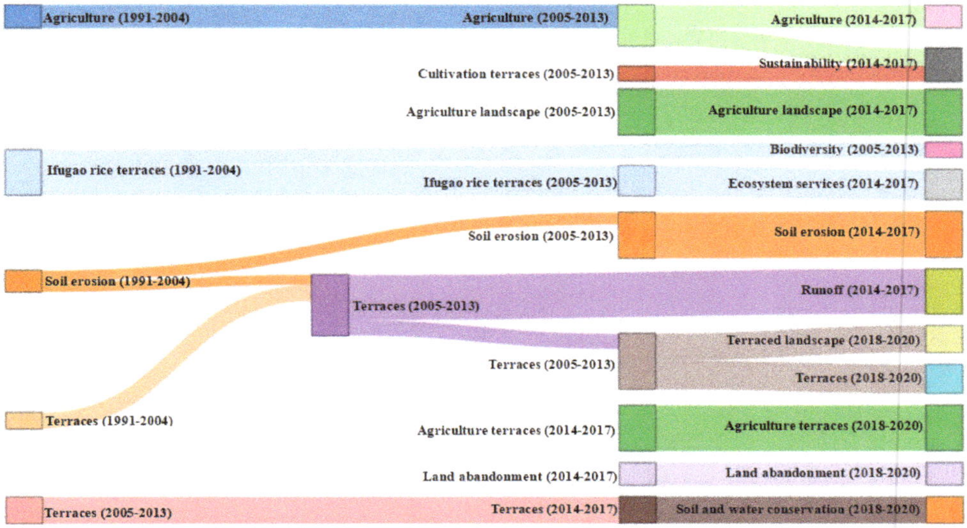

Figure 9. Thematic evolution of agricultural-terrace research (1991–2020).

The four main single research paths are as follows:

(1) Agriculture (1991–2017);
(2) Soil erosion (1991–2017);
(3) Agricultural landscape (2005–2017);
(4) Land abandonment (2014–2020).

According to the study period, paths 1, 2, and 3 are traditional paths, while path 4 is an emerging path. The temporal information shows that early terrace research (1991–2017) focused on agricultural and soil erosion. Traditional terraces were generally built on hillsides to develop cultivated land, regulate water circulation, and maintain soil. The early research paths were consistent with the original development needs for terraced fields. Namely, to control soil and water loss of production land by slope treatment, and to improve soil physical and chemical properties through terrace cultivation to accumulate nutrients needed for crop growth, thus improving agricultural production capacity. Terraced fields play an important role in maintaining slope landscape pattern and tourism development by changing land landscape and increasing regional landscape heterogeneity. Therefore, with the development and improvement of terrace engineering technology, scholars pay more attention to the agricultural landscape value of terraced fields, including landscape pattern change, typical watershed landscape patterns, terraced cultural landscapes, comprehensive evaluation of landscape multifunctional value, and landscape protection. In recent years, due to the decline of agricultural comparative income, the rise of agricultural opportunity cost, and the constraints in terrace engineering design, abandoned terrace fields gradually expanded and became an emerging research focus. The advancement of industrialization and urbanization highlights the intensifying terrace abandonment. Therefore, study of abandoned terraced fields will continue to be a hotspot in future.

The expanded research paths are as follows:

(1) Cultivation terraces (2005–2013) → sustainability (2014–2017);
(2) Ifugao rice terraces (1991–2004) → biodiversity (2005–2013);
(3) Ifugao rice terraces (1991–2013) → ecosystem services (2014–2017);
(4) Soil erosion (1991~2004) → terraces (2005–2013) → runoff (2014–2017);
(5) Soil erosion (1991~2004) → terraces (2005–2017) → terraces landscape (2018–2020);
(6) Terraces (2005~2017) → soil and water conservation (2018–2020).

The evolutionary process of expanded research paths is more abundant than the single research paths. Path (1) reflects the sustainability goals and research trends of terrace cultivation, including the necessity of terrace sustainable development [88], sustainable terrace management [89], and sustainable terrace practice mode and its evaluation [90,91]. For example, Ni (2014) believed that the Hani terrace in China should develop small-scale sustainable and ecological agriculture, and take this as an opportunity to develop ecological agricultural tourism [92]. Paths (2) and (3) reflect the importance of the studies on the ecosystem service value of rice terraces. Among them, the literature on the Ifugao rice terrace in the Philippines is the most abundant and representative. For example, the study on the resilience of the Ifugao terraced agricultural system provides a reference for evaluating the biodiversity and sustainable development of terraced ecosystems. More than food supply, the functions of a rice terrace ecosystem with rice as the main crop also include climate regulation [93], soil and water conservation [94], pest regulation, tourism [95,96], and aesthetics [97]. The cultivation and development of rice terrace is of great significance in maintaining ecological security and food security in mountainous areas. Biodiversity maintenance is a research hotspot of terrace ecosystem service. For example, Drechsler and Settele (2001) studied the predator–prey interactions in rice ecosystems of Philippine rice terraces, thus exploring the effects of guild composition, trophic relationships, and land use changes on terrace biodiversity [98]. These studies provide important support for the realization of ecosystem services. Paths (4), (5), and (6) are similar in research content regarding the effective management of water and soil resources in terraced fields, including terrace soil erosion and runoff and its countermeasures. Currently, soil and water conservation measures have also become an important part of the terraced landscape, that is conducive to the planning and stability of landscape pattern. As the role of soil and water conservation in maintaining terrace landscape ecology highlights, the theme of path (5) develops from soil erosion (1991–2004) to terrace landscapes (2018–2020).

The theme evolution pathways present a terrace research trend from single to diversified and from simple to complex in the past 30 years. Themes have evolved from the initial topics of agriculture, terraces, and soil erosion to sustainable utilization, biodiversity, and ecosystem services. However, research on the classified management of terrace fields is still deficient, and research on terrace historical evolution, value evaluation, eco-environmental effect analysis, and specific management strategies is expected to become popular in the future.

5. Conclusions and Suggestions

This paper conducted quantitative and qualitative analysis on literature of terraced fields from 1991–2020 from the WoS database. The quantitative analysis included article, author, and regional analysis, and the main conclusions were as follows: (1) terrace research presents an increasing trend with fluctuations during 1991–2020; (2) terrace research has attracted scholars from various fields, and is expected to remain a hotspot in the future; and (3) agricultural production, agricultural engineering, cultural tourism, and ecological environment are the main research aspects in China. Qualitative studies included keyword co-occurrence and thematic evolution analysis. The main conclusions were as follows: (1) keyword co-occurrence analysis formed five research clusters focused on terrace geological archaeology, terrace landscape and cultural value, the effectiveness of terrace soil and water conservation, and the utilization and change of terraces, reflecting the main modules of current terrace studies and (2) thematic evolution analysis presented 10 main evolutionary paths of terrace research in a temporal dynamic perspective, which were characterized by single and expanded research paths and included "land abandonment (2014–2020)" and "soil erosion (1991–2004) $\rightarrow$ terraces (2005–2017) $\rightarrow$ terraces landscape (2018–2020)", reflecting the trend of terrace research towards sustainable terrace development of ecological agriculture and ecosystem service. By reflecting terrace research emphases in different periods, thematic evolution analysis provides ideas for analyzing the evolutionary process and mechanism of terraced field research, thus providing a basis for understanding the future research direction of terraced fields. Based on the existing literature, this paper puts forward the key content and direction of terrace research in future.

(1) As important agricultural landscapes and cultural heritages, terraced fields have become a model of the coupling between man and nature, and also a unique way for humans to adapt to the mountain ecosystem during their long-term struggle against harsh natural conditions. Terrace ecosystems provide a variety of services for mankind, and their ecosystem structure and processes affect the service supply. However, there is little research on the spatial–temporal matching of terrace ecosystem services in current studies, thus failing to reveal the interaction between multiple services from the service formation mechanism, and research on the evaluation of terrace ecosystem services is also lacking. Therefore, based on the structure and processes of terrace ecosystems, future studies may focus on terrace ecosystem services and human well-being, and carry out quantitative and qualitative evaluation of terrace ecosystem services according to the research environment and purpose.

(2) Terrace research shows a comprehensive and diversified trend in terms of research content, research scope, and research scale. However, studies on the driving forces and risk assessment of terrace abandonment are few. In view of the complexity and coupling effect of terrace abandonment, it is necessary to give full play to the multi-disciplinary advantages of natural ecology, sociology, economics, geography, and human policies and to take into account the demands of stakeholders including scientists, decision makers, grass-roots managers, and farmers, thus comprehensively analyzing the interaction and effect of natural process and humanistic driving forces on terrace abandonment.

(3) There is a need to study countermeasures to strengthen ecological protection and restoration of terraced fields, and to explore the multi-functional benefits and sustainable management of high-quality terraced landscapes. In fact, under the impact of the modern scientific and technological revolution and the new ways of production and life, terrace ecosystems in most parts of the world are threatened by abandonment and extinction. It is necessary to establish a long-term self-sustaining mechanism of terrace ecosystems through dynamic protection and adaptive management of terrace landscapes, so as to realize the sustainable utilization and development of terraced multi-functional landscapes.

Author Contributions: Conceptualization, Q.C. and Y.W.; methodology, Y.W.; software, Z.Z.; validation, X.Z.; formal analysis, Q.C.; investigation, Z.Z.; resources, Y.W.; data curation, Z.Z.; writing—original draft preparation, Y.W.; writing—review and editing, Q.C.; visualization, Y.W.; supervision, Q.C.; project administration, Q.C. All authors have read and agreed to the published version of the manuscript.

Funding: This research was funded by the National Natural Science Foundation of China, grant numbers 41971243 and 41930757.

Institutional Review Board Statement: Not applicable.

Informed Consent Statement: Not applicable.

Data Availability Statement: Not applicable.

Acknowledgments: The authors would like to thank the reviewers and editors whose suggestions greatly improved the manuscript.

Conflicts of Interest: The authors declare no conflict of interest.

References

1. Ni, G. *Liang Jiamian's Collection of Agricultural History*; China Agricultural Publishing House: Beijing, China, 2002; pp. 218–236. (In Chinese)
2. Branch, N.P.; Kemp, R.A.; Silva, B.; Meddens, F.M.; Williams, A.; Kendall, A.; Pomacanchari, C.V. Testing the sustainability and sensitivity to climatic change of terrace agricultural systems in the Peruvian Andes: A pilot study. *J. Archaeol. Sci.* **2007**, *34*, 1–9. [CrossRef]
3. Barker, G.W.; Adams, R.; Creighton, O.H.; Daly, P.; Gilbertson, D.D.; Grattan, J.P.; Hunt, C.O.; Mattingly, D.J.; Mclaren, S.J.; Newson, P.L. Archaeology and Desertification in the Wadi Faynan: The Fourth (1999) Season of the Wadi Faynan Landscape Survey. *Levant* **2000**, *32*, 27–52. [CrossRef]

4. Chen, D.; Wei, W.; Chen, L. Historical distribution of terrace landscape and typical international case analysis. *Chin. J. Appl. Ecol.* **2017**, *2*, 326–335. (In Chinese)
5. Mishra, P.K.; Rai, A.; Rai, S.C. Indigenous knowledge of terrace management for soil and water conservation in the Sikkim Himalaya, India. *Indian J. Tradit. Knowl.* **2020**, *19*, 475–485.
6. Xu, Q.X.; Pan, W.; Dai, J.F.; Wan, T.W. The effects of rainfall regimes and terracing on runoff and erosion in the Three Gorges area, China. *Environ. Sci. Pollut. Res.* **2018**, *25*, 9474–9484. [CrossRef] [PubMed]
7. Zhao, X.; Zhu, H.S.; Dong, K.H.; Li, D.Y. Plant Community and Succession in Lowland Grasslands under Saline–Alkali Conditions with Grazing Exclusion. *Agron. J.* **2017**, *109*, 2428–2437. [CrossRef]
8. Gebreslase, M.S.; Oliveira, L.A.A.; Yazew, E.; Bresci, E.; Castelli, G. Spatial Variability of Soil Moisture in Newly Implemented Agricultural Bench Terraces in the Ethiopian Plateau. *Water* **2019**, *11*, 2134. [CrossRef]
9. Londero, A.L.; Minella, J.P.G.; Deuschle, D.; Schneider, F.J.A.; Boeni, M.; Merten, G.H. Impact of broad-based terraces on water and sediment losses in no-till (paired zero-order) catchments in southern Brazil. *J. Soils Sediments* **2017**, *18*, 1159–1175. [CrossRef]
10. Bary, A.A.; Pierzgalski, E. Ridged terraces—Functions, construction and use. *J. Environ. Eng. Landsc. Manag.* **2008**, *16*, 1–6.
11. Xu, G.C.; Zhang, T.G.; Li, Z.B.; Li, P.; Cheng, Y.T.; Cheng, S.D. Temporal and spatial characteristics of soil water content in diverse soil layers on land terraces of the Loess Plateau, China. *Catena* **2017**, *158*, 20–29. [CrossRef]
12. Blecourt, M.; Hansel, V.M.; Brumme, R.; Corre, M.D.; Veldkamp, E. Soil redistribution by terracing alleviates soil organic carbon losses caused by forest conversion to rubber plantation. *For. Ecol. Manag.* **2014**, *313*, 26–33. [CrossRef]
13. Ni, S.J.; Zhang, J.H. Variation of chemical properties as affected by soil erosion on hillslopes and terraces. *Eur. J. Soil Sci.* **2007**, *58*, 1285–1292. [CrossRef]
14. Koyanagi, T.F.; Yamada, S.; Yonezawa, K.I.; Kitagawa, Y.; Ichikawa, K.; Ohlemuller, R. Plant species richness and composition under different disturbance regimes in marginal grasslands of a Japanese terraced paddy field landscape. *Appl. Veg. Sci.* **2015**, *17*, 636–644. [CrossRef]
15. Chen, S.K.; Chen, R.S.; Yang, T.Y. Application of a tank model to assess the flood-control function of a terraced paddy field. *Hydrol. Sci. J.-J. Des Sci. Hydrol.* **2014**, *59*, 1020–1031. [CrossRef]
16. Agnoletti, M.; Conti, L.; Frezza, L.; Santor, A. Territorial Analysis of the Agricultural Terraced Landscapes of Tuscany (Italy): Preliminary Results. *Sustainability* **2015**, *7*, 4564–4581. [CrossRef]
17. Pietsch, D.; Mabit, L. Terrace soils in the Yemen Highlands: Using physical, chemical and radiometric data to assess their suitability for agriculture and their vulnerability to degradation. *Geoderma* **2012**, *185–186*, 48–60. [CrossRef]
18. Xie, H.L.; Chen, Q.R. Land Use and Ecological Civilization: A Collection of Empirical Studies. *J. Resour. Ecol.* **2021**, *12*, 137–142.
19. Lasanta, T.; Arnaez, J.; Ruiz-Flaño, P.; Lana-Renault, N. Agricultural terraces in the Spanish mountains: An abandoned landscape and a potential resource. *South Asian Stud.* **2013**, *63*, 487–491.
20. Caga, K.A.D. Mixed Views on the Philippines' Ifugao Rice Terraces: 'Good' versus 'Beautiful' in the Management of a UNESCO World Heritage Site. *J. Southeast Asian Stud.* **2018**, *49*, 84–104. [CrossRef]
21. Fukamachi, K. Sustainability of terraced paddy fields in traditional satoyama landscapes of Japan. *J. Environ. Manag.* **2016**, *202*, 543–549. [CrossRef]
22. Mori, Y.; Sasaki, M.; Morioka, E.; Tsujimoto, K. When do rice terraces become rice terraces? *Paddy Water Environ.* **2019**, *17*, 323–330. [CrossRef]
23. Kladnik, D.; Geršič, M.; Pipan, P.; Bahun, M.V. Land-use changes in Slovenian terraced landscapes. *Acta Geogr. Slov.* **2019**, *59*, 119–141. [CrossRef]
24. Delius, P.; Schirmer, S. Order, openness, and economic change in precolonial southern Africa: A perspective from the bokoni terraces. *J. Afr. Hist.* **2014**, *55*, 37–54. [CrossRef]
25. Bizoza, A.R.; Graaff, J.D. Financial cost–benefit analysis of bench terraces in Rwanda. *Land Degrad. Dev.* **2012**, *23*, 103–115. [CrossRef]
26. Posthumus, H.; Graaff, J.D. Cost-benefit of bench terraces, a case study in Peru. *Land Degrad. Dev.* **2010**, *16*, 1–11. [CrossRef]
27. Pepe, G.; Mandarino, A.; Raso, E.; Scarpellini, P.; Brandolini, P.; Cevasco, A. Investigation on Farmland Abandonment of Terraced Slopes Using Multitemporal Data Sources Comparison and Its Implication on Hydro-Geomorphological Processes. *Water* **2019**, *11*, 1552. [CrossRef]
28. Soriano, M.A.; Herath, S. Quantifying the role of traditional rice terraces in regulating water resources: Implications for management and conservation efforts. *Agroecol. Sustain. Food Syst.* **2018**, *42*, 885–910. [CrossRef]
29. Ubeda, X.; Alcaniz, M.; Borges, G.; Outeiro, L.; Francos, M. Soil Quality of Abandoned Agricultural Terraces Managed with Prescribed Fires and Livestock in the Municipality of Capafonts, Catalonia, Spain (2000–2017). *Agronomy* **2019**, *9*, 340. [CrossRef]
30. Lucas-Borja, M.E.; Calsamiglia, A.; Fortesa, J.; García-Comendador, J.; Guardiola, E.L.; García-Orenes, F.; Gago, J.; Estrany, J. The role of wildfire on soil quality in abandoned terraces of three Mediterranean micro-catchments. *Catena* **2018**, *170*, 256. [CrossRef]
31. Brandolini, P.; Cevasco, A.; Capolongo, D.; Pepe, G.; Lovergine, F.; Del Monte, M. Response of Terraced Slopes to a Very Intense Rainfall Event and Relationships with Land Abandonment: A Case Study from Cinque Terre (Italy). *Land Degrad. Dev.* **2018**, *29*, 630–642. [CrossRef]
32. Chen, X.L.; Zou, D.; Kohnke, L.; Xie, H.R.; Cheng, G. Affective states in digital game-based learning: Thematic evolution and social network analysis. *PLoS ONE* **2021**, *16*, e0255184. [CrossRef] [PubMed]

33. Li, H.; Pan, Y.G.; Wang, L. Biblimetrix: Introduction and evaluation of a new bibliometrics software based on R language. *J. Librariansh. Inf. Sci.* **2018**, *36*, 93–104.
34. Xie, H.L.; Zhang, Y.W.; Choi, Y.R.; Li, F.Q. A Scientometrics Review on Land Ecosystem Service Research. *Sustainability* **2020**, *12*, 2959. [CrossRef]
35. Huang, Y.Q.; Li, F.Q.; Xie, H.L.A. Scientometrics Review on Farmland Abandonment Research. *Land* **2020**, *9*, 263. [CrossRef]
36. Xie, H.L.; Zhang, Y.W.; Zeng, X.J.; He, Y.F. Sustainable land use and management research: A scientometric review. *Landsc. Ecol.* **2020**, *35*, 2381–2411. [CrossRef]
37. Zong, Q.; Yuan, Q.; Shen, H. Analysis of hot spots of library science research in China in 2010 based on VOSviewer. *Library* **2012**, *4*, 88–90. (In Chinese)
38. Li, J.; Hale, A. Output distributions and topic maps of safety related journals. *Saf. Sci.* **2016**, *82*, 236–244. [CrossRef]
39. Jeong, D.; Koo, Y. Analysis of Trend and Convergence for Science and Technology using the VOSviewer. *Int. J. Contents* **2016**, *12*, 54–58. [CrossRef]
40. Lu, X.; Peng, W.L.; Huang, X.J.; Fu, Q.Q.; Zhang, Q.J. Homestead management in China from the "separation of two rights" to the "separation of three rights": Visualization and analysis of hot topics and trends by mapping knowledge domains of academic papers in China National Knowledge Infrastructure (CNKI). *Land Use Policy* **2020**, *97*, 104670. [CrossRef]
41. Xie, H.L.; Wen, Y.Y.; Choi, Y.R.; Zhang, X.M. Global Trends on Food Security Research: A Bibliometric Analysis. *Land* **2021**, *10*, 119. [CrossRef]
42. Gallart, F.; Llorens, P.; Latron, J. Studying the role of old agricultural terraces on runoff generation in a small Mediterranean mountainous basin. *J. Hydrol.* **1994**, *159*, 291–303. [CrossRef]
43. Dunjo, G.; Pardini, G.; Gispert, M. Land use change effects on abandoned terraced soils in a Mediterranean catchment, NE Spain. *Catena* **2003**, *52*, 23–37. [CrossRef]
44. Koulouri, M.; Giourga, C. Land abandonment and slope gradient as key factors of soil erosion in Mediterranean terraced lands. *Catena* **2007**, *69*, 274–281. [CrossRef]
45. Zhao, H.Q.; Fang, X.; Ding, H.; Strobl, J. Extraction of Terraces on the Loess Plateau from High-Resolution DEMs and Imagery Utilizing Object-Based Image Analysis. *ISPRS Int. J. Geo-Inf.* **2017**, *6*, 157. [CrossRef]
46. Liu, M.C.; Xiong, Y.; Zheng, Y.; Min, Q.W.; Sun, Y.H.; Fuller, A.M. Standards of ecological compensation for traditional eco-agriculture: Taking rice-fish system in Hani terrace as an example. *J. Mt. Sci.* **2014**, *11*, 1049–1059. [CrossRef]
47. He, Y.; Zhan, F.; Zu, Y.; Liu, C.; Yuan, L. Effect of Elevated UV-B Radiation on the Antioxidant System of Two Rice Landraces in Paddy Fields on Yuanyang Terrace. *Int. J. Agric. Biol.* **2014**, *16*, 585–590.
48. Li, X.; He, Y.M.; Xie, C.M.; Zu, Y.Q.; Zhan, F.D.; Mei, X.Y.; Xia, Y.; Li, Y. Correction: Effects of UV-B radiation on the infectivity of Magnaporthe oryzae and rice disease-resistant physiology in Yuanyang terraces. *Photochem. Photobiol. Sci.* **2017**, *17*, 7393. [CrossRef]
49. Li, Y.; Lindstrom, M.J. Evaluating Soil Quality–Soil Redistribution Relationship on Terraces and Steep Hillslope. *Soil Sci. Soc. Am. J.* **2001**, *65*, 1500–1508. [CrossRef]
50. Yang, H.; Shao, X.X.; Wu, M.A. Review on Ecosystem Health Research: A Visualization Based on CiteSpace. *Sustainability* **2019**, *11*, 4908. [CrossRef]
51. Ran, Q.H.; Chen, X.X.; Hong, Y.Y.; Ye, S.; Gao, J.H. Impacts of terracing on hydrological processes: A case study from the Loess Plateau of China. *J. Hydrol.* **2020**, *588*, 125045. [CrossRef]
52. Xu, G.C.; Ren, Z.P.; Li, P.; Li, Z.B.; Yuan, S.L.; Zhang, H.; Wang, D.; Zhang, Z.Y. Temporal persistence and stability of soil water storage after rainfall on terrace land. *Environ. Earth Sci.* **2016**, *75*, 966. [CrossRef]
53. Shi, Y.Z.; Wang, Y.H.; Cui, Y.L.; Wang, S.W.; Zhang, Y.S. A new rainwater harvesting and recycling system for transforming sloping land into terraced farmland. *J. Mt. Sci.* **2014**, *11*, 205–214. [CrossRef]
54. Pipan, P.; Kokalj, Z. Transformation of the Jeruzalem Hills cultural landscape with modern vineyard terraces. *Acta Geogr. Slov.* **2017**, *57*, 149–162. [CrossRef]
55. Zhou, X.; Zhang, C. The relationship evolution and interactive mechanism between heritage and tourism of Hani Terrace in Yuanyang. *Hum. Geogr.* **2019**, *34*, 154–160. (In Chinese)
56. Zhang, A. Evolution and differentiation of agricultural land use behavior of different types of farmers in Agricultural Cultural Heritage Tourism Destination: A case study of Hani Terrace. *Tour. Sci.* **2020**, *35*, 51–63. (In Chinese)
57. Feng, Z. The mutual benefit and symbiosis of terrace sightseeing, rice farming and national culture. *J. Hubei Minzu Univ. (Philos. Soc. Sci.)* **2020**, *38*, 96–103. (In Chinese)
58. Miao, J.Q.; Wang, Z.Q.; Yang, W.T.; Sun, S.; Huang, G.Q. Ecosystem service function and value evaluation of Chongyi Hakka terrace area. *J. Nat. Resour.* **2016**, *31*, 1817–1831. (In Chinese)
59. Chen, Y. The cultural beauty of Hani's terrace Tourism. *Heilongjiang Natl. Ser.* **2008**, *2*, 154–158. (In Chinese)
60. Beach, T.; Luzzadder-Beach, S.; Terry, R.; Dunning, N.; Houston, S.; Garrison, T. Carbon isotopic ratios of wetland and terrace soil sequences in the Maya Lowlands of Belize and Guatemala. *Catena* **2011**, *85*, 109–118. [CrossRef]
61. Borisov, A.V.; Chernysheva, E.V.; Korobov, D.S. Buried Paleoanthrosols of the Bronze Age agricultural terraces in the Kislovodsk basin (Northern Caucasus, Russia). *Quat. Int.* **2015**, *418*, 28–36. [CrossRef]
62. Ferro-Vazquez, C.; Lang, C.; Kaal, J.; Stump, D. When is a terrace not a terrace? The importance of understanding landscape evolution in studies of terraced agriculture. *J. Environ. Manag.* **2017**, *202*, 500–513. [CrossRef] [PubMed]

63. Jiang, Y.G.; Li, S.J.; Cai, D.S.; Chen, W.; Liu, Y.; Yu, Z. The genesis and paleoenvironmental records of Longji agricultural terraces, southern China: A pilot study of human–environment interaction. *Quat. Int.* **2014**, *321*, 12–21. [CrossRef]
64. Lansing, J.S.; Thurner, S.; Ning, N.C.; Coudurier-Curveur, A.; Chew, L.Y. Adaptive self-organization of Bali's ancient rice terraces. *Proc. Natl. Acad. Sci. USA* **2017**, *114*, 6504–6509. [CrossRef]
65. Jiao, Y.M.; Li, X.Z.; Liang, L.H.; Takeuchi, K.; Okuro, T.; Zhang, D.D.; Sun, L.F. Indigenous Ecological Knowledge and Natural Resource Management in the Cultural Landscape of China's Hani Terraces. *Ecol. Res.* **2011**, *27*, 247–263. [CrossRef]
66. Chen, Q.R.; Xie, H.L. Research Progress and Discoveries Related to Cultivated Land Abandonment. *J. Resour. Ecol.* **2021**, *12*, 165–174.
67. Hribar, M.M.; Geri, M.; Pipan, P.; Repolusk, P.; Cigli, R. Cultivated terraces in Slovenian landscapes. *Acta Geogr. Slov.* **2017**, *57*, 83–97.
68. Chen, B.; Qiu, Z.; Nakamura, K. Tourist preferences for agricultural landscapes: A case study of terraced paddy fields in Noto Peninsula, Japan. *J. Mt. Sci.* **2016**, *13*, 1880–1892. [CrossRef]
69. Kizos, T.; Koulouri, M.; Vakoufaris, H.; Psarrou, M. Preserving Characteristics of the Agricultural Landscape through Agri-Environmental Policies: The Case of Cultivation Terraces in Greece. *Landsc. Res.* **2010**, *35*, 577–593. [CrossRef]
70. Zhang, J.H.; Su, Z.A.; Liu, G.C. Effects of terracing and agroforestry on soil and water loss in hilly areas of the Sichuan Basin, China. *J. Mt. Sci.* **2008**, *5*, 241–248. [CrossRef]
71. Chen, D.; Wei, W.; Chen, L.D. Effects of terracing practices on water erosion control in China: A meta-analysis. *Earth-Sci. Rev.* **2017**, *173*, 109–121. [CrossRef]
72. Liu, S.L.; Dong, Y.H.; Li, D.; Liu, Q.; Wang, J.; Zhang, X.L. Effects of different terrace protection measures in a sloping land consolidation project targeting soil erosion at the slope scale. *Ecol. Eng.* **2013**, *53*, 46–53. [CrossRef]
73. Chen, S.; Chen, Y.; Peng, Y. Experimental study on soil erosion characteristics in flooded terraced paddy fields. *Paddy Water Environ.* **2013**, *11*, 433–444. [CrossRef]
74. Lesschen, J.P.; Cammeraat, L.H.; Nieman, T. Erosion and terrace failure due to agricultural land abandonment in a semi-arid environment. *Earth Surf. Process. Landf.* **2010**, *33*, 1574–1584. [CrossRef]
75. Boselli, V.; Ouallali, A.; Briak, H.; Houssni, M.; Kassout, J.; El Ouahrani, A.; Michailidi, E.M. System Dynamics Applied to Terraced Agroecosystems: The Case Study of Assaragh (Anti Atlas Mountains, Morocco) (Special Issue: Terraced Landscapes and Hydrological-Geological Hazards). *Water* **2020**, *12*, 1693. [CrossRef]
76. Paliaga, G.; Luino, F.; Turconi, L.; Graff, J.V.D.; Faccini, F. Terraced Landscapes on Portofino Promontory (Italy): Identification, Geo-Hydrological Hazard and Management. *Water* **2020**, *12*, 435. [CrossRef]
77. José, R.A.; Marisa, T.; Concepción, J.; Jorge, A.R.B.; Francisco, J.D. Plant species composition and richness in abandoned agricultural terraces vs. natural soils on Lanzarote (Canary Islands). *J. Arid. Environ.* **2016**, *124*, 165–171.
78. Arévalo, J.R.; Fernández-Lugo, S.; Reyes-Betancort, J.A.; Tejedor, M.; Jiménez, C.; Díaz, F.J. Relationships between soil parameters and vegetation in abandoned terrace fields vs. non-terraced fields in arid lands (Lanzarote, Spain): An opportunity for restoration. *Acta Oecologica* **2017**, *85*, 77–84. [CrossRef]
79. Bantayan, N.C.; Calderon, M.M.; Dizon, J.T.; Sajise, A.J.U.; Salvador, M. Estimating the Extent and Damage of the UNESCO World Heritage Sites of the Ifugao Rice Terraces, Philippines. *J. Environ. Manag.* **2012**, *15*, 1–5.
80. Joshi, R.C.; Matchoc, O.R.O.; Bahatan, R.G.; Pena, F.A.D. Farmers' knowledge, attitudes and practices of rice crop and pest management at Ifugao Rice Terraces, Philippines. *Pans Pest Artic. News Summ.* **2000**, *46*, 43–48. [CrossRef]
81. Joshi, R.C.; Delacruz, M.S.; Martin, E.C.; Cabigat, J.C.; Bahatan, R.G.; Bahatan, A.D.; Abayao, E.H.; Choy-Awon, J.; Chilagan, N.P.; Cayong, A.B. Current Status of the Golden Apple Snail in the Ifugao Rice Terraces, Philippines. *J. Sustain. Agric.* **2001**, *18*, 71–90. [CrossRef]
82. Pascal, A.; Laurence, B.; Pierre, G.; Frédéric, M.; Hervé, S.; Romain, F.; Denis, F.; Eric, D.; Frédérique, C.; Jin, B.H.; et al. Heterogeneity of the rice microbial community of the Chinese centuries-old Honghe Hani rice terraces system. *Environ. Microbiol.* **2020**, *22*, 3429–3445. [CrossRef]
83. Yang, L.; Liu, M.C.; Lun, F.; Min, Q.W.; Zhang, C.Q.; Li, H.Y. Livelihood Assets and Strategies among Rural Households: Comparative Analysis of Rice and Dryland Terrace Systems in China. *Sustainability* **2018**, *10*, 2525. [CrossRef]
84. Dizon, J.T.; Calderon, M.M.; Sajise, A.J.U.; Andrada, R.T.; Salvador, M.G. Youths' Perceptions of and Attitudes towards the Ifugao Rice Terraces. *J. Environ. Manag.* **2012**, *15*, 52–58.
85. Liu, M.C.; Liu, W.W.; Yang, L.; Jiao, W.J.; He, S.Y.; Min, Q.W. A dynamic eco-compensation standard for Hani Rice Terraces System in southwest China. *Ecosyst. Serv.* **2019**, *36*, 100897. [CrossRef]
86. Gu, H.; Jiao, Y.; Liang, L. Strengthening the socio-ecological resilience of forest-dependent communities: The case of the Hani Rice Terraces in Yunnan, China. *For. Policy Econ.* **2012**, *22*, 53–59. [CrossRef]
87. Benz, U.C.; Hofmann, P.; Willhauck, G.; Lingenfelder, I.; Heynen, M. Multi-resolution, object-oriented fuzzy analysis of remote sensing data for GIS-ready information. *ISPRS J. Photogramm. Remote Sens.* **2004**, *58*, 239–258. [CrossRef]
88. Wang, Y.N.; Sun, H.J. On the importance of environmental protection and sustainable development from the collapse of terrace. *City Geogr.* **2017**, *10*, 189–190. (In Chinese)
89. Miao, J.Q. Ecosystem Services of Chongyi Hakka Rice Terrace and Its Sustainable Management in Jiangxi Province. Ph.D. Thesis, Jiangxi Agricultural University, Nanchang, China, 2012.

90. Jiang, Z.S.; Chen, H.L.; Jiang, G.Z. Combinative Practice of Sustainable Protection of World Cultural Heritage and Targeted Poverty Alleviation: A Case Study of "Rice-fish Culture" Mode in Hani Terraced Fields of Honghe, Yunnan. *Chin. Agric. Sci. Bull.* **2018**, *34*, 160–164.
91. Miao, J.Q.; Wang, Z.Q.; Ma, Y.Q.; Yang, W.T.; Yang, B.J. Comprehensive evaluation of sustainable development of the Hakka Terrace Ecosystem of Chongyi. *Acta Ecol. Sin.* **2018**, *38*, 6326–6336.
92. Ni, H. The concept and path of sustainable development of Hani Terrace Agriculture from the perspective of ecological civilization. *World Agric.* **2014**, *11*, 156–159. (In Chinese)
93. Agus, F.; Irawan, I.; Suganda, H.; Wahyunto, W.; Setiyanto, A.; Kundarto, M. Environmental multi-functionality of Indonesian agriculture. *Paddy Water Environ.* **2006**, *4*, 181–188. [CrossRef]
94. Gan, D.X.; Long, Y.L.; Huang, H.; Xiong, X.Y. The disaster prevention mechanism and benefit analysis of the mountain terraced field landscape—taking the Ziquejie terraced field as an example. *J. Nat. Disasters* **2006**, *15*, 106–108. (In Chinese)
95. Bulilan, C.M.R. Experiencing cultural heritage and indigenous tourism in Banaue. *Philipp. Q. Cult. Soc.* **2007**, *35*, 100–128.
96. Li, Z.M.; Zou, H.X.; Yi, Q.Q.; Zhou, Q. Research on the evaluation of tourist resources potential of terraced agricultural cultural heritage. *Econ. Geogr.* **2015**, *35*, 198–208. (In Chinese)
97. Jiao, Y.M.; Yang, Y.J.; Hu, W.Y.; Su, S.H. Analysis of spatial pattern and aesthetic features of Hani Terraced Fields. *Geogr. Res.* **2006**, *25*, 624–632. (in Chinese).
98. Drechsler, M.; Settele, J. Predator-prey interactions in rice ecosystems: Effects of guild composition, trophic relationships, and land use changes—A model study exemplified for Philippine rice terraces. *Ecol. Model.* **2001**, *137*, 135–159. [CrossRef]

International Journal of
Environmental Research and Public Health

Review

Evolutionary Overview of Land Consolidation Based on Bibliometric Analysis in Web of Science from 2000 to 2020

Xin Xu, Qianru Chen * and Zhenhong Zhu

Institute of Ecological Civilization, Jiangxi University of Finance and Economics, Nanchang 330013, China; 2202021937@stu.jxufe.edu.cn (X.X.); 2202010105@stu.jxufe.edu.cn (Z.Z.)
* Correspondence: cqrjufe@163.com; Tel.: +86-791-8382-0732

Abstract: Land consolidation is widely used as a powerful tool for land use management in many countries. In order to objectively reveal the current research status in the field of land consolidation, this paper uses the Bibliometrix and Biblioshiny software packages, and VOSviewer to analyze the literature in the field of land consolidation in the last 20 years of the Web of Science Core Collection Database. The results show that: (1) In the past two decades, the annual publication of papers on land consolidation rose. It can be divided into three stages: 2000–2007 for the embryonic period, 2008–2012 for the long-term, and 2013–2020 for the high-yield period. (2) Land consolidation studies covered 68 countries or regions. The top three countries were China, Poland, and the United States. China and the United States played an important role in international cooperation in the field of land consolidation, and Turkey mainly conducted independent research in the field of land consolidation. (3) Land consolidation, reclamation, China, remote sensing, and land fragmentation were the high-frequency keywords in the field of land consolidation in recent years. (4) The research focusing on the field of land consolidation involved its development course, its impact on ecosystem services, and the evaluation of its benefits. (5) The theme of land consolidation studies was shunted and evolved over time, and nine evolution paths could be summarized in the studies of cultivated land fragmentation, development course of land consolidation, and impacts of land consolidation on soil. Finally, this paper predicted the future research directions of land consolidation: exploring new methods for evaluating the benefits of land consolidation, the scale effects of the impact of land consolidation on ecosystem services, research on the mechanism and comprehensive effects of land consolidation on soil, research on land consolidation and rural revitalization, and land consolidation theory research.

Keywords: Bibliometrix; land consolidation; thematic evolution; reclamation; Biblioshiny

Citation: Xu, X.; Chen, Q.; Zhu, Z. Evolutionary Overview of Land Consolidation Based on Bibliometric Analysis in Web of Science from 2000 to 2020. *Int. J. Environ. Res. Public Health* **2022**, *19*, 3218. https://doi.org/10.3390/ijerph19063218

Academic Editors: Wei Song, Hualin Xie and Paul B. Tchounwou

Received: 26 January 2022
Accepted: 4 March 2022
Published: 9 March 2022

1. Introduction

Land is the basic resource for human survival and development [1,2]. The rapid development of the social economy and population growth increase the demand for land. However, the limited land cannot provide more space and resources for human beings, and the contradiction between man and land becomes prominent [2,3]. Currently, more than half (54%) of the world's population lives in urban areas, and this proportion is expected to reach 66% by 2050 [4]. Food production is expected to increase by 70% (by 2050) to satisfy the growing population, and land requires a more sustainable management mode [5]. It demonstrates the need for the rational and efficient use of the world's limited land resources in order to maximize the provision of products and services to the growing food needs of humankind.

Land consolidation refers to the activities of integrated management of unused, inefficient and idle, damaged, and degraded land to meet the needs of human production, life, and ecological functions. It is the general term for land development, consolidation, reclamation, and restoration [6]. The function of land consolidation is reflected in many ways. As an effective tool to supplement cultivated land, land consolidation plays an active role in achieving the balance of arable land [7], ensuring national food security [8], and promoting rural revitalization [9]. It has been widely used all over the world in recent decades. Land consolidation policies have gradually shifted from initial agricultural production objectives to a means of supporting rural development [10]. Moreover, land consolidation has far-reaching effects on promoting the scale of agricultural production and improving the competitiveness of agricultural products, adjusting the structure of land use, developing modern agriculture, alleviating ecological risks, and improving agricultural production efficiency [11–15].

Bibliometrics is a quantitative analysis method that uses mathematical and statistical tools to measure the interrelationship and impact of publications in a particular research area [16]. It enables researchers to sketch out complex knowledge maps that represent the structure of knowledge in a field of study and study their properties through statistical and mathematical methods [17–19]. As a powerful tool for analyzing the field of knowledge and revealing its cognitive–epistemological structure [20], it provides a macro-overview of a large number of academic studies and reliably identifies influential research, authors, journals, organizations, and countries [21].

Scholars have used large-scale data sets and bibliometric methods to carry out bibliometric analysis in the field of land consolidation, and obtained valuable research results. However, most of the research focuses on keyword co-occurrence analysis, journal sources, and author publications, while studies on the context of historical citation, high-frequency keyword clustering analysis, subject evolution, future development direction prediction, etc. are rare. Therefore, the Bibliometrix and Biblioshiny software packages in the *R* tool are adopted in this paper to systematically measure the literature of land consolidation in the Web of Science core collection database set during 2000–2020. This study aims to solve the following scientific questions.

(1) How are the keywords in the field of land consolidation clustered?
(2) What countries have cooperated in the field of land consolidation?
(3) How did the history of citations in the field of land consolidation develop?
(4) What is the focus and direction of future research in the field of land consolidation?

2. Data Sources and Research Methods

2.1. Data Sources

Web of Science is the world's largest and most subject-covering comprehensive academic information resources, including more than 8700 core academic journals in various fields of natural sciences, engineering, biomedicine, social sciences, and arts and humanities. This paper took the core collection in the Web of Science database as the data source. The search term included TI = "land consolidation" or TI = "land reclamation". The document type was limited to "Article", and the retrieval time was 2000–2020. The languages were English and Chinese. After pre-processing, such as de-duplication and the removal of irrelevant data, a total of 599 papers in the field of land consolidation were obtained. Among them, 594 papers were in English, accounting for 99.17% of the total number of articles. The downloaded data were saved in a text format.

2.2. Research Methods

Bibliometric analysis provides a comprehensive overview of a large body of research literature and further developed previously unevaluated insights by allowing quantitative and objective identification of past and present research topics [22].

The Bibliometrix *R* package provides a set of tools for quantitative research in scientific metrology. It is written in *R*, an open-source environment and ecosystem. The existence of a large number of effective statistical algorithms, access to high-quality numerical routines, and integrated data visualization tools may be the strongest qualities of *R* languages in scientific computing over other languages [23].

Biblioshiny was developed by Massimo Aria in a secondary development of the Bibliometrix-based Shiny package in the *R* language, encapsulating Bibliometrix's core code and creating a web-based online data analysis framework. Users can carry out relevant scientific measurement and visual analysis work on the interactive web interface, which reduces the user's threshold of use and the intensity of information input, to a certain extent [24].

Developed by Nis Jan van Eck and Ludo Waltman, VOSviewer is also widely used in bibliometric analysis due to its more aesthetically pleasing visualization processing, in particular, keyword co-occurrence analysis [25,26]. The specific Bibliometrix and science-mapping workflow is shown in Figure 1.

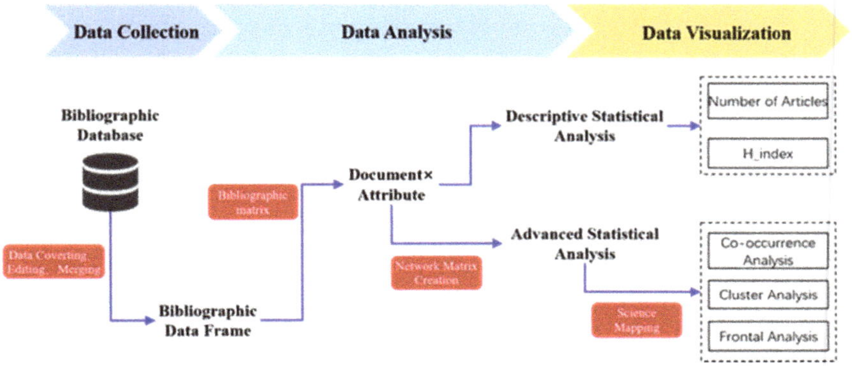

Figure 1. Bibliometrix and the recommended science-mapping workflow.

3. Results Analysis

3.1. Distribution of Annual Documents

The analysis of the distribution of literation volume from the time series can reflect the trend of study [24]. Figure 2 shows that, from 2000 to 2020, the number of research studies published in the field of land consolidation fluctuated slightly, but is on an upward trend overall. Combined with the macro-policy changes in land consolidation, it was divided into three research stages: 2000–2007, 2008–2012, 2013–2020. The 2000–2007 period was the germination period of land consolidation research. The annual publication volume was very small with few differences. The 2008–2012 period was the long term. Although the growth rate of literature publication was not obvious, but compared with the previous stage, the annual publication volume was more stable, indicating that the function of land consolidation attracted the attention of scholars. The 2013–2020 period was the high-yield period with an obvious growth rate. The number of publications reached a peak of 92 in 2020. Combined with China's comprehensive land consolidation policy formulated in 2013, it shows that the research on land consolidation in recent years is of important practical significance.

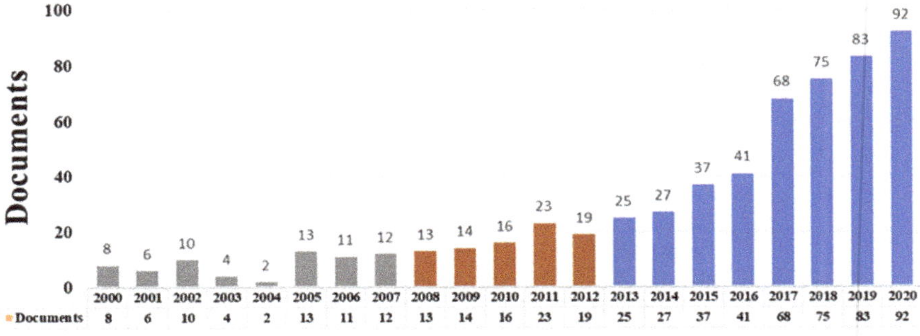

Figure 2. Number of land consolidation research documents published from 2000 to 2020.

3.2. Analysis of Cited Papers in Land Consolidation Research

3.2.1. Annual Development Trend of Citations

As can be seen from the average citation distribution of papers per year (Figure 3), the citations were low in 2000–2004. Especially in 2004, the citation was lowest, showing that the research period was in its infancy. The cited frequency was at an average of 3.0 in 2006–2012 and 2014–2016. It reached its highest level in 2014 at 5.18. At the same time, the fluctuation was greater than the annual change in paper yield. In 2018–2020, the average citation of papers on land consolidation was in a declining stage, while the volume of published papers was on a steady upward trend. This indicates that, despite the various research directions on land consolidation, the influence of papers was decreasing.

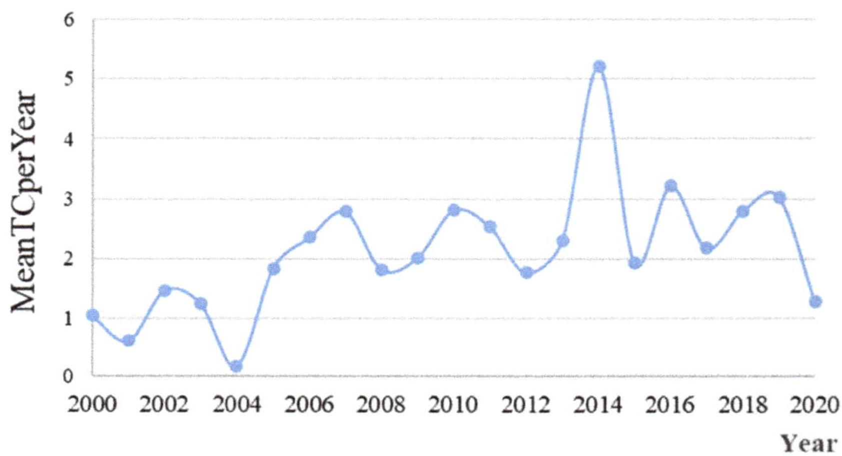

Figure 3. The number of annual cited papers.

The highest yield of cited papers was in 2014 at 5.20. From 2012 to 2014, the average citation of highly cited papers increased explosively. Most scholars in this period mainly analyzed the effects of land consolidation based on multi-type land policies. For example, Li analyzed the negative impacts of land consolidation on China's coastal ecosystem and its services, and called on China to strengthen the construction of laws and regulations, improve marine spatial planning, and fully assess the negative impacts of land consolidation [27].

3.2.2. Historical Cited Papers of Land Consolidation Research

Using the historical citation visual analysis in the Bibliometrix installation package in *R* Studio, 20 nodes were selected, and the pioneering works and some classic studies in this field were found. CiteScore is the youngest indicator, which was released in December 2016 [28]. In this paper, LCS and GCS indicators were used to analyze the research methods and contents of classical literature. LCS refers to the reference score in the downloaded paper dataset, and GSC refers to the reference score in the Web of Science core collection database.

As can be seen from Figure 4, the earliest node in the literature on land consolidation was an article published in *Ground Water* in 2001 entitled *"Analytical Studies on The Impact of Land Reclamation on The Water Flow"*. On the basis of the negative impacts that land reclamation will have on the ecosystem and its services, this paper analyzed the influence of land reclamation on groundwater flow. The results showed that the larger the scale of land reclamation, the more significant the increase in the horizontal line of the coast [29].

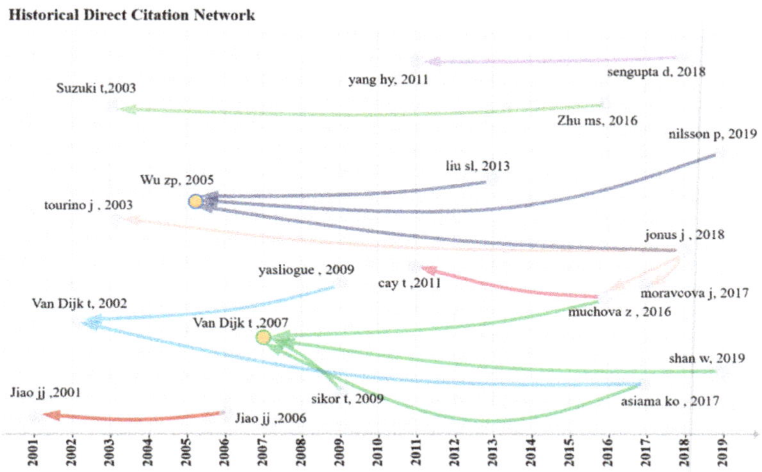

Figure 4. Historical direct citation network of top-cited papers in the field of land consolidation during 2000–2020.

Several classic articles emerged between 2005 and 2007 (Figure 4 clearly cites the relationship). Terry's article in 2007 in *Geoforum* entitled "*Complications for Traditional Land Consolidation in Central Europe*" had four distinct chains of citations, and its citation frequency of 39 was also the highest of the LCS. This article described the complexity of implementing traditional land consolidation in Central Europe from factors such as unfavorable macro conditions, absentee landowners, land ties, and unfinished privatization [30]. Terry first published a paper on land consolidation in 2002 in *European Planning Research*, entitled "*Export of Planning Knowledge Needs Comparative Analysis: The Case of Applying Western Land Consolidation Experience in Central*". Its innovation lies in putting forward cross-border knowledge transfer research comparative analysis by demonstrating land consolidation as a case [31].

Wu's article in *China Economic Review* in 2005 entitled "*Land Consolidation and Productivity in Chinese Household Crop Production*" also presented three significant chains of citation relationships. As shown in Table 1, both LCS and GCS ranked second, which were 17 and 81 respectively. This indicates that the literature is not only a classic document in the field of land consolidation, but also favored by scholars in many fields with a strong cross-cutting with other disciplines. Based on raw data from 227 Chinese households, this paper evaluated the effectiveness of land consolidation projects in the agricultural comprehensive development plan. The results showed that the land consolidation project improved the land quality and contributed 1.52% to the crop yield [32]. The LCS and GCS of Hoeksema's article in *Irrigation and Drainage* in 2007 entitled "*Three Stages in the History of Land Reclamation in the Netherlands*" are at a high level. This paper briefly introduced the three stages of land reclamation in the history of the Netherlands and laid a foundation for other scholars' research [33].

Table 1. Top 10 local citation scores (LCS) in land consolidation research.

Documents	DOI	Year	LCS	GCS
TERRY VD, 2007, GEOFORUM	10.1016/j.geoforum.2006.11.010	2007	39	78
WU ZP, 2005, CHINA ECON REV	10.1016/j.chieco.2004.06.010	2005	17	81
CAY T, 2011, EXPERT SYST APPL	10.1016/j.eswa.2011.02.150	2011	16	30
YASLIOGLU E, 2009, EUR PLAN STUD	10.1080/09654310802553639	2009	14	20
SUZUKI T, 2003, MAR POLLUT BULL	10.1016/S0025-326X(02)00405-8	2003	12	48

Table 1. *Cont.*

Documents	DOI	Year	LCS	GCS
ASLAN STA, 2007, SPAN J AGRIC RES		2007	11	21
ASIAMA KO, 2017, J RURAL STUD	10.1016/j.jrurstud.2017.09.007	2017	11	25
HOEKSEMA RJ, 2007, IRRIG DRAIN	10.1002/ird.340	2007	10	71
JIAO JJ, 2001, GROUND WATER	10.1111/j.1745-6584.2001.tb02479.x	2001	9	25
TOURINO J, 2003, INT J GEOGR INF SCI	10.1080/1365881031000072636	2003	9	26

Yang et al. [34] published an article in *Bird Conservation International* in 2011 entitled "Impacts of Tidal Land Reclamation in Bohai Bay, China: Ongoing Losses of Critical Yellow Sea Waterbird Staging and Wintering Sites". As shown in Table 2, the GCS is as high as 90, while the LCS is not ideal, which is 6, and only one lead chain appeared. Based on remote sensing technology to collect data, this paper analyzed the influence of wetland reclamation on water birds on the Bohai coast of China.

Table 2. Top 10 global citation scores (GCS) in land consolidation research.

Documents	DOI	Year	LCS	GCS
YANG HY, 2011, BIRD CONSERV INT	10.1017/S0959270911000086	2011	6	90
WU ZP, 2005, CHINA ECON REV	10.1016/j.chieco.2004.06.010	2005	17	81
TERRY VD, 2007, GEOFORUM	10.1016/j.geoforum.2006.11.010	2007	39	78
HOEKSEMA RJ, 2007, IRRIG DRAIN	10.1002/ird.340	2007	10	71
SIKOR T, 2009, WORLD DEV	10.1016/j.worlddev.2008.08.013	2009	6	63
LIU SL, 2013, ECOL ENG	10.1016/j.ecoleng.2012.12.001	2013	9	54
SUZUKI T, 2003, MAR POLLUT BULL	10.1016/S0025-326X(02)00405-8	2003	12	48
ADRIANSEN HK, 2009, GEOFORUM	10.1016/j.geoforum.2009.05.006	2009	5	37
MUCHOVA Z, 2016, ECOL ENG	10.1016/j.ecoleng.2016.01.018	2016	8	31
CAY T, 2011, EXPERT SYST APPL	10.1016/j.eswa.2011.02.150	2011	16	30

3.3. Analysis of Main Researchers

A total of 1585 authors were involved in the paper data set on land consolidation, of whom Liu, Zhou, and Jin were the top three authors, with 12, 12, and 11 publications, respectively (Table 3). In this field, the largest number of publications was from China's Liu, with a h-index of 7, g-index of 12, and total number of citations of 349. It indicates the high quality and great influence of Liu's papers. As can be seen from Figure 5, Liu's most frequently cited paper appeared in 2019 (the darkest color of the graph), which was cited 28 times. Liu's article entitled *"Land consolidation boosting poverty alleviation in China: Theory and practice"*, published in *Land Use Policy* in 2019, analyzed the mechanisms and dynamics of land consolidation to alleviate poverty in China. The results showed that land consolidation plays an active role in increasing the cultivated land area, promoting the agricultural production scale, improving the conditions of rural production, and reducing ecological risks [35].

Zhou's most frequently cited paper also appeared in 2019. His article entitled *"Land consolidation boosting poverty alleviation in China: Theory and practice"* revealed the link between land consolidation and poverty alleviation [35]. The total number of citations of Li from China was the highest (Table 3), which was 455, and his most frequently cited year was 2014.

Hirsch believes that the h-index is not only an acceptable tool for measuring the importance, significance, and broad impact of authors' cumulative research contributions, but also for assessing current paper volumes and predicting authors' future performance, as this indicator combines productivity and impact [36,37]. However, comparison of the h-index alone may be misleading due to the loss of citation information [38]. To compare the influence of authors in this field at different times, the m-index can be introduced. To be specific, $m = h/n$, where n indicates the age of the author's publication in the field [28].

Len's m-index was the highest, which was 1.167. Len has been publishing papers in the field of land consolidation since 2016, while the most frequently cited year was 2018. For example, Len's article in *Computers and Electronics in Agriculture* in 2018 entitled *"An algorithm for selecting groups of factors for prioritization of land consolidation in rural areas"* presented a general algorithm for factor group selection. Based on similar areas of two regions, but varied significantly in spatial structure in Poland, this paper proved that the algorithm could prioritize land consolidation activities in these areas [39].

Table 3. Top 10 influential authors in the field of land consolidation.

Author	h-Index	g-Index	m-Index	TC	NP	PY-Start
LIU Y	7	12	0.7	349	12	2012
ZHOU Y	7	12	1	197	12	2015
JIN X	7	11	1	161	11	2015
LI Y	7	10	0.875	455	10	2014
LEN P	7	9	1.167	114	9	2016
ARULRAJAH A	6	7	0.333	170	7	2004
BO MW	6	7	0.333	170	7	2004
DEMETRIOU D	6	6	0.6	180	6	2012
ZHANG Y	5	10	0.833	104	11	2016
CAY T	5	8	0.417	196	8	2010

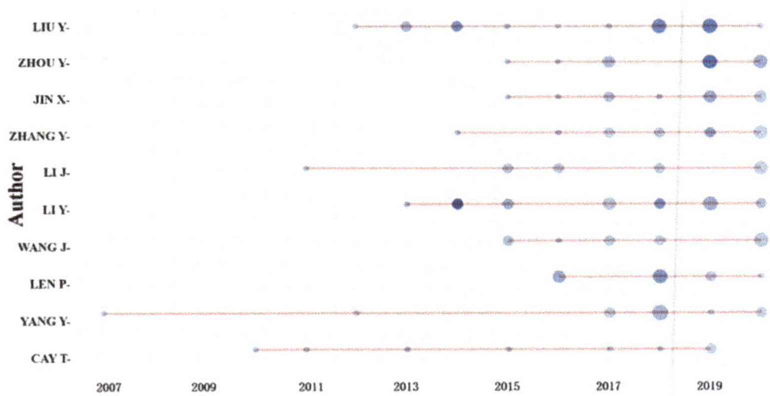

Figure 5. Authors' production over time in the field of land consolidation.

3.4. Analysis of Distribution Characteristics of Major Research Countries/Regions

The distribution characteristics of major research countries/regions reflect each country's influence in the field of land consolidation and provide conditions for further exploitation on the reasons for different degrees of influence. The dataset used in this article was published in 68 countries or regions. The top 25 published papers were distributed as follows: nine Asian countries (China, Turkey, Korea, Japan, India, Singapore, Iran, Malaysia, and Thailand), three American countries (the United States, Canada, and Brazil), two Oceania countries (Poland and Australia), and eleven European countries (Netherlands, United Kingdom, Spain, Czech Republic, Italy, Russia, Germany, France, Serbia, Slovakia, and Romania).

As can be seen from Figure 6, papers on land consolidation were published mainly in Asia and Europe. Specifically, China is the only developing country in the top three, with several times the number of studies of that of other countries, accounting for about 50% of the total production. The reasons for such high production may be China's far-reaching history in land consolidation and the evolution of related policies (Figure 7). In 2003, the Chinese "National Land Development and Consolidation Plan" specifically elaborated the objectives, principles, key areas, etc. of land consolidation, providing a guiding role for the nationwide implementation of land consolidation. Before 2008, the prevailing idea of "emphasized quantity over quality" in land consolidation increased the cultivated land area in China [40]. However, grain production at this stage still remains stagnant. In 2011, the Ministry of Natural Resources issued the *"High-standard Basic Farmland Construction Specification"*, aiming at improving the quality and productivity of arable land. It meant that the quality of arable land was as important as quantity. In March 2012, the Chinese "National Land Consolidation Plan (2011–2015)" was implemented [41]. It meant the beginning of comprehensive land consolidation in China, which included agricultural land, construction land, and unused land, and more attention was paid to the protection of the ecological environment. However, the average citation for each paper was 13.44. It was lower than that of the United States of 17.90, while its paper volume ranked the third in the world. This indicates that the influence of literature on land consolidation in China needs to be improved. Thailand leads the list, with an average of 57.75 citations per article. Niroula and Thapa [42], for example, analyzed the structural problems of land consolidation that did not address fragmentation, and concluded with an overview of a wide range of sustainable land consolidation policies and legal measures.

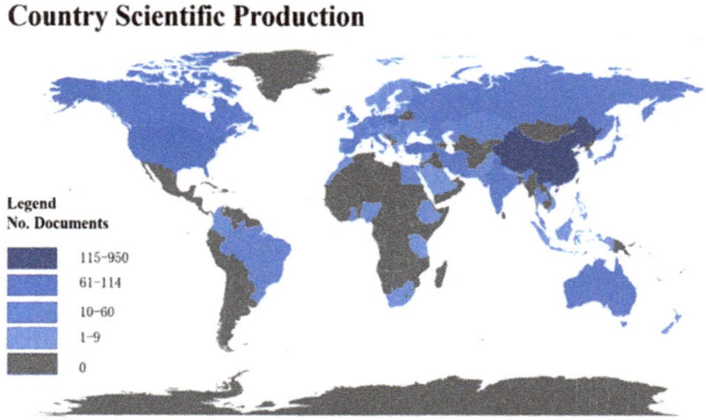

Figure 6. Scientific production distribution in the field of land consolidation.

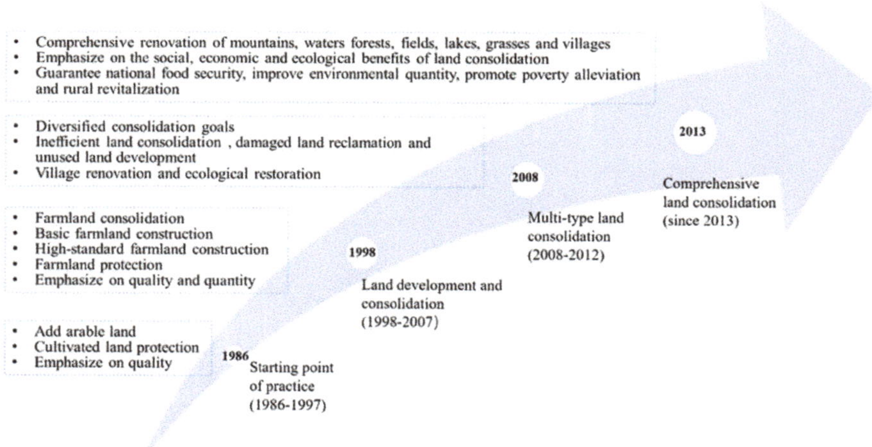

Figure 7. The history of land consolidation in China. Source: Land consolidation and rural revitalization in China: Mechanisms and paths [43].

Based on the downloaded studies from the Web of Science database, we used VOSviewer software to screen out the cooperation between 18 countries (Figure 8). Country was set as the analysis unit, full counting was set as the counting method, and the minimum number of documents in each country was set to 7. The size of the circle in the figure indicates the number of papers published by each country. The finer the line between the two labels, the weaker the cooperation intensity between countries, and vice versa.

Figure 8. Country collaboration map in the field of land consolidation.

As can be seen from Figure 8, China is at the center of international cooperation, followed by the United States, Australia, and Canada, with 31, 11, and 6 respectively. In total, 244 Chinese documents were sourced from state cooperation, accounting for about 22% of all published papers in China (Figure 9). China's main cooperative countries were Australia and Germany, accounting for 73% and 66%, respectively. Among them, Australia and China, Singapore, and Canada constituted a cooperative, with frequencies of 11, 6, and 6, respectively. Germany published a total of nine documents in the field of land consolidation, mainly in cooperation with China and the Netherlands, with frequencies of three and two, respectively. In particular, the 29 papers published in Turkey were all independent research. The above indicates that international cooperation should be strengthened in the field of land consolidation.

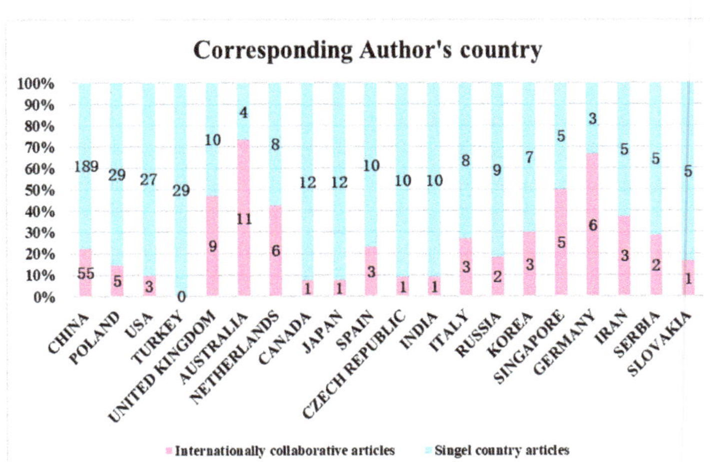

Figure 9. Corresponding authors' nationalities in the 20 most prolific countries in the field of land consolidation.

3.5. Analysis of Keywords

3.5.1. Analysis of High-Frequency Keywords

Keywords are the high generalization of a research topic and content. Analysis of high-frequency keywords reflects the hotspots in the field of land consolidation in a straightforward way. The Bibliometrix and Biblioshiny installation packages in the *R* tool were adopted to count the author keywords and draw a Word TreeMap of the top 20 keywords in this area. Figure 10 shows that land consolidation, land reclamation, and reclamation accounted for over half of the total keywords, with proportions of 27%, 19%, and 9% respectively. The keyword "China" in the bright red box of Figure 10 accounted for 8% of the keywords, indicating that China was one of the main study areas of land consolidation studies. It corresponded to the total paper number and the state of international cooperation in China analyzed above. Remote sensing and GIS accounted for 2% of the keywords. This indicates that they are two major study tools in this area. For example, Karan et al. [44] used remote sensing and GIS to monitor the land degradation and reclamation of coal mines through the pioneering combination of ratio vegetation index (RVI), enhanced vegetation index (EVI), normalized vegetation index (NDVI), and normalized moisture index (NDMI).

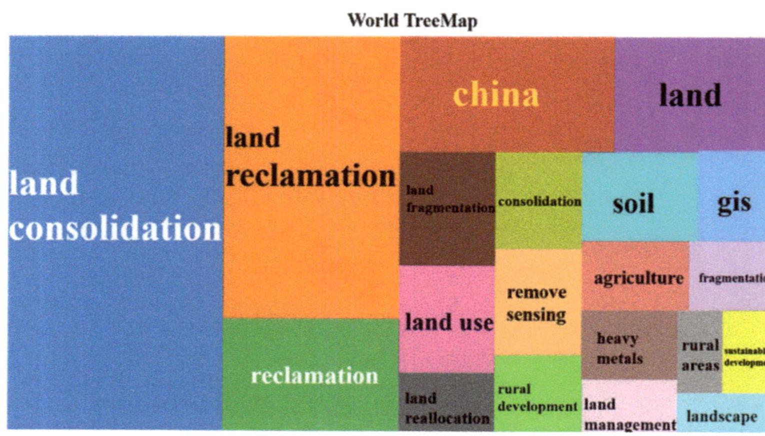

Figure 10. Word treemap of high-frequency keywords in the field of land consolidation.

Land fragmentation and fragmentation were the major problems to be solved by land consolidation, accounting for 4% and 2% respectively. Land fragmentation limits agricultural production and development in many countries, so it is urgent to address the problem of land fragmentation [45]. For example, Liu et al. [46] took Jiangsu Province of China as an example to analyze the distribution characteristics, influencing factors, and classification of arable land, and finally suggested that land fragmentation should be incorporated into the land consolidation plan so as to achieve high-quality and sustainable cultivated land development.

Finally, the proportions of land, land use, land reallocation, rural development, and sustainable development were lower, accounting for 1−2%. Even so, they have become an integral part of land consolidation policy, as there is a strong correlation between rural development and sustainable development and the issue of land fragmentation.

3.5.2. Cluster Analysis and Multiple Correspondence Analysis of High-Frequency Keywords

Clustering analysis in literature metrology, based on the frequency of two or two key words appearing at the same time, uses statistical methods to simplify the complex keyword mesh relationship into a few relatively small groups of classes [47]. It is designed to detect the natural division of network groupings (clusters) based on similarity and to minimize similarity between clusters [48]. This study used hierarchical clustering to treat each clustered keyword as a category, then merged it among the clusters with the highest degree of similarity, and finally grouped all of the individuals into one category and demonstrated the similarity of the key words in the field of land consolidation research in the form of treemaps (Figure 11).

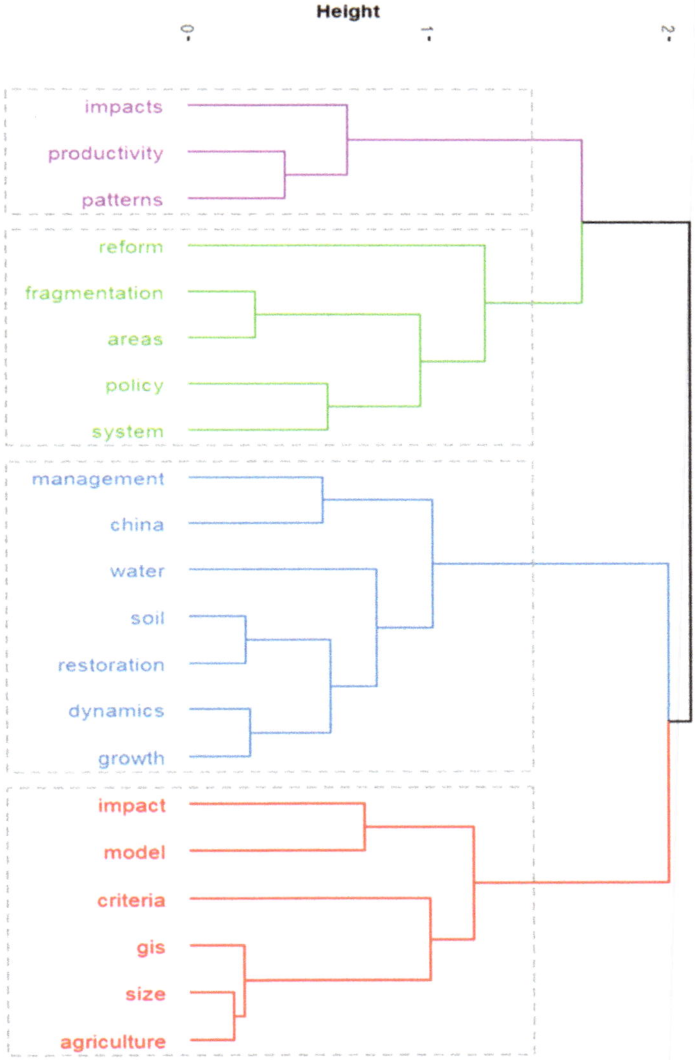

Figure 11. Tree dendrogram of hierarchical cluster analysis of keywords in the field of land consolidation.

Correspondence analysis first appeared in the 1960s, and has had a long and varied history [49,50]. It represents an exploratory approach of graphically representing the associations between variables in large, classified data sets to explore their relationships [51]. The corresponding analysis is designed to reveal the correspondence between different variables or categories of the same variable in qualitative data by lowering the dimension. Figures 11 and 12 show the clustering results of multiple corresponding analyses in the field of land consolidation. It can be classified into four categories.

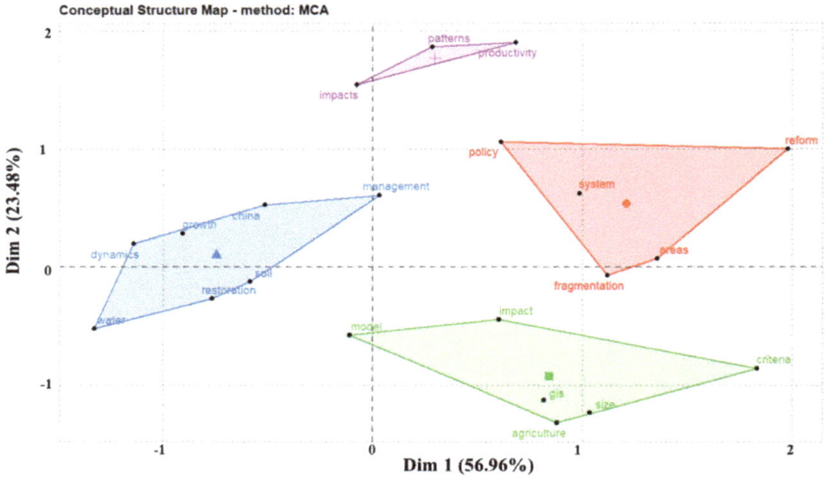

Figure 12. Multiple correspondence analysis of high-frequency keywords in the field of land consolidation.

(1) The first category of cluster analysis: this category is mainly related to crop productivity, technical efficiency, arable land patterns, and landscape types. For example, Zeng et al. [52] used random cutting-edge analytical methods to calculate the efficiency of agricultural techniques for land consolidation. The results showed that the overall agricultural technology efficiency of producers was greatly improved after land consolidation, with an average technical efficiency of 0.924. Using the revised ecological connectivity index, the study of Wang et al. [53] in Da'an City, Jilin Province of China during 2008–2014 verified the negative impacts of land consolidation on ecosystem services, reflecting the problems in the implementation of the land consolidation project.

(2) The second category of cluster analysis: this category is mainly related to the impacts of land consolidation policies, reform, and land fragmentation. The review and prospects of land consolidation development, as well as the study of the influencing factors of land fragmentation, help the government to formulate relevant land consolidation policies. Based on the regional data of Jiangsu Province, Liu et al. [46] explored the characteristics, influencing factors, and classification of the spatial distribution of cultivated land, which is of great significance to the improvement of agricultural production capacity on the regional scale. At the same time, the fragmentation degree of cultivated land in the construction area is higher than that outside the construction area. The awareness of this situation helps the government to formulate relevant land consolidation policies.

(3) The third category of cluster analysis: based on population growth, this category mainly studies the impacts of land consolidation on ecosystem services through land-use change under the background of accelerated urbanization. It is represented by studies in China. For instance, using GIS-RS technology, Hao et al. [54] selected typical farmlands to analyze the change in the cropland ecosystem service value with land-use change in northeast China.

(4) The fourth category of cluster analysis: this category is mainly related to the benefit evaluation of land consolidation and the adoption of models. The benefit evaluation of land consolidation includes three aspects: economic benefit, social benefit, and ecological benefit. Based on land-use patch data, Shi et al. [55] combined landscape pattern analysis with production, life, and benefit evaluation to overcome the shortcomings of previous single-benefit evaluation and conduct comprehensive research

on land consolidation projects. The results showed that land consolidation directly or indirectly improved the landscape ecological pattern of the project area, and land consolidation obviously improved the balanced distribution of cultivated land and the centralized distribution of construction land in the project area.

3.6. Evolution Analysis of Themes in the Field of Land Consolidation

According to Weismayer and Pezenka [56], it is important to study research development in a field in terms of themes and thematic evolution. Sankey diagrams, or Sankey energy shunts, are also known as Sankey energy shunts. Sankey diagrams describe the flow of different nodes in a network, and they are most typically used to analyze the flow of energy or matter. Arrows or direction lines are used to represent these flows, and the thickness of the arrows or direction lines is proportional to the flow size. These diagrams are commonly used in industrial ecology to describe product life cycle assessments and for rapidly visualizing energy efficiency in engineering [57]. Sankey diagrams emphasize the size and direction of traffic within the system, which, due to their wide practicality, have been applied to many geographic or human environmental research environments. Based on the Sankey diagram's visual presentation of the changes in the subject matter of the land consolidation research field over time, this study can see the diversion of different topics in the field of land consolidation, and clarify the quantitative information, such as subject flow and conversion relationship [58].

Referring to Zhou et al. [43], this paper divides the subject diversion of land consolidation development process into three stages, with 2007 and 2012 as breakpoints. Over the past 20 years, the research topics in the field of land consolidation have shown several evolutionary paths in three directions (Figure 13).

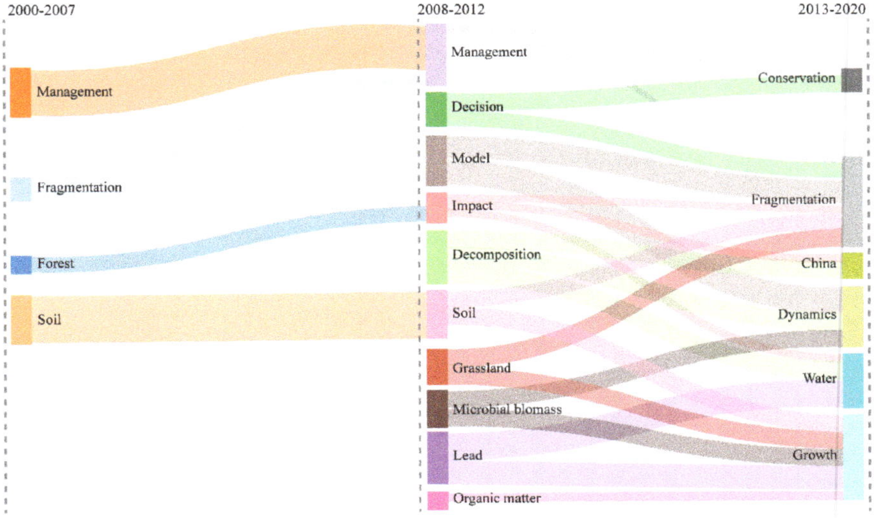

Figure 13. Thematic evolution of the land consolidation research field (2000–2020).

(1) Studies on land fragmentation related to land consolidation. ① Fragmentation→management→conservation, fragmentation. ② Fragmentation→impact→fragmentation, China, water. While enriching the structure and dispersing the risk of agricultural cultivation, and increasing farmers' income, the fine fragmentation of cultivated land has also caused the waste of land resources and the increase of the agricultural production cost to a certain extent, thus reducing agricultural production efficiency and hindering the development of agricultural mechanization. Land consolidation

is an effective way to reduce cultivated land fragmentation. For example, Ela [59] took Turkey as an example, and analyzed the impacts of land consolidation projects on agricultural land fragmentation. The proportions of agricultural enterprises with an index value of less than 0.40 were 1.17% and 3.7%, respectively, before land consolidation, and decreased to 0.6% and 2.3%, respectively, after land consolidation. The resulting values indicate a decrease in the degree of plot fragmentation in the area. Moreover, land consolidation projects have brought great economic benefits to the owners of agricultural enterprises in the region.

(2) Research on the development process of land consolidation. ① Management→management→conservation. ② Management→management→fragmentation. As time passed, the number of publications decreased significantly after 2012 with evolved themes. The development stages and priorities of land consolidation differed between countries. The study of the development process is conducive to the formulation of relevant policies. The focus of land consolidation in Germany has changed from adjusting farmland ecology and improving agricultural supporting facilities to the current stage of focusing on the transformation of rural infrastructure construction and regional planning, and the protection of natural landscapes; in addition, special attention is paid to public participation in land consolidation projects [60]. The focus of the current land consolidation work in the Netherlands is to enhance the comprehensive role of the land and strengthen the protection and improvement of the ecological environment in the process of renovation [61]. Land consolidation in Japan was initially aimed at the treatment of farmland salinization and the reclamation of land from the sea, but later developed into an overall plan of the region. Since the 1980s, land consolidation has been an important means to realize the symbiosis of residential space and agricultural land. The preparation of a comprehensive village development plan that maintains agricultural characteristics has fully demonstrated the important position and huge role of land consolidation in rural development [62]. Portugal believes that comprehensive land consolidation projects are geographically defined as rural land development activities [43].

(3) Research on the impacts of land consolidation on soil. ① Soil→soil→fragmentation, growth. ② Forest→impact→fragmentation, China, water. ③ Microbial biomass→dynamics, growth. ④ Decomposition→China, dynamics, water. ⑤ Organic matter→growth. Soil is an important resource basis for agricultural production, as well as an important object of land consolidation. The direction of the thematic evolution is mainly between the effects of land consolidation behavior on soil physical structure, nutrient cycling, and microbial functions. This research is of great significance for avoiding the negative effects of land consolidation and improving the benefits of land consolidation. For example, He et al. [63] characterized the soil microbial communities under five land-use patterns, and used DNA fingerprinting and metabolic analysis as the characteristics, revealing that land reclamation has severely affected the population size, composition, and structure of soil microbial communities, as well as bacteria. In addition, Hou et al. [64] investigated soil samples of cinder-reclaimed land after reclamation periods of 1, 6, and 15 years, compared the characteristics of various soil microbial communities in the reclamation area, and compared areas not affected by coal mining. Soil sample analysis revealed that the application of microbial remediation technology can be used to adjust the structure of the soil microbial community, improve soil quality, and shorten the soil recovery cycle.

4. Conclusions and Discussion

Based on the Web of Science database to retrieve studies in the field of land consolidation during 2000–2020 and adopting the Bibliometrix and Biblioshiny software packages for data mining and analysis, the conclusions of this study are as follows.

(1) In view of year distribution and the number of publications, the development of land consolidation studies can be divided into three stages: 2000–2007 is the germination

period, 2008–2012 is the growth period, and 2013–2020 is the high-yield period. The number of publications reached its peak in 2020, indicating that more and more attention was paid to this field.

(2) In terms of the distribution of research countries, the published papers were mainly from Asia and Europe. Among them, China had the largest number of publications, while Thailand had the greatest influence. The analysis of cooperation between countries showed that most studies were independent research, which was not conducive to the globalization of scientific research forces in land consolidation.
(3) The high-frequency keywords in land consolidation mainly included land consolidation, land reclamation, China, remote sensing, and land fragmentation. The future research directions included land consolidation, its impacts on ecosystem services, benefit evaluation of land consolidation, research combining land fragmentation, etc.
(4) The research on land consolidation presents three evolutionary paths, namely the study of land fragmentation related to land consolidation, the study of the developing process of land consolidation, and the study of the impacts of land consolidation on soil.

Based on existing literature, future studies on land consolidation can be carried out from the following aspects.

(1) Exploring new methods for benefit evaluation of land consolidation. At present, the evaluation method of land consolidation benefits is relatively single, and the standard for index selection is not uniform. The determination of index weight values is mostly qualitative. Future research should explore new benefit evaluation methods for land consolidation, improve the evaluation index system, and strengthen quantitative research on the benefit evaluation of land consolidation [65].
(2) The scale effect of the impacts of land consolidation on ecosystem services. Under the influence of land consolidation, the spatial heterogeneity and temporal dynamics of ecosystem services will change in time and space [66]. In view of the characteristics of land consolidation implementation and research, the coupling mechanism analysis of ecosystem services at different temporal–spatial scales and the construction of a multi-scale coupling model to achieve the integration of multi-scale land consolidation management and regulation is of important scientific significance and practical value.
(3) Research on the mechanism and comprehensive effects of land consolidation on soil. Future research is expected to focus on exploring the morphological characteristics, evolutionary process, and causative mechanism of the soil system under different remediation modes, types of areas, and remediation years [67]; continue an in-depth study of the impact of land consolidation on the overall soil environment; and establish a quantitative model of the impact of land consolidation on the soil environment.
(4) Multidisciplinary integrated system research on land consolidation. Land consolidation is a comprehensive system engineering involving natural resources, the economy, society, ecological environment, technology, etc. [68]. Current studies on land consolidation are mostly based on independent studies. It is necessary to strengthen the cooperation between experts and scholars in different fields and countries, which is of great significance for condensing innovative ideas, sharing research resources, and promoting the development of land consolidation.
(5) Research on land consolidation and rural revitalization. The Chinese Rural Revitalization strategy aims to establish a sound urban–rural integration development system, mechanism, and policy system, and accelerate the modernization of agriculture and rural areas. Based on this perspective, land consolidation should activate key development factors, such as the rural population, land, and industry, and coordinate the revitalization of material space and the promotion of a spiritual core. Under the unified spatial planning system, coordinate land consolidation planning and rural revitalization planning, and vigorously develop a new model that combines land consolidation and multi-functional agriculture [8].

(6) Strengthen theoretical research on land consolidation. Land consolidation has been practiced in many countries for many years, and there is an urgent need for theoretical improvement and systematic summary. It is necessary to make full use of the theoretical fruits of related disciplines to carry out solid internal theoretical research and promptly introduce innovative methods and research concepts from other disciplines so as to realize the continuous development of land consolidation [69].

(7) Promote 3S technology research and the application of land consolidation. Due to the complex process and high technical requirements, on-site investigation of land consolidation is difficult [70,71]. The time-consuming and laborious traditional survey methods perform poorly in locating, surveying, and recording tasks [72]. Applying 3S technology in land consolidation will help to solve these problems, and refers to RS, GIS, and GPS. Remote Sensing (RS) can be used to obtain various ground feature elements. The Global Positioning System (GPS) can be used for the spatial positioning of important features. The Geographic Information System (GIS) can be used for comprehensive processing and integrated management of land consolidation data [73]. As an efficient means of acquiring and managing spatial information, 3S technology has a broad application prospect in the field of land consolidation.

(8) Improve the supervision mechanism of land consolidation. A scientific and reasonable supervision and management mechanism is not only a guarantee for the smooth completion of land consolidation projects, but also has irreplaceable practical significance for promoting the harmonious development of localities [74]. It is essential to speed up the improvement of land consolidation supervision and management mechanisms, and implement joint responsibilities for land consolidation. Combine legal, administrative, technological, and other management methods to reasonably establish a government-led, land-based, departmental-collaborative, and public-participation working mechanism, effectively implement land consolidation objectives and responsibilities, and guarantee the completion quality and implementation level of land consolidation projects.

Author Contributions: Conceptualization, Q.C. and X.X.; methodology, X.X.; software, Z.Z.; validation, Q.C., X.X. and Z.Z.; formal analysis, Q.C.; investigation, X.X.; resources, Z.Z.; data curation, X.X.; writing—original draft preparation, X.X.; writing—review and editing, Q.C.; visualization, Z.Z.; supervision, Z.Z.; project administration, Q.C. All authors have read and agreed to the published version of the manuscript.

Funding: This research was funded by the National Natural Science Foundation of China, grant numbers 41971243 and 41930757.

Institutional Review Board Statement: Not applicable.

Informed Consent Statement: Not applicable.

Data Availability Statement: Not applicable.

Acknowledgments: The authors would like to thank the reviewers and the editor whose suggestions greatly improved the manuscript.

Conflicts of Interest: The authors declare no conflict of interest.

References

1. Liu, Y.S. The basic theory and methodology of rural revitalization planning in China. *Acta Geogr. Sin.* **2020**, *75*, 1120–1133. (In Chinese)
2. Liu, Y.; Fang, F.; Li, Y. Key issues of land use in China and implications for policy making. *Land Use Policy* **2014**, *40*, 6–12. [CrossRef]
3. Long, H.L. Land use policy in China: Introduction. *Land Use Policy* **2014**, *40*, 1–5. [CrossRef]
4. Masini, E.; Tomao, A.; Barbati, A.; Corona, P.; Serra, P.; Salvati, L. Urban growth, land-use efficiency and local socioeconomic context: A comparative analysis of 417 metropolitan regions in Europe. *Environ. Manag.* **2019**, *63*, 322–337. [CrossRef]
5. Long, H.L. Land consolidation: An indispensable way of spatial restructuring in rural China. *J. Geogr. Sci.* **2014**, *24*, 211–225. [CrossRef]

6. Wang, M.; Bai, Z.; Dong, X. Land Consolidation Zoning in Shaanxi Province based on the Supply and Demand of Ecosystem Services. *China Land Sci.* **2018**, *32*, 73–80. (In Chinese) [CrossRef]
7. Zhang, H.; Zhou, S.; Wu, S.; Li, Z.; Li, L. Provincial scale spatial variation of cultivated land production capacity and its impact factors. *Nongye Gongcheng Xuebao Trans. Chin. Soc. Agric. Eng.* **2010**, *26*, 308–314. (In Chinese)
8. Long, H.L.; Zhang, Y.N.; Tu, S.S. Rural vitalization in China: A perspective of land consolidation. *J. Geogr. Sci.* **2019**, *29*, 517–530. [CrossRef]
9. Jiang, Y.F.; Long, H.L.; Tang, Y.T. Land consolidation and rural vitalization: A perspective of land use multifunctionality. *Prog. Geogr.* **2021**, *40*, 487–497. [CrossRef]
10. Crecente, R.; Alvarez, C.; Fra, U. Economic, social and environmental impact of land consolidation in Galicia. *Land Use Policy* **2002**, *19*, 135–147. [CrossRef]
11. Yan, J.M.; Xia, F.Z.; Bao, H.X. Strategic planning framework for land consolidation in China: A top-level design based on SWOT analysis. *Habitat Int.* **2015**, *48*, 46–54. [CrossRef]
12. Wang, Q.; Zhang, M.; Cheong, K.C. Stakeholder perspectives of China's land consolidation program: A case study of Dongnan Village, Shandong Province. *Habitat Int.* **2014**, *43*, 172–180. [CrossRef]
13. Jin, X.; Xu, X.; Xiang, X.; Bai, Q.; Zhou, Y. System-dynamic analysis on socio-economic impacts of land consolidation in China. *Habitat Int.* **2016**, *56*, 166–175. [CrossRef]
14. Colombo, S.; Perujo-Villanueva, M. A practical method for the ex-ante evaluation of land consolidation initiatives: Fully connected parcels with the same value. *Land Use Policy* **2019**, *81*, 463–471. [CrossRef]
15. Allahyari, M.S.; Damalas, C.A.; Masouleh, Z.D.; Ghorbani, M. Land consolidation success in paddy fields of northern Iran: An assessment based on farmers' satisfaction. *Land Use Policy* **2018**, *73*, 95–101. [CrossRef]
16. Lee, I.S.; Lee, H.; Chen, Y.H.; Chae, Y. Bibliometric analysis of research assessing the use of acupuncture for pain treatment over the past 20 years. *J. Pain Res.* **2020**, *13*, 367–376. [CrossRef]
17. De Bellis, N. *Bibliometrics and Citation Analysis: From the Science Citation Index to Cybermetrics*; The Scarecrow Press: Lanham, MD, USA, 2009.
18. Ying, D.; Gobinda, G.C.; Schubert, S.B.F. Bibliometric cartography of information retrieval research by using co-word analysis. *Inf. Process. Manag.* **2001**, *37*, 817–842.
19. Godin, B. On the origins of bibliometrics. *Scientometrics* **2006**, *68*, 109–133. [CrossRef]
20. Mora, L.; Deakin, M.; Reid, A. Combining co-citation clustering and text-based analysis to reveal the main development paths of smart cities. *Technol. Forecast. Soc. Chang.* **2019**, *142*, 56–69. [CrossRef]
21. Van Eck, N.J.; Waltman, L. Software survey: VOSviewer, a computer program for bibliometric mapping. *Scientometrics* **2010**, *84*, 523–538. [CrossRef]
22. Chen, X.L.; Zou, D.; Kohnke, L.; Xie, H.R.; Cheng, G. Affective states in digital game-based learning: Thematic evolution and social network analysis. *PLoS ONE* **2021**, *16*, e0255184. [CrossRef] [PubMed]
23. Aria, M.; Cuccurullo, C. Bibliometrix: An R-Tool for Comprehensive Science Mapping Analysis. *J. Informetr.* **2018**, *11*, 959–975. [CrossRef]
24. Xie, H.L.; Zhang, Y.W.; Wu, Z.L.; Lv, T.G. A Bibliometric Analysis on Land Degradation: Current Status, Development, and Future Directions. *Land* **2020**, *9*, 28. [CrossRef]
25. Iwona, G.; Beata, B.; Marzena, T.; Danuta, K.K. Mapping the structure of food waste management research: A Co-keyword analysis. *Int. J. Environ. Res. Public Health* **2020**, *17*, 4798.
26. Jeong, D.H.; Koo, Y. Analysis of Trend and Convergence for Science and Technology using the VOSviewer. *Int. J. Contents* **2016**, *12*, 54–58. [CrossRef]
27. Wang, W.; Liu, H.; Li, Y.Q.; Su, J.L. Development and management of land reclamation in China. *Ocean Coast. Manag.* **2014**, *102*, 415–425. [CrossRef]
28. Roldan-Valadez, E.; Salazar-Ruiz, S.Y.; Ibarra-Contreras, R.; Rios, C. Current concepts on bibliometrics: A brief review about impact factor, Eigenfactor score, CiteScore, SCImago Journal Rank, Source-Normalised Impact per Paper, H-index, and alternative metrics. *Ir. J. Med. Sci.* **2019**, *188*, 939–951. [CrossRef]
29. Jiao, J.J.; Nandy, S.; Li, H. Analytical studies on the impact of land reclamation on ground water flow. *Groundwater* **2001**, *39*, 912–920. [CrossRef]
30. Van Dijk, T. Complications for traditional land consolidation in Central Europe. *Geoforum* **2006**, *38*, 505–511. [CrossRef]
31. Van Dijk, T. Export of Planning Knowledge Needs Comparative Analysis: The Case of Applying Western Land Consolidation Experience in Central Europe. *Eur. Plan. Stud.* **2002**, *10*, 911–922. [CrossRef]
32. Wu, Z.P.; Liu, M.Q.; Davis, J. Land consolidation and productivity in Chinese household crop production. *China Econ. Rev.* **2005**, *16*, 28–49. [CrossRef]
33. Robert, J.H. Three stages in the history of land reclamation in the Netherlands. *Irrigation Drainage* **2007**, *56* (Suppl. 1), S113–S126.
34. Yang, H.Y.; Chen, B.; Barter, M.; Piersma, T.; Zhou, C.F.; Li, F.S.; Zhang, Z.W. Impacts of tidal land reclamation in Bohai Bay, China: Ongoing losses of critical Yellow Sea waterbird staging and wintering sites. *Bird Conserv. Int.* **2011**, *21*, 241–259. [CrossRef]
35. Zhou, Y.; Guo, L.Y.; Liu, Y.S. Land consolidation boosting poverty alleviation in China: Theory and practice. *Land Use Policy* **2019**, *82*, 339–348. [CrossRef]

36. Hirsch, J.E. An index to quantify an individual's scientific research output. *Proc. Natl. Acad. Sci. USA* **2005**, *102*, 16569–16572. [CrossRef]
37. Hirsch, J.E. Does the h index have predictive power? *Proc. Natl. Acad. Sci. USA* **2007**, *104*, 19193–19198. [CrossRef]
38. Zhang, C.T.; Etienne, J. The e-Index, Complementing the h-Index for Excess Citations. *PLoS ONE* **2009**, *4*, e5429. [CrossRef]
39. Leń, P. An algorithm for selecting groups of factors for prioritization of land consolidation in rural areas. *Electron. Agric.* **2018**, *144*, 216–221. [CrossRef]
40. Tang, H.; Yun, W.; Liu, W.; Sang, L. Structural changes in the development of China's farmland consolidation in 1998–2017: Changing ideas and future framework. *Land Use Policy* **2019**, *89*, 104212. [CrossRef]
41. Ministry of Land and Resources (MLR) and National Development and Reform Commission (NDRC). National Land Consolidation Plan (2011–2015). 2012. Available online: http://www.tdzyw.com/2012/0704/17747.html (accessed on 3 March 2022).
42. Niroula, G.S.; Thapa, G.B. Impacts and causes of land fragmentation, and lessons learned from land consolidation in South Asia. *Land Use Policy* **2005**, *22*, 358–372. [CrossRef]
43. Zhou, Y.; Li, Y.; Xu, C. Land consolidation and rural revitalization in china: Mechanisms and paths. *Land Use Policy* **2020**, *91*, 104379. [CrossRef]
44. Karan, S.K.; Samadder, S.R.; Maiti, S.K. Assessment of the capability of remote sensing and GIS techniques for monitoring reclamation success in coal mine degraded lands. *J. Environ. Manag.* **2016**, *182*, 272–283. [CrossRef]
45. Jürgenson, E. Land reform, land fragmentation and perspectives for future land consolidation in Estonia. *Land Use Policy* **2016**, *57*, 34–43. [CrossRef]
46. Liu, J.; Jin, X.B.; Xu, W.Y.; Sun, R.; Han, B.; Yang, X.H.; Gu, Z.M.; Xu, C.L.; Sui, X.Y.; Zhou, Y.K. Influential factors and classification of cultivated land fragmentation, and implications for future land consolidation: A case study of Jiangsu Province in eastern China. *Land Use Policy* **2019**, *88*, 104185. [CrossRef]
47. Ding, Y. Scientific collaboration and endorsement: Network analysis of co-authorship and citation networks. *J. Informetr.* **2010**, *5*, 187–203. [CrossRef] [PubMed]
48. Zakaria, N.N.; Convey, P.; Gomezfuentes, C.; Zulkharnain, A.; Sabri, S.; Shaharuddin, N.A.; Ahmad, S.A. Oil Bioremediation in the Marine Environment of Antarctica: A Review and Bibliometric Keyword Cluster Analysis. *Microorganisms* **2021**, *9*, 419. [CrossRef]
49. Blasius, J.; Greenacre, M.; Groenen, P.J.F.; van de Velden, M. Special issue on correspondence analysis and related methods. *Comput. Stat. Data Anal.* **2008**, *53*, 3103–3106. [CrossRef]
50. Le Roux, B.; Rouanet, H. *Multiple Correspondence Analysis*; SAGE Publications: Thousand Oaks, CA, USA, 2010.
51. Clausen, S.-E. *Applied Correspondence Analysis: An Introduction (Quantitative Applications in the Social Science)*; SAGE Publications: Thousand Oaks, CA, USA, 1998.
52. Zeng, S.Y.; Zhu, F.W.; Chen, F.; Yu, M.; Zhang, S.L.; Yang, Y.J. Assessing the Impacts of Land Consolidation on Agricultural Technical Efficiency of Producers: A Survey from Jiangsu Province, China. *Sustainability* **2018**, *10*, 2490. [CrossRef]
53. Wang, J.; Yan, S.C.; Guo, Y.Q.; Li, J.R.; Sun, G.Q. The effects of land consolidation on the ecological connectivity based on ecosystem service value: A case study of Da'an land consolidation project in Jilin province. *J. Geogr. Sci.* **2015**, *25*, 603–616. [CrossRef]
54. Hao, F.H.; Lai, X.H.; Ouyang, W.; Xu, Y.M.; Wei, X.F.; Song, K.Y. Effects of Land Use Changes on the Ecosystem Service Values of a Reclamation Farm in Northeast China. *Environ. Manag.* **2012**, *50*, 888–899. [CrossRef]
55. Shi, Y.; Cao, X.; Fu, D.; Wang, Y. Comprehensive Value Discovery of Land Consolidation Projects: An Empirical Analysis of Shanghai, China. *Sustainability* **2018**, *10*, 2039. [CrossRef]
56. Weismayer, C.; Pezenka, I. Identifying emerging research fields: A longitudinal latent semantic keyword analysis. *Scientometrics* **2017**, *113*, 1757–1785. [CrossRef]
57. Schmidt, M. The Sankey diagram in energy and material flow management. *J. Ind. Ecol.* **2008**, *12*, 82–94. [CrossRef]
58. Kamal, S.; Hiang, K.H.; Bin, S. Sankey diagram framework for energy and exergy flows. *Appl. Energy* **2014**, *136*, 1035–1042.
59. Ela, E. Analysis of the effect of land consolidation projects in terms of land fragmentation and parcel shapes: The case of Konya, Turkey. *Arab. J. Geosci.* **2020**, *13*, 350. [CrossRef]
60. Swartjes, F.A.; Rutgers, M.; Lijzen, J.; Janssen, P.; Otte, P.F.; Wintersen, A.; Brand, E.; Posthuma, L. State of the art of contaminated site management in The Netherlands: Policyframework and risk assessment tools. *Sci. Total Environ.* **2012**, *427*, 1–10. [CrossRef] [PubMed]
61. Asiama, K.O.; Bennett, R.M.; Zevenbergen, J.A. Land consolidation on Ghana's rural customarylands: Drawing from the Dutch, Lithuanian and Rwandan Experiences. *J. Rural Stud.* **2017**, *56*, 87–99. [CrossRef]
62. Abubakari, Z.; Van der Molen, P.; Bennett, R.M.; Kuusaana, E.D. Land consolidation, customary lands, and Ghana's Northern Savannah Ecological Zone: An evaluation of the possibilities and pitfalls. *Land Use Policy* **2016**, *54*, 386–398. [CrossRef]
63. He, X.Y.; Su, Y.R.; Liang, Y.M.Y.; Chen, X.B.; Zhu, H.H.; Wang, K.L. Land reclamation and short-term cultivation change soil microbial communities and bacterial metabolic profiles. *J. Sci. Food Agric.* **2012**, *92*, 1103–1111. [CrossRef]
64. Hou, H.P.; Wang, C.; Ding, Z.Y.; Zhang, S.L.; Yang, Y.J.; Ma, J.; Chen, F.; Li, J.R. Variation in the Soil Microbial Community of Reclaimed Land over Different Reclamation Periods. *Sustainability* **2018**, *10*, 2286. [CrossRef]
65. Wang, J.; Zhong, L.N. Literature Analysis on Land Consolidation Research in China. *China Land Sci.* **2016**, *30*, 88–97.

66. Wang, J.; Zhong, L.N.; Ying, L.X. Review on the Study of the Impacts of Land Consolidation on Ecosystem Services. *J. Ecol. Rural Environ.* **2018**, *34*, 803–812.
67. Bu, C.L.; Li, Y.R.; Wang, Y.S.; Yuan, Y.F. Effects of land consolidation on soil: Research progress and prospect. *Land Dev. Eng. Res.* **2020**, *5*, 32–39. (In Chinese)
68. Yang, Y.J.; Zhang, S.L.; Bian, Z.F.; Peter Erskine Li, X.S. Spatial-temporal pattern differentiation and influencing mechanism of land reclamation in China. *Trans. Chin. Soc. Agric. Eng. Trans. CSAE* **2016**, *32*, 206–214.
69. Hu, Z.Q.; Zhao, Y.L.; Miao, H.L.; Yu, Y. Comments on Land Sciences Research Priorities in 2012 and Perspective for 2013: Sub-report of Land Consolidation and Readjustment. *China Land Sci.* **2013**, *27*, 89–96.
70. Jia, W.; Liu, J.; Yu, L.; Wang, M. Development and application of field survey technology-based GPS and GIS for land consolidation. *Trans. Chin. Soc. Agric. Eng.* **2009**, *25*, 197–201.
71. Xi, K.; Yang, X.; Zhao, J.; Han, T. PDA + 3S technology application in land change survey. Geomatics Spatial Inform. *Technology* **2013**, *36*, 91–94.
72. Wang, J.; Ge, A.; Hu, Y.; Li, C.; Wang, L. A fuzzy intelligent system for land consolidation—A case study in Shunde, China. *Solid Earth Discuss* **2015**, *7*, 1347–1374. [CrossRef]
73. Bao, H.J.; Wu, C.F.; Ye, Y.M. Preliminary discussion on cropland consolidation under the supporting system of 3S technology. *Econ. Geogr.* **2001**, *S1*, 229–232.
74. Hu, Z.Q.; Yu, Y.; Long, J.H. Comments on Land Sciences Research Priorities in 2013 and Perspective for 2014: Sub-report of Land Consolidation and Readjustment. *China Land Sci.* **2014**, *28*, 13–21.

MDPI
St. Alban-Anlage 66
4052 Basel
Switzerland
www.mdpi.com

International Journal of Environmental Research and Public Health Editorial Office
E-mail: ijerph@mdpi.com
www.mdpi.com/journal/ijerph

www.ingramcontent.com/pod-product-compliance
Lightning Source LLC
Chambersburg PA
CBHW061447220926
51850CB00004B/37
9783036588421